Aging
Exploring a Complex Phenomenon

Aging
Exploring a Complex Phenomenon

Edited by
Shamim I. Ahmad

CRC Press
Taylor & Francis Group
Boca Raton London New York

CRC Press is an imprint of the
Taylor & Francis Group, an **informa** business

CRC Press
Taylor & Francis Group
6000 Broken Sound Parkway NW, Suite 300
Boca Raton, FL 33487-2742

First issued in paperback 2020

© 2018 by Taylor & Francis Group, LLC
CRC Press is an imprint of Taylor & Francis Group, an Informa business

No claim to original U.S. Government works

ISBN 13: 978-0-367-65756-7 (pbk)
ISBN 13: 978-1-138-19697-1 (hbk)

Library of Congress Cataloging-in-Publication Data

Names: Ahmad, Shamim I., editor.
Title: Aging : exploring a complex phenomenon / editor, Shamim I. Ahmad.
Other titles: Aging (Ahmad)
Description: Boca Raton : Taylor & Francis, 2018. | Includes bibliographical references.
Identifiers: LCCN 2017028774 | ISBN 9781138196971 (hardback : alk. paper)
Subjects: | MESH: Aging | Age Factors | Geriatrics--methods
Classification: LCC QP86 | NLM WT 104 | DDC 612.6/7--dc23
LC record available at https://lccn.loc.gov/2017028774

Visit the Taylor & Francis Web site at
http://www.taylorandfrancis.com

and the CRC Press Web site at
http://www.crcpress.com

For the aging book

The editor dedicates this book to his late father, Abdul Nasir, and mother, Anjuman Ara, who played very important roles to bring him to this stage of academic achievements with their esteemed love, sound care, and sacrifice. Dedication also goes to his wife, Riasat Jan, for her patience and persistent encouragement to produce this book, as well as to his children, Farhin, Mahrin, Tamsin, Alisha, and Arsalan, especially the latter two for providing great pleasure with their innocent interruptions, leading to his energy revitalization. Finally, his best wishes go to the aged ailing patients for their remaining life to run smoothly.

Contents

SECTION I   Introduction to Aging

SECTION II   Aging Hypothesis

SECTION III Diseases Associated with Aging and Treatment

SECTION IV Mechanisms of Aging

SECTION VII   Anti-Aging Drugs

SECTION VIII   Aging in Caenorhabditis elegans

SECTION IX   Hibernation and Aging

SECTION X   Mathematical Modeling of Aging

Preface

AGING: AN UNEXPLAINED PHENOMENON

Since the time mankind has developed a "thinking brain," there has been an everlasting curiosity to know what aging is and whether it will ever be possible to increase the human life span or achieve immortality? The first book on this subject was written as early as 1582 by Mohammad Ibn Yusuf Al-Harawi titled *Ainul Hayat*. The book discussed the diet, lifestyle, and environmental factors influencing aging. According to PubMed, serious research on ageing and its first publication in scientific journals was recorded in 1925. Since then, 395,708 research and review articles have been published in various journals on this subject. From this, it is clear that the subject is not only extremely important but also very difficult to reach to the bottom of its molecular mechanisms and control, and hence remains an unexplained phenomenon.

The mechanism of aging, in fact, appears to be very complex and, according to Chapter 6 of this book, over 300 theories alone on aging have been proposed. Among those, some are more acceptable than others, but none stands clear that can bring all the available research data on one platform. Among the more accepted theories on aging are the adaptive theory, cellular senescence theory, DNA damage theory, free radical and oxidative stress theory, genetic program theory, immunological theory, mitochondrial mutation theory, and the telomerase shortening theory. Less important are the antagonistic pleiotropy and stress theory, biogerontologic theory and psychosocial theories, chaos theory, codon restriction theory, continuity theory of normal aging, cross-linkage of macromolecule theory, developmental theory, ecological stress theory, evolutionary theory, gene regulation theory, general theory of aging, kinetic theory, melatonin theory, metabolic causes of aging, modern evolutionary mechanisms theories, neuroendocrine theory, parametabolic theory, participation theory, programmed and non-programmed theory, rate of metabolic theory, redundant message theory, somatic mutation theory, reliability theory of aging and longevity, error theory, thermodynamic theory, transcriptional event theory, unifying theory of aging, united theory of aging, and many more.

The research publication existing in such a large number, especially covering the areas such as mechanisms of aging, age-related diseases, healthy aging and antiaging agents, and devices to control or slow down the aging process points toward a continued deep interest in this subject.

It is almost impossible to detail all these aging theories and only a selected number of them have been presented in this book. Sections I and II in this book present the indispensable Soma hypothesis programmed aging of perennial neurons, the genetic program theory, aging epigenetics, and a couple of rare and novel approaches on aging affected by urbanization and associated culture as well on the role of multicellularity, speciation, and ecosystem on aging.

Section III addresses a number of diseases and the malfunctioning of body organs and systems specifically associated with aging. Most important of these are premature aging in which a number of RecQ helicases play roles in maintaining the genome stability leading to progeria and some other diseases. Thus, a mutation in the *WRN* gene leads to Werner's syndrome in which, besides rapid aging phototypes, the subjects are prone to cancer susceptibility. Another progeroid disease, known as Hutchinson–Gilford progeria, ironically could not be included in this book. Bloom syndrome (*BLM* gene mutation and another RecQ helicase) gives rise to a variety of cancers and multiple malignancies including several other physical deformities. Mutation in *RecQ4* can lead to skin abnormalities and skeletal defects. These have been intricately highlighted in Chapter 10. The other children's disease associated with aging is Cockayne syndrome, which is a disorder with premature aging and short life expectancy due to a mutation in one of the two genes, *ERCC8* or *ERCC6*, known as CSA and CSB, respectively. It is a complex and progressive disease leading to developmental and cognitive delay apart from several other physiological and biochemical impairments. This has been comprehensively described in Chapter 11.

The advancement of age and the probability of cancer susceptibility have been linked with the decreased gene repair activities and the accumulation of mutation contributing to multiple deleterious biological and molecular events; Chapter 12 highlights this issue. Other less important diseases linked with old age are osteoporosis due to immunological problems (Chapter 17), nodular thyroid diseases (Chapter 13), and breakdown of blood–brain barrier (Chapter 15). These are intricately presented to enrich the knowledge of our readers. Chapter 16 presents a generalized view of age-related diseases, and Chapters 18 and 19 highlight the manipulation of aging to treat age-related diseases and the therapeutic options available to enhance the post-stroke recovery in aged humans.

Section IV covers the mechanisms of aging as in Chapter 20 is described the significance of increased mitochondrial DNA fragments inside nuclear DNA, which seems to occur throughout life and plays a role in the aging process. These studies have mostly been carried out on yeast and in mice and implied to long living animals, which have low rates of mitochondrial reactive oxygen species production and thus less damage to their DNA. Chapters 21 and 23 address the important roles of oxidative stress and oxidation of ion channels in aging. The involvement of immune systems in aging and cancer development is described in Chapter 22 and the modulation of aging by autophagy is described in Chapter 25. The role of lipid raft alteration and impairment of age neural membrane have been addressed in Chapter 24.

Section V focuses on the skin's aging clock (Chapter 27) and on the grounds and determents of aging (Chapter 26).

Section VI is equally valuable in making readers aware on the topics of healthy and successful aging. This issue has been addressed by different authors differently, such as in Chapter 28 the social structure and its influence on healthy aging is explained and in Chapter 29 the author describes that cognition, personality, and the whole person is responsible for determining the wellness and extension of age. Physical activities also have major impacts not only on age longevity but also can be helpful in maintaining good health and wellness.

Section VII covers valuable chapters on finding anti-aging drugs. The fact is that the desire to live longer is fairly powerful and although ample funds are being spent on the aging research, a satisfactory outcome is still awaited. Certain vitamins (such as vitamin E) and biochemical agents, plant extracts, and so on have been listed to have antiaging ability but none so far has clearly and categorically been shown to have any significant effect on the human aging process. Interestingly, however, success has been achieved in producing mutant lower organisms, flies, and small animals through genetic manipulation that have a longer life than their non-mutant counterparts. Mutant mice living longer were reported by Migliaccio et al. (*Nature*, 402, 309–313, 1999) in which the mice had mutations in gene p66 shc protein and lived almost one third longer. Resistance to oxidative damage was considered to be the prime reason. Since then, a large number of studies have been carried out to find other reasons for the longevity in mutant mice, but a clear, full-length picture is still awaited. Research on long-lived Drosophila melanogaster revealed that evolutionary conserved function for the mitochondrial electron transport chain in the modulation of life span was responsible (Linford and Pletcher, *Curr Biol*, 19, 2009) and in another report on *Caenorhabditis elegans* age-1 (hx546) mutant was shown to have an increased mean life span averaging 40% and a maximal life span averaging 60% at 20°C (Friedman and Johnson, *Genetics*, 118, 75–86, 1988). These results are providing promising information to extend the work on finding the magic potion(s) for human beings to increase their aging life.

Sections VIII and IX address aging in lower organisms and those that having longer life due to hibernating in adverse environmental conditions. The authors of Section X uniquely presents their own devised mathematical models of aging.

From the materials presented in this book, it is evident that the interest in aging, especially discovering antiaging drugs and craving for a longer life, is not waning; hence, it is anticipated that the information presented will stimulate further research among both specialists and novices in the field with excellent overviews of the current status of research and pointers to future research goals. However, an interesting situation remains to imagine how the world is going to look when human beings will live for 150 years or more.

Shamim I. Ahmad, BSc, MSc, PhD
School of Science and Technology
Nottingham Trent University
Nottingham, United Kingdom

One World, One Humanity

AN APPEAL TO THE UNITED NATIONS SECURITY COUNCIL AND THE HEAD OF THE STATES

It is yet to be determined when the desire beings originated in humans to have eternal life or at least to extend it as much as possible. Whenever, the desire must be very strong and hence the thinking brain must have been extensively used to find the ways to achieve them. According to the research record of PubMed although 395,708 research papers (May 2017) on various aspects of aging have been published, reflecting the public demands and the tireless efforts of the scientists and the industries to achieve it, yet the goal remains fairly aloof. The only achievement we can see is the increased number of geriatrics and centenarians, which has been due to the massive development in medical science, and the knowledge gained through research about the better quality of life leading to age longevity.

Recent developments in gene technology, however, have managed to prolong the life of certain microbes, flies, and small animals, and these have been addressed in this book. The next question therefore asked is whether this technology could ever be applied to humans to prolong their life span. It is an open-ended question and, in the editor's opinion, it is highly unlikely for many years to come due to stringent ethical issues and controls. The alternative, therefore, is to discover the most active and safe antiaging agents, which can be tested on animals and later applied on humans beings. In the editor's opinion, it may not be too long that such magic potion(s) will be found and, if safely used from the early age, it may slow down the aging process throughout the life and may culminate in the childhood age to stay say up to 20 years, the teen age may extend up to 60 years, youth up to 120 years, old age up to 150 years, and then the death at the ripe age 160 or over. It is purely assumptive but not impossible.

Now, if we look at the other side of the coin, we can see that the present world, especially the developed and the developing countries, are extensively busy in developing more and more super-powered war weapons of various kinds, competing with each other, which can fairly rapidly kill the human species at massive scales. Also, it is apparent that most often innocent people, including children, women, and the fragile and old people, will become victims of such actions. There is no doubt that, with the passage of time, biological, chemical, and nuclear weapons are being manufactured and stored in increasing quantities in several countries; the sensible and crunchy question being asked is why. If there are weapons, there is use, and if there is use it has only one consequence: death of the human species and massive infrastructure destruction besides the collateral heartbreaking results. Here I must add that in this world there still exist very many kindhearted people and organizations who are risking their lives and their resources to reduce this destruction, and I salute them.

I recall someone asking of the very famous and super intelligent Einstein his opinion about the Third World War; the answer he gave was, "I do not know about the third war but I know about the fourth that it will be fought by bows and arrows." The editor sincerely hopes that the destruction of mankind should never reach that level, and that the human species does not deserve to see the Third World War causing their massive elimination and other destructions predicted by this ingenious scientist and great philosopher. *Hence, I am appealing to the United Nations Security Council and the Head of every State in the world to think seriously about the prediction presented by this highly respected scientist, and put all their efforts possible to minimize the production, storage, and especially the use of the weapons of mass destruction. Minimizing weapons production and storage will have the ripple effect and the number of wars should reduce, thus saving this world and especially our human race. Thinks we humans are tied up with one slogan and one philosophy – the*

Philosophy of Humanity and this should be applied with full strength – please adapt this slogan, **One World, One Humanity.**

Shamim I. Ahmad, BSc, MSc, PhD
School of Science and Technology
Nottingham Trent University
Nottingham, United Kingdom

Acknowledgments

The editor cordially acknowledges the authors of this book for their contribution and in-depth knowledge, high skill, and professional presentation. Without their input, it would not have been possible to bring out this book on this topic. He would also like to acknowledge the staff, especially Ms. Hillary Lafoe, Ms. Jennifer Blaise, and Mr. Chuck Crumly of CRC Press, for their hard work, friendly approach, and patience and also CRC Press/Taylor & Francis Group for their efficient and highly professional handling of this project. Finally, he wishes to acknowledge his university for providing him this platform and especially the IT service staff members for helping him on every IT-associated problem with their super-skills. Finally acknowledgement presented to Ms. Teena Lawrence, Mr. Todd Perry, and the staff of the printing division for their most professional input to this excellent production.

Editor

Shamim I. Ahmad, after obtaining his master's in botany from Patna University, Bihar, India, and his PhD in molecular genetics from Leicester University, England, joined Nottingham Polytechnic as a grade 1 lecturer and was subsequently promoted as a senior lecturer. After serving for about 35 years in Nottingham Polytechnic (which subsequently became Nottingham Trent University), he volunteered for early retirement, although yet serving as a part-time senior lecturer. He is now involved with writing medical books. For more than three decades, he has researched on different areas of molecular biology/genetics including thymineless death in bacteria, genetic control of nucleotide catabolism, development of anti-AIDs drug, control of microbial infection of burns, phages of thermophilic bacteria, and microbial flora of Chernobyl after the accident at the nuclear power station. However, his main interest, which started about 30 years ago, is DNA damage and repair, specifically by near ultraviolet light, mostly through the photolysis of biological compounds, production of reactive oxygen species, and their implications on human health including skin cancer. He is also investigating near ultraviolet photolysis of nonbiological compounds such as 8-methoxypsoralen and mitomycin C and their importance in psoriasis treatment and in Fanconi anemia. Collaborating with the University of Osaka, Japan, in his latest research publication, he and his colleagues were able to show that a number of naturally occurring enzymes were able to scavenge the reactive oxygen species.

In 2003 he received a prestigious "Asian Jewel Award" in Central Britain for *Excellence in Education*. His longtime ambition to produce medical books started in 2007 and since then has published *Molecular Mechanisms of Fanconi Anemia*; *Molecular Mechanisms of Xeroderma Pigmentosum*; *Molecular Mechanisms of Cockayne Syndrome*; *Molecular Mechanisms of Ataxia Telangiectasia*; *Diseases of DNA Repair*; *Neurodegenerative Diseases*; *and Diabetes: An Old Disease a New Insight*. As a co-author, he has also published *Obesity: A Practical Guide*; *Thyroid: Basic Science and Clinical Practice*; *and Diabetes: A Comprehensive Treatise for Patients and Caregivers*, published by Landes Bioscience/Springer. Recently CRC Press/Taylor & Francis Group has published his book on *Reactive Oxygen Species in Biology and Human Health* and a book on *Ultraviolet Light in Human Health, Diseases and Environment* will be out soon by Springer.

Contributors

María de la Luz Arenas-Sordo
Servicio de Genética
Instituto Nacional de Rehabilitación
Secretaria de Salud
México City, Mexico

Vasily V. Ashapkin
Belozersky Institute of Physico-Chemical
    Biology
Lomonosov Moscow State University
Moscow, Russia

Yousun Baek
Department of Human Development and
    Family Studies
Iowa State University
Ames, Iowa

Ágnes Bajza
Faculty of Information Technology and Bionics
Pázmány Péter Catholic University
Budapest, Hungary

Enke Baldini
Department of Surgical Sciences, "Sapienza"
    University of Rome
Rome, Italy

Gustavo Barja
Department of Animal Physiology-II
Faculty of Biological Sciences
Complutense University of Madrid (UCM)
Madrid, Spain

Betr Brož
Catholic Theology Faculty
Charles University
Prague, Czech Republic

R. Glen Calderhead
Clinique L Dermatology
Goyang-si, Gyeonggi-do, South Korea

Antonio Catania
Department of Surgical Sciences
"Sapienza" University of Rome
Rome, Italy

Julie Colin
UR AFPA (INRA USC 340, EA 3998)
Équipe Biodisponibilité et Fonctionnalités des
    Lipides Alimentaires (BFLA)
Université de Lorraine
Vandœuvre-lès-Nancy, France

Attila Csorba
Department of Pharmacognosy
Faculty of Pharmacy
University of Szeged
Szeged, Hungary

Dao-Fu Dai
Department of Pathology
University of Iowa
Carver College of Medicine
Iowa City, Iowa

Dumbrava Danut
Department of Functional Sciences
Center of Clinical and Experimental
    Medicine
University of Medicine and Pharmacy
    of Craiova
Craiova, Romania

Elizabeth de Lange
Translational Pharmacology
Cluster of Systems Pharmacology
Leiden Academic Center for Drug
    Research
Leiden University
Leiden, The Netherlands

Massimo De Martinis
Department of Life, Health and
    Environmental Sciences
University of L'Aquila
L'Aquila, Italy

László Dénes
Faculty of Information Technology
    and Bionics
Pázmány Péter Catholic University
Budapest, Hungary

Taylor N. Dennis
Department of Biochemistry
University of California
Riverside, California

Daniela Di Silvestre
Department of Life, Health and Environmental
    Sciences
University of L'Aquila
L'Aquila, Italy

G. Dupuis
Department of Biochemistry
Graduate Program in Immunology
Faculty of Medicine and Health Sciences
University of Sherbrooke
Sherbrooke, Quebec, Canada

Barbora Dvořánková
First Faculty of Medicine
Institute of Anatomy and BIOCEV
Charles University
Prague, Czech Republic

Raluca Sandu Elena
Department of Functional Sciences, Center of
    Clinical and Experimental Medicine
University of Medicine and Pharmacy of Craiova
Craiova, Romania

Preetham Elumalai
Department of Processing Technology
    (Biochemistry)
Kerala University of Fisheries and Ocean Studies
Kochi, Kerala, India

Franciska Erdő
Faculty of Information Technology and Bionics
Pázmány Péter Catholic University
Budapest, Hungary

Angelo Filippini
Department of Surgical Sciences
"Sapienza" University of Rome
Rome, Italy

T. Fulop
Research Center on Aging
Graduate Program in Immunology
Faculty of Medicine and Health Sciences
University of Sherbrooke
Sherbrooke, Quebec, Canada

Lia Ginaldi
Department of Life, Health and Environmental
    Sciences
University of L'Aquila
L'Aquila, Italy

Daniela-Gabriela Glavan
2nd Psychiatry Clinic Hospital
University of Medicine and Pharmacy of
    Craiova
Craiova, Romania

Lynn Gregory-Pauron
UR AFPA (INRA USC 340, EA 3998)
Équipe Biodisponibilité et Fonctionnalités des
    Lipides Alimentaires (BFLA)
Université de Lorraine
Vandœuvre-lès-Nancy, France

Péter Imre
Faculty of Information Technology and
    Bionics
Pázmány Péter Catholic University
Budapest, Hungary

Priit Kaasik
Institute of Sport Sciences and Physiotherapy
University of Tartu
Tartu, Estonia

Alexander N. Khokhlov
Evolutionary Cytogerontology Sector
School of Biology
Lomonosov Moscow State University
Moscow, Russia

Elise A. Kikis
Biology Department
The University of the South
Sewanee, Tennessee

Gukbin Kim
Global Management of Natural Resources
University College London (UCL)
London, United Kingdom

Jong In Kim
Division of Social Welfare and Health
    Administration
and
Institute for Longevity Sciences
Wonkwang University
Iksan, Republic of Korea

Alexander A. Klebanov
Evolutionary Cytogerontology Sector
School of Biology
Lomonosov Moscow State University
Moscow, Russia

Lyudmila I. Kutueva
Belozersky Institute of Physico-Chemical
  Biology
Lomonosov Moscow State University
Moscow, Russia

Marios Kyriazis
ELPIs Foundation for Indefinite Lifespans
London, United Kingdom

Lukáš Lacina
First Faculty of Medicine
Institute of Dermatovenerology and BIOCEV
Charles University
Prague, Czech Republic

Sreeja Lakshmi
Department of Processing Technology
  (Biochemistry)
Kerala University of Fisheries and Ocean Studies
Kochi, Kerala, India

A. Larbi
Singapore Immunology Network (SIgN)
Biopolis
Agency for Science Technology and Research
  (A*STAR)
Singapore

Kyuho Lee
Department of Human Development and
  Family Studies
Iowa State University
Ames, Iowa

A. Le Page
Research Center on Aging
Graduate Program in Immunology
Faculty of Medicine and Health Sciences
University of Sherbrooke
Sherbrooke, Quebec, Canada

Giacinto Libertini
Department of Translational Medical Sciences
Federico II University
Naples, Italy

Zhanjun Lv
Department of Genetics
Hebei Medical University
Hebei Key Lab of Laboratory Animal
Shijiazhuang, Hebei Province, China

Catherine Malaplate-Armand
UR AFPA (INRA USC 340, EA 3998)
Équipe Biodisponibilité et Fonctionnalités des
  Lipides Alimentaires (BFLA)
Université de Lorraine
Vandœuvre-lès-Nancy, France

V. Mallikarjun
Wellcome Trust Centre for Cell-Matrix
  Research
Division of Cell Matrix Biology and
  Regenerative Medicine
School of Biological Sciences
Medicine and Health
Manchester Academic Health Science
  Centre
University of Manchester
Manchester, United Kingdom

Jennifer A. Margrett
Department of Human Development and
  Family Studies
Iowa State University
Ames, Iowa

Cielo Mae D. Marquez
Institute of Biology
University of the Philippines Diliman
Quezon City, Philippines

Peter Martin
Department of Human Development and
  Family Studies
Iowa State University
Ames, Iowa

Mark Mc Auley
Faculty of Science and Engineering
University of Chester
Chester, United Kingdom

Massimo Monti
Department of Surgical Sciences
"Sapienza" University of Rome
Rome, Italy

Kathleen Mooney
Faculty of Health and Social Care
Edge Hill University
Lancashire, United Kingdom

Amy Morgan
Faculty of Science and Engineering
University of Chester
Chester, United Kingdom

Galina V. Morgunova
Evolutionary Cytogerontology Sector
School of Biology
Lomonosov Moscow State University
Moscow, Russia

Denissa-Greta Olaru
Department of Ophthalmology
Medlife Clinic
Craiova, Romania

Christ Ordookhanian
Department of Biochemistry
University of California
Riverside, California

Thierry Oster
UR AFPA (INRA USC 340, EA 3998)
Équipe Biodisponibilité et Fonctionnalités des
    Lipides Alimentaires (BFLA)
Université de Lorraine
Vandœuvre-lès-Nancy, France

G. Pawelec
Center for Medical Research
Second Department of Internal Medicine
University of Tübingen
Tübingen, Germany

J. Jefferson P. Perry
Department of Biochemistry
University of California
Riverside, California

and

Amrita University
Kerala, India

and

Universidad Francisco de Vitoria
Madrid, Spain

Eugen Petcu
Griffith University School of Medicine
Queensland, Australia

and

Queensland Eye Institute
Brisbane, Queensland, Australia

João Pinto da Costa
CESAM and Department of
    Chemistry
University of Aveiro
Aveiro, Portugal

Daniele Pironi
Department of Surgical Sciences
"Sapienza" University of Rome
Rome, Italy

Leonard W. Poon
Institute of Gerontology
University of Georgia
Athens, Georgia

Aurel Popa-Wagner
Griffith University School of
    Medicine
Southport, Queensland, Australia

and

Queensland Eye Institute
Brisbane, Queensland, Australia

and

Department of Functional Sciences
Center of Clinical and Experimental
    Medicine
University of Medicine and Pharmacy of
    Craiova
Craiova, Romania

Aleksi Šedo
First Faculty of Medicine
Institute of Biochemistry and Experimental
    Oncology
Charles University
Prague, Czech Republic

Teet Seene
Institute of Sport Sciences and Physiotherapy
University of Tartu
Tartu, Estonia

Federico Sesti
Department of Neuroscience and
    Cell Biology
Robert Wood Johnson Medical School
Rutgers University
Piscataway, New Jersey

Michael A. Singer
Faculty of Health Sciences
Queen's University
Kingston, Ontario, Canada

Maria Maddalena Sirufo
Department of Life
Health and Environmental Sciences
University of L'Aquila
L'Aquila, Italy

Karel Smetana Jr.
First Faculty of Medicine
Institute of Anatomy and BIOCEV
Charles University
Prague, Czech Republic

Salvatore Sorrenti
Department of Surgical Sciences
"Sapienza" University of Rome
Rome, Italy

Kenneth B. Storey
Department of Biology
Carleton University
Ottawa, Canada

Libo Su
Department of Genetics
Hebei Medical University
Hebei Key Lab of Laboratory Animal
Shijiazhuang, Hebei Province, China

Roxana Surugiu
Department of Functional Sciences
Center of Clinical and Experimental
    Medicine
University of Medicine and Pharmacy
    of Craiova
Craiova, Romania

J. Swift
Wellcome Trust Centre for Cell-Matrix Research
Division of Cell Matrix Biology and
    Regenerative Medicine
School of Biological Sciences
Medicine and Health
Manchester Academic Health Science Centre
University of Manchester
Manchester, United Kingdom

Pavol Szabo
First Faculty of Medicine
Institute of Anatomy and BIOCEV
Charles University
Prague, Czech Republic

Francesco Tartaglia
Department of Surgical Sciences
"Sapienza" University of Rome
Rome, Italy

Salvatore Ulisse
Department of Surgical Sciences
"Sapienza" University of Rome
Rome, Italy

Quentin Vanhaelen
Insilico Medicine Inc.
Johns Hopkins University
Baltimore, Maryland

Boris F. Vanyushin
Belozersky Institute of Physico-Chemical Biology
Lomonosov Moscow State University
Moscow, Russia

Michael C. Velarde
Institute of Biology,
University of the Philippines
Diliman, Quezon City, Philippines

Massimo Vergine
Department of Surgical Sciences
"Sapienza" University of Rome
Rome, Italy

Xiufang Wang
Department of Genetics
Hebei Medical University
Hebei Key Lab of Laboratory Animal
Shijiazhuang, Hebei Province, China

Edward J. Wing
Department of Medicine
Warren Alpert Medical School of Brown
   University
Providence, Rhode Island

J. M. Witkowski
Department of Pathophysiology
Medical University of Gdańsk
Gdańsk, Poland

Cheng-Wei Wu
Department of Biology
University of Florida
Gainesville, Florida

Frances T. Yen
UR AFPA (INRA USC 340, EA 3998)
Équipe Biodisponibilité et
   Fonctionnalités des Lipides
   Alimentaires (BFLA)
Université de Lorraine
Vandœuvre-lès-Nancy, France

Huanling Zhang
Department of Genetics
Hebei Medical University
Hebei Key Lab of Laboratory Animal
Shijiazhuang, Hebei Province, China

Section I

Introduction to Aging

1 A Synopsis on Aging

João Pinto da Costa

CONTENTS

INTRODUCTION

Aging is an inherently fascinating topic that has captivated scientists and philosophers throughout history. For Plato (428–347 BC), those who lived longer reached a "philosophical understanding of mortal life," which led to the desire to understand everlasting ideas and truths, beyond the mortal world (Baars 2012): "for wisdom and assured true conviction, a man is fortunate if he acquires them even on the verge of old age" (Cary, Davis, and Burges 1852). However, the most accurate description of the general human perception of aging comes from the Italian poet, philosopher, and essayist Giacomo Leopardi (1798–1837): "Old age is the supreme evil, because it deprives us of all pleasures, leaving us only the appetite for them and it brings with it all sufferings. Nevertheless, we fear death, and we desire old age" (Leopardi, Thomson, and Dobell 1905).

What, then, constitutes aging? In its broadest sense, aging broadly encompasses the general changes that occur during an organism's life span, though the rate at which these take place, as well as the order and mode in which they occur, varies widely (Kirkwood 2005). Hence, such a definition comprises changes that are not necessarily deleterious, such as wrinkles and graying hair in humans, which do not affect the individual's viability (although some might disagree with such an assertion). Anton and coworkers summarize it as an equation (1.1) (Anton et al. 2005), in which the phenotype is the end result of the interaction between genotype and external factors, namely, diet, lifestyle, and the surrounding environment

$$[\text{phenotype}] = [\text{genotype}] + [(\text{diet, lifestyle, and environment})]. \tag{1.1}$$

These changes that may be classified as innocuous must then be differentiated from those that may lead to an increased risk of disease, disability, or death. A more precise term was therefore coined by biogerontologists, scientists who work in the subfield of gerontology concerned with the biological process of aging, its evolutionary origins, and potential interventions in the process: senescence (Dollemore and Aging 2002). Senescence is therefore the progressive deterioration of bodily functions over time and normal human aging has been associated with a loss of complexity

3

in multiple physiological processes and anatomic structures (Goldberger, Peng, and Lipsitz 2002), including blood pressure (Kaplan et al. 1991), stride intervals (Hausdorff et al. 1997, Terrier and Dériaz 2011), respiratory cycles (Peng et al. 2002, Schumann et al. 2010), and vision (Azemin et al. 2012), among others, such as postural dynamics (Manor et al. 2010). All these detrimental consequences of senescence ultimately lead to decreased fertility and increased risk of mortality (Lopez-Otin et al. 2013, Chesser 2015). In this chapter, the more inclusive term "aging" will be used, however, due to its extensive use in both the scientific and nonscientific literatures. This describes the breakdown of self-organizing systems and the subsequent reduced ability to adapt to the environment (Vasto et al. 2010), although aging is a rather complex biological process with still poorly understood mechanism(s) of regulation.

Since the knowledge about aging throughout the years has unexpectedly become increasingly complicated, the initial quest for one overall encompassing theory explaining both the reasons and the intrinsic mechanisms of aging has given place to numerous processes, which may interact at multiple levels, for explaining this phenomenon (Dollemore and Aging 2002, Guarente 2014). Hence, the relatively young science of aging is now becoming of age, and with it has come the understanding of some of the underlying biochemical mechanisms at the core of the aging process (Yin and Chen 2005). These are the result of advanced analytical studies aimed at the observation and identification of the "subtle, quiet" age-related changes that occur in living organisms. In addition, new synthetic and medicinal chemistry methods are being developed, resulting in the availability of small molecule tools that may actively contribute to the extended elucidation of complex biological pathways, and, perhaps, pave the way for potential life span extending therapeutics (Ostler 2012).

In order to better understand how such experimental tools may help to extend the knowledge of the mechanisms of aging, it is necessary to first examine the basis on which these are developed. Hence, the prevalent theories of aging will be discussed in the following sections, focusing on the major biological, chemical, and pathological aspects of aging. This will hopefully make clear the current need to comprehend the different models of senescence and how only a systems approach will actively yield a new, integrative view of the aging process.

THEORIES OF AGING

What is aging? There have been multiple theories as to why most organisms (but not all, such as hydras) age. In 1990, Medvedev compiled, classified, and published the then existing theories of aging (Medvedev 1990). The number of theories exceeded 300 and was classified in seven distinct groups. Many of these theories have since been discredited or have undergone modifications and adaptations, while others have stayed. Broadly speaking, aging has been ascribed to free radical-induced damages (Harman 1993), telomere shortening (Kruk, Rampino, and Bohr 1995), molecular cross-linking (Bjorksten 1968), changes in immunological functions (Effros 2005), and senescence genes in the chromosomes (Warner et al. 1987). More recently, there have been attempts to develop a new, unified theory encompassing genes, the performance of genetic maintenance and repair systems and the surrounding environment, as well as chance (Rattan 2006), highlighting the need for a systematic and integrative view and understanding of the aging process. The prevailing theories of aging are discussed in detail in the following chapters.

It is almost impossible to provide an exhaustive list of all proposed theories of aging. Most of these can, however, be classified into three main categories: error theories, program hypotheses, and combined theories, the latter of which contains elements of both groups, as shown in Figure 1.1. Nonetheless, it should be emphasized that this classification is subjective and many others have been suggested (Weinert and Timiras 2003, de Magalhães 2005, Vina, Borras, and Miquel 2007, Jin 2010, Baltes, Rudolph, and Bal 2012).

Ultimately, all these theories, despite whatever described mechanism or their concomitant classifications are, aim at answering two questions: what is aging and why does it occur? The following sections, as well as the concurrent chapters in this book, clearly denote the assumption that no one

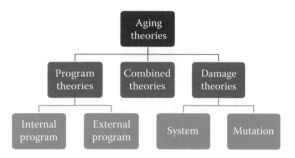

FIGURE 1.1 Categorization of the main theories of aging. (Classification based on the work developed by Semsei, I. 2000. *Mech Ageing Dev* 117 (1–3):93–108. http://dx.doi.org/10.1016/S0047-6374(00)00147-0 and de Magalhães, J. P. 2015. Senescence.info 2013 [cited October 2015]. Available from http://www.senescence.info.)

cause for aging is correct and, alas, there may not be one underlying mechanism or set of mechanisms for aging across species.

THE CHEMICAL INTERPLAY

The process of aging has been described as a raging war between chemical and biochemical processes (Clarke 2003), although a more suitable and accurate description might be that of a complex and rather interconnected gear mechanism (da Costa et al. 2016). On the basis of this perspective, aging is, at its core, the outcome of multiple unwanted chemical processes that result in spontaneous side products of normal metabolism, including less active, mutated, and perhaps toxified biomolecules such as lipids, proteins, DNA, RNA, and other small molecules (Clarke 2003). Ultimately, an organism's endurance reflects the extent to which it is capable of minimizing the accumulation of these modified biomolecules (Yin and Chen 2005). Minimization strategies are based on enzyme-mediated reactions, which are at the core of the metabolic pathways involved in biosynthesis, energy generation, and signal transduction (Vogel et al. 2004). Theoretically, optimizing these strategies could potentially make life considerably longer and, if sufficiently perfected, indefinite.

What does seem to work against biochemistry is chemistry itself (da Costa et al. 2016). Enzymes act as catalysts to speed up the biochemical reactions and it is difficult to slow them down (unless mutated or genetically modified). Hence, side reactions take place, and these result in the build-up of undesirable side products (Clarke 2003). Since most biomolecules are thermodynamically unstable (Ross and Subramanian 1981), they become susceptible to nonenzymatic conversion, which can impact orderly biochemical process, the fundamental reason at the heart of the damage-based theories of aging. There are reparation mechanisms, though such mechanisms are seldom 100% effective (Yin and Chen 2005).

Such pathways are summarized in Figure 1.2, which highlights the routes for degradation, but also for repair and replacement of aged proteins (Schiene and Fischer 2000, Grimaud et al. 2001, Ruan et al. 2002, Clarke 2003).

An example of the aforesaid processes is described in Figure 1.3, detailing the spontaneous chemical degradation of the aspartyl and asparaginyl residues in proteins and the methyltransferase-mediated repair mechanism.

Understanding the overall aging mechanisms and processes has become much more complicated than ever before. However, it seems clear that phenomena such as oxidative stress and associated damages are neither linked to the changes observed during aging nor correlated with maximum life span, due to the existence of the aforementioned repair and defense mechanisms (Sohal, Mockett, and Orr 2002). What may define the extension of the age-related modifications and, ultimately, the extension of life are the alterations that remain after repair, as well as the interactions of these with

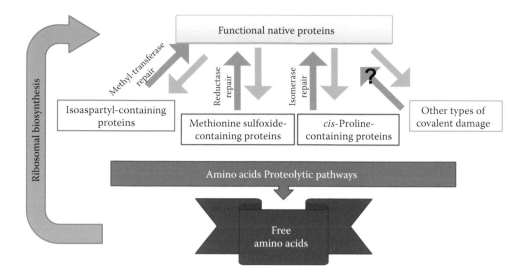

FIGURE 1.2 Pathways of nonenzymatic degradation, repair, and replacement of aged proteins. Functional proteins can be covalently altered by a number of pathways (light grey arrows). Enzymatic mechanisms exist that are capable of directly repairing, at least partially, this damage (dark grey arrows), though, so far, no repair mechanisms have been described for many other types of damage. Altered proteins can be proteolytically digested to free amino acids and these can be used for synthesizing new functional proteins ("proteolytic pathways" arrow). (Adapted from Clarke, S. 2003. *Ageing Res Rev* 2 (3):263–85.)

other biomolecules (da Costa et al. 2016). These appear to result in complex interconnected processes which scientists are continuously trying to unravel and fully comprehend.

SENESCENCE: WHAT CHANGES?

As already noted, aging is intrinsically complex. It encompasses multiple changes that take place at numerous levels of the biological hierarchy. There is no clear, undisputed evidence of which changes (molecular, cellular, or physiological) are the most important drivers of senescence and/or how they influence one another (da Costa et al. 2016). Each mechanism tends to be, in total or in part, supported by data indicating that they may play a role in the overall process. Nonetheless, the magnitude of an isolated mechanism is usually modest (Kirkwood 2011). Consequently, such restricted approaches may thwart the full appreciation of how different physiological, cellular, and/or molecular components interact with one another. An important effort to circumvent this limitation has been the development of the Digital Aging Atlas (http://ageing-map.org), which aims at integrating the multiplicity of reported age-related changes into a unified, freely accessible resource (Craig et al. 2015). Ultimately, the goal of an integrative approach will be the compilation of the acquired knowledge into a single depiction of how the aging process takes place (da Costa et al. 2016), ideally capable of characterizing the phenotype at a systemic/organism level (Cevenini et al. 2010). This goal can only be achieved through a multidisciplinary approach relying on the overall identification of key genes, biochemical pathways, and interactions that are involved in the aging process. Furthermore, it is also necessary to study inheritable genetic diseases that affect the "speed" of the aging process, often resulting in premature aging (e.g., Hutchinson–Gilford syndrome, commonly referred as progeria), and physiological experiments targeted at effectively decreasing this rate of aging, such as caloric intake restriction. Molecular and cellular biology will be decisive contributors in unveiling the changes that organisms undergo during senescence and the underlying causes. The multitude of available data, especially from high-throughput studies (de Magalhães, Curado, and Church 2009), will require a systems biology approach, making

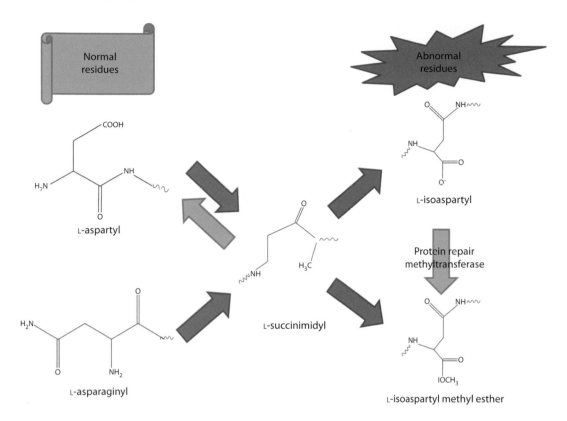

FIGURE 1.3 Pathways of nonenzymatic chemical degradation of aspartyl and asparaginyl residues in proteins and of the methyltransferase-mediated repair mechanism. Spontaneous degradation of normal L-aspartyl and L-asparaginyl residues leads to the formation of a ring succinimidyl intermediate. This can spontaneously hydrolyze to either the L-aspartyl residue or the abnormal L-isoaspartyl residue. The L-isoaspartyl residue is specifically recognized by the protein L-isoaspartate (D-aspartate) O-methyltransferase. The result is the formation of an unstable methyl ester that is converted back to L-succinimidyl. Net repair occurs when the L-succinimidyl residue is hydrolyzed to the L-aspartyl form. With the exception of the repair methyltransferase step, all the reactions are nonenzymatic. (Adapted from Clarke, S. 2003. *Ageing Res Rev* 2 (3):263–85.)

use of computational and mathematical modeling (for more details, see Chapter X in this book). Considering all these, it may become possible to decisively understand the "old problem of aging" (Hou et al. 2012).

In the next sections, some of the most notorious changes associated with aging that take place at the physiological and molecular levels, as well at the pathological level, will be discussed. It should be emphasized that others exist, though their exhaustive exploration would be unfeasible.

MOLECULAR CHANGES

Between 1/5th and 1/3rd of the total variation in the adult life span may be attributed to genetic variation, which makes a key feature for survival, particularly at advanced age (Hjelmborg et al. 2006). Consequently, multiple studies have focused on the elucidation of the genetic basis of senescence, looking for distinct "signatures" of the aging process. Many gene-centric studies have resulted in the identification of genes whose expression leads to significant alteration in senescent cells (Zhang 2007). Table 1.1 summarizes a selection of these studies, highlighting some of the most dramatic age-related changes that occur at the molecular level. Nonetheless, in spite of the clearly identified changes in gene expression, these studies have so far failed to unequivocally demonstrate if such

TABLE 1.1
Molecular Age-Related Changes

Measured Variable	Variation (%) p Value	Observations	Ref.
Calcium binding protein A8	↑1228.2% $p < 0.01$	Level of expression in the parietal lobe from those aged 69–99 years compared to cells from younger individuals (aged 20–52 years)	Cribbs et al. (2012)
Major histocompatibility complex, class II, DQ alpha 1	↑496.7% $p < 0.01$		
CD163 molecule	↑515.7% $p < 0.01$		
miR-320b	↑1049.9% $p = 1.33 \times 10^{-8}$	Level of expression from blood samples of the German elderly (mean age: 98.9 years) compared to younger controls (mean: 43.8 years)	ElSharawy et al. (2012)
miR-320d	↑529.9% $p = 6.93 \times 10^{-7}$		
miR-106a	↓98.7% $p = 1.20 \times 10^{-10}$		
Zinc-finger RNA binding protein	↑150.0% $p = 5.68 \times 10^{-3}$	Level of expression of TMEM33 in oocytes donated from women aged 37–39 years decreased compared to those from women aged 25–35 years	Grondahl et al. (2010)
Transmembrane protein 33	↓63.9% $p = 8.62 \times 10^{-3}$		
Zinc-finger RNA binding protein	↑251.4% $p = 5.67 \times 10^{-12}$	Significantly increased expression in foreskin cells from the elderly when compared with younger controls	Hackl et al. (2010)
miR-144	↑390.2% $p = 1.00 \times 10^{-25}$		
miR-100	↓49.1% $p = 1.28 \times 10^{-16}$		

Source: Data compiled based on the information available at the *Aging Digital Atlas* from Craig et al. 2015. The Digital Ageing Atlas: Integrating the diversity of age-related changes into a unified resource. *Nucleic Acids Res* 43 (Database issue):D873–8. doi: 10.1093/nar/gku843.

Note: The level of variation is also indicated, as well as the statistical significance.

alterations are unique and causal to senescence or if they are a nonspecific consequence of reduced or nonexistent cell proliferation. In addition, another limitation associated with such types of studies is that these are often carried out in animal models, which may have a limited contribution to the elucidation of the underlying aging mechanisms in humans, as senescence pathways are significantly different among cells from different species. For example, mouse fibroblasts express telomerase and display very long telomeres, in contrast to human fibroblasts (Kipling and Cooke 1990, Greenberg et al. 1998). When cultured, mouse fibroblasts undergo senescence, which takes place independent of telomere shortening (Sherr and DePinho 2000, Banito and Lowe 2013). Even within the same species, cells can exhibit significant differences in their senescence pathways (Zhang 2007). For instance, human fibroblasts undergo senescence after a finite number of divisions, although telomerase expression has been demonstrated to avoid this halt in proliferation (Yamashita et al. 2012). Contrarily, however, human mammary epithelial cells reach a growth arrest state that is not related to telomere shortening; rather, they are mediated by a tumor suppressing protein, p16 (Stampfer, LaBarge, and Garbe 2013). The importance of p16 in growth arrest was confirmed by immortalization of these cells by resorting to short hairpin RNA (shRNA), targeted at p16 (Stampfer et al. 2013). Taken as a whole, these data strongly suggest that there exist various pathways to senescence (Zhang 2007). In a post-genome era, where established and emerging *–omics* strategies and technologies allow for a more detailed and comprehensive characterization of the molecular changes that take place during aging (da Costa, Rocha-Santos, and Duarte 2016), it will be possible to link these changes to specific cellular and physiological processes (Craig et al. 2015), thus improving our understanding of aging. To date, different platforms and/or methodologies have yielded disparate results, and even the distinct nomenclature and definitions used can have an impact on the final conclusions (da Costa, Rocha-Santos, and Duarte 2016). This underscores the current need for creating and developing standardized methods for data acquisition and analysis. Despite many attempts (Weis 2005, Sun et al. 2013, Kohl et al. 2014, Zheng et al. 2015), these have so far failed to be universally implementable (da Costa et al. 2016). The volume of generated data also presents a challenge in the recent technological advances observed in *–omics* research. These allow for the simultaneous measurement of millions of biochemical entities (Zierer et al. 2015), which has led to reductionist association studies showing not only a high degree of correlation between *–omics* data and the aging process, but also with age-related diseases as well. Hence, it is becoming evident that integrated networks and *–omics* analyses focusing on a given biological process, such as aging, must be conducted at a systems level, resulting in what could be previously unattainable information, namely, the pathways involved and their interactions with both internal and external factors and variables (Valdes, Glass, and Spector 2013, Van Assche et al. 2015, Zierer et al. 2015).

Lastly, *–omics* research heavily relies on bioinformatics tools. Such software packages and databases are continuously updated, meaning that results should be constantly "revisited" (Zhou, Sha, and Guo 2016). Updated studies and results can benefit from new annotated information, and the original data could take into account previously unreported and valuable results, as evidenced by some studies disclosing new findings from previously published data (Matic, Ahel, and Hay 2012, Mann and Edsinger 2014). To date, no standard guidelines for "revisiting" are available, as well as no protocols for their comprehensive reanalysis exist, although some recent attempts have been carried out (Zhou, Sha, and Guo 2016). Establishing such procedural rules, guidelines, and protocols and successfully implementing them could yield potential key discoveries toward the understanding of relevant biological processes, namely, aging.

PHYSIOLOGICAL CHANGES

Physiological changes occur in all organ systems with aging. For example, aging is usually accompanied by a decrease in cardiac output and an increase in blood pressure, which often leads to arteriosclerosis. Movement impairment is also frequent, attributable to the degenerative changes that take place in joints combined with the loss of muscle mass (Boss and Seegmiller 1981). These observable changes have been the focus of various studies, and an attempt to summarize some of the most prominent works can be found in Table 1.2.

However, the compiled results found in Table 1.2 should be systematically analyzed and the described findings must be looked at critically. For example, Arking described a positive correlation between the hypodermal layer atrophy with age (Arking 2006); it should be emphasized that this was a "regional change," commonly affecting the face and back of hands but not other parts of the anatomy, such as the waist, leading to the possibility of being a symptom that may, at least to some extent, correlate with exposure. Another example is the generally held belief that there is a global neuron loss with age when, in fact, the difference in total neuron number between the ages of 20 and 90 is less than 10% (Pakkenberg et al. 2003, Pannese 2011). However, there are some morphological changes, namely, the significant decrease of synapses (Mostany et al. 2013), axon demyelination (Adamo 2014), or loss of dendritic spines (Dickstein et al. 2013).

PATHOLOGICAL CHANGES

Pathological changes, that is, the structural and functional changes produced during specific biological processes, are not always easily identifiable. What distinguishes pathological from "normal" age-related changes is, consequently, somewhat fleeting and not clearly defined. One example is the

TABLE 1.2
Physiological Age-Related Changes

Measured Variable	Tissue/Organ	Observations	References
Function of epithelial barriers	Lung; oral cavity; pharynx; esophagus; stomach; intestine; epidermis	Decreased epithelial barrier function; increased pathogenic invasion of mucosal tissues	Weiskopf, Weinberger, and Grubeck-Loebenstein (2009)
Modifications to proteins and membrane components	Lenses	Increased incidence of presbyopia and age-related nuclear cataract	Truscott and Zhu (2010)
Expiratory volume	Lung	Lung forced expiratory volume decreased with age	Klocke (1977)
Rates of neuronal and astroglial tricarboxylic acid cycles and neuroglial glutamate–glutamine cycling	Brain (mitochondria)	Neuronal mitochondrial metabolism and glutamate–glutamine cycle decreased ~30% in the elderly	Boumezbeur et al. (2010)
Protein level	Arterial intima	Collagen content of human arterial intima showed an average increase of 100%. Large increase in intimal embrittlement	Johnson et al. (1986)
Cell proliferation		Increase in intima cell proliferation	Chisolm and Steinberg (2000)
Prevalence of arteriosclerosis	Artery	Arteriosclerosis incidence increased with age	Wilkinson and McEniery (2012)
Cholesterol level	Plasma	Incidence increased in age brackets 45–64 and 65–74	CDC (2015)
Clonal mosaic abnormalities	Blood; oral cavity	Detectable clonal mosaic events increased with age	Jacobs et al. (2012)
Hematopoietic bone marrow volume	Bone marrow	Volume of hematopoietic bone marrow decreased with age	Sharma, Hanania, and Shim (2009)
Atrophy of hypodermal layer	Subcutaneous skin	Hypodermal layer underwent atrophy with age	Arking (2006)
White matter volume	Brain white matter	White matter volume decreased in individuals aged 59–85 years	Resnick, Lamar, and Driscoll (2007)

The affected tissue(s)/organ(s), as well as the measured variables, are listed.

normal, mild changes in neurologic functions that go with aging, and which do not substantially interfere with everyday activities, unless disease prevails (Morris and McManus 1991). There are, nonetheless, macroscopic, clearly visible changes in the aging brain, including the thickening of the arachnoid, the increased ventricular volume, and variable degrees of white and cortical matter atrophy, that are almost universal (Donahue 2012). It is also necessary to critically evaluate what may sometimes be reported as age-related pathological incidences. For example, a positive correlation between hip fracture incidence and age among postmenopausal women has been reported (Banks et al. 2009), although this may be a mere consequence of the reduced movement coordination and visual acuity observed in older ages. In other words, these impairments may result in a higher rate of falls and collisions, which in turn may have resulted in the reported high frequency of hip (and other) fractures. From these simple yet compelling examples, it is possible to infer that the age-related pathological changes may not be the result of single, one-time measurements, but rather a continuous observation, adequately including incidence and serial reports, has been highlighted in Table 1.3.

TABLE 1.3
Pathological Age-Related Changes

Measured Variable	Tissue/Organ	Observations	References
Cancer incidence	Multiple	Morbidity per 100,000 was >370× higher in 85 years old compared to individuals aged 18–24	CDC (2006)
Incidence of acute rheumatic fever and chronic rheumatic heart diseases	Heart; skin; brain	Morbidity per 100,000 was >165× higher in 85 years old than in individuals aged 25–44	
Coronary artery disease incidence	Heart; artery	Prevalence of coronary artery disease markedly increased with age	CDC (2006), Odden et al. (2011)
Chronic obstructive pulmonary disease (COPD) and small airway obstruction incidence	Lung	COPD increased with age, as well as small airway obstruction	Sharma, Hanania, and Shim (2009)
Incidence of presbyscusis (hearing loss)	Cochlea (inner ear)	Positive correlation of presbyscusis with age	Albert and Knoefel (2011)
Renal arteriosclerosis incidence	Kidney	Renal arteriosclerosis increase with age was noted	Bolignano et al. (2014), Glassock and Rule (2012)
Gastroesophageal reflux disease incidence	Esophagus; stomach	Incidence and severity of gastroesophageal reflux disease increased with age, particularly >50	Becher and Dent (2011)
Asthma incidence	Lung	Morbidity per 100,000 was >40× higher in individuals aged 85 years than in those aged 18–24 years	CDC (2006)
Clinical presentation and pathological staging in colorectal cancer	Colon	Older patients exhibited lower frequency of abdominal pain; time from onset to diagnosis and pathological staging were similar	Paganini Piazzolla et al. (2015)

PSYCHOLOGICAL CHANGES

Considerations pertaining to the psychology of aging inevitably lead to sociological considerations (Tischler 2013). Although concrete analyses yielding tangible results may be carried out, such as evaluating altered sleep patterns and measurements of cognitive deficits, psychological age-related changes are also intrinsically interlaced with the physiological changes and concomitant stress dynamics and coping mechanisms. Individually, as well as collectively, aging members of society "must learn to age" (da Costa et al. 2016). Such a need derives from the current mixed feelings toward the senior population demonstrated by Western societies. Although generally respected and appreciated, the popular culture is youth oriented: we strive to preserve our younger self, resorting to a wide variety of hyped age-delaying strategies, and are sometimes willing to undergo surgical procedures to look younger and healthy. As a result, older people are often considered to have some physical and/or mental degree of limitation, stereotypes that the ever-popular television does little to contradict (Lee, Carpenter, and Meyers 2007). There are, nonetheless, defined and quantifiable variations that take place at a psychological level with aging, such changes being observed in the functional neuroanatomy that result in alterations in speech production, as reported by Soros and coworkers (Soros et al. 2011). In Table 1.4, some of these quantifiable variations are listed.

CURING OF AGING

Before looking for a cure for aging, it is necessary to determine whether aging is in fact a disease. Aging is a process—no matter the cause or mechanism—that is characterized by various

TABLE 1.4
Psychological Age-Related Changes

Measured Variable	Observations	Ref.
Speech production	Speech production problems and reduced speech rate increased with aging	Soros et al. (2011)
Alterations in sleep patterns	Older individuals reported a higher number of awakenings and modifications in sleep duration and patterns	Feinberg, Koresko, and Heller (1967), Crowley (2011)
Subjective memory	Normal aging was found to be accompanied by memory impairment	Gazzaley et al. (2005)
Long-term depression	Individuals aged 65+ showed increased incidence of depression	Roblin (2015)
Cognitive decline	Cognitive decline was found to be almost universal in the general elderly population and increases with age	Park, O'Connell, and Thomson (2003), Schönknecht et al. (2014)
Cognitive processing speed	Processing speed decreases with age	Eckert (2011)
Cognitive executive functions	Executive functions (e.g., planning), decreased with age	Glisky (2007)
Verbal memory	Age-related differences were found in eight verbal span tasks	Bopp and Verhaeghen (2005)
Visual memory	Interaction of deficits in inhibition and processing speed was found to contribute to age-related cognitive impairment	Gazzaley et al. (2008)
Long-term potentiation	Greater and longer stimulation was necessary for long-term potentiation in older subjects	Kumar (2011)

pathologies, which inevitably lead to death, by the loss of homeostasis and the accumulation of molecular damage (Vijg and de Grey 2014). Disease, on the other hand, is defined as a "disorder or abnormality of structure and/or function" (Scully 2004), meaning that aging is not a disease, due to the fact that everyone suffers from it, though disease and aging often overlap (da Costa et al. 2016). The better suited question is perhaps "*should* we cure aging"? Many authors have expressed opposing opinions (Anton et al. 2005, Caplan 2005, Baars 2012, de Magalhães 2013, 2014, Vijg and de Grey 2014, Aledo and Blanco 2015). Those opposed often emphasize the obvious concerns about overpopulation and inequality, economic collapse due to increasing healthcare needs, and a perhaps "purist" view of biology, noting that aging is natural and should not be tampered with (de Magalhães 2013, 2014). On the other side of the spectrum, life extension research advocates underline that curing aging is not scientifically implausible and that we may soon reach the "longevity escape velocity" (de Grey 2004), a stage of medical progress at which the delay in age-related degeneration overcomes death to an extent that allows for additional research seeking more effective therapies later on (Vijg and de Grey 2014). These authors also dispute the alarms raised by others by noting the failed Malthusian predictions, which foresaw severe disasters caused by overpopulation (Trewavas 2002, Sethe and de Magalhães 2013).

Although such views are inherently personal and a matter of opinion, there should be little disagreement regarding the need to fight illnesses, including those that are age-related/caused and comorbidities (Longo et al. 2015). As detailed in subsequent chapters, great endeavors are being undertaken to prolong life, in spite of personal beliefs. Owing to the current knowledge on some of the well-studied mechanisms of aging, there has been considerable research and significant advances in delaying age have been achieved. For example, it has been known since the 1930s that restricting calories (caloric restriction [CR]) resulted in life extension in rodents (McCay 1935). Also related to CR, it has been shown that modern diets are largely based on heat-processed foods, which may result in the consumption of high quantities of advanced glycation end-products (AGEs) (Uribarri et al. 2015). These are proteins and lipids that covalently bond to sugars, without the controlling mechanisms of enzymes. AGEs can affect almost every type of cell in the organism and are thought to play a pivotal role in aging and in age-related illnesses.

Stem cells therapies have also been heralded as potential treatments of age-related diseases and rejuvenation. Recently, platelet-rich plasma was used for the recovery of stem cell senescence in mice (Liu et al. 2014) and the authors conjectured that the transplantation of restored stem cells in aged individuals could be achieved, which could be applied in the treatment of age-related diseases (Feng et al. 2007, Poulsen et al. 2013, Van Puyvelde et al. 2014). It is uncertain whether a low-calorie diet or a reduced intake of AGEs has a major effect in aging, though animal models suggest that the high levels of AGEs in the CR-high diet compete with the benefits of CR, although the mechanisms of action of these is still uncertain (Cai et al. 2008, da Costa et al. 2016). Consequently, there is the potential for using pharmacological agents that act as blockers of the cross-linking reactions leading to AGEs, or as blockers of their actions, namely, aminoguanidine (Thornalley 2003), benfotiamine (Stirban et al. 2006), metformin (Ishibashi et al. 2012), and inhibitors of the renin–angiotensin system (Zhenda et al. 2014), as well as aspirin (Urios, Grigorova-Borsos, and Sternberg 2007), to fight AGEs and their potentially deleterious consequences. Perhaps more notoriously, ALT-711 has received much attention as a next-generation antiaging agent. This compound acts by catalytically breaking AGE crosslinks and it has been demonstrated to have a beneficial effect in heart failure (Little et al. 2005), diabetic nephropathy (Thallas-Bonke et al. 2004), type II diabetes (Freidja et al. 2012), and age-associated ventricular and vascular stiffness (Steppan et al. 2012), among others (da Costa et al. 2016). Nonetheless, there is still the need to adequately and exhaustively determine the potential effects and side effects of these drugs. Hence, there may be a long time before such compounds emerge as safe and efficient agents with pharmacological and therapeutic actions against AGEs and, ultimately, aging itself (Luevano-Contreras and Chapman-Novakofski 2010).

Initially considered as only harmful to organisms, reactive oxygen species (ROS) have been recognized to contribute to cellular signaling and homeostasis (Lushchak 2014). These highly reactive chemical species, formed as a natural by-product of the normal metabolism of oxygen, do exert numerous damaging effects over lipids (Sharma et al. 2012), proteins (Youle and Van Der Bliek 2012), and nucleic acids (Ray, Huang, and Tsuji 2012). These are counteracted by a diverse array of endogenous cellular antioxidant systems, whose action may be enhanced by the ingestion of exogenous antioxidants (Rahman 2007). The most commonly recognized antioxidants by the general public are vitamins A, C, and E, as well as the coenzyme Q_{10}, the latter widely advertised in beauty creams (Prahl et al. 2008), though it has also been described as having a positive effect in the preservation of mitochondrial respiratory function in aged rat skeletal (Sugiyama, Yamada, and Ozawa 1995) and cardiac muscles (Park and Prolla 2005). Nonetheless, antioxidants do not delay or halt the aging process; rather, they contribute to increasing longevity (Holloszy 1998). More worrisome is the increasing commercialization of dietary supplements containing high concentrations of these compounds, as some have been implicated in the accelerated cancer development in mice (Sayin et al. 2014). Cumulatively, the intake of high-dose antioxidant supplements may in fact be more harmful than good (Bjelakovic et al. 2004, 2008, Combet and Buckton 2014), due to the fact that ROS play a role in cell signal and homeostasis, and, at some level, may have a positive role in life span (Lee, Hwang, and Kenyon 2010).

Telomere-based therapies have also been considered as potential avenues of research in antiaging therapies. It has been clearly demonstrated that telomere extension increased cellular proliferative capacity *in vitro* (Ramunas et al. 2015) and positively contributed to the reversal of tissue degeneration in mice (Jaskelioff et al. 2011). The hyped potential of such strategies, particularly by the media (Pollack 2011, Geddes and Macrae 2015, Knight 2015), is remaining unproven. This has not barred pharmaceutical companies from actively looking for viable age-prolonging telomerase-based therapies, such as TA-65®, a telomerase activator, already available (Harley et al. 2011). In spite of failing to increase life span, it has resulted in positive immune remodeling effects and beneficial outcomes over bone, cardiovascular, and metabolic health (Harley et al. 2013). Again, some precaution is required, as enhanced telomerase expression has been closely associated with cell proliferation and tumor growth (Peterson, Mok, and Au 2015).

Another potential direction for fighting the aging process is resorting to hormonal-based therapies, as patients with growth hormone (GH) and insulin-like growth factor 1 (IGF-1) deficiencies exhibit signs of early aging (Anisimov and Bartke 2013, Vanhooren and Libert 2013). There are some experimental data reporting on the benefits of human GH having beneficial effects in the elderly (Taub, Murphy, and Longo 2010) and hGH supplements have had a positive effect over muscle mass and the strengthening of the immune system (de Magalhães 2013). However, there have also been reports detailing the effects of such supplements over body composition and metabolism (Carroll et al. 1998), also resulting in high blood and intracranial pressure (Malozowski et al. 1993) and the development of diabetes (Lewis et al. 2013).

Certain additional approaches are also presently being studied as viable alternatives for delaying the aging process. These include the use of rapamycin, an immunosuppressant which has been shown to extend maximal life span in male mice (Neff et al. 2013), although with limited effects on the consequences of aging. However, rapamycin exhibits serious side effects, including nephrotoxicity (Murgia, Jordan, and Kahan 1996), a severe decrease in platelet numbers (thrombocytopenia) (Sacks 1999), and a steep elevation in the levels of lipids (hyperdyslipidemia) (Stallone et al. 2009).

Another potential route of intervention is through the manipulation of the *klotho* gene. This gene codes for one membrane protein and one secreted transcript, which acts as a circulating hormone (da Costa et al. 2016). Mutations in the *klotho* gene have resulted in accelerated aging in mice (Kuro-o et al. 1997) and the overexpression of *klotho* has been accompanied by an extension in life span by about 30% (Kurosu et al. 2005). The action mechanism of this gene remains unclear (the insulin/IGF-1 signaling pathways may be involved [Tsujikawa et al. 2003]), though more research is needed to confirm the role of the *klotho* gene in the aging process, which also holds true for other genes, which have also been described as having a role in the aging process(es) (Hackl et al. 2010, ElSharawy et al. 2012, Klement et al. 2012, Zhong et al. 2015).

However, the most futuristic antiaging therapy is that based on the use of nanotechnology. In spite of the many promises that nanotechnology still holds in a vast array of applications, the nanotechnology-based biomedical therapies remain elusive, as they entail a level of technological advances that will be a reality in the near future, but are not yet fully available. There are, nonetheless, promising works, such as those by Agostini and coworkers who reported the development of a nanodevice consisting of capped silica nanoparticles devised to selectively release drugs in aged human cells (Agostini et al. 2012), with an enormous potential in the treatment of, for example, cancer and Alzheimer's. Hence, such nanostructures may, in the future, be able to drive chemical reactions that can ultimately result in the slowing down of the aging process or even completely revert senescence. For now, however, the science of antiaging is very much in its infancy and the road to longevity is still long (da Costa et al. 2016). Nonetheless, the multifactorial nature of the aging process suggests that the long searched Fountain of Youth (Grene 2010) will remain out of our collective reach for quite some time.

CONCLUSIONS

Aging is one of the most complex biological mechanisms, and theories for this process abound. They can generally be grouped into program theories, damage theories, or combined theories, the latter of which encompass elements of both program and damage theories. Despite the ongoing discussion, the debate still continues over not only the prevailing mechanism(s), but also the reasons for aging. It is perhaps worth noting that the most effective way of disproving a theory is based on finding negative evidences for it. In other words, when there are facts contrary to the proposed theory, then the theory is no longer valid. However, for aging, this may prove to be a flawed approach.

Aging is intrinsically complex, as it not only depends on multiple aspects, such as lifestyle, diet, and genetic factors, among others, but may very well rely on multiple biological and chemical processes, as well as on their interactions.

During the past decades, there have been increasing reports demonstrating that aging is not an irreversible process. Some of the underlying mechanisms of aging have been elucidated and this has led to the development of promising antiaging therapies. However, most of these life extension mechanisms have been observed in simpler organisms and their translation into viable mammalian and, ultimately, human therapies have yet to be demonstrated. Biogerontologists are now aware of the inherent interconnectivity of senescence, and research into aging is a thriving field of research. Eventually, we will fully understand the aging process, which may help us to answer a larger question: what is life?

ACKNOWLEDGMENTS

This work was funded by the Portuguese Science Foundation (FCT) through the SFRH/BPD/122538/2016 research fellowship under POCH funds, co-financed by the European Social Fund and Portuguese National Funds from MEC. This work was also funded by national funds through FCT/MEC (PIDDAC) under project IF/00407/2013/CP1162/CT0023. Thanks are also due for the financial support to CESAM (UID/AMB/50017), and to co-funding by the FEDER, within the PT2020 Partnership Agreement and Compete 2020.

The author declares no conflict of interest.

REFERENCES

Adamo, A. M. 2014. Nutritional factors and aging in demyelinating diseases. *Genes Nutr* 9 (1):360. doi: 10.1007/s12263-013-0360-8.

Agostini, A., L. Mondragón, A. Bernardos, R. Martínez-Máñez, M. Dolores Marcos, F. Sancenón, J. Soto et al. 2012. Targeted cargo delivery in senescent cells using capped mesoporous silica nanoparticles. *Angew Chem Int Ed* 51 (42):10556–60. doi: 10.1002/anie.201204663.

Albert, M. L., and J. E. Knoefel. 2011. *Clinical Neurology of Aging*. Oxford University Press, Oxford, UK.

Aledo, J. C., and J. M. Blanco. 2015. Aging is neither a failure nor an achievement of natural selection. *Curr Aging Sci* 8 (1):4–10.

Anisimov, V. N., and A. Bartke. 2013. The key role of growth hormone–insulin–IGF-1 signaling in aging and cancer. *Crit Rev Oncol Hematol* 87 (3):201–23. http://dx.doi.org/10.1016/j.critrevonc.2013.01.005.

Anton, B., L. Vitetta, F. Cortizo, and A. Sali. 2005. Can we delay aging? The biology and science of aging. *Ann N Y Acad Sci* 1057 (1):525–35. http://dx.doi.org/10.1196/annals.1356.040.

Arking, R. 2006. *Biology of Aging: Observations and Principles*. Oxford University Press, Oxford, UK.

Azemin, M. Z. C., D. K. Kumar, T. Y. Wong, J. J. Wang, P. Mitchell, R. Kawasaki, and H. Wu. 2012. Age-related rarefaction in the fractal dimension of retinal vessel. *Neurobiol Aging* 33 (1):194.e1–e4.

Baars, J. 2012. *Aging and the Art of Living*. Johns Hopkins University Press, Baltimore, MD, USA.

Baltes, B. B., C. W. Rudolph, and A. C. Bal. 2012. A review of aging theories and modern work perspectives. In *The Oxford Handbook of Work and Aging*, edited by W. C. Borman and J. W. Hedge, 117–36. doi: 10.1093/oxfordhb/9780195385052.013.0069.

Banito, A., and S. W. Lowe. 2013. A new development in senescence. *Cell* 155 (5):977–8. http://dx.doi.org/10.1016/j.cell.2013.10.050.

Banks, E., G. K. Reeves, V. Beral, A. Balkwill, B. Liu, and A. Roddam. 2009. Hip fracture incidence in relation to age, menopausal status, and age at menopause: Prospective analysis. *PLoS Med* 6 (11):e1000181. doi: 10.1371/journal.pmed.1000181.

Becher, A., and J. Dent. 2011. Systematic review: Ageing and gastro-oesophageal reflux disease symptoms, oesophageal function and reflux oesophagitis. *Aliment Pharmacol Ther* 33 (4):442–54.

Bjelakovic, G., D. Nikolova, R. G. Simonetti, and C. Gluud. 2004. Antioxidant supplements for prevention of gastrointestinal cancers: A systematic review and meta-analysis. *The Lancet* 364 (9441):1219–28. http://dx.doi.org/10.1016/S0140-6736(04)17138-9.

Bjelakovic, G., D. Nikolova, R. G. Simonetti, and C. Gluud. 2008. Systematic review: Primary and secondary prevention of gastrointestinal cancers with antioxidant supplements. *Aliment Pharmacol Ther* 28 (6):689–703. doi: 10.1111/j.1365-2036.2008.03785.x.

Bjorksten, J. 1968. The crosslinkage theory of aging. *J Am Geriatr Soc* 16 (4):408–27.

Bolignano, D., F. Mattace-Raso, E. J. G. Sijbrands, and C. Zoccali. 2014. The aging kidney revisited: A systematic review. *Ageing Res Rev* 14:65–80.

Bopp, K. L., and P. Verhaeghen. 2005. Aging and verbal memory span: A meta-analysis. *J Gerontol B Psychol Sci Soc Sci* 60 (5):P223–33.

Boss, G. R., and J. E. Seegmiller. 1981. Age-related physiological changes and their clinical significance. *West J Med* 135 (6):434–40.

Boumezbeur, F., G. F. Mason, R. A. de Graaf, K. L. Behar, G. W. Cline, G. I. Shulman, D. L. Rothman, and K. F. Petersen. 2010. Altered brain mitochondrial metabolism in healthy aging as assessed by in vivo magnetic resonance spectroscopy. *J Cereb Blood Flow Metab* 30 (1):211–21. doi: 10.1038/jcbfm.2009.197.

Cai, W., J. C. He, L. Zhu, X. Chen, F. Zheng, G. E. Striker, and H. Vlassara. 2008. Oral glycotoxins determine the effects of calorie restriction on oxidant stress, age-related diseases, and lifespan. *Am J Pathol* 173 (2):327–36. doi: 10.2353/ajpath.2008.080152.

Caplan, A. L. 2005. Death as an unnatural process. Why is it wrong to seek a cure for aging? *EMBO Rep* 6 (Spec No):S72–5. doi: 10.1038/sj.embor.7400435.

Carroll, P. V., E. R. Christ (the members of Growth Hormone Research Society Scientific Committee), B. Å. Bengtsson, L. Carlsson, J. S. Christiansen, D. Clemmons, R. Hintz et al. 1998. Growth hormone deficiency in adulthood and the effects of growth hormone replacement: A review. *J Clin Endocrinol Metab* 83 (2):382–95. doi: 10.1210/jcem.83.2.4594.

Cary, H., H. Davis, and G. Burges. 1852. *The Works of Plato: The Laws*. H.G. Bohn, Charlottesville, VA, USA.

CDC. 2006. Mortality by underlying and multiple cause, ages 18+: US, 1981–2006. edited by Center for Disease Control and Prevention. Center for Disease Control and Prevention.

CDC. 2015. Cholesterol Level, ages 20+: US, 1988–2012. edited by CDC. Centers for Disease Control and Prevention.

Cevenini, E., E. Bellavista, P. Tieri, G. Castellani, F. Lescai, M. Francesconi, M. Mishto et al. 2010. Systems biology and longevity: An emerging approach to identify innovative anti-aging targets and strategies. *Curr Pharm Des* 16 (7):802–13.

Chesser, B. 2015. *Senescence in Humans*. M L Books International, New Delhi, India.

Chisolm, G. M., and D. Steinberg. 2000. The oxidative modification hypothesis of atherogenesis: An overview. *Free Radic Biol Med* 28 (12):1815–26.

Clarke, S. 2003. Aging as war between chemical and biochemical processes: Protein methylation and the recognition of age-damaged proteins for repair. *Ageing Res Rev* 2 (3):263–85.

Combet, E., and C. Buckton. 2014. Micronutrient deficiencies, vitamin pills and nutritional supplements. *Medicine (United Kingdom)* 43 (2):66–72. doi: 10.1016/j.mpmed.2014.11.002.

Craig, T., C. Smelick, R. Tacutu, D. Wuttke, S. H. Wood, H. Stanley, G. Janssens et al. 2015. The Digital Ageing Atlas: Integrating the diversity of age-related changes into a unified resource. *Nucleic Acids Res* 43 (Database issue):D873–8. doi: 10.1093/nar/gku843.

Cribbs, D. H., N. C. Berchtold, V. Perreau, P. D. Coleman, J. Rogers, A. J. Tenner, and C. W. Cotman. 2012. Extensive innate immune gene activation accompanies brain aging, increasing vulnerability to cognitive decline and neurodegeneration: A microarray study. *J Neuroinflammation* 9:179. doi: 10.1186/1742-2094-9-179.

Crowley, K. 2011. Sleep and sleep disorders in older adults. *Neuropsychol Rev* 21 (1):41–53.

da Costa, J. P., T. Rocha-Santos, and A. C. Duarte. 2016. Analytical tools to assess aging in humans: The rise of geri-omics. *TRAC Trends Anal Chem* 80:204–12. http://dx.doi.org/10.1016/j.trac.2015.09.011.

da Costa, J. P., R. Vitorino, G. M. Silva, C. Vogel, A. C. Duarte, and T. Rocha-Santos. 2016. A synopsis on aging—Theories, mechanisms and future prospects. *Ageing Res Rev* 29:90–112. http://dx.doi.org/10.1016/j.arr.2016.06.005.

de Grey, A. D. N. J. 2004. Escape velocity: Why the prospect of extreme human life extension matters now. *PLoS Biol* 2 (6):e187. doi: 10.1371/journal.pbio.0020187.

de Magalhães, J. P. 2005. Open-minded scepticism: Inferring the causal mechanisms of human ageing from genetic perturbations. *Ageing Res Rev* 4 (1):1–22. http://dx.doi.org/10.1016/j.arr.2004.05.003.

de Magalhaes, J. P. 2014. The scientific quest for lasting youth: Prospects for curing aging. *Rejuvenation Res* 17 (5):458–67. doi: 10.1089/rej.2014.1580.

de Magalhães, J. P. 2015. Senescence.info 2013 [cited October 2015]. Available from http://www.senescence.info.

de Magalhães, J. P., J. Curado, and G. M. Church. 2009. Meta-analysis of age-related gene expression profiles identifies common signatures of aging. *Bioinformatics* 25 (7):875–81. doi: 10.1093/bioinformatics/btp073.

Dickstein, D. L., C. M. Weaver, J. I. Luebke, and P. R. Hof. 2013. Dendritic spine changes associated with normal aging. *Neuroscience* 251:21–32.

Dollemore, D., and National Institute on Aging. 2002. *Aging under the Microscope: A Biological Quest*. National Institutes of Health, National Institute on Aging, Office of Communications and Public Liaison, Ithaca, NY, USA.

Donahue, J. E. 2012. "Normal" and pathological changes with age in the brain. *Med Health R I* 95 (3):75–6.

Eckert, M. A. 2011. Slowing down: Age-related neurobiological predictors of processing speed. *Front Neurosci* 5:25.

Effros, R. B. 2005. Roy Walford and the immunologic theory of aging. *Immun Ageing* 2 (1):7.

ElSharawy, A., A. Keller, F. Flachsbart, A. Wendschlag, G. Jacobs, N. Kefer, T. Brefort et al. 2012. Genome-wide miRNA signatures of human longevity. *Aging Cell* 11 (4):607–16. doi: 10.1111/j.1474-9726.2012.00824.x.

Feinberg, I., R. L. Koresko, and N. Heller. 1967. EEG sleep patterns as a function of normal and pathological aging in man. *J Psychiatr Res* 5 (2):107–44.

Feng, J. X., F. F. Hou, M. Liang, G. B. Wang, X. Zhang, H. Y. Li, D. Xie, J. W. Tian, and Z. Q. Liu. 2007. Restricted intake of dietary advanced glycation end products retards renal progression in the remnant kidney model. *Kidney Int* 71 (9):901–911.

Freidja, M. L., K. Tarhouni, B. Toutain, C. Fassot, L. Loufrani, and D. Henrion. 2012. The AGE-breaker ALT-711 restores high blood flow-dependent remodeling in mesenteric resistance arteries in a rat model of type 2 diabetes. *Diabetes* 61 (6):1562–72.

Gazzaley, A., W. Clapp, J. Kelley, K. McEvoy, R. T. Knight, and M. D'Esposito. 2008. Age-related top-down suppression deficit in the early stages of cortical visual memory processing. *Proc Natl Acad Sci USA* 105 (35):13122–6. doi: 10.1073/pnas.0806074105.

Gazzaley, A., J. W. Cooney, J. Rissman, and M. D'Esposito. 2005. Top-down suppression deficit underlies working memory impairment in normal aging. *Nat Neurosci* 8 (10):1298–300.

Geddes, L., and F. Macrae. 2015. Why stress of divorce could make you age more quickly: Breakups, bereavements and unemployment can make body's genetic material deteriorate prematurely. *Daily Mail* http://www.dailymail.co.uk/health/article-3214862/Why-stress-divorce-make-age-quickly-Breakups-bereavements-unemployment-make-body-s-genetic-material-deteriorate-prematurely.html.

Glassock, R. J., and A. D. Rule. 2012. The implications of anatomical and functional changes of the aging kidney: With an emphasis on the glomeruli. *Kidney Int* 82 (3):270–7.

Glisky, E. L. 2007. Changes in cognitive function in human aging. In *Brain Aging: Models, Methods, and Mechanisms*, edited by D. R. Riddle, 3–20. Boca Raton (FL): CRC Press/Taylor & Francis.

Goldberger, A. L., C. K. Peng, and L. A. Lipsitz. 2002. What is physiologic complexity and how does it change with aging and disease? *Neurobiol Aging* 23 (1):23–6.

Greenberg, R. A., R. C. Allsopp, L. Chin, G. B. Morin, and R. A. DePinho. 1998. Expression of mouse telomerase reverse transcriptase during development, differentiation and proliferation. *Oncogene* 16 (13):1723–30.

Grene, D. 2010. *The History*. University of Chicago Press, Chicago, IL, USA.

Grimaud, R., B. Ezraty, J. K. Mitchell, D. Lafitte, C. Briand, P. J. Derrick, and F. Barras. 2001. Repair of oxidized proteins identification of a new methionine sulfoxide reductase. *J Biol Chem* 276 (52):48915–20.

Grondahl, M. L., C. Yding Andersen, J. Bogstad, F. C. Nielsen, H. Meinertz, and R. Borup. 2010. Gene expression profiles of single human mature oocytes in relation to age. *Hum Reprod* 25 (4):957–68. doi: 10.1093/humrep/deq014.

Guarente, L. 2014. Aging research—Where do we stand and where are we going? *Cell* 159 (1):15–9. http://dx.doi.org/10.1016/j.cell.2014.08.041.

Hackl, M., S. Brunner, K. Fortschegger, C. Schreiner, L. Micutkova, C. Muck, G. T. Laschober et al. 2010. miR-17, miR-19b, miR-20a, and miR-106a are down-regulated in human aging. *Aging Cell* 9 (2):291–6. doi: 10.1111/j.1474-9726.2010.00549.x.

Harley, C. B., W. Liu, M. Blasco, E. Vera, W. H. Andrews, L. A. Briggs, and J. M. Raffaele. 2011. A natural product telomerase activator as part of a health maintenance program. *Rejuvenation Res* 14 (1):45–56. doi: 10.1089/rej.2010.1085.

Harley, C. B., W. Liu, P. L. Flom, and J. M. Raffaele. 2013. A natural product telomerase activator as part of a health maintenance program: Metabolic and cardiovascular response. *Rejuvenation Res* 16 (5):386–95.

Harman, D. 1993. Free radical involvement in aging. *Drugs Aging* 3 (1):60–80. doi: 10.2165/00002512-199303010-00006.

Hausdorff, J. M., S. L. Mitchell, R. Firtion, C. K. Peng, M. E. Cudkowicz, J. Y. Wei, and A. L. Goldberger. 1997. Altered fractal dynamics of gait: Reduced stride-interval correlations with aging and Huntington's disease. *J Appl Physiol (1985)* 82 (1):262–9.

Hjelmborg, J. B., I. Iachine, A. Skytthe, J. W. Vaupel, M. McGue, M. Koskenvuo, J. Kaprio, N. L. Pedersen, and K. Christensen. 2006. Genetic influence on human lifespan and longevity. *Hum Genet* 119 (3):312–21. doi: 10.1007/s00439-006-0144-y.

Holloszy, J. O. 1998. Longevity of exercising male rats: Effect of an antioxidant supplemented diet. *Mech Ageing Dev* 100 (3):211–9.

Hou, L., J. Huang, C. D. Green, J. Boyd-Kirkup, W. Zhang, X. Yu, W. Gong, B. Zhou, and J. D. Han. 2012. Systems biology in aging: Linking the old and the young. *Curr Genomics* 13 (7):558–65. doi: 10.2174/138920212803251418.

Ishibashi, Y., T. Matsui, M. Takeuchi, and S. Yamagishi. 2012. Metformin inhibits advanced glycation end products (AGEs)-induced renal tubular cell injury by suppressing reactive oxygen species generation via reducing receptor for AGEs (RAGE) expression. *Horm Metab Res* 44 (12):891–5.

Jacobs, K. B., M. Yeager, W. Zhou, S. Wacholder, Z. Wang, B. Rodriguez-Santiago, A. Hutchinson et al. 2012. Detectable clonal mosaicism and its relationship to aging and cancer. *Nat Genet* 44 (6):651–8. doi: 10.1038/ng.2270.

Jaskelioff, M., F. L. Muller, J.-H. Paik, E. Thomas, S. Jiang, A. Adams, E. Sahin et al. 2011. Telomerase reactivation reverses tissue degeneration in aged telomerase deficient mice. *Nature* 469 (7328):102–6. doi: 10.1038/nature09603.

Jin, K. 2010. Modern biological theories of aging. *Aging Dis* 1 (2):72–4.

Johnson, W. T., G. Salanga, W. Lee, G. A. Marshall, A. L. Himelstein, S. J. Wall, and O. Horwitz. 1986. Arterial intimal embrittlement. A possible factor in atherogenesis. *Atherosclerosis* 59 (2):161–71.

Kaplan, D. T., M. I. Furman, S. M. Pincus, S. M. Ryan, L. A. Lipsitz, and A. L. Goldberger. 1991. Aging and the complexity of cardiovascular dynamics. *Biophys J* 59 (4):945–9.

Kipling, D., and H. J. Cooke. 1990. Hypervariable ultra-long telomeres in mice. *Nature* 347 (6291):400–402. doi:10.1038/347400a0.

Kirkwood, T. B. 2011. Systems biology of ageing and longevity. *Philos Trans R Soc Lond B Biol Sci* 366 (1561):64–70. doi: 10.1098/rstb.2010.0275.

Kirkwood, T. B. L. 2005. Understanding the odd science of aging. *Cell* 120 (4):437–47. http://dx.doi.org/10.1016/j.cell.2005.01.027.

Klement, K., C. Melle, U. Murzik, S. Diekmann, J. Norgauer, and P. Hemmerich. 2012. Accumulation of annexin A5 at the nuclear envelope is a biomarker of cellular aging. *Mech Ageing Dev* 133 (7):508–22. http://dx.doi.org/10.1016/j.mad.2012.06.003.

Klocke, R. A. 1977. Influence of aging in the lung. In *Handbook of the Biology of Aging*, edited by C. E. Finch and L. Hayflick, 432–44. New York: Van Nostrand Reinhold.

Knight, M. 2015. Buy your telomere testing kit here! Evidence based or pseudo-science? Genetic Literacy Project.

Kohl, M., D. A. Megger, M. Trippler, H. Meckel, M. Ahrens, T. Bracht, F. Weber et al. 2014. A practical data processing workflow for multi-OMICS projects. *Biochim Biophys Acta (BBA)—Proteins Proteom* 1844 (1, Part A):52–62. http://dx.doi.org/10.1016/j.bbapap.2013.02.029.

Kruk, P. A., N. J. Rampino, and V. A. Bohr. 1995. DNA damage and repair in telomeres: relation to aging. *Proc Natl Acad Sci USA* 92 (1):258–62.

Kumar, A. 2011. Long-term potentiation at CA3–CA1 hippocampal synapses with special emphasis on aging, disease, and stress. *Front Aging Neurosci* 3:7. doi: 10.3389/fnagi.2011.00007.

Kuro-o, M., Y. Matsumura, H. Aizawa, H. Kawaguchi, T. Suga, T. Utsugi, Y. Ohyama et al. 1997. Mutation of the mouse klotho gene leads to a syndrome resembling ageing. *Nature* 390 (6655):45–51. doi: 10.1038/36285.

Kurosu, H., M. Yamamoto, J. D. Clark, J. V. Pastor, A. Nandi, P. Gurnani, O. P. McGuinness et al. 2005. Suppression of aging in mice by the hormone klotho. *Science* 309 (5742):1829–33. doi: 10.1126/science.1112766.

Lee, M. M., B. Carpenter, and L. S. Meyers. 2007. Representations of older adults in television advertisements. *J Aging Stud* 21 (1):23–30.

Lee, S.-J., A. B. Hwang, and C. Kenyon. 2010. Inhibition of respiration extends *C. elegans* life span via reactive oxygen species that increase HIF-1 activity. *Curr Biol* 20 (23):2131–6. http://dx.doi.org/10.1016/j.cub.2010.10.057.

Leopardi, G., J. Thomson, and B. Dobell. 1905. *Essays, Dialogues and Thoughts: (Operette Morali and Pensieri) of Giacomo Leopardi*. G. Routledge & Sons, Limited, London, UK.

Lewis, U. J., R. N. P. Singh, G. F. Tutwiler, M. B. Sigel, E. F. Vander-Laan, and W. P. VanderLaan. 2013. Human Growth Hormone: A Complex of Proteins, In *Recent Progress in Hormone Research*, edited by R. O. Greep, 36:477–508. *Proceedings of the 1979 Laurentian Hormone Conference (Mont-Tremblant Canada)*, Academic Press: Boston. https://doi.org/10.1016/B978-0-12-571136-4.50019-X.

Little, W. C., M. R. Zile, D. W. Kitzman, W. Gregory Hundley, and T. X. O'Brien. 2005. The effect of alagebrium chloride (ALT-711), a novel glucose cross-link breaker, in the treatment of elderly patients with diastolic heart failure. *J Card Fail* 11 (3):191–5.

Liu, H.-Y., C.-F. Huang, T.-C. Lin, C.-Y. Tsai, S.-Y. T. Chen, A. Liu, W.-H. Chen et al. 2014. Delayed animal aging through the recovery of stem cell senescence by platelet rich plasma. *Biomaterials* 35 (37):9767–76. http://dx.doi.org/10.1016/j.biomaterials.2014.08.034.

Longo, V. D., A. Antebi, A. Bartke, N. Barzilai, H. M. Brown-Borg, C. Caruso, T. J. Curiel et al. 2015. Interventions to slow aging in humans: Are we ready? *Aging Cell* 14 (4):497–510. doi: 10.1111/acel.12338.

Lopez-Otin, C., M. A. Blasco, L. Partridge, M. Serrano, and G. Kroemer. 2013. The hallmarks of aging. *Cell* 153 (6):1194–217. doi: 10.1016/j.cell.2013.05.039.

Lucvano-Contreras, C., and K. Chapman-Novakofski. 2010. Dietary advanced glycation end products and aging. *Nutrients* 2 (12):1247–65.

Lushchak, V. I. 2014. Free radicals, reactive oxygen species, oxidative stress and its classification. *Chem Biol Interact* 224:164–75. http://dx.doi.org/10.1016/j.cbi.2014.10.016.

Malozowski, S., L. A. Tanner, D. Wysowski, and G. A. Fleming. 1993. Growth hormone, insulin-like growth factor I, and benign intracranial hypertension. *N Engl J Med* 329 (9):665–6. doi: 10.1056/nejm199308263290917.

Mann, K., and E. Edsinger. 2014. The *Lottia gigantea* shell matrix proteome: Re-analysis including MaxQuant iBAQ quantitation and phosphoproteome analysis. *Proteome Sci* 12 (1):28.

Manor, B., M. D. Costa, K. Hu, E. Newton, O. Starobinets, H. G. Kang, C. K. Peng, V. Novak, and L. A. Lipsitz. 2010. Physiological complexity and system adaptability: Evidence from postural control dynamics of older adults. *J Appl Physiol* 109 (6):1786–91.

Matic, I., I. Ahel, and R. T. Hay. 2012. Reanalysis of phosphoproteomics data uncovers ADP-ribosylation sites. *Nat Meth* 9 (8):771–2. http://www.nature.com/nmeth/journal/v9/n8/abs/nmeth.2106.html#supplementary-information.

McCay, C. M. 1935. Iodized salt a hundred years ago. *Science* 82 (2128):350–1. doi: 10.1126/science.82.2128.350-a.

Medvedev, Z. A. 1990. An attempt at a rational classification of theories of ageing. *Biol Rev Camb Philos Soc* 65 (3):375–98.

Morris, J. C., and D. Q. McManus. 1991. The neurology of aging: Normal versus pathologic change. *Geriatrics* 46 (8):47–8, 51–4.

Mostany, R., J. E. Anstey, K. L. Crump, B. Maco, G. Knott, and C. Portera-Cailliau. 2013. Altered synaptic dynamics during normal brain aging. *J Neurosci* 33 (9):4094–104.

Murgia, M. G., S. Jordan, and B. D. Kahan. 1996. The side effect profile of sirolimus: A phase I study in quiescent cyclosporine-prednisone-treated renal transplant patients. *Kidney Int* 49 (1):209–16.

Neff, F., D. Flores-Dominguez, D. P. Ryan, M. Horsch, S. Schröder, T. Adler, L. C. Afonso et al. 2013. Rapamycin extends murine lifespan but has limited effects on aging. *J Clin Invest* 123 (8):3272–91. doi: 10.1172/JCI67674.

Odden, M. C., P. G. Coxson, A. Moran, J. M. Lightwood, L. Goldman, and K. Bibbins-Domingo. 2011. The impact of the aging population on coronary heart disease in the United States. *Am J Med* 124 (9):827–33. e5. doi: 10.1016/j.amjmed.2011.04.010.

Ostler, E. 2012. Chemistry of ageing. *Chemistry Central*, 2 August 2012 [cited October 2015]. Available from https://www.biomedcentral.com/series/Chemistry_of_Ageing.

Paganini Piazzolla, L., R. Medeiros de Almeida, A. C. Nóbrega dos Santos, P. Gonçalves de Oliveira, E. Freitas da Silva, and J. Batista de Sousa. 2015. Does aging influence clinical presentation and pathological staging in colorectal cancer? *Eur Geriatr Med* 6 (5):433–6. http://dx.doi.org/10.1016/j.eurger.2015.04.007.

Pakkenberg, B., D. Pelvig, L. Marner, M. J. Bundgaard, H. J. Gundersen, J. R. Nyengaard, and L. Regeur. 2003. Aging and the human neocortex. *Exp Gerontol* 38 (1–2):95–9.

Pannese, E. 2011. Morphological changes in nerve cells during normal aging. *Brain Struct Funct* 216 (2):85–9. doi: 10.1007/s00429-011-0308-y.

Park, H. L., J. E. O'Connell, and R. G. Thomson. 2003. A systematic review of cognitive decline in the general elderly population. *Int J Geriatr Psychiatry* 18 (12):1121–34. doi: 10.1002/gps.1023.

Park, S.-K., and T. A. Prolla. 2005. Gene expression profiling studies of aging in cardiac and skeletal muscles. *Cardiovasc Res* 66 (2):205–12.

Peng, C.-K., J. E. Mietus, Y. Liu, C. Lee, J. M. Hausdorff, H. Eugene Stanley, A. L. Goldberger, and L. A Lipsitz. 2002. Quantifying fractal dynamics of human respiration: Age and gender effects. *Ann Biomed Eng* 30 (5):683–92.

Peterson, D. R., H. O. Mok, and D. W. Au. 2015. Modulation of telomerase activity in fish muscle by biological and environmental factors. *Comp Biochem Physiol C Toxicol Pharmacol* 178:51–9. doi: 10.1016/j.cbpc.2015.09.004.

Pollack, A. 2011. A blood test offers clues to longevity. *New York Times*. http://www.nytimes.com/2011/05/19/business/19life.html.

Poulsen, M. W., R. V. Hedegaard, J. M. Andersen, B. de Courten, S. Bügel, J. Nielsen, L. H. Skibsted, and L. O. Dragsted. 2013. Advanced glycation endproducts in food and their effects on health. *Food Chem Toxicol* 60:10–37.

Prahl, S., T. Kueper, T. Biernoth, Y. Wöhrmann, A. Münster, M. Fürstenau, M. Schmidt, C. Schulze, K.-P. Wittern, and H. Wenck. 2008. Aging skin is functionally anaerobic: Importance of coenzyme Q10 for anti aging skin care. *Biofactors* 32 (1–4):245–55.

Rahman, K. 2007. Studies on free radicals, antioxidants, and co-factors. *Clin Interv Aging* 2 (2):219–36.

Ramunas, J., E. Yakubov, J. J. Brady, S. Y. Corbel, C. Holbrook, M. Brandt, J. Stein, J. G. Santiago, J. P. Cooke, and H. M. Blau. 2015. Transient delivery of modified mRNA encoding TERT rapidly extends telomeres in human cells. *FASEB J* 29 (5):1930–9.

Rattan, S. I. S. 2006. Theories of biological aging: Genes, proteins, and free radicals. *Free Radic Res* 40 (12):1230–8. doi: 10.1080/10715760600911303.

Ray, P. D., B.-W. Huang, and Y. Tsuji. 2012. Reactive oxygen species (ROS) homeostasis and redox regulation in cellular signaling. *Cell Signal* 24 (5):981–90.

Resnick, S. M., M. Lamar, and I. Driscoll. 2007. Vulnerability of the orbitofrontal cortex to age-associated structural and functional brain changes. *Ann N Y Acad Sci* 1121:562–75. doi: 10.1196/annals.1401.027.

Roblin, J. 2015. Les dépressions du sujet âgé: du diagnostic à la prise en charge. *NPG Neurologie—Psychiatrie—Gériatrie* 15 (88):206–18. http://dx.doi.org/10.1016/j.npg.2014.11.001.

Ross, P. D., and S. Subramanian. 1981. Thermodynamics of protein association reactions: Forces contributing to stability. *Biochemistry* 20 (11):3096–102.

Ruan, H., X. D. Tang, M.-L. Chen, M. A. Joiner, G. Sun, N. Brot, H. Weissbach, S. H. Heinemann, L. Iverson, and C.-F. Wu. 2002. High-quality life extension by the enzyme peptide methionine sulfoxide reductase. *Proc Natl Acad Sci USA* 99 (5):2748–53.

Sacks, S. H. 1999. Rapamycin on trial. *Nephrol Dial Transplant* 14 (9):2087–9. doi: 10.1093/ndt/14.9.2087.

Sayin, V. I., M. X. Ibrahim, E. Larsson, J. A. Nilsson, P. Lindahl, and M. O. Bergo. 2014. Antioxidants accelerate lung cancer progression in mice. *Sci Transl Med* 6 (221):221ra15. doi: 10.1126/scitranslmed.3007653.

Schiene, C., and G. Fischer. 2000. Enzymes that catalyse the restructuring of proteins. *Curr Opin Struct Biol* 10 (1):40–5.

Schönknecht, P., J. Pantel, A. Kruse, and J. Schröder. 2014. Prevalence and natural course of aging-associated cognitive decline in a population-based sample of young-old subjects. *Am J Psychiatry* 162 (11):2071–2077.

Schumann, A. Y., R. P. Bartsch, T. Penzel, P. Ch Ivanov, and J. W. Kantelhardt. 2010. Aging effects on cardiac and respiratory dynamics in healthy subjects across sleep stages. *Sleep* 33 (7):943.

Scully, J. L. 2004. What is a disease? *EMBO Rep* 5 (7):650–3. doi: 10.1038/sj.embor.7400195.

Semsei, I. 2000. On the nature of aging. *Mech Ageing Dev* 117 (1–3):93–108. http://dx.doi.org/10.1016/S0047-6374(00)00147-0.

Sethe, S., and J. P. de Magalhães. 2013. Ethical perspectives in biogerontology. In *Ethics, Health Policy and (Anti-) Aging: Mixed Blessings*, edited by M. Schermer and W. Pinxten, 173–88. Springer, Dordrecht, The Netherlands.

Sharma, G., N. A. Hanania, and Y. M. Shim. 2009. The aging immune system and its relationship to the development of chronic obstructive pulmonary disease. *Proc Am Thorac Soc* 6 (7):573–80. doi: 10.1513/pats.200904-022RM.

Sharma, P., A. B. Jha, R. S. Dubey, and M. Pessarakli. 2012. Reactive oxygen species, oxidative damage, and antioxidative defense mechanism in plants under stressful conditions. *J Bot*. https://www.hindawi.com/journals/jb/2012/217037/.

Sherr, C. J., and R. A. DePinho. 2000. Cellular senescence: Minireview mitotic clock or culture shock? *Cell* 102 (4):407–10.

Sohal, R. S., R. J. Mockett, and W. C. Orr. 2002. Mechanisms of aging: An appraisal of the oxidative stress hypothesis. *Free Radic Biol Med* 33 (5):575–86. http://dx.doi.org/10.1016/S0891-5849(02)00886-9.

Soros, P., A. Bose, L. G. Sokoloff, S. J. Graham, and D. T. Stuss. 2011. Age-related changes in the functional neuroanatomy of overt speech production. *Neurobiol Aging* 32 (8):1505–13. doi: 10.1016/j.neurobiolaging.2009.08.015.

Stallone, G., B. Infante, G. Grandaliano, and L. Gesualdo. 2009. Management of side effects of sirolimus therapy. *Transplantation* 87 (8S):S23–6.

Stampfer, M., L. Vrba, L. Fuchs, A. Brothman, M. LaBarge, B. Futscher, and J. Garbe. 2013. Abstract B008: Efficient immortalization of normal human mammary epithelial cells using two pathologically relevant agents does not require gross genomic alterations. *Mol Cancer Res* 11 (10 Supplement):B008–B008.

Stampfer, M. R., M. A. LaBarge, and J. C. Garbe. 2013. An integrated human mammary epithelial cell culture system for studying carcinogenesis and aging. In *Cell and Molecular Biology of Breast Cancer*, edited by H. Schatten. Humana Press. doi: 10.1007/978-1-62703-634-4, 323–61. Springer.

Steppan, J., H. Tran, A. M. Benjo, L. Pellakuru, V. Barodka, S. Ryoo, S. M. Nyhan, C. Lussman, G. Gupta, and A. R. White. 2012. Alagebrium in combination with exercise ameliorates age-associated ventricular and vascular stiffness. *Exp Gerontol* 47 (8):565–72.

Stirban, A., M. Negrean, B. Stratmann, T. Gawlowski, T. Horstmann, C. Götting, K. Kleesiek, M. Mueller-Roesel, T. Koschinsky, and J. Uribarri. 2006. Benfotiamine prevents macro-and microvascular endothelial dysfunction and oxidative stress following a meal rich in advanced glycation end products in individuals with type 2 diabetes. *Diabetes Care* 29 (9):2064–71.

Sugiyama, S., K. Yamada, and T. Ozawa. 1995. Preservation of mitochondrial respiratory function by coenzyme Q10 in aged rat skeletal muscle. *Biochem. Mol. Biol. Int.* 37 (6):1111–20.

Sun, H., H. Wang, R. Zhu, K. Tang, Q. Gong, J. Cui, Z. Cao, and Q. Liu. 2013. iPEAP: Integrating multiple omics and genetic data for pathway enrichment analysis. *Bioinformatics* 30 (5):737–739. doi: 10.1093/bioinformatics/btt576.

Taub, D. D., W. J. Murphy, and D. L. Longo. 2010. Rejuvenation of the aging thymus: Growth hormone-mediated and ghrelin-mediated signaling pathways. *Curr Opin Pharmacol* 10 (4):408–24.

Terrier, P., and O. Dériaz. 2011. Kinematic variability, fractal dynamics and local dynamic stability of treadmill walking. *J Neuroeng Rehabil* 8 (1):12.

Thallas-Bonke, V., C. Lindschau, B. Rizkalla, L. A. Bach, G. Boner, M. Meier, H. Haller, M. E. Cooper, and J. M. Forbes. 2004. Attenuation of extracellular matrix accumulation in diabetic nephropathy by the advanced glycation end product cross-link breaker ALT-711 via a protein kinase C-α-dependent pathway. *Diabetes* 53 (11):2921–30.

Thornalley, P. J. 2003. Use of aminoguanidine (Pimagedine) to prevent the formation of advanced glycation endproducts. *Arch Biochem Biophys* 419 (1):31–40.

Tischler, H. 2013. *Cengage Advantage Books: Introduction to Sociology*. Cengage Learning, Independence, KY, USA.

Trewavas, A. 2002. Malthus foiled again and again. *Nature* 418 (6898):668–70. doi: 10.1038/nature01013.

Truscott, R. J. W., and X. Zhu. 2010. Presbyopia and cataract: A question of heat and time. *Prog Retin Eye Res* 29 (6):487–99. http://dx.doi.org/10.1016/j.preteyeres.2010.05.002

Tsujikawa, H., Y. Kurotaki, T. Fujimori, K. Fukuda, and Y. Nabeshima. 2003. Klotho, a gene related to a syndrome resembling human premature aging, functions in a negative regulatory circuit of vitamin D endocrine system. *Mol Endocrinol* 17 (12):2393–403. doi: 10.1210/me.2003-0048.

Uribarri, J., M. D. del Castillo, M. P. de la Maza, R. Filip, A. Gugliucci, C. Luevano-Contreras, M. H. Macias-Cervantes et al. 2015. Dietary advanced glycation end products and their role in health and disease. *Adv Nutr* 6 (4):461–73. doi: 10.3945/an.115.008433.

Urios, P., A.-M. Grigorova-Borsos, and M. Sternberg. 2007. Aspirin inhibits the formation of pentosidine, a cross-linking advanced glycation end product, in collagen. *Diabetes Res Clin Pract* 77 (2):337–40.

Valdes, A. M., D. Glass, and T. D. Spector. 2013. Omics technologies and the study of human ageing. *Nat Rev Genet* 14 (9):601–7.

Van Assche, R., V. Broeckx, K. Boonen, E. Maes, W. De Haes, L. Schoofs, and L. Temmerman. 2015. Integrating-omics: Systems biology as explored through *C. elegans* research. *J Mol Biol* 427 (21):3441–51. http://dx.doi.org/10.1016/j.jmb.2015.03.015.

Vanhooren, V., and C. Libert. 2013. The mouse as a model organism in aging research: Usefulness, pitfalls and possibilities. *Ageing Res Rev* 12 (1):8–21. http://dx.doi.org/10.1016/j.arr.2012.03.010.

Van Puyvelde, K., T. Mets, R. Njemini, I. Beyer, and I. Bautmans. 2014. Effect of advanced glycation end product intake on inflammation and aging: A systematic review. *Nutr Rev* 72 (10):638–50.

Vasto, S., G. Scapagnini, M. Bulati, G. Candore, L. Castiglia, G. Colonna-Romano, D. Lio et al. 2010. Biomarkes of aging. *Front Biosci (Schol Ed)* 2:392–402.

Vijg, J., and A. D. N. J. de Grey. 2014. Innovating aging: Promises and pitfalls on the road to life extension. *Gerontology* 60 (4):373–80.

Vina, J., C. Borras, and J. Miquel. 2007. Theories of ageing. *IUBMB Life* 59 (4–5):249–54. doi: 10.1080/15216540601178067.

Vogel, C., M. Bashton, N. D. Kerrison, C. Chothia, and S. A. Teichmann. 2004. Structure, function and evolution of multidomain proteins. *Curr Opin Struct Biol* 14 (2):208–16.

Warner, H. R., R. L. Sprott, E. L. Schneider, and R. N. Butler. 1987. Modern biological theories of aging. In: *Related Information Volume 31. Aging.*

Weinert, B. T., and P. S. Timiras. 2003. Invited review: Theories of aging. *J Appl Physiol* 95 (4):1706–16.

Weis, B. K. 2005. Standardizing global gene expression analysis between laboratories and across platforms. *Nat Methods* 2 (5):351–6.

Weiskopf, D., B. Weinberger, and B. Grubeck-Loebenstein. 2009. The aging of the immune system. *Transpl Int* 22 (11):1041–50. doi: 10.1111/j.1432-2277.2009.00927.x.

Wilkinson, I. B., and C. M. McEniery. 2012. Arteriosclerosis: Inevitable or self-inflicted? *Hypertension* 60 (1):3–5. doi: 10.1161/hypertensionaha.112.193029.

Yamashita, S., K. Ogawa, T. Ikei, M. Udono, T. Fujiki, and Y. Katakura. 2012. SIRT1 prevents replicative senescence of normal human umbilical cord fibroblast through potentiating the transcription of human telomerase reverse transcriptase gene. *Biochem Biophys Res Commun* 417 (1):630–4.

Yin, D., and K. Chen. 2005. The essential mechanisms of aging: Irreparable damage accumulation of biochemical side-reactions. *Exp Gerontol* 40 (6):455–65.

Youle, R. J., and A. M. Van Der Bliek. 2012. Mitochondrial fission, fusion, and stress. *Science* 337 (6098):1062–5.

Zhang, H. 2007. Molecular signaling and genetic pathways of senescence: Its role in tumorigenesis and aging. *J Cell Physiol* 210 (3):567–74. doi: 10.1002/jcp.20919.

Zhenda, Z., C. Cailian, D. Ruimin, Q. Xiaoxian, and C. Lin. 2014. GW25-e3403 Advanced glycation end products upregulated the expression of angiopoietin-like protein 4 via activation the renin-angiotensin system in endothelials. *J Am Coll Cardiol* 64 (16_S).

Zheng, C. L., V. Ratnakar, Y. Gil, and S. K. McWeeney. 2015. Use of semantic workflows to enhance transparency and reproducibility in clinical omics. *Genome Med* 7 (1):73. doi: 10.1186/s13073-015-0202-y.

Zhong, Y., J. Zhao, Y. J. Gu, Y.-F. Zhao, Y.-W. Zhou, and G.-X. Fu. 2015. Differential levels of cathepsin B and L in serum between young and aged healthy people and their association with matrix metalloproteinase 2. *Arch Gerontol Geriatr* 61 (2):285–8. http://dx.doi.org/10.1016/j.archger.2015.04.010.

Zhou, T., J. Sha, and X. Guo. 2016. The need to revisit published data: A concept and framework for complementary proteomics. *Proteomics* 16 (1):6–11. doi: 10.1002/pmic.201500170.

Zierer, J., C. Menni, G. Kastenmüller, and T. D. Spector. 2015. Integration of 'omics' data in aging research: From biomarkers to systems biology. *Aging Cell* 14 (6):933–44. doi: 10.1111/acel.12386.

2 Understanding Aging after Darwin

Michael A. Singer

CONTENTS

INTRODUCTION

When Darwin published his seminal work in 1859, he laid the foundations of the evolutionary process and in so doing outlined a mechanism that could account for the diversity of animal species in both form and function. Darwin was an empiricist and his proposal was based upon innumerable measurements and observations; the cornerstone of his theory was the principle of natural selection acting on variation in individual phenotypes within a species. To quote Darwin (1859, p. 61): "Owing to this struggle for life, any variation, however slight and from whatever cause proceeding, if it be in any degree profitable to an individual of any species, in its infinitely complex relations to other organic beings and to external nature, will tend to the preservation of that individual, and will generally be inherited by its offspring…. I have called this principle, by which even slight variation, if useful, is preserved, by the term of Natural Selection, in order to mark its relation to man's power of selection."

The fitness of an individual organism in the context of its struggle for life is generally defined as survival to the age of reproductive maturation with production of offspring.

The key ingredients of Darwin's evolutionary recipe were: variation in individual traits with some variants conferring a fitness advantage in the competition with other individuals and with respect to environmental pressures, heritability of advantageous variants making them transmissible across successive generations, and the action of natural selection. Gould (1980) noted that evolution works on several levels. At the foundational level, genes are the building blocks of species since they supply the variation upon which natural selection operates. Individual organisms are the unit of selection but individuals do not evolve. Individuals only grow, reproduce, and die. The actual unit of evolution is a group of interacting organisms defined as a species.

In Darwin's construct, the target of natural selection was the individual organism or, more specifically, phenotypic variation between individual organisms. Multilevel selection was not considered. Interactions between species were not included; a species evolved independently of the presence or absence of other species or the nature of coexisting species. Changes in the environment altered the "choices" natural selection made as to what constituted an advantageous phenotype, but the species–environment interaction was treated as being unidirectional. There was no evolutionary role attributed to species-mediated environmental modifications. However, as discussed in Chapter 7, organisms and species reside in ecosystems which are characterized by mutually advantageous interactions between diverse species across multiple trophic levels. Ecosystems evolve under the action of multilevel selection, that is, selection acting at the level of the individual, species, and ecosystem. It has also been demonstrated that species-mediated environmental modifications play a significant role in the subsequent evolution of that species.

There are several biological situations that are not readily accounted for by Darwin's concepts: evolution of eusocial species and emergence of an aging phenotype.

Eusocial species, such as those comprising the genus *Apis* (honey bee), display a division of labor and a division of reproductive function. The queen is the sole source of offspring and other members of the colony perform sustaining functions such as foraging, nursing, and hive guarding. How does one explain the evolution of individuals within a community who are incapable of reproduction if natural selection strictly acts at the level of the individual organism?

All organisms age but the aging phenotype does not confer an obvious fitness advantage to an individual organism. How then has the aging phenotype evolved?

In the remainder of this chapter, different theories will be reviewed which have been proposed to account for the evolution of the aging phenotype. I have classified these theories into two main categories: evolutionary–mechanistic theories and developmental theories.

EVOLUTIONARY–MECHANISTIC THEORIES OF AGING

Weismann examined the evolutionary basis of aging and life span in the late nineteenth century (Weismann 1891; Rose et al. 2008; Ljubuncic and Reznick 2009). Weismann proposed that aging and life span were determined by the needs of the species. Quoting from an essay by Weismann published in 1891, p. 24: "When one or more individuals have provided a sufficient number of successors they themselves as consumers of nourishment in a constantly increasing degree, are an injury to those successors. Natural selection therefore weeds them out and in many cases favors such races as die almost immediately after they have left successors." Weismann also proposed several other interesting theories (Ljubuncic and Reznick 2009). One theory focused on programmed death, which he believed evolved because it conferred an evolutionary advantage to the species, not the individual. He also developed the idea that somatic cells could undergo only a predetermined limited number of divisions. Weismann appears to be one of the first biologists to examine aging and death of individual organisms within the context of the needs of the species. Although the principle of group selection was not referred to by Weismann, this concept is implicit in his ideas.

Haldane was probably the first to propose that the strength of natural selection declined with age (Rose et al. 2008), although this idea was developed more fully by Medawar (Rose et al. 2008) and

Fisher (Charlesworth 2000). This insight, which allowed the aging phenotype to be accommodated within Darwin's evolutionary framework, resulted in several expanded theories of aging.

MUTATION ACCUMULATION THEORY

The declining force of selection with age was used by Medawar to postulate that deleterious alleles (mutations) could accumulate in the late stages of life (Charlesworth 2000; Ljubuncic and Reznick 2009). In essence, deleterious mutations in young individuals would be strongly selected against since they would reduce that individual's fitness. Deleterious mutations acting in late life would accumulate because of the reduced strength of natural selection. Aging becomes the manifestation of age-related accumulation of deleterious mutations.

ANTAGONISTIC PLEIOTROPY THEORY

This theory was outlined by Williams (1957) and was based upon four assumptions: soma and germline were distinct within the organism, natural selection of alternative alleles in a population, existence of pleiotropic genes which had opposite effects on fitness at different ages, and decreasing probability of reproduction with increasing age. Williams (1957) believed that senescence was a characteristic of the soma, hence his inclusion of assumption one. According to Williams (1957), senescence would not be a feature of unicellular organisms. However, since there is evidence that senescence is ubiquitous in unicellular organisms (Singer 2016), a distinct soma and germline are not a necessary condition for senescence to occur. The essence of William's proposal was that a pleiotropic gene that increased fitness at an early age would be selected even though that same gene decreased fitness at a later post reproductive age. As stated by Williams (1957, pp. 408 and 410), "Natural selection will frequently maximize vigor in youth at the expense of vigor later on and thereby produce a declining vigor (senescence) during adult life. The time of reproductive maturation should mark the onset of senescence."

Charlesworth (2000) summarized the relationship between these two evolutionary theories: mutation accumulation and antagonistic pleiotropy. Quoting Charlesworth (2000, p. 930), "Our understanding of the evolution of senescence is at one level, very complete; we know that senescence is an evolutionary response to the diminishing effectiveness of selection with age and that this explains many aspects of the comparative biology of senescence. On the other hand, it is at present hard to be sure which of the two most likely important mechanisms by which this property of selection influences senescence (accumulation of late-acting deleterious mutations or fixation of mutations with favorable early effects and deleterious late effects) plays the more important role, especially as these are not mutually exclusive possibilities."

DISPOSABLE SOMA THEORY

The basis of this theory was outlined by Kirkwood in 1977. Kirkwood pointed out that errors in the synthesis of proteins which performed vital cellular functions would have serious consequences for organismal fitness. If the occurrence of errors in protein synthesis was an ongoing phenomenon, the result would be the accumulation of defective proteins and ultimately death of the organism. To avoid such a situation, mechanisms would have to be in place to detect and correct errors which occurred during protein synthesis. Kirkwood (1977) drew a distinction between unicellular and multicellular organisms. He only considered aging of multicellular organisms since they had distinct germline and somatic cells. Germline transmits information to succeeding generations but somatic cells do not. Since accuracy in germline protein synthesis was critical for successful reproductive function and given the finite resources of the organism, more of these resources would be invested into error regulation and repair in germline and less into error regulation and repair in somatic cells. Aging resulted from the progressive accumulation of defective proteins in somatic cells and this accumulation would be greater post reproductively when the force of selection was weaker.

In summary, these three evolutionary theories have much in common. In all three, aging is the result of accumulation of somatic damage in later life, primarily because of diminished natural selection. The theories differ primarily in the nature of the somatic damage: late life accumulation of deleterious mutations, late life injurious actions of pleiotropic genes which act favorable in early life, or the incomplete regulation and repair of errors in somatic cell protein synthesis.

In these evolutionary theories, fitness is defined in terms of survival to reproductive maturation, and therefore organisms preferentially invest resources into somatic maintenance and repair during this reproductive period. Strong selection at this stage weeds out deleterious genetic variants. A corollary is that species subject to high extrinsic mortality due, for example, to predation select for short life spans with early and frequent reproduction while species subject to low extrinsic mortality generally have longer life spans and delayed and prolonged reproductive periods (Healy et al. 2014).

Within the framework of these evolutionary theories, investigators have begun to define the actual biological processes, dysregulation of which are responsible for the aging process. I refer to these studies as mechanistic extensions of evolutionary theories.

MECHANISTIC EXTENSIONS

Mechanistic extensions are based on the premise that aging is the result of time-dependent accumulation of cellular damage (Lopez-Otin et al. 2013) and that aging is a nonadaptive phenotype. Lopez-Otin et al. (2013) categorized molecular processes associated with the aging phenotype into nine categories which they termed the "Hallmarks of Aging." Their three criteria for inclusion of a hallmark were: it should manifest during normal aging, its experimental aggravation should accelerate aging, and its experimental amelioration should retard the normal aging process. However, as discussed by the investigators, these criteria, particularly the third one, were only met to a varying degree by the proposed hallmarks. The nine hallmarks identified were: genomic instability, telomere attrition, loss of proteostasis, deregulated nutrient sensing, mitochondrial dysfunction (generation of reactive oxygen species), cellular senescence, stem cell exhaustion, altered intercellular communication, and epigenetic alterations. The investigators pointed out that even though these hallmarks were treated as separate, they display extensive interconnectedness. The biological processes included in these nine categories are quite far-reaching, which is not surprising considering the complexity of the aging process. More recently, additional molecular processes have been proposed to account for the aging process; one such proposal is the retrotransposon activation theory. Since retrotransposon activation is one of the most recently proposed "Hallmarks of Aging," I will review some of the experimental evidence in support of this hallmark.

Wood et al. (2016) looked at the relationship between activation of transposable elements and aging/life span using *Drosophila melanogaster* as the test subject. Transposable elements are mobile genes that constitute a large fraction of the eukaryotic genome and are enriched in heterochromatic regions. Heterochromatin is a form of densely packed DNA which is associated with gene and transposable element silencing. Increased expression of transposable elements promotes increased transposition (mobilization) which can lead to genomic damage.

Wood et al. (2016) examined heterochromatic regions in the adult female fly head and fat body. The Drosophila fat body is the homolog of the mammalian adipose tissue and liver. Mobile genes in these heterochromatic regions of the fly are chiefly retrotransposons (RTEs) which transpose between DNA sites through a copy and paste mechanism. Wood et al. (2016) found that many in a set of 250 genes and RTEs located within head/fat body heterochromatin showed an increased expression in old (40 days) compared to young (10 days) flies. Caloric restriction, a life-extension/antiaging regimen, reversed the age-related increase in gene and RTE expression.

These investigators also tracked the transposition (mobilization) of RTEs in the fly fat body using the gypsy-TRAP reporter system. Longitudinal fly cohorts were used and the investigators documented an age-related increase in transposition which began at about midlife (20 days of age). Caloric restriction delayed the age of onset as well as the magnitude of this increase. Using flies with two different genetic backgrounds and exposure to normal or caloric restricted diets, Wood et al. (2016) observed a correlation between the age at which transposition increased and fly life span.

The investigators looked at the effects of genetic manipulations that altered heterochromatin structure on RTE expression. Interventions that stabilized heterochromatin structure reduced the age-related increase in RTE expression and extended fly life span.

These experimental observations indicate that the age-related increase in expression and transposition of RTEs is primarily due to a breakdown in the integrity of heterochromatin structure with aging. In summary, Wood et al. (2016) found that an increase in RTE expression and transposition appears to be a hallmark of somatic aging and interventions that stabilize heterochromatin and thereby suppress RTE activity lead to life span extension.

Mechanistic extensions have defined many biological processes that appear to be involved in the aging process. Although most studies focus on specific individual biological processes, it is important to remember that, as noted by Lopez-Otin et al. (2013), the various processes that make up the hallmarks of aging have extensive interconnections.

NEWER COMPLEXITIES IN POST DARWIN THEORY OF EVOLUTION

Since Darwin's original formulation of his theory of evolution, there have been a number of modifications. For simplicity, I will consider two categories of modifications: advances in understanding genetic mechanisms and expansion of the scope of the evolutionary process. Since these two categories are inter-related, this division is clearly arbitrary. The next two sections highlight the extent of post Darwin modifications in the scope of evolutionary theory as well as the extent of advances in genomics. Despite these modifications/advances, the three major theories of aging (mutation accumulation, antagonistic pleiotropy, and disposable soma) have remained in vogue for over 40 years. These three evolutionary theories were formulated at a time when our understanding of genomics and our concepts of evolutionary processes were much more rudimentary. A good discussion of current controversies in evolutionary theory can be found in the parallel papers of Laland et al. (2014) and Wray et al. (2014).

Genetic Mechanisms

In the past half century, there has been an exponential increase in our understanding of genetic mechanisms. A review of this topic is clearly beyond the scope of this chapter, but a few observations will underscore the complexity of the genome and, in particular, regulation of gene expression.

Approximately 15 years ago, the human genome was sequenced (Chi 2016) and the results raised a whole series of new questions. The genome contained a smaller number of protein coding genes than predicted, about 19,000. These genes accounted for only about 2% of genomic DNA, suggesting that the bulk of DNA was involved in regulatory functions (Chi 2016). The function of much of noncoding DNA is still unknown, and in fact the estimate is that perhaps only 10%–20% of noncoding DNA actually has a vital function. These observations indicate that the evolution of diversity in animal form and function is principally the result of changes in the regulation of gene expression patterns rather than the emergence of new protein coding genes. In support of this contention, Kachroo et al. (2015) found that almost half of the essential genes in yeast could be replaced by orthologous human genes. Even though yeast and humans are separated by a vast evolutionary distance, a significant number of yeast/human protein coding genes perform similar functions.

Another interesting observation has emerged from studies of the human exome, the protein coding region of the genome (Hayden 2016). Many mutations thought to be pathogenic have turned out not to be harmful. For example, the gene PRNP encodes for the major prion protein PrP. Mutations in this gene are associated with prion diseases. The human exome data bank contains exon sequencing from 60,706 individuals. The expected number of individuals with PRNP mutations based upon the incidence of prion diseases would be 1.7. The actual number of people with PRNP mutations in the exome database was 52. Hence, the great majority of PRNP mutations detected in these 60,706 individuals were not pathogenic for prion diseases.

The role of the epigenome in regulating gene expression is a relatively new development. The epigenome is viewed as a link between environmental cues and gene expression patterns, and it has been proposed that epigenetic chemical modifications can be passed down to succeeding generations. In plants, inheritance of DNA methylation tags throughout many generations has been well documented (Miska and Ferguson-Smith 2016), but transgenerational epigenetic inheritance is still very controversial in vertebrates. Wray et al. (2014) explicitly state that there has been no documented case in which a new trait has evolved based strictly on an epigenetic mechanism divorced from DNA sequence.

Newer technologies have yielded much information about genome spatial organization and regulation of gene expression. The nucleosome is the basic repeating unit of chromatin, consisting of an octamer of histone proteins around which is wrapped DNA (Gross et al. 2015). Epigenetic marks involve chemical modifications to histone tails (e.g., methylation and acetylation) and methylation of DNA at CpG sites and these chemical modifications are involved in regulation of gene expression. Nucleosomes are packed together in a filament arrangement to form the primary structure of chromatin. Within the nucleus, chromatin is organized into three-dimensional structures (Gross et al. 2015). The largest compartments are termed A (active; euchromatic) and B (inactive; heterochromatic) where activity refers to the expression activity of the contained genes. Within the A and B compartments are smaller topologically associating domains (TADs) which contain preferentially interacting gene clusters. Within TADs, regulation of gene expression involves specific DNA sequences known as promoters and enhancers. It is now known that at the single gene level transcription is sporadic rather than continuous, consisting of bursts of mRNA production followed by refractory periods with little or no transcription activity (Muerdter and Stark 2016). Enhancers act by regulating burst frequency and in the current model, enhancer-bound protein molecules directly contact promoter-attached protein molecules via DNA looping. However, new evidence suggests that a single enhancer can simultaneously activate two promoters (Muerdter and Stark 2016). This observation is inconsistent with a DNA loop arrangement and suggests a model in which a single enhancer creates a micro-environment in which diffusible activators can regulate two genes at the same time.

These few examples underline several important points. Since most of the genome is involved in regulation of gene expression, the evolution of species is directed primarily via variation in gene expression patterns rather than via creation of new protein coding genes. Regulation of gene expression is a complex process involving enhancers, promoters, epigenetic chemical modifications to histones/DNA, and the three-dimensional spatial organization of chromatin within the cell nucleus. In essence, gene expression is regulated by a multidimensional system which at the single gene level is subject to stochastic influences. How stochasticity at the molecular level gives rise to stability and order at the level of the organism is unknown. These very complex features of the genome need to be kept in mind when considering the developmental programs that underlie the evolution of species.

Scope of the Evolutionary Process

Darwin was well aware that organisms change their environmental space, a process referred to as niche construction. In his study of earthworms, Darwin discussed how the activities of the earthworm create topsoil and improve soil quality and plant growth (Feller et al. 2003). However, evolutionary biologists have expanded the scope of interactions between organisms and their environmental space (Laland et al. 2014). Living organisms do not evolve to fit into preexisting environments but co-construct and co-evolve with their environments. Niche construction not only modifies an organism's environmental space, but in so doing alters the selective forces acting on that organism. Such a feedback system will change the evolutionary course of that organism and through this mechanism the organism becomes an active player in its own evolution.

Evolutionary theory now includes nongenetic forms of inheritance. Culture represents a body of knowledge and practices that is based upon social learning, is modified by each generation, and then passed down with high fidelity to succeeding generations, that is, cultural inheritance. Modified environmental space (e.g., cities, roadways) is transmitted across generations and can be considered as ecological inheritance. These nongenetic forms of inheritance play major roles in the evolution of species and, in fact, gene-culture co-evolution is now considered the principal driver of human evolution. The role of nongenetic forms of inheritance is discussed more fully in Chapter 7.

Recently, another theory of evolution, termed the plasticity-first hypothesis, has been proposed. In this construct, an environmentally induced phenotypic change (phenotypic plasticity) precedes and even facilitates evolutionary adaptation (Levis and Pfennig 2016). The genetic basis of such a phenotypic-driven process is not known but may involve cryptic genetic variation or epistasis. Two examples will illustrate this type of evolutionary process.

Allf et al. (2016) explored the premise that behavioral plasticity could promote morphological novelty using the evolution of the rattlesnake rattle. This morphologic feature is only present in species of rattlesnake and therefore has evolved only once. The rattle consists of a series of loose fitting interlocking segments of keratin located at the tip of the snake tail. Specialized muscles cause the segments to vibrate rapidly against one another generating the distinctive sound. The rattle sound functions as a deterrent signal to predators, a cue that a bite with extrusion of venom is imminent.

Tail vibration is a ubiquitous behavior of snakes which is expressed when the snake is threatened; it is therefore a form of behavioral plasticity being elicited by an environmental cue, that is, presence of a predator. Allf et al. (2016) measured tail vibration in 155 individual snakes from 56 species: 38 species of rattlesnake and 18 other species. Tail vibration, which is a defensive behavioral response, was induced under standardized conditions and the snake's response was filmed. The rate and duration of tail vibration were the measured parameters. A phylogenetic tree was constructed from the experimental observations. The more closely related a species was to rattlesnakes, the more similar its rate and duration of tail vibration was to that of rattlesnakes. The phylogenetic data were consistent with the proposal that an ancestral defensive behavior of tail vibration preceded the evolution of the rattle. The shorter the phylogenetic distance between a species and the rattlesnake clade, the stronger that species rate and duration of defensive tail vibration. The actual sequence of events by which a plastic defensive behavior, tail vibration, drove the evolution of a morphological novelty, the rattle, remains speculative. The genetic basis for this behavioral driven evolution of the rattle is also unknown but could involve cryptic variation or genetic epistasis.

The second example involves domestication experiments in the silver fox (Singer 2015). The Russian biologist D.K. Belyaev used farmed silver foxes as his test subject. The behavior he selected was tameability in fox pups based upon their responses to hand feeding and to attempts to touch or pet them. Less than 10% of the tamest animals of every generation were used as parents of the next generation. By the sixth generation, some pups sought out human contact by whining, whimpering, and licking in a doglike manner. He referred to these foxes as the elite of domestication and by the fortieth year of the experiment he had a complete population of domesticated foxes. Although these foxes were selectively bred using tameability, a behavioral trait, as the only criterion, many of them also developed morphological and other phenotypic changes. Some of these domesticated foxes showed altered reproductive patterns and some showed morphological changes such as floppy ears, curly tails, and white spotting of the coat. The results of this domestication experiment in which selection for a behavioral phenotype promoted co-evolution of morphological changes and alterations in reproductive behavior would constitute an example of the plasticity-first hypothesis.

When Darwin outlined his theory of evolution, he considered the principle of natural selection to act at the level of the individual (Darwin 1859). However, Darwin did accept the possibility of group selection. He apparently postulated that moral men might not do any better than immoral men but that tribes of moral men would certainly have an immense advantage over fractious bands of pirates (Mirsky 2009). Group or multilevel selection has had a checkered history but I believe the evidence for multilevel selection is quite strong (Mirsky 2009; Singer 2016). Nature is characterized by a hierarchy of biological organizations: individual organisms, species occupying specific ecological niches and ecosystems. Diversity and abundance of species are the prime determinants of ecosystem functionality and regulation of these parameters requires the action of multilevel selection (Singer 2016; Chapter 7).

CRITIQUE OF EVOLUTIONARY–MECHANISTIC THEORIES OF AGING

Jones et al. (2014) collected mortality and reproductive data from a number of multicellular species: 11 mammals, 12 other vertebrates, 10 invertebrates, 12 vascular plants, and a green alga. These various species have a wide range of life spans so as to make comparisons possible, the investigators standardized the age range to begin at the mean age of reproductive maturation and to end when only 5% of adults were still alive. Fertility (related to number of offspring) and mortality were mean standardized by dividing age-specific fertility and mortality by the respective weighted average levels of fertility and mortality for all adults alive from reproductive maturation to the terminal age (only 5% of adults alive). The two parameters of mortality and fertility correspond to a measure of Darwinian fitness. The data set reveals an incredible variation in the trajectories of mortality and fertility across these 46 diverse species; each species appears to have evolved its own unique pattern. In a separate figure, Jones et al. (2014) highlighted the significant intraspecific variation in the trajectory of standardized mortality in strains of two laboratory model species: rat (*Rattus norvegicus*) and mouse (*Mus musculus*). These data illustrated two important features of aging: each species had evolved a unique mortality/fertility trajectory over its life span and there were considerable between-individual variations within a species in mortality/fertility trajectories.

Do evolutionary–mechanistic theories of aging provide plausible mechanism(s) to explain these two features? Neither evolutionary theories nor their mechanistic extensions explain species-specific aging/life span. Evolutionary theories with their focus on early-life investment in somatic maintenance and age-related weakening of selection provide a general framework for organismal aging but not a species-specific mechanism. Neither do the mechanistic extensions. To explain species specificity, one would have to postulate that each species has evolved its own unique mix of "hallmarks of aging." For example, the mix of "hallmarks" underlying aging in the chimpanzee would be different from that underlying aging in the human. If the mix of "hallmarks" underlying aging is species-specific, then this mechanism becomes indistinguishable from aging due to evolved species-specific developmental programs. Evolutionary theories and their mechanistic extensions could account for intraspecies variation in aging. For example, individuals within a species could have different amounts of accumulated deleterious mutation in late life or "hallmarks" could have different levels of dysregulation between individuals.

There are several other features of the aging process that are not accommodated by evolutionary–mechanistic theories of aging. Speciation takes place and species evolve within the context of ecological niches and ecosystems. That is to say, there is a direct relationship between organismal evolution and the environmental space in which that organism resides. This relationship was discussed in the section titled "Newer Complexities in Post Darwin Theory of Evolution." This direct relationship between environment and evolution is not included in evolutionary–mechanistic theories of aging. Nongenetic forms of inheritance, that is, cultural and ecological inheritance, are important drivers of human evolution but are not part of evolutionary–mechanistic theories of aging. When aging is viewed in terms of the ecological context in which a species evolves, it is an adaptive phenotype (Singer 2016). There is a biochemical basis to this assertion. Aging/death programs such as cellular senescence and apoptosis are adaptive within the context of a multicellular organism. These programs play critical roles in embryogenesis and tissue repair/maintenance (Singer 2016). None of these features are included in evolutionary–mechanistic theories of aging. These theories view aging as a nonadaptive pathological process.

Several recent studies have documented the important role of ecological and behavioral factors in determining the maximum life spans of species.

Healy et al. (2014) looked at the relationships between maximum life spans, mode-of-life and ecological factors in species of mammals and birds. The factors collected included: flight capability (volant or non-volant), activity period (nocturnal, daytime, or dusk and dawn), foraging environment (terrestrial, arboreal, aerial, or aquatic) and fossoriality (adapted to digging and burrowing). The data set used included 589 bird species (579 volant and 10 non-volant) and 779 species of mammal (83 volant and 696 non-volant).

It is well known that maximum life span correlates strongly with body mass such that in general larger species live longer than smaller species. The purpose of this study was to estimate the impact of the collected mode-of-life and ecological factors on maximum life span when body size was controlled. Volant species lived longer than non-volant species of similar body mass and for volant species, those active during the daytime or nighttime had longer life spans than those active at dusk and dawn. For non-volant species, activity period had no impact on longevity, arboreal foragers lived longer than terrestrial foragers, and fossorial species had longer life spans than non-fossorial species. The investigators noted that the most important factor associated with a longer life span in both birds and mammals after controlling for body size and phylogeny was the ability to fly.

Healy et al. (2014) interpreted their results in terms of how mode-of-life/ecological factors altered extrinsic mortality, primarily predation. Species that fly and non-volant species that forage in trees or burrow underground (fossorial) will be less exposed to predators. According to the evolutionary theories of aging, organisms subject to low extrinsic mortality will select for longer life spans than organisms that are subject to high extrinsic mortality. However, there are several lines of evidence that this interpretation is probably incorrect.

First, as discussed in Chapter 7, the predator–prey relationship is not simply an evolutionary arms war. The relationship between predator biomass and prey biomass follows the same scaling rule across diverse terrestrial and aquatic ecosystems. For example, apex predators regulate directly or indirectly the population density of prey species and smaller predators, abundance and diversity of plant species, river/stream morphology as well as nutrient recycling. The universality of this relationship underscores that the relative abundance of predators and prey within ecosystems is a strongly controlled parameter. Given this observation, it becomes very unlikely that the mode-of-life/ecological features that Healy et al. (2014) found to alter species life span evolved simply to allow species to escape predation. These features evolved under the action of multilevel selection to allow species to adapt to their ecological niches and ecosystems.

Second, Williams and Shattuck (2015) looked at the effect of fossoriality and sociality on life span in mammals using a larger data set than that used by Healy et al. (2014). Some mammalian fossorial species are eusocial, eusociality being an extreme form of social behavior characterized by overlapping generations, reproductive division of labor, and cooperative breeding. One well-known mammalian eusocial fossorial species is the naked mole-rat, *Heterocephalus glaber*, which has an exceptionally long life span (Buffenstein 2005).

Williams and Shattuck (2015) used as their data set ground dwelling mammals (excluding those that were arboreal or aquatic) that were fossorial (101 species) or non-fossorial (339 species). Species were also categorized into strongly eusocial (17 species) and non-eusocial (solitary and social, 423 species). The investigators found that with control of body mass and correction for phylogenetic relatedness, species that were both fossorial and eusocial had a greater life span than non-fossorial and non-eusocial species. Hence, the exceptional life span of the naked mole-rat can be accounted for by its elaborate tunnel system ecological niche and its eusocial behavior.

Naked mole-rats live in large colonies of, on average, 75 individuals (Buffenstein 2005). Their ecological niche consists of a maze of burrows several meters below the soil surface. A colony contains a single breeding female and one to three breeding males. Breeders and nonbreeders have similar life spans. As noted by Buffenstein (2005) and supported by the data of Williams and Shattuck (2015), extended longevity is only observed in fossorial animals that are also eusocial. Furthermore, the data for the naked mole-rat indicate that the extended longevity associated with eusociality is conferred on nonbreeders as well as breeders.

These observations are not consistent with evolutionary theories of evolution. If natural selection acts at the level of the individual and "chooses" phenotypes that confer a fitness (survival to reproductive maturation) advantage, how do we explain how eusociality extends the life span of nonbreeding fossorial animals like the naked mole-rat? The observations of Buffenstein (2005) and Williams and Shattuck (2015) are more consistent with the evolution of eusocial behavior as an adaptive phenotype through the action of multilevel selection.

Eusocial behavior has been well documented in certain insect groups: ants, termites, and the honeybee (Keller and Genoud 1997). These eusocial groups show a morphological/functional caste system with a division of labor including reproductive function. For example, the honeybee colony has a queen who functions as the reproductive unit and has an extended life span measured in years and sterile worker bees that function as foragers, nurses, and guards and have a more limited life span. How did such a eusocial community and its sterile worker bees evolve if natural selection strictly acts at the level of the individual organism? Worker bees cannot reproduce and so, by definition, the individuals in this caste are unfit. Clearly, these eusocial insect communities evolved under the action of multilevel selection such that the community as a whole is adaptive and so are the constituent castes within the community. It is interesting that eusociality in these insect groups differs from that of the naked mole-rat. In the insect community, only the reproductive queen enjoys an extended life span while worker castes are relatively short-lived. In the eusocial naked mole-rat community, both breeders and nonbreeders have equivalent extended life spans. Hence, the manifestations of eusociality differ across animal groups.

In summary, although evolutionary theories of aging and their mechanistic extensions (hallmarks of aging) can account for intraspecies variation in aging, they do not explain the aging phenotype itself. All of the observations discussed in this section point to aging being more than the accumulation of late life deleterious mutations or the late life deleterious action of pleiotropic genes or the sum total of dysregulated biological processes. The aging of a species is best accounted for as the expression of a species-specific developmental program that evolved within the structure of the specific ecosystem inhabited by that species.

AGING AS A SPECIES-SPECIFIC DEVELOPMENTAL PROGRAM

INTRODUCTION

What is a developmental program? There is no agreed upon definition of what constitutes a developmental program and defining this biological process is easiest done by example. Embryogenesis is considered the expression of a species-specific developmental program and one that has a genetic basis. Embryogenesis is also a plastic process; there is a considerable between-individual variation in function and form at the time of birth. This variation stems from the stochasticity of the involved genetic mechanisms as well as the influence of nongenetic environmental factors. For example, reproduction in the nine-banded armadillo involves the regular production of monozygous quadruplets, sets of animals with identical genomes (Storrs and Williams 1968). Storrs and Williams (1968) measured 20 parameters in 16 sets of quadruplets. These parameters included body weight, different organ weights, catecholamine levels, and brain neurotransmitter levels. All of the parameters showed quantitative differences within each set and these differences ranged from 1% to as high as 43%. The important observation is that, despite identical genomes, the four armadillo embryos within the same set develop differently due to influences imparted by the uterine environment. The observations of Storrs and Williams (1968) demonstrate that environmental cues can modify a developmental program and that cues begin to act within the intrauterine environment.

There have been some attempts to incorporate developmental programs into the aging process. de Magalhaes (2012) proposed a model in which some aspects of aging were based on developmental programs, but the investigator was quite explicit that aging did not evolve for a purpose, that is, aging was not adaptive. The premise of the model was as follows. If developmental programs which enhance function during the early reproductive period were to continue beyond this stage because of weaker selection, they could become detrimental in late life. In this model, aging becomes the manifestation of developmental programs that continue to operate beyond the lifestage during which they are beneficial. This proposal is similar to the antagonistic pleiotropy theory, the difference being that developmental programs are substituted for genes; developmental programs exert beneficial effects in early life but deleterious effects in late life.

Mitteldorf (2012) has proposed that aging had an adaptive origin and evolved under the action of group selection. Control of reproduction through aging and death prevents populations from increasing in number such as to exceed the abundance of producer species in their ecosystem. Another adaptive feature of aging is that it creates vacancies in the ecological niche for younger individuals. The population turns over faster; aging promotes diversity. On the basis of the conservation of genetic pathways across a wide range of species that mediate longevity, Longo, Mitteldorf, and Skulachev (2005) proposed a programmed longevity theory which they use as a mechanism to explain examples of apparent altruistic aging in plants and animals (Longo et al. 2005).

In contrast, Vijg and Kennedy (2016) conclude that the evidence for adaptive programmed aging is weak at best. They emphasize that the target of selection is first and foremost the individual and that even though a developmental program creates an organism, this program is interrupted by a random, aimless process that leads to death. Vijg and Kennedy (2016) conclude that evolution optimizes for fitness (reproduction) but not for longevity.

Neither of these two groups of investigators, Mitteldorf and colleagues or Vijg and Kennedy, has included the rich multidimensional nature of evolution in their arguments (see discussion in the section, "Newer Complexities in Post Darwin Theory of Evolution" and Chapter 7).

Multidimensional Evolutionary Framework

Evolution of a species takes place through a two-way interaction between individual members of a species and their environmental space. The environmental space imposes constraints on the evolution of a species, but at the same time individuals of that species modify their environmental space and those modifications themselves will shape the evolution of that species. The environmental space includes overlapping ecological niches occupied by various species, all of which make up a larger organization known as an ecosystem. An ecosystem is a highly regulated structure characterized by mutually advantageous interactions between its diverse member species. These between-species interactions shape the evolution of each species; a species does not evolve independently of the other species within its ecosystem. Species diversity and relative abundance are the primary determinants of ecosystem functionality and these determinants are regulated parameters. As proposed by Darwin, species evolution is directed by the action of natural selection but the complex ecosystem structure in which a species evolves requires that natural selection acts at multiple levels: individuals, species, and groups of species, that is, multilevel selection.

Aging and Defined Life Spans Are Adaptive Traits

Within the inherent multidimensional nature of the evolutionary framework, species-specific aging trajectories and maximum life spans evolved as adaptive phenotypes. The origin of aging and maximum life spans as adaptive traits has deep evolutionary roots. In communities of unicellular organisms, cells undergoing programmed death release substances that enhance the growth of neighboring single cell organisms. The manner in which phytoplankton, a foundational organism within marine ecosystems, dies largely determines the flow of nutrients and the fate of marine organisms within the ecosystem. Since the relative biomass of predators and prey is a controlled feature of marine and terrestrial ecosystems, predation cannot be the primary mechanism regulating ecosystem biodiversity. As a general principle, species-specific aging and maximum life spans are the instruments that regulate the diversity and relative abundance of species within an ecosystem. These examples are discussed more fully in Chapter 7.

Within multicellular organisms, programmed aging and death of individual cells are adaptive, conferring a benefit to the whole organism. Cellular senescence, apoptosis, and autophagy play critical roles in morphogenesis, tissue repair, and tissue maintenance (Singer 2016). The aging and mortality pattern of humans is associated with the intergenerational transfer of information and resources, including the care of grand offspring (Singer 2015). These intergenerational transfers between parents and children are considered adaptive since they enhance the reproductive function of the children.

All of these examples support the contention that post reproductive organismal lifestages serve an adaptive function.

EVOLUTION IS DRIVEN BY CHANGES IN GENE EXPRESSION

In Darwin's era, there was little understanding of the genetic mechanisms underlying phenotypic diversity. It is now recognized that protein coding genes make up only a small fraction of the genome and that protein coding genes are highly conserved across vast evolutionary distances. The evolution of species is driven primarily by changes in the spatial/temporal regulation of gene expression. This regulatory process is complex and made up of many elements: DNA sequences known as enhancers and promoters, various proteins that bind to these sequences, chromatin which is composed of nucleosomes (a core of histone proteins around which DNA is wrapped), chemical modifications to histone tails and DNA nucleotides (epigenetic marks), and the three-dimensional packing of the chromatin within the cell nucleus. All of these elements have been documented to be involved in the regulation of gene expression.

SPECIES-SPECIFIC DEVELOPMENTAL PROGRAMS ARE KEY

Let us consider different mammals. When an elephant, giraffe, human, or chimpanzee is born, that individual animal will have a distinctive morphology and repertoire of behaviors. In addition, that animal will be projected to have a characteristic lifecycle including time to reach reproductive maturation, pattern of offspring production, aging, and maximum life span. That individual animal's morphology and repertoire of behaviors at birth as well as its projected lifestages (including aging and maximum life span) would all be recognized as characteristic of a particular species, be it an elephant, giraffe, human, or chimpanzee. These species characteristics are encoded in species-specific developmental programs. However, species-specific developmental programs are not "set in stone" since there is considerable variation between individuals within a species. This variation between individuals has as its basis at least two mechanisms: the stochasticity of the developmental program itself and the modifying influence of environmental cues, starting with the intrauterine environment. It has been proposed that environmental cues alter organismal phenotypes through epigenetic reprogramming. The important point is that the specific aging trajectory and maximum life span of a species is encoded in a species-specific developmental program.

As an aside, it is clearly impossible to define how many developmental programs characterize a given species. For example, is the developmental program that determines a species morphology and behaviors, that is, embryogenesis, different from the developmental program that determines that species aging process and maximum life span? This question is at present unanswerable. If a species does have multiple developmental programs, they would no doubt all be interconnected.

Divergence of the human lineage from its common ancestry with the chimpanzee/bonobo was due to the evolution of a new human lineage-specific developmental program. This new program would have encoded a changed aging trajectory and changed the maximum life span for this early hominin compared to its common ancestry with the chimpanzee/bonobo. Further evolution of the human lineage with the emergence of Homo sapiens about 200,000 years ago reflected ongoing evolution of the human lineage-specific developmental program. In essence, the aging trajectory and maximum life span of a species evolves because the developmental program that underwrites these traits is subject to continuing evolutionary change. The key to understanding speciation (emergence of a new species) and evolution of an existing species is in understanding the nature of developmental programs. I propose that species-specific developmental programs encode all life-cycle stages, including aging and maximum life span. Hence, understanding what constitutes such a developmental program will help unravel the intrinsic nature of the aging process and the basis of species-specific maximum life spans.

The topic of developmental programs was discussed in Chapter 7. In that discussion, some of the general features of developmental programs were described but the point was made that our understanding of species-specific aspects of these programs is still very lacking. In this section, the focus will be more directed toward species-specific aspects. Three examples are discussed.

Example 1

In an elegant series of experiments, Kvon et al. (2016) looked at the genetic mechanisms underpinning limb development in vertebrates, focusing on snakes. One of the best characterized enhancers involved in limb-specific development is known as the zone of polarizing activity regulatory sequence (ZRS). ZRS is a long-range enhancer that controls the expression of the Sonic hedgehog (Shh) gene in developing limb buds (Villar and Odom 2016). Basal snakes such as the boa constrictor and Burmese python retain a vestigial pelvic girdle and rudimentary hindlimbs, whereas advanced snakes such as the viper, rattlesnake, and king cobra have lost all skeletal limb structures.

In nine finned or limbed vertebrates, orthologous ZRS enhancers displayed spatial/temporal activity patterns indistinguishable from that of the mouse enhancer. In four species of snakes, the boa constrictor, a basal snake, showed a low level of ZRS activity, whereas the other snake species had no enhancer activity. Using gene editing techniques, the investigators replaced the mouse enhancer with human or coelacanth orthologs. Coelacanths are a taxon of cartilaginous fish phylogenetically separated from the mouse by 400 million years. Mouse, human, and coelacanth enhancers were largely interchangeable functionally. In contrast, replacing the mouse ZRS with the orthologous cobra enhancer resulted in loss of Shh gene expression and development of very truncated limbs.

Sequencing of snake enhancers identified a 17 bp sequence that was conserved across all examined tetrapods and fish but deleted in the snake ZRS. To test the functional significance of this deletion, the 17 bp sequence was reintroduced into the python ZRS to create an "ancestral" ZRS. When the investigators replaced the endogenous mouse enhancer with this "ancestral" modified python ZRS, mice developed normal limbs.

These experimental findings underscore three important points.

First, the process of speciation and the evolution of existing species are driven primarily by changes in the spatial/temporal regulation of gene expression patterns.

Second, linking phenotypes to genotypes is very difficult. A minor sequence modification in a specific enhancer that regulates limb development is associated with extensive phenotypic changes ranging from normal limb development to absence of limb development.

Finally, "mother nature" is a tinkerer and generally relies on minor modifications to existing successful body plans. The ZRS enhancer is highly conserved and regulates limb/fin development in a wide range of vertebrates (Lettice et al. 2003; Kvon et al. 2016). Minor modifications in this enhancer result in major changes in limb/fin development programs. Hence, unraveling what gives developmental programs species specificity will be a very daunting task.

Example 2

A second example involves the admixture of genetic material between two diverging species, a process referred to as reticulate evolution (Hoelzel 2016). There is DNA evidence that a low level of interbreeding occurred between Homo sapiens and two other human-lineage species, Neanderthal and Denisovan. In fact, the genetic underpinnings of the Tibetan population adaptation to high altitude are due to introgression of Denisovan genomic DNA (Fan et al. 2016). In addition, de Manuel et al. (2016) have documented gene flow between subspecies of chimpanzees as well as gene flow from bonobos to chimpanzees that occurred after divergence of these two species 1.5 to 2 million years ago. These examples indicate that natural hybridization may be an important contributor to speciation and evolution of existing species.

Example 3

The third example involves organisms that have undergone dramatic simplifications of their body plan with associated reductions in genome size and complexity. The phylum Cnidaria contains multiple taxa which have been difficult to classify phylogenetically (Chang et al. 2015; Okamura and Gruhl 2016). Cnidaria have a relatively simple body plan consisting of mesoglea between two epithelial layers. The data of Chang et al. (2015) indicate that obligatory endoparasitism evolved as a single event within free living Cnidaria giving rise to a clade known as myxozoa. At a later time, myxozoa underwent a further simplification of body plan and genomes compared to free living cnidaria. The lifecycle of myxozoa involves fish and worms as successive hosts and its body plan consists of several cell types only. The simplified body plan of myxozoa is associated with massive simplification of its genome (Chang et al. 2015). The myxozoa genome is 20-fold smaller than the genome of free living cnidaria and contains less than one third of the complement of protein coding genes. Myxozoan genes are depleted in biological categories related to development, cell differentiation, and cell–cell communication. Genes involved in Wnt and Hedgehog signaling pathways have been lost; these pathways are important mediators of developmental processes.

This example illustrates a parallelism between body plan complexity and genome complexity. A change from a free-living existence to one of obligatory parasitism was associated with parallel simplifications in body morphology/complexity and simplification in the associated genome encoded developmental program.

These three examples give glimpses into the complexities and subtleties of developmental programs. The evolution of a species is driven by changes in the spatial/temporal regulation of gene expression patterns. Subtle modifications in gene regulatory mechanisms can result in large changes in morphology and simpler body plans are associated with simpler developmental programs. These observations only touch the surface of what constitutes a developmental program and what imparts species specificity to these programs.

WHY STUDY AGING?

The human lineage has a long history of treating death as a special event. There is good evidence that the Neanderthal practiced purposeful burial between 57,000 and 71,000 years ago (Rendu et al. 2014). Perhaps the Neanderthal viewed death as a lifestage that needed to be commemorated by the process of burial. In ancient polytheistic religions, burial of the dead, particularly the elite, was considered the start of a journey into an afterlife. Successful entry into the afterlife required specific burial practices.

When Darwin formulated his theory of evolution, the course of biological research was irrevocably changed. Research efforts became focused on determining the mechanisms underpinning aging and life span.

Why study aging? The answer to this question has at least two parts. First, as an important stage in the lifecycle of an organism, the study of aging represents an important area of biological research. The second part of the answer is more diffuse and hence more difficult to define.

From a human perspective, aging and maximum life span are "charged" topics and easily related to personal experiences since aging and life span touch on our own mortality and wellbeing. Why study aging? The second part of the answer to this question reflects the human desire or perhaps need to understand the aging process with a view toward modifying this process. Believing that it is possible to modify the aging process will depend on what are the underlying mechanisms. Furthermore, I suspect that we humans will be biased in favor of mechanisms that are susceptible to modifications. This bias is in harmony with our highly technological culture which promotes solutions to various environmental and biological problems through technological strategies.

In this post Darwin era, the most widely held view of what accounts for an organism's aging trajectory and life span is based on evolutionary–mechanistic theories of aging. The basic features of these theories include selection for early-life traits that advance survival/reproduction and after the reproductive period a weakened natural selection that allows organisms to accumulate biological damage. Natural selection is considered to act only at the level of the individual organism. Within this framework, aging reflects accumulated molecular/tissue damage due to dysregulated biological processes, the so called "Hallmarks of Aging" (Lopez-Otin et al. 2013).

However, the scientific community has extended the scope of the aging process to include a direct relationship between aging in humans and diseases that generally occur in late life such as atherosclerotic cardiovascular disease, neurodegenerative processes, diabetes, and cancer (Vijg and Campisi 2008). As a result of this proposed relationship, the boundary between aging and chronic degenerative diseases has become blurred. In humans, aging is now considered the principal risk factor for such diseases (Kennedy et al. 2014) and since diseases are the chief causes of death in the elderly (Vijg and Campisi 2008), the logic holds that interventions that correct the disordered biology of the aging process will reduce the prevalence of these diseases.

A countervailing theory is that species-specific developmental programs underlie the aging process and determine species maximum life span. Such species-specific developmental programs evolved within the context of the ecological niche and ecosystem that a species inhabits. The evolution of each species is influenced by the evolution of other species within the ecosystem with which it interacts. No species evolves independently of other species and each species is characterized by its own unique aging trajectory and life span. These traits have evolved under the action of multi-level selection and are adaptive when viewed through the lens of the ecosystem's functionality. The aging and maximum life span of a species are encoded in a developmental program much the same way that the morphology and behaviors of a species are encoded in a developmental program.

However species-specific developmental programs are quite plastic: morphology, behaviors, aging, and life span all display considerable between-individual variation within a species. Singer (2013, 2015) has argued that what we have designated as chronic degenerative diseases in humans are in reality between-individual variations in the (human) aging process. Within the developmental model, significant modifications to the human aging process or maximum life span would require significant alterations in our species developmental programs. Such alterations would modify not only our aging trajectory and life span but also our morphology and behaviors. In essence, we would cease to be humans as we now know them. We cannot simply engraft the mechanisms responsible for longevity in another species onto the human-specific developmental program. Since the developmental model does not allow significant modifications to a species aging process or life span, it will be less appealing to our anthropocentric viewpoint than the evolutionary–mechanistic model.

In a recent study, Dong et al. (2016) used epidemiological data to estimate maximum human life span. This is an appropriate study with which to end this chapter. Dong et al. (2016) reported new observations about human life span but this study highlights some of the peculiarities of aging research.

Since life expectancy in humans has shown recent dramatic improvements, Dong et al. (2016) used mortality data to determine whether there was a limit to the human life span; does our species have a maximum life span? These investigators used two databases: the Human Mortality Database and the International Database on Longevity. They posited that if human life span had no fixed upper limit, then with advancing calendar years ever-older age groups should experience the largest gains in survival years. A plot of the age group showing the greatest gains in survival years versus calendar year showed this expected relationship between the years 1920 and 1980. For example, in

1920, the 85-year-old age group had the largest gains in survival years, whereas by 1980 the largest gain in survival years was experienced by the 100-year-old age group. However, this relationship appeared to plateau after 1980. In the calendar year 2000, the age group showing the greatest gains in survival was still the 100-year-old age group. In a second analysis, Dong et al. (2016) plotted the ages of the oldest persons to die in a given year versus calendar years. Between the calendar years 1970 and 1995, there was a steady increase in the maximum reported age of death from 110 years to 115 years. In 1995, there was a trend break in this relationship and, in fact, between 1995 and 2006 there was actually a small decrease in the maximum reported age of death. On the basis of their analysis, Dong et al. (2016) concluded that human longevity has a fixed ceiling and that the natural limit was about 115 years.

Several features of this report reflect peculiarities in the field of aging research. The concept of maximum longevity or life span is well accepted in biology and in fact a database listing maximum life spans for over 4000 species is available (de Magalhaes and Costa 2009). As a background to their study, Dong et al. (2016) remarked that since technological advances had dramatically increased human life expectancy over the recent past, perhaps human life span had no upper limit. Determining whether human longevity had a fixed natural limit was the purpose of their study. Does this study reflect an anthropocentric view that our species is not part of the evolutionary continuum? If all other species have a maximum life span, why should our species be an exception? There is increasing evidence that traits that were once considered to be uniquely human are actually present in other species (de Waal 2016; Gross 2016; Krupenye et al. 2016).

A second feature relates to the explanation advanced by Dong et al. (2016) as to the biological mechanism accounting for a maximum life span in humans. First, Dong et al. (2016, p. 258) state, "The idea that ageing is a purposeful, programmed series of events that evolved under the direct force of natural selection to cause death has now been all but discredited." They postulate that the natural limit to human longevity is an inadvertent by-product of early-life fixed genetic programs such as development, growth, and reproduction. In a commentary on the study by Dong et al. (2016), Olshansky (2016) proposed a similar biological explanation. Aging is an accidental consequence of fixed genetic programs for growth, development, maturation, and reproduction. This proposal seems to be a hybrid explanation. A fixed maximum life span in humans is not due to its own genetic program but a consequence of early-life developmental programs. Is not the proposal that a maximum life span for humans is a by-product of early-life genetic programs the same as ascribing human maximum life span to being encoded within a human species-specific developmental program?

CONCLUSION

Research into aging, particularly human aging, appears to have an "intensity" over and above the intrinsic research value of studying aging and maximum life span as simply interesting lifestages of an organism. A quick survey of Google Scholar makes that point. Searching the words "aging process" yields 2,620,000 citations while searching the words "human aging process" yields 4,210,000 citations, almost double.

Post Darwin, a number of theories have been proposed to account for an organism's aging and maximum life span. These theories fall into two principal categories: evolutionary–mechanistic and species specific developmental programs. Unfortunately, since human scientists are studying the aging and life span of their own species, it is difficult to avoid biases. The human lineage has had a special interest in death for thousands of years. Now, within the context of our highly technological culture, we view aging and a fixed life span as pathological processes. In addition, we have constructed a direct link between the aging process and a host of "chronic diseases." A primary research goal has become the development of interventions that attenuate or even eliminate human aging. Hence, will we tend to bias our view of the mechanisms underpinning the aging process in a direction that presupposes the development of such interventions?

REFERENCES

Allf, J., Durst, P. A. P. and D. W. Pfennig. 2016. Behavioral plasticity and the origins of novelty: The evolution of the rattlesnake rattle. *The American Naturalist* 188: 475–483.

Buffenstein, R. 2005. The naked mole-rat: A new long-living model for human aging research. *The Journals of Gerontology Series A: Biological Sciences and Medical Sciences* 60: 1369–1377.

Chang, E. S., Neuhof, M., Rubinstein, N. D. et al. 2015. Genomic insights into the evolutionary origin of Myxozoa within Cnidaria. *Proceedings of the National Academy of Sciences, USA* 112: 14912–14917.

Charlesworth, B. 2000. Fisher, Medawar, Hamilton and the evolution of aging. *Genetics* 156: 927–931.

Chi, K. R. 2016. The dark side of the human genome. *Nature* 538: 275–277.

Darwin, C. 1859. *The Origin of Species, by Means of Natural Selection.* John Murray, London.

de Magalhaes, J. P. 2012. Programmatic features of aging originating in development: Aging mechanisms beyond molecular damage? *The FASEB Journal* 26: 4821–4826.

de Magalhaes, J. P. and J. Costa. 2009. A database of vertebrate longevity records and their relation to other life-history traits. *Journal of Evolutionary Biology* 22: 1770–1774.

de Manuel, M., Kuhlwilm, M., Frandsen, P. et al. 2016. Chimpanzee genomic diversity reveals ancient admixture with bonobos. *Science* 354: 477–481.

de Waal, F. B. M. 2016. Apes know what others believe. *Science* 354: 39–40.

Dong, X., Milholland, B. and J. Vijg. 2016. Evidence for a limit to human lifespan. *Nature* 538: 257–259. doi:10.1038/nature19793.

Fan, S., Hansen, M. E. B., Lo, Y. and S. A. Tishkoff. 2016. Going global by adapting local: A review of recent human adaptation. *Science* 354: 54–59.

Feller, C., Brown, G. G., Blanchart, E., Deleporte, P. and S. S. Chernyanskii. 2003. Charles Darwin, earthworms and the natural sciences: Various lessons from past to future. *Agriculture, Ecosystems and Environment* 99: 29–49.

Gould, S. J. 1980. *The Panda's Thumb. More Reflections in Natural History.* W. W. Norton, New York and London.

Gross, D. S., Chowdhary, S., Anandhakumar, J. and A. S. Kainth. 2015. Chromatin. *Current Biology* 25: R1158–R1163.

Gross, M. 2016. Chimpanzees, our cultured cousins. *Current Biology* 26: R83–R85.

Hayden, E. C. 2016. Seeing deadly mutations in a new light. *Nature* 538: 154–157.

Healy, K., Guillerme, T., Finlay, S. et al. 2014. Ecology and mode-of-life explain lifespan variation in birds and mammals. *Proceedings of the Royal Society B: Biological Sciences* 281: 20140298.

Hoelzel, A. R. 2016. The road to speciation runs in both ways. *Science* 354: 414–415.

Jones, O. R., Scheuerlein, A., Salguero-Gomez, R. et al. 2014. Diversity of ageing across the tree of life. *Nature* 505: 169–174.

Kachroo, A. H., Laurent, J. M., Yellman, C. M., Meyer, A. G., Wilke, C. O. and E. M. Marcotte. 2015. Systematic humanization of yeast genes reveals conserved functions and genetic modularity. *Science* 348: 921–925.

Keller, L. and M. Genoud. 1997. Extraordinary lifespans in ants: A test of evolutionary theories of ageing. *Nature* 389: 958–960.

Kennedy, B. K., Berger, S. L., Brunet, A. et al. 2014. Geroscience: Linking aging to chronic disease. *Cell* 159: 709–713.

Kirkwood, T. B. L. 1977. Evolution of ageing. *Nature* 270: 301–304.

Krupenye, C., Kano, F., Hirata, S., Call, J. and M. Tomasello. 2016. Great apes anticipate that other individuals will act according to false beliefs. *Science* 354: 110–114.

Kvon, E. Z., Kamneva, O. K., Melo, U. S. et al. 2016. Progressive loss of function in a limb enhancer during snake evolution. *Cell* 167: 633–642.

Laland, K., Uller, T., Feldman, M. et al. 2014. Does evolutionary theory need a rethink? Yes, urgently. *Nature* 514: 161–162, 164.

Lettice, L. A., Heaney, S. J. H., Purdie, L. A. et al. 2003. A long range Shh enhancer regulates expression in the developing limb and fin and is associated with preaxial polydactyly. *Human Molecular Genetics* 12: 1725–1735.

Levis, N. A. and D. W. Pfennig. 2016. Evaluating "plasticity-first" evolution in nature: Key criteria and empirical approaches. *Trends in Ecology and Evolution* 31: 563–574.

Ljubuncic, P. and A. Z. Reznick. 2009. The evolutionary theories of aging revisited—A mini-review. *Gerontology* 55: 205–216.

Longo, V. D., Mitteldorf, J. and V. P. Skulachev. 2005. Programmed and altruistic ageing. *Nature Reviews Genetics* 6: 866–872.

Lopez-Otin, C., Blasco, M. A., Partridge, L., Serrano, M. and G. Kroemer. 2013. The hallmarks of aging. *Cell* 153: 1194–1217.

Mirsky, S. 2009. What's good for the group. *Scientific American* 300: 51.

Miska, E. A. and A. C. Ferguson-Smith. 2016. Transgenerational inheritance: Models and mechanisms of non-DNA sequence-based inheritance. *Science* 354: 59–63.

Mitteldorf, J. J. 2012. Adaptive aging in the context of evolutionary theory. *Biochemistry (Moscow)* 77: 716–725.

Muerdter, F. and A. Stark. 2016. Gene regulation: Activation through space. *Current Biology* 26: R895–R898.

Okamura, B. and A. Gruhl. 2016. Myxozoa + Polypodium: A common route to endoparsitism. *Trends in Parasitology* 32: 268–271.

Olshansky, S. 2016. Measuring our strip of life. *Nature* doi:10.1038/nature19475.

Rendu, W., Beauval, C., Crevecoeur, I. et al. 2014. Evidence supporting an intentional Neanderthal burial at LaChapelle-aux-saints. *Proceedings of the National Academy of Sciences, USA* 111: 81–86.

Rose, M. R., Burke, M., Shahrestani, P. and L. D. Mueller. 2008. Evolution of ageing since Darwin. *Journal of Genetics, Indian Academy of Sciences* 87: 363–371.

Singer, M. A. 2013. *Are Chronic Degenerative Diseases Part of the Ageing Process? Insights from Comparative Biology.* Nova Science Publishers, New York.

Singer, M. A. 2015. *Human Ageing, a Unique Experience.* World Scientific Publishing, Singapore.

Singer, M. A. 2016. The origins of aging: Evidence that aging is an adaptive phenotype. *Current Aging Science* 9: 99–115.

Storrs, E. E. and R. J. Williams. 1968. A study of monozygous quadruplet armadillos in relation to mammalian inheritance. *Proceedings of the National Academy of Sciences, USA* 60: 910–914.

Vijg, J. and J. Campisi. 2008. Puzzles, promises and a cure for ageing. *Nature Reviews* 454: 1065–1071.

Vijg, J. and B. K. Kennedy. 2016. The essence of aging. *Gerontology* 62: 381–385.

Villar, D. and D. T. Odom. 2016. Unwinding limb development. *Cell* 167: 598–600.

Weismann, A. 1891. *Essays Upon Heredity and Kindred Biological Problems.* Vol. 1, edited by E. D. Poulton, S. Schonland and A. E. Shipley. Clarendon Press, Oxford.

Williams, G. C. 1957. Pleiotropy, natural selection, and the evolution of senescence. *Evolution* 11: 398–411.

Williams, S. A. and M. R. Shattuck. 2015. Ecology, longevity and naked mole-rats: Confounding effects of sociality? *Proceedings of the Royal Society B* 282: 20141664.

Wood, J. G., B. C. Jones, N. Jiang et al. 2016. Chromatin-modifying genetic interventions suppress age-associated transposable element activation and extend life span in Drosophila. *Proceedings of the National Academy of Sciences, USA* 113: 11277–11282.

Wray, G. A., H. E. Hoekstra, D. J. Futuyma et al. 2014. Does evolutionary theory need a rethink? No, all is well. *Nature* 514: 161, 163–164.

Section II

Aging Hypothesis

3 Evolutionary Theories of Aging
A Systemic and Mechanistic Perspective

Quentin Vanhaelen

CONTENTS

INTRODUCTION

On our earth, there exist a tremendous number of different species. Since the seminal work of Darwin, this diversity is explained in part as being the result of a succession of adaptations occurring at various timescales as a response to environmental or internal changes. Understanding how and why these adaptations, called evolutionary changes, occur is one of the most fundamental questions in life science. Owing to the inherent complexity of the problematic and the accumulation of various experimental data demonstrating the variety of evolutionary patterns encountered within the tree of life, the initial version of the theory of evolution proposed by Darwin has continuously undergone modifications and improvements. Its most recent version, known as the Modern Synthesis, has been shaped upon the experimental discoveries and theoretical progress made in the different fields of life sciences and was formalized by Ronald Fisher, J.B.S. Haldane, Sewall Wright, and Julian Huxley [1,2]. Broadly speaking, Modern Synthesis is based on the mechanics of evolution and the formalism describing the effects of the driving forces of evolution. As its name suggests, the main achievement of this evolutionary theory is to provide a conceptual framework able to unify the point of view of geneticists, naturalists, and paleontologists about the mechanisms behind evolutionary changes.

Experimental observations show that each species can be characterized by a set of specific treats. Nevertheless, the development of an individual of any species follows a common sequence, the life cycle, which contains two main stages. The first one is growth which includes birth, childhood, and maturation, and the second one is the reproduction stage which is followed by aging and death. Evolution is thought to act on the different steps of this sequence by progressively modifying specific traits through different types of mechanisms. Historically, natural selection was considered as the only driving force of evolution and for that reason studies in that field were essentially restricted at the level of the phenotypic traits. The development of population genetics and molecular genetics allowed the analysis of biological changes induced by evolution from a molecular and a cellular perspective and it is now admitted that other processes such as random genetic drifts act as a part of the global driving force of the evolution [3–5]. These studies showed, for example, that modifications at the genomic level can arise through mutations affecting the expression of specific genes, which in turn affects one or several phenotypic traits. Modifications affecting some traits, called changing life history traits, have a bigger impact than modifications on other minor traits of the species. Furthermore, the modifications can have beneficial, neutral, or deleterious effects and modifications which are beneficial for the species on a long-term basis are not necessarily beneficial for a single individual on the timescale of a life span.

Globally, it is assumed that evolution acts by eliminating detrimental mutations through natural selection while keeping neutral and beneficial ones unaffected. Currently, determining which process between natural selection, genetic drift, and other suggested mechanisms should be considered as the dominant term of the forces of evolution is still a matter of debate. For example, while theories based on the Selectionist Hypothesis still see selection as the main force of evolution [6,7], the neutral theory proposed by Kimura and based on the neutralist hypothesis supports that evolutionary forces are ruled by random genetic drift [8–10]. This debate should actually be considered from a broader perceptive. Indeed, several hypotheses behind the theoretical foundations of Modern Synthesis are under strong pressure [11] or were discarded [12]. This implies that although the current version of Modern Synthesis is still the main theoretical framework to analyze the various aspects of the evolution, one can expect that new conceptual developments could emerge [13,14].

Independently of the equilibrium between the various known or still unknown processes contributing to the forces of evolution, the purpose of the evolutionary mechanism is to ensure that, on a more or less long-term basis, a species is able to adapt to changes occurring in its environment, a necessary condition to guarantee the perpetuation of the species. Thus, a successful adaptation will result in an improved equilibrium between mortality and reproduction rates in order to allow each generation to produce a larger progeny. Indeed, more descendants imply that the beneficial traits are more easily carried on into future generations and can contribute to strengthening the species. The number of descendants of a species is associated with its fitness function. By modifying the mortality and reproduction rates, evolution works toward increasing the fitness and the fitness function is usually expressed in terms of reproductive and mortality rates. Thus, from an evolutionary perspective, the changing life history traits of a species are the ones whose adaptation induces the most significant difference in the level of fitness.

Aging and its associated biological process of organismic decay called senescence is a treat shared by most, if not all, living species. Historically, it is Leonard Hayflick and Paul Moorhead who discovered that normal human fibroblasts have a finite proliferative capacity in culture [15]. They named this process cellular senescence, and assumed that it could be associated with the onset of aging. From a molecular point of view, cellular senescence can be defined as a process in which cells cease dividing and undergo distinctive phenotypic alterations, including profound chromatin changes and tumor-suppressor activation. Nevertheless, from an evolutionary perspective, the potential functions of senescence and aging with respect to survival, adaptability, or natural selection are still not well understood.

During the past decades, many theories were suggested to explain from various perspectives the role of aging and its associated regulatory mechanisms. From an evolutionary perceptive, aging can be analyzed from different complementary points of view. First, as a trait, does aging, despite having obvious deleterious effects at the level of individuals, provide any significant advantages in terms of probability of survival for the kin? Is it possible that these advantages could depend on the context and environment faced by the species? To answer this question, most of the evolutionary theories of aging are based on the evolutionary mechanics and suggest mechanisms of aging using the interplay between the processes of mutation and selection. Nevertheless, these theories diverge on many key aspects. For example, they provide different answers when it comes to defining aging as a beneficial, neutral, or detrimental trait for individuals and/or kin. Furthermore, they often support different mechanisms as ultimate causes of aging [16–18]. These different mechanisms are most of the time complementary and represent incomplete descriptions of aging from different perspectives which should be unified in a more integrated framework. To that end, it has been recently suggested in Reference [19] to categorize the main theories of aging depending on how aging is supposed to affect the fitness function. The four main categories obtained can be summarized as follows. The first category is called maladaptive aging. An example is given by the mutation accumulation theory proposed by Medawar and Charlesworth which defines aging as a by-product of natural selection with an accumulation of genes with negative effects at old age from one generation to another. This accumulation of deleterious genetic material is caused by the lack of selection pressure in long-lived animals [20]. The second category is called secondary aging. The antagonistic pleiotropy theory belongs to this category. This theory proposes that some genes can have positive effects at young ages but become harmful when the organism gets older. Thus, both mutation accumulation and antagonistic pleiotropy predict that specific mutations in particular genes will cause senescence; however, antagonistic pleiotropy adds an adaptive aspect in the sense that mutations that are damaging for the organism later in life could actually be favored by natural selection if they are advantageous early in life, resulting in increased reproductive success. Another example of this category is the disposable soma theory. The disposable soma theory is actually an improved version of the antagonistic pleiotropy in the sense that it provides a more detailed explanation about how a gene could have both deleterious and positive effects. The hypothesis supported by the disposable soma theory is that organisms face a trade-off between dedicating energy to reproduction or investing it in the maintenance and growth of their somas. This ultimately results in disruption of repair mechanisms, leading to an accumulation of cellular damage. The third category is called assisted death. These theories consider aging as a beneficial trait for the kin. Finally, the category of senemorphic aging considers that aging can be beneficial or detrimental depending on the environmental conditions [21].

Another purpose of evolutionary theories of aging is to understand how aging patterns and mortality curves obtained from demographic data are shaped. Indeed, aging patterns strongly vary between species [22] and in order to explain these observations, it is necessary to identify what are the mechanisms whose effects can result in the shapes of survival curves. Currently, there are several of these causal models which are able to describe several important topological features of these mortality curves, but generally these evolutionary demographic models of aging do not provide us with an integrated framework for understanding the origin of the diversity of aging patterns observed through the tree of life. The difficulty comes from the fact that the mortality curves are the result of complex relationships between living style, effects of natural selection, environmental conditions, and fine-tuning of cellular mechanisms, all of these parameters being at various degrees specific to each species. Demographic models have been built to describe the effects of natural selection together with mechanisms such as accumulation of mutations. They usually also integrate other variables such as resources availability and reproduction rate and can take into account the effects of competition between species. However, a more challenging task is to include the effects of the fine-tuning of cellular mechanisms. Indeed, incorporating such complex information within these models requires having a detailed description of these mechanisms, that is, a description of how aging occurs and propagates with the living system.

There are currently many recognized mechanisms of aging including inflammation, apoptosis, oxidative stress, accumulation of DNA damage, cell-cycle deregulation, mitochondrial dysfunction, and telomere shortening, just to name a few. Each of these mechanisms can be associated with specific disparate damages and pathologies, called aging-related diseases (ARDs), which commonly appear when individuals get older. For example, various types of cancers are identified as ARDs and their origin is assumed to be connected to genomic instability and decreased capacity for DNA repair, two characteristics of both cancer and aging [23]. Telomere length and telomerase activity are also involved in aging and diseases like Alzheimer's dementia [24].

Mitochondrial dysfunction is another common property of aging and cardiovascular diseases [25,26]. Chronic inflammation also appears in old individuals and could also contribute to cardiovascular diseases [27] and neurodegenerative diseases [28]. Hence, during the past decades, many of these mechanisms triggering ARDs have been used to elaborate specific theories of aging. There are currently an impressive number of theories trying to explain the onset and propagation of aging, usually from a restricted set of molecular mechanisms associated with ARDs. Nevertheless, according to Trindade et al. [19], these causality theories of aging can be classified into two classes. First, there are the entropy-based theories. These theories see death as the result of a relatively long period of degeneration. From a mechanistic point of view, aging is then described as a collection of cumulative changes to the molecular and cellular structure of the adult organism, which result from essential metabolic processes, but which also, once they progress far enough, increasingly disrupt metabolism, resulting in pathology and death [29,30]. Second, there are the sudden death theories. The main hypothesis behind these theories is that death follows either a relatively short period of degeneration or is an almost instantaneous process. Altogether, these theories provide different insights of aging and are helpful to develop adapted therapies to cure ARDs. Nevertheless, they remain based on one or a few mechanisms connected to a subset of disparate damages and ARDs, and consequently, they do not provide the comprehensive and unified mechanistic description required to fully integrate the systemic nature of aging.

The origin of the systemic nature of aging is a consequence of the hierarchical organization of living systems such as the human body. Indeed, the human body is a multilevel complex system firstly constituted of billions of independent cells which form different types of tissues. These tissues are the main blocks used to assemble organs and these organs are themselves organized in different systems such as lymphatic, respiratory, digestive, urinary, and reproductive systems to achieve specific tasks. As a result, dysfunctions affecting even a restricted number of biological processes within the cells of one or several organs can propagate to all parts of the body. This explains why aging cannot be fully understood or controlled when monitoring only a restricted number of physiological processes.

The study of this systemic organization of the living system, which has its equivalent inside any single cell, has been made possible by two main technological trends. First of all, there is the accumulation of various high-throughput data generated from different research areas such as proteomics, genomics, chemoproteomics, and phenomics which provides a better insight of the main components of the living cell [31–33]. Second, there is the progress made in computational and mathematical sciences [34]. These progresses, combined with the availability of increasingly powerful computational resources, allowed the development of software for retrospective analysis as well as the maintenance of web-based databases which are required for the gathering, classification, and efficient use of these experimental data.

These trends have also been accompanied by a conceptual change within biology with the transition from a qualitative, structural, and most of the time static description of the cell to a more systemic description in terms of functional but also dynamical properties [35–37]. For example, in addition to the traditional approaches of biochemistry and biophysics, studies focused on the hierarchical organization of the cellular environment and, on the dynamics of its components, have identified dynamical motifs and cycles [38,39] as key elements involved in the regulation of the cellular behavior. Inside the cytoplasm, proteins interacting together are organized as structured modules such as the signaling pathways which are well known for being involved in the transmission of all signals received from the external environment to any concerned cellular components. Inside the nucleus, genes and transcription factors also form structured dynamical patterns called gene regulatory networks (GRN). GRN can be visualized in the form of a directed graph whose nodes represent the genes. An edge between two nodes represents an interaction which can be either an inhibition or activation. These structures are deeply involved in many regulatory processes including the regulation of genes expression. The regulation of gene expression is a complex process also involving mechanisms such as mRNA splicing [40,41], chromatin remodeling [42], and epigenetics modifications. Epigenetics refers to how transcription, DNA replication, and other aspects of genome function are regulated in a manner that is independent of DNA sequence.

Epigenetic modifications are dynamical adaptations of the structure of the chromatin which contribute to the regulation of gene transcription. The structure of the chromatin can be modified at the level of the histones, the predominant protein components of chromatin, via different molecular processes [43–46]. It has been demonstrated that the histone modifications are regulated by conserved protein modules and follow a well-organized dynamical scheme known as the histone code [47,48]. The apparent complexity of these dynamical structures is progressively better understood with the establishment of relationships between their topology and their dynamical behavior [49]. For example, the complexity of the regulatory machinery can be characterized in terms of generic properties such as robustness [50], modularity, and evolvability [51–53]. Within this framework, it is intuitively easier to understand how components of the proteome or genome participate together in many different processes which occur in different cellular compartments. As a consequence, a dysfunction of a small set of molecules affecting a restricted number of defined epigenetic [54] and metabolic processes [55,56] may propagate to all parts of the cell, leading to a progressive disruption of the general homeostasis. Hence, the systemic description of the organism naturally leads to the definition of aging as a dynamical and systemic process whose external symptoms are the disparate damages and ARDs described above [57,58].

This chapter is organized as follows. We begin with a presentation of an example of an evolutionary experiment performed on mutated colonies of *Escherichia coli*. *E. coli* is a simple organism with a short life span which can be easily cultured in various experimental conditions. It can also be modified through genetic engineering to introduce mutation if necessary. This example illustrates how it is possible to follow the effects of selected mutations over generations by selecting species with appropriate traits. This type of experiment allows to observe in real time the effects of the evolutionary dynamics and to identify regulators of such mechanisms. In this experiment, one is interested in understanding how stochasticity in gene expression can play a role in the evolutionary process. Indeed, it has been shown that the magnitude of fluctuations in protein abundance among bacterial species such as *E. coli* is large and is reflected in the diversity of phenotypes that can be encountered.

Fluctuations in gene expression have an important impact on the selection and robustness of biological functions and traits. We will discuss how phenotypic fluctuations increase the ability of a species to quickly adapt to various types of environments including severe conditions. Furthermore, to better understand the quantitative connections between phenotypic fluctuations, effects of mutations, robustness, selection, and fitness, we discuss several theoretical works which show how these relationships can be modeled and studied. These simple models show how systems eliciting stochastic fluctuations are shaped through evolution to be more robust with respect to noise and mutations during development. Another model illustrates how a combination of selection driven by fluctuations is accompanied by optimal growth rate selection to improve fitness and robustness. In the last example, the possibility that the growth rate could be an alternative mechanism for the adaptation to changes in environment is discussed.

In the following section, we begin our journey through the evolutionary theories of aging by clarifying the fundamental difference between the common chronological age and its biological counterpart. Biological age is intrinsically connected to health status, which is itself dependent on the ability to maintain homeostasis, which decreases with age. The systemic nature of aging makes any measurement of biological age a technical and conceptual challenge. Modern computational techniques which offer great promises for addressing these issues are shortly described. In the next section, we discuss the relationships between aging and its most obvious symptoms, the disparate damages. It is well known that these disparate damages are a consequence of the onset of age-related diseases (ARDs). Moreover, the role played by specific genes which undergo change in expression with age, the so-called age-related genes (ARG), is also investigated. Using computational analysis, it is found that these ARGs are strongly connected to genes known to be related to ARDs. Furthermore, the results tend to demonstrate that aging could be the consequence of local topological changes in the vicinity of ARGs which ultimately affect the stability of the complete biological system.

In the following section, the parametric models and associated demographic models used to model the aging patterns obtained from life tables are introduced. We discuss how such functions can be helpful to identify generic properties of these aging patterns. The limitations of these simple models are emphasized and the possibilities offered by more complex models recently published are described. These models integrate explicit descriptions of key biological mechanisms such as senescence or mutations and, as a result, they are able to suggest mechanistic explanations of the aging patterns observed. Furthermore, by reinforcing the description of the connections between the apparition of mutations, the selection process, and the topological properties of GRN, these models are able to suggest mechanisms to explain several experimental findings. Interestingly, these models suggest that the various theories of aging, far from being mutually exclusive, could be unified within a single framework.

The next section focuses on mathematical models based on population dynamics which includes environmental constraints. The results show how changes in environment can affect the behavior and the survival capacity of individuals. This confirms the fact that evolutionary changes are triggered in order to adapt to the characteristics of the environment. This can be seen as a continuous optimization of the fitness function. These models also clarify the role played by aging in terms of mechanism of evolution.

In the last section, following a bottom-up approach, we develop a succinct mechanistic and systemic description of central cellular components involved in the regulation of cellular maintenance. Recent results regarding our understanding of senescence, telomere attrition, regulation of ROS, and epigenetic regulation demonstrate that aging should be considered as the consequence of the disruption of the cellular homeostasis. These findings illustrate that it is important to consider a living organism as a dynamical system made of connected components organized through different levels of organization with genomic and proteomic levels being interconnected [59–62]. This feature has a strong implication in the sense that any biochemical process, including the most complex ones, is always a component of a wider set of connected processes acting on each other in continuous interaction with the external environment of the living organism. We end up with a conclusion emphasizing that aging is essentially a systemic process that is a consequence of not only local malfunctions of specific components but also a result of a disruption of interactions occurring between different subsystems that are critical for the maintenance of the homeostasis within the body. Within this framework, cellular maintenance is mainly a continuous search for a dynamical homeostasis between the constraints inherent to the biochemical cellular processes, the evolutionary forces and the external environment. Thus, aging should be considered as a disruption of this dynamical homeostasis which could also serve as an evolutionary mechanism favoring adaptation to strong changes in living conditions.

PHENOTYPIC FLUCTUATIONS: AN EVOLUTIONARY MECHANISM TO IMPROVE FITNESS

The behavior of living organisms is regulated by biochemical reactions. As biochemical reactions occur, concentrations of metabolites, proteins, and mRNA vary over time. It is well known that biochemical compounds, especially mRNA, can be present at very low levels. Low concentrations have strong implications on the temporal behavior of these compounds because it induces stochasticity, that is, rather than switching from one specific concentration corresponding to a given steady state to another fixed one, the concentration of these compounds continuously fluctuates around the equilibrium. These fluctuations are caused not only by the stochasticity in intracellular reaction processes but also by external noise. Strong fluctuations can affect the level of proteins; the various processes there are involved in and lead ultimately to different phenotypes. Since these fluctuations are a characteristic shared by all species, one may ask whether they serve a specific purpose from an evolutionary point of view. Are these fluctuations providing an advantage in terms of adaptation abilities, improved fitness, or robustness? The impact of these phenotypic fluctuations has been examined in relation to biological processes, differentiation, and also evolution. It has been suggested that these fluctuations prevent the maintenance of a state with higher function in the sense that, under fixed environmental conditions, fluctuations around an optimal state could reduce fitness. Thus, a decrease in fluctuation during the evolutionary process is advantageous as it reduces the fluctuation around the fittest state, which in turn contributes to maintenance of optimal traits and functions. Experiments have shown that the enhancement of a trait is accompanied by a decrease in both the fluctuation and rate of evolution, and other studies made additional findings by showing that the fitness improvement with generation is accompanied by a reduction of the phenotypic fluctuation. Other experiments have shown that the magnitude of phenotypic fluctuations actually increases with the rate of evolution under fixed environmental conditions.

Another example of an evolutionary experiment is presented in Reference [63]. In this case, cycles of mutation and selection for higher GFP fluorescence were carried out in *E. coli*, cultured in a severe environment allowing a small number of individuals to survive, in order to better understand how natural selection could affect these phenotypic fluctuations and what effects these fluctuations could have on the fitness. Interestingly, when the selection from one generation to the next one is based on the average phenotype of each genotype, the fluctuation decreased gradually. On the other hand, when selection is based on the individual phenotype, the fluctuations tend to increase. These findings support the hypothesis that under severe selection happening at the individual levels, the increase in phenotypic fluctuation could be an evolutionary strategy as these fluctuations together with a genetic diversity allow maintaining a diverse phenotype which is necessary to adapt to a severe environment when it is encountered. From this point of view, the phenotypic fluctuations may be a mechanism maintained in the evolutionary history of facing severe environment.

Modeling approaches can be helpful for formulating and testing hypotheses to explain the connections between the fitness function, the effect of natural selection, and dynamical characteristics of the biological system such as robustness and the presence of stochastic fluctuations. Such models can help understand why a large amount of phenotypic noise has been preserved through evolution and how it can intervene in the evolution of biological functions. An example of such a modeling approach is presented in Reference [64]. The mathematical model is based on a GRN which provides the genotype of the living system and hence different genotypes can be obtained by simply varying the topology of the GRN. The effects of natural selection and the fitness function associated with a given phenotype are computed as follows. First, a set of Boolean-based dynamical equations is established to simulate the change of the gene expression level over time. These equations form the developmental dynamical systems. A state variable is associated with each gene and is set to zero if the gene is inhibited or to one otherwise. The set of values associated with these variables defines the gene expression pattern. Second, starting from an initial condition, one computes the evolution of the dynamical system. The set of values of the state variables changes in time according to the dynamical equations and eventually reaches a stationary pattern which is the phenotype. Third, the fitness function is computed using these state variables. Concretely, the fitness is determined by setting a target gene expression pattern. The fitness is at its maximum if a given set of genes is activated after a transient time and at its minimum if all are inhibited. The fitness is set to zero, which is the optimal value, if all the target genes are on. Finally, during reproduction, mutations are introduced by modifying the topology of the network and, as a result, there is a redistribution of dynamical systems at each generation. Selection is applied after the introduction of mutation at each generation. Among the mutated networks, those with higher fitness values are selected. Since the model contains a noise term, the fitness can fluctuate at each run, which leads to a distribution of the fitness, even among individuals sharing the same network. For each individual network, the average fitness over a given number of runs is taken and networks with the highest fitness are selected from one generation to the next. If evolution serves to increase fitness, dynamical systems with a higher function should be shaped through this selection–mutation process.

Robustness is defined as the ability to function against changes in the parameters of a given system. In the case of the system considered here, one needs to consider changes of genetic and epigenetic origin. The former concerns structural robustness of the phenotype, that is, rigidity of the phenotype against the genetic changes produced by mutations which are represented by changes in the network topology. The latter concerns the robustness against the stochasticity that can appear as a result of fluctuations in initial states and stochasticity occurring during developmental dynamics or in the environment. In terms of dynamical systems, these two types of robustness are the stability of a state, also called attractor, to external noise and the structural stability of the state against changes in the underlying equations.

The results obtained with this model are in agreement and complete the experimental observations discussed in Reference [63]. The average fitness computed using this model exhibits three distinct behaviors depending on the level of noise. For a very low level of noise, the average fitness stays lower than the fittest value. On the other hand, for very large values of noise, the distribution of the average fitness is sharp and concentrated around the top value. Nevertheless, the top fittest value is never achieved. However, for a middle range of noise level, the distribution is not only sharp but also concentrated at the fittest value. Even individuals with the lowest fitness approach the fittest value. This range of value, called the robust concentration region, is reduced as the mutation rate increases. Interestingly, individuals evolved at very low noise do not have robustness against mutation, whereas individuals evolved at higher noise do. In other words, the fitness landscape has an almost neutral region where it is insensitive to mutation, demonstrating the evolution of mutational robustness.

To summarize, under high noise conditions, the selection process favors a developmental process that is robust against noise. This robustness to noise is then embedded into robustness against mutations. Hence, a dynamical system that is robust both to noise and to structural variation is shaped through evolution under noise. This establishes a correlation between developmental robustness to noise and genetic robustness to mutation with the former leading to the latter.

Cells adapt to a variety of environmental conditions by changing the pattern of gene expression and metabolic flux distribution. As discussed in the introduction, adaptive responses can be explained by signal transduction mechanisms, where extracellular events are translated into intracellular events through regulatory molecules. However, considering that the huge number of various environmental conditions encountered by a living species such as *E. coli* is higher than the set of regulatory gene mechanisms available, one may ask whether an alternative adaptation process could exist in addition to adaptation through gene regulation by signal transduction mechanisms. Actually, it has been shown that species such as *E. coli* cells are able to select appropriate intracellular state according to environmental conditions without the help of signal transduction.

To better explain this alternative mechanism of adaptation, a more elaborated model combining a GRN together with a metabolic network has been developed in Reference [65]. In this model, the regulation of gene expression affects the metabolic network which is also affected by changes in external conditions. The mathematical model takes into account the change in gene expression over time and includes effects linked to the synthesis of proteins, dilution of proteins by cell volume growth, and the molecular fluctuations arising from stochasticity in chemical reactions. In order to include the effects of the environment, temporal changes in concentrations of metabolic substrates are also considered. They are given by metabolic reactions and transportation of substrates from the outside of the cell and some nutrient substrates are supplied from the environment by diffusion though the cell membrane, to ensure the growth of the cell. The temporal evolution of both expression levels and growth rate are analyzed for various levels of noise and one notices that without noise or for a very small level of noise, the cellular state rapidly converges deterministically to an attractor and the final growth rates are broadly distributed. This broad distribution is explained by the fact that, in this model, the selection of the more adapted state is only allowed by the stochasticity of gene expression. Thus, in the absence of noise, selection cannot happen and cells simply remain trapped in the first attracting state they encounter. When the level of noise is very large, the cellular state continues to change without being able to settle into any attractor and a similar broad distribution of the growth rates is also obtained. On the other hand, for a range of average values of noise, one observes that the selected states are always associated with a significantly higher growth rate.

From a dynamical point of view, these observations are explained by the fact that growth rate is the deterministic part of the protein expression dynamics. If the noise is very small, this deterministic part determines completely the selection of the state. As noise level increases, there is a competition between the deterministic and stochastic parts which result in the selection of an optimal growth rate according to the level of noise. It is worth mentioning that selecting a state with a high growth rate is also a way to increase the robustness of the chosen state and connected biological functions as a higher growth rate reduces the probability to switch to another state because of noise. It can also be noted that any perturbation of the environment leads to a perturbation of the state, but ultimately the system still chooses the available state with the optimal growth rate.

DEFINITION AND MEASUREMENT OF BIOLOGICAL AGING

The age commonly attributed to individuals, called chronological age, measures how long a human has been alive. This measurement is based on arbitrary units from the calendar time and is disconnected from clinical measurements of the physiological status of the individuals. The actual physiological health of the individual is better represented by the biological age, also called the physiological age. The biological age can be broadly defined as a measure of how well the different organs, physiological processes, and regulatory systems of the body perform and at what extent they are being maintained. Thus, in theory, monitoring the biological age can provide a better estimate of the health status than the chronological one. Consequently, it could be a quantity of great interest for building theories of aging or demographic models and for testing their predictions using real data. However, aging is a systemic process which involves and affects many different physiological systems at the same time and, consequently, obtaining a correct estimate of the biological age is not straightforward. In practice, it is possible to measure separately any physiological process inside the body with clinical procedures based on the use of predefined biomarkers. A biomarker is a characteristic that is objectively measured and evaluated as an indicator of normal biological processes or pathogenic processes. Generally, biomarkers are developed with the purpose of measuring a very well-defined functionality within the body and, as such, they are not necessarily adapted for measuring the effects of a systemic process such as aging.

So far, there have been several attempts to develop markers of aging. These methods monitor not only one but a restricted set of physiological functionalities whose disruptions are known to trigger the onset of specific diseases and malfunctions correlated with aging. Nevertheless, these biomarkers of aging consider a restricted number of cellular mechanisms involved in aging and, as a result, they are unable to represent the health state with enough accuracy. Furthermore, many of these biomarkers, such as biomarkers based on the measurement of epigenetic mechanisms, are not easily measured or targeted with already known clinical interventions. Thus, an accurate and practical measurement of the biological age requires a set of biomarkers preferably selected from standard clinical biomarkers. Taken together, these biomarkers should not only be an objective quantifiable and easily measurable characteristics of biological aging but should also be able to take into account that aging is not a single specific process, but rather a suite of changes that are felt across multiple physiological systems.

From an experimental point of view, the development of biomarkers is a time-consuming and tedious multistep process which includes proof of concept, experimental validation, and analytical performance validation and more effective approaches can be helpful for developing the complex biomarkers required to measure biological age.

As discussed above, the increase in throughput technologies has generated a massive accumulation of various types of -omics data which can be used in many ways for the development of adapted biomarkers. In practice, the use of these data can actually be complicated because they are highly variable, high-dimensional, and sourced from multiple often incompatible data platforms.

Nevertheless, several types of appropriate in silico methods can be applied to exploit this huge amount of information in order to identify potential biomarkers and contribute to accelerating the development process. For example, machine learning (ML) techniques are already routinely used in biomarker development. Among them, deep learning (DL) methods are the latest generation of ML techniques and they have already shown promising results in different applications such as predicting various physical and chemical properties, modeling drug–target interactions using structural data, or predicting toxicity issues [66–68]. Owing to this flexibility and adaptability of DL, these methods are now considered as interesting computational approaches for tackling many current biomedical related issues [69–71].

DL techniques are based on the use of deep architecture, called Deep Neural Networks (DNNs). DNNs are collections of units, also called neurons, connected in an acyclic graph. DNN-based models are often organized into distinct layers of units. For regular DNNs, the most common layer type is the fully connected layer in which units between two adjacent layers are fully pairwise connected, but units within a single layer share no connections. One of the main features of these DNNs is that units are controlled by nonlinear activation functions. This nonlinearity, combined with the deep architecture, makes it possible to take into account complex combinations of the input features leading ultimately to a wider understanding of the relationships between them. Once correctly trained and optimized, DNNs are capable of providing more reliable final output than standard ML approaches. Recently, Putin et al. [72] have published a study demonstrating the capacity of transcriptomic-based DNN methods to accurately predict biological age. They were able to identify a set of relevant biomarkers which can be used for tracking physiological processes related to aging. The features used as inputs, a set of 41 biomarkers for each sample, were extracted from tens of thousands of blood samples from patients undergoing routine physical examinations. Although being highly variable in nature, the blood biochemistry test has the advantage of being very simple to perform in practice. Furthermore, it is approved for clinical use and, as a consequence, commonly used by physicians. In the study, identified biomarkers and, as a consequence, the associated physiological processes were ordered according to their importance with respect to the aging process itself. The five most important biomarkers identified were albumin whose low level is associated with increased risk for heart failure in the elderly, glucose which is linked to metabolic health, alkaline phosphatase whose level in blood increases with age, erythrocytes which are known to be damaged by oxidative stress, and urea which is known to increase oxidative stress. These five biomarkers monitor the physiological status of renal, liver, and metabolic systems as well as respiratory function.

This kind of computational approach demonstrates interesting performances and further improvements have already been suggested. For example, adding other sources of features including transcriptomic and metabolomics markers from blood, urine, individual organ biopsies, and even imaging data. Furthermore, one should take into account genetic determinants, environmental conditions, and living styles which can be very different between human communities and can affect the aging rates across countries. From a more practical point of view, specific care should be taken when collecting the samples because there might be specific biases coming from the methods used to collect and analyze clinical samples which differ substantially across health systems. Furthermore, in order to better represent the diversity of the aging patterns observed through the different populations, one should probably implement population-specific algorithms. Indeed, it is unlikely that a single algorithm could accurately predict biological age for all populations. The measurements of the biological age being more representative of the health of individuals, it could be interesting to analyze the difference between mortality curves obtained from measurement based on chronological or biological age measurements and for what conditions they strongly differ from each other.

RELATIONSHIPS BETWEEN ARGS, ARDS, AND AGING

We have discussed in the introduction that several evolutionary theories explain the onset of aging by the fact that specific genes, called ARGs, which are beneficial at a young age can, for whatever reason, become harmful at older age. Also, we have mentioned that various sources support that it is indisputable that many diseases appear or have a higher probability to appear as individuals get older, a concrete example, well known in developed societies, being the dramatic increase in the onset of cancers within the population aged 65 years and older compared to the population aged between 20 and 44 years [73,74]. We have already mentioned that these ARDs could be interpreted as being the external symptoms of systemic disruptions of the homeostasis within the organism.

However, to better define the connection between ARGs, ARDs, and aging (senescence), it is necessary to clearly identify what distinguishes ARGs from other genes and also whether they have specific functions or occupy a specific location within GRNs. Currently, the establishment of mechanistic relationships between the ARDs, ARGs, and the aging process itself is still a matter of intensive research but several studies have provided interesting results in recent years.

To begin with, it is worth mentioning that attempts to identify ARGs and related ARDs using gene expression data alone have provided us with results revealing that from as few as 442 [75] to as many as 8277 [76] human genes can be related to aging. This significant variation can be explained by the fact that, in many studies, the interactions between genes of the associated network are taken into consideration with poorly designed topology measures and that ARGs are identified mainly using the fact that these genes demonstrate a significant change in expression with age.

In order to overcome these limitations, more complex approaches have been developed. For example, in Reference [77], ARGs are identified by investigating how the topology of the corresponding protein–protein interactions (PPI) networks evolves with age. To perform this analysis, they firstly designed a set of dynamic age-specific PPI networks by selecting all proteins that correspond to actively expressed genes at different ages and all PPIs involving these active proteins. Hence, each age-specific network is the network that is active at a given age. These networks were then analyzed with various constraining measures to study how the topology can evolve at the global and local levels. The authors have formulated three interesting conclusions. First, global network topologies do not change with age and the overlap of age-specific networks is large. The age-specific networks share on average 92% of the nodes and 89% of the edges, depending on age, whereas every pair of the networks shares at least 82% of the nodes and 74% of the edges. Second, local topologies of PPIs show significant modifications with age. More precisely, local topologies around only a subset of proteins in the networks do change with age and aging-related information remains stored only locally within the networks. This feature can explain why other global network analyses were unable to uncover any aging-related information. Third, the nodes of the network which undergo major modifications in their interactions are related to the ARGs and ARDs. This conclusion is based on the observation of significant overlaps between predicted and well-known ARGs. Furthermore, there is an overlap between functions and diseases that are enriched in their aging-related predictions and those that are enriched in well-known aging-related data.

In their study, the authors were also able to provide experimental evidences that diseases which are enriched in their aging-related predictions are also related to human aging. Other studies have been performed to analyze these specific relationships. These works, based on a top-bottom strategy, are also based on systemic approaches in the sense that they analyze the interactions between biological processes and related genes rather than considering them separately. A common characteristic of these strategies is that they use various kinds of rather large interaction networks (including PPIs, gene interactions, pathway maps, etc.) where, on the one hand, one identifies genes undergoing strong changes in their regulation with age (ARGs) and, on the other hand, one identifies genes that are included in the transcriptomic profiles associated with ARDs or non-ARDs. The core of the statistical analysis which follows relies firstly on applying topology-based criteria to determine whether ARDs-related genes are closely connected (directly or indirectly) to ARGs. Second, one identifies overlapping sets of ARGs and ARD-genes that could clarify the mechanistic connections between them. By following this strategy, it has been shown that cellular senescence was interconnected with the aging process and ARDs, either by sharing common genes and regulators or common PPIs and signaling pathways [78]. In another work, on the basis of a network of ARGs and disease genes only, it has been found that ARGs were topologically closer to disease genes than by random chance. Furthermore, the results also supported that diseases having their transcriptomic signatures significantly close to ARGs can be clearly distinguished from non-ARDs [79].

Recently, Yang et al. [80] presented the results of a similar study made on a larger scale. By using an elaborated combination of statistical and topology-based analysis of gene–gene interactions within modularized networks, they have been able to identify not only meaningful biological processes involved in the diseases and their connections to ARGs but also key genes and modules related to biological processes that potentially mediate disease-aging connections. This includes, for example, modules involved in cell-cycle regulation and modules controlling the response to decreased oxygen levels. These results are corroborated by the fact that decreased oxygen levels are usually associated with aging [81] and, more generally, hypoxia condition is known for being involved in cancers [82], tumor survival, and inflammatory response [83]. Furthermore, the complexity of the observed interactions suggests that connections between ARDs and ARGs are mediated by different numbers of sub-networks. They also conclude, in agreement with [79], that ARDs and non-ARDs have significantly different interaction patterns with ARGs in modularized networks. These important topological features confirm that the onset of ARDs is strongly connected to the aging process and, as a result, these features could be used to differentiate ARDs from non-ARDs. Not surprisingly, various types of cancers including lung cancer, melanoma, bladder cancer, prostate cancer, leukemia, and breast cancer are found to be closely connected to ARGs. These findings are supported by previous studies emphasizing the significant increase in the onset of these diseases with age after maturity [74]. More interestingly, the functional modules associated with these ARDs include pathways well known for their involvement in various mechanisms of aging. Among them, one can cite the intrinsic apoptotic pathway activated in response to DNA damage and various regulatory pathways of apoptosis, the signal transduction by the p53 class mediator and several modules related to cell cycle.

These findings demonstrate the role played by the evolution of the topology of the PPI network in the onset of aging. However, it remains to investigate what mechanisms drive these topological changes within the PPI networks during aging and how these changes occur. As will be discussed in the next section, the interplay between the different levels of organization of the regulatory systems and the interactions with the external environment plays a central role in this process.

To conclude, one can describe similar findings of a computational data analysis published by Rodriguez et al. [84]. The aim of this study was to identify connections between ARDs, ARGs, and senescence. Interestingly, the analysis was performed by combining human genome-wide association studies (GWAS) and senescence data from various human databases. By making use of the relationship between senescence and ARDs, together with the abundance of information on the effects of genetic variants associated with complex disease, they were able to identify evidences supporting that specific mutations in particular genes cause senescence. These findings are in agreement with the main hypothesis of the mutation accumulation and the antagonistic pleiotropy theories. The authors ended up with three interesting conclusions. First, risk alleles associated with diseases which appear at a late age have higher frequencies than risk alleles associated with diseases that manifest themselves earlier in life. From these observations, one can deduce that natural selection allows late-onset genetic variants with large effects on disease risk to reach higher frequencies, another observation consistent with the mutation accumulation theory. Second, the analysis of the patterns of pleiotropy shows that, when age thresholds from 40 to 50 years are considered, there is a significant excess of antagonistic pleiotropies with a maximum at ages 46–50 years. Third, in all data-sets used, they observed a high level of association between senescence (ARGs) and pleiotropic genes. Furthermore, using molecular evolution techniques to screen for the signature of natural selection in genes and genomic regions involved in early–late antagonistic pleiotropies, they have identified several genes that increase survival or fertility at early ages but have deleterious effects at older ages.

AGING AS A RESULT OF DYNAMICAL INSTABILITIES
CAUSED BY MUTATION ACCUMULATION

Demographic data (life tables) such as the ones available from the Human Mortality Database (http://www.mortality.org) are the primary source of information from which demographic trends and key parameters including mortality rates and fertility rates can be analyzed. Using life tables, it is possible to draw graphs, called survival curves, showing the number or proportion of individuals surviving to each age for a given species. According to Demetrius [85], the various patterns of survival curve can be classified into three types. Type-I survival curves which change at early and middle ages and then decline at late ages, as seen for humans. Type-II curves almost linearly decrease with age, as seen for short lived birds. Type-III curves quickly decrease at early ages, as seen for most plants. On these curves, the onset of processes such as senescence, for example, can be identified as an increase in mortality and a decrease in fertility with age or by the sharpness or abruptness of the increase in mortality. Furthermore, when analyzing the patterns of these survival (or alternatively mortality) curves, one observes that they elicit specific topological features which provide meaningful information about the aging patterns of the species.

More precisely, three important characteristics can be identified. First, trajectories of life and, as a consequence, mortality and fertility rates strongly vary from one species to another [22]. The reason for this variety is still a matter of debate and its understanding is one of the main motivations of current aging research. Nevertheless, a possible explanation is that the number of contributions to the mortality rate could be higher than initially expected. Indeed, various analyses have shown that the age-trajectory of mortality can be decomposed into three parts: a first part due to the accumulation of unfavorable mutations, a second part which is the result of the selection processes that optimize the trade-offs necessitated by resource limitations, and a last part attributed to unavoidable external risks of death. Second, as thoroughly discussed in Reference [86], human survival curves show a tendency to evolve toward the slowest aging rates. This dynamics is called rectangularization [87] and it formally refers to the tendency of survival curves of the human population of developed countries to evolve over time, over the past several decades, universally toward a rectangular shape. This observation could be explained by the fact that human mortality curves evolve toward the slowest aging rates, also a synonym of healthy aging. Third, although mortality curves show a clear increase in mortality with age, this trend is not applicable to very old age. Indeed, after growing exponentially, the risk of death saturates at a constant level at very old ages. The formation of this plateau is a phenomenon known as mortality deceleration. The reason explaining why mortality rate slows down for this specific period of life is not clearly understood.

There are various parametric mathematical models which have been elaborated to represent the shape of the survival curves [88,89]. One widely used model is the Gompertz law [90] which represents the mortality curve as an exponential function of time. The model contains only two parameters: One to capture the initial level of adult mortality, also called initial mortality rate (IMR), and the Gompertz exponent slope which is inversely proportional to the mortality rate doubling time (MRDT). MRDT represents the rate of aging, that is, the relative change in mortality with a given age. In practice, these parameters can be inferred from life tables, although this operation is not necessarily trivial [91]. The Gompertz law is adapted to describe the survival curves of individuals living in a protected environment where external causes of death are rare. When this assumption is not valid, the Gompertz–Makeham law can be used. This model adds to the age-dependent component, represented by the standard Gompertz law, an age-independent component, called the Makeham term, which takes into account external factors affecting the mortality rate. This generalized law of mortality is able to describe the age dynamics of human mortality rather accurately in the age window from about 30 to 80 years of age. Another commonly used function to represent the survival curves is the Kohlrausch–Williams–Watts (KWW) function. This function is more complex but, interestingly, it is possible to model the three different kinds of survival curves cited above by using the same KWW function with various parameter values. For a specific range of ages, the predictions made using these functions can be useful for the quantification of survival dynamics to scientists, such as demographers, biologists, and gerontologists. Nevertheless, since mortality is by far more complex than a two- or three-parameter process, these functions taken alone do not suffice and difficulties arise for explaining the mortality rate at a very old age, and alternative mathematical curves such as the Heligman–Pollard, the Kannisto, the quadratic, and the logistic models share similar issues [89,92].

The difficulties encountered by these parametric functions to accurately describe the survival curves are in part explained by the fact that they do not include a detailed description of the major processes that intervene in the survival rate of all species. Thus, more elaborated models are required for explaining the aging patterns and the topology of the survival curves in terms of the combined effects of accumulation of mutations, senescence, effects of natural selection, or any other parameter which is supposed to intervene in the onset and propagation of aging. Conceptually, there is a need for models able to provide formal representations of demographic processes which are coherent with the current evolutionary framework. From a historical perspective, Hamilton was the first to establish a mathematical description of the forces of natural selection [93,94]. His mathematical model expressed the survival rate in terms of the mortality rate and other parameters describing the demographic evolution of a population. The derivation of this model is based on the hypothesis that mutation accumulation is the main process behind the onset and progression of senescence through cellular tissues. In his work, Hamilton concluded that senescence was an inevitable outcome of the evolutionary process. However, other analysis performed with a similar model but with another choice of values for the parameters provided different results [95], and the extension of the mathematical formulation of this model to include nonlinear effects also challenged the traditional established scenario of aging [96].

An example of a more recent model is presented in Reference [97]. The authors have computed a mortality function to describe the survival and senescence processes. The idea is to investigate the biochemical mechanisms responsible for the shape of the Gompertz law of mortality as well as the effect of the temperature. The main hypothesis of the model is that there is a living energy which is used to protect the chemical substance that is critical for life from being impaired by damaging energy. This living energy is proportional to the quantity of a vital molecular unit and linearly decreases with time. Thus, changes in certain substances and inherent energy are involved in the aging process [98–101] and senescence is assumed to result from the imbalance between damaging energy and protecting energy for the critical chemical substance in the body. Here, telomere length is taken as the only vital molecular unit representing this living energy. The dynamical equation for the global survival function is obtained from a first-order kinetic equation for the evolution of a molecular component and contains a time-dependent coefficient rate. Furthermore, a term is added to take into account that the living energy can be reduced and enlarged by the onset of diseases or effects of medical treatments, respectively. The parameters of the function were estimated by fitting survival curves of several countries. The results show that the mortality function is similar to the Gompertz mortality function for young and middle-aged individuals. However, the rate of senescence increases when the protecting energy decreases in a temperature-dependent manner. In agreement with previous experimental findings, the analytical expression of the function predicts that reduced temperatures induce an increased life span. The model is able to explain the plateau observed on the mortality curves at late ages. This plateau is reached when the protecting energy decreases to its minimal levels.

Interestingly, the curves are in agreement with the previously observed tendency of rectangularization [102,103] of the patterns of human survival curves. Regarding the rectangularization, it is worth mentioning that a new mathematical formulation of the survival curve based on the KWW function presented in Reference [86] provides a quantitative criterion for the rectangularization tendency. The authors hypothesized that evolution toward the slowest aging rates could be a species-independent, scale invariant, universal aspect in survival dynamics of living systems. Thus, common features of mortality curves such as the Gompertz law and the mortality plateau could spontaneously emerge from the age-dependent shaping exponents that dynamically evolve toward the slowest aging rates of living systems. From an evolutionary point of view, the slowest aging rates could be a consequence of the fact that living systems tend to evolve to optimize their capabilities and strategies for survival.

Other models recently published combine an explicit mathematical formulation describing the effects of natural selection and the propagation of mutations together with several systemic properties such as the topology of GRNs associated with the onset of aging. For example, the study described in Reference [104] proposes a quantitative model for explaining aging as a consequence of the critical dynamical behavior elicited by GRNs. This specific behavior can in turn induce dynamical instabilities which are interpreted as a cause of aging. Interestingly, the presence of this instable behavior is the result of specific topological and connectivity properties of the GRNs. This is coherent with the experimental findings described in Reference [77] regarding the connections between the topology of networks associated with ARGs, ARDs, and the onset of aging itself. The mathematical model is built on the assumption that mutation accumulation is the main cause of aging and other potential causes and how they could affect each other are not taken into account. Nevertheless, this model is able to establish relationships between the mortality rate and the increasing number of regulatory errors. Furthermore, using experimental data, the model is also able to recover the main properties of the Gompertz law. Interestingly, several characteristic timescales such as MRDT and IMR were explicitly related to generic parameters of the network such as translation, gene repair, and protein turnover rates. Another interesting point is that the model suggests that the dynamics of aging could be a combination of a stochastic component, directly related to the accumulation of regulatory errors and a deterministic component, linked to some development program.

This has at least two strong implications. First, it supports the hypothesis discussed earlier that current theories of aging (antagonistic pleiotropy and theories based on damage/error accumulation) describe only a restricted aspect of aging and that the driving mechanism suggested by these theories could be complementary. This illustrates that establishing connections between the standard formulation of the forces of selection and the mechanistic description of aging could lead to a merging of these theories. This situation is also encountered when comparing from a mechanistic point of view the description of aging proposed by the mutation accumulation and disposable soma theories. These mechanisms are not only complementary but also provide a conceptual link to the biochemical nature of aging. Second, the results obtained in Reference [104] show that the dynamics of these networks contain characteristics that can be interpreted from a non-programmed as well as programmed point of view [105–108]. Thus, this is a step to establish a bridge between two (apparently) opposed interpretations of aging. This does not only suggest that these theories could be compatible but also that programmed and non-programmed theories of aging could be two aspects of a deeper mechanistic process that requires a more detailed description.

Finally, the work presented in Reference [109] investigates other quantitative aspects of the relationships between aging and stability of GRNs involved in aging which were developed in Reference [104]. The dynamical model is based on a set of linearized equations describing the temporal evolution of the improperly copied proteins and improperly expressed genes occurring as a result of mutation accumulation. Although simple, these equations include a coupling rate constant characterizing the regulation of gene expression by the proteins and a DNA repair rate for representing the action of the repair system which can prevent harmful effects arising from the accumulation of genetic mutations. Furthermore, a constant to measure the overall connectivity of the GRN and a generic constant to take into account the combined efficiency of proteolysis and heat shock response systems, mediating degradation, and refolding of misfolded proteins are also included. Using appropriate approximations, the authors were able to derive the corresponding effective equation for the age-dependent changes in the number of regulatory errors and, consequently, the mortality. The solution of this equation gives a time-dependent analytical expression of the mortality rate which includes all parameters described above. By performing a stability analysis of this solution, they identified a parameter characterizing the stability of the associated GRN. This parameter is itself dependent on the propagation rate of gene-expression-level perturbations and it can be directly interpreted as the Gompertz coefficient. Depending on the value of this parameter, the mortality rate shows two very different behaviors. For one range of values, the system is instable and the number of errors in gene expression defects grows exponentially with mortality. In that case, one recovers the expression of the well-known Gompertz law. However, for another range of values, the network stability is conserved and mortality rate is shown to increase at a smaller rate. These analytical predictions were successfully tested using several experimental data. Interestingly, this model not only suggests quantitative and experimentally verifiable links between the life span of a species and the stability of its most vulnerable GRNs related to aging but also several approaches to slow down the aging process. Unsurprisingly, acting at the level of the repair and maintenance system appears to be the most straightforward solution to slow down aging but other suggestions are discussed. The description of aging provided by this model also emphasizes the role of the GRNs in the regulation of the levels of somatic maintenance and repair functions.

The results obtained by the last two models are interesting in the sense that they do not only provide us with mechanistic explanations regarding the various patterns of mortality rates observed but are also able to connect the dynamics of GRNs, especially its instable character, to the process of aging itself and suggest testable hypotheses and mechanisms to explain the experimental data from which the relationships between aging, ARDs and the change in the topology of these networks were initially deduced. Of course, it must be stressed that these models are based on a simplified view of aging in the sense that the sole driving force generating instability in the network is the accumulation of transcription errors and mutations and this mechanistic description based on the regulation of genes is incomplete. Indeed, GRNs are controlled by various signaling pathways and other mechanisms taking place in the cytoplasm where proteins are involved in many processes which can in turn regulate the transcription in a positive or negative fashion. These details are important and other evolutionary demographic models which include an explicit description of the mutation accumulation and epistasis between genes, such as the one developed in Reference [110], could be improved with a more detailed description of the regulation of the mutated genes.

AGING AS AN EVOLUTIONARY MECHANISM TO ADAPT TO ENVIRONMENTAL PRESSURES

Individuals are under the constant pressure of their direct environment. One can expect that the environment and, more precisely, characteristics such as resource availability and competition between individuals, which can result from resource scarcity, may affect the survival rate of a group of individuals. From this point of view, evolutionary changes can act as adaptation mechanisms for improving the survival rate. From an evolutionary perceptive, aging being also a major evolutionary trait, it is of interest to ask whether it provides an advantage to individuals facing environmental changes and competition to access vital resources. In what follows, we describe two different population models which incorporate various types of environmental changes.

The model presented in Reference [111] simulates the temporal evolution of two populations of species which are in competition for supremacy in a spatial grid. These two populations are initially identical but one can die of senescence (the aging population), whereas the other can only die due to competition and incidents. Individuals of the population suffering the effects of senescence die at the same programmed age. The model takes into account the availability of the resources and allows competition between individuals. The interactions between individuals include only individual competition but a small viscosity term is also introduced to explore the effects of group selection. Members of each population can produce offspring and since the progeny is created near the parents' locations, competition can happen between the parents and their progeny. Moreover, although the initial conditions of the simulations are set to provide enough resources for both populations, these optimal conditions are not stationary. Finally, the model includes the effects of mutations which can appear from one generation to another and are assumed to help each species to remain competitive. The survival ability of individuals is characterized by a fitness function whose value decreases by a small amount at each time step of the simulation. This simple implementation allows capturing the influence of the environment including the changing conditions. In practice, the fitness function of the parents and their progeny is not necessarily the same and can be modified by a small random value. This allows taking into account the effects of mutations that can occur from one generation to another.

The results obtained can be classified into three distinct cases. First, in the absence of mutation and changing conditions, the senescent population is driven to extinction very quickly while the other survives and dominates. Second, when mutations are introduced, senescence causes the extinction of the aging population in most simulations but, nevertheless, this population manages to survive longer and it is even able to lead the non-aging population to extinction in some cases. Interestingly, this trend increases as the mutation rate increases. When analyzing the combined effects of both random mutations and environmental changes, one observes that the average fitness of a new generation is a little larger than that of the previous ones. It means that, on average, each generation is a little better adapted than the previous one. The evolution of several simulations shows that the average fitness of the senescent population becomes larger than that of the non-senescent ones. Interestingly, the age of death set for the senescent population plays an important role in its ability to survive. When this value is very high, it is unlikely that any individual could reach it and any difference that could exist between both groups becomes negligible but, on the other hand, if the aging population dies too soon, the price of senescence is far too large to be overcome by any fitness advantage. However, as soon as the age of death is large enough to overcome that cost, the aging population is able to dominate the non-aging one.

Thus, when no mutations or environmental changes are present, this model suggests that death by senescence appears to have a too important evolutionary cost for a species that adopts it [112,113]. However, in the presence of mutations, the aging population has an improved survival rate because it can adapt faster to new conditions. Furthermore, when introducing changes in the environment, the extinction of the non-aging population becomes a systematic outcome and this result holds for a large range of parameter values. It is worth mentioning that these conclusions are in agreement with the hypothesis supporting the senemorphic aging theory mentioned in the introduction. The fact that the age of death or senescence rate plays a critical role in the ability of the aging population to survive is also worth mentioning. These results are in agreement with another evolutionary demographic model of aging which supports that "evolution favors sustenance over senescence if the sacrifice in reproduction to achieve sustenance is smaller than the sacrifice in life expectancy resulting from senescence" [114]. These results support that senescence seems to be a well-adapted answer to changes that can be adopted by evolutionary dynamics as aging produces a pruning effect on the populations, eliminating older, less adapted individuals who had managed to survive only by chance.

In what fellows, another model based on the simulation of population dynamics is described. Its purpose is to analyze the relationships between aging, mutation accumulation, and effects of natural selection to understand how these interactions affect the fitness of individuals on a long-term basis [115]. In this model, each individual is represented by its own genome that defines the probabilities to reproduce and survive. The exact impact of this genome on survival and the reproduction rate at different life stages depends on how it evolves due to mutation and selection. The probability of survival through each step depends on three main factors: the status of the genome, the difference between the resources available and the size of population, and an extrinsic death crisis parameter. At the beginning, the population is composed of genetically heterogeneous individuals, whose genomes are represented by randomly generated bit arrays. Each individual from the initial population has a chronological age, and becomes one time-unit older at each step of the simulation. Under these conditions, individuals evolve through a sequence of discrete time intervals and one may analyze the evolution of this genetically heterogeneous population under various external constraints including the size of the initial population, the effects of limited resources (which are fixed at the beginning of the simulation), and whether reproduction is sexual or asexual.

The results obtained agree with several characteristics of real population aging. For example, the populations develop a longer reproductive life span and survival under stable environmental conditions. Faster aging appears in more unstable populations because they have large portions of the genome that accumulate mutations which have deleterious effects on the individuals. The model predicts an increased early survival and a rapid increase in the death rate and decrease in fertility after sexual maturation, a scenario which is coherent with the already known aging patterns. Interestingly, results also show that the time-dependent decrease in individual survival and reproduction is more pronounced in the sexual model than in the asexual model. Finally, the fitness effect of beneficial mutations acting in late life is negligible compared to mutations affecting survival and reproduction at earlier ages and one can observe an age-dependent increased genetic variance following sexual maturation.

Together, the models discussed in this section provide another point of view describing the multifactorial nature of aging. It might not be surprising that the different mechanisms which intervene in the onset of aging can lead to different outcomes depending on whether they are considered together in a single model or taken into account separately. From a dynamical perspective, the evolutionary mechanisms, the environmental constraints, and the dynamical cellular equilibriums which must be maintained to sustain the living organism act on very different timescales and they have their own dynamics. Nevertheless, they are obviously also under the continuous influence of each other. This fact reinforces the importance of assembling demographic models of aging which are capable of including the various causes of aging altogether in one single framework.

Furthermore, the development presented here also supports the hypothesis discussed earlier that the disruptions of the dynamical equilibriums described in the previous sections could also be a consequence of modifications in the allocation strategy of resources to maintenance and repair [114], a hypothesis supported by well-established evolutionary theories of aging such as the disposable soma theory. The rate, schedule, and nature of these modifications should be considered as the product of an evolutionary optimization constrained by the necessity to respond to individual variations and environmental changes. The continuously evolving constraints faced by any species could explain why, rather than creating a perfect organism, natural selection promotes survival by ensuring that an individual can maintain the efficiency of key biological functions while maintaining mechanisms allowing adaptation whenever necessary. The phenotypic fluctuations discussed at the beginning of this chapter were suggested to be an example of such mechanism. Exposure to severe environmental changes but also limited availability of resources can explain why natural selection prefers to act through a continuous search for an equilibrium using optimization (resulting in individuals performing all functional obligations reasonably well simultaneously but not perfectly in terms of each individual task) rather than maximization. Indeed, maximization reduces adaptability because it creates individuals which could perfectly maintain a set of functionalities in detriment to some others. From this point of view, aging could be considered as a consequence of the constant search for an equilibrium established at and between different levels of organization and defined as the optimal solution of an evolutionary process with respect to the constraints encountered by any species within their environment [12,116].

AGING AS A RESULT OF THE DISRUPTION OF CELLULAR DYNAMICAL EQUILIBRIUMS

Self-renewal ability and pluripotency maintenance are two characteristics of pluripotent stem cells (PSCs) which attract much interest. At the genomic level, pluripotency maintenance is regulated by transcription factors [117–119] which act as master regulators of GRNs [120–122]. GRNs are organized following various dynamical motifs which act together and contribute to improving the adaptation abilities and robustness of the system [123]. From a dynamical point of view, transcription factors have been shown to continually attempt to specify differentiation to their own lineage. Consequently, direct external interventions, through activation or inhibition of one or several signaling pathways, are necessary to reinforce the pluripotency state or to control the differentiation to a specific lineage [124–126]. From this point of view, the pluripotency state could be considered as a metastable state whose maintenance depends on the properties of the external environment of the cell. Consequently, the most accurate approaches to model the dynamics of such open system exchanging continuously with its external environment do not only take into account the dynamics of the gene regulatory network but also include the network formed by the signaling pathways which are responsible for receiving and transmitting signals from and to the environment [127,128].

Major epigenetic events intervene during the transition from the pluripotency state to the differentiated state. Indeed, differentiation is also characterized by a large remodeling of the chromatin structure [129,130]. The chromatin remodeling occurring during early differentiation induces the silencing of hundreds of genes while others become available for expression [131–133]. hTERT is one of the genes which is inhibited during the early differentiation process [134]. It is well known for being a catalytic subunit with reverse transcriptase activity identified as part of the telomerase complex [135], although it is also involved in many other cellular functionalities [136,137]. Telomerase is a cellular enzyme which, through the *de novo* addition of TTAGGG repeats to the chromosome ends, is capable of compensating the progressive telomere attrition which occurs at each cell division [138]. Telomerase activity is also controlled by regulators of the apoptosis pathways [139] and when telomeres shorten down to a critical length, they are identified as DNA damage. As a result, a DNA damage signaling response mediated by p53, an important check point whose activation prevents the proliferation of cells when DNA has undergone damages, is activated and cells ultimately become senescent. Thus, inhibition of hTERT induces the decrease in telomerase activity and the loss of self-renewal capacities. These findings lead to the identification of telomerase regulation as a key regulator of aging [140] and a stem cell theory of telomere mediated aging has been formalized on this basis [141].

The fate of hTERT is an interesting illustration of how maintenance and disruption of homeostasis may affect key biological functionalities. Indeed, although silencing of hTERT occurs during stem cell differentiation, it is not clear whether this silencing is required for proper differentiation or if it is a collateral damage of the chromatin remodeling [142]. It should be noticed that many different genes coding for an identical protein are usually found in different chromosome locations. This guarantees that the requested promoter will be available for transcription when needed even after modifications of the chromatin structure. The absence of this feature in the case of hTERT could mean that, from an evolutionary point of view, maintaining a significant level of telomerase activity is not favored. On the other hand, the fact that hTERT may be expressed via exogenous activation means that there is no sustained pressure to continuously prevent hTERT expression. The fact that the activity of a pro-growth regulator such as hTERT requires the maintenance of a set of elaborated mechanisms to protect the organism against uncontrolled proliferation and cancer could be an explanation.

Although PSCs are not present in mature organisms, other types of stem cells such as unipotent, bipotent, or multipotent stem cells exist in most adult tissues. They are involved in the maintenance of tissue homeostasis and also contribute to tissue repair and regeneration. Adult tissue stem cells generally reside within specific compartments called stem cell niches. The size of these specific stem cell populations depends on the balance between self-renewal and differentiation. As for PSCs, this balance is a function of the external environment of the cell and the role of the stem cell niches is to maintain the specific conditions required for the preservation of self-renewal capacities. Inside the niches, stem cells are organized according to their telomere length [143]. The longest telomeres are located in the most primitive adult stem cell compartments while the shortest telomeres belong to the more differentiated compartments. Similar to the case of PSCs, differentiation and lineage commitment of adult stem cells is determined by a complex epigenetic program which is itself controlled by several signaling mechanisms [144,145] and pathways such as Wnt [146], Notch [147], and Hedgehog [148] which react to the influence of environmental factors whose effects are transmitted to stem cells by their niches [149]. The niches themselves can undergo dynamical changes to regulate stem cell function according to the external signals received. These changes can affect the stem cell fate in various ways and when the condition are not favorable, stem cells loss their ability to self-renew and in some circumstances the activation of senescence pathways can occur, resulting in the depletion of the stem cell pool [150]. Thus, as discussed in References [144,151], the external environment has an important effect with regard to stem cell aging. Moreover, owing to the p53 checkpoint discussed above, only those stem cells with sufficiently long telomeres will be allowed to regenerate tissues [152,153].

These observations support the hypothesis that the maintenance of a sufficient telomere length is a direct way to prevent tissue degeneration and aging. Moreover, the situation is more complex because even if decreased stem cell mobilization leads ultimately to tissue degeneration, it also provides the organism with a mechanism for cancer protection. Once again, the maintenance of a dynamical equilibrium between antagonistic processes is the key. Several important works performed on mice [154–157] have proven that telomerase activation may be used to slow down aging only if the extra proliferation afforded by lengthened telomeres is controlled by other mechanisms. The conclusion is that the optimal situation relies on the maintenance of a balance between lengthening and shortening activities which contributes to a positive maintenance of the telomere length. This systemic mechanism is called telomere homeostasis and is discussed in further detail in Reference [158] and more recently in Reference [159].

The different signals received from the environment by the signaling pathways are centralized by mTOR (mechanistic target of rapamycin) [160]. mTOR is a large serine/threonine protein kinase which regulates cellular and organismal homeostasis as well as various metabolic cues involved in the regulation of stem cell self-renewal [161]. Broadly speaking, mTOR is responsible for managing cell growth, proliferation, and self-renewal maintenance and, as such, it acts as a sensor which interprets the received signals and adapts the behavior of the cell accordingly [162,163]. Among the regulatory pathways connected to mTOR, one can cite the pathways regulating the level of reactive oxidative species (ROS), the glucose and amino acid metabolism or the LKB1–AMPK pathway which restricts cell growth under energetically unfavorable conditions such as increased cellular AMP/ATP ratios. It is worth mentioning that all of them are known for being involved in the onset of aging.

From a mechanistic point of view, the mTOR pathway is composed of two distinct parts organized around the mTORC1 and mTORC2 complexes. The formation and initial activation of these complexes is dependent on the availability of essential amino acids and other nutrients such as glucose [160,164]. The concentration of fully activated mTORC1 is a function of the signals received from upstream pathways such as the LKB1–AMPK pathway [165] and several key growth receptors involved in the self-renewal maintenance and differentiation of stem cells [163]. One of the major downstream targets of mTORC1 is the set of processes controlling the formation of ribosomal proteins and translation of proteins [166]. The implication of mTORC1 in the regulation of mRNA translation also establishes the mechanistic connection between mTORC1 and the second dynamical complex mTORC2. Indeed, recent experimental findings have shown that the collaboration between activated ribosomes by mTORC1 and effectors of the PI3 K pathway is required for the activation of mTORC2 [167,168]. mTORC2 regulates pro and anti-apoptotic molecules such as p53, BAX, and BCL-XL [169,170]. The fact that p53 is a downstream target of mTORC2 may also explain why the mTORC2 signaling cascade could contribute to resistance or survival in the face of DNA damage in cancer cells [171]. From a dynamical point of view, the interaction between the regulators of apoptosis, mTORC2 and mTORC1, is an example of a negative feedback loop which increases the robustness of the system by allowing an accurate cellular response when changes occur at the genomic level or in the external environment of the cell. Considering the central implication of mTORC1 and mTORC2 in the regulation of key cellular processes, it has been emphasized that a comprehensive molecular description of aging requires an in-depth understanding of the dynamics of the mTOR pathway [172].

It is well known that nutrient and energy availability as well as their correct management by mTOR is critical for the development and maintenance of the organism. The various components involved in the mTOR pathway mediate complex mechanistic relationships between, on the one hand, the external environment and, on the other hand, cell proliferation and development as well as cellular and energy metabolism. Donohoe and Bultman [173] have extended these mechanistic connections by analyzing the metaboloepigenetics, that is, how energy metabolism and the epigenetic control of gene expression interact directly with each other. To that end, detailed evidences supporting that diet and energy metabolism can directly affect gene expression through the regulation of epigenetic mechanisms are provided. From a mechanistic point of view, it is emphasized that energy metabolites such as SAM, acetyl-CoA, or ATP can act as cofactors of epigenetic enzymes regulating DNA methylation, posttranslational histone modifications, etc. Since energy metabolites are themselves dependent on bioactive food components and nutrients, they play the role of a rheostat able to maintain the homeostasis by regulating the level of activity of epigenetic enzymes which work to control gene transcription. This mechanistic relationship between energy, nutrients availability, and epigenetic enzymes has many implications. For example, maternal diet can affect gene expression in a mother's offspring in utero, and these changes could persist after the birth and for the entire life of the offspring. The regulation of the cell-cycle progression is also dependent on the nutrient availability and cofactor abundance. Indeed, if cells are in an external environment lacking enough energy to properly proceed to the next cell division, they are stopped at the G1 checkpoint and become quiescent in order to prevent genomic instability and onset of cancer.

The DNA-response damage triggered by the failure to maintain a minimal telomere length is called replicative senescence. Replicative senescence is one of the different subtypes of DNA-damage induced senescence. Another one is the stress-induced senescence which is caused by high intracellular levels of ROS induced by the RAS–RAFMEK–ERK cascade which activates the p38 MAPK–p16 pathway leading to an increased transcriptional activity of p53. The increased level of ROS is often attributed to higher oxidative stress and, as such, it is considered as a factor of aging. The effects of increased level of ROS observed in older individuals are the subject of intensive research [174,175]. This increase has been shown to have strong consequences on the bioenergetics and metabolism which can be partially explained by the mechanistic connections established between ROS, stress-induced senescence, and p53 activation through mTOR signaling. However, the role of ROS and associated oxidative stress could be more complex as initially expected. For example, ROS appears to be an important physiological regulator of several intracellular signaling pathways [176] and oxidative stress plays an important role in the regulation of stem cell self-renewal [161]. Indeed, although it was shown that an increase in ROS reduces the adult stem cells' self-renewal abilities and induces stem cell senescence and tissue damage, data also suggest that a minimal level of intracellular ROS is essential to maintain quiescence of several types of stem cells such as hematopoietic stem cells. Thus, taken together, these evidences show that ROS is involved in various metabolism processes and that its effects are complex and probably depend on its concentration. In order to describe this dual role of ROS, a conceptual model called the ROS rheostat was presented in Reference [161]. The fundamental hypothesis behind this model is that the level of intracellular ROS monitors stem cell fate decisions. With this aspect taken into account, the regulation of the homeostasis within stem cell niches seems to be more dependent on the establishment of an optimal level of ROS rather than on a complete elimination of ROS inside the intracellular compartments. The elaboration of such a model contributes to the building of a more refined description of the mechanisms behind the maintenance of the quiescent state of long-term adult stem cells.

Although senescence was historically defined and is still largely considered as an irreversible cell-cycle arrest mechanism responsible for the onset and propagation of aging, recent findings suggest that senescence could actually have both negative and positive effects. This dual role has been recently formalized within a unified model of senescence called the senescence–clearance–regeneration model [184]. The model describes the mechanism which functions in the case of adult somatic damages. The purpose of this mechanism is to restore the damaged tissue by elimination of the damaged cells. In the ideal situation, that is, in a healthy organism, senescence initiates tissue repair by recruiting immune cells through the senescence-associated secretory phenotype. In that case, disposal of senescent cells by immune-mediated clearance is efficient and under control: macrophages clear the senescent cells, and progenitor cells repopulate and regenerate the damaged tissue. However, this programmed sequence may be impaired upon persistent damage or aging. Indeed, during aging-related senescence, the switch from temporal to persistent cell-cycle arrest appears unscheduled and the program of senescence–clearance–regeneration is not finalized. The disruption of this program could come from the fact that in aging organisms the immune system undergoes several changes in both the innate and adaptive immunity that culminate in age-associated immunodeficiency. As this evolution occurs, the immune system becomes unable to efficiently recruit macrophages. As a result, the efficiency of senescent cell clearance is strongly reduced and leads to an accumulation of senescent cells in tissues. Thus, the negative effects of senescence on tissues are essentially caused by a disruption of a collaborative program involving the immune system. These findings support the idea of an extended role of senescence that should be taken into account by the current demographic evolutionary aging theory in order to build a new conceptual framework to explain the diversity of patterns within the life table, the unexpected lack of association between the length of life and the degree of senescence, and why senescence has evolved in some species and not in others.

Critical biological functionalities such as senescence, telomere attrition, or the regulation of ROS appear to be more complex than initially expected. These mechanistic models provide a more comprehensive overview of how a single cellular process can have both positive and negative effects. Taken together, these findings support the hypothesis that aging should be interpreted as the long-term result of the disruption of different dynamical equilibriums established between antagonistic processes rather than as the result of a sudden appearance of isolated molecular processes or components with intrinsic negative effects. Of course, considering the systemic organization of the cellular environment, it is obvious that this progressive disruption of dynamical equilibriums which was described here only at the proteomic level has also strong implications at the genomic level and could probably play a central role in the local topological changes characterizing the emergence of ARGs and related ARDs. The effects of the apparition and accumulation of mutations which appear randomly at the genomic level as a result, for example, of dysfunctions in the mechanisms regulating DNA duplication cannot be neglected, but they must be considered within a larger framework which should include the dynamical description of biological processes such as the ones described here. This will allow integrating the fact that the tuning of dynamical equilibriums established between different molecular processes is specific to each species or individual.

The need for improvement was pointed out previously when the pace of life and the shape of mortality patterns have been related to the metabolic rate of an organism and to its capacity for repair, regeneration, and growth, respectively [177]. Interestingly, there is an increasing amount of information and number of dynamical models available for many of these fundamental processes. For example, computational modeling has already been used to study the mechanisms behind histones modifications and their effects on aging of stem cells [178]. One can also mention the successful elaboration of complete models of individual living organisms [179], the realistic descriptions of some dynamical subsystems of the human body such as the ones involved in metabolism [180,181], and the development of more efficient methods to simulate population dynamics [182,183]. On a long-term basis, advances made in the understanding of the systemic biochemical mechanisms involved directly or indirectly in the onset and progression of aging will be used to build more realistic models. For example, an in silico population whose members are described by a biochemical model similar to the one in Reference [179] could be implemented as part of a demographic model. In that case, the evolution of key parameters such as mortality rate and fertility could be directly obtained from the biochemical processes occurring inside the living organisms.

CONCLUSION

In this chapter, we have emphasized that the understanding of aging from biological, molecular, and evolutionary perspectives can be greatly improved by taking into account the latest developments made in the field of cellular, systemic, and molecular biology. These approaches that mainly rely on building reduced mathematical models sometimes combined with transcriptomic data to capture key features of the aging dynamics have proven to provide in-lighting information. For example, we have seen that large-scale analysis of biological data using top-bottom approaches identified functional modules as responsible for the onset of various ARDs. ARGs were identified in sub-networks undergoing topological modifications. Interestingly, mechanistic models based on a bottom-up approach were able to suggest hypotheses to explain how these topological changes caused by mutation accumulation could make GRNs instable and trigger the onset and progression of aging. These connections established between the disparate damages of aging and dynamical and topological properties of the cellular environment illustrate with the other dynamical models described here that aging appears to be caused by disruptions of dynamical equilibriums established between cellular biochemical processes, although other external constraints can also intervene. These disruptions can affect the ability of various components to interact together. This can induce changes in the topology of core GRNs, which in turn could affect other key cellular functionalities. Interestingly, we have seen that the mechanisms behind the ROS rheostat model, the telomere homeostasis model, or the senescence–clearance–regeneration model are in agreement with the fundamental basis of several evolutionary theories of aging such as the antagonistic pleiotropy theory and the disposable soma theory. Thus, more detailed descriptions of molecular processes involved in aging do not only improve evolutionary theories of aging individually but also show that these theories simply describe different aspects of the same fundamental process.

Obviously, it is important to underline that the models described here are always the result of a consensus between simplicity and complexity. Indeed, in any modeling process, if one wants to make possible the extraction of meaningful and interpretable results, simplifications are always necessary to reduce the system to a set of parameters and variables which can be more easily interpreted. One can expect that more detailed models will be elaborated in the near future. Other fields of sciences such as system biology will continue to contribute to building a more integrated and detailed description of aging. This might be achieved by implementing more detailed mathematical models for the forces of natural selection and senescence and for dynamics of mutation accumulation [184,185]. Hence, these models will continue to be integrated step by step in demographic models of aging. Furthermore, these models should take into account the fine-tuning of the molecular processes occurring among species as well as the environmental constraints experienced by the individuals.

From a mechanistic perspective, the interplay between the environment and the evolutionary processes, including aging, has also been discussed. We have seen that stochastic fluctuations of gene concentrations, a pure mechanistic property, were thought to be a potential evolutionary mechanism. Phenotypic fluctuations could work to improve the adaptability of the species facing severe environmental conditions. Mechanistic models were also used to study the relationships between these fluctuations, the evolution of the fitness over generations, and another dynamical property called robustness. From an evolutionary point of view, models simulating the dynamics of populations suggest that aging could provide an advantage to adapt to environmental changes. Thus, whenever necessary, aging could act as a mechanism working to optimize the fitness function across the generations. However, this evolutionary mechanism might not be of primary importance for species evolving in a safe and controlled environment. This fact is illustrated by the rectangularization of the survival curves observed for the populations with an easy access to all necessary resources. Of course, when analyzing the effects of environment in developed human societies, other factors must be considered such as diet, quality, and access to health care systems or working and social conditions.

These results obtained from different kind of methods also illustrate that multidisciplinary approaches provide new insights and contribute to improve our understanding of aging. Aging is one of the most complex biological processes to be addressed by the scientific community. The purpose of this work was obviously not to cover an extensive review of all aspects of the subject. The general guideline was rather to emphasize several current methodological and conceptual trends which show promising perspectives. The conclusion is that aging research is in need of multidisciplinary and global approaches. We have illustrated this point with several interesting progresses made regarding the mechanistic description of aging. These results show that the contributions of scientists from various fields and domains of expertise can provide new insights and unexpected new directions of investigation. Aging is a phenomenon for which contributions from all fields of sciences will certainly continue to give promising results in the future.

REFERENCES

1. Haldane JBS. *The Causes of Evolution*. Princeton University Press, Princeton, 1990.
2. Huxley J. *Evolution: The Modern Synthesis*. The Definitive Edition. MIT Press, Cambridge, 2010.
3. Cann RL, Stoneking M, Wilson AC. Mitochondrial DNA and human evolution. *Nature* 1987; 325: 31–5.
4. Harrison GA, Tanner JM, Pilbeam DR, Baker PT. In: *Human Biology. An Introduction to Human Evolution, Variation, Growth and Adaptability*. Harrison, GA, Tanner, JM, Pilbeam, DR, Baker, PT (eds). 3rd ed. Oxford University Press, Oxford 1988, pp. 214–5.
5. Lande R. Natural selection and random genetic drift in phenotypic evolution. *Evolution* 1976; 30(2): 314–34.
6. Hahn MW. Toward a selection theory of molecular evolution. *Evolution* 2008; 62(2): 255–65.
7. Hershberg R, Petrov DA. Selection on codon bias. *Annu Rev Genet* 2008; 42(1): 287–99.
8. Kimura M. The neutral theory of molecular evolution: A review of recent evidence. *Jpn J Genet* 1991; 66: 367–86.
9. Ohta T. The nearly neutral theory of molecular evolution. *Annu Rev Ecol Syst* 1992; 23: 263–86.
10. Nei M, Suzuki Y, Nozawa M. The neutral theory of molecular evolution in the genomic era. *Annu Rev Genomics Hum Genet* 2010; 11: 265–89.
11. Laland K, Uller T, Feldman M et al. Does evolutionary theory need a rethink? *Nature* 2014; 514: 161–4.
12. Rose MR, Oakley TH. The new biology: Beyond the modern synthesis. *Biol Direct* 2007; 2: 30.
13. Kutschera U, Niklas KJ. The modern theory of biological evolution: An expanded synthesis. *Naturwissenschaften* 2004; 91: 255–76.
14. Pigliucci M. An extended synthesis for evolutionary biology. *Ann N Y Acad Sci* 2009; 1168: 218–28.
15. Hayflick L, Moorhead PS. The serial cultivation of human diploid cell strains. *Exp Cell Res* 1961; 25: 585–621.
16. Weinert BT, Timiras PS. Physiology of aging invited review: Theories of aging. *J Appl Physiol* 2003; 95: 1706–16.
17. Gavrilov LA, Gavrilova NS. Evolutionary theories of aging and longevity. *The Scientific World J* 2002; 2: 339–56.
18. Rose MR, Burke MK, Shahrestani P, Mueller LD. Evolution of ageing since Darwin. *J Genet* 2008; 87: 363–71.
19. Trindade LS, Aigaki T, Peixoto AA et al. A novel classification system for evolutionary aging theories. *Front Genet* 2013; 4 article 25. doi:10.3389/fgene.2013.00025.
20. Ljubuncic P, Reznick AZ. The evolutionary theories of aging revisited. A mini-review. *Gerontology* 2009; 55: 205–16.
21. Kirkwood TB. Understanding ageing from an evolutionary perspective. *J Intern Med* 2008; 263: 117–27.
22. Jones OR, Scheuerlein A, Salguero-Gomez R et al. Diversity of ageing across the tree of life. *Nature* 2014; 505: 169–73.
23. Maslov AY, Vijg J. Genome instability, cancer and aging. *Biochim Biophys Acta (BBA)-Gen Subjects* 2009; 1790: 963–9.
24. von Zglinicki T, Martin-Ruiz CM. Telomeres as biomarkers for ageing and age-related diseases. *Curr Mol Med* 2005; 5: 197–203.
25. Dai DF, Chiao YA, Marcinek DJ, Szeto HH, Rabinovitch PS. Mitochondrial oxidative stress in aging and healthspan. *Longevity Healthspan* 2014; 3: 6. doi:10.1186/2046-2395-3-6.
26. Lpez-Otin C, Blasco MA, Partridge L, Serrano M, Kroemer G. The hallmarks of aging. *Cell* 2013; 153: 1194–217. doi:10.1016/j.cell.2013.05.039.
27. Guarner V, Rubio-Ruiz ME. Low-grade systemic inflammation connects aging, metabolic syndrome and cardiovascular disease. *Interdiscip Top Gerontol* 2015; 40: 99–106. doi:10.1159/000364934.
28. Singhal G, Jaehne EJ, Corrigan F, Toben C, Baune BT. Inflammasomes in neuroinflammation and changes in brain function: A focused review. *Front Neurosci* 2014; 8: 315. doi:10.3389/fnins.2014.00315.

29. de Grey A. A strategy for postponing aging indefinitely. *Stud Health Technol Inform* 2005; 118: 209–19.
30. Liu JP. Molecular mechanisms of ageing and related diseases. *Clin Exp Pharmacol Physiol* 2014; 41: 445–58.
31. Altelaar AF, Munoz J, Heck AJ. Next-generation proteomics: Towards an integrative view of proteome dynamics. *Nat Rev Genet* 2012; 14(1): 35–48.
32. Lowell S. Getting the measure of things: The physical biology of stem cells. *Development* 2013; 140: 4125–8.
33. Gruebele M, Thirumalai D. Perspective: Reaches of chemical physics in biology. *J Chem Phys* 2013; 139(12): 121701.
34. Alberghina L, Hans VW eds. *Systems Biology: Definitions and Perspectives. Topics in Current Genetics 13.* Springer-Verlag, Berlin, 2005.
35. Barabasi AL, Oltvai ZN. Network biology: Understanding the cells functional organization. *Nat Rev Genet* 2004; 5: 101–13.
36. Bruggeman FJ, Westerhoff HV. The nature of systems biology. *Trends Microbiol* 2006; 15(1): 45–50.
37. Liang J, Luo Y, Zhao H. Synthetic biology: Putting synthesis into biology. *Wiley Interdiscip Rev Syst Biol Med* 2011; 3(1): 7–20.
38. Rajapakse I, Scalzo D, Tapscott SJ, Kosak ST, Groudine M. Networking the nucleus. *Mol Syst Biol* 2010; 6: 395.
39. Alon U. Network motifs: Theory and experimental approaches. *Nat Rev Genet* 2007; 8: 450–61.
40. Matera AG, Wang Z. A day in the life of the spliceosome. *Nat Rev Mol Cell Biol* 2014; 15: 108–21.
41. Bentley DL. Coupling mRNA processing with transcription in time and space. *Nat Rev Genet* 2014; 15: 163–75.
42. Varga-Weisz PD. Chromatin remodeling: A collaborative effort. *Nat Struct Mol Biol* 2014; 21(1): 14–6.
43. Marmorstein R. Protein modules that manipulate histone tails for chromatin regulation. *Nat Rev Mol Cell Biol* 2001; 2: 422–32.
44. Grummt I, Pikaard CS. Epigenetic silencing of RNA polymerase I transcription. *Nat Rev Mol Cell Biol* 2003; 4: 641–9.
45. Klose RJ, Zhang Y. Regulation of histone methylation by demethylimination and demethylation. *Nat Rev Mol Cell Biol* 2007; 8: 307–18.
46. Wagner EJ, Carpenter PB. Understanding the language of Lys36 methylation at histone H3. *Nat Rev Mol Cell Biol* 2012; 13: 115–26.
47. Jenuwein T, Allis CD. Translating the histone code. *Science* 2001; 293: 1074–80.
48. Richards EJ, Elgin SCR. Epigenetic codes for heterochromatin formation and silencing: Rounding up the usual suspects. *Cell* 2002; 108: 489–500.
49. Bhardwaja N, Yana KK, Gerstein MB. Analysis of diverse regulatory networks in a hierarchical context shows consistent tendencies for collaboration in the middle levels. *Proc Natl Acad Sci USA* 2010; 107(15): 6841–6.
50. Pechenick DA, Payne JL, Moore JH. Phenotypic robustness and the assortativity signature of human transcription factor networks. *PLoS Comput Biol* 2014; 10(8): e1003780.
51. Ibez-Marcelo E, Alarcn T. The topology of robustness and evolvability in evolutionary systems with genotype-phenotype map. *J Theor Biol* 2014; 356: 144–62.
52. Marashi SA, Tefagh M. A mathematical approach to emergent properties of metabolic networks: Partial coupling relations, hyperarcs and flux ratios. *J Theor Biol* 2014; 355: 185–93.
53. Kuwahara H, Soyer OS. Bistability in feedback circuits as a byproduct of evolution of evolvability. *Mol Syst Biol* 2012; 8: 564.
54. Gentilini D, Mari D, Castaldi D et al. Role of epigenetics in human aging and longevity: Genome-wide DNA methylation profile in centenarians and centenarians offspring. *Age (Dordr)* 2013; 35: 1961–73.
55. Tavernarakis N. Ageing and the regulation of protein synthesis: A balancing act? *Trends Cell Biol* 2008; 18(5): 228–35.
56. Shimizu I, Yoshida Y, Suda M, Minamino T. DNA damage response and metabolic disease. *Cell Metab* 2014; 20: 967–77.
57. Pawelec G, Goldeck D, Derhovanessian E. Inflammation, ageing and chronic disease. *Curr Opin Immunol* 2014; 29: 23–8.
58. Kennedy BK, Berger SL, Brunet A et al. Geroscience: Linking aging to chronic disease. *Cell* 2014; 159: 709–13.
59. Han JDJ, Bertin N, Hao T et al. Evidence for dynamically organized modularity in the yeast protein-protein interaction network. *Nature* 2004; 430: 88–93.

60. Trott J, Hayashi K, Surani A, Babu MM, Martinez-Arias A. Dissecting ensemble networks in ES cell populations reveals microheterogeneity underlying pluripotency. *Mol Biosyst* 2011; 8(3): 744–52.

61. Buescher JM, Liebermeister W, Jules M et al. Global network reorganization during dynamic adaptations of *Bacillus subtilis* metabolism. *Science* 2012; 335: 1099–103.

62. Nicolas P, Mader U, Dervyn E, Rochat T, Leduc A, Pigeonneau N. Condition-dependent transcriptome reveals high-level regulatory architecture in *Bacillus subtilis*. *Science* 2012; 335: 1103–6.

63. Ito Y, Toyota H, Kaneko K, Yomo T. How selection affects phenotypic fluctuation. *Mol Syst Biol* 2009; 5 Article number 264, 28 April, 2009. doi:10.1038/msb.2009.23.

64. Kaneko K. Shaping robust system through evolution. *Chaos: Interdiscip J Nonlinear Sci* 2008; 18: 026112. doi:10.1063/1.2912458.

65. Furusawa C, Kaneko K. A generic mechanism for adaptive growth rate regulation. *PLoS Comput Biol* 2008; 4(1): e3. doi:10.1371/journal.pcbi.0040003.

66. Mayr A, Klambauer G, Unterthiner T. et al. DeepTox: Toxicity prediction using deep learning. *Front Environ Sci*. 2016. doi:10.3389/fenvs.2015.00080.

67. Ma J et al. Deep neural nets as a method for quantitative structure-activity relationships. *J Chem Inf Model* 2015; 55(2): 263–74.

68. Wang C. et al. Pairwise input neural network for target-ligand interaction prediction. *Bioinformatics and Biomedicine (BIBM), 2014 IEEE International Conference*. Tara Lodge, Belfast, United Kingdom. 67–70.

69. Xu Y. et al. Deep learning for drug-induced liver injury. *J Chem Inf Model* 2015; 55(10): 2085–93. doi:10.1021/acs.jcim.5b00238.

70. Mamoshina P et al. 2016 Applications of deep learning in biomedicine. *Mol Pharm* 2016; 13(5): 1445–54. doi:10.1021/acs.molpharmaceut.5b00982.

71. Hughes TB et al. Modeling epoxidation of drug-like molecules with a deep machine learning network. *ACS Cent Sci* 2015; 1(4): 168–80. doi:abs/10.1021/acscentsci.5b00131.

72. Putin E et al. Deep biomarkers of human aging: Application of deep neural networks to biomarker development. *Aging* 2016; 8(5): 1021–33.

73. Geller AC et al. Melanoma incidence and mortality among US whites, 1969–1999. *JAMA: J Am Med Assoc* 2002; 288: 719–1720.

74. de Magalhaes, JP. How ageing processes influence cancer. *Nat Rev Cancer* 2013; 13: 357–65.

75. Lu, T. et al. Gene regulation and DNA damage in the ageing human brain. *Nature* 2004; 429: 883–91.

76. Berchtold, NC et al. Gene expression changes in the course of normal brain aging are sexually dimorphic. *Proc Natl Acad Sci USA* 2008; 105: 15605–10.

77. Faisal FE, Milenkovic T. Dynamic networks reveal key players in aging. *Bioinformatics* 2014; 30(12):1721–9. doi:10.1093/bioinformatics/btu089.

78. Tacutu, R, Budovsky, A, Yanai, H, Fraifeld, VE. Molecular links between cellular senescence, longevity and age-related diseases—A systems biology perspective. *Aging (Albany NY)* 2011; 3: 1178–91.

79. Wang J, Zhang S, Wang Y, Chen, L, Zhang, XS. Disease-aging network reveals significant roles of aging genes in connecting genetic diseases. *PLoS Comput Biol* 2009; 5: e1000521. doi:10.1371/journal.pcbi.1000521.

80. Yang J, Huang T, Song WM, Petralia F, Mobbs CV, Zhang B et al. Discover the network mechanisms underlying the connections between aging and age-related diseases. *Sci Rep* 2016; 6: 32566. doi:10.1038/srep32566.

81. Wyczalkowska-Tomasik A, Czarkowska-Paczek B, Zielenkiewicz M, Paczek L. Inflammatory markers change with age, but do not fall beyond reported normal ranges. *Arch Immunol Ther Exp*. 2015; doi:10.1007/s00005-015-0357-7.

82. Wilson WR, Hay MP. Targeting hypoxia in cancer therapy. *Nat Rev Cancer* 2011; 11: 393–410.

83. Ziello JE, Jovin IS, Huang Y. Hypoxia-inducible factor (HIF)-1 regulatory pathway and its potential for therapeutic intervention in malignancy and ischemia. *Yale J Biol Med* 2007; 80: 51–60.

84. Rodrguez JA, Marigorta UM, Hughes DA, Spataro N, Bosch E, Navarro A. Antagonistic pleiotropy and mutation accumulation influence human senescence and disease. *Nat Ecol Evol* 2017; 1: 0055. doi:10.1038/s41559-016-0055.

85. Demetrius L. Adaptative value, entropy and survivorship curves. *Nature* 1988; 275: 213–4.

86. Weon BM, Je JH. Plasticity and rectangularity in survival curves. *Sci Rep* 2011; 1: 104. doi:10.1038/srep00104.

87. Fries JF. Aging, natural death, and the compression of morbidity. *N Engl J Med* 1980; 303: 130–6.

88. Wachter KW, Finch C eds. *Between Zeus and the Salmon: The Biodemography of Longevity*. National Academic Press, Washington, DC, 1997.

89. Thatcher A R, Kannisto V, Vaupel JW. *The Force of Mortality at Ages 80–120. (Odense Monographs on Population Aging 5.* Odense University Press, Odense, 1998.

90. Gompertz B. On the nature of the function expressive of the law of human mortality. *Philos Trans R Soc Lond A* 1825; 115: 513–80.

91. Tarkhov AE, Menshikova LI, Fedicheva PO. Strehler-Mildvan correlation is a degenerate manifold of Gompertz fit. *J Theor Biol* 2017; 416: 180–9. doi:10.1016/j.jtbi.2017.01.017.

92. Robine JM, Vaupel JW. Emergence of supercentenarians in low mortality countries. *N Am Actuarial J* 2002; 6: 54–63.

93. Hamilton WD. The moulding of senescence by natural selection. *J Theor Biol* 1966; 12: 12–45.

94. Rose MR, Rauser CL, Benford G, Matos M, Mueller LD. Hamiltons forces of natural selection after forty years. *Evolution* 2007; 61(6): 1265–76.

95. Baudisch A. Hamiltons indicators of the force of selection. *Proc Natl Acad Sci USA* 2005; 102(23): 8263–8.

96. Wachtera KW, Evans SN, Steinsaltz D. The age-specific force of natural selection and biodemographic walls of death. *Proc Natl Acad Sci USA* 2013; 110(25): 10141–6.

97. Liu X. Life equations for the senescence process. *Biochem Biophys Rep* 2015; 4: 228–33.

98. Brownlee J. Notes on the biology of a life-table. *J R Stat Soc* 1919; 82: 34–77.

99. Loeb J, Northrop JH. Is there a temperature coefficient for the duration of life? *Proc Natl Acad Sci USA* 1916; 2: 456–7.

100. Gavrilov LA, Gavrilova NS. The reliability theory of aging and longevity. *J Theor Biol* 2001; 213: 527–45.

101. Bebbington M, Lai CD, Zitikis R. Modelling deceleration in senescent mortality. Math. *Popul Stud* 2011; 18: 18–37.

102. Vaupel WJ. Biodemography of human ageing. *Nature* 2010; 464: 536–42.

103. Conti B, Sanchez-Alavez M, Winsky-Sommerer R, Morale MC, Lucero J, Brownell S et al. Transgenic mice with a reduced core body temperature have an increased lifespan. *Science* 2006; 314: 825–8.

104. Podolskiy D, Molodtcov I, Zenin A, Kogan V, Menshikov LI, Reis RJS, Fedichev PO. Critical dynamics of gene networks is a mechanism behind ageing and Gompertz law. arXiv:1502.04307v1 [q-bio.MN] 15 Feb 2015.

105. Austad SN. Is aging programed? *Aging Cell* 2004; 3: 249–51.

106. Goldsmith TC. Aging, evolvability, and the individual benefit requirement; medical implications of aging theory controversies. *J Theor Biol* 2008; 252: 764–8.

107. Goldsmith TC. On the programmed/non-programmed aging controversy. *Biochem (Mosc)* 2012; 77(7): 729–32.

108. Kirkwood TBL, Melov S. On the programmed/non-programmed nature of ageing within the life history. *Curr Biol* 2011; 21: R701–7.

109. Kogan V, Molodtsov I, Menshikov LI, Reis RJS, Fedichev P. Stability analysis of a model gene network links aging, stress resistance, and negligible senescence. *Sci Rep* 2016; 5: 13589. doi:10.1038/srep13589.

110. Steinsaltz D, Evans SN, Wachter KW. A generalized model of mutation-selection balance with applications to aging. *Adv Appl Math* 2005; 35: 16–33.

111. Martins ACR. Change and aging senescence as an adaptation. *PLoS ONE* 2011; 6(9): e24328. doi:10.1371/journal.pone.0024328.

112. Medawar PB. *An Unsolved Problem of Biology.* Lewis, London, 1952.

113. Ricklefs RE. Evolutionary theories of aging: Confirmation of a fundamental prediction, with implications for the genetic basis and evolution of life span. *Am Natural* 1998; 152: 24–44.

114. Baudisch A, Vaupel JW. Senescence vs. sustenance: Evolutionary demographic models of aging. *Demographic Res* 2010; 23(23): 655–68.

115. Sajina A, Valenzano DR. An In Silico Model to Simulate the Evolution of Biological Aging. arXiv:1602.00723v1 [q-bio.PE] 1 Feb 2016.

116. Mitteldorf JJ. Adaptive aging in the context of evolutionary theory. *Biochem (Mosc)* 2012; 77(7): 716–25.

117. Thomson M, Liu SJ, Zou LN, Smith Z, Meissner A, Ramanathan S. Pluripotency factors in embryonic stem cells regulate differentiation into germ layers. *Cell* 2011; 145: 875–89.

118. Walker E, Ohishi M, Davey RE et al. Prediction and testing of novel transcriptional networks regulating embryonic stem cell self-renewal and commitment. *Cell Stem Cell* 2007; 1: 71–86.

119. Tantin D. Oct transcription factors in development and stem cells: Insights and mechanisms. *Development* 2013; 140: 2857–66.

120. Iglesias-Bartolome R, Gutkind JS. Signaling circuitries controlling stem cell fate: To be or not to be. *Curr Opin Cell Biol* 2011; 23: 716–23.

121. Ng HH, Surani MA. The transcriptional and signalling networks of pluripotency. *Nat Cell Biol* 2011; 13(5): 490–6.
122. Dalton S. Signaling networks in human pluripotent stem cells. *Curr Opin Cell Biol* 2013; 25(2): 241–6.
123. Walker E, Stanford WL. Transcriptional networks regulating embryonic stem cell fate decisions. *Regulatory Networks in Stem Cells, Stem Cell Biology and Regenerative Medicine.* Rajasekhar VK, Vemuri MC (eds). Humana Press: USA, 2009; pp. 87–100.
124. Silva J, Smith A. Capturing pluripotency. *Cell* 2008; 132: 532–6.
125. Nowick K, Stubbs L. Lineage-specific transcription factors and the evolution of gene regulatory networks. *Brief Funct Genomic* 2010; 9(1): 65–78.
126. Loh KM, Lim B. A precarious balance: Pluripotency factors as lineage specifiers. *Cell Stem Cell* 2011; 8: 363–9.
127. Andrieux G, Le Borgne M, Thret N. An integrative modeling framework reveals plasticity of TGF- signaling. *BMC Syst Biol* 2014; 8(30).
128. Liu JK, OBrien EJ, Lerman JA, Zengler K, O Palsson B, Feist AM. Reconstruction and modeling protein translocation and compartmentalization in *Escherichia coli* at the genome-scale. *BMC Syst Biol* 2014; 8(110).
129. Chen T, Dent SYR. Chromatin modifiers and remodellers: Regulators of cellular differentiation. *Nat Rev Genet* 2014; 15: 93–106.
130. Hemberger M, Dean W, Reik W. Epigenetic dynamics of stem cells and cell lineage commitment: Digging Waddingtons canal. *Nat Rev Mol Cell Biol* 2009; 10: 526–37.
131. Jung I, Kim D. Histone modification profiles characterize function specific gene regulation. *J Theor Biol* 2012; 310: 132–42.
132. Xie W, Schultz MD, Lister R et al. Epigenomic analysis of multilineage differentiation of human embryonic stem cells. *Cell* 2013; 153: 1134–48.
133. Gifford CA, Ziller MJ, Gu H et al. Transcriptional and epigenetic dynamics during specification of human embryonic stem cells. *Cell* 2013; 153: 1149–63.
134. Blasco MA. The epigenetic regulation of mammalian telomeres. *Nat Rev Genet* 2007; 8: 299–309.
135. Zvereva MI, Shcherbakova DM, Dontsova OA. Telomerase: Structure, functions, and activity regulation. *Biochem (Mosc)* 2010; 75(13): 1563–83.
136. Ye J, Renault VM, Jamet K, Gilson E. Transcriptional outcome of telomere signalling. *Nat Rev Genet* 2014; 15: 491–503.
137. Ale-Agha N, Dyballa-Rukes N, Jakob S, Altschmied J, Haendeler J. Cellular functions of the dual-targeted catalytic subunit of telomerase, telomerase reverse transcriptase. Potential role in senescence and aging. *Exp Gerontol* 2014; 56: 189193.
138. Lu W, Zhang Y, Liu D, Songyang Z, Wan M. Telomeres structure, function, and regulation. *Exp Cell Res* 2013; 319: 133–41.
139. Oh W, Lee EW, Lee D et al. Hdm2 negatively regulates telomerase activity by functioning as an E3 ligase of hTERT. *Oncogene* 2010; 29: 4101–12.
140. Boccardi V, Paolisso G. Telomerase activation: A potential key modulator for human health span and longevity. *Ageing Res Rev* 2014; 15: 1–5.
141. Mikhelson VM, Gamaley IA. Telomere shortening is the sole mechanism of aging. *Open Longevity Sci* 2008; 2: 23–8.
142. Park J, Venteicher AS, Hong JY, Choi J, Jun S, Shkreli M. Telomerase modulates Wnt signalling by association with target gene chromatin. *Nature* 2009; 460: 66–73.
143. Flores I, Canela A, Vera E et al. The longest telomeres: A general signature of adult stem cell compartments. *Genes Dev* 2008; 22: 654–67.
144. Gopinath SD, Rando TA. Stem cell review series: Aging of the skeletal muscle stem cell niche. *Aging Cell* 2008; 7: 590–8.
145. Brizzi MF, Tarone G, Defilippi P. Extracellular matrix, integrins, and growth factors as tailors of the stem cell niche. *Curr Opin Cell Biol* 2012; 24: 645–51.
146. Sugimura R, He XC, Venkatraman A et al. Noncanonical Wnt signaling maintains hematopoietic stem cells in the niche. *Cell* 2012; 150: 351–65.
147. Weber JM, Calvi LM. Notch signaling and the bone marrow hematopoietic stem cell niche. *Bone* 2010; 46(2): 281–5.
148. Chotinantakul K, Leeanansaksiri W. Hematopoietic stem cell development, niches, and signaling pathways. *Bone Marrow Res* 2012; 2012: 270425.
149. Kuang S, Gillespie MA, Rudnicki MA. Niche regulation of muscle satellite cell self-renewal and differentiation. *Cell Stem Cell* 2008; 2: 22–31.

150. Armstrong L, Al-Aama J, Stojkovic M, Lako M. The epigenetic contribution to stem cell ageing can we rejuvenate our older cells? *Stem Cells* 2014; 32(9): 2291–8.
151. Carlson ME, Conboy IM. Loss of stem cell regenerative capacity within aged niches. *Aging Cell* 2007; 6: 371–82.
152. Sahin E, DePinho RA. Axis of ageing: Telomeres, p53 and mitochondria. *Nat Mol Cell Bio* 2012; 13: 397–404.
153. Flores I, Blasco MA. A p53-dependent response limits epidermal stem cell functionality and organismal size in mice with short telomeres. *PLoS One* 2009; 4(3): e4934.
154. Kelland LR. Overcoming the immortality of tumour cells by telomere and telomerase based cancer therapeutics current status and future prospects. *Eur J Cancer* 2005; 41: 971–9.
155. Corey DR. Telomeres and telomerase: From discovery to clinical trials. *Chem Biol* 2009; 16(12): 1219–23.
156. Reichert S, Bize P, Arriv M, Zahn S, Massemin S, Criscuolo F. Experimental increase in telomere length leads to faster feather regeneration. *Exp Gerontol* 2014; 52: 36–8.
157. Bernardes de Jesus B, Vera E, Schneeberger K, Tejera AM, Ayuso E, Bosch F. Telomerase gene therapy in adult and old mice delays aging and increases longevity without increasing cancer. *EMBO Mol Med* 2012; 4: 691–704.
158. Donate LE, Blasco MA. Telomeres in cancer and ageing. *Phil Trans R Soc B* 2011; 366: 76–84.
159. Zhao Z, Pan X, Liu L, Liu N. Telomere length maintenance, shortening, and lengthening. *J Cell Physiol* 2014; 229: 1323–9.
160. Sengupta S, Peterson TR, Sabatini DM. Regulation of the mTOR complex 1 pathway by nutrients, growth factors, and stress. *Mol Cell* 2010; 40: 310–22.
161. Ito K, Suda T. Metabolic requirements for the maintenance of self-renewing stem cells. *Mol Cell Biol* 2014; 15: 243–56.
162. Caron E, Ghosh S, Matsuoka Y et al. A comprehensive map of the mTOR signaling network. *Mol Syst Biol* 2010; 6: 453.
163. Foster KG, Fingar DC. Mammalian target of rapamycin (mTOR): Conducting the cellular signaling symphony. *J Biol Chem* 2010; 285(19): 14071–7.
164. Buller CL, Loberg RD, Fan MH et al. A GSK-3/TSC2/mTOR pathway regulates glucose uptake and GLUT1. *Am J Physiol Cell Physiol* 2008; 295: C836–43.
165. Inoki K, Ouyang H, Zhu T et al. TSC2 integrates Wnt and energy signals via a coordinated phosphorylation by AMPK and GSK3 to regulate cell growth. *Cell* 2006; 126: 955–68.
166. Thoreen CC, Chantranupong L, Keys HR, Wang T, Gray NS, Sabatini DM. A unifying model for mTORC1-mediated regulation of mRNA translation. *Nature* 2012; 485: 109–13.
167. Zinzalla V, Stracka D, Oppliger W, Hall MN. Activation of mTORC2 by association with the ribosome. *Cell* 2011; 144: 757–68.
168. Dalle Pezze P, Sonntag AG, Thien A et al. A dynamic network model of mTOR signaling reveals TSC independent mTORC2 regulation. *Sci Signal* 2012; 5(217): ra25.
169. Fingar DC, Inoki K. Deconvolution of mTORC2 "in Silico". *Sci Signal* 2012; 5(217): pe12.
170. Feng Z. p53 regulation of the IGF-1/AKT/mTOR pathways and the endosomal compartment. *Cold Spring Harb Perspect Biol* 2010; 2(2): a001057.
171. Weisman R, Cohen A, Gasser SM. TORC2-a new player in genome stability. *EMBO Mol Med* 2014; 6(8): 995–1002.
172. McCormick MA, Tsai SY, Kennedy BK. TOR and ageing: A complex pathway for a complex process. *Phil Trans R Soc B* 2011; 366: 17–27.
173. Donohoe DR, Bultman SJ. Metaboloepigenetics: Interrelationships between energy metabolism and epigenetic control of gene expression. *J Cell Physiol* 2012; 227: 3169–77.
174. Darzynkiewicz Z, Zhao H, Halicka HD et al. In search of antiaging modalities: Evaluation of mTOR- and ROS/DNA damage signaling by cytometry. *Cytometry Part A* 2014; 85A: 386–99.
175. Trubitsyn AG. The joined aging theory. *Adv Gerontol* 2013; 3(3): 155–72.
176. Finkel T. Signal transduction by reactive oxygen species. *JCB* 2011; 194(1): 7–15.
177. Baudisch, A. The pace and shape of ageing. *Methods Ecol Evol* 2011; 2: 375–82.
178. Przybilla J, Rohlf T, Loeffler M, Galle J. Understanding epigenetic changes in aging stem cells—A computational model approach. *Aging Cell* 2014; 13: 320–8.
179. Thiele I, Swainston N, Fleming RM et al. A community-driven global reconstruction of human metabolism. *Nat Biotechnol* 2013; 31(5): 419–25.
180. Karr JR, Sanghvi JC, Macklin DN et al. A whole-cell computational model predicts phenotype from genotype. *Cell* 2012; 150: 389–401.

181. Duarte NC, Becker SA, Jamshidi N et al. Global reconstruction of the human metabolic network based on genomic and bibliomic data. *Proc Natl Acad Sci USA* 2007; 104(6): 1777–82.

182. You L, Cox RS3rd, Weiss R, Arnold FH. Programmed population control by cell-cell communication and regulated killing. *Nature* 2004; 428: 868–71.

183. Charlebois DA, Kaern M. An accelerated method for simulating population dynamics. *Commun Comput Phys* 2013; 14(2): 461–76.

184. Munoz-Espin D, Serrano M. Cellular senescence: From physiology to pathology. *Nat Mol Cell Biol* 2014; 15: 482–96.

185. van Deursen JM. The role of senescent cells in ageing. *Nature* 2014; 509: 439–46.

4 The Indispensable Soma Hypothesis in Aging

Marios Kyriazis

CONTENTS

INTRODUCTION

In Chapter 31, I discussed in some detail the shortcomings of reductionist thinking and expanded on the suitability of using a *systems thinking* approach in aging. Here, I am going to give a specific example where "systems thinking" can be applied specifically in aging research and I will provide evidence that such an approach may lead to significant insights regarding reduction of age-related degeneration. There is no doubt that the increasing progress of technology is now significantly affecting our lives. For the first time in human history, we are experiencing a fusion between the biological (physical) and the digital (cognitive) world [1–4], which bears the characteristics of an evolutionary transition [5–7]. Therefore it is legitimate, even necessary, to initiate a learned debate examining the consequences of inhabiting this self-organizing, adaptive, bio-technological eco-system [8]. With regard to aging, established natural laws have hitherto assured continual survival through an emphasis on germline repair, with trade-offs against somatic repair and survival [9]. This was an essential and appropriate evolutionary strategy in a world where organisms were at risk from early death due to infections, predation, famine, and other natural events [10]. However, this situation is now changing [11]. Many humans live in relatively secure and safe societies, where technology increasingly buffers against any external or internal risks to survival. As a result, the very reason for the presence of aging (the disproportionate allocation of repair resources, leading to an increased time-related dysfunction) [12] is now gradually becoming less relevant. Consider this concept under a different perspective: It is frequently found both in the natural and the artificial realm that any agents* who contribute to the overall function and to the successful remodeling of a system are maintained for longer within that system [13–17].

---

* In cybernetics, an *"agent"* is any entity which acts on its own environment [25]. This entity could be organic (including human), digital, cultural, or abstract.

When an agent transmits a challenge or a stimulus via a specific link, and it senses that the challenge has been sufficiently acted upon, then this agent will be more likely to use the same link in the future to transmit similar challenges [18]. However, if the challenge has not been successfully dealt with, then the probability of using the same link in the future will be decreased [19]. Extrapolating this to human agents, it can be assumed that when we successfully deal with a task (and thus help humanity at large to improve, adapt, and evolve), the probability of our being retained within this system will increase [20,21]. On the basis of this, we can now speculate that humans who become essential to the function of a global techno-cognitive entity (an ecosystem defined by technology and cognition) will experience new biological realities which must ensure their continuing survival [22]. It is also possible to begin the study of certain biological mechanisms which may underpin such a situation. One such mechanism may be based on conserved biological functions and pathways [23,24] which are currently under-expressed but could be made functional through exposure to a cognitive/digital environment. The information input would require a continual repair and good function of neurons, and thus their continual survival. Below, I describe a mechanism (the Indispensable Soma Hypothesis) by which participating humans may experience elimination of age-related degeneration, based on a reversal of the repair resource trade-offs between the germline and the soma, specifically the neuron.

The notion that there is an underlying power struggle between the neuron and the germline is encountered deep into prehistoric times. It is interesting to note that ancient wisdom believed that the brain and the reproductive system shared common frontiers. Even in the prehistoric Stone Age, the brain was linked to the genital reproductive function. Homer and the ancient Greeks thought that the brain contained the immortal psyche (symbolizing the persistence of *self*, that is, the continual survival of an individual human), and this was directly connected to the function of semen (the continual survival of mankind). Basically, the ancients symbolized the penis (germline) as representing the immortality of the human race versus the brain (soma) symbolizing the immortality of the individual. Clearly, certain instinctive concepts have deep roots within the human understanding of nature. Human intuition was able to grasp an inherent association between living longer as a cognitive individual (somatic neuron), against living longer as a species (germline).

INDISPENSABLE AGENTS

According to the hypothesis under consideration here, I am suggesting that cognitive humans who place more emphasis on cognition [26], who are well integrated within a digital ecosystem [27], and who are indispensable to the overall adaptability and remodeling of such environment will be retained for longer within that environment. These humans may be able to survive longer because their functional usefulness to the network is more important and less thermodynamically costly than their elimination. I will discuss the biological basis for this later, but here I need to clarify the notion of *indispensability* or *usefulness* of agents within any system. The term "usefulness" is meant in the wider sense and refers to the value of an agent in facilitating the adaptation and evolution of the entire network. It has no ethical or moral connotations. A *useful* agent is a creative individual who improves the quality of function of the system and helps it to regulate its control tasks. It makes the system more effective at coping quickly with several challenges. It improves the performance and productivity of the system, and helps it to accomplish work which would not have been possible without that agent.

Drawing inspiration from the digital realm, we see that any computer nodes (agents) that are "useful" are retained for longer within the network system [28]. This is mirrored in neurobiology, where neurons which are well integrated and well functioning last longer than other cells [29] (see apoptosis section below). This effect is also encountered in social networks where links which are associated with positivism and approval have an increased tendency to connect with similar ones, whereas links carrying negative connotations (such as disapproval, rejection, dislike, and refusal) are characterized by a decreased propensity to connect and are thus eliminated [30]. If this general principle is applied to humans, it could suggest that each one of us has an inherent *value* to the overall evolution and adaptability of our ecosystem. For a significant section of humanity, the current ecosystem is now becoming considerably different compared to previous ones. It is no longer a community of interacting species, but a combination of just two principal elements: humans and computers (artificial intelligence/Internet). Digitally *hyperconnected* humans help maintain the resilience [31,32] and antifragility [33] of this ecosystem by absorbing disturbances and inhibiting propagation of damaging stress through the entire system.

We are now in a position to formalize this general principle, The Law of Requisite Usefulness, which states that *the duration of retention of an agent within a complex adaptive system is proportional to the contribution of that agent to the overall adaptability of the system* [21]. Systems tend to increase both the fitness and the life span of their individual components if these components provide useful functional feedback to the system [34]. Can "usefulness" be calculated? For the purposes of this hypothesis, the Law of Requisite Usefulness is represented by

$$LR(t) = U + C \times CC - Mr$$

where $LR(t)$ = length of retention of an agent within a system (in humans, this represents how long repair resources are being diverted to the soma for)

U = Usefulness value of the agent, given by

$$U = \frac{b - h}{t}$$

where
b = the probability of causing benefits to the system
h = the probability of causing harm to the system
t = time elapsed since birth
C = a buffering constant
Mr = Maintenance resources needed for basic repair and maintenance of the agent

CC = The probability that the agent will capture the benefits of their actions. Being unable to capture the consequences (in this case, benefits) of an action, an agent could still perish even though it may be useful to the system. For example, somatic cells currently invest resources in cooperation with the germline but fail to capture these benefits; thus, they perish and are outcompeted [35].

So, in more detail:

$$LR(t) = \frac{b - h}{t} + C \times Cc - Mr$$

The cybernetic concept of *Selective Reinforcement* provides additional theoretical support for the above submission [36]. The concept holds that an appropriate agent (or action) is selected and retained if its content of information is of a sufficient and appropriate magnitude. For instance, research has shown [37] that an online avatar can *live longer* (through selective user retention) within a virtual society if it is integrated and well connected within that specific society. User disengagement and short retention of users have been correlated with low quality content [38]. The ability to earn online reputation and intrinsic motivation for participating content are also factors that affect retention [39]. This is a universal theme also encountered in the retention (survival) of neurons in the brain [40]. The concept can be extrapolated to humans who act as autonomous creative agents within a technological environment.

These creative, cognitively biased humans act within a global technological society and help to invent novel solutions for continually emerging problems, integrating the network in response to unexpected environmental circumstances, and determining the system's potential for quick adaptation [41]. Such creative humans improve the flexibility of the system and provide many degrees of freedom, enabling the system to store more useful information and thus enhance functionality, adaptation, and evolvability (i.e., longer survival).

Attempts have already been made to elaborate on some possible biological mechanisms which may potentially underpin this increased physical longevity [42]. One such mechanism involves information-sharing processes which force a reallocation of resources from germline cells to somatic cells [43]. This will be discussed in detail below. In addition, it has been shown that appropriate external stimulation may cause epigenetic reprogramming, where an adult cell (old) can revert to a pluripotent (young) stem cell [44], and even that this epigenetic information may be heritable [45]. This provides additional support in principle to the arguments discussed here, showing that epigenetic influence (caused by continual exposure to cognitively challenging information) may result in improved somatic repair and thus survival [46].

Apoptosis in Mature Neurons

The importance of being well integrated within a fully functioning system (and how this integration may help prolong individual survival) is clarified through the example of mature neurons. These are post-mitotic cells which need to consistently operate efficiently until their host (us) dies, and perhaps even longer [47]. Adult neurons are very resistant to apoptosis [48]. This may be achieved through a variety of antiapoptotic measures deployed by mature neurons:

1. Pre-mitochondrial apoptotic breaks. One such break is mediated by nerve growth factor (NGF) which binds to its tyrosine kinase receptor, TrkA and phosphorylates it. This initiates signaling that promotes neuronal survival [49,50].
2. Post-mitochondrial apoptotic brakes, which may involve resistance to cytochrome c (which plays a part in initiating apoptosis) [51] and also near total loss of sensitivity to Apaf-1 (apoptotic protease activating factor 1) which is essential in the apoptotic cascade [52–54].

It is known that deprivation of NGF activates Bax, a pro-apoptotic member of the Bcl-2 protein family (thus triggering apoptotic cascades) [52], but mature neurons which have sufficient availability of NGF through being continually exposed to external stimulation are resistant to this apoptotic risk [55].

Developing neurons may undergo cell death through apoptosis in response to deprivation of sensory inputs [56,57]. This is also true in the case of mature neurons [58]. In mature neurons, survival after injury is correlated with neuronal age: the more mature the neuron is, the less likely it is that it will die [59–61]. The importance here lies in the fact that being mature (i.e., well integrated and well functioning) may provide resistance to apoptotic death [62], and this resistance may be dependent upon maintaining a sufficient level of cognitive sensory input and output.

SOMATIC TISSUE REPAIR: A POSSIBLE MECHANISM FOR HUMAN SURVIVAL

During the co-evolution of the genomic repair mechanisms deployed by somatic and germline cells, there was a fierce race for survival, whereby germline elements have succeeded in modifying the somatic control systems [63]. Even though the somatic elements deployed a vast array of countermeasures in order to acquire sufficient repair resources, there was a relatively rapid divergence of the functionality of the control and regulation systems, which clearly favored the robustness of replicative fidelity of the germline [64]. For the purposes of this discussion, it must be made clear that by *soma* I essentially refer to a human neuron, or to any other organic agent which stores, transmits, or elaborates information. "Soma" is also any other non-germ cell or tissue which is necessary for the survival of this primary neuron. Therefore, the fundamental trade-off is not just between the soma and the germline as suggested by the disposable soma theory, but, more precisely, between the neuron and the germline. Kirkwood and others [65–67] have suggested that although efficient repair mechanisms are conserved in somatic cells, these are significantly downgraded, as a result of trade-offs with germline cells. In this respect, the process of aging (i.e., the time-dependent loss of function) can be conceived as an ongoing conflict, a *war of trade-offs* between the germline and the soma, specifically the neuron [68,69]. During this conflict, any fidelity-assuring factors or processes are preferentially diverted toward the germline in order to safeguard the survival of the given species, instead of being used by the somatic cells [70–72]. This process results in a continual and successful repair of the germline. As a result, the germline exhibits preferential and energy-efficient robust anti-senescence traits, to the detriment of somatic cells which eventually become subjugated by time-related damage.

In this chapter, I am positing that this "war of trade-offs" is not necessarily *a priori* won by the germline. It may be possible to examine fitness landscapes which favor somatic (neuronal) instead of germline survival [73]. Inhabiting a highly cognitive ecosystem exposes us to information that requires action, and this places a continual energetic burden upon the neuron (energetic burden needed *both* for the assimilation of this information *and* for successful neuronal repair). The information has to be incorporated within the neuron in the physical sense in order to cause a change in a biological status from a less useful to a more useful one [74], that is, to cause *adaptation*. This process demands a continual flow of repair resources to the somatic neuron, resources that must be reallocated from the germline [43]. This reallocation may take place if the survival of somatic cells becomes a priority over that of the germline [75].

We know that there is cross talk between the neuron and the germline [76]. Gracida and Eckmann [63] have reported an evolutionarily conserved soma-to-germline communication pathway whereby somatic nuclear receptors (such as the *nhr-114* receptor) buffer against toxic dietary metabolites and actively protect germline stem cells. The *nhr-114* somatic nuclear receptor acts as a detoxifier and shields germline stem cells from damaging environmental stress [77]. Evidence is now gathering to suggest that this process is not unidirectional, with resources flowing from the soma to the germline only. Instead, under certain circumstances, resources can flow from the germline back to the soma [78,79]. In a paper by Ermolaeva et al. [80], the possibility was raised that genetic injury in the germline may initiate protective effects and upregulate stress resistance pathways in somatic cells, at least in *C. elegans*. For instance, the ubiquitin–proteasome system in somatic cells may be upregulated through agents generated in the germline [81], such as the MAP (mitogen activated protein)-kinase homologue MPK-1 [82]. This germ-initiated somatic protective response may reflect a conserved propensity to reverse the trade-offs between germ cell and somatic cell repair [83]. An example of the continual conflict between germline and soma [84] is also encountered in a study by Labbadia and Morimoto [85]. They found that removing germline stem cells in *C. elegans* preserves expression of factors such as *jmjd-3.1* and this maintains an effective response to stress in somatic lines.

This conflict for resource acquisition may have originated from elementary principles where two essential for life processes, namely storing hereditary information and performing enzymatic activities, had to be separated. In what has been described as a *division of labor* by Bozza et al. [69], the conflict between these two functions was initiated because enzymatic activities of the ribozymes were in a trade-off with their replication rates. The result of these trade-offs was the development of strong asymmetry between enzymatic and genetic processes. This suggests that the schism between replication (in the germline) and enzymatic efficiency (in the soma) has ancient origins buried in the earliest stages of evolution of life. In addition, this example may help drive the point that evolution preferentially selects situations which are beneficial to the continuation of life at a low thermodynamic cost and, if (as I am arguing here) the balance of benefits is shifted in favor of somatic repair, then it would be logical to expect that this situation would be preferentially selected.

Other authors confirm that germline functions affect somatic life span [86]. These authors showed that removal of *the C. elegans* germline triggers changes in regulatory functions (some of which involve microRNA activities) which result in increased longevity [86]. It is possible to encounter soma-to-germline transformation of gene expression [87], such as the improved function of the FOXO transcription factor Daf-16, which is normally encountered only in the germline [88]. The maintenance of the functional stability of the germline may be dependent on factors which may also be present in somatic cells, but are significantly downregulated. Such factors are [66]:

- Both general and specific cellular repair and maintenance mechanisms. Certain DNA repair systems including GG-NER (global genome nucleotide excision repair) are specifically active in germline cells, but these are also active in somatic cells [89].
- Efficient selection of fully functional germ cells which are allowed to propagate at the expense of less efficient germ cells [90].
- Nonautonomous contributions of the soma toward germline rejuvenation [72].

The above facts raise the possibility that somatic cells lines may, under certain circumstances, withhold any *immortality* contributions for their own needs. This may happen when somatic cells are under intense pressure to maintain themselves (following, e.g., unremitting cognitive information inputs and challenges). Fontana and Wrobel [91] have speculated that certain efficient repair sequences may be created in the somatic cells and subsequently migrate to the germline (germline penetration). It can be argued that this process can be arrested, and therefore encourage somatic cells to use these repair sequences for their own repairs instead.

It was shown that neurons may interfere with germline survival. Levi-Ferber et al. [92] have shown that neuronal stress induces apoptosis in the germline. This process is mediated by the IRE-1 (inositol-requiring enzyme 1) factor, an endoplasmic reticulum stress response sensor, which then activates p53 and initiates the apoptotic cascade in the germline [93]. Phosphorylated IRE-1 also activates tumor-necrosis factor (TNF)-receptor-associated factor 2 (TRAF2) which is another apoptosis-initiating factor [94].

The suggestion that stress response factors initiated in neurons may cause apoptosis in germline cells is astounding. Stress response pathways may be activated in neurons through cognitive stimulation of the magnitude and type we encounter in our modern technological society [95] and then cause apoptotic germline cell death. The result is that neurons are better maintained and continue to survive in tandem with diminished germ cell survival (Figure 4.1). This is not to say that the matter

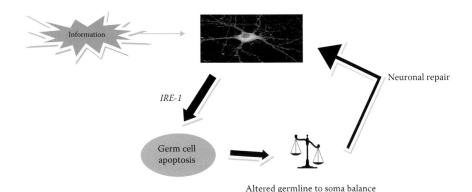

FIGURE 4.1 Cognitively challenging information results in damaged germline cells. Environmental high-quality information, which creates a tendency for action, up-regulates human neuronal stress responses, which generates IRE-1 (among many other stress response factors) which results in germline cell apoptosis, an event which subsequently releases signals to upgrade the repair of somatic cells.

is simple. It was recently shown that germline cells may *retaliate* in order to increase degeneration of the neurons. Wu et al. [96] found that germline loss resulted in resistance to neurodegeneration following infection. In other words, germline factors initiate degenerative sequences in neurons, and when these germline factors are lost, neural degeneration decreases. These effects are mediated by the maternal sterile gene *mes-1* which encodes a receptor protein that is needed for the essential (from the germline point of view) unequal cell division in embryonic germ line stem cells.

Role of Noncoding RNAs

Information from the environment is transmitted to the genome via microRNAs and other noncoding RNAs, and it can cause epigenetic modifications [97]. This is encountered in plants, worms, and also mammals [98]. microRNAs also transmit information between the soma and the germline [99] and at least in *C. elegans*, neurons transmit double-stranded RNA to the germline and cause transgenerational gene silencing [100]. Sharma [101] has discussed how the transfer of heritable epigenetic information from the soma to the germline depends upon microRNAs and exosome function.

Intracellular clearance systems are upregulated following signals from the germline [83]. This is significant because it indicates that germ cells have direct control over the health and longevity of somatic cells. In addition, protein homoeostasis in somatic cells is well maintained when germ cells are damaged, and it is significantly downgraded when germ cell function improves [81]. There exist mechanisms in germ cells which may induce somatic cell reprogramming and somatic stem cell pluripotency [102,103]. Boulias and Horvitz [104] have claimed that the microRNA *mir-71* mediates the effects of germline cell loss in *C. elegans*. Overexpression of *mir-71* extends the life span when germline cells are missing or damaged.

piRNAs in the Germline

piRNAs (PIWI-Interacting RNAs) are noncoding RNAs but distinct from microRNAs in some characteristics. They are involved in gene silencing of transposons among others, so they protect the germline. These are also expressed in tumor cells and in certain somatic cells which exhibit biological immortality, such as those in hydra. It is well known that genomic instability in somatic cells increases with age. Evidence indicates that the disintegration of somatic genomes (progressive chromatin decondensation for instance) is accompanied by the mobilization of transposable elements (TEs) which can be mutagenic. However, since TEs are silenced in the germline, there is less risk of adverse mutations [105].

piRNAs are defined by their specific binding to the PIWI (P-element Induced WImpy testis) subfamily of the Argonaute/PIWI family proteins [106]. One of the fidelity preservation mechanisms deployed by the germline involves the piRNA and retrotransposon defense mechanisms [107,108]. piRNA silencing may have an impact (through very intricate mechanisms) on the maintenance of the stability of the germ cell genome [109]. In somatic cells, PIWI may act in association with the conserved flamenco piRNA cluster and prevent retroviral elements from propagating and from damaging germ cells [110].

Even though piRNAs were thought to be germline-specific, and have a germline-restricted expression [111], we now know that there is an expression of at least 28 types of piRNA in the brain [112], including the hippocampus [113]. piRNAs in the brain may regulate the function of transposons within the neuron [114] and protect the brain from mutation accumulation, thus playing a role in maintaining cognitive function [115]. Noncoding RNAs in general, and piRNAs specifically, are expressed within dendrites in the mammalian neurons [116,117] and are especially active near synapses [118]. They participate in plasticity processes which define learning and memory [119]. We also know that developing neurons and germline cells share similar polarization morphologies and both depend on piRNA activity [120].

Therefore, if piRNAs are present in the germline and are pivotal in assuring germline perpetual repair, and if they are also present in neurons [121], it is tempting to speculate that they may also be pivotal in continual neuronal repair. This provides another biological mechanism for the soma versus germline conflict argument. It can be surmised that cognitive information has a direct effect on piRNAs, which then improve neuronal repair and are also in direct cross talk/conflict with the germline.

This supports the notion that mechanisms which ensure the integrity of the germline and somatic nervous system share common frontiers and are mediated through common elements which are dependent on external epigenetic signals. These examples show that mechanisms making possible a germline-to-soma reallocation of resources exist in nature, are likely to be widespread, and can be made to operate by manipulating the environment to make it less dangerous, albeit cognitively challenging [95].

STEM CELLS

An important communication pathway between the neuron and the germline is one that involves stem cells. A source of both *neuron-like* cells [122] and definitive neural stem cells [123] could be from germline (spermatogonial) stem cells. Multipotent neural and glial precursors can be derived from multipotent adult germ line stem cells [124]. Following this, the resulting neurons can achieve full maturation and efficient integration within the existing neural network. This possibility (that the germline may act as a source of fully functional neurons [125–127]) is extraordinary. However, research is lacking in this area and further questions need to be answered. For instance, it is necessary to establish under what circumstances the germline can facilitate the provision of functioning neural stem cells, what the interrelationship between these two elements are, and if external cognitive challenges which activate the stress response in neurons may have an impact on spermatogonial germ stem cells. If this is explained, it will then lend more impact on my general speculation that the trade-offs are specifically relevant between the neuron and the germline. A short video of a direct conversion of germ stem cells into a specific neuron is available from Hobert lab [128] here https://vimeo.com/60414246.

It is also possible to examine the reprogramming of somatic cells into induced pluripotent stem cells (iPSCs). This is one way whereby somatic cells can be reprogrammed and so be used when needed by somatic tissues [129]. An intriguing possibility is the reprogramming of somatic cells into specifically neural stem cells via microRNAs (which provide the interface between external stimulation, such as cognitive information, and the cellular milieu) [130]. Although there are many still unanswered questions regarding this process, such as the stress response triggered by the reprogramming process (the reprogramming-induced senescence (RIS) [131], this is a useful concept to consider as it shows that the aged soma is not necessarily irreparable. Epigenetic regulation (via chromatin modification for instance) can assure a successful repair sequence and thus increase the life span [129]. We know that chromatin modifiers play a positive role in neural health [132]. One such chromatin modulator is the barrier-to-autointegration factor (BAF) which plays a role both in gonad development [133] and in neurodevelopment [134]. Stimuli originating from the environment constantly influence the epigenome and remodel our responses to these stimuli. The process of aging is subjected to epigenetic regulations [135]. Taking these facts a little more into the realm of speculation, we may assume that aging somatic cells may be reprogrammed through the action of microRNAs (which operate following suitable external, cognitive, technology-assisted stimulation) into functioning neurons [136] and thus improve overall brain function. This must result in a reduction of age-related degeneration, and also represents an example where germline repair resources have been diverted to the neuron [137].

TESTABILITY, PREDICTIONS, AND FUTURE PROSPECTS

The hypothesis provides biological and evolutionary interpretations concerning the near to midterm future integration of humans with technology [138]. If the prediction of the hypothesis is correct, we may expect to encounter upregulated somatic survival mechanisms in those people who are digitally hyperconnected in a meaningful fashion (those who *continually share meaningful information-that-requires-action, and receive constructive instant feedback on their actions*). This can be tested in a variety of ways, and three examples are mentioned below:

TELOMERE LENGTH

It would be relatively easy to devise a clinical study of telomere length (or telomerase expression) in people who are hyperconnected (and thus are more cognitive than physical), compared to those who are not connected, perhaps living in a totally agricultural setting (and are thus more physical than cognitive), matched for age, sex, education, health status, and other parameters. This observational case-control study, which can also be a longitudinal study, may support the hypothesis if it shows that participants who are digitally hyperconnected (as defined above) experience a lengthening of their somatic telomeres and, perhaps, a selective shortening of their germline telomeres. The opposite finding would be true in those who are not hyperconnected.

Several studies found a correlation between a cognitive environment and telomere length [139,140] to the point where some authors suggest that telomere length may be used as a biological marker for age-related cognitive decline [141]. It is also known that short telomeres are associated with reduced cognitive function [142]. These studies provide support in principle that cognitive activity is indeed positively associated with telomere length [143] and that a cognitive, digital environment may enhance this association.

APOPTOSIS

It is also possible to examine apoptotic markers (such as microRNA-29, Apaf-1, Bax/Bcl-2) [144] in the two study groups mentioned above. Here, we can expect to find a decreased expression of pro-apoptotic markers in neurons and an increased rate of apoptosis in the germline [48].

Although, of course, there could be various other elements affecting apoptosis, the findings, if taken in association with other parameters, may provide a general support to the hypothesis under consideration. We know that mature neurons deploy a vast array of antiapoptotic measures [48]; it would be necessary to also show that intense cognitive activity upregulates neuronal antiapoptotic factors to an even higher degree, compared to those encountered in ordinary, relatively *cognition-poor* individuals.

STRESS RESPONSE

In addition to the above studies, it could be possible to evaluate biomarkers which indicate if the stress response has been activated. Following exposure to meaningful information, there is neural expression of factors such as the endoplasmic reticulum stress response factor IRE-1 [93] mentioned above, which may then interfere with normal germline function, as in Figure 4.1. Thus, evaluating this and similar biomarkers may prove an effective way of studying some basic premises of the Indispensable Soma Hypothesis. Other stress response factors in neurons are PERK (protein kinase RNA-like endoplasmic reticulum kinase [145], ATF6 (activating-transcription-factor-6) [146], c-Jun N-terminal kinase 1 (JNK) [147], and ATF4 [148]. These may initiate apoptosis depending on the level of sensed damage [149] and can thus be used as biomarkers of a cognitive stress response, which may then affect germline function.

Highly sensitive techniques, such as Western blot or qRT-PCR (reverse transcription polymerase chain reaction), have already been developed to measure stress response sensors such as IRE-1 [150], PERK phosphorylation, and ATF6 cleavage by specific antibodies [151]. However, it may be difficult, costly, and impractical to test all of these sensor molecules in humans. Nevertheless, it remains a possibility which can be exploited at a later stage with improved technology.

CAVEATS AND LIMITATIONS

When considering the finer mechanisms of the hypothesis, there is a considerable degree of informed speculation. For instance, I assume that humans will continue to exploit technology at the current predicted rates [152] and that no new unforeseen elements will disrupt this line of development.

More research is needed in order to confirm and corroborate some biological assumptions, although the soma-to-germline conflict argument which is central to this hypothesis seems to be well substantiated [63,76–80]. It is also possible that there are other mechanisms involved in addition to the germline versus soma argument, which I have not considered or discussed. Emergent research may contradict certain details, although it is unlikely that it will invalidate the general central tenet. This is because the hypothesis does not compete with any other major evolutionary hypotheses, theories of aging, principles of nature, or laws of physics. It fits well within nearly all existing and accepted knowledge systems. The hypothesis is likely to help explain emergent and new thinking regarding the interactions of humans with their technological environment, and it is merely an extension of current evolutionary mechanisms which must ensure adaptation and survival within a rapidly changing cognitive environment.

One problem is that the space between beneficial integration within a human–computer ecosystem and the emergence of Internet addiction (due to overuse of this technology) is very narrow [153], so it would be necessary to establish suitable ways of exploiting this human–computer relationship without risking addiction or overuse disorders. I have discussed this elsewhere [154].

Another narrow space is encountered between the achievement of healthy fidelity of genomic repair in somatic cells and the risk of immortalization in cancer cells. The relationship between somatic immortalization and cancers is still poorly understood. It is known that ectopic expression of genes, which are normally necessary for germline longevity, may result in unregulated immortalization of somatic cells, that is, cancer, at least in drosophila [155]. In this case, the key (between cancer development and normal somatic repair) appears to be the ectopic expression of the genes. Germline traits which were ectopically expressed may result in cancer, whereas inactivation of these genes may result in cancer-suppressing benefits. This is an area which requires further research and consideration.

CONCLUSION

A continual input of cognitive information makes such energetic demands on the brain that it may cause a situation whereby the energy available to the neuron will be proportional to the amount of information which the cell is required to process [156]. This energy may be used not only for the mechanisms associated with information processing but, necessarily, also for neuronal repair in order to make this information processing possible. I quote from one of my papers [157]:

> It is clearly unlikely that a high level of efficient information processing takes place in a damaged neuron. The neuron needs to be repaired (this includes age-related damage repair) in order to be in a position to process the information. Indeed, studies have estimated that 50%–80% of energy consumption is allocated to signalling, and the remainder is allocated to neuronal maintenance and repairs [158]. Here a balance must be found between the energy costs of information processing and the contribution to overall fitness made by this information. If the fitness of the organism increases as a result of the new information, then it means that the cost of processing that information was worth it.

Upregulated neuronal repair results in healthy neurons, which in turn improve homeostasis and affect the aging of other somatic cells through a variety of mechanisms [159].

It is obvious that the biological mechanisms associated with this hypothesis need additional elucidation, and the speculative inferences need further grounding. However, on the basis of the above discussion, it seems likely that the environment, via epigenetic factors such as noncoding RNAs (among others), can influence the balance between somatic versus germline allocation of repair resources. In order to reverse these trade-offs, it is necessary to place a critical mass of humanity under such a load of information processing (e.g., through digital hyperconnection and information sharing) [8] that conserved evolutionary mechanisms are activated in favor of somatic (neuronal) cells which are required in order to process this information. When such a critical mass is reached, efficient *network effects* [160] may emerge and facilitate (through mutual feedback loops and reciprocal selective reinforcement) the transition. In other words, the process of assimilating new information may initiate a sequestration sequence that diverts resources from germ cells to neuronal cells. It is important to emphasize that this is taken in the evolutionary context of increased fitness within a technological, information-laden cognitive (and less physical) niche. Thus, information processing carries an adaptive value within this specific niche.

The Indispensable Soma Hypothesis posits that: *humans who are suitably integrated within a technological, information-rich, cognitive environment, and who are indispensable for the adaptability of such environment, may experience a reversal of their resource allocation priorities from the germline to the soma, resulting in improved somatic repair and a progressive reduction of age-related degeneration.* In other words, a neuron makes itself indispensable by acting positively within the general process of successful cognitive adaptation, may experience a reversal of the principles that govern the disposable soma theory, and thus experiences a progressive reduction of age-related degeneration.

REFERENCES

1. Gillings MR, Hilbert M, Kemp DJ. 2015. Information in the biosphere: Biological and digital worlds. *Trends Ecol Evol*. http://dx.doi.org/10.1016/j.tree.2015.12.013 (accessed 10 January 2017).
2. Lifton J, Paradiso JA. 2010. *Dual Reality: Merging the Real and Virtual. Facets of Virtual Environments*. Springer, Berlin Heidelberg. 12–28. http://resenv.media.mit.edu/pubs/papers/2009-07-fave2009.pdf (accessed 4 January 2017).
3. Applin S, Fischer M. 2011. A cultural perspective on mixed, dual and blended reality. *IUI-Workshop on Location Awareness for Mixed and Dual Reality (LAMDa'11)*.
4. Gershon N. Wearables, Humans, and Things as a Single Ecosystem! November 9, 2015 http://iot.ieee.org/newsletter/november-2015/wearables-humans-and-things-as-a-single-ecosystem.html (accessed 2 January 2017).
5. Jablonka E, Lamb MJ. 2006. The evolution of information in the major transitions. *J Theor Biol* 239:236–246.
6. Szathmáry E. 2015. Toward major evolutionary transitions theory 2.0. *Proc Natl Acad Sci USA* 112:10104–10111.
7. Ball P. The Strange Inevitability of Evolution, Creativity 020, January 8, 2015. http://nautil.us/issue/20/creativity/the-strange-inevitability-of-evolution (accessed 15 January 2017).
8. Kyriazis M. 2016. A cognitive-cultural segregation and the three stages of aging. *Curr Aging Sci* 9(1):1–6.
9. Flatt T, Heyland A. 2011. *Mechanisms of Life History Evolution: The Genetics and Physiology of Life History Traits and Trade-offs*. Oxford, UK: Oxford University Press.
10. Stearns SC. 2008. Life history evolution: Successes, limitations, and prospects. *Naturwissenschaften* 87(11):476–486.
11. Heyes C. 2012. New thinking: The evolution of human cognition. *Philos Trans R Soc Lond B Biol Sci* 367(1599):2091–2096.
12. Kirkwood TBL. 2002. Evolution of ageing. *Mech Ageing Dev* 123(7):737–745.
13. Garcia D, Mavrodiev P, Schweitzer F. 2013. Social resilience in online communities: The autopsy of Friendster. *COSN '13 Proceedings of the First ACM Conference on Online Social Networks*. 39–50. Boston, MA.
14. Garcia D, Tanase D. 2013. Measuring cultural dynamics through the Eurovision song contest. *Advs Complex Syst* 16:350037.
15. Iandoli L, Klein M, Zollo G. 2009. Enabling on-line deliberation and collective decision-making through large-scale argumentation: A new approach to the design of an Internet-based mass collaboration platform. *Int J Decis Support Syst Technol* 1(1):69–92.

16. Garas A, Garcia D, Skowron M, Schweitzer F. 2012. Emotional persistence in online chatting communities. *Sci Rep* 2:402.

17. Gupte M, Eliassi-Rad T. 2012. Measuring tie strength in implicit social networks. *Proceedings of the 3rd Annual ACM Web Science Conference.* 109–118. Evanston, Illinois.

18. Gontis V, Kononovicius A. 2014. Consentaneous agent-based and stochastic model of the financial markets. *PLoS One* 9(7):e102201.

19. Zhu J, Hong J, Hughes JG. 2002. Using Markov Chains for link prediction in adaptive web sites. In: Bustard, D, Sterritt, W. and Liu, R. (Eds) *Soft-Ware: Computing in an Imperfect World: First International Conference, Soft-Ware 2002 Belfast.* Lecture Notes in Computer Science, 2311. Springer, 60–73.

20. Kitsak M, Gallos LK, Havlin S et al. 2010. Identification of influential spreaders in complex networks. *Nat Phys* 6(11):888–893.

21. Kyriazis M. 2015. Technological integration and hyper-connectivity: Tools for promoting extreme human lifespans. *Complexity* 20(6):15–24.

22. Kyriazis M. 2015. Environmental challenges may impact on somatic repair. *Peer J Perprint.* https://peerj.com/preprints/1560/ (accessed January 2, 2017).

23. Aguirre A, Montserrat N, Zacchigna S et al. 2014. In vivo activation of a conserved MicroRNA program induces mammalian heart regeneration. *Cell Stem Cell* 15(5):589–604.

24. Hodgkinson CP, Dzau VJ. 2015. Conserved microRNA program as key to mammalian cardiac regeneration: Insights from zebrafish. *Circ Res* 116(7):1109–1111.

25. Franklin S, Graesser A. 2005. Is It an agent, or just a program? A taxonomy for autonomous agents. *Intell Agents III Agent Theor, Archit, Lang* 1193:21–35.

26. Zacarias M, Valente de Oliveira J (Eds) 2012. *Human-Computer Interaction: The Agency Perspective. Studies in Computational Intelligence.* Springer-Verlag, Berlin.

27. Briscoe G, De Wilde P. 2006. Digital ecosystems: Evolving service-oriented architectures. In *Conference on Bio Inspired Models of Network, Information and Computing Systems.* IEEE Press, Cambridge, UK

28. Bader DA, Pennington R. 2001. Cluster computing: Applications. *Int J High Performance Comp* 15(2):181–185.

29. Alvarez-Buylla A, Kohwi M, Nguyen TM, Merkle FT. 2008. The heterogeneity of adult neural stem cells and the emerging complexity of their niche. *Cold Spring Harb Symp Quant Biol* 73:357–365.

30. Ciotti V, Bianconi G, Capocci A, Colaiori F, Panzarasa P. 2014. Degree correlations in signed social networks. arXiv:1412.1024 [physics.soc-ph].

31. Gunderson LH, Holling CS (Eds) 2002. *Panarchy: Understanding Transformations in Systems of Humans and Nature.* Island Press, Washington DC.

32. Walker B, Holling CS, Carpenter SR, Kinzig A. 2014. Adaptability and transformability in social-ecological systems. *Ecol Soc* 9:5.

33. Taleb NN. 2012. *Antifragile: Things That Gain from Disorder.* Random House, NY, USA.

34. Gershenson C. 2011. The sigma profile: A formal tool to study organization and its evolution at multiple scales. *Complexity* 16(5):37–44.

35. Stewart JE. 2014. The direction of evolution: The rise of cooperative organisation. *BioSystems* 123:27–36.

36. Pack Kaelbling L, Littman M, Moore A. 1996. Reinforcement learning: A survey. *J Artif Intell Res* 4:237–285.

37. Teng CY, Adamic L. 2010. Longevity in second life. *Proceedings of ICWSM,* https://www.aaai.org/ocs/index.php/ICWSM/ICWSM10/paper/viewFile/1520/1893 (accessed January 8, 2017).

38. Brandtzaeg PB, Heim J. 2008. User loyalty and online communities: Why members of online communities are not faithful. In *Intetain* 1–10. http://dl.acm.org/citation.cfm?id=1363215 (Accessed January 4, 2017).

39. Bryant SL, Forte A, Bruckman A. 2005. Becoming wikipedian: Transformation of participation in a collaborative online encyclopedia. In *Proceedings of GROUP: International Conference on Supporting Group Work,* 1–10, ACM Sanibel Island, FL.

40. Ramírez-Rodríguez G, Ocaña-Fernández MA, Vega-Rivera NM et al. 2014. Environmental enrichment induces neuroplastic changes in middle age female BalbC mice and increases the hippocampal levels of BDNF, p-Akt and p-MAPK1/2. *Neuroscience* 260:158–170.

41. Csermely PC. 2008. Reactive elements: Network-based predictions of active centres in proteins and cellular and social networks. *Cell* 33(12):569–577.

42. Heylighen F. 2014. Cybernetic principles of aging and rejuvenation: The buffering-challenging strategy for life extension. *Curr Aging Sci* 7(1), 60–75.

43. Kyriazis M. 2014. Reversal of informational entropy and the acquisition of germ-like immortality by somatic cells. *Curr Aging Sci* 7(1):9–16.
44. Obokata H, Wakayama T, Sasai, Y et al. 2014. Stimulus-triggered fate conversion of somatic cells into pluripotency. *Nature* 505:641–647.
45. Dias BG, Ressler KJ. 2014. Parental olfactory experience influences behavior and neural structure in subsequent generations. *Nat Neurosci* 17(1):89–96.
46. Kyriazis, M. 2014. Information-sharing, adaptive epigenetics and human longevity. http://arxiv.org/abs/1407.6030.
47. Magrassi L, Leto K, Rossi F. 2013. Lifespan of neurons is uncoupled from organismal lifespan. *Proc Natl Acad Sci USA* 110(11):4374–4379.
48. Kole AJ, Annis RP, Deshmukh M. 2013. Mature neurons: Equipped for survival. *Cell Death Disease* 4:e689.
49. Snider WD. 1994. Functions of the neurotrophins during nervous system development: What the knock-outs are teaching us. *Cell* 77:627–638.
50. Kaplan DR, Miller FD. 2000. Eurotrophin signal transduction in the nervous system. *Curr Opin Neurobiol* 10:381–391.
51. Putcha GV, Deshmukh M, Johnson EM Jr. 2000. Inhibition of apoptotic signaling cascades causes loss of trophic factor dependence during neuronal maturation. *J Cell Biol* 149:1011–1018.
52. Wright KM, Smith MI, Farrag L, Deshmukh M. 2007. Chromatin modification of Apaf-1 restricts the apoptotic pathway in mature neurons. *J Cell Biol* 179:825–832.
53. Yakovlev AG, Ota K, Wang G et al. 2001. Differential expression of apoptotic protease-activating factor-1 and caspase-3 genes and susceptibility to apoptosis during brain development and after traumatic brain injury. *J Neurosci* 21:7439–7446.
54. Johnson CE, Huang YY, Parrish AB et al. 2007. Differential Apaf-1 levels allow cytochrome c to induce apoptosis in brain tumors but not in normal neural tissues. *Proc Natl Acad Sci USA* 104:20820–20825.
55. Sofroniew MV, Howe CL, Mobley WC. 2001. Nerve growth factor signaling, neuroprotection, and neural repair. *Annu Rev Neurosci* 24:1217–1281.
56. Harris JA, Rubel EW. 2006. Afferent regulation of neuron number in the cochlear nucleus: Cellular and molecular analyses of a critical period. *Hear Res* 216–217:127–137.
57. Mandairon N, Jourdan F, Didier A. 2003. Deprivation of sensory inputs to the olfactory bulb up-regulates cell death and proliferation in the subventricular zone of adult mice. *Neuroscience* 119(2):507–516.
58. Kikuta S, Sakamoto T, Nagayama S, Kanaya K, Kinoshita M, Kondo K, Tsunoda K, Mori K, Yamasoba T. 2015. Sensory deprivation disrupts homeostatic regeneration of newly generated olfactory sensory neurons after injury in adult mice. *J Neurosci* 35(6):2657–2673.
59. McKernan DP, Caplis C, Donovan M, O'Brien CJ, Cotter TG. 2006. Age-dependent susceptibility of the retinal ganglion cell layer to cell death. *Invest Ophthalmol Vis Sci* 47:807–814.
60. Guerin MB, McKernan DP, O'Brien CJ, Cotter TG. 2006. Retinal ganglion cells: Dying to survive. *Int J Dev Biol* 50:665–674.
61. Bittigau P, Sifringer M, Pohl D et al. 1999. Apoptotic neurodegeneration following trauma is markedly enhanced in the immature brain. *Ann Neurol* 45:724.
62. Liu CL, Siesjo BK, Hu BR. 2004. Pathogenesis of hippocampal neuronal death after hypoxia-ischemia changes during brain development. *Neuroscience* 127:113–123.
63. Gracida X, Eckmann CR. 2013. Fertility and germline stem cell maintenance under different diets requires nhr-114/HNF4 in *C. elegans*. *Curr Biol* 23(7):607–613.
64. Heininger K. 2002. Aging is a deprivation syndrome driven by a germ-soma conflict. *Ageing Res Rev* 1(3):481–536.
65. Westendorp RG, Kirkwood TB. 1998. Human longevity at the cost of reproductive success. *Nature* 396(6713):743–746.
66. Smelick C, Ahmed S. 2005. Achieving immortality in the *C. elegans* germline. *Ageing Res Rev* 4:67–82.
67. Drenos F, Kirkwood TBL. 2005. Modelling the disposable soma theory of ageing. *Mech Ageing Develop* 126(1):99–103.
68. Charlesworth, B. 2000. Fisher, Medawar, Hamilton and the evolution of aging. *Genetics* 156:927–931.
69. Bozza G, Szilagyi A, Kun A, Santos M, Szathmary E. 2014. Evolution of the division of labor between genes and enzymes in the RNA world. *PLOS Comput Biol*. Doi: 10.1371/journal.pcbi.1003936.
70. Cossetti C, Lugini L, Astrologo L, Saggio I, Fais S, Spadafora C. 2014. Soma-to-germline transmission of RNA in mice xenografted with human tumour cells: Possible transport by exosomes. *PLoS One* 9(7):e101629.
71. Baudisch A, Vaupel JW. 2012. Getting to the root of aging. *Science* 338(6107):618–619.

72. Kirkwood TB. 1987. Immortality of the germ-line versus disposability of the soma. *Basic Life Sci* 42:209–218.

73. Chastain E, Antia R, Bergstrom CT. 2012. Defensive complexity and the phylogenetic conservation of immune control. arXiv: 1211.2878 [q-bio.PE].

74. Joshi NJ, Tononi G, Koch C. 2013. The minimal complexity of adapting agents increases with fitness. *PLoS Comput Biol* 9:e1003111.

75. Heininger K. 2001. The deprivation syndrome is the driving force of phylogeny, ontogeny and oncogeny. *Rev Neurosci* 12(3):217–218.

76. Avise JC. 1993. The evolutionary biology of aging, sexual reproduction and DNA repair. *Evolution* 47:1293–1301.

77. Gracida X, Eckmann CR. 2013. Mind the gut: Dietary impact on germline stem cells and fertility. *Commun Integr Biol* 6(6):e260040.

78. Douglas PM, Dillin A. 2014. The disposable soma theory of aging in reverse. *Cell Res* 24:7–8.

79. Qian Y, Ng CL, Schulz C. 2015. CSN maintains the germline cellular microenvironment and controls the level of stem cell genes via distinct CRLs in testes of *Drosophila melanogaster*. *Dev Biol* 398(1):68–79.

80. Ermolaeva MA, Segref A, Dakhovnik A et al. 2013. DNA damage in germ cells induces an innate immune response that triggers systemic stress resistance. *Nature* 501(7467):416–420.

81. Shemesh N, Shai N, Ben-Zvi A. 2013. Germline stem cell arrest inhibits the collapse of somatic proteostasis early in *Caenorhabditis elegans* adulthood. *Aging Cell* 12:814–822.

82. Ermolaeva M, Schumacher B. 2013. The innate immune system as mediator of systemic DNA damage responses. *Commun Integr Biol* 6(6):e26926.

83. Khodakarami A, Saez I, Mels J, Vilchez D. 2015. Mediation of organismal aging and somatic proteostasis by the germline. *Front Mol Biosci*. http://dx.doi.org/10.3389/fmolb.2015.00003.

84. Smendziuk CM, Messenberg A, Vogl AW, Tanentzapf G. 2015. Bi-directional gap junction-mediated soma-germline communication is essential for spermatogenesis. *Development* 142(15):2598–2609.

85. Labbadia J, Morimoto RI. 2015. Repression of the heat shock response is a programmed event at the onset of reproduction. *Mol Cell* 59(4):639–650.

86. Shen Y, Wollam J, Magner D, Karalay O, Antebi A. 2012. A steroid receptor-microRNA switch regulates life span in response to signals from the gonad. *Science* 338(6113):1472–1476.

87. Kenyon C. 2010. A pathway that links reproductive status to lifespan in *Caenorhabditis elegans*. *Ann N Y Acad Sci* 1204:156–162.

88. Curran SP, Wu X, Riedel CG, Ruvkun G. 2009. A soma-to-germline transformation in long-lived *Caenorhabditis elegans* mutants. *Nature* 459:1079–1084.

89. Petruseva IO, Evdokimov AN, Lavrik OI. 2014. Molecular mechanism of global genome nucleotide excision repair. *Acta Naturae* 6(1):23–34.

90. Murphey P, McLean DJ, McMahan CA, Walter CA, McCarrey JR. 2013. Enhanced genetic integrity in mouse germ cells. *Biol Reprod* 88(1):6.

91. Fontana A, Wróbel B. 2012. A model of evolution of development based on germline penetration of new "no-junk" DNA. *Genes (Basel)* 3(3):492–504.

92. Levi-Ferber M, Salzberg Y, Safra M, Haviv-Chesner A, Bülow HE, Henis-Korenblit S. 2014. It's all in your mind: Determining germ cell fate by neuronal IRE-1 in *C. elegans*. *PLoS Genet* 10(10):e1004747.

93. Levi-Ferber M, Gian H, Dudkevich R, Henis-Korenblit S. 2015. Transdifferentiation mediated tumor suppression by the endoplasmic reticulum stress sensor IRE-1 in *C. elegans*. *Elife*. 4. doi: 10.7554/eLife.08005.

94. Prause MC, Berchtold LA, Urizar AI, Hyldgaard Trauelsen AM, Billestrup N, Mandrup-Poulsen T, Størling J. 2015. TRAF2 mediates JNK and STAT3 activation in response to IL-1β and IFNγ and facilitates apoptotic death of insulin-producing β-cells. *Mol Cell Endocrinol* 420:24–36. S0303-7207(15)30146-5.

95. Kyriazis M. 2016. Epigenetic regulation and adaptation to stimuli. In *Challenging Ageing: The Anti-Senescence Effects of Hormesis, Environmental Enrichment, and Information Exposure*. Bentham Science Publishers, UAE.

96. Wu Q, Cao X, Yan D, Wang D, Aballay A. 2015. Genetic screen reveals link between maternal-effect sterile gene mes-1 and *P. aeruginosa*-induced neurodegeneration in *C. elegans*. *J Biol Chem* Oct 16. pii: jbc.M115.674259. [Epub ahead of print].

97. Yoshioka Y, Katsuda T, Ochiya T. 2015. Circulating microRNAs as hormones: Intercellular and inter-organ conveyors of epigenetic information? *EXS*. 106:255–267.

98. Sharma A. 2014. Novel transcriptome data analysis implicates circulating microRNAs in epigenetic inheritance in mammals. *Gene* 538(2):366–372.

99. Toledano H. 2013. The role of the heterochronic microRNA let-7 in the progression of aging. *Exp Gerontol* 48(7):667–670.

100. Devanapally S, Ravikumar S, Jose AM. 2015. Double-stranded RNA made in C. *elegans* neurons can enter the germline and cause transgenerational gene silencing. *Proc Natl Acad Sci USA* 112(7):2133–2138.

101. Sharma A. 2015. Transgenerational epigenetic inheritance: Resolving uncertainty and evolving biology. *Biomol Concepts* 6(2):87–103.

102. Bazley FA, Liu CF, Yuan X et al. 2015. Direct reprogramming of human primordial germ cells into induced pluripotent stem cells. *Stem Cells Dev* 24(22):2634–2648.

103. Nagamatsu G, Kosaka T, Saito S, Honda H, Takubo K, Kinoshita T. 2013. Induction of pluripotent stem cells from primordial germ cells by single reprogramming factors. *Stem Cells* 32:3.

104. Boulias K, Horvitz HR. 2012. The *C. elegans* MicroRNA mir-71 acts in neurons to promote germline-mediated longevity through regulation of DAF-16/FOXO. *Cell Metab* 15(4):439–450.

105. Sturm Á, Ivics Z, Vellai T. 2015. The mechanism of ageing: Primary role of transposable elements in genome disintegration. *Cell Mol Life Sci* 72:1839. doi: 10.1007/s00018-015-1896-0.

106. Seto AG, Kingston RE, Lau NC. 2007. The coming of age for Piwi proteins. *Mol Cell* 26(5):603–609.

107. Crichton JH, Dunican DS, MacLennan M, Meehan RR, Adams IR. 2014. Defending the genome from the enemy within: Mechanisms of retrotransposon suppression in the mouse germline. *Cell Mol Life Sci* 71(9):1581–1605.

108. Bao J, Yan W. 2012. Male germline control of transposable elements. *Biol Reprod* 86(5):162.

109. Kibanov MV, Egorova KS, Ryazansky SS et al. 2011. A novel organelle, the piNG-body, in the nuage of Drosophila male germ cells is associated with piRNA-mediated gene silencing. In: Brill JA ed. Garland Science, New York. *Molecular Biology of the Cell*. 22(18):3410–3419.

110. Malone CD, Brennecke J, Dus M, Stark A, McCombie WR, Sachidanandam R, Hannon GJ. 2009. Specialized piRNA pathways act in germline and somatic tissues of the Drosophila ovary. *Cell* 137(3):522–535.

111. Ishizu H, Siomi H, Siomi MC. 2012. Biology of PIWI-interacting RNAs: New insights into biogenesis and function inside and outside of germlines. *Genes Dev* 26:2361–2373.

112. Rajasethupathy P, Antonov I, Sheridan R, Frey S, Sander C, Tuschl T, Kandel ER. 2012. A role for neuronal piRNAs in the epigenetic control of memory-related synaptic plasticity. *Cell* 149(3):693–707.

113. Lee EJ, Banerjee S, Zhou H, Jammalamadaka A, Arcila M, Manjunath BS, Kosik KS. 2011. Identification of piRNAs in the central nervous system. *RNA* 17(6):1090–1099.

114. Iyengar BR, Choudhary A, Sarangdhar MA, Venkatesh KV, Gadgil CJ, Pillai B. 2014. Non-coding RNA interact to regulate neuronal development and function Front. *Cell. Neurosci* 8:47.

115. Butler AA, Webb WM, Lubin FD. 2015. Regulatory RNAs and control of epigenetic mechanisms: Expectations for cognition and cognitive dysfunction. *Epigenomics* 8(1):135–51.

116. Weiss K, Antoniou A, Schratt G. 2015. Non-coding mechanisms of local mRNA translation in neuronal dendrites. *Eur J Cell Biol* 94(7–9):363–367.

117. Bredy TW, Lin Q, Wei W, Baker-Andresen D, Mattick JS. 2011. MicroRNA regulation of neural plasticity and memory. *Neurobiol Learn Mem* 96(1):89–94.

118. Earls LR, Westmoreland JJ, Zakharenko SS. 2014. Non-coding RNA regulation of synaptic plasticity and memory: Implications for aging. *Ageing Res Rev* 17:34–42.

119. Smalheiser NR. 2014. The RNA-centred view of the synapse: Non-coding RNAs and synaptic plasticity. *Philos Trans R Soc Lond B Biol Sci* 369:1652.

120. Zhao P, Yao M, Chang S et al. 2015. Novel function of PIWIL1 in neuronal polarization and migration via regulation of microtubule-associated proteins. *Mol Brain* 8:39.

121. Ku HY, Lin H. 2014. PIWI proteins and their interactors in piRNA biogenesis, germline development and gene expression. *Natl Sci Rev* 1(2):205–218.

122. Wang X, Chen T, Zhang Y, Li B, Xu Q, Song C. 2015. Isolation and culture of pig spermatogonial stem cells and their in vitro differentiation into neuron-like cells and adipocytes. *Int J Mol Sc* 16(11):26333–26346.

123. Teichert AM, Pereira S, Coles B et al. 2014. The neural stem cell lineage reveals novel relationships among spermatogonial germ stem cells and other pluripotent stem cells. *Stem Cells Dev* 23(7):767–778.

124. Glaser T, Opitz T, Kischlat T et al. 2008. Adult germ line stem cells as a source of functional neurons and glia. *Stem Cells* 26(9):2434–2443.

125. Kim BJ, Lee YA, Kim KJ et al. 2015. Effects of paracrine factors on CD24 expression and neural differentiation of male germline stem cells. *Int J Mol Med* 36(1):255–262.

126. Streckfuss-Bömeke K, Vlasov A, Hülsmann S. et al. 2009. Generation of functional neurons and glia from multipotent adult mouse germ-line stem cells. *Stem Cell Res* 2(2):139–154.

127. Yang H, Liu Y, Hai Y et al. 2015. Efficient conversion of spermatogonial stem cells to phenotypic and functional dopaminergic neurons via the PI3 K/Akt and P21/Smurf2/Nolz1 pathway. *Mol Neurobiol* 52(3):1654–1669.

128. Tursun B, Patel T, Kratsios P, Hobert O. 2011. Direct conversion of *C. elegans* germ cells into specific neuron types. *Science* 331(6015):304–8.

129. Mahmoudi, S, Brunet, A. 2012. Aging and reprogramming: A two-way street. *Curr Opinion Cell Biol* 24(6):744–756.

130. Yang, H, Zhang, L, An, J et al. 2016. MicroRNA-mediated reprogramming of somatic cells into neural stem cells or neurons. *Mol Neurobiol* 54(2):1587–1600.

131. Banito, A, Rashid, ST, Acosta, JC et al. 2009. Senescence impairs successful reprogramming to pluripotent stem cells. *Genes Dev* 23:2134–2139.

132. Ronan, JL, Wu, W, Crabtree, GR. 2013. From neural development to cognition: Unexpected roles for chromatin. *Nat Rev Gen* 14(5):347–359.

133. Tifft, KE, Segura-Totten, M, Lee, KK, Wilson, KL. 2006. Barrier-to-autointegration factor-like (BAF-L): A proposed regulator of BAF. *Exp Cell Res* 312(4):478–487.

134. Ooi, L, Belyaev, ND, Miyake, K et al. 2006. BRG1 chromatin remodeling activity is required for efficient chromatin binding by repressor element 1-silencing transcription factor (REST) and facilitates REST-mediated repression. *J Biol Chem* 281(51):38974–38980.

135. Benayoun, BA, Pollina, EA, Brunet, A. 2015. Epigenetic regulation of ageing: Linking environmental inputs to genomic stability. *Nat Rev Mol Cell Biol* 16(10):593–610.

136. Adlakha, YK, Seth, P. 2017. The expanding horizon of MicroRNAs in cellular reprogramming. *Prog Neurobiol*. 2017 Jan;148:21–39. doi: 10.1016/j.pneurobio.2016.11.003.

137. Käser-Pébernard, S, Müller, F, Wicky, C. 2014. LET-418/Mi2 and SPR-5/LSD1 cooperatively prevent somatic reprogramming of *C. elegans* germline stem cells. *Stem Cell Rep* 2(4):547–559.

138. Kyriazis M. 2015. Systems neuroscience in focus: From the human brain to the global brain? *Front Syst Neurosci* 9:7 http://dx.doi.org/10.3389/fnsys.2015.00007.

139. Ma SL, Lau ES, Suen EW, Lam LC, Leung PC, Woo J, Tang NL. 2013. Telomere length and cognitive function in southern Chinese community-dwelling male elders. *Age Ageing* 42(4):450–455.

140. Vaez-Azizi LM, Ruby E, Dracxler R et al. 2015. Telomere length variability is related to symptoms and cognition in schizophrenia. *Schizophr Res* 164(1–3):268–269.

141. Yaffe K, Lindquist K, Kluse M et al. 2011. Health ABC Study. Telomere length and cognitive function in community-dwelling elders: Findings from the Health ABC Study. *Neurobiol Aging* 32(11):2055–2060.

142. Cohen-Manheim I, Doniger GM, Sinnreich R et al. 2015. Increased attrition of leukocyte telomere length in young adults is associated with poorer cognitive function in midlife. *Eur J Epidemiol* 31(2):147–57.

143. Carlson LE, Beattie TL, Giese-Davis J et al. 2014. Mindfulness-based cancer recovery and supportive-expressive therapy maintain telomere length relative to controls in distressed breast cancer survivors. *Cancer* 121(3):476–84. doi: 10.1002/cncr.29063.

144. Jafarinejad-Farsangi S, Farazmand A, Mahmoudi M et al. 2015. MicroRNA-29a induces apoptosis via increasing the Bax:Bcl-2 ratio in dermal fibroblasts of patients with systemic sclerosis. *Autoimmunity* 48(6):369–378.

145. Chavez-Valdez R, Flock DL, Martin LJ, Northington FJ. 2016. Endoplasmic reticulum pathology and stress response in neurons precede programmed necrosis after neonatal hypoxia-ischemia. *Int J Dev Neurosci* 48:58–70.

146. Yoshikawa A, Kamide T, Hashida K et al. 2015. Deletion of Atf6α impairs astroglial activation and enhances neuronal death following brain ischemia in mice. *J Neurochem* 132(3):342–353.

147. Riches JJ, Reynolds K. 2014. Jnk1 activity is indispensable for appropriate cortical interneuron migration in the developing cerebral cortex. *J Neurosci* 34(43):14165–14166.

148. Nolan K, Walter F, Tuffy LP et al. 2016. ER stress-mediated upregulation of miR-29a enhances sensitivity to neuronal apoptosis. *Eur J Neurosci* 43(5):640–52. doi: 10.1111/ejn.13160.

149. Iurlaro R, Muñoz-Pinedo C. 2015. Cell death induced by endoplasmic reticulum stress. *FEBS J* 283(14):2640–52. doi: 10.1111/febs.13598.

150. Hikiji T, Norisada J, Hirata Y et al. 2015. A highly sensitive assay of IRE-1 activity using the small luciferase NanoLuc: Evaluation of ALS-related genetic and pathological factors. *Biochem Biophys Res Commun* 463(4):881–887.

151. Oslowski CM, Urano F. 2011. Measuring ER stress and the unfolded protein response using mammalian tissue culture system. *Methods Enzymol* 490:71–92.

152. Smart J. 2005. Measuring innovation in an accelerating world. *Technological Forecasting & Social Change*, V72N8 72:980–986.
153. Frölich J, Lehmkuhl G, Döpfner M. 2009. Computer games in childhood and adolescence: Relations to addictive behavior, ADHD, and aggression. *Z Kinder Jugendpsychiatr Psychother* 37(5):393–402.
154. Kyriazis M. 2016. Hormesis and adaptation. In *Challenging Ageing: The Anti-Senescence Effects of Hormesis, Environmental Enrichment, and Information Exposure*. Bentham Science Publishers, UAE.
155. Janic A, Mendizabal L, Llamazares S, Rossell D, Gonzalez C. 2010. Ectopic expression of germline genes drives malignant brain tumor growth in Drosophila. *Science* 330(6012):1824–1827.
156. Rajan K, Bialek W. 2013. Maximally informative "stimulus energies" in the analysis of neural responses to natural signals. *PLoS One* 8(11):e71959.
157. Kyriazis M. 2016. Engagement with a technological environment for ongoing homoeostasis maintenance. In *Challenging Ageing: The Anti-Senescence Effects of Hormesis, Environmental Enrichment, and Information Exposure*. Bentham Science Publishers, UAE.
158. Engl E, Attwell D. 2015. Non-signalling energy use in the brain. *J Physiol* 593(16):3417–3429.
159. Alcedo J, Flatt T, Pasyukova EG. 2013. Neuronal inputs and outputs of aging and longevity. *Front Genet* 4:71. doi: 10.3389/fgene.2013.00071.
160. Sundararajan A. Network Effects 2006. http://oz.stern.nyu.edu/io/network.html (accessed 3 January 2017).

5 Programmed Aging Paradigm and Aging of Perennial Neurons

Giacinto Libertini

CONTENTS

INTRODUCTION

THE PROBLEM

Alzheimer's disease (AD), Parkinson's disease (PD), and age-related macular degeneration (AMD) are degenerative diseases of the nervous system that manifest in the accumulation of specific substances (β-amyloid and tau protein for AD, α-synuclein for PD, A2E for AMD) and progressive impairment of psychomotor (AD, PD) or visual (AMD) functions (Weiner and Lipton 2009, Holz et al. 2013, Lim 2013, Pahwa and Lyons 2013, Husain and Schott 2016).

These diseases primarily affect elderly people and, as a result of the increase in life expectancy, there is a proportionate increase in the percentage and number of affected people. In the United States, approximately 96% of the 5.3 million Americans suffering from AD are over 65 years of age (Alzheimer's Association 2015). The worldwide absolute number of people with dementia and over 65 years old was estimated or predicted at 24.3 million in 2001, 42.3 in 2020, and 81.1 in 2040 (Rizzi et al. 2014). AD frequency increases progressively with age so that the frequency is 1.5% at the age of 65, which becomes 30% at 80 (Gorelick 2004), leaving a very high probability that a centenarian will suffer from AD.

Excluding AD, PD is the most frequent of the neurodegenerative disorders (de Lau and Breteler 2006, Yao et al. 2013). In industrialized countries, its frequency is approximately 0.3% of the whole population, 1% of the population older than 60, and 4% of the individuals older than 80 (de Lau and Breteler 2006). The mean age of onset is around 60; however, PD manifestations begin before 50 years of age in 5%–10% of cases (Samii et al. 2004). "Meta-analysis of the worldwide data showed a rising prevalence of PD with age (all per 100,000): 41 in 40 to 49 years; 107 in 50 to

59 years; 173 in 55 to 64 years; 428 in 60 to 69 years; 425 in 65 to 74 years; 1087 in 70 to 79 years; and 1903 in older than age 80" (Pringsheim et al. 2014).

The estimated frequency of AMD is 1.4% at 70 years of age, 5.6% at age 80, and 20% at age 90 (Rudnicka et al. 2012), which implies that a centenarian is a probable AMD sufferer. The projected number of AMD patients in 2020 is 196 million, increasing to 288 million in 2040 (Wong et al. 2014).

In general, there are only symptomatic or palliative cures for these illnesses that do not appear able to block their basic pathogenetic mechanisms:

1. AD is treated by cholinesterase inhibitors (donepezil, galantamine, rivastigmine), memantine, souvenaid (Waite 2015), psychotropic drugs, etc., which aim to relieve its neurological and psychiatric manifestations, and by anti-amyloid or anti-tau drugs, which try to limit the accumulation of the substances that are considered to cause the disease. However, "Current therapies for Alzheimer's disease do not modify the course of disease and are not universally beneficial" (Waite 2015).
2. PD is treated by pharmacotherapy, functional stereotaxic neurosurgery, physiotherapy, etc., all symptomatic cures, but "All treatments available until 2016 are of symptomatic nature. No therapy is currently available that slows down the progression of PD or even to prevent its manifestation" (Oertel and Schulz 2016).
3. There is no approved therapy for the atrophic (dry) form of AMD, while the costly treatment with anti-vascular endothelial growth factors (anti-VEGFs) contrast effectively the rapid evolution of the neovascular (wet) form (Azad et al. 2007, Nowak 2014), but do not block or revert it.

For the growing number of people affected and for the impairments caused by these diseases, the human, social, and economic costs are progressively growing and becoming unsustainable, especially for the cases of dementia. In the United States: "Total payments in 2015 for health care, long-term care and hospice services for people age 65 years with dementia are expected to be $226 billion" (Alzheimer's Association 2015).

In general, these diseases are described as without a known cause or origin, although many details of their pathogenetic mechanisms have been elucidated. AMD is defined as a "retinal disease with an unprecise etiopathogenesis" (Nowak 2014). PD "is a chronic, progressive neurological disease… The molecular mechanisms underlying the loss of these neurons still remain elusive" (Blesa et al. 2015). For AD: "The etiological mechanisms underlying these neuropathological changes remain unclear, but are probably caused by both environmental and genetic factors" (Reitz and Mayeux 2014).

These data indicate the gravity of the problem and the absence of an effective strategy to prevent and treat these illnesses. This serious and disastrous situation is a strong incentive to seek a completely new approach that could enable the understanding of the etiology of these diseases in order to develop and implement effective measures of prevention and treatment. In this work, a different interpretation of these illnesses is discussed to obtain a necessary premise for the achievement of such an ambitious and seemingly unrealistic goal.

NON-PROGRAMMED AND PROGRAMMED AGING PARADIGM

AD, PD, AMD, and other age-related illnesses of the nervous system (presbycusis, age-related hyposmia, and age-related deficits of other sensory abilities), disregarding the early cases due to "risk factors" (see below for the definition) or to genetic defects, are diseases whose frequency and severity are clearly related to age. Therefore, it appears logical that the nature of their connections with aging must be clarified for their proper understanding (and vice versa a better understanding of their origins may lead to a better understanding of aging mechanisms).

Therefore, it is an imperative preliminary to understand what aging is and the relationship between the mechanisms of aging and the mechanisms underlying these diseases.

It is necessary to specify that

- A comprehensive discussion about aging is outside the scope of this work. Only a few main points will be outlined while the appropriate papers will be indicated for a more detailed exposition.
- The interpretation here will be reaffirmed that the nature and mechanisms of aging are different and in clear contrast with the contradictory ideas generally accepted as a coherent (!) explanation for aging. However, such a different interpretation allows a unified and consistent view of the aforesaid diseases. Moreover, it will clarify why the current therapeutic approaches fail and will allow the proposition of a rationale for the development of effective treatment.

A necessary premise is to state a definition of aging that is descriptive and not vitiated by a preconceived explanation of this phenomenon. Therefore, aging will be defined precisely, in a neutral descriptive way, as "increasing mortality [= decreasing fitness] with increasing chronological age in populations in the wild" (Libertini 1988). This phenomenon, which has been documented for 175 different animal species (Nussey et al. 2013), may be also more concisely defined as "actuarial senescence" in reference to animals studied "in the wild" (Holmes and Austad 1995).

Aging is explained in two general ways, which are completely different and mutually incompatible and so deserve the definition of opposite "paradigms" (Libertini 2015a) according to the meaning that was proposed by Kuhn (1962).

The first paradigm, which could be defined as the "non-programmed aging paradigm," or briefly "old paradigm," explains aging as the effect of many degenerative phenomena, insufficiently contrasted by natural selection, which inexorably and inevitably cause the accumulation of random damages over time, as proposed by mutation accumulation (Medawar 1952, Hamilton 1966), antagonistic pleiotropy (Williams 1957, Rose 1991), disposable soma (Kirkwood 1977, Kirkwood and Holliday 1979), and other theories (Libertini 2015b). According to this paradigm, in principle, we could slow down and fight aging manifestations by using appropriate drugs and measures without the possibility to stop or reverse them (Libertini 2015a).

The second paradigm, which could be defined as the "programmed aging paradigm," or briefly "new paradigm," interprets aging as a physiological phenomenon that:

- Is evolutionarily favored in terms of supra-individual natural selection (Libertini 2015a).
- Is a particular type of phenoptosis (Skulachev 1997, Libertini 2012), that is, "programmed death of an individual" (Skulachev 1999).
- Is the result of specific mechanisms that are genetically determined and regulated, and therefore, in principle, might be modified up to a complete control of aging (Libertini 2009a,b).
- Has its specific pathological forms (Libertini 2009a,b, 2014).
- Has its specific phylogeny (Libertini 2015b).

A discussion about the arguments and the evidence that support or contrast each of the two paradigms has been proposed in another study where arguments and evidence appear to be against the old paradigm and support the new paradigm (Libertini 2015a).

Although the old paradigm is still the prevalent idea (Kirkwood and Melov 2011), the new paradigm will be considered a working hypothesis for the aims of this work. Accepting the interpretation of the new paradigm as a working hypothesis will provide an easy and unifying interpretation of AD, PD, AMD, and other diseases, which may be profitable for the goal of the formulation of appropriate therapies. By contrast, the current concepts based on the old paradigm have led to the current situation of an insufficient understanding of these diseases and of ineffective therapies. The reader will be able to judge whether this approach is more consistent and efficacious.

SHORT DESCRIPTION OF THE TELOMERE THEORY

The new paradigm absolutely requires the existence of specific mechanisms that determine a progressive fitness reduction. These mechanisms are documented by a long series of authoritative studies and can be summarized under the term "telomere theory," which for the sake of brevity cannot be explained here in detail. Therefore, only a brief description will be provided here, referring to a more detailed description from additional references (Libertini 2009a, 2014, Libertini and Ferrara 2016a).

During every replication cycle, the DNA molecule requires the telomerase enzyme in order to complete the duplication of a repetitive nucleotide sequence at its terminal regions, the telomere. In cells where the telomerase is inactive or partially active, any duplication results in telomere shortening. Since the telomere is covered by a heterochromatin hood with a fixed length, the gradual telomere shortening causes the sliding of the hood on the adjacent part of the DNA molecule, the subtelomere, which so is progressively inhibited. This inhibition has two main effects:

1. The progressive alteration of the regulatory functions of the subtelomere that determines a progressive alteration of numberless cell functions, a phenomenon here defined as "gradual cell senescence" or just as "gradual senescence" (Fossel 2004, Libertini 2014, 2015b).
2. An increase in the probability of activation of a particular cell program, the cell senescence, which is characterized by the blockage of cell replicative capacities and by the alterations of gradual senescence to its maximum degree (Ben-Porath and Weinberg 2005). The activation of this program is a random function with a probability that grows in every cell with the shortening of the telomere (Blackburn 2000) and so is likely related to the progressive inhibition of the subtelomere (Libertini and Ferrara 2016a) (Figure 5.1).

It has to be highlighted that, in plain contrast with the tenets of the old paradigm, the aforesaid cell alterations are not caused by irreversible degenerative phenomena, as it has been well demonstrated by experiments in which telomerase reactivation determined the regression of all the

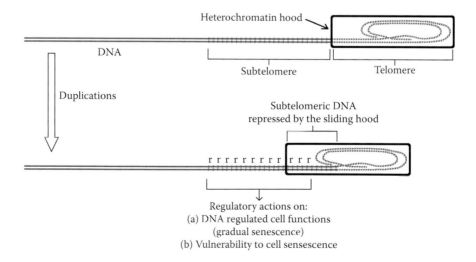

FIGURE 5.1 A scheme of the progressive subtelomere inhibition caused by the sliding of the telomeric hood. (Modified and redrawn from Figure 5.4 Libertini G. *Biochem. (Mosc.)* 2015b, 80(12):1529–46.)

aforesaid phenomena (Bodnar et al. 1998, Counter et al. 1998, Vaziri 1998, Vaziri and Benchimol 1998).

The majority of cell types in vertebrates undergo a continuous cell turnover. In a young individual, there is a perfect balance between the effects of various types of programmed cell death (apoptosis, keratinization of hair or epidermal cells, detachment of intestine cells, osteocytes phagocytized by osteoclasts, etc.) and the duplication of the opportune stem cells. As telomeres shorten, there is an increase in the number of cells in the cell senescence state, resulting in a gradual decrease in cell renewal capacity. This deficit, combined with the growing number of cells in various degrees of gradual senescence, leads to what has been defined as "atrophic syndrome" (Libertini 2009a), characterized by:

"1. Reduced mean cell duplication capacity and slackened cell turnover
2. Reduced number of cells (atrophy)
3. Substitution of missing specific cells with nonspecific cells
4. Hypertrophy of the remaining specific cells
5. Altered functions of cells with shortened telomeres or definitively in a noncycling state
6. Alterations of the surrounding milieu and of the cells depending on the functionality of the senescent or missing cells
7. Vulnerability to cancer because of the dysfunctional telomere-induced instability ..." (Libertini 2014)

Factors such as unhealthy lifestyles (which cause diabetes mellitus, hypertension, obesity, etc.) and toxic substances (e.g., cigarette smoke and alcohol abuse), which may be defined on the whole as "risk factors," increase cell duplication necessities and so accelerate the aging process. By contrast, factors such as healthy lifestyles and drugs that have organ protection qualities ("protective drugs"),

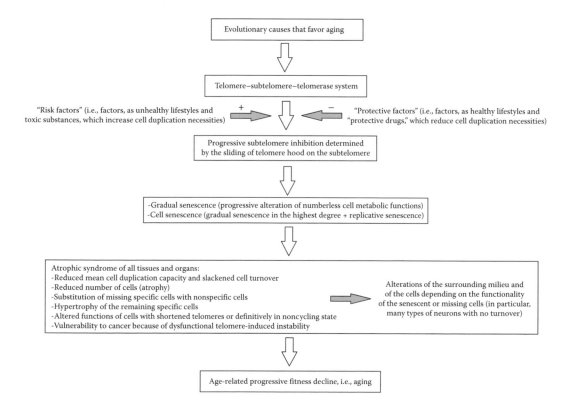

FIGURE 5.2   A scheme of the aging process. The discussion of the evolutionary causes is omitted in this work.

which may be defined on the whole as "protective factors," reduce cell duplication capacities and neutralize the negative effects of the "risk factors" (Libertini 2009b) (Figure 5.2).

These mechanisms appear readily applicable to all cell types that are subject to cell turnover, and this could easily explain a large part of the aging phenomenon (Libertini 2009b). This, however, would not seem to relate to cells not subject to turnover, such as the majority of the neurons. However, a careful reading of the synthetic description of the atrophic syndrome shows (see point 6 in the description of the atrophic syndrome) that if a perennial cell is dependent for its functionality on other cells subject to turnover, the numerical and functional decline in these "satellite" cells may well explain the concomitant numerical and functional decline in the perennial cells (Libertini and Ferrara 2016b).

This discussion is deepened in the present analysis.

AGING OF NEURONS

TYPES OF NEURONS

There are many different neuronal cell types. For the purpose of this work, a preliminary distinction between two opposite extremes is necessary:

1. Certain types of specialized neurosensory neurons, such as olfactory receptor cells (ORCs), receive a signal from the external world by a single dendrite and are connected by a single axon to other neurons that process the signal (Bermingham-McDonogh and Reh 2011). For these neurons, the connections are quite simple and stereotyped and, in principle, the turnover of these cells should not entail any particular difficulty.

2. By contrast, each neuron in the central nervous system (CNS) is connected to a large number of other neurons. According to an estimate, in the human brain, there are 10^{11} neurons connected by 10^{14} synapses (Williams and Herrup 1988), that is, there is an estimated average of 1000 synapses for each neuron. Given the large number of synapses and the expected variability in the connections between neurons, the possible replacement of a neuron with a new neuron should also restore any previous connection through an unlikely mechanism that would be exceedingly complex. If brain functions are dependent on the net of synapses among neurons that have been established over time, and the connections are different for each individual neuron and, on the whole, are specific for each individual, the failure to restore the exact synapses would create unbearable harm caused by any turnover of these neurons. This would justify the absence of cell turnover for CNS neurons, with some exceptions (Horner and Gage 2000, Zhao et al. 2008), for example, "for certain forms of brain function involving the olfactory bulb and the hippocampus, which is important for some forms of learning and memory" (Zhao et al. 2008). This absence is likely for functions that require new connections and for which new neurons are not a problem but a necessity.

Different strategies are necessary for these two opposite neuronal cell types. For the first type, as will be expounded in more detail for the ORCs, the strategy of cell turnover can be implemented as well as it is implemented for nonneuronal cells.

For the second type of neuron, the practical impossibility of replacing the neurons without damaging brain functions and, at the same time, the need to ensure their cell functionality even after many years are solved, as we will see in the following sections, by a different strategy that is very well documented for a neuronal cell type, the retinal photoreceptor cells, and is quite likely for other types of CNS neurons.

For the following discussion, endothelial cells will be used as an important example of nonneuronal cells that undergo turnover. Risk factors for cardiovascular diseases, such as age, diabetes, smoking, body mass index (i.e., overweight and obesity), and hypertension (Wilson et al. 1987), are associated with a reduced number of endothelial progenitor cells (Hill et al. 2003), likely caused by a quicker turnover of endothelial cells: "continuous endothelial damage or dysfunction leads to an eventual depletion or exhaustion of a presumed finite supply of endothelial progenitor cells … continuous risk-factor-induced injury may lead to eventual depletion of circulating endothelial cells" (Hill et al. 2003). This would be analogous to what happens in patients with muscular dystrophy, where there is an exhaustion of skeletal muscle stem cells (Webster and Blau 1990, Decary et al. 2000, Seale et al. 2001) as well as in a number of age-related conditions (Tyner et al. 2002, Geiger and Van Zant 2002).

ACE inhibitors, AT1 blockers (sartans), and statins, which may be considered protective drugs for cardiovascular diseases, improve endothelial function (Su 2015). Statin therapy accelerates reendothelialization through endothelial progenitor cells (Walter et al. 2002), and this could explain the effects of statins, and likely of other "protective drugs," on the prevention and treatment of cardiovascular diseases.

OLFACTORY RECEPTOR CELLS

ORCs are specialized neurons, which are present in the upper part of the nasal cavity and allow the perception of smell. They "have a single dendrite that extends to the apical surface of the epithelium and ends in a terminal knob, which has many small cilia extending into the mucosa. A single axon projects through the basal side of the epithelium through the lamina cribrosa to terminate in the olfactory bulb. Each of the receptor neurons expresses one of a family of over 1000 olfactory receptor proteins … The neurons are surrounded by glial-like cells, called sustentacular cells. Other cells in the epithelium contribute to the continual production of the new receptor neurons …" (Bermingham-McDonogh and Reh 2011) (Figure 5.3).

The continuous turnover of the ORCs in normal individuals is well documented (Maier et al. 2014). "The ongoing genesis of olfactory receptor cells is common to all vertebrates (see Graziadei and Monti Graziadei, 1978, for a review) and the rate of production is quite high. The production of new olfactory receptor cells is critical to the maintenance of this system, as the olfactory receptor cells only last a few months. The rate of production of new olfactory receptor cells is balanced by their loss so that a relatively stable population of these receptors is maintained" (Bermingham-McDonogh and Reh 2011).

Analyses of the healthy olfactory epithelium show that the turnover of ORCs is enabled by some slow cycling stem cells and by transit-amplifying progenitor cells (globose basal cells and horizontal basal cells, respectively) (Caggiano et al. 1994, Huard et al. 1998, Chen et al. 2004, Leung et al. 2007, Iwai et al. 2008). This two-stage modality of cell reproduction to allow cell turnover is similar to that of the epidermis and other cell types (Watt et al. 2006).

Owing to their position, ORCs are highly exposed to damage by external factors but have simple connections as each cell has a single dendrite, with many small ramifications in the mucosa, and a single axon. Therefore, the turnover of the ORCs is both necessary and simple. If this turnover follows the same patterns as other cell types, one can predict an age-related slowing of cell substitution with the consequent impairment of olfactory function. Other factors, perfectly compatible with the aforesaid mechanism, which could contribute to this progressive impairment, are: (i) an increase in the proportion of ORCs with various degrees of gradual senescence; (ii) age-related slowing of cell turnover of the gliocytes that are satellites of ORCs and of olfactory bulb neurons; and (iii) age-related decay of other CNS areas that contribute to olfactory function.

As a matter of fact, the impairment of olfactory function is age-related (Schubert et al. 2012, Doty and Kamath 2014, Gouveri et al. 2014), and approximately 50% of 65–80-year-old individuals suffer from clear olfactory dysfunction (Doty et al. 1984, Duffy et al. 1995, Murphy et al. 2002).

If AD, PD, AMD, and other diseases of the CNS are caused (as discussed below) by the failure in cell turnover of the gliocytes that are essential neuronal satellite cells, a clear relation between olfactory dysfunction and these diseases is expected. In fact: (i) olfactory dysfunction has been

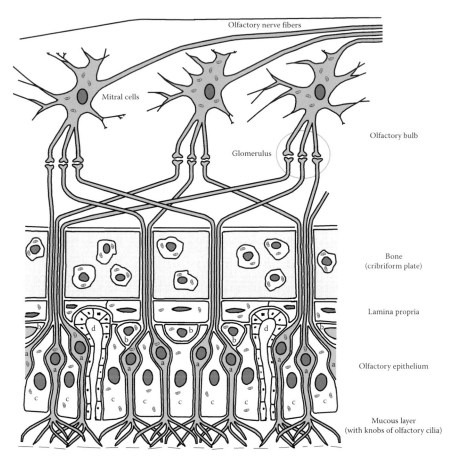

FIGURE 5.3 A scheme of the peripheral olfactory system. In the olfactory epithelium: a = olfactory receptor cells, b = regenerative basal cells; c = sustentacular or supporting cells (differentiated gliocytes); d = olfactory (Bowman's) glands.

estimated to be as high as 90% of early-stage PD cases (Doty 2012) and 100% in AD (Duff et al. 2002); (ii) hyposmia is a common symptom in PD dementia, in AD, and in other forms of dementias (Barresi et al. 2012); (iii) it has been reported as an early sign of PD and a precocious and constant characteristic of AD and of dementia with Lewy bodies (DLB) (Factor and Weiner 2008); and (iv) there is a correlation between olfactory dysfunction and AMD (Kar et al. 2015).

There is a possible association between olfactory dysfunction and unhealthy lifestyles: (i) "Olfactory dysfunction is a known complication of diabetes" (Mehdizadeh et al. 2015); (ii) there is an association between olfactory dysfunction and type 2 diabetes or hypertension (Gouveri et al. 2014); (iii) obesity is associated with the risk of olfactory dysfunction (Richardson et al. 2004, Patel et al. 2015); (iv) in the rat, ethanol and tobacco smoke damage olfactory function, and this "could explain the decreased olfactory ability seen in patients who use these products" (Vent et al. 2003); (v) "alcoholism appears to be associated with a variety of disturbances in olfactory processing" (Rupp et al. 2004); and (vi) prevention and treatment of such diseases indicated as risk factors should also be effective for the olfactory dysfunctions. It was not possible to find studies on the specific use of drugs for such purposes. However, in a mouse model, anosmia caused by a toxic substance was successfully treated by a statin (Kim et al. 2012).

Given that ORC turnover is essential and simple for the aforesaid reasons, the turnover of other sensory neurons and, consequently, an age-related functional decline may also be expected. For example, in humans: (i) taste buds turn over rapidly, with an average life span of 8–12 days, and an age-related decline in function is documented (Feng et al. 2014); (ii) "Reported changes include reduced total number of taste buds, reduced taste bud density in the epithelium, and reduced number of taste cells per taste bud" (Feng et al. 2014); (iii) there is a decline in gustatory capacities of patients with AD (Aliani et al. 2013). A negative correlation between age and the number of Meissner's corpuscles per mm^2 was observed (Iwasaki et al. 2003) and "there is an approximate two-thirds reduction in numbers of Pacinian and Meissner's corpuscles with age" (Griffiths 1998).

RETINAL PHOTORECEPTORS

Retinal photoreceptors (cones and rods) are very specialized nervous cells and have complex connections with other neurons of the retina, where the initial processing of data perceived by the eye occurs. Therefore, photoreceptors belong to the aforesaid second type of neurons (as do almost all neurons) and have no turnover. However, photoreceptors depend on other cells with turnover, the cells of the so-called retinal pigment epithelium (RPE), which consist of specialized gliocytes that turn over. The heads of the photoreceptors are associated with RPE cells. Photoreceptors have particular membranes covered by photopsin molecules, which allow light reception but suffer from high oxidative damage, and each day about one-tenth of these membranes is phagocytized and metabolized by RPE cells. At the same time, photoreceptors synthesize an equal quantity of membrane so that cell structure and function remain stable. An RPE cell serves approximately 50 photoreceptors and, therefore, each day metabolizes the membranes of approximately five photoreceptors, showing an exceptional metabolic activity. RPE function is essential for the vitality of the photoreceptors (Fine et al. 2000, Jager et al. 2008) (Figure 5.4).

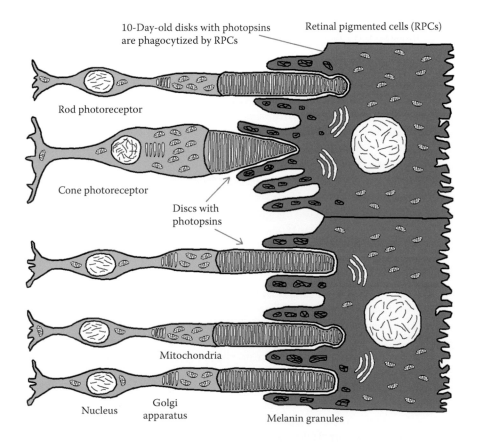

FIGURE 5.4 A scheme of some photoreceptor cells and retinal pigmented cells.

The limits in the duplication capacities of RPE stem cells do not allow for an unlimited turnover of these cells. The age-related decline in RPE cell turnover determines the enlargement of the remaining RPE cells and the accumulation of A2E, a breakdown product derived from vitamin A, and of other damaging substances (Sparrow 2003). The continued decline in turnover causes the formation of holes in the RPE, and the photoreceptors die at this point (Berger et al. (1999)).

This decline is more precociously evident in the pivotal part of the retina, the macula, where photoreceptors are denser and A2E accumulation is most abundant (Sparrow 2003, Ablonczy et al. 2012). For this reason, the trouble is defined as "age-related macular degeneration," although the entire retina is affected (Fine et al. 2000).

AMD may arise at precocious ages depending on the particular genetic defects affecting RPE cells or from the effects of toxic substances or metabolic stresses caused by unhealthy lifestyles. Disregarding these cases, the frequency of AMD increases exponentially with age (Rudnicka et al. 2012) and for its genesis and frequency, it should be considered a standard feature of aging.

Risk factors for cardiovascular diseases (Wilson et al. 1987) can have detrimental effects on RPE turnover that are analogous to those on endothelial progenitor cells (Hill et al. 2003), and this could justify their association with AMD (Klein et al. 2007) and between unhealthy lifestyles and AMD (Mares et al. 2011). "Smoking is the risk factor most consistently associated with AMD. Current smokers are exposed to a two to three times higher risk of AMD than non-smokers and the risk increases with intensity of smoking" (Armstrong and Mousavi 2015).

About the possible prevention or treatment of AMD by drugs, "… epidemiologic, genetic, and pathological evidence has shown AMD shares a number of risk factors with atherosclerosis, leading to the hypothesis that statins may exert protective effects in AMD. … Evidence from currently available randomized controlled trials is insufficient to conclude that statins have a role in preventing or delaying the onset or progression of AMD" (Gehlbach et al. 2015).

As regards alcohol consumption and the risk of vascular diseases, low alcohol consumption appears to have a beneficial effect (Roerecke and Rehm 2014, Gardner and Mouton 2015, de Gaetano et al. 2016). "On the other hand, ethanol chronically consumed in large amounts acts as a toxin to the heart and vasculature" (Gardner and Mouton 2015). "Evidence consistently suggests a J-shaped relationship between alcohol consumption … and all-cause mortality, with lower risk for moderate alcohol consumers than for abstainers or heavy drinkers" (de Gaetano et al. 2016). For AMD: "Moderate alcohol consumption is unlikely to increase the risk of AMD" (Armstrong and Mousavi 2015). However, in the Beaver Dam Offspring Study, no association between a history of heavy drinking and early AMD was observed (Klein et al. 2010).

Since the retina is constantly exposed to light and has high oxygen content, it is clearly susceptible to strong oxidative damage. However, in evident contrast with the predictions of the old paradigm, antioxidant supplements did not prevent early AMD as shown by the results of the meta-analysis of 12 studies (Chong et al. 2007).

Brain Neurons and the Genesis of AD

The preceding section has shown that specialized neurons without cell turnover (retina photoreceptors) depend for their function and vitality on specialized gliocytes with cell turnover (RPE cells). This suggests that other types of neurons without cell turnover (mostly CNS neurons) could also be dependent on specific gliocytes with cell turnover (Figure 5.5). Therefore, the gradual senescence and cell senescence of these specialized gliocytes may cause pathologies equivalent to AMD for their genesis.

Even without considering the proposed AMD genesis, AD was hypothesized to have originated from the functional decline in particular gliocytes caused by telomere shortening (Fossel 1996, 2004): "One function of the microglia … is degradation of β-amyloid through insulin-degrading enzyme (IDE), a function known to falter in Alzheimer's disease …" (Fossel 2004, p. 233).

The role of microglial cells in the degradation of β-amyloid protein is well known (Qiu et al. 1998, Vekrellis et al. 2000, Miners et al. 2008), as is the fact that this function is altered in AD (Bertram

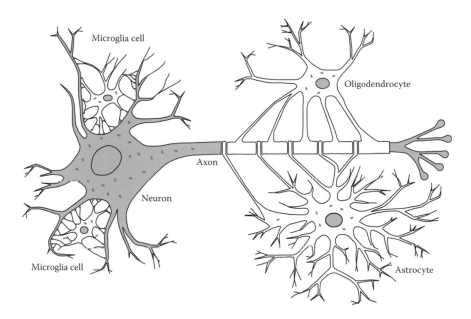

FIGURE 5.5 A scheme of a neuron and its auxiliary gliocytes.

et al. 2000) with the consequent accumulation of the substance. As regards the relation between the decline in microglial function and telomere length, a significantly reduced telomere length in circulating monocytes is associated with at least vascular dementia (von Zglinicki et al. 2000).

The hypothesis that AD is determined by the decline in microglial cells was proposed again without mentioning the association between satellite gliocyte failure and AMD (Flanary 2009) or, with stronger arguments, considering it (Libertini 2009a, 2009b).

"A cell senescence model might explain Alzheimer dementia without primary vascular involvement" (Fossel 2004, p. 235) and it is likely that many AD cases have in part a vascular etiology caused by age-related endothelial dysfunction (Fossel 2004). The similarities or identities between the symptoms of a "pure" AD and those of a "pure" vascular dementia and the cases where the two conditions overlap to varying extents must not obscure an affinity that is deeper and not linked only to some identical symptoms.

If it is true that AD neuronal decay results from the decline in satellite cells and that vascular dementia is a neuronal decay caused by vascular dysfunction, there is a common pathogenetic mechanism in their origins. The vascular diseases, including those that affect the brain, are caused by endothelial dysfunction originating from the depletion of their turnover capacity marked by an insufficient number of endothelial progenitor cells (Hill et al. 2003). Similarly, AD could originate from microglial dysfunction caused by the slowing or exhaustion of microglia cell turnover capacity. In both cases, telomere shortening in the stem cells of the endothelial cells or of the microglial cells is the primary cause.

Therefore, the risk factors for cardiovascular diseases and for AD, in general, should both accelerate telomere failure, whereas protective factors should counter these effects. Moreover, risk factors for cardiovascular diseases and for AD should coincide, and analogously protective factors for the same diseases should coincide.

In support of these claims: (i) an association between cardiovascular risk factors and AD has been shown (Vogel et al. 2006, Rosendorff et al. 2007); (ii) cigarette smoking is a risk factor for AD (Durazzo et al. 2014); (iii) many reviews, with some exceptions (e.g., Li et al. 2016), maintain a positive correlation between diabetes mellitus and the risk of developing AD (Baglietto-Vargas et al. 2016, Rani et al. 2016, Saedi et al. 2016, Vicente Miranda et al. 2016); (iv) there is a positive relation between hypertension and AD risk (Qiu et al. 2005, Michel 2016); (v) statins, ACE inhibitors, and sartans, "protective drugs" for cardiovascular diseases, are considered effective against AD (Vogel et al. 2006, Ellul et al. 2007); and (vi) in a recent meta-analysis, ACE inhibitors and sartans appear to reduce the risk of developing AD (Yasar et al. 2016); (vii) "Mid-life dyslipidemia appears to play an important role in the development of AD amongst a host of other risk factors that affect vascular health. Results from observational cohorts have been mixed, though many of the highest-quality studies have found a protective effect for statins" (Wanamaker et al. 2015).

As regards alcohol consumption, "Light to moderate drink may decrease the risk of Alzheimer's disease" (Ilomaki et al. 2015). A study has highlighted "weak evidence" for the association between a higher risk of AD and excessive alcohol consumption while "a lower risk of AD is associated with moderate alcohol consumption" (Campdelacreu 2014).

The traditional hypothesis that the decisive factor for AD pathogenesis is the damage accumulation of substances such as β-amyloid and tau protein has been widely contradicted by the failures of the pharmaceutical industry in the trials based on this thesis. Drugs or vaccines used with the aim to counter the formation of β-amyloid plaques have been disappointing, not for their ability to eliminate the plaques but for the purpose of obtaining positive clinical effects (Abbott 2008). In particular, in 2008, a study showed that an experimental amyloid peptide vaccine was very effective in eliminating the plaques or in avoiding their formation, but: "Seven of the eight immunised patients who underwent post-mortem assessment, including those with virtually complete plaque removal, had severe end stage dementia before death" (Holmes et al. 2008) and the authors observed "Although immunization with Abeta42 resulted in clearance of amyloid plaques in patients with Alzheimer's disease, this clearance did not prevent progressive neurodegeneration" (Holmes et al. 2008). A recent authoritative study (Sevigny et al. 2016) has used a human monoclonal antibody (aducanumab) to eliminate the amyloid-β plaques. The technique has shown significant and dose-dependent effectiveness in eliminating the plaques, while the clinical results (for which one must observe that the study was not specifically designed) remain uncertain. After 1 year of treatment, one of the two clinical evaluation parameters (Mini Mental State Examination) showed positive results for the doses of antibody 3 and 10 mg/kg, but for the intermediate dose of 6 mg/kg, the result was similar to that of placebo.

The hypothesis that AD originates from the accumulation of β-amyloid (amyloid cascade hypothesis) has been strongly criticized (Herrup 2010, Mondragón-Rodríguez et al. 2010, Reitz 2012), but there is a reluctance to drop it, perhaps for the false idea that there is no plausible alternative such as the one discussed in the current work.

N-Methyl-D-aspartate receptor antagonist (e.g., memantine), acetylcholinesterase inhibitors (e.g., galantamine, donepezil, tacrine, and rivastigmine), and other drugs are used to treat AD (Mendiola-Precoma et al. 2016). However, "The only drugs available for Alzheimer's patients aim to treat symptoms … They are marginally effective at best" (Abbott 2008). Moreover, the treatment of AD cognitive alterations by antipsychotic drugs increased the long-term risk of mortality (Ballard et al. 2009).

BRAIN NEURONS AND THE GENESIS OF PD

PD is characterized by the accumulation of the protein α-synuclein (AS) inside CNS neurons, which forms particular inclusions known as Lewy bodies (Davie 2008, Schulz-Schaeffer 2010). A disease that is akin to PD, classified as a Parkinson-plus syndrome, is the DLB (Nuytemans et al. 2010). It has the signs of a primary parkinsonism but shows some additional features (Samii et al. 2004). In addition, multiple system atrophy is a rare genetic disease in which AS accumulates in oligodendrocytes (Sturm and Stefanova 2014).

While AD is referred to as a tauopathy due to the accumulation of tau protein and the formation of neurofibrillary tangles, PD, DLB, and multiple system atrophy are referred to as synucleinopathies for the accumulation of AS (Galpern and Lang 2006). Despite the difference between the accumulated substances, many symptoms and pathological manifestations are similar or identical in these neuropathies (Aarsland et al. 2009).

PD is mainly a disease of the CNS motor system because its more common and typical symptoms include disorders of movement caused by extrapyramidal motor dysfunction. However, nonmotor manifestations such as dementia and sensory deficits are common (Barnett-Cowan et al. 2010). In PD cases without dementia, behavior and mood alterations (e.g., depression, anxiety, apathy, etc.) are more frequent than in the general population and are common symptoms in cases with dementia (Jankovic 2008).

Dementia is frequent in PD patients, in particular at advanced stages, and AD patients often manifest parkinsonism (Galpern and Lang 2006). In PD patients, the risk of dementia is two to six times that of the whole population and is related to disease duration (Caballol et al. 2007). However, for AD patients, dementia is more precocious and senile plaques and neurofibrillary tangles are characteristic of AD, while they are uncommon in PD without dementia (Dickson 2007).

As regards the distinction between the clinical manifestations of PD and DLB, "PD and DLB are common neurodegenerative diseases in the population over the age of 65. About 3% of the general population develops PD after the age of 65, whereas about 20% of all diagnosed dementia patients have DLB … In both disorders movement and cognition, as well as mood and autonomic function are severely affected. Diagnosis to distinguish PD and DLB is very difficult, because of the overlap of symptoms and signs …" (Brück et al. 2016).

In PD pathogenesis, Lewy bodies are found in the olfactory bulb, medulla oblongata and pontine tegmentum before clear symptoms appear. In the following phases, they are present in the substantia nigra, in some regions of the basal forebrain and of the midbrain, and finally in the neocortex (Davie 2008). PD motor symptoms are interpreted as a consequence of neuronal cell death in a particular area of the brain (the pars compacta region of the substantia nigra), which causes a reduction in dopamine secretion (Obeso et al. 2008).

In the aforesaid areas, there is neuronal degeneration and loss, for which Lewy bodies have been proposed not to be the cause of neuron death but as a protection against other factors (Obeso et al. 2010, Schulz-Schaeffer 2010). In demented AD patients, Lewy bodies are observed in particular in cortical areas in the brain, and it has been suggested that "reduced AS clearance is involved in the generation of AS inclusions in DLB and PD" (Brück et al. 2016), but this does not clarify if Lewy bodies are the cause of neuronal degeneration or a simple consequence of the cause.

About this, some evidence must be considered:

> Glial cells are important in supporting neuronal survival, synaptic functions and local immunity … However, glial cells might be crucial for the initiation and progression of different neurodegenerative diseases, including ASP [α-Synucleinopathies] … Microglial cells contribute to the clearance of debris, dead cells and AS thereby supporting neuronal survival. But on the other hand, microglial cells can get over-activated in the course of the disease and might contribute to disease initiation and progression by enhancing neurodegeneration through elevated oxidative stress and inflammatory processes.
>
> **Brück et al. 2016**

In analogy with AMD and AD genesis, these elements suggest that the pivotal event in PD genesis is the loss of essential functions of specific gliocytes (astrocytes), mainly dedicated to axon trophism, and the activation of astroglial and microglial cells could be an event caused by the consequent AS accumulation (Morales et al. 2015), while the decline in function of these specific gliocytes would be caused by the decline in their turnover.

As regards the associations of PD with "risk factors," there is the following evidence: (i) A study has shown that, in middle age, high skinfold thickness is associated with PD (Abbott et al. 2002); (ii) Metabolic syndrome (i.e., in short, an unhealthy lifestyle that causes dyslipidemia, obesity, glucose intolerance, hypertension, etc.) has been indicated as an important risk factor for PD (Zhang and Tian 2014); (iii) Independently of other risk factors, body mass index appears to be related to an increased risk of PD (Hu et al. 2006), and obesity in midlife increases the risk of dementia (Whitmer et al. 2005); (iv) Hyperglycemia in aging subjects is associated with PD (Hu et al. 2007, Tomlinson and Gardiner 2008); (v) Type 2 diabetes is a risk factor for PD (Vicente Miranda et al. 2016); (vi) Alcohol (light to moderate) use has been found to have an inverse association with PD (Ishihara and Brayne 2005) while alcohol use disorder appears to increase the risk of PD (Eriksson et al. 2013).

The risk of PD appears to be lowered by statins (Gao et al. 2012, Friedman et al. 2013, Undela et al. 2013, Sheng et al. 2016). Captopril, an angiotensin-converting enzyme inhibitor, has been shown to protect nigrostriatal dopamine neurons in animal models for PD (Lopez-Real et al. 2005, Sonsalla et al. 2013). A strange finding is that cigarette smoking, a risk factor for AD (Durazzo et al. 2014), AMD (Klein et al. 2007), hearing loss (Fransen et al. 2008, Chang et al. 2016), and olfactory dysfunction (Vent et al. 2003), appears to lower PD risk (Li et al. 2015). A possible explanation is that nicotine has a neuroprotective effect on dopaminergic neurons and could even be used as a medicine to contrast PD progression or symptoms (Thiriez et al. 2011, Quik et al. 2015).

Hyperhomocysteinemia (HHcy), which is considered a risk factor for endothelial dysfunction and for vascular diseases (Woo et al. 1997), has been shown to be associated with AD and PD (Kruman et al. 2000). However, limiting the discussion to AD, "it is still controversial if HHcy is an AD risk factor or merely a biomarker" (Zhuo et al. 2011).

Hearing Neurons

The organ of Corti is a sensory receptor inside the cochlea of the inner ear, which has a hearing function. It has differentiated neurons, the auditory or hair cells, divided into two groups (inner and outer hair cells) and connected to specific neurons (spiral ganglion neurons). In a newborn, for each organ of Corti, there are approximately 35,000 neurons and 15,500 hair cells, and both of these cell types are perennial (Wong and Ryan 2015): "… no epithelial maintenance has been described for

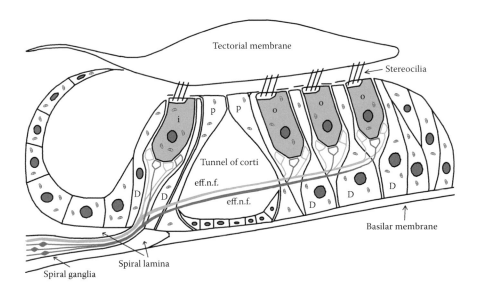

FIGURE 5.6 Organ of Corti. i = inner hair cells; o = outer hair cells; D = Deiters' cells; p = pillar cells; aff.n.f. = afferent nerve fibers of cochlear nerve; eff.n.f. = afferent nerve fibers of cochlear nerve. Hair cells, specialized neurons that constitute the sensorineural cells of the auditive organ, are supported by Deiters' cells, which are specialized gliocytes.

the hair cells of the cochlea of mammals, though hair cell addition and repair occur in lower verte-brates ..." (Maier et al. 2014) (Figure 5.6).

It is well known that there is an age-related progressive reduction of hearing ability (Zhan et al. 2010). In the United States, hearing loss of more than 25 dB in speech frequency pure tone aver-age has been reported to be 45.6%, 67.6%, 78.2%, and 80.6% in groups aged 70–74, 75–79, 80–84, and >84 years, respectively (Lin et al. 2011b). Noise exposure is not an indispensable cause for the development of presbycusis, as it has been observed in healthy animals that were reared in silence (Sergeyenko et al. 2013, Yan et al. 2013).

Cardiovascular and cerebrovascular diseases, diabetes, and cigarette smoking have been associ-ated with increased risk of hearing loss (Yamasoba et al. 2013). In a study with 3,753 participants, "current smokers were 1.69 times as likely to have a hearing loss as nonsmokers" (Cruickshanks et al. 1998) and in another study with 12,935 participants, "Current smoking was associated with hearing impairment in both speech-relevant frequency and high frequency across all ages" (Chang et al. 2016). Interestingly, "a remarkable parallelism between the risk factors for ARHI [age-related hearing impairment] and those for CVD [cardiovascular disease] with similar roles for smoking, high BMI, and regular moderate alcohol consumption" (Fransen et al. 2008) has been observed.

Cardiovascular risk factors "adversely affect hearing acuity" (Oron et al. 2014). Type 2 diabetes mellitus is associated with alterations in hearing (Akinpelu et al. 2014, Calvin and Watley 2015, Helzner and Contrera 2016), and hypertension has been found to be associated with cochlear hear-ing loss (Przewoźny et al. 2015). In a highly endogamous population, "adults with DM [diabetes mellitus] and hypertension associated showed greater hearing impairment" (Bener et al. 2016).

There is an association between chronic alcohol abuse and hearing impairment (Rosenhall et al. 1993), while a protective effect of moderate alcohol consumption has been shown (Fransen et al. 2008).

Hearing impairment is common in idiopathic PD (Vitale et al. 2012), and there are 77% more cases of PD in patients with hearing loss than in those without hearing loss (Lai et al. 2014). An association between incident dementia and hypacusia has been shown (Lin et al. 2011a).

In rats, low doses of a statin (atorvastatin) appear to prevent noise-induced hearing loss (Jahani et al. 2016). In patients with hyperlipidemia, tinnitus may be successfully treated by atorvastatin (Hameed et al. 2014). In mice, another statin (pravastatin) attenuates cochlear injury caused by noise (Park et al. 2012). In a mouse strain with accelerated aging, atorvastatin slows down the dete-rioration of inner ear function with age and this "suggest[s] that statins could also slow down the age-related deterioration of hearing in man" (Syka et al. 2007). In a rat model for type 2 diabetes, losartan treatment effectively contrasts the hearing dysfunction that is typical in these animals (Meyer zum Gottesberge et al. 2015).

CONCLUSION

The traditional interpretation of aging, based on the tenets of the old paradigm, leads the research-ers to justify the many ailments that afflict the elderly as diseases caused by various and differ-ent degenerative processes. According to the old paradigm, this implies that the disorders may well have common characteristics when there are identical or similar degenerative factors, but the similarities do not allow at all to consider them as a single disorder with various manifestations. Therefore, aging is explained as the sum and the overlapping of many different diseases and the term "aging" is only a useful term that summarizes them all without indicating a unique and distinct entity. This theoretical conception, among other things, has a practical consequence so that, in the International Classification of Diseases (ICD-10 2016, ICD-9-CM 2016), a code to define aging is nonexistent and so, in the international statistics of the World Health Organization, aging is ignored as a cause of death (World Ranking Total Deaths 2014).

In complete contrast to this classical view, in the interpretation of the new paradigm, the term "aging" indicates a precise and distinct physiological mechanism, genetically determined and regulated, which manifests itself in all tissues and organs, and in the organism as a whole, as a consequence of a single, specific, and well-defined mechanism. As briefly discussed above, this mechanism originates in the telomere–telomerase–subtelomere system and causes a progressive increase in the fraction of cells in replicative senescence and in various degrees of gradual senescence, with a parallel decline in cell turnover rates and in the functions of tissues and organ, which leads to a condition defined as "atrophic syndrome."

For all cell types, as well as for the tissues and organs they form and for the functions resulting from them, this decline has two general modes:

i. Cells subjected to turnover (e.g., nonneuronal cells, gliocytes included, certain types of neuronal cells such as the ORCs) undergo a direct decline

ii. Cells not subjected to turnover (e.g., most neurons) but dependent on satellite cells that show turnover suffer from a decline in the cells upon which they depend

However, it is useful and necessary to add further considerations that are based on the evidence:

1. Unhealthy lifestyles which cause diabetes, obesity, hypertension, and exposure to toxic substances and other "risk factors" damage cells and accelerate their turnover, causing alterations and anticipation of physiological aging.

2. By avoiding or reducing these "risk factors" and/or by using "protective drugs," such as ACE inhibitors, sartans, and statins, cellular damage and the associated disorders, caused by slackened or exhausted cell turnover, are avoided or reduced. This occurs both for the decline in cells with turnover (e.g., emphysema caused by alveolocyte turnover failure and arteriosclerosis due to endothelial cell turnover failure) and for the decline in cells without turnover but depending on cells with turnover (e.g., AMD, AD, PD, etc. caused by the decline in their trophic cells).

3. The full avoidance of the risk factors annuls the anticipation or increased probability of the aforementioned diseases but does not cancel the physiological rhythm and characteristics of aging.

4. The alterations of the characteristics of physiological or "normal" aging are definable as diseases if "disease" is conceived as something that causes sufferings but not if "disease" is used in the meaning of alterations shown in a normal condition. Perhaps a more neutral term (e.g., "trouble") should be used for alterations that are associated with physiological aging.

5. For the troubles of aging, if we disregard the precocious cases due to genetic anomalies, it is difficult to distinguish between the physiological forms (which we must consider as part of physiological aging and are not prevented or modified by a healthy lifestyle and/or protective drugs) and the precocious form, which are due to some risk factor and may be undoubtedly defined as diseases.

These concepts are summarized in Figures 5.1, 5.2, and 5.7 and in Table 5.1.
Now, some important general concepts may be expressed:

1. The telomere theory allows a unified and coherent vision of various diseases of the nervous system (e.g., AD, PD, AMD, presbycusis, and age-related hyposmia), which are apparently quite distinct and without a common root.

2. The telomere theory allows a unified and consistent view of these diseases in the framework of a wide range of diseases that are not pertaining to the nervous system. Through this unitary conception, it is possible to have a rational explanation of the noteworthy commonality of factors that increase the risks (risk factors) or reduce them (protective factors) for all the aforementioned diseases.

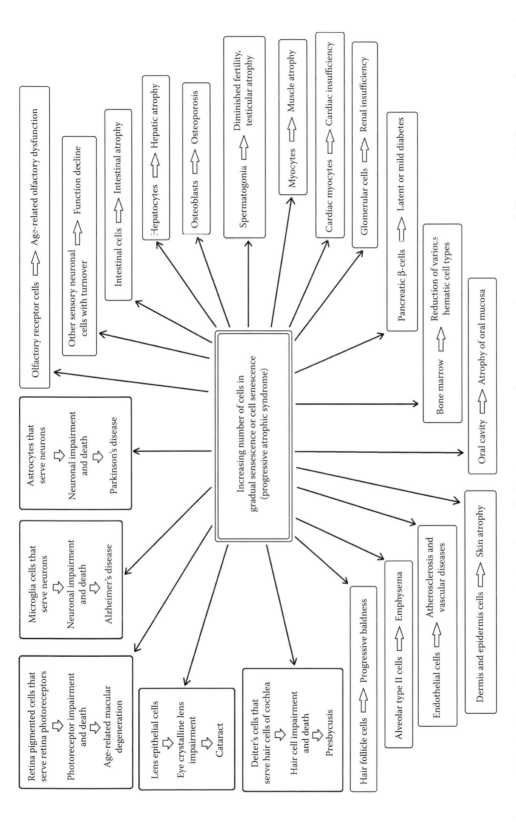

FIGURE 5.7 A scheme of the aging process. Most cell types, including some types of neurons, are subject to turnover. Cell turnover decline causes the progressive impairment of the functions depending on these cells. Perennial neurons are compromised by turnover decline and gradual senescence of essential satellite gliocytes.

TABLE 5.1
Relations between Some Troubles and Some "Risk Factors" or "Protective Drugs"

Troubles	Cell Turnover	Risk increased (+) or Perhaps Increased (+?) or Lowered (−) or Unaltered (/) by							Protective Effect by	
		Age	Diabetes	Obesity/ Dyslipidemia	Hypertension	Smoke	Alcohol Moderate Use	Alcohol Abuse	Statins	ACE-i; Sartans
Endothelial dysfunction	Yes	+[1]	+[2]	+[3]	+[4]	+[5]	−[6]	+[7]	+[8]	+[9]
Olfactory dysfunction	Yes	+[10]	+[11]	+[12]	+[13]	+[14]	[15]	+[16]	+?[17]	[18]
AMD	No	+[19]	+[20]	+[21]	+[22]	+[23]	/[24]	/[25]	+?[26]	[27]
Ad	No	+[28]	+[29]	+[30]	+[31]	+[32]	−[33]	+[34]	+[35]	+[36]
PD	No	+[37]	+[38]	+[39]	+[40]	−[41]	−[42]	+[43]	+[44]	+?[45]
Hearing impairment	No	+[46]	+[47]	+[48]	+[49]	+[50]	−[51]	+[52]	+?[53]	+?[54]

Notes: (1–5) Wilson et al. (1987), Hill et al. (2003); (6) Roerecke and Rehm (2014), Gardner and Mouton (2015), de Gaetano et al. (2016), (7) Gardner and Mouton (2015), de Gaetano et al. (2016), (8) Walter et al. (2002), Su (2015), (9) Su (2015), (10) Schubert et al. (2012), Doty and Kamath (2014), Gouveri et al. (2014), Mehdizadeh et al. (2015), (12) Richardson et al. (2004), Patel et al. (2015), (13) Gouveri et al. (2014), (11) Gouveri et al. (2014), (14) Vent et al. (2003), (15) No specific study; (16) Rupp et al. (2004), (17) Kim et al. (2012), (18) No specific study; (19) Rudnicka et al. (2012), (20–22) Klein et al. (2007), Mares et al. (2011), (23) Klein et al. (2007), Armstrong and Mousavi (2015), (24) Armstrong and Mousavi (2015), (25) Klein et al. (2010), (26) Gehlbach et al. (2015), (27) No specific study; (28) Gorelick (2004), (29) Rosendorff et al. (2007), Baglietto-Vargas et al. (2016), Rani et al. (2016), Saedi et al. (2016), Vicente Miranda et al. (2016), (30) Rosendorff et al. (2007), Wanamaker et al. (2015), (31) Rosendorff et al. (2007), Qiu et al. (2005), Michel (2016), (32) Rosendorff et al. (2007), Durazzo et al. (2014), (33) Campdelacreu (2014), Ilomaki et al. (2015), (34) Rosendorff et al. (2007), Campdelacreu (2014), (35) Vogel et al. (2006), Ellul et al. (2007), Wanamaker et al. (2015), (36) Vogel et al. (2006), Ellul et al. (2007), Yasar et al. (2016), (37) De Lau and Breteler (2006), Pringsheim et al. (2014), (38) Hu et al. (2007), Tomlinson and Gardiner (2008), Zhang and Tian (2014), Vicente Miranda et al. (2016), (39) Abbott et al. (2002), Hu et al. (2006), Zhang and Tian (2014), (40) Zhang and Tian (2014), (41) Li et al. (2015), (42) Ishihara and Brayne (2005), (43) Eriksson et al. (2013), (44) Gao et al. (2012), Friedman et al. (2013), Undela et al. (2013), Sheng et al. (2016), (45) Lopez-Real et al. (2005), Sonsalla et al. (2013), (46) Zhan et al. (2010), Lin et al. (2011b), (47) Akinpelu et al. (2014), Oron et al. (2014), Calvin and Watley (2015), Helzner and Contrera (2016), Bener et al. (2016), (48) Fransen et al. (2008), Oron et al. (2014), (49) Oron et al. (2014), Przewóźny et al. (2015), Bener et al. (2016), (50) Cruickshanks et al. (1998), Fransen et al. (2008), Oron et al. (2014), Chang et al. (2016), (51) Fransen et al. (2008), (52) Rosenhall et al. (1993), (53) Syka et al. (2007), Park et al. (2012), Hameed et al. (2014), Jahani et al. (2016), (54) Meyer zum Gottesberge et al. (2015).

3. The possible criticism against the telomere theory, and consequently against the programmed aging paradigm, caused by the evidence of tissue and organs largely composed of cells without turnover and that age, is overcome by the evidence previously expounded. Conversely, the telomere theory and the programmed aging paradigm emerge strengthened from the overcoming of such criticism.

4. If aging is a physiological phenomenon, genetically determined and regulated, in principle it could be regulated, slackened, or even cancelled. This statement, which might seem unrealistic and, for the supporters of the non-programmed aging paradigm, the declaration of something clearly utopian, that is, impossible, on the contrary, is something that has been proved: (i) since 1998, it is known that in cultivated normal cells, telomerase activation leads to longer telomeres and cancels all the manifestations of cell senescence, both the biochemical alterations and the incapacity to duplicate (Bodnar et al. 1998; Counter et al. 1998; Vaziri 1998; Vaziri and Benchimol 1998); (ii) telomerase reactivation in aged mice with blocked telomerase determined the clear reversal of all the manifestations of aging, those of the nervous system included (Jaskelioff et al. 2011); (iii) in one- and two-year-old normal mice, telomerase reactivation delayed the aging manifestations and increased the life span (Bernardes de Jesus et al. 2012); (iv) human skin obtained from aged fibroblasts with reactivated telomerase was not distinguishable from skin obtained from young fibroblasts (Funk et al. 2000).

5. In the search for a treatment to slacken or reverse aging, attention should be focused on the telomere–telomerase–subtelomere system. Telomerase reactivation has been proposed as treatment for AD (Fossel 1996, 2004) and for AMD (Libertini 2009b). As an intermediate step for the goal of a complete control of aging, the effective treatment of diseases that are strongly invalidating and expensive, such as AD, PD, and AMD, by actions on the telomere–telomerase–subtelomere system has been proposed (Libertini and Ferrara 2016a).

REFERENCES

Aarsland D, Londos E, Ballard C. Parkinson's disease dementia and dementia with Lewy bodies: Different aspects of one entity. *Int. Psychogeriatr.* 2009, 21(2):216–9.

Abbott A. Neuroscience: The plaque plan. *Nature* 2008, 456:161–4.

Abbott RD, Ross GW, White LR, Nelson JS, Masaki KH, Tanner CM et al. Midlife adiposity and the future risk of Parkinson's disease. *Neurology* 2002, 59(7):1051–7.

Ablonczy Z, Gutierrez DB, Grey AC, Schey KL, Crouch RK. Molecule-specific imaging and quantitation of A2E in the RPE. *Adv. Exp. Med. Biol.* 2012, 723:75–81.

Akinpelu OV, Mujica-Mota M, Daniel SJ. Is type 2 diabetes mellitus associated with alterations in hearing? A systematic review and meta-analysis. *Laryngoscope.* 2014, 124(3):767–76.

Aliani M, Udenigwe CC, Girgih AT, Pownall TL, Bugera JL, Eskin MNa. Aroma and taste perceptions with Alzheimer disease and stroke. *Crit. Rev. Food Sci. Nutr.* 2013, 53(7):760–9.

Alzheimer's Association. Alzheimer's disease facts and figures. *Alzheimers Dement.* 2015, 11(3):332–84.

Armstrong RA, Mousavi M. Overview of risk factors for age-related macular degeneration (AMD). *J. Stem Cells.* 2015, 10(3):171–91.

Azad R, Chandra P, Gupta R. The economic implications of the use of anti-vascular endothelial growth factor drugs in age-related macular degeneration. *Indian J. Ophthalmol.* 2007, 55(6):441–3.

Baglietto-Vargas D, Shi J, Yaeger DM, Ager R, LaFerla FM. Diabetes and Alzheimer's disease crosstalk. *Neurosci Biobehav. Rev.* 2016, 64:272–87.

Ballard C, Hanney ML, Theodoulou M, Douglas S, McShane R, Kossakowski K et al. The dementia antipsychotic withdrawal trial (DART-ad): Long-term follow-up of a randomised placebo-controlled trial. *Lancet Neurol.* 2009, 8(2):151–7.

Barnett-Cowan M, Dyde RT, Fox SH, Moro E, Hutchison WD, Harris LR. Multisensory determinants of orientation perception in Parkinson's disease. *Neuroscience* 2010, 167(4):1138–50.

Barresi M, Ciurleo R, Giacoppo S, Foti Cuzzola V, Celi D, Bramanti P, Marino S. Evaluation of olfactory dysfunction in neurodegenerative diseases. *J. Neurol. Sci.* 2012, 323(1-2):16–24.

Ben-Porath I, Weinberg RA. The signals and pathways activating cellular senescence. *Int. J. Biochem. Cell Biol.* 2005, 37(5):961–76.

Bener A, Al-Hamaq AO, Abdulhadi K, Salahaldin AH, Gansan L. Interaction between diabetes mellitus and hypertension on risk of hearing loss in highly endogamous population. *Diabetes Metab. Syndr.* 2016. pii: S1871-4021(16)30203-X. doi: 10.1016/j.dsx.2016.09.004.

Berger JM, Fine SL, Maguire MG. *Age-related Macular Degeneration.* Mosby, USA 1999.

Bermingham-McDonogh O, Reh TA. Regulated reprogramming in the regeneration of sensory receptor cells. *Neuron* 2011, 71(3):389–405.

Bertram L, Blacker D, Mullin K, Keeney D, Jones J, Basu S et al. Evidence for genetic linkage of Alzheimer's disease to chromosome 10q. *Science* 2000, 290:2302–3.

Blackburn EH. Telomere states and cell fates. *Nature* 2000, 408:53–6.

Blesa J, Trigo-Damas I, Quiroga-Varela A, Jackson-Lewis VR. Oxidative stress and Parkinson's disease. *Front. Neuroanat.* 2015, 9:91. doi: 10.3389/fnana.2015.00091.

Bodnar AG, Ouellette M, Frolkis M, Holt SE, Chiu CP, Morin GB et al. Extension of life-span by introduction of telomerase into normal human cells. *Science* 1998, 279:349–52.

Brück D, Wenning GK, Stefanova N, Fellner L. Glia and alpha-synuclein in neurodegeneration: A complex interaction. *Neurobiol. Dis.* 2016, 85:262–74.

Caballol N, Martí MJ, Tolosa E. Cognitive dysfunction and dementia in Parkinson disease. *Mov. Disord.* 2007, 22(Suppl 17):S358–66.

Caggiano M, Kauer JS, Hunter DD. Globose basal cells are neuronal progenitors in the olfactory epithelium: A lineage analysis using a replication-incompetent retrovirus. *Neuron* 1994, 13(2):339–52.

Calvin D, Watley SR. Diabetes and hearing loss among underserved populations. *Nurs. Clin. North Am.* 2015, 50(3):449–56.

Campdelacreu J. Parkinson disease and Alzheimer disease: Environmental risk factors. *[Article in English, Spanish] Neurologia.* 2014, 29(9):541–9.

Chang J, Ryou N, Jun HJ, Hwang SY, Song JJ, Chae SW. Effect of cigarette smoking and passive smoking on hearing impairment: Data from a population-based study. *PLoS One.* 2016, 11(1):e0146608 doi: 10.1371/journal.pone.0146608.

Chen X, Fang H, Schwob JE. Multipotency of purified, transplanted globose basal cells in olfactory epithelium. *J. Comp. Neurol.* 2004, 469(4):457–74.

Chong EWT, Wong TY, Kreis AJ, Simpson JA, Guymer RH. Dietary antioxidants and primary prevention of age related macular degeneration: Systematic review and meta-analysis. *BMJ* 2007, 335(7623):755, doi: 10.1136/bmj.39350.500428.47.

Counter CM, Hahn WC, Wei W, Caddle SD, Beijersbergen RL, Lansdorp PM, Sedivy JM, Weinberg RA. Dissociation among *in vitro* telomerase activity, telomere maintenance, and cellular immortalization. *Proc. Natl. Acad. Sci. USA* 1998, 95(25):14723–8.

Cruickshanks KJ, Klein R, Klein BE, Wiley TL, Nondahl DM, Tweed TS. Cigarette smoking and hearing loss: The epidemiology of hearing loss study. *JAMA* 1998, 279(21):1715–9.

Davie CA. A review of Parkinson's disease. *Br. Med. Bull.* 2008, 86:109–27.

Decary S, Hamida CB, Mouly V, Barbet JP, Hentati F, Butler-Browne GS. Shorter telomeres in dystrophic muscle consistent with extensive regeneration in younger children. *Neuromuscul. Disord.* 2000, 10(2):113–20.

Dickson DW. Neuropathology of Parkinsonian disorders. In: Jankovic JJ, Tolosa E (eds). *Parkinson's Disease and Movement Disorders.* Lippincott Williams and Wilkins, Philadelphia, 2007, pp 271–83.

Doty RL. Olfactory dysfunction in Parkinson disease. *Nat. Rev. Neurol.* 2012, 8(6):329–39.

Doty RL, Kamath V. The influences of age on olfaction: A review. *Front Psychol.* 2014, 5:20. doi: 10.3389/fpsyg.2014.00020.

Doty RL, Shaman P, Applebaum SL, Giberson R, Siksorski L, Rosenberg L. Smell identification ability: Changes with age. *Science* 1984, 226:1441–3.

Duff K, McCaffrey RJ, Solomon GS. The pocket smell test: Successfully discriminating probable Alzheimer's dementia from vascular dementia and major depression. *J. Neuropsychiatry Clin. Neurosci.* 2002, 14(2):197–201.

Duffy VB, Backstrand JR, Ferris AM. Olfactory dysfunction and related nutritional risk in free-living, elderly women. *J. Am. Diet. Assoc.* 1995, 95(8):879–84.

Durazzo TC, Mattsson N, Weiner MW. Smoking and increased Alzheimer's disease risk: A review of potential mechanisms. *Alzheimers Dement.* 2014, 10(3 Suppl):S122–45.

Ellul J, Archer N, Foy CM, Poppe M, Boothby H, Nicholas H, Brown RG, Lovestone S. The effects of commonly prescribed drugs in patients with Alzheimer's disease on the rate of deterioration. *J. Neurol. Neurosurg. Psychiatry* 2007, 78(3):233–9.

Eriksson AK, Löfving S, Callaghan RC, Allebeck P. Alcohol use disorders and risk of Parkinson's disease: Findings from a Swedish national cohort study 1972–2008. BMC *Neurol.* 2013, 13:190. doi: 10.1186/1471-2377-13-190.

Factor SA, Weiner WJ (eds). *Parkinson's Disease: Diagnosis and Clinical Management*, 2nd edition. Demos Medical Publishing, New York, 2008, pp. 72–3.

Feng P, Huang L, Wang H. Taste bud homeostasis in health, disease, and aging. *Chem. Senses.* 2014, 39(1):3–16.

Fine SL, Berger JW, Maguire MG, Ho AC. Age-related macular degeneration. *N. Engl. J. Med.* 2000, 342:483–92.

Flanary B. Telomeres: Function, Shortening, and lengthening. In: Mancini L. (ed.), *Telomeres: Function, Shortening and Lengthening*. Nova Science Publ. Inc., New York, 2009, pp 379–86.

Fossel MB. *Reversing Human Aging*. William Morrow and Company, New York, 1996.

Fossel MB. *Cells, Aging and Human Disease*. Oxford University Press, New York, 2004.

Fransen E, Topsakal V, Hendrickx JJ, Van Laer L, Huyghe JR, Van Eyken E et al. Occupational noise, smoking, and a high body mass index are risk factors for age-related hearing impairment and moderate alcohol consumption is protective: A European population-based multicenter study. *J. Assoc. Res. Otolaryngol.* 2008, 9(3):264–76.

Friedman B, Lahad A, Dresner Y, Vinker S. Long-term statin use and the risk of Parkinson's disease. *Am. J. Manag. Care* 2013, 19(8):626–32.

Funk WD, Wang CK, Shelton DN, Harley CB, Pagon GD, Hoeffler WK. Telomerase expression restores dermal integrity to *in vitro*-aged fibroblasts in a reconstituted skin model. *Exp. Cell Res.* 2000, 258(2):270–8.

de Gaetano G, Costanzo S, Di Castelnuovo A, Badimon L, Bejko D, Alkerwi A et al. Effects of moderate beer consumption on health and disease: A consensus document. *Nutr. Metab. Cardiovasc. Dis.* 2016, 26(6):443–67.

Galpern WR, Lang AE. Interface between tauopathies and synucleinopathies: A tale of two proteins. *Ann. Neurol.* 2006, 59(3):449–58.

Gao X, Simon KC, Schwarzschild MA, Ascherio A. A prospective study of statin use and risk of Parkinson disease. *Arch. Neurol.* 2012, 69(3):380–4.

Gardner JD, Mouton AJ. Alcohol effects on cardiac function. *Compr. Physiol.* 2015, (2):791–802.

Gehlbach P, Li T, Hatef E. Statins for age-related macular degeneration. *Cochrane Database Syst. Rev.* 2015, (2):CD006927. doi: 10.1002/14651858.CD006927.pub4.

Geiger H, Van Zant G. The aging of lympho-hematopoietic stem cells. *Nat. Immunol.* 2002, 3:329–33.

Gorelick PB. Risk factors for vascular dementia and Alzheimer disease. *Stroke* 2004, 35:2620–2.

Gouveri E, Katotomichelakis M, Gouveris H, Danielides V, Maltezos E, Papanas N. Olfactory dysfunction in type 2 diabetes mellitus: An additional manifestation of microvascular disease? *Angiology.* 2014, 65(10):869–76.

Griffiths CEM. Aging of the skin. In: Tallis RC, Fillit H, Brocklehurst JCl. (eds), *Brocklehurst's Textbook of Geriatric Medicine and Gerontology*, 5th edition. Churchill Livingstone, New York, 1998.

Hameed MK, Sheikh ZA, Ahmed A, Najam A. Atorvastatin in the management of tinnitus with hyperlipidemias. *J. Coll. Physicians Surg. Pak.* 2014, 24(12):927–30.

Hamilton WD. The moulding of senescence by natural selection. *J. Theor. Biol.* 1966, 12:12–45.

Helzner EP, Contrera KJ. Type 2 diabetes and hearing impairment. *Curr. Diab. Rep.* 2016, 16(1):3.

Herrup K. Reimagining Alzheimer's disease—An age-based hypothesis. *J. Neurosci.* 2010, 30(50):16755–62.

Hill JM, Zalos G, Halcox JPJ, Schenke WH, Waclawiw MA, Quyyumi AA, Finkel T. Circulating endothelial progenitor cells, vascular function, and cardiovascular risk. *N. Engl. J. Med.* 2003, 348:593–600.

Holmes C, Boche D, Wilkinson D, Yadegarfar G, Hopkins V, Bayer A et al. Long-term effects of Abeta42 immunisation in Alzheimer's disease: Follow-up of a randomised, placebo-controlled phase I trial. *Lancet* 2008, 372(9634):216–23.

Holmes DJ, Austad SN. Birds as animal models for the comparative biology of aging: A prospectus. *J. Gerontol. A Biol. Sci.* 1995, 50(2):B59–66.

Holz FG, Pauleikhoff D, Spaide RF, Bird AC (eds). *Age-related Macular Degeneration*, 2nd edition. Springer-Verlag, Berlin, 2013.

Horner PJ, Gage FH. Regenerating the damaged central nervous system. *Nature* 2000, 407:963–70.

Hu G, Jousilahti P, Bidel S, Antikainen R, Tuomilehto J. Type 2 diabetes and the risk of Parkinson's disease. *Diabetes Care* 2007, 30(4):842–7.

Hu G, Jousilahti P, Nissinen A, Antikainen R, Kivipelto M, Tuomilehto J. Body mass index and the risk of Parkinson disease. *Neurology* 2006, 67(11):1955–9.

Huard JMT, Youngentob SL, Goldstein BJ, Luskin MB, Schwob JE. Adult olfactory epithelium contains multipotent progenitors that give rise to neurons and non-neural cells. *J. Comp. Neurol.* 1998, 400(4):469–86.

Husain M, Schott JM (eds.). *Oxford Textbook of Cognitive Neurology and Dementia*, 1st edition. Oxford University Press, Oxford 2016.

ICD-10, 2016. Available: http://www.who.int/classifications/apps/icd/icd10online.

ICD-9-CM, 2016. Available: http://www.cdc.gov/nchs/icd/icd9cm.htm.

Ilomaki J, Jokanovic N, Tan EC, Lonnroos E. Alcohol consumption, dementia and cognitive decline: An overview of systematic reviews. *Curr. Clin. Pharmacol.* 2015, 10(3):204–12.

Ishihara L, Brayne C. A systematic review of nutritional risk factors of Parkinson's disease. *Nutr. Res. Rev.* 2005, 18(2):259–82.

Iwai N, Zhou Z, Roop DR, Behringer RR. Horizontal basal cells are multipotent progenitors in normal and injured adult olfactory epithelium. *Stem Cells* 2008, 26(5):1298–306.

Iwasaki T, Goto N, Goto J, Ezure H, Moriyama H. The aging of human Meissner's corpuscles as evidenced by parallel sectioning. *Okajimas Folia Anat. Jpn.* 2003, 79(6):185–9.

Jager RD, Mieler WF, Miller JW. Age-related macular degeneration. *N. Engl. J. Med.* 2008, 358(24):2606–17.

Jahani L, Mehrparvar AH, Esmailidehaj M, Rezvani ME, Moghbelolhossein B, Razmjooei Z. The effect of atorvastatin on preventing noise-induced hearing loss: An experimental study. *Int. J. Occup. Environ. Med.* 2016, 7(1):15–21.

Jankovic J. Parkinson's disease: Clinical features and diagnosis. *J. Neurol. Neurosurg. Psychiatry* 2008, 79:368–76.

Jaskelioff M, Muller FL, Paik JH, Thomas E, Jiang S, Adams AC et al. Telomerase reactivation reverses tissue degeneration in aged telomerase-deficient mice. *Nature* 2011, 469:102–6.

Bernardes de Jesus B, Vera E, Schneeberger K, Tejera AM, Ayuso E, Bosch F, Blasco MA. Telomerase gene therapy in adult and old mice delays aging and increases longevity without increasing cancer. *EMBO Mol. Med.* 2012, 4(8):691–704.

Kar T, Yildirim Y, Altundağ A, Sonmez M, Kaya A, Colakoglu K et al. The relationship between age-related macular degeneration and olfactory function. *Neurodegener. Dis.* 2015, 15(4):219–24.

Kim HY, Kim JH, Dhong HJ, Kim KR, Chung SK, Chung SC et al. Effects of statins on the recovery of olfactory function in a 3-methylindole-induced anosmia mouse model. *Am. J. Rhinol. Allergy.* 2012, 26(2):e81–4. doi: 10.2500/ajra.2012.26.3719.

Kirkwood TBL. Evolution of ageing. *Nature* 1977, 270:301–4.

Kirkwood TBL, Holliday R. The evolution of ageing and longevity. *Proc. R. Soc. Lond. B Biol. Sci.* 1979, 205:531–46.

Kirkwood TBL, Melov S. On the programmed/non-programmed nature of ageing within the life history. *Curr. Biol.* 2011, 21(18):R701–7.

Klein R, Cruickshanks KJ, Nash SD, Krantz EM, Nieto FJ, Huang GH, Pankow JS, Klein BE. The prevalence of age-related macular degeneration and associated risk factors. *Arch. Ophthalmol.* 2010, 128(6):750–8.

Klein R, Deng Y, Klein BE, Hyman L, Seddon J, Frank RN et al. Cardiovascular disease, its risk factors and treatment, and age-related macular degeneration: Women's Health Initiative Sight Exam ancillary study. *Am. J. Ophthalmol.* 2007, 143(3):473–83.

Kruman II, Culmsee C, Chan SL, Kruman Y, Guo Z, Penix L, Mattson MP. Homocysteine elicits a DNA damage response in neurons that promotes apoptosis and hypersensitivity to excitotoxicity. *J. Neurosci.* 2000, 20(18):6920–6.

Kuhn TS. *The Structure of Scientific Revolutions*. The University of Chicago Press, Chicago, 1962.

Lai SW, Liao KF, Lin CL, Lin CC, Sung FC. Hearing loss may be a non-motor feature of Parkinson's disease in older people in Taiwan. *Eur. J. Neurol.* 2014, 21(5):752–7.

de Lau LM, Breteler MM. Epidemiology of Parkinson's disease. *Lancet Neurol.* 2006, 5(6):525–35.

Leung CT, Coulombe PA, Reed RR. Contribution of olfactory neural stem cells to tissue maintenance and regeneration. *Nat. Neurosci.* 2007, 10:720–6.

Li J, Cesari M, Liu F, Dong B, Vellas B. Effects of diabetes mellitus on cognitive decline in patients with Alzheimer disease: A systematic review. *Can. J. Diabetes* 2016, pii: S1499-2671(15)30064-2. doi: 10.1016/j.jcjd.2016.07.003.

Li X, Li W, Liu G, Shen X, Tang Y. Association between cigarette smoking and Parkinson's disease: A meta-analysis. *Arch. Gerontol. Geriatr.* 2015, 61(3):510–6.

Libertini G. An adaptive theory of the increasing mortality with increasing chronological age in populations in the wild. *J. Theor. Biol.* 1988, 132:145–62.

Libertini G. The role of telomere-telomerase system in age-related fitness decline, a tameable process. In: Mancini L (ed.), *Telomeres: Function, Shortening and Lengthening.* Nova Science Publ. Inc., New York, 2009a, pp 77–132.

Libertini G. Prospects of a longer life span beyond the beneficial effects of a healthy lifestyle. In: Bently JV, Keller MA (eds.), *Handbook on Longevity: Genetics, Diet and Disease.* Nova Science Publ. Inc., New York, 2009b, pp 35–95.

Libertini G. Classification of phenoptotic phenomena. *Biochem. (Mosc).* 2012, 77(7):707–15.

Libertini G. The programmed aging paradigm: How we get old. *Biochem. (Mosc.)* 2014, 79(10):1004–16.

Libertini G. Non-programmed versus programmed aging paradigm. *Curr. Aging Sci.* 2015a, 8(1):56–68.

Libertini G. Phylogeny of aging and related phenoptotic phenomena. *Biochem. (Mosc.)* 2015b, 80(12):1529–46.

Libertini G, Ferrara N. Possible interventions to modify aging. *Biochem. (Mosc.)* 2016a, 81(12):1413–28.

Libertini G, Ferrara N. Aging of perennial cells and organ parts according to the programmed aging paradigm. *Age (Dordr)* 2016b, 38(2):35 doi: 10.1007/s11357-016-9895-0.

Lim JI (ed.). *Age-related Macular Degeneration*, 3rd edition. CRC Press, Taylor & Francis Group, London and New York, 2013.

Lin FR, Metter EJ, O'Brien RJ, Resnick SM, Zonderman AB, Ferrucci L. Hearing loss and incident dementia. *Arch. Neurol.* 2011a, 68(2):214–20.

Lin FR, Thorpe R, Gordon-Salant S, Ferrucci L. Hearing loss prevalence and risk factors among older adults in the United States. *J. Gerontol. A Biol. Sci. Med. Sci.* 2011b, 66A(5):582–90.

Lopez-Real A, Rey P, Soto-Otero R, Mendez-Alvarez E, Labandeira-Garcia JL. Angiotensin-converting enzyme inhibition reduces oxidative stress and protects dopaminergic neurons in a 6-hydroxydopamine rat model of Parkinsonism. *J. Neurosci. Res.* 2005, 81(6):865–73.

Maier EC, Saxena A, Alsina B, Bronner ME, Whitfield TT. Sensational placodes: Neurogenesis in the otic and olfactory systems. *Dev. Biol.* 2014, 389(1):50–67.

Mares JA, Voland RP, Sondel SA, Millen AE, Larowe T, Moeller SM et al. Healthy lifestyles related to subsequent prevalence of age-related macular degeneration. *Arch. Ophthalmol.* 2011, 129(4):470–80.

Medawar PB. *An Unsolved Problem in Biology.* H. K. Lewis, London, 1952. Reprinted in: Medawar PB. *The Uniqueness of the Individual.* Methuen, London, 1957.

Mehdizadeh Seraj J, Mehdizadeh Seraj S, Zakeri H, Bidar Z, Hashemi S, Mahdavi Parsa F, Yazdani N. Olfactory dysfunction in Iranian diabetic patients. *Acta Med. Iran.* 2015, 53(4):204–6.

Mendiola-Precoma J, Berumen LC, Padilla K, Garcia-Alcocer G. Therapies for prevention and treatment of Alzheimer's disease. *Biomed Res. Int.* 2016; 2016:2589276. doi: 10.1155/2016/2589276.

Meyer zum Gottesberge AM, Massing T, Sasse A, Palma S, Hansen S. Zucker diabetic fatty rats, a model for type 2 diabetes, develop an inner ear dysfunction that can be attenuated by losartan treatment. *Cell Tissue Res.* 2015, 362(2):307–15.

Michel JP. Is it possible to delay or prevent age-related cognitive decline? *Korean J. Fam. Med.* 2016, 37(5):263–6.

Miners JS, Baig S, Palmer J, Palmer LE, Kehoe PG, Love S. Aβ-degrading enzymes in Alzheimer's disease. *Brain Pathol.* 2008, 18(2):240–52.

Mondragón-Rodríguez S, Basurto-Islas G, Lee HG, Perry G, Zhu X, Castellani RJ, Smith MA. Causes versus effects: The increasing complexities of Alzheimer's disease pathogenesis. *Expert Rev. Neurother.* 2010, 10(5):683–91.

Morales I, Sanchez A, Rodriguez-Sabate C, Rodriguez M. The degeneration of dopaminergic synapses in Parkinson's disease: A selective animal model. *Behav. Brain Res.* 2015, 289:19–28.

Murphy C, Schubert CR, Cruickshanks KJ, Klein BE, Klein R, Nondahl DM. Prevalence of olfactory impairment in older adults. *JAMA* 2002, 288(18):2307–12.

Nowak JZ. AMD--the retinal disease with an unprecised etiopathogenesis: In search of effective therapeutics. *Acta Pol. Pharm.* 2014, 71(6):900–16.

Nussey DH, Froy H, Lemaitre JF, Gaillard JM, Austad SN. Senescence in natural populations of animals: Widespread evidence and its implications for bio-gerontology. *Ageing Res. Rev.* 2013, 12(1):214–25.

Nuytemans K, Theuns J, Cruts M, Van Broeckhoven C. Genetic etiology of Parkinson disease associated with mutations in the SNCA, PARK2, PINK1, PARK7, and LRRK2 genes: A mutation update. *Hum. Mutat.* 2010, 31(7):763–80.

Obeso JA, Rodríguez-Oroz MC, Benitez-Temino B, Blesa FJ, Guridi J, Marin C, Rodriguez M. Functional organization of the basal ganglia: Therapeutic implications for Parkinson's disease. *Mov. Disord.* 2008, 23(S3):S548–59.

Obeso JA, Rodriguez-Oroz MC, Goetz CG, Marin C, Kordower JH, Rodriguez M et al. Missing pieces in the Parkinson's disease puzzle. *Nat. Med.* 2010, 16(6):653–61.

Oertel W, Schulz JB. Current and experimental treatments of Parkinson disease: A guide for neuroscientists. *J. Neurochem.* 2016, doi: 10.1111/jnc.13750 [Epub ahead of print].

Oron Y, Elgart K, Marom T, Roth Y. Cardiovascular risk factors as causes for hearing impairment. *Audiol. Neurootol.* 2014, 19(4):256–60.

Pahwa R, Lyons KE. *Handbook of Parkinson's Disease*, 5th edition. CRC Press, Taylor & Francis Group, London and New York, 2013.

Park JS, Kim SW, Park K, Choung YH, Jou I, Park SM. Pravastatin attenuates noise-induced cochlear injury in mice. *Neuroscience.* 2012, 208:123–32.

Patel ZM, DelGaudio JM, Wise SK. Higher body mass index is associated with subjective olfactory dysfunction. *Behav. Neurol.* 2015, 2015:675635. doi: 10.1155/2015/675635.

Pringsheim T, Jette N, Frolkis A, Steeves TD. The prevalence of Parkinson's disease: A systematic review and meta-analysis. *Mov. Disord.* 2014, 29(13):1583–90.

Przewoźny T, Gójska-Grymajło A, Kwarciany M, Gąsecki D, Narkiewicz K. Hypertension and cochlear hearing loss. *Blood Press.* 2015, 24(4):199–205.

Qiu C, Winblad B, Fratiglioni L. The age-dependent relation of blood pressure to cognitive function and dementia. *Lancet Neurol.* 2005, 4(8):487–99.

Qiu WQ, Walsh DM, Ye Z, Vekrellis K, Zhang J, Podlisny MB et al. Insulin-degrading enzyme regulates extracellular levels of amyloid beta-protein by degradation. *J. Biol. Chem.* 1998, 273(49):32730–8.

Quik M, Bordia T, Zhang D, Perez XA. Nicotine and nicotinic receptor drugs: Potential for Parkinson's disease and drug-induced movement disorders. *Int. Rev. Neurobiol.* 2015; 124:247–71.

Rani V, Deshmukh R, Jaswal P, Kumar P, Bariwal J. Alzheimer's disease: Is this a brain specific diabetic condition? *Physiol. Behav.* 2016, 164(Pt A):259–67.

Reitz C. Alzheimer's disease and the amyloid cascade hypothesis: A critical review. *Int. J. Alzheimers Dis.* 2012, 2012:369808. doi: 10.1155/2012/369808.

Reitz C, Mayeux R. Alzheimer disease: Epidemiology, diagnostic criteria, risk factors and biomarkers. *Biochem. Pharmacol.* 2014, 88(4):640–51.

Richardson BE, Vander Woude EA, Sudan R, Thompson JS, Leopold DA. Altered olfactory acuity in the morbidly obese. *Obes. Surg.* 2004, 14(7):967–9.

Rizzi L, Rosset I, Roriz-Cruz M. Global epidemiology of dementia: Alzheimer's and vascular types. *BioMed. Res. Int.* 2014, 2014:908915, doi: 10.1155/2014/908915.

Roerecke M, Rehm J. Alcohol consumption, drinking patterns, and ischemic heart disease: A narrative review of meta-analyses and a systematic review and meta-analysis of the impact of heavy drinking occasions on risk for moderate drinkers. *BMC Med.* 2014, 12:182. doi: 10.1186/s12916-014-0182-6.

Rose MR. *Evolutionary Biology of Aging.* Oxford University Press, New York, 1991.

Rosendorff C, Beeri MS, Silverman JM. Cardiovascular risk factors for Alzheimer's disease. *Am. J. Geriatr. Cardiol.* 2007, 16(3):143–9.

Rosenhall U, Sixt E, Sundh V, Svanborg A. Correlations between presbyacusis and extrinsic noxious factors. *Audiology* 1993, 32(4):234–43.

Rudnicka AR, Jarrar Z, Wormald R, Cook DG, Fletcher A, Owen CG. Age and gender variations in age-related macular degeneration prevalence in populations of European ancestry: A meta-analysis. *Ophthalmology.* 2012, 119(3):571–80.

Rupp CI, Fleischhacker WW, Hausmann A, Mair D, Hinterhuber H, Kurz M. Olfactory functioning in patients with alcohol dependence: Impairments in odor judgements. *Alcohol Alcohol.* 2004, 39(6):514–9.

Saedi E, Gheini MR, Faiz F, Arami MA. Diabetes mellitus and cognitive impairments. *World J. Diabetes.* 2016, 7(17):412–22.

Samii A, Nutt JG, Ransom BR. Parkinson's disease. *Lancet.* 2004, 363:1783–93.

Schubert CR, Cruickshanks KJ, Fischer ME, Huang GH, Klein BE, Klein R et al. Olfactory impairment in an adult population: The Beaver Dam Offspring Study. *Chem. Senses* 2012, 37(4):325–34.

Schulz-Schaeffer WJ. The synaptic pathology of α-synuclein aggregation in dementia with Lewy bodies, Parkinson's disease and Parkinson's disease dementia. *Acta Neuropathol.* 2010, 120(2):131–43.

Seale P, Asakura A, Rudnicki MA. The potential of muscle stem cells. *Dev. Cell* 2001, 1(3):333–42.

Sergeyenko Y, Lall K, Liberman MC, Kujawa SG. Age-related cochlear synaptopathy: An early-onset contributor to auditory functional decline. *J. Neurosci.* 2013, 33(34):13686–94.

Sevigny J, Chiao P, Bussière T, Weinreb PH, Williams L, Maier M et al. The antibody aducanumab reduces Aβ plaques in Alzheimer's disease. *Nature* 2016, 537:50–6.

Sheng Z, Jia X, Kang M. Statin use and risk of Parkinson's disease: A meta-analysis. *Behav. Brain Res.* 2016, 309:29–34.

Skulachev VP. Aging is a specific biological function rather than the result of a disorder in complex living systems: Biochemical evidence in support of Weismann's hypothesis. *Biochem. (Mosc.)* 1997, 62(11):1191–5.

Skulachev VP. Phenoptosis: Programmed death of an organism. *Biochem. (Mosc.)* 1999, 64(12):1418–26.

Sonsalla PK, Coleman C, Wong LY, Harris SL, Richardson JR, Gadad BS, Li W, German DC. The angiotensin converting enzyme inhibitor captopril protects nigrostriatal dopamine neurons in animal models of parkinsonism. *Exp. Neurol.* 2013, 250:376–83.

Sparrow JR. Therapy for macular degeneration: Insights from acne. *Proc. Natl. Acad. Sci. USA* 2003, 100(8): 4353–4.

Sturm E, Stefanova N. Multiple system atrophy: Genetic or epigenetic? *Exp. Neurobiol.* 2014, 23(4):277–91.

Su JB. Vascular endothelial dysfunction and pharmacological treatment. *World J. Cardiol.* 2015, 7(11):719–41.

Syka J, Ouda L, Nachtigal P, Solichová D, Semecký V. Atorvastatin slows down the deterioration of inner ear function with age in mice. *Neurosci. Lett.* 2007, 411(2):112–6.

Thiriez C, Villafane G, Grapin F, Fenelon G, Remy P, Cesaro P. Can nicotine be used medicinally in Parkinson's disease? *Expert Rev. Clin. Pharmacol.* 2011, 4(4):429–36.

Tomlinson DR, Gardiner NJ. Glucose neurotoxicity. *Nat. Rev. Neurosci.* 2008, 9(1):36–45.

Tyner SD, Venkatachalam S, Choi J, Jones S, Ghebranious N, Igelmann H et al. p53 mutant mice that display early ageing-associated phenotypes. *Nature* 2002, 415:45–53.

Undela K, Gudala K, Malla S, Bansal D. Statin use and risk of Parkinson's disease: A meta-analysis of observational studies. *J. Neurol.* 2013, 260(1):158–65.

Vaziri H. Extension of life span in normal human cells by telomerase activation: A revolution in cultural senescence. *J. Anti-Aging Med.* 1998, 1(2):125–30.

Vaziri H, Benchimol S. Reconstitution of telomerase activity in human normal cells leads to elongation of telomeres and extended replicative life span. *Curr. Biol.* 1998, 8:279–82.

Vekrellis K, Ye Z, Qiu WQ, Walsh D, Hartley D, Chesneau V, Rosner MR, Selkoe DJ. Neurons regulate extracellular levels of amyloid β-protein via proteolysis by insulin-degrading enzyme. *J. Neurosci.* 2000, 20(5):1657–65.

Vent J, Bartels S, Haynatzki G, Gentry-Nielsen MJ, Leopold DA, Hallworth R. The impact of ethanol and tobacco smoke on intranasal epithelium in the rat. *Am. J. Rhinol.* 2003, 17(4):241–7.

Vicente Miranda H, El-Agnaf OM, Outeiro TF. Glycation in Parkinson's disease and Alzheimer's disease. *Mov. Disord.* 2016, 31(6):782–90.

Vitale C, Marcelli V, Allocca R, Santangelo G, Riccardi P, Erro R et al. Hearing impairment in Parkinson's disease: Expanding the nonmotor phenotype. *Mov. Disord.* 2012, 27(12):1530–5.

Vogel T, Benetos A, Verreault R, Kaltenbach G, Kiesmann M, Berthel M. Risk factors for Alzheimer: Towards prevention? [Article in French]. *Presse Med.* 2006, 35:1309–16.

Waite LM. Treatment for Alzheimer's disease: Has anything changed? *Aust. Prescr.* 2015, 38(2):60–3.

Walter DH, Rittig K, Bahlmann FH, Kirchmair R, Silver M, Murayama T et al. Statin therapy accelerates reendothelialization: A novel effect involving mobilization and incorporation of bone marrow-derived endothelial progenitor cells. *Circulation* 2002, 105(25):3017–24.

Wanamaker BL, Swiger KJ, Blumenthal RS, Martin SS. Cholesterol, statins, and dementia: What the cardiologist should know. *Clin. Cardiol.* 2015, 38(4):243–50.

Watt FM, Lo Celso C, Silva-Vargas V. Epidermal stem cells: An update. *Curr. Opin. Genet. Dev.* 2006, 16(5):518–24.

Webster C, Blau HM. Accelerated age-related decline in replicative life-span of Duchenne muscular dystrophy myoblasts: Implications for cell and gene therapy. *Somat. Cell Mol. Genet.* 1990, 16(6):557–65.

Weiner MF, Lipton AM. *The American Psychiatric Publishing Textbook of Alzheimer Disease and Other Dementias.* American Psychiatric Publishing Inc, Washington and London 2009.

Whitmer RA, Gunderson EP, Barrett-Connor E, Quesenberry CP Jr., Yaffe K. Obesity in middle age and future risk of dementia: A 27 year longitudinal population based study. *BMJ* 2005, 330(7504):1360, doi: 10.1136/bmj.38446.466238.E0.

Williams GC. Pleiotropy, natural selection and the evolution of senescence. *Evolution* 1957, 11:398–411.

Williams RW, Herrup K. The control of neuron number. *Annu. Rev. Neurosci.* 1988, 11(1):423–53.

Wilson PW, Castelli WP, Kannel WB. Coronary risk prediction in adults (the Framingham Heart Study). *Am. J. Cardiol.* 1987, 59(14):91–4G. [Erratum, *Am. J. Cardiol.* 1987; 60(13):A11.].

Wong ACY, Ryan AF. Mechanisms of sensorineural cell damage, death and survival in the cochlea. *Front. Aging Neurosci.* 2015, 7:58 doi: 10.3389/fnagi.2015.00058.

Wong WL, Su X, Li X, Cheung CMG, Klein R, Cheng CY, Wong TY. Global prevalence of age-related macular degeneration and disease burden projection for 2020 and 2040: A systematic review and meta-analysis. *Lancet Glob. Health.* 2014, 2(2):e106–16.

Woo KS, Chook P, Lolin YI, Cheung AS, Chan LT, Sun YY et al. Hyperhomocyst(e)inemia is a risk factor for arterial endothelial dysfunction in humans. *Circulation* 1997, 96(8):2542–4.

World Ranking Total Deaths, 2014. Available: http://www.worldlifeexpectancy.com/world-rankings-total-deaths. See also: http://www.worldlifeexpectancy.com/sitemap.

Yamasoba T, Lin FR, Someya S, Kashio A, Sakamoto T, Kondo K. Current concepts in age-related hearing loss: Epidemiology and mechanistic pathways. *Hear Res.* 2013, 303:30–8.

Yan D, Zhu Y, Walsh T, Xie D, Yuan H, Sirmaci A et al. Mutation of the ATP-gated P2X(2) receptor leads to progressive hearing loss and increased susceptibility to noise. *Proc. Natl. Acad. Sci. USA* 2013, 110(6):2228–33.

Yao SC, Hart ad, Terzella MJ. An evidence-based osteopathic approach to Parkinson disease. *Osteopathic Family Physician* 2013, 5(3):96–101.

Yasar S, Schuchman M, Peters J, Anstey KJ, Carlson MC, Peters R. Relationship between antihypertensive medications and cognitive impairment: Part I. Review of human studies and clinical trials. *Curr. Hypertens. Rep.* 2016, 18(8):67. doi: 10.1007/s11906-016-0674-1.

von Zglinicki T, Serra V, Lorenz M, Saretzki G, Lenzen-Grossimlighaus R, Gessner R, Risch A, Steinhagen-Thiessen E. Short telomeres in patients with vascular dementia: An indicator of low antioxidative capacity and a possible risk factor? *Lab. Invest.* 2000, 80:1739–47.

Zhan W, Cruickshanks KJ, Klein BE, Klein R, Huang GH, Pankow JS, Gangnon RE, Tweed TS. Generational differences in the prevalence of hearing impairment in older adults. *Am. J. Epidemiol.* 2010, 171(2): 260–6.

Zhang P, Tian B. Metabolic syndrome: An important risk factor for Parkinson's disease. *Oxid. Med. Cell Longev.* 2014, 2014:729194 doi: 10.1155/2014/729194.

Zhao C, Deng W, Gage FH. Mechanisms and functional implications of adult neurogenesis. *Cell* 2008, 132(4):645–60.

Zhuo JM, Wang H, Praticò D. Is hyperhomocysteinemia an Alzheimer's disease (AD) risk factor, an AD marker, or neither? *Trends Pharmacol. Sci.* 2011, 32(9):562–71.

6 The Genetic Program of Aging

Xiufang Wang, Huanling Zhang, Libo Su, and Zhanjun Lv

CONTENTS

INTRODUCTION

Aged individuals can no longer maintain homeostasis in response to physiologic and environmental changes as easily as they once could (Wolden-Hanson et al., 2002). Through the years, many theories have been proposed to explain the aging mechanisms (Salmon et al., 2010). These theories include two main types (Sergiev et al., 2015): one is a stochastic event resulting from the accumulation of random errors and the other is an orderly, genetically programmed event that is the consequence of differentiation, growth, and maturation. Wear and tear effects, the waste accumulation theory, the free radical theory (Harman, 2006), the error catastrophe theory (Holliday, 1996), and the network theory of aging (Kowald and Kirkwood, 1996) are the representatives of the former. The other type of theories postulates that genetically programming changes that occur with increasing age are responsible for the deleterious changes that accompany aging (Wang et al., 2014). It is well known that development is genetically programmed, so logic dictates that aging changes might also be programmed, which means that aging starts from zygotes, the programmed changes during the early period of life are favorable to individual survival and the programmed changes during the later period of life prejudice individual survival. The life span of an organism includes a stage of development that culminates in sexual maturity, a stage of maximal fitness and fertility (maturity stage), and a senescent stage that is characterized by functional decline and an increased probability of death. Figure 6.1 shows the abridged general views of the human aging process.

Although more than 300 theories of aging were proposed (Viña et al., 2007), until recently the very definition of aging—senescence—still remains uncertain (da Costa et al., 2016). In this chapter, the early and middle periods of life are called the development–maturity stage and the later period of life is called the senescence stage. These two stages are called aging (Figure 6.1). The aging process starts from a zygote and refers to the changes that occur during an organism's life span (Kirkwood, 2002), which means that aging is consistent with life span or longevity. The senescence stage starts after completing the process of development and maturity; therefore, senescence is the progressive deterioration of bodily functions with time, which means that function decline is consistent with increasing age during the senescence stage (Wang et al., 2014).

After completing the process of development and maturity, aging leads to loss of homeostasis that predisposes to the many diseases and frailty characteristic of older age, and to changes of many molecular and cellular hallmarks (López-Otín et al., 2013; Ocampo et al., 2016). During the aging process, though the curves of different survival-related functions have variation, the tendencies of changing during the whole life are the same, so we used a curve that represents all the functions (Figure 6.1).

The senescence stage in a species' lifetime could be defined by the increased mortality rate caused by the increased number of emerged illnesses. In humans, the turning point of the mortality curve is at the age of 45 (Tepp et al., 2016). In Figure 6.1, we accept the viewpoint that the senescent stage starts from 45 years of age, but, in fact, many human survival functions decrease after 20 years of age. Here, the survival functions refers to the functions that are related to postnatal survival, for example, audition, vision, immunity, and so on.

THE IMPORTANT PHENOMENA OF AGING

General characteristics: The aging process starts from the zygote stage (Figure 6.1). Aging is a general phenomenon in all multicellular eukaryotes (Zs-Nagy, 1987; Holmes and Ottinger, 2003).

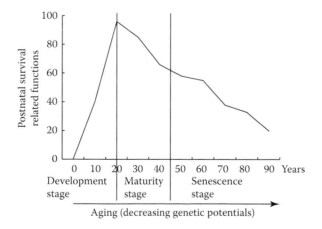

FIGURE 6.1   The abridged general view of the human aging process. We name the early period of life (e.g., from the zygote stage to 20 years in humans) as the development stage. Human postnatal survival-related functions increase during this stage, but the differentiation potential decreases. Therefore, not all functions of humans increase during this stage. The next stage is the middle period of life named as the maturity stage when fitness and fertility display maximally. The late period of life (e.g., after 45 years in humans) is called the senescent stage. The early and middle periods of life are called the development-maturity stage and the late period of life is called the senescence stage. Aging includes all these stages.

One challenge is to explain why aging occurs despite its apparent detrimental effects. Since natural selection favors the survival of the fittest individuals, this poses the question why natural selection did not eliminate something as detrimental to an organism as aging (Lipsky and King, 2015). Senescent cells accumulate with age in organisms, albeit at different rates in the various organs (Paradis et al., 2001; Erusalimsky and Kurz, 2005; Herbig et al., 2006; Jeyapalan et al., 2007; Bialystok et al., 2016).

Genetic control: It has been proven that the aging process is genetically controlled via intraspecific and interspecific comparative studies (de Magalhães, 2003; Prinzinger, 2005). In fact, the first described mutation to yield a significant extension in the life span of *Caenorhabditis elegans* was in the age-I gene, which was shown to result in a 65% increase in mean life span and a 110% increase in maximum life span of this organism (Johnson, 1990). Measuring the replicative life spans for cells from a series of mammals of different life spans, the decline in replicative capacity was seen and the species' maximum life span potential is expressed in fibroblast culture in terms of proliferative capacity (Macieira-Coelho, 1976; Röhme, 1981). Pairs of monozygotic twins showed no significant difference in replicative life span within each of the twin pairs but did show such differences among pairs (Ryan et al., 1981).

Influence factors: Aging is intrinsic because internal factors are the decisive ones; however, external factors play important roles in the aging process. Dietary restriction, an intervention where the consumption of food is decreased but without malnutrition, is a well-established mechanism that has a wide range of important outcomes including improved health span and delayed aging (Hadem and Sharma, 2016). Environmental temperature greatly affects life span in a wide variety of animals; a moderate temperature decrease from 22°C to 16°C extends the life span of the monogonont rotifer *Brachionus manjavacas* by up to 163% (Johnston and Snell, 2016). As we can see, antioxidant supplementation led in most cases to the life extension (Vaiserman et al., 2016). The salmon of the genus *Oncorhynchus* are semelparous and usually die immediately after spawning. In the period leading up to sexual maturation and spawning, they undergo a series of degenerative changes that resemble the effects of aging in other vertebrates (Allard and Duan, 2011; de Magalhães, 2012). Ice water is in favor of the survival of Atlantic salmon (Hedger et al., 2013).

Aging occurring at the cellular level: It is now well established that populations of normal human diploid fibroblasts (Swim and Parker, 1957; Hayflick and Moorhead, 1961; Schneider and Mitsui, 1976; Balin et al., 2002), lymphocytes (Effros, 1996; Chou and Effros, 2013), hematopoietic stem cells (Rossi et al., 2007), and stem cells (Oh et al., 2014) can proliferate in culture for only finite periods. The serial transplantation of normal somatic tissues to new, young inbred hosts each time the recipient approaches old age. Normal cells seem to show a decline in proliferative capacity and cannot survive indefinitely after they were serially transplanted to the inbred host (Daniel et al., 1968; Albright and Makinodan, 1976).

Although death happens suddenly, the cell senescence in culture is a gradual process. Loss of proliferative capacity is characterized by a gradually declining fraction of cells which synthesize DNA, increasing cell cycle time and declining saturation density (Cristofalo, 1988; Yamamoto et al., 2000). Cell culture passage numbers are inversely proportional to the donor age, approximately 0.20 cell doublings/year from ages 0 to 90 years (Martin et al., 1970; Schneider and Mitsui, 1976; Pignolo et al., 1992).

Immortality: The new individuals that develop from the zygote are young and not affected by parental ages though their genomes are from parents. In this sense, the germ line is normally "immortal" (Katz et al., 2009; Petralia et al., 2014). Under appropriate *in vitro* culture conditions, embryonic stem cells (ESCs) proliferate indefinitely without differentiation, and retain the developing potential to generate cells of all three primary germ layers, termed "pluripotency." Mouse ESCs were firstly established in 1981 and rat ESCs in 2008 (Evans and Kaufman, 1981; Martin, 1981; Buehr, et al., 2008; Li et al., 2008). Some plants obtain immortally through clonal reproduction (de Witte and Stöcklin, 2010; Myles et al., 2011) and certain transforming cell lines exhibited immortalized characteristics (Hosaka et al., 1992; Benraiss et al., 1996; Liu et al., 2010).

TABLE 6.1
Theories of Aging are Described and Discussed in This Chapter

Theories of Aging	The Central Themes of Theories
Biological clock theory	An internal biological clock regulates development, growth, maturity, and aging by sequentially switching genes on and off
The evolutionary theory	Aging results from a decline in the force of natural selection
Redundant message theory	The selective repetitions of some definite genes, cistrons, operons, and other linear structures on the DNA molecule, the bulk of which are repressed, behave as redundant messages to be called into action when active genome messages become faulty
Codon-restriction theory	The kinds of proteins synthesized by cells are controlled by the set of code words that a cell can decode
Vital essence loss theory	Humans have two kinds of vital essences: one is innate that is inherited from parents and one that is acquired that comes from diet. The innate essences will decrease with age since they cannot receive supplement. The decrease of the innate essences is the reason of leading to aging
Immune functions and aging	Impairment of immune function results in disease and death. Aging may represent deterioration in the immune ability
Hormones and aging	The levels of hormones may change with age
Telomere and aging	Telomere shortening is related to aging. Telomere length determines the multiplying capacity of human cells
Chromatin modification and aging	Chromatin is altered during the aging process, which interfere with chromatin regulatory complexes, impacts the life span
Aging genes	Genes undoubtedly contribute to longevity and several gene mutations affecting the life span
The gene regulation theory	Senescence results from changes in gene expression
RNA population model	Gene transcription can be altered by changes in the interactions between the RNA population and DNA. Such changes are the foundation of aging and differentiation

Aging-related genes: The aging process is associated with the expression changes of many genes (Baskerville et al., 2008; Kim et al., 2009; Mun et al., 2009; Lui et al., 2010). Cell fusion experiments proved that aging was a dominant phenotype and cellular immortality was a recessive phenotype in hybrids (Pereira-Smith and Ning, 1992; Leung and Pereira-Smith, 2001). Four complementation groups (A, B, C, D) for infinite divisions have been identified from extensive studies by fusing different immortal human cell lines with each other (Pereira-Smith and Smith, 1988). By microcell-mediated chromosome transfer, the senescence-related genes for three of the complementation groups B–D have been identified on human chromosomes 4, 1, and 7, respectively (Ogata et al., 1993; Tominaga et al., 2002).

Most of the aging theories are monistic in nature; they omit numerous key factors of senescence during the process of model creation. It is possible that all the cell components could contribute to the process of aging, and roles of some cell components are weak and others are strong (Semsei, 2000). Therefore, it is necessary that putting forward a theory of aging should consider the above entire phenomena of aging. Here, we review the current main genetic program theories of aging. Table 6.1 shows the central themes of theories of aging that are described and discussed in this chapter.

BIOLOGICAL CLOCK THEORY OF AGING

The biological clock theory of aging argues that an internal biological clock regulates development, growth, maturity, and senescence by sequentially switching genes on and off (Weinert and Timiras, 2003; Plikus et al., 2015). The biological clocks are the internal appliances that in some ways measure the time and control aging process. Bird migration and plant flowering rhythms are controlled by biological clocks or rhythm clocks (Kumar et al., 2010; Kondratova and Kondratov, 2012; Mizuno and Yamashino, 2015). The biological clocks in this chapter refer to the timing devices that control the aging process. The question then is what makes this clock "tick." There are a large number of theories on what controls aging processes, but none of these conclude whether the phenomenon is the clock itself or a subsidiary mechanism that is controlled by the clock.

THE EVOLUTIONARY THEORY OF AGING

Evolutionary theories argue that aging results from a decline in the force of natural selection (Weinert and Timiras, 2003). Since evolution acts mainly to maximize reproductive fitness in an individual, longevity is a characteristic that is selected only if it is in favour of the reproductive fitness. Life span is the result of selective pressures and has a large degree of plasticity within a species and among different species. The evolutionary theory was first proposed in the 1940s based on the study of Huntington's disease that remained in the population even though it should be strongly selected against (Haldane, 1941). The late age of onset of Huntington's disease (30–40 years) allows a carrier to reproduce before dying and allows the disease avoiding the force of natural selection. The basic concept that aging results from a lack of selection has many experimental support. Long-lived *Drosophila* strains can be bred by selecting the offspring of older adults, suggesting that life span can be changed directly by selective pressure (Rose and Charlesworth, 1980; Partridge et al., 1999). Life span is species specific because it is a function of survivability and reproductive strategy in the competitive environment. Living organisms that die primarily from environmental hazards and predation will increase a life span optimized for their own particular environments. The viewpoint was experimented with in a natural environment by comparing mainland opossums that are subject to predation to a population of opossums living on an island free of predators (Austad, 1993). The evolutionary theory predicted that the island opossums would have the opportunity to evolve a longer life span, if it were conducive to fitness. Observation indeed showed that island opossums did live longer and aged more slowly than did their mainland counterparts (Austad, 1993). From many corners of the field, many biologists are finding the fundamental mechanisms of natural selection. Aging has universal, intrinsic, deleterious, and progressive characteristics. In the biology field, the universal characteristics are not deleterious and vice versa. Aging is the only exception.

REDUNDANT MESSAGE THEORY OF AGING

The redundant message theory of aging argues that the selective repetitions of some definite genes, cistrons, operons, and other linear structures on the DNA molecule, the bulk of which are repressed, behave as redundant messages to be called into action when active genome messages become faulty (Hayflick, 1975). The genomes of mammals are made up of not less than 10^5 structural genes or cistrons, but in each cell hardly more than 0.2%–0.4% of this number is expressed during development. If 1/500 of all genes is active and 499/500 are repressed, and if mutagenic factors act equally on the repressed and active cistrons, the mutation rate of repressed genes must yield more mutations than those occurring in active genes. The different species' life spans may be a function of the degree of repeated sequences. Long-lived species should then have more redundant message than short-lived species. As errors accumulate in functioning genes, reserve sequences containing the same information take over until the redundancy in the system is exhausted, which leads to biological age changes. The degree of gene repetition in different species may be a determiner of species' life spans.

CODON-RESTRICTION THEORY

Strehler et al. (1971) proposed a mechanism of codon modulation. They suggested that the kinds of proteins synthesized by cells are controlled by the set of code words that a cell can decode. Most biological aging theories can be divided into two concepts: one of them proposes that age-dependent deterioration is the result of an active "self-destruct" program; another argues that it is the result of passive "wearing out" processes. The former aging theories impute aging to a sequential program of events, which in turn is ultimately specified by the order of DNA nucleotides in each species. The operation of this program leads to a predetermined series of events which cause aging and the eventual death of the individual. Strehler et al. (1971) studied the phenomenon of programmed cell death as observed during the embryogeny of limbs and the observations on the finite lifetime of cultured normal cells as providing support for these theories.

VITAL ESSENCE LOSS THEORY

Traditional Chinese Medicine suggests that humans survive because of vital essences. Humans have two kinds of vital essences: one is the innate vital essence that is inherited from parents and the other is the acquired vital essence that mainly comes from diets. Humans do not lack the acquired vital essence during the whole life because of the diet supplement. The innate vital essence will decrease with age since it cannot receive the supplement, which leads to aging.

IMMUNE FUNCTIONS AND AGING

The immune function impairment theory of aging argues that impairment of immune function results in disease and death and aging may represent a deterioration in immune capacity (Prinzinger, 2005; Cardinali et al., 2008). It has been easy to demonstrate that the immune system changes with age. Recognizing foreign biological entities and inactivating them either by tagging them with very specific antibodies or by directly killing them is the main function of the immune system. Mammals produce lymphocytes in the thymus (T-lymphocytes) or the bone marrow (B-lymphocytes) for this purpose. The thymus peaks in both size and function during puberty and then progressively atrophies with age, producing fewer mature T cells. The immune system can distinguish between foreign antigens and indigenous antigens. The response of the immune system to indigenous antigens is called autoimmunity, and the frequency of autoimmune interactions increases with aging. In fact, these inappropriate autoimmune responses can lead to many age-related diseases. The apparently programmed changes in the immune system could be important factors in aging. As the immune system declines, a living organism becomes more susceptible to infection and immune dysfunction is linked to an increased risk of cancer and Alzheimer's disease (Lipsky and King, 2015).

HORMONES AND AGING

The levels of hormones may change with age. This is particularly true for growth hormones, dehydroepiandrosterone (DHEA) (Chatterjee and Mondal, 2014), and melatonin (Devore et al., 2016). It is not clear whether the decreases observed are developmentally programmed to benefit the organism, or whether they are simply another example of dysregulation with aging. An example is provided by estrogen. Menopause results in a relatively rapid loss of estradiol and progesterone, and is programmed to occur at about the age of 50 in women (Barron and Pike, 2012). The decline of estrogen production in women is one of the risk factors of age-related diseases, for example, osteoporosis and cardiovascular disease and Alzheimer's disease. The late-life programmed changes may result in a variety of effects, many of which are detrimental.

Somatopause, or the age-related decrease of growth hormone and the insulin-like signaling pathway, appears to play roles in regulating life span. Insulin-like signaling activity and the expression of insulin-like peptides are reduced in long-lived nematodes, mice, and humans. Centenarians are generally more sensitive to insulin. Mutations in insulin-like growth factor-1 (IGF-1) receptors are overrepresented in many Ashkenazi Jewish centenarians (Kenyon, 2010; Moskalev et al., 2014). daf-2 is a gene that encodes for a hormone receptor similar to mammalian insulin and IGF-1 in nematodes. In nematodes, mutations that decrease the activity of daf-2 doubled a nematode's life span (Kenyon, 2010). In humans, insulin resistance, changes in body composition, physiologic declines in growth hormone IGF-1, and sex steroids feature the process of aging. The enhanced levels of corticosteroid hormones led to degenerative changes in salmon. The life span of salmon was extended when the gonad was removed from salmon before maturation via prevention of interrenal hyperplasia (Allard and Duan, 2011).

TELOMERE AND CELL AGING

Telomeres consist of TTAGGG repeats that are extended 9–15 kb and are located at the ends of chromosomes in eukaryotes. They can protect the chromosome ends against degradation, fusion, and recombination (Lin and Yan, 2005) and make cells distinguish the appropriate chromosome ends from the double-strand DNA breaks induced by exogenous factors such as radiation. Olovnikov (1971) first proved that linear arrangement of chromosomes was one cause of the end replication. DNA replication needs an RNA primer to initiate synthesis, followed by its removal, which progressively shortens chromosome ends with each cell cycle. Telomerase is a ribonucleoprotein reverse transcriptase and can stabilize telomere length by adding TTAGGG repeats to the ends of the chromosomes. It consists of two main components: one is telomere RNA, which acts as a template for telomere synthesis, and the other is telomere reverse transcriptase, which catalyzes the elongation (Mason et al., 2011).

Allsopp et al. (1992) put forward evidence to support the theory that telomere shortening is related to aging. Three possible hypotheses by which cell aging due to telomere shortening have been proposed. The first hypothesis suggests that telomeres are characterized by heterochromatin. After the telomere length is shortened, it will attack the nearby DNA and thus affect the expression of genes that are related to growth regulation. It has been proved that gene transferation near telomere DNA has been inhibited in low-level eukaryotes. The second hypothesis argues that the total loss of telomere DNA produces a damaging signal, which will induce growth inhibition and DNA damage via activation of p53 gene expression and PI3-kinase-like kinase activation (Karlseder et al., 1999; di Fagagna et al., 2003; Greenberg, 2005). The third hypothesis suggests that it is not the damaging signal but the shortened telomere itself that activates p53 and causes permanent growth suppression. P53 is the key mediator of the response to dysfunctional telomere (Artandi and DePinho, 2010).

Many experiments have proved that telomere shortening is a kind of molecular clock that counts cell divisions (Allsopp and Harley, 1995; Takubo et al., 2010). Telomere length positively relates to years of healthy life in health, aging, and body composition in humans (Hunt et al., 2008; Njajou et al., 2009; Ren et al., 2009; Oeseburg et al., 2010). The immortal human cells were treated using widowed nucleotide acid in order to extend telomere length and integrated with fatal cells (Wright et al., 1996). Telomerase lengthened the proliferative life span and prevented the replicative senescence of cells when it was introduced into human lens epithelial cells and fibroblasts (Bodnar et al., 1998; Huang et al., 2005; Wieser et al., 2008).

In addition, telomere sequences may be lost from human chromosomes as a result of the end replication problem, exonuclease degradation, or oxidative damage (Olovnikov, 1973; von Zglinicki et al., 1995; Makarov et al., 1997; Munro et al., 2001; Sato et al., 2012).

CHROMATIN MODIFICATIONS AND AGING

Chromatin is altered during the aging process, which interfering with chromatin regulatory complexes impacts life span. The accumulating evidence so far highlights the importance of many chromatin features during aging (Feser and Tyler, 2011).

DNA METHYLATION AND AGING

Methylation of the 5-carbon of cytosine in CpG dinucleotide sites is a conserved epigenetic modification, which is classically linked to transcriptional silencing in vertebrates. During the aging process, DNA methylation experiences remodeling in many different tissues in mice and humans and then contributes to misregulation of gene expression. So DNA methylation is regarded as an "aging clock."

In human cells, compared to actively cycling cells, the global levels of 5-methylcytosine (5-mC) are reduced in senescent cells (Wilson and Jones, 1983; Cruickshanks et al., 2013). Studies probing the methylation status of specific CpG sites through the genome illustrated that the methylation of such sites outside of promoter CpG "islands" tended to decrease in a variety of human tissues (Day et al., 2013). In contrast, in both mice and humans, CpG "islands" near promoters are typically hypermethylated with age increasing most notably on genes related in differentiation or development (Maegawa et al., 2010; Day et al., 2013). Regions with changed DNA methylation overlap binding sites of specific sets of transcription factors (Sun et al., 2014). In addition, the discovery of other forms of DNA methylation, such as 5-hydroxymethylcytosine (5-hmC), cytosine methylation at non-CpG dinucleotides, and N6-methylation of adenines (6-mA), deserve study in the context of aging (Lister et al., 2013; Fu et al., 2015; Greer et al., 2015; Zhang et al., 2015a). Traditional model organisms for aging are not well suited for studying the role of 5-mC methylation in organismal life span, as this modification has not been observed in *Saccharomyces cerevisiae* (Capuano et al., 2014) or *C. elegans* (Simpson et al., 1986). Although DNA methylation is present in *Drosophila melanogaster*, its intensity is low (Capuano et al., 2014).

HISTONE METHYLATION AND AGING

Histone methylation is related to active or repressed genome regions, depending on the residue affected and the level of methylation, which induces heterochromatin remodeling. Although the mechanisms underlying heterochromatin remodeling are not fully understood, they may involve interactions between the chromatin machinery and nuclear periphery proteins including nuclear lamins. Such interactions can set up nuclear microdomains that adumbrate regions of active and inhibited gene expression (Dechat et al., 2008; Benayoun et al., 2015).

Histone methyltransferases and histone demethylases can dynamically regulate histone methylation. The global level or genomic distribution of a lot of histone methylations alters in organismal and cellular models of aging. The manipulation of histone methyltransferases and histone demethylases can modulate longevity of model organisms. Widespread changes in heterochromatin organization are found in mesenchymal stem cells derived from a Werner syndrome ESC model, including a generalized reduction of H3K9me3 (Zhang et al., 2015b). Targeted RNAi screens probing the effects of histone methyltransferases and demethylases on longevity in worms and flies have shown that H3K4me3 regulators can modulate life span (Jin, et al., 2011; Maures et al., 2011; Ni et al., 2012). Keeping the levels of another active histone methylation, H3K36me3, which is linked to transcriptional elongation, is required for healthy aging of worms and yeast. The mutation of the yeast *RPH1* gene, which encodes a H3K36 demethylase, prolongs the yeast replicative life span, and yeast cells carrying H3 mutant forms that cannot be H3K36-methylated are short lived (Sen et al., 2015). In *C. elegans*, somatic levels of H3K36me3 moderately reduce with increasing age (Ni et al., 2012), and appear to be particularly decreased at genes that are deregulated with increasing age. Knock down of *met-1* encodes the putative *C. elegans* enzyme depositing the H3K36me3 mark, shortening the life span in worms (Pu et al., 2015). These results suggest that the correct maintenance of H3K36me3 may be a key process during the senescent stage.

Changing specific histone methyltransferases or demethylases regulates life spans of yeast, worms, and flies. However, apart from studies in yeast, it is yet well understood whether alterations in histone methylation directly affect aging. Future studies should confirm whether the role of histone methylation complexes in life span modulation is conserved throughout evolution. It will be interesting to determine the mechanisms that trigger large-scale alterations in suppressive chromatin with aging.

HISTONE ACETYLATION AND AGING

Histone acetylation directly affects the physical association of histones and DNA. Evidence suggests that the pattern of histone acetylation changes during normal aging. The global levels of H3K56ac decrease during replicative aging in yeast, while those of H4K16ac increase, resulting in de-silencing of telomeric repeats (Dang et al., 2009). Global H4K16ac levels reduce during normal aging and in a mouse model of Hutchinson–Gilford progeria syndrome (HGPS) and may be linked, at least in the progeroid model, to a decreased association of histone acetyltransferases (HAT) with the nuclear periphery (Krishnan et al., 2011). Following contextual fear conditioning, older mice cannot upregulate H4K12ac, a mark that accelerates transcriptional elongation (Hargreaves et al., 2009), in their hippocampus, and this relates to changed gene expression and memory impairment (Peleg et al., 2010). Alterations in histone acetylation may be a result and a cause of the failure of older cells to transduce external stimuli to downstream transcriptional responses, a process that is detrimental for rapid cell-to-cell signaling in the brain. Both HAT and deacetylases (HDAC) regulate life span and metabolic health. For example, H4K16ac is deacetylated by the sirtuin SIR263, and reduced *SIR2* dosage prolongs life span in *S. cerevisiae* (Kaeberlein et al., 1999) by limiting aberrant recombination at the ribosomal DNA locus. More generally, sirtuins may have a pro-longevity role by promoting enhanced genomic stability (Mostoslavsky et al., 2006; Toiber et al., 2013; Van Meter et al., 2014).

General speaking, deregulation of both HATs and HDACs via genetic manipulation or targeting by drugs has been related to significant changes in longevity across taxa. However, in most cases, whether modulation of these enzymes regulates life span by affecting chromatin or also by modulating non-histone substrates needs further studies.

HETEROCHROMATIN LOSS AND AGING

The "loss of heterochromatin" model of aging was first put forth by Villeponteau (1997), who proposed that heterochromatin domains established early in embryogenesis are broken down (decreased heterochromatin and/or the inappropriate redistribution of heterochromatin) during the aging process, contributing to the derepression of silenced genes and resulting in abnormal gene expression patterns (Villeponteau, 1997).

Consistent with this proposal, these findings were verified in studies of natural aging in *C. elegans*, *Drosophila*, and humans. A similar alteration in nuclear architecture and a loss of peripheral heterochromatin were found in nonneuronal cells during aging in *C. elegans* (Haithcock et al., 2005). In *Drosophila*, a gradual loss of heterochromatin correlates with increasing age and lamin proteins were found to have an important role in abnormal nuclear morphology and decreased life span (Brandt et al., 2008; Larson et al., 2012). In humans, skin fibroblast cell lines established from old individuals showed nuclear defects similar to those of cells from an HGPS patient, including changes of histone modifications and reduced DNA damage (Scaffidi and Misteli, 2006). Interestingly, in cultured cells from healthy individuals, repressing the sporadic use of a cryptic splice site in lamin A that produces progerin reverses the nuclear defects related to senescence, further proving that the global loss of heterochromatin markers and dissociation of heterochromatin proteins seen in HGPS also occur in physiological aging (Scaffidi and Misteli, 2006).

It appears that aging is related to an overall net decrease in heterochromatin, but accompanied by an increase in heterochromatin at specific loci, which leads to the changed gene expression patterns related to aging.

AGING GENES

"Aging genes" induce their effects by slowing or stopping biochemical metabolic pathways, which are important to the aging process, include the insulin/IGF-1 and targets of rapamycin (rTOR) pathways, although the mechanisms of how these pathways control aging remain unclear (Park and Yo, 2013). Genes undoubtedly contribute to longevity, though the extents of genetic versus nongenetic factor contributions to aging are unclear (Cournil and Kirkwood, 2001). The human's twin observations estimate that genes contribute to about 20%–30% of aging (Barzilai et al., 2012); however, it is not clear how genes are responsible for this variability. Researchers have confirmed several gene mutations affecting life span in many model organisms. In roundworms, genetic mutations that double the life span provide clear evidence that genes influence aging and longevity. However, there is no single identified gene completely controlling aging, which is consistent with the concept that the genetic control of aging is multifactorial.

THE GENE REGULATION THEORY OF AGING

This theory proposes that senescence results from changes in gene expression (Kanungo, 1975). Though it is clear that the expression of many genes change with age (Zou et al., 2000; Pletcher et al., 2002; Weindruch et al., 2002), it is unlikely that selection could act on genes that facilitate aging directly (Kirkwood, 2002). Rather, life span is affected by the selection of genes that enhance longevity. The genome-wide transcriptional changes with age in several model organisms have been analyzed using DNA microarrays (Pletcher et al., 2002; Weindruch et al., 2002). Genome-level analysis allows researchers to compile a transcriptional fingerprint of "normal" senescence. These data can be compared with interventions that slow or facilitate aging, perhaps enabling the identification of gene expression changes that are related to the senescence process (Zou et al., 2000; Weindruch et al., 2001).

RNA POPULATION MODEL

OVERVIEW

In 2014, Wang et al. (2014) proposed the RNA population model of aging. The RNA population in a cell comprises all of its transcriptional RNAs. The RNAs produced from a single transcription site (including multiple genes) make up an RNA subpopulation that forms a local network via RNA repetitive sequence complementation. In addition, an RNA subpopulation may also include a small number of RNAs produced from other transcription sites. Interactions between DNA and RNA in the local network disturb the tight packing of chromatin and maintain gene activation. DNA transcription resulting from the interaction between DNA and the RNA network produces an RNA population, which in turn accelerates DNA transcription via changing chromatin packing. DNA transcription can be altered by changes in the interactions between the RNA population and DNA. Such changes are the foundation of aging and differentiation. If the interaction between the RNA population and DNA runs a cyclical course, it would result in immortal cells. Figure 6.2 shows the RNA population model. Simply speaking, the core of the RNA population model is that DNAs transcribe into RNAs, which in turn promote DNA transcription.

THE BIOLOGICAL PHENOMENA EXPLAINED BY THE RNA POPULATION MODEL

Figure 6.3 shows that the RNA population model accounts for aging, differentiation, reproduction, and epigenetics. For ease of description, the RNA population model divides genes into grades A–D, based on gene expression changes and chromatin remodeling within cell cycles. This gene activating process is illustrated from the zygote (grade A genes), to the triploblast (grade B), to various primordiums (grade C), and to formation of an individual (grade D) (Figure 6.3a). Gene expression is the process of continuous change (activation and silent) during individual development, maturation, and senescence. Activation of each grade of genes requires specific activating material (mainly the RNA population), which is produced from higher-grade genes. The expression of lower-grade

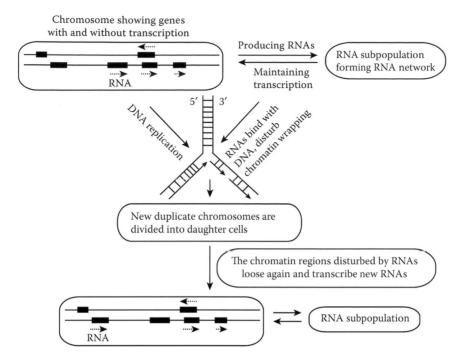

FIGURE 6.2 The RNA population model. The RNA population comprises all transcriptional RNA in a cell and can be divided into many different RNA subpopulations that interact with DNA and disturb chromatin tight packing to maintain gene activation. DNA transcription produces the RNA population, which in turn affects DNA transcription. The RNA population alters chromatin packing, thereby exerting an epigenetic influence on the process of cell division.

genes will stop or change without the RNA population produced by higher-grade genes because the changing RNA population results in changes in chromatin conformation and gene expression. For example, grade A genes produce the RNA population that will activate grade B genes. In this process, cellular internal and external environmental factors can also affect the expression of the grade B genes, which in turn may drive cells into differentiation (B1, B2, B3, etc.). The resulting differentiated cells produce a different RNA population that activates lower-grade genes. Thus, gene expression in differentiated cells is due to activation of higher-grade genes. If higher-grade genes are still expressed when lower-grade genes are activated, differentiation cannot be completed. Alternatively, termination of high-grade gene expression decreases cellular proliferation and differentiation, leading to cellular senescence. Likewise, in the process of cell proliferation, epigenetic information is maintained because the RNA population affects chromatin conformation to produce another RNA population.

Postnatal survival-associated functions (e.g., intelligence, audition, vision, immunocompetence, etc.) increase in early life stages and decrease with age (Figure 6.3b). Many of the changes in gene expression that occur during the aging process originate during the period of juvenile growth deceleration (Terao et al., 2002; Zhao et al., 2002; Ihm et al., 2007; Chang et al., 2008; Finkielstain et al., 2009; Lui et al., 2010). According to the RNA population model, from a zygote to an aged individual, changes in gene expression occur continuously, allotting a specific period of time that is suitable for survival.

When the RNA population produced as a result of certain chromatin conformation results in production of an RNA population identical to itself, the cells become immortal and oncogenic (Figure 6.3c). Aging of multicellular eukaryotes is compulsory. In the chain of evolution, the early organisms (prokaryotes) are not aging, yet aging occurs in multicellular eukaryotic organisms where differentiation has emerged. Thus, aging should have an important biological significance.

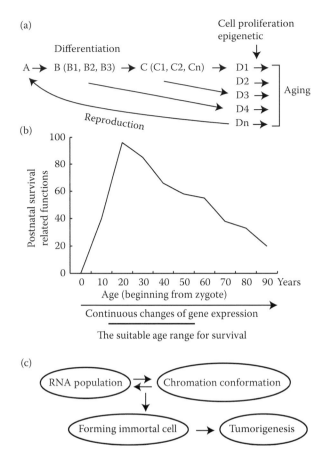

FIGURE 6.3 The RNA population model accounts for aging, differentiation, reproduction, epigenetics, and tumorigenesis. (a) All eukaryotic genes can be divided into several grades such as A–D. Activation of each grade requires a specific activating RNA population that is produced by the higher-grade genes. Internal and external environmental factors promote cellular differentiation (e.g., B1, B2, and B3). Differentiation requires that expression of higher-grade genes decreases after lower-grade genes are activated. However, after the expression of higher-grade genes decreases, the expression of lower-grade genes eventually decreases and changes, which results in aging. In the process of cellular proliferation, epigenetic information is maintained because the RNA population affects chromatin conformation that results in production of another RNA population. For example, sperm–egg binding activates grade A genes; in turn, the production of grade A genes activates grade B genes and then the expression of grade A genes will stop. In a similar way, grade B genes will stop after they activate grade C genes and grade C genes will stop after they activate grade D genes. The expression of grade D genes produces postnatal survival-related functions. Since the activation of grade D genes is the result of grade C genes activating, so the stopping of grade C genes will result in the expression of grade D genes stopping or changing (aging). The sperm and egg binding in adult individuals will activate grade A genes again. Life process is a cycle. (b) Postnatal survival related functions (e.g., intelligence, audition, vision) alter with age. Abscissa, age and ordinate, functions. The functions increase during the early life stage with age and later decrease with age. According to the RNA population model, from a zygote to an aged individual, change is a continuous process during which there is a set period of time when overall gene expression is conducive to survival. (c) During the process of development, the RNA population regulates chromatin conformation, which results in production of another RNA population. Immortalization of the cell and subsequent tumorigenesis occurs when an RNA population induces a chromatin conformation that results in the production of an RNA population identical to that of the initial population.

ANTIAGING STRATEGIES

Multicellular organisms attain immortality via reproduction. However, each individual of the organisms appears aging and death (see "The Important Phenomena of Aging" and the "RNA Population Model"). Although the transformed cells escape aging, this phenomenon does not belong to the normal physiological process (see "The Important Phenomena of Aging" and the "RNA Population Model"). Changing life conditions of organisms or cells can affect their life span (see "The Important Phenomena of Aging"). We proposed the antiaging strategies based on the genetic program theories of aging as below:

CHANGING DNA

The individuals either with strong or weak functions can survive since selective pressure decreases and intraspecific variation increases in today's society. In evolutionary history, humans are at a stage of low mortality. A new notion was proposed according to evolutionary theory: selective pressure will increase with environmental pollution, climate change, etc., which leads to increasing mortality of individuals with low functions and shortening of the human life span. The authors in this chapter offer proposals to solve this issue: (1) eugenics; (2) single-celled sequencing of embryos (keep advantageous genome); and (3) DNA editing (repair detrimental genes or add the advantageous DNA sequences). So, the humans will regain their evolution and weed out individuals carrying detrimental genes and increase the number of individuals with advantageous genes no longer through natural selection.

CELLULAR REPROGRAMMING

Recent advances in cellular reprogramming technologies have enabled detailed analyses of the aging process, often involving cell types derived from aged individuals, or patients with premature aging syndromes (Ocampo et al., 2016).

In the RNA population model, local transcription sites produce RNA subpopulations that decide chromatin conformation. So, cell reprogramming requires that specific RNAs (noncoding RNAs) act on the local transcription site. Owing to the regional distribution of the RNA subpopulation, cell reprogramming needs to develop new methods of changing artificially specific RNA subpopulations. Aging (including age-related diseases) is the result of gene expression changes, which in turn are due to a change in RNA subpopulations. Modulating artificially RNA subpopulations of aged cells may result in antiaging.

CONCLUSION

Aging is a continuous process in which genetic potentials persistently reduce. The survival functions increase with an increase in age during the development stage, and decline with an increase in age during the senescent stage. The aging process can prevent tumorigenesis and is an indispensable step for differentiation, which suggests that aging has an important biological significance. There are two kinds of theories about aging timers—genetic program and damage theories, and also theories integrating the two types. To explore the aging timers, scientists need to distinguish between the influencing factors of aging and the nature of aging (Chen et al., 2007). In the development-maturity stage, the genetic program plays a main role, and in the later period of the senescent stage, owing to declining physiological functions, damage plays a main role.

ACKNOWLEDGMENTS

This work was supported by grants from the National Natural Sciences Foundation of China (grant number 30873001), the Hebei Province Natural Science Foundation of China (H2013206101 and

C2011206043), and the Key Project of Hebei Province (08276101D-90) and Scientific Research Funding from the Department of Public Health of Hebei Province (grant No. 20100242).

REFERENCES

Albright JW, Makinodan T. Decline in the growth potential of spleen cog bone marrow stem cells of long lived aging mice. *J Exp Med* 1976; 144: 1204–1213.

Allard JB, Duan C. Comparative endocrinology of aging and longevity regulation. *Front Endocrinol (Lausanne)* 2011; 2: 75.

Allsopp RC, Harley CB. Evidence for a critical telomere length in senescent human fibroblasts. *Exp Cell Res* 1995; 219: 130–136.

Allsopp RC, Vaziri H, Patterson C et al. Telomere length predicts replicative capacity of human fibroblasts. *Proc Natl Acad Sci USA* 1992; 89: 10114–10118.

Artandi SE, DePinho RA. Telomeres and telomerase in cancer. *Carcinogenesis* 2010; 31: 9–18.

Austad SN. Retarded senescence in an insular population of opossums. *J Zool* 1993; 229: 695–708.

Balin AK, Fisher AJ, Anzelone M et al. Effects of establishing cell cultures and cell culture conditions on the proliferative life span of human fibroblasts isolated from different tissues and donors of different ages. *Exp Cell Res* 2002; 274(2): 275–287.

Barron AM, Pike CJ. Sex hormones, aging, and Alzheimer's disease. *Front Biosci (Elite Ed)* 2012; 4: 976–997.

Barzilai N, Guarente dL, Kirkhope TB et al. The place of genetics in ageing research. *Genetics* 2012; 13: 589–592.

Baskerville KA, Kent C, Personett D et al. Aging elevates metabolic gene expression in brain cholinergic neurons. *Neurobiol Aging* 2008; 29: 1874–1893.

Benayoun BA, Pollina EA, Brunet A. Epigenetic regulation of ageing: Linking environmental inputs to genomic stability. *Nat Rev Mol Cell Biol* 2015; 16(10): 593–610.

Benraiss A, Caubit X, Arsanto JP et al. Clonal cell cultures from adult spinal cord of the amphibian urodele *Pleurodeles waltl* to study the identity and potentialities of cells during tail regeneration. *Dev Dyn* 1996; 205(2): 135–149.

Bialystok E, Abutalebi J, Bak TH et al. Aging in two languages: Implications for public health. *Ageing Res Rev* 2016; 27: 56–60.

Bodnar AG, Ouellette M, Frolkis M et al. Extension of life-span by introduction of telomerase into normal human cells. *Science* 1998; 279: 349–352.

Brandt A, Krohne G, Grosshans J. The farnesylated nuclear proteins KUGELKERN and LAMIN B promote aging-like phenotypes in *Drosophila* flies. *Aging Cell* 2008; 7: 541–551.

Buehr M, Meek S, Blair K et al. Capture of authentic embryonic stem cells from rat blastocysts. *Cell* 2008; 135: 1287–1298.

Capuano F, Mulleder M, Kok R et al. Cytosine DNA methylation is found in *Drosophila melanogaster* but absent in *Saccharomyces cerevisiae*, *Schizosaccharomyces pombe*, and other yeast species. *Anal chem* 2014; 86: 3697–3702.

Cardinali DP, Esquifino AI, Srinivasan V et al. Melatonin and the immune system in aging. *Neuroimmunomodulation* 2008; 15(4–6): 272–278.

Chang M, Parker EA, Muller TJ et al. Changes in cell-cycle kinetics responsible for limiting somatic growth in mice. *Pediatr Res* 2008; 64: 240–245.

Chatterjee S, Mondal S. Effect of regular yogic training on growth hormone and dehydroepiandrosterone sulfate as an endocrine marker of aging. *Evid Based Complement Alternat Med* 2014; 2014: 240581.

Chen JH, Hales CN, Ozanne SE. DNA damage, cellular senescence and organismal ageing: Causal or correlative? *Nucleic Acids Res* 2007; 35(22): 7417–7428.

Chou JP, Effros RB. T cell replicative senescence in human aging. *Curr Pharm Des* 2013; 19(9): 1680–1698.

Cournil A, Kirkwood TB. If you would live long, choose your parents well. *Trends Genet* 2001; 15(5): 233–235.

Cristofalo VJ. Cellular biomarkers of aging. *Exp Gerontol* 1988; 23(4–5): 297–307.

Cruickshanks HA, McBryan T, Nelson DM et al. Senescent cells harbour features of the cancer epigenome. *Nat Cell Biol* 2013; 15: 1495–1506.

da Costa JP, Vitorino R, Silva GM et al. A synopsis on aging—Theories, mechanisms and future prospects. *Ageing Res Reviews* 2016; 29: 90–112.

Dang W, Steffen KK, Perry R et al. Histone H4 lysine 16 acetylation regulates cellular lifespan. *Nature* 2009; 459: 802–807.

Daniel CW, Daniel CW, De Ome KB et al. The in vivo life span of normal and preneoplastic mouse mammary glands: A serial transplantation study. *Proc Natl Acad Sci USA* 1968; 61(1): 53–60.

Day K, Waite LL, Thalacker-Mercer A et al. Differential DNA methylation with age displays both common and dynamic features across human tissues that are influenced by CpG landscape. *Genome Biol* 2013; 14: R102.

Dechat T, Pfleghaar K, Sengupta K et al. Nuclear lamins: Major factors in the structural organization and function of the nucleus and chromatin. *Genes Dev* 2008; 22: 832–853.

de Magalhães JP. Is mammalian aging genetically controlled? *Biogerontology* 2003; 4(2): 119–120.

de Magalhães JP. Programmatic features of aging originating in development: Aging mechanisms beyond molecular damage? *FASEB J* 2012; 26(12): 4821–4826.

Devore EE, Harrison SL, Stone KL et al. Association of urinary melatonin levels and aging-related outcomes in older men. *Sleep Med* 2016; 23: 73–80.

de Witte LC, Stöcklin J. Longevity of clonal plants: Why it matters and how to measure it. *Ann Bot* 2010; 106(6): 859–870.

di Fagagna FD, Reaper PM, Clay-Farrace L et al. A DNA damage checkpoint response in telomere-initiated senescence. *Nature* 2003; 426: 194–198.

Effros RB. Insights on immunological aging derived from the T lymphocyte cellular senescence model. *Exp Gerontol* 1996; 31(1–2): 21–27.

Erusalimsky JD, Kurz DJ. Cellular senescence in vivo: Its relevance in ageing and cardiovascular disease. *Exp. Gerontol* 2005; 40: 634–642.

Evans MJ, Kaufman MH. Establishment in culture of pluripotential cells from mouse embryos. *Nature* 1981; 292: 154–156.

Feser J, Tyler J. Chromatin structure as a mediator of aging. *FEBS Lett* 2011; 585(13): 2041–2048.

Finkielstain GP, Forcinito P, Lui JC et al. An extensive genetic program occurring during postnatal growth in multiple tissues. *Endocrinology* 2009; 150: 1791–1800.

Fu Y, Luo GZ, Chen K et al. N6-methyldeoxyadenosine marks active transcription start sites in *Chlamydomonas*. *Cell* 2015; 161: 879–892.

Greenberg RA. Telomeres, crisis and cancer. *Curr Mol Med* 2005; 5: 213–218.

Greer EL, Blanco MA, Gu L et al. DNA methylation on N^6-adenine in *C. elegans*. *Cell* 2015; 161: 868–878.

Hadem IK, Sharma R. Differential regulation of hippocampal IGF-1-associated signaling proteins by dietary restriction in aging mouse. *Cell Mol Neurobiol* 2017; 37(6): 985–993.

Haithcock E, Dayani Y, Neufeld E et al. Age-related changes of nuclear architecture in *Caenorhabditis elegans*. *Proc Natl Acad Sci USA* 2005; 102: 16690–16695.

Haldane JBS. *New Paths in Genetics*. London: Allen & Unwin, 1941.

Hargreaves DC, Horng T, Medzhitov R. Control of inducible gene expression by signal-dependent transcriptional elongation. *Cell* 2009; 138: 129–145.

Harman D. Free radical theory of aging: An update. *Ann NY Acad Sci* 2006; 1067: 10–21.

Hayflick L. Current theories of biological aging. *Fed Proc* 1975; 34(1): 9–13.

Hayflick L, Moorhead PS. The serial cultivation of human diploid cell strains. *Exp Cell Res* 1961; 25: 585–621.

Hedger RD, Næsje TF, Fiske P et al. Ice-dependent winter survival of juvenile Atlantic salmon. *Ecol Evol* 2013; 3(3): 523–535.

Herbig U, Ferreira M, Condel L et al. Cellular senescence in aging primates. *Science* 2006; 311: 1257.

Holliday R. The current status of the protein error theory of aging. *Exp Gerontol* 1996; 31: 449–452.

Holmes DJ, Ottinger MA. Birds as long-lived animal models for the study of aging. *Exp Gerontol* 2003; 38(11–12): 1365–1375.

Hosaka Y, Kitamoto A, Shimojo M et al. Generation of microglial cell lines by transfection with simian virus 40 large T gene. *Neurosci Lett* 1992; 141(2): 139–142.

Huang XQ, Wang J, Liu JP et al. hTERT extends proliferative lifespan and prevents oxidative stress-induced apoptosis in human lens epithelial cells. *Invest Ophthalmol Vis Sci* 2005; 46: 2503–2513.

Hunt SC, Chen W, Gardner JP et al. Leukocyte telomeres are longer in African Americans than in whites: The National Heart, Lung, and Blood Institute Family Heart Study and the Bogalusa Heart Study. *Aging Cell* 2008; 7: 451–458.

Ihm SH, Moon HJ, Kang JG et al. Effect of aging on insulin secretory function and expression of beta cell function-related genes of islets. *Diabetes Res Clin Pract* 2007; 77(Suppl 1): S150–S154.

Jeyapalan JC, Ferreira M, Sedivy JM et al. Accumulation of senescent cells in mitotic tissue of aging primates. *Mech Ageing Dev* 2007; 128: 36–44.

Jin C, Li J, Green CD et al. Histone demethylase UTX-1 regulates *C. elegans* life span by targeting the insulin/IGF-1 signaling pathway. *Cell Metab* 2011; 14: 161–172.

Johnson TE. Increased life-span of age-1 mutants in *Caenorhabditis elegans* and lower Gompertz rate of aging. *Science* 1990; 249: 908–912.

Johnston RK, Snell TW. Moderately lower temperatures greatly extend the lifespan of *Brachionus manjava-cas* (Rotifera): Thermodynamics or gene regulation? *Exp Gerontol* 2016; 78: 12–22.

Kaeberlein M, McVey M, Guarente L. The SIR2/3/4 complex and SIR2 alone promote longevity in *Saccharomyces cerevisiae* by two different mechanisms. *Genes Dev* 1999; 13: 2570–2580.

Kanungo MS. A model for ageing. *J Theor Biol* 1975; 53: 253–261.

Karlseder J, Broccoli D, Dai YM et al. p53- and ATM-dependent apoptosis induced by telomeres lacking TRF2. *Science* 1999; 283: 1321–1325.

Katz DJ, Edwards TM, Reinke V et al. A *C. elegans* LSD1 demethylase contributes to germline immortality by reprogramming epigenetic memory. *Cell* 2009; 137(2): 308–320.

Kenyon C. The genetics of ageing. *Nature* 2010; 464(25): 504–512.

Kim KS, Kang KW, Seu YB et al. Interferon-gamma induces cellular senescence through p53-dependent DNA damage signaling in human endothelial cells. *Mech Ageing Dev* 2009; 130: 179–188.

Kirkwood TB. New science for an old problem. *Trends Genet* 2002; 18: 441–442.

Kondratova AA, Kondratov RV. Circadian clock and pathology of the ageing brain. *Nat Rev Neurosci* 2012; 13(5): 325–335.

Kowald A, Kirkwood TB. A network theory of ageing: The interactions of defective mitochondria, aberrant proteins, free radicals and scavengers in the ageing process. *Mutat Res* 1996; 316(5–6): 209–236.

Krishnan V, Chow MZ, Wang Z et al. Histone H4 lysine 16 hypoacetylation is associated with defective DNA repair and premature senescence in Zmpste24-deficient mice. *Proc Natl Acad Sci USA* 2011; 108: 12325–12330.

Kumar V, Wingfield JC, Dawson A et al. Biological clocks and regulation of seasonal reproduction and migration in birds. *Physiol Biochem Zool* 2010; 83(5): 827–835.

Larson K, Yan SJ, Tsurumi A et al. Heterochromatin formation promotes longevity and represses ribosomal RNA synthesis. *PLoS Genet* 2012; 8: 1002473.

Leung JK, Pereira-Smith OM. Identification of genes involved in cell senescence and immortalization: Potential implications for tissue ageing. *Novartis Found Symp* 2001; 235: 105–110.

Li P, Tong C, Mehrian-Shai R et al. Germline competent embryonic stem cells derived from rat blastocysts. *Cell* 2008; 135: 1299–1310.

Lin KW, Yan J. The telomere length dynamic and methods of its assessment. *J Cell Mol Med* 2005; 9: 977–989.

Lipsky MS, King M. Biological theories of aging. *Dis Mon* 2015; 61(11): 460–466.

Lister R, Mukamel EA, Nery JR et al. Global epigenomic reconfiguration during mammalian brain development. *Science* 2013; 341: 1237905.

Liu M, Gu Y, Liu Y et al. Establishment and characterization of two cell lines derived from primary cultures of *Gekko japonicus* cerebral cortex. *Cell Biol Int* 2010; 34(2): 153–161.

López-Otín C, Blasco MA, Partridge L et al. The hallmarks of aging. *Cell* 2013; 153: 1194–1217.

Lui JC, Chen W, Barnes KM et al. Changes in gene expression associated with aging commonly originate during juvenile growth. *Mech Ageing Dev* 2010; 131: 641–649.

Macieira-Coelho A. Metabolism of aging cells in culture. *Gerontology* 1976; 22: 3–8.

Maegawa S, Hinkal G, Kim HS et al. Widespread and tissue specific age-related DNA methylation changes in mice. *Genome Res* 2010; 20: 332–340.

Makarov VL, Hirose Y, Langmore JP. Long G tails at both ends of human chromosomes suggest a C strand degradation mechanism for telomere shortening. *Cell* 1997; 88: 657–666.

Martin GM, Sprague CA, Epstein GJ. Replicative lifespan of cultivated human cells effects of donor age, tissue, and genotype. *Lab Invest* 1970; 23: 86–92.

Martin GR. Isolation of a pluripotent cell line from early mouse embryos cultured in medium conditioned by teratocarcinoma stem cells. *Proc Natl Acad Sci USA* 1981; 78: 7634–7638.

Mason M, Schuller A, Skordalakes E. Telomerase structure function. *Curr Opin Struct Biol* 2011; 21: 92–100.

Maures TJ, Greer EL, Hauswirth AG et al. The H3K27 demethylase UTX-1 regulates *C. elegans* lifespan in a germline-independent, insulin-dependent manner. *Aging Cell* 2011; 10: 980–990.

Mizuno T, Yamashino T. The plant circadian clock looks like a traditional Japanese clock rather than a modern Western clock. *Plant Signal Behav* 2015; 10(12): e1087630.

Moskalev AA, Aliper AM, Smit-McBride Z et al. Genetics and epigenetics of aging and longevity. *Cell Cycle* 2014; 13(7): 1063–1077.

Mostoslavsky R, Chua KF, Lombard DB et al. Genomic instability and aging-like phenotype in the absence of mammalian SIRT6. *Cell* 2006; 124: 315–329.

Mun GI, Lee SJ, An SM et al. Differential gene expression in young and senescent endothelial cells under static and laminar shear stress conditions. *Free Radic Bio Med* 2009; 47: 291–299.

Munro J, Steeghs K, Morrison V et al. Human fibroblast replicative senescence can occur in the absence of extensive cell division and short telomeres. *Oncogene* 2001; 20(27): 3541–3552.

Myles S, Boyko AR, Owens CL et al. Genetic structure and domestication history of the grape. *Proc Natl Acad Sci USA* 2011; 108(9): 3530–3535.

Ni Z, Ebata A, Alipanahiramandi E et al. Two SET domain containing genes link epigenetic changes and aging in *Caenorhabditis elegans*. *Aging Cell* 2012; 11: 315–325.

Njajou OT, Hsueh WC, Blackburn EH et al. Association between telomere length, specific causes of death, and years of healthy life in health, aging, and body composition, a population-based cohort study. *J Gerontol Ser A Biol Sci Med Sci* 2009; 64: 860–864.

Ocampo A, Reddy P, Izpisua Belmonte JC. Anti-aging strategies based on cellular reprogramming. *Trends Mol Med* 2016; 22(8): 725–738.

Oeseburg H, de Boer RA, van Gilst WH et al. Telomere biology in healthy aging and disease. *Pflugers Arch* 2010; 459: 259–268.

Ogata T, Ayusawa D, Namba M et al. Chromosome 7 suppresses indefinite division of nontumorigenic immortalized human fibroblast cell lines KMST-6 and SUSM-1. *Mol Cell Biol* 1993; 13(10): 6036–6043.

Oh J, Lee YD, Wagers AJ. Stem cell aging: Mechanisms, regulators and therapeutic opportunities. *Nat Med* 2014; 20(8): 870–880.

Olovnikov AM. Principle of marginotomy in template synthesis of polynucleotides. *Dokl Akad Nauk SSSR* 1971; 201(6): 1496–1499.

Olovnikov AM. A theory of marginotomy. The incomplete copying of template margin in enzymic synthesis of polynucleotides and biological significance of the phenomenon. *J Theor Biol* 1973; 41(1): 181–190.

Paradis V, Youssef N, Dargère D et al. Replicative senescence in normal liver, chronic hepatitis C, and hepatocellular carcinomas. *Hum Pathol* 2001; 32: 327–332.

Park DC, Yo SG. Aging. *Korean J Audiol* 2013; 17: 39–44.

Partridge L, Prowse N, Pignatelli P. Another set of responses and correlated responses to selection on age at reproduction in *Drosophila melanogaster*. *Proc R Soc Lond B Biol Sci* 1999; 266: 255–261.

Peleg S, Sananbenesi F, Zovoilis A et al. Altered histone acetylation is associated with age-dependent memory impairment in mice. *Science* 2010; 328: 753–756.

Pereira-Smith OM, Ning Y. Molecular genetic studies of cellular senescence. *Exp Gerontol* 1992; 27(5–6): 519–522.

Pereira-Smith OM, Smith JR. Genetic analysis of indefinite division in human cells: Identification of four complementation groups. *Proc Natl Acad Sci USA* 1988; 85(16): 6042–6046.

Petralia RS, Mattson MP, Yao PJ. Aging and longevity in the simplest animals and the quest for immortality. *Ageing Res Rev* 2014; 16: 66–82.

Pignolo RL, Masoro EJ, Nichols WW. Skin fibroblasts from aged Fischer 344 rats undergo similar changed in replicative life span but not immortalization with caloric restriction of donors. *Exp Cell Res* 1992; 201: 16–22.

Pletcher SD, Macdonald SJ, Marguerie R et al. Genome-wide transcript profiles in aging and calorically restricted *Drosophila melanogaster*. *Curr Biol* 2002; 12: 712–723.

Plikus MV, Van Spyk EN, Pham K et al. The circadian clock in skin: Implications for adult stem cells, tissue regeneration, cancer, aging, and immunity. *J Biol Rhythms* 2015; 30(3): 163–182.

Prinzinger R. Programmed aging: The theory of maximal metabolic scope. How does the biological clock tick? *EMBO Rep* 2005; 6(Spec No): S14–19.

Pu M, Ni Z, Wang M et al. Trimethylation of Lys36 on H3 restricts gene expression change during aging and impacts life span. *Genes Dev* 2015; 29: 718–731.

Ren F, Li CY, Xi HJ et al. Estimation of human age according to telomere shortening in peripheral blood leukocytes of Tibetan. *Am J Forensic Med Pathol* 2009; 30: 252–255.

Röhme D. Evidence for a relationship between longevity of mammalian species and life spans of normal fibroblasts in vitro and erythrocytes in vivo. *Proc Natl Acad Sci USA* 1981; 78(8): 5009–5013.

Rose M, Charlesworth B. A test of evolutionary theories of senescence. *Nature* 1980; 287: 141–142.

Rossi DJ, Bryder D, Weissman IL. Hematopoietic stem cell aging: Mechanism and consequence. *Exp Gerontol* 2007; 42(5): 385–390.

Ryan JM, Ostrow DG, Breakefield XO et al. A comparison of the proliferative and replicative lifespan kinetics of cell cultures derived from monozygotic twins. *In Vitro* 1981; 17: 20–27.

Salmon AB, Richardson A, Pérez VI. Update on the oxidative stress theory of aging: Does oxidative stress play a role in aging or healthy aging? *Free Radic Biol Med* 2010; 48, 642–655.

Sato M, Shin-ya K, Lee JI et al. Human telomerase reverse transcriptase and glucose-regulated protein 78 increase the life span of articular chondrocytes and their repair potential. *BMC Musculoskelet Disord* 2012; 13: 51.

Scaffidi P, Misteli T. Lamin A-dependent nuclear defects in human aging. *Science* 2006; 312: 1059–1063.

Schneider EL, Mitsui Y. The relationship between in vitro cellular aging and in vivo human age. *Proc Natl Acad Sci USA* 1976; 73(10): 3584–3588.

Semsei I. On the nature of aging. *Mech Ageing Dev* 2000; 117(1–3): 93–108.

Sen P, Dang W, Donahue G et al. H3K36 methylation promotes longevity by enhancing transcriptional fidelity. *Genes Dev* 2015; 29: 1362–1376.

Sergiev PV, Dontsova OA, Berezkin GV. Theories of aging: An ever-evolving field. *Acta Naturae* 2015; 7(1): 9–18.

Simpson VJ, Johnson TE, Hammen RF. *Caenorhabditis elegans* DNA does not contain 5-methylcytosine at any time during development or aging. *Nucleic Acids Res* 1986; 14: 6711–6719.

Strehler B, Hirsch G, Gusseck D et al. Codon-restriction theory by aging and development. *J Theor Biol* 1971; 33(3): 429–474.

Sun D, Luo M, Jeong M et al. Epigenomic profiling of young and aged HSCs reveals concerted changes during aging that reinforce self-renewal. *Cell Stem Cell* 2014; 14: 673–688.

Swim HE, Parker RF. Culture characteristics of human fibroblasts propagated serially. *Am J Hyg* 1957; 66: 235–243.

Takubo K, Aida J, Izumiyama-Shimomura N et al. Changes of telomere length with aging. *Geriatr Gerontol Int* 2010; 10: S197–S206.

Tepp K, Timohhina N, Puurand M et al. Bioenergetics of the aging heart and skeletal muscles: Modern concepts and controversies. *Ageing Res Rev* 2016; 28: 1–14.

Terao A, Apte-Deshpande A, Dousman L et al. Immune response gene expression increases in the aging murine hippocampus. *J Neuroimmunol* 2002; 132: 99–112.

Toiber D, Erdel F, Bouazoune K et al. SIRT6 recruits SNF2H to DNA break sites, preventing genomic instability through chromatin remodeling. *Mol Cell* 2013; 51: 454–468.

Tominaga K, Olgun A, Smith JR et al. Genetics of cellular senescence. *Mech Ageing Dev* 2002; 123(8): 927–936.

Vaiserman AM, Lushchak OV, Koliada AK. Anti-aging pharmacology: Promises and pitfalls. *Ageing Res Rev* 2016; 31: 9–35.

Van Meter M, Kashyap M, Rezazadeh S et al. SIRT6 represses LINE1 retrotransposons by ribosylating KAP1 but this repression fails with stress and age. *Nat Commun.* 2014; 5: 5011.

Villeponteau B. The heterochromatin loss model of aging. *Exp Gerontol* 1997; 32: 383–394.

Viña J, Borrás C, Miquel J. Theories of ageing. *IUBMB Life* 2007; 59(4–5): 249–254.

von Zglinicki T, Saretzki G, Docke W et al. Mild hyperoxia shortens telomeres and inhibits proliferation of fibroblasts: A model for senescence? *Exp Cell Res* 1995; 220: 186–193.

Wang X, Ma Z, Cheng J, Lv Z. A genetic program theory of aging using an RNA population model. *Ageing Res Rev* 2014; 13: 46–54.

Weindruch R, Kayo T, Lee CK et al. Microarray profiling of gene expression in aging and its alteration by caloric restriction in mice. *J Nutr* 2001; 131: 918S–923S.

Weindruch R, Kayo T, Lee CK et al. Gene expression profiling of aging using DNA microarrays. *Mech Ageing Dev* 2002; 123: 177–193.

Weinert BT, Timiras PS. Invited review: Theories of aging. *J Appl Physiol (1985)* 2003; 95(4): 1706–1716.

Wieser M, Stadler G, Jennings P et al. hTERT alone immortalizes epithelial cells of renal proximal tubules without changing their functional characteristics. *Am J Physiol Renal Physiol* 2008; 295: F1365–F1375.

Wilson VL, Jones PA. DNA methylation decreases in aging but not in immortal cells. *Science.* 1983; 220: 1055–1057.

Wolden-Hanson T, Marck BT, Matsumoto AM. Troglitazone treatment of aging Brown Norway rats improves food intake and weight gain after fasting without increasing hypothalamic NPY gene expression. *Exp Gerontol* 2002; 37: 679–691.

Wright WE, Brasiskyte D, Piatyszek MA et al. Experimental elongation of telomeres extends the lifespan of immortal x normal cell hybrids. *EMBO J* 1996; 15: 1734–1741.

Yamamoto M, Akazawa K, Aoyagi M et al. Changes in biological characteristics during the cellular aging of ligament fibroblasts derived from patients with prolapsus uteri. *Mech Ageing Dev* 2000; 115(3): 175–187.

Zhang G, Huang H, Liu D et al. N^6-methyladenine DNA modification in *Drosophila*. *Cell* 2015a; 161: 893–906.

Zhang W, Li J, Suzuki K et al. A Werner syndrome stem cell model unveils heterochromatin alterations as a driver of human aging. *Science* 2015b; 348(6239): 1160–1163.

Zhao, H., Patra, A., Yeh, CC et al. Effects of aging on growth factors gene and protein expression in the dorsal and ventrallobes of rat prostate. *Biochem Biophys Res Commun* 2002; 292: 482–491.

Zou S, Meadows S, Sharp L et al. Genome-wide study of aging and oxidative stress response in *Drosophila melanogaster*. *Proc Natl Acad Sci USA* 2000; 97: 13726–13731.

Zs-Nagy I. An attempt to answer the questions of theoretical gerontology on the basis of the membrane hypothesis of aging. *Adv Biosci* 1987; 64: 393–413.

7 The Origins of Aging
Multicellularity, Speciation, and Ecosystems

Michael A. Singer

CONTENTS

INTRODUCTION

The biology of aging is an area of intense research interest. This interest is driven by the proposition that aging represents the principal risk factor for the development of chronic degenerative diseases (e.g., neurodegenerative, metabolic syndromes, most cancers, and cardiovascular disease) that have almost reached epidemic proportions in modern humans. This view was recently summarized in an article by Kennedy et al. (2014). These investigators stated that since aging was the predominant risk factor for most chronic diseases, understanding the link between aging and pathological processes would provide scientists the opportunity to develop therapies which would be effective for multiple chronic diseases rather than targeting a single disease. However, this view is based upon a specific construct as to the nature of the aging process.

Within the evolutionary framework, natural selection acting at the level of the individual "chooses" phenotypes that maximize "Darwinian fitness" in the context of the environmental constraints experienced by that individual. Darwinian fitness is defined in terms of an organism's survival to reproductive maturity. If a selected adaptive phenotype has a heritable basis it is passed on to subsequent generations. Within this framework, the environmental context is primarily a passive player in the evolutionary process and heritability is understood to be through genetic (Mendelian) inheritance.

Once, reproductive maturity is attained, the strength of natural selection declines. There is less investment into somatic repair mechanisms and genetic and biological damage accumulate. The organism ages and ultimately dies. The mechanistic extension of this construct has focused on the specific biological processes, dysregulation of which is responsible for age-associated pathologies (Lopez-Otin et al. 2013; Kennedy et al. 2014). Accumulation of age-associated pathologies is the basis for chronic degenerative diseases.

However, organisms, both unicellular and multicellular, do not live in isolation; organisms live within communities of their own species as well as in ecosystems composed of mixtures of diverse species. Communities and ecosystems are acted upon by multilevel natural selection and it is through the action of multilevel selection that aging has evolved as an adaptive species-specific developmental life stage. Furthermore, the environment is not a passive player in the evolutionary process. Organisms modify their environments through a process of niche construction (Laland et al. 2015) and the modified environment feeds back and influences the evolutionary process itself. It is now recognized that culture represents an important nongenetic form of inheritance and for humans gene–culture coevolution is considered the primary driver of evolutionary change. Within this larger view, aging is not the stochastic accumulation of biological damage, but an adaptive species-specific developmental program. Within a given species, there is considerable variation in the aging trajectory between individuals and in the human species, chronic diseases are not "pathologies" but just manifestations of this variation in aging across individuals (Singer 2013, 2015).

Although aging is a familiar concept, it lacks a singular definition. In terms of Darwinian fitness, aging is defined using a measure of age-specific mortality and/or a measure of age-specific reproduction, for example, number of offspring. In this chapter, aging will be defined as a progressive time-dependent decline in the physiological functions of an organism ultimately resulting in death of that organism. I have also taken the liberty of using programmed non-predatory, noninfectious death as a proxy for aging.

UNICELLULAR ORGANISMS AND THE EMERGENCE OF MULTICELLULARITY

UNICELLULAR ORGANISMS

Aging and programmed death have been well described in unicellular organisms (Durand et al. 2016; Singer 2016) and as pointed out by Durand et al. (2016), programmed death in single cell organisms can increase the fitness of others in the community. For example, the growth rate of *Dunalielia salina* (a unicellular chlorophyte) was greater when cultured with the bacterium *Halobacterium salinarum* than when cultured alone. The growth enhancement of the mixed culture was due to the processing of nutrients released from dying *D. salina* cells by the bacterium and then the reutilization of these processed nutrients by the growing *D. salina* population. This example describes how programmed cell death can forge an interaction between two different species which then become physiologically connected. The evidence reviewed by Durand et al. (2016) strongly suggests that programmed cell death made possible complex interactions between individual organisms and that the development of such complex interactions was a necessary step in the evolution of multicellularity (Durand et al. 2016).

Experimental studies have underscored the adaptive nature of programmed death in the unicellular world. Durand et al. (2011) used the green alga *Chlamydomonas reinhardtii* to compare the effects of programmed and non-programmed death on population growth. Programmed death was induced by heat stress (heating cells to 50° C) while non-programmed death was induced by sonication with cell lysis. Supernatants were collected from control cultures as well as from cultures of *C. reinhardtii* exposed to heat stress or sonication. The investigators measured the effects of these different supernatants on the growth characteristics of *C. reinhardtii* populations. Population growth was greater when *C. reinhardtii* was cultured in a medium supplemented with supernatant from cultured "programmed death" cells compared to controls. In contrast, supernatant from cultures exposed to sonication significantly decreased population growth.

In summary, substances released by cells undergoing programmed death were beneficial to the growth of neighboring cells, whereas substances released by cells undergoing non-programmed death (sonication induced lysis) were detrimental to the growth of neighboring cells. As Durand et al. (2011) concluded, the manner in which an organism dies can have beneficial or detrimental effects on the fitness characteristics of its neighbors.

As an extension of this study, Durand et al. (2014) examined whether the fitness effects of programmed death were species specific. Programmed death was induced in *C. reinhardtii* using heat stress. The supernatant from the culture medium was collected and tested for its effects on the population growth of two other species of *Chlamydomonas*. As in the previous experiments, supernatant from cultures exposed to heat stress increased the population growth of the same strain of *C. reinhardtii*. In contrast, supernatant added to the culture medium of two other species of *Chlamydomonas* caused an inhibition of population growth. Supernatant added to the culture medium of *C. reinhardtii*, same species but a different strain, had no effect (positive or negative) on population growth. Hence, it appears that the fitness effects of programmed death are most apparent in unicellular organisms that are genetically related, that is, the same species and strain.

Programmed cell death is recognized as an important component of the ecology of phytoplankton (Bidle 2015, 2016). Phytoplankton refers to a diverse group of marine unicellular organisms that engage in photosynthesis and form the foundation of the aquatic food web. These organisms are responsible for nearly half of global carbon-based primary production and about 2.2 billion years ago were influential in oxygenating the Earth's atmosphere (Bidle 2015, 2016). They have helped maintain the Earth's biogeochemical cycles and such maintenance requires growth and death cycles. Recently, it has been determined that one of the modes of death for these organisms is programmed death (Bidle 2015, 2016).

As summarized by Bidle (2016), the manner in which phytoplankton die largely determines the flow of nutrients and the fate of marine organisms. There are three main ecosystem pathways. One pathway involves ingestion of phytoplankton by zooplankton, which in turn are eaten by larger animals. A second pathway involves vertical sinking of aggregates of phytoplankton, zooplankton, and other polymeric material which transports nutrients to the deeper levels of the oceans. A third pathway involves programmed death or death by viral infestation. (However, viral infestation can also trigger programmed death.) Nutrients and other material released by the programmed death of phytoplankton are utilized by bacteria, which facilitates the rapid recycling of organic/inorganic elements in the upper layers of the ocean. In addition, there is evidence that the production of extracellular polysaccharide particles by phytoplankton is enhanced by programmed death (Berman-Frank et al. 2007). These particles are extremely sticky and can aggregate with bacteria, phytoplankton, and other matter to enhance the downward flux of organic material. Hence, the process of programmed death can influence the movement of nutrients and other elements through several mechanisms; by facilitating bacterial recycling of released material in the upper ocean layers and by facilitating particle/aggregate-mediated downward movement of organic material to deeper ocean layers. The third pathway is thought to be the most important of the three.

Although our understanding of the full adaptive nature of programmed death in phytoplankton is incomplete, the fact that phytoplankton have maintained the genes underwriting programmed death for billions of years indicates that programmed death must confer a selective advantage. Otherwise, the genes underwriting programmed death would have been eliminated a long time ago (Bidle 2016).

Programmed death is ubiquitous in unicellular organisms and probably emerged about 2.8 billion years ago (Bidle 2016); this phenotype has deep evolutionary roots. Programmed death evolved under the action of multilevel selection since this phenotype conferred a fitness advantage to conspecifics within the same community; a form of kin selection. Within more complex hierarchical structures such as marine ecosystems, programmed death within phytoplankton helps regulate the movement of nutrients within the different ocean layers.

The few studies described in the previous section underscore that the adaptive nature of non-predatory, noninfectious organismal death can only be appreciated when a species is viewed in the context of the community and ecosystem in which it lives.

MULTICELLULARITY

There is an extensive literature on the evolution of multicellular organisms and as reviewed by Durand et al. (2016), there is evidence that the emergence of programmed death in unicellular organisms was a prerequisite to the evolution of multicellular structures.

Ratcliff et al. (2012) developed an experimental model for the evolution of multicellularity. The test organism was the yeast *Saccharomyces cerevisiae* and the action of gravity was used to select for multicellular structures. Clusters of cells settle in liquid medium more rapidly than single cells and centrifugation was used to enhance the settling process. After about 60 settling-transfer steps, roughly spherical snowflake-like clusters of cells were obtained. These clusters arose through post-division adhesion of cells, that is, mother and daughter cells remain attached, rather than by aggregation of previously separate cells. The post-division adhesion mode of formation means that there is high relatedness between individual cells within the cluster. Snowflake clusters are capable of reproduction via the production of smaller daughter clusters, that is, multicellular propagules. Programmed cell death (apoptosis) was involved in the reproductive process. A small number of cells within the cluster undergo apoptosis and these dying/dead cells act as a breakpoint for the release of smaller multicellular propagules.

The experimental model of Ratcliff et al. (2012) represents a plausible representation of how multicellular structures might have evolved and presents an example of how functional specificity could have emerged in a multicellular structure. In the yeast snowflake cluster, a small number of cells undergo apoptosis and thereby facilitate the generation of daughter multicellular clusters. In essence, the cells that undergo programmed death perform a specific reproductive function. This role of programmed death in the reproduction of snowflake yeast foreshadows the well-documented functions of programmed cell death in complex multicellular organisms (Singer 2016). In complex multicellular organisms, programmed death is a critical component of embryogenesis, and tissue remodeling/repair. Programmed death is obviously destructive to the cell undergoing this process but clearly adaptive to the functioning of the community of cells that makes up a complex multicellular organism.

SPECIATION

There has been extensive research into the process of speciation beginning with the work of Darwin. The focus in this section is on the links between speciation and aging/death phenotypes as well as the role of developmental programs. What is a species? Perhaps, it is best to say that there is not one definition but several. Examples will suffice. A species can be defined as a group of organisms that can reproduce with other members of the group but are reproductively isolated from other groups. Another definition based upon phylogenetic principles defines a species as a group of organisms that share a common evolutionary history and ancestry.

AGING/DEATH PROGRAMS

As Darwin concluded, speciation is a slow process directed by natural selection acting on a mixture of heritable phenotypic traits and environmental constraints. Speciation has been divided into allopatric in which the diverging population is geographically isolated from other populations, that is, gene exchange is physically impossible and sympatric in which gene flow can occur between the diverging population and other populations (Via 2009). Allopatric speciation can be driven by strong or weak (divergent) selection since there is a physical barrier to gene exchange whereas sympatric speciation requires much stronger selection since gene flow is possible. However, this conceptualization assumes that interactions between species are so weak as to not exert selection on the process of speciation. In this construct, species should evolve in parallel in response to environmental pressures irrespective of which other species are present. We could term this construct the "single-species" theory.

Barraclough (2015) has reviewed the validity of the single-species theory and explored the role of species interactions in the evolutionary process. One form of species interaction is ecological sorting. Species within an ecosystem vary in their capacity to respond to changes in environmental conditions. When there is fluctuation in a variable, for example, temperature, those species with a standing trait variation relevant to the new conditions will increase in abundance while species less able to tolerate the new conditions will decline in abundance. However, there is evidence that other types of species interactions, even in the absence of abiotic changes, can influence evolution across communities. An example of biotic species interactions has been described by Whitam et al. (2006). These authors documented how phenotypic variation in a foundational species within an ecosystem can have significant effects on community and ecosystem phenotypes. Two examples illustrate this point. Poplar trees are a foundational species within an ecosystem. The investigators used crosses of poplar trees to create variation in the quantitative trait locus underlying the concentration of condensed tannins within different poplar types. Tannins are complex chemical substances found throughout the structures of trees and plants which prevent infestations by bacteria and fungi. In a series of studies performed in the field and in experimental gardens, the investigators found correlations between variation in poplar concentration of condensed tannins and various community phenotypes; composition of arthropod species living in poplar canopies, endophytic fungal species living in poplar bark, species composition of aquatic organisms feeding on leaves falling from the poplars into an adjacent stream. A change in arthropod species in the tree canopy can affect the abundance of avian predators. There were (negative) correlations between poplar condensed tannin concentrations and ecosystem phenotypes; microbial conversion of organic nitrogen to plant-preferred inorganic nitrogen and decomposition of leaves that had fallen into an adjacent stream.

In a second example, the investigators documented a negative correlation between the tannin concentration in the poplar bark and felling of poplars by the beaver *Castor canadensis*. Beavers selectively took down trees with a low concentration of tannins in the bark. After about five years, the ecosystem had about triple the number of poplars with a high concentration of tannins. The abundance of high-tannin trees inhibited the microbial conversion of organic to inorganic nitrogen. This effect resulted in reduced soil nutrients which led to enhanced fine root production by the poplars as compensation.

These studies underscore that speciation does not occur in a vacuum. The "single-species" model is too simplistic. Different species within a community and within the larger ecosystem interact and not simply through changes in environmental conditions. Variation in tannin production within the poplar tree population leads to variations in nutrient availability, composition of insect species living in the tree canopy, and even the abundance of avian predators feeding on those insects. Hence, biotic interactions between species can have widespread community and ecosystem effects across multiple trophic levels. These species interactions have emerged under the action of multilevel selection. In addition, the lifecycles (including aging/death programs) of species have evolved to promote fitness at the level of the individual, community, and ecosystem. Unfortunately, these types of genomic studies which integrate interactions within an ecosystem have only been attempted in ecosystems containing foundational plant species (Whitam et al. 2008).

Jones et al. (2014) collected data on the aging pattern of numerous species; 11 mammals, 12 non-mammalian vertebrates, 10 invertebrates, 12 vascular plants, and a green alga. For each species, relative mortality and fertility were calculated by dividing age-specific rates by the weighted mean for the whole adult population sampled. The age range used for these calculations was the adult life span defined as starting at the mean age of reproductive maturation and ending when only 5% of adults were still alive. This method allowed the investigators to compare widely different life spans across species. Their graphed data show that each species has its own unique aging trajectory, that is, mortality and fertility pattern over the life span of the species.

Ecosystems are highly regulated biological structures (see section on *Ecosystems*) and speciation takes place within ecosystems under the action of multilevel selection. The data presented by Jones et al. (2014) reflect species-specific aging patterns that have evolved within constraints presented by the organization of the ecosystem. Hence, speciation and aging/death programs are linked. As a new species emerges, the features of its specific lifecycle, including aging/death, will be shaped to fit the ecosystem it inhabits.

DEVELOPMENTAL PROGRAMS

Speciation is driven by the evolution of species-specific developmental programs. Members of a species share a repertoire of behaviors, a common morphology, and a similar lifecycle. For example, it is easy to distinguish a human from other primate species in all of these features. It is also clear that within a species there is extensive behavioral and morphologic variation; although humans differ from other primates, no two humans look or behave alike. Each species has a set of specific developmental programs different from those of other species, although the nature of these differences is not well understood. Thousands of species have had their genomes sequenced yet much remains unanswered (Kruglyak 2016). Which specific genomic sequences underlie the unique phenotypic traits of each species? If we are handed over a specific genome, can we tell whether it encodes a mouse, elephant, or a whale? Conversely, given the features of a specific species, can we design a genome that encodes these features? Although humans and chimpanzees look and behave quite differently, the genomes of these two species differ by only 4% in terms of nucleotide variations (Varki and Altheide 2005).

Another issue compounding the difficulty defining the genomic basis of species-specific developmental programs is the conservative nature of evolution. The evolutionary process proceeds by tinkering; building upon successful body designs already in existence. Two examples underscore this feature: Kachroo et al. (2015) found that almost half of essential yeast genes surveyed could be successfully replaced by orthologous human genes. Hence, despite the vast evolutionary distance between yeast and humans, a significant fraction of yeast/human protein coding genes perform much the same function in both organisms. In a second example, evolution of the wrist and digits in tetrapod vertebrates occurred by hijacking developmental processes used in fish fin formation (Saxena and Cooper 2016). In particular, the genes common to fin and digit formation are members of the Hox family of transcription factors.

Although much is known about developmental programs in general, much less is known about the species-specific aspects of these programs. It is these unknowns that will be highlighted in this section. A few examples will be presented; the hourglass model, preimplantation studies in mammals, and the stochastic nature of stem cell fate decisions.

An interesting aspect of developmental programs is the so-called hourglass feature. An hourglass model of embryonic development was proposed based upon the observation that morphological patterns during embryogenesis across species within a phylum diverged extensively early and late compared to the middle of the developmental process (Kalinka et al. 2010; Prud'homme and Gompel 2010). The middle period or waist was characterized by a conserved morphology such that species within each animal phylum showed a maximum similarity at this stage of development. The waist was named the phylotypic period and at this stage of development, the basic body plan of the phylum was laid down. The significance of this model was largely unknown primarily because it was based upon subjective comparisons of animal morphological likeness. However, two recent papers report molecular signatures that support the hourglass model.

Kalinka et al. (2010) measured differences in gene expression throughout embryogenesis across six distinct species of *Drosophila* separated by up to 40 million years of evolution. They compared at two-hour time intervals the expression of 3019 genes known to be involved in embryonic development. They applied a sophisticated statistical analysis of the data to extract two measures of divergence of gene expression; divergence (differences) in gene expression across the whole time course and temporal divergence which reflects differences in gene expression at specific time points. Temporal divergence of gene expression across the six species followed an hourglass pattern with maximum conservation (minimum divergence) at the embryonic stage corresponding to the arthropod phylotypic period. Genes that followed the hourglass pattern of expression were enriched for core organismal and cellular development whereas genes that did not follow an hourglass pattern were enriched for metabolic processes and immune functions. The findings of Kalinka et al. (2010) indicated that the expression of genes involved in key developmental processes was most evolutionarily constrained during mid-embryogenesis (phylotypic stage) by which time the "common" core body plan of species within the phylum was established. As development proceeded after the phylotypic stage, species within the phylum displayed significant divergence in the expression of these genes as their morphologies began to assume species-specific recognizable forms.

A complementary study was performed by Domazet-Loso and Tautz (2010). These investigators used phylostratigraphy to determine the evolutionary age of genes active during zebrafish embryogenesis. They developed a transcriptome age index (TAI) by combining at a given developmental stage, the evolutionary age of a gene with its expression level and summing these values over all genes expressed at the respective stage. Higher values of the TAI denoted younger transcriptomes.

The TAI profile across 60 stages of zebrafish development (unfertilized egg to aged animal) showed that genes of different evolutionary ages were expressed at different time points. The oldest genes were expressed at the embryonic stages equated with the zebrafish phylotypic period. Evolutionarily younger genes were expressed before and after this phylotypic period. This pattern of gene evolutionary age across developmental stages faithfully mirrored the hourglass model of morphological divergence. The evolutionary youngest gene set was observed in the young adult. However, as animals became older, they began to express evolutionary older genes.

Domazet-Loso and Tautz (2010) did a similar analysis for *Drosophila*, *Caenorhabditis elegans*, and the mosquito *Anopheles*. For all three organisms, a TAI pattern similar to that of zebrafish was found across development stages and into adulthood. The evolutionarily oldest genes were expressed during the phylotypic stage and with aging the three organisms expressed increasingly evolutionary older genes.

These two studies (Domazet-Loso and Tautz 2010; Kalinka et al. 2010) give legitimacy to the hourglass model and furnish a molecular mechanism. The phylotypic stage is a developmental period involving the expression of genes with a very old evolutionary origin and under tight evolutionary control. During the phylotypic stage, the shared core body plan characteristic of the phylum is established. Once this basic body plan is assembled, post-phylotypic development is characterized by extensive species-specific divergence in morphology. This divergence is driven by the expression of genes of more recent evolutionary origin and under less evolutionary constraint. As adult organisms age, they again begin to express genes with an older evolutionary origin. The findings in these two studies suggest some type of link or sharing between the set of genes which drive core morphological development during the phylotypic embryonic stage and the set of genes associated with aging in the adult organism.

Although these two studies provide a genomic foundation for the hourglass nature of development, we still do not understand how this hourglass phenomenon evolved or what selective advantage it provides.

Early stages of mammalian embryogenesis have been examined using the mouse as the model (Maitre et al. 2016; Plusa and Hadjantonakis 2016). The fertilized mouse egg (zygote) undergoes a series of divisions giving rise to blastomeres. At the eight cell stage, blastomeres become polarized by forming an apical domain. With asymmetric division, one daughter cell inherits the apical domain and remains polarized while the other daughter cell becomes apolar. Hence, the 16 cell stage will have both polarized and unpolarized blastomeres. The apical domain has lower levels of myosin than non-apical cortical areas and displays reduced contractility. In addition, unpolarized blastomeres show greater contractility than polarized blastomeres. This heterogeneity in blastomere contractility translates into cell fate decisions at the 32 cell stage. In the polar cells, the transcription factor yes-associated protein (YAP) enters the nucleus and specifies blastomeres to become trophectoderm (TE); the origins of the placenta. Apolar highly contractile blastomeres lack nuclear YAP, become internalized, and give rise to the inner cell mass (ICM); origin of the embryo proper. Hence, the contractile state of the cell regulates its location within the blastocyst, the subcellular localization of YAP, and the fate of the cell, TE or ICM. The experimental data of Maitre et al. (2016) tie together the polarity of the blastomere, its contractile state, and the sublocalization of YAP to account for cell specification. However, these experimental results may be specific to the mouse in view of the observations of Berg et al. (2011).

Berg et al. (2011) examined the time course of TE and ICM cell fate commitment in cattle and mice. By early blastocyst stages, cattle and mouse embryos are characterized by these two morphologically distinct cell populations; the ICM and outer TE. In the mouse, these two lineages are committed to their fate by the mid-blastocyst stage whereas in cattle, TE cells of the late blastocyst are still not fully committed to their fate. A number of regulatory factors are involved in cell fate "decisions" but two homeodomain DNA binding motif-containing transcription factors were particularly prominent, Oct4 and Cdx2. Berg et al. (2011) found that in mice, compared to cattle, the Oct4/Cdx2 regulatory network has been modified to allow earlier TE commitment and differentiation and hence earlier blastocyst implantation than in other mammals.

The observations of Berg et al. (2011) underscore that studies related to early developmental stages performed in the mouse appear to be species specific and not representative of early embryogenesis in other mammals.

Developmental programs rely on stem cells. The fertilized egg and its immediate progeny represent totipotential stem cells with the capacity to generate all the tissues of the fully formed organism. The so-called somatic stem cells have two general characteristics; self-renewal and the capacity to generate tissue-specific differentiated cells. Gurevich et al. (2016) and Rompolas et al. (2016) described two different mechanisms by which dividing stem cells can both renew the stem cell pool and give rise to committed progenitor cells. Muscle stem cells divide asymmetrically, that is, daughter cells have predetermined cell fates at birth with one remaining a stem cell while the other is committed to differentiation (Gurevich et al. 2016). Epidermal stem cells divide symmetrically, that is, at birth both daughter cells are initially uncommitted stem cells (Rompolas et al. 2016). Post-division, stochastic events determine the fate of each daughter cell, that is, whether the cell remains a stem cell or commits to differentiation. Asymmetric divisions result in direct coupling between the processes of self-renewal and lineage differentiation. Such is not the case for symmetric divisions. In this case, coupling between the processes of self-renewal and differentiation is indirect and mediated by post-mitotic stochastic events that determine daughter cell fates. Symmetrical compared to asymmetrical divisions reflect a plastic process that allows the balance between self-renewal and lineage differentiation to change in response to fluctuations in microenvironmental conditions.

Evidence is accumulating that the outcomes of somatic stem cell divisions are not pre-programmed, but are primarily stochastic. Stochasticity can arise in several ways. First, stochasticity might derive from the possibility that committed and uncommitted stem cells represent reversible interconvertible states (Greulich and Simons 2016). Alternatively, whether a stem cell divides symmetrically or asymmetrically is most likely a stochastic event rather than a preestablished cell autonomous process. However, one of the important elements regulating cell fate decisions is the cell cycle. The evidence is strong that the conditions necessary for a stem cell to commit to lineage differentiation are only present during G1 phase of the cell cycle (Julian et al. 2016).

Developmental programs are at the core of speciation. These programs are species specific and determine the behavioral repertoire, morphology, and lifecycle characteristic of a given species. They have evolved under the action of multilevel selection such as to balance the fitness of individuals, species, and the whole ecosystem in which diverse species reside (see section on *Ecosystems*).

However, there is still a very poor understanding of what makes developmental programs species specific. For example, we are still unable to define the genomic features that results in the formation of a chimpanzee versus a human. The hourglass model gives a broad view of embryogenesis and the existence of the phylotypic stage indicates that there is an underlying commonality to developmental programs across different species. In many experimental studies, the mouse has been used as the test mammal. However, the limited data available suggest that results obtained with the mouse may be specific to that species and not generalizable to other mammalian species. Deciphering the details of species-specific developmental programs is a key to understanding speciation since these programs encode the lifecycle characteristics, including aging/death, of a given species. Given that evolution is a tinkerer, the mechanisms that confer species specificity to developmental programs will probably be quite subtle.

Another important observation is that developmental programs are plastic; they only set the boundaries for the phenotypic characteristics of a given species. Within a species, there is considerable between-individual variation in morphology, behavior, and lifecycle stages. The molecular basis of this phenotypic plasticity is not understood although epigenetic mechanisms have been proposed as a possibility. However, there is still controversy as to whether epigenetic marks can be transmitted across generations (van Otterdijk and Michels 2016). Between-individual phenotypic variation is critical since it allows the species to adapt to and survive short-term environmental perturbations. Interestingly, large phenotypic plasticity has been described in clonal marine plants even though these organisms share a common genotype and have a low somatic mutation rate (Arnaud-Haond et al. 2012). It has been suggested that the particular genotypes retained by clonal organisms are ones that "encode" large phenotypic plasticity. Using this observation from clonal organisms, one could propose that during the evolution of species-specific developmental programs, only those programs that confer significant phenotypic plasticity are retained.

ECOSYSTEMS

Ecosystems represent a high order of biological organization comprising diverse interacting species and the ecological niches they inhabit. Ecosystems contain multiple trophic layers with plants and algae, so-called producers, being the foundational layer. The apical trophic layer is occupied by predators such as the lion. Species diversity and abundance are primary determinants of ecosystem functionality (Cardinale et al. 2012) and under the action of multilevel selection species found in the ecosystem have evolved lifecycles (including aging and death) which optimize the fitness of each species as well as the "fitness" of the ecosystem as a whole. Defining the fitness of an ecosystem is somewhat problematical but would include such attributes as species abundance and diversity, sustainability and resilience, that is, capacity of an ecosystem to restore itself following an environmental perturbation. Ecosystem fitness requires a balance in the relative population sizes of the constituent species and one tool for regulating such balance would be evolved species-specific aging and death programs. Some of the biotic interactions that have been documented between different species within an ecosystem were discussed in the section on *Speciation*.

Coral reefs represent the most biologically diverse ecosystems on Earth and provide coastal protection to millions of people worldwide (Quistad et al. 2014). The biological foundation of coral reefs is the mutualistic symbiosis (living together of two or more organisms) between two organisms; one is a multicellular organism of the phylum *Cnidaria* and the other is a photosynthetic unicellular dinoflagellate alga referred to as zooxanthellae (Davy et al. 2012). *Cnidaria* have a simple body plan consisting of two cell layers, an endoderm and ectoderm, held together by a jelly-like mesoglea. The algal symbiont resides within the cells of the gastrodermis of the host cnidarian where they are bound by a membrane complex derived from both the algal cell and the host cell. In general, symbionts are taken up by host cells through phagocytosis followed by the development of a stable relationship (Davy et al. 2012). Each host cell generally harbors one to two symbionts although this number can change through proliferation of intracellular dinoflagellate cells.

The photosynthetic activity of the intracellular algae supports the metabolic needs of the cnidarian host and supplies the energy for the host cells to form the calcium carbonate exoskeleton from which the name coral is derived.

The number or abundance of symbionts is a prime determinant of coral reef physiology and depending upon the specifics of the host–symbiont relationship, optimal symbiont abundance can be defined (Cunning and Baker 2014). For example, if each symbiont provides the same photosynthetic activity, then increasing the symbiont number will increase the overall photosynthesis. However, there is a cost to the host cell of harboring more symbionts. This cost includes creating and maintaining host-derived membrane to enclose the symbiont, providing carbon dioxide for symbiont photosynthesis, and detoxifying reactive oxygen species (ROS). As described by Cunning and Baker (2014), a "cost-benefit" calculation can be made to determine an optimal symbiont number or abundance.

The existence of an optimal symbiont abundance means that mechanisms must be in place to regulate the density of symbionts (number of intracellular algal cells per host cell) in order to optimize metabolic/energy transfers between symbiont and host cells. One mechanism for regulating symbiont numbers is through expulsion of symbiont cells (Dimond and Carrington 2008; Davy et al. 2012) and it has been suggested that expulsion takes place via host cell programmed death (apoptosis) (Davy et al. 2012). Quistad et al. (2014) have demonstrated that the coral cnidarian *Acropora digitifera* possessed an intact tumor necrosis factor (TNF) receptor–ligand superfamily which is a known mediator of programmed cell death (apoptosis). This superfamily showed remarkable homology with human TNF receptor–ligands. Hence, cnidarians have the genetic tool kit for programmed death.

Programmed death has been well described in the phenomenon of coral bleaching. Bleaching involves the breakdown of the cnidarian–dinoflagellate symbiosis when the coral reef is exposed to stressors such as an increase in seawater temperature and increased ultraviolet radiation (Edmunds and Gates 2003). The symbiotic algal cells are expelled from the host cells which results in a loss of color (bleaching) of the reef and reduced fitness of the host (reduced growth rates and fecundity) (Paxton et al. 2013). The prevailing model is that stressors such as increased water temperatures lead to induction of programmed death in the host cell followed by expulsion and loss of symbiotic algal cells (Tchernov et al. 2011). This model was experimentally supported by Quistad et al. (2014) who demonstrated that treating coral tissue with TNF ligand-induced host cell programmed death and expulsion of intracellular algae.

The coral reef ecosystem is built upon a mutually beneficial interaction between two organisms; a multicellular cnidarian host and an intracellular algal cell (symbiont). Optimal functioning of the ecosystem requires that the ratio of symbionts per host cell be kept within a certain range. One of the mechanisms regulating this ratio is programmed death of host cells with expulsion of symbionts. This same mechanism also occurs as part of a stress response known as coral reef bleaching when coral reefs are exposed to environmental perturbations such as elevated sea temperatures. The important point is that programmed death of the cnidarian (host) organism is an adaptive phenotype, beneficial to the overall functioning of the entire ecosystem.

The notion that ecosystem organization is highly regulated is supported by the analysis of Hatton et al. (2015). These investigators examined for systematic patterns using a database consisting of over 2200 ecosystems in 1512 locations globally. The predator–prey relationship was defined in both terrestrial and aquatic ecosystems. Land predators included large species of carnivores (e.g., lion) and their herbivore prey and in lakes and oceans, zooplankton was considered a predator and phytoplankton the prey. Across multiple terrestrial and aquatic ecosystems, a log–log plot of predator biomass (Y-axis) versus prey biomass (X-axis) gave a straight line with a slope of 0.74 for land systems and 0.71 for water systems. Similarly, ecosystem production, the increase in biomass per unit time (Y-axis) and ecosystem population, biomass summed over all individuals (X-axis) gave a slope of 0.67–0.76 on a log–log plot. Hence, an increase in number of prey was not associated with a proportionate increase in the number of predators and an increase in ecosystem biomass was not associated with a proportionate increase in biomass production.

For both terrestrial and aquatic ecosystems, the relationship between predator–prey biomass and production–biomass followed similar sublinear power law scaling. These scaling patterns, which appear to be a feature of ecosystems in general, indicate that ecosystems are highly organized and that the relative population sizes of different species within the ecosystem must be regulated. Although the regulatory mechanisms are unknown, it is likely that organismal aging and death programs represent one of the tools used to control species population densities.

The universality of the predator–prey biomass scaling relationship in diverse ecosystems underscores the essential role this relationship plays in trophic cascades within ecosystems. Ripple et al. (2014) reviewed the ecological effects of large mammalian carnivore species. These apex predators, despite their low population densities, regulate both directly and indirectly such ecosystem properties as population density of prey species and smaller predators, abundance and diversity of plant species, and river/stream morphology, as well as nutrient recycling. An earlier article by Ripple and Beschta (2005) summarized the multiple trophic cascades that were regulated by wolves in ecosystems within the western hemisphere. These regulatory functions performed by wolves were amply demonstrated in ecosystems in which wolves had been eliminated by purposeful human actions. As noted in a previous section, Hatton et al. (2015) documented that the relationship between predator and prey biomass across numerous ecosystems, both terrestrial and aquatic, follows the same scaling rule. Such a universal feature points to a level of ecosystem structure/function regulation that would necessitate species-specific aging and death programs (Singer 2016). In fact, optimal multifunctionality of an ecosystem requires regulated interactions across the multiple trophic groups comprising that ecosystem (Soliveres et al. 2016).

The human species is also an apex predator. However, as documented by Darimont et al. (2015), the predatory behavior of humans is very different from that of nonhuman terrestrial and marine predators. Unlike other predators, the predatory behavior of humans is global in scope. The fishing activities of humans are supported by large-scale industrial type technologies and unlike other mammalian carnivores, humans tend to hunt adult forms of other species rather than juveniles. In addition, humans often hunt for sport-type reasons and not for the necessity of obtaining food. Humans have domesticated many animals and plants and through agricultural practices produce food on a large industrial scale.

Currently, almost two-thirds of humans live in cities and cities can be considered an urban ecosystem. However, urban ecosystems and natural ecosystems are fundamentally different in organization. Natural ecosystems show sublinear scaling between biomass production and total ecosystem population or biomass; scaling exponent 0.67–0.76 (Hatton et al. 2015). Cities show a different scaling relationship (Bettencourt 2013). Gross domestic product, a measure of ecosystem production, shows supra-linear scaling with population, a measure of urban ecosystem biomass. In this case, the scaling exponent is 1.12. The reasons underlying the different scaling exponents between cities and natural ecosystems are not understood, but the implications for human evolution are clear. The natural ecosystem from which modern humans emerged was a much different structure than the human-created urban ecosystem in which modern humans now reside.

Carr et al. (2002) experimentally highlighted the complex interactions among species that characterize ecosystems. The test system consisted of experimental reefs of living coral and the investigators measured density-dependent mortality in reef fish. Early larvae from reef fish disperse in the plankton but late stage larvae settle to the reef habitat where they become site attached. This process is known as recruitment and the recruits rapidly develop into juveniles and then adults. The investigators studied the density-dependent mortality of recently attached recruits in experimental reefs that had been modified as follows; removal of predators and competitors, competitors or predators present, both competitors and predators present. When both predators and competitors were present, the natural state of the ecosystem, settled recruits showed a significant density-dependent mortality rate with mortality increasing linearly with the density of the recruit population. In the other experimental situations, absence of competitors and predators or presence of either predators or competitors, settled recruit mortality was highly variable and showed no relationship to the population density of the settled recruits.

Field studies have demonstrated the importance of density-dependent mortality in regulating species population sizes in marine ecosystems (Carr et al. 2002) and the experiments of Carr et al. (2002) established that within coral reef ecosystems, density-dependent regulation of settled fish population mortality required the presence of both competitors and predators. Hence, complex interactions among multiple species within an ecosystem are necessary to maintain stability of constituent species population sizes.

In summary, natural ecosystems are highly structured and regulated hierarchical biological organizations. One of the mechanisms regulating species diversity and abundance within ecosystems is evolved species-specific lifecycles including aging and death. As discussed in previous sections, the regulatory role of species-specific aging and death programs has its evolutionary origins in the emergence of programmed cell death in communities of unicellular organisms.

HUMANS

EVOLUTION

The human lineage originated in Africa and diverged from a common ancestor with chimpanzees and bonobos about four and a half million years ago (Prufer et al. 2012). The evolutionary pattern leading to modern humans is quite complex and our understanding of this pattern is still in a state of flux. Initially, human evolution was viewed in terms of a linear pattern. The evolutionary trajectory was represented by a single branch of a tree with this single branch incorporating all early human species (Anton et al. 2014). However, based upon accumulated fossil and genomic data, this linear pattern has been replaced by a pattern in which human evolution is depicted as a tree with many branches. A number of early human species coexisted, but a more dominant species *Homo erectus* emerged about two million years ago. *H. erectus* was probably the first human species to leave Africa for Eurasia about 1.8 million years ago (Anton et al. 2014). All of the branches of the evolutionary tree were dead-ends except for the branch leading to modern humans, *Homo sapiens*. Modern humans first appeared about 200,000 years ago in a part of present day Ethiopia (Harcourt 2016) and initially coexisted with other human species including Neanderthal and Denisovan. In fact, there is DNA evidence indicating that gene flow via a low level of interbreeding occurred between modern humans and Neanderthal/Denisovan (Sankararaman et al. 2016). The early human lineage inhabited a small ecological niche in eastern Africa but beginning with migration of *H. erectus* out of Africa, this ecological niche expanded and today is global in extent. Within their ecosystem, early human species occupied a high trophic level but were not necessarily the apex predator. Present day humans are not only an apex predator but display predatory behaviors and practices very different from other marine or terrestrial apex predators (see section on *Ecosystems*).

Researchers have tried to reconstruct environmental conditions that were present in eastern Africa during early human evolution, but environmental pressures alone cannot account for evolution of the human lineage since speciation is driven by natural selection acting at multiple levels of biological organization; the individual organism, species, and the ecosystem in which that species lives (see sections *Speciation* and *Ecosystems*).

Data are available that give insights into the evolution of aging and life span patterns in early human species. Caspari and Lee (2004) used fossil dentition samples to estimate ages in four human groups. The earliest group contained the species *Australopithecus* and *Paranthropus* from a time period about 2.5 million years ago and the fourth group consisted of modern humans from about 11,000 years ago. The total number of individual dentition samples was 768. They used the third molar, which generally erupts at the time of reproductive maturation, as a dating tool. Dental specimens were divided categorically into young adult (Y) when there was evidence of molar eruption and older adult (O) when the molar had erupted and there were also signs of molar wear. The investigators then calculated an O/Y ratio for the dental remains from the four human groups. The ratios for the four groups were; 0.12, 0.25, 0.39, and 2.08.

Caspari and Lee (2004) drew several conclusions from their data. First, the increasing O/Y ratio across all four groups indicated a trend for increased survival of older individuals throughout the course of human evolution. Second, the greatest increase in longevity occurred in the most recent group which lived prior to the emergence of agriculture, about 11,000 years ago. Preagricultural humans showed a greater number of older than younger adults in the death distribution. The survival of older adults made possible the development of connections between generations since an older adult could theoretically be a grandparent. Hence, social groupings in the preagricultural time period most likely comprised more than one generation and such an arrangement would have allowed for the transfer of resources between generations including the care of grand-offspring by the older generation.

Gurven and Kaplan (2007) have performed an extensive review of life span and aging patterns in preagricultural hunter-gatherers. These investigators amassed a data base for the mortality patterns of extant human populations living a lifestyle most similar to that of our foraging ancestors. Some of their results for the pooled data base were as follows. About 60% of children born reached the age of 15 years. Of children reaching the age of 15 years, about 60% will live to be 45 years. The mean life expectancy for those who reached 45 years was about 20 years. All populations showed a significant post-reproductive life span in women. In general, these populations experienced a lengthy period of prime adulthood, delayed senescence, and an overall extended life span. The modal age of death in these extant hunter-gatherers was 72 years with a range of 68–78 years. Modal was defined as the age at which most deaths occur. Gurven and Kaplan (2007) noted that the mean mortality pattern for these forager populations was similar to that of the Swedish population in 1751.

The important findings in these hunter-gatherer populations as summarized by Gurven and Kaplan (2007) were; post-reproductive longevity was a robust feature, adult mortality rates were low from maturity (age 15 years) to age 40 years, and thereafter mortality rate increased exponentially and the modal age of death was 72 years. This general mortality pattern of foragers is similar to that of modern humans which means that this pattern was established at least 10,000–12,000 years ago. A major difference between present day human populations and hunter-gatherers were the causes of death. In the forager populations illnesses accounted for 70% of deaths, with 20% of deaths due to violence and accidents and only 9% of deaths due to degenerative diseases. In present day populations, degenerative diseases are the predominant cause of death.

While I have focused on the evolutionary course of human aging and life span, new morphologic and behavioral traits were also evolving in the human lineage over the same time frame (Anton et al. 2014). Bipedal locomotion emerged about 3.5 million years ago. The bipedal gait and erect posture led to changes in pelvic anatomy which had consequences for delivery of infants with proportionately large heads. A large increase in brain size and complexity evolved over the past two million years and, in modern humans, brain maturation is not completed until aged to the early 20s (Lebel et al. 2008). The consistent use of stone tools emerged about 2 million years ago and, about 10,000–11,000 years ago, humans domesticated plants and animals and introduced agricultural practices. Agriculture was apparently acquired by a number of groups independently and in addition was spread by migration of groups using agriculture (Gibbons 2016). These practices afforded humans access to reliable and adequate food supplies. Humans became less migratory in their behavior and began to establish communities in fixed locations. The first cities were constructed about five and a half thousand years ago (Department of Ancient Near Eastern Art 2000).

As an aside, the predominant view has been that agriculture replaced hunting and gathering because agricultural practices made a higher population growth rate possible. However, newer data do not support this view (Bettinger 2016). Hunter-gatherers from North America and Australia had a similar population growth rate as that of early agriculture-based populations in North America and Europe. The question becomes; if mobile hunter-gatherer populations grew as fast as fixed location agriculture-based populations, why then did agricultural practices steadily replace hunting and gathering during the Holocene period? Although the answer is not known, the implication is that agriculture conveyed a selective advantage other than that of increased population growth rate.

This brief summary illustrates the parallel evolution of morphologic/behavioral traits and cultural practices that occurred in the human lineage. The emergence of culture as a nongenetic inherited system has been pivotal in the evolution of the human species and is discussed in the next section.

CULTURE

The human species has altered its own evolutionary trajectory and that of many other species through the development of a parallel system of nongenetic inheritance known as culture. There are many definitions of culture but for our purposes culture can be considered a body of knowledge and practices that is socially learned, that is modified and added to by each generation, and which can then be transmitted from generation to generation with high fidelity. Culture has made it possible for our species to expand the physical/functional space, that is, the ecological niche that it occupies to include the entire globe and potentially extraterrestrial space. Culture is clearly a group attribute and its underpinning is the process of social learning. Social learning is not unique to humans and has been documented in a number of species (Galef and Laland 2005). For example, Slagsvold and Wiebe (2011) experimentally documented social learning in birds.

The test subjects for the study by Slagsvold and Wiebe (2011) were two species of passerine birds; blue tits and great tits. The foraging behavior of these two species differs in that blue tits forage high in trees on twigs and buds whereas great tits feed mainly on the ground. In an early study, the investigators found that nestlings raised by parents of the other species (blue tit by great tit and vice versa), that is, cross-fostered birds, adopted the foraging behavior of the foster species and this shift in learned behavior of nestlings lasted for life (Slagsvold and Wiebe 2007). Hence, foraging behavior was shaped through a process of social learning between young birds and parents and the type of foraging behavior adopted was strongly influenced by the species of the parents. The investigators then questioned whether birds raised by a different species (cross-fostered) and which had learned foraging behavior from their foster parents would differ in the prey items they delivered to their nestlings compared to birds reared by their own species (Slagsvold and Weibe 2011). In the wild, great tits provide a larger prey volume (larvae, spiders, and flies) than blue tits and the cross-fostering experiment gave the expected results. Cross-fostered great tits provided smaller prey volumes than controls and cross-fostered blue tits provided larger volumes than controls.

In summary, the experimental observations of Slagsvold and Weibe (2007, 2011) document that social learning occurs in birds and that this learning can be transferred across generations. Young birds learn their foraging behavior from their parents and then pass that learned foraging behavior to their offspring. Since social learning is the basis of culture, it is reasonable to conclude that culture, in a more restricted form than that in humans, will be present in many species. In fact, Whiten (2011) has surveyed data from nonhuman primates and it is apparent that chimpanzees do have a rudimentary form of culture. Hence, social learning and culture are not unique to humans. What makes human culture unique are the layers of complexity and sophistication that have been added on by the human lineage since it diverged from the common ancestor with chimpanzees.

There is an extensive literature on various aspects of human culture. In this section, the focus will be on the impact culture has had on the evolution of humans and other species as well as the impact of human culture on our planet.

The human lineage began its evolutionary course occupying a small ecological niche and living in a natural ecosystem. Initially, human culture was rudimentary, but the evolution of a large complex brain over the past two million years made possible the emergence of a complex and intricate culture. The life span of our ancestors progressively increased such that social groupings began to comprise multiple generations. The general aging pattern of modern humans was established about 10,000–11,000 years ago and at that time the modal age of death was 72 years. Human cultural practices became highly sophisticated and the size of our ecological niche has now reached global proportions.

As defined at the beginning of this section, culture comprises a body of knowledge and practices that is inherited across generations through nongenetic mechanisms (cultural inheritance). Cultural knowledge and practices allow organisms to modify their environmental space, for example, the building of a city or a dam (a process known as niche construction) and this modified environment is passed on to subsequent generations. The transgenerational transmission of modified environmental space can be considered ecological inheritance. A further complexity is that organismal modification of its environment changes the selection pressures on the organism itself, its descendants, and other organisms within the same environmental space thereby influencing their evolutionary course (Kendal et al. 2011; Singer 2015). Through these processes, organisms become active participants in their own evolution and in the evolution of other organisms sharing their environmental space. Cultural and ecological inheritances as well as niche construction are most highly developed in the human species.

Perhaps, the clearest example of the action of these processes is the evolution of lactase persistence (LP) in humans. Lactose is the main sugar found in mammalian milk and mammals have the enzyme lactase to break down this sugar. After weaning, the expression of this enzyme is generally downregulated. In humans, a number of populations do not downregulate lactase expression and this enzyme can be found in adults, a trait known as LP (Gerbault et al. 2011). LP has a patchy geographical distribution with a high frequency in northern Europe, the United Kingdom, and Scandinavia and a low frequency in East Asia. The LP trait has been associated with single nucleotide polymorphisms (SNPs) located in the vicinity of the lactase gene. The estimated dates of origin of these SNPs correspond to the dates at which domesticated animals and dairying emerged in Europe and pastoralism in Africa. Hence, the data indicate that these SNPs spread rapidly through certain human populations reaching high frequencies and that this phenomenon was temporally linked to the emergence of the cultural practice of dairying.

Although there is not a consensus as to why LP should confer such a selective advantage, the simplest explanation is that LP allowed humans to use fresh milk from domesticated animals as an important food source in the post-weaning period and into adulthood. This example illustrates the link between culture and human evolution. The introduction of domesticated animals and dairying gave a selective advantage to those individuals who through natural variation had LP and who were therefore able to use fresh milk as a food source (Singer 2015). Under these new circumstances, the adaptive value of the LP phenotype became so high that SNPs responsible for LP were subject to strong positive selection and spread through populations practicing dairying. A cultural practice, dairying, modified the ecological niche of certain human populations (niche construction) and this modified ecological niche changed the selective pressures on those populations such that alleles (SNPs) that were previously non-adaptive became highly adaptive.

A second example relating the effects of cultural practices to human evolution was described by Ryan and Shaw (2015). These investigators found that bone strength in the hip joint of mobile hunter-gatherers was comparable to that of wild nonhuman primates and greater than that of more sedentary agriculturalists. The emergence of agriculture modified the lifestyle of human populations engaged in this cultural practice. The physically demanding lifestyle of the hunter-gatherer, necessary for the hunting of prey, was replaced by the less physically demanding lifestyle of the agriculturist. This finding can be extended to present day humans who by virtue of the use of technology have a very sedentary life style. Cultural practices such as agriculture and the present day widespread use of technology have modified the selective pressures in our ecological niche. Physical strength and the ability to engage in strenuous physical activity, traits that were adaptive in preagricultural societies, no longer have a selective advantage. Reduced physical activity has resulted in reduced bone strength and a high prevalence of osteoporosis.

These two examples illustrate the complex relationships between cultural practices and evolution. An emerging cultural practice will generally serve a certain purpose for our species. At the same time, that practice modifies the environmental space which we inhabit and this ecological modification can lead to a change in our evolutionary course. This reciprocal process linking culture and population genetics has been termed gene–culture coevolution. In addition, by virtue of the widespread interactions that exist between species occupying the same ecosystem, this human-mediated ecological modification could potentially alter the evolutionary course of these other species as well.

As reviewed in the sections *Speciation* and *Ecosystems*, the evolution of a species is driven by multilevel selection acting within the context of the ecosystem in which that species lives. The important contextual features of the ecosystem that influence the evolutionary course of that species include abiotic factors characterizing the environmental space, interactions that exist between the given species and other species inhabiting the ecosystem, and modifications made by the given species as well as other species to the shared environmental space. For a species to evolve, phenotypes targeted by natural selection must be ones underwritten by a heritable developmental program. Morphology, behavioral repertoire, and lifecycle stages are the phenotypic categories that characterize a species. Lifecycle stages would include aging/death programs.

The extent of the modifications a species makes to its environmental space will be determined by the cultural practices of that species. All species modify their environmental space by, for example, building nests, hives, or burrows. These types of structures create a microclimate for that species. Humans have developed the most sophisticated culture of all species as discussed in the beginning of this section. The fact that the present geological era has been termed the Anthropocene is a testament to the enormous impact that human culture has had on Earth. The power of human cultural practices to modify our species environmental space is so profound that culture has become the major force driving human evolution through the process of gene–culture coevolution. In the next sections, some of the impacts that humans have made to Earth will be reviewed. Although we tend to view these cultural-mediated environmental changes as generally beneficial, we must not forget that because of the reciprocal nature of gene–culture coevolution, these cultural-mediated changes in our environmental space will alter the course of human evolution (including our aging/death programs) in ways over which we have no direct control.

As an aside, data collected by Beauchamp (2016) give a measure of the difference between the action of natural selection when only Darwinian fitness is considered and when cultural practices are also included. Beauchamp (2016) used as his subjects the approximately 20,000 Americans enrolled in a longitudinal health and retirement study. He chose subjects born between the years 1931 and 1953. Since modern humans have a low mortality, fitness was based upon reproductive success; the total number of children an individual ever gave birth to or fathered. He measured seven phenotypes for his subject population but the only one I will consider is educational attainment (EA); years of education. He surveyed genome wide association studies done with respect to EA and calculated a polygenic score for EA. He then did a regression of reproductive success on the phenotype EA and on the polygenic score for EA. Both regressions showed a negative correlation. Individuals with higher EA (more years of school) or individuals with the genomic background favoring higher EA had reduced reproductive fitness (fewer lifetime children). Beauchamp (2016) estimated that selection directed toward increasing reproductive fitness would result in a decrease in EA of 1.5 months per generation. The cost of enhancing Darwinian fitness was reduced education attainment. However, Americans born between 1876 and 1951 achieved a mean level of EA of two years per generation. In reality, EA was not driven by selection acting on reproductive fitness but by gene–culture coevolution.

Anthropocene

The effects of human cultural practices have become so profound that the current geological era has been named the Anthropocene (Lewis and Maslin 2015) although there is still no consensus as to when to date the start of this era (Voosen 2016). For example, there is evidence that the impact of human activities on the organization of ecosystems became significant about 6000 years ago (Pennisi 2015; Lyons et al. 2016).

The major processes responsible for anthropogenic changes are accelerating technological development, rapid growth of the human population (was estimated to be about 2 million 12,000–14,000 years ago), and increasing consumption of resources (Waters et al. 2016). Until recently, the doctrine of uniformitarianism has been a guiding principle to predict the extent of future geological changes. This principle holds that the processes operating today to cause geophysical changes are just a linear extension of the same forces that have been acting throughout geological time. However, the evidence is that human cultural practices have moved geophysical forces away from linearity (Knight and Harrison 2014). Human-induced climate and land surface changes have entered a stage of nonlinear dynamics; the rate and extent of past geophysical changes cannot be extrapolated to predict the rate and extent of future changes.

In this section, I have arbitrarily divided human cultural impacts into four domains: (1) climate change, (2) environmental space, (3) biodiversity, and (4) economics. An important point that needs emphasizing is that the effects of human cultural practices are not unidirectional. Cultural practices will modify the environmental space that humans occupy. These modifications will alter the direction and strength of multilevel natural selection and thereby change the evolutionary course of the human species (and other species) in unpredictable ways. This culture-induced change in human evolution would include modifications of our species lifecycle stages such as aging and longevity. Hence, understanding the impacts of human activities embodied within the Anthropocene is crucial to understanding the course of human evolution.

CLIMATE CHANGE

There is now a strong scientific consensus that human cultural practices (primarily fossil-fuel burning and large-scale changes in land use) have significantly altered the Earth's climate system. Quoting from a press release from the Intergovernmental Panel on Climate Change (IPCC), November 2, 2014; "Human influence on the climate system is clear and growing, with impacts observed on all continents. If left unchecked, climate change will increase the likelihood of severe, pervasive and irreversible impacts for people and ecosystems." The IPCC is a scientific and intergovernmental agency under the auspices of the United Nations.

Carleton and Hsiang (2016) have reviewed the social and economic impacts of climate change. Analyzing the relationship between climate events and social outcomes requires translating weather data into a socially meaningful form and then developing a dose–response function between specific climate events and the social outcome of interest. In addition, studies can be done longitudinally by following an individual population and examining how that specific population responds to changes in climatic conditions; the population right before the climate event becomes the "control" for that same population after the event. A few of the examples presented by Carleton and Hsiang (2016), depicting dose–response functions for pairings of social outcomes and climate events are; a U-shaped relationship between relative risk of human mortality and daily temperatures with risk of mortality highest at very hot and very cold temperatures, a decrease in agricultural income with both low and very high total seasonal rainfall levels, and an increase in interpersonal and intergroup violence with increasing daily maximum temperatures. The quantitative influence of climate change (temperature and rainfall anomalies) on personal and intergroup conflict was also explored by Hsiang et al. (2013). These dose–response functions represent the quantitative change in a social outcome for unit changes in the climate event.

The following two examples illustrate the social impacts of extreme climate events. In the first example involving the collapse of Mayan society, the climate event was probably not human-induced whereas in the second example, an extreme drought in Syria, the climate event had a significant anthropogenic component.

The collapse of Mayan society presents a case study illustrating the social impacts of climatic events (Kennett and Beach 2013). Classic Mayan society was organized around the institution of kingship and consisted of a collection of low density urban centers, ruled by these kings, which were spread across Mesoamerica. Total population is estimated to have been between 3 and 10 million individuals. Agricultural food production varied from region to region but primary foods included maize, beans, and squash. Domesticated animals played a limited role in Mayan society and in particular herd animals (sheep, goats, and cows) were absent. Agricultural practices included terracing to stabilize the landscape and wetland systems. Food and by extension labor were the foundation of hierarchical Mayan society with kings and a small group of nonfood producing elite at the top of the pyramid. Production of an adequate supply of food allowed for population growth.

The widespread collapse of classic Mayan society began about year 800 and had multiple causes. One cause was the environmental impacts of extensive deforestation and soil erosion which were a consequence of intense agricultural practices necessary to support an expanding population. Another cause was climate events. The early classic period was a time of high rainfall which favored agricultural food production and population expansion. However, in the late classic period, there was less rainfall and extended periods of severe drought. These climate events had a very negative impact on food production and the primary response of kings during this period of climate-induced instability was to wage wars. The combination of sustained conflict and war led to a breakdown in Mayan society with population decline and dispersal. Another factor was the "rigidity" of the hierarchical organization of Mayan society. The lack of societal resilience amplified the negative social impacts of repeated severe periods of drought and prevented Mayan society from responding to the climate events in an adaptive manner. In fact, there is now evidence that fragile societies, those that lack resilience, are at particular risk to breakdown or collapse Scheffer (2016).

Although the long periods of drought during the late classic Mayan period were not human induced, the negative social outcome of this event played a major role in the collapse of that society.

The second example involves modern day Syria (Kelley et al. 2015). Syria has been subject to repeated droughts but a severe drought began in the winter of 2006/2007 and lasted for three years. The more recent droughts are believed to have a significant anthropogenic basis. Just prior to this last drought the Syrian government instituted unsustainable policies to increase agricultural production. One consequence of these policies was a reduction in groundwater which increased Syria's vulnerability to drought. When the drought occurred, the agricultural system collapsed and wheat, for example, had to be imported. A mass migration of rural farming families to urban areas ensued. It is estimated that almost 1.5 million people were internally displaced. Between 2002 and 2010, Syria's urban population had increased by 50% mostly due to migration of rural inhabitants plus refugees from other countries such as Iraq. These communities on the urban outskirts were overcrowded and lacked proper infrastructure. The population living in these communities was unemployed and crime/violence became widespread. The Syrian "uprising" occurred in 2011.

These two examples illustrate a chain of events which begins with an extreme climate event, such as a severe and prolonged drought, followed by mass migration of people from rural to urban areas, overtaxing of urban housing and infrastructure, breakdown in civil society, and emergence of widespread personal and intergroup violence. This chain of events is much more likely to occur in rigid societies that lack resilience.

ENVIRONMENTAL SPACE

The human lineage began its evolutionary path millions of years ago, occupying a small ecological niche within a natural ecosystem. An ecological niche represents a functional/physical space occupied by a specific species and an ecosystem is an organization containing multiple overlapping ecological niches. The overlapping of these niches results in extensive interactions between diverse species inhabiting the ecosystem. Culture has allowed modern humans to expand their ecological niche to being global in extent and to develop a new type of ecosystem referred to as an urban ecosystem. For the human species, the boundaries between its ecological niche and its urban ecosystem have become blurred. The emergence of the urban ecosystem has had extensive impacts on the Earth's environment and a sampling of these impacts is discussed in this section.

One of the hallmarks of the Anthropocene has been the accumulation of materials that are either man-made or have been extracted in significant amounts (Waters et al. 2016). These materials include pottery, glass, bricks, copper alloys and more recently concrete, aluminum, and various types of plastics. Many of these materials are not biodegradable.

There has been a decline in wilderness areas of almost 10% in the last two decades due to an expanding human footprint, that is, widespread agricultural land use, construction of large urban centers with roadways, navigable waterways, and rail lines to connect these centers to name just a few components of this footprint (Pennisi 2016; Watson et al. 2016). Wilderness is defined as biologically and ecologically intact landscapes that are mostly free of human disturbance. Currently, only 23% of land areas remain as wilderness. Although there have been efforts to protect wilderness areas, losses due to human activities exceed gains due to protection. Wilderness areas sustain many endangered species and biomes such as boreal and tropical forests that store a considerable fraction of total global carbon.

The first cities were built about five and a half thousand years ago and today humans have constructed extensive urban settings inhabited by millions of people. Almost two out of three humans live in an urban center. However, the scaling rules for urban ecosystems (see section on **Ecosystems**) are different than that of natural ecosystems indicating that these two structures are fundamentally different.

Light pollution is a byproduct of the global collection of large urban centers (Panko 2016). Estimates are that more than 80% of humans experience light-polluted night skies. The consequences of nighttime artificial light are many, for example, interference with bird migratory patterns, deterrence of nighttime pollinators like bats, and decreased melatonin production in humans with dysregulation of day–night cycles.

Urbanization has multiple ecological impacts which are best defined in terms of an urban–rural gradient (McKinney 2002). Changes in physical properties and plant/animal diversity occur along the transect from inner city to surrounding areas. As one approaches the urban core, there are increases in population density, road density, soil compaction, average ambient temperature, and percentage of surface area covered by an impervious material such as concrete or asphalt. Natural habitat is reduced and fragmented in the inner core and there is a corresponding decrease in species biodiversity along the rural to urban gradient. There is also a change in the types of plants and animals found in the heart of the urban ecosystem. The urban core generally contains nonnative trees and vegetation as well as animals such as mice, rats, raccoons, and certain bird species that can exploit the urban environment. In essence, the urban ecosystem is a habitat built by and dominated by the human species. Natural habitat areas within the urban ecosystem are small and disconnected and only certain species of plants and animals can adapt to such an environmental space. The extensive breadth and high density of urbanization means that in general humans living in core urban areas are disconnected from nature.

A new discipline has emerged known as urban ecology and projects have been instituted to make large urban areas more ecologically sustainable (Pickett et al. 2008; Wachsmuth et al. 2016).

Two examples illustrate the evolutionary consequences of urbanization. Harris et al. (2016) looked at the effect of urbanization within the New York City area on the distribution of the white-footed mouse, *Peromyscus leucopus*. In the post-glacial period about 13,000 years ago, the population of *P. leucopus* became segregated, by natural barriers made up of the East River and Long Island (LI) Sound, into two main groups termed mainland and Manhattan (MM) and LI. Europeans entered the area about the year 1600 and over the ensuing 400 years urbanized about 96% of natural green spaces. Urbanization resulted in fragmenting the landscape and created isolated green spaces which were surrounded by dense urban development. Genetic analysis was done on 191 individual mice from 23 sampling sites both on the mainland and on LI. During the process of urbanization, small populations of *P. leucopus* became "trapped" within these isolated green parks. For five of the larger parks, the separated subpopulations showed genetic differentiation and little evidence of admixture. These observations imply that buildings, roads, and other urban structures which surround these green parks present sufficient physical barriers to maintain isolation of the small contained populations. In the case of the species *P. leucopus*, there was evidence that small subpopulations isolated in these parks had evolved genetic differentiation and perhaps were on a trajectory to form new species. Regardless, the data of Harris et al. (2016) indicate that physical isolation of subpopulations of animals within urban green parks has a strong enough allopatric impact to alter the evolutionary history of those animals.

A second example involves the phenomenon of industrial melanism in the peppered moth *Biston betulana*. This phenomenon refers to the emergence of a previously unknown black form (carbonaria) and its replacement of the common pale type (typica) during the industrial revolution in Britain (van't Hof et al. 2011). This evolutionary event was driven by the interaction between bird predation and coal pollution. Black moths would be less visible to avian predators in an environment polluted by coal fumes. The first reported sighting of the carbonaria form was in Manchester in 1848 and by the end of the 1800s, this variant outnumbered the original phenotype. Genetic studies indicated that the evolution of the carbonaria phenotype was the result of a single mutation which arose about 1819 (van't Hof et al. 2011, 2016). Within 20–30 moth generations, this genomic variant had become widespread. The responsible mutation was the insertion of a 22 kb transposable element into the gene, cortex, and the carbonaria phenotype resulted from altered expression of this gene. The gene, cortex controls pigmentation pattern through its regulation of wing-scale development (Nadeau et al. 2016) although the exact mechanism is unknown.

In these examples, alterations of environmental space by the process of urbanization directly changed the evolutionary course of two species; genetic differentiation between subpopulations of the mouse *P. leucopus* due to isolation of these subpopulations in physically separated urban parks and emergence of a new pigmented phenotype in the peppered moth as a consequence of environmental pollution by coal-driven industrial processes.

BIODIVERSITY

From a human perspective, ecosystems provide a number of benefits which have been termed ecosystem services. In fact, the total economic value of global ecosystem services has been estimated to be in the range of tens of trillions of dollars (Oliver 2016). However, this concept is clearly artificial since it defines the value of ecosystems with reference to the human species only. In reality, an ecosystem is an ensemble of overlapping ecological niches in which a diverse mixture of species interact. Each ecosystem has evolved under the action of multilevel selection as a structure which is beneficial to its entire constituent species. However, the notion of ecosystem services has become a metric for measuring the "health" of ecosystems. Ecosystem services are categorized under four broad domains; supporting services (nutrient recycling, synthesis of organic substances through photosynthesis, soil formation), provisioning services (food, raw materials, energy), regulating services (carbon sequestration, climate regulation, waste decomposition), and cultural services (nonmaterial human benefits of ecosystems).

One of the most important attributes determining ecosystem "health" is biodiversity which is a combination of the abundance of each species and the richness of species diversity. These two parameters can be combined into a function called the biodiversity intactness index (BII) which allows for the quantification of biodiversity loss (Newbold et al. 2016). Newbold et al. (2016) estimated the effect of human land use on the biodiversity intactness of terrestrial biomes by comparing the BII of present day biomes to the BII of habitats undisturbed by man. A biome is a large geographical area containing multiple ecosystems, for example, montane grasslands and shrub lands. Across 14 biomes, 9 showed a decrease in biodiversity intactness of 10%–25%. Grasslands were the most affected and tundra and boreal forests the least. On a global scale, 58% of land surface shows a reduction of local species abundance of 10% or more and 62% of land surface shows a reduction in local species richness of 10% or more both due to human land use. These global losses in biodiversity will reduce ecosystem services, which reduction can be assigned a monetary value. However, a more important consequence of human land use is widespread fragmentation of the structural/ functional integrity of ecosystems with loss of interaction webs between species. This breakdown of ecosystem integrity and functional vitality will have significant evolutionary ramifications.

During past geological eras, there have been five mass extinctions and the loss of biodiversity during the present Anthropocene era has been considered a sixth mass extinction. Hull et al. (2015) looked at this possibility. Human cultural behaviors have resulted in a significant reduction in the abundance of many species largely due to loss of habitat through human use of land for purposes such as agriculture, cities, and roadways to name a few. Hull et al. (2015) note that to date the number of species extinctions is small compared to the number of species mass rarities across the globe. Hull et al. (2015) argue that relying on the fossil record has made it difficult to collect unequivocal evidence as to past mass extinctions. In fact, a mass rarity due to a profound reduction in the abundance of a species can masquerade as extinction since the small number of remaining individuals in the species will be invisible in the fossil record. Past mass extinctions may have actually been widespread mass rarity events and if so the present global reduction in the abundance of many species would be on a par with past mass rarities/extinctions.

One feature specific to human-mediated reduction in species abundance has been the loss of larger animals in general and apex predators in particular (Estes et al. 2011). Estes et al. (2011) review the consequences to ecosystems of removing apex predators. Because ecosystems are built around a web of interconnections between constituent species, loss of the apex predator results in a trophic cascade which is a change in the relative numbers of predators and prey down through trophic levels to the foundational layer of primary producers; plants and phytoplankton. The documented indirect consequences of apex predator loss in various marine and terrestrial ecosystems include a reduction in biodiversity, influx of invasive species, and alteration in soil quality to name a few of the consequences reviewed by Estes et al. (2011).

Boivin et al. (2016) examined the historical roots of culturally mediated global changes in species abundance and distribution and identified four key stages. *H. sapiens* emerged about 200,000 years ago in the horn of Africa and by about 12,000 years ago, had radiated across the globe. This geographical expansion was accompanied by an anthropogenic footprint on species populations and distribution. For example, early humans used fire to promote growth of specific plants and transported different species of plants and animals to new locations. A second stage was the emergence and spread of agriculture and domestication of plants and animals. Both these cultural activities had impacts on species abundance and distribution: domestication generated new species and extensive land use for agricultural food production reduced species habitats. A third stage involved colonization of islands with the translocation of plants and animals not native to these islands by colonists. A final stage involved urbanization and the rapid growth of the human population. These processes resulted in loss of natural habitats through deforestation, intensification of agricultural food production, and the building of large urban sites including connecting roadways, waterways, and rail lines. In summary, over thousands of years, modern humans have changed the abundance and distribution of wild species primarily through the expropriation and exploitation of large areas of wilderness for intensive agricultural food production and the creation of widespread urban sites with connecting links. As discussed in the section *Ecosystems*, large cities are now the new ecosystem for the majority of humans. The historical record of human cultural activities presented by Boivin et al. (2016) underscores that culture has now become a major evolutionary force on the planet.

ECONOMICS

Humans have established an elaborate system of economics and trade that is global in extent. Money has now become the primary material resource largely replacing land and domesticated animals. Money allows individuals, companies, and countries to purchase various goods and services. Humans primarily use an economic metric to measure progress and that metric is the monetary value of goods and services produced. In addition, we have adopted a model of unlimited economic growth (Martin et al. 2016). A corollary of this model is that successful companies and countries are those that show sustained increases in this metric, that is, increasing production of goods and services. However, the ever increasing production of goods and services must be fueled by a society geared to unrestrained consumption (Martin et al. 2016). Such unbridled production/consumption is made possible by our species stripping the planet of resources. The rate at which humans use up natural resources generally exceeds their renewal. The outcomes of this mismatch between resource use and renewal are the various anthropogenic impacts discussed in the sections; climate change, environmental space, and biodiversity.

The current era has been named the Anthropocene in recognition of the profound impacts that the cultural behavior of our species has had on the geophysical properties of our planet. We have become the Earth's dominant species. Of course other invertebrate species are more abundant, but our species is unique in that we have actually altered Earth's ecological path. We may think that we have control of the direction of these ecological modifications but such is not the case. For example, made-made solutions to mitigate the effects of climate change such as planting forests to increase carbon storage or protecting coastal wetlands to adapt to rising sea levels can also lead to unplanned changes in local ecological selections and cause unpredictable shifts in evolutionary trajectories (Sarrazin and Lecomte 2016). Through the feedback loop of gene–culture coevolution, human-mediated ecological modifications will alter the course of our own evolution in ways that cannot be predicted. Gene–culture coevolution will determine the phenotypes of future generations including their aging pattern and life span.

CONCLUSIONS

Aging/death programs have very deep evolutionary roots. All organisms live in ecosystems comprised of conspecifics and other species and aging/death programs emerged as a means of transferring resources and information between organisms and as a "tool" for regulating species populations. The importance of the ecosystem organization in driving the evolution of species-specific aging/death programs has been obscured by the study of aging/death programs in model laboratory animals isolated from their natural ecosystem. A case in point is the hydra (Singer 2016). When studied in a controlled laboratory setting, the hydra appears to be immortal. However, field studies have shown that in their natural habitat hydra have a life span of only a few months. In their natural ecosystem, fluctuations in food supply, water temperature, and organism density modulate the hydra stem cell program; the rate of asexual budding declines and may completely cease. Triggered by changes in water temperature and food supply, certain species of hydra revert to sexual reproduction and show signs of aging.

The human lineage evolved over millions of years within the constraints imposed by natural ecosystems. Modern humans (*H. sapiens*) emerged about 200,000 years ago and the characteristic aging pattern and longevity of the human species was established about 10,000–12,000 years ago. The impacts of human cultural practices have become much more dominant over the past 10,000 years and our species has now created its own ecosystem, large cities. Culture has allowed humans to be largely free of constraints imposed by our habitat since we are capable of modifying our urban ecosystem through our mastery of technology. Humans are now an active player in their own evolution through the process of gene–culture coevolution and indirectly a player in the evolution of other species through cultural practices. The process of gene–culture coevolution will shape future human phenotypes: morphology, behaviors, and aging/life span. Gene–culture coevolution is a relatively recent process, but more important it is a process specific to the human species. This point is pivotal. The Anthropocene represents a unique epoch, one in which the human species through culture has significantly altered the evolutionary process. In essence, our species is the pilot of its own boat floating on the river of evolution. However, we have no map charting the course of the river and more so we lack a defined destination.

REFERENCES

Anton, S. C., Potts, R. and L. C. Aiello. 2014. Evolution of early *Homo*: An integrated biological perspective. *Science* 345 (6192). doi:10.1126/science.1236828.

Arnaud-Haond, S., Duarte, C. M., Diaz-Almela, E., Marba, N., Sintes, T. and E. A. Serrao. 2012. Implications of extreme life span in clonal organisms: Millenary clones in meadows of the threatened seagrass *Posidonia oceanica*. *PLoS ONE* 7(2): e30454. doi:10.1371/journal.pone.0030454.

Barraclough, T. G. 2015. How do species interactions affect evolutionary dynamics across whole communities? *Annual Review of Ecology, Evolution and Systematics* 46: 25–48.

Beauchamp, J. P. 2016. Genetic evidence for natural selection in humans in the contemporary United States. *Proceedings of the National Academy of Sciences, USA* 113: 7774–7779.

Berg, D. K., Smith, C. S., Pearton, D. J. et al. 2011. Trophectoderm lineage determination in cattle. *Developmental Cell* 20: 244–255.

Berman-Frank, I., Rosenberg, G., Levitan, O., Haramaty, L. and X. Mari. 2007. Coupling between autocatalytic cell death and transparent exopoly particle production in the marine cyanobacterium *Trichodesmium*. *Environmental Microbiology* 9: 1415–1422.

Bettencourt, L. M. A. 2013. The origins of scaling in cities. *Science* 340: 1438–1441.

Bettinger, R. L. 2016. Prehistoric hunter-gatherer population growth rates rival those of agriculturalists. *Proceedings of the National Academy of Sciences, USA* 113: 812–814.

Bidle, K. D. 2015. The molecular ecophysiology of programmed cell death in marine phytoplankton. *Annual Review o Marine Science* 7: 341–375.

Bidle K. D. 2016. Programmed cell death in unicellular phytoplankton. *Current Biology* 26: R594–R607.

Boivin, N. L., Zeder, M. A., Fuller, D. Q. et al. 2016. Ecological consequences of human niche construction: Examining long term anthropogenic shaping of global species distributions. *Proceedings of the National Academy of Sciences, USA* 113: 6388–6396.

Cardinale, B. J., Duffy, J. E. Gonzalez, A. et al. 2012. Biodiversity loss and its impact on humanity. *Nature* 486: 59–67.

Carleton, T. A. and S. M. Hsiang. 2016. Social and economic impacts of climate. *Science* 353 (6304). doi:10.1126/science.aad9837.

Carr, M. H., Anderson, T. W. and M. A. Hixon. 2002. Biodiversity, population regulation, and the stability of coral-reef fish communities. *Proceedings of the National Academy of Sciences, USA* 99: 11241–11245.

Caspari, R. and S.-H. Lee. 2004. Older age becomes common late in human evolution. *Proceedings of the National Academy of Sciences, USA* 101: 10895–10900.

Cunning, R. and A. C. Baker. 2014. Not just who, but how many: The importance of partner abundance in reef coral symbioses. *Frontiers in Microbiology*. http://dx.doi.org/10.3389/fmicb.2014.00400.

Darimont, C. T., Fox, C. H., Bryan, H. M. and T. E. Reimchen. 2015. The unique ecology of human predators. *Science* 349: 858–860.

Davy, S. K., Allemand, D. and V. M. Weis. 2012. Cell biology of cnidarian-dinoflagellate symbiosis. *Microbiology and Molecular Biology Reviews* 76: 229–261.

Department of Ancient Near Eastern Art. Uruk: The first city. In *Heilbrunn Timeline of Art History*. New York: The Metropolitan Museum of Art, 2000. htpp://www.metmuseum.org/toah/hd_uruk.htm (October 2003).

Dimond, J. and E. Carrington. 2008. Symbiosis regulation in a facultatively symbiotic temperate coral: Zooxanthellae division and expulsion. *Coral Reefs* 27: 601–604.

Domazet-Loso, T. and D. Tautz. 2010. A phylogenetically based transcriptome age index mirrors ontogenetic divergence patterns. *Nature* 468: 815–818.

Durand, P. M., Choughury, R., Rashidi, A. and R. E. Michod. 2014. Programmed death in a unicellular organism has species-specific fitness effects. *Biology Letters* 10: 20131088. http://dx.doiorg/10.1098/rsbl.2013.1088.

Durand, P. M., Rashidi, A. and R. E. Michod. 2011. How an organism dies affects the fitness of its neighbors. *American Naturalist* 177: 224–232.

Durand, P. M., Sym, S. and R. E. Michod. 2016. Programmed cell death and complexity in microbial systems. *Current Biology* 26: R587–R593.

Edmunds, P. J. and R. D. Gates. 2003. Has coral bleaching delayed our understanding of fundamental aspects of coral-dinoflagellate symbioses? *BioScience* 53: 976–980.

Estes, J. A., Terborgh, J., Brashares, J. S. et al. 2011. Trophic downgrading of planet Earth. *Science* 353: 301–306.

Galef, B. G. Jr. and K. N. Laland. 2005. Social learning in animals: Empirical studies and theoretical models. *BioScience* 55: 489–499.

Gerbault, P., Liebert, A., Itan, Y. et al. 2011. Evolution of lactase persistence: An example of human niche construction. *Philosophical Transactions of the Royal Society B* 366: 863–877.

Gibbons, A. 2016. First farmers' motley roots. *Science* 353: 207–208.

Greulich, P. and B. D. Simons. 2016. Dynamic heterogeneity as a strategy of stem cell self-renewal. *Proceedings of the National Academy of Sciences, USA* 113: 7509–7514.

Gurevich, D. B., Nguyen, P. D., Siegel, A. L. et al. 2016. Asymmetric division of clonal muscle stem cells coordinates muscle regeneration *in vivo*. *Science* 353 (6295). doi:10.1126/science.aad9969.

Gurven, M. and H. Kaplan. 2007. Longevity among hunter-gatherers: A cross-cultural examination. *Population and Development Review* 33: 321–365.

Harcourt, A. H. 2016. Human phylogeography and diversity. *Proceedings of the National Academy of Sciences, USA* 113: 8072–8078.

Harris, S. E., Xue, A. T., Alvarado-Serrano, D. et al. 2016. Urbanization shapes the demographic history of a native rodent (the white-footed mouse, *Peromyscus leucopus*) in New York City. *Biology Letters* 12: 20150983. http://dx.doi.org/10.1098/rsbl.2015.0893.

Hatton, I. A., McCann, K. S., Fryxell, J. M. et al. 2015. The predator–prey power law: Biomass scaling across terrestrial and aquatic biomes. *Science* 349 (6252). doi:10.1126/science. aac6284.

Hsiang, S. M., Burke, M. and E. Miguel. 2013. Quantifying the influence of climate on human conflict. *Science* 341 (6151). doi:10.1126/science.1235367.

Hull, P. M. Darroch, S. A. F. and D. H. Erwin. 2015. Rarity in mass extinctions and the future of ecosystems. *Nature* 528: 345–351.

IPCC *Intergovernmental Panel on Climate Change (United Nations)* press release November 2, 2014.

Jones, O. R., Scheuerlein, A., Salguero-Gomez, R. et al. 2014. Diversity of ageing across the tree of life. *Nature* 505: 169–174.

Julian, L. M., Carpenedo, R. L., Rothberg, J. L. M. and W. L. Stanford. 2016. Formula G1: Cell cycle in the driver's seat of stem cell fate determination. *Bioessays* 38: 325–332.

Kachroo, A. H., Laurent, J. M., Yellman, C. M., Meyer, A. G. Wilke, C. O. and E. M. Marcotte. 2015. Systematic humanization of yeast genes reveals conserved functions and genetic modularity. *Science* 348: 921–925.

Kalinka, A. T., Varga, K. M. Gerrard, D. T. et al. 2010. Gene expression divergence recapitulates the developmental hourglass model. *Nature* 468: 811–814.

Kelley, C. P., Mohtadi, S., Cane, M. A., Seager, R. and Y. Kushnir. 2015. Climate change in the fertile crescent and implications of the recent Syrian drought. *Proceedings of the National Academy of Sciences, USA* 112: 3241–3246.

Kendal, J., Tehrani, J. J. and J. Odling-Smee. 2011. Human niche construction in interdisciplinary focus. *Philosophical Transactions of the Royal Society B* 366: 785–792.

Kennedy, B. K., Berger, S. L., Brunet, A. et al. 2014. Geroscience: Linking aging to chronic disease. *Cell* 159: 709–713.

Kennett, D. J. and T. P. Beach. 2013. Archeological and environmental lessons for the Anthropocene from the Classic Maya Collapse. *Anthropocene* 4: 88–100.

Knight, J. and S. Harrison. 2014. Limitations of uniformitarianism in the Anthropocene. *Anthropocene* 5: 71–75.

Kruglyak, L. 2016. Big questions in evolution. *Cell* 166: 529.

Laland, K. N., Uller, T., Feldman, M. W. et al. 2015. The extended evolutionary synthesis: Its structure, assumptions and predictions. *Philosophical Transactions of the Royal Society B* 282: 20151019. http://dx.doi.org/10.1098/rspb.2015.1019.

Lebel, C., Walker, L., Leemans, A., Phillips, L. and C. Beaulieu. 2008. Microstructural maturation of the human brain from childhood to adulthood. *NeuroImage* 40: 1044–1055.

Lewis, S. L. and M. A. Maslin. 2015. Defining the Anthropocene. *Nature* 519: 171–180.

Lopez-Otin, C., Blasco, M., Partridge, L., Serrano, M. and G. Kroemer. 2013. The hallmarks of aging. *Cell* 153: 1194–1217.

Lyons, S. K., Amatangelo, K. L., Behrensmeyer, A. K. et al. 2016. Holocene shifts in the assembly of plant and animal communities implicate human impacts. *Nature* 529: 80–83.

Maitre, J.-L., Turlier, H., Illukkumbura, R. et al. 2016. Asymmetric division of contractile domains couples cell positioning and fate specification. *Nature* 536: 344–348.

Martin, J.-L., Maris, V. and Simberloff, S. 2016. The need to respect nature and its limits challenges society and conservation science. *Proceedings of the National Academy of Sciences, USA* 113: 6105–6112.

McKinney, M. L. 2002. Urbanization, biodiversity, and conservation. *BioScience* 52: 883–890.

Nadeau, N. J., Pardo-Diaz, C., Whibley, A. et al. 2016. The gene cortex controls mimicry and crypsis in butterflies and moths. *Nature* 534: 106–110.

Newbold, T., Hudson, L. N., Arnell, A. P. et al. 2016. Has land use pushed terrestrial biodiversity beyond the planetary boundary? A global assessment. *Science* 353: 288–291.

Oliver, T. H. 2016. How much biodiversity loss is too much? *Science* 353: 220–221.

Panko, B. 2016. Nighttime light pollution covers nearly 80% of the globe. *Science News* June 10. doi:10.1126/science.aaf5777.

Paxton, C. W., Davy, S. K. and V. M. Weis. 2013. Stress and death of cnidarians host cells play a role in cnidarians bleaching. *Journal of Experimental Biology* 216: 2813–2820.

Pennisi, E. 2015. Human impacts on ecosystems began thousands of years ago. *Science* 350: 1452.

Pennisi, E. 2016. We've destroyed one-tenth of Earth's wilderness in just 2 decades. *Science News* September 8. doi:10.1126/science.aah7279.

Pickett, S. T. A., Cadenasso, M. L., Grove, J. M. et al. 2008. Beyond urban legends: An emerging framework of urban ecology, as illustrated by the Baltimore ecosystem study. *BioScience* 58: 139–150.

Plusa, B. and A.-K. Hadjantonakis. 2016. Mechanics drives cell differentiation. *Nature* 536: 281–282.

Prud'homme, B. and N. Gompel. 2010. Genomics hourglass. *Nature* 468: 768–769.

Prufer, K., Munch, K., Hellmann, I. et al. 2012. The bonobo genome compared with the chimpanzee and human genomes. *Nature* 486: 527–531.

Quistad, S. D., Stotland, A., Barott, K. L. et al. 2014. Evolution of TNF-induced apoptosis reveals 550 My of functional conservation. *Proceedings of the National Academy of Sciences, USA* 111: 9567–9572.

Ratcliff, W. C., Denison, R. F., Borrello, M. and M. Travisano. 2012. Experimental evolution of multicellularity. *Proceedings of the National Academy of Sciences, USA* 109: 1595–1600.

Ripple, W. J. and R. L. Beschta. 2005. Linking wolves and plants; Aldo Leopold on trophic cascades. *Bioscience* 44: 613–621.

Ripple, W. J., Estes, J. A., Beschta, R. L. et al. 2014. Status and ecological effects of the world's largest carnivores. *Science* 343: 151–162.

Rompolas, P., Mesa, K. R., Kawaguchi, K. et al. 2016. Spatiotemporal coordination of stem cell commitment during epidermal homeostasis. *Science* 352 (6292): 1471–1474. doi:10.1126/science.aaf7012.

Ryan, T. M. and C. N. Shaw. 2015. Gracility of the moder *Homo sapiens* skeleton is the result of decreased biomechanical loading. *Proceedings of the National Academy of Sciences, USA* 112: 372–377.

Sankararaman, S., Malllick, S., Patterson, N. and D. Reich. 2016. The combined landscape of Denisovan and Neanderthal ancestry in present-day humans. *Current Biology* 26: 1241–1247.

Sarrazin, F. and J. Lecomte. 2016. Evolution in the Anthropocene. *Science* 351: 922–923.

Saxena, A. and K. L. Cooper. 2016. Fin to limb within our grasp. *Nature* 537: 176–177.

Scheffer, M. 2016. Anticipating societal collapse, hints from the stone age. *Proceedings of the National Academy of Sciences, USA* V113: 10733–10735. doi:10.1073/pnas.1612728113.

Singer, M. A. 2013. *Are Chronic Degenerative Diseases Part of the Ageing Process? Insights from Comparative Biology.* Nova Science Publishers, New York.

Singer, M. A. 2015. *Human Ageing: A Unique Experience—Implications for the Disease Concept.* World Scientific Publishing, Singapore.

Singer, M. A. 2016. The origins of aging: Evidence that aging is an adaptive phenotype. *Current Aging Science* 9: 99–115.

Slagsvold, T. and K. L. Wiebe. 2007. Learning the ecological niche. *Philosophical Transactions of the Royal Society B* 274: 19–23.

Slagsvold, T. and K. L. Wiebe. 2011. Social learning in birds and its role in shaping a foraging niche. *Philosophical Transactions of the Royal Society B* 366: 969–977.

Soliveres, S., van der Plas, F., Manning, P. et al. 2016. Biodiversity at multiple trophic levels is needed for ecosystem multifunctionality. *Nature* 536: 456–459.

Tchernov, D., Kvitt, H., Haramaty, L. et al. 2011. Apoptosis and the selective survival of host animals following thermal bleaching in zooxanthellate corals. *Proceedings of the National Academy of Sciences, USA* 108: 9905–9909.

van Otterdijk, S. D. and K. B. Michels. 2016. Transgenerational epigenetic inheritance in mammals: How good is the evidence? *FASEB Journal* 30: 2457–2465.

van't Hof, A. E., Campagne, P., Rigden, D. et al. 2016. The industrial melanism mutation in British peppered moths is a transposable element. *Nature* 534: 102–105.

van't Hof, A. E., Edmonds, N., Dalikova, M., Marec, F. and I. J. Saccheri. 2011. Industrial melanism in British peppered moths has a singular and recent mutational origin. *Science* 332: 958–960.

Varki, A. and T. K. Altheide. 2005. Comparing the human and chimpanzee genomes: Searching for needles in a haystack. *Genome Research* 15: 1746–1758.

Via, S. 2009. Natural selection in action during speciation. *Proceedings of the National Academy of Sciences, USA* 106: 9939–9946.

Voosen, P. 2016. Anthropocene pinned to postwar period. *Science* 353: 852–853.

Wachsmuth, D., Cohen, D. A. and H. Angelo. 2016. Expand the frontiers of urban sustainability. *Nature* 536: 391–393.

Waters, C. N., Zalasiewicz, J., Summerhayes, C. et al. 2016. The Anthropocene is functionally and stratigraphically distinct from the Holocene. *Science* 351: 137–147.

Watson, J. E. M., Shanahan, D. F., Di Marco, M. et al. 2016. Catastrophic declines in wilderness areas undermine global environment targets. *Current Biology* 26: 1–6.

Whitam, T. G., Bailey, J. K., Schweitzer, J. A. et al. 2006. A framework for community and ecosystem genetics: From genes to ecosystems. *Nature Reviews* 7: 510–523.

Whitam, T. G., Difazio, S. P., Schweitzer, J. A. et al. 2008. Extending genomics to natural communities and ecosystems. *Science* 320: 492–495.

Whiten, A. 2011. The scope of culture in chimpanzees, humans and ancestral apes. *Philosophical Transactions of the Royal Society B* 366: 997–1007.

8 Human Culture
Urbanization and Human Aging

Michael A. Singer

CONTENTS

INTRODUCTION

As a starting point, what is culture? In this chapter, culture will be defined as an acquired collection of practices and knowledge modified by each generation and passed down to succeeding generations with high fidelity. Culture is an inherited system transmitted across generations via nongenetic mechanisms and is considered the principal force in human evolution (Singer 2015). Although culture is not unique to our species (Whiten 2011; Gross 2016a), humans have expanded the scope of this inherited system much more extensively than any other species. One very significant human cultural activity has been the development of a unique urban environment known as the city and currently more than half of all humans reside in cities. The development of a unique urban environment or ecosystem can be defined as the process of urbanization and urbanization is the specific cultural activity discussed in this chapter. This specific cultural activity, like all aspects of culture, will be modified by each generation and then that modified version passed down to the next generation.

This chapter has been structured as follows. In the first sections, the nature of cities is explored with a focus on features that appear to be common to all cities. These common features can be quantified through the technique of allometric scaling. Cities represent a novel urban ecosystem and evidence is presented that the urban environment is a powerful driver of the evolution of species. In the final section, the role of the urban ecosystem in shaping the aging and life span trajectory of humans is examined.

The premise of this chapter is that cities are now the principal vehicle of human evolution, that is, transmission of heritable phenotypes to succeeding generations and the focus will be on the aging/life span phenotype.

CITIES, SCALING

The process of urbanization probably began with the introduction of agricultural practices, including plant and animal domestication, about 10,000 years ago (Gross 2016b). One of the earliest cities was Uruk about 5200 years ago in southern Mesopotamia (Department of Ancient Near Eastern Art 2000). The growth of the urban population, using the largest 50 cities as the metric, has been exponential since about 1850 and by 2008 more than 50% of the world's population lived in cities (Batty 2011). It has been estimated that by 2050, the proportion of the global population residing in cities will reach 70% (Batty 2011). *Hence, for our species the urban landscape will be the environmental context in which current and future evolution will take place.*

As cities expand and coalesce, it will become increasingly difficult to delineate precise geographical boundaries for individual cities (Batty 2013). In the future, cities will have to be considered within a framework of clusters or systems. As Batty (2011, 2013) pointed out, with the growth of cities network effects will supervene and cities will begin to coalesce functionally. Clearly, the ecological effects of an expanding globalized network of cities will be extensive and will be exacerbated by the rapid pace at which this process is occurring.

Despite obvious differences in cities across both time and space, they appear to share a common set of structural and functional features as defined by allometric scaling analysis. Cities that have been studied include; pre-Hispanic Aztec settlements in Mexico (Ortman et al. 2014), medieval European cities (Cesaretti et al. 2016), present day European cities (Bettencourt and Lobo 2016), and US metropolitan areas (Bettencourt 2013). The primary structural feature scaled was urbanized land area whereas the functional features scaled included gross metropolitan domestic product, number of patents as a measure of urban innovation, and employment.

The mathematical relations between urban functional/structural characteristics and urban population can be described by power law scaling equations; the scaling exponent is less than one for urbanized land area, approximately one for employment, and greater than one for urban gross domestic product and number of patents. Hence, as the urban population grows, productivity (gross domestic product, number of patents) increases at a greater rate and urban area increases at a slower rate than the rate of population growth. Larger cities are proportionately more productive and proportionately denser than smaller cities. These scaling relationships appear to be fundamental relationships in that they hold across a variety of urban areas both present day and historic.

Scaling analysis has been applied to a cluster of cities in the United Kingdom that are interconnected directly and indirectly through the road network. The data set included 283 towns and cities with a population of more than 10,000. The parameter calculated was accessibility, a measure of flows into and out of each urban area from and/or to all other urban areas in the data set. Accessibility scaled to population by a power law equation with a scaling coefficient greater than one. Hence, larger cities are not only proportionately more productive and denser than smaller cities but they are also characterized by a proportionately greater movement of goods and people to and from other connected cities. This example demonstrates that scaling rules appear to apply to interactions between cities and not just to features of the city itself.

Bettencourt and colleagues have developed a model consistent with the scaling relationships observed for cities (Bettencourt 2013; Ortman et al. 2014). Small settlements are characterized by a low population density and a lack of spatial organization; so-called amorphous settlements. As a small settlement grows through a process of agglomeration, a richer mix of social and economic interactions develop between individuals and groups of individuals. In addition, not only do socioeconomic interactions occur but also populations within the city mix with each other. This mixing requires that individuals engage in different types of transport processes depending upon distances traveled. As cities grow, population density increases and land use within the city becomes structured into settled areas, public spaces, and transecting roads. The scaling relationships indicate that population growth exceeds growth in urban space so the city becomes denser. Within Bettencourt's model, socioeconomic interactions within the city underlie the determination of productivity

(Cesaretti et al. 2016) and growth of productivity exceeds growth in number of city inhabitants; large cities are proportionately more productive than small cities.

The framework developed by Bettencourt (2013) and Ortman et al. (2014) can be considered a socioeconomic model since the main variables included were social interactions and mixing of individuals in a structured physical space. In this model, land use and productivity were determined by these socioeconomic interactions. But cities are also considered ecosystems (Grimm et al. 2000) since ecological influences play a major role in shaping city structure and function. A more complete framework for modeling the complexities of cities must include both socioeconomic and ecological factors as discussed by McPhearson et al. (2016).

Cities as Ecosystems

For most of human evolution, our ancestors lived in "natural" ecosystems and occupied one of the higher trophic levels. Major impacts of human cultural practices on the environment have only occurred in the past 10,000–15,000 years (Grimm et al. 2000). Urbanization is one of the most significant effects of human cultural practices. Cities are now the habitat for more than half of the human population and the rapid pace of urbanization means that an even larger proportion of the future global human population will reside in cities.

Natural ecosystems are composed of overlapping ecological niches. An ecological niche has both physical and functional attributes and can be defined as the environmental space occupied by a species and within which that species can carry out its evolved behaviors/functions. Since an ecosystem is a collection of overlapping ecological niches, it will contain a diverse mixture of interacting species organized in a hierarchical manner according to trophic levels. Ecosystems are highly structured and regulated biotic/abiotic organizations which have evolved to optimize the functioning of both individual species as well as the ecosystem as a whole.

Natural ecosystems, like urban ecosystems (cities), can be described by power law scaling equations. Across a variety of aquatic and terrestrial ecosystems, the relationships between predator–prey biomass and production-population biomass followed power law equations with scaling exponents of about 0.7 (Hatton et al. 2015). The observation that, for a variety of cities, productivity scales supralinear with population while for natural ecosystems productivity scales sublinear with population biomass implies that these two types of ecosystems are fundamentally different (Kerkhoff and Enquist 2006; Hatton et al. 2015).

The development of cities as a novel ecosystem is an obvious example of the inventiveness of human cultural practices. However, human activities have also had profound negative effects on "natural" ecosystems worldwide; examples include loss of biodiversity, loss of wilderness areas, and deforestation to name a few (Estes et al. 2011; Newbold et al. 2016; Oliver 2016; Watson et al. 2016). However, in all of the current discussions related to the impact of human activities on the biological and physical properties of our planet, an aspect that has received little attention is the effect of human culture on the evolution of the human species itself. Since most humans now reside in cities, the urban ecosystem must be considered the principal environmental space that will shape the evolution of humans and other species as well. As we design the cities of tomorrow (Groffman et al. 2017), we are actually determining our own evolution in ways over which we have no direct control. The next section reviews some example of the influence of cities on the evolution of its own inhabitants.

Effects of Cities on Evolution of Its Inhabitants

Plants

Thompson et al. (2016) examined the effect of an urban environment on the cyanogenesis phenotype of the plant white clover (*Trifolium repens*). The cyanogenic phenotype is underwritten by two genetic loci, CYP79D15 (Ac) and Li. One gene, Ac, codes for the plant's ability to produce cyanogenic glucosides which are cyanide containing sugars while the second gene Li codes for the enzyme that catalyzes hydrolysis of the cyanogenic glucoside. The two components of the cyanide-generating system are kept separate; glucosides are stored in the vacuoles of leaf and stem cells and the enzyme is stored in the cell wall. When herbivore predators such as grasshoppers, slugs, snails, and voles "eat" the clover, the tissue damage brings the two components together and hydrogen cyanide, a toxin to the predator, is released. For clover to produce cyanide, the plant must possess a dominant allele at both loci. Cyanogenesis not only acts as a deterrent to predators but also reduces plant freezing tolerance.

Thompson et al. (2016) surveyed over 5000 plants from 266 populations of white clover along 50 km transects radiating out from the core of each of four large metropolitan areas. They observed that the proportion of plants with the cyanogenesis phenotype decreased with increasing proximity to the downtown urban core. The frequency of dominant alleles at the two loci, Ac and Li, also showed the same distribution pattern. The investigators observed that ground surface temperatures showed a gradient with decreasing temperatures toward the urban core. This temperature gradient inversely correlated with the gradient in snow cover thickness; snow cover thickness was less in the urban core than in rural areas. Since cyanogenesis reduces the plant's freezing tolerance, plants subject to colder ground temperatures in the urban core would be susceptible to frost-induced tissue damage resulting in autocatalytic generation of self-toxic hydrogen cyanide. Hence, the clover plant has adapted to the different micro climate in the urban core compared to more rural areas by altering the expression of the cyanogenic phenotype.

Tropical Lizard

Winchell et al. (2016) looked at the effect of urbanization on morphological phenotypes in the tropical lizard *Anolis cristatellus* living in Puerto Rico. In the nineteenth and early twentieth century, Puerto Rico was subject to intense agriculturalization mainly for cash crops. Subsequently after the middle of the twentieth century, Puerto Rico underwent rapid urbanization and now about 94% of the population resides in cities. *A. cristatellus* is an arboreal lizard mainly living in the trunk-ground zone. It has long limbs and a stocky build and clings to broad surfaces. It has an island wide distribution and is found in both natural and urban areas. Winchell et al. (2016) compared morphologic traits between adult male lizards residing in urban and natural habitats. Natural sites had a diverse collection of native and nonnative plant species while urban sites were chosen from the three largest cities. In general, urban areas were warmer and had a higher relative humidity, much reduced canopy cover, and much more extensive impervious surface areas.

Urban lizards tended to perch on wide artificial surfaces such as walls, glass windows, metal fence posts, and the sides of buildings. These artificial perches were much broader than the perches used in natural areas. Hindlimb and forelimb lengths were longer in urban lizards and urban lizards had a greater number of subdigital lamellae (adapted for climbing and adhesion) in both fore- and hindlimbs. To test the inheritance of these morphological differences, Winchell et al. (2016) mated female–male pairs from urban and natural habitats. Morphologic differences between urban and natural populations were maintained in captive-reared offspring; offspring from urban populations had more lamellae and longer front and rear limbs than offspring from natural populations.

The longer limbs and increased number of subdigital lamellae in urban lizards is consistent with the differences between the two habitats. Urban areas are more open with less canopy cover and longer limbs facilitate locomotion in this type of terrain. Artificial surfaces in urban areas are smoother, harder, and broader than the natural surfaces in the rural areas. Perching on these artificial surfaces would be facilitated by an increased number of lamellae. Hence, the urban terrain has selected for certain morphological variants and the mating experiments suggest that these differences have a genetic basis.

Small Mammals

Snell-Rood and Wick (2013) tested the hypothesis that since the urban ecosystem represented a novel environment, it would select for increased behavioral plasticity. They used small mammals such as voles, bats, mice, shrew, gophers, and squirrels as test subjects and rural and urban areas in Minnesota as the test sites. In this state, about 75% of the population lives in one urban area, Minneapolis-Saint Paul. They used cognition (measured as cranial capacity) as a proxy for behavioral plasticity. Of the 10 species measured, two showed a significant difference in mean cranial capacity between urban and rural populations of each of the species with the urban population having a larger mean size. When relative differences in cranial capacity between rural and urban populations were calculated, all 10 species displayed a significant trend for the urban population of a species to have a relatively larger cranial capacity than the rural population of the same species. Fertility as measured by litter size showed a positive correlation with percentage increase in cranial capacity for urban over rural populations.

The investigators drew several conclusions. First, behavioral plasticity as measured by cranial capacity appeared to be positively selected for by an urban environment compared to a rural one. Second, species with a high reproductive rate and hence the potential for rapid population growth had a larger cranial capacity in urban populations compared to rural ones. In general, larger populations of a species are characterized as having more phenotypic plasticity than smaller populations of that species.

As discussed in Chapter 7, Harris et al. (2016) studied the effect of urbanization on the evolution of the white-footed mouse (*Peromyscus leucopus*) in the New York City metropolitan area. Over the past 400 years, this area has undergone intense urbanization which resulted in a fragmentation of the landscape into isolated green spaces (parks). During this process of urbanization, small populations of *P. leucopus* became trapped within these physically isolated green parks and Harris et al. (2016) documented that these isolated subpopulations evolved genetic differentiation. This case study illustrates an important evolutionary timeline; in less than 400 years, this small mammal had genetically evolved within the context of an urban ecosystem.

These examples, plant, lizard, and small mammals, underscore the powerful influence of the urban environment on species evolution. Since the urban ecosystem is the principal habitat of humans, the implication is clear that the urban landscape will be a powerful influence on our species evolution. In the next section, I review the limited data available relating urbanization to human aging and life span.

URBANIZATION; HUMAN AGING AND LIFE SPAN

Colchero et al. (2016) studied what they refer to as longevous populations. They define a longevous population as having two characteristics; a long average life span and a low variation in life span. Variation in life span can be understood as follows (Smits and Monden 2009). At the individual level, length of life varies between individuals, that is, there is an inequality in length of life across individuals. At the population level, length of life is closely related to life expectancy, the expected average length of life based upon the current mortality pattern. For a population, length of life inequality is the variation or splay of length of life distribution. A population with a high level of life span inequality will have a wide age distribution for life expectancy, whereas a population with a low level of life span inequality will have a very narrow age distribution for life expectancy.

Since the ultimate outcome of the aging process is death, at the level of the individual, an early age at death indicates a more rapid aging trajectory. At the population level, life span inequality can be used as a proxy measure for aging, that is, a population with a high level of inequality is composed of individuals with very different rates of aging whereas a population with a low level of inequality is composed of individuals with very similar rates of aging.

Colchero et al. (2016) used published life history data for human populations (hunter-gatherers; Hadza and Ache, preindustrial Sweden [1751–1759], modern Sweden [2000–2009], modern Japan [2012], modern Nigeria, India, Russia, China, and the United States; all from 2013 and England [1600–1725]), six species of nonhuman primates and nine additional mammalian species. For the six nonhuman primates, life expectancy, for example, for females varied from 5.58 to 28.1 years and for the human populations (e.g., females) life expectancy varied from 37 (hunter-gatherers) to 86.4 years (modern Japan).

Although there were differences between males and females for both human and nonhuman primate populations, I will not refer to these differences since they do not change the general conclusions.

The investigators plotted the logarithm of life span equality (Y-axis) versus life expectancy (X-axis) for both human populations and the six nonhuman primate species. The human data set spanned a time period from hunter-gatherers through to 2013. There was a highly significant positive correlation between life expectancy and life span equality for both humans and nonhuman primates; greater life expectancy correlated with a lower level of life span inequality. The data fell into two separate regression lines; one for nonhuman primates with a slope of 0.014 and one for human populations with a slope of 0.037. These two lines intersected at the level of preindustrial hunter-gatherer populations. For the nine additional mammalian species studied, there was no correlation between life expectancy and life span inequality.

What is the interpretation of these observations? Primates, both humans and nonhuman species, form social groupings. Nonhuman primates with longer life expectancies showed a reduced level of life span inequality compared to nonhuman primates with shorter life expectancies. Human hunter-gatherer populations fell on this nonhuman primate line and hence nonhuman primates and hunter-gatherer populations form a continuum.

Group size in nonhuman primates is small and varies from three for Sifaka to 200 for baboons while hunter-gatherers form groups varying from 6 to 80 (Hill et al. 2011). The hunter-gatherer populations are at the intersection of the two lines. This finding indicates that the period when humans were hunter-gatherers was the time at which the slope of the relationship between life expectancy and life span equality changed for subsequent human populations; the slope of the relationship became steeper with a value of 0.037 compared to a value of 0.014 for the nonhuman primate-hunter-gatherer line. This change in slope corresponds to the introduction of agriculture about 10,000 years ago; social groupings became larger and the process of urbanization, as we know it, began. For the Swedish population between the years 1751–1759, life expectancy was less than 40 years and life span inequality was high; values not much different from those for hunter-gatherers. It is about this time, circa 1850, that the pace of urbanization rapidly accelerated (Batty 2011). For the Japanese population in 2012, life expectancy was in the range of 80 years for males and 86 years for females. For both genders, life span inequality was very low.

The data of Colchero et al. (2016) can be linked to the process of urbanization via the following hypothesis. I would propose that urbanization is the primary factor responsible for shifting our species off the nonhuman primate-hunter-gatherer trajectory and placing us on a course characterized by comparatively greater increases in life expectancy and greater reductions in rate of aging (decreasing life span inequality).

This proposal is consistent with earlier data presented by Smits and Monden (2009). These investigators looked at life expectancy and length of life inequality for the total human population and for specific geographical areas. The data set for the global population was for the year 2000 and included life tables for 191 countries; over 99% of the world population was covered. A plot of length of life inequality (Y-axis) versus life expectancy (X-axis) gave a highly significant inverse relationship for both men and women (plotted separately); high life expectancy was associated with low level of length of life inequality. For the global population, variation in length of life inequality was due principally to within country variation (91.3%) as opposed to between country differences (8.7%).

The investigators illustrated plots of length of life inequality (Y-axis) versus life expectancy (X-axis) for four geographical regions covering approximately the last 200 years; England and

Wales 1841–2003, France 1806–2004, United States 1850–2003, and Sweden 1751–2005. During these time periods, these geographical areas would have been undergoing rapid urbanization. In all four regions, there was a strong inverse relationship between these two parameters and a progressive increase in life expectancy and reduction in length of life inequality over the time spans examined. For example, in the United States over the period 1850–2003, life expectancy increased from about 57 to 75 years and length of life inequality decreased from 0.20 to 0.10 (Theil index). In the year 1790, 5.1% of the US population lived in cities, but by 2010, the proportion of the population living in cities had increased to 80.7% (Jenkins et al. 2016). In other words, a significant increase in life expectancy, a significant reduction in the rate of aging, and a rapid rate of urbanization occurred in the population of the United States over the time period spanning 1850–2003.

Although the hypothesis implicating urbanization as the causal factor responsible for an increased life expectancy and reduced rate of aging is based on data which show only a temporal association, known mechanisms that could account for such an association (discussed in the next section) suggest that the relationship is most likely causal.

Mechanisms Linking Urbanization and Aging and Life Span

There are probably multiple mechanisms that underlie modifications to human aging and life span by the process of urbanization. Several possibilities are presented in this section.

Alteration in Natural Selection

Evolution proceeds by natural selection acting on phenotypic variation in a population constrained by certain environmental pressures. If the selected phenotype has a heritable basis, it can be passed down to successive generations. Humans have developed a novel environment, the urban ecosystem which is now the specific habitat for the global majority of humans. This novel ecosystem will present different environmental pressures than those experienced by a human population in a natural ecosystem. In this manner, the direction and action of natural selection will be altered and the evolution of humans and other species inhabiting this urban environmental space modified. However, this is not the only possible mechanism linking the process of urbanization to alterations of human aging and life span.

Genomic Diversity and the Rural–Urban Divide

There is a wealth of studies documenting differences in life expectancy and the incidence of so-called chronic diseases between rural and urban environments (e.g., Singh and Siahpush 2014). In the United States by 2005–2009, the absolute gap in life expectancy for men and women combined between metropolitan and rural environments had increased to 2.0 years (greater in urban compared to rural settings). In that same period, cardiovascular and pulmonary diseases as well as lung cancer accounted for 70% of the urban–rural gap in life expectancy. Generally, these rural–urban discrepancies have been attributed to the differences in social determinants of health including accessibility to medical care. However, could there also be a genetic component to urban–rural disparities?

Jenkins et al. (2016) have discussed possible urban–rural disparities in genomic diversity. They point out that small rural communities in the United States are subject to founder effects and experience reduced admixture with other outside communities. Both these processes will lead to decreased genetic diversity compared to large metropolitan areas and to a higher frequency of rare possibly deleterious alleles. In addition, the process of genetic drift has a greater impact in small isolated communities. Unfortunately, as Jenkins et al. (2016) point out, there are no studies comparing the genetic diversity of urban versus rural populations. However, it is reasonable to speculate that reduced genetic diversity, due to founder effects and lack of admixture with outside communities, will make rural residents more susceptible to adverse health outcomes thereby lowering life expectancy in rural environments. It has been well documented that genetically diverse populations are much less susceptible to the spread of infectious diseases compared to small isolated more geneti-cally homogeneous populations (Lively 2010).

Hence, reduced genetic diversity of rural compared to large urban populations plus differences in social determinants in health between the two settings appear to play a significant role in explaining the disparity in rural–urban aging and life span.

Changes in Gut Microbiota

The human gastrointestinal tract harbors a large and diverse collection of microbial species and there is accumulating evidence that the gut microbial ecosystem is very important for the health of the host human (Saraswati and Sitaraman 2015). For example, frailty scores in elderly individuals were associated with a decrease in microbial diversity (Saraswati and Sitaraman 2015). Hornef and Pabst (2016) in a recent commentary reviewed the capacity of specific commensal gut microbes such as *Faecalibacterium* to modulate the immune/inflammatory response through the production of various factors. Depleted gastrointestinal levels of this bacterium have been associated with development of Crohn's disease, a form of inflammatory bowel disease (Velasquez-Manoff 2015).

Moeller et al. (2014) detailed the changes that have occurred in the gut microbiota during the course of human evolution. These investigators used a phylogenetic approach with sequencing of the microbial V4 region of 16S rRNA to characterize microbial taxa in fecal samples of wild nonhuman primates (chimpanzees, bonobos, and gorillas), urban human populations (USA, Europe), rural human populations (Malawi), hunter-gatherers (Tanzania), and a preindustrial population (rainforest southern Amazon). All human and ape populations shared a core set of bacterial genera consistent with human and nonhuman primates having a common ancestor.

Humans diverged from the common ancestor of chimpanzee and bonobo about 7–13 million years ago. Since the divergence, the human microbiota has shown a number of changes; an increase in microbial taxa associated with diets rich in animal fat and protein and a decrease in taxa associated with plant-fermentation and degradation of complex plant polysaccharides. The human gut microbiota displayed a lower level of diversity, measured as the number of bacterial genera per individual, than any of the nonhuman primate species tested. Among the human populations studied, urban dwellers had the lowest levels of microbial diversity while the preindustrial Amazon population had the highest level of diversity although still lower than that of nonhuman primates.

This study detailed the coevolution of primates and their gastrointestinal collection of bacterial species. Compared to nonhuman primates, humans have evolved a gut microbiota characterized by much less biodiversity as measured by the number of bacterial taxa per individual. Among human populations, those living in an urban ecosystem have the lowest level of microbiota diversity.

In an interesting study, Martinez et al. (2015) compared the biodiversity and composition of gut microbiota between preindustrial and urbanized humans and proposed a mechanism to account for the differences. Study participants included populations from two traditional communities in Papua New Guinea (PNG) and a US urban community composed of male and female students attending the University of Nebraska. Fecal bacterial taxa were identified using 16S rRNA sequencing.

The whole fecal data set from the PNG population contained a higher number of bacterial taxa than the whole fecal data set from the urban US population indicating a greater total biodiversity in the PNG compared to the US populations. On an individual level, PNG residents showed a higher level of microbiota diversity (more bacterial taxa per individual) than urban individuals in the United States while between individual differences were lower in the PNG population. Hence, a traditional preindustrial PNG community compared to an urban US society has a greater number of bacterial taxa per individual but is more homogenous in terms of less variation in biodiversity within the population. Martinez et al. (2015) noted that these results were consistent with previous rural–urban comparative studies.

The investigators documented the presence of a core set of bacterial species common to both PNG and US populations, a finding previously documented for US and European populations. Of the total of 1520 bacterial taxa identified in the total fecal data set, 664 were detected in both PNG and the United States. This observation implies that despite large geographical distances and cultural differences the human gut microbiota is dominated by globally distributed microbial species (Martinez et al. 2015).

The PNG population shared a core group of 186 taxa of which 47 could not be detected in the United States. The US population shared a core group of 169 taxa of which only four could not be detected in PNG. These observations underscore some of the compositional differences between the US and PNG populations.

The investigators proposed the following model to explain the differences in biodiversity between the PNG and US populations. PNG settlements are characterized by a lower level of community health practices such as sanitation and water treatment which results in transmission or dispersal of microbes between individuals as well as greater environmental exposure of individuals to microbes. This process leads to greater individual and total levels of biodiversity but reduced between individual differences in biodiversity. Within an urban population, community health measures decrease the dispersal and transmission of microbes between individuals as well as reduce environmental exposure to microbes. The result is less individual biodiversity but more variation in biodiversity between individuals.

Within this model, dispersal or transmission of microbes is the key feature underlying the differences in microbiota diversity between traditional and urban populations. In addition, microbial dispersal predisposes traditional societies to epidemics of infectious diseases whereas reduced microbial dispersion in urban societies might preclude the acquisition of certain microbial taxa that protect against noncommunicable immune/inflammatory diseases (Martinez et al. 2015).

Biagi et al. (2016) looked at the relationship between gut microbiota and human longevity. The study populations included males and females in the following age categories; 105–109, 99–104, 65–75, and 22–48 years. These subjects all came from the Emilia Romagna, a region in northeast Italy containing the city of Bologna. This geographical area has a high prevalence of long living individuals. In Italy, the prevalence of 100+ year old individuals is 31.4 per 100,000 inhabitants; in Emilia Romagna it is 31.9 and in the city of Bologna, the prevalence is 64.0. In this particular region of Italy, an urban setting appears to be associated with a higher prevalence of very elderly individuals compared to a rural environment. For all subjects, fecal analysis was done by sequencing the V3–V4 region of the bacterial 16S rRNA gene.

The fecal microbiota of all age groups was dominated by three taxa: Bacteroidaceae, Lachnospiraceae, and Ruminococcaceae. The cumulative relative abundance decreased with aging from 77.8% in 22–48 year olds to 57.7% in 105–109 year olds. With the age-related decrease in these three taxa, subdominant taxa contributed a greater relative proportion. Individual bacterial families showed variations in age-related abundance with some showing a decrease in abundance with age and others showing an increase.

The investigators looked at the co-occurrence of pairs of bacterial taxa and on the basis of the co-occurrence pattern were able to identify four co-occurrence groups. One of these named the *Bacteroides* co-occurrence group contained the three dominant taxa referred to in the previous paragraph. This group appeared to form a core microbiota accounting for about 70% of relative abundance in all samples. The abundance of this core group fell with age from 75.3% in the youngest group to 64.9% in the oldest. The other three co-occurrence groups did not show any consistent changes with age. The investigators did note that in the oldest age group, there was an increase in several bacterial taxa, specifically *Christensenellaceae*, *Akkermansia*, and *Bifidobacterium*. Everard et al. (2013) demonstrated the occurrence of decreased abundance of *Akkermansia* in obese and type 2 diabetic mice. Restoring *Akkermansia* abundance in these mice improved their metabolic abnormalities. On the other hand, the abundance of the taxon *Faecalibacterium* decreased with age and as previously noted, members of this taxon produce factors which modify the immune/inflammatory response.

In summary, with increasing age, there is a decrease in the abundance of a core microbiota formed predominantly by three taxa, Ruminococcaceae, Lachnospiraceae, and Bacteroidaceae and a concomitant increase in the abundance of subdominant taxa. In addition, the very elderly show an increase in relative abundance of specific bacterial taxa that have been associated with healthy aging.

The human gut microbiota represents a co-evolving collection of symbionts which plays a very important role in the regulation of a number of metabolic processes as well as in modulating the immune/inflammatory process. The gut microbiota is also involved in the aging process since the microbiota of the very elderly differs from that of younger adults (Biagi et al. 2016). It is safe to say that our understanding of the effects of host-gut microbiota interactions on host metabolic phenotypes is still very much in its infancy.

Humans have a core gut microbiota that has been derived from our common ancestor with chimpanzees and bonobos. During our evolutionary course, gut microbiota diversity has decreased compared to nonhuman primates and there have been compositional changes consistent with a diet oriented toward animal fat and protein. The process of urbanization has further decreased gut microbiota diversity compared to traditional preindustrial human populations. However, urbanization has been associated with a lengthening of human life span and a reduction in the rate of aging and the very elderly display some specific changes in the composition and biodiversity of the gut microbiota that could be partially responsible for these changes in aging and life span.

CONCLUSIONS

Human cultural activities have had deep effects on our planet, so much so that there is a proposal to rename the current geological era the Anthropocene. When the Anthropocene began is still up for debate (Ruddiman et al. 2015). One of the more visible aspects of our cultural activities is the development of a novel ecosystem, the city. Currently, cities are the specific environmental space which more than half of all humans inhabit.

However, the city as an ecosystem differs in fundamental ways from natural ecosystems. As we design future sustainable cities, we should not be under the illusion that such designs will affect only the convenience and "quality" aspects of our lives. When we modify the structure of our cities, we will be unknowingly changing the evolutionary trajectory of our species. This is an important principle. Humans continue to evolve, but there is evidence for "evolutionary override" caused by cultural, economic, and social factors (Courtiol et al. 2016). *The premise of this discussion is that the human-developed urban ecosystem, that is, the city is now the principal cultural innovation driving the evolution of our species.* However, since evolution is by nature a random directionless process, predicting future effects of human cultural activities (in particular urbanization) on the course of our evolution will be an almost impossible challenge.

In this chapter, the focus has been on human aging and life span. Since the introduction of agricultural practices including plant and animal domestication, there has been a significant increase in global life expectancy and a significant decrease in the rate of aging. I would propose that the principal cultural activity responsible for these global changes in aging and life span is the process of urbanization which over the past several hundred years has shown an exponential increase in rate. An obvious question is how might the process of urbanization promote these phenotypic changes? In the previous sections, several mechanisms were considered likely candidates to account for the influence of an urban landscape on the trajectory of human aging/life span.

First, the urban ecosystem, which is now the habitat of most humans, represents a novel environmental space and this novel environment will modify the action and direction of natural selection.

Second, cities are home to large numbers of humans and many of these individuals have originated from geographical areas outside of the urban site. The city thus becomes a cauldron in which the mixing of diverse humans occurs. As such, cities represent regions of genetic diversity and genetic diversity is an important ingredient of the evolutionary recipe.

Third, urbanization has had profound effects on the human gut microbiota. Recent research has implicated the gut microbiota as playing an important role in healthy aging.

Urbanization has been associated with a reduction in microbiota diversity as well as with specific compositional changes. Community health and hygienic practices within the city have reduced the dispersal and transmission of microbes between individuals. These practices have limited the spread of infectious diseases within the modern city but urbanization-mediated changes in the human gut microbiota appear to have increased the occurrence of autoimmune-type disorders. In northeast Italy, the urban environment is associated with a higher prevalence of very elderly individuals and these aged humans appear to have a specific gut microbiota signature.

No doubt the interplay between the urban ecosystem and human evolution is much more complex than discussed in this chapter. However, an important starting point is to recognize the pivotal role the urban setting plays in the evolution of our species and to somehow incorporate this functional aspect of the urban ecosystem into our future planning.

As a corollary, we will not truly understand human aging as long as this phenotype is studied outside our natural environment which for the majority of our species is the urban ecosystem.

ADDENDUM TO THIS CHAPTER

Recent studies on comparative biology have revealed interesting correlations between life-history traits across species. The data of Smits and Monden (2009) and Colchero et al. (2016), discussed in this chapter, showed that in Homo sapiens life expectancy and between- individual variation in life span (life span inequality) were inversely and tightly correlated. Increases in life expectancy were associated with reductions in life span inequality. In addition, the data of Colchero et al. (2016) and Smits and Monden (2009) showed a temporal pattern; since agricultural practices were introduced and cities developed, there has been a progressive extension of human life expectancy and a corresponding reduction in life span inequality. Simply put, with the human population living longer, between-individual variation in age of death has diminished; the great majority of human deaths now occur within a narrow age band compressed into the extreme right end of the life span. Is this inverse relationship between length and inequality of life span a life-history trait specific to our species?

The data of Colchero et al. (2016) and Smits and Monden (2009) were based upon a single species, Homo sapiens, considered a long lived species. Observations of Peron et al. (2016) showed that long lived non-human species appeared to have a low level of life span inequality comparable to humans. The data of Colchero et al. (2016) and Smits and Monden (2009) also indicated that the human species had recently evolved a progressive extension of life expectancy and reduction in life span inequality over a time period that correlated with the introduction of agriculture and subsequent urbanization of the human population (discussed in this chapter). This observation underscores the role of human cultural activities as principal driver of human evolution.

Data in vertebrates indicate that a number of life- history traits co-evolved in a complex manner. Larger sized animals live longer and show much less individual variations in age of death than smaller sized animals. In addition, animals with a long generation time (slow pace of life) show much less between individual variation in body size than animals with a short generation time (fast pace of life) (Hamel et al. 2016). Humans as a species are large sized, have a long generation time and a long life span with a low level of between individual variations in the age of death.

However, there is a further dimension to the links between these life- history traits; large sized species are more "homogeneous" in their life history traits than smaller sized species. The data indicate that long lived species have a slower pace of life (longer generation time) and much less between individual variation in body size and life span compared to smaller sized species. What does this observation imply about species-specific developmental programs? Large sized species with a slow life pace appear to have developmental programs that are less plastic than those of smaller sized species with a faster pace of life (short generation time). Does a less plastic developmental program make large sized species less resilient in terms of coping with drastic changes in environmental conditions? Is this prediction supported by observations as to what life history traits characterized species that have become extinct?

McKinney (1997) reviewed traits that render species vulnerable to extinction. In general, extinction prone species are "specialized" or narrowly adapted. McKinney (1997) listed a series of individual traits which correlated with increased extinction risk; high trophic level, large body size, low fecundity, long-lived, slow growth/development and complex morphology are some of the traits on the list. It is obvious that all of these features are characteristic of primates and in particular the human lineage. In addition, McKinney (1997) listed "abundance" traits that are associated with risk of extinction; restricted geographical range and low population density. According to Huff et al. (2010) the population size of human ancestors about 0.9 to 1.5 million years ago was between 14,500 and 26,000. This is a relatively small number and yet by 1.8 million years ago Homo erectus had migrated out of east Africa and spread into Eurasia (**Chapter 7, The Origins of Aging: Multicellularity, Speciation and Ecosystems**). In other words, our human ancestors had already developed a sufficient level of cultural expertise to be able to expand their geographical range from a small area in Eastern Africa to a wide area encompassing both Africa and Eurasia. This nascent form of culture allowed our ancestors to avoid extinction despite a very limited population size.

What about non-human primates, our closest cousins on the phylogenetic tree. Morris et al. (2011) looked at survival and fertility in wild primate populations under variable environmental conditions i.e. temporal variability. The investigators found that survival did show temporal variability but the very small magnitude of this variation in survival produced only weak effects on long term fitness.

How do non-human primates maintain near constant adult survival rate despite fluctuations in environmental conditions? Morris et al. (2011) proposed that the answer lay in the cognitive abilities of primates and their ability to share information within complex social groupings i.e. cultural activities.

I believe that if our lineage had not developed culture our ancestors would have become extinct. Cultural practices allowed a small population of hominids to expand their ecological niche from a restricted geographical area to global in extent. Agricultural practices including domestication as well as urbanization resulted in a 60% increase in human life expectancy over the past 200 years (Smits and Monden 2009). Humans are living longer and individual ages at death are now compressed into the right end of the life span arc. The process of urbanization required adaptations that made it possible for thousands of humans to live together in a functional and safe city (Miller 2016). However a cultural activity such as urbanization is a two way process; we may design cities but the cities we design will inevitably direct our future evolutionary path.

REFERENCES

Batty, M. 2011. Commentary: When all the world's a city. *Environment and Planning A* 43: 765–772.

Batty, M. 2013. A theory of city size. *Science* 340: 1418–1419.

Bettencourt, L. M. A. 2013. The origins of scaling in cities. *Science* 340: 1438–1419.

Bettencourt, L. M. A. and J. Lobo. 2016. Urban scaling in Europe. *Journal of the Royal Society Interface* 13: 20160005. http://dx.doi.org/10.1098/rsif.2016.005

Biagi, E., Franceschi, C., Rampelli, S. et al. 2016. Gut microbiota and extreme longevity. *Current Biology* 26: 1480–1485.

Cesaretti, R., Lobo, J., Bettencourt, L. M. A., Ortman, S. G. and M. E. Smith. 2016. Population-area relationship for medieval European cities. *Plos One* 11(10): e0162678, doi:10.1371/journal.pone.0162678.

Colchero, F., Rau, R., Jones, O. R. et al. 2016. The emergence of longevous populations. *Proceedings of the National Academy of Sciences, USA* 113: E7681–E7690.

Courtiol, A., Tropf, F. C. and M. C. Mills. 2016. When genes and environment disagree: Making sense of trends in recent human evolution. *Proceedings of the National Academy of Sciences, USA* 113: 7693–7695.

Department of Ancient Near Eastern Art. 2000. Uruk: The first city. In *Heilbrunn Timeline of Art History*. New York: The Metropolitan Museum of Art.

Estes, J. A., Terborgh, J., Brashares, J. S. et al. 2011. Trophic downgrading of planet Earth. *Science* 333: 301–306.

Everard, A., Belzer, C., Geurts, L. et al. 2013. Cross-talk between *Akkermansia muciniphila* and intestinal epithelium controls diet-induced obesity. *Proceedings of the National Academy of Sciences, USA* 110: 9066–9071.

Grimm, N. B., Grove, J. M., Pickett, S. T. A. and C. L. Redman. 2000. Integrated approaches to long-term studies of urban ecological systems. *Bioscience* 50: 571–584.

Groffman, P. M., Cadenasso, M. L., Cavender-Bares, J. et al. 2017. Moving towards a new urban systems science. *Ecosystems* 20: 38–43. doi:10.1007/s10021-016-0053-4.

Gross, M. 2016a. Chimpanzees, our cultured cousins. *Current Biology* 26: R83–R85.

Gross, M. 2016b. The urbanization of our species. *Current Biology* 26: R1205–R1208.

Harris, S. E., Xue, A. T., Alvarado-Serrano, D. et al. 2016. Urbanization shapes the demographic history of a native rodent (the white-footed mouse, *Peromyscus leucopus*) in New York City. *Biology Letters* 12: 20150983. http://dx.doi:org/10.1098/rsbl.2015.0983.

Hatton, I. A., McCann, K. S., Fryxell, J. M. et al. 2015. The predator–prey power law: Biomass scaling across terrestrial and aquatic biomes. *Science* 349. doi:10.1126/science.aac6284.

Hill, K. R., Walker, R. S., Bozicevic, M. et al. 2011. Co-residence patterns in hunter-gatherer societies show unique human social structure. *Science* 331: 1286–1289.

Hornef, M. W. and O. Pabst. 2016. Real friends: *Faecalibacterium prausnitzii* supports mucosal immune homeostasis. *British Medical Journal* 65: 365–367.

Jenkins, W. D., Lipka, A. E., Fogleman, A. J., Delfino, K. R., Malhi, R. S. and B. Hendricks. 2016. Variance in disease risk: Rural populations and genetic diversity. *Genome* 59: 519–525.

Kerkhoff, A. J. and B. J. Enquist. 2006. Ecosystem allometry: The scaling of nutrient stocks and primary productivity across plant communities. *Ecology Letters* 9: 419–427.

Lively, C. M. 2010. The effect of host genetic diversity on disease spread. *The American Naturalist* 175: E-Note, E149–E152.

Martinez, I., Stegen, J. C., Maldonado-Gomez, M. X. et al. 2015. The gut microbiota of rural Papua New Guineans: Composition, diversity patterns and ecological processes. *Cell Reports* 11: 527–538.

McPhearson, T., Pickett, S. T. A., Grimm, N. B. et al. 2016. Advancing urban ecology toward a science of cities. *Bioscience* 66: 198–212.

Moeller, A. H., Li, Y., Ngole, E. M. et al. 2014. Rapid changes in the gut microbiome during human evolution. *Proceedings of the National Academy of Sciences, USA* 111: 16431–16435.

Newbold, T., Hudson, L. N., Arnell, A. P. et al. 2016. Has land use pushed terrestrial biodiversity beyond the planetary boundary? A global assessment. *Science* 353: 288–291.

Oliver, T. H. 2016. How much biodiversity loss is too much? *Science* 353: 220–221.

Ortman, S. G., Cabaniss, A. H. F., Sturm, J. O. and L. M. A. Bettencourt. 2014. The pre-history of urban scaling. *Plos One* 9(2): e87902. doi:10.1371/journal.pone.0087902.

Ruddiman, W. F., Ellis, E. C., Kaplan, J. O. and D. Q. Fuller. 2015. Defining the epoch we live in. *Science* 348: 38–39.

Saraswati, S. and R. Sitaraman. 2015. Aging and the human gut microbiota—From correlation to causality. *Frontiers in Microbiology* 5: 1–4.

Singer, M. A. 2015. *Human Ageing: A Unique Experience—Implications for the Disease Concept.* World Scientific Publishing, Singapore.

Singh, G. K. and M. Siahpush. 2014. Widening rural–urban disparaties in life expectancy, U.S., 1969–2009. *American Journal of Preventative Medicine* 46: e19–e29.

Smits, J. and C. Monden. 2009. Length of life inequality around the globe. *Social Science and Medicine* 68: 1114–1123.

Snell-Rood, E. C. and N. Wick. 2013. Anthropogenic environments exert variable selection on cranial capacity in mammals. *Proceedings of the Royal Society B* 280: 20131384. http://dx.doi:org/10.1098/rspb.2013.1384.

Thompson, K. A., Renaudin, M. and M. T. J. Johnson. 2016. Urbanization drives the evolution of parallel clines in plant populations. *Proceedings of the Royal Society B* 283: 20162180. http://dx.doi:org/10.1098/rspb.2016.2180.

Velasquez-Manoff, M. 2015. Gut microbiome, the peace-keepers. *Nature* 518: S4–S11.

Watson, J. E. M., Shanahan, D. F., Di Marco, M. et al. 2016. Catastrophic declines in wilderness areas undermine global environment targets. *Current Biology* 26: 1–6.

Whiten, A. 2011. The scope of culture in chimpanzees, humans and ancestral apes. *Philosophical Transactions of the Royal Society B* 366: 997–1007.

Winchell, K. M., Reynolds, R. G., Prado-Irwin, S. R., Puente-Rolon, A. R. and L. J. Revell. 2016. Phenotypic shifts in urban areas in the tropical lizard *Anolis cristatellus*. *Evolution* 70–5: 1009–1022.

9 Aging Epigenetics
Accumulation of Errors and More

*Vasily V. Ashapkin, Lyudmila I. Kutueva,
and Boris F. Vanyushin*

CONTENTS

INTRODUCTION

Why we age has been a question of heated debate for a very long time. Now, it is quite clear that aging has multiple causes [1]. Nine hallmarks of aging proposed in a recent review are genomic instability, telomere attrition, epigenetic alterations, loss of proteostasis, deregulated nutrient sensing, mitochondrial dysfunction, cellular senescence, stem cell exhaustion, and altered intercellular communication [2]. Epigenetic systems control gene activity and thus, directly or indirectly, affect all other hallmarks. Once differentiated, every cell in metazoan organisms must "remember" its appropriate pattern of gene expression for the organism to survive and function normally. Studies of developmental biology have established a role for epigenetic systems in establishing and maintaining these differentiated states of cells. For quite a long time, the epigenome was thought of as a static entity; once a cell becomes differentiated and its genome is appropriately methylated and chromatin configured, no further changes in the epigenome are supposed to occur. Unexpectedly, multiple studies in the past few years have shown that, in fact, an epigenome is a dynamically regulated system involved in aging, or at least affected by it, and responsive to various external and internal factors.

An ability of epigenetic systems to affect all other drivers of aging puts them in a key position to affect aging per se. On the other hand, the epigenetic systems, like any others in the living organisms, are subjected to aging-related deterioration. Active components of the epigenetic systems are encoded in genomes and thus are themselves under epigenetic control. Besides, there are numerous links, both feedback and feed-forward ones, between different elements of the epigenome. Transcripts of genes encoding proteins of the DNA methylation and histone modification systems could be targeted and controlled by miRNAs, whereas expression of genes encoding miRNAs is controlled by cytosine methylation and histone modification. DNA methylation and histone modification systems are known to be coupled at both gene expression and target recognition levels. Which alterations in DNA methylation and other epigenetic marks play a causal role in aging and which are just manifestations of aging has yet to be elucidated.

GENE EXPRESSION

A comparative study of the transcription profiles between middle-aged and young adult individuals in *Caenorhabditis elegans* and *Drosophila melanogaster* showed that most aging-related changes in gene expression are species specific [3]. Nevertheless, there is a conserved part in these expression profiles that includes several hundred *C. elegans–D. melanogaster* ortholog gene pairs. Aging in both *D. melanogaster* and *C. elegans* is accompanied by repression of genes in Gene Ontology (GO) categories for mitochondrial membranes, including many components of the respiratory chain, the adenosine triphosphate (ATP) synthase complex, and the citric acid cycle. Conserved patterns of regulation of genes encoding peptidases, and proteins for catabolism and DNA repair were also found. Both the conserved global changes in gene expression and the conserved repression of oxidative metabolism genes occur rather abruptly early in adulthood. These results show that changes in gene expression with adult age are not solely implemented in response to cumulative damage. Instead, the timing of these conserved features of aging suggests developmentally timed transcriptional regulation in young adults.

Genes that change expression with aging, both downregulated and upregulated ones, were found in transcriptional profiles of the human frontal cortex from individuals ranging from 26 to 106 years of age [4]. The groups of younger (<42 years) and older (>73 years) individuals showed the most homogeneous pattern of gene expression, whereas the middle age group (45–71 years) exhibited much greater heterogeneity, with some individuals resembling the younger group and others resembling the older group. The genetic signature of human cortical aging seems to be defined starting in young adult life, and the rate of age-related changes may be variable among middle-aged individuals. About 4% of the approximately 11,000 genes analyzed were significantly ($\geq$1.5-fold) different between <42 year and >73 year aged groups. Genes that play a role in synaptic function and the plasticity that underlies learning and memory were among those most significantly affected. Several neurotransmitter receptors that are centrally involved in synaptic plasticity showed significantly reduced expression after 40 years age. Members of the major signal transduction systems that mediate long-term potentiation (LTP) and memory storage were downregulated with age. Furthermore, multiple members of the protein kinase C (PKC) and Ras-MAP kinase signaling pathways showed decreased expression. Thus, neuronal signaling may be affected in the aged cortex. Aging of the human frontal cortex was also associated with increased expression of genes that mediate stress response and repair functions, such as protein folding, antioxidant defense, and inflammatory or immune responses. Increased expression of the genes encoding base excision repair (BER) enzymes 8-oxoguanine DNA glycosylase and uracil DNA glycosylase was observed, consistent with increased oxidative DNA damage in the aged cortex.

The effects of aging and caloric restriction (CR), which extends life span in laboratory rodents, on gene expression were studied in the heart, liver, and hypothalamus of young (4–6 months age), old (26–28 months age), and young CR (4–6 months age with 2.5–4.5 months of CR) mice [5]. Aging significantly altered expression of 309, 1819, and 1085 genes in the heart, liver, and hypothalamus, respectively. Only nine genes altered expression with aging across all three tissues, some of them in opposite directions. CR significantly affected the expressions of 192, 839, and 100 genes in the heart, liver, and hypothalamus, respectively. Only seven genes altered expression across all three tissues (three upregulated and four downregulated). The upregulation of antigen processing/presentation genes by aging and downregulation of stress response genes by CR were observed in all three tissues. The immune response category was significantly upregulated, whereas the biosynthesis category was significantly downregulated, by aging in the heart and liver but not in the hypothalamus. Comparison between aging and CR showed that there are 389 genes, 18 biological processes, and 20 molecular functions in common. Partial overlaps of age-related gene expression profiles between different tissues of the same animal were observed. Generally, aging has been found to affect gene expression more significantly and broadly compared with CR in all tissues. Tissue-specific gene expression changes predominated in responses both to aging and CR.

In a transcriptome-wide investigation of aging-dependent gene expression in mice organs (liver, kidney, spleen, lung, and brain), mainly organ-specific differentially expressed genes (DEGs) were found: 6973 in the liver, 2325 in the kidney, 925 in the spleen, 1025 in the lung, and 15 in the brain [6]. Most DEGs (88%) were exclusively found in one organ. The number of overlapping DEGs rapidly decreased when more tissues were compared at the same time. Only one gene was found to be differentially expressed in all organs tested, namely *Lilrb4*. It encodes an immunoglobulin-like receptor that is involved in regulation of immune tolerance. Expression of *Lilrb4* has been found to increase with aging. Five of the 11 other DEGs that are found in at least four organs are immunoglobulin lambda and kappa complex related, and all these DEGs showed increased gene expression during aging. Many functionally related gene sets altered their expression with aging in each organ, most of them exclusively in one organ. In all organs, except the liver, many gene sets involved in immunological processes were changed during aging. Generally, each organ seems to have a specific aging course.

One of the consequences of the age-associated DNA damage accumulation in somatic cells could be stochastic deregulation of gene expression manifested as increased cell-to-cell variations in gene expression. When the expression levels of seven housekeeping genes, three heart-specific genes, two protease-encoding genes, and three mitochondrial genes were measured in individual cardiomyocytes isolated from the ventricular heart tissue of young (6 months) and old (27 months) male mice, a highly significant increase in cell-to-cell variations of the expression levels of all nuclear genes was observed in old animals compared with young ones [7]. In contrast, no significant cell-to-cell variations were observed in the expression levels of mitochondrial genes. Noise is inherent in the basic process of transcription, especially in genes expressed at low levels. When cultured mouse embryonic fibroblasts were treated at early passages with 0.1 mM H_2O_2, a known generator of oxidative damage, cell-to-cell variations in gene expression were greatly increased. The effect was absent at 6 h after treatment, but was significant at 48 h, when the cell populations started to show an increased number of senescent cells. At nine days after treatment, almost all cells were senescent, and the increased cell-to-cell variations in gene expression were still highly significant. It is conceivable that a variety of persistent forms of damage to biological macromolecules, including random changes to DNA methylation and chromatin remodeling, could initiate a gradual increase in transcriptional noise introducing phenotypic variation among cells. Such variation may become detrimental to tissue functioning, which would explain many etiological characteristics of the aging process, most notably the highly variable progressive decline in organ functions.

A special form of progressive transcription deregulation with aging, *transcriptional drift*, has been recently described in *C. elegans* [8]. It was shown that as animals age, different genes of the same functional groups change expression in opposing directions, thus disrupting relative mRNA ratios within respective pathways, when compared to young adults. In many cases, the fractions of genes that increased, decreased, or did not change in expression showed no consistent pattern. In the superoxide detoxification pathway, a well-defined cellular function that is widely believed to decline with age, the expression levels of some superoxide detoxification genes increased with age (*sod-4*, *sod-5*), while others decreased (*sod-1*, *sod-2*, *prdx-2*, *prdx-3*, *prdx-6*) or did not change (*ctl-1*, *ctl-2*, *ctl-3*). Interestingly, treatments that increase *C. elegans* longevity, such as mianserin or *daf-2* RNAi, attenuated the effect of aging across the whole transcriptome and preserved the co-expression patterns observed in young adults into later ages, whereas the treatments that decrease longevity, such as *daf-16* RNAi, increased the transcription drift to values beyond those seen in control animals. A reanalysis of data obtained from aging mouse tissues [6] and human brains [4] showed that the progressive transcription drift with age also occurs in various organs of mammals.

In a study of normal muscle samples from variously aged humans, a molecular profile for aging consisting of 250 age-regulated genes was established [8]. A common signature for aging in diverse human tissues was defined by comparing these data with transcriptional profiles of aging in the kidney and brain. The common aging signature consists of six genetic pathways, four of them (extracellular matrix, cell growth, complement activation, and cytosolic ribosome) increasing expression with age, while two others (chloride transport and mitochondrial electron transport chain) decreasing. A great heterogeneity in the transcriptional changes with age was found in different mouse tissues [9]. Some tissues display large transcriptional differences in old mice, suggesting that these tissues may contribute strongly to organismal decline. Other tissues show few or no changes in expression with age, indicating strong levels of homeostasis throughout life. Nevertheless, different tissues of individual mice age in a coordinated fashion, such that certain mice exhibit rapid aging, whereas others exhibit slow aging for multiple tissues. In the transcriptional profiles for aging in mice, humans, flies, and worms, the electron transport chain genes show common age regulation in all four species, indicating that these genes may be exceptionally good markers of aging.

Chronological aging is associated with diverse and widespread changes in gene expression that reflect the history and physiologic functions of the aged tissues [10]. Comparison of the age-regulated gene sets across diverse tissues, physiologic states, and species showed the NF-κB family binding sites to be the most prevalent upstream motifs in gene sets significantly induced with age in various human and murine tissues. Coordinated induction of gene sets defined by the NF-κB motif alone or by NF-κB in combination with other transcription factor (TF) motifs tended to occur in humans after 40 years age and was even more prevalent after 70 years. Multiple NF-κB motif genes were strongly induced with age in the mouse liver, heart, and hematopoietic stem cells (HSCs). Treatment of mice with high fat diet induced NF-κB motif gene sets in the liver, while concomitant treatment with resveratrol, an agent that reverses the effects of high fat diet and prolongs life span, repressed these gene sets. Thus, NF-κB seems to be a master regulator of gene expression programs in mammalian aging.

A meta-analysis of age-related gene expression profiles using 27 datasets from mice, rats, and humans revealed several common signatures of aging, including 56 genes consistently overexpressed with age, the most significant of which was *APOD*, and 17 genes underexpressed with age [11]. Of the biological processes, inflammation, immune response, and lysosome genes were most consistently overexpressed on aging, whereas collagen genes, those associated with energy metabolism, particularly mitochondrial genes, and genes related to apoptosis, cell cycle, and cellular senescence were most consistently underexpressed. These gene sets supposedly reflect not only a mix of degenerative processes accompanying aging, but also adaptive responses of healthy cells to the aging-associated derangements.

Only a very small proportion (2%) of transcripts demonstrated robust age-related differences in the human peripheral blood leukocytes (PBL) [12]. The pathways most disrupted by aging include genes involved in messenger RNA splicing, polyadenylation, and other posttranscriptional events. Deregulation of mRNA processing may be one of the mechanisms of human cellular aging. As was predicted, genes showing the largest variations in expression with age are involved in inflammatory responses or immune function [11]. Among the top 100 age-dependent ones were genes encoding components of nutrient sensing pathways, such as insulin or target of rapamycin (TOR) signaling, oxidative stress, DNA repair, inhibition of respiration, reproductive system signaling, and telomere-related mechanisms. Only seven of the GO molecular or biological function pathways are robustly associated with age. Four of these pathways are involved in mRNA processing. The remaining pathways relate to the chromatin accessibility and to the processes associated with mRNA translation. Since disruptions to the proteins involved in mRNA processing appear to occur without widespread alterations in gene expression levels, it may be suggested that, while aged leukocytes *in vivo* may be expressing most genes at levels comparable to those found in younger cells, there may be differences in the relative balance of splice products produced or increases in the occurrence of aberrantly spliced transcripts. Disruptions to the patterns of isoform expression with increasing age were found in 7 of 10 alternatively spliced transcripts studied.

A comparison of the cerebral cortex transcriptomes between 6-, 12-, and 28-month-old rats revealed differential expression of genes related to MHC II presentation and serotonin biosynthesis, as well as a wide group of noncoding genes [13]. Across the three age groups, differential expression was found for 136 transcripts, 37 of which do not map to known exons. Fourteen of these transcripts were identified as novel long noncoding RNAs (lncRNAs). Evidence of isoform switching was also found. Therefore, in addition to changes in the expression of protein-coding genes, changes in transcript splicing, isoform usage, and noncoding RNAs occur with age.

DNA METHYLATION

CHANGES IN DNA METHYLATION LEVELS DURING AGING

The first experimental data concerning age-dependent changes in DNA methylation were obtained in our lab nearly half a century ago. DNA methylation levels in different organs of humpback salmon and rat have been found to gradually decrease with age [14,15]. The age-dependent loss of methylation mainly affects the heavily methylated repeat fraction of the mammalian genome [16], possibly leading to a general genome destabilization, since this genome fraction contains transposable elements (TEs). The repression of such elements is widely believed to be one of the main defensive functions of DNA methylation.

The DNA methylation levels in different mice organs showed a correlation with chronological age [17]. The rates of 5mC (5-methyl-cytosine) loss upon aging in DNA of two mouse species were inversely correlated with their maximal life spans. A progressive decrease in the DNA methylation levels was observed in the cultured mouse, hamster, and human fibroblasts [18]. Again, the rates of this hypomethylation were inversely correlated with the cell life spans (maximal number of the cell population doublings before senescence). It could be suggested that the progressive 5mC loss in dividing cells is mainly caused by errors in the maintenance DNA methylation, which has a fidelity of approximately 95% per cell generation [19,20]. Such loss of 5mC, essentially stochastic as it seems to be, could affect aging at both cellular and organismal levels by increasing the transcriptional noise and eventually leading to aberrant transcription of various genes. No changes in DNA methylation levels were observed in dividing immortal cell lines [18]. Since the fidelity of the maintenance DNA methylation could hardly be higher in these immortal cell lines compared with their normal counterparts, the age-related loss of 5mC cannot be explained by the maintenance methylation errors alone. Indeed, similar losses of 5mC upon aging occur in mice tissues widely different in proliferative activity (liver, brain, and small intestine mucosa) [17]. Moreover, it was found that selective hypermethylation of some genes occurs upon aging, along with the global DNA hypomethylation [21]. The tumor-suppressor genes, known to be hypermethylated in tumor cells, are often among the aging-hypermethylated ones. The promoter CpG island (CGI) sequences of the *LOX*, *CDKN2A* (also known as *p16*, *INK4a* or *p16^{INK4a}*), *RUNX3*, and *TIG1* genes were found to be essentially unmethylated in normal stomach epithelial cells before the 50-year patient age but progressively methylated between the 50-year and 80-year ages [22]. This progressive methylation could explain the increase in tumor occurrences in aged people. Since the age-dependent increase in DNA methylation levels is nonlinear in this case, it evidently must be caused by a specific process of some kind, not just accumulation of stochastic errors.

CHANGES IN GENOME METHYLATION PATTERNS UPON AGING

DNA methylation patterns in human solid tissues were found to be dependent both on tissue and age [23]. A rather distinct correlation between age and methylation was observed, with loci in CGIs gaining methylation with age and loci not in CGIs losing it. This pattern was consistent across tissues and in blood-derived DNA. Gene loci that earlier have been reported to be associated with aging, *ESR1*, *GSTP1*, *IGF2*, *MGMT*, *MYOD1*, *RARB*, and *RASSF1*, displayed significant methylation alterations (mainly increases) with age. Loci in epigenetic regulatory genes (*LAMB1*, *DNMT1*, *DNMT3B*, *HDAC1*, and *HDAC7*), telomere maintenance genes (*TERT*, *ERCC1*, and *RAD50*), and a premature aging syndrome gene (*WRN*), all showed age-related methylation alterations. In contrast to the predominantly increased age-associated methylation at other gene loci, there was a significant age-related decrease in CpG methylation of *DNMT3B* that, unlike the vast majority of hypermethylated CpG loci tested, was not located in a CGI.

Of about 3600 mouse genes studied, 21% displayed hypermethylation and 13% displayed hypomethylation with age in the small intestine cells [24]. Among genes devoid of the promoter CGIs, 7% displayed hypermethylation and 11% hypomethylation. The genes known to be hypermethylated in colon cancer (*Cdh13*, *Dok5*, *Esr1*, *Igf2*, *Myod1*, *Nkx2-5*, *Cdkn2a*, *Pgr*, and *Tmeff2*) showed age-related hypermethylation in the normal small intestine. Globally, the hypermethylated gene group was enriched for genes involved in development and differentiation, whereas the hypomethylated gene group was not enriched for any specific functional category. These age-dependent gene methylation patterns were partially conserved between humans and mice, and the conserved genes were mainly among the hypermethylated ones.

Of the approximately 350 CpG loci differently methylated with age (aDMRs) in human PBL, approximately 200 were hypermethylated (hyper-aDMRs) and approximately 150 hypomethylated (hypo-aDMRs) [25]. More than 95% of these aDMRs were located within 500 bp of the transcriptional start site (TSS). An essential share ($>60\%$) of the hyper-aDMRs found were also present in purified fractions of CD14+ monocytes (short-life span cells of the myeloid lineage) and CD4+ T cells (long-life span cells of the lymphoid lineage). The hypo-aDMRs set was significantly reproduced in T cells, but not in monocytes. Thus, it could be suggested that most hyper-aDMRs represent epigenetic perturbations inherent to aging per se, whereas the hypo-aDMRs may reflect modifications associated both with aging per se and with age-dependent changes in relative proportions of the blood cell subtypes. The conservative hyper-aDMRs group has been found to be enriched for genes with tissue-specific functions, including neural cell-related processes. Most of these genes are inactive or moderately active in the peripheral blood cells. These hyper-aDMRs significantly overlap genes hypermethylated in various human cancers. Similar to cancer cells, their methylation could decrease the cell differentiation potential and increase the stem cell self-renewal. Possibly, this is one of the mechanisms that enhances cancer frequency at advanced ages.

In contrast to total human PBL, 12,275 and 48,876 differentially methylated (dm) with age CpG sites (aDMRs) were identified in the purified CD4+ and CD8+ T cell samples [26]. A majority of the hypermethylated CpG sites were located in CGIs, at silent gene promoter regions that were enriched for repressive histone methylation marks. Hypomethylated CpG sites were located more at the borders of the CGIs and in gene bodies. In a subset of genes expressed in CD8+ T cells, a negative correlation between DNA methylation and transcription levels was observed. Many genes that are essential for the differentiation and function of T cells were among these last ones, explaining the possible causes of the age-related decline in immune response.

In variously aged human donors, 589 CpGs showed age-dependent methylation in one brain division, 167 in two, 86 in three, and only 10 in all four divisions studied [27]. Of all age-related CpGs found, approximately 82% were located within CGIs, approximately 11% were not within CGIs, and approximately 7% were in regions that could not be unequivocally defined as CGIs or non-CGIs. All 10 CpG sites that showed significant correlation with age in four brain regions were located within CGIs and their methylation levels were increased with age. Four of these sites were earlier identified as dm with age in other tissues [23,25]. A positive correlation of the methylation level with age was observed also in a majority ($>95\%$) of the remaining CpG sites. Of the age-related CpG sites located within CGIs, 98% were hypermethylated at advanced ages. Compared with other tissues, the brain contains mainly sites hypermethylated with age, most of them located in the promoter-associated

CGIs of the transcription factor encoding genes. Apparently, the age-dependent hypermethylation of specific genes could be a part of an epigenetic program affecting genome expression in aging cells, whereas the age-dependent loss of total DNA methylation seems to be a result of stochastic methylation errors. Indeed, intergenic regions and CpG poor promoters were found most likely to undergo changes in methylation during the early life in monozygotic (MZ) twin pairs, whereas the methylation of CGIs and CpG-rich promoters was most stable [28]. Since environmental and lifestyle differences between MZ twins could be excluded at these ages, their discordance in DNA methylation should be a result of stochastic methylation errors.

The sites located within CGIs were found to predominate among those hypermethylated with age in various human organs (brain, blood, kidney, and skeletal muscle), whereas hypomethylation with age was more characteristic of CpG sites located beyond CGIs, and their methylation levels were more variable between tissues [29]. Hence, age-related variations in methylation common for different tissues are mainly observed in CGIs and are usually represented by increases in methylation levels with age. On the other hand, tissue-specific variations in methylation are more characteristic of the CpG sites located beyond CGIs and are often represented by decreases in methylation levels with age. Gene loci, functionally connected with regulation of transcription and control of morphogenesis, predominate among those hypermethylated with age in various tissues. Skeletal muscle showed the least overlap in total age-dependent CpG sites with other tissues, and had the strongest association of age-dependent CpGs related to tissue-specific gene expression. In a larger-scale study on human skeletal muscle, approximately 6000 CpG sites were identified to be dm between the young (18–27 years) and older (68–89 years age) male adults [30]. Of these, 92% (5518 dmCpG sites) were hypermethylated in the older subjects, while the remaining 8% were hypomethylated. Interestingly, in aging human skeletal muscle, a strong preference for the dmCpG sites to localize within gene bodies, especially their central and 3′ end parts, was found, and surprisingly, no preference was observed for the promoter.

Discordance in DNA methylation profiles between MZ twins can be used to discriminate between the age-related changes in DNA methylation, which represent a cumulative result of stochastic errors, and those that are a part of the hypothetical epigenetic program of aging. The methylation levels of total lymphocyte DNA were found to be practically identical in 65% of MZ twin pairs and significantly different in 35% of MZ twin pairs [31]. Identical DNA methylation levels were usually observed in young pairs, whereas aged pairs had most different ones. Thus, DNA methylation difference between MZ twins gradually increases with age. Nearly identical methylation patterns were characteristic of the young twin pairs that lived together for the most part of their life and had similar lifestyles, whereas most discordant methylation patterns were characteristic of the older twin pairs that lived separately and had different lifestyles. Thus, large phenotypic discordance in MZ twin pairs may be caused by the accumulated epigenetic differences. These differences could be due both to effects of external and internal factors (smoking, physical activity, dietary preferences, etc.), and stochastic methylation errors, "epigenetic drift," accompanying aging.

DNA METHYLATION AS AGE PREDICTORS

The aging rates are non-equal in different persons. Women are known to have a longer average life span compared with men. Nearly, all supercentenarians who have reached the 110-year age are women. Aging can be accelerated by unhealthy life habits, such as smoking, or slowed down by good ones, such as physical training. Molecular markers of biological, rather than passport, age are needed to evaluate more precisely the degree of age-dependent deterioration in physical welfare. The age-related variance in DNA methylation seems to be a good contender for this role. Unfortunately, the age-related DNA methylation loci are masked by a plethora of methylation variations caused by other factors or stochastic errors. In saliva samples of MZ twin pairs, 88 CpG sites were found to have methylation levels significantly correlated with age, 69 of them positively correlated and 19 negatively correlated [32]. Most (83%) age-correlated sites were within promoter CGIs.

Three genes that showed the most clear correlation with age and had the widest distribution of the methylation values, *Edaradd*, *NPTX2*, and *Tom1L1*, showed a clear correlation with age in men, but only two of them (*Edaradd* and *Tom1L1*) in women. The methylation levels of *Edaradd* and *Tom1L1* were linearly decreasing with age, whereas the methylation level of *NPTX2* was increasing. On the basis of the methylation levels of just two CpG sites (located in *Edaradd* and *NPTX2* genes), the age of the test subjects could be predicted with a 5–6-year accuracy, whereas the addition of one more site (located in the *ELN* gene) reduced the average error to approximately 3.5 years. No epigenetic drift was detected in the promoter CpG sites studied. This finding corroborates the view that the stochastic methylation errors are mainly accumulated in repeat sequences and intergenic regions, whereas the gene and promoter methylation is under a more robust control.

Comparative studies of age-related methylation patterns in various tissues showed these patterns to be highly tissue specific. Nevertheless, there are some loci that have methylation levels significantly correlated with age in various tissues. Obviously, these common methylated loci are of the highest relevance to the mechanisms of aging per se, and their methylation status could be used as an epigenetic signature to estimate the biological age. The first non-cell-type-dependent epigenetic aging signature was elaborated based on the DNA methylation datasets from several independent studies that used the Illumina HumanMethylation27 BeadChip platform [33]. Of more than 450 age-correlated CpG sites found, most were hypermethylated with age and only 25 were hypomethylated. This is in accord with the view noted above that hypermethylation at specific sites is a predominant trend upon aging, whereas hypomethylation seems to be less stringently regulated. Most accurate age predictions were obtained when a set of four hypermethylated loci, *TRIM58*, *KCNQ1DN*, *NPTX2*, and *GRIA2*, has been used. To further enhance the prediction accuracy, a hypomethylated locus *BIRC4BP* was added to the set. When all five loci were used, the average prediction accuracy across all datasets was ±12.7 years, whereas the use of only the three most reliable of them (*NPTX2*, *GRIA2*, and *KCNQ1DN*) enhanced the accuracy to ±11.4 years. It should be noted that, in the work described, the age prediction was applicable to various tissues and was gender independent, whereas in the previous study described above [32], the prediction was based only on the saliva samples. When the blood samples were investigated, the set of CpG loci with a high predictive capability could be narrowed down to just three (*ITGA2B*, *ASPA*, and *PDE4C*), and the accuracy of age prediction was ±4.5 years [34].

In a larger-scale investigation using the Illumina HumanMethylation450 BeadChip platform, methylation levels of 485,577 CpG sites were analyzed in blood DNA samples from more than 650 volunteers of 19- to 101-year ages [35]. An age predictive model was built using a set of 71 methylation markers. The mean error of age prediction by this model was ±3.9 years (96% correlation between the passport age and the predicted age). Nearly all markers in the model were located within or near genes with known functions in age-related conditions, such as Alzheimer's disease, cancer, tissue degradation, DNA damage, and oxidative stress. Two sites chosen were within the gene of somatostatin, a regulator of endocrine and nervous system function, and six sites within the gene encoding KLF14, a "master regulator" of obesity and other metabolic diseases. The model was capable not only of predicting the age, but also of revealing the factors that affect the personal rate of aging. For example, the gender was found to affect it very significantly, DNA methylation "aging" in men being approximately 4% faster than in women. The body mass index (BMI) was found not to affect the aging rate of blood, adipose, and muscle cells, whereas age acceleration by approximately 2.2 years per each additional 10 BMI units was observed in the liver [35,36]. When the model was used to estimate the age of tumors, these appeared approximately 40% more aged than respective normal cells of the same person. The age prediction model worked with DNA samples from other organs (breast, lung, kidney, and skin) with the same accuracy as with blood samples, when a linear offset specific for each organ was used. When the epigenetic predictive models were constructed using the same algorithm but based on the age-related methylation data from other organs (breast, lung, and kidney), the main differences were in the sets of the most informative CpG sites chosen. Only two CpGs near *ELOVL2*, a gene involved in the skin cell aging, appeared to be common. Not only were the methylation levels of age-related CpG sites changing with age, but also the variation limits of these methylation levels between different persons became larger for most sites. For any specific person, the extent of deviation in these values from the population averages seems to be a fairly accurate measure of the individual aging rate.

An "ultimate" biological age predictor has been built via bioinformatics analysis of all publicly available datasets concerning age-related variations of DNA methylation in various tissues and cell lines (nearly 8000 samples, 51 tissue and cell types) [37]. A total of 353 methylation sites were chosen, which allowed of most reliable age prediction for various tissues and cells (96% correlation to the passport age, ±3.6 years accuracy). Blood cells that have very different life spans, CD14+ monocytes (myeloid lineage) living several weeks at the most, and CD4+ T cells (lymphoid lineage) living for months to years, were shown to have identical epigenetic ages in blood samples from healthy male subjects. Hence, epigenetic age reflects some internal methylome features, related to chronological age of the person, not just the age-dependent peculiarities of respective blood cells. The mean epigenetic age is highly correlated with the chronological age in most tissues. The variations in epigenetic age between different tissues of the same person are rather small. The notable exceptions are breast tissue in women (epigenetically older compared with other tissues) and sperm in men (epigenetically younger compared with other tissues). Surprisingly, though all brain regions of the same person have similar epigenetic ages in subjects younger than 80 years, the cerebellum and, to a lesser extent, the occipital cortex exhibit progressively negative epigenetic age acceleration in the older persons, that is, these brain regions are younger than others [38]. This relative resistance of cerebellum to aging correlates with overexpression of 1239 genes compared with other brain regions, two RNA helicase superfamilies (SF1 and SF2) being most enriched among these genes.

As could be expected, the epigenetic age of embryonic stem cells (ESCs) is close to zero [37]. Interestingly, induced pluripotent stem cells (iPSCs) do not differ from ESCs by their epigenetic age. Therefore, the epigenetic age is reset when iPSCs are produced from differentiated somatic cells. When cells are maintained in culture, ESCs and iPSCs included, their epigenetic age increases with each passage. The epigenetic age is clearly not a reflection of the mitotic age, since it tracks chronological age in tissues widely different in proliferative potential, including postmitotic neurons. It is also not related to cell senescence, since its correlation with chronological age is observed in immortal cell lines, such as ESCs.

Obviously, the epigenetic age could serve as a convenient marker in assessment of the rejuvenation treatments efficiency. A comparative analysis of DNA methylation levels of peripheral blood mononuclear cells between semi-supercentenarians (mean age: 105.6 years), their offspring (mean age: 71.8 years), and age-matched controls (mean age: 69.8 years) showed that the offspring of semi-supercentenarians have a lower epigenetic age than age-matched controls (age difference=5.1 years) and that the epigenetic age of centenarians is less by 8.6 years than their chronological age [39]. The aging rates of different tissues of the same person can be used to identify those with the evidence of disease, uppermost cancer. An interesting example of application of the epigenetic age concept was estimation of the mortality risk [40]. It was found that a 5-year acceleration of the epigenetic aging compared with the chronological aging results in a 16% increase in mortality risk, irrespective of general health, lifestyle, and genetic factors. Similar results were obtained in another study; a 5-year higher epigenetic age compared with chronological age has been associated with 15%–20% increases in all-case, cancer, and cardiovascular disease mortalities [41].

The predictive accuracy of the epigenetic age can be enhanced when less universal models are used. For example, for the blood cells, it reached a value of 2.6 years when 17 marker CpG sites were used [42]. In a follow-up study of the same persons eight years later, the predicted increases in methylation levels of hypermethylated sites and decreases in methylation levels of hypomethylated sites were observed.

The human life span is widely believed to be genetically determined by 20%–30% [43]. The searches for gene loci affecting longevity were essentially unsuccessful. The possible significance of DNA methylation as a factor controlling life span in humans was studied in leukocytes of female centenarians (>100-year ages), their daughters of about 70 years age selected from pairs, where the father was also long-lived (died at the age of >77 years), females of about 70 years age, whose parents were both non-long-lived (mothers died at ≤72 years age, fathers at ≤67 years age), and a control group of young (17–34 years) women [44]. Clinical histories showed that, in general, the centenarians' daughters have a much better health status (age-related diseases, permanent usage of prescribed drugs, etc.), than daughters of non-long-lived parents. Evidently, the probability to become long-lived is inheritable to a very considerable degree. Global DNA methylation levels were significantly decreased in all three aged groups compared with the control group, but to different extents. Maximum hypomethylation was observed in daughters of non-long-lived parents, minimal in centenarians' daughters, and intermediate in centenarians themselves. About 700 CpG sites (located in ~600 genes) were hypermethylated in all three aged groups—to similar extents in both daughters' groups and to a greater extent in centenarians. This set of hypermethylated loci was enriched for genes functionally related to organ development, cell differentiation, and transcription regulation. A total of 330 CpG sites (located in 326 genes) were hypomethylated in aged groups—to similar extents in both daughters' groups and to a greater extent in centenarians. This set of hypomethylated loci was enriched for genes involved in defense responses, acute inflammation, and signal transduction. A total of approximately 150 CpG sites (located in 124 genes) were significantly hypermethylated in centenarians' daughters compared with daughters of non-long-lived parents. This set was enriched for genes functionally related to nucleotide metabolism, nucleic acids synthesis, and cellular signaling. On the other hand, 67 CpG sites (located in 65 genes) were significantly hypomethylated in centenarians' daughters compared with daughters of non-long-lived parents. This set was enriched for genes related to downstream processes of the signal transmission. Most strongly pronounced differences in methylation levels between two daughters' groups were found for 12 CpG sites (10 hypermethylated and 2 hypomethylated in centenarians' daughters) located in 9 genes. Six of the hypermethylated genes (*SLC38A4*, *SLC22A18*, *MGC3207*, *ECRG4*, *ATP13A4*, and *AGPAT2*) are involved in metabolic processes, one hypermethylated gene (*DUSP22*) is a tumor-suppressor gene, and yet another hypermethylated gene (*ZNF169*) encodes a zinc finger DNA binding protein with unknown function. The function of the only hypomethylated gene (*FLJ32569*) is also unknown. Obviously, genome methylation in centenarians' daughters is more stable compared with that in daughters of non-long-lived parents. Thus, epigenome stability and a more robust epigenetic control for nucleotide metabolism, nucleic acids synthesis, and signal transduction may contribute to an increase in life span and healthy aging in centenarians.

HISTONE MODIFICATIONS

HISTONE METHYLATION

A genome-wide RNAi screen for genes that regulate life span in *C. elegans* resulted in a number of genes encoding the SET domain containing histone methyltransferases [45]. Knockdown of *set-2*, *set-4*, *set-9*, *set-15*, and *ash-2* extended the worm life span, *ash-2* having the most significant effect. The encoded protein ASH-2 is a member of H3K4 trimethylation (H3K4me3) complex in yeast, flies, and mammals. In *C. elegans*, *ash-2* knockdown decreases global H3K4me3 levels. WDR-5 is a protein that interacts with ASH-2 in mammals, and is important for the mono-, di-, and trimethylation of H3K4 both in *C. elegans* and mammals. The *wdr-5* knockdown also decreases H3K4me3 levels and significantly (by ~30%) extends life span in *C. elegans*. Thus, ASH-2 and WDR-5 mediated H3K4 trimethylation seems to promote aging and limit the life span in *C. elegans*. In mammals, ASH-2 and WDR-5 form a complex with several H3K4me3 methyltransferases of the SET1/ MLL family. Of the four SET1/MLL orthologues in *C. elegans*, SET-1, SET-2, SET-12, and SET-16, only SET-2 affects the life span. The *set-2* knockdown worms have reduced H3K4me3 levels. On the other hand, neither *set-9* nor *set-15* knockdowns affect global H3K4me3 levels, even though they both regulate the life span. The bacterially expressed SET-2 methyltransferase methylates histone

H3 at lysine 4 *in vitro* to generate H3K4me2, whereas ASH-2 converts H3K4me2 to H3K4me3. Analysis of the life span-extending effects of combined mutations showed that ASH-2, WDR-5, and SET-2 act in the same pathway to limit the life span. RBR-2 is an H3K4me3 demethylase homologous to the human KDM5A and KDM5B—H3K4me3 demethylases of the JARID family. The *rbr-2* mutant worms show increased H3K4me3 levels and a significantly decreased life span, indicating that RBR-2 activity is necessary for normal longevity. The *ash-2* knockdown leads to changes in the expression of 220 genes at the larval stage L3 and of 847 genes at D5 (day 5) of adulthood. This set of ASH-2-controlled genes is most enriched for genes known to affect the life span and to change expression during aging. These results show that members of the H3K4me3 methyltransferase complex ASH-2 and of the H3K4me3 demethylase complex RBR-2 regulate aging by controlling the expression of a specific subset of genes.

RNAi inhibition of expression of the *utx-1* gene encoding an H3K27me3 demethylase in young adult *C. elegans* extends the mean life span by approximately 30% [46,47]. Thus, H3K27me3 may be a regulatory mark in aging. Compared with control worms, *utx-1* RNAi worms were more resistant to heat (35°C), DNA damage (UV), and oxidative (paraquat) stresses [47]. By an analysis of *utx-1* RNAi effects on mutant backgrounds, the life span extension was shown to be mediated by activity of the insulin/IGF-1 signaling (IIS) pathway genes [46,47]. The exact target of *utx-1* regulation seems to be the *daf-2* gene, encoding an insulin-like growth factor receptor (IGF1R). The expression level of *daf-2* increases dramatically, lagging behind the increase in *utx-1* expression during the aging time course [47]. Apparently, this transcriptional regulation of the *daf-2* gene might be through its H3K27me3-demethylation. UTX-1 regulation of the IIS pathway genes and the life span may well be relevant to mammals, since both UTX and IIS are highly conserved from invertebrates to mammals. Indeed, when the UTX mRNA in 3T3-L1 cells was downregulated by a complementary shRNA, a remarkable decrease in *Igf1r* and downstream genes expression was detected [47]. Thus, *utx-1* regulation of the IIS pathway is conserved between worms and mice. A significant decrease in the *IGF1R*-associated H3K27me3 levels was observed in macaque muscle during aging. Interestingly, a genome-wide decrease in the promoter H3K27me3 levels with age has been observed that can be restored by *utx-1* RNAi. In synchronized worms, H3K27me3 levels were not significantly affected between youth and middle age (up to D11), but drastically dropped to almost undetectable levels in older worms (D14-20) [46]. Thus, increased H3K27me3 seems to be associated with youthfulness. Obviously, *utx-1* deficiency extends life span by maintaining high levels of H3K27me3, perhaps allowing a better control of chromatin repression. Therefore, UTX-1 may not specifically target IIS genes, but instead IIS genes may serve as sensors for the global changes in epigenetic status and tune downstream cell growth and stress resistance functions accordingly [47].

In a focused RNAi screen of putative histone methyltransferases (38 SET domain containing genes) and histone demethylases (6 LSD1/amine oxidase genes, and 14 jmjC-domain containing genes), knockdown of six genes reproducibly extended life span in *C. elegans*, namely *set-9*, *set-26*, *rbr-2*, *utx-1*, *mes-2*, and *jmjd-2* [48]. Neither aging nor the *set-26* mutation had a significant effect on the global levels of the two active gene expression marks H3K4me3 and H3K36me3, whereas the two repressive chromatin marks, H3K9me3 and H3K27me3, were decreased with aging by 4.17-fold and 3-fold, respectively, whereas no decrease was observed in *set-26* deletion mutants. Thus, inactivation of *set-26* prevents the age-dependent loss of histone modifications that are normally associated with highly compacted heterochromatin. No gross changes in quantities and distribution of transcription activating marks (RNA pol II, H3K4me3, and H3K36me3) were found between young (D10) and old (D40) *D. melanogaster*, whereas a striking redistribution of repressive marks (H3K9me3 and HP1) between heterochromatin and euchromatin regions occurred in old flies [49].

In *Drosophila*, E(Z) is the catalytic subunit of Polycomb repressive complex 2 (PRC2) and its H3K27 trimethylation activity is essential for the establishment and maintenance of Polycomb silencing. Flies heterozygous for the catalytically inactive *E(z)* mutant alleles exhibit a substantially greater life span than the wild-type (wt) controls [50]. Similar effects on longevity have mutations of *ESCL*, encoding another component of PRC2 that is required for H3K27 trimethylation by E(Z).

Trithorax (TRX) is a methyltransferase that specifically trimethylates histone H3 on lysine 4 and antagonizes PRC2-induced silencing. It also interacts with the histone acetyltransferase CREB-binding protein (CBP) and is required for CBP-mediated acetylation of H3K27, which directly blocks H3K27 trimethylation by E(Z) because the two modifications are mutually exclusive. A *trx* null mutation partially (by ~75%) suppresses the long life span phenotype of *E(z)* mutants. Obviously, the increased longevity of heterozygous PRC2 mutants is most likely due to reduced H3K27 trimethylation and consequent perturbation of PRC2-induced silencing. The *E(z)* and *escl* heterozygotes displayed substantially greater resistance to acute oxidative stress (paraquat) and to starvation than controls. Both of these phenotypes were suppressed by *trx* mutation, suggesting that they are also due to impaired Polycomb silencing. Several hundred putative direct targets of PRC2 were found in *Drosophila* genome. One candidate that might play a direct role in the increased longevity and stress resistance of *E(z)* mutants is *Odc1*, encoding ornithine decarboxylase, an evolutionarily conserved enzyme that catalyzes the first rate-limiting step in polyamine biosynthesis. Polyamines have been implicated in resistance to oxidative stress and were reported to increase longevity in *C. elegans*, *Drosophila*, and mice. Increased expression of ornithine decarboxylase (ODC) in mammalian cells protects them from apoptosis induced by oxidative damage and other stresses. *Odc1* expression was elevated more than 2-fold in *E(z)* heterozygotes, and this increase was partially suppressed by the *trx* mutation. PRC2 and TRX play key roles in promoting epigenetically stable states of various genome loci, and thus modulate aging and affect the life span.

In aging brain cells of rhesus macaque, the H3K4me2 mark was found to be concentrated at gene promoter and enhancer sequences and to positively correlate with gene expression levels [51]. H3K4me2 peaks showed age-related increases both in intensity and breadth. Thus, a global opening of the chromatin seems to occur in brain cells upon aging. The genes that showed an increase in H3K4me2 at the promoter with age overlap significantly with those showing an increase in expression with age. On the other hand, no significant overlap was found between gene sets showing decreased promoter H3K4me2 and decreased expression with age. Thus, not all expression changes are caused by *cis* changes in promoter H3K4me2, and this is particularly true for aging-downregulated genes. The genes that show age-related increase both in expression and in promoter H3K4me2 in rhesus macaque significantly overlap with genes that show age-related increase in H3K4me3 at promoters in human brain neurons, whereas the age-related H3K4me2-downregulated macaque genes do not significantly overlap with the H3K4me3-downregulated human genes. Thus, promoter H3K4me2 upregulation seems to be more biologically relevant to the aging process than downregulation. The H3K4me2-upregulated gene set is significantly enriched for aging-related functions, such as chromatin regulation, oxidative stress, DNA damage and repair, inflammation, and metabolism, whereas the H3K4me2-downregulated gene set is not enriched for any specific biological functions. The chromatin regulatory functions were significantly enriched among genes proximal to H3K4me2-upregulated enhancers. Thus, chromatin modifiers themselves are among the major targets of the age-related gene expression upregulation by H3K4me2 at both promoters and enhancers, implicating extensive cross talk, feedback, or feed-forward controls of different chromatin modification events. Interestingly, of the various specific stresses, oxidative stress-related genes are the most overrepresented in the H3K4me2-upregulated gene sct, followed by UV stress-related genes. This, together with the enriched DNA repair functions, points to the possibility that oxidative stress and UV-induced DNA damage might trigger H3K4 methylation. At the whole genome level, H3K4me2 marks seem to induce a progressive opening of chromatin structure during development and aging, as a consequence of stress responses to various insults from the cellular environment, such as DNA damage and oxidative stress [48]. Histone modifications play important roles in shaping the epigenetic landscape and chromatin structure. It would be interesting to study the possible cross talk between H3K4 methylation and other epigenetic modifications and their net effect on chromatin structure.

H3K36me3 is widely known to be enriched on the gene body regions of actively transcribed genes [52]. Interestingly, no large changes in genome-wide levels and distributions of H3K36me3 were found during *C. elegans* aging [53] confirming previous observations [48]. A still more intriguing finding was that the H3K36me3 levels do not strictly correlate with expression levels of respective genes. An inverse correlation between H3K36me3 levels and mRNA expression changes during aging was observed. Genes with a dramatic expression difference between young (D2) and old (D12) worms were marked with minimal levels of H3K36me3 in their gene bodies, irrespective of their corresponding mRNA abundance. A similar correlation was observed in *D. melanogaster*, suggesting a conserved mechanism for H3K36me3 in suppressing age-dependent mRNA expression variability. A global reduction in H3K36me3 levels via inactivation of the respective methyltransferase gene *met-1* caused an increase in mRNA expression variability with age and shortened the life span. Thus, H3K36me3 functions to restrain gene expression variability during aging and positively affects longevity.

In ycast, H3K36me3 methyltransferase mutants (*set2Δ*) were found to have shortened life spans, whereas the H3K36me3 demethylase mutants (*rph1Δ*) increased life spans [54]. These effects were correlated with changes in transcription fidelity. An increase in cryptic transcription from intragenic promoter-like sequences was observed in old yeast cells, whereas H3K36me3 demethylase mutation suppressed such cryptic transcription. Both age-dependent cryptic transcription and H3K36me3 loss were most obvious in the same set of long and infrequently transcribed genes, and both were suppressed by *rph1Δ* mutation. An age-dependent increase in cryptic transcription was also observed in a large (>400) set of *C. elegans* genes. Thus, a decrease in transcription fidelity may be an evolutionarily conserved aging mechanism.

Current evidence supports a popular view of progressive breakdown of repressive chromatin structure via accumulation of activating epigenetic marks and loss of inhibiting ones that initiates spurious transcriptional events and promotes aging. TEs are particularly enriched in heterochromatin regions of eukaryotic genomes. Since TEs activity involves their insertion into new genomic locations, loss of heterochromatin repression during aging may well lead to genome destabilization. Indeed, a number of genes and TEs located in heterochromatic regions increased expression with age in *D. melanogaster* [55]. Calorie restriction (CR), an intervention known to delay aging in various species, diminished or even prevented age-dependent increase in expression levels of most heterochromatin genes and TEs. Activation of TEs expression has been correlated with an increase in frequency of transposition events that was diminished by CR. A correlation between the life span and the timing of increased transposition was observed in different genetic backgrounds. Lines exhibiting earlier transposition showed a shorter life span, and those with later transposition a longer life span. Super-expression of *Sir2*, *Su(var)3–9*, and *Dicer-2*, known to stabilize heterochromatin, decreased the age-dependent TEs activation and extended the life span. Thus, prevention of TEs activation increases genomic stability and homeostasis during aging and extends life span.

H3.3 is an evolutionarily conserved histone variant involved in epigenetic regulation of the differentiated cell plasticity [56]. Two genes encoding histone H3.3 (*his-71* and *his-72*) have been shown to increase expression in *C. elegans* postmitotic cells [57]. H3.3-deficiency (*his-72;his-71* double mutation) and expression of human H3.3 or HIS-72::GFP transgenes did not affect the *C. elegans* life span in wt background. On the other hand, *his-72;his-71* double mutation significantly decreased the life span extension induced by *daf-2* mutation. Deletion of *daf-16* completely abolished the long-lived phenotype of *daf-2* single mutants, and *his-72;his-71* double mutation did not affect the short life span of *daf2;daf-16* double mutants. Thus, H3.3 seems to be necessary for the proper action of pro-survival processes in *daf-2* animals. Less than 1000 genes were dysregulated (up- or downregulated) in *his-72;his-71* double mutants compared with wt worms, whereas more than 8000 genes were dysregulated in *his-72;his-71;daf-2* triple mutants compared with *daf-2* single mutant worms. Thus, H3.3 loss leads to much severe changes of gene expression in long-lived *daf-2* animals compared with wt ones. Of about 8000 putative direct DAF-16 target genes, 1208 were significantly upregulated in *daf-2* mutants, and 722 of these last were further affected in *his-72;his-71;daf-2* triple mutants compared with *daf-2* single mutant worms. Interestingly, 65% of 722 genes affected by *his-72;his-71* double mutation in *daf-2* background were downregulated back to wt expression levels. Thus, H3.3 appears to be profoundly involved in pro-longevity DAF-16-mediated transcriptional programs. A mild decrease in mitochondrial activity in *nuo-6* mutant nematodes promotes metabolic changes that lead to an increased life span independent of the DAF-2-DAF-16 pathway activity. Surprisingly, H3.3 loss abolished the life span extension induced by *nuo-6* mutation and caused dysregulation of 6305 genes in *nuo-6* background. Obviously, H3.3 has a general role in life span regulation by pro-longevity transcriptional programs.

Mild mitochondrial dysfunction positively affects longevity of *C. elegans* [58]. Massive changes in gene expression in response to such mitochondrial stress involves an upregulation of genes involved in mitochondrial unfolded protein response (UPR^{mt}), approximately half of these genes via the activity of transcription factor ATFS-1 [59]. In a large-scale screen for genes that are required for induced UPR^{mt}, mutations of the gene encoding nuclear cofactor *lin-65* have been shown to be as effective in UPR^{mt} suppression as *atfs-1* deletions [60]. Mitochondrial stress leads to LIN-65 nuclear accumulation that depends on activity of histone H3K9me1/2 methyltransferase MET-2 (a homologue of mammalian SETDB1). Thus, epigenetic modifications seem to be involved in UPR^{mt} regulation. Activity of both *met-2* and *lin-65* is needed for H3K9me2 mark establishment. Striking changes in chromatin structure were observed on mitochondrial stress. Nuclei appeared smaller and much more condensed compared with those in unstressed animals. In contrast, *met-2* and *lin-65* mutant animals had enlarged nuclei and loose chromatin structure even after mitochondrial stress. Thus, mitochondrial stress induces global chromatin reorganization and decreased nuclear size. This correlates with altered expression of more than 1300 genes. In the absence of *lin-65* and *met-2*, a majority of these genes exhibited wide variance in expression patterns but did not significantly change compared with untreated animals. The proper regulation of gene expression in response to mitochondrial stress appears to be mediated by the chromatin reorganization that depends on H3K9me2 methylation. Both *lin-65/met-2* and *atfs-1* mutations separately only partially suppressed the life span extension by UPR^{mt}. On the other hand, this life span extension was fully abolished in combined *met-2;atfs-1* mutants. Thus, H3K9me2 methylation and ATFS-1 act synergistically to mediate mitochondrial stress-induced longevity. In a similar screen for UPR^{mt} suppressors, two histone lysine demethylases, *jmjd-1.2* and *jmjd-3.1*, were identified [61]. These demethylases differ in their substrate specificity, *jmjd-3.1* being mainly if not exclusively an H3K27me2/3-specific demethylase, whereas *jmjd-1.2* possesses a wider range of activities toward H3K9me1/2, H3K27me2, and H4K20me1. Both *jmjd-1.2* and *jmjd-3.1* are necessary for a UPR^{mt} induced increase in longevity. Interestingly, *jmjd-1.2* has no effect on the life span extension by other pathways, whereas *jmjd-3.1* does not affect CR-mediated longevity but partially suppresses life span extension in *daf-2* mutant and germline deficient *glp-1* worms. Intriguingly, overexpression of either *jmjd-1.2* or *jmjd-3.1* appeared to be sufficient to induce the UPR^{mt} and to increase the life span, and a majority of the gene expression changes induced by UPR^{mt} were recapitulated by overexpression of *jmjd-1.2* or *jmjd-3.1*. Furthermore, of all nine JmjC domain-encoding genes, *jmjd-1.2* and *jmjd-3.1* were specifically upregulated upon UPR^{mt} induction. Thus, *jmjd-1.2* and *jmjd-3.1* act downstream of mitochondrial stress to mediate the majority of its transcriptional response.

Sirtuins

Silent information regulator (Sir) proteins regulate life span in multiple model organisms [62]. In yeast, an extra copy of the *Sir2* gene extends replicative life span by 50%, while deleting *Sir2* shortens the life span. Sir2 exerts its effects via modulation of chromatin structure by catalyzing NAD$^+$-dependent deacetylation of histone H4K16ac [63]. Sir2 homologues are required for the life span extension by CR in various model organisms. Mammals have seven Sir2 homologues, sirtuins (SIRT1–7) [62]. Unlike yeast Sir2, mammalian sirtuins target multiple substrates and affect a broad range of cellular functions. SIRT1 has multiple substrates, known to be involved in aging (p53, NF-κB, FOXO family proteins, and others), and is generally regarded as a guardian against cellular oxidative stress and DNA damage. It also regulates the activities of the nuclear receptor PPARγ and its co-activator protein PGC-α to influence differentiation of muscle cells, adipogenesis, fat storage, and metabolism. SIRT1 has been shown to modify chromatin and silence transcription via deacetylation of histones and some other proteins (H4K16Ac, H3K9Ac, H1K26Ac, p300, and multiple TFs) [64], recruitment of histone H1b, and promotion of the loss of H3K79me2 (a mark associated with transcriptionally active chromatin) and the gain of H3K9me3 and H4K20me1 (marks associated with repressed chromatin) [65]. SIRT1 has been shown to interact directly with the major mammalian H3K9me3-specific histone methyltransferase SUV39H1, recruit it to the respective chromatin loci, and stimulate its activity via deacetylation of K266 residue in its catalytic SET domain [66]. SIRT1 deficiency results in a complete loss of SUV39H1-dependent H3K9me3 and impairs localization of HP1 (heterochromatin protein 1) to heterochromatic regions. The involvement of SIRT1 in DNA methylation at damaged CGI-containing promoters, occasionally causing their aberrant silencing—events typical of aging and cancer—has also been demonstrated [67]. On the genome-wide scale, genes involved in chromatin assembly and transcriptional repression, ubiquitin-regulated protein degradation, and the cell cycle regulation were significantly overrepresented among the SIRT1-bound promoters in mouse ESC cells [68]. Oxidative stress (treatment with non-cytotoxic levels of H_2O_2) caused a major redistribution of SIRT1, such that less than 10% of the SIRT1-associated promoters overlapped between untreated and H_2O_2-treated cells. Multiple genes significantly increased their expression upon H_2O_2 treatment, coincident with SIRT1 release, including regulators of metabolism, apoptosis, ion transport, cell motility, and G-protein signaling. Of these genes, 85% were repressed by a modest overexpression of SIRT1. Interestingly, more than two-thirds of the SIRT1-bound genes that were derepressed by oxidative stress *in vitro* were also derepressed during aging in mice brain. Moreover, the SIRT1 target genes were significantly overrepresented among the age-upregulated genes, and SIRT1 overexpression completely abolished this age-dependent upregulation. Thus, SIRT1 seems to be a participant in oxidative stress response that may provide a direct link between DNA damage and gene expression changes that occur during aging. Lower levels of DNA damage, decreased expression of the senescence-associated gene *p16^Ink4a*, better general health, and fewer spontaneous tumors were observed at old ages in transgenic mice moderately overexpressing *Sirt1* compared with wt controls [69].

SIRT6 is a histone H3K9Ac deacetylase required for normal BER maintenance of genomic integrity and telomeric chromatin modulation [70,71]. Its deficiency in mice leads to the development of an acute degenerative aging-like phenotype. SIRT6-depleted cells exhibit abnormal telomere structures that resemble defects observed in Werner syndrome, a premature aging disorder. At telomeric chromatin, SIRT6 activity is required for stable association of the *WRN* encoded helicase RecQ that is mutated in Werner syndrome. SIRT6 has been shown to physically interact with an NF-κB subunit, RELA [72]. Upon cell stimulation with TNF-α, a known activator of NF-κB, SIRT6 is recruited to the promoters of several NF-κB target genes (*IAP2, MnSOD, ICAM,* and *NFKBIA*), leading to their repression. The SIRT6-repressed NF-κB target gene set overlaps with genes that are known to be increasingly expressed with age in human tissues [10]. The SIRT6-dependent transcription repression has been shown to occur via H3K9Ac deacetylation that promotes RELA destabilization from the target gene chromatin loci and termination of NF-κB signaling. Not all putative NF-κB target genes are affected by SIRT6; rather, the SIRT6-regulated NF-κB target genes are selectively enriched for those related to immune response, cell signaling, and metabolism. In transgenic male mice overexpressing *Sirt6*, a significant (~15%) increase in the average life span was observed, whereas the life span of transgenic female mice was not significantly changed [73]. A whole genome microarray analysis uncovered a subset of genes whose expression differed significantly between transgenic and control males. This subset was significantly enriched for functional categories related to metabolism and cellular responses. Of these genes, 50% were also differentially expressed between male and female wt mice, and approximately 30% showed a similar expression pattern in male mice fed a CR diet. Specifically, the upregulated genes *Lpin1, Lpin2, Gadd45g, Fkbp5, Dusp1,* and *Cebpd,* and the downregulated genes *Vnn1, Vnn3, Pctp, Vldlr, Car3,* and *G0s2,* in the expression profile of male *Sirt6*-transgenic mice were also differentially expressed in the livers of mice fed a CR diet. One of the genes that was highly upregulated in *Sirt6*-transgenic males, to the same levels as in control or *Sirt6*-transgenic females, was the gene encoding IGF-binding protein 1 (IGFBP1), known to be the main short-term modulator of IGF1 bioavailability. CR increases the expression of IGFBP1, and high levels of IGFBP1 correlate with protection against metabolic disorders. *Sirt6* overexpression extends life span only in male mice, potentially by reducing IGF1 signaling in white adipose tissue. *Sirt6* deficiency has been demonstrated to cause a cell-autonomous upregulation of glucose uptake, triggering a nutrient-stress response and a switch in glucose metabolism toward glycolysis and away from mitochondrial respiration [74]. It was proposed that SIRT6 functions as a corepressor of Hif1a-dependent transcriptional activity, deacetylating H3K9Ac at Hif1a target gene promoters. In this way, SIRT6 maintains an efficient glucose flux in the tricarboxylic acid (TCA) cycle under normal nutrient conditions.

NONCODING RNAS

SMALL NONCODING RNAS

Of the 114 *C. elegans* miRNAs cataloged in miRBase version 5.0, 34 have been found to change their expression levels during the post-young adult aging (from D4 to D15) [75]. Several of these miRNAs exhibit very large changes in relative abundance over adult life. In particular, miR-231 is noteworthy for its high level of expression and its great increase in abundance over time (>100-fold higher at D15 than at D4). A substantial increase in miR-34 level and decrease in levels of five other miRNAs occur between D6 and D8. These miRNAs were suggested to influence aging and life span. Level of miRNA lin-4, known to affect life span [76], gradually declines from D6 to D11 and remains essentially constant thereafter. Level of let-7 is highest early in adulthood, rapidly decreases between D6 and D11 to minimal values and remains at these values through the post-reproductive period. The mRNA transcript for nuclear hormone receptor DAF-12 is among the validated targets of let-7. Since the loss of function *daf-12* mutations extends life span in *C. elegans*, the diminishing levels of let-7 with age could influence life span similar to lin-4. Concentration of miR-1, a highly conserved miRNA expressed in somatic muscle of different species, has been found to exhibit a rapid decline during the adult life of *C. elegans*. This pattern parallels the progressive decline of the body wall muscle integrity during aging. Of more than 200 *C. elegans* genes known to impact life span, about 40 encode mRNAs that could be direct targets of age-regulated miRNAs; 10 of them could be targeted by multiple age-regulated miRNAs.

In a deep-sequencing survey of small RNA fractions from aged tissue in wt and in long-lived *daf-2* mutant *C. elegans*, mature forms of 120 out of the 155 miRNAs annotated in miRBase 13.0 were detected, as well as 10 sequences that probably represented new miRNAs [77]. In wt animals, 8.6% of miRNAs increased and 28.4% decreased in expression more than 2-fold from D0 to D10. A similar pattern was observed in *daf-2* animals, where 9.8% of miRNAs increased and 25.6% miRNAs decreased with age. The most strong increase in expression during aging in wt worms was observed for miR-246, miR-71, miR-34, miR-253, miR-238, and miR-239a/b. Conversely, let-7, miR-41, miR-70, and miR-252 were the most downregulated miRNAs with aging. The insulin/IGF-1 receptor DAF-2 is a key component of the IIS pathway in *C. elegans* that responds to environmental signals to regulate stress response and life span. The vast majority of miRNAs exhibited similar changes in expression with aging between wt and *daf-2* animals. Some of the few specific miRNAs that exhibited altered expression in *daf-2* mutants compared with wt also exhibited altered expression with aging. The miRNAs with the greatest increase in expression in *daf-2* compared with the wt animals, miR-237, miR-62, and miR-252, were among the most downregulated with aging. Conversely, miR-239b, which was downregulated in *daf-2* animals, was among the most upregulated with aging miRNAs. Collectively, these *daf-2*-dependent miRNAs reveal the intersection of the IIS pathway with miRNA-regulated genes that impact life span. Deletions of *mir-71*, *mir-238*, and *mir-246* genes decreased life span, whereas deletion of *mir-239* increased life span compared with wt animals. Since miR-71, miR-238, and miR-246 are highly upregulated during aging, these results suggest that these miRNAs promote longevity. Indeed, overexpression of both miR-71 and miR-246 significantly increased life span, whereas overexpression of miR-239 reduced it. It has been found by genetic criteria that both miR-71 and miR-239 function through the IIS pathway, and that miR-71 also interacts with the DNA damage response pathway; miR-71 may serve as a possible link between the IIS and the DNA checkpoint pathways. No discernible phenotypic effects were observed for deletion mutants of *mir-71*, *mir-239*, and *mir-246*, besides their effects on longevity. Thus, *mir-71*, *mir-239*, and *mir-246* may be aging-specific genes, functioning specifically during adulthood to regulate genetic pathways that affect life span.

Retention of "youthful" miR-71 states at D3 to D7 (beginning of the reproductive period to the onset of mortality), both in terms of high levels of expression and of maintenance of these levels through mid-adulthood, has been found to correlate with longevity [78]. Animals with higher and/ or longer-lasting miR-71 expression tend to live longer, consistent with the known role of miR-71 in promoting life span [77]. In the *daf-16* mutant background, miR-71 expression differences were no longer apparent between animals with different life spans. The expression of miR-246 increased over time, and showed a gradual plateauing in late adulthood. Animals in which miR-246 levels plateaued more slowly were relatively longer lived. The mean level of miR-246 between D3 and D7, however, did not clearly predict longevity. Unlike miR-71 and miR-246, miR-239 antagonized longevity. The ability of expression states to predict later longevity quite early in life suggested that some fraction of the longevity variance is indeed a result of developmentally determined epigenetic states of "robustness." The degree of strong miR-71 expression appeared to be the most powerful biomarker identified, explaining 47% of future longevity, and also the earliest indicator of future longevity. Together with earlier findings on knockout animals [77], the data described could indicate that the individual wt variability in the expression of these miRNAs not only affects longevity but also likely determines it. In the case of miR-71 and miR-239, the mechanism of individual life span determination may be via the well-known IIS pathway. The *mir-71* knockout leads to increased expression of components of the IIS transduction machinery [77], whereas *daf-16* null animals have increased levels of miR-71 expression [78]. These observations suggest that DAF-16 and miR-71 levels are held in homeostatic balance via a mutual regulatory feedback loop. Thus, it is possible that interindividual fluctuations in miR-71 levels may directly determine individual levels of tonic IIS activity and hence individual life spans.

Of 174 *C. elegans* miRNAs annotated in miRBase release 14, 39 were found to change in abundance more than 2-fold between D0 (young adult) and D8 (old adult) ages, 23 of them upregulated and 16 downregulated [79]. When miRNA maturation was inhibited in adult worms by an RNAi against *alg-1*, encoding an Argonaute protein, a significant decrease in life span was observed. Thus, age-regulated miRNAs could play a significant role in aging and life span regulation. Compared with control animals, which were kept at 23°C during all life stages, animals shifted to lower temperature (15°C) at D0 of adulthood showed a significantly longer life span, while those shifted to higher temperature (27°C) showed a shorter life span. Age-regulated miRNAs that increased expression during aging showed a significant delay in expression changes at 15°C, while at 27°C their expression changes were accelerated. As examples, miR-34 and miR-239a, both of which increased in expression approximately 10−15-fold between D0 and D8 in control worms, reached a similar fold increase at D15 in the long-lived (15°C) worms and at D3 in the short-lived (27°C) worms. These miRNAs seem to be tightly regulated during aging, and their expression at lower levels might contribute to a delay in aging. In contrast, miRNAs that decreased in expression during aging did not show a delay when shifted to long-lived condition at 15°C. Their downregulation could be possibly programed at earlier stages of development or controlled independently from the life span. In many cases of miRNAs upregulated with age, at later stages of development they seem to be expressed in tissues or cells in which they are not expressed at earlier stages, suggesting an age-dependent loss of proper regulation of tissue-specific expression. Such ectopic activation of miRNA expression might trigger unfavorable age-related decline, since miRNAs affect expression of many target genes. Age-associated alterations of miRNA expression levels may cause reciprocal changes in the expression of their target genes during aging. A total of 354 protein-coding genes were predicted to be the targets of at least one age-associated miRNA. Of these genes, 58 encode proteins that are involved in aging-related pathways (phagosome and lysosome function, protein processing, splicing, oxidative phosphorylation, methionine metabolism, and others) and affect life span.

A temperature-sensitive allele of the miRNA pathway gene *pash-1* in *C. elegans* allows rapid and reversible inactivation of miRNA synthesis [80]. The mean life span of *pash-1ts* adults shifted to the restrictive temperature was shorter by about 30% compared with similarly temperature-shifted wt animals. This defect was rescued by the wt *pash-1* transgene, confirming that it was due to the *pash-1ts* lesion. Hence, miRNAs synthesized post-developmentally must have functions that affect life span. Decline in locomotion is a well-known marker of physiological aging in *C. elegans*. It was noted that locomotion declines more rapidly in *pash-1ts* adult worms shifted to the restrictive temperature, suggesting that these animals age more rapidly. Adult *pash-1ts* animals placed at the restrictive temperature frequently display defects associated with advanced age much earlier than rescued animals. Very few genes were misregulated in *pash-1ts* mutants at the permissive temperature, whereas hundreds of genes became rapidly misregulated upon upshift to a restrictive temperature, indicating that ongoing miRNA synthesis during adulthood is required to maintain normal patterns of gene expression. Genes that are known to be upregulated during normal aging in wt animals are rapidly upregulated in *pash-1ts* animals at the restrictive temperature. Conversely, genes normally downregulated in old animals are rapidly repressed. Genes that are upregulated, then downregulated, over the first 15 days of normal aging are upregulated, then downregulated over the course of 24 h in *pash-1ts* animals at the restrictive temperature. These data suggest that the reduced life span and the age-associated defects of adult animals unable to synthesize miRNAs result from accelerated aging. Surprisingly, inactivation of the DAF-2 insulin/IGF-1 receptor extended life span even in animals unable to synthesize miRNAs. This life span extension was *daf-16* dependent, indicating that FOXO can promote longevity via mechanisms that do not require synthesis of miRNAs. The FOXO-mediated effects of the IIS pathway activity on life span are known to involve regulation of a large set of target genes. The data described suggest that this program of transcriptional regulation is sufficient to extend life span even in the absence of post-transcriptional regulation by miRNAs. Removal of miRNA expression reduced life span in a *daf-16*–null mutant, in which IIS cannot regulate life span. These data suggest that miRNAs and IIS determine life span in parallel.

The PHA-4 and SKN-1 transcription factors are known to play key roles in mediating the pro-longevity effects of CR in *C. elegans*. A network analysis of aging-associated miRNAs that could target the *pha-4* and *skn-1* genes, and whose expression could be regulated by PHA-4 and SKN-1, revealed extensive regulatory interactions between PHA-4, SKN-1, and miRNAs, especially two aging-associated miRNAs, miR-71 and miR-228 [81]. Both miR-71 and miR-228 appeared to be necessary for the life span extension by CR. It was shown that CR induces the expression of miR-71 and miR-228, which are in feedback loop regulation relationships with PHA-4 and SKN-1. Specifically, miR-228 targets both *pha-4* and *skn-1*. On the other hand, PHA-4 promotes miR-228 expression, whereas SKN-1 antagonizes it. Conversely, SKN-1 promotes miR-71 expression, and miR-71 targets *pha-4*. These feedback loops fine-tune the CR effect to extend the life span.

In *Drosophila*, a hypomorphic mutation in *loquacious* (*loqs*), a key gene in fly miRNA processing, was found to lead to a significantly shortened life span and late-onset brain morphological deterioration [82]. Thus, some miRNAs may be critically involved in age-associated events impacting long-term brain integrity. Of the 29 miRNAs expressed in the adult brain, most maintained a steady level or decreased, while one, namely miR-34, increased with age. This last miRNA is markedly conserved, with identical seed sequences in fly, *C. elegans*, mouse, and human orthologues. Flies null mutant for *mir-34* have no obvious developmental defects, showing normal adult appearance and early survival, but display a catastrophic decline in viability just after reaching mid-age (30-day). Older wt flies show sporadic, age-correlated vacuoles in the brain—a morphological hallmark of neural deterioration. Flies null mutant for *mir-34* are born with normal brain morphology, but show dramatic vacuolization with age, indicative of loss of brain integrity. Transcription profiling of the fly brain at various ages showed that the majority of transcripts positively correlated with age display a faster pace of increase in *mir-34* mutants compared with controls. Thus, loss of miR-34 seems to accelerate brain aging. One of the putative targets of miR-34 is the *E74EF* gene encoding a protein of steroid hormone signaling pathways. In the wild type, E74EF protein is highly expressed in young flies, but undergoes a dramatic decrease within a 24 h time window, the temporal pattern appearing to be opposite to that of miR-34. Moreover, in flies lacking miR-34, downregulation of the protein during this critical period is dampened. Thus, adult-onset expression of miR-34 seems to function, at least in part, to attenuate E74EF expression in the young adults, and maintain that repression through adulthood. When a hypomorphic mutation of the *E74EF* gene has been introduced into *mir-34* mutant flies, proper regulation of the E74EF protein was partially restored, and age-associated defects of miR 34 loss, including the shortened life span and the brain vacuolization, were mitigated. When the adult activity of E74EF was upregulated with an *E74EF* transgene that lacked miR-34 binding sites, animals showed late-onset brain degeneration and significantly shortened life span. Obviously, miR-34 functions to suppress the *E74EF* gene expression at the post-transcriptional level, thus preventing its adult stage-specific deleterious effects on brain integrity and viability. E74EF appears to have opposite effects on animal fitness at different life stages, being essential during preadult development, but harmful to the adults during aging. This biological property is known as antagonistic pleiotropy.

Genes associated with antagonistic pleiotropy are likely to be evolutionarily retained due to their earlier beneficial functions, whereas miRNA pathways provide a mechanism to suppress their potentially deleterious age-related activities, in order to prolong healthy life span and longevity. It has been noted that miR-34 is elevated with age in *C. elegans*, and its orthologues in mammals are highly expressed in the adult brain, increase with age, and are misregulated in degenerative diseases in humans. Furthermore, miR-34a has been shown to function as an important mediator of aging and cardiac and endothelial cell dysfunctions due to direct downregulation of its target, SIRT1 [83]. An aging-associated increase in miR-34a levels in vascular smooth muscle cells leads to downregulation of SIRT1 and induction of the senescent-associated secretory factors, thereby provoking cell senescence and inflammation that characterize arterial dysfunctions, such as vascular calcification and atherosclerosis.

An analysis of miRNA expression profiles between old and young murine liver specimens revealed four miRNAs upregulated with aging, miR-669c (9.9-fold upregulation in 33-month-old mice compared with 4-month-old ones), miR-709 (7.6-fold), miR-214 (3.4-fold), and miR-93 (4.3-fold) [84]. Only miR-669c and miR-709 showed a gradual increase with age, whereas miR-93 and miR-214 showed biphasic expression at 10- and 18–33-month ages. Several miRNAs were found to decrease at 33 months compared with 10 months (miR-375, let-7i and let-7g). At least 24% of proteins significantly downregulated at 33 months appeared to be targets of the four upregulated miRNAs. Importantly, several of these proteins (Acsl1, Gstz1, Uqcrc1, and Mgst1) are linked with mitochondria. Several downregulated glutathione S-transferases are also among the predicted targets. Obviously, these miRNAs may be implicated in an aging mechanism related to oxidative stress. Specifically, a decline in oxidative protection is correlated with miR-93 targeted Mgst1, and the failure of the mitochondrial respiratory chain is correlated with miR-709 targeted Uqcrc1. Insulin-like growth factor 1 (Igf1) is among the predicted targets of miR-93; it may have a role in the IIS pathway similar to lin-4 in *C. elegans*. Decreased IIS pathway activity is intrinsically related to conditions of CR and may influence longevity. The gene for miR-669c is located in an intron of the gene encoding Sfmbt2, a Polycomb group (PcG) protein, known to functionally interact with TF Yinyang1 (YY1), possibly forming a PcG silencing complex [85]. The gene for miR-709 is found in an intron of the gene encoding TF Rfx1, a known activator of the virus gene expression and suppressor of cellular genes, such as *c-myc* and *PCNA*. One of the miR-709 targets is Brother of the Regulator of Imprinted Sites (BORIS), known to play an important role in epigenetic reprogramming during the male germ cell differentiation [86]. Its downregulation by miR-709 may be the cause of age-dependent decline in spermatogenesis.

Stress factors can induce upregulated miRNA expression [85]. Permanent growth arrest, seen either in replicative senescence via serial passaging or in premature senescence induced by peroxide treatment, is associated with significant upregulated expression of miRNAs. This stress-induced activation of miRNAs may promote tissue aging. Certain miRNAs, including miR-210 and miR-373, dampen the expression of key DNA repair proteins [87]. Interestingly, in a screen of over 800 miRNAs in human peripheral blood mononuclear cells, 16.5% of the miRNAs declined in abundance with age, while only 2.5% increased [88]. It has been shown that miR-24, significantly downregulated with age, can target E2F2, p16^{INK4a}, MKK4, and H2AX mRNAs. Several miRNAs downregulated with age are known to be associated with cancers (miR-103, -107, -128, -155, and -221). These findings underscore the importance of miRNA expression in age-related diseases, such as cancer. PI3K (phosphatidylinositol-3-kinase) is a known integrator of multiple signaling pathways that promote tumorigenesis. Downregulation of miR-221 with age leads to increased PI3K mRNA and protein. Thus, miR-221 may be a modulator of pathways important in aging and tumorigenesis.

Numerous proteins exhibited age-dependent variations in mouse brain, including 13 subunits of the mitochondrial electron transport chain complexes or ATPase, of which 10 were downregulated [89]. Of the 367 miRNAs studied, 70 showed significant age-dependent variations. Among them, 31 miRNAs exhibited consistent upregulation, and 17 were consistently downregulated with age. Of the 70 age-variable miRNAs, 27 targeted the 10 mitochondrial subunits that exhibited decreased expression in elderly animals. Among them, 16 miRNAs exhibited consistent upregulation with age, whereas 4 were consistently downregulated. These data show that multiple miRNAs are upregulated during normal aging. Expression levels of miR-22, -101a, -720, and -721 were inversely correlated with those of their predicted targets, Uqcrc2, Cox7a1, Atp5b, and Cox5b. Two main groups of miRNAs involved in aging could be discerned. The first consisted of crucial miRNAs that carry out tissue-specific targeting of genes, such as oxidative phosphorylation in the brain or detoxification in the liver, which leads to tissue-specific aging dysfunction. The other consisted of miRNAs that take part in regulating the common aging process. Thus, upregulated miRNAs miR-30d, -34a, -468, -669b, and -709 were found both in aging liver and brain, whereas miR-22, -101a, -720, and -721 were found specifically in aging brains, and miR-669c, -712, -214, and -93 were specific to aging livers. Interestingly, brains of the extremely old (33 months) mice had less upregulated miRNAs than those of 24-month-old mice. The proportion of mice surviving to 33 months of age being about 4%, one may suggest that these mice had a more stable epigenome, leading to reduced levels of deregulation of miRNA expression and extended life span.

In a genome-wide screen of the expression profiles of 863 miRNAs in the whole blood samples from 15 long-lived (mean age 96.4 years) and 55 younger (mean age 45.9 years) human individuals, 80 differentially regulated miRNAs were found, 16 of them upregulated and 64 downregulated with age [90]. A total of 4957 mRNAs could be potentially targeted by these miRNAs, 1233 by the upregulated miRNAs and 4229 by the downregulated ones. There are 505 mRNAs that could be potentially targeted both by upregulated and downregulated miRNAs. No significantly enriched cellular pathways were found for the set of targets of upregulated miRNAs. On the other hand, four significantly enriched target pathways were detected for the downregulated miRNAs, including the p53 signaling cascade, cancer pathways, citrate acid cycle, and the hedgehog signaling pathway. A total of 14 diseases were correlated with 64 downregulated miRNAs, mainly cancers. In contrast, no enrichment for specific diseases was found for 16 upregulated miRNAs. This observation implies that changes in the expression levels of miRNAs during physiological aging may suppress the development of tumors and other age-related pathologies. Differential regulation of miRNAs has been shown to influence senescence, a mechanism that leads to cancer resistance at the price of aging. The same miRNAs described to be downregulated in senescent cells have also been found to be downregulated in the long-lived individuals, for example, miR-17, miR-20a, miR-106a, and miR-21. Interestingly, miR-21 has been shown to be differentially regulated in approximately 40 different diseases. It is likely that either the deregulation of miR-21 plays a key role in all of these diseases by a common mechanism or its regulation is controlled by various cellular signals specific for each disease pathway. Downregulated miRNAs might be involved in regulation of the pro-aging versus pro-longevity effects of p53, and their effects on aging and longevity could be, at least partly, mediated by p53. In the long-lived individuals, activities of the aging-related miRNA and p53 networks could be balanced such as to prevent tumorigenesis and maintain genomic integrity.

Single-nucleotide polymorphisms (SNPs) in the 3′-untranslated regions (3′-UTRs) targeted by miRNAs can alter the strength of miRNA binding and, hence, the regulation of target gene expression. In a screen for such SNPs (miRSNPs) in the 3′-UTRs of 140 human aging-related genes, 24 miRSNPs with the high difference of binding energy between the two alleles were found [91]. These miRSNPs were then investigated for their possible association with longevity. Two of them, SIRT2-rs45592833 G/T and DRD2-rs6276 A/G, showed a significant correlation with longevity. In both cases, the minor allele was associated with a significantly decreased chance to become long-lived. Thus, individual variability in the gene regulation by specific miRNAs could modulate the aging phenotype.

LONG NONCODING RNAS

A quantitative proteomic study of long-lived *daf-2* mutants of *C. elegans* revealed a remarkable decrease in proteins involved in mRNA processing and transport, the translational machinery, and protein metabolism compared with wt and short-lived *daf-16;daf-2* double mutant animals [92]. Polyribosome profiling in the *daf-2* and *daf-16;daf-2* double mutants confirmed that a DAF-16-dependent reduction in overall translation activity occurs in *daf-2* mutant worms. Since the total amount of protein in the *daf-2* mutants does not significantly differ from that in wt animals, concurrent reduction in protein metabolism probably occurs. The RNA sets associated with ribosomal fractions were significantly different between *daf-2* mutants, on the one side, and wt and *daf-2;daf-16* double mutants on the other [93]. These sets in *daf-2* animals were enriched for mRNAs that encode proteins involved in aging and stress response, whereas respective sets in both wt and *daf-2;daf-16* worms were enriched for mRNAs coding for proteins involved in general biological activities, such as growth, development, the cell cycle, and reproduction. One striking difference between these sets was a long noncoding RNA (lncRNA), transcribed telomeric sequence 1 (*tts-1*), present at a high level in ribosomal fractions of *daf-2* cells, but nearly absent from those of wt and *daf-2;daf-16* cells. A double-stranded small interfering RNA (siRNA) construct against *tts-1* reduced the levels of *tts-1* in *daf-2* mutants by >10-fold. Importantly, this siRNA significantly shortened the extended life span of *daf-2* mutants, but did not affect the short life span of wt and *daf-2;daf-16* double mutants. Increase in *tts-1* expression has been observed in two other life span-extending mutations, *clk-1* (a mitochondrial pathway mutation that reduces respiration and decreases ubiquinone biosynthesis) and *eat-2* (a model of CR with defective feeding behavior), with much higher levels of *tts-1* found in the *clk-1* mutants compared with *eat-2*. In good accordance with these results, the siRNA depletion of *tts-1* resulted in a substantial shortening of life span in *clk-1* mutants, but only a marginal shortening in *eat-2*. It could be suggested that *tts-1* lncRNA is able to reduce ribosome levels in a manner that is necessary for the life span extension.

REJUVENATION: IS IT POSSIBLE?

A natural example of the full reversal of aging is fertilization, a process where the fusion of the haploid oocyte and sperm cells results in a diploid cell (zygote) that has a zero age. Since there are no reasons to believe the germinal cells to be specifically protected from chronological aging, the species would progressively age with each generation, if there is no such age reset. A similar age reset occurs upon somatic cell nuclear transfer (SCNT) when the nucleus of a somatic cell is transferred to the cytoplasm of an enucleated oocyte, resulting in development of a new individual. The oocyte cytoplasm seems to possess a capability to "erase" all the aging features accumulated in the nucleus of the somatic cell. Thus, age-related features of the cell nuclei, whatever their nature, are principally reversible. Regarding the epigenetic aging marks, the main difficulty in their resetting is probably the "needle in a haystack" problem. The epigenome of any cell is a complex mosaic of epigenetic marks, where age-related ones are intermixed with a plethora of others. A most easy way to reset the age-related epigenetic changes is to fully erase all existing epigenetic information and then rebuild it from scratch in a form corresponding to the zero age. Such erasure occurs during the first hours following fertilization [94–96]. Accumulation of mutations and some other irreversible changes in germinal cells during chronological aging could hardly be avoided. The resetting of age upon fertilization proves these changes not to be the specific causes of aging. Evidently, gradual accumulation of mutations increases the genetic variability of humans as a biological species, and most of these mutations are not directly related to aging.

Yet another example of age resetting is production of the iPSCs [97]. Recent observations showed that iPSCs generated from senescent cells or centenarian donor cells rejuvenate telomeres, gene expression profiles, oxidative stress, and mitochondrial metabolism to the levels characteristic of ESCs [98,99], and are able to redifferentiate into fully rejuvenated cells [99]. In addition, as was noted earlier, such iPSCs have an epigenetic age close to zero [37]. In all the cases described above, resetting of the aging clock was coupled to cell dedifferentiation. Is it possible to epigenetically reprogram cells to a more youthful state without disturbing its differentiation status? Two distinct transcriptional waves or phases were distinguished during cellular reprogramming, the first (stochastic) driven by c-Myc and Klf4, and the second (deterministic) driven by Oct4, Sox2, and Klf4 [100,101]. Importantly, changes in active and repressive histone marks, such as H3K4me3 and H3K27me3, occur during the first phase, changes in miRNA expression are observed during both phases, whereas alterations in DNA methylation take place during the second phase. Early in the reprogramming process, the four factors induce epigenetic changes by a stochastic mechanism, leading to an intermediate, partially reprogrammed state. Activation of endogenous Sox2 that occurs eventually in a small fraction of intermediate state cells triggers a gene activation cascade that drives these cells to the pluripotent state. Since epigenetic reprogramming of somatic cells to iPSCs involves an intermediate state, one may suggest that a partial rejuvenation could be achieved at this stage without complete reprogramming to iPSCs. Indeed, short-term expression of the Yamanaka factors (Oct4, Sox2, Klf4, and c-Myc) in fibroblasts derived from a premature aging mouse model has been found to ameliorate multiple age-associated hallmarks of aging, including the accumulation of DNA damage, cellular senescence markers (p16[INK4a], p21[CIP1]), epigenetic dysregulation (changed levels of H3K9me3 and H4K20me3), and nuclear envelope defects, without loss of their cellular identity [102]. Although expression of the Yamanaka factors *in vivo* has been shown to lead to cancer development, their short-term cyclic induction ameliorated hallmarks of aging and extended the life span in a mouse model of premature aging. In addition, it improved the regenerative capacity of the pancreas and muscle following injury in physiologically aged mice. Thus, reprogramming of aging without affecting cellular identity is principally possible. An opposite possibility of reprogramming cellular identity without rejuvenation has been demonstrated by induced direct conversion of human fibroblasts to neurons [103]. Unlike neurons obtained via intermediate iPSCs, these directly induced neurons retained the epigenetic signature of the donor age. Interestingly, of more than 200 genes differentially expressed between neurons obtained from young and old donors, only seven overlapped between fibroblasts and neurons, whereas 49 overlapped between induced neurons and human prefrontal cortex samples. Thus, both cell types showed cell-type-specific age-dependent expression profiles. Only three aging genes, *LAMA3*, *PCDH10*, and *RANBP17*, were shared between aging fibroblasts, induced neurons, and the brain. Since the fibroblast-specific epigenetic signature of donor age was somehow transformed into a neuron-specific one during direct cell conversion, a logical suggestion was that putative master regulators of aging are among these shared genes. The nuclear pore-associated transport receptor RanBP17, which has been found to be significantly downregulated with age, deserves special attention. The decrease in RanBP17 with age could indicate a progressive functional impairment of the nuclear pore function. Indeed, induced neurons from the oldest donors showed significant nucleo-cytoplasmic compartmentalization defects compared with both middle-aged (29 year–50 year) and young donor-derived neurons; the extent of these defects correlated with donor age and RanBP17 downregulation. Furthermore, an shRNA-mediated knockdown of *RanBP17* in young fibroblasts (one year) changed the expression of 68% of the 78 identified fibroblast aging genes in the same direction, as was observed in progressive aging, confirming the idea that RanBP17 plays a causative role in cellular aging. As expected, iPSCs derived neurons showed essentially rejuvenated transcriptome signatures, and no detectable age-dependent impairment of nucleocytoplasmic compartmentalization was detectable. The accelerated aging phenotype in the Hutchinson–Gilford progeria syndrome is known to be a consequence of progerin-impaired nuclear envelope structure that might lead to nuclear leakiness and disruption of nucleocytoplasmic compartmentalization. Thus, a decrease of the RanBP17 levels in aging cells of different organs can contribute to their aging-related phenotypes.

In an experimental procedure, known as hetcrochronic parabiosis, a shared circulatory system between young and old mice is established, thus exposing old mice to factors present in young serum. Heterochronic parabiosis was found to restore activity of the Notch signaling pathway, and proliferation and regenerative capacity of the aged skeletal muscle satellite cells [104]. Furthermore, heterochronic parabiosis increased proliferation of the aged hepatocytes and restored their cEBP-α level to values seen in young animals. Thus, age-related decline of the progenitor cell activity could be modulated by systemic factors that change with age. The number of newly born neurons, proliferating cells, and neural progenitors in the dentate gyrus of hippocampus decreased in the young heterochronic parabiont mice and increased in the old ones [105]. A systemic environment appears to affect the biological age of cells. Most likely, its effects are caused by changes in activity of the main signaling pathways (Notch, Wnt, and TGFβ) due to changed concentrations of respective cytokines. In the adult brain, neural stem cells (NSCs) reside in a heterogeneous niche where they are in direct contact with blood vessels and the cerebrospinal fluid. The vasculature influences NSC proliferation and differentiation by providing signaling molecules secreted from endothelial cells and by delivering systemic regulatory factors. In the aging niche, the vasculature deteriorates with a consequent reduction in blood flow, and the neurogenic potential of NSCs declines, leading to reduced neuroplasticity and cognition. Systemic factors can affect these aging-associated events. In a mouse heterochronic parabiosis model, remodeling of the aged cerebral vasculature in response to young systemic factors was observed, producing a noticeably higher blood flow [106]. NSCs proliferation in the subventricular zone and olfactory neurogenesis were activated and an improvement in the olfactory function occurred. GDF11, a circulating TGF-β family member that has been reported to reverse cardiac hypertrophy in aged mice [107], also stimulated vascular remodeling and increased neurogenesis in aging mice. Thus, circulating factors have diverse positive effects in aging mice, including enhancing neurogenesis and improving the vasculature in the cortex and other parts of the brain. As regards the specific rejuvenating role of GDF11, it has been questioned in subsequent studies. First, it was shown that GDF11 levels in the serum and muscle of rats do not decrease, but rather increase with age [108]. Furthermore, GDF11 has been shown not to stimulate muscle regeneration, but rather to inhibit it in a dose-dependent manner. In humans, GDF11 levels have been found to be not statistically different between variously aged (21–93 years) subjects of both genders [109]. Furthermore, in older adults with cardiovascular disease, increased circulating GDF11 has been shown to correlate with a higher prevalence of comorbid conditions, frailty, and a larger number of adverse health outcomes following aortic valve replacement surgery.

A genome-wide microarray analysis of hippocampi showed a distinct gene expression profile between aged isochronic (aged–aged) and aged heterochronic (aged–young) parabiont mice [110]. Furthermore, genes related to synaptic plasticity signaling pathways, including *Creb*, were among the top GO enrichment categories associated with heterochronic parabiosis. These changes in gene expression were correlated with increased numbers of cells expressing the immediate-early genes *Egr1* and *c-Fos* and with increased dendritic spine number on granule cell neurons in the dentate gyrus of heterochronic compared with isochronic parabionts. Collectively, the data described show that exposure to young blood counteracts aging at the molecular, structural, functional, and cognitive levels in the aged brain.

Effects of the temporal inhibition of NF-κB activity on cell aging were tested in chronologically aged transgenic mice with *NF-κB* gene conditionally repressed in epidermal skin cells [10]. Comparison of the global expression profiles between young and old skin samples revealed 414 genes that were significantly altered (mostly upregulated) in old skin. About 50% of these age-dependent genes were putative direct targets of NF-κB, confirming the proposed role of NF-κB in aging. Upon 2-wk *NF-κB* blockade in old skin, the expression levels of 225 of the 414 age-related genes returned to values indistinguishable from those in the young skin samples. Globally, the expression profile of the NF-κB-quenched aged skin was more similar to that of the young skin than to that of the control aged skin. Thus, NF-κB activity is required to maintain a substantial portion of the global gene expression program induced with age in murine skin. In addition to alterations in gene expression, aged skin is characterized by epidermal atrophy, decreased proliferative capacity, and increased frequency of cellular senescence. Aged murine tissues also exhibit increased expression of SA-β-gal and p16^{INK4A}, two markers of cellular senescence. The *NF-κB* blockade increased epidermal thickness in old skin to a degree intermediate between young and old skin, increased cell proliferation, and significantly decreased expression of SA-β-gal and p16^{INK4A}. Skin constitution and general condition improved. Importantly, reversal of skin cell senescence and increased proliferative capacity induced by *NF-κB* blockade occurred with preservation of normal tissue homeostasis and differentiation. Cell proliferation occurred predominantly in the basal layer of the epidermis, the normal proliferative compartment, and the spatial organization of the mature epidermal stratification was intact. Thus, NF-κB activity is continually required to maintain cellular senescence associated with chronological aging in murine skin. These data suggest that many features of mammalian aging may not be due to the passive accumulation of stochastic cellular damages and errors but rather are actively enforced by special gene expression programs and thus can be substantially reversed by selective gene expression interventions. The NF-κB action in skin aging appears to be cell-autonomous since reversion of the aging features was possible in limited patches of the epidermis in otherwise old animals. As a TF that is responsive to oxidative stress, DNA damage, growth signals, and immune activation, on the one hand, and acting on a great number of various target genes, on the other hand, NF-κB seems to be in an ideal position to transduce diverse extracellular signals to adaptive changes in gene expression and tissue homeostasis. The contribution of specific NF-κB target genes to aging remains unknown. The biological effects of NF-κB may be mediated by the combined effects of large numbers of the NF-κB responsive genes, individual target genes having at best a very modest influence on aging.

HSCs activity decreases with age, manifesting in reduced self-renewal, hematopoiesis, and lymphopoiesis [111]. In mammals, the target of the rapamycin (mTOR) pathway integrates multiple signals from nutrients, growth factors, and oxygen to regulate cell growth, proliferation, and survival [112,113]. Conditional deletion of the *Tsc1* gene in HSCs of young adult mice drives them from quiescence into rapid cycling, increased mitochondrial biogenesis, and elevated levels of reactive oxygen species [114]. Importantly, this deletion dramatically reduces both hematopoiesis and self-renewal of HSCs and leads to constitutive activation of mTOR. In murine HSCs, the mTOR activity increases with age [115]. Treatment of the old mice with an mTOR inhibitor, rapamycin, significantly increases their life span. Moreover, the treatment causes a significant increase in the proliferative activity of HSCs and a decrease in the expression of *p16^{Ink4a}* and *p19^{Arf}*, known markers of cell aging. Old mice have impaired B cell generation due to decreased numbers of pre-B cells. Rapamycin treatment enhances the generation of B cells due to a 4-fold increase in the number of pre-B cells.

Thus, rejuvenation of differentiated or committed cells can be achieved without disturbing their differentiation programs. These results suggest the possibility of targeted therapies to reverse individual features of aging and to alleviate age-related pathologies in the elderly.

CONCLUSION

A number of studies described above has shown that aging is accompanied by and probably caused to a significant extent by epigenetic changes. Epigenetic perturbations are seen in progeroid syndromes and provoke progeroid phenotypes in model organisms. Small noncoding RNAs, in particular miRNAs, have been implicated in various aspects of animal development. The targeting of single genes by multiple miRNAs and of multiple genes by single miRNAs makes this epigenetic system very versatile and efficient. Ironically, these very features could cause serious derangement of gene expression programs when an improper activity of miRNA systems occurs, such as changed miRNA levels observed during aging.

Whether epigenetic changes accompanying aging are the drivers of the aging process or are just molecular consequences of aging is a crucial question. Studies that have shown the elimination of histone methylation–demethylation enzymes to affect the life span in *C. elegans* and *D. melanogaster* demonstrate a causal role of epigenetic systems as specific drivers of aging. It is worth noting that all other hallmarks of aging (genomic instability, telomere attrition, loss of proteostasis, deregulated nutrient sensing, mitochondrial dysfunction, cellular senescence, stem cell exhaustion, and altered intercellular communication) are potentially regulated by epigenetic mechanisms. Thus, epigenetic remodeling to a more youthful state could correspondingly ameliorate these other hallmarks of aging. Hence, targeted interventions in epigenetic mechanisms of aging appear to be a most promising and effective strategy for developing antiaging therapies. The rejuvenation of the aged HSCs to a more youthful state by rapamycin could serve as a convincing illustration.

The possibility of the aging clock being reset suggests that epigenetic signals are not mere correlates of the aging process, but rather a substantial part of the aging mechanism. The links between epigenome and aging can be mutual; an epigenome is changed by age factors, but it also affects aging. An epigenome can be viewed as a general sensor of cellular dysfunction, responsive to any changes in the genome and internal milieu, including those related to aging. On the other hand, the epigenome determines changes in gene expression patterns that underlie the aging phenotype. Similar to other cell systems, the epigenome is prone to gradual degradation due to genome damage, stressful agents, and other aging factors. However, unlike mutations and other kinds of genome damage, age-related epigenetic changes could be fully or partially reversed to a "young" state.

Aging is generally believed to be an inevitable and essentially irreversible process in all living organisms. However, multiple studies described above have shown that it could be completely reversed at the cellular level and reversed to a considerable extent at the organismal level. First, cellular reprogramming has been shown to reset the age of somatic cells to zero. Second, on the whole organism level, manipulations of specific signaling pathways (insulin/IGF-1, mTOR, AMPK, and sirtuins) and external interventions (CR and physical activity) were shown to extend the life span of model animals, probably by slowing down the aging process. Third, the heterochronic parabiosis experiments have shown that, at the organismal level, aging not only could be slowed down but also reversed to a significant extent, at least in some organs. It would be interesting to determine whether these rejuvenating effects of parabiosis are caused by epigenetic changes.

Studies of the aging details using predictive epigenetic models could have many practical implications, from health assessment to forensic analysis. Similar to the analysis of the influence of gender on the aging rates described above, the effects of various environmental and life-style factors, such as smoking, alcohol consumption, dietary preferences, and many others, could be assessed. As the predictive accuracy of the models improves, it seems quite probable that the biological age, measured by epigenetic markers, might become more useful in clinical practice than the passport age. Of course, the epigenetic age could become indispensable in evaluating efficiency of the new rejuvenation procedures.

CONFLICT OF INTEREST

The authors confirm that this article content has no conflict of interest.

ACKNOWLEDGMENT

This work was supported by grant No 14-50-00029 from the Russian Science Foundation (RSF).

REFERENCES

1. Holliday, R. Aging is no longer an unsolved problem in biology. *Ann. N.Y. Acad. Sci.*, 2006, 1067, 1–9. Doi: 10.1196/annals.1354.002.
2. López-Otín, C., Blasco, M.A., Partridge, L., Serrano, M., Kroemer, G. The hallmarks of aging. *Cell*, 2013, 153, 1194–1217. Doi: 10.1016/j.cell.2013.05.039.
3. McCarroll, S.A., Murphy, C.T., Zou, S., Pletcher, S.D., Chin, Ch.-Sh., Jan, Y.N., Kenyon, C., Bargmann, C.I., Li, H. Comparing genomic expression patterns across species identifies shared transcriptional profile in aging. *Nat. Genet.*, 2004, 36, 197–204. Doi: 10.1038/ng1291.
4. Lu, T., Pan, Y., Kao, Sh.-Y., Li, Ch., Kohane, I., Chan, J., Yankner, B.A. Gene regulation and DNA damage in the ageing human brain. *Nature*, 2004, 429, 883–891. Doi: 10.1038/nature02661.
5. Fu, Ch., Hickey, M., Morrison, M., McCarter, R., Han, E.-S. Tissue specific and non-specific changes in gene expression by aging and by early stage CR. *Mech. Ageing Dev.*, 2006, 127, 905–916. Doi: 10.1016/j.mad.2006.09.006.
6. Jonker, M.J., Melis, J.P.M., Kuiper, R.V., van der Hoeven, T.V., Wackers, P.F.K., Robinson, J., van der Horst, G.T.J. et al. Life spanning murine gene expression profiles in relation to chronological and pathological aging in multiple organs. *Aging Cell*, 2013, 12, 901–909. Doi: 10.1111/acel.12118.
7. Bahar, R., Hartmann, C.H., Rodriguez, K.A., Denny, A.D., Busuttil, R.A., Dollé, M.E.T., Calder, R.B. et al. Increased cell-to-cell variation in gene expression in ageing mouse heart. *Nature*, 2006, 441, 1011–1014. Doi: 10.1038/nature04844.
8. Zahn, J.M., Sonu, R., Vogel, H., Crane, E., Mazan-Mamczarz, K., Rabkin, R., Davis, R.W., Becker, K.G., Owen, A.B., Kim, S.K. Transcriptional profiling of aging in human muscle reveals a common aging signature. *PLoS Genet.*, 2006, 2(7), e115. Doi: 10.1371/journal.pgen.0020115.
9. Zahn, J.M., Poosala, S., Owen, A.B., Ingram, D.K., Lustig, A., Carter, A., Weeraratna, A.T. et al. AGEMAP: A gene expression database for aging in mice. *PLoS Genet.*, 2007, 3(11), e201. Doi: 10.1371/journal.pgen.0030201.
10. Adler, A.S., Sinha, S., Kawahara, T.L.A., Zhang, J.Y., Segal, E., Chang, H.Y. Motif module map reveals enforcement of aging by continual NF-κB activity. *Genes Dev.*, 2007, 21, 3244–3257. Doi: 10.1101/gad.1588507.
11. de Magalhães, J.P., Curado, J., Church, G.M. Meta-analysis of age-related gene expression profiles identifies common signatures of aging. *Bioinformatics*, 2009, 25, 875–881. Doi: 10.1093/bioinformatics/btp073.
12. Harries, L.W., Hernandez, D., Henley, W., Wood, A.R., Holly, A.C., Bradley-Smith, R.M., Yaghootkar, H. et al. Human aging is characterized by focused changes in gene expression and deregulation of alternative splicing. *Aging Cell*, 2011, 10, 868–878. Doi: 10.1111/j.1474-9726.2011.00726.x.
13. Wood, Sh.H., Craig, Th., Li, Y., Merry, B., de Magalhães, J.P. Whole transcriptome sequencing of the aging rat brain reveals dynamic RNA changes in the dark matter of the genome. *Age*, 2013, 35, 763–776. Doi: 10.1007/s11357-012-9410-1.
14. Berdyshev, G.D., Korotaev, G.K., Boyarskikh, G.V., Vanyushin, B.F. Nucleotide composition of DNA and RNA from somatic tissues of humpback salmon and its changes during spawning. *Biokhimiya*, 1967, 32, 988–993.
15. Vanyushin, B.F., Nemirovsky, L.E., Klimenko, V.V., Vasiliev, V.K., Belozersky, A.N. The 5-methylcytosine in DNA of rats: Tissue and age specificity and the changes induced by hydrocortisone and other agents. *Gerontologia (Basel)*, 1973, 19, 138–152.
16. Romanov, G.A., Vanyushin, B.F. Methylation of reiterated sequences in mammalian DNAs: Effects of the tissue type, age, malignancy and hormonal induction. *Biochim. Biophys. Acta*, 1981, 653, 204–218. Doi: 10.1016/0005-2787(81)90156-8.
17. Wilson, V.L., Smith, R.A., Mag, S., Cutler, R.G. Genomic 5-methyldeoxycytidine decreases with age. *J. Biol. Chem.*, 1987, 262, 9948–9951.
18. Wilson, V.L., Jones, P.A. DNA methylation decreases in aging but not in immortal cells. *Science*, 1983, 220, 1055–1057. Doi: 10.1126/science.6844925.
19. Wigler, M., Levy, D., Perucho, M. The somatic replication of DNA methylation. *Cell*, 1981, 24, 33–40. Doi: 10.1016/0092-8674(81)90498-0.
20. Stein, R., Gruenbaum, Y., Pollack, Y., Razin, A., Cedar, H. Clonal inheritance of the pattern of DNA methylation in mouse cells. *Proc. Natl. Acad. Sci. U.S.A.*, 1982, 79, 61–65.
21. Boyd-Kirkup, J.D., Green, C.D., Wu, G., Wang, D., Han, J.-D.J. Epigenomics and the regulation of aging. *Epigenomics*, 2013, 5, 205–227. Doi: 10.2217/EPI.13.5.

22. So, K., Tamura, G., Honda, T., Homma, N., Waki, T., Togawa, N., Nishizuka, S., Motoyama, T. Multiple tumor suppressor genes are increasingly methylated with age in non-neoplastic gastric epithelia. *Cancer Sci.*, 2006, 97, 1155–1158. Doi: 10.1111/j.1349-7006.2006.00302.x.

23. Christensen, B.C., Houseman, E.A., Marsit, C.J., Zheng, S., Wrensch, M.R., Wiemels, J.L., Nelson, H.H. et al. Aging and environmental exposures alter tissue-specific DNA methylation dependent upon CpG island context. *PLoS Genet.*, 2009, 5, e1000602. Doi: 10.1371/journal.pgen.1000602.

24. Maegawa, S., Hinka, G., Kim, H.S., Shen, L., Zhang, L., Zhang, J., Zhang, N., Liang, S., Donehower, L.A., Issa, J.-P.J. Widespread and tissue specific age-related DNA methylation changes in mice. *Genome Res.*, 2010, 20, 332–340. Doi: 10.1101/gr.096826.109.

25. Rakyan, V.K., Down, T.A., Maslau, S., Andrew, T., Yang, T.-P., Beyan, H., Whittaker, P. et al. Human aging-associated DNA hypermethylation occurs preferentially at bivalent chromatin domains. *Genome Res.*, 2010, 20, 434–439. Doi: 10.1101/gr.103101.109.

26. Tserel, L., Kolde, R., Limbach, M., Tretyakov, K., Kasela, S., Kisand, K., Saare, M. et al. Age-related profiling of DNA methylation in CD8+ T cells reveals changes in immune response and transcriptional regulator genes. *Sci. Rep.*, 2015, 5, 13107. Doi: 10.1038/srep13107.

27. Hernandez, D.G., Nalls, M.A., Gibbs, J., Arepalli, S., van der Brug, M., Chong, S., Moore, M. et al. Distinct DNA methylation changes highly correlated with chronological age in the human brain. *Hum. Mol. Genet.*, 2011, 20, 1164–1172. Doi: 10.1093/hmg/ddq561.

28. Martino, D., Loke, Y.J., Gordon, L., Ollikainen, M., Cruickshank, M.N., Saffery, R., Craig, J.M. Longitudinal, genome-scale analysis of DNA methylation in twins from birth to 18 months of age reveals rapid epigenetic change in early life and pair-specific effects of discordance. *Genome Biol.*, 2013, 14, R42. Doi: 10.1186/gb-2013-14-5-r42.

29. Day, K., Waite, L.L., Thalacker-Mercer, A., West, A., Bamman, M.M., Brooks, J.D., Myers, R.M., Absher, D. Differential DNA methylation with age displays both common and dynamic features across human tissues that are influenced by CpG landscape. *Genome Biol.*, 2013, 14, R102. Doi: 10.1186/gb-2013-14-9-r102.

30. Zykovich, A., Hubbard, A., Flynn, J.M., Tarnopolsky, M., Fraga, M.F., Kerksick, C., Ogborn, D., MacNeil, L., Mooney, S.D., Melov, S. Genome-wide DNA methylation changes with age in disease-free human skeletal muscle. *Aging Cell*, 2014, 13, 360–366. Doi: 10.1111/acel.12180.

31. Fraga, M.F., Ballestar, E., Paz, M.F., Ropero, S., Setien, F., Ballestar, M.L., Heine-Suner, D. et al. Epigenetic differences arise during the lifetime of monozygotic twins. *Proc. Natl. Acad. Sci. U.S.A.*, 2005, 102, 10604–10609. Doi: 10.1073/pnas.0500398102.

32. Bocklandt, S., Lin, W., Seh, M.E., Sanchez, F.J., Sinsheimer, J.S., Horvath, S., Vilain, E. Epigenetic predictor of age. *PLoS ONE*, 2011, 6, e14821. Doi: 10.1371/journal.pone.0014821.

33. Koch, C.M., Wagner, W. Epigenetic aging signature to determine age in different tissues. *Aging*, 2011, 3, 1018–1027. Doi: 10.18632/aging.100395.

34. Weidner, C.I., Lin, Q., Koch, C.M., Eisele, L., Beier, F., Ziegler, P., Bauerschlag, D.O. et al. Aging of blood can be tracked by DNA methylation changes at just three CpG sites. *Genome Biol.*, 2014, 15, R24. Doi: 10.1186/gb-2014-15-2-r24.

35. Hannum, G., Guinney, J., Zhao, L., Zhang, L., Hughes, G., Sadda, S., Klotzle, B. et al. Genome-wide methylation profiles reveal quantitative views of human aging rates. *Mol. Cell*, 2013, 49, 359–367. Doi: 10.1016/j.molcel.2012.10.016.

36. Horvath, S., Erhart, W., Brosch, M., Ammerpohl, O., von Schönfels, W., Ahrens, M., Heits, N. et al. Obesity accelerates epigenetic aging of human liver. *Proc. Natl. Acad. Sci. U.S.A.*, 2014, 111, 15538–15543. Doi: 10.1073/pnas.1412759111.

37. Horvath, S. DNA methylation age of human tissues and cell types. *Genome Biol.*, 2013, 14, R115. Doi: 10.1186/gb-2013-14-10-r115.

38. Horvath, S., Mah, V., Lu, A.T., Woo, J.S., Choi, O.-W., Jasinska, A.J. et al. The cerebellum ages slowly according to the epigenetic clock. *Aging*, 2015, 7, 294–305. Doi: 10.18632/aging.100742.

39. Horvath, S., Pirazzini, C., Bacalini, M.G., Gentilini, D., Di Blasio, A.M., Delledonne, M., Mari, D. et al. Decreased epigenetic age of PBMCs from Italian semi-supercentenarians and their offspring. *Aging*, 2015, 7, 1159–1170. Doi: 10.18632/aging.100861.

40. Marioni, R.E., Shah, S., McRae, A.F., Chen, B.H., Colicino, E., Harris, S.E., Gibson, J. et al. DNA methylation age of blood predicts all-cause mortality in later life. *Genome Biol.*, 2015, 16, 25. Doi: 10.1186/s13059-015-0584-6.

41. Perna, L., Zhang, Y., Mons, U., Holleczek, B., Saum, K.-U., Brenner, H. Epigenetic age acceleration predicts cancer, cardiovascular, and all-cause mortality in a German case cohort. *Clin. Epigenet.*, 2016, 8, 64. Doi: 10.1186/s13148-016-0228-z.

42. Florath, I., Butterbach, K., Müller, H., Bewerunge-Hudler, M., Brenner, H. Cross-sectional and longitudinal changes in DNA methylation with age: An epigenome-wide analysis revealing over 60 novel age-associated CpG sites. *Hum. Mol. Genet.*, 2014, 23, 1186–1201. Doi: 10.1093/hmg/ddt531.

43. Hjelmborg, J.vB., Iachine, I., Skytthe, A., Vaupel, J.W., McGue, M., Koskenvuo, M., Kaprio, J., Pedersen, N.L., Christensen, K. Genetic influence on human lifespan and longevity. *Hum. Genet.*, 2006, 119, 312–321. Doi: 10.1007/s00439-006-0144-y.

44. Gentilini, D., Mari, D., Castaldi, D., Remondini, D., Ogliari, G., Ostan, R., Bucci, L., Sirchia, S.M., Tabano, S., Cavagnini, F., Monti, D., Franceschi, C., Di Blasio, A.M., Vitale, G. Role of epigenetics in human aging and longevity: Genome-wide DNA methylation profile in centenarians and centenarians' offspring. *Age*, 2013, 35, 1961–1973. Doi: 10.1007/s11357-012-9463-1.

45. Greer, E.L., Maures, T.J., Hauswirth, A.G., Green, E.M., Leeman, D.S., Maro, G.S., Han, S., Banko, M.R., Gozani, O., Brunet, A. Members of the H3K4 trimethylation complex regulate lifespan in a germline-dependent manner in *C. elegans*. *Nature*, 2010, 466, 383–387. Doi: 10.1038/nature09195.

46. Maures, T.J., Greer, E.L., Hauswirth, A.G., Brunet, A. The H3K27 demethylase UTX-1 regulates *C. elegans* lifespan in a germline-independent, insulin-dependent manner. *Aging Cell*, 2011, 10, 980–990. Doi: 10.1111/j.1474-9726.2011.00738.x.

47. Jin, C., Li, J., Green, C.D., Yu, X., Tang, X., Han, D., Xian, B. et al. Histone demethylase UTX-1 regulates *C. elegans* life span by targeting the insulin/IGF-1 signaling pathway. *Cell Metab.*, 2011, 14, 161–172. Doi: 10.1016/j.cmet.2011.07.001.

48. Ni, Zh., Ebata, A., Alipanahiramandi, E., Lee, S.S. Two SET domain containing genes link epigenetic changes and aging in *Caenorhabditis elegans*. *Aging Cell*, 2012, 11, 315–325. Doi: 10.1111/j.1474-9726.2011.00785.x.

49. Wood, J.G., Hillenmeyer, S., Lawrence, Ch., Chang, Ch., Hosier, S., Lightfoot, W., Mukherjee, E. et al. Chromatin remodeling in the aging genome of *Drosophila*. *Aging Cell*, 2010, 9, 971–978. Doi: 10.1111/j.1474-9726.2010.00624.x.

50. Siebold, A.P., Banerjee, R., Tie, F., Kiss, D.L., Moskowitz, J., Harte, P.J. Polycomb repressive complex 2 and trithorax modulate *Drosophila* longevity and stress resistance. *Proc. Natl. Acad. Sci. U.S.A.*, 2010, 107, 169–174. Doi: 10.1073/pnas.0907739107.

51. Han, Y., Han, D., Yan, Zh., Boyd-Kirkup, J.D., Green, Ch.D., Khaitovich, Ph., Han, J.-D.J. Stress-associated H3K4 methylation accumulates during postnatal development and aging of rhesus macaque brain. *Aging Cell*, 2012, 11, 1055–1064. Doi: 10.1111/acel.12007.

52. Black, J.C., Van Rechem, C., Whetstine, J.R. Histone lysine methylation dynamics: Establishment, regulation, and biological impact. *Mol. Cell*, 2012, 48, 491–507. Doi: 10.1016/j.molcel.2012.11.006.

53. Pu, M., Ni, Zh., Wang, M., Wang, X., Wood, J.G., Helfand, S.L., Yu, H., Lee, S.S. Trimethylation of Lys36 on H3 restricts gene expression change during aging and impacts life span. *Genes Develop.*, 2015, 29, 718–731. Doi: 10.1101/gad.254144.

54. Sen, P., Dang, W., Donahue, G., Dai, J., Dorsey, J., Cao, X., Liu, W. et al. H3K36 methylation promotes longevity by enhancing transcriptional fidelity. *Genes Develop.*, 2015, 29, 1362–1376. Doi: 10.1101/gad.263707.

55. Wood, J.G., Jones, B.C., Jiang, N., Chang, C., Hosier, S., Wickremesinghe, P., Garcia, M. et al. Chromatin-modifying genetic interventions suppress age-associated transposable element activation and extend life span in Drosophila. *Proc. Natl. Acad. Sci. USA*, 2016, 113, 11277–11282. Doi: 10.1073/pnas.1604621113.

56. Maze, I., Noh, K.-M., Soshnev, A.A., Allis, C.D. Every amino acid matters: Essential contributions of histone variants to mammalian development and disease. *Nat. Rev. Genet.*, 2014, 15, 259–271. Doi: 10.1038/nrg3673.

57. Piazzesi, A., Papic, D., Bertan, F., Salomoni, P., Nicotera, P., Bano, D. Replication-independent histone variant H3.3 controls animal lifespan through the regulation of pro-longevity transcriptional programs. *Cell Rep.*, 2016, 17, 987–996. Doi: 10.1016/j.celrep.2016.09.074.

58. Lee, S.S., Lee, R.Y.N., Fraser, A.G., Kamath, R.S., Ahringer, J., Ruvkun, G. A systematic RNAi screen identifies a critical role for mitochondria in *C. elegans* longevity. *Nat. Genet.*, 2003, 33, 40–48. Doi: 10.1038/ng1056.

59. Nargund, A.M., Pellegrino, M.W., Fiorese, C.J., Bakcr, B.M., Haynes, C.M. Mitochondrial import efficiency of ATFS-1 regulates mitochondrial UPR activation. *Science*, 2012, 337, 587–590. Doi: 10.1126/science.1223560.

60. Tian, Y., Garcia, G., Bian, Q., Steffen, K.K., Joe, L., Wolff, S., Meyer, B.J., Dillin, A. Mitochondrial stress induces chromatin reorganization to promote longevity and UPR^{mt}. *Cell*, 2016, 165, 1197–1208. Doi: 10.1016/j.cell.2016.04.011.

61. Merkwirth, C., Jovaisaite, V., Durieux, J., Matilainen, O., Jordan, S.D., Quiros, P.M., Steffen, K.K., Williams, E.G., Mouchiroud, L., Tronnes, S.U., Murillo, V., Wolff, S.C., Shaw, R.J., Auwerx, J., Dillin, A. Two conserved histone demethylases regulate mitochondrial stress-induced longevity. *Cell*, 2016, 165, 1209–1223. Doi: 10.1016/j.cell.2016.04.012.

62. Haigis, M.C., Guarente, L.P. Mammalian sirtuins—Emerging roles in physiology, aging, and calorie restriction. *Genes Dev.*, 2006, 20, 2913–2921. Doi: 10.1101/gad.1467506.

63. Dang, W., Steffen, K.K., Perry, R., Dorsey, J.A., Johnson, F.B., Shilatifard, A., Kaeberlein, M., Kennedy, B.K., Berger, S.L. Histone H4 lysine 16 acetylation regulates cellular lifespan. *Nature*, 2009, 459, 802–808. Doi: 10.1038/nature08085.

64. Vaquero, A., Reinberg, D. Calorie restriction and the exercise of chromatin. *Genes Dev.*, 2009, 23, 1849–1869. Doi: 10.1101/gad.1807009.

65. Vaquero, A., Scher, M., Lee, D., Erdjument-Bromage, H., Tempst, P., Reinberg, D. Human SirT1 interacts with histone H1 and promotes formation of facultative heterochromatin. *Mol. Cell*, 2004, 16, 93–105. Doi: 10.1016/j.molcel.2004.08.031.

66. Vaquero, A., Scher, M., Erdjument-Bromage, H., Tempst, P., Serrano, L., Reinberg, D. SIRT1 regulates the histone methyl-transferase SUV39H1 during heterochromatin formation. *Nature*, 2007, 450, 440–444. Doi: 10.1038/nature06268.

67. O'Hagan, H.M., Mohammad, H.P., Baylin, S.B. Double strand breaks can initiate gene silencing and SIRT1-dependent onset of DNA methylation in an exogenous promoter CpG island. *PLoS Genet.*, 2008, 4(8), e1000155. Doi: 10.1371/journal.pgen.1000155.

68. Oberdoerffer, Ph., Michan, Sh., McVay, M., Mostoslavsky, R., Vann, J., Park, S.-K., Hartlerode, A., Stegmuller, J., Hafner, A., Loerch, P., Wright, S.M., Mills, K.D., Bonni, A., Yankner, B.A., Scully, R., Prolla, T.A., Alt, F.W., Sinclair, D.A. SIRT1 redistribution on chromatin promotes genomic stability but alters gene expression during aging. *Cell*, 2008, 135, 907–918. Doi: 10.1016/j.cell.2008.10.025.

69. Herranz, D., Muñoz-Martin, M., Cañamero, M., Mulero, F., Martinez-Pastor, B., Fernandez-Capetillo, O., Serrano, M. Sirt1 improves healthy ageing and protects from metabolic syndrome-associated cancer. *Nat. Commun.*, 2010, 1, 3. Doi: 10.1038/ncomms1001.

70. Mostoslavsky, R., Chua, K.F., Lombard, D.B., Pang, W.W., Fischer, M.R., Gellon, L., Liu, P. et al. Genomic instability and aging-like phenotype in the absence of mammalian SIRT6. *Cell*, 2006, 124, 315–329. Doi: 10.1016/j.cell.2005.11.044.

71. Michishita, E., McCord, R.A., Berber, E., Kioi, M., Padilla-Nash, H., Damian, M., Cheung, P. et al. SIRT6 is a histone H3 lysine 9 deacetylase that modulates telomeric chromatin. *Nature*, 2008, 452, 492–496. Doi: 10.1038/nature06736.

72. Kawahara, T.L., Michishita, E., Adler, A.S., Damian, M., Berber, E., Lin, M., McCord, R.A. et al. SIRT6 links histone H3 lysine 9 deacetylation to NF-κB-dependent gene expression and organismal life span. *Cell*, 2009, 136, 62–74. Doi: 10.1016/j.cell.2008.10.052.

73. Kanfi, Y., Naiman, Sh., Amir, G., Peshti, V., Zinman, G., Nahum, L., Bar-Joseph, Z., Cohen, H.Y. The sirtuin SIRT6 regulates lifespan in male mice. *Nature*, 2012, 483, 218–221. Doi: 10.1038/nature10815.

74. Zhong, L., D'Urso, A., Toiber, D., Sebastian, C., Henry, R.E., Vadysirisack, D.D., Guimaraes, A. et al. The histone deacetylase Sirt6 regulates glucose homeostasis via Hif1a. *Cell*, 2010, 140, 280–293. Doi: 10.1016/j.cell.2009.12.041.

75. Ibáñez-Ventoso, C., Yang, M., Guo, S., Robins, H., Padgett, R.W., Driscoll, M. Modulated microRNA expression during adult lifespan in *Caenorhabditis elegans*. *Aging Cell*, 2006, 5, 235–246. Doi: 10.1111/j.1474-9726.2006.00210.x.

76. Boehm, M., Slack, F. A developmental timing microRNA and its target regulate life span in *C. elegans*. *Science*, 2005, 310, 1954–1957. Doi: 10.1126/science.1115596.

77. de Lencastre, A., Pincus, Z., Zhou, K., Kato, M., Lee, S.S., Slack, F.J. MicroRNAs both promote and antagonize longevity in *C. elegans*. *Curr. Biol.*, 2010, 20, 2159–2168. Doi: 10.1016/j.cub.2010.11.015.

78. Pincus, Z., Smith-Vikos, T., Slack, F.J. MicroRNA predictors of longevity in *Caenorhabditis elegans*. *PLoS Genet.*, 2011, 7(9), e1002306. Doi: 10.1371/journal.pgen.1002306.

79. Kato, M., Chen, X., Inukai, S., Zhao, H., Slack, F.J. Age-associated changes in expression of small, non-coding RNAs, including microRNAs, in *C. elegans*. *RNA*, 2011, 17, 1804–1820. Doi: 10.1261/rna.2714411.

80. Lehrbach, N.J., Castro, C., Murfitt, K.J., Abreu-Goodger, C., Griffin, J.L., Miska, E.A. Post-developmental microRNA expression is required for normal physiology, and regulates aging in parallel to insulin/IGF-1 signaling in *C. elegans*. *RNA*, 2012, 18, 2220–2235. Doi: 10.1261/rna.035402.112.

81. Smith-Vikos, T., de Lencastre, A., Inukai, S., Shlomchik, M., Holtrup, B., Slack, F.J. MicroRNAs mediate dietary-restriction-induced longevity through PHA-4/FOXA and SKN-1/Nrf transcription factors. *Curr. Biol.*, 2014, 24, 2238–2246. Doi: 10.1016/j.cub.2014.08.013.

82. Liu, N., Landreh, M., Cao, K., Abe, M., Hendriks, G.J., Kennerdell, J.R., Zhu, Y., Wang, L.S., Bonini, N.M. The microRNA miR-34 modulates aging and neurodegeneration in *Drosophila*. *Nature*, 2012, 482, 519–523. Doi: 10.1038/nature10810.

83. Badi, I., Burba, I., Ruggeri, C., Zeni, F., Bertolotti, M., Scopece, A., Pompilio, G., Raucci, A. MicroRNA-34a induces vascular smooth muscle cells senescence by SIRT1 downregulation and promotes the expression of age-associated pro-inflammatory secretory factors. *J. Gerontol. A. Biol. Sci. Med. Sci.*, 2015, 70, 1304–1311. Doi: 10.1093/gerona/glu180.

84. Maes, O.C., An, J., Sarojini, H., Wang, E. Murine microRNAs implicated in liver functions and aging process. *Mech. Ageing Dev.*, 2008, 129, 534–541. Doi: 10.1016/j.mad.2008.05.004.

85. Liang, R., Bates, D.J., Wang, E. Epigenetic control of microRNA expression and aging. *Curr. Genomics*, 2009, 10, 184–193. Doi: 10.2174/138920209788185225.

86. Tamminga, J., Kathiria, P., Koturbash, I., Kovalchuk, O. DNA damage-induced upregulation of miR-709 in the germline downregulates BORIS to counteract aberrant DNA hypomethylation. *Cell Cycle*, 2008, 7, 3731–3736. Doi: 10.4161/cc.7.23.7186.

87. Crosby, M.E., Kulshreshtha, R., Ivan, M., Glazer, P.M. MicroRNA regulation of DNA repair gene expression in hypoxic stress. *Cancer Res.*, 2009, 69, 1221–1229. Doi: 10.1158/0008-5472.CAN-08-2516.

88. Noren Hooten, N., Abdelmohsen, K., Gorospe, M., Ejiogu, N., Zonderman, A.B. Evans, M.K. MicroRNA expression patterns reveal differential expression of target genes with age. *PLoS ONE*, 2010, 5(5), e10724. Doi: 10.1371/journal.pone.0010724.

89. Li, N., Bates, D.J., An, J., Terry, D.A., Wang, E. Up-regulation of key microRNAs, and inverse downregulation of their predicted oxidative phosphorylation target genes, during aging in mouse brain. *Neurobiol. Aging*, 2011, 32, 944–955. Doi: 10.1016/j.neurobiolaging.2009.04.020.

90. ElSharawy, A., Keller, A., Flachsbart, F., Wendschlag, A., Jacobs, G., Kefer, N., Brefort, T., Leidinger, P., Backes, C., Meese, E., Schreiber, S., Rosenstiel, P., Franke, A., Nebel, A. Genome-wide miRNA signatures of human longevity. *Aging Cell*, 2012, 11, 607–616. Doi: 10.1111/j.1474-9726.2012.00824.x.

91. Crocco, P., Montesanto, A., Passarino, G., Rose, G. Polymorphisms falling within putative miRNA target sites in the 3′UTR region of *SIRT2* and *DRD2* genes are correlated with human longevity. *J. Gerontol. A Biol. Sci. Med. Sci.*, 2016, 71, 586–592. Doi: 10.1093/gerona/glv058.

92. Stout, G.J., Stigter, E.C., Essers, P.B., Mulder, K.W., Kolkman, A., Snijders, D.S., van den Broek, N.J. et al. Insulin/IGF-1-mediated longevity is marked by reduced protein metabolism. *Mol. Syst. Biol.*, 2013, 9, 679. Doi: 10.1038/msb.2013.35.

93. Essers, P.B., Nonnekens, J., Goos, Y.J., Betist, M.C., Viester, M.D., Mossink, B., Lansu, N. et al. A long noncoding RNA on the ribosome is required for lifespan extension. *Cell Rep.*, 2015, 10, 339–345. Doi: 10.1016/j.celrep.2014.12.029.

94. Surani, M.A., Hayashi, K., Hajkova, P. Genetic and epigenetic regulators of pluripotency. *Cell*, 2007, 128, 747–762. Doi: 10.1016/j.cell.2007.02.010.

95. Albert, M., Peters, A.H.F.M. Genetic and epigenetic control of early mouse development. *Curr. Opin. Genet. Dev.*, 2009, 19, 113–121. Doi: 10.1016/j.gde.2009.03.004.

96. Zhou, L., Dean, J. Reprogramming the genome to totipotency in mouse embryos. *Trends Cell Biol.*, 2015, 25, 82–91. Doi: 10.1016/j.tcb.2014.09.006.

97. Yamanaka, S., Blau, H.M. Nuclear reprogramming to a pluripotent state by three approaches. *Nature*, 2010, 465, 704–712. Doi: 10.1038/nature09229.

98. Marión, R.M., Blasco, M.A. Telomere rejuvenation during nuclear reprogramming. *Curr. Opin. Genet. Dev.*, 2010, 20, 190–196. Doi: 10.1016/j.gde.2010.01.005.

99. Lapasset, L., Milhavet, O., Prieur, A., Besnard, E., Babled, A., Aït-Hamou, N., Leschik, J. et al. Rejuvenating senescent and centenarian human cells by reprogramming through the pluripotent state. *Genes Dev.*, 2011, 25, 2248–2253. Doi: 10.1101/gad.173922.111.

100. Buganim, Y., Faddah, D.A., Cheng, A.W., Itskovich, E., Markoulaki, S., Ganz, K., Klemm, S.L., van Oudenaarden, A., Jaenisch, R. Single-cell expression analyses during cellular reprogramming reveal an early stochastic and a late hierarchic phase. *Cell*, 2012, 150, 1209–1222. Doi: 10.1016/j.cell.2012.08.023.

101. Polo, J.M., Anderssen, E., Walsh, R.M., Schwarz, B.A., Nefzger, C.M., Lim, S.M., Borkent, M. et al. A molecular roadmap of reprogramming somatic cells into iPS cells. *Cell*, 2012, 151, 1617–1632. Doi: 10.1016/j.cell.2012.11.039.

102. Ocampo, A., Reddy, P., Martinez-Redondo, P., Platero-Luengo, A., Hatanaka, F., Hishida, T., Li, M. et al. In vivo amelioration of age-associated hallmarks by partial reprogramming. *Cell*, 2016, 167, 1719–1733. Doi: 10.1016/j.cell.2016.11.052.

103. Mertens, J., Paquola, A.C.M., Ku, M., Hatch, E., Bohnke, L., Ladjevardi, S., McGrath, S. et al. Directly reprogrammed human neurons retain aging-associated transcriptomic signatures and reveal age-related nucleocytoplasmic defects. *Cell Stem Cell*, 2015, 17, 705–718. Doi: 10.1016/j.stem.2015.09.001.

104. Conboy, I.M., Conboy, M.J., Wagers, A.J., Girma, E.R., Weissman, I.L., Rando, T.A. Rejuvenation of aged progenitor cells by exposure to a young systemic environment. *Nature*, 2005, 433, 760–764. Doi: 10.1038/nature03260.

105. Villeda, S.A., Luo, J., Mosher, K.I., Zou, B., Britschgi, M., Bieri, G., Stan, T.M. et al. The ageing systemic milieu negatively regulates neurogenesis and cognitive function. *Nature*, 2011, 477, 90–94. Doi: 10.1038/nature10357.

106. Katsimpardi, L., Litterman, N.K., Schein, P.A., Miller, Ch.M., Loffredo, F.S., Wojtkiewicz, G.R., Chen, J.W., Lee, R.T., Wagers, A.J., Rubin, L.L. Vascular and neurogenic rejuvenation of the aging mouse brain by young systemic factors. *Science*, 2014, 344, 630–634. Doi: 10.1126/science.1251141.

107. Loffredo, F.S., Steinhauser, M.L., Jay, S.M., Gannon, J., Pancoast, J.R., Yalamanchi, P., Sinha, M. et al. Growth differentiation factor 11 is a circulating factor that reverses age-related cardiac hypertrophy. *Cell*, 2013, 153, 828–839. Doi: 10.1016/j.cell.2013.04.015.

108. Egerman, M.A., Cadena, S.M., Gilbert, J.A., Meyer, A., Nelson, H.N., Swalley, S.E., Mallozzi, C. et al. GDF11 increases with age and inhibits skeletal muscle regeneration. *Cell Metab.*, 2015, 22, 164–174. Doi: 10.1016/j.cmet.2015.05.010.

109. Schafer, M.J., Atkinson, E.J., Vanderboom, P.M., Kotajarvi, B., White, T.A., Moore, M.M., Bruce, C.J. et al. Quantification of GDF11 and myostatin in human aging and cardiovascular disease. *Cell Metab.*, 2016, 23, 1207–1215. Doi: 10.1016/j.cmet.2016.05.023.

110. Villeda, S.A., Plambeck, K.E., Middeldorp, J., Castellano, J.M., Mosherm K.I., Luo, J., Smith, L.K. et al. Young blood reverses age-related impairments in cognitive function and synaptic plasticity in mice. *Nat. Med.*, 2014, 20, 659–663. Doi: 10.1038/nm.3569.

111. Rossi, D.J., Bryder, D., Zahn, J.M., Ahlenius, H., Sonu, R., Wagers, A.J., Weissman, I.L. Cell intrinsic alterations underlie hematopoietic stem cell aging. *Proc. Natl. Acad. Sci. U.S.A.*, 2005, 102, 9194–9199. Doi: 10.1073/pnas.0503280102.

112. Wullschleger, S., Loewith, R., Hall, M.N. TOR signaling in growth and metabolism. *Cell*, 2006, 124, 471–484. Doi: 10.1016/j.cell.2006.01.016.

113. Gough, N.R. Focus issue: TOR signaling, a tale of two complexes. *Sci. Signal.*, 2012, 5(2170), eg4. Doi: 10.1126/scisignal.2003044.

114. Chen, Ch., Liu, Y., Liu, R., Ikenoue, T., Guan, K.-L., Liu, Y., Zheng, P. TSC—mTOR maintains quiescence and function of hematopoietic stem cells by repressing mitochondrial biogenesis and reactive oxygen species. *J. Exp. Med.*, 2008, 205, 2397–2408. Doi: 10.1084/jem.20081297.

115. Chen, C., Liu, Y., Liu, Y., Zheng, P. mTOR regulation and therapeutic rejuvenation of aging hematopoietic stem cells. *Sci. Signal.*, 2009, 2, ra75. Doi: 10.1126/scisignal.2000559.

Section III

Diseases Associated with Aging and Treatment

10 The Premature Aging Characteristics of RecQ Helicases

Christ Ordookhanian, Taylor N. Dennis, and J. Jefferson P. Perry

CONTENTS

INTRODUCTION

The faithful replication of DNA and its maintenance and correct repair is critical to perturbing genetic changes that would otherwise drive increased aging, neurodegenerative disease, and cancer [1]. The RecQ helicase family of proteins function as key cellular mediators of genomic integrity, through their various roles in DNA metabolism, which includes functions in replication, recombination, and repair [2,3]. The RecQ helicase family is named after its founding member from *E. coli* [4], and RecQ proteins are distributed across the domains of life [5]. Where characterized, the RecQ helicases function through ATP hydrolysis to unwind and translocate along double-stranded (ds)DNA substrate in a 3′–5′ direction. This is in addition to having single-strand annealing properties. A number of alternate, structure-specific substrates have been defined for RecQ helicases, which contain similarities to intermediates of recombination and repair, providing further support for their functions in these DNA metabolism processes [6].

The human genome contains five RecQ helicases, RecQL1, BLM, WRN, *RECQL4*, and RecQL5, and their importance to human health is highlighted by mutations in WRN, BLM, and *RECQL4* resulting in rare, distinct autosomal recessive disease syndromes [7]. So far, diseases caused by mutations in RecQL1 or RecQ5 have yet to be described [8]. The inherited RecQ disease syndrome share commonalities of increased cancer risk, dwarfism/a short stature, and have aging-related phenotypes to various degrees [9]. Hereditary mutations in BLM result in Bloom syndrome (BS) (OMIM #210900) that is unusual for a hereditary defect in having a broad-spectrum and marked cancer predisposition, with cancers that occur early and often in childhood [10]. Mutations in WRN give rise to the segmental progeroid disorder Werner syndrome (WS) (OMIM #277700), which has an early onset of aging that most closely resembles the normal aging process among the progerias [11–13]. Mutations in RecQ4 can give rise to three distinct syndromes with overlapping phenotypes that include skin abnormalities and skeletal defects [14]. The observed differences between the pathologies of these syndromes suggest that WRN, BLM, and *RECQL4* have discrete functions within the cell.

WS is considered a prototypical progeroid syndrome [15], and is the focus of this chapter together with RecQ4 that also exhibits certain aging phenotypes. In terms of aging-related symptoms, WS and RecQ4 disease syndromes both have cataract formation and osteoporosis in common, but only WS has an observed predisposition to cardiovascular disease and diabetes mellitus. Also, the cancers observed between the WS and the RecQ4 spectra of disorders differ, and these also differ from BS. All cancer types occur in BS; sarcomas are common to WS patients; and osteosarcoma [16], squamous cell and basal cell carcinomas of the skin, and lymphoma are the most prevalent in individuals with RecQL4 mutations. BS-related studies could suggest some aging-related characteristics, such as hypogonadism and short stature, but the cancer clinical phenotype predominates. The broad spectrum of cancers observed in BS individuals includes leukemia, lymphomas, and carcinomas, there is the presence of synchronous or metachronous cancers, the average age of cancer diagnoses of a BLM cohort is approximately 26 years old, and the average life span for an individual with BS is 27 years. The driving force behind these cancers is elevated genome instability in BS cells, driven through sister chromatid exchanges (SCEs) and excessive crossovers between homologous chromosomes.

RECQ HELICASE DOMAIN ARCHITECTURE

The central component of RecQ helicases is an architecture that belongs to the larger Superfamily 2 (SF2) helicases, with some features unique to RecQ proteins, consisting of helicase domains 1 and 2 (HD1, HD2) and a RecQ C-terminal (RQC) region (Figure 10.1). Within HD1 & 2 reside the classical seven-helicase sequence motifs (I, Ia, II, III, IV, V, and VI) that are known to couple ATP binding and hydrolysis with DNA unwinding. Mutation of these motifs can disrupt RecQ function [17] where a mutation within these motifs of WRN protein resulted in a Werner phenotype in mice tail-derived fibroblasts [18]. The Walker A and B Motifs (Motifs I and II) directly bind to ATP (or ATP analogs in the RecQ crystal structures) [19], as observed in the SF1 & 2 helicases, but there is also an extra region conserved in RecQs termed "Motif 0" that is a pocket that preferentially accommodates the adenine base of ATP. This pocket is similar to a pocket in RNA DEAD-box helicases known as the Q motif, suggesting a potential evolutionary link between these two classes of helicases. X-ray crystallography studies on several RecQ helicase structures, including human RecQ1 [20] and human BLM [21,22], have revealed that the RQC region is integral to the helicase core. This region is composed of a Zn^{2+}-binding region and a winged-helix domain. The Zn^{2+} ion is chelated by four conserved cysteine residues, the mutation of which disrupts RecQ helicase function, and causes BS when mutated in *BLM* [23]. The winged-helix domain has high affinity for DNA likely providing substrate specificities, in addition to unique protein partner interactions for the RecQs.

Regions outside of these conserved central helicase region are important for function, as their divergence between the RecQ family members likely provides distinct functionalities through

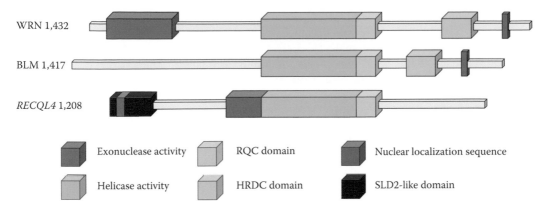

FIGURE 10.1 Domain organization of three disease-related human RecQ helicases. All the RecQ homologous enzymes share a highly conserved helicase domain (teal). The nuclear localization sequence (blue) is present in the C-termini of WRN and BLM; *RECQL4* has two NLSs which are both located in the N-terminus. The RQC domain (lime green) and the RNase domain C-terminal of the helicase (HRDC) (light blue) are located only in BLM and WRN following the helicase domain; an RQC domain has recently been identified in *RECQL4* as well. Unique features include the N-terminal exonuclease domain (green) of WRN, and the N-terminal SLD2-like domain (purple) of *RECQL4*.

differing DNA substrate specificities and protein partner interactions [24]. WRN, along with BLM, contains an RNase D-like C-terminal domain (HRDC) (Figure 10.1) that likely imparts specific functions to WRN, as it plays key functions in localizing WRN to sites of DNA damage and laser-induced dsDNA breaks. WRN is unique among the RecQ helicases in containing an N-terminal exonuclease domain (WRN-exo) (Figure 10.1). WRN-exo degrades DNA substrates in a 3′–5′ direction and X-ray crystallographic analyses revealed that it is structurally homologous to the *E. coli* DNA Pol I proofreading domain. WRN-exo substrates include bubble and forked dsDNA structures that have a 3′-recessed ends [25,26], indicating potential roles in replication, recombination, and repair. The N-terminus of RecQ4 is unique in containing an SLD2 domain (Figure 10.1), which has not been fully structurally characterized, but is known to play key roles in DNA substrate recognition and in DNA replication [27–29].

WRN RECQ HELICASE

WERNER SYNDROME CLINICAL PHENOTYPE

Hereditary mutations in the WRN RecQ helicase gene give rise to WS [30], with some of the highest frequencies of WS being observed in the Japanese population, with a heterozygote frequency estimated at 1 in 166 [31] and also in the Sardinian population with a heterozygote frequency estimated at 1 in 120 [32]. Clinical manifestation of WS includes type II diabetes, increased incidence of cancer, hypogonadism, and aging of mesodermal tissues giving rise to atherosclerosis, osteoporosis, bilateral cataracts, graying of the hair, wrinkled skin, and loss of subcutaneous fat [12,13]. Increased aging in the central nervous system is under debate, but support for some effects include an analysis of WS individuals that revealed brain atrophy in 40% of individuals [33]. Also, sensitive magnetic resonance imaging methods have revealed diffuse structural and metabolic tissue damage in the brains of WS individuals [34]. Schizophrenia has been noted in 10% of WS individuals, while there is no evidence for increased amyloidosis, and there are only a few noted cases of senile dementia that were not linked to Alzheimer's [33]. Also, mental retardation is not a common feature is WS, and neither are birth defects or skeletal abnormalities that are observed in REQ4-related disorders.

The German medical student Otto Werner described this syndrome in 1904, based on a family of four siblings, and the term Werner's syndrome was first used by Oppenheimer and Kugel in 1934, when describing a new case of WS [35]. WS is unusual for a rapid-aging syndrome in that individuals appear to have a relatively normal development until adolescence. The lack of a growth spurt gives WS individuals a relatively small stature, and they begin to display the rapid-aging phenotype in the third decade of life, with surgery being required for bilateral cataracts that occur in nearly all cases at this stage. Other common occurrences are a high-pitched hoarse voice, a bird-like facial appearance, thin limbs, and flat feet. Extensive subcutaneous calcifications are present that result in deep ulcerations around the Achilles and at the elbows to a lesser extent, and this can lead to amputation of the lower extremities. Gonadal atrophy occurs in the thirties for WS individuals, but 30%–40% of WS individuals are likely to have had children by this time. There are differences between WS and the normal aging process, as the cataracts observed in WS individuals are almost always posterior sub-capsular, rather than nuclear cataracts. Sarcomas occur much more frequently than for an equivalent age-matched cohort, and the increased incidence tumor risk is 2–60-fold higher, with thyroid follicular carcinomas being the most common. Hypertension is not typically present in WS, but severe forms of atherosclerosis and arteriolosclerosis do occur, along with medial calcinosis. Cancer and myocardial infarction are the most common cause of death, with a mean life expectancy of 54 years for a WS individual, which is seven years higher than that reported in the mid-1990s, and this can likely be attributed to an overall improvement in medical care [36].

WRN GENE AND MUTATIONS

The WRN gene was identified in 1996 through positional cloning and either homozygous or compound heterozygous loss of function mutations are the cause of WS, numbering greater than 80 distinct mutations in the WRN gene [37]. WRN encodes a 160-kDa protein, WRN, of 1432 amino acids that contains a central ATP-dependent $3'$–$5'$ RecQ DNA helicase domain that includes the RQC region, an N-terminal $3'$–$5'$ exonuclease domain, as well as a C-terminal HRDC domain and a nuclear localization signal (NLS). Most of the disease-causing mutations are either stop codons, splice variants, or small indels that promote nonsense-mediated mutant mRNA decay [38] and/or result in a truncated WRN. The truncated forms of WRN protein have lost the C-terminal NLS and therefore cannot reach the nucleus and are subsequently degraded in the cytoplasm [39–41]. There are a few mutations that instead cause amino acid substitutions. Lys125Asn and Lys135Glu both occur in the exonuclease domain, and both of these result in an unstable form of WRN protein that functionally acts as a null mutation. Further, two amino acid substitution mutations, Gly574Arg and Arg637Trp, have been characterized in the central WRN helicase domain and are observed as compound heterozygotes associated with null mutations in WS individuals [42,43]. The Gly574Arg mutation ablates the WRN helicase activity, while Arg637Trp is predicted to inactivate helicase function [44], and the clinical phenotypes of individuals with either of these mutations appear identical to those carrying null mutations [43,44].

OVERVIEW OF WRN CELLULAR FUNCTIONS

WRN has multiple roles in DNA metabolism that include DNA transcription, replication, recombination, and repair, in addition to telomere maintenance, suggesting WS pathogenesis relates to genomic instability [45,46]. However, precise roles for WRN remain incompletely defined [13]. Analysis of WS cells has shown a shortened replicative life span [47–49], and genomic instability traits that include an increased frequency of chromosomal rearrangements [46], an unusually high rate of spontaneous deletions [50], and a slightly increased sensitivity to X-ray radiation [51–54]. Recent research has also highlighted a role for WRN in heterochromatin maintenance, suggesting that heterochromatin disorganization that is linked to a lack of/decline in WRN activity within the cell could be a key determinant in human aging [55]. Biochemical analyses have revealed that WRN protein displays an ATPase activity and unwinds partial-duplex DNA substrates with $3'$–$5'$ polarity [56]. Alternate complex DNA conformations are preferred over double-stranded DNA (dsDNA), and these substrate specificities, including Holliday junctions, bubble DNA, and G4 quadraplexes, further suggestive roles in DNA replication, recombination, and repair, in addition to telomere maintenance [26,57–60].

WRN EXONUCLEASE

Unique to WRN among the RecQ family is an N-terminal extension containing a 3′–5′ exonuclease domain [61,62]. This domain was first identified through sequence analysis [63–65], and was subsequently demonstrated to degrade dsDNA with 3′ recessed termini with a 3′–5′ directionality [25]. WRN exonuclease also functions on a variety of alternate, structured DNA substrates that include bubbles, stem-loops, forks, and Holliday junctions, in addition to RNA–DNA duplexes, further implicating WRN roles in replication, recombination, and repair [26]. These similarities in substrate specificity between the WRN 3′–5′ helicase and exonuclease activities, indicates that these two enzymatic activities may have coordinated functions on several classes of DNA structures [66,67]. Structural biochemistry studies determined that the core WRN exonuclease domain (WRN-exo) has the same fold and two-metal ion-mediated molecular mechanism of that of the *E. coli* DNA polymerase I proofreading domain (Klenow fragment exonuclease) [68]. The active site accommodates Zn^{2+}, Mg^{2+}, and/or Mn^{2+} divalent cations to support activity [68,69], while the lanthanide Eu^{3+} inhibits exonuclease activity [68], probably due to either a greater charge state or larger radii or both, causing misalignment of the scissile phosphate of the DNA substrate [68]. WRN-exo has evolved to have distinct functions compared to DNA polymerase I proofreading exonuclease, which includes an increased enzymatic activity in the presence of the Ku70/80 subunit of DNA-dependent protein kinase (DNA–PK) as compared to the Klenow fragment exonuclease that was inhibited by Ku70/80 [68]. WRN-exo activity is required to fully compliment a Werner syndrome DNA-end joining phenotype in an *in vivo* plasmid based assay [68]. While this data does not define a specific cellular pathway, the elevated microhomology-mediated repair observed in WRN-exo deficient cells is similar to the phenotypes observed with essential NHEJ proteins [70,71], suggesting possible links to this this pathway. However, WS cells have only mild radiation sensitivity, suggesting a more limited function likely involved in resolving a subset of DNA DSBs repaired by NHEJ.

WRN MULTIMERIZATION

RecQ helicases are indicated to oligomerize, which may help control their activities. The WRN-exo domain as defined by crystallography studies was monomeric in solution. However, a similar WRN exonuclease construct forms homo-hexamers upon interaction with DNA or PCNA [72] and a larger WRN N-terminal construct, containing extensions to the N- and C-termini, was observed to form a stable homomultimer [68,73]. More recently, a multimerization region was characterized that is immediately C-terminal to the WRN-exo domain and it contains a coiled-coil sequence motif, which is a motif common to protein–protein interactions and oligomerization [45]. Inclusion of this coiled-coil motif region provided increased processivity and reduced pausing/terminating double-stranded DNA substrate to WRN-exo *in vitro*. Also, the expression of this coiled-coil motif region of WRN was sufficient to assemble hetero-multimers with full-length WRN in human cells, and this hetero-multimer formation disrupted WRN function by causing sensitivity to camptothecin and 4-nitroquinoline 1-oxide, similar to that observed in WS cells. A structural homolog of WRN-exo from *A. thaliana* (PDB code: 1VK0) was used as a template to build a hexameric ring model of the WRN-exo multimer [68]. The WRN-exo ring model contains a central cavity that is positively charged, with the exonuclease domain facing the center, and a ring structure that is large enough to accommodate dsDNA. Such a ring structure could potentially allow for an increased regulation of DNA-end processing, thus avoiding the risks of releasing broken strands of DNA. Support for a RecQ ring like homo-multimeric structure includes the human homolog BLM that has been previously observed to form hexameric and/or tetrameric rings [74]. Furthermore, a second region, between the WRN RQC and HRDC domains, may also promote full-length WRN multimerization, and this region additionally promotes WRN ssDNA annealing activity.

RQC AND HRDC DOMAIN FUNCTIONS

Structural and biochemical studies have revealed that WRN contains RQC and HRDC domains, which have key functions in DNA substrate and protein partner interactions. The WRN winged-helix domain of the RQC region facilitates targeting of WRN to the nucleolus [40] and it interacts with alternate DNA substructures, including forks, holding junctions, 3′ recessed dsDNA, and binds blunt-ended dsDNA, or ssDNA [26]. The WRN HRDC domain has been determined to preferentially bind with high affinity to forked duplex DNA and Holliday junction DNA, and also 3′ recessed DNA to a somewhat lesser extent [26], indicating potential replication and recombination roles. The C-terminal region is also the site of interaction of several of the potential protein partners of WRN [75], regulating the activity of WRN or of the partner protein. A WRN C-terminal interaction with p53 regulates WRN-exo activity [76], as does interactions with Ku70/80 [77–79] and PARP-1 [80]. The WRN helicase activity is regulated by C-terminal interactions that include TRF2 [81], RAD52 [82], and PARP-1 [80]. An example of alteration of partner protein activity is the WRN:FEN-1 interaction, which occurs in the WRN winged-helix domain and stimulates FEN-1 nucleolytic activity by more than 80-fold [83]. How the protein partner interactions control distal WRN activities is unknown, and this may involve changes in conformation that translate through the polypeptide chain, changes in oligomerization states and/or that distal domains in sequence are closer in three-dimensional space in the *holo*-enzyme. How these key DNA and/or protein interactions potentially allow for WRN mediated pathway progression, and thus how pathway progression breaks down in the absence of functioning WRN and so likely giving rise to the disease phenotype, remains to be defined.

WRN DNA REPAIR FUNCTIONS

The physical and functional interactions of WRN with proteins in the DNA repair pathways, such as FEN-1, Ku70/80, RAD52, and PARP-1, strongly indicate key repair roles for WRN that may impact the clinical phenotype of WS. Links to the base excision repair (BER) pathway include physical and functional interactions with multiple BER components that include: replication protein A (RPA) [84], PCNA [85] polδ [86,87], flap endonuclease 1 (FEN-1) [83], polβ [88], and poly(ADP-ribose) polymerase 1 (PARP-1) [80]. A potential link to the nucleotide excision repair (NER) pathway is indicated by the interaction of WRN and XPG [89], while links to DNA DSB repair pathways are perhaps the most numerous. An interaction of WRN with the NHEJ-essential protein kinase DNA–PK that also has capping functions at mammalian telomeres, provided a first link to NHEJ pathway [52,90,91]. WRN is only one of the four known *in vivo* substrates of DNA–PK, and full-length WRN activity is regulated by both the Ku70/80 subunit and *holo*–DNA–PK [52,91] and WRN exonuclease activity is stimulated *in vitro* the DNA–PK subunit Ku70/80 [77,78,91–93]. WRN has been observed in an endogenous complex with the Ku70/80 subunit and poly(ADP-ribose) polymerase-1 (PARP-1) [94] that binds sites of SSBs and DSBs, and is also implicated in the control of genomic integrity and mammalian life span [95]. XRCC4/Ligase IV that functions in NHEJ also stimulates the exonuclease, but not the helicase activity of WRN [78,92]. Interestingly, NHEJ-mediated repair in WS cells displays extensive deletions, which suggests that another, less regulated exonuclease may substitute for WRN in these cells [96].

Possible roles in the HRR of DNA DSBs are suggested by the observed co-localizing of WRN with Rad51 in camptothecin-treated cells [97] as well as interactions with Rad52 [82] and the MRE11:RAD50:NBS1 (MRN) complex [53]. WRN studies using integrated recombination reporter substrates identified WRN functions in promoting the resolution of recombinant DNA molecules, during the postsynaptic phase of HRR [98]. More recently, WRN has been observed to participate in 5′–3′ DNA-end resection, and the WRN helicase activity likely provides substrates the HRR-related DNA2 endonuclease [99]. In such a role, WRN may overlap with BLM functions in the cell, with potentially different cell lineages preferring one of these RecQ helicases to the other. Support for this idea includes a finding that cells with a heterozygous RAD51 T131P mutation use WRN:DNA2 (DNA replication ATP-dependent helicase/nuclease DNA2), rather than BLM:DNA2, for DNA resection at mitomycin C-mediated crosslinks [100].

The HRR pathway also functions to resolve stalled or broken replication forks [101,102], and WRN may have a nonenzymatic role by preventing unproductive nascent DNA resection occurring at replication forks that formed double strand breaks via camptothecin-induced topoisomerase I-DNA adducts [103]. However, the WRN exonuclease and helicase activities are both observed to be required for preserving forks, to allow for replication restart, and in this capacity WRN likely functions with DNA2 for the limited resection of stalled forks [104–109]. WRN may also have a function in promoting DNA synthesis in the initial moments during fork restart, due to interactions with polymerase eta (Pol η) [110] and potentially other polymerases [87,107,108].

Interestingly, an increased expression of common fragile sites in WRN-deficient cells is observed, suggesting that chronic but low-grade fork demise could be a driving force behind the genome instability phenotype of WS [111,112]. This is further supported by studies on forks that have not been significantly disrupted by genotoxic compounds. Here, increased stalling of forks was observed during the S-phase in WRN-deficient primary fibroblasts [113], while WRN-depleted transformed human fibroblasts have been noted to have a subtly reduced rate of fork progression [108].

WRN Telomeric Maintenance Functions

Telomeres are hotspots for DNA damage, which if not correctly repaired can lead to telomere instability and shortening [47–49] that is associated with aging phenotypes. The ends of telomeres are also bound and protected from aberrant activation of the DNA DSB repair machinery by a group of proteins that form the shelterin complex that prevent the fusion chromosomal ends, which would otherwise lead to chromosomal abnormalities. The shelterin proteins TRF1, TRF2, and POT1 can all interact and alter the enzymatic activities of WRN [114,115], suggesting that dysfunctions in telomere maintenance could contribute to the aging phenotypes associated with WS. G4 quadruplexes are present in telomeres and four-way junctions are required to be resolved during telomeric replication, and both these from part of the preferred, alternate DNA substrates for WRN helicase [58,116]. A dominant negative WRN helicase mutant resulted in significant loss of telomeric DNA [117], while immortalization of WS fibroblasts by human telomerase expression can eliminate both the proliferative defect of these cells, as well as somewhat limit their hypersensitivity to genotoxic agents [118].

WRN Roles in Supporting Heterochromatin

A recent key finding supporting heterochromatin functions was uncovered through the generation of a model of WS that used human embryonic stem cells (ESCs) [55]. Differentiation of WRN-null ESCs to mesenchymal stem cells (MSCs) was observed to cause premature cellular aging, together with a global loss of the histone modification H3K9me3 and changes in heterochromatin architecture. The authors of this study further revealed that WRN was associated with two known heterochromatin proteins SUV39H1 and HP1α together with the nuclear lamina-heterochromatin anchoring protein LAP2β. Knock-in of a catalytically inactive SUV39H1 in wild-type MSCs also caused an accelerated cellular senescence, which resembled the phenotype for the WRN-deficient MSCs, providing further support for an *in vivo* functional interaction. Analysis of MSCs from older individuals revealed a decrease in WRN protein levels together with decreased SUV39H1, HP1α, LAP2β, and H3K9me3 modification, suggesting potential heterochromatin deregulation as a key driver of human aging.

RECQL4 HELICASE

RECQL4 Disease Syndromes

Mutations in *RECQL4* can give rise to three syndromes, Baller–Gerold (BGS) (OMIM #218600), RAPADILINO (OMIM #266280), and Rothmund–Thomson (RTS) (OMIM #268400) [2,3], which are clinically distinct but show overlap in their signature of developmental abnormalities, chromosome instability, and cancer predisposition phenotypes [119–122]. As such, these diseases have been noted as part of the *RECQL4* spectrum of disorders [121,123]. Two thirds of the RTS cases are due to either homozygous or compound heterozygous mutations in *RECQL4*, known as Type 2 RTS, while the remaining third, Type 1 RTS, belong to mutations of an unknown gene/set of genes. The clinical phenotypes of RTS are highly variable, causing mild-to-severe forms of RTS. What is common to all RTS individuals is a sun-sensitive skin rash with prominent poikiloderma, consisting of both hyper and hypo-pigmentation, telangiectases and atrophy. Other clinical phenotypes include small stature, juvenile cataracts, and skeletal defect spathologies which consist of osteopenia, hypoplastic or aplastic bones, missing thumbs, bone fusions, and lack of or sparse hair on the scalp, eyebrows, and eye lashes. There is also an increased cancer risk, particularly osteosarcoma, and basal cell and squamous cell skin carcinomas. Thus, although RTS patients do not display a preponderance of rapid-aging characteristics as associated with WS, they do display certain aspects of an aging phenotype, such as the presence of juvenile cataracts, sparse hair, and cancer predisposition.

The rare congenital disorder RAPADILINO syndrome has so far been observed to be most prevalent in Finland, and its name is an acronym for the clinical phenotypes that characterize it. This phenotypes are radial ray defects (RA) that can include absence of thumbs, patellae hypoplasia/aplasia, as well as a high arched or cleft palate (PA), limb malformation, dislocated joints, and diarrhea in infancy (DI), a small stature (LI, little size), a long slender nose, and a normal level of intelligence (NO). Thus, apart from the lack of poikiloderma, many of these RAPADILINO phenotypes are common to RTS, as is an increased cancer predisposition, observed at 40% among a group of 15 Finnish patients, with two individuals having osteosarcoma and four having lymphoma [121]. The cardinal features of the BGS are craniosynostosis, radial aplasia [124] with absent or hypoplastic thumbs, in addition to poikiloderma. A short stature is also highly prevalent, as are skeletal anomalies of the spine and pelvis, along with abnormalies of the urogenital system, and mental and/or motor retardation has also been noted in some BGS individuals. Phenotypic overlap of BGS has been noted with certain Fanconi anemia cases and in particular with the autosomal dominant Saethre–Chotzen syndrome.

RecQ4 Biochemical Activities

Human *RECQL4* (RecQ4) is a 1208 amino acid protein that has both nuclear and mitochondrial targeting sequences, with RecQ4 being the only member of the RecQ family of helicases that is noted to be in the mitochondria, cytoplasm, and nucleus [125–127]. The intracellular localization of RecQ4 could be specific to cell type, as a difference in distribution has been observed between the nucleus and cytoplasm in varying cell types [125]. In the nucleus, RecQ4 is largely in the nucleoplasm, with smaller amounts being observed to co-localize to the nucleolus, as well as telomeres [125,128,129]. Studies on RecQ4 mRNA levels have revealed both tissue and cell-cycle-specific expression. High-level expression of RecQ4 was observed in the human thymus, placenta, and testis, while more moderate expression levels occurred in the brain, heart, colon, and small intestines [130]. In the cell cycle, RecQ4 mRNA expression is at its maximum in the S-phase, which is similar to BLM RecQ helicase expression [130]. Notably, certain cancers also have high levels of RecQ4 expression, in particular osteosarcomas, cervical, and prostate cancers [131–134].

Biochemical characterizations have revealed a potentially smaller range of substrates for RecQ4, as compared to the other RecQ helicases [135], but studies have suggested multiple cellular roles that include functions in DNA replication, DNA repair events such as homologous recombination, and maintenance of both mitochondrial DNA and telomeres. Dysfunction of RecQ4 causes an increased rate of chromosome aneuploidy and chromatid cohesion defects [136–138]. *RECQL4* is also observed to dynamically associate with foci that have been produced through the use of DNA damaging agents, such as UV, topoisomerase inhibitors, or hydrogen peroxide [128,129,139–141]. This observation, together with the observed co-localization and physical interaction with DNA repair factors, such as Rad51 and XPA, strongly imply roles for RecQ4 in the DNA damage response [128,139,140,142,143].

RecQ4 helicase contains two functional domains conserved in higher eukaryotes, an N-terminal Sld2-like domain that has essential roles in DNA replication initiation, and the RecQ helicase core that contains most of the disease-associated mutations [14,27–29,144]. Detailed biochemical analyses of RecQ4 initially revealed a DNA-stimulated ATPase activity [125], followed by a DNA annealing activity [145]. The $3'$–$5'$ helicase activity in common with other RecQ family members was confirmed later, as it appears that the DNA annealing function of *RECQL4* can predominate and mask helicase activity [29]. Studies on RecQ4 recombinant material from *E. coli* indicate that RecQ4 has a helicase activity that can additionally unwind blunt-ended substrates, unlike other RecQ helicases [29]. This study also defined a helicase activity in the N-terminal Sld2 domain [29] although due to the assay used, it may also instead be a strand-exchange activity that was being observed. A second study using recombinant human protein expressed in insect cells, which could potentially improve protein folding, revealed a $3'$–$5'$ helicase activity that could unwind 17 base pair long oligonucleotide annealed to single-stranded circular DNA [146]. This recombinant protein could not unwind a larger 37 base pair region, suggesting a limited processivity for RecQ4 [146]. Studies on recombinant Drosophila RecQ4 expressed in insect cells also revealed DNA-stimulated ATP hydrolysis, DNA annealing, and a $3'$–$5'$ helicase activity, with substrates including duplex DNA with a single-stranded $3'$ extension and fork substrates. Helicase activity was not observed in these studies on the Sld2 domain of Drosophila RecQ4 [147].

The DNA-binding activities of RecQ4 have been further characterized in recent studies, revealing that the N-terminus of human RecQ4 contains regions of intrinsic disorder, but this disorder is sufficient for strand annealing activity [148]. Multiple DNA-binding sites are observed to be present, and this N-terminal region has the highest affinity of G4 quadruplex DNA, estimated at 60-fold over other substrates that include single-stranded, double-stranded, and Y-structured DNA [148]. G4 quadruplex elements are present near origins of replication, likely highlighting RecQ4 functions in this process. A separate study also indicated the presence of two key sites in the N-terminus for DNA binding, and the helicase core contained a third that has affinity for branched substrates and in particular, Holliday junctions [149].

Bioinformatics analysis initially indicated that the RecQ4 N-terminus contains a novel "Zn-knuckle" domain as well as the Sld2 homology region, and that the helicase core also likely contains a RQC region [150], which had not been previously noted. Structural and biochemical analysis of the N-terminal region revealed that Xenopus RecQ4 contains a Zn-knuckle fold, while the human region is disordered due to mutation of one of the Zn-binding amino acids [151]. Both Xenopus and human Zn-knuckle regions bind to oligonucleotides, and both were shown to have a mild preference to RNA substrate. A region upstream of the Zn-knuckle was also studied, which forms the C-terminal part of the Sld2 domain, and a nucleic acid-binding propensity was observed for this part of RecQ4 [151]. The putative RQC domain has also been analyzed by the same laboratory, using inductively coupled plasma-atomic emission spectrometry that detected the presence of two zinc clusters, within the zinc-binding domain of the RQC, as indicated by bioinformatics studies. Studies on the potential winged-helix domain included mutational analyses and low resolution in solution small angle X-ray scattering studies, indicating the presence of a winged-helix domain that likely adopts alternate conformations in binding DNA [152].

RecQ4 DNA Replication Functions

The indication that RecQ4 functions in DNA replication was first suggested by two independent research groups, both noting the sequence similarity between the N-terminus of the Xenopus

RecQ4 and the replication-associated Sld2 domain from budding yeast [27,28], and which is known as DRC1 in fission yeast. This RecQ4 N-terminal Sld2 domain is the only known metazoan homolog of the yeast Sld2 replication factor. Yeast Sld2 functions alongside Sld3 to interact with Dpb11, where the interaction first requiring that Sld2 and Sld3 are phosphorylated by the S-phase CDK, and this interaction is essential for the initiation of DNA replication in yeast [153]. The Sld2 containing N-terminal region of the Xenopus RecQ4 homolog was observed to be essential for the initiation of replication in Xenopus oocyte extracts, in addition to maintaining the chromatin binding ability of DNA Polymerase α [27]. Depleting Xenopus oocyte extract of Xenopus RecQ4 and adding in the Sld2 containing N-terminus of human RecQ4 also allowed for replication to re-occur, although at much lower levels than the wild type [28]. Interestingly, an inactivating point mutation within the RecQ helicase domain was also unable to restore replication, indicating potential replication-associated roles for the helicase core [28]. In human cells, depletion of RecQ4 was observed to inhibit DNA synthesis and cell proliferation, and RecQ4 was found to associate with origins of replication during G1 and S-phase [154]. Studies in Drosophila also further support these findings, revealing that cells with either null RecQ4 or having hypomorphic expression of RecQ4 are defective in the initiation of pre-mitotic replication [155]. Also, replication origin firing and nascent DNA synthesis are reduced after depletion of *RECQL4* and of RecQL1 [154]. Mass spectroscopy studies on chromatin-bound RecQ4 have revealed interactions with key DNA replication factors, namely CDC45, the GINS complex, MCM2-7 replicative DNA helicase, MCM10, and SLD5 [156]. The N-terminal region of RecQ4 interacts with MCM10 to form a larger complex with the GINS complex and MCM2-7 helicase, and this occurs in a cell-cycle-dependent manner [156], and where *RECQL4*, MCM10, as well as CTF4 are all necessary for the proper loading and assembly of CDC45, MCM10, and the GINS complex onto the origins of replication [157].

RecQ4 Functions in DNA Repair

RecQ4 has been implicated in having functions across many of the DNA repair pathways, with RTS cells or cells with RecQ4 knockdown observed to fail to start DNA synthesis after UV irradiation or hydroxyurea treatment [158,159]. RecQ4 has also been indicated to directly interact with the DNA damage sensor PARP-1, where PARP-1 was observed to PARylate the C-terminus of *RECQL4* [129]. The range of sensitivities to DNA damaging agents varies, with RTS fibroblasts having hypersensitivity to hydroxyurea, camptothecin, and doxorubicin [36], they have mild sensitivity to UV light, ionizing radiation, or cisplatin [21], and have limited sensitivity to NER-related 4-nitroquinolone oxide damage (4-NQO) [36]. RecQ4 functions in repair of DNA DSBs were initially supported due to RecQ4 co-immunoprecipitating with HRR protein RAD51. More recently, functions for RecQ4 have been observed in DNA-end resection, which is an initial and essential step of HRR [160]. RecQ4 depletion was observed to dramatically reduce cellular 5′-end resection and HRR rates. RecQ4 was further shown to physically interact with the MRE11–RAD50–NBS1 (MRN) complex that functions to initiate DNA-end resection. The N-terminus of RecQ4 was determined to bind to the MRN partner CtIP, which functions with the MRN complex to sense DSBs and to initiate DNA-end resection, while the RccQ4 helicase activity was shown to be critical for promoting DNA-end processing and efficient HRR. RecQ4 may also have functions in the restart of replication forks, as RTS cells are sensitive to agents that interfere with replication [142]. Potential NHEJ functions are indicated by RecQ4 co-immunoprecipitating with KU70/80, and that depletion of RecQ4 was also observed to reduce cellular NHEJ activity.

PARP-1 also has key functions in long patch BER, indicating potential roles in this repair mechanism. RecQ4 re-localizes to the nucleolus under oxidative stress [129], and this re-localization results in RecQ4 interacting with key BER pathway proteins, which includes APE1, FEN1, and DNA polymerase β, stimulating the activities of all three partner proteins [140]. RTS cells and cells with transient RecQ4 knockdown are noted to be hypersensitive to oxidative damage [140]. RTS cells have also been observed to contain elevated levels of 8-oxoG and of formamidopyrimidine, which are DNA lesions that are known markers of oxidative damage [161]. Microarray analyses also showed that key BER pathway genes were upregulated in RTS patient fibroblasts, as compared to normal cell controls [161]. Finally, some potential NER functions are supported by an interaction with Xeroderma pigmentosum group A (XPA) protein, with this interaction observed to increase after UV irradiation of human cells [139], suggesting perhaps lesion-type specific functions in DNA repair pathway considering an observed insensitivity to 4-NQO.

Studies in yeast were critical to unravel many key functions of BLM RecQ helicase, and a recent bioinformatics analysis identified potential orthologues of RecQ4 in fungi and plants, called the Hrq1 family [162]. This discovery potentially provides the power of yeast genetics to uncover key functions of RecQ4 family members, which can hopefully help unravel the complexity of RecQ4-associated disorders. Studies on the fission yeast Hrq1 protein first determined that this protein plays fundamental roles in genome stability and DNA repair, similar to other RecQ helicases [135], and has been further corroborated by studies in budding yeast Hrq1 [163–168]. Fission yeast Hrq1 demonstrated a 3′–5′ DNA helicase activity, similar to human RecQ4 [135], and these yeast cells lacking Hrq1 have hypersensitivity to the chemotherapeutic agent cisplatin, have constitutive activation of the DNA damage checkpoint, and they exhibit spontaneous genome instability. Hrq1 was observed to facilitate the processing of a subset of NER by interacting with XPA, in keeping with human RecQ4 functions [139], and functioning in a pathway that runs parallel to post-replication repair.

RecQ4 Functions in Telomere and Mitochondrial Maintenance

Dysfunction in telomere and mitochondrial maintenance can contribute to increased rates of aging, which could explain at least in part the clinical phenotypes observed with RecQ4 mutations. Some phenotypic similarities of REQ4 syndromes to Dyskeratosis congenita, which is characterized by shortened telomeres, also support the concept of aging-related phenotypes arising due to disrupted RecQ4 functions in telomere maintenance. Such a telomeric function for RecQ4 was observed by studies on RTS patients cells and human cells depleted of RecQ4, both of which showed elevated fragile telomeres [169]. RecQ4 protein was also observed to co-localize to telomeres and to co-immunoprecipitate with the shelterin complex protein TRF2 [169]. RecQ4 can unwind the telomeric D-loops, as well as oxidatively damaged D-loops, and provide support for a role in telomere maintenance [170]. Three of the shelterin proteins, TRF1, TRF2, and POT1, can stimulate RecQ4 helicase activity on telomeric substrates [169]. Also, RecQ4 functions cooperatively with WRN on telomeric D-loops *in vitro*, stimulating WRN helicase activity and suggesting a potential cooperative mode of action between the two RecQ helicases [169].

Depletion of RecQ4 is observed to result in increased mitochondrial (mt)DNA damage, to lessen the mitochondrial reserve capacity, and to decreased mtDNA copy number as well as reduced mtDNA replication [126,127,171]. Mitochondrial dysfunction is a known factor in aging, being observed in aged cells and those from individuals with age-related diseases, including Alzheimer's [172]. RecQ4 was demonstrated to localize to the mitochondria, initially identified through proteomics-based analyses [173] and followed by immunocytochemistry studies [126,127,171,174,175]. RecQ4 can modulate mitochondrial DNA polymerase (Pol A) activity [174], increasing both the polymerase and proofreading activities of Pol A. RecQ4 also interacts with the mitochondrial helicase PEO1. This interaction was shown to be increased in a RecQ4 mutant, with a 44 residue deletion present N-terminally adjacent to the helicase core, resulting in an elevation of mtDNA copy number [175] that may have deleterious effects to the mitochondria, potentially resulting in increased rates of aging.

CONCLUSIONS

RecQ helicase function appears important for promoting DNA repair that includes responses to oxidative stress, and telomere maintenance, both of which are known to have important links to the aging process. There are some similarities in the RecQ diseases, and WRN, BLM, and RecQ4 are known to have certain physical and functional interactions within the cell, and so one question remains is that are these clinical similarities potentially related to such interactions and/or overlapping functions between the RecQs. WRN is regarded as being a caretaker of the genome to promote genome stability, and the functions and protein partner interactions that have been defined for WRN are numerous, but further research is needed. For example, is there a particular mechanism of action that is the key component behind the WS rapid-aging phenotype, such as dysfunctional telomere maintenance or because of disruptions to heterochromatin. Or is it that WRN has a role at the center of various key cellular processes, and loss of WRN activity promotes a more general disruption within the cell to drive it toward senescence. The effects of single nucleotide polymorphisms and epigenetics of WRN and RecQ4 function, which have been under study, are worth exploring further, to define any impacts rates of aging and cancer in the general population. Also, studies on the molecular mechanisms of WRN and RecQ4 have revealed many insights into genome stability functions, but questions still remain. WRN has an exonuclease domain within the polypeptide chain, but critical functions of this domain within the cell remain elusive. Perhaps a lack of function of this unique exonuclease domain in mutant WRN helps underlie the unique aging phenotype. Also, other RecQs are known to function with nuclease partners, but why is it that WRN has evolved to include the exonuclease domain within its sequence?

RecQ4 studies are perhaps less thorough than for either WRN or BLM, but very recent research has begun to clarify its functions within the cell. Intriguing questions remain for RECQ4, including which functions during development give rise to the spectrum of disease phenotypes when *RECQL4* is mutated. Also, it appears that RecQ4 functions as a guardian of both the mitochondrial and nuclear genomes, including protecting telomeres. This is of much interest, especially as new theories of aging suggest potential interactions between mitochondrial and telomere dysfunction that could help drive cellular aging. However, RecQ4 only displays limited aging phenotypes compared to WRN, leading to questions of why the aging phenotype of the other RecQs are limited, in spite of functional overlaps with WRN. Finally, is there a mechanism of action for WRN and/or RecQ4 that gives rise to the specific cancers that are observed in mutations of *WRN* or *RECQL4*, considering the broad spectrum of cancers readily observed in *BLM* mutations?

ACKNOWLEDGMENTS

We would like to thank Dr. Anna Travesa Centrich for critical reading of the manuscript. J.J.P.P. acknowledges funding from the University of California Office of the President Lab Fees Grant No. LFR-17-476732.

REFERENCES

1. Heyer WD, Ehmsen KT, Liu J. Regulation of homologous recombination in eukaryotes. *Annu Rev Genet*. 2010;44:113–39. doi: 10.1146/annurev-genet-051710-150955. PubMed PMID: 20690856; PMCID: PMC4114321.
2. Bernstein KA, Gangloff S, Rothstein R. The RecQ DNA helicases in DNA repair. *Annu Rev Genet*. 2010;44:393–417. doi: 10.1146/annurev-genet-102209-163602. PubMed PMID: 21047263; PMCID: PMC4038414.
3. Chu WK, Hickson ID. RecQ helicases: Multifunctional genome caretakers. *Nat Rev Cancer*. 2009;9(9):644–54. doi: 10.1038/nrc2682. PubMed PMID: 19657341.
4. Nakayama H, Nakayama K, Nakayama R, Irino N, Nakayama Y, Hanawalt PC. Isolation and genetic characterization of a thymineless death-resistant mutant of *Escherichia coli* K12: Identification of a new mutation (recQ1) that blocks the RecF recombination pathway. *Mol Gen Genet*. 1984;195(3):474–80. PubMed PMID: 6381965.
5. Hickson ID. RecQ helicases: Caretakers of the genome. *Nat Rev Cancer*. 2003;3(3):169–78. doi: 10.1038/nrc1012. PubMed PMID: 12612652.

6. Croteau DL, Popuri V, Opresko PL, Bohr VA. Human RecQ helicases in DNA repair, recombination, and replication. *Annu Rev Biochem*. 2014;83:519–52. doi: 10.1146/annurev-biochem-060713-035428. PubMed PMID: 24606147; PMCID: PMC4586249.

7. Suhasini AN, Brosh RM, Jr. Disease-causing missense mutations in human DNA helicase disorders. *Mutat Res*. 2013;752(2):138–52. doi: 10.1016/j.mrrev.2012.12.004. PubMed PMID: 23276657; PMCID: PMC3640642.

8. Fu W, Ligabue A, Rogers KJ, Akey JM, Monnat RJ, Jr. Human RECQ helicase pathogenic variants, population variation and "Missing" diseases. *Hum Mutat*. 2017;38(2):193–203. doi: 10.1002/humu.23148. PubMed PMID: 27859906.

9. Suhasini AN, Brosh RM, Jr. DNA helicases associated with genetic instability, cancer, and aging. *Adv Exp Med Biol*. 2013;767:123–44. doi: 10.1007/978-1-4614-5037-5_6. PubMed PMID: 23161009; PMCID: PMC4538701.

10. Cunniff C, Bassetti JA, Ellis NA. Bloom's syndrome: Clinical spectrum, molecular pathogenesis, and cancer predisposition. *Mol Syndromol*. 2017;8(1):4–23. doi: 10.1159/000452082. PubMed PMID: 28232778; PMCID: PMC5260600.

11. Oshima J, Sidorova JM, Monnat RJ, Jr. Werner syndrome: Clinical features, pathogenesis and potential therapeutic interventions. *Ageing Res Rev*. 2017;33:105–14. doi: 10.1016/j.arr.2016.03.002. PubMed PMID: 26993153; PMCID: PMC5025328.

12. Muftuoglu M, Oshima J, von Kobbe C, Cheng WH, Leistritz DF, Bohr VA. The clinical characteristics of Werner syndrome: Molecular and biochemical diagnosis. *Hum Genet*. 2008;124(4):369–77. doi: 10.1007/s00439-008-0562-0. PubMed PMID: 18810497; PMCID: PMC4586253.

13. Ozgenc A, Loeb LA. Werner Syndrome, aging and cancer. *Genome Dyn*. 2006;1:206–17. doi: 10.1159/000092509. PubMed PMID: 18724062.

14. Larizza L, Roversi G, Volpi L. Rothmund–Thomson syndrome. *Orphanet J Rare Dis*. 2010;5:2. doi: 10.1186/1750-1172-5-2. PubMed PMID: 20113479; PMCID: PMC2826297.

15. Epstein CJ, Martin GM, Schultz AL, Motulsky AG. Werner's syndrome a review of its symptomatology, natural history, pathologic features, genetics and relationship to the natural aging process. *Medicine (Baltimore)*. 1966;45(3):177–221. PubMed PMID: 5327241.

16. Goto M, Miller RW, Ishikawa Y, Sugano H. Excess of rare cancers in Werner syndrome (adult progeria). *Cancer Epidemiol Biomarkers Prev*. 1996;5(4):239–46. PubMed PMID: 8722214.

17. Killoran MP, Keck JL. Sit down, relax and unwind: Structural insights into RecQ helicase mechanisms. *Nucleic Acids Res*. 2006;34(15):4098–105. doi: 10.1093/nar/gkl538. PubMed PMID: 16935877; PMCID: PMC1616949.

18. Wang L, Ogburn CE, Ware CB, Ladiges WC, Youssoufian H, Martin GM, Oshima J. Cellular Werner phenotypes in mice expressing a putative dominant-negative human WRN gene. *Genetics*. 2000;154(1):357–62. PubMed PMID: 10628995; PMCID: PMC1460888.

19. Bernstein DA, Zittel MC, Keck JL. High-resolution structure of the *E. coli* RecQ helicase catalytic core. *EMBO J*. 2003;22(19):4910–21. doi: 10.1093/emboj/cdg500. PubMed PMID: 14517231; PMCID: PMC204483.

20. Pike AC, Shrestha B, Popuri V, Burgess-Brown N, Muzzolini L, Costantini S, Vindigni A, Gileadi O. Structure of the human RECQ1 helicase reveals a putative strand-separation pin. *Proc Natl Acad Sci U S A* 2009;106(4):1039–44. doi: 10.1073/pnas.0806908106. PubMed PMID: 19151156; PMCID: PMC2628305.

21. Newman JA, Savitsky P, Allerston CK, Bizard AH, Ozer O, Sarlos K, Liu Y et al. Crystal structure of the Bloom's syndrome helicase indicates a role for the HRDC domain in conformational changes. *Nucleic Acids Res*. 2015;43(10):5221–35. doi: 10.1093/nar/gkv373. PubMed PMID: 25901030; PMCID: PMC4446433.

22. Swan MK, Legris V, Tanner A, Reaper PM, Vial S, Bordas R, Pollard JR, Charlton PA, Golec JM, Bertrand JA. Structure of human Bloom's syndrome helicase in complex with ADP and duplex DNA. *Acta Crystallogr D Biol Crystallogr*. 2014;70(Pt 5):1465–75. doi: 10.1107/S139900471400501X. PubMed PMID: 24816114.

23. Ellis NA, Groden J, Ye TZ, Straughen J, Lennon DJ, Ciocci S, Proytcheva M, German J. The Bloom's syndrome gene product is homologous to RecQ helicases. *Cell*. 1995;83(4):655–66. PubMed PMID: 7585968.

24. Vindigni A, Hickson ID. RecQ helicases: Multiple structures for multiple functions? *HFSP J*. 2009;3(3):153–64. doi: 10.2976/1.3079540. PubMed PMID: 19949442; PMCID: PMC2714954.

25. Huang S, Li B, Gray MD, Oshima J, Mian IS, Campisi J. The premature ageing syndrome protein, WRN, is a $3'$–$5'$ exonuclease. *Nat Genet*. 1998;20(2):114–6. doi: 10.1038/2410. PubMed PMID: 9771700; PMCID: PMC4940158.

26. von Kobbe C, Thoma NH, Czyzewski BK, Pavletich NP, Bohr VA. Werner syndrome protein contains three structure-specific DNA binding domains. *J Biol Chem*. 2003;278(52):52997–3006. doi: 10.1074/jbc.M308338200. PubMed PMID: 14534320.

27. Matsuno K, Kumano M, Kubota Y, Hashimoto Y, Takisawa H. The N-terminal noncatalytic region of Xenopus RecQ4 is required for chromatin binding of DNA polymerase alpha in the initiation of DNA replication. *Mol Cell Biol*. 2006;26(13):4843–52. doi: 10.1128/MCB.02267-05. PubMed PMID: 16782873; PMCID: PMC1489170.

28. Sangrithi MN, Bernal JA, Madine M, Philpott A, Lee J, Dunphy WG, Venkitaraman AR. Initiation of DNA replication requires the RECQL4 protein mutated in Rothmund-Thomson syndrome. *Cell*. 2005;121(6):887–98. doi: 10.1016/j.cell.2005.05.015. PubMed PMID: 15960976.

29. Xu X, Liu Y. Dual DNA unwinding activities of the Rothmund-Thomson syndrome protein, RECQ4. *EMBO J*. 2009;28(5):568–77. doi: 10.1038/emboj.2009.13. PubMed PMID: 19177149; PMCID: PMC2657580.

30. Yu CE, Oshima J, Fu YH, Wijsman EM, Hisama F, Alisch R, Matthews S et al. Positional cloning of the Werner's syndrome gene. *Science*. 1996;272(5259):258–62. PubMed PMID: 8602509.

31. Satoh M, Imai M, Sugimoto M, Goto M, Furuichi Y. Prevalence of Werner's syndrome heterozygotes in Japan. *Lancet* 1999;353(9166):1766. doi: 10.1016/S0140-6736(98)05869-3. PubMed PMID: 10347997.

32. Masala MV, Scapaticci S, Olivieri C, Pirodda C, Montesu MA, Cuccuru MA, Pruneddu S, Danesino C, Cerimele D. Epidemiology and clinical aspects of Werner's syndrome in North Sardinia: Description of a cluster. *Eur J Dermatol*. 2007;17(3):213–6. doi: 10.1684/ejd.2007.0155. PubMed PMID: 17478382.

33. Goto M. Hierarchical deterioration of body systems in Werner's syndrome: Implications for normal ageing. *Mech Ageing Dev*. 1997;98(3):239–54. PubMed PMID: 9352493.

34. De Stefano N, Dotti MT, Battisti C, Sicurelli F, Stromillo ML, Mortilla M, Federico A. MR evidence of structural and metabolic changes in brains of patients with Werner's syndrome. *J Neurol*. 2003;250(10):1169–73. doi: 10.1007/s00415-003-0167-4. PubMed PMID: 14586596.

35. Oppenheimer BSaK, V.H. Werner's syndrome. *Trans Ass Amer Physicians*. 1934;49:358–70.

36. Huang S, Lee L, Hanson NB, Lenaerts C, Hoehn H, Poot M, Rubin CD et al. The spectrum of WRN mutations in Werner syndrome patients. *Hum Mutat*. 2006;27(6):558–67. doi: 10.1002/humu.20337. PubMed PMID: 16673358; PMCID: PMC1868417.

37. Yokote K, Chanprasert S, Lee L, Eirich K, Takemoto M, Watanabe A, Koizumi N et al. WRN mutation update: Mutation spectrum, patient registries, and translational prospects. *Hum Mutat*. 2017;38(1):7–15. doi: 10.1002/humu.23128. PubMed PMID: 27667302; PMCID: PMC5237432.

38. Yamabe Y, Sugimoto M, Satoh M, Suzuki N, Sugawara M, Goto M, Furuichi Y. Down-regulation of the defective transcripts of the Werner's syndrome gene in the cells of patients. *Biochem Biophys Res Commun*. 1997;236(1):151–4. doi: 10.1006/bbrc.1997.6919. PubMed PMID: 9223443.

39. Matsumoto T, Shimamoto A, Goto M, Furuichi Y. Impaired nuclear localization of defective DNA helicases in Werner's syndrome. *Nat Genet*. 1997;16(4):335–6. doi: 10.1038/ng0897-335. PubMed PMID: 9241267.

40. von Kobbe C, Bohr VA. A nucleolar targeting sequence in the Werner syndrome protein resides within residues 949–1092. *J Cell Sci*. 2002;115(Pt 20):3901–7. PubMed PMID: 12244128.

41. Moser MJ, Kamath-Loeb AS, Jacob JE, Bennett SE, Oshima J, Monnat RJ, Jr. WRN helicase expression in Werner syndrome cell lines. *Nucleic Acids Res*. 2000;28(2):648–54. PubMed PMID: 10606667; PMCID: PMC102521.

42. Friedrich K, Lee L, Leistritz DF, Nurnberg G, Saha B, Hisama FM, Eyman DK et al. WRN mutations in Werner syndrome patients: Genomic rearrangements, unusual intronic mutations and ethnic-specific alterations. *Hum Genet*. 2010;128(1):103–11. doi: 10.1007/s00439-010-0832-5. PubMed PMID: 20443122; PMCID: PMC4686336.

43. Uhrhammer NA, Lafarge L, Dos Santos L, Domaszewska A, Lange M, Yang Y, Aractingi S, Bessis D, Bignon YJ. Werner syndrome and mutations of the WRN and LMNA genes in France. *Hum Mutat*. 2006;27(7):718–9. doi: 10.1002/humu.9435. PubMed PMID: 16786514.

44. Tadokoro T, Rybanska-Spaeder I, Kulikowicz T, Dawut L, Oshima J, Croteau DL, Bohr VA. Functional deficit associated with a missense Werner syndrome mutation. *DNA Repair (Amst)*. 2013;12(6):414–21. doi: 10.1016/j.dnarep.2013.03.004. PubMed PMID: 23583337; PMCID: PMC3660515.

45. Perry JJ, Asaithamby A, Barnebey A, Kiamanesch F, Chen DJ, Han S, Tainer JA, Yannone SM. Identification of a coiled coil in Werner syndrome protein that facilitates multimerization and promotes exonuclease processivity. *J Biol Chem*. 2010;285(33):25699–707. doi: 10.1074/jbc.M110.124941. PubMed PMID: 20516064; PMCID: PMC2919133.

46. Salk D, Au K, Hoehn H, Martin GM. Cytogenetics of Werner's syndrome cultured skin fibroblasts: Variegated translocation mosaicism. *Cytogenet Cell Genet.* 1981;30(2):92–107. PubMed PMID: 7273860.

47. Fujiwara Y, Higashikawa T, Tatsumi M. A retarded rate of DNA replication and normal level of DNA repair in Werner's syndrome fibroblasts in culture. *J Cell Physiol.* 1977;92(3):365–74. doi: 10.1002/jcp.1040920305. PubMed PMID: 903377.

48. Martin GM, Sprague CA, Epstein CJ. Replicative life-span of cultivated human cells. Effects of donor's age, tissue, and genotype. *Lab Invest.* 1970;23(1):86–92. PubMed PMID: 5431223.

49. Faragher RG, Kill IR, Hunter JA, Pope FM, Tannock C, Shall S. The gene responsible for Werner syndrome may be a cell division "counting" gene. *Proc Natl Acad Sci U S A* 1993;90(24):12030–4. PubMed PMID: 8265666; PMCID: PMC48119.

50. Fukuchi K, Martin GM, Monnat RJ, Jr. Mutator phenotype of Werner syndrome is characterized by extensive deletions. *Proc Natl Acad Sci U S A* 1989;86(15):5893–7. PubMed PMID: 2762303; PMCID: PMC297737.

51. Grigorova M, Balajee AS, Natarajan AT. Spontaneous and X-ray-induced chromosomal aberrations in Werner syndrome cells detected by FISH using chromosome-specific painting probes. *Mutagenesis* 2000;15(4):303–10. PubMed PMID: 10887208.

52. Yannone SM, Roy S, Chan DW, Murphy MB, Huang S, Campisi J, Chen DJ. Werner syndrome protein is regulated and phosphorylated by DNA-dependent protein kinase. *J Biol Chem.* 2001;276(41):38242–8. doi: 10.1074/jbc.M101913200. PubMed PMID: 11477099.

53. Cheng WH, von Kobbe C, Opresko PL, Arthur LM, Komatsu K, Seidman MM, Carney JP, Bohr VA. Linkage between Werner syndrome protein and the Mre11 complex via Nbs1. *J Biol Chem.* 2004;279(20):21169–76. doi: 10.1074/jbc.M312770200. PubMed PMID: 15026416.

54. Comai L, Li B. The Werner syndrome protein at the crossroads of DNA repair and apoptosis. *Mech Ageing Dev.* 2004;125(8):521–8. doi: 10.1016/j.mad.2004.06.004. PubMed PMID: 15336909.

55. Zhang W, Li J, Suzuki K, Qu J, Wang P, Zhou J, Liu X et al. Aging stem cells. A Werner syndrome stem cell model unveils heterochromatin alterations as a driver of human aging. *Science.* 2015;348(6239):1160–3. doi: 10.1126/science.aaa1356. PubMed PMID: 25931448; PMCID: PMC4494668.

56. Gray MD, Shen JC, Kamath-Loeb AS, Blank A, Sopher BL, Martin GM, Oshima J, Loeb LA. The Werner syndrome protein is a DNA helicase. *Nat Genet.* 1997;17(1):100–3. doi: 10.1038/ng0997-100. PubMed PMID: 9288107.

57. Shen JC, Gray MD, Oshima J, Loeb LA. Characterization of Werner syndrome protein DNA helicase activity: Directionality, substrate dependence and stimulation by replication protein A. *Nucleic Acids Res.* 1998;26(12):2879–85. PubMed PMID: 9611231; PMCID: PMC147646.

58. Brosh RM, Jr., Opresko PL, Bohr VA. Enzymatic mechanism of the WRN helicase/nuclease. *Methods Enzymol.* 2006;409:52–85. doi: 10.1016/S0076-6879(05)09004-X. PubMed PMID: 16793395.

59. Compton SA, Tolun G, Kamath-Loeb AS, Loeb LA, Griffith JD. The Werner syndrome protein binds replication fork and Holliday junction DNAs as an oligomer. *J Biol Chem.* 2008;283(36):24478–83. doi: 10.1074/jbc.M803370200. PubMed PMID: 18596042; PMCID: PMC2528990.

60. Kamath-Loeb A, Loeb LA, Fry M. The Werner syndrome protein is distinguished from the Bloom syndrome protein by its capacity to tightly bind diverse DNA structures. *PLoS One.* 2012;7(1):e30189. doi: 10.1371/journal.pone.0030189. PubMed PMID: 22272300; PMCID: PMC3260238.

61. Shen JC, Gray MD, Oshima J, Kamath-Loeb AS, Fry M, Loeb LA. Werner syndrome protein. I. DNA helicase and DNA exonuclease reside on the same polypeptide. *J Biol Chem.* 1998;273(51):34139–44. PubMed PMID: 9852073.

62. Kamath-Loeb AS, Shen JC, Loeb LA, Fry M. Werner syndrome protein. II. Characterization of the integral 3′–5′ DNA exonuclease. *J Biol Chem.* 1998;273(51):34145–50. PubMed PMID: 9852074.

63. Mian IS. Comparative sequence analysis of ribonucleases HII, III, II PH and D. *Nucleic Acids Res.* 1997;25(16):3187–95. PubMed PMID: 9241229; PMCID: PMC146874.

64. Mushegian AR, Bassett DE, Jr., Boguski MS, Bork P, Koonin EV. Positionally cloned human disease genes: Patterns of evolutionary conservation and functional motifs. *Proc Natl Acad Sci U S A* 1997;94(11):5831–6. PubMed PMID: 9159160; PMCID: PMC20866.

65. Moser MJ, Holley WR, Chatterjee A, Mian IS. The proofreading domain of *Escherichia coli* DNA polymerase I and other DNA and/or RNA exonuclease domains. *Nucleic Acids Res.* 1997;25(24):5110–8. PubMed PMID: 9396823; PMCID: PMC147149.

66. Opresko PL, Laine JP, Brosh RM, Jr., Seidman MM, Bohr VA. Coordinate action of the helicase and 3′–5′ exonuclease of Werner syndrome protein. *J Biol Chem.* 2001;276(48):44677–87. doi: 10.1074/jbc.M107548200. PubMed PMID: 11572872.

67. Opresko PL, Cheng WH, von Kobbe C, Harrigan JA, Bohr VA. Werner syndrome and the function of the Werner protein; what they can teach us about the molecular aging process. *Carcinogenesis.* 2003;24(5):791–802. PubMed PMID: 12771022.

68. Perry JJ, Yannone SM, Holden LG, Hitomi C, Asaithamby A, Han S, Cooper PK, Chen DJ, Tainer JA. WRN exonuclease structure and molecular mechanism imply an editing role in DNA end processing. *Nat Struct Mol Biol.* 2006;13(5):414–22. doi: 10.1038/nsmb1088. PubMed PMID: 16622405.

69. Choi JM, Kang SY, Bae WJ, Jin KS, Ree M, Cho Y. Probing the roles of active site residues in the 3′–5′ exonuclease of the Werner syndrome protein. *J Biol Chem.* 2007;282(13):9941–51. doi: 10.1074/jbc. M609657200. PubMed PMID: 17229737.

70. Melek M, Gellert M, van Gent DC. Rejoining of DNA by the RAG1 and RAG2 proteins. *Science.* 1998;280(5361):301–3. PubMed PMID: 9535663.

71. Verkaik NS, Esveldt-van Lange RE, van Heemst D, Bruggenwirth HT, Hoeijmakers JH, Zdzienicka MZ, van Gent DC. Different types of V(D)J recombination and end-joining defects in DNA double-strand break repair mutant mammalian cells. *Eur J Immunol.* 2002;32(3):701–9. doi: 10.1002/1521-4141(200203) 32:3<701::AID-IMMU701>3.0.CO;2-T. PubMed PMID: 11870614.

72. Xue Y, Ratcliff GC, Wang H, Davis-Searles PR, Gray MD, Erie DA, Redinbo MR. A minimal exonuclease domain of WRN forms a hexamer on DNA and possesses both 3′–5′ exonuclease and 5′-protruding strand endonuclease activities. *Biochemistry.* 2002;41(9):2901–12. PubMed PMID: 11863428.

73. Huang S, Beresten S, Li B, Oshima J, Ellis NA, Campisi J. Characterization of the human and mouse WRN 3′–5′ exonuclease. *Nucleic Acids Res.* 2000;28(12):2396–405. PubMed PMID: 10871373; PMCID: PMC102739.

74. Karow JK, Newman RH, Freemont PS, Hickson ID. Oligomeric ring structure of the Bloom's syndrome helicase. *Curr Biol.* 1999;9(11):597–600. PubMed PMID: 10359700.

75. Lee JW, Harrigan J, Opresko PL, Bohr VA. Pathways and functions of the Werner syndrome protein. *Mech Ageing Dev.* 2005;126(1):79–86. doi: 10.1016/j.mad.2004.09.011. PubMed PMID: 15610765.

76. Brosh RM, Jr., Karmakar P, Sommers JA, Yang Q, Wang XW, Spillare EA, Harris CC, Bohr VA. p53 Modulates the exonuclease activity of Werner syndrome protein. *J Biol Chem.* 2001;276(37):35093–102. doi: 10.1074/jbc.M103332200. PubMed PMID: 11427532.

77. Li B, Comai L. Functional interaction between Ku and the Werner syndrome protein in DNA end processing. *J Biol Chem.* 2000;275(37):28349–52. doi: 10.1074/jbc.C000289200. PubMed PMID: 10880505.

78. Cooper MP, Machwe A, Orren DK, Brosh RM, Ramsden D, Bohr VA. Ku complex interacts with and stimulates the Werner protein. *Genes Dev.* 2000;14(8):907–12. PubMed PMID: 10783163; PMCID: PMC316545.

79. Karmakar P, Snowden CM, Ramsden DA, Bohr VA. Ku heterodimer binds to both ends of the Werner protein and functional interaction occurs at the Werner N-terminus. *Nucleic Acids Res.* 2002;30(16):3583–91. PubMed PMID: 12177300; PMCID: PMC134248.

80. von Kobbe C, Harrigan JA, Schreiber V, Stiegler P, Piotrowski J, Dawut L, Bohr VA. Poly(ADP-ribose) polymerase 1 regulates both the exonuclease and helicase activities of the Werner syndrome protein. *Nucleic Acids Res.* 2004;32(13):4003–14. doi: 10.1093/nar/gkh721. PubMed PMID: 15292449; PMCID: PMC506806.

81. Opresko PL, von Kobbe C, Laine JP, Harrigan J, Hickson ID, Bohr VA. Telomere-binding protein TRF2 binds to and stimulates the Werner and Bloom syndrome helicases. *J Biol Chem.* 2002;277(43):41110–9. doi: 10.1074/jbc.M205396200. PubMed PMID: 12181313.

82. Baynton K, Otterlei M, Bjoras M, von Kobbe C, Bohr VA, Seeberg E. WRN interacts physically and functionally with the recombination mediator protein RAD52. *J Biol Chem.* 2003;278(38):36476–86. doi: 10.1074/jbc.M303885200. PubMed PMID: 12750383.

83. Brosh RM, Jr., von Kobbe C, Sommers JA, Karmakar P, Opresko PL, Piotrowski J, Dianova I, Dianov GL, Bohr VA. Werner syndrome protein interacts with human flap endonuclease 1 and stimulates its cleavage activity. *EMBO J.* 2001;20(20):5791–801. doi: 10.1093/emboj/20.20.5791. PubMed PMID: 11598021; PMCID: PMC125684.

84. Brosh RM, Jr., Orren DK, Nehlin JO, Ravn PH, Kenny MK, Machwe A, Bohr VA. Functional and physical interaction between WRN helicase and human replication protein A. *J Biol Chem.* 1999;274(26):18341–50. PubMed PMID: 10373438.

85. Lebel M, Spillare EA, Harris CC, Leder P. The Werner syndrome gene product co-purifies with the DNA replication complex and interacts with PCNA and topoisomerase I. *J Biol Chem.* 1999;274(53):37795–9. PubMed PMID: 10608841.

86. Szekely AM, Chen YH, Zhang C, Oshima J, Weissman SM. Werner protein recruits DNA polymerase delta to the nucleolus. *Proc Natl Acad Sci U S A*. 2000;97(21):11365–70. doi: 10.1073/pnas.97.21.11365. PubMed PMID: 11027336; PMCID: PMC17206.

87. Kamath-Loeb AS, Shen JC, Schmitt MW, Loeb LA. The Werner syndrome exonuclease facilitates DNA degradation and high fidelity DNA polymerization by human DNA polymerase delta. *J Biol Chem*. 2012;287(15):12480–90. doi: 10.1074/jbc.M111.332577. PubMed PMID: 22351772; PMCID: PMC3320997.

88. Harrigan JA, Opresko PL, von Kobbe C, Kedar PS, Prasad R, Wilson SH, Bohr VA. The Werner syndrome protein stimulates DNA polymerase beta strand displacement synthesis via its helicase activity. *J Biol Chem*. 2003;278(25):22686–95. doi: 10.1074/jbc.M213103200. PubMed PMID: 12665521.

89. Trego KS, Chernikova SB, Davalos AR, Perry JJ, Finger LD, Ng C, Tsai MS et al. The DNA repair endonuclease XPG interacts directly and functionally with the WRN helicase defective in Werner syndrome. *Cell Cycle*. 2011;10(12):1998–2007. doi: 10.4161/cc.10.12.15878. PubMed PMID: 21558802; PMCID: PMC3154418.

90. Li B, Comai L. Displacement of DNA-PKcs from DNA ends by the Werner syndrome protein. *Nucleic Acids Res*. 2002;30(17):3653–61. PubMed PMID: 12202749; PMCID: PMC137412.

91. Karmakar P, Piotrowski J, Brosh RM, Jr., Sommers JA, Miller SP, Cheng WH, Snowden CM, Ramsden DA, Bohr VA. Werner protein is a target of DNA-dependent protein kinase *in vivo* and *in vitro*, and its catalytic activities are regulated by phosphorylation. *J Biol Chem*. 2002;277(21):18291–302. doi: 10.1074/jbc.M111523200. PubMed PMID: 11889123.

92. Li B, Comai L. Requirements for the nucleolytic processing of DNA ends by the Werner syndrome protein-Ku70/80 complex. *J Biol Chem*. 2001;276(13):9896–902. doi: 10.1074/jbc.M008575200. PubMed PMID: 11152456.

93. Orren DK, Machwe A, Karmakar P, Piotrowski J, Cooper MP, Bohr VA. A functional interaction of Ku with Werner exonuclease facilitates digestion of damaged DNA. *Nucleic Acids Res*. 2001;29(9):1926–34. PubMed PMID: 11328876; PMCID: PMC37248.

94. Li B, Navarro S, Kasahara N, Comai L. Identification and biochemical characterization of a Werner's syndrome protein complex with Ku70/80 and poly(ADP-ribose) polymerase-1. *J Biol Chem*. 2004;279(14):13659–67. doi: 10.1074/jbc.M311606200. PubMed PMID: 14734561.

95. Burkle A, Diefenbach J, Brabeck C, Beneke S. Ageing and PARP. *Pharmacol Res*. 2005;52(1):93–9. doi: 10.1016/j.phrs.2005.02.008. PubMed PMID: 15911337.

96. Oshima J, Huang S, Pae C, Campisi J, Schiestl RH. Lack of WRN results in extensive deletion at non-homologous joining ends. *Cancer Res*. 2002;62(2):547–51. PubMed PMID: 11809708.

97. Sakamoto S, Nishikawa K, Heo SJ, Goto M, Furuichi Y, Shimamoto A. Werner helicase relocates into nuclear foci in response to DNA damaging agents and co-localizes with RPA and Rad51. *Genes Cells*. 2001;6(5):421–30. PubMed PMID: 11380620.

98. Saintigny Y, Makienko K, Swanson C, Emond MJ, Monnat RJ, Jr. Homologous recombination resolution defect in Werner syndrome. *Mol Cell Biol*. 2002;22(20):6971–8. PubMed PMID: 12242278; PMCID: PMC139822.

99. Sturzenegger A, Burdova K, Kanagaraj R, Levikova M, Pinto C, Cejka P, Janscak P. DNA2 cooperates with the WRN and BLM RecQ helicases to mediate long-range DNA end resection in human cells. *J Biol Chem*. 2014;289(39):27314–26. doi: 10.1074/jbc.M114.578823. PubMed PMID: 25122754; PMCID: PMC4175362.

100. Wang AT, Kim T, Wagner JE, Conti BA, Lach FP, Huang AL, Molina H et al. A dominant mutation in human RAD51 reveals its function in DNA interstrand crosslink repair independent of homologous recombination. *Mol Cell*. 2015;59(3):478–90. doi: 10.1016/j.molcel.2015.07.009. PubMed PMID: 26253028; PMCID: PMC4529964.

101. Kakarougkas A, Jeggo PA. DNA DSB repair pathway choice: An orchestrated handover mechanism. *Br J Radiol*. 2014;87(1035):20130685. doi: 10.1259/bjr.20130685. PubMed PMID: 24363387; PMCID: PMC4064598.

102. Petermann E, Helleday T. Pathways of mammalian replication fork restart. *Nat Rev Mol Cell Biol*. 2010;11(10):683–7. doi: 10.1038/nrm2974. PubMed PMID: 20842177.

103. Su F, Mukherjee S, Yang Y, Mori E, Bhattacharya S, Kobayashi J, Yannone SM, Chen DJ, Asaithamby A. Nonenzymatic role for WRN in preserving nascent DNA strands after replication stress. *Cell Rep*. 2014;9(4):1387–401. doi: 10.1016/j.celrep.2014.10.025. PubMed PMID: 25456133; PMCID: PMC4782925.

104. Ammazzalorso F, Pirzio LM, Bignami M, Franchitto A, Pichierri P. ATR and ATM differently regulate WRN to prevent DSBs at stalled replication forks and promote replication fork recovery. *EMBO J*. 2010;29(18):3156–69. doi: 10.1038/emboj.2010.205. PubMed PMID: 20802463; PMCID: PMC2944071.

105. Franchitto A, Pirzio LM, Prosperi E, Sapora O, Bignami M, Pichierri P. Replication fork stalling in WRN-deficient cells is overcome by prompt activation of a MUS81-dependent pathway. *J Cell Biol.* 2008;183(2):241–52. doi: 10.1083/jcb.200803173. PubMed PMID: 18852298; PMCID: PMC2568021.

106. Iannascoli C, Palermo V, Murfuni I, Franchitto A, Pichierri P. The WRN exonuclease domain protects nascent strands from pathological MRE11/EXO1-dependent degradation. *Nucleic Acids Res.* 2015;43(20):9788–803. doi: 10.1093/nar/gkv836. PubMed PMID: 26275776; PMCID: PMC4787784.

107. Sidorova JM. Roles of the Werner syndrome RecQ helicase in DNA replication. *DNA Repair (Amst).* 2008;7(11):1776–86. doi: 10.1016/j.dnarep.2008.07.017. PubMed PMID: 18722555; PMCID: PMC2659608.

108. Sidorova JM, Kehrli K, Mao F, Monnat R, Jr. Distinct functions of human RECQ helicases WRN and BLM in replication fork recovery and progression after hydroxyurea-induced stalling. *DNA Repair (Amst).* 2013;12(2):128–39. doi: 10.1016/j.dnarep.2012.11.005. PubMed PMID: 23253856; PMCID: PMC3551992.

109. Thangavel S, Berti M, Levikova M, Pinto C, Gomathinayagam S, Vujanovic M, Zellweger R et al. DNA2 drives processing and restart of reversed replication forks in human cells. *J Cell Biol.* 2015;208(5):545–62. doi: 10.1083/jcb.201406100. PubMed PMID: 25733713; PMCID: PMC4347643.

110. Kamath-Loeb AS, Lan L, Nakajima S, Yasui A, Loeb LA. Werner syndrome protein interacts functionally with translesion DNA polymerases. *Proc Natl Acad Sci U S A.* 2007;104(25):10394–9. doi: 10.1073/pnas.0702513104. PubMed PMID: 17563354; PMCID: PMC1965524.

111. Murfuni I, De Santis A, Federico M, Bignami M, Pichierri P, Franchitto A. Perturbed replication induced genome wide or at common fragile sites is differently managed in the absence of WRN. *Carcinogenesis.* 2012;33(9):1655–63. doi: 10.1093/carcin/bgs206. PubMed PMID: 22689923.

112. Pirzio LM, Pichierri P, Bignami M, Franchitto A. Werner syndrome helicase activity is essential in maintaining fragile site stability. *J Cell Biol.* 2008;180(2):305–14. doi: 10.1083/jcb.200705126. PubMed PMID: 18209099; PMCID: PMC2213598.

113. Rodriguez-Lopez AM, Jackson DA, Iborra F, Cox LS. Asymmetry of DNA replication fork progression in Werner's syndrome. *Aging Cell.* 2002;1(1):30–9. PubMed PMID: 12882351.

114. Edwards DN, Orren DK, Machwe A. Strand exchange of telomeric DNA catalyzed by the Werner syndrome protein (WRN) is specifically stimulated by TRF2. *Nucleic Acids Res.* 2014;42(12):7748–61. doi: 10.1093/nar/gku454. PubMed PMID: 24880691; PMCID: PMC4081078.

115. Opresko PL. Telomere ResQue and preservation–roles for the Werner syndrome protein and other RecQ helicases. *Mech Ageing Dev.* 2008;129(1-2):79–90. doi: 10.1016/j.mad.2007.10.007. PubMed PMID: 18054793.

116. Opresko PL, Sowd G, Wang H. The Werner syndrome helicase/exonuclease processes mobile D-loops through branch migration and degradation. *PLoS One.* 2009;4(3):e4825. doi: 10.1371/journal.pone.0004825. PubMed PMID: 19283071; PMCID: PMC2653227.

117. Crabbe L, Jauch A, Naeger CM, Holtgreve-Grez H, Karlseder J. Telomere dysfunction as a cause of genomic instability in Werner syndrome. *Proc Natl Acad Sci U S A.* 2007;104(7):2205–10. doi: 10.1073/pnas.0609410104. PubMed PMID: 17284601; PMCID: PMC1794219.

118. Hisama FM, Chen YH, Meyn MS, Oshima J, Weissman SM. WRN or telomerase constructs reverse 4-nitroquinoline 1-oxide sensitivity in transformed Werner syndrome fibroblasts. *Cancer Res.* 2000;60(9):2372–6. PubMed PMID: 10811112.

119. Van Maldergem L, Siitonen HA, Jalkh N, Chouery E, De Roy M, Delague V, Muenke M et al. Revisiting the craniosynostosis-radial ray hypoplasia association: Baller–Gerold syndrome caused by mutations in the RECQL4 gene. *J Med Genet.* 2006;43(2):148–52. doi: 10.1136/jmg.2005.031781. PubMed PMID: 15964893; PMCID: PMC2564634.

120. Kitao S, Shimamoto A, Goto M, Miller RW, Smithson WA, Lindor NM, Furuichi Y. Mutations in RECQL4 cause a subset of cases of Rothmund–Thomson syndrome. *Nat Genet.* 1999;22(1):82–4. doi: 10.1038/8788. PubMed PMID: 10319867.

121. Siitonen HA, Sotkasiira J, Biervliet M, Benmansour A, Capri Y, Cormier-Daire V, Crandall B et al. The mutation spectrum in RECQL4 diseases. *Eur J Hum Genet.* 2009;17(2):151–8. doi: 10.1038/ejhg.2008.154. PubMed PMID: 18716613; PMCID: PMC2986053.

122. Vennos EM, James WD. Rothmund–Thomson syndrome. *Dermatol Clin.* 1995;13(1):143–50. PubMed PMID: 7712640.

123. Lu L, Jin W, Wang LL. Aging in Rothmund–Thomson syndrome and related RECQL4 genetic disorders. *Ageing Res Rev.* 2017;33:30–5. doi: 10.1016/j.arr.2016.06.002. PubMed PMID: 27287744.

124. Galea P, Tolmie JL. Normal growth and development in a child with Baller–Gerold syndrome (craniosynostosis and radial aplasia). *J Med Genet.* 1990;27(12):784–7. PubMed PMID: 2074565; PMCID: PMC1017284.

125. Yin J, Kwon YT, Varshavsky A, Wang W. RECQL4, mutated in the Rothmund–Thomson and RAPADILINO syndromes, interacts with ubiquitin ligases UBR1 and UBR2 of the N-end rule pathway. *Hum Mol Genet*. 2004;13(20):2421–30. doi: 10.1093/hmg/ddh269. PubMed PMID: 15317757.
126. Croteau DL, Rossi ML, Canugovi C, Tian J, Sykora P, Ramamoorthy M, Wang ZM et al. RECQL4 localizes to mitochondria and preserves mitochondrial DNA integrity. *Aging Cell*. 2012;11(3):456–66. doi: 10.1111/j.1474-9726.2012.00803.x. PubMed PMID: 22296597; PMCID: PMC3350572.
127. De S, Kumari J, Mudgal R, Modi P, Gupta S, Futami K, Goto H et al. RECQL4 is essential for the transport of p53 to mitochondria in normal human cells in the absence of exogenous stress. *J Cell Sci*. 2012;125(Pt 10):2509–22. doi: 10.1242/jcs.101501. PubMed PMID: 22357944.
128. Petkovic M, Dietschy T, Freire R, Jiao R, Stagljar I. The human Rothmund–Thomson syndrome gene product, RECQL4, localizes to distinct nuclear foci that coincide with proteins involved in the maintenance of genome stability. *J Cell Sci*. 2005;118(Pt 18):4261–9. doi: 10.1242/jcs.02556. PubMed PMID: 16141230.
129. Woo LL, Futami K, Shimamoto A, Furuichi Y, Frank KM. The Rothmund–Thomson gene product RECQL4 localizes to the nucleolus in response to oxidative stress. *Exp Cell Res*. 2006;312(17):3443–57. doi: 10.1016/j.yexcr.2006.07.023. PubMed PMID: 16949575.
130. Kitao S, Ohsugi I, Ichikawa K, Goto M, Furuichi Y, Shimamoto A. Cloning of two new human helicase genes of the RecQ family: Biological significance of multiple species in higher eukaryotes. *Genomics*. 1998;54(3):443–52. doi: 10.1006/geno.1998.5595. PubMed PMID: 9878247.
131. Choi YW, Bae SM, Kim YW, Lee HN, Kim YW, Park TC, Ro DY et al. Gene expression profiles in squamous cell cervical carcinoma using array-based comparative genomic hybridization analysis. *Int J Gynecol Cancer*. 2007;17(3):687–96. doi: 10.1111/j.1525-1438.2007.00834.x. PubMed PMID: 17504382.
132. Saglam O, Shah V, Worsham MJ. Molecular differentiation of early and late stage laryngeal squamous cell carcinoma: An exploratory analysis. *Diagn Mol Pathol*. 2007;16(4):218–21. doi: 10.1097/PDM.0b013e3180d0aab5. PubMed PMID: 18043285.
133. Maire G, Yoshimoto M, Chilton-MacNeill S, Thorner PS, Zielenska M, Squire JA. Recurrent RECQL4 imbalance and increased gene expression levels are associated with structural chromosomal instability in sporadic osteosarcoma. *Neoplasia*. 2009;11(3):260–8, 3p following 8. PubMed PMID: 19242607; PMCID: PMC2647728.
134. Su Y, Meador JA, Calaf GM, Proietti De-Santis L, Zhao Y, Bohr VA, Balajee AS. Human RecQL4 helicase plays critical roles in prostate carcinogenesis. *Cancer Res*. 2010;70(22):9207–17. doi: 10.1158/0008-5472.CAN-10-1743. PubMed PMID: 21045146; PMCID: PMC3058916.
135. Rossi ML, Ghosh AK, Kulikowicz T, Croteau DL, Bohr VA. Conserved helicase domain of human RecQ4 is required for strand annealing-independent DNA unwinding. *DNA Repair (Amst)*. 2010;9(7):796–804. doi: 10.1016/j.dnarep.2010.04.003. PubMed PMID: 20451470; PMCID: PMC2893255.
136. Durand F, Castorina P, Morant C, Delobel B, Barouk E, Modiano P. Rothmund–Thomson syndrome, trisomy 8 mosaicism and RECQ4 gene mutation. *Ann Dermatol Venereol*. 2002;129(6–7):892–5. PubMed PMID: 12218919.
137. Mann MB, Hodges CA, Barnes E, Vogel H, Hassold TJ, Luo G. Defective sister-chromatid cohesion, aneuploidy and cancer predisposition in a mouse model of type II Rothmund–Thomson syndrome. *Hum Mol Genet*. 2005;14(6):813–25. doi: 10.1093/hmg/ddi075. PubMed PMID: 15703196.
138. Miozzo M, Castorina P, Riva P, Dalpra L, Fuhrman Conti AM, Volpi L, Hoe TS et al. Chromosomal instability in fibroblasts and mesenchymal tumors from 2 sibs with Rothmund–Thomson syndrome. *Int J Cancer*. 1998;77(4):504–10. PubMed PMID: 9679749.
139. Fan W, Luo J. RecQ4 facilitates UV light-induced DNA damage repair through interaction with nucleotide excision repair factor xeroderma pigmentosum group A (XPA). *J Biol Chem*. 2008;283(43):29037–44. doi: 10.1074/jbc.M801928200. PubMed PMID: 18693251; PMCID: PMC2570873.
140. Schurman SH, Hedayati M, Wang Z, Singh DK, Speina E, Zhang Y, Becker K et al. Direct and indirect roles of RECQL4 in modulating base excision repair capacity. *Hum Mol Genet*. 2009;18(18):3470–83. doi: 10.1093/hmg/ddp291. PubMed PMID: 19567405; PMCID: PMC2729667.
141. Singh DK, Karmakar P, Aamann M, Schurman SH, May A, Croteau DL, Burks L, Plon SE, Bohr VA. The involvement of human RECQL4 in DNA double-strand break repair. *Aging Cell*. 2010;9(3):358–71. doi: 10.1111/j.1474-9726.2010.00562.x. PubMed PMID: 20222902; PMCID: PMC4624395.
142. Jin W, Liu H, Zhang Y, Otta SK, Plon SE, Wang LL. Sensitivity of RECQL4-deficient fibroblasts from Rothmund–Thomson syndrome patients to genotoxic agents. *Hum Genet*. 2008;123(6):643–53. doi: 10.1007/s00439-008-0518-4. PubMed PMID: 18504617; PMCID: PMC2585174.
143. Smith PJ, Paterson MC. Enhanced radiosensitivity and defective DNA repair in cultured fibroblasts derived from Rothmund Thomson syndrome patients. *Mutat Res*. 1982;94(1):213–28. PubMed PMID: 7099192.

144. Larizza L, Magnani I, Roversi G. Rothmund–Thomson syndrome and RECQL4 defect: Splitting and lumping. *Cancer Lett.* 2006;232(1):107–20. doi: 10.1016/j.canlet.2005.07.042. PubMed PMID: 16271439.

145. Macris MA, Krejci L, Bussen W, Shimamoto A, Sung P. Biochemical characterization of the RECQ4 protein, mutated in Rothmund–Thomson syndrome. *DNA Repair (Amst).* 2006;5(2):172–80. doi: 10.1016/j.dnarep.2005.09.005. PubMed PMID: 16214424.

146. Suzuki T, Kohno T, Ishimi Y. DNA helicase activity in purified human RECQL4 protein. *J Biochem.* 2009;146(3):327–35. doi: 10.1093/jb/mvp074. PubMed PMID: 19451148.

147. Capp C, Wu J, Hsieh TS. Drosophila RecQ4 has a 3'–5' DNA helicase activity that is essential for viability. *J Biol Chem.* 2009;284(45):30845–52. doi: 10.1074/jbc.M109.008052. PubMed PMID: 19759018; PMCID: PMC2781483.

148. Keller H, Kiosze K, Sachsenweger J, Haumann S, Ohlenschlager O, Nuutinen T, Syvaoja JE, Gorlach M, Grosse F, Pospiech H. The intrinsically disordered amino-terminal region of human RecQL4: Multiple DNA-binding domains confer annealing, strand exchange and G4 DNA binding. *Nucleic Acids Res.* 2014;42(20):12614–27. doi: 10.1093/nar/gku993. PubMed PMID: 25336622; PMCID: PMC4227796.

149. Sedlackova H, Cechova B, Mlcouskova J, Krejci L. RECQ4 selectively recognizes Holliday junctions. *DNA Repair (Amst).* 2015;30:80–9. doi: 10.1016/j.dnarep.2015.02.020. PubMed PMID: 25769792.

150. Marino F, Vindigni A, Onesti S. Bioinformatic analysis of RecQ4 helicases reveals the presence of a RQC domain and a Zn knuckle. *Biophys Chem.* 2013;177-178:34–9. doi: 10.1016/j.bpc.2013.02.009. PubMed PMID: 23624328.

151. Marino F, Mojumdar A, Zucchelli C, Bhardwaj A, Buratti E, Vindigni A, Musco G, Onesti S. Structural and biochemical characterization of an RNA/DNA binding motif in the N-terminal domain of RecQ4 helicases. *Sci Rep.* 2016;6:21501. doi: 10.1038/srep21501. PubMed PMID: 26888063; PMCID: PMC4757822.

152. Mojumdar A, De March M, Marino F, Onesti S. The Human RecQ4 helicase contains a functional RecQ C-terminal region (RQC) that is essential for activity. *J Biol Chem.* 2017;292(10):4176–84. doi: 10.1074/jbc.M116.767954. PubMed PMID: 27998982; PMCID: PMC5354486.

153. Sclafani RA, Holzen TM. Cell cycle regulation of DNA replication. *Annu Rev Genet.* 2007;41:237–80. doi: 10.1146/annurev.genet.41.110306.130308. PubMed PMID: 17630848; PMCID: PMC2292467.

154. Thangavel S, Mendoza-Maldonado R, Tissino E, Sidorova JM, Yin J, Wang W, Monnat RJ, Jr., Falaschi A, Vindigni A. Human RECQ1 and RECQ4 helicases play distinct roles in DNA replication initiation. *Mol Cell Biol.* 2010;30(6):1382–96. doi: 10.1128/MCB.01290-09. PubMed PMID: 20065033; PMCID: PMC2832491.

155. Wu J, Capp C, Feng L, Hsieh TS. Drosophila homologue of the Rothmund–Thomson syndrome gene: Essential function in DNA replication during development. *Dev Biol.* 2008;323(1):130–42. doi: 10.1016/j.ydbio.2008.08.006. PubMed PMID: 18755177; PMCID: PMC2600506.

156. Xu X, Rochette PJ, Feyissa EA, Su TV, Liu Y. MCM10 mediates RECQ4 association with MCM2-7 helicase complex during DNA replication. *EMBO J.* 2009;28(19):3005–14. doi: 10.1038/emboj.2009.235. PubMed PMID: 19696745; PMCID: PMC2760112.

157. Im JS, Ki SH, Farina A, Jung DS, Hurwitz J, Lee JK. Assembly of the Cdc45-Mcm2-7-GINS complex in human cells requires the Ctf4/And-1, RecQL4, and Mcm10 proteins. *Proc Natl Acad Sci U S A.* 2009;106(37):15628–32. doi: 10.1073/pnas.0908039106. PubMed PMID: 19805216; PMCID: PMC2747170.

158. Park SJ, Lee YJ, Beck BD, Lee SH. A positive involvement of RecQL4 in UV-induced S-phase arrest. *DNA Cell Biol.* 2006;25(12):696–703. doi: 10.1089/dna.2006.25.696. PubMed PMID: 17184169.

159. Werner SR, Prahalad AK, Yang J, Hock JM. RECQL4-deficient cells are hypersensitive to oxidative stress/damage: Insights for osteosarcoma prevalence and heterogeneity in Rothmund–Thomson syndrome. *Biochem Biophys Res Commun.* 2006;345(1):403–9. doi: 10.1016/j.bbrc.2006.04.093. PubMed PMID: 16678792.

160. Lu H, Shamanna RA, Keijzers G, Anand R, Rasmussen LJ, Cejka P, Croteau DL, Bohr VA. RECQL4 Promotes DNA end resection in repair of DNA double-strand breaks. *Cell Rep.* 2016;16(1):161–73. doi: 10.1016/j.celrep.2016.05.079. PubMed PMID: 27320928.

161. Kohzaki M, Chiourea M, Versini G, Adachi N, Takeda S, Gagos S, Halazonetis TD. The helicase domain and C-terminus of human RecQL4 facilitate replication elongation on DNA templates damaged by ionizing radiation. *Carcinogenesis.* 2012;33(6):1203–10. doi: 10.1093/carcin/bgs149. PubMed PMID: 22508716.

162. Barea F, Tessaro S, Bonatto D. In silico analyses of a new group of fungal and plant RecQ4-homologous proteins. *Comput Biol Chem.* 2008;32(5):349–58. doi: 10.1016/j.compbiolchem.2008.07.005. PubMed PMID: 18701350.

163. Kwon SH, Choi DH, Lee R, Bae SH. *Saccharomyces cerevisiae* Hrq1 requires a long 3′-tailed DNA substrate for helicase activity. *Biochem Biophys Res Commun.* 2012;427(3):623–8. doi: 10.1016/j. bbrc.2012.09.109. PubMed PMID: 23026052.

164. Choi DH, Lee R, Kwon SH, Bae SH. Hrq1 functions independently of Sgs1 to preserve genome integrity in *Saccharomyces cerevisiae. J Microbiol.* 2013;51(1):105–12. doi: 10.1007/s12275-013-3048-2. PubMed PMID: 23456718.

165. Bochman ML, Paeschke K, Chan A, Zakian VA. Hrq1, a homolog of the human RecQ4 helicase, acts catalytically and structurally to promote genome integrity. *Cell Rep.* 2014;6(2):346–56. doi: 10.1016/j. celrep.2013.12.037. PubMed PMID: 24440721; PMCID: PMC3933191.

166. Choi DH, Min MH, Kim MJ, Lee R, Kwon SH, Bae SH. Hrq1 facilitates nucleotide excision repair of DNA damage induced by 4-nitroquinoline-1-oxide and cisplatin in *Saccharomyces cerevisiae. J Microbiol.* 2014;52(4):292–8. doi: 10.1007/s12275-014-4018-z. PubMed PMID: 24682993.

167. Rogers CM, Bochman ML. *Saccharomyces cerevisiae* Hrq1 helicase activity is affected by the sequence but not the length of single-stranded DNA. *Biochem Biophys Res Commun.* 2017;486(4):1116–21. doi: 10.1016/j.bbrc.2017.04.003. PubMed PMID: 28385527.

168. Rogers CM, Wang JC, Noguchi H, Imasaki T, Takagi Y, Bochman ML. Yeast Hrq1 shares structural and functional homology with the disease-linked human RecQ4 helicase. *Nucleic Acids Res.* 2017. doi: 10.1093/nar/gkx151. PubMed PMID: 28334827.

169. Ghosh AK, Rossi ML, Singh DK, Dunn C, Ramamoorthy M, Croteau DL, Liu Y, Bohr VA. RECQL4, the protein mutated in Rothmund–Thomson syndrome, functions in telomere maintenance. *J Biol Chem.* 2012;287(1):196–209. doi: 10.1074/jbc.M111.295063. PubMed PMID: 22039056; PMCID: PMC3249070.

170. Ferrarelli LK, Popuri V, Ghosh AK, Tadokoro T, Canugovi C, Hsu JK, Croteau DL, Bohr VA. The RECQL4 protein, deficient in Rothmund–Thomson syndrome is active on telomeric D-loops containing DNA metabolism blocking lesions. *DNA Repair (Amst).* 2013;12(7):518–28. doi: 10.1016/j. dnarep.2013.04.005. PubMed PMID: 23683351; PMCID: PMC3710707.

171. Chi Z, Nie L, Peng Z, Yang Q, Yang K, Tao J, Mi Y, Fang X, Balajee AS, Zhao Y. RecQL4 cytoplasmic localization: Implications in mitochondrial DNA oxidative damage repair. *Int J Biochem Cell Biol.* 2012;44(11):1942–51. doi: 10.1016/j.biocel.2012.07.016. PubMed PMID: 22824301; PMCID: PMC3461334.

172. Coskun P, Wyrembak J, Schriner SE, Chen HW, Marciniack C, Laferla F, Wallace DC. A mitochondrial etiology of Alzheimer and Parkinson disease. *Biochim Biophys Acta.* 2012;1820(5):553–64. doi: 10.1016/j.bbagen.2011.08.008. PubMed PMID: 21871538; PMCID: PMC3270155.

173. Jiang Y, Liu X, Fang X, Wang X. Proteomic analysis of mitochondria in Raji cells following exposure to radiation: Implications for radiotherapy response. *Protein Pept Lett.* 2009;16(11):1350–9. PubMed PMID: 20001925.

174. Gupta S, De S, Srivastava V, Hussain M, Kumari J, Muniyappa K, Sengupta S. RECQL4 and p53 potentiate the activity of polymerase gamma and maintain the integrity of the human mitochondrial genome. *Carcinogenesis.* 2014;35(1):34–45. doi: 10.1093/carcin/bgt315. PubMed PMID: 24067899.

175. Wang JT, Xu X, Alontaga AY, Chen Y, Liu Y. Impaired p32 regulation caused by the lymphoma-prone RECQ4 mutation drives mitochondrial dysfunction. *Cell Rep.* 2014;7(3):848–58. doi: 10.1016/j. celrep.2014.03.037. PubMed PMID: 24746816; PMCID: PMC4029353.

11 Cockayne Syndrome and the Aging Process

María de la Luz Arenas-Sordo

CONTENTS

INTRODUCTION

Cockayne syndrome (CS) is a disorder with premature aging and short-life expectancy. It is a very complex disease, with progressive degeneration in almost all the organs. The loss of the self-renewal capacity of the stem cells that are responsible to maintain the organs' tissues is one of the causes of premature aging (Kenyon and Gerson, 2007).

It was Edward Alfred Cockayne who, in 1936, described a new syndrome: dwarfism with retinal atrophy and deafness. Later, Neill and Dingwall (1950) described the case of two brothers with Cockayne, and for this reason the syndrome was called Neill–Dingwall. They claimed that it was the same disease described by Cockayne, and these women contributed to the study of the disease through their discovery of the presence of calcifications in the brain, and then around 150 cases have been described. In these cases progressive failure to thrive, neurological deterioration with mental retardation, facial and skeletal changes, photodermatitis, and visual and hearing loss were described (Gorlin et al., 1995).

DEFINITION

CS is a genetic disease with an autosomal recessive inheritance. It is a multi-organic and progressive disease which causes developmental and cognitive delay. It is secondary to a DNA repair failure in the NER (nucleotide excision repair) pathway. There are also other situations related to the failure of proteins produced by the mutated genes responsible for CS (Arenas-Sordo et al., 2006; Baez et al., 2013).

CLASSIFICATION

Depending on its clinical findings, it has been classified in three or four types:

    CS 1: Moderate or classical
    CS 2: Severe, early-onset
    CS 3: Mild, late-onset

There is also other CS which shares findings from Xeroderma pigmentosum (XP); in these, other genes are involved: *XPB, XPD,* and *XPG.* They are considered variants of classical CS and are named complex Cockayne–Xeroderma pigmentosum (CS–XP). This one is considered the fourth type (Calmels et al., 2016).

CLINICAL FINDINGS

In general all patients, regardless of the type (1, 2, and 3), have the same symptoms, but the time of onset and the rate of progression vary significantly. The patients have important postnatal failure to thrive and of brain growth. They also have cachexia, dementia, loss of hearing and of vision, and an important hypersensitivity to ultraviolet (UV) light. It has been considered a progressive disorder. They can have several other signs and symptoms than those described above, such as: spasticity, ataxia, peripheral neuropathy, weakness, osteopenia, joint contractures, thin, dry hair, dental caries, endocrinopathies, pigmentary retinopathy, cataracts and their progeroid appearance (Ahmad, 2009) (Table 11.1).

The classical CS is designated Type 1, the congenital form is Type 2, and the form that begins later in life and with normal development is Type 3. Finally, Type 4 is the group with patients having the complex CS–XP. The clinical findings have some changes, for example, they have the XP classical skin lesions. These patients share manifestations of both diseases. The life expectancy of the patients is reduced, the mean age at death is around 12.5 years; however, some patients have lived into their late 20s, which depends on the CS type they suffer from (Arenas-Sordo et al., 2006; Calmels et al., 2016).

ETIOPATHOLOGY

From the etiological point of view CS originates due to mutation in one or both the genes, the *ERCC8* and the *ERCC6.* Depending on the gene mutation causing the disease, it is classified in

TABLE 11.1
Clinical Findings in CS

Growth and developmental	Intrauterine and postnatal growth retardation, the second one more severe; short stature; cachectic body
Head	Microcephaly, prognatism, thickened calvarium, loss of adipose tissue in the cheeks, slender nose, sunken eyes, pigmentary retinopathy, optic atrophy, strabismus, cataracts, nystagmus, corneal opacity, malocclusion, hypodontia, partial macrodontia (central incisors), dental caries, deep palate, dental hypoplasia, short roots, condylar hypoplasia, sensorineural hearing loss, less frequently malformed ears
Thorax	Cardiac arrhythmias, hypertension, kyphosis and vertebral body abnormalities, small clavicles, pectus carinatum
Abdomen	Hepatomegaly and splenomegaly
Limbs	Very thin, joint limitation, sclerotic phalangeal epiphyses, Pes valgus, short second toes
Genitourinary	Cryptorchidism, micropenis, proteinuria, and renal failure
Pelvis	Squared shape, small, hypoplastic iliac wings, coxa valga
Skin	Photosensitivity, atrophy, pigmentation, anhidrosis, dry, loss of subcutaneous adipose tissue
Endocrine system	Hypogonadism, irregular menstrual cycles, thymic hormone decreased. Osteoporosis
Neurologic	Abnormalities of brainstem evoked responses
	Mental retardation, delayed psychomotor skills, calcifications of basal ganglia, areas of demyelination, seizures, dysarthric speech (when there is a speech), gait trouble, ataxia, tremor, weakness, peripheral neuropathy

CSA and CSB. The first one is linked to *ERCC8* and the second to *ERCC6*. In the CS population about two thirds are linked to mutations in *ERCC6* and one third to *ERCC8* (Kondo et al., 2016). Now around 30 mutations have been identified in the *ERCC8* gene and 78 in the *ERCC6*. CSA corresponds to a CS Type I and CSB to Type II (Laugel, 2013). There is no molecular defect found yet in the CS Type III. In the complex CS–XP the involved genes are *XPB*, *XPD*, and *XPG* (Calmels et al., 2016).

DNA REPAIR SYSTEMS AND ITS DEFICIENCY IN CS

Any event that brings about changes from the usual double-helical structure of DNA is a threat to the genetic constituent of the cell. For this reason a number of different repair systems exist which try to repair those changes (Modrich, 2016). However, the repair may not be always perfect and a misrepair can lead to long lasting problems in DNA including mutagenesis and cancer (Larrea et al., 2010)

When there is only single-strand DNA damage (such as UV-induced cyclobutane pyrimidine dimer or CPD), the other strand can be used as a template to rectify the error. The damages are usually corrected by one of the three types of repair systems as shown below (Larrea et al., 2010).

1. BER: Base-excision repair involving ~15 genes
2. NER: Nucleotide excision repair involving ~30 genes
3. MR: Mismatch repair involving ~15 genes

In order to repair DNA damage one of the endonucleases binds to the damaged site of the strand and with the help of an exonuclease removes a number of nucleotides around the damaged section and the repair is carried out by filling the gap by DNA polymerase; instruction is taken from the complementary strand of the DNA, and finally DNA ligase binds the newly synthesized strand with the old strand. When there is mutation in any of the DNA repair gene(s), cells become exceedingly sensitive to the agent causing damage such as UV irradiation (Park and Gerson, 2005).

Out of these three repair systems in CS, NER is the one which is mostly involved. This is because *ERCC6* gene product (excision repair 6, chromatin remodeling factor) and *ERCC8* product (excision repair 8, chromatin remodeling factor) are part of the NER pathway. This pathway acts on a wide variety of DNA damage, especially those involved in the distortion of the double helical structure. The most studied are the CPDs that result from UV radiation exposure (Calmels et al., 2016).

The mis-repaired or unrepaired DNA damage leaving an excessive amount of DNA damage can lead to the loss of genome integrity and can induce apoptosis, senescence, and influence the aging process. The DNA damage also may inhibit the stem cells in their role as the regenerator of tissues (Park and Gerson, 2005).

GENES IN CS: *ERCC6* AND *EERC8*

ERCC6 gene (excision repair 6, cross-complementing protein): its cytogenetic location is 10q11.23. Its product is a protein (ERCC6) which is a polypeptide of 1493 amino acids and has a molecular weight of 168 kDa. This protein belongs to a wsi/snf superfamily and is a part of the NER pathway. This system is very complex and its function is to eliminate a variety of DNA lesions, including UV-induced CPD, and DNA inter-strand cross-links. NER is subdivided into two pathways: global genome NER (GG-NER) and transcription-coupled NER (TC-NER) (Hanawalt, 2001; Kashiyama et al., 2013).

In the CS the specific defect is in TC-NER. GG-NER is involved in the global surveillance and genome repair, while TC-NER repairs lesions that block elongation of the complexes of transcription. The *ERCC6* gene function is the processing of the elongating form of RNA polymerase II that is stalled at the site of UV-induced DNA damage (Hashimoto et al., 2016). The TC-NER is a very specific and efficient system (Figure 11.1). The defect in any one of the repair steps of the DNA

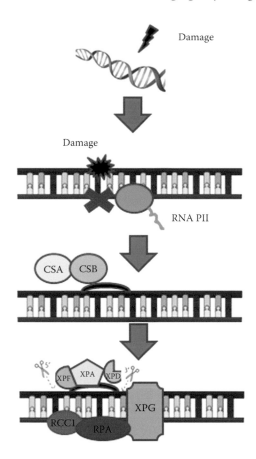

FIGURE 11.1   CSA and CSB function in NER pathway.

repair system can cause cell death and for this reason there is no cancer observed in CS patients but there is premature aging (Hoeijmakers, 2009; Wang et al., 2016).

ERCC8 gene: (excision repair 8, cross-complementing protein): its cytogenetic location is 5q12.1. The protein is a polypeptide of 396 amino acids with a molecular weight of 44 kDa. The function of this gene product is similar to the one of the *ERCC6* gene. It has domains with repeated sequences, a WD40, that joins the ERCC6 protein and forms a complex and also joins with the p44 subunit from TFIIH from the RNA polymerase (Kashiyama et al., 2013).

It seems that at least in the patients with CS Type A, it is possible to have mutations in homozygosity or in compound heterozygosity. There is no clear evidence of genotype/phenotype correlation (Bertola et al., 2006).

In the CS–XP complex the genes involved are related to GG–NER pathway. The *XPC* gene product is the one which recognizes the damage in addition to protein HR23B, after that, it forms a complex with the D complex (DDB1 and DDB2 (XPE)) (Table 11.2).

In cases of DNA double-strand breaks (DSB), there are other repair systems:

1. Nonhomologous end joining (NHEJ)
2. Microhomology-mediated end joining (MMEJ)
3. Homologous recombination (HR)

Besides the mutation in genes that causes CS, mutation in other genes (Table 11.3) of the NER pathway can cause other diseases such as XP A–G (Xeroderma pigmentosum A–G), ERCC-1, and trichothiodystrophy (Cleaver et al., 2009).

TABLE 11.2
Genes Related to CS

Type/Gene	Meaning	Function
CS I/ERCC8	Excision repair 8, chromatin remodeling factor	Protein that works in the NER pathway
CS II/ERCC6	Excision repair 6, chromatin remodeling factor	Protein that works in the NER pathway
CS IV/*XPB, XPD* and *XPG*.	Xeroderma pigmentosum B, D and G	Protein that works in the NER pathway

TABLE 11.3
Diseases with DNA Repair Systems Failure

Disease	Gene	Affected Pathway
Cockayne	ERCC8/ERCC6	NER
Xeroderma pigmentosum	XP (A–G)	NER
Trichothiodystrophy	XP/CS	NER
Rothmund–Thomson syndrome	RECQ[4]	HR
Werner syndrome	WRN (a RECQ family)	HR
Bloom	BLM (a RECQ family)	HR
Ataxia teleangiectasia	ATM (a PI3K)	HR

NER, nucleotide excision repair. *HR*, homologous recombination.

STEM CELLS AND MITOCHONDRIAL DNA

Stem cells in adulthood are responsible for the long-term maintenance of all body tissues. They regenerate cells after damage and replace differentiated senescent cells that no longer function. That means the cells have to have precise balance between proliferation and apoptosis. The cells' longevity depends on gene expression, cell cycle, and DNA repair. To avoid the premature aging process and prevention of diseases a robust DNA repair is necessary. Thus the process of aging is linked to proper function of stem cells and to DNA damage repair processes (Kenyon and Gerson, 2007).

The repair processes seem to become less effective with age and then, besides aging, other problems like cancers of various types appear. In CS, the mutation in genes responsible to drive the NER pathway induces the premature aging of tissues (Hoeijmakers, 2009).

Recently, besides the role of stem cells and DNA repair, the importance of normal integrity of mtDNA has been identified to avoid age-related phenotypes. In CS it is especially important, because as Chatre et al. found that the impaired function of mitochondria is due to a depletion of the mitochondrial DNA polymerase (POLG1) which depends on the CSA (gene *ERCC8*)/CSB (gene *ERCC6*) deficiency in these patients (Chatre et al., 2015; Pascucci et al., 2016).

CONCLUSION

CS is a complex autosomal recessive genetic disease, in which there is premature aging. These genes, *ERCC8, ERCC6, XPB, XPD*, and XPG, are responsible in DNA repair and mutation in any of them can cause loss of repair function and mitochondrial dysfunction leading to a number of human diseases and an enhanced aging process.

REFERENCES

Ahmad, SI (ed.). *Molecular Mechanisms of Cockayne Syndrome*, Landes Bioscience Publication, Madam Curie Bioscience Database, Austin, Texas, 2009.

Arenas-Sordo ML, Hernández-Zamora E, Montoya-Pérez LA, Aldape-Barrios BC. Cockayne's syndrome: A case report. Literature review. *Med Oral Patol Oral Cir Bucal.* 2006; 11: E236–8.

Baez S, Couto B, Herrera E, Bocanegra Y, Trujillo-Orrego N, Madrigal-Zapata L, Cardona JF, Manes F, Ibanez A, Villegas A. Tracking the cognitive, social, and neuroanatomical profile in early neurodegeneration: Type III Cockayne syndrome. *Front Aging Neurosci.* 2013; 5: 80. doi: 10.3389/fnagi.2013.00080

Bertola DR, Cao H, Albano LM, Oliveira DP, Kok F, Marques-Dias MJ, Kim CA, Hegele RA. Cockayne syndrome type A: Novel mutations in eight typical patients. *J Hum Genet.* 2006; 51(8): 701–5.

Calmels N, Greff G, Obringer C, Kempf N, Gasnier C, Tarabeux J, Miguet M et al. Uncommon nucleotide excision repair phenotypes revealed by targeted high-throughput sequencing. *Orphanet J Rare Dis.* 2016 Mar 22; 11: 26. doi: 10.1186/s13023-016-0408-0

Chatre L, Biard DS, Sarasin A, Ricchetti M. Reversal of mitochondrial defects with CSB-dependent serine protease inhibitors in patient cells of the progeroid Cockayne syndrome. *Proc Natl Acad Sci U S A.* 2015; 112(22): E2910–9. doi: 10.1073/pnas.1422264112

Cleaver JE, Lam ET, Revet I. Disorders of nucleotide excision repair: The genetic and molecular basis of heterogeneity. *Nat Rev Genet* 2009; 10: 756–768. doi: 10.1038/nrg2663

Cockayne EA. Dwarfism with retinal atrophy and deafness. *Arch Dis Child* 1936; 11(61): 1–8.

Gorlin R, Toriello H, Cohen M. *Hereditary Hearing Loss and Its Syndromes.* Oxford University Press, USA, 1995.

Hanawalt PC. Controlling the efficiency of excision repair. *Mut Res* 2001; 485(1): 3–13.

Hashimoto S, Anai H, Hanada K. Mechanisms of interstrand DNA crosslink repair and human disorders. *Genes Environ.* 2016; 38: 9. doi: 10.1186/s41021-016-0037-9

Hoeijmakers, J.H.J. DNA damage, aging, and cancer. *N Engl J Med* 2009; 361: 1475–85.

Kashiyama K, Nakazawa Y, Pilz DT, Guo C, Shimada M, Sasaki K, Fawcett H et al. Malfunction of nuclease ERCC1-XPF results in diverse clinical manifestations and causes Cockayne syndrome, xeroderma pigmentosum, and Fanconi anemia. *Am J Hum Genet.* 2013; 92(5): 807–19. doi: 10.1016/j.ajhg.2013.04.007

Kenyon J, Gerson SL. The role of DNA damage repair in aging of adult stem cells. *Nucl Acids Res*, 2007; 35(22): 7557–7565. doi: 10.1093/nar/gkm1064

Kondo D, Noguchi A, Tamura H, Tsuchida S, Takahashi I, Kubota H, Yano T et al. Elevated urinary levels of 8-hydroxy-2'-deoxyguanosine in a Japanese child of Xeroderma pigmentosum/Cockayne syndrome complex with infantile onset of nephrotic syndrome. *Tohoku J. Exp. Med.*, 2016; 239: 231–235.

Larrea AA, Lujan SA, Kunke TA. SnapShot: DNA mismatch repair. *Cell* 2010: 141. doi: 10.1016/j.cell.2010.05.002

Laugel V. Cockayne syndrome: The expanding clinical and mutational spectrum. *Mech Ageing Dev.* 2013; 134(5–6): 161–170. doi: 10.1016/j.mad.2013.02.006

Modrich P. Mechanisms in eukaryotic mismatch repair. *J Biol Chem*, 2006; 281(41): 30305–30309.

Neill CA, Dingwall MM. A syndrome resembling progeria. A review of two case. *Arch Dis Child* 1950; 25(123): 213–223.

Park Y, Gerson SL. DNA repair defects in stem cell function and aging. *Annu. Rev. Med.* 2005; 56: 495–508. doi: 10.1146/annurev.med.56.082103.104546

Pascucci B, D'Errico M, Romagnoli A, De Nuccio C, Savino M, Pietraforte D, Lanzafame M et al. Overexpression of parkin rescues the defective mitochondrial phenotype and the increased apoptosis of Cockayne Syndrome A cells. *Oncotarget.* 2016. doi: 10.18632/oncotarget.9913

Wang Y, Li F, Zhang G, Kang L, Guan H. Ultraviolet-B induces ERCC6 repression in lens epithelium cells of age-related nuclear cataract through coordinated DNA hypermethylation and histone deacetylation. *Clin Epigenetics.* 2016; 8: 62. doi: 10.1186/s13148-016-0229-y

12 Cancer
The Price for Longevity

Karel Smetana Jr., Barbora Dvořánková, Lukáš Lacina,
Pavol Szabo, Betr Brož, and Aleksi Šedo

CONTENTS

INTRODUCTION: AGING—ITS IMPORTANCE AND DEFINITION

Aging is an essential aspect of the life of all organisms including humans. The populations of developed countries have a higher chance of prolonged life span than their grandparents. This becomes even more apparent in comparison to developing regions. It is necessary to prepare society for the adoption of daily practices reflecting various specific needs of these elderly citizens ranging from public transportation to barrier-free public buildings and accommodation and the necessity for available senior-specific leisure time facilities. Governmental, social, and medical authorities, as well as families, must accumulate adequate funds to cover expenses for retired people (frequently over a period longer than 20 years), provide housing, nursing services, and medical care for old individuals, because of their polymorbidity. This perspective needs to define aging, and understand its biology and the future directions of the demographic development of society.

The explanation of aging is, usually, not concrete. According to Encyclopedia Britannica, aging is defined as a number of progressive physiological changes in the organism that lead to senescence, or a decline of biological functions and of the organism's ability to adapt to metabolic stress (http://www.britannica.com/science/aging-life-process). This and similar definitions are of limited help with regard to the above-listed tasks.

LIMITATION OF HUMAN LIFE SPAN

Life expectancy has increased throughout human history. Data from ancient Egypt predominantly based on paleoanthropological inspection of skeletons indicated that people died between 20 and 30 years of age [1]. Life expectancy of people in Papua New Guinea is between 50 and 60 years of age and it is directly related to distance from the city, where they can seek medical care. Communicable diseases prevail as the main cause of death in these societies [2]. Of note, the situation in western/central Europe was very similar up to the mid-twentieth century [3]. Further socioeconomic development enabled the remarkable increase of life expectancy up to the age of around 80 years with somewhat higher life span in females than in males. It is generally accepted that in almost every country the proportion of people aged over 60 years is growing faster than any other age group. It is interpreted as a result of both longer life expectancy and declining fertility rates. This extensive prolongation of human life raises the question about the limits of our life span. The age over 110 years is extremely rare and mathematical and demographical modeling defined the borderline of our life expectancy at 125 years [4]. The explanation of this trend is not easy and it is influenced by many factors. However, it seems to be that progress in medical technology such as sulfonamides and anti-biotics as well as preventive and interventional cardiology in combination with the wide availability of medical care greatly has participated in this success [5,6]. This medical technology revolution significantly influenced the epidemiological situation in relevant countries. The high incidence of infectious diseases including tuberculosis was reduced and the tendency to decrease also manifests in the mortality of cardiovascular diseases. On the other hand, the outburst of cancer incidence has been detected worldwide, namely in developed countries [6].

AGING AND CANCER

As mentioned above, aging is a physiological process that includes many changes of the human organism such as: neurodegeneration, sensual eye disease (presbyopia, cataract, and macular degen-eration), hearing loss, cognitive decline, dementia, atherosclerosis, sarcopenia, osteoarthritis, osteo-penia, diabetes mellitus, and reduced production of extracellular matrix [7]. It demonstrates that aging is associated with a well-recognized panel of disorders. However, there are also numerous other comorbidities such as cancer associated with the elderly. It is generally understood that while the incidence of malignant diseases is quite low until 50 years of age, their occurrence reaches a peak in old individuals above 75 years of age. After this age, the incidence of cancer is somewhat reduced [8]. For example, every third or even second person in the Czech Republic runs the risk of suffering from cancer during their life [6]. The interpretation of this situation is rather complicated but the high percentage of citizens in developed countries worldwide now a good chance of life expectancy beyond 75 years. This long life expectancy is dependent on multiple aspects of human life, ranging from lifestyle to climate and accessibility of efficient medical care [6]. Collectively, many people run the risk to develop cancer due to their prolonged life span.

LESSON FROM THE FRUIT FLY AND CHIMPANZEE

The phenomenon of aging and its course is genetically determined and it is also environmentally modulated. *Drosophila melanogaster*, a common fruit fly, emerged as an efficient model to eluci-date essential events in the genetic and cellular pathways of aging. It became highly popular due to its short generation time, availability of powerful genetic tools, and functionally conserved anatomy and physiology. Several important cellular aging-related mechanisms and signaling cascades have been deciphered by utilizing this model. In early times [9], for example, several experiments demon-strated the effects of temperature and food on fly longevity. On the other hand, multiple genes were found more recently to expand life spans in *Drosophila melanogaster* [10]. To address the problems of aging in this model, multiple assays were proposed to describe, for example, life expectancy, age-related behavioral changes, reproductive strategy, oxidative stress, etc. Despite notable achieve-ments in this field, this model organism is too far from human in aging various significant aspects.

Mammals are seemingly in many aspects more applicable models for human aging. However, mammals have also evolved an outstanding diversity of aging rates. In rodents [11], highly popu-lar models in biomedical research, maximum life spans range from 4 years in mice to 32 years in naked mole rats. Notably, the incidence of sporadic malignant tumors [12] also differs substan-tially between cancer-prone mice, almost cancer-proof naked mole rats, and blind mole rats. It is recently well-recognized fact that tumors occurring in humans often differ from those in mice. This was clearly documented in the example of Ras-mediated oncogenesis suggesting great variation between different species [13,14].

Further, chimpanzees are genetically extremely similar, almost identical, to humans. Both species have a common ancestor approximately 6 million years ago before they were evolutionarily split [15,16]. The estimation of the life span of a chimpanzee in the wilderness is not easy. One of the rare studies in this field, performed in Gombe, calculated their life expectancy sometime between 15 and 30 years of age [17,18]. In the wilderness, the main causes of their death were trauma and infections. On the other hand, the life span of captive chimpanzees was significantly higher, but never exceeding 60 years of age [19]. These captive chimpanzees also usually die because of infections, but they also suffer from cardiovascular diseases and tumors, more frequently benign than malignant [20]. The incidence of malignant tumors is lower in captive old chimpanzees than in humans and can be influenced by many factors including genetics, although the genetic difference between both species is minimal as mentioned above [21–23]. However, it is possible to speculate that medical care of captive chimpanzees is better than in the jungle but because of its specificity, it is less than that for humans with the longest life span in developed countries. The malignant tumor incidence in 50s, when life expectancy was also between 50 and 60 years of age was quite low as in captive apes. Considering the primary importance of life span in aging research in this model, it is also important to mention that multiple veterinary medical interventions have developed and become broadly available in last two or three decades and thus, their impact cannot be evaluated yet.

LIFE SPAN AS A DEVELOPMENTAL STRATEGY

The process of reproduction and its timing in life represents the fundamental strategy of each organism. Selection experiments [24] in *Drosophila* have revealed that long-lived flies decrease their early reproduction. Using a different selection strategy approach, late-life reproduction was used for identification of lines with increased life span. Notably, long-lived mutant females have reduced fecundity or fertility. However, several long-lived lines, including Indy gene mutants, do not exhibit reduced reproduction in general [25]. In humans, the relationship between longevity and reproduction has been addressed in numerous studies, many of them based on historiographic data. Even in humans, the results have yielded similarly conflicting conclusions.

It was demonstrated that the ability of agametic reproduction can be associated with extreme longevity and high regenerative potential [26]. On the other hand, the increasing complexity of the body construction plane during evolution is generally associated with prolonged life span, but it negatively influences the regenerative potential of the complex organism. This can be well depicted in a comparison of regeneration in flatworms, newts, and humans.

Flatworms [27] are an important model presenting a unique insight into a plethora of reproductive/regenerative strategies. Some species switch seasonally between the two reproductive modes— asexual and sexual. The mechanisms underlying the determination of reproductive strategy and sexual switching in these simple organisms remain enigmatic. However, this can be also correlated to their regenerative capacity.

Considering these phylogenetic and longevity aspects, the incidence of malignant tumors is negligible in flatworms and extremely high in humans [28]. However, these observations require explanation at the molecular level.

BIOLOGY OF AGING RELATION TO CANCER

The human population in Okinawa (Japan) has one of the highest prevalence of centenarians worldwide [29]. These people are genetically quite homogeneously presenting a low expression of anti-inflammatory HLA genes. Widely spread active genes *APOE* and *FOXO3* suggest some genetic background of their longevity. Similar observation was also confirmed in other countries [30]. Interestingly, also long-lived Drosophila lines were prepared by overexpressing the transcription factor *dFOXO* [31] in the adult head fat body.

Wrinkles represent the conspicuous but medically unimportant sign of aging associated with the aberrant function of dermal fibroblasts. *In vitro* experiments with fibroblasts clearly demonstrated that the function of these cells is age-dependent and reflects their genetic alteration [32–35]. At the cellular level, aging is associated with numerous processes that include genomic instability, telomere attrition, epigenetic alterations, qualitative and quantitative changes of protein spectra, cellular senescence-associated secretory phenotype, deregulated nutrient sensing, mitochondrial dysfunction, stem cell exhaustion, and altered intercellular communications [6,36,37].

It is generally assumed that the number of mitotic divisions of each cell is limited by the length of telomeres [38]. However, this limit can be circumvented by a special reprogramming occurring in immortalized stem cells [38]. Biostatistical analysis demonstrated that the number of previous mitotic divisions of adult tissue stem cells reflects the risk of cancer [39]. This hypothesis harmonizes with the assumed role of cancer stem cells in the majority of types of malignant tumors [40]. The increased number of stem cell divisions during the course of a long life span also increases the risk of mutations in the course of the S-phase of the mitotic cycle. However, another type of genetic damage such as DNA breaks also increases with the age of the cell donor [41]. The eukaryotic cells have a powerful protective mechanism responsible for DNA mistake repair. Beside this, cell death is induced in cells once extensive damage repair is not possible. This repair system seems to be less efficient in humans of age approximately 50 years and more [42,43]. The number of mutations acquired during a lifetime is extremely high. It is expected that 3000–13,000 genes per genome can be affected by 5000–50,000 mutations [44]. This number of errors is remarkably high, as the total number of genes in the human genome is estimated to 20,000. But the number of mutations is probably not crucial because even clinically asymptomatic humans accumulate leukemia-associated gene mutation in the elderly [44]. As mentioned above, gene *FOXO3* is active in quite healthy very old individuals including centenarians. This transcription factor is an important component of activation of p53 cascade stimulating the apoptosis in cells with serious DNA damage [45]. Genes from the *FOXO* family also somewhat reduce proliferation of cancer cells and their activity can be minimized by specific micro-RNA. It consequently can stimulate cancer cells growth and migration [46]. The role of FOXO transcription factors in longevity with diminished cancer risk seems to be therefore possible. Furthermore, this concept is supported by the observation that *FOXO* genes activity influence the biorhythms in insects [47]. It is included in the regulation of crosstalk between circadian and metabolic activities in cooperation with the *Clock* gene product [48,49]. From this aspect, the positive effect of caloric restriction for longevity can be explained [50].

The importance of gene repair and proapoptotic machinery for cancer formation in aging and their gradual decrease can be therefore clinically relevant [51–53]. As depicted in Figure 12.1, the harmonization between the decrease of the gene repair activity, accumulation of genetic mutations accompanied by genetic instability, and reduction of fertility is present in normal human ontogeny. These age-dependent alterations clearly preclude the age of high incidence of malignant tumors as also shown in Figure 12.1 [6]. It can be possible that this reduction of gene repair capacity at the beginning of our elderly can be an integral part of our ontogenetic program.

This hypothesis provokes questions about the measurements of personal biological time and the role of a clockmaker. Sensory neurone deprivation influences not only circadian rhythms but also the life span of *Caenorhabditis elegans* [54]. The position of a central chronometer in vertebrates including mammals can be predicted to the suprachiasmatic nucleus of the hypothalamus [55,56] where light exposition influences the expression profile of neurones of this nucleus [57]. However, it must be noted here that the suprachiasmatic nucleus is also sensitive to aging that results in many aging-related disorders [58].

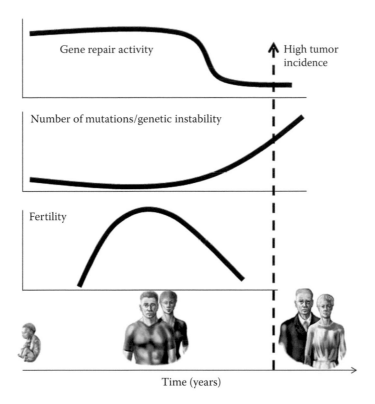

Time (years)

FIGURE 12.1 Comparison of gene repair activity, genetic instability, fertility, and cancer incidence as a function of human aging. (From Smetana K Jr., Lacina L, Szabo P, Dvořánková B, Brož P, Šedo A. *Anticancer Res.* 2016; 36: 5009–5018. With permission of the International Institute of Anticancer Research.)

DISEASES WITH ALTERED GENE REPAIR, AGING, AND CANCER

Gene defect repair is very important for the maintenance of the integrity of the genome. Genetically based failure of this mechanism is usually associated with serious medical problems. Well-recognized examples of this type of disorders are Xeroderma pigmentosum (XP), Werner syndrome, Lynch syndrome, and Fanconi anemia. For example, XP is frequently associated with premature (photo) aging and the occurrence of malignant tumors is significantly higher compared to the normal population [59–62]. Partial failure of gene repair mechanism was also noted in trisomy of the 21st chromosome in Down syndrome. This is associated with increased incidence of myeloid leukemia and also symptoms of accelerated aging [63–65]. The most paradigmatic disease with extremely exaggerated aging is the Hutchinson–Gilford progeria syndrome. Although this disease is caused by a defect of a structural protein, lamin A, it is associated with typical signs of aging including genetic instability also seen in normal not-accelerated aging [66]. The existence of these diseases supports the presented idea about the role of age-dependent failure of gene repair and consequent cancer formation.

Considering all above-presented data in an anthropocentric manner, it seems possible to interpret cancer as a "biologically logical" consequence of the regular aging process. Reaching the culmination point of aging, cancer as a disease of tissue growth regulation and homeostasis becomes the instrument finally executing a programmed termination of its host. On the other hand, specific mutations of particular genes significantly participating in the regulation of aging processes could not only accelerate the senescence, as we see in syndromes of accelerated aging but also, in opposite manner, it could provide an "individual advantage" causing longevity of "successful old people." Biological mechanisms of life span regulations on the individual level thus might have a critical impact on population phenomena and may raise questions for social and even philosophic disciplines, which are more competent to think in categories as a "sense" or "destiny."

FUTURE PERSPECTIVES

Developmental biology traditionally presents the human ontogeny with emphasis on the birth of a child or on body growth. However, aging is another integral component of human ontogeny. This process seems to be influenced by many endogenous as well as exogenous factors. It was described in monozygotic twins that the age of onset, as well as disease occurrence and course, can be quite discordant [67,68]. However, our life span seems to be genetically determined including reduced activity of gene repair that can reflect the high incidence of malignant tumors in the elderly. The activity of genes such as transcription factors of the FOXO family can affect the extreme longevity of centenarians in a pleiotropic manner, influencing several cell-regulated activities such as stress resistance, metabolism, cell cycle arrest, and apoptosis, and probably minimize cancer incidence in the elderly [69].

CONCLUSION

The increased incidence of cancer worldwide, but predominantly in developed countries is associated with aging. Based on demographic data we can expect a tsunami of cancer in the near future and society must be prepared to provide sufficient care for these sick and therapeutically fragile old people.

ACKNOWLEDGMENTS

Part of results presented in this article was obtained through research supported by Charles University (PROGRES Q28 and UNCE 204013), Ministry of Education, Youth and Sports of CR within the National Sustainability Program II (project BIOCEV-FAR Reg. No. LQ1604), and project BIOCEV (CZ.1.05/1.1.00/02.0109).

REFERENCES

1. Parsche F. Paleodemographic and cultural historical studies of skeletal remains of the pre- and early dynasty necropole in Minshat Abu Omar (east Nile delta). *Anthropol Anz.* 1991; 49: 49–64.
2. Kakazo M, Lehmann D, Coakley K, Gratten H, Saleu G, Taime J, Riley ID, Alpers MP. Mortality rates and the utilization of health services during terminal illness in the Asaro Valley, Eastern Highlands Province, Papua New Guinea. *P N G Med J.* 1999; 42: 13–26.
3. Mackenbach JP. Political conditions and life expectancy in Europe, 1900–2008. *Soc Sci Med.* 2013; 82: 134–146.
4. Dong X, Milholland B, Vijg J. Evidence for a limit to human life span. *Nature* 2016; 538: 257–259.
5. Mackenbach JP. Convergence and divergence of life expectancy in Europe: A centennial view. *Eur J Epidemiol.* 2013; 28: 229–240.
6. Smetana K Jr., Lacina L, Szabo P, Dvořánková B, Brož P, Šedo A. Ageing as an important risk factor for cancer. *Anticancer Res.* 2016; 36: 5009–5018.
7. Pitt JN, Kaeberlein M. Why is aging conserved and what can we do about it? *PLoS Biol.* 2015; 13: e1002176.
8. Global Burden of Disease Cancer Collaboration. The Global Burden of Cancer 2013. *JAMA Oncol.* 2015; 1: 505–527.
9. Loeb J, Northrop JH. On the influence of food and temperature upon the duration of life. *J Biol Chem.* 1917; 32:103–121.

10. Rogina B, Reenan RA, Nilsen SP, Helfand SL. Extended life-span conferred by cotransporter gene mutations in Drosophila. *Science* 2000; 290: 2137–2140.
11. Gorbunova V, Seluanov A, Zhang Z, Gladyshev VN, Vijg J. Comparative genetics of longevity and cancer: Insights from long-lived rodents. *Nat Rev Genet.* 2014; 15: 531–540.
12. Azpurua J, Seluanov A. Long-lived cancer-resistant rodents as new model species for cancer research. *Front Genet.* 2013; 3: 319.
13. Balasubramanian P, Longo VD. Aging, nutrient signaling, hematopoietic senescence, and cancer. *Crit Rev Oncog.* 2013; 18: 559–571.
14. Lees H, Walters H, Cox LS. Animal and human models to understand ageing. *Maturitas* 2016; 93: 18–27.
15. Disotell TR. 'Chimpanzee' evolution: The urge to diverge and merge. *Genome Biol.* 2006; 7: 240.
16. Patterson N, Richter DJ, Gnerre S, Lander ES, Reich D. Genetic evidence for complex speciation of humans and chimpanzees. *Nature* 2006; 441: 1103–1108.
17. Hill K, Boesch C, Goodall J, Pusey A, Williams J, Wrangham R. Mortality rates among wild chimpanzees. *J Human Evol.* 2001; 40: 437–450.
18. Terio KS, Kinsel MJ, Raphael J, Mlengeya T, Lipende I, Kirchhoff CA, Gilagiza B, Wilson ML, et al. Pathologic lesions in chimpanzees (*Pan Troglodytes* Schweinfurthii) from Gombe Nationa Park, Tanzania, 2004–2010. *J Zoo Wildlife Med.* 2011; 42: 597–607.
19. Bronikowski AM, Cords M, Alberts SC, Altmann J, Brockman DK, Fedigan LM, Pusey A, Stoinski T, et al. Female and male life tables for seven wild primate species. *Sci Data* 2016; 3: 160006.
20. Strong VJ, Grindlay D Redrobe S, Cobb M, White K. A systematic review of the literature relating to captive great ape morbidity and mortality. *J Zoo Wildlife Med.* 2016; 47: 697–710.
21. Puente XS, Velasco G, Gutiérrez-Fernández A, Bertranpetit J, King M-C, López-Otín C. Comparative analysis of cancer genes in the human and chimpanzee genomes. *BMC Genomics* 2006; 7: 15.
22. Brown SL, Anderson DC, Dick EJ Jr, Guardado-Mendoza R, Garcia AP, Hubbard GB. Neoplasia in the Chimpanzee (*Pan* spp.). *J Med Primatol.* 2009; 38: 137–144.
23. Varki NV, Varki A. On the apparent rarity of epithelial cancers in captive chimpanzees. *Phil. Trans. R. Soc. B* 2016; 370: 20140225.
24. Iliadi KG, Knight D, Boulianne GL. Healthy aging—Insights from *Drosophila. Front Physiol.* 2012; 3: 106.
25. Rogina B, Reenan RA, Nilsen SP, Helfand SL. Extended life-span conferred by cotransporter gene mutations in Drosophila. *Science* 2000; 290: 2137–2140.
26. Bilinski T, Bylak A, Zadrag-Tecza R. Principles of alternative gerontology. *Aging* 2016; 8: 589–602.
27. Nodono H, Ishino Y, Hoshi M, Matsumoto M. Stem cells from innate sexual but not acquired sexual planarians have the capability to form a sexual individual. *Mol Reprod Dev* 2012; 79: 757–766.
28. Smetana K, Jr, Dvořánková B, Lacina L. Phylogeny, regeneration, ageing and cancer: Role of microenvironment and possibility of its therapeutic manipulation. *Folia Biol.* 2013; 59: 207–216.
29. Willcox BJ, Willcox DC, Suzuki M. Demographic, phenotypic, and genetic characteristics of centenarians in Okinawa and Japan: Part 1—centenarians in Okinawa. *Mech Ageing Develop.* 2016; in press.
30. Shadyab AH, LaCroixc AZ. Genetic factors associated with longevity: A review of recent findings. *Ageing Res Rev.* 2015; 19: 1–7.
31. Hwangbo DS, Gershman B, Tu MP, Palmer M, Tatar M. Drosophila dFOXO controls life span and regulates insulin signalling in brain and fat body. *Nature* 2004; 429: 562–566.
32. Kalfalah F, Seggewiß S, Walter R, Tigges J, Moreno-Villanueva M, Bürkle A, Ohse S, Busch H, et al. Structural chromosome abnormalities, increased DNA strand breaks and DNA strand break repair deficiency in dermal fibroblasts from old female human donors. *Aging* 2015; 7: 110–122.
33. Krejčí E, Kodet O, Szabo P, Borský J, Smetana K Jr, Grim M, Dvořánková B. In vitro differences of neonatal and later postnatal keratinocytes and dermal fibroblasts. *Physiol. Res.* 2015; 64: 561–569.
34. Brun C, Jean-Louis F, Oddos T, Bagot M, Bensussan A, Michel L. Phenotypic and functional changes in dermal primary fibroblasts isolated from intrinsically aged human skin. *Exp Dermatol.* 2016; 25: 113–119.
35. Mateu R, Živicová R, Drobná Krejčí E, Grim M, Strnad H, Vlček Č, Kolář M, Lacina L, et al. Functional differences between neonatal and adult fibroblasts and keratinocytes: Donor age affects epithelial-mesenchymal crosstalk in vitro. *Int J Mol Med.* 2016; 38: 1063–1074.
36. López-Otín C, Blasco MA, Partridge L, Serrano M, Kroemer G. The hallmarks of aging. *Cell* 2013; 153: 1194–1217.
37. Burkhalter MD, Rudolph KL, Sperka T. Genome instability of ageing stem cells—Induction and defence mechanisms. *Ageing Res Rev.* 2015; 23: 29–36.

38. MacNeil DE, Bensoussan HJ, Autexier H. Telomerase regulation from beginning to the end. *Genes* 2016; 7: 64. Liu L: Linking telomere regulation to stem cell pluripotency. *Trends Genet* 2017; 33: 16–33.

39. Tomasetti C, Vogelstein B. Cancer etiology. Variation in cancer risk among tissues can be explained by the number of stem cell divisions. *Science* 2015; 347: 78–81.

40. Sell S. Stem cell origin of cancer and differentiation therapy. *Crit Rev Oncol Hematol.* 2004; 51: 1–28.

41. Wang JL, Guo HL, Wang PC, Liu CG. Age-dependent down-regulation of DNA polymerase δ1 in human lymphocytes. *Mol Cell Biochem.* 2012; 371: 157–163.

42. Løhr M, Jensen A, Eriksen L, Grønbæk M, Loft S, Møller P. Association between age and repair of oxidatively damaged DNA in human peripheral blood mononuclear cells. *Mutagenesis* 2015; 30: 695–700.

43. Bavarva JH, Tae H, McIver L, Karunasena E, Garner HR. The dynamic exome: Acquired variants as individuals age. *Aging* 2014; 6: 511–521.

44. McKerrell T, Park N, Moreno T, Grove CS, Ponstingl H, Stephens J, Understanding Society Scientific Group, Crawley C, et al. Leukemia-associated somatic mutations drive distinct patterns of age-related clonal hemopoiesis. *Cell Rep.* 2015; 10: 1239–1245.

45. Matt S, Hofmann TG. The DNA damage-induced cell death response: A roadmap to kill cancer cells. *Cell Mol Life Sci.* 2016; 73: 2829–2850.

46. Coomans de Brachene A, Demoulin J-B. FOXO transcription factors in cancer development and therapy. *Cell Mol Life Sci.* 2016; 73: 1159–1172.

47. Sim C, Denlinger DL. Insulin signaling and FOXO regulate the overwintering diapause of the mosquito Culex pipiens. *Proc Natl Acad Sci U S A.* 2008; 105: 6777–6781.

48. Asher G, Schibler U. Crosstalk between components of circadian and metabolic cycles in mammals. *Cell Metab.* 2011; 13: 125–137.

49. Chaves I, van der Horst GTJ, Schellevis R, Nijman RM, Groot Koerkamp M, Holstege FCP, Smidt MP, Hoekman MFM. Insulin-FOXO3 signaling modulates circadian rhythms via regulation of clock transcription. *Curr Biol.* 2015; 24: 1248–1255.

50. Biliński T, Paszkiewicz T, Zadrag-Tecza R. Energy excess is the main cause of accelerated aging of mammals. *Oncotarget* 2015; 6: 12909–12919.

51. Moraes MC, Neto JB, Menck CF. DNA repair mechanisms protect our genome from carcinogenesis. *Front Biosci (Landmark Ed)* 2012; 17: 1362–1388.

52. Suhasini AN, Brosh RM Jr. DNA helicases associated with genetic instability, cancer, and aging. *Adv Exp Med Biol.* 2013; 767: 123–144.

53. Edifizi D, Schumacher B. Genome instability in development and aging: Insights from nucleotide excision repair in humans, mice, and worms. *Biomolecules* 2015; 5: 1855–1869.

54. Gaglia MM, Jeong DE, Ryu EA, Lee D, Kenyon C, Lee SJ. Genes that act downstream of sensory neurons to influence longevity, dauer formation, and pathogen responses in *Caenorhabditis elegans. PLoS Genet.* 2012; 8: e1003133.

55. Chiang CK, Mehta N, Patel A, Zhang P, Ning Z, Mayne J, Sun WY, Cheng HY, Figeys D. The proteomic landscape of the suprachiasmatic nucleus clock reveals large-scale coordination of key biological processes. *PLoS Genet.* 2014; 10: e1004695.

56. Cornelissen G, Otsuka K. Chronobiology of aging: A mini-review. *Gerontology* 2016; doi: 10.1159/000450945.

57. Park J, Zhu H, O'Sullivan S, Ogunnaike BA, Weaver DR, Schwaber JS, Vadigepalli R. Single-cell transcriptional analysis reveals novel neuronal phenotypes and interaction networks involved in the central circadian clock. *Front Neurosci.* 2016; 10: 481.

58. Farajnia S, Deboer T, Rohling JH, Meijer JH, Michel S. Aging of the suprachiasmatic clock. *Neuroscientist* 2014; 20: 44–55.

59. Cleaver JE. Common pathways for ultraviolet skin carcinogenesis in the repair and replication defective groups of xeroderma pigmentosum. *J Dermatol Sci.* 2000; 23: 1–11.

60. Knoch J, Kamenisch Y, Kubisch C, Berneburg M. Rare hereditary diseases with defects in DNA-repair. *Eur J Dermatol.* 2012; 22: 443–455.

61. Romick-Rosendale LE, Lui VW, Grandis JR, Wells SI. The Fanconi anemia pathway: Repairing the link between DNA damage and squamous cell carcinoma. *Mutat Res.* 2013; 743–744: 78–88.

62. Oshima J, Sidorova JM, Monnat RJ Jr. Werner syndrome: Clinical features pathogenesis and potential therapeutic interventions. *Ageing Res Rev.* 2017; 33: 105–114.

63. Xavier AC, Ge Y, Taub JW. Down syndrome and malignancies: A unique clinical relationship: A paper from the 2008 William Beaumont hospital symposium on molecular pathology. *J Mol Diagn.* 2009; 11: 371–380.

64. Patterson D, Cabelof DC. Down syndrome as a model of DNA polymerase beta haploinsufficiency and accelerated aging. *Mech Ageing Dev.* 2012; 133: 133–137.
65. Necchi D, Pinto A, Tillhon M, Dutto I, Serafini MM, Lanni C, Govoni S, Racchi M, Prosperi E. Defective DNA repair and increased chromatin binding of DNA repair factors in Down syndrome fibroblasts. *Mutat Res.* 2015; 780: 15–23.
66. Gonzalo S, Kreienkamp R. DNA repair defects and genome instability in Hutchinson–Gilford progeria syndrome. *Curr Opin Cell Biol.* 2015; 34: 75–83.
67. Tan T, Christiansen L, Thomassen M, Kruse TA, Christensen K. Twins for epigenetic studies of human aging and development. *Ageing Res Rev.* 2013; 12: 182–187.
68. Castillo-Fernandez JE, Spector TD, Bell JT. Epigenetics of discordant monozygotic twins: Implications for disease. *Genome Med.* 2014; 6: 60.
69. Martins R, Lithgow GJ, Link W. Long live FOXO: Unravelling the role of FOXO proteins in aging and longevity. *Aging Cell* 2016; 15: 196–207.

13 Nodular Thyroid Disease with Aging

Enke Baldini, Salvatore Sorrenti, Antonio Catania,
Francesco Tartaglia, Daniele Pironi, Massimo Vergine,
Massimo Monti, Angelo Filippini, and Salvatore Ulisse

CONTENTS

INTRODUCTION

Thyroid cancer (TC) represents the most common endocrine malignancy and accounts for 2.5% of all malignancies [1,2]. TC incidence has been increasing over the last few decades, mainly due to the improved ability to diagnose malignant transformation in small non-palpable thyroid nodules [3,4]. More than 90% of thyroid carcinomas are represented by the differentiated TC (DTC), papillary TC (PTC), and follicular TC (FTC) histotypes, which following dedifferentiation are thought to generate the poorly DTC (PDTC) and the highly aggressive and invariably fatal anaplastic TC (ATC) accounting for about 1%–2% of all TC (Figure 13.1) [5,6]. Although derived from the same cell type, the different epithelial thyroid tumors show specific histological features, biological behavior, and degree of differentiation because of their different genetic mutations [7].

TCs manifest themselves as thyroid nodules, which are present, on ultrasonography examination, in about 50%–70% of the adult population [8]. However, only about 5% of thyroid nodules harbor malignant lesions. Thus, the first aim in their clinical evaluation is to exclude malignancy. Fine-needle aspiration cytology (FNAC) represents the main diagnostic tool for the evaluation of thyroid nodules showing an overall accuracy of 84%–95% [9]. However, the method shows a major limitation leading to indeterminate results in 10%–15% of cases, in which cytopathologists cannot discriminate between benign and malignant lesions [4,5]. Indetermined cytology remains a gray zone also for clinicians, who have a choice between surgery and medical follow-up. In clinical practice, most of these patients undergo total or subtotal thyroidectomy, even though only about 20% of them harbor a malignant lesion. Over the last few years, cytological classifications have been modified in order to improve the management of patients with indeterminate cytology, and new molecular and ultrasound (US) approaches, detailed below, are making their way into clinical practice.

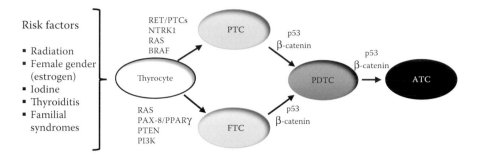

FIGURE 13.1 Risk factors and step model of thyroid carcinogenesis. The model is based on histological, molecular, and clinical features of the different TC histotypes.

RISK FACTORS FOR THYROID CANCER

Established risk factors for TC include radiation exposure, family history of TC, lymphocytic thyroiditis, reduced intake of iodine, and hormonal factors (Figure 13.1) [10].

A clear association between PTC and radiation exposure has been demonstrated following the fallout consequent to the launch of atomic bombs in Hiroshima and Nagasaki at the end of the World War II, or to the nuclear testing in Marshall Islands and in Nevada in the 1950s, as well as following the accident at the nuclear plant of Chernobyl [10]. External beam radiation exposure of the head and neck also increases the risk of PTC development [11].

Although the molecular mechanism remains to be identified, a role for iodine in thyroid carcinogenesis is strongly suggested by epidemiological and experimental observations [10]. Epidemiological data, in fact, demonstrated that in areas characterized by iodine deficiency the incidence of FTC is higher compared to iodine-sufficient regions, where PTC represents the most frequent TC [10,12,13]. Of interest are the observations showing that experimental-induced TC following the administration of *N*-nitrosobis(2-hydroxypropyl)amine (BHP) caused the formation of PTC in animals receiving an adequate iodine diet, while FTC was observed in animals receiving an iodine-deficient diet [14]. This would indicate that iodine might affect TC progression by affecting tumor morphology.

It is well known that chronic inflammation may promote tumor growth [15]. In particular, the presence of infiltrating cells of the immune system is now recognized to represent a generic constituent of human tumors, including TC [16–18]. In this context, it is worth to note that patients affected by chronic lymphocytic thyroiditis present premalignant lesions [19]. In particular, inflammatory cells, including macrophages, mast cells, neutrophils, as well as T and B lymphocytes may have tumor-promoting effects by releasing growth factors (i.e., epidermal growth factor), angiogenic growth factors (i.e., vascular endothelial growth factors and fibroblast growth factor 2), and matrix-degrading enzymes (i.e., matrix metalloproteinase 9, cysteine cathepsin protease, and heparanase) [17]. Consequently, tumor-infiltrating inflammatory cells may support tumor expansion in the primary site and invasion of adjacent tissue and metastatization to distant sites.

The observation that both benign and malignant thyroid tumors is 3–4 times more frequent in females than males have led to hypothesize the role of estrogens in TC promotion [20]. In fact, the estrogen receptor is expressed in thyroid cells, and estrogen has been shown to be a potent growth factor for benign and malignant thyroid cells. Its effects are exerted through the classical genomic and non-genomic pathways. In the latter, the estrogen receptor has been shown to affect the mitogen-activated protein kinases (MAPK) that, as we will see below, plays a primary role in TC pathogenesis [20]. It has to be mentioned, however, that despite the lower frequency, TC in men is usually more aggressive than in women showing a higher mortality, recurrences, extrathyroidal invasion, and distant metastases rates [21].

DTC is generally sporadic. Familial forms are responsible for 3%–6% of cases, and are observed in familial syndromes including: (i) familial adenomatous polyposis, characterized by mutations in the APC (adenomatosis polyposis coli) gene; (ii) Cowden disease, characterized by mutations in the PTEN (phosphatase and tensin homolog) gene; (iii) Werner syndrome, associated with mutations in the WRN (Werner syndrome, RecQ helicase-like) gene; (iv) Carney complex, showing mutations in the PRKAR1A (protein kinase, cAMP-dependent, regulatory subunit type I alpha) gene. In addition, susceptibility gene loci have been identified in other familial syndromes in which PTC is associated with papillary renal carcinomas (1q21), clear-cell renal-cell carcinoma (p14.2;q24.1) [3,10], and multinodular goiter (19p13.2) [10,22,23].

MOLECULAR ALTERATIONS UNDERLYING THYROID CANCER PROGRESSION

Human cancer progression is characterized by the acquisition of molecular alterations capable of conferring novel functional competences to the malignant cells [17]. Genomic instability is thought to be a driving force in the transforming process [17,24–26]. In fact, the number and the frequency of chromosomal abnormalities encountered during TC progression increases from DTC to PDTC and ATC [24,25]. Aberrant activation of the MAPK pathway, radiation, and estrogens are held responsible for the observed genomic instability in TC cells [10,27,28]. Furthermore, dysregulation of the expression and/or activity of key cell cycle regulators are thought to be involved in TC genomic instability [10]. In particular, malignant TC cells are characterized by a deregulated control of the G1/S transition of the cell cycle, due to either an increased expression of promoting factors (i.e., cyclin D1 and E2F), and/or a downregulation or loss-of-function mutations of inhibiting factors (i.e., retinoblastoma, $p^{16INK4A}$, p^{21CIP1}, p^{27KIP1}, and p53) [10]. Abnormal expression of mitotic kinases, such as the polo-like kinase and the Aurora kinase family members (Aurora-A, -B, and -C), regulating the G2/M phase transition and several mitotic processes, are thought to contribute to anomalous cell divisions with consequent generation of aneuploid cancer cells [29–32].

In PTC activating somatic mutations of genes encoding key players of the MAPK signaling pathway, including RET (rearranged during transformation) (RET/PTCs) and NTRK1 (neurotrophic tyrosine kinase receptor 1) gene rearrangements, activating point mutations of the three RAS oncogenes (HRAS, KRAS, and NRAS) and BRAF have been shown to occur with high frequency (70%–80%) (Figure 13.1) [33–41]. In addition, mutations of genes involved in the phosphoinositide 3-kinase (PI3K) pathway (i.e., PTEN, PIK3CA, and AKT1) have been demonstrated to take place at lower frequencies in PTC [41].

Regarding FTC, activating point mutations of the RAS genes are observed in about 45% of cases; rearrangements of the paired-box gene 8 (PAX-8) with the peroxisome proliferator-activated receptor-γ (PAX8-PPARγ) are observed in 35% of FTC; loss of function mutations of the tumor suppressor PTEN gene are encountered in about 10% of FTC; activating mutations or amplification of the PI3KCA gene are found in about 10% of FTC (Figure 13.1) [10,42,43]. PTEN loss-of-function mutations, as occurring in Cowden disease, or activating mutations of RAS or PI3K strongly promote cell proliferation and survival [43].

DTC progression to PDTC and ATC implies TC cell acquisition of novel genetic alterations, which are either absent or present at low frequency in DTC cells (Figure 13.1). These include mutations of the tumor suppressor gene p53, which restrains DTC progression into the more aggressive cancers [26,44–46]. A similar trend is observed with the CTNNB1 gene, which encodes the β-catenin involved in cell adhesion and the wingless signaling pathway [47,48].

The conversion of early-stage TC into more aggressive and invasive malignancies implies epithelial-to-mesenchymal transition (EMT), characterized by the loss of cell–cell contacts, remodeling of cytoskeleton, and the acquisition of a migratory phenotype [49,50]. In this context, reduced expression of E-cadherin and deregulated expression of integrins, Notch, MET, TGFβ, NF-κB, PI3K, TWIST1, matrix metalloproteinases (MMP), components of the urokinase plasminogen-activating system (uPAS) all involved in the EMT, have been demonstrated during TC progression [10,49–54].

EPIDEMIOLOGICAL AND CLINICAL FEATURES OF THYROID CANCERS

As mentioned above, thyroid nodules are present in about 50%–70% of the adult population observed on ultrasonography examination [8]. Thyroid nodules are approximately 4–5 times more common in women than in men, and their frequency increases with age and with decreasing iodine intake [55,56]. In particular, it has been estimated that about 90% of women over 60 years of age and 60% of men over 80 years are affected by nodular thyroid disease [9,57–60]. However, only about 5% of thyroid nodules harbor a malignant lesion [55]; and this percentage increases in geriatric patients. In particular, a retrospective study on 3629 Taiwanese subjects who underwent thyroid surgery showed one peak of TC in patients aged 20–29 years, and a second one in patients older than 65 years [57]. In another study, it was shown that the proportion of malignant nodules was smaller for patients in the fourth decade of life and was greater in patients under 30 or over 60 years [58]. In the latter study, it was also reported that in men of 64 years or over, the odds of cold thyroid nodules harboring a malignant lesion is increased four fold and that in patients over 70 years, more than 50% of nodules are malignant [58]. Moreover, histological types of DTC with less favorable diagnosis were more frequently encountered in the geriatric population [61]. In a recent study, it has been shown that nearly all thyroid malignancies in younger patients were well differentiated, while older patients were more likely to have high-risk PTC variants, PDTC, or ATC [62]. Actually, in patients older than 60 years tumors show increased mitotic activity and likelihood of distant metastasis [63]. Thus, not surprisingly, the TC-specific mortality rate has been shown to be higher in aging patients [64–66]. For this reasons early and accurate diagnosis of thyroid nodules and urgent and aggressive treatment of diagnosed TC are of crucial importance in aging patients.

The diagnostic approaches for thyroid nodules are the same irrespective of a patient's age. As mentioned, FNAC represents the main diagnostic tool for the evaluation of both palpable and non-palpable thyroid nodules [9]. In particular, FNAC-based diagnosis of thyroid nodules shows a sensitivity of 65%–98%, specificity of 72%–100%, and accuracy of 84%–95% [60,66,68]. However, the management of patients with indeterminate or suspicious FNAC specimens remains problematic, the main challenge being the differentiation of nodules requiring surgical treatment from benign ones that can simply be monitored over time [60,66,68]. Over the last few years cytological classifications have been developed in a bid to further improve the management of patients with indeterminate cytology.

The Bethesda System for reporting thyroid cytopathology (BSRTC) classifies the FNAC outcome in six diagnostic categories comprising: (I) non-diagnostic; (II) benign; (III) atypia/follicular lesion of undetermined significance (AUS/FLUS); (IV) follicular neoplasm or suspicious for follicular neoplasm (FN/SFN); (V) suspicious for malignancy (SUSP); (VI) malignant [69,70]. In 2009, the classification of the British Thyroid Association–Royal College of Pathologists (BTA–RCPath) was modified along the lines of the BSRTC [71]. In 2014 the Italian Society for Anatomic Pathology and Cytology together with the Italian Thyroid Association, the Italian Society of Endocrinology, and the Italian Association of Clinical Endocrinologists (SIAPEC 2014) modified the cytological classification of thyroid nodule, replacing its TIR3 class with two new classes (TIR3A and TIR3B), which differ for the risk of malignancy and in the clinical actions required [72]. TIR3A and TIR3B are comparable both to the BSRTC classes III and IV, and to the BTA–RCPath classes Thy 3a and Thy 3f. However, the diagnostic categories 3 and 4 of the BSRTC, Thy 3a and 3f of the BTA–RCPath, and TIR 3A and 3B of the SIAPEC still represent a gray area, in which the encountered cellular atypias are of indeterminate significance and do not permit the discrimination of malignant (i.e., follicular carcinoma and follicular variant of papillary carcinoma) from benign (i.e., follicular adenoma and nodular adenomatous goiter) tumors [60,68]. Below, we will focus on new molecular and US approaches claimed to improve thyroid nodule diagnostic accuracy in these patients.

Total thyroidectomy followed by adjuvant therapy with ^{131}I is the treatment of choice for most patients affected by DTC [9]. Thyroid surgery at an advanced age is usually considered risky for complications yet several studies demonstrated that it can be safely performed in aged patients and that the presence of aggressive forms may justify an aggressive approach [73–79]. Following thyroidectomy and ^{131}I treatment, patients' follow-up includes radioiodine scanning 6–12 months after surgery, periodic US of the thyroid bed and cervical lymph node compartments, measurement of basal and recombinant human TSH-stimulated thyroglobulin serum level [9].

In general, the prognosis of patients with DTC is favorable, with 10-year-survival rate of nearly 90%. However, about 20% of patients face the morbidity of disease recurrence and TC-related deaths [9,80]. It is worthwhile to note that although the overall outcome of women with PTC is similar to men, differences exist in patients younger than 55 years compared to patients older than that [81]. In fact, women under 55 years show a better outcome than men, while among older patients women and men have similar outcomes [82].The worst upshots are usually observed in patients with PDTC or ATC that, as above mentioned, are more frequent in aging patients [26,80]. In particular, ATC has a mean survival time of few months from the diagnosis, which is not influenced by current anticancer treatments, including chemotherapy and radiotherapy. In the majority of these patients, death occurs due to tumor airway obstruction.

EMERGING DIAGNOSTIC APPROACHES FOR THYROID NODULE DIAGNOSIS

As previously stated, FNAC represents the gold standard in the diagnosis of thyroid nodules. According to the BSRTC, the diagnostic categories 3–5 connote a gray zone, in which morphological parameters of the cells do not allow the differentiation of malignant from benign lesions [60,68]. Consequently, the diagnostic category AUS/FLUS has an average cancer risk of 15.9%, FN/SFN of 26.1%, and SUSP of 75.2% [82]. Thus, despite the revision of the diagnostic terminology and morphological criteria adopted in the Bethesda system, a consistent number of patients are still facing needless thyroid surgery. Therefore, the identification of alternative diagnostic procedures to avoid morbidity and costs associated to unnecessary surgery is really needed. This is particularly evident in aged patients where surgery could be problematic due to the frequent presence of comorbidities. In the next sections, new molecular tests and the thyroid imaging reporting and data system (TIRADS) scores, claimed to overcome the diagnostic limits of FNAC, are described.

MOLECULAR-BASED DIAGNOSIS OF THYROID NODULES

Over the last few years, several molecular tests conceived to predict the histological outcome in patients with indeterminate FNAC have been developed. They were designed with the aim to either increase the preoperative detection of malignant nodules (ruling-in approach) or to reduce the number of patients with benign nodules undergoing needless thyroidectomy (ruling-out approach) [83–88]. Among the former are mutational testing and the ThyroSeq v2 next-generation assay. The first, proposed by Nikiforov et al., was aimed to differentiate benign from malignant thyroid nodules based on the detection in FNA samples of the seven most common proto-oncogene mutations found in DTC (RET/PTC1, RET/PTC3, PAX8/PPARγ, BRAF[V600E], HRAS, KRAS, NRAS) [89–92]. In fact, with the exclusion of RAS, mutation tests for each of these genes possess a very high predictive value for malignancy. When verified on a prospective study, the reported positive predictive value ranged from 87% to 95% in the three diagnostic categories with indeterminate cytology of the BSRTC system [87,90]. The false positive results were found for RAS mutation positive samples, showed to be benign lesions on histological examination. Nodules carrying RAS mutations could be, however, considered premalignant lesions and these false positives cases may be considered acceptable. Less acceptable in this test is the high percentage of false-negative results, responsible for a low sensitivity, ranging from 57% to 68% in the BSRTC categories III to IV. However, such findings could be expected since the analyzed mutations are held responsible for only 60%–70% of all DTC. More recently, a new mutational test allowing the simultaneous assay of multiple genetic markers involved in DTC progression has been realized (ThyroSeq v2 from CBLPATH, NY, USA) [93,94]. The test has been demonstrated to possess a sensitivity of 90%, specificity of 93%, positive and negative predictive values of 83% and 96%, respectively, with an overall accuracy of 92% [94]. A similar study performed on FNA samples with indeterminate cytology for the presence of hotspot mutations in 50 genes involved in TC progression showed a sensitivity of 71%, specificity of 89%, positive predictive value of 63%, negative predictive value of 92%, and overall accuracy of 85% [95]. If confirmed in large-scale studies, these findings may lead to a comprehensive genotyping through the next-generation sequencing of FNA samples with indeterminate cytology, which may dramatically improve the diagnostic accuracy and management of the disease.

The rule-out gene-expression classifier (GEC) assay, termed Afirma (from Veracyte Co., San Francisco, USA) was designed to identify benign, rather than malignant, thyroid nodules, with the main aim to avoid useless thyroid surgery in patients with indeterminate cytology [96–98]. The GEC consists of a microarray analyzing the expression of 167 different genes in the total RNA extracted from FNA samples [96–98]. The test employs a multidimensional algorithm to analyze gene-expression data, and divides thyroid FNA into benign or suspicious of malignancy [99]. In a prospective multicenter study, the test showed a high negative predictive value (about 95%) [97]. Therefore, thyroidectomy in patients with benign GEC results may be avoided, and the patients followed in a watchful waiting schedule. On the other hand, a suspicious result from the test is less informative due to the low positive predictive value observed in the AUS/FLUS and FN/SFN categories, respectively, 38% and 37% [97]. However, in the last few years, independent studies were not able to reproduce the abovementioned results using the GEC raising some doubts regarding clinical utility [100–104].

THYROID IMAGING REPORTING AND DATA SYSTEM SCORE

US is of critical importance in thyroid nodules evaluation and for the selection of nodules to submit to FNAC [9]. Despite the presence of several US parameters suggestive or suspicious of malignancy (i.e., hypoechogenicity, presence of microcalcifications, and border irregularity) used to select nodules for FNAC, the need of a better US classification of thyroid nodules emerged [9,105]. In 2009, based on the concepts of the Breast Imaging Reporting Data System (BI-RADS) of the American College of Radiology, two different studies were introduced for the TIRADS score to better stratify the risk of thyroid nodule malignancy [106–108]. In particular, Horvath et al. described 10 US patterns of thyroid nodules associated with different rates of malignancy [107]. In a second study, Park et al. [108] proposed an equation for predicting the probability of malignancy in thyroid nodules based on 12 US features. However, both systems were found difficult to translate into clinical practice. Later on, Kwak et al. proposed a more practical and convenient TIRADS score for the management of thyroid nodules in which risk stratification of thyroid malignancy was based on the number of suspicious US features such as hypoechogenicity, marked hypoechogenicity, microlobulated or irregular margins, microcalcifications, and taller-than-wide shape regarded as independent US features of malignancy [109]. In particular, in the absence of US features of malignancy a TIRADS score of 3 is assigned which corresponds to a 1.7% risk of malignancy; when one suspicious US feature is present (TIRADS score 4a) a 3.3% risk of malignancy is present; when two suspicious US features are present (TIRADS score 4b) the risk of malignancy is 9.2%; in the presence of three or four suspicious US features (TIRADS score 4c) the risk of malignancy increases to 44%–72%; finally when five US features suspicious of malignancy are detected, the risk of malignancy rises up to 88%. Even if the TIRADS score is now considered a powerful tool to select high-risk nodules for FNAC, it was more recently employed in combination with the BSRTC to further stratify the malignancy risk of thyroid nodules with indeterminate results on cytology [110–113]. Maia et al. demonstrated that in patients with diagnosis falling in the III, IV, or V categories, based on the Bethesda classification system, TIRADS assessment may help the clinicians decide between repeating FNAC or submitting patients to surgery. For instance, in case of Bethesda AUS/FLUS cytology and a TIRADS score 3 or 4A, a conservative approach may be suggested, while in those with higher TIRADS scores, thyroidectomy should be considered [110]. Similarly, we recently reported that dichotomizing the TIR 3A and TIR 3B categories of the Italian SIAPEC classification, based on low risk TIRADS score (3, 4a, and 4b) and high-risk TIRADS score (4c and 5), the risk of malignancy for indeterminate lesions could be stratified in three classes of low (below 10%), intermediate (about 20%), and high (about 80%) risk of malignancy [114]. This may point to the use of a conservative approach for the low-risk class and a surgical approach for the high-risk class. For patients with intermediate risk a careful evaluation of risk factors for thyroid malignancy and a close follow-up is recommended. Similar results were also reported by Chng et al. [111] using the BTA–RCPath system. Thus, TIRADS combined with the new cytological classification systems could be used to better stratify TC risk in patients with indeterminate thyroid lesions.

CONCLUSIONS

The prevalence of nodular thyroid disease increases with age. In the elderly, the occurrence of more aggressive histological variants of thyroid carcinoma and higher disease-specific mortality impose an accurate evaluation of thyroid nodules in order to exclude malignancy. At present, our comprehension of the molecular pathways involved in TC progression has considerably increased and this begins now to be translated into clinical practice influencing the clinical management of these patients. In addition, the newly proposed TIRADS score may prove useful in reducing the diagnostic limit still present in the revised cytological classifications of thyroid nodules.

REFERENCES

1. Jemal, A., Siegel, R., Ward, E., Hao, Y., Xu, J., Thun, M.J. 2009. Cancer statistics, 2009. *Ca-A Cancer J Clin* 59:225–49.
2. British Thyroid Association and Royal College of Physician. 2007. Guidelines for the management of thyroid cancer. In *Report of the Thyroid Cancer Guidelines Update Group*, ed. P. Perros. London: Royal College of Physicians. http://www.british-thyroid-association.org/news/Docs/Thyroid_cancer_guidelines_2007.pdf.
3. Davies, L., Welch, H.G.2006. Increasing incidence of thyroid cancer in the United States, 1973–2002. *JAMA* 295:2164–7.
4. SEER Stat Fact Sheets: Thyroid Cancer. Surveillance Epidemiology and End Results (SEER) Program. http://seer.cancer.gov/statfacts/html/thyro.html.
5. Kinder, B.K. 2003. Well differentiated thyroid cancer. *Curr Opin Oncol* 15:71–7.
6. Pasieka, J.L. 2003. Anaplastic thyroid cancer. *Curr Opin Oncol* 15:78–83.
7. Nikiforov, Y.E., Biddinger, P.W., Thompson, L.D.R. 2009. *Diagnostic Pathology and Molecular Genetics of the Thyroid*. Philadelphia: Lippincott Williams & Wilkins.
8. Mehanna, H.M., Jain, A., Morton, R.P., Watkinson, J., Shaha, A. 2009. Investigating the thyroid nodule. *BMJ* 338:b733.
9. American Thyroid Association (ATA) Guidelines Taskforce on Thyroid Nodules and Differentiated Thyroid Cancer, Cooper, D.S., Doherty, G.M. et al. 2009. Revised American Thyroid Association management guidelines for patients with thyroid nodules and differentiated thyroid cancer. *Thyroid* 19:1167–1214.
10. Kondo, T., Ezzat, S., Asa, S.L. 2006. Pathogenic mechanisms in thyroid follicular-cell neoplasia. *Nat Rev Cancer* 6:292–306.
11. Ron, E., Lubin, J.H., Shore, R.E. et al. 1995. Thyroid cancer after exposure to external radiation: A pooled analysis of seven studies. *Radiat Res* 141:259–77.
12. DeLellis, R.A., Lloyd, R.V., Heitz, P.U., Eng, C. 2004. *World Health Organization Classification of Tumors. Pathology and Genetics of Tumor of the Endocrine Organs*. Lyon: IARC Press.
13. Harach, H.R., Escalante, D.A., Day, E.S. 2002. Thyroid cancer and thyroiditis in Salta, Argentina: A 40-yr study in relation to iodine prophylaxis. *Endocr Pathol* 13:175–81.
14. Yamashita, H., Noguchi, S., Murakami, N. et al. 1990. Effects of dietary iodine on chemical induction of thyroid carcinoma. *Acta Pathol Jpn* 40:705–12.
15. Grivennikov, S.I., Greten, F.R., Karin, M. 2010. Immunity, inflammation, and cancer. *Cell* 140:883–9.
16. Prasad, M.L., Huang, Y., Pellegata, N.S., de la Chapelle, A., Kloos, R.T. 2004. Hashimoto's thyroiditis with papillary thyroid carcinoma (PTC)-like nuclear alterations express molecular markers of PTC. *Histopathology* 45:39–46.

17. Hanahan, D., Weinberg, R.A. 2011. Hallmarks of cancer: The next generation. *Cell* 144:646–74.

18. Schreiber, R.D., Old, L.J., Smyth, M.J. 2011. Cancer immunoediting: Integrating immunity's role in cancer suppression and promotion. *Science* 331:1565–8.

19. Gasbarri, A., Sciacchitano, S., Marasco, A. et al. 2004. Detection and molecular characterisation of thyroid cancer precursor lesions in a specific subset of Hashimoto's thyroiditis. *Br J Cancer* 91:1096–104.

20. Derwahl, M., Nicula, D. 2014. Estrogen and its role in thyroid cancers. *Endocr Relat Cancer* 21:T273–83.

21. Sipos, J.A., Mazzaferri, E.L. 2010. Thyroid cancer epidemiology and prognostic variables. *Clin Oncol* 22:395–4.4.

22. Son, E.J., Nosé, V. 2012. Familial follicular cell-derived thyroid carcinoma. *Front Endocrinol* 3:61.

23. Mazeh, H., Benavidez, J., Poehls, J.L., Youngwirth, L., Chen, H., Sippel, R.S. 2012. In patients with thyroid cancer of follicular cell origin, a family history of nonmedullary thyroid cancer in one first-degree relative is associated with more aggressive disease. *Thyroid* 22:3–8.

24. Shahedian, B., Shi, Y., Zou, M. et al. 2001. Thyroid carcinoma is characterized by genomic instability: Evidence from p53 mutations. *Mol Gen Metab* 72:155–63.

25. Wressmann, V.B., Ghossein, R.A., Patel, S.G. et al. 2002. Genome-wide appraisal of thyroid cancer progression. *Am J Pathol* 161:1549–56.

26. Patel, K.N., Shaha, A.R. 2006. Poorly differentiated and anaplastic thyroid cancer. *Cancer Control* 3:119–28.

27. Saavedra, H.I., Knauf, J.A., Shirokawa, J.M. et al. 2000. The RAS oncogene induces genomic instability in thyroid PCCL3 cells via the MAPK pathway. *Oncogene* 19:3948–54.

28. Morgan, W.F., Sowa, M.B.2005. Effects of ionizing radiation in nonirradiated cells. *Proc Natl Acad Sci USA* 102:14127–8.

29. Ito, Y., Miyoshi, E., Sasaki, N. et al. 2004. Polo-like kinase 1 overexpression is an early event in the progression of papillary carcinoma. *Br J Cancer* 90:414–8.

30. Salvatore, G., Nappi, T.C., Salerno, P. et al. 2007. A cell proliferation and chromosomal instability signature in anaplastic thyroid carcinoma. *Cancer Res* 67:10148–58.

31. Ulisse, S., Delcros, J.G., Baldini, E. et al. 2006. Expression of Aurora kinases in human thyroid carcinoma cell lines and tissues. *Int J Cancer* 119:275–82.

32. Baldini, E., D'Armiento, M., Ulisse, S. 2014. A new aurora in anaplastic thyroid cancer therapy. *Int J Endocrinol* 816430.

33. Kimura, E.T., Nikiforova, M.N., Zhu, Z. et al. 2003. High prevalence of BRAF mutations in thyroid cancer: Genetic evidence for constitutive activation of the RET/PTC-RAS-BRAF signaling pathway in papillary thyroid carcinoma. *Cancer Res* 63:1454–7.

34. Cohen, Y., Xing, M., Mambo, E. et al. 2003. BRAF mutation in papillary thyroid carcinoma. *J Natl Cancer Inst* 95:625–7.

35. Suárez, H.G., Du Villard, J.A., Caillou, B. 1988. Detection of activated ras oncogenes in human thyroid carcinomas. *Oncogene* 2:403–6.

36. Lemoine, N.R., Mayall, E.S., Wyllie, F.S. et al. 1988. Activated ras oncogenes in human thyroid cancers. *Cancer Res* 48:4459–63.

37. Pierotti, M.A., Bongarzone, I., Borrello, M.G. et al. 1995. Rearrangements of TRK proto-oncogene in papillary thyroid carcinomas. *J Endocrinol Invest* 18:130–3.

38. Grieco, M., Santoro, M., Berlingieri, M.T. et al. 1990. PTC is a novel rearranged form of the ret proto-oncogene and is frequently detected in vivo in human thyroid papillary carcinomas. *Cell* 60:557–63.

39. Soares, P., Trovisco, V., Rocha, A.S. et al. 2003. BRAF mutations and RET/PTC rearrangements are alternative events in the etiopathogenesis of PTC. *Oncogene* 22:4578–80.

40. Xing, M. 2013. Molecular pathogenesis and mechanisms of thyroid cancer. *Nat Rev Cancer* 13:184–99.

41. The Cancer Genome Atlas Research Network. 2014. Integrated genomic characterization of papillary thyroid carcinoma. *Cell* 159:676–90.

42. Giordano, T.J., Au, A.Y., Kuick, R. et al. 2006. Delineation, functional validation, and bioinformatic evaluation of gene expression in thyroid follicular carcinomas with the PAX8-PPARG translocation. *Clin Cancer Res* 12:1983–93.

43. Omur, O., Baran, Y. 2014. An update on molecular biology of thyroid cancer. *Crit Rev Oncol Hematol* 90:233–52.

44. Pita, J.M., Figueiredo, I.F., Moura, M.M., Leite, V., Cavaco, B.M. 2014. Cell cycle deregulation and TP53 and RAS mutations are major events in poorly differentiated and undifferentiated thyroid carcinomas. *J Clin Endocrinol Metab* 99: E497–507.

45. Ito, T., Seyama, T., Mizuno, T. et al. 1992. Unique association of p53 mutations with undifferentiated but not with differentiated carcinomas of the thyroid gland. *Cancer Res* 52:1369–71.

46. Donghi, R., Longoni, A., Pilotti, S. et al. 1993. Gene p53 mutations are restricted to poorly differentiated and undifferentiated carcinomas of the thyroid gland. *J Clin Invest* 91:1753–60.
47. Garcia-Rostan, G., Camp, R.L., Herrero, A. et al. 2001. Beta-catenin dysregulation in thyroid neoplasms: Down-regulation, aberrant nuclear expression, and CTNNB1 exon 3 mutations are markers for aggressive tumor phenotypes and poor prognosis. *Am J Pathol* 158: 987–96.
48. Miyake, N., Maeta, H., Horie, S. et al. 2001. Absence of mutations in the beta-catenin and adenomatous polyposis coli genes in papillary and follicular thyroid carcinomas. *Pathol Int* 51:680–5.
49. Huber, M.A., Kraut, N., Beug, H. 2005. Molecular requirements for epithelial-mesenchymal transition during tumor progression. *Curr Opin Cell Biol* 17:548–58.
50. Vasko, V., Espinosa, A.V., Scouten, W. et al. 2007. Gene expression and functional evidence of epithelial-to-mesenchymal transition in papillary thyroid carcinoma invasion. *Proc Natl Acad Sci USA* 104:2803–8.
51. Baldini, E., Toller, M., Graziano, F.M. et al. 2004. Expression of matrix metalloproteinases and their specific inhibitors (TIMPs) in normal and different human thyroid tumor cell lines. *Thyroid* 14:881–8.
52. Ulisse, S., Baldini, E., Toller, M. et al. 2006. Differential expression of the components of the plasminogen activating system in human thyroid tumour derived cell lines and papillary carcinomas. *Eur J Cancer* 42:2631–8.
53. Ulisse, S., Baldini, E., Sorrenti, S. et al. 2011. High expression of the urokinase plasminogen activator and its cognate 1 receptor associates with advanced stages and reduced disease-free interval in papillary thyroid carcinoma. *J Clin Endocrinol Metab* 96:504–8.
54. Ulisse, S., Baldini, E., Sorrenti, S. et al. 2012. In papillary thyroid carcinoma BRAFV600E is associated with increased expression of the urokinase plasminogen activator and its cognate receptor, but not with disease-free interval. *Clin Endocrinol* 77:780–6.
55. Burman, K.D., Wartofsky, L. 2015. Thyroid nodules. *N Engl J Med* 373:2347–56.
56. Hegedus, L. 2015. The thyroid nodule. *N Engl J Med* 373:2347–56.
57. Lin, J.D., Chao, T.C., Huang, B.Y. et al. 2005. Thyroid cancer in the thyroid nodules evaluated by ultrasonography and fine-needle aspiration cytology. *Thyroid* 15:708–17.
58. Belfiore, A., La Rosa, G.L., La Porta, G.A. et al. 1992. Cancer risk in patients with cold thyroid nodules: Relevance of iodine intake, sex, age, and multinodularity. *Am J Med* 93:363–9.
59. Cantisani, V., Grazhdani, H., Ricci, P. et al. 2014. Q-elastosonography of solid thyroid nodules: Assessment of diagnostic efficacy and interobserver variability in a large patient cohort. *Eur Radiol* 24:143–50.
60. Gharib, H. 1994. Fine-needle aspiration biopsy of thyroid nodules: Advantages, limitations, and effect. *Mayo Clin Proceed* 69:44–9.
61. Biliotti, G.C., Martini, F., Vezzosi, V. et al. 2006. Specific features of differentiated thyroid carcinoma in patients over 70 years of age. *J Surg Oncol* 93:194–8.
62. Kwong, N., Medici, M., Angell, T.E. et al. 2015. The influence of patient age on thyroid nodule formation, multinodularity, and thyroid cancer risk. *J Clin Endocrinol Metab* 100:4434–40.
63. Toniato, A., Bernardi, C., Piotto, A., Rubello, D., Pelizzo, M.R. 2011. Features of papillary thyroid carcinoma in patients older than 75 years. *Updates Surg* 63:115–8.
64. Park, H.S., Roman, S.A., Sosa, J.A. 2010. Treatment patterns of aging Americans with differentiated thyroid cancer. *Cancer* 116:20–30.
65. Orosco, R.K., Hussain, T., Brumund, K.T. et al. 2015. Analysis of age and disease status as predictor of thyroid cancer-specific mortality using the surveillance, epidemiology, and end results database. *Thyroid* 25:125–32.
66. Rukhman, N., Silverberg, A. 2011. Thyroid cancer in older men. *Aging Male* 14:91–8.
67. Hamburger, J.I. 1994. Diagnosis thyroid nodules by fine needle biopsy: Use and abuse. *J Clin Endocrinol Metab* 79:335–9.
68. Baloch, Z.W., Sack, M.J., Yu, G.H. et al. 1998. Fine-needle aspiration of thyroid: An institutional experience. *Thyroid* 8:565–569.
69. Cibas, E.S., Ali, S.Z. 2009. The Bethesda system for reporting thyroid cytopathology. *Am J Clin Pathol* 132:658–65.
70. Baloch, Z.W., LiVolsi, V.A., Asa, S.L. et al. 2008. Diagnostic terminology and morphologic criteria for cytologic diagnosis of thyroid lesions: A synopsis of the National Cancer Institute Thyroid Fine-Needle Aspiration State of the Science Conference. *Diagn Cytopathol* 36:425–37.
71. Perros, P., Boelaert, K., Colley, S. et al. 2014. British Thyroid Association guidelines for the management of thyroid cancer. *Clin Endocrinol* 81 (Suppl. 1):1–122.

72. Nardi, F., Basolo, F., Crescenzi, A. et al. 2014. Italian consensus for the classification and reporting of thyroid cytology. *J Endocrinol Invest* 37:593–9.

73. Del Rio, P., Sommaruga, L., Bezer, L., Arcuri M.F., Cataldo, S. 2009. Thyroidectomy for differentiated thyroid carcinoma in older patients on a short stay basis. *Acta Biomed* 80:65–8.

74. Gervasi, R. Orlando, G., Lerose, M.A. et al. 2012. Thyroid surgery in geriatric patients: A literature review. *BMC Surg* 12(Suppl. 1): S16.

75. Raffaelli, M., Bellantone, R., Princi, P. et al. 2010. Surgical treatment of thyroid diseases in elderly patients. *Am J Surg* 200:467–472.

76. Mekel, M., Stephen, A.E., Gaz, R.G., Perry, Z.H., Hodin, R.A., Parangi, S. 2009. Thyroid surgery in octogenarians is associated with higher complication rates. *Surg* 146:913–21.

77. Seybt, M.W., Khichi, S., Terris, D.J. 2009. Geriatric thyroidectomy. *Arch Otolaryngol Head Neck Surg* 135:1041–4.

78. Sosa, J.A., Mehta, P.J., Wang, T.S., Boudourakis, L., Roman, S.A. 2008. A population-based study of outcomes from thyroidectomy in aging Americans: At what cost? *J Am Coll Surg* 206:1097–105.

79. Falvo, L., Catania, A., Sorrenti, S. et al. 2004. Prognostic significance of the age factor in thyroid cancer: Statistical analysis. *J Surg Oncol* 88:217–22.

80. Eustatia-Rutten, C.F., Corssmit, E.P., Biermasz, N.R., Pereira, A.M., Romijn, J.A., Smit, J.W. 2006. Survival and death causes in differentiated thyroid carcinoma. *J Clin Endocrinol Metab* 91:313–9.

81. Jonklaas, J., Noguera-Gonzalez, G., Munsell, M. et al. 2012. The impact of age and gender on papillary thyroid cancer survival. *J Clin Endocrinol Metab* 97:E878–87.

82. Bongiovanni, M., Spitale, A., Faquin, W.C., Mazzucchelli, L., Baloch, Z.W. 2012. The Bethesda system for reporting thyroid cytopathology: A meta-analysis. *Acta Cytol* 56:333–9.

83. Xing, M., Haugen, B., Schlumberger, M. 2013. Progress in molecular-based management of differentiated thyroid cancer. *Lancet* 381:1058–69.

84. Eszlinger, M., Hegedüs, L., Paschke, R. 2014. Ruling in or ruling out thyroid malignancy by molecular diagnostics of thyroid nodule. *Best Pract Res Clin Endocrinol Metab* 28:545–57.

85. Nishino, M. 2016. Molecular cytopathology for thyroid nodules: A review of methodology and test performance. *Cancer Cytopathol* 124:14–27.

86. Bernet, V., Hupart, K.H., Parangi, S. et al. 2015. Molecular diagnostic testing of thyroid nodule with indeterminate cytology. *Endocr Pract* 20:360–3.

87. Baldini, E., Tuccilli, C., Prinzi, N. et al. 2013. New molecular approaches in the diagnosis and prognosis of thyroid cancer patients. *Global J Oncol* 1:20–9.

88. Ferris, R.L., Baloch, Z., Bernet, V. et al. 2015. American thyroid association statement on surgical application of molecular profiling for thyroid nodules: Current impact on preoperative decision making. *Thyroid* 25:760–8.

89. Nikiforov, Y.E., Steward, D.L., Robinson-Smith, T.M. et al. 2009. Molecular testing for mutations in improving the fine-needle aspiration diagnosis of thyroid nodules. *J Clin Endocrinol Metab* 94:2092–8.

90. Nikiforov, Y.E., Ohori, N.P., Hodak, S.P. et al. 2011. Impact of mutational testing on the diagnosis and management of patients with cytologically indeterminate thyroid nodules: A prospective analysis of 1056 FNA samples. *J Clin Endocrinol Metab* 96:3390–7.

91. Moses, W., Weng, J., Sansano, I. et al. 2010. Molecular testing for somatic mutations improves the accuracy of thyroid fine-needle aspiration biopsy. *World J Surg* 34:2589–94.

92. Cantara, S., Capezzone, M., Marchisotta, S. et al. 2010. Impact of proto-oncogene mutation detection in cytological specimens from thyroid nodules improves the diagnostic accuracy of cytology. *J Clin Endocrinol Metab* 95:1365–9.

93. Nikiforova, M.N., Wald, A.I., Roy, S. et al. 2013. Targeted next-generation sequencing panel (RhyroSeq) for detection of mutation in thyroid cancer. *J Clin Endocrinol Metab* 98:E1852–60.

94. Nikiforov, Y.E., Carty, S.E., Chiosea, S.I. et al. 2014. Highly accurate diagnosis of cancer in thyroid nodule with follicular neoplasm/suspicious for follicular neoplasm cytology by ThroSeq v2 next-generation sequencing assay. *Cancer* 120:3627–34.

95. Le Mercier, M., D'Haene, N., De Nève, N. et al. 2015.Next-generation sequencing improves the diagnosis of thyroid FNA specimens with indeterminate cytology. *Histopathology* 66:215–24.

96. Chudova, D., Wild, J.I., Wang, E.T. et al. 2010. Molecular classification of thyroid nodules using high-dimensionality genomic data. *J Clin Endocrinol Metab* 95:5296–304.

97. Alexander, E.K., Kennedy, G.C., Baloch, Z.W. et al. 2012. Preoperative diagnosis of benign thyroid nodules with indeterminate cytology. *N Engl J Med* 367:705–15.

98. Walsh, P.S., Wild, J.I., Tom, E.Y. et al. 2012. Analytical performance verification of a molecular diagnostic for cytology-indeterminate thyroid nodules. *J Clin Endocrinol Metab* 97:E2297–306.

99. Faquin, W.C. 2013. Can a gene-expression classifier with high negative predictive value solve the indeterminate thyroid fine-needle aspiration dilemma? *Cancer Cytopathol* 121:116–9.
100. Harrel, R.M., Bimston, D.N. 2014. Surgical utility of Afirma: Effects of high cancer prevalence and oncocytic cell types in patients with indeterminate thyroid cytology. *Endocr Pract* 20:364–369.
101. McIver, B., Castro, M.R., Morris, J.C. et al. 2014. An independent study of a gene expression classifier (Afirma) in the evaluation of cytologically indeterminate thyroid nodules. *J Clin Endocrinol Metab* 99:4069–77.
102. Krane, J.F. 2014. Lessons from early clinical experience with the Afirma gene expression classifier. *Cancer Cytopathol* 122:715–9.
103. Marti, J.L., Avadhani, V., Donatelli, L.A. et al. 2015. Wide inter-institutional variation in performance of a molecular classifier for indeterminate thyroid nodules. *Ann Surg Oncol* 22:3996–4001.
104. Noureldine, S.I., Olson, M.T., Agrawal, N. et al. 2015. Effect of gene expression classifier molecular testing on the surgical decision-making process for patients with thyroid nodules. *JAMA Otolaryngol Head Neck Surg* 141:1082–8.
105. Cantisani, V., Ulisse, S., Guaitoli, E. et al. 2012. Q-elastography in the presurgical diagnosis of thyroid nodules with indeterminate cytology. *PloS One* 7:e50725.
106. American College of Radiology. 2003. *Breast Imaging Reporting and Data System: BI-RADS Atlas.* 4th ed. Reston, VA.
107. Horvath, E., Majlis, S., Rossi, R. et al. 2009. An ultrasonogram reporting system for thyroid nodules stratifying cancer risk for clinical management. *J Clin Endocrinol Metab* 94:1748–51.
108. Park, J.Y., Lee, H.J., Jang, H.W. et al. 2009. A proposal for a thyroid imaging reporting and data system for ultrasound features of thyroid carcinoma. *Thyroid* 19:1257–64.
109. Kwak, J.Y., Han, K.H., Yoon, J.H. 2011. Thyroid imaging reporting and data system for US features: A step in establishing better stratification of cancer risk. *Radiology* 260:892–9.
110. Maia, F.F.R., Matos, P.S., Pavin, E.J., Zantut-Wittmann, D.E. 2015. Thyroid imaging reporting and data system score combined with Bethesda system for malignancy risk stratification in thyroid nodules with indeterminate results on cytology. *Clin Endocrinol* 82:439–44.
111. Chng, C.L., Kurzawinski, T.R., Beale, T. 2015. Value of sonographic features in predicting malignancy in thyroid nodules diagnosed as follicular neoplasm on cytology. *Clin Endocrinol* 83:711–6.
112. Kim, D.W., Lee, E.J., Jung, S.J., Ryu, J.H., Kim, Y.M. 2011. Role of sonographic diagnosis in managing Bethesda class III nodules. *AJNR Am J Neuroradiol* 32:2136–41.
113. Russ, G., Royer, B., Bigorgne, C., Rouxel, A., Bienvenu-Perrard, M., Leenhardt, L. 2013. Prospective evaluation of thyroid imaging reporting and data system on 4550 nodules with and without elastography. *Eur J Endocrinol* 168:649–55.
114. Ulisse, S., Bosco, D., Nardi, F. et al. 2017. Thyroid imaging reporting and data system score combined with the new Italian classification for thyroid cytology improves the clinical management of indeterminate nodules. *Int J Endocrinol* 9692304.

14 HIV and Aging
A Multifaceted Relationship

Edward J. Wing

CONTENTS

INTRODUCTION

Patients living with the human immunodeficiency virus (HIV) no longer die from opportunistic infections and tumors but live successfully with their disease for decades. This stunning medical success is due almost entirely to the development of antiretroviral drugs that suppress but do not eliminate the virus. As a result, the HIV-infected population of approximately 1.3 million people in the United States is aging. Estimates are that 50% of HIV-infected people are now over the age of 50 and by the year 2030, 70% will be over 50. Worldwide, the 35 million people infected with HIV, 80% of whom live in sub-Saharan Africa, are now living longer with the widespread availability of antiretroviral drugs, and will be aging as well.

As people living with HIV age, they face a variety of new challenges, including greater rates of comorbidities such as cardiovascular disease, metabolic conditions, and neurodegenerative disease. They also face the earlier development of geriatric syndromes and frailty, problems with polypharmacy, as well as increasing social isolation, stigma, and depression. Thus, the challenges facing patients with HIV are no longer life-threatening acute illnesses but unique issues related to the aging process [1].

HISTORY

HIV was probably first transmitted to humans in West Africa in the 1930s, and the first case identified in retrospect occurred in 1959. The early history is tragically outlined in The Band Played On by Randy Shilts published in 1987. The disease was first recognized in the United States in 1981 as a fatal disease of young gay men and IV drug abusers; the virus was first identified as the cause of HIV in 1983, and by 1985 there was a serum test to identify infected individuals. In March 1987, the first drug, zidovudine or AZT, was approved by the federal government for treatment of HIV and by 1996 there was widespread use of effective combinations of antiretroviral drugs now termed cART. In the mid-1990s, the incidence of HIV and the death rate from HIV began to fall. Application of cART worldwide saw fewer cases in the world for the first time in 2012. This remarkable reversal of the prognosis of HIV has caused a progressive increase in the age of people living with HIV.

EPIDEMIOLOGY

The average age of the first 1000 cases reported in the United States was 34 [2]. Most patients were classified as intravenous drug users, men who have sex with men, hemophiliacs, or Haitians. Patients typically presented with unusual opportunistic infections such as *Pneumocystis carinii* pneumonia, *Toxoplasma gondii* encephalitis, *Cryptococcus neoformans* meningitis and *Candida albicans* pharyngitis and esophagitis, or tumors such as Kaposi's sarcoma and non-Hodgkin lymphoma. In countries endemic for tuberculosis, HIV-infected patients often presented with severe pulmonary and extra pulmonary disease. While the infections and tumors could be treated in the short term, there was a very high recurrence rate and resulting mortality.

In the mid-1990s, cART became widely available in the United States and other high income countries, and as a result both the incidence and mortality fell dramatically (Figure 14.1). Consequently, the age of people living with HIV rose dramatically. For example, in 1990 in San Francisco, one of the epicenters of the epidemic in the United States, only 10% of HIV+ individuals were over the age of 50. By 2010, 50% were over the age of 50 [3]. Similarly, in New York City, in 2013, 49% of people living with HIV were over the age of 50, and 17% were over the age of 60 (New York City HIV/AIDS Annual Surveillance Statistics 2013. NYC Department of Health and Mental Hygiene). A similar trend was observed in Europe [4]. Less appreciated are data indicating that 18% of newly diagnosed patients are over the age of 50 [5]. Both cART and new diagnoses in the elderly have pushed up the average age of people living with HIV.

The same trends, albeit at lower rates, have occurred in low- and middle- income countries. Since 1995, cART has saved an estimated 14 million lives in these countries, including 9 million in

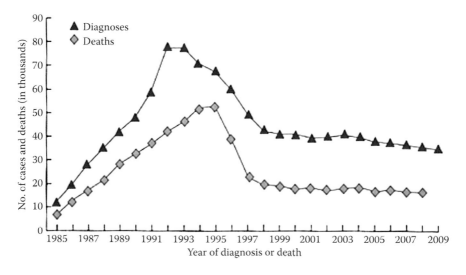

FIGURE 14.1 HIV Diagnoses from 1985 to 2011 in the United States. (Adapted from Dan L. Longo et al., (eds.) 2012, *Harrison's Principles of Internal Medicine*, 18th ed., The McGraw Hill Companies, Inc., 1518.)

sub-Saharan Africa. As a result, an estimated 3.6 million people living with HIV out of 35.6 million people worldwide are over the age of 50 [6].

MORTALITY AND LIFE SPAN

Despite the effectiveness of cART and the increasing longevity of the HIV-infected population, the mortality rate remains 2–7 times that of the general uninfected population [7–11]. A number of factors influence the heterogeneity seen in the HIV mortality rates including chronic viral infections such as hepatitis C, illicit drug and alcohol use, smoking, lifestyle and behavioral factors (e.g., obesity), mental illness, and socioeconomic factors. In addition, adherence to cART, viral suppression, and nadir and current CD4 T cell count influence mortality. Mortality rates appear to be lowest among those individuals who are adherent to cART and have suppressed viral loads and CD4 T cell counts greater than 350–500/mm³. Similarly, calculated life expectancy for HIV+ populations remains only two-thirds that of the uninfected population. As with mortality rates, life expectancy of a subset of patients— HIV+ individuals with suppressed viral loads and preserved CD4 counts above 350—approaches that of the uninfected population [7,12–17].

An important recent article from the Kaiser Permanent Health System using data collected from 1996 to 2011 examined the life expectancy of HIV+ individuals compared to matched uninfected individuals [18]. Although calculated life expectancy rose for HIV+ individuals during this time, it remained significantly less than that for the uninfected population. Even after controlling for risk factors such as hepatitis virus infection and smoking, the HIV+ population had a significantly shorter life expectancy than uninfected individuals. Thus, HIV infection itself, independent of other risk factors, appears to shorten life expectancy.

CHANGES IN CAUSES OF DEATH

With the introduction of cART, the rates of AIDS-associated opportunistic infections including pneumocystis pneumonia, cryptococcal meningitis, toxoplasma encephalitis, and candida infections as well as rates for Kaposi's sarcoma and non-Hodgkin lymphoma fell dramatically; however, the rates never reached the rates of the uninfected population. Data on the causes of death from numerous studies reflected the decreases in the rates of infection and tumors [19–24]. AIDS-related infections and tumors continue to occur primarily in individuals who have not been treated with cART or are non-adherent to medications. Of concern as the patient population ages are the increasing rates of morbidity and mortality caused by comorbidities such as cardiovascular disease, renal disease, liver disease, and non-HIV-related malignancies. Unfortunately, other causes of death, such as suicide, substance abuse, and violence, which reflect the socioeconomic and behavioral characteristics of some HIV+ populations, continue at unacceptably high rates [25].

MORBIDITY IN THE AGING HIV POPULATION

Chronic comorbidities, such as cardiovascular disease, increase steadily in the general population with increasing age. During the cART era, numerous studies have documented that the rates of chronic comorbidities are increased in HIV individuals compared to age-matched controls (Table 14.1). For example, in a cross-sectional study from the VA, published in 2013, the rates of acute myocardial infarction for HIV+ individuals was 1.5 times that of uninfected individuals controlled for common risk factors such as hypertension and smoking [26]. This risk was found in all age groups. Similarly, increased risk for stroke was found in a similar VA population [27].

Chronic renal disease [28], osteoporosis with resulting fractures [29,30], and obesity-related metabolic syndrome and diabetes mellitus [31,32] are also found at increased rates in HIV-infected individuals compared with age-matched controls. Furthermore, HIV-associated neurodegenerative disease (HAND) is commonly identified on neuropsychiatric testing in HIV patients. HAND is usually asymptomatic [33], but approximately 12% of patients have mild and 2% severe symptomatic

TABLE 14.1
Chronic Comorbidities Increased in HIV+ Populations

Cardiovascular disease including stroke
Chronic renal disease
Osteoporosis and fractures
Metabolic syndrome and diabetes mellitus
Neurodegenerative disease and dementia
Malignancies, e.g., liver, lung, and cervical
Liver disease
Bacterial pneumonia and emphysema
Mental illness including depression
Geriatric syndromes including slow gait, difficulties with daily living
  (ADLs), falls, urinary incontinence, sensory deficits, and
  neurocognitive deficits
Frailty
Polypharmacy

neurocognitive disease. As patients with HIV age, the normal age-related decline in neurological function will be compounded by the effect of HIV.

While the rates of HIV-related malignancies such as Kaposi's sarcoma have declined, the rates of lung cancer, related to smoking, and the rates of virally related malignancies such as cervical cancer and hepatocellular carcinoma have increased [28,34,35]. Patients with HIV also are at increased risk for liver disease due to chronic viral hepatitis, alcohol abuse, and fatty liver leading to cirrhosis. Rates of pulmonary disease including pneumococcal pneumonia and emphysema are also associated with chronic HIV infection [36–38]. Finally, patients with HIV also have increased rates of depression and social isolation as they age [39,40].

An important concept in geroscience is geriatric syndromes which are clusters of aging-related illnesses and disabilities that are predictive of morbidity, hospitalization, and mortality. While definitions vary, geriatric syndromes usually include slow gait, difficulties with activities of daily living (ADLs), falls, urinary incontinence, sensory deficits, and neurocognitive deficits. In general, geriatric syndromes occur at earlier ages in HIV+ populations compared to HIV-uninfected populations [41]. A similar and related concept, frailty, is defined by unexplained weight loss, exhaustion, decreased muscle strength, slow gait, and decreased activities. This syndrome predicts poor outcomes and is found at higher rates in HIV+ populations than age-matched controls [42–44].

An important but underappreciated risk factor for aging HIV+ individuals is polypharmacy which can be defined as taking five or more medications. Polypharmacy, while it may not directly affect the aging process, increases the risk of drug side effects and interactions as well as poor adherence. cART consists of at least three drugs, which, when combined with other medications for comorbidities listed above, often results in polypharmacy. In addition to polypharmacy in the elderly, factors such as renal impairment and sarcopenia increase the risk for adverse drug effects. Finally, components of cART can have direct adverse effects. Older nucleoside reverse transcriptase inhibitors and protease inhibitors are associated with hyperlipidemia, lipodystrophy, and insulin resistance. Abacavir has been associated in some studies with increased risk for cardiovascular disease. Tenofovir disoproxil is associated with both renal tubular dysfunction and osteoporosis. Efavirenz, a non-nucleoside reverse transcriptase inhibitor, has significant central nervous system side effects.

MECHANISMS AND CORRELATES OF AGING IN HIV

Aging has been defined as the functional decline over time that occurs in all living things [45]. The mechanisms underlying the process of aging have been intensely investigated in recent years in many different animal models from yeasts to primates, but there is not yet a unifying framework for the process. A number of physiological changes occur, described by Lopez-Otin [45], including altered intercellular communication which includes causes of inflammation, telomere shortening, epigenetic changes, mitochondrial dysfunction, cellular senescence, altered nutrient sensing (GH/IGF-1 pathway), genomic instability, stem cell exhaustion, and abnormal proteostasis. The following discussion will highlight the important features of certain mechanisms as they relate to aging and HIV (Table 14.2).

INFLAMMATION

The aging process is accompanied by increasing levels of inflammation resulting from immune dysregulation, chronic viral infections, and loss of intestinal integrity. During acute HIV infection, inflammatory markers are broadly increased, including markers for T cell and monocyte activation. Chronically infected HIV+ individuals, even when virally suppressed with cART, continue to express evidence of low-level inflammation, similar to aging noninfected individuals. The level of chronic inflammation, however, occurs at much earlier ages in HIV+ compared to aging uninfected individuals [46]. For example, IL-6 has been found to be elevated in HIV+ chronically infected patients on cART and also been linked in many studies to non-AIDS comorbidities and mortality [47–51]. Chronic inflammation is also evidenced by other markers of inflammation, for example, C-reactive protein (CRP) and lipopolysaccharide-induced monocyte activation (soluble CD14 [sCD14]) [52].

HIV results in progressive T cell dysfunction due to destruction of CD4 T cells over time and continuous stimulation of both CD8 T cells and CD4 T cells. During early HIV infection, there is widespread CD4 T cell destruction, including a massive depletion of gut mucosal immune T cells. Partial recovery occurs in the peripheral blood and lymphoid tissue, but gut T cells remain profoundly depleted [53]. With continued turnover and destruction of the CD4 T cell population in HIV+ patients, there is a gradual decrease in absolute numbers and a loss of antigen reactivity. CD8 T cell expansion occurs, even with viral suppression from cART, resulting in an increased CD8:CD4 ratio. With continuing immune stimulation, destruction, and massive cell turnover, there is eventual inability to respond to new antigens. In both aging HIV infected and uninfected individuals, there is an increase in a CD8 T cell phenotype (CD28–, CD57+) that is often directed toward cytomegalovirus (CMV). This phenotype is termed senescent, resists apoptosis, and is proinflammatory (producing inflammatory cytokines such as interferon gamma, tumor necrosis factor alpha, interleukin 1 beta, and interleukin 6) [54]. The expansion and proinflammatory effect of these cells appears to be greater in HIV+ individuals compared to uninfected individuals.

There is increasing evidence for the role of immune dysfunction of monocytes during HIV infection [55]. Chronic immune activation of monocytes can be indicated by the biomarkers IL-6, soluble

TABLE 14.2

Potential Mechanisms of Accelerated Aging in HIV-Infected Patients

Inflammation

Epigenetic changes

Telomere shortening

Mitochondrial dysfunction

Metabolic dysregulation

Abnormal proteostasis

Disease drivers of aging

Resiliency and stress

CD14, and soluble CD163 in serum. Activation is believed to be caused by increased microbial translocation, residual HIV replication, and reactivation of other viruses including CMV. Monocyte activation has been associated with cardiovascular disease and HIV-associated neurocognitive disorder.

One of the principal drivers of immune inflammation in HIV+ patients is low-level HIV replication even when serum viremia is suppressed by cART. Evidence for viral replication in cART-treated patients has recently been based on detecting HIV viral genetic mutations over time [56–58]. More direct evidence at this point is lacking. In addition, reactivation of chronic viral infections, particularly CMV, but also including Epstein–Barr virus and chronic hepatitis virus, can drive the inflammatory response [59,60].

Of considerable importance and scientific investigation is microbial translocation. Owing to the dysfunction in mucosal gut immunity (see above), gut microbes and their products such as lipopolysaccharide are able to penetrate the gut wall and enter the systemic circulation. Stimulation of components of innate immunity, particularly monocytes and macrophages, produces activated cells that secrete inflammatory cytokines [61].

Thus, chronic inflammation is one of the signature mechanisms that potentially drives progressive and accelerated aging in HIV+ patients.

EPIGENETICS

There is increasing interest in identifying specific markers for the aging process, in addition to chronological age. Epigenetic alterations including DNA methylation potentially affect DNA function and have been used to determine physiological age. This technology has been applied recently to HIV populations [62,63]. For example, in one study using six novel DNA methylation data sets, Horvath and Levine measured epigenetic changes in tissue from HIV+ patients [64]. They estimated that the epigenetic age in HIV+ patients was increased by 7.4 years in brain tissue and by 5.2 years in blood tissue compared to uninfected patients.

TELOMERE SHORTENING

Telomeres are tandem repeats at the end of chromosomes necessary for replication. Telomeres shorten with each cell division [65] and telomere length, therefore, has been used as a measure of biological age. Telomere shortening has been criticized as a biomarker for aging because of its modest accuracy and because numerous factors can affect telomere length including stress, smoking, obesity, and other conditions. However, it has been demonstrated in HIV+ patients and the effect appears to occur primarily during acute infection [66]. Length of HIV infection does not appear to affect telomere length.

MITOCHONDRIAL DYSFUNCTION

Mitochondrial function is abnormal in HIV infection and can manifest as a decreased efficiency of oxidative phosphorylation to produce energy (ATP). Direct measurement of oxygen utilization as a measure of cardiovascular fitness in HIV patients was shown to be impaired by 40% at all ages compared to uninfected patients [67]. One potential mechanism for this dysfunction, in addition to HIV infection, is the inhibition of mitochondrial DNA polymerase by cART drugs, particularly a nucleotide reverse transcriptase, stavudine, used early in the epidemic. Other drugs of this class produce a similar but lessened effect on mitochondrial function.

METABOLIC DYSREGULATION

Perturbations in the growth hormone/insulin-like growth factor-1/insulin (GH/IGF-1/I) pathway, including metabolic syndrome and diabetes mellitus, are common in HIV patients [68] and may contribute to accelerated aging [45,69]. Disorders in the GH/IGF-1/I pathway have been associated with traditional risk factors such as obesity (especially visceral obesity) as well as cART (particularly older nucleoside reverse transcriptase inhibitors and protease inhibitors associated with lipodystrophy). Accumulating evidence suggests that HIV itself and the resulting inflammation also contribute to disorders in this pathway, and consequently its effect on aging [68].

ABNORMAL PROTEOSTASIS

HIV may affect proteostasis, the ability of an organism to maintain the balance of cellular and extracellular proteins and peptides by correcting abnormal protein folding or degrading dysfunctional proteins by processes such as autophagy. There is evidence that HIV itself can both inhibit and enhance autophagy in immune cells (T cells and macrophages), depending on the cell type [70,71]. These effects on autophagy are a direct result of HIV viral proteins such as env, vif, and tat. This disruption of the normal protein degradation processes theoretically could affect cellular function and accelerate aging.

DISEASE DRIVERS OF AGING

Aging is strongly correlated with the development of chronic diseases such as cardiovascular disease, renal disease, cancer, and type 2 diabetes mellitus. There is increasing interest in the reverse relationship, that is the effect of chronic disease on the acceleration of the aging process [72,73]. For example, cancer and its treatment have complex and often deleterious effects on the normal aging process. Traditional cancer chemotherapy can affect genomic stability, epigenetic stability, and cellular senescence [73], each of which may adversely affect aging. Cancer chemotherapy has also correlated with decreased telomere length, a marker of advancing age. In addition, there may be inflammatory effects of cancer itself as well as cancer chemotherapy that could accelerate aging. Diabetes mellitus may adversely affect the aging process as discussed above. Drugs used to treat diabetes have been particularly interesting because of their antiaging effects For example, metformin, the standard initial medication for type 2 diabetes, has multiple effects on aging mechanisms and is currently being proposed for clinical trials testing its antiaging potential. Finally, HIV disease itself affects numerous aging processes as discussed above that potentially accelerate the normal aging process.

RESILIENCY

As HIV patients age, many are under increasing stress due to physical disabilities, chronic diseases, social isolation with the loss of partners and friends, and stigma [39,74,75]. In addition, some are faced with financial difficulties and suboptimal living environments with the risk of violence and crime. These multiple factors may result in depression and adverse health outcomes in HIV patients. Stress has been proposed as an aging mechanism in HIV patients. However, in a review of 23 articles examining stress and clinical outcomes in HIV patients, Weinstein and Li found that only a minority of studies demonstrated a significant relationship [76]. Measures of stress varied widely. Additional complexity was added when acute stress, chronic stress, and stress varying over time were taken into account. Other studies have shown that while many factors potentially increase stress on aging HIV+ individuals, those who score high on measures of resiliency (positive adaptation to adverse events) and mastery (cognitive or affective resources that help one develop a sense of self control) deal with both emotional and physical stress more successfully than those who have lower scores. Thus, resiliency and mastery in the face of stress seem to be important factors in successful aging and may be modifiable [77,78].

DOES HIV ACCELERATE AGING?

Whether HIV infection itself results in accelerated aging remains controversial (References 38, 39 IJID) [46,79]. The biological changes, including inflammatory and immunological, increased rates of comorbidities at younger ages, and the shortened calculated life span in HIV populations, even on cART, support accelerated aging. Nonetheless, the lack of accelerated rates of comorbidities with increasing time of HIV infection suggests merely an increased risk, not an acceleration of the underlying biological processes of aging. Clarification on this issue awaits further research into the mechanisms of aging and accumulating epidemiological data

CONCLUSION

The prognosis of patients with HIV infection has gone from dismal to excellent over a span of nearly 40 years. The challenges facing HIV-infected patients have changed from acute infections and tumors to the problems of aging including chronic illnesses associated with aging, geriatric syndromes, and frailty. Accumulating data indicate that underlying biological processes of aging seen in all humans are accelerated in HIV patients. One of the most exciting areas of research will be illuminating the mechanisms of aging in HIV patients and with that understanding developing interventions to slow the processes.

REFERENCES

1. Wing EJ. HIV and aging. *Int J Infect Dis* 2016, 53:61–68.
2. Jaffe HW, Bregman DJ, Selik RM. Acquired immune deficiency syndrome in the United States: The first 1,000 cases. *J Infect Dis* 1983, 148:339–345.
3. O'Keefe KJ, Scheer S, Chen MJ, Hughes AJ, Pipkin S. People fifty years or older now account for the majority of AIDS cases in San Francisco, California, 2010. *AIDS Care* 2013, 25:1145–1148.
4. Mary-Krause M, Grabar S, Lievre L, Abgrall S, Billaud E, Boue F et al. Cohort profile: French hospital database on HIV (FHDH-ANRS CO4). *Int J Epidemiol* 2014, 43:1425–1436.
5. Centers for Disease Control and Prevention. *HIV Surveillance Report* 2015, 27. http://www.cdc.gov/hiv/library/reports/hiv-surveillance.html
6. Global report: UNAIDS report on the global AIDS epidemic 2013. "UNAIDS/JC2502/1/E" – revised and reissued, November 2013.
7. Antiretroviral Therapy Cohort C. Life expectancy of individuals on combination antiretroviral therapy in high-income countries: A collaborative analysis of 14 cohort studies. *Lancet* 2008, 372:293–299.
8. Collaboration of Observational HIVEREiE, Lewden C, Bouteloup V, De Wit S, Sabin C, Mocroft A et al. All-cause mortality in treated HIV-infected adults with CD4 >/=500/mm3 compared with the general population: Evidence from a large European observational cohort collaboration. *Int J Epidemiol* 2012, 41:433–445.
9. Legarth RA, Ahlstrom MG, Kronborg G, Larsen CS, Pedersen C, Pedersen G et al. Long-term mortality in HIV-infected individuals 50 years or older: A nationwide, population-based cohort study. *J Acquir Immune Defic Syndr* 2016, 71:213–218.
10. Lewden C, Chene G, Morlat P, Raffi F, Dupon M, Dellamonica P et al. HIV-infected adults with a CD4 cell count greater than 500 cells/mm3 on long-term combination antiretroviral therapy reach same mortality rates as the general population. *J Acquir Immune Defic Syndr* 2007, 46:72–77.
11. Rodger AJ, Lodwick R, Schechter M, Deeks S, Amin J, Gilson R et al. Mortality in well controlled HIV in the continuous antiretroviral therapy arms of the SMART and ESPRIT trials compared with the general population. *AIDS* 2013, 27:973–979.
12. Lohse N, Hansen AB, Pedersen G, Kronborg G, Gerstoft J, Sorensen HT et al. Survival of persons with and without HIV infection in Denmark, 1995–2005. *Ann Intern Med* 2007, 146:87–95.
13. Losina E, Schackman BR, Sadownik SN, Gebo KA, Walensky RP, Chiosi JJ et al. Racial and sex disparities in life expectancy losses among HIV-infected persons in the United States: Impact of risk behavior, late initiation, and early discontinuation of antiretroviral therapy. *Clin Infect Dis* 2009, 49:1570–1578.

14. May M, Gompels M, Delpech V, Porter K, Post F, Johnson M et al. Impact of late diagnosis and treatment on life expectancy in people with HIV-1: UK Collaborative HIV Cohort (UK CHIC) Study. *Br Med J* 2011, 343:d6016.

15. May MT, Gompels M, Delpech V, Porter K, Orkin C, Kegg S et al. Impact on life expectancy of HIV-1 positive individuals of CD4+ cell count and viral load response to antiretroviral therapy. *AIDS* 2014, 28:1193–1202.

16. Sabin CA. Do people with HIV infection have a normal life expectancy in the era of combination antiretroviral therapy? *BMC Med* 2013, 11:251.

17. Samji H, Cescon A, Hogg RS, Modur SP, Althoff KN, Buchacz K et al. Closing the gap: Increases in life expectancy among treated HIV-positive individuals in the United States and Canada. *PLoS One* 2013, 8:e81355.

18. Marcus JL, Chao CR, Leyden WA, Xu L, Quesenberry CP, Jr., Klein DB et al. Narrowing the gap in life expectancy between HIV-infected and HIV-uninfected individuals with access to care. *J Acquir Immune Defic Syndr* 2016, 73:39–46.

19. Weber R, Ruppik M, Rickenbach M, Spoerri A, Furrer H, Battegay M et al. Decreasing mortality and changing patterns of causes of death in the Swiss HIV Cohort Study. *HIV Med* 2013, 14:195–207.

20. Antiretroviral Therapy Cohort C. Causes of death in HIV-1-infected patients treated with antiretroviral therapy, 1996–2006: Collaborative analysis of 13 HIV cohort studies. *Clin Infect Dis* 2010, 50:1387–1396.

21. French AL, Gawel SH, Hershow R, Benning L, Hessol NA, Levine AM et al. Trends in mortality and causes of death among women with HIV in the United States: A 10-year study. *J Acquir Immune Defic Syndr* 2009, 51:399–406.

22. Lewden C, Salmon D, Morlat P, Bevilacqua S, Jougla E, Bonnet F et al. Causes of death among human immunodeficiency virus (HIV)-infected adults in the era of potent antiretroviral therapy: Emerging role of hepatitis and cancers, persistent role of AIDS. *Int J Epidemiol* 2005, 34:121–130.

23. Morlat P, Roussillon C, Henard S, Salmon D, Bonnet F, Cacoub P et al. Causes of death among HIV-infected patients in France in 2010 (national survey): Trends since 2000. *AIDS* 2014, 28:1181–1191.

24. Smith CJ, Ryom L, Weber R, Morlat P, Pradier C, Reiss P et al. Trends in underlying causes of death in people with HIV from 1999 to 2011 (D:A:D): A multicohort collaboration. *Lancet* 2014, 384:241–248.

25. Goehringer F, Bonnet F, Salmon D, Cacoub P, Paye A, Chene G et al. Causes of death in HIV-infected individuals with immunovirologic success in a National prospective survey. *AIDS Res Hum Retroviruses* 2017, 33:187–193.

26. Freiberg MS, Chang CC, Kuller LH, Skanderson M, Lowy E, Kraemer KL et al. HIV infection and the risk of acute myocardial infarction. *JAMA Intern Med* 2013, 173:614–622.

27. Sico JJ, Chang CC, So-Armah K, Justice AC, Hylek E, Skanderson M et al. HIV status and the risk of ischemic stroke among men. *Neurology* 2015, 84:1933–1940.

28. Rasmussen LD, May MT, Kronborg G, Larsen CS, Pedersen C, Gerstoft J et al. Time trends for risk of severe age-related diseases in individuals with and without HIV infection in Denmark: A nationwide population-based cohort study. *Lancet HIV* 2015, 2:e288–e298.

29. Kooij KW, Wit FW, Bisschop PH, Schouten J, Stolte IG, Prins M et al. Low bone mineral density in patients with well-suppressed HIV infection: Association with body weight, smoking, and prior advanced HIV disease. *J Infect Dis* 2015, 211:539–548.

30. Prieto-Alhambra D, Guerri-Fernandez R, De Vries F, Lalmohamed A, Bazelier M, Starup-Linde J et al. HIV infection and its association with an excess risk of clinical fractures: A nationwide case-control study. *J Acquir Immune Defic Syndr* 2014, 66:90–95.

31. Becofsky KM, Wing RR, Richards KE, Gillani FS. Obesity prevalence and related risk of comorbidities among HIV+ patients attending a New England ambulatory centre. *Obes Sci Prac* 2016, 2:123–127.

32. Hernandez-Romieu AC, Garg S, Rosenberg ES, Thompson-Paul AM, Skarbinski J. Is diabetes prevalence higher among HIV-infected individuals compared with the general population? Evidence from MMP and NHANES 2009–2010. *BMJ Open Diabetes Res Care* 2017, 5:e000304.

33. Heaton RK, Clifford DB, Franklin DR, Jr., Woods SP, Ake C, Vaida F et al. HIV-associated neurocognitive disorders persist in the era of potent antiretroviral therapy: CHARTER study. *Neurology* 2010, 75:2087–2096.

34. Silverberg MJ, Lau B, Achenbach CJ, Jing Y, Althoff KN, D'Souza G et al. Cumulative incidence of cancer among persons with HIV in North America: A cohort study. *Ann Intern Med* 2015, 163:507–518.

35. Yanik EL, Katki HA, Engels EA. Cancer risk among the HIV-infected elderly in the United States. *AIDS* 2016, 30:1663–1668.

36. Attia EF, Akgun KM, Wongtrakool C, Goetz MB, Rodriguez-Barradas MC, Rimland D et al. Increased risk of radiographic emphysema in HIV is associated with elevated soluble CD14 and nadir CD4. *Chest* 2014, 146:1543–1553.

37. Marcus JL, Baxter R, Leyden WA, Muthulingam D, Yee A, Horberg MA et al. Invasive pneumococcal disease among HIV-infected and HIV-uninfected adults in a large integrated healthcare system. *AIDS Patient Care STDS* 2016, 30:463–470.

38. Almeida A, Boattini M. Community-acquired pneumonia in HIV-positive patients: An update on etiologies, epidemiology and management. *Curr Infect Dis Rep* 2017, 19:2.

39. Masten J. 'A shrinking kind of life': Gay Men's experience of aging with HIV. *J Gerontol Soc Work* 2015, 58:319–337.

40. Pinquart M, Duberstein PR, Lyness JM. Effects of psychotherapy and other behavioral interventions on clinically depressed older adults: A meta-analysis. *Aging Ment Health* 2007, 11:645–657.

41. Greene M, Covinsky KE, Valcour V, Miao Y, Madamba J, Lampiris H et al. Geriatric syndromes in older HIV-infected adults. *J Acquir Immune Defic Syndr* 2015, 69:161–167.

42. Brothers TD, Kirkland S, Guaraldi G, Falutz J, Theou O, Johnston BL et al. Frailty in people aging with human immunodeficiency virus (HIV) infection. *J Infect Dis* 2014, 210:1170–1179.

43. Desquilbet L, Jacobson LP, Fried LP, Phair JP, Jamieson BD, Holloway M et al. HIV-1 infection is associated with an earlier occurrence of a phenotype related to frailty. *J Gerontol A Biol Sci Med Sci* 2007, 62:1279–1286.

44. Levett TJ, Cresswell FV, Malik MA, Fisher M, Wright J. Systematic review of prevalence and predictors of frailty in individuals with human immunodeficiency virus. *J Am Geriatr Soc* 2016, 64:1006–1014.

45. Lopez-Otin C, Blasco MA, Partridge L, Serrano M, Kroemer G. The hallmarks of aging. *Cell* 2013, 153:1194–1217.

46. Pathai S, Bajillan H, Landay AL, High KP. Is HIV a model of accelerated or accentuated aging? *J Gerontol A Biol Sci Med Sci* 2014, 69:833–842.

47. Borges AH, O'Connor JL, Phillips AN, Ronsholt FF, Pett S, Vjecha MJ et al. Factors associated with plasma IL-6 levels during HIV infection. *J Infect Dis* 2015, 212:585–595.

48. Nixon DE, Landay AL. Biomarkers of immune dysfunction in HIV. *Curr Opin HIV AIDS* 2010, 5:498–503.

49. Nordell AD, McKenna M, Borges AH, Duprez D, Neuhaus J, Neaton JD et al. Severity of cardiovascular disease outcomes among patients with HIV is related to markers of inflammation and coagulation. *J Am Heart Assoc* 2014, 3:e000844.

50. So-Armah KA, Tate JP, Chang CC, Butt AA, Gerschenson M, Gibert CL et al. Do biomarkers of inflammation, monocyte activation, and altered coagulation explain excess mortality between HIV infected and uninfected people? *J Acquir Immune Defic Syndr* 2016, 72:206–213.

51. Hunt PW, Lee SA, Siedner MJ. Immunologic biomarkers, morbidity, and mortality in treated HIV infection. *J Infect Dis* 2016, 214(Suppl 2):S44–S50.

52. Sereti I, Krebs SJ, Phanuphak N, Fletcher JL, Slike B, Pinyakorn S et al. Persistent, albeit reduced, chronic inflammation in persons starting antiretroviral therapy in acute HIV infection. *Clin Infect Dis* 2017, 64:124–131.

53. Mudd JC, Brenchley JM. Gut mucosal barrier dysfunction, microbial dysbiosis, and their role in HIV-1 disease progression. *J Infect Dis* 2016, 214(Suppl 2):S58–S66.

54. Gianella S, Letendre S. Cytomegalovirus and HIV: A dangerous Pas de Deux. *J Infect Dis* 2016, 214(Suppl 2):S67–S74.

55. Anzinger JJ, Butterfield TR, Angelovich TA, Crowe SM, Palmer CS. Monocytes as regulators of inflammation and HIV-related comorbidities during cART. *J Immunol Res* 2014, 2014:569819.

56. Dampier W, Nonnemacher MR, Mell J, Earl J, Ehrlich GD, Pirrone V et al. HIV-1 genetic variation resulting in the development of new quasispecies continues to be encountered in the peripheral blood of well-suppressed patients. *PLoS One* 2016, 11:e0155382.

57. Lorenzo-Redondo R, Fryer HR, Bedford T, Kim EY, Archer J, Kosakovsky Pond SL et al. Persistent HIV-1 replication maintains the tissue reservoir during therapy. *Nature* 2016, 530:51–56.

58. Martinez-Picado J, Deeks SG. Persistent HIV-1 replication during antiretroviral therapy. *Curr Opin HIV AIDS* 2016, 11:417–423.

59. Naeger DM, Martin JN, Sinclair E, Hunt PW, Bangsberg DR, Hecht F et al. Cytomegalovirus-specific T cells persist at very high levels during long-term antiretroviral treatment of HIV disease. *PLoS One* 2010, 5:e8886.

60. Freeman ML, Lederman MM, Gianella S. Partners in crime: The role of CMV in immune dysregulation and clinical outcome during HIV infection. *Curr HIV/AIDS Rep* 2016, 13:10–19.

61. Dinh DM, Volpe GE, Duffalo C, Bhalchandra S, Tai AK, Kane AV et al. Intestinal microbiota, microbial translocation, and systemic inflammation in chronic HIV infection. *J Infect Dis* 2015, 211:19–27.
62. Gross AM, Jaeger PA, Kreisberg JF, Licon K, Jepsen KL, Khosroheidari M et al. Methylome-wide analysis of chronic HIV infection reveals five-year increase in biological age and epigenetic targeting of HLA. *Mol Cell* 2016, 62:157–168.
63. Rickabaugh TM, Baxter RM, Sehl M, Sinsheimer JS, Hultin PM, Hultin LE et al. Acceleration of age-associated methylation patterns in HIV-1-infected adults. *PLoS One* 2015, 10:e0119201.
64. Horvath S, Levine AJ. HIV-1 infection accelerates age according to the epigenetic clock. *J Infect Dis* 2015, 212:1563–1573.
65. Mason PJ, Perdigones N. Telomere biology and translational research. *Transl Res* 2013, 162:333–342.
66. Zanet DL, Thorne A, Singer J, Maan EJ, Sattha B, Le Campion A et al. Association between short leukocyte telomere length and HIV infection in a cohort study: No evidence of a relationship with antiretroviral therapy. *Clin Infect Dis* 2014, 58:1322–1332.
67. Oursler KK, Sorkin JD, Smith BA, Katzel LI. Reduced aerobic capacity and physical functioning in older HIV-infected men. *AIDS Res Hum Retroviruses* 2006, 22:1113–1121.
68. Lake JE, Currier JS. Metabolic disease in HIV infection. *Lancet Infect Dis* 2013, 13:964–975.
69. Lopez-Otin C, Galluzzi L, Freije JM, Madeo F, Kroemer G. Metabolic control of longevity. *Cell* 2016, 166:802–821.
70. Dinkins C, Pilli M, Kehrl JH. Roles of autophagy in HIV infection. *Immunol Cell Biol* 2015, 93:11–17.
71. Espert L, Beaumelle B, Vergne I. Autophagy in *Mycobacterium tuberculosis* and HIV infections. *Front Cell Infect Microbiol* 2015, 5:49.
72. Hodes RJ, Sierra F, Austad SN, Epel E, Neigh GN, Erlandson KM et al. Disease drivers of aging. *Ann N Y Acad Sci* 2016, 1386:45–68.
73. Kohanski RA, Deeks SG, Gravekamp C, Halter JB, High K, Hurria A et al. Reverse geroscience: How does exposure to early diseases accelerate the age-related decline in health? *Ann N Y Acad Sci* 2016, 1386:30–44.
74. Vincent W, Fang X, Calabrese SK, Heckman TG, Sikkema KJ, Hansen NB. HIV-related shame and health-related quality of life among older, HIV-positive adults. *J Behav Med* 2016, 40:434–444.
75. Moore RC, Moore DJ, Thompson WK, Vahia IV, Grant I, Jeste DV. A case-controlled study of successful aging in older HIV-infected adults. *J Clin Psychiatry* 2013, 74:e417–e423.
76. Weinstein TL, Li X. The relationship between stress and clinical outcomes for persons living with HIV/AIDS: A systematic review of the global literature. *AIDS Care* 2016, 28:160–169.
77. Fang X, Vincent W, Calabrese SK, Heckman TG, Sikkema KJ, Humphries DL et al. Resilience, stress, and life quality in older adults living with HIV/AIDS. *Aging Ment Health* 2015, 19:1015–1021.
78. Emlet CA, Shiu C, Kim HJ, Fredriksen-Goldsen K. Bouncing back: Resilience and mastery among HIV-positive older gay and bisexual men. *Gerontologist* 2017, 57:S40–S49.
79. Justice A, Falutz J. Aging and HIV: An evolving understanding. *Curr Opin HIV AIDS* 2014, 9:291–293.

15 Maturation, Barrier Function, Aging, and Breakdown of the Blood–Brain Barrier

Elizabeth de Lange, Ágnes Bajza, Péter Imre,
Attila Csorba, László Dénes, and Franciska Erdő

CONTENTS

BLOOD–BRAIN BARRIER DEVELOPMENT DURING EMBRYOGENESIS

The blood–brain barrier (BBB) develops early during brain development (Saunders et al. 2008). The genetic program of embryogenesis coordinates the development of the central nervous system (CNS) including the development, maturation, and complex composition of BBB. This main stream of the process is characterized by a strong construction of the brain with neuronal cells surrounded with vasculature. The development of vasculature is critical for establishment of the vessel network and maturation events. Nutrition is a part of embryogenesis and neonatal development of CNS and BBB. It is very difficult to separate the different phases of embryogenesis, because the processes run in parallel. Early prenatal period can be divided into four stages: blastulation, gastrulation, neurulation, and articulation (Erich Blechschmit 2004, Mason and Price 2016). The vascular progenitor cells appear at a very early stage of neurulation in the prenatal period (E7.5 in mice and E20 in humans) (Table 15.1).

The extra- and endo-embryonic tissues are separated first in the blastulation phase, and the germ layers (ectoderm, mesoderm, and endoderm) are then formed in the gastrulation period. The next step is the development of the ectoderm–neuro–endoderm surfaces (neurulation), and finally the articulation of tissues can be observed.

Following gastrulation, the ectoderm gives rise to epithelial and neural tissues. Neurulation was detected in mice on day E7.5 and in humans on E18, while on day E8 in mice and E20 in human tissue the separation occurs.

The population of early dividing neuroepithelial cells quickly transforms and diversifies. The new cell types are radial glial cells (RGCs). RGCs also generate other types of progenitors, neurons, and glial cells. In mice, the production of cortical neurons begins at about 10 days and continues for

TABLE 15.1
Early Prenatal Development after Fertilization

Stages	Zygote	Blastocyst	Histogenesis			Neurulation
			Gastrulation			
			Ectoderm	Mesoderm	Endoderm	
			Skin	Heart	Lungs	
			Nervous system	Circulation	Intestine	
			Connective tissue	Bones	Thyroid	
				Muscles	Pancreas	
				Kidneys	Bladder	
			Implantation			
Mice/day			6			7.5–8
Human/day			16			18–20

about 8 days. In humans, cortical neurogenesis occurs over many weeks, starting at about 35 days, and the cell cycle times can be up to five times longer than in rodents (Erich Blechschmit 2004).

The permeability of the BBB is rather high in the early developmental phases, but later on it becomes a more compact structure protecting the brain against xenobiotics and toxic agents. The timeline of BBB development during embryonic prenatal phases and postnatal days in mice is shown in Figure 15.1.

Restriction of paracellular and transcellular transport of agents is accomplished by elimination of endothelial fenestrae and pinocytosis. This is accompanied by the formation of a continuous endothelial monolayer connected by tight junctions (TJs) and creation of highly selective endothelial transport systems in the perinatal period during embryogenesis. Establishment of specialized perivascular structures, including the basement membrane (BM) and the coverage of the endothelial capillary wall by pericytes and astrocytic end feet is the main stream of the postnatal period of BBB embryogenesis (E = embryonic days; P = postnatal days).

MATURATION IN THE PRENATAL PERIOD

Endothelial cells (ECs) and pericytes both contribute to the BM development and induction of the expression of integrins. The EC-derived factors, platelet-derived growth factor-BB (PDGF-BB) and heparin binding-epidermal growth factor (HB-EGF) support growth, and they are critical in tube maturation. The new vessels are formed from preexisting vessels which sprout into the embryonic neuroectoderm. The early vessels have TJs, transporters, transcytotic vesicles, and leukocyte adhesion molecules. The mature BBB vessels come into close contact with cells of the neurovascular units (NVUs) (pericytes, astroglia, neurons), transcytosis is decreased, the regulated efflux transport increased (Obermeier et al. 2013), and maturation is completed.

Vascular endothelial growth factor (VEGF) has a fundamental role in embryonic angiogenesis including VEGF receptor 2 (VEGFr-2); fetal liver kinase 1 (Flk-1); and kinase insert domain receptor (Kdr). Activation of the PI3K-AKT/PKB and similarly the p38/MAPK-HSP27 pathways supports EC survival and promotes EC migration (Olsson et al. 2006, Jiang and Nardelli 2016).

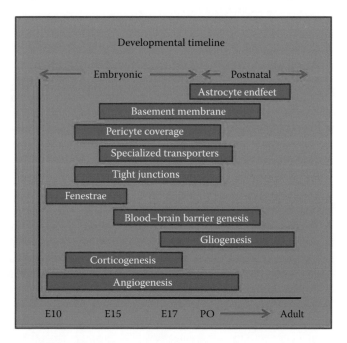

FIGURE 15.1 The timeline of development of the BBB during the embryonic prenatal and postnatal days in mice. (Modified from Zhao, Z. et al. 2015. *Cell* 163(5): 1064–1078.)

During embryogenesis, the Wnt/β-catenin pathway is activated in the CNS ECs and therefore, it drives angiogenesis specifically in the CNS. Wnt ligands inhibit the degradation of β-catenin in the proteasome. Downstream signaling element of β-catenin in ECs results in failed vascularization of the CNS (Daneman et al. 2009). Wnt induces the expression of different genes, including nutrient transporters. The same signal that drives EC migration into the CNS also induces BBB functions. Endothelial β-catenin also regulates the formation of TJs so it has a key role in embryonic and postnatal BBB maturation (Liebner et al. 2008).

GPR124, a member of the G protein-coupled receptor family, known as tumor endothelial marker 5 (TEM5), has a pivotal role in the brain-specific angiogenesis (Anderson et al. 2011). The phenotype is characterized by impaired EC survival, growth, and migration, which result in an eligible vascular sprout to embryonic neuroectoderm. GPR124 seems to act independently from VEGF in vessel sprouting. Expression of VEGFr was unaffected in the absence of GPR124 (Cullen et al. 2011).

BBB formation in the embryonic stage is regulated together with other factors by an important protein, Sonic hedgehog (SHH). SHH knockout mice exhibit embryonic lethality between E11 and E13.5. Their phenotype is associated with abnormalities in BBB formation with decreased expression of TJ proteins, occludin and claudin-5. Moreover, when smoothened (Smo), a downstream signaling protein was selectively deleted from ECs this resulted in lowering the TJ protein expression associated with vessel leakage. These data suggest that SHH is required for the maturation of the BBB vessels. SHH upregulates TJ protein expression in human BBB ECs and decreases the permeability, indicating that SHH could also have a role in maintaining BBB functions (Alvarez et al. 2011a, b). SHH can induce VEGF and angiopoietin (Ang) expression, which are strong angiogenic factors (Pola et al. 2001).

The infant brain is vulnerable to neurotoxic substances connecting partly to the immature BBB. Neonatal BBB cells have lower barrier and p-glycoprotein (P-gp) functions than adult BBB cells and are well associated with the lower expressions of barrier-related proteins and the age-dependent BBB permeability of drugs (Takata et al. 2013).

INTERACTIONS WITHIN THE NEUROVASCULAR UNIT

Two astrocyte markers, the glial water channel aquaporin-4 (AQP4) and the glial fibrillary acidic protein (GFAP), have been implicated in several physiological and pathological conditions in the CNS as well as in BBB breakdown. AQP4 and GFAP increase expression in the cerebellum of neonate (14-day-old) and adult (8-week-old) rats regulating the age- and time-related water/electrolyte balance (Stavale et al. 2013). The permeability of water and low/high molecular-mass tracers is connected to the activity of pericytes and regulated by the endothelial and astrocyte collaborations (Obermeier et al. 2013). The main activity of the postnatal period remains development and reproduction. Among the capillary network of the neural tissue shows the highest pericyte coverage (Armulik et al. 2011), suggesting that pericytes have an important role in the BBB maturation and functions. They are also responsible for performing vessel stability, regulation of blood flow, and control BBB integrity and function (Winkler et al. 2011). Pericytes are a heterogeneous and dynamic cell population whose expression of surface markers varies according to cell differentiation and the tissue where they reside (Sa-Pereira et al. 2012). When pericytes have proliferated and directed to sprouting vessels, adhesion between ECs and pericytes is mediated by the transforming growth factor-β (TGF-β). Both cell types secrete TGF-β and express its receptor TGF-βR2 (Winkler et al. 2011). TGF-β signaling in pericytes initiates production of extracellular matrix (ECM) molecules whereas TGF-β signaling in ECs promotes pericyte adhesion by upregulating Cadherin-2 (N-cadherin) (Winkler et al. 2011). Pericytes may induce astrocytic foot processes to the endothelial tube, which subsequently initiates proper end-foot polarization.

Astroglial perivascular connectivity occurs and develops during postnatal BBB maturation (days 2–20). The absence of astroglial connexins (Cx30 and Cx43) weakens the BBB, which opens upon increased hydrostatic vascular pressure and shear stress, demonstrating that astroglial connexins are necessary to maintain BBB integrity (Ezan et al. 2012, Elahy et al. 2015). Perivascular astrocyte end feet have intramembranous particles as water channel AQP4 and the adenosine triphosphate (ATP)-sensitive inward rectifier potassium channel Kir4 (Kostic et al. 2013). The astrocyte lineages influence the BBB phenotype of the cerebral endothelium. Astrocyte precursors release soluble factors (IL-6, glial cell-derived neurotrophic factor [GDNF], FGF-2) that determine the fate of cerebral vascular ECs and define an elevated expression of transporters and increased TJ formation (Obermeier et al. 2013, Elahy et al. 2015). Astrocytes also produce the cholesterol and phospholipid transporter molecule apolipoprotein E (APOE), which mediates regulatory processes related to brain homeostasis (Gee and Keller 2005).

BM is essential for proper BBB and astrocytes and pericytes contribute to the formation of BM. The BM is mostly made up of collagen type IV, laminin, and fibronectin, and also contains cell adhesion molecules and immobilized signaling proteins (Obermeier et al. 2013). At the level of the post-capillary venue, there are two distinct BMs: endothelial and parenchymal. They have distinct compositions of ECM molecules, which determine their different functions and regulate their intercellular crosstalk.

Interactions between the BMs and their associated cells are enabled by two types of matrix transmembrane receptors: dystroglycan and integrins and their extracellular ECM ligands: laminin, fibronectin, collagen type IV, nidogen, osteonectin, glycosaminoglycans (GAGs), agrin, and perlecan (Baeten and Akassoglou 2011). Ligand binding leads to the activation of various growth factors and signaling cascades that control cell growth, differentiation, migration, and survival during BBB development and maintenance. During brain vascularization, angiogenic ECs express $\alpha4\beta1$ and $\alpha5\beta1$ integrins, and binding of the ECM ligand fibronectin induces cell proliferation through mitogen-activated protein kinase (MAPK) signaling (Milner and Campbell 2002, Wang and Milner 2006). In the adult mouse, EC differentiation and vessel stabilization is promoted via laminin binding to $\alpha1\beta1$ and $\alpha6\beta1$ integrins (Wang and Milner 2006). The $\beta1$ integrin interaction with laminin directly affects cerebrovascular integrity because blocking this receptor *in vitro* increased vascular permeability due to decreased expression of the TJ protein claudin-5 (Osada et al. 2011).

The Main Contributors of the Mature Blood–Brain Barrier

The mature BBB is composed of capillary ECs tightly connected with TJs and adherens junctions (AJs) that prevent paracellular transport (Abbott et al. 2010) and have a low pinocytotic activity (Sedlakova et al. 1999, Redzic 2011), although limited transcellular transport does occur (De Bock et al. 2016). In addition, the BBB is influenced by closely associated perivascular astrocytic end feet, pericytes, and microglia (De Bock et al. 2014) (Figure 15.2). These different cell types play essential roles in BBB induction and maintenance by regulating the proliferation, migration, and vascular branching of the brain ECs. In addition, the BM provides structural support around the pericytes and ECs. The basal lamina (BL) is contiguous with the plasma membranes of the astrocyte end feet (Hawkins and Davis 2005). The BBB provides strong resistance to movement of ions, with transendothelial electrical resistance (TEER) around $1500\times$ cm^2 (Redzic 2011), 100 times higher than that for peripheral microvessels (Crone and Christensen 1981, Gorle et al. 2016).

Endothelial Cells

Compared to peripheral vasculature, BBB ECs are characterized by increased mitochondrial content, exhibit minimal pinocytotic activity, and lack fenestrations (Oldendorf et al. 1977, Fenstermacher et al. 1988, Takakura et al. 1991, Sanchez-Covarrubias et al. 2014). Increased mitochondrial content is essential for these cells to maintain various active transport mechanisms such as those utilized to transport ions, nutrients, and waste products into and out of brain parenchyma, thus contributing to precise regulation of the CNS microenvironment and ensuring proper neuronal function. The high concentration of mitochondria in cerebrovascular ECs might account for the sensitivity of the BBB to oxidant stressors. Furthermore, the physiology and pathophysiology of ECs are closely linked to

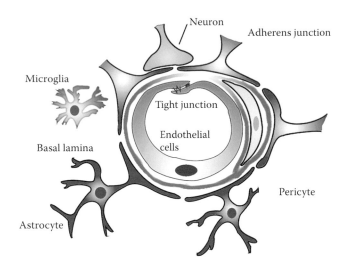

FIGURE 15.2 Schematic structure of the BBB. The brain capillary ECs connected to each other by TJs and AJs. They are surrounded by basal membrane which also covers the connecting pericytes. Around the brain microvessels astrocyte end feet are also essential providers of the barrier function. Additional supporting cell types are microglia cells and connecting neurons.

the functioning of their mitochondria, and mitochondrial dysfunction is another important mediator of disease pathology in the brain (Grammas et al. 2011).

The brain endothelium is highly reactive because it serves as both a source of, and a target for, inflammatory proteins and reactive oxygen species (ROS). BBB breakdown thus leads to neuro-inflammation and oxidative stress, which are implicated in the pathogenesis of CNS disease. Cell polarity of ECs is ascribed to differing functional expression of transporter proteins and metabolic enzymes that are differentially expressed on the luminal (or apical) and abluminal (or basolateral) membranes, which further contribute to the high selectivity of the BBB (Betz et al. 1980, Sanchez et al. 1995, Vorbrodt and Dobrogowska 2003). Of the many transporters expressed at the BBB endo-thelium, several have been implicated in influx and/or efflux of drugs into the CNS. Breakdown of the BBB as a result of disruption of transporters, leads to increased leukocyte transmigration and is an early event in the pathology of several disorders. The restricted paracellular permeability of capillary EC layer is warranted by two intercellular molecular binding systems: the AJs and TJs.

Tight Junctions

Although disruption of AJs can result in increased BBB permeability, TJs are primarily responsible for restricting paracellular permeability at the BBB (Hawkins and Davis 2005, Zlokovic 2008). TJs form the primary physical barrier component of the BBB and function to greatly restrict paracel-lular entry of various endogenous and exogenous substances that can potentially be neurotoxic. TJs are dynamic complexes of multiple protein constituents including junctional adhesion molecules (JAMs), occludin, claudins (i.e., claudin-1, -3, and -5), and membrane-associated guanylate kinase (MAGUK)-like proteins (i.e., ZO-1, -2, and -3) (Hawkins and Davis 2005) (Figure 15.3).

Adherens Junctions

AJs are found throughout the CNS microvasculature and are responsible for intercellular adherence between adjacent ECs (Hawkins and Davis 2005, Sanchez-Covarrubias et al. 2014). AJs are com-posed of multiple protein components including vascular endothelium (VE) cadherin, actin, and catenin (Vorbrodt and Dobrogowska 2003). Cell–cell adhesion is mediated by homophilic interac-tions of VE-cadherin expressed on adjacent ECs. Such interactions mediate calcium-dependent cell adhesion by binding to the actin cytoskeleton. Cytoskeletal binding occurs via catenin accessory

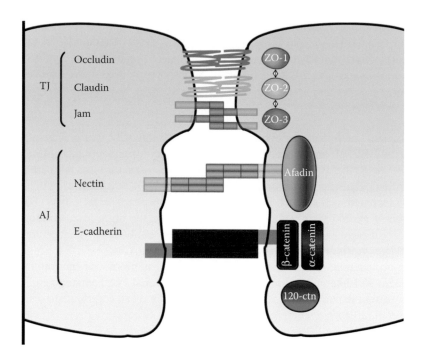

FIGURE 15.3 Connective surface of ECs in cerebral microvasculature. The TJ proteins are expressed close to the apical membrane of EC, while the AJ proteins are located more basolaterally.

proteins. Specifically, β-catenin links VE-cadherin to α-catenin, an interaction that induces the direct binding to actin (Oldendorf et al. 1977, Sanchez del Pino et al. 1995) (Figure 15.3).

Membrane Transporters

Before entering the brain, the molecules have to cross the EC membrane. There are different routes for the molecules to go through this barrier. Simple diffusion transports molecules through an aqueous channel formed within the membrane. Several channels are expressed at high levels at the CNS barriers, for example, aquaporin (AQP)-1 (Nazari et al. 2015) and ion channels. The presence of specific ion channels at the BBB and blood-cerebrospinal fluid barrier (BCSFB) ensures that the ionic composition is optimal for synaptic signaling. The ions are also important for the formation of the brain fluids: interstitial fluid (ISF) and cerebrospinal fluid (CSF). ECs are an important source of ISF (Abbott 2004), while ion transport across the choroid plexus epithelium (CPE) drives CSF secretion (Redzic 2011).

In carrier-mediated diffusion, solute molecules bind to specific membrane protein carriers. This process is energy independent and plays an essential role in transport of, for example, monocarboxylates, hexoses, amines, amino acids, nucleosides, glutathione, and small peptides (Abbott et al. 2010). Examples are glucose transporter (GLUT) (Simpson et al. 2007), monocarboxylate transporter (MCT) (Roberts et al. 2008), equilibrative nucleoside transporter (ENT), and organic anion transporter (OAT) (Redzic 2011). Carrier-mediated endocytosis is important for the delivery of nutrients to the brain. For example, glucose is the main energy source for the brain and a continuous supply is essential (Simpson et al. 2007). Glucose is constantly catabolized in the brain, which creates a gradient for hexose transport into the brain mediated by glucose transporters. Both the BBB and BCSFB express GLUT1 (Redzic 2011).

Active transport also occurs through protein carriers with a specific binding site for the solute. These active transporters require ATP hydrolysis and allow movement against the concentration gradient. Efflux transporters in the brain at the blood–brain interfaces include P-glycoprotein (P-gp), breast cancer resistance protein (BCRP in humans; Bcrp in rodents) and multidrug resistance proteins (MRPs in humans; Mrps in rodents). Transporters that facilitate drug entry into the brain (uptake or influx transporters) include organic anion transporting polypeptides (OATPs in humans, Oatps in rodents), OATs (OATs in humans; Oats in rodents), organic cation transporters (OCTs in humans, Octs in rodents), nucleoside transporters, MCTs (MCTs in humans, Mcts in rodents), and putative transport systems for peptide transport (Erdo et al. 2016, Nagy et al. 2016, Erdo et al. 2017).

Pericytes

Under physiological conditions, pericytes regulate (1) BBB integrity, that is, TJs or AJs and trans-cytosis across the BBB; (2) angiogenesis, that is, microvascular remodeling, stability, and architecture; (3) phagocytosis, that is, clearance of toxic metabolites from the CNS; (4) cerebral blood flow (CBF) and capillary diameter; (5) neuroinflammation, that is, leukocyte trafficking into the brain; and (6) multipotent stem cell activity. Pericyte dysfunction is characterized by (1) BBB breakdown causing leakage of neurotoxic blood-derived molecules into the brain (e.g., fibrinogen, thrombin, plasminogen, erythrocyte-derived free iron, and anti-brain antibodies); (2) aberrant angiogenesis; (3) impaired phagocytosis causing CNS accumulation of neurotoxins; (4) CBF dysfunction and ischemic capillary obstruction; (5) increased leukocyte trafficking promoting neuroinflammation; and (6) impaired stem cell-like ability to differentiate into neuronal and hematopoietic cells. Pericyte dysfunction is present in numerous neurological conditions and can contribute to disease pathogenesis (Sweeney et al. 2016).

Astrocytes

Astrocytes are glial cells that help support and protect neurons by controlling neurotransmitter and ion concentrations to maintain the homeostatic balance of the neural microenvironment, modulating synaptic transmission, and regulating immune reactions (Rodriguez-Arellano et al. 2016). Astrocytes are also known to interact with ECs through their end feet projections that encircle the abluminal side of cerebral capillaries (Abbott et al. 2006). In the adult brain, such interactions are important in synchronizing metabolite levels with cerebral blood flow and vasodilation, and regulating brain water content (Zlokovic 2008). For example, the most abundant water channel protein, aquaporin 4, is predominantly expressed in astrocytic end feet surrounding CNS vessels (Tait et al. 2008). Astrocytes are also a key cellular support of BBB integrity. Recent *in vitro* and *in vivo* molecular studies have revealed several effector molecules released by astrocytes that function to enhance and maintain barrier tightness (Keaney and Campbell 2015).

Supportive Cell Types

Neurons

There is considerable evidence for direct innervation of both brain microvessel ECs and associated astrocyte processes via distinct connections with noradrenergic (Ben-Menachem et al. 1982, Cohen et al. 1997, Sanchez-Covarrubias et al. 2014), serotonergic (Cohen et al. 1996), cholinergic (Vaucher and Hamel 1995, Tong and Hamel 1999), and GABAergic (Vaucher et al. 2000) neurons. For example, studies have shown that loss of direct noradrenergic input from the locus coeruleus results in increased BBB susceptibility to the effects of acute hypertension, resulting in significantly increased permeability to 125-I labeled albumin (Berezowski et al. 2004, Erdo et al. 2017).

Microglia

Microglia, the primary immune cells of the brain, are ubiquitously distributed in the CNS and are activated in response to systemic inflammation, trauma, and several CNS pathophysiologies (Kettenmann et al. 2011, Kofler and Wiley 2011, Harry 2013, Sanchez-Covarrubias et al. 2014). Microglia present with a ramified morphology that is characterized by a small soma and fine cellular processes during their "resting state." Microglial activation in response to pathophysiological stressors can trigger changes in cell morphology, which include reduced complexity of cellular processes and transition from a ramified morphology to an amoeboid appearance (Kettenmann et al. 2011). Microglia can exist in one of two active states: in the classically activated M1 pathway, microglia primarily release proinflammatory cytokines like interleukin-1β and tumor necrosis factor-α, whereas in the alternative M2 pathway microglia are involved in tissue repair, phagocytosing damaged neurons and foreign material, releasing chemokines and VEGF, and activating neurotrophic pathways (Keaney and Campbell 2015). Activated microglia produce high levels of neurotoxic and proinflammatory mediators and proteases, all of which result in cell injury and neuronal death (Ronaldson and Davis 2012), As immune cells, microglia scavenge apoptotic cells, tissue debris after trauma, or microbes (Ronaldson and Davis 2012). They can also act as scavengers of extracellular molecules such as amyloid-β (Kettenmann et al. 2011, Kofler and Wiley, 2011). Activation of microglia is associated with altered TJ protein expression and increased BBB permeability (Huber et al. 2006).

Basal Lamina

BMs (also called as BL), are considered to be uniform, approximately 100 nm-thin extracellular matrix sheets that serve as a substrate for ECs, epithelial cells, and myotubes. To find out whether BMs maintain their ultrastructure, protein composition and biophysical properties throughout life the natural aging history of the human inner limiting membranes (ILMs) was investigated by Candiello et al. (2010). Transmission electron microscopy showed that the ILM steadily increases in thickness from 70 nm at fetal stages to several microns at age 90. Furthermore, the relative concentrations of collagen IV and agrin increase, and the concentration of laminin decreases with age. Force-indentation measurements by atomic force microscopy also showed that ILMs become increasingly stiffer with advancing age (Candiello et al. 2010).

Extracellular Matrix

The extracellular matrix of the BL serves as an anchor for the cerebral microvascular endothelium. The anchoring function of the extracellular matrix is mediated via interactions between endothelial integrin receptors, lamin, and other matrix proteins. Disruption of extracellular matrix is associated with the loss of barrier function, resulting in increased permeability. In addition, matrix proteins have been shown to influence the expression of TJ proteins, such as occludin, suggesting that the extracellular matrix plays a role in maintaining TJ protein integrity (Hawkins and Davis 2005, Ronaldson and Davis 2012, Sanchez-Covarrubias et al. 2014).

EFFECT OF PHYSIOLOGICAL AGING ON THE BLOOD–BRAIN BARRIER

Although it has been controversial whether BBB permeability significantly increases with aging in human brains, a large-scale meta-analysis study including 31 BBB permeability studies demonstrated that BBB permeability, evaluated by CSF/serum albumin ratios, increased with normal aging, and further increased in patients with dementia, and with accumulation of white matter lesions (Farrall et al. 2009). Recently, BBB breakdown was shown to be an early event in the aging brain, beginning in the hippocampus, and may contribute to cognitive impairment (Montagne et al. 2015). Similarly, in experimental animals, BBB permeability to serum albumin increased with aging in three different strains of mice (Ueno et al. 1993). This increase in permeability was accelerated in aged mice showing cognitive impairment, such as senescence-accelerated prone mice (SAMP8) (Ueno et al. 1996, 1997, 2016, Vorbrodt and Dobrogowska 2003).

MORPHOLOGICAL CHANGES AT THE BLOOD–BRAIN BARRIER

Electron Microscopy

Many of the modern techniques for brain imaging such as enhanced computed tomography (CT), enhanced magnetic resonance imaging (MRI), single-photon emission computed tomography

(SPECT), and positron emission tomography (PET) are able to study damage of the normal BBB. It became possible to define the ultrastructural features of blood vessels by electron microscopy (Sage and Wilson 1994).

By electron microscopy, Burns et al. (1981) were able to show changes at the BBB in a total of 25 rats and 18 *Macaca nemestrina* of both male and female genders. Changes in the numerical density of rat cerebral cortical capillaries, during the life span, probably reflect the rapid increase in volume of neural tissue occurring during the brain "growth spurt" and the decreasing volume of neural tissue frequently observed during aging as well as the changes occurring in the temporal pattern of brain vascularization. The changes were similar in monkeys and rats. The thinning of capillary walls in the monkey brain was attributable almost entirely to a decrease in the cross-sectional area of its endothelial component. This change is consistent with EC loss with increasing age in rats. A significant increase in cerebral capillary BL thickness was observed in rats with aging, whereas in the monkey a marked increase in this parameter was observed only between 4 and 10 years of age. Thickened BL in cortical capillaries was thought to be associated with altered BBB permeability. The authors observed aberrant interendothelial TJs in cerebral capillary profiles from one out of five 20 year old monkeys and this was associated with thickened BL and with an attenuated endothelial component. They found a decline in the number of endothelial mitochondria per cerebral capillary profile with increasing age in the monkey. A significant decrease in the number of ECs in the rat was also found. The ratio of the total cross-sectional area of endothelial mitochondria to that of EC-BL did not change significantly with age in either monkey or rat; however, both parameters decreased with age (Burns et al. 1981).

Our electron microscopy studies also demonstrated the age-dependent ultrastructural changes at the BBB as shown in Figure 15.4.

Confocal Microscopy

Confocal microscopy is suitable for the examination of fluorescently labeled samples. DiNapoli and coworkers (2008) examined the effects of age on stroke progression and outcome in order to explore the association between BBB disruption, neuronal damage, and functional recovery by confocal microscopy and other methods. They were using middle cerebral artery occlusion. Young adult (3 months) and aged (18 months) rats were assessed for BBB disruption. Results indicated that aged rats suffer larger infarctions, reduced functional recovery, and increased BBB disruption preceding observable neuronal injury. Infarction volumes were significantly affected by animal age in both the cortical and striatal regions. The volume of infarcted cortex relative to contralateral hemisphere was significantly higher in the aged rats when compared to the young. Striatal infarction volumes were also significantly greater in the aged group. BBB permeability in young and aged rats was assessed by quantifying extravasation of albumin and visualization of fluorescent vascular markers by confocal microscopy. Aged animals demonstrated a 175% increase in BBB permeability to albumin in the infarcted hemisphere as compared to young animals (DiNapoli et al. 2008). Confocal microscopy has many advantages; it is suitable for the transport experiments too (Markoutsa et al. 2011) or can even be used in association with a microfluidic platform (Griep et al. 2013).

FUNCTIONAL CHANGES AT THE BLOOD–BRAIN BARRIER

The homeostasis of the CNS is essential for the proper functioning of brain cells. The BBB participates in CNS homeostasis by the regulation of influx/uptake and efflux transport, protection from harm, preventing the brain from neurotoxins, and by transporting nutrients and products from brain metabolism. The barrier function results from a combination of: *physical barrier*—is formed by the TJs between ECs of the capillaries reducing flux through the intercellular cleft (paracellular pathway); *transport barrier*—a specific transport mechanism mediating transcellular solute flux; and *metabolic barrier*—enzymes that may degrade or alter substances prior to passage. The barrier function is not fixed, but can be modulated and regulated, both in physiology and in pathology (Abbott et al. 2010).

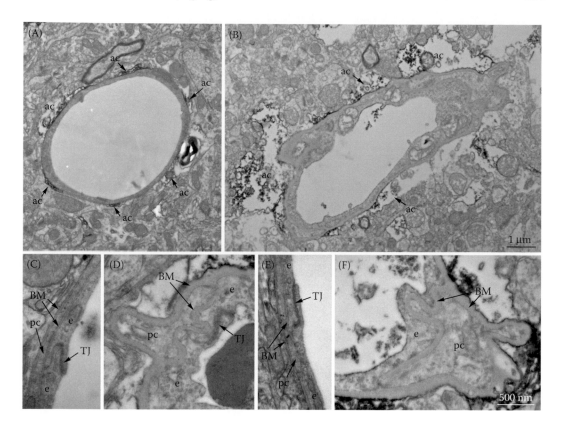

FIGURE 15.4 (A) In young animal the capillary wall is thin, surrounded by some astrocyte endfeet (ac). (B) In aged animal the capillary wall is thicker and the astrocyte endfeet are much wider. (C, E) Capillary wall in young rat by 30,000 fold magnification. Endothelial cells (e) tight junction (TJ), and on the basolateral side the basal membrane (BM) and a perycite (pc) are visisble. (D, F) Capillary wall in aged rat by 30,000 fold magnification. Less tight junction can be found between the endothelial cells than in young rat. The basal membrane is remarkably thicker and covers pericytes. (Scale: A–B: 1 um, C–F: 500 nm). This study was performed by Kinga Tóth, PhD at the Institute of Cognitive Neuroscience and Psychology, Budapest.

Permeability

The BBB is permeable to small lipophilic gases (O_2 and CO_2) and other gaseous molecules. These molecules can pass the barrier freely by diffusion across ECs along their concentration gradient. Moreover, the BBB is also permeable to water however solute carriers of the basal and apical membranes together with enzymes regulate entry and efflux. Nutrients (amino acids, glucose, etc.) enter the brain through specific transporters but larger molecules (insulin, transferrin, leptin) are transported by receptor-mediated endocytosis.

Transfer of some larger molecules through the BBB requires the presence of receptor systems or specific transporters that can limit their concentration within the CNS (e.g., multidrug transporters and P-gp-like proteins that limit access of drugs to the brain, thereby may prevent their accumulation) (Abbott and Friedman 2012, Serlin et al. 2015).

The metabolites can be eliminated from the brain by passive diffusion and with transporters/receptors located on the brain side of the ECs (e.g., glutamine, Aβ peptides). Moreover, two important efflux transporters (MDR1 and BCRP) play important roles in preventing xenobiotics from crossing the BBB. Both transporters are on the apical part of the endothelial membrane, and pump their substrates from the cell into the bloodstream.

In summary, the BBB or more precisely the CNS barriers attend the maintenance of the stable microenvironment and also protect the CNS from chemical and physical damage. The BBB dysfunction can be expressed as a mild or transient TJ opening and also as a chronic barrier break-down (Forster 2008).

Cognitive Decline

During aging, the permeability of the brain barriers increase (leakages in the BBB), tiny ruptures of blood vessels may occur, some substances that are normally kept out of the brain, gain entry and can potentially cause neuronal damage and cognitive dysfunction.

Cognitive decline in aging presents an increasing challenge. Age-related vascular pathologies (stroke, microinfarcts, microhemorrhages, etc.) are promoted, the brain homeostasis changes, and cognitive skills decline which enhances the development of dementia or neurodegenerative diseases (NDs) in elderly people (Toth et al. 2017).

Processes and vascular pathological consequences of cerebral microvascular dysregula-tion that correlate with age-related declines in cognitive function are neuroinflammation, microglia dysfunction, reduced synaptic densities, BBB disruption, followed by blood-to-brain extravasation of circulating neuroinflammatory molecules and low levels of neurogenesis. The age-related disruption of BBB may potentiate the progress of some cerebrovascular-based NDs such the Alzheimer's disease (AD), vascular dementia, and sclerosis multiplex (Marschallinger et al. 2015).

There is also a direct association between cognitive impairment, hypercholesterolemia, and diabetes. Both risk factors, diabetes and hypercholesterolemia increase risk of neurodegeneration, cognitive impairment, and dementia, as a consequence of changes in the brain macrovasculature and microvasculature. Dysfunction and the breakdown of the BBB in AD are associated with chronic influx of plasma components (such as amyloid-β peptides, serum albumin, immunoglobu-lins, complements) into the brain tissue (Acharya et al. 2013).

CHANGES AT THE BLOOD–BRAIN BARRIER ASSOCIATED WITH AGE-RELATED NEURODEGENERATIVE PATHOLOGIES

An important component of age-related pathology is neurodegeneration. It can be defined as pro-gressive loss of neuronal structure and function, leading to neuronal cell death. Most NDs have their onset in mid-life and can be characterized by motor and/or cognitive symptoms that progressively worsen with age and may reduce life expectancy.

The general characteristics of selected NDs are presented below, followed by the main neurode-generative processes and disorders, and finally more specific information on BBB dysfunction in these diseases will be discussed.

Neurodegenerative Diseases

AD: is characterized by brain processes that include formation of amyloid (senile) plaques; neurofi-brillary tangles in the intracellular space of neurons, with a high content of the hyperphosphorylated protein "tau," as well as brain atrophy and shrinkage (Finder 2010). In this disease the neurons, typically of the temporal lobe and parietal lobe, and parts of the frontal cortex and cingulate gyrus degenerate, leading to gross atrophy of these brain areas (Wenk 2003). Amyloid-β (Aβ) is produced from the amyloid-β precursor protein (APP) both in the brain and in peripheral tissues. Clearance of amyloid-β from the brain normally maintains its low levels in the brain. Aβ accumulation in the AD affected brain is likely due to its faulty clearance from the brain (Holtzman and Bitterman 1956, Zlokovic et al. 2000, Tanzi ct al. 2004, Selkoe 2011).

Multiple sclerosis (MS): is a typical inflammatory disease, but axonal loss and neurodegenera-tion have been observed even in its earliest stages (Engelhardt and Ransohoff 2005, Hendriks et al. 2005, Minagar et al. 2012, Stangel 2012). EC stress and apoptosis are characteristic hallmarks of MS. The inflammatory demyelinating disease processes in early MS triggers a cascade of events that lead to neurodegeneration, which are amplified by pathogenic mechanisms related to brain aging and accumulated disease burden.

Parkinson's disease (PD): is a chronic, progressive neurological disorder. The roles of oxidative stress, apoptosis, mitochondrial dysfunction, inflammation, and impairment of the protein degradation pathways have been highlighted in animal models (Grunblatt et al. 2000, Bove and Perier 2012). The mechanism by which the brain cells in PD are lost may consist of an abnormal accumulation of the protein alpha-synuclein bound to ubiquitin in the damaged cells. The alpha-synuclein–ubiquitin complex cannot be directed to the proteosome. This protein accumulation forms proteinaceous cytoplasmic inclusions called Lewy bodies, which are one of the hallmarks of PD (De Vos et al. 2008).

Pharmacoresistant epilepsy: In very broad and general terms, pharmacoresistance is the failure of seizures to come under complete control or acceptable control in response to antiepileptic drugs. About 30%–40% of all people with epilepsy do not become fully seizure free with present medication, even when treated at the maximal tolerated dose. This pharmacoresistance is particularly prominent in partial epilepsies and some severe syndromes in infants, but essentially it can occur in nearly all types of epilepsies and epileptic syndromes. In addition, unresponsiveness in these patients is not limited to a specific drug or drug class, but occurs with the complete range of antiepileptic drugs (Regesta and Tanganelli 1999, Schmidt and Loscher 2005).

Neurodegenerative Processes

NDs are characterized by a continuous progress of neuronal death and progressive nervous system dysfunction. The NDs have many processes in common, though these processes may be qualitatively, quantitatively, temporally, and spatially distinct. The main processes include gene defects and variants, oxidative stress, protein misfolding, and accumulation, which will affect different cellular and intercellular signaling pathways to cause neuronal cell death by necrosis or apoptosis. Here, we briefly describe these main processes with references for further reading (Ahmad 2012), as this chapter focuses on changes at the level of the BBB, which will be discussed afterward in more detail for AD, MS, PD, and pharmacoresistant epilepsy.

Gene Defects and Variants

Over the years, many genetic defects related to human neurodegeneration have been identified, which in the presence of environmental factors determine the course of progressive neurodegenerative processes (Bertram and Tanzi 2005, Coppede et al. 2006). In Table 15.2, the genes that have been found to be associated with AD, MS, PD, and pharmacoresistant epilepsies are shown.

It should be noted that genome-wide association studies have successfully revealed numerous susceptibility genes for NDs, but that it has been found that odds ratios (a statistical number, based on the link between absence/presence of property A and absence/presence of property B in a given population) associated with risk alleles are generally low. This indicates that it is mere the "common

TABLE 15.2

Disease Condition and Associated Changes in Genes

Disease Condition	Genes	References
AD	APOE, APP, PSEN1, PSEN2	Genetics home reference (of the National Institutes of Health)
MS	CYP27B1, HLA-DRB1, IL2RA, IL7R, TNFRSF1A	Genetics home reference (of the National Institutes of Health)
Parkinson's disease	ATP13A2, GBA, LRRK2, PARK2, PARK7, PINK1, SNCA, UCHL1, VPS35	Takakura, Audus et al. (1991)
Pharmacoresistant epilepsies	ABCB1, CSTB, EMPA, GABRG2, NHRC1, SCN1a, SCN9A, SLC2A1, TSC1, TSC2	Martínez-Juárez (2013)

disease-multiple rare gene variants" than the "common disease-common gene variants" determine the susceptibility for a given ND (Tsuji 2010).

Oxidative Stress

Disturbances in the normal redox state of cells can cause toxic effects via the production of peroxides and free radicals, ROS, and reactive nitrogen species (RNS) which may damage most components of the cell, such as proteins, lipids, and nucleic acids. Such oxidative stress can cause disruptions in normal cellular signaling and seems to have a ubiquitous role in neurodegeneration via the induction of cell death (Sayre et al. 2008, Navarro and Boveris 2010, Arnold 2012, Perez-Pinzon et al. 2012). In response to oxidative stress, cells increase and activate their cellular antioxidant mechanisms. Glutathione (GSH) is the major antioxidant in the brain, and as such plays a pivotal role in the detoxification of reactive oxidants.

For AD oxidation/dysfunction of a number of enzymes specifically involved in energy metabolism have been found, that support the view that reduced glucose metabolism and loss of ATP are crucial events triggering neurodegeneration and progression of AD (Tramutola et al. 2016).

For MS, oxidative stress has been strongly implicated in both the inflammatory and neurodegenerative pathological mechanisms. The onset of MS is characterized by inflammation-mediated demyelination due to lymphocyte infiltration from peripheral blood and microglial cell activation. In different disease stages accumulation of ROS and RNS has been observed. Also, it has been shown that GSH homeostasis is altered in MS (Carvalho et al. 2014, Ibitoye et al. 2016, Lepka et al. 2016).

For PD, oxidative stress is generally recognized as one of the main causes of this disease, and excessive ROS can lead to dopamine neuron vulnerability and eventual death (Munoz et al. 2016, Xie and Chen 2016). Finally, no information is available on the role of oxidative stress in specifically pharmacoresistant epilepsies.

Protein Misfolding and Accumulation

Protein misfolding is a naturally occurring process in living cells. Under healthy conditions, the so-called protein quality control systems dispose of misfolded proteins. When the capacity of these systems becomes limiting, misfolded proteins may accumulate and aggregate. It seems that proteins that have repetitive amino acid motifs are mostly prone to changing into a misfolded state. Such a state is often toxic and can be viewed as "infective" when the misfolded protein is able to induce the conversion of other, normally folded proteins into the toxic configuration, and may lead to an amplification loop. Accumulation and aggregation of misfolded proteins further lead to impairment of proper cellular functioning and ultimate cell killing (Figure 15.5).

Prion proteins are an important and well-known example of this catastrophal cascade (Stokin and Goldstein 2006, De Vos et al. 2008, Soto and Estrada 2008, Jellinger 2012). For AD, the accumulation proteins are amyloid-β and hyperphosphorylated tau (Holtzman and Bitterman 1956, Zlokovic et al. 2000, Tanzi et al. 2004, Selkoe 2011), and for PD the accumulating protein is α-synuclein (Recasens and Dehay 2014, Chu and Kordower 2015).

Cell Death

Inappropriate death of cells in the nervous system is the cause of multiple NDs and pathological neuronal death can occur by apoptosis, by necrosis, or by a combination of both. Elevated intracellular calcium is the most ubiquitous feature of neuronal death with the concomitant activation of cysteine calcium-dependent proteases, calpains. Calpains and lysosomal, catabolic aspartyl proteases, play key roles in the necrotic death of neurons (Artal-Sanz and Tavernarakis 2005).

Neuronal loss is almost invariably accompanied by abnormal insoluble protein aggregates, either intra- or extracellular. Ambegaokar et al. (2010) have reviewed methods by which *Drosophila melanogaster* has been used to model aspects of PD and AD.

In the pathogenesis of neurodegeneration, an aberrant regulation of apoptosis is known to play a role. For example, ROS are known to be able to initiate apoptosis via the mitochondrial and death

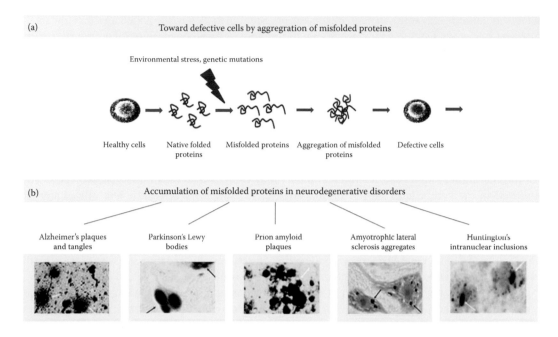

FIGURE 15.5 (a) The cascade from normal cells to defective cells due to aggregation of misfolded proteins. (b) Accumulation of misfolded proteins is a common characteristic of many NDs. The location and type of protein is specific to the ND. (Redrawn from Soto, C. 2003. Unfolding the role of protein misfolding in neurodegenerative diseases. *Nat Rev Neurosci* 4(1): 49–60.)

receptor pathways (Okouchi et al. 2007). Abnormality in mitochondrial Ca(2+) handling has been detected in a range of NDs, and emerging evidence from disease models suggests that mitochondrial Ca(2+) may play a role in disease pathogenesis (Abeti and Abramov 2015).

Excess cells are eliminated through programmed cell death or apoptosis. The initiation of neuronal apoptosis in response to numerous extracellular agents has been widely reported and lethal functions for the lipid effectors ceramide and sphingosine have been identified in both normal and pathophysiological conditions. Inappropriate initiation of apoptosis by deregulated ceramide and sphingosine has been proposed to underlie the progressive neuronal attrition associated with, amongst other NDs, for AD and PD (Ariga et al. 1998).

Bains and Shaw (1997) proposed a GSH-depletion model of neurodegenerative disorders. Oxidative stress-mediated neuronal loss may be initiated by a decline in GSH which plays multiple roles in the nervous system including that of free radical scavenger, redox modulator of ionotropic receptor activity, and possible neurotransmitter. GSH depletion can enhance oxidative stress and may also increase the levels of excitotoxic molecules; both types of action can initiate cell death in distinct neuronal populations. Evidence for a role of oxidative stress and diminished GSH status is present, among other NDs, for PD and AD (Bains and Shaw 1997).

Autophagy is a highly conserved intracellular pathway involved in the elimination of proteins and organelles by lysosomes. Known originally as an adaptive response to nutrient deprivation in mitotic cells, autophagy is now recognized as an arbiter of neuronal survival and death decisions in NDs. Studies using postmortem human tissue, genetic and toxin-induced animal and cellular models indicate that many of the etiological factors associated with NDs can perturb the autophagy process. Emerging data support the view that dysregulation of autophagy might play a critical role in the pathogenesis of NDs, such as AD and PD (Banerjee et al. 2010).

The DNA damage response is a key factor in the maintenance of genome stability. As such, it is a central axis in sustaining cellular homeostasis in a variety of contexts: development, growth, differentiation, and maintenance of the normal life cycle of the cell. The DNA damage response defects in neurons may result in neurodegeneration. Barzilai et al. have provided an overview on the potential role of the DNA damage response in the etiology and pathogenesis of NDs (Barzilai 2010).

Macroautophagy: is a cellular process by which cytosolic components and organelles are degraded in double-membrane bound structures upon fusion with lysosomes. It is a pathway for selective degradation of mitochondria by autophagy, known as mitophagy, and is of particular importance to neurons. It appears that the regulation of mitophagy shares key steps with the macro-autophagy pathway, while exhibiting distinct regulatory steps specific for mitochondrial autophagic turnover. Mitophagy has been linked to the pathogenesis of PD through the study of recessively inherited forms of this disorder, involving PINK1 and Parkin. Recent work indicates that PINK1 and Parkin together maintain mitochondrial quality control by regulating mitophagy. In the Purkinje cell degeneration (pcd) mouse, altered mitophagy may contribute to the dramatic neuron cell death observed in the cerebellum, suggesting that overactive mitophagy or insufficient mitophagy can both be deleterious (Batlevi and La Spada 2011).

Exposure of hippocampal neuronal/glial cocultures to β-amyloid peptides activates the glial nicotinamide adenine dinucleotide phosphate oxidase, followed by predominantly neuronal cell death, which is mediated by poly(ADP-ribose) polymerase in response to oxidative stress generated by the astrocytic nicotinamide adenine dinucleotide phosphate oxidase (Abeti et al. 2011).

Aneuploidy is an abnormal number of copies of a genomic region. Genomic instability has been associated with aneuploidy, however, can also lead to developmental abnormalities and decreased cellular fitness. Arendt et al. (2010) have shown that neurons with a more-than-diploid content of DNA are increased in preclinical stages of AD and are selectively affected by cell death during progression of the disease. Present findings show that neuronal hyperploidy in AD is associated with a decreased viability. Hyperploidy of neurons thus represents a direct molecular signature of cells prone to death in AD and indicates that a failure of neuronal differentiation is a critical pathogenetic event in AD (Arendt et al. 2010).

The metabolic turnover of sphingolipids produces several signaling molecules that profoundly affect the proliferation, differentiation, and death of cells. It is well known that specifically ceramide and sphingosine-1-phosphate play an important role in the so-called cell death pathways. A wide body of evidence indicates that ceramide and amyloid beta protein plays a key role in attacking mitochondria to set in the pathways of cell death in AD (Chakrabarti et al. 2016).

Mechanisms by which dopaminergic neurons die in PD seems to be related to, (i) defects in ubiquitin–proteasome pathway and protein misfolding and aggregation caused by α-synuclein and Parkin gene defects; (ii) defects in mitochondrial morphology and function in PINK1/Parkin and DJ-1 mutations; and (iii) increased susceptibility to cellular oxidative stress which appear to underlie defects in α-synuclein, Parkin and DJ-1 genes (Abdel-Salam 2014).

It is known that PD is characterized by the progressive loss of select neuronal populations, but prodeath genes mediating the neurodegenerative processes is a novel proposed concept by Aimé et al. (2015). They have proposed a pathway involving tribbles pseudokinase (Trib3), which is a stress-induced gene with proapoptotic activity, to have a role in neuronal death associated with PD (Aimé et al. 2015). For PD, furthermore, Michel et al. (2016) have provided an overview of cell-autonomous mechanisms that are likely to participate in dopamine (DA) cell death in both sporadic and inherited forms of PD. Damage to vulnerable DA neurons may arise from cellular disturbances produced by protein misfolding and aggregation, disruption of autophagic catabolism (a conserved catabolic process that degrades cytoplasmic constituents and organelles in the lysosome), endoplasmic reticulum stress, mitochondrial dysfunction, or loss of calcium homeostasis and where pertinent show how these mechanisms may mutually cooperate to promote neuronal death (Michel et al. 2016).

For MS, it has been postulated that glutamate excitotoxicity, as a result of an excessive amount of glutamate that overactivates its cellular receptors and induces cell death, could be a missing link between inflammatory and neurodegenerative processes evident in MS (Kostic et al. 2013). Glutamate-mediated excitotoxicity is the principal mechanism driving neuronal death after status epilepticus, whereby excessive glutamate release leads to intracellular calcium overload, oxidative stress, organelle swelling and rupture of intracellular membranes, activation of proteases and necrosis. Based on the work of Meldrum et al., it has generally been accepted that status epilepticus, even in the absence of systemic complications, causes neuronal death in vulnerable brain regions, often but not necessarily including the hippocampus (Meldrum 2002, Thom et al. 2005, Nobili et al. 2015).

Finally, it can be said that in different NDs a selective vulnerability of particular neuronal groups or brain structures exists. What makes such specific vulnerability to be associated with a particular ND remains to be elucidated.

NEURODEGENERATIVE DISORDERS AND CHANGES IN BLOOD–BRAIN BARRIER FUNCTIONALITY

In neurodegeneration changes in functionality of the BBB and supporting cells (NVU) occur (Sandoval and Witt 2008, Zlokovic 2008, 2010) that lead to BBB dysfunction in a more progressed state of the disease process (Abbott et al. 2006, Zlokovic 2008, Zlokovic 2010, 2011, Al Ahmad et al. 2012, Freeman and Keller 2012) (Figure 15.6).

Zhao et al. (2015) reviewed the establishment and dysfunction of the BBB, with associations between BBB breakdown and pathogenesis of inherited monogenic neurological disorders and complex multifactorial diseases, including AD. Brain vascular damage can cause BBB dysfunction and/or reduced brain–blood perfusion and hypoxia. Both processes lead to neuronal injury and neurodegeneration. Changes in BBB functioning in AD, MS, PD, and pharmacoresistant epilepsies are presented below in a more disease specific manner.

BLOOD–BRAIN BARRIER IN ALZHEIMER'S DISEASE

Available evidence suggests that alteration of the BBB plays an important role in AD (Miyakawa 2010, Zlokovic 2011, Baloyannis 2015). While BBB breakdown is an early event in the aging human brain that begins in the hippocampus and may contribute to cognitive impairment (Montagne et al.

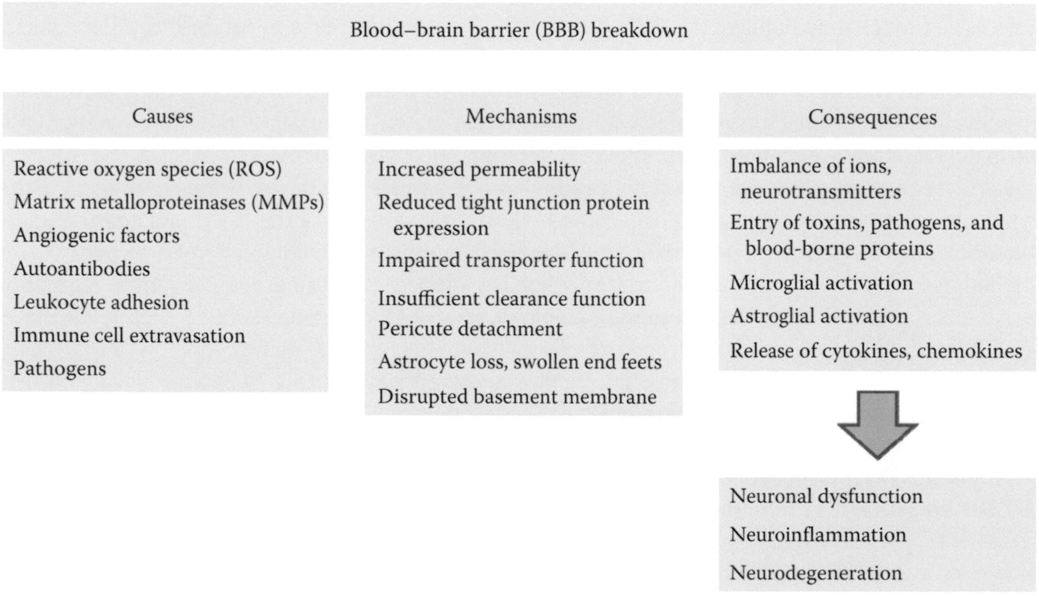

Blood–brain barrier (BBB) breakdown		
Causes	Mechanisms	Consequences
Reactive oxygen species (ROS)	Increased permeability	Imbalance of ions, neurotransmitters
Matrix metalloproteinases (MMPs)	Reduced tight junction protein expression	Entry of toxins, pathogens, and blood-borne proteins
Angiogenic factors	Impaired transporter function	Microglial activation
Autoantibodies	Insufficient clearance function	Astroglial activation
Leukocyte adhesion	Pericute detachment	Release of cytokines, chemokines
Immune cell extravasation	Astrocyte loss, swollen end feets	
Pathogens	Disrupted basement membrane	
		Neuronal dysfunction
		Neuroinflammation
		Neurodegeneration

FIGURE 15.6 Causes, characteristics, and consequences of the BBB breakdown. (Redrawn from Obermeier, B., R. Daneman and R. M. Ransohoff. 2013. *Nat Med* 19(12): 1584–1596.)

2015), the BBB breakdown in the hippocampus worsened with mild cognitive impairment and was correlated with injury to BBB-associated pericytes, as shown by the cerebrospinal fluid analysis (Montagne et al. 2015).

Cellular and molecular mechanisms in cerebral blood vessels and the pathophysiological events leading to cerebral blood flow dysregulation and disruption of the NVU and the BBB, may contribute to the onset and progression of dementia and AD. There is a link between neurovascular dysfunction and neurodegeneration. This includes the effects of AD genetic risk factors on cerebrovascular functions and clearance of Alzheimer's amyloid-β peptide toxin. Vascular risk factors, environment, and lifestyle have impact on cerebral blood vessels, which in turn may affect synaptic, neuronal, and cognitive functions, as reviewed by Zhao et al. (2015).

Tight Junctions

Both the components and functioning of TJs proteins are affected by neurodegenerative processes (Zlokovic 2011). TJs proteins include occludin and claudins (e.g., claudin-3, -5, -12). Occludin is vulnerable to be attacked by matrix metalloproteinases (MMPs) and MMPs seem to have implications in AD (Rosenberg and Yang 2007, Yang and Rosenberg 2011). Furthermore, connection of AJs and TJs to the actin cytoskeleton is influenced by tau and may result in tau-induced neurotoxicity (Fulga et al. 2007).

Without knowing the exact mechanisms, TJs seem to be involved in receptor for advanced glycation end-products (RAGE)-mediated Aβ cytotoxicity to the brain microvascular ECs, resulting in damaged BBB structural integrity. RAGE is a multiligand membrane receptor and is the main factor mediating Aβ cytotoxicity in AD. It has interaction with Aβ stimulating activation of proinflammatory cytokines, release of ROS, which leads to neuron damage and BBB dysfunction (Wan et al. 2014).

Astrocytes

Astrocytes normally regulate the BBB and abnormal astrocytic activity coupled to vascular instability has been observed in AD models (Takano et al. 2006). It seems that astrocyte properties may be affected upon development of amyloid deposits (Yang et al. 2011, Zlokovic 2011).

Pericytes

Loss of pericytes may damage the BBB due to an associated decrease in cerebral capillary perfusion, blood flow, and blood flow responses to brain activation. This will lead to more chronic perfusion problems like hypoxia, while BBB breakdown may further lead to brain accumulation of blood proteins and several macromolecules with toxic effects on the vasculature and brain parenchyma, ultimately leading to secondary neuronal degeneration (Bell and Zlokovic 2009, Zlokovic 2011).

Pericytes are uniquely positioned within the NVU between the ECs of brain capillaries, astrocytes, and neurons. Winkler et al. (2014) have reviewed the concept of the NVU and neurovascular functions of CNS pericytes, discussing vascular contributions to AD and new roles of pericytes in the pathogenesis of AD such as vascular-mediated Aβ-independent neurodegeneration, regulation of Aβ clearance and contributions to tau pathology, neuronal loss and cognitive decline (Winkler et al. 2014).

The key signaling pathways between pericytes and their neighboring ECs, astrocytes, and neurons that control neurovascular functions have been reviewed by Sweeney et al. (2016). Halliday et al. (2016) have shown that accelerated pericyte degeneration in AD APOE4 carriers is the highest, high in AD APOE3 carriers and lower in non-AD controls. This correlates with the magnitude of BBB breakdown to immunoglobulin G and fibrin. Also, accumulation of the proinflammatory cytokine CypA and MMP-9 in pericytes and ECs in AD (APOE4 higher than APOE3), have been shown to lead to BBB breakdown in transgenic APOE4 mice. This indicates that APOE4 leads to accelerated pericyte loss and enhanced activation of LRP1-dependent CypA–MMP-9 BBB-degrading pathway in pericytes and ECs, which can mediate a greater BBB damage in AD APOE4 compared with AD APOE3 carriers (Halliday et al. 2016)

In the brain, excess cholesterol is metabolized into 24S-hydroxycholesterol (24S-OH-chol) and eliminated into the circulation across the BBB. 24S-OH-chol is a natural agonist of the nuclear liver X receptors (LXRs) involved in peripheral cholesterol homeostasis. The effects of this oxysterol on the pericytes have demonstrated that pericytes express LXR nuclear receptors and their target gene ATP-binding cassette, subfamily A, member 1 (ABCA1), known to be one of the major transporters involved in peripheral lipid homeostasis. Furthermore, pericytes are able to internalize the amyloid-β peptides which accumulate in the brain of AD patients (Saint-Pol et al. 2012).

Shimizu et al. (2012) demonstrated that the GDNF secreted from the brain and peripheral nerve pericytes was one of the key molecules responsible for the upregulation of claudin-5 expression and permeability in the BBB. Amyloid deposits detected within degenerating pericytes in the brains of patients with AD suggest that pericyte dysfunction may play a role in cerebral hypoperfusion and impaired amyloid β-peptide clearance in AD (Dalkara et al. 2011).

Transport and Elimination of Amyloid-β (Aβ) at the Blood–Brain Barrier

Active transport of Aβ across the BBB seems to occur by a number of transporters that control the level of the soluble isoform of Aβ in brain.

P-glycoprotein: P-glycoprotein (P-gp, MDR1, ABCB1) contributes to the efflux of brain-derived Aβ into blood (Kuhnke et al. 2007, Bell and Zlokovic 2009, Hartz et al. 2010, Brenn et al. 2011, Vogelgesang et al. 2011, Sagare et al. 2012, Sharma et al. 2012). It seems that, in addition to the age-related decrease of P-gp expression, Aβ1-42 itself downregulates the expression of P-gp and other Aβ-transporters, which could exacerbate the intracerebral accumulation of Aβ and thereby accelerate neurodegeneration in AD and cerebral β-amyloid angiopathy. Defects in P-gp-mediated Aβ clearance from the brain are thought be triggered by systemic inflammation by lipopolysaccharide, leading to increased brain accumulation of Aβ. Recently, reduction of P-gp expression and transport activity has been found in isolated capillaries, as a result of Aβ40 mediated P-gp ubiquitination, internalization, and proteasome-dependent degradation (Hartz et al. 2016).

Low-density lipoprotein receptor-related protein 1 (LRP1): Aβ is produced from the APP, both in the brain and in peripheral tissues. In plasma, a soluble form of LRP1 (sLRP1) is the major transport protein for peripheral Aβ. sLRP1 maintains a plasma "sink" activity for Aβ through binding of peripheral amyloid-β which in turn inhibits reentry of free plasma amyloid-β into the brain. LRP1 in the liver mediates systemic clearance of amyloid-β. LRP1 at the BBB and contributes to the clearance of amyloid-β from the brain. LRP1 mediates rapid efflux of a free, unbound form of amyloid-β and of amyloid-β bound to apolipoprotein E2 (APOE2), APOE3 or α2-macroglobulin from the brain's ISF into the blood, and APOE4 inhibits such transport. In AD, LRP1 expression at the BBB is reduced and amyloid-β binding to circulating sLRP1 is compromised by oxidation (Sagare et al. 2012). Moreover, amyloid-β damages its own LRP1-mediated transport by oxidating LRP1 (Owen et al. 2010). Defects in LRP-1-mediated Aβ clearance from the brain are thought be triggered by systemic inflammation by lipopolysaccharide, leading to increased brain accumulation of amyloid-β (Bulbarelli et al. 2012, Erickson et al. 2012).

Low-density lipoprotein receptor-related protein 2 (LRP2)-mediated transcytosis (Chun et al. 2000) eliminates amyloid-β that is bound to clusterin (also known as apolipoprotein J) by transport across the BBB, and shows a preference for the 42-amino acid form of this peptide.

Receptor for advanced glycation end products (RAGE) provides the key mechanism for influx of peripheral amyloid-β into the brain across the BBB either as a free, unbound plasma-derived peptide and/or by amyloid-β-laden monocytes. Faulty vascular clearance of amyloid-β from the brain and/or an increased reentry of peripheral amyloid-β across the blood vessels into the brain can elevate amyloid-β levels in the brain parenchyma and around cerebral blood vessels. At pathophysiological concentrations, amyloid-β forms neurotoxic oligomers and also self-aggregates, which leads to the development of cerebral β-amyloidosis and cerebral amyloid angiopathy. RAGE provides the key mechanism for the influx of peripheral amyloid-β into the brain across the BBB either as a free, unbound plasma-derived peptide and/or by amyloid-β-laden monocytes. Faulty vascular clearance of amyloid-β from the brain and/or an increased reentry of peripheral amyloid-β across the blood vessels into the brain can elevate amyloid-β levels in the brain parenchyma and around cerebral blood vessels. At pathophysiological concentrations, amyloid-β forms neurotoxic oligomers and also self-aggregates, which leads to the development of cerebral β-amyloidosis and cerebral amyloid angiopathy (Zlokovic 2011).

More insight into the molecular mechanisms underlying amyloid-β-RAGE interaction-induced alterations in the BBB have been provided by Kook et al. (2012). They found that $A\beta_{1-42}$ induces enhanced permeability, disruption of zonula occludin-1 (ZO-1) expression in the plasma membrane, and increased intracellular calcium and MMPs secretion in cultured ECs *in vitro*, and disrupted microvessels near amyloid-β plaque-deposited areas, elevated RAGE expression, and enhanced MMP secretion in microvessels of the brains of 5XFAD mice, an animal model for AD.

Cellular prion protein: There are also indications of transcytosis by cellular prion protein (PrP(c)) that binds amyloid-β (1-40) (Pflanzner et al. 2012).

Amyloid-β-degrading enzymes: Cerebral ECs, pericytes, vascular smooth muscle cells, astrocytes, microglia, and neurons express different amyloid-β-degrading enzymes, including neprilysin, insulin-degrading enzyme, tissue plasminogen activator, and MMPs, which contribute to amyloid-β clearance. In the circulation, amyloid-β is bound mainly to soluble LRP1 (sLRP1), which normally prevents its entry into the brain. Systemic clearance of amyloid-β is mediated by its removal by the liver and kidneys.

Other Transporters at the Blood–Brain Barrier

Glutamate transporters: The transporters, EAAT1, EAAT2, and EAAT3, at the BBB determine the levels of brain extracellular glutamate and are essential to prevent excitotoxicity (Lipton 2005), prompting the question whether changes in these transporters may contribute to glutamate excess and excitotoxicity. It has been suggested that glutamate excitotoxicity plays a role in the neurodegenerative processes in AD (Lipton 2005). Strict control L-glutamate concentration in the brain ISF is important to maintain neurotransmission and avoid excitotoxicity. The role of astrocytes in handling L-glutamate transport and metabolism is well known, however, ECs may also play an important role through mediating brain-to-blood L-glutamate efflux. These can account for high affinity concentrative uptake of L-glutamate from the brain ISF into the capillary ECs. The mechanisms in between L-glutamate uptake in the ECs and L-glutamate appearing in the blood may involve a luminal transporter for L-glutamate, metabolism of L-glutamate, and transport of metabolites, or a combination of the two (Cederberg et al. 2014).

Na(+)-dependent transporters for glutamate exist on astrocytes (EAAT1 and EAAT2) and neurons (EAAT3). These transporters presumably assist in keeping the glutamate concentration low in the extracellular fluid of the brain. Recently, Na(+)-dependent glutamate transport on the abluminal membrane of the BBB was described (O'Kane et al. 1999).

Glucose transporters: Also, facilitative glucose transport in the brain is affected in different pathophysiological conditions including AD (Guo et al. 2005). Protein expression of the glucose transporter GLUT1 is reduced in brain capillaries in AD, without changes in GLUT1 mRNA structure (Mooradian et al. 1997) or levels of GLUT1 mRNA transcripts (Wu et al. 2005). Furthermore, a reduction in CNS energy metabolites has been seen in several PET scanning studies of AD's patients using fluoro-deoxy-glucose (FDG) (Mosconi et al. 2006, Samuraki et al. 2007, Mosconi et al. 2008), likely because the surface area at the BBB available for glucose transport is substantially reduced in AD (Bailey et al. 2004, Wu et al. 2005). Furthermore, GLUT1 deficiency in mice overexpressing amyloid β-peptide precursor protein leads to early cerebral microvascular degeneration, blood flow reductions, dysregulation, and BBB breakdown. Also, it leads to accelerated amyloid β-peptide pathology, reduced amyloid β clearance, diminished neuronal activity, behavioral deficits, progressive neuronal loss, and neurodegeneration that develop after initial cerebrovascular degenerative changes. Moreover, GLUT1 deficiency in endothelium, but not in astrocytes, initiates the vascular phenotype as shown by BBB breakdown. This indicates that reduced BBB GLUT1 expression worsens AD cerebrovascular degeneration, neuropathology, and cognitive function (Winkler et al. 2015).

Miscellaneous

MMP-2 and *MMP-9.* MMPs are a family of enzymes able to degrade components of the extracellular matrix, which is important for normal BBB function. MMP-2 and MMP-9 have been implicated in the physiological catabolism of AD's amyloid-β. Conversely, their association with vascular amyloid deposits, BBB disruption, and hemorrhagic transformations after ischemic stroke also highlights their involvement in pathological processes (Hernandez-Guillamon et al. 2015).

MMP function is regulated by tissue inhibitors of matrix metalloproteinases (TIMPs). Specifically, the metalloproteinases MMP-2 and MMP9 have been associated in BBB breakdown in NDs, including AD (Duits et al. 2015, Weekman and Wilcock 2015).

Chemokines in the brain can recruit immune cells from the blood or from within the brain (Britschgi and Wyss-Coray 2007) to secrete MMP-2 and MMP-9 that increase BBB permeability (Feng et al. 2001). Inhibition of this process is linked to more rapid disease progression (Dimitrijevic et al. 2007).

APOE4 Homozygosity

While it is well appreciated that APOE4 homozygosity is associated with an increased risk of sporadic AD, its effects on the brain microvasculature and BBB have been less appreciated. Interestingly, APOE(4,4) is associated with thinning of the microvascular BM in AD (Bell et al. 2012). In APOE4 transgenic mice, a high fat diet induced deleterious effects on BBB permeability (Mulder et al. 2001). A recent study by Bell et al. (2012) suggested that CypA is a key target for treating APOE4-mediated neurovascular injury and the resulting neuronal dysfunction and degeneration; indeed, activating a proinflammatory CypA-nuclear factor-κB-MMP-9 pathway in pericytes is associated with increased susceptibility of the BBB to injury in APOE4 conditions (Zhao et al. 2015).

Cerebral microbleeds. Cerebral microbleeds as consequences of BBB breakdown in AD are thought to represent cerebral amyloid angiopathy. Cerebral microbleeds are seen as possible predictors of intracerebral hemorrhage. It was found that the incidence of brain microbleeds positively correlates with age (Yates et al. 2014, Shams et al. 2015). Their increasing prevalence with an increasing number of risk factors (hypertension, hyperlipidemia, diabetes, male sex, and advanced age) (Yates et al. 2014, Shams et al. 2015) and lower levels of TIMPs in AD patients with microbleeds suggest less MMP inhibition in patients with concurrent cerebral microbleeds (Duits et al. 2015).

BLOOD–BRAIN BARRIER IN MULTIPLE SCLEROSIS

Formation of MS focal lesions follows extravasation of activated leukocytes from blood through the BBB into the CNS. Once the activated leukocytes enter the CNS environment, they propagate massive destruction to finally result in the loss of both myelin/oligodendrocyte complex and neurodegeneration. Also, the activated leukocytes locally release inflammatory cytokines and chemokines leading to focal immune activation of the brain ECs, and loss of normal functioning of the BBB (Bartholomaus et al. 2009, Greenwood et al. 2011).

While peripheral blood leukocyte infiltration plays an essential role in lesion development, there is also evidence suggesting that BBB dysfunction precedes immune cell infiltration. Recent evidence suggests that immune-mediated activation (or damage) of the various BBB cellular components significantly contributes to lesion development and progression (McQuaid et al. 2009, Alvarez et al. 2011a, b). Chemokines seem to play an important role in the cascade of leukocyte extravasation. Chemokines displayed along the endothelial lumen bind chemokine receptors on circulating leukocytes, initiating intracellular signaling that culminates in integrin activation, leukocyte arrest, and extravasation (Holman et al. 2011).

Tight Junctions

Increased permeability of the BBB is associated with decreased expression of TJ and AJ proteins in the brain capillary ECs. Dephosphorylation of occludin in a MS mouse model precedes visible signs of disease, before changes in the BBB permeability were observed (Morgan et al. 2007). Tight junctional abnormalities are most common in active lesions, but are present in inactive lesions in normal-appearing white matter. Tight junctional abnormalities are positively associated with leakage of the serum protein fibrinogen which has recently been shown to be an activator of microglia (Persidsky et al. 2006, McQuaid et al. 2009).

While the complex network of molecular players that leads to BBB dysfunction in MS is yet to be fully elucidated, recent studies indicated a critical role for microRNAs (miRNAs) in controlling the function of the barrier endothelium in the brain (Kamphuis, 2015).

Multiple Matrix Proteins (MMPs)

In MS MMPs seem to have implications (Rosenberg and Yang 2007, Yang and Rosenberg 2011), as increased activity of MMPs may attack the TJ protein occludin. An increase in MMP-9 activity has been demonstrated at sites of BBB disruption showing leukocyte infiltration. Moreover, the timing of MMP activity in pial and parenchymal vessels correlated with the timing of increased BBB permeability (Persidsky et al. 2006).

Active Transporters

P-glycoprotein (p-gp), multidrug resistance-related protein-1 (MRP1), BCRP: P-gp functionality seems to be impaired in neuroinflammation and may play a role in immunomodulation (Kooij et al. 2009, Kooij et al. 2010). P-gp expression and function are strongly decreased during neuroinflammation. *In vivo*, the expression and function of brain endothelial P-gp in experimental allergic encephalomyelitis (EAE), an animal model for MS, were significantly impaired. Strikingly, vascular P-gp expression was decreased in both MS and EAE lesions and its disappearance coincided with the presence of perivascular infiltrates consisting of lymphocytes (Kooij et al. 2010).

Cerebrovascular expression of P-glycoprotein was decreased in both active and chronic inactive MS lesions. Moreover, foamy macrophages in active MS lesions showed enhanced expression of MRP-1 and BCRP (Kooij et al. 2011). Also, CD8 (+) T lymphocyte trafficking into the brain is dependent on the activity of P-gp (Kooij et al. 2014).

Blood–Brain Barrier in Parkinson's Disease

Using histologic markers of serum protein, iron, and erythrocyte extravasation, increased permeability of the BBB in part of the caudate putamen of PD patients has been significantly shown (Gray and Woulfe 2015). It has been established that the process of reactive gliosis is a common feature of astrocytes during BBB disruption, which may have implications of astrocyte functions in the protection of the BBB, and in the development of PD (Cabezas et al. 2014).

It seems that alterations in BBB permeability can also be reflected by changes in CSF/blood albumin, although changes in CSF turnover might also be responsible for such findings. Significant differences in albumin CSF/serum ratios (AR), based on samples obtained from non-demented subjects with idiopathic PD and age-matched control subjects, were found between patients with advanced disease, and both early stage and unaffected groups. Conversely, early-phase patients did not differ from healthy subjects. In addition, dopaminergic treatment seems to exert a possible effect on AR values. This indicates possible dysfunction of the BBB (and/or blood-CSF-barrier) in PD progression (Pisani et al. 2012).

Tight Junctions

In a intracerebral rotenone model of PD in rats, using fluorescein as a tight junctional BBB permeability marker, no changes were detected (Ravenstijn et al. 2008).

Active Transporters

P-glycoprotein: Brain distribution studies of paradigm compounds selected, based upon their differential transport across the BBB (l-3,4-dihydroxyphenylalanine, carbamazepine, quinidine, lovastatin, and simvastatin) was analyzed in healthy and 1-methyl-4-phenyl-1,2,3,6-tetrahydropiridine (MPTP)-treated macaques. Only changes in brain distribution of quinidine were found, indicating changes in P-gp functionality (Thiollier et al. 2016). In contrast, studies carried out by Hou et al. (2014) suggest that P-gp inhibition increases BBB permeability to *N*-[2-(4-hydroxy-phenyl)-ethyl]-2-(2,5-dimethoxy-phenyl)-3-(3-methoxy-4-hydroxy-phenyl)-acrylamide (FLZ), a novel synthetic squamosamide derivative and potential anti-PD agent. However, no significant differences were observed in the brain distribution of FLZ between normal and PD model rats, suggesting no significant change in P-gp in PD (Hou et al. 2014).

Bartels et al. (2008a, b) investigated *in vivo* BBB P-gp function in patients with parkinsonian neurodegenerative syndromes, using [11C]-verapamil PET in PD patients. Advanced PD patients had increased [11C]-verapamil uptake in frontal white matter regions compared to controls; while *de novo* PD patients. The authors concluded that lower [11C]-verapamil uptake in midbrain and frontal regions of *de novo* PD patients could indicate a regional upregulation of P-gp function (Bartels et al. 2008a, b). However, in a later study of this group, the decreased BBB P-gp function in early stage PD patients could not be confirmed (Bartels et al. 2008a, b).

LRP1: Alpha-synuclein (α-Syn), being one of the dominant proteins found in Lewy bodies in brains of PD, has been found in body fluids, including blood and CSF, and is likely produced by both peripheral tissues and the CNS. Radioactively labeled α-Syn has been found to cross the BBB in both the brain-to-blood and the blood-to-brain directions at rates consistent with saturable mechanisms. LRP-1, (but not P-gp) seems to be involved in α-Syn efflux (Bates and Zheng 2014, Sui et al. 2014).

Large neutral amino acid transporter (LNAA): BBB transport of L-DOPA (L-dihydroxy-phanyl-alanin) transport in conjunction with its intra-brain conversion was studied in both control and diseased cerebral hemispheres in the unilateral rat rotenone model of PD. PD-like pathology, indicated by a huge reduction of tyrosine hydroxylase as well as by substantially reduced levels and higher elimination rates of dihydoxy-phenyl-acetatic acid (DOPAC) and homovanillic acid (HVA), does not result in changes in BBB transport of L-DOPA (Ravenstijn et al. 2012).

Matrix Metalloproteinases

Like for AD and MS, MMPs seem to have implications in PD (Rosenberg and Yang 2007, Yang and Rosenberg 2011), and are associated in neurodegeneration of dopaminergic neurons (Rosenberg 2009). Altogether it can be concluded that there is much controversy on the role of the BBB in PD.

Blood–Brain Barrier in Pharmacoresistant Epilepsy

Older studies already indicated that seizures induce BBB transport changes (Padou et al. 1995, Sahin et al. 2003). Focal epilepsies are often associated with BBB leakage. For example, BBB leakage to albumin-bound Evans blue has been found in pentylenetetrazol (PTZ)-induced epilepsy, with the location and pattern depending on the rat strain (Ates et al. 1999). When a normal brain develops epilepsy (epileptogenesis), the immunoglobulin G (IgG) leaks across the BBB and neuronal IgG uptake increases concomitantly with the occurrence of seizures. IgG-positive neurons show signs of neurodegeneration, such as shrinkage and eosinophilia. This may suggest that IgG leakage across the BBB is related to neuronal impairment and may be a pathogenic mechanism in epileptogenesis and chronic epilepsy (Ndode-Ekane et al. 2010, Michalak et al. 2012). The information on pharmacoresistant epilepsy associated changes in the BBB is relatively sparse.

Tight Junctions

Claudin-8: Selective downregulation of claudin-8 by kindling epilepsy (Lamas et al. 2002) suggests that selective modulation of claudin expression in response to abnormal neuronal synchronization may lead to BBB breakdown and brain edema.

Transport Systems

Facilitative glucose transport: Facilitative glucose transport in the brain is affected in different pathophysiological conditions including epilepsy. It has been shown that GLUT1 mediates BBB transport of some neuroactive drugs, such as glycosylated neuropeptides, low molecular weight heparin, and D-glucose derivatives (Guo et al. 2005).

P-glycoprotein: A role for ATP-binding cassette (ABC) transporters in the pathogenesis and treatment of pharmacoresisant epilepsy has been proposed (Marchi et al. 2004, Bankstahl et al. 2011, Loscher et al. 2011). A positive association between the polymorphism in the MDR1 gene encoding P-gp (/ABCB1) and pharmacoresistant epilepsy has been reported in a subset of epilepsy patients (Siddiqui et al. 2003). However, the follow-up association genetics studies did not support a major role for this polymorphism (Tate and Sisodiya 2007, Sisodiya and Mefford 2011). Then, an increased expression of P-gp at the BBB has been reported, which was determined in epileptogenic brain tissue of patients with pharmacoresistant epilepsy (Dombrowski et al. 2001, Desai et al. 2007) as well as in rodent models of temporal lobe epilepsy, including the pilocarpine model. Other studies point to a profound role of seizure-induced neuronal cyclooxygenase-2 (COX-2) expression in neuropathologies that accompany epileptogenesis (Serrano et al. 2011) and it is thought that epileptic seizures drive expression of the BBB efflux transporter P-gp via a glutamate/COX-2-mediated signaling pathway.

CONCLUSION

In conclusion, we can say that the breakdown and leakage of the BBB are among the initiating factors in the process of physiological brain aging and also in the different neurodegenerative pathologies associated with advanced aged. The investigation of risk factors (genetic factors, oxidative stress, protein missfolding, and accumulation, leukocyte adhesion, immune cell extravasation, pathogens, etc.) inducing BBB disruption is the focus of the research. The fundamental role of membrane transporters located at the cerebral microvasculature is also revealed. This chapter intends to give an overview of the scientific literature on the structural and functional consequences of aging for the BBB including embryonic development, brain maturation, postnatal changes, and senescence. We hope that this summary will generate new ideas and help in the orientation of new projects for drug design against chronic neurodegenerative disorders with unmet needs.

ACKNOWLEDGMENTS

The authors sincerely appreciate Kinga Tóth for electron microscopy, Szimonetta Tamás for preparation of figures, and Tímea Rosta for editing the manuscript.

REFERENCES

Abbott, N. J. 2004. Evidence for bulk flow of brain interstitial fluid: Significance for physiology and pathology. *Neurochem Int* 45(4): 545–552.

Abbott, N. J. and A. Friedman 2012. Overview and introduction: The blood–brain barrier in health and disease. *Epilepsia* 53 (Suppl 6): 1–6.

Abbott, N. J., A. A. Patabendige, D. E. Dolman, S. R. Yusof and D. J. Begley 2010. Structure and function of the blood–brain barrier. *Neurobiol Dis* 37(1): 13–25.

Abbott, N. J., L. Ronnback and E. Hansson 2006. Astrocyte–endothelial interactions at the blood–brain barrier. *Nat Rev Neurosci* 7(1): 41–53.

Abdel-Salam, O. M. 2014. The paths to neurodegeneration in genetic Parkinson's disease. *CNS Neurol Disord Drug Targets* 13(9): 1485–1512.

Abeti, R. and A. Y. Abramov 2015. Mitochondrial Ca(2+) in neurodegenerative disorders. *Pharmacol Res* 99: 377–381.

Abeti, R., A. Y. Abramov and M. R. Duchen 2011. Beta-amyloid activates PARP causing astrocytic metabolic failure and neuronal death. *Brain* 134(Pt 6): 1658–1672.

Acharya, N. K., E. C. Levin, P. M. Clifford, M. Han, R. Tourtellotte, D. Chamberlain, M. Pollaro, N. J. Coretti, M. C. Kosciuk, E. P. Nagele, C. Demarshall, T. Freeman, Y. Shi, C. Guan, C. H. Macphee, R. L. Wilensky and R. G. Nagele 2013. Diabetes and hypercholesterolemia increase blood–brain barrier permeability and brain amyloid deposition: Beneficial effects of the LpPLA2 inhibitor darapladib. *J Alzheimers Dis* 35(1): 179–198.

Ahmad, S. I. ed. 2012. *Neurodegenerative Diseases. Advances in Experimental Medicine and Biology*, vol. 724. Springer and Landes Bioscience, New York, ISBN 978-1-4614-0652-5.

Aimé, P., X. Sun, N. Zareen, A. Rao, Z. Berman, L. Volpicelli-Daley, P. Bernd, J. F. Crary, O. A. Levy and L. A. Greene 2015. Trib3 is elevated in Parkinson's disease and mediates death in Parkinson's disease models. *J Neurosci* 35(30): 10731–10749.

Al Ahmad, A., M. Gassmann and O. O. Ogunshola 2012. Involvement of oxidative stress in hypoxia-induced blood–brain barrier breakdown. *Microvasc Res* 84(2): 222–225.

Alvarez, J. I., R. Cayrol and A. Prat 2011a. Disruption of central nervous system barriers in multiple sclerosis. *Biochim Biophys Acta* 1812(2): 252–264.

Alvarez, J. I., A. Dodelet-Devillers, H. Kebir, I. Ifergan, P. J. Fabre, S. Terouz, M. Sabbagh, K. Wosik, L. Bourbonniere, M. Bernard, J. van Horssen, H. E. de Vries, F. Charron and A. Prat 2011b. The Hedgehog pathway promotes blood–brain barrier integrity and CNS immune quiescence. *Science* 334(6063): 1727–1731.

Ambegaokar, S. S., B. Roy and G. R. Jackson 2010. Neurodegenerative models in *Drosophila*: Polyglutamine disorders, Parkinson disease, and amyotrophic lateral sclerosis. *Neurobiol Dis* 40(1): 29–39.

Anderson, K. D., L. Pan, X. M. Yang, V. C. Hughes, J. R. Walls, M. G. Dominguez, M. V. Simmons, P. Burfeind, Y. Xue, Y. Wei, L. E. Macdonald, G. Thurston, C. Daly, H. C. Lin, A. N. Economides, D. M. Valenzuela, A. J. Murphy, G. D. Yancopoulos and N. W. Gale 2011. Angiogenic sprouting into neural tissue requires Gpr124, an orphan G protein-coupled receptor. *Proc Natl Acad Sci USA* 108(7): 2807–2812.

Arendt, T., M. K. Bruckner, B. Mosch and A. Losche 2010. Selective cell death of hyperploid neurons in Alzheimer's disease. *Am J Pathol* 177(1): 15–20.

Ariga, T., W. D. Jarvis and R. K. Yu 1998. Role of sphingolipid-mediated cell death in neurodegenerative diseases. *J Lipid Res* 39(1): 1–16.

Armulik, A., G. Genove and C. Betsholtz 2011. Pericytes: Developmental, physiological, and pathological perspectives, problems, and promises. *Dev Cell* 21(2): 193–215.

Arnold, S. 2012. The power of life—Cytochrome c oxidase takes center stage in metabolic control, cell signalling and survival. *Mitochondrion* 12(1): 46–56.

Artal-Sanz, M. and N. Tavernarakis 2005. Proteolytic mechanisms in necrotic cell death and neurodegeneration. *FEBS Lett* 579(15): 3287–3296.

Ates, N., N. Esen and G. Ilbay 1999. Absence epilepsy and regional blood–brain barrier permeability: The effects of pentylenetetrazole-induced convulsions. *Pharmacol Res* 39(4): 305–310.

Baeten, K. M. and K. Akassoglou 2011. Extracellular matrix and matrix receptors in blood–brain barrier formation and stroke. *Dev Neurobiol* 71(11): 1018–1039.

Bailey, T. L., C. B. Rivara, A. B. Rocher and P. R. Hof 2004. The nature and effects of cortical microvascular pathology in aging and Alzheimer's disease. *Neurol Res* 26(5): 573–578.

Bains, J. S. and C. A. Shaw 1997. Neurodegenerative disorders in humans: The role of glutathione in oxidative stress-mediated neuronal death. *Brain Res Brain Res Rev* 25(3): 335–358.

Baloyannis, S. J. 2015. Brain capillaries in Alzheimer's disease. *Hell J Nucl Med* 18(Suppl 1): 152.

Banerjee, R., M. F. Beal and B. Thomas 2010. Autophagy in neurodegenerative disorders: Pathogenic roles and therapeutic implications. *Trends Neurosci* 33(12): 541–549.

Bankstahl, J. P., M. Bankstahl, C. Kuntner, J. Stanek, T. Wanek, M. Meier, X. Q. Ding, M. Muller, O. Langer and W. Loscher 2011. A novel positron emission tomography imaging protocol identifies seizure-induced regional overactivity of P-glycoprotein at the blood–brain barrier. *J Neurosci* 31(24): 8803–8811.

Bartels, A. L., B. N. van Berckel, M. Lubberink, G. Luurtsema, A. A. Lammertsma and K. L. Leenders 2008a. Blood–brain barrier P-glycoprotein function is not impaired in early Parkinson's disease. *Parkinsonism Relat Disord* 14(6): 505–508.

Bartels, A. L., A. T. Willemsen, R. Kortekaas, B. M. de Jong, R. de Vries, O. de Klerk, J. C. van Oostrom, A. Portman and K. L. Leenders 2008b. Decreased blood–brain barrier P-glycoprotein function in the progression of Parkinson's disease, PSP and MSA. *J Neural Transm (Vienna)* 115(7): 1001–1009.

Bartholomaus, I., N. Kawakami, F. Odoardi, C. Schlager, D. Miljkovic, J. W. Ellwart, W. E. Klinkert, C. Flugel-Koch, T. B. Issekutz, H. Wekerle and A. Flugel 2009. Effector T cell interactions with meningeal vascular structures in nascent autoimmune CNS lesions. *Nature* 462(7269): 94–98.

Barzilai, A. 2010. DNA damage, neuronal and glial cell death and neurodegeneration. *Apoptosis* 15(11): 1371–1381.

Bates, C. A. and W. Zheng 2014. Brain disposition of alpha-Synuclein: Roles of brain barrier systems and implications for Parkinson's disease. *Fluids Barriers CNS* 11: 17.

Batlevi, Y. and A. R. La Spada 2011. Mitochondrial autophagy in neural function, neurodegenerative disease, neuron cell death, and aging. *Neurobiol Dis* 43(1): 46–51.

Bell, R. D., E. A. Winkler, I. Singh, A. P. Sagare, R. Deane, Z. Wu, D. M. Holtzman, C. Betsholtz, A. Armulik, J. Sallstrom, B. C. Berk and B. V. Zlokovic 2012. Apolipoprotein E controls cerebrovascular integrity via cyclophilin A. *Nature* 485(7399): 512–516.

Bell, R. D. and B. V. Zlokovic 2009. Neurovascular mechanisms and blood–brain barrier disorder in Alzheimer's disease. *Acta Neuropathol* 118(1): 103–113.

Ben-Menachem, E., B. B. Johansson and T. H. Svensson 1982. Increased vulnerability of the blood–brain barrier to acute hypertension following depletion of brain noradrenaline. *J Neural Transm* 53(2–3): 159–167.

Berezowski, V., C. Landry, S. Lundquist, L. Dehouck, R. Cecchelli, M. P. Dehouck and L. Fenart 2004. Transport screening of drug cocktails through an in vitro blood–brain barrier: Is it a good strategy for increasing the throughput of the discovery pipeline? *Pharm Res* 21(5): 756–760.

Bertram, L. and R. E. Tanzi 2005. The genetic epidemiology of neurodegenerative disease. *J Clin Invest* 115(6): 1449–1457.

Betz, A. L., J. A. Firth and G. W. Goldstein 1980. Polarity of the blood–brain barrier: Distribution of enzymes between the luminal and antiluminal membranes of brain capillary endothelial cells. *Brain Res* 192(1): 17–28.

Bove, J. and C. Perier 2012. Neurotoxin-based models of Parkinson's disease. *Neuroscience* 211: 51–76.

Brenn, A., M. Grube, M. Peters, A. Fischer, G. Jedlitschky, H. K. Kroemer, R. W. Warzok and S. Vogelgesang 2011. Beta-amyloid downregulates MDR1-P-glycoprotein (Abcb1) expression at the blood–brain barrier in mice. *Int J Alzheimers Dis* 2011: 690121.

Britschgi, M. and T. Wyss-Coray 2007. Immune cells may fend off Alzheimer disease. *Nat Med* 13(4): 408–409.

Bulbarelli, A., E. Lonati, A. Brambilla, A. Orlando, E. Cazzaniga, F. Piazza, C. Ferrarese, M. Masserini and G. Sancini 2012. Abeta42 production in brain capillary endothelial cells after oxygen and glucose deprivation. *Mol Cell Neurosci* 49(4): 415–422.

Burns, E. M., T. W. Kruckeberg and P. K. Gaetano 1981. Changes with age in cerebral capillary morphology. *Neurobiol Aging* 2(4): 283–291.

Cabezas, R., M. Avila, J. Gonzalez, R. S. El-Bacha, E. Baez, L. M. Garcia-Segura, J. C. Jurado Coronel, F. Capani, G. P. Cardona-Gomez and G. E. Barreto 2014. Astrocytic modulation of blood brain barrier: Perspectives on Parkinson's disease. *Front Cell Neurosci* 8: 211.

Candiello, J., G. J. Cole and W. Halfter 2010. Age-dependent changes in the structure, composition and biophysical properties of a human basement membrane. *Matrix Biol* 29(5): 402–410.

Carvalho, A. N., J. L. Lim, P. G. Nijland, M. E. Witte and J. Van Horssen 2014. Glutathione in multiple sclerosis: More than just an antioxidant? *Mult Scler* 20(11): 1425–1431.

Cederberg, H. H., N. C. Uhd and B. Brodin 2014. Glutamate efflux at the blood–brain barrier: Cellular mechanisms and potential clinical relevance. *Arch Med Res* 45(8): 639–645.

Chakrabarti, S. S., A. Bir, J. Poddar, M. Sinha, A. Ganguly and S. Chakrabarti 2016. Ceramide and sphingosine-1-phosphate in cell death pathways: Relevance to the pathogenesis of Alzheimer's disease. *Curr Alzheimer Res* 13(11): 1232–1248.

Chu, Y. and J. H. Kordower 2015. The prion hypothesis of Parkinson's disease. *Curr Neurol Neurosci Rep* 15(5): 28.

Chun, H. S., J. J. Son and J. H. Son 2000. Identification of potential compounds promoting BDNF production in nigral dopaminergic neurons: Clinical implication in Parkinson's disease. *Neuroreport* 11(3): 511–514.

Cohen, Z., G. Bonvento, P. Lacombe and E. Hamel 1996. Serotonin in the regulation of brain microcirculation. *Prog Neurobiol* 50(4): 335–362.

Cohen, Z., G. Molinatti and E. Hamel 1997. Astroglial and vascular interactions of noradrenaline terminals in the rat cerebral cortex. *J Cereb Blood Flow Metab* 17(8): 894–904.

Coppede, F., M. Mancuso, G. Siciliano, L. Migliore and L. Murri 2006. Genes and the environment in neuro-degeneration. *Biosci Rep* 26(5): 341–367.

Crone, C. and O. Christensen 1981. Electrical resistance of a capillary endothelium. *J Gen Physiol* 77(4): 349–371.

Cullen, M., M. K. Elzarrad, S. Seaman, E. Zudaire, J. Stevens, M. Y. Yang, X. Li, A. Chaudhary, L. Xu, M. B. Hilton, D. Logsdon, E. Hsiao, E. V. Stein, F. Cuttitta, D. C. Haines, K. Nagashima, L. Tessarollo and B. St Croix 2011. GPR124, an orphan G protein-coupled receptor, is required for CNS-specific vascularization and establishment of the blood–brain barrier. *Proc Natl Acad Sci USA* 108(14): 5759–5764.

Dalkara, T., Y. Gursoy-Ozdemir and M. Yemisci 2011. Brain microvascular pericytes in health and disease. *Acta Neuropathol* 122(1): 1–9.

Daneman, R., D. Agalliu, L. Zhou, F. Kuhnert, C. J. Kuo and B. A. Barres 2009. Wnt/beta-catenin signaling is required for CNS, but not non-CNS, angiogenesis. *Proc Natl Acad Sci USA* 106(2): 641–646.

De Bock, M., V. Van Haver, R. E. Vandenbroucke, E. Decrock, N. Wang and L. Leybaert 2016. Into rather unexplored terrain-transcellular transport across the blood–brain barrier. *Glia* 64(7): 1097–1123.

De Bock, M., R. E. Vandenbroucke, E. Decrock, M. Culot, R. Cecchelli and L. Leybaert 2014. A new angle on blood-CNS interfaces: A role for connexins? *FEBS Lett* 588(8): 1259–1270.

De Vos, K. J., A. J. Grierson, S. Ackerley and C. C. Miller 2008. Role of axonal transport in neurodegenerative diseases. *Annu Rev Neurosci* 31: 151–173.

Desai, B. S., A. J. Monahan, P. M. Carvey and B. Hendey 2007. Blood–brain barrier pathology in Alzheimer's and Parkinson's disease: Implications for drug therapy. *Cell Transplant* 16(3): 285–299.

Dimitrijevic, O. B., S. M. Stamatovic, R. F. Keep and A. V. Andjelkovic 2007. Absence of the chemokine receptor CCR2 protects against cerebral ischemia/reperfusion injury in mice. *Stroke* 38(4): 1345–1353.

DiNapoli, V. A., J. D. Huber, K. Houser, X. Li and C. L. Rosen 2008. Early disruptions of the blood–brain barrier may contribute to exacerbated neuronal damage and prolonged functional recovery following stroke in aged rats. *Neurobiol Aging* 29(5): 753–764.

Dombrowski, S. M., S. Y. Desai, M. Marroni, L. Cucullo, K. Goodrich, W. Bingaman, M. R. Mayberg, L. Bengez and D. Janigro 2001. Overexpression of multiple drug resistance genes in endothelial cells from patients with refractory epilepsy. *Epilepsia* 42(12): 1501–1506.

Duits, F. H., M. Hernandez-Guillamon, J. Montaner, J. D. Goos, A. Montanola, M. P. Wattjes, F. Barkhof, P. Scheltens, C. E. Teunissen and W. M. van der Flier 2015. Matrix metalloproteinases in Alzheimer's disease and concurrent cerebral microbleeds. *J Alzheimers Dis* 48(3): 711–720.

Elahy, M., C. Jackaman, J. C. Mamo, V. Lam, S. S. Dhaliwal, C. Giles, D. Nelson and R. Takechi 2015. Blood–brain barrier dysfunction developed during normal aging is associated with inflammation and loss of tight junctions but not with leukocyte recruitment. *Immun Ageing* 12: 2.

Engelhardt, B. and R. M. Ransohoff 2005. The ins and outs of T-lymphocyte trafficking to the CNS: Anatomical sites and molecular mechanisms. *Trends Immunol* 26(9): 485–495.

Erdo, F., L. Denes and E. de Lange 2017. Age-associated physiological and pathological changes at the blood–brain barrier: A review. *J Cereb Blood Flow Metab* 37(1): 4–24.

Erdo, F., B. Hutka and L. Denes 2016. Function, aging and dysfunction of blood–brain barrier. Crossing the barrier. *Orv Hetil* 157(51): 2019–2027.

Erich Blechschmit, B. F. 2004. *A Biodynamic Approach to Development from Conception to Birth. The Ontogenic Basis of Human Anatomy*. North Atlantic Books, Berkeley, California. pp. 43, 65–67.

Erickson, M. A., P. E. Hartvigson, Y. Morofuji, J. B. Owen, D. A. Butterfield and W. A. Banks 2012. Lipopolysaccharide impairs amyloid beta efflux from brain: Altered vascular sequestration, cerebro-spinal fluid reabsorption, peripheral clearance and transporter function at the blood–brain barrier. *J Neuroinflammation* 9: 150.

Ezan, P., P. Andre, S. Cisternino, B. Saubamea, A. C. Boulay, S. Doutremer, M. A. Thomas, N. Quenech'du, C. Giaume and M. Cohen-Salmon 2012. Deletion of astroglial connexins weakens the blood–brain barrier. *J Cereb Blood Flow Metab* 32(8): 1457–1467.

Farrall, A. J. and J. M. Wardlaw 2009. Blood-brain barrier: Ageing and microvascular disease--systematic review and meta-analysis. *Neurobiol Aging* 30(3): 337–352.

Feng, M. R., D. Turluck, J. Burleigh, R. Lister, C. Fan, A. Middlebrook, C. Taylor and T. Su 2001. Brain microdialysis and PK/PD correlation of pregabalin in rats. *Eur J Drug Metab Pharmacokinet* 26(1–2): 123–128.

Fenstermacher, J., P. Gross, N. Sposito, V. Acuff, S. Pettersen and K. Gruber 1988. Structural and functional variations in capillary systems within the brain. *Ann N Y Acad Sci* 529: 21–30.

Finder, V. H. 2010. Alzheimer's disease: A general introduction and pathomechanism. *J Alzheimers Dis* 22 (Suppl 3): 5–19.

Forster, C. 2008. Tight junctions and the modulation of barrier function in disease. *Histochem Cell Biol* 130(1): 55–70.

Freeman, L. R. and J. N. Keller 2012. Oxidative stress and cerebral endothelial cells: Regulation of the blood–brain-barrier and antioxidant based interventions. *Biochim Biophys Acta* 1822(5): 822–829.

Fulga, T. A., I. Elson-Schwab, V. Khurana, M. L. Steinhilb, T. L. Spires, B. T. Hyman and M. B. Feany 2007. Abnormal bundling and accumulation of F-actin mediates tau-induced neuronal degeneration in vivo. *Nat Cell Biol* 9(2): 139–148.

Gee, J. R. and J. N. Keller 2005. Astrocytes: Regulation of brain homeostasis via apolipoprotein E. *Int J Biochem Cell Biol* 37(6): 1145–1150.

Genetics home reference (of the National Institutes of Health, U). https://ghr.nlm.nih.gov/condition

Gorle, N., C. Van Cauwenberghe, C. Libert and R. E. Vandenbroucke 2016. The effect of aging on brain barriers and the consequences for Alzheimer's disease development. *Mamm Genome* 27(7–8): 407–420.

Grammas, P., J. Martinez and B. Miller 2011. Cerebral microvascular endothelium and the pathogenesis of neurodegenerative diseases. *Expert Rev Mol Med* 13: e19.

Gray, M. T. and J. M. Woulfe 2015. Striatal blood–brain barrier permeability in Parkinson's disease. *J Cereb Blood Flow Metab* 35(5): 747–750.

Greenwood, J., S. J. Heasman, J. I. Alvarez, A. Prat, R. Lyck and B. Engelhardt 2011. Review: Leucocyte-endothelial cell crosstalk at the blood–brain barrier: A prerequisite for successful immune cell entry to the brain. *Neuropathol Appl Neurobiol* 37(1): 24–39.

Griep, L. M., F. Wolbers, B. de Wagenaar, P. M. ter Braak, B. B. Weksler, I. A. Romero, P. O. Couraud, I. Vermes, A. D. van der Meer and A. van den Berg 2013. BBB on chip: Microfluidic platform to mechanically and biochemically modulate blood–brain barrier function. *Biomed Microdevices* 15(1): 145–150.

Grunblatt, E., S. Mandel and M. B. Youdim 2000. MPTP and 6-hydroxydopamine-induced neurodegeneration as models for Parkinson's disease: Neuroprotective strategies. *J Neurol* 247 (Suppl 2): II95–102.

Guo, X., M. Geng and G. Du 2005. Glucose transporter 1, distribution in the brain and in neural disorders: Its relationship with transport of neuroactive drugs through the blood–brain barrier. *Biochem Genet* 43(3–4): 175–187.

Halliday, M. R., S. V. Rege, Q. Ma, Z. Zhao, C. A. Miller, E. A. Winkler and B. V. Zlokovic 2016. Accelerated pericyte degeneration and blood–brain barrier breakdown in apolipoprotein E4 carriers with Alzheimer's disease. *J Cereb Blood Flow Metab* 36(1): 216–227.

Harry, G. J. 2013. Microglia during development and aging. *Pharmacol Ther* 139(3): 313–326.

Hartz, A. M., D. S. Miller and B. Bauer 2010. Restoring blood–brain barrier P-glycoprotein reduces brain amyloid-beta in a mouse model of Alzheimer's disease. *Mol Pharmacol* 77(5): 715–723.

Hartz, A. M., Y. Zhong, A. Wolf, H. LeVine, 3rd, D. S. Miller and B. Bauer 2016. Abeta40 reduces P-glycoprotein at the blood–brain barrier through the ubiquitin-proteasome pathway. *J Neurosci* 36(6): 1930–1941.

Hawkins, B. T. and T. P. Davis 2005. The blood–brain barrier/neurovascular unit in health and disease. *Pharmacol Rev* 57(2): 173–185.

Hendriks, J. J., C. E. Teunissen, H. E. de Vries and C. D. Dijkstra 2005. Macrophages and neurodegeneration. *Brain Res Brain Res Rev* 48(2): 185–195.

Hernandez-Guillamon, M., S. Mawhirt, S. Blais, J. Montaner, T. A. Neubert, A. Rostagno and J. Ghiso 2015. Sequential amyloid-beta degradation by the matrix metalloproteases MMP-2 and MMP-9. *J Biol Chem* 290(24): 15078–15091.

Holman, D. W., R. S. Klein and R. M. Ransohoff 2011. The blood–brain barrier, chemokines and multiple sclerosis. *Biochim Biophys Acta* 1812(2): 220–230.

Holtzman, W. H. and M. E. Bitterman 1956. A factorial study of adjustment to stress. *J Abnorm Psychol* 52(2): 179–185.

Hou, J., Q. Liu, Y. Li, H. Sun and J. Zhang 2014. An in vivo microdialysis study of FLZ penetration through the blood–brain barrier in normal and 6-hydroxydopamine induced Parkinson's disease model rats. *Biomed Res Int* 2014: 850493.

Huber, J. D., C. R. Campos, K. S. Mark and T. P. Davis 2006. Alterations in blood–brain barrier ICAM-1 expression and brain microglial activation after lambda-carrageenan-induced inflammatory pain. *Am J Physiol Heart Circ Physiol* 290(2): H732–H740.

Ibitoye, R., K. Kemp, C. Rice, K. Hares, N. Scolding and A. Wilkins 2016. Oxidative stress-related biomarkers in multiple sclerosis: A review. *Biomark Med* 10(8): 889–902.

Jellinger, K. A. 2012. Interaction between pathogenic proteins in neurodegenerative disorders. *J Cell Mol Med* 16(6): 1166–1183.

Jiang, X. and J. Nardelli 2016. Cellular and molecular introduction to brain development. *Neurobiol Dis* 92(Pt A): 3–17.

Kamphuis, W. W., C. Derada Troletti, A. Reijerkerk, I. A. Romero and H. E. de Vries 2015. The blood-brain barrier in multiple sclerosis: MicroRNAs as key regulators. *CNS Neurol Disord Drug Targets* 14(2): 157–167.

Keaney, J. and M. Campbell 2015. The dynamic blood–brain barrier. *FEBS J* 282(21): 4067–4079.

Kettenmann, H., U. K. Hanisch, M. Noda and A. Verkhratsky 2011. Physiology of microglia. *Physiol Rev* 91(2): 461–553.

Kofler, J. and C. A. Wiley 2011. Microglia: Key innate immune cells of the brain. *Toxicol Pathol* 39(1): 103–114.

Kooij, G., R. Backer, J. J. Koning, A. Reijerkerk, J. van Horssen, S. M. van der Pol, J. Drexhage et al. 2009. P-glycoprotein acts as an immunomodulator during neuroinflammation. *PLoS One* 4(12): e8212.

Kooij, G., J. Kroon, D. Paul, A. Reijerkerk, D. Geerts, S. M. van der Pol, B. van Het Hof et al. 2014. P-glycoprotein regulates trafficking of CD8(+) T cells to the brain parenchyma. *Acta Neuropathol* 127(5): 699–711.

Kooij, G., M. R. Mizee, J. van Horssen, A. Reijerkerk, M. E. Witte, J. A. Drexhage, S. M. van der Pol et al. 2011. Adenosine triphosphate-binding cassette transporters mediate chemokine (C-C motif) ligand 2 secretion from reactive astrocytes: Relevance to multiple sclerosis pathogenesis. *Brain* 134(Pt 2): 555–570.

Kooij, G., J. van Horssen, E. C. de Lange, A. Reijerkerk, S. M. van der Pol, B. van Het Hof, J. Drexhage et al. 2010. T lymphocytes impair P-glycoprotein function during neuroinflammation. *J Autoimmun* 34(4): 416–425.

Kook, S. Y., H. S. Hong, M. Moon, C. M. Ha, S. Chang and I. Mook-Jung 2012. Abeta(1)(-)(4)(2)-RAGE interaction disrupts tight junctions of the blood–brain barrier via Ca(2)(+)-calcineurin signaling. *J Neurosci* 32(26): 8845–8854.

Kostic, M., N. Zivkovic and I. Stojanovic 2013. Multiple sclerosis and glutamate excitotoxicity. *Rev Neurosci* 24(1): 71–88.

Kuhnke, D., G. Jedlitschky, M. Grube, M. Krohn, M. Jucker, I. Mosyagin, I. Cascorbi et al. 2007. MDR1-P-glycoprotein (ABCB1) mediates transport of Alzheimer's amyloid-beta peptides—Implications for the mechanisms of abeta clearance at the blood–brain barrier. *Brain Pathol* 17(4): 347–353.

Lamas, M., L. Gonzalez-Mariscal and R. Gutierrez 2002. Presence of claudins mRNA in the brain. Selective modulation of expression by kindling epilepsy. *Brain Res Mol Brain Res* 104(2): 250–254.

Lepka, K., C. Berndt, H. P. Hartung and O. Aktas 2016. Redox events as modulators of pathology and therapy of neuroinflammatory diseases. *Front Cell Dev Biol* 4: 63.

Liebner, S., M. Corada, T. Bangsow, J. Babbage, A. Taddei, C. J. Czupalla, M. Reis et al. 2008. Wnt/beta-catenin signaling controls development of the blood–brain barrier. *J Cell Biol* 183(3): 409–417.

Lipton, S. A. 2005. The molecular basis of memantine action in Alzheimer's disease and other neurologic disorders: Low-affinity, uncompetitive antagonism. *Curr Alzheimer Res* 2(2): 155–165.

Loscher, W., C. Luna-Tortos, K. Romermann and M. Fedrowitz 2011. Do ATP-binding cassette transporters cause pharmacoresistance in epilepsy? Problems and approaches in determining which antiepileptic drugs are affected. *Curr Pharm Des* 17(26): 2808–2828.

Marchi, N., K. L. Hallene, K. M. Kight, L. Cucullo, G. Moddel, W. Bingaman, G. Dini, A. Vezzani and D. Janigro 2004. Significance of MDR1 and multiple drug resistance in refractory human epileptic brain. *BMC Med* 2: 37.

Markoutsa, E., G. Pampalakis, A. Niarakis, I. A. Romero, B. Weksler, P. O. Couraud and S. G. Antimisiaris 2011. Uptake and permeability studies of BBB-targeting immunoliposomes using the hCMEC/D3 cell line. *Eur J Pharm Biopharm* 77(2): 265–274.

Marschallinger, J., I. Schaffner, B. Klein, R. Gelfert, F. J. Rivera, S. Illes, L. Grassner et al. 2015. Structural and functional rejuvenation of the aged brain by an approved anti-asthmatic drug. *Nat Commun* 6: 8466.

Martínez-Juárez, I. E., L. Hernández-VanegasN. Rodríguez y Rodríguez,, J. A. León-Aldana and A. V. Delgado-Escueta 2013. Genes involved in pharmacoresistant epilepsy. In: *Pharmacoresistance in Epilepsy—From Genes and Molecules to Promising Therapies*, E. A. Cavalheiro and L. Rocha (eds.). Springer Science+Business Media, 1–25.

Mason, J. O. and D. J. Price 2016. Building brains in a dish: Prospects for growing cerebral organoids from stem cells. *Neuroscience* 334: 105–118.

McQuaid, S., P. Cunnea, J. McMahon and U. Fitzgerald 2009. The effects of blood–brain barrier disruption on glial cell function in multiple sclerosis. *Biochem Soc Trans* 37(Pt 1): 329–331.

Meldrum, B. S. 2002. Concept of activity-induced cell death in epilepsy: Historical and contemporary perspectives. *Prog Brain Res* 135: 3–11.

Michalak, Z., A. Lebrun, M. Di Miceli, M. C. Rousset, A. Crespel, P. Coubes, D. C. Henshall, M. Lerner-Natoli and V. Rigau 2012. IgG leakage may contribute to neuronal dysfunction in drug-refractory epilepsies with blood–brain barrier disruption. *J Neuropathol Exp Neurol* 71(9): 826–838.

Michel, P. P., E. C. Hirsch and S. Hunot 2016. Understanding dopaminergic cell death pathways in Parkinson disease. *Neuron* 90(4): 675–691.

Milner, R. and I. L. Campbell 2002. Developmental regulation of beta1 integrins during angiogenesis in the central nervous system. *Mol Cell Neurosci* 20(4): 616–626.

Minagar, A., A. H. Maghzi, J. C. McGee and J. S. Alexander 2012. Emerging roles of endothelial cells in multiple sclerosis pathophysiology and therapy. *Neurol Res* 34(8): 738–745.

Miyakawa, T. 2010. Vascular pathology in Alzheimer's disease. *Psychogeriatrics* 10(1): 39–44.

Montagne, A., S. R. Barnes, M. D. Sweeney, M. R. Halliday, A. P. Sagare, Z. Zhao, A. W. Toga et al. 2015. Blood–brain barrier breakdown in the aging human hippocampus. *Neuron* 85(2): 296–302.

Mooradian, A. D., H. C. Chung and G. N. Shah 1997. GLUT-1 expression in the cerebra of patients with Alzheimer's disease. *Neurobiol Aging* 18(5): 469–474.

Morgan, L., B. Shah, L. E. Rivers, L. Barden, A. J. Groom, R. Chung, D. Higazi, H. Desmond, T. Smith and J. M. Staddon 2007. Inflammation and dephosphorylation of the tight junction protein occludin in an experimental model of multiple sclerosis. *Neuroscience* 147(3): 664–673.

Mosconi, L., S. De Santi, J. Li, W. H. Tsui, Y. Li, M. Boppana, E. Laska, H. Rusinek and M. J. de Leon 2008. Hippocampal hypometabolism predicts cognitive decline from normal aging. *Neurobiol Aging* 29(5): 676–692.

Mosconi, L., S. Sorbi, M. J. de Leon, Y. Li, B. Nacmias, P. S. Myoung, W. Tsui et al. 2006. Hypometabolism exceeds atrophy in presymptomatic early-onset familial Alzheimer's disease. *J Nucl Med* 47(11): 1778–1786.

Mulder, M., A. Blokland, D. J. van den Berg, H. Schulten, A. H. Bakker, D. Terwel, W. Honig et al. 2001. Apolipoprotein E protects against neuropathology induced by a high-fat diet and maintains the integrity of the blood–brain barrier during aging. *Lab Invest* 81(7): 953–960.

Munoz, Y., C. M. Carrasco, J. D. Campos, P. Aguirre and M. T. Nunez 2016. Parkinson's disease: The mitochondria-iron link. *Parkinsons Dis* 2016: 7049108.

Nagy, I., B. Toth, Z. Gaborik, F. Erdo and P. Krajcsi 2016. Membrane transporters in physiological barriers of pharmacological importance. *Curr Pharm Des* 22(35): 5347–5372.

Navarro, A. and A. Boveris 2010. Brain mitochondrial dysfunction in aging, neurodegeneration, and Parkinson's disease. *Front Aging Neurosci* 2: 1–11.

Nazari, Z., N. M., Safaei Nejad, Z., Delfan, B. and Irian, S. 2015. Expression of aquaporins in the rat choroid plexus. *Arch Neurosci* 2: e17312.

Ndode-Ekane, X. E., N. Hayward, O. Grohn and A. Pitkanen 2010. Vascular changes in epilepsy: Functional consequences and association with network plasticity in pilocarpine-induced experimental epilepsy. *Neuroscience* 166(1): 312–332.

Nobili, P., F. Colciaghi, A. Finardi, S. Zambon, D. Locatelli and G. S. Battaglia 2015. Continuous neurodegeneration and death pathway activation in neurons and glia in an experimental model of severe chronic epilepsy. *Neurobiol Dis* 83: 54–66.

O'Kane, R. L., I. Martinez-Lopez, M. R. DeJoseph, J. R. Vina and R. A. Hawkins 1999. Na(+)-dependent glutamate transporters (EAAT1, EAAT2, and EAAT3) of the blood–brain barrier. A mechanism for glutamate removal. *J Biol Chem* 274(45): 31891–31895.

Obermeier, B., R. Daneman and R. M. Ransohoff 2013. Development, maintenance and disruption of the blood–brain barrier. *Nat Med* 19(12): 1584–1596.

Okouchi, M., O. Ekshyyan, M. Maracine and T. Y. Aw 2007. Neuronal apoptosis in neurodegeneration. *Antioxid Redox Signal* 9(8): 1059–1096.

Oldendorf, W. H., M. E. Cornford and W. J. Brown 1977. The large apparent work capability of the blood–brain barrier: A study of the mitochondrial content of capillary endothelial cells in brain and other tissues of the rat. *Ann Neurol* 1(5): 409–417.

Olsson, A. K., A. Dimberg, J. Kreuger and L. Claesson-Welsh 2006. VEGF receptor signalling—In control of vascular function. *Nat Rev Mol Cell Biol* 7(5): 359–371.

Osada, T., Y. H. Gu, M. Kanazawa, Y. Tsubota, B. T. Hawkins, M. Spatz, R. Milner and G. J. del Zoppo 2011. Interendothelial claudin-5 expression depends on cerebral endothelial cell-matrix adhesion by beta(1)-integrins. *J Cereb Blood Flow Metab* 31(10): 1972–1985.

Owen, J. B., R. Sultana, C. D. Aluise, M. A. Erickson, T. O. Price, G. Bu, W. A. Banks and D. A. Butterfield 2010. Oxidative modification to LDL receptor-related protein 1 in hippocampus from subjects with Alzheimer disease: Implications for Abeta accumulation in AD brain. *Free Radic Biol Med* 49(11): 1798–1803.

Padou, V., S. Boyet and A. Nehlig 1995. Changes in transport of [14C] alpha-aminoisobutyric acid across the blood–brain barrier during pentylenetetrazol-induced status epilepticus in the immature rat. *Epilepsy Res* 22(3): 175–183.

Perez-Pinzon, M. A., R. A. Stetler and G. Fiskum 2012. Novel mitochondrial targets for neuroprotection. *J Cereb Blood Flow Metab* 32(7): 1362–1376.

Persidsky, Y., S. H. Ramirez, J. Haorah and G. D. Kanmogne 2006. Blood–brain barrier: Structural components and function under physiologic and pathologic conditions. *J Neuroimmune Pharmacol* 1(3): 223–236.

Pflanzner, T., B. Petsch, B. Andre-Dohmen, A. Muller-Schiffmann, S. Tschickardt, S. Weggen, L. Stitz, C. Korth and C. U. Pietrzik 2012. Cellular prion protein participates in amyloid-beta transcytosis across the blood–brain barrier. *J Cereb Blood Flow Metab* 32(4): 628–632.

Pisani, V., A. Stefani, M. Pierantozzi, S. Natoli, P. Stanzione, D. Franciotta and A. Pisani 2012. Increased blood-cerebrospinal fluid transfer of albumin in advanced Parkinson's disease. *J Neuroinflammation* 9: 188.

Pola, R., L. E. Ling, M. Silver, M. J. Corbley, M. Kearney, R. Blake Pepinsky, R. Shapiro et al. 2001. The morphogen Sonic hedgehog is an indirect angiogenic agent upregulating two families of angiogenic growth factors. *Nat Med* 7(6): 706–711.

Ravenstijn, P. G., H. J. Drenth, M. J. O'Neill, M. Danhof and E. C. de Lange 2012. Evaluation of blood–brain barrier transport and CNS drug metabolism in diseased and control brain after intravenous L-DOPA in a unilateral rat model of Parkinson's disease. *Fluids Barriers CNS* 9: 4.

Ravenstijn, P. G., M. Merlini, M. Hameetman, T. K. Murray, M. A. Ward, H. Lewis, G. Ball et al. 2008. The exploration of rotenone as a toxin for inducing Parkinson's disease in rats, for application in BBB transport and PK-PD experiments. *J Pharmacol Toxicol Methods* 57(2): 114–130.

Recasens, A. and B. Dehay 2014. Alpha-synuclein spreading in Parkinson's disease. *Front Neuroanat* 8: 159.

Redzic, Z. 2011. Molecular biology of the blood–brain and the blood-cerebrospinal fluid barriers: Similarities and differences. *Fluids Barriers CNS* 8(1): 3.

Regesta, G. and P. Tanganelli 1999. Clinical aspects and biological bases of drug-resistant epilepsies. *Epilepsy Res* 34(2–3): 109–122.

Roberts, L. M., K. Woodford, M. Zhou, D. S. Black, J. E. Haggerty, E. H. Tate, K. K. Grindstaff, W. Mengesha, C. Raman and N. Zerangue 2008. Expression of the thyroid hormone transporters monocarboxylate transporter-8 (SLC16A2) and organic ion transporter-14 (SLCO1C1) at the blood–brain barrier. *Endocrinology* 149(12): 6251–6261.

Rodriguez-Arellano, J. J., V. Parpura, R. Zorec and A. Verkhratsky 2016. Astrocytes in physiological aging and Alzheimer's disease. *Neuroscience* 323: 170–182.

Ronaldson, P. T. and T. P. Davis 2012. Blood–brain barrier integrity and glial support: Mechanisms that can be targeted for novel therapeutic approaches in stroke. *Curr Pharm Des* 18(25): 3624–3644.

Rosenberg, G. A. 2009. Matrix metalloproteinases and their multiple roles in neurodegenerative diseases. *Lancet Neurol* 8(2): 205–216.

Rosenberg, G. A. and Y. Yang 2007. Vasogenic edema due to tight junction disruption by matrix metalloproteinases in cerebral ischemia. *Neurosurg Focus* 22(5): E4.

Sa-Pereira, I., D. Brites and M. A. Brito 2012. Neurovascular unit: A focus on pericytes. *Mol Neurobiol* 45(2): 327–347.

Sagare, A. P., R. Deane and B. V. Zlokovic 2012. Low-density lipoprotein receptor-related protein 1: A physiological Abeta homeostatic mechanism with multiple therapeutic opportunities. *Pharmacol Ther* 136(1): 94–105.

Sage, M. R. and A. J. Wilson 1994. The blood–brain barrier: An important concept in neuroimaging. *AJNR Am J Neuroradiol* 15(4): 601–622.

Sahin, D., G. Ilbay and N. Ates 2003. Changes in the blood–brain barrier permeability and in the brain tissue trace element concentrations after single and repeated pentylenetetrazole-induced seizures in rats. *Pharmacol Res* 48(1): 69–73.

Saint-Pol, J., E. Vandenhaute, M. C. Boucau, P. Candela, L. Dehouck, R. Cecchelli, M. P. Dehouck, L. Fenart and F. Gosselet 2012. Brain pericytes ABCA1 expression mediates cholesterol efflux but not cellular amyloid-beta peptide accumulation. *J Alzheimers Dis* 30(3): 489–503.

Samuraki, M., I. Matsunari, W. P. Chen, K. Yajima, D. Yanase, A. Fujikawa, N. Takeda, S. Nishimura, H. Matsuda and M. Yamada 2007. Partial volume effect-corrected FDG PET and grey matter volume loss in patients with mild Alzheimer's disease. *Eur J Nucl Med Mol Imaging* 34(10): 1658–1669.

Sanchez-Covarrubias, L., L. M. Slosky, B. J. Thompson, T. P. Davis and P. T. Ronaldson 2014. Transporters at CNS barrier sites: Obstacles or opportunities for drug delivery? *Curr Pharm Des* 20(10): 1422–1449.

Sanchez del Pino, M. M., D. R. Peterson and R. A. Hawkins 1995. Neutral amino acid transport characterization of isolated luminal and abluminal membranes of the blood–brain barrier. *J Biol Chem* 270(25): 14913–14918.

Sandoval, K. E. and K. A. Witt 2008. Blood–brain barrier tight junction permeability and ischemic stroke. *Neurobiol Dis* 32(2): 200–219.

Saunders, N. R., C. J. Ek, M. D. Habgood and K. M. Dziegielewska 2008. Barriers in the brain: A renaissance? *Trends Neurosci* 31(6): 279–286.

Sayre, L. M., G. Perry and M. A. Smith 2008. Oxidative stress and neurotoxicity. *Chem Res Toxicol* 21(1): 172–188.

Schmidt, D. and W. Loscher 2005. Drug resistance in epilepsy: Putative neurobiologic and clinical mechanisms. *Epilepsia* 46(6): 858–877.

Sedlakova, R., R. R. Shivers and R. F. Del Maestro 1999. Ultrastructure of the blood–brain barrier in the rabbit. *J Submicrosc Cytol Pathol* 31(1): 149–161.

Selkoe, D. J. 2011. Alzheimer's disease. *Cold Spring Harb Perspect Biol* 3(7).

Serlin, Y., I. Shelef, B. Knyazer and A. Friedman 2015. Anatomy and physiology of the blood–brain barrier. *Semin Cell Dev Biol* 38: 2–6.

Serrano, G. E., N. Lelutiu, A. Rojas, S. Cochi, R. Shaw, C. D. Makinson, D. Wang, G. A. FitzGerald and R. Dingledine 2011. Ablation of cyclooxygenase-2 in forebrain neurons is neuroprotective and dampens brain inflammation after status epilepticus. *J Neurosci* 31(42): 14850–14860.

Shams, S., J. Martola, T. Granberg, X. Li, M. Shams, S. M. Fereshtehnejad, L. Cavallin, P. Aspelin, M. Kristoffersen-Wiberg and L. O. Wahlund 2015. Cerebral microbleeds: Different prevalence, topography, and risk factors depending on dementia diagnosis—The Karolinska Imaging Dementia Study. *AJNR Am J Neuroradiol* 36(4): 661–666.

Sharma, H. S., R. J. Castellani, M. A. Smith and A. Sharma 2012. The blood–brain barrier in Alzheimer's disease: Novel therapeutic targets and nanodrug delivery. *Int Rev Neurobiol* 102: 47–90.

Shimizu, F., Y. Sano, K. Saito, M. A. Abe, T. Maeda, H. Haruki and T. Kanda 2012. Pericyte-derived glial cell line-derived neurotrophic factor increase the expression of claudin-5 in the blood–brain barrier and the blood–nerve barrier. *Neurochem Res* 37(2): 401–409.

Siddiqui, A., R. Kerb, M. E. Weale, U. Brinkmann, A. Smith, D. B. Goldstein, N. W. Wood and S. M. Sisodiya 2003. Association of multidrug resistance in epilepsy with a polymorphism in the drug-transporter gene ABCB1. *N Engl J Med* 348(15): 1442–1448.

Simpson, I. A., A. Carruthers and S. J. Vannucci 2007. Supply and demand in cerebral energy metabolism: The role of nutrient transporters. *J Cereb Blood Flow Metab* 27(11): 1766–1791.

Sisodiya, S. M. and H. C. Mefford 2011. Genetic contribution to common epilepsies. *Curr Opin Neurol* 24(2): 140–145.

Soto, C. 2003. Unfolding the role of protein misfolding in neurodegenerative diseases. *Nat Rev Neurosci* 4(1): 49–60.

Soto, C. and L. D. Estrada 2008. Protein misfolding and neurodegeneration. *Arch Neurol* 65(2): 184–189.

Stangel, M. 2012. Neurodegeneration and neuroprotection in multiple sclerosis. *Curr Pharm Des* 18(29): 4471–4474.

Stavale, L. M., E. S. Soares, M. C. Mendonca, S. P. Irazusta and M. A. da Cruz Hofling 2013. Temporal relationship between aquaporin-4 and glial fibrillary acidic protein in cerebellum of neonate and adult rats administered a BBB disrupting spider venom. *Toxicon* 66: 37–46.

Stokin, G. B. and L. S. Goldstein 2006. Axonal transport and Alzheimer's disease. *Annu Rev Biochem* 75: 607–627.

Sui, Y. T., K. M. Bullock, M. A. Erickson, J. Zhang and W. A. Banks 2014. Alpha synuclein is transported into and out of the brain by the blood–brain barrier. *Peptides* 62: 197–202.

Sweeney, M. D., S. Ayyadurai and B. V. Zlokovic 2016. Pericytes of the neurovascular unit: Key functions and signaling pathways. *Nat Neurosci* 19(6): 771–783.

Tait, M. J., S. Saadoun, B. A. Bell and M. C. Papadopoulos 2008. Water movements in the brain: Role of aquaporins. *Trends Neurosci* 31(1): 37–43.

Takakura, Y., K. L. Audus and R. T. Borchardt 1991. Blood–brain barrier: Transport studies in isolated brain capillaries and in cultured brain endothelial cells. *Adv Pharmacol* 22: 137–165.

Takano, T., G. F. Tian, W. Peng, N. Lou, W. Libionka, X. Han and M. Nedergaard 2006. Astrocyte-mediated control of cerebral blood flow. *Nat Neurosci* 9(2): 260–267.

Takata, F., S. Dohgu, A. Yamauchi, J. Matsumoto, T. Machida, K. Fujishita, K. Shibata et al. 2013. In vitro blood–brain barrier models using brain capillary endothelial cells isolated from neonatal and adult rats retain age-related barrier properties. *PLoS One* 8(1): e55166.

Tanzi, R. E., R. D. Moir and S. L. Wagner 2004. Clearance of Alzheimer's Abeta peptide: The many roads to perdition. *Neuron* 43(5): 605–608.

Tate, S. K. and S. M. Sisodiya 2007. Multidrug resistance in epilepsy: A pharmacogenomic update. *Expert Opin Pharmacother* 8(10): 1441–1449.

Thiollier, T., C. Wu, H. Contamin, Q. Li, J. Zhang and E. Bezard 2016. Permeability of blood–brain barrier in macaque model of 1-methyl-4-phenyl-1,2,3,6-tetrahydropyridine-induced Parkinson disease. *Synapse* 70(6). 231–239.

Thom, M., J. Zhou, L. Martinian and S. Sisodiya 2005. Quantitative post-mortem study of the hippocampus in chronic epilepsy: Seizures do not inevitably cause neuronal loss. *Brain* 128(Pt 6): 1344–1357.

Tong, X. K. and E. Hamel 1999. Regional cholinergic denervation of cortical microvessels and nitric oxide synthase-containing neurons in Alzheimer's disease. *Neuroscience* 92(1): 163–175.

Toth, P., S. Tarantini, A. Csiszar and Z. Ungvari 2017. Functional vascular contributions to cognitive impairment and dementia: Mechanisms and consequences of cerebral autoregulatory dysfunction, endothelial impairment, and neurovascular uncoupling in aging. *Am J Physiol Heart Circ Physiol* 312(1): H1–H20.

Tramutola, A., C. Lanzillotta, M. Perluigi and D. A. Butterfield 2016. Oxidative stress, protein modification and Alzheimer disease. *Brain Res Bull*.

Tsuji, S. 2010. Genetics of neurodegenerative diseases: Insights from high-throughput resequencing. *Hum Mol Genet* 19(R1): R65–70.

Ueno, M., I. Akiguchi, H. Yagi, H. Naiki, Y. Fujibayashi, J. Kimura and T. Takeda 1993. Age-related changes in barrier function in mouse brain I. Accelerated age-related increase of brain transfer of serum albumin in accelerated senescence prone SAM-P/8 mice with deficits in learning and memory. *Arch Gerontol Geriatr* 16(3): 233–248.

Ueno, M., I. Akiguchi, M. Hosokawa, M. Shinnou, H. Sakamoto, M. Takemura and K. Higuchi 1997. Age-related changes in the brain transfer of blood-borne horseradish peroxidase in the hippocampus of senescence-accelerated mouse. *Acta Neuropathol* 93(3): 233–240.

Ueno, M., Y. Chiba, R. Murakami, K. Matsumoto, M. Kawauchi and R. Fujihara 2016. Blood–brain barrier and blood–cerebrospinal fluid barrier in normal and pathological conditions. *Brain Tumor Pathol* 33(2): 89–96.

Ueno, M., D. H. Dobrogowska and A. W. Vorbrodt 1996. Immunocytochemical evaluation of the blood–brain barrier to endogenous albumin in the olfactory bulb and pons of senescence-accelerated mice (SAM). *Histochem Cell Biol* 105(3): 203–212.

Vaucher, E. and E. Hamel 1995. Cholinergic basal forebrain neurons project to cortical microvessels in the rat: Electron microscopic study with anterogradely transported *Phaseolus vulgaris* leucoagglutinin and choline acetyltransferase immunocytochemistry. *J Neurosci* 15(11): 7427–7441.

Vaucher, E., X. K. Tong, N. Cholet, S. Lantin and E. Hamel 2000. GABA neurons provide a rich input to microvessels but not nitric oxide neurons in the rat cerebral cortex: A means for direct regulation of local cerebral blood flow. *J Comp Neurol* 421(2): 161–171.

Vogelgesang, S., G. Jedlitschky, A. Brenn and L. C. Walker 2011. The role of the ATP-binding cassette transporter P-glycoprotein in the transport of beta-amyloid across the blood–brain barrier. *Curr Pharm Des* 17(26): 2778–2786.

Vorbrodt, A. W. and D. H. Dobrogowska 2003. Molecular anatomy of intercellular junctions in brain endothelial and epithelial barriers: Electron microscopist's view. *Brain Res Brain Res Rev* 42(3): 221–242.

Wan, W., H. Chen and Y. Li 2014. The potential mechanisms of Abeta-receptor for advanced glycation endproducts interaction disrupting tight junctions of the blood–brain barrier in Alzheimer's disease. *Int J Neurosci* 124(2): 75–81.

Wang, J. and R. Milner 2006. Fibronectin promotes brain capillary endothelial cell survival and proliferation through alpha5beta1 and alphavbeta3 integrins via MAP kinase signalling. *J Neurochem* 96(1): 148–159.

Weekman, E. M. and D. M. Wilcock 2015. Matrix metalloproteinase in blood–brain barrier breakdown in dementia. *J Alzheimers Dis* 49(4): 893–903.

Wenk, G. L. 2003. Neuropathologic changes in Alzheimer's disease. *J Clin Psychiatry* 64(Suppl 9): 7–10.

Winkler, E. A., R. D. Bell and B. V. Zlokovic 2011. Central nervous system pericytes in health and disease. *Nat Neurosci* 14(11): 1398–1405.

Winkler, E. A., Y. Nishida, A. P. Sagare, S. V. Rege, R. D. Bell, D. Perlmutter, J. D. Sengillo et al. 2015. GLUT1 reductions exacerbate Alzheimer's disease vasculo-neuronal dysfunction and degeneration. *Nat Neurosci* 18(4): 521–530.

Winkler, E. A., A. P. Sagare and B. V. Zlokovic 2014. The pericyte: A forgotten cell type with important implications for Alzheimer's disease? *Brain Pathol* 24(4): 371–386.

Wu, Z., H. Guo, N. Chow, J. Sallstrom, R. D. Bell, R. Deane, A. I. Brooks et al. 2005. Role of the MEOX2 homeobox gene in neurovascular dysfunction in Alzheimer disease. *Nat Med* 11(9): 959–965.

Xie, Y. and Y. Chen 2016. microRNAs: Emerging targets regulating oxidative stress in the models of Parkinson's disease. *Front Neurosci* 10: 298.

Yang, J. Lunde, L. K. Nuntagij, P. Oguchi, T. Camassa, L. M. Nilsson, L. N. Lannfelt, L. et al. 2011. Loss of astrocyte polarization in the tg-ArcSwe mouse model of Alzheimer's disease. *J Alzheimers Dis* 27(4): 711–722.

Yang, Y. and G. A. Rosenberg 2011. MMP-mediated disruption of claudin-5 in the blood–brain barrier of rat brain after cerebral ischemia. *Methods Mol Biol* 762: 333–345.

Yates, P. A., P. M. Desmond, P. M. Phal, C. Steward, C. Szoeke, O. Salvado, K. A. Ellis et al. A. R. Group 2014. Incidence of cerebral microbleeds in preclinical Alzheimer disease. *Neurology* 82(14): 1266–1273.

Zhao, Z., A. R. Nelson, C. Betsholtz and B. V. Zlokovic 2015. Establishment and dysfunction of the blood–brain barrier. *Cell* 163(5): 1064–1078.

Zlokovic, B. V. 2008. The blood–brain barrier in health and chronic neurodegenerative disorders. *Neuron* 57(2): 178–201.

Zlokovic, B. V. 2010. Neurodegeneration and the neurovascular unit. *Nat Med* 16(12): 1370–1371.

Zlokovic, B. V. 2011. Neurovascular pathways to neurodegeneration in Alzheimer's disease and other disorders. *Nat Rev Neurosci* 12(12): 723–738.

Zlokovic, B. V., S. Yamada, D. Holtzman, J. Ghiso and B. Frangione 2000. Clearance of amyloid beta-peptide from brain: Transport or metabolism? *Nat Med* 6(7): 718–719.

16 Senescent Cells as Drivers of Age-Related Diseases

Cielo Mae D. Marquez and Michael C. Velarde

CONTENTS

INTRODUCTION: THE AGING CELL

Somatic cells have a limited capacity to proliferate. They reach a state of permanent growth arrest after a certain number of cell divisions. This irreversible cell cycle arrest was first described by Hayflick and Moorhead, who showed that normal human fibroblasts can only undergo up to about 60 cell divisions in their lifetime before they cease to divide [1]. The phenomenon is proposed to be part of the aging process that leads to senescence in somatic cells. Hence, it is called replicative senescence or cellular senescence [1]. However, unlike organismal senescence which leads to death, cells reaching the Hayflick limit continue to be viable in culture despite their inability to replicate. They are able to resist apoptotic signals and remain metabolically active [2–4]. Unlike quiescent cells, the senescent state is permanent and cannot be reversed even in the presence of sufficient proliferative inducers, nutrients, and other growth factors [1,5].

Replicative senescence is found in the cells of humans and other animals [6]. It is observed in multiple cell types such as keratinocytes, lymphocytes, endothelial cells, vascular smooth muscle cells (VSMC), chondrocytes, and several embryonic-derived cells [7–12]. Stem cells may also undergo cellular senescence, leading to their proliferative arrest and impaired differentiation [13,14]. Decline in stem cell function with age is associated with decreased levels of *Bmi1*, a member of the *Polycomb* group of transcription repressors, which maintain adult stem cells in tissues partly through the repression of genes responsible for inducing cellular senescence [15]. Expression of *Bmi1* extends life span in stem cells, but its downregulation causes the enhanced expression of the senescence marker p16 [16]. Other types of somatic stem cells, specifically hematopoietic stem cells and neuronal stem cells, also undergo senescence [16–18].

In this chapter, we will introduce the concept of cellular senescence, including the features that distinguish them from normal cells. We will investigate the factors which trigger cells to undergo senescence and provide evidence that link cellular senescence to aging and several age-related pathologies. We will further explore some potential interventions that may delay and attenuate the effects of cellular senescence in humans and other model organisms.

FEATURES AND MARKERS OF SENESCENT CELLS

Senescent cells are marked by their inability to divide and by their dramatic changes in morphology and metabolic processes. Senescent cells are characterized by their enlarged cell size and flattened cytoplasm compared to the normal cell morphology in proliferating, quiescent, or terminally differentiated cells (Figure 16.1). Senescent phenotypes also include development of nuclear blebbing, increase in lysosomal mass, increased activity of the pH-dependent senescence-associated β-galactosidase (SA β-gal), downregulation of lamin B1, formation of senescence-associated heterochromatin foci (SAHF), establishment of DNA segments with chromatin alterations reinforcing senescence (DNA-SCARS), activation of p53/p21 and p16/pRb pathways, and secretion of high mobility group box 1 (HMGB1) (Figure 16.2) [19–23]. Cells that undergo senescence develop resistance to apoptosis partly due to the absence of insulin-like growth factor-binding protein 3 (IGFBP-3) within the nucleus [4]. Senescent cells also express many cytokines, chemokines, growth factors, and other proteins collectively termed as the senescence-associated secretory phenotype (SASP) [24,25].

PERMANENT CELL CYCLE ARREST

Progression through the cell cycle requires cells to pass cell cycle checkpoints, namely the G1-S transition, the S-phase checkpoint, the G2 to M transition, and the mitotic spindle checkpoint [26]. Replicative senescent cells become permanently arrested at G1 and G2 phases of the cell cycle [27,28]. Cellular senescence caused by telomere erosion evokes a DNA damage response (DDR), prompting the persistent activation of p53 via the ataxia telangiectasia mutated (ATM)/ATR DNA damage signaling pathway [29]. Upregulation of p21, a downstream effector of p53, is necessary to

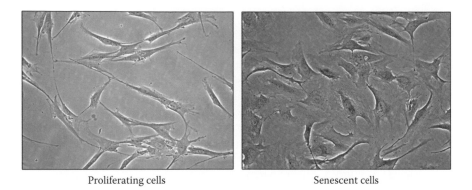

Proliferating cells   Senescent cells

FIGURE 16.1   Morphology of proliferating cells versus senescent cells. Primary human foreskin fibroblasts (HCA2) were treated for seven days with vehicle control (left) or 200 nM rotenone (right) to induce cellular senescence by mitochondrial dysfunction. Based on their morphology, senescent cells display marked increase in cell size and flattened cytoplasm, in contrast to the elongated and narrow-shaped proliferating cells. *Photomicrographs were taken at 200× magnification.*

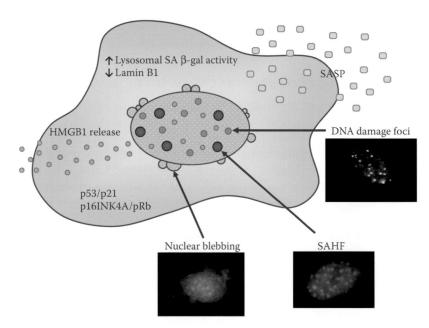

FIGURE 16.2   Distinct features of senescent cells. Senescent cells display one or more features that distinguish them from normal cells. Senescent cells have altered nuclear structures such as enlarged nuclear size, pronounced nuclear blebbing, increased DNA damage foci, and marked SAHF. Senescent cells display increased activity of SA β-gal, decreased lamin B1 expression, reduced nuclear HMGB1, and persistent upregulation of p53/p21 and p16INK4A/pRb. A robust secretion of numerous cytokines, growth factors, and other proteins (SASP) may also be observed in senescent cells.

induce G1 cell cycle arrest following exposure to DNA-damaging agents [30]. Aside from triggering proliferative arrest in G1, p21 may also induce permanent G2 arrest in a number of senescent cells [31]. In contrast, loss of p21 causes DNA damaged cells to go through additional S-phases, resulting in polyploidy and subsequent apoptosis [32].

The inability of senescent cells to progress through the cell cycle even after appropriate proliferative stimuli may be used as a marker for cellular senescence. One conventional method of measuring cell cycle is through radioactive tagging with tritiated thymidine (^{3}H-dT). ^{3}H-dT is a radioactive isotope which incorporates within the genomic DNA during DNA synthesis and is detected by autoradiographic technique [33]. Actively dividing cells will acquire the radioactive ^{3}H-dT, while senescent cells will remain untagged. Because the ^{3}H-dT technique often lacks cytologic resolution and requires laborious, costly handling of radiolabeled substances, the use of nonradioactive bromodeoxyuridine (BrdU) labeling to distinguish senescent from proliferating cells has increasingly gained popularity [34]. BrdU is a thymidine analog that is integrated within the DNA during the S-phase of the cell cycle, allowing the detection of proliferating cells in culture and in tissues [33]. This method makes use of antibodies that detect BrdU, which were incorporated into the DNA of actively dividing cells [35]. BrdU labeling can be detected in histological sections, cell cultures, and flow cytometry preparations [34,36].

Cell cycle analysis may also be performed using flow cytometry. Flow cytometric analysis uses DNA stains that bind proportionally to the amount of DNA present within the cell [37]. The more dyes taken up, the brighter the fluorescence. This helps to distinguish senescent cells (arrested at G1 or G2/M phases) versus non-senescent cells (quiescent at G0 or undergoing G1/S phase transitions) [27,28]. DNA-binding dyes which are frequently used include propidium iodide (PI), 4′,6-diamidino-2-phenylindole (DAPI), 7-aminoactinomycin-D (7-AAD), and Hoechst 33342 [38]. Generally, cells must be fixed and/or permeabilized to allow passage of the dyes, which is otherwise excluded by live cells. Common fixatives applied are alcohol and aldehyde which may be used in combination with a detergent or other permeabilizing reagents. Alcohols fix cells by dehydration and protein denaturation, while aldehydes fix cells by cross-linking proteins and other macromolecules [39].

PERSISTENT UPREGULATION OF CYCLIN-DEPENDENT KINASE INHIBITORS

Senescent cells show prominent perturbations in gene and protein expression, including altered expression of several cell cycle activators and inhibitors [40,41]. Two examples of cell cycle inhibitors that are typically elevated in senescent cells are the cyclin-dependent kinase inhibitors (CDKIs) p21 (also p21Cip1) and p16 (also p16INK4a) as (Figure 16.3). The p21 (CIP1/WAF1) protein is a potent CDKI that binds to and inhibits the activity of cyclin E/CDK2 and cyclin D/CDK4/6 complexes, and thus prevents cell cycle progression at G1 and G2 [42]. The expression of this gene is upregulated by the tumor suppressor protein p53 [30]. Sustained activity in p53 maintains elevated expression of p21, which subsequently suppresses the phosphorylation and activation of the retinoblastoma protein (pRB) pathway, leading to growth arrest [5,40,43]. p16 is also important in commencing growth arrest in cells. Once activated, p16 inhibits the phosphorylation of pRb which consequently prevents the release of transcription factor E2F, thus culminating cell cycle arrest [20,44]. Mutation or epigenetic silencing of at least one crucial regulator of these tumor suppressor pathways allows cancer cells to bypass the senescence checkpoint and progress toward tumorigenesis [45]. Hence, persistent activation of the p53/p21 and p16/Rb pathway is a widely used marker for senescent cells. Activation of p53/p21 and p16/Rb pathway can be identified by measuring the expression levels of p21 and p16 through Western blot, immunohistochemical analysis, and qRT-PCR [20,42,45–49]. Hyperphosphorylation of p53 in senescent cells is correlated with persistent p53 activation, while hypophosphorylation of Rb is consistent with growth arrest in senescent cells.

INCREASED SENESCENCE-ASSOCIATED β-GALACTOSIDASE (SA β-GAL) ACTIVITY

Identifying senescent cells from quiescent and terminally differentiated cells was a challenge in the early years, because markers used for cell growth experiments are not specific and are difficult to manipulate *in vivo*. However, in 1995, Dimri et al. developed a biomarker which detects human senescent cells by measuring the activity of an endogenous β-galactosidase (β-gal) enzyme [50]. Cultures of senescent human fibroblasts develop a blue–green staining at pH 6.0 when exposed to the X-gal substrate (Figure 16.4) [50]. The staining recognizes senescent cells but not proliferating, quiescent, and terminally differentiated cells, hence the term senescence-associated β-galactosidase (SA

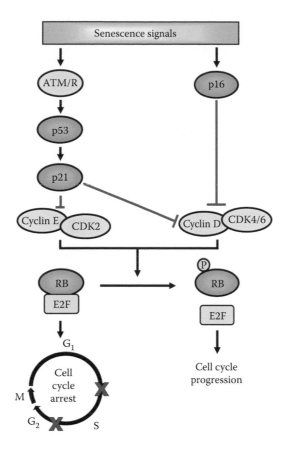

FIGURE 16.3 Cell cycle arrest is mediated by CDKIs. During senescence, there is a persistent upregulation of CDKI p21 and p16. The p21 protein inhibits cell cycle by binding to the cyclin E/CDK2 and cyclin D/CDK4/6 complexes, which halt proliferation at G1 or G2. At the same time, p16 may bind to cyclin D/CDK4/6 complex and inhibit the phosphorylation of pRb to prevent subsequent release of the E2F transcription factor and promote cell cycle arrest.

β-gal) was described. From then on, β-galactosidase activity has become a widely used biomarker for cellular senescence and has been greatly relied on due to its convenience and simplicity [51,52].

Positive staining by senescent cells at pH 6.0 is due in part by the increase in lysosomal content of cells undergoing cellular senescence [53]. Lysosomal size and mass increase in late passage cultures, thus, larger lysosomes are detected in senescent versus normal cells [53]. In addition, the positive staining of senescent cells with SA β-gal is due to the overexpression of the endogenous lysosomal β-galactosidase protein called galactosidase beta 1 (GLB1), a gene encoding the beta-D-galactosidase protein localized in the mammalian cell [54]. While GLB1 may be responsible for increased SA β-gal activity in senescent cells, this gene is not necessary for the establishment of cellular senescence [54].

Despite widespread relevance of SA β-gal as a biomarker, disputes regarding the specificity and selectivity of the assay are being raised, especially when identification of senescent cells is performed solely through this assay. Several reports indicate that SA β-gal may not be limited to senescent cells as this can also be observed in non-senescent cells upon extended culture at low serum conditions, treatment with high concentrations of nonspecific growth inhibitor (i.e., heparin), and incubation at high cell density [55–57]. Nevertheless, SA β-gal activity still remains to be a valid marker for cellular senescence. Care should be taken in assessing whether certain cells in culture or tissue are senescent based on this method [53].

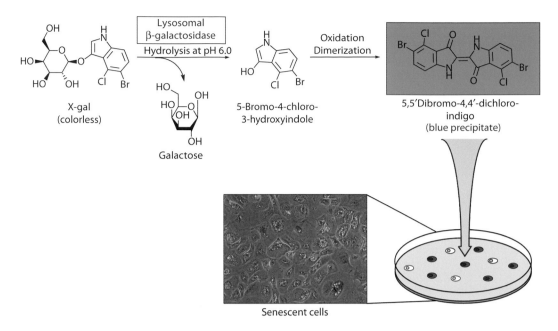

FIGURE 16.4 Chemical reaction involving lysosomal SA β-galactosidase enzyme. In the presence of senescent cells, the colorless substrate called X-gal becomes hydrolyzed by the lysosomal SA β-galactosidase enzyme at pH 6.0. Upon hydrolysis at the β-glycosidic bond, X-gal yields galactose and 5-bromo-4-chloro-3-hydroxyindole, which undergoes oxidation dimerization to become 5,5′ dibromo-4.4′-dichloro-indigo, an intensely blue precipitate. Senescent but not proliferating cells will stain blue with the assay.

INCREASED CELL SIZE AND ALTERED MORPHOLOGY

Cellular senescence is associated with cellular changes, including protein aggregation in endoplasmic reticulum, enlarged and dysfunctional mitochondria, increased number of nonfunctional lysosomes, and structural alterations of Golgi apparatus [58–60]. Senescent cells have an enlarged and flattened appearance, with single prominent nucleoli and cytoplasmic granules [20,61]. Senescent cells show a decrease in the number of gap junctions and an accumulation of surface glycosaminoglycans and glycoproteins [62]. Senescent cells are also associated with increased incidence of polyploidy [63]. Replicative and premature senescence have spatial rearrangement of cytoskeleton and focal adhesion proteins [64,65]. Specifically, senescent cells display enhanced actin stress fiber with accompanying increase and sporadic redistribution of focal proteins vinculin and paxillin. Cytoskeletal rearrangement and protein redistribution develop gradually over time and require the presence of hypophosphorylated Rb and *de novo* protein synthesis [65]. Senescent cells may also exhibit strong autofluorescence due to accumulation of lipofuscin within the cells [66,67].

PRESENCE OF DNA DAMAGE FOCI (MARKER OF GENOTOXIC STRESS)

Cells experiencing telomere uncapping due to telomere erosion and those undergoing DNA strand breaks due to genotoxic stress can activate a robust DDR [29]. This response subsequently triggers the production of DDR proteins that are associated with the induction of DNA damage foci at the sites of DNA strand breaks. DNA damage foci can be detected using microscopic imaging, immunofluorescence, histochemical detection, or by fluorescent protein tagging of DDR proteins (i.e., γH2AX, 53BP1, MDC1, and NBS1) involved in DNA strand break repair [29,68]. Presence of DNA damage foci in senescent cells has been initially described using the γH2AX, a variant of histone H2AX that is phosphorylated by kinases ATM, ATM-Rad3-related (ATR), and DNA-dependent protein kinase (DNA-PK) [69–72]. Staining senescent cells with antibodies against γH2AX, which aggregates at DNA damage sites, display a distinct bright foci or spot in the nucleus under a fluorescence microscope [73]. Chromogenic substrates for γH2AX with hematoxylin counterstaining may also be employed to detect DNA damage from paraffin-embedded tissue sections [74].

Senescent cells with activated DDR signaling pathway harbor persistent nuclear foci called DNA-SCARS. DNA-SCARS contain dysfunctional telomeres as well as phosphorylated DDR proteins including ATM, ATR, Rad3, and other related substrates [29,47,75,76]. DNA-SCARS are easily distinguished from transient damage foci because of their association with promyelocytic leukemia (PML) nuclear bodies and 53BP1 [77].

Presence of SAHF

Chromatin in the nucleus exists in euchromatin and heterochromatin states, which dictate transcriptional activity in a cell. Euchromatin is lightly packed and contains gene-rich regions that are under active transcription [78]. In contrast, heterochromatin is densely packed and is usually found in many transcriptionally repressed regions and relatively low gene concentrations [79]. In senescent cells, reorganization of chromatin structure generates discrete subnuclear heterochromatin clusters called SAHF [20,80]. SAHF is found in senescent human primary fibroblasts induced by replicative exhaustion and oncogene activation, but it is notably absent in quiescent cells [20]. DAPI may be used to detect SAHF formation in senescent cells, as demonstrated by brightly stained foci in the nucleus of these cells [81].

SAHF formation is triggered by several histone chaperones such as histone cell cycle regulator (HIRA) and anti-silencing function protein 1a (ASF1A) known to mediate gene silencing [81]. At the onset of senescence, these proteins are initially recruited within PML bodies, the punctate structures of the nucleus. This subsequently enriches the histone H2A variant macroH2A at chromatin regions to initiate chromatin condensation. HIRA and ASF1A stably translocate to the SAHF and causes deposition of several proteins such as heterochromatin protein 1 (HP1), methylated lysine 9 of histone H3 (H3K9Me), and macroH2A at the SAHF regions when cells become senescent [81–83]. Efficient SAHF formation is dependent on functional pRB and p53 tumor suppressor pathways [20,84].

Decreased Lamin B1 Expression

The nuclear envelope is lined by the nuclear lamina, a dense fibrillary network which provides mechanical support and regulates size, shape, and stability of the nucleus [85,86]. Nuclear lamina also partakes in a number of other functions including regulation of DNA synthesis, RNA transcription, and chromatin organization [87]. In mammals, the lamina contains major structural proteins categorized as type A (lamin A and C) and type B (lamin B1 and B2) lamins based on their isoelectric points [85]. Nuclear lamins are dynamic structures that are assembled and disassembled throughout the cell cycle. Lamin A and C are derived from the gene *LMNA* by alternative splicing and are expressed by non-proliferating and differentiated cells. Lamin B1 and B2 are encoded by separate genes *LMNB1* and *LMNB2* respectively and are strongly expressed by dividing and undifferentiated cells [88–90]. While there are two types of lamin B, expression of one or the other is sufficient for cell survival [86,91]. Expression of type A and B lamins are differentially regulated in specific tissues during embryogenesis [91,92].

Lamin B1 expression decreases during cellular senescence [22]. Significant loss of Lamin B1 is observed in senescent cells but not in normal and non-senescent cells, regardless of cell strain and type of senescence inducer employed [22,93]. Lamin B1 loss is triggered by the activation of p53 and pRb tumor suppressor pathways and is independent of p38MAPK, NF-κB, ATM, or reactive oxygen species (ROS) signaling pathways [22]. Senescent cells also display striking chromatin rearrangement that is associated with gene repression [94]. Knockdown of Lamin B1 facilitates the spatial relocalization of perinuclear H3K9me3-positive heterochromatin and promotes the formation of SAHF during cellular senescence [95]. Decreased Lamin B1 expression also contributes to the development of nuclear blebs observed in senescent cells (Figure 16.2) [93,96]. In contrast to other senescence biomarkers such as SA β-gal and SASP, lamin B1 develops shortly after treating with ionizing radiation (IR), suggesting that lamin B1 may be an early marker of the senescence program [22].

DECREASED NUCLEAR HMGB1

HMGB1 is a nuclear non-histone protein that controls gene expression by bending chromatin structures to allow transcription factors to gain access to promoter regions of DNA [97,98]. HMGB1 is also involved in DNA repair pathways [99]. It attaches to DNA lesions caused by irradiation and chemotherapeutic reagents, and it is involved in nucleotide excision repair [100–102]. Aside from its role in activating transcription and DNA repair, HMGB1 may also function by inducing inflammation. As a member of the alarmin family, HMGB1 is secreted intracellularly in response to cell and tissue damage caused by trauma and pathogens [103–105].

HMGB1 is also implicated in cellular senescence. While HMGB1 is initially found in the nucleus, it becomes redistributed to the extracellular milieu as a secreted alarmin, shortly after DNA damage and activation of a p53-dependent growth arrest [23,106]. HMGB1 secretion occurs in a p53-dependent manner and does not require ATM or p16 activation. HMGB1 release stimulates cytokine secretion and promotes senescence-associated inflammation by signaling through the toll-like receptor-4 (TLR-4) [23].

EXPRESSION OF THE SASP

One of the key features that distinguishes senescent cells from quiescent and terminally differentiated cells is their ability to produce the SASP. These SASP components consist of a variety of factors such as cytokines, chemokines, growth factors, secreted proteases, and extracellular matrix (ECM) proteins [107]. The SASP regulates various cellular processes such as cell growth, motility, inflammation, differentiation, tissue repair, vascularization, and angiogenesis [107–109]. Several SASP factors that have been described include interleukin 6 (IL6), IL1A, IL1B, chemokine C-X-C motif ligand 1 (CXCL1, also GRO-α), CXCL8, vascular endothelial growth factor (VEGF), and matrix metalloproteinase-1 (MMP1) [24,76,107,109–112].

SASP factors reinforce senescence in a paracrine and autocrine manner [113]. They reinforce growth arrest of neighboring non-senescent cells through activation of TGFβ signaling in a paracrine manner [114]. Senescent cells also cause surrounding normal, proliferation-competent cells to senesce through a bystander effect [115]. Hence, cellular senescence may act as a defense mechanism to lock cancer cells into a permanent state of cell cycle arrest [116]. Paradoxically, while cellular senescence may act as a strong tumor suppressor mechanism, some cancer cells may be resistant to senescence signals and may even proliferate in response to the SASP secreted by surrounding senescent cells. The proinflammatory arm of the SASP may induce an epithelial–mesenchymal transition and an invasiveness phenotype in premalignant epithelial cells [24]. In addition to cancer, incessant accumulation of the SASP also promotes persistent inflammation within the tissue microenvironment that exacerbates the pathogenesis of several chronic diseases [117].

Senescent cells develop the SASP gradually over time rather than immediately after sustained DNA damage. In cell culture studies, fibroblasts develop SASPs several days after exposure to stressors like genotoxic stress, irradiation, and oncogenic retrovirus-associated DNA sequences (RAS) [110]. It is also worth mentioning that senescent features are not universally expressed nor are they exclusive for a particular senescent cell type. SASP factors are not concomitantly produced by senescent cells, and not all senescent cells produce the same levels and types of SASP. Studies using various senescence inducers reveal that specific types of cells develop certain phenotype and may not engage in the production of SASPs, particularly those cells induced to senesce via p16INK4A pathway [110]. Senescent cells induced by mitochondrial dysfunction also elicit a distinct SASP phenotype [106].

TRIGGERS OF CELLULAR SENESCENCE

Replicative senescence is caused by shortening of telomeres after several cycles of DNA replication. While repetitive culture of somatic cells is a strong inducer of cellular senescence, several other factors can also trigger cellular senescence. Proliferating cells that experience excessive stress from various exogenous and endogenous sources also undergo cellular senescence [5]. This premature senescence or stress-induced premature senescence is caused by DNA damage from IR, cytotoxic drugs, oxidative stress, and oncogene activation. It may also be caused by non-DNA-damaging effects, such as tumor suppressor gene overexpression and mitochondrial dysfunction.

TELOMERE ATTRITION AND REPLICATIVE EXHAUSTION

One of the best known mechanisms that drive cellular senescence is telomere shortening. Telomeres are regions comprised of proteins and nucleotides of TTAGGG repeats found at the ends of each DNA strand that serve as caps to protect the chromosome [118]. Whether *in vivo* or *in vitro*, progressive shortening of telomeres occurs because DNA polymerase incompletely replicates the DNA ends over the course of cell division. Due to the nature and mechanism of replication, telomere exhaustion ultimately results in deletion of essential genes situated at the end-replicons [119]. This occurrence can be rescued by telomerase which catalyzes the addition of short telomere repeats at chromosomal ends. However, humans do not have sufficient telomerase that would maintain and counteract the losses [118]. Hence, telomere ends continue to erode, thereby limiting the cell cycles and life span of somatic cells. Shortened telomeres decrease proliferative potential of cells until such time that they no longer divide. This growth arrest entered by cells is called replicative senescence [1,120].

IONIZING RADIATION

Cells are constantly exposed to various stressors that generate DNA damage. Cells are equipped with DNA repair mechanisms to negate the effects of this damage. These repair mechanisms play pivotal roles in keeping the integrity of the genome, and hence, the normal physiology of the cells. However, repair mechanisms are not always fool-proof processes and could be compromised, resulting in chromosomal instability [121]. Repair machineries may fail to correct the damage DNA, which could be passed on to the DNA of the daughter cells [122]. Excessive unrepaired damaged DNA causes cells to enter into the senescence state.

IR generates high levels of DNA damage and induces senescence [123]. Charged particles from IR target DNA directly by ionizing it or indirectly by reacting with the surrounding water molecules to produce hydroxyl OH• radicals [124]. Chemical reactions in the DNA by direct ionization or indirectly through OH• radicals consequently lead to DNA strand breaks [124]. Minor base damage or single strand breaks are essentially repaired through base excision repair [125,126]. However, in more severe cases, DNA damage can no longer be repaired and may lead to senescence [123].

Stress-induced premature senescence via IR occurs when cells detect damage, triggering the DDR and p53 activation. p53 induces subsequent gene transcription of p21, which leads to permanent G1 arrest and induction of senescence. Senescence by IR is heavily dependent on the p53 pathway but IR may also increase p16 expression during the latter stages of the establishment of senescence [5].

CYTOTOXIC DRUGS

For decades, cytotoxic drugs served as the backbone for cancer treatment. Cancer undergoes perpetual cell division and can spread and invade neighboring tissues [127]. Previously, it was a widely held view that cancers do not undergo growth arrest. However, through various manipulation studies, it is now apparent that cancer may also be stimulated to undergo growth arrest by cyto-toxic drug treatments [127–129]. Despite lacking functional tumor suppressor proteins, cancer cells exposed to low to moderate doses of cytotoxic drugs will still undergo a terminally arrested state with accompanying morphologic and enzymatic changes reminiscent of the senescence phenotype [130]. This senescence response after treatment with cytotoxic drugs is observed in a subset of cells, specifically in colon carcinoma (HCT116), breast cancer (MCF7), osteosarcoma (SAOS-2), and prostate (DU145, LNCaP) cancer cells using chemotherapeutic drugs, such as doxorubicin, docetaxel, cisplatin, fluorouracil, etoposide, camptothecin, and bleomycin [130–134]. Cytotoxic drugs can also induce senescence in tumors expressing wild-type p53 [135]. Treatment with clini-cally relevant concentrations of topoisomerase inhibitors induces senescence in several cancer cell lines via activation of p53, p21, and p16 pathways. [130,133]. Taken altogether, cellular senescence induced by cytotoxic drug treatments may serve as a cancer therapy to inhibit cancer progression and malignant transformations [130,136].

OXIDATIVE STRESS

Oxidative stress is essentially the imbalance between the production of reactive oxygen species and antioxidant defenses of the body [137]. Oxidative stress may arise from increased oxidant expo-sure, decreased oxidant protection, or both [138]. Oxygen plays a major contributing role in oxida-tive stress. Oxygen molecule (O_2) undergoes a series of reduction steps when it is metabolized *in vivo*. During the process, several reactive metabolites (i.e., superoxide, hydrogen peroxide, hydroxyl radical, singlet oxygen) are produced through the excitation of electrons or interaction with transi-tion elements [138]. Other related endogenous and exogenous oxidants also contribute to oxidative stress. These oxidative stress-inducing agents can attack molecules such as proteins, carbohydrates, lipids, and nucleic acids (i.e., DNA) [137].

Exposure to low concentrations of oxidants may be beneficial because of its mitotic effect on cells [139,140]. However, increasing levels of oxidants causes oxidative stress that is capable of damaging the cells and causing them to enter a permanent state of growth arrest [141]. Oxidative stress may enhance telomere dysfunction and promote development of the senescent phenotype. In the presence of oxidative stress, telomeric damage is less repaired than elsewhere in the chromo-some region [141].

Caveolin-1, a scaffolding protein within the plasma membrane, is involved in oxidative stress-induced senescence in fibroblasts [142]. Overexpression of caveolin-1 causes mitotic arrest in mouse embryonic fibroblasts (MEFs) in G_0–G_1 phase of the cell cycle and promotes premature senescence in a p53/p21-dependent pathway [143,144]. Cells may also undergo senescence via p38MAPK path-way following oxidative stress.

ONCOGENE ACTIVATION

Cells undergo senescence when there is an overexpression of oncogenes. This phenomenon has been first described when oncogenic RAS promoted premature cellular senescence in normal human fibroblasts through the increased expression and activation of two tumor suppressor pathways p53 and p16INK4a [45]. Expression of p53 protein activates transcription genes responsible for cell cycle arrest. p53 transcriptionally activates its downstream target, p21, which also promotes cell cycle arrest. p16INK4a is a potent cell cycle inhibitor that blocks the activity of the cyclin-dependent kinase CDK4. When these kinases are inhibited, they can no longer phosphorylate their target reti-noblastoma protein (Rb), which is responsible for promoting cell cycle progression. Hence, oncogenic activation culminates in premature senescence that involves p53 and p16INK4a expression [45].

An oncogene also represses ribonucleotide reductase subunit M2 (RRM2), a rate-limiting pro-tein in dNTP synthesis, to decrease dNTP levels and promote senescence-associated cell cycle exit [145]. Because dNTP is necessary for DNA synthesis during the S-phase of the cell cycle, loss of dNTP can prevent cells from progressing through the cell cycle. Consistently, RRM2 downregulation is both necessary and sufficient for senescence [145].

TUMOR SUPPRESSOR GENE OVEREXPRESSION

Several different signaling pathways that induce cellular senescence seem to converge toward the activation of Rb and its family members (p107 and p130) [146]. Activation or overexpression of Rb-family proteins is sufficient to induce a senescence-like irreversible cell cycle arrest [147]. Ablation of all three Rb-family genes renders MEFs completely insensitive to senescence-inducing signals [148,149]. Importantly, once the Rb-family proteins are fully activated, the cell cycle arrest becomes irreversible and is no longer revoked by subsequent inactivation of the Rb pathway [150–152].

Several intracellular signaling pathways may influence the ability of p16/Rb to induce cellular senescence. Activation of phosphoinositide 3-kinase/protein kinase B (PI3K/AKT) pathway is important to establish cellular senescence and not quiescence upon overexpression of nuclear p16 [153]. This mitogenic signal in the presence of p16 decreases SOD2 expression, increases ROS production, and increases DNA damage foci, resulting in cellular senescence [153]. Induction of ROS by the combination of the p16/Rb pathway and mitogenic signaling cascade activates protein kinase C delta (PKC delta) [152]. This in turn, enhances ROS production to create a positive feedback loop between ROS and PKC delta, leading to an irreversible cell cycle arrest [152]. In contrast, serum starvation and contact inhibition can increase FoxO3a expression, increase SOD2 expression, decrease ROS production, and decrease DNA damage foci, resulting in quiescence instead of cellular senescence [153].

MITOCHONDRIAL DYSFUNCTION

Mitochondria are dubbed as the "powerhouses of the cells" because of their ability to produce the adenosine triphosphate (ATP) needed to support the proper functioning of the cell. However, genetic mutations and environmental factors may affect mitochondria and may cause their inability to generate energy for the cells. Mitochondrial DNA (mtDNA) and other proteins are in close proximity to the site of electron transport chain and may become oxidatively damaged. When mitochondrial function is compromised in normal proliferating cells, it induces a senescence state termed mitochondrial dysfunction-associated senescence (MiDAS), which may occur *in vivo* and *in vitro* [106,154,155]. MiDAS is marked by decreased levels of NAD+/NADH and increased activation of AMPK and p53, which are important for the establishment of permanent growth arrest in these cells [106].

CELLULAR SENESCENCE AND ORGANISMAL AGING

Multiple theories were proposed to describe the aging process, but explanations were not completely satisfactory because of the myriad of factors that came into play [156]. Great strides were made to understand the phenomenon of aging at a molecular and cellular level. Cellular senescence brought about by replicative exhaustion or excessive cellular stress was proposed to be one of the underlying causes of aging [157]. Much of our understanding on cellular senescence were based on cell culture observations. However, cellular senescence is also now described *in vivo*. Senescent cells are identified in several somatic tissues like stroma, epithelial organs, and hematopoietic systems [50,158–160]. For instance, dermal samples from human donors show an age-dependent increase in activity of the senescent marker SA β-gal [50]. Dermal fibroblasts from aging baboons also display several senescent markers, including increased telomere erosion, heterochromatin foci formation, ATM activation, and p16INK4A expression [159].

Cellular senescence is also detected in aging rodents. Various organs (i.e., heart, kidney) in aging rats demonstrate upregulation of p16INK4A [160]. Hippocampal pyramidal cells show increased cellular senescence with age, but no SA β-gal expression is detected in neurons of hippocampal dental gyrus across all rat age groups [161]. Aging phenotype in the skin is also evident in *Sod2*-deficient mice [162]. Lungs, spleen, dermis, liver, and intestinal epithelium of C57BL/6 mice show a positive correlation between age and induction of DNA damage foci with γH2AX [163].

CELLULAR SENESCENCE AND AGING MOUSE MODEL

Mouse models extensively contribute to our understanding of many important biological processes. Using these models, researchers show that clearance of senescent cells prevents or delays the progression of age-related diseases. For instance, in a study conducted by Jaskelioff et al. (2011) on telomerase-deficient adult mice, reactivation of the endogenous telomerase results in the elongation of telomere ends, reduction in DNA damage signals, resumption of cell proliferation, and rescue of degenerative phenotypes across multiple organs [164]. Studies on BubR1 progeroid mice support the role of senescent cells in developing age-associated pathologies in certain tissues, such as eye, adipose, and skeletal tissues [1,165,166]. Baker et al. (2011) further developed a novel transgenic mouse model called INK-ATTAC, which allows selective elimination of p16INK4A senescent cells in the animal upon administration of the drug AP20187 [166]. Lifelong clearance of senescent cells in various tissues and organs delays and attenuates the progression of age-associated degenerative pathologies without notable side effects. Elimination of these senescent cells also extends the life span of BubR1 progeroid mice [166,167]. Another mouse model that can eliminate senescent cells is the 3MR mouse model, which can detect senescent cells *in vivo* by bioluminescence or fluorescence

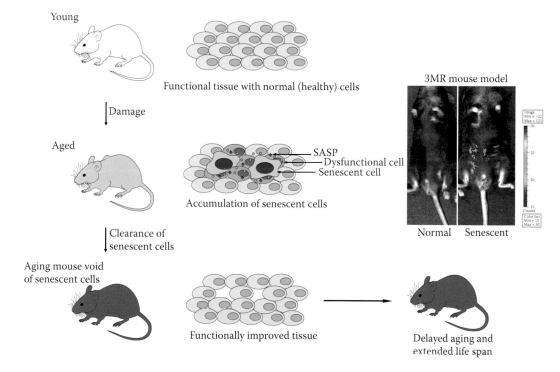

FIGURE 16.5 Clearance of senescent cells in a novel transgenic 3MR mouse model. This mouse model has a senescence-sensitive promoter p16INK4A that drives the expression of trimodality reporter (3MR) fusion protein. The 3MR contains luciferase and red fluorescent protein reporters, along with herpes simplex virus-1 thymidine kinase. The fluorescent reporters allow the detection of 3MR-expressing cells through luminescence and permits sorting of these cells from tissues. (*Photo inset courtesy of Marco Demaria, European Research Institute for the Biology of Aging (ERIBA), University Medical Center Groningen (UMCG), Groningen, the Netherlands.*)

as seen in Figure 16.5. Senescent cells in these mice may be efficiently removed by treating with ganciclovir [168]. However, it is yet to be shown whether elimination of senescent cells in these mice will reverse or delay the onset of aging pathologies.

CELLULAR SENESCENCE AND AGE-RELATED PATHOLOGIES IN HUMANS

Despite beneficial aspects of senescent cells in wound healing and tissue repair [168,169], the presence of senescent cells in a microenvironment may negatively impact the behavior of neighboring cells by altering their normal physiological function. Senescent cells are implicated in many chronic diseases and age-related diseases, which are largely associated with inflammation. Examples of these age-associated diseases include neurodegenerative diseases, diabetes, cataract, macular degeneration, cardiovascular diseases (CVD), osteoporosis, sarcopenia, and arthritis.

ALZHEIMER'S DISEASE

Alzheimer's disease (AD) is the most prevalent form of dementia, a disorder associated with cognitive impairment that afflicts the elderly [170]. AD is marked by considerable amount of amyloid plaques and neurofibrillary tangles within the hippocampus and neocortex of the brain [171]. Individuals affected by this disease show early clinical symptoms such as memory loss and inability to learn new things. As the disease progresses, behavioral changes, disorientation, delusion, and impaired judgment may occur [172]. AD disrupts signal transduction in the brain and damages nerve connections, resulting in neuronal cell death and decreased brain size [173].

One major risk factor for AD is aging, and it is thought that mitochondrial dysfunction contributes to this disease [174]. Early onset may also occur due to mutations in the amyloid precursor protein (APP) [175]. Mutation in APP produces a cleaved form, amyloid-β1–42 (Aβ) protein, which forms aggregates, responsible for plaque formation [176]. Aβ protein promotes premature senescence in adult hippocampal neural stem/progenitor cells [177]. Pyramidal neurons of the hippocampus, as well as neurofibrillary and neritic components of the amyloid plaques also show increased expression of p16INK4A, which are not observed in terminally differentiated neurons [178]. Astrocytes have also been described as a component in the development of AD. Astrocytes execute diverse functions and are responsible for maintaining neuronal health. However, these cells are highly susceptible to oxidative stress and may respond by undergoing premature senescence [173]. Indeed, astrocytes from brain tissues of aged individuals and patients with AD show a profound increase in the expression of p16INK4A and MMP, both of which are markers of cellular senescence [173]. Hence, cellular senescence is proposed to play a contributory role in the development of AD.

PARKINSON'S DISEASE

Parkinson's disease (PD) is the second most common type of age-related neurodegenerative disease, which is marked by tremors, rigidity, postural instability, and bradykinesia [179,180]. PD is characterized by the preferential loss of dopaminergic neurons in the substantia nigra of the brain, largely responsible for the motor control of the organism [181]. Several treatment options are employed in patients with PD, such as the introduction of dopamine neurotransmitters. While parts of the brain are capable of neurogenesis, this regenerative ability is reduced with age [182]. In PD, dopaminergic neurons continue to degenerate, and treatments employed in PD patients are only able to delay the symptoms and slightly improve the condition of patients inflicted with the disease [179].

Similar to AD, several stressors that lead to the pathogenesis of PD are mitochondrial dysfunction and oxidative stress [181,183–185]. These stressors may cause neuronal cells to senesce and elicit SASP factors, which later reduces neuronal integrity in patients with PD. Indeed, expression of senescence markers is positively correlated with the incidence of PD upon examination of postmortem brain tissues from patients [186].

DIABETES

Diabetes mellitus type II (also Type 2 diabetes or D2M) is a metabolic disorder characterized by insulin resistance or insulin deficiency [187]. D2M due to insulin resistance is the most common cause of diabetes in adults, resulting in the inability of cells to absorb glucose and consequently maintain elevated levels of blood glucose. However, D2M due to insulin deficiency may also be a factor in adults with diabetes. This is caused by gradual depletion and exhaustion of insulin production by pancreatic beta cells [187,188]. D2M is the most common form of diabetes prevalent among middle-aged to elderly individuals and is quickly becoming a global health concern worldwide [189].

Senescence is implicated in diabetes [190]. Cellular senescence permits pathogenesis of diabetes through its contribution to adipose tissue dysfunction, SASP-induced chronic or sterile inflammation, and pancreatic beta cell senescence [190,191]. Adipose tissues from obese mice express high levels of p53 that is involved in development of insulin resistance [192]. On the contrary, inhibition of p53 expression in adipose tissues significantly ameliorates senescence features and improves insulin sensitivity in mice [192].

CATARACT

Several eye disorders are associated with increasing age. One of the most prevalent age-associated eye diseases is cataract, a clouding of the lens within the eye [193]. The lens of the eye consists of water and proteins. The lens should stay clear and transparent to properly focus the light at various distances, thereby allowing formation of sharp images on the retina. Once protein aggregates, this forms a lump over a small area of the lens and obstructs the passage of light, hence the fuzzy vision and eventual blindness. If left untreated, patients with cataracts may undergo surgery to treat the disorder. The procedure involves the removal of the cloudy lens and replacement with an artificial lens to correct their impaired vision.

Studies on BubR1 mouse models are conducted to demonstrate putative roles of senescent cells in cataract formation. For instance, progressive bilateral cataracts developed at an early stage in the BubR1 hypomorphic mouse model. A positive correlation in senescence and cataract development is also observed in aging BubR1 hypomorphic mice. Aging mice display overt cataracts that have very similar features with age-associated human cataracts [194]. Moreover, specific mouse tissues respond to BubR1 hypomorphism by inducing the expression of p16INK4A, an important effector of cellular senescence. Ablation of p16INK4A reduces the incidence and latency of cataract formation in BubR1 holomorphic mice [165]. In addition, long-term clearance of p16INK4A-positive senescent cells in BubR1 progeroid mice also delays onset of cataract development [166].

Human epithelial lens cells (HLEC) also undergo senescence. With increasing serial passage, HLEC cultures from old donors display enlarged cytoplasm and nucleus and increased DNA fragmentation compared to their embryonic counterpart [11]. Depletion of lens stem/progenitor cells (LCS) has also been implicated in age-associated cataracts. Proportion of LCS within HLEC from cataract donors declines as a function of age as evidenced by the decreased expression of LCS markers Sox2, Ki67, and Abcg2. Furthermore, an accompanying increase in positive staining for SA β-gal in HLEC cultures is also observed from older patients [193]. Other senescent features such as increased cell size, flattened cytoplasm, and large vacuoles are also evident in adult HLEC. Lastly, increase in senescent cells within HLECs is associated with the severity of cortical cataracts experienced by older patients [193].

MACULAR DEGENERATION

Macular degeneration or age-related macular degeneration (AMD) is one of the leading causes of vision loss worldwide [195]. AMD incidence increases with age and has become a public health concern, especially in growing populations [196,197]. AMD is considered an incurable eye condition caused by the deterioration of macula, a small central area of the retina that provides visual acuity [196]. Two main types of macular degeneration exist: dry and wet form [198,199]. Dry from is characterized by the presence of small yellow crystalline deposits called drusen in the macula [200]. Typically, few small drusen do not significantly cause changes in vision, but over time, these drusen may increase in size and number and could lead to distorted vision. In more advanced stages, retinal pigment epithelial (RPE) cells undergo atrophy, reducing the central vision, and affecting color perception [201,202]. On the other hand, the wet form is the more severe type of AMD. In the wet form, blood vessels develop at the layer behind the retina called the choroid. These blood vessels eventually leak blood and fluid into the retina, causing the macula to swell and rapidly damage the central vision [199,200].

Several biological pathways have been implicated in the etiology and pathogenesis of macular degeneration [199]. Oxidative stress contributes to pathogenesis of AMD. The retina is particularly susceptible to oxidative damage due to its increased oxygen consumption and exposure to visible light [203]. RPE cells undergo senescence and the loss of normal physiological function in aging RPE cells contributes to accumulation of lipofuscin and genesis of drusen in the macula [203–206].

CARDIOVASCULAR DISEASES

CVD contribute to the biggest mortality worldwide [207]. Risk factors contributing to heart diseases include gender, physical inactivity, tobacco use, excessive alcohol intake, unhealthy diet, genetic predisposition, hypertension, and diabetes [207–211]. In addition, age is by far the most important risk factor for CVD, likely through hypercholesterolemia and changes in structural properties of the vascular walls causing reduced arterial elasticity and arterial compliance [211,212].

Atherosclerosis is the most common cause of CVD. Atherosclerosis is a multifactorial disease involving chronic inflammation that causes subsequent cardiovascular conditions, such as myocardial infarction, stroke, and ischemic heart failure [213,214]. Atherosclerosis is initiated by oxidative modification of low-density lipoprotein in the arterial intima, which in turn triggers local inflammation and atherosclerotic plaque development [213]. Persistent inflammation ruptures the plaque and subsequent thrombus formation, resulting in ischemia and myocardial infarction [213,215].

Cellular senescence have a direct association with atherosclerosis. Atherosclerotic plaque consists of VSMC, collagen, elastin, inflammatory cells, and extracellular and intracellular lipids and debris [216]. Atherosclerotic plaques exhibit senescence markers such as SA β-galactosidase and CDKIs p16 and p21 that are not observed in normal vessels [217]. Pronounced expression of CDKIs is mediated by oxidative stress, thereby initiating the senescent state in VSMC [217]. SASP components secreted by senescent VSMC may also amplify the atherosclerotic inflammatory environment generated by macrophages [214,216]. Moreover, reduced telomere length in plaques is commensurate to the severity of the disease [217].

OSTEOPOROSIS

Osteoporosis is a condition characterized by bone loss, which leads to increased risk of bone fractures [218]. A number of risk factors underlying senile osteoporosis include endocrine, metabolic, and mechanical factors [219,220]. Inflammation also influences bone turnover by fostering the progression of osteoporosis [221,222]. Indeed, clinical findings show coincidence of osteoporosis with the period of systemic inflammation [223]. Cytokines associated with systemic inflammation display similar profiles with the cytokines that regulate bone resorption [224,225].

Bone remodeling is regulated by several proinflammatory cytokines such as IL6 [226–228]. However, alteration in cytokine levels may tip the balance between bone formation and resorption. Inflammation driven by immunosenescence, a state of gradual deterioration of immune functions associated with aging, may play a role in pathogenesis of osteoporosis [229]. Immune cells, particularly T cells, affect bone remodeling through the secretion of soluble factors that activate osteoclast and osteoblast functions [219]. At the same time, NF-κB and receptor activator of nuclear factor kappa B ligand (RANKL), which are responsible for immune responses, are upregulated during T cell replicative senescence [230]. RANKL expressed by T cells activate osteoclasts and initiate bone resorption. In addition, IL6 can also heighten osteoclast activity in a RANKL-independent manner by inhibiting apoptosis and increasing cell survival in osteoclasts [230].

SARCOPENIA

Sarcopenia is a gradual reduction of skeletal muscle mass and strength that occurs in concert with aging [231]. Sarcopenic muscles show decreased regenerative capacity. They fail to replace damaged myofibers, thereby causing incapacitation and loss of independence among geriatric individuals [232]. Multiple factors that lead to the development of age-related sarcopenia include physical inactivity, decreased hormonal levels, anorexia, decreased anabolic stimuli, and increased catabolism [231]. Sarcopenia involves denervation of neuromuscular junctions and skeletal muscles [233].

Cellular senescence is implicated in sarcopenia. Mice with an increased number of senescent cells in skeletal muscles show a strong incidence of sarcopenia [165]. Cellular senescence due to mitochondrial dysfunction is also associated with the development of sarcopenia [234,235]. Cellular senescence may promote sarcopenia by depleting the pool of satellite cells, which are responsible for regenerating and repairing damaged tissues in muscles [235,236]. In geriatric mice, resting satellite cells fail to switch from quiescence to proliferative state. These cells proceed to senesce and subsequently lose their ability to rejuvenate despite having a youthful milieu. The switch from quiescent to senescent state is attributed to the derepression of p16INK4A [237]. Hence, ablation of p16INK4A attenuates sarcopenia and extends the healthspan in mice. This result is further supported in a study involving the BubR1-insufficient mouse model [165]. Overall, results suggest that senescent cells may have vital roles in sarcopenia development.

ARTHRITIS

Arthritis is a chronic multifactorial disease where the immune system attacks and gradually degrades the joints [238]. Arthritis inflicts individuals from all age group but occurs more frequently in the elderly. This disease comes in many forms, specifically osteoarthritis (OA) and rheumatoid arthritis (RA). Together with aging, other risk factors such as gender, genetic predisposition, joint overuse, and obesity seem to promote pathogenesis of arthritis [211,239].

Chronic inflammation is the main driver for OA and RA [240]. Numerous inflammatory cytokines are found in chondrocytes of deteriorating cartilage from patients with OA [241]. Many of these cytokines, including IL6, IL8, IL1B, and MMPs, are components of the SASP [107]. Reduced telomere lengths, increased SA β-gal activity, and elevated p16 and p21 expression levels are observed in damaged cartilage of OA patients [241]. Hence, cellular senescence may play a causal role in promoting OA by manifesting the SASP.

POTENTIAL AGING INTERVENTIONS

The accumulation of senescent cells with age has a negative impact on cell function. Hence, eliminating these cells and attenuating their negative effects seem to be an attractive strategy to improve healthspan in humans. Several attempts are proposed to translate the positive results of research from mouse models to humans.

SENOLYTIC THERAPIES

Due to the enhanced healthy life span observed in mice after elimination of senescent cells, subsequent experiments have been undertaken to develop small molecule drugs that could mimic the same effects in human [166,167,242]. Senolytics are a new class of drugs that selectively target senescent cells for clearance. Senescent cells show a marked increase in pro-survival genes, responsible for resistance to apoptosis [243]. Transcript analysis reveals the identities of several upregulated pro-survival genes, which include ephrins (EFNB1), p21, PI3Kγ, BCL-xL, and plasminogen-activated inhibitor-2.

Several drugs that eliminate senescent cells are being investigated. The first senolytic drugs, namely dasatinib and quercetin, have been effective in removing senescent cells, but not quiescent or proliferating cells. In combination, these two drugs act synergistically to eliminate senescent murine embryonic fibroblasts [242]. Intermittent treatment with dasatinib and quercetin via oral gavage also improve vascular aging and hypercholesterolemia in aged mice [244]. Another recently developed senolytic drug termed ABT-263 (also Navitoclax) is an inhibitor of the Bcl-2 family of anti-apoptotic factors [131]. Oral administration of ABT-263 drug effectively clears senescent cells from irradiated and aged transgenic mouse models, and is able to rejuvenate aged tissue stem cells [131]. In an atherosclerotic mouse model, ABT-263 also suppresses growth of atherosclerotic plaque, reduces inflammation, and modifies the structural features of plaque, making it less susceptible to lesions.

Further studies demonstrate the benefits of senolytics in human samples. For instance, ABT-263 effectively reduces viability of senescent human umbilical vein cells (HUVECS) and lung fibroblasts (IMR90) [243]. The same drug also induces apoptosis in senescent cells from cultures of various human cells, irrespective of cell type and senescence inducer [245]. Dasatinib causes selective killing of senescent adipocyte progenitors, whereas quercetin effectively kills human endothelial cells [242]. The Mayo Clinic is currently recruiting participants for their phase 2 clinical trial on senolytic drugs dasatinib and quercetin. This particular study will assess the effectivity of dasatinib and quercetin to eliminate senescent cells in chronic kidney diseases [246].

In 2009, a biotech start-up company called Unity Biotechnology was founded with the goal of clearing senescent cells as a means of rejuvenation therapy to extend and promote healthy life span. Unity Biotechnology aims to design and develop senolytic drugs that could be used to ease chronic diseases caused by senescence. Unity Biotechnology was able to move 91 therapeutic candidates into human clinical trials and they have gained 13 food and drug administration (FDA) approved medicines in the market [247]. It will be very interesting to see whether the company and researchers in the field will be able to successfully create medicines to delay the onset of age-related diseases.

DAMPENING OF THE SASP

The SASP is thought to play a role in persistent inflammation during the progression of age-related chronic diseases. Hence, besides eliminating senescent cells, another therapeutic strategy is to reduce adverse effects by dampening the expression of SASP factors [248]. Several examples of natural compounds that target the SASP include resveratrol, apigenin, kaempferol, and wogonin [249,250]. Resveratrol is a phenolic compound that increases the activity of adenosine monophosphate (AMP)-activated kinase (AMPK) and sirtuins and consequently ameliorates the effects of hydrogen peroxide-induced senescence [251]. Moreover, chronic treatment with resveratrol ameliorates the SASP, possibly though nuclear factor kappa B (NF-κB) modulation [249,252]. Apigenin, kaempferol, and wogonin are flavonoids that exert anti-inflammatory effects by reducing NF-κB activity through downregulation of IκBζ expression [250]. However, it still remains to be seen whether these compounds can treat age-related diseases. *In vivo* studies need to be done to demonstrate the significant contribution of the SASP in various age-related diseases.

CONCLUSION

Cellular senescence is a process of establishing a permanent growth arrest in cells, induced by a myriad of factors. It is proposed to contribute to the pathogenesis of many age-related disorders.

Because of the implications of senescent cells in several diseases, it is suggested that these cells be eliminated in the body to improve and prolong healthspan in humans. Extensive research is being conducted to dampen the negative effects of senescence in humans. Although much progress has been made to understand the role of cellular senescence in several age-related diseases, there is still a lot of work to be done to produce clinically relevant drugs that will target senescent cells. It will take some time before senolytic drugs will be made available in the market. Nonetheless, it is with high hopes that we expect there will be major breakthroughs in the treatment of age-related pathologies in the coming decades.

ACKNOWLEDGMENTS

The authors thank Marco Demaria, European Research Institute for the Biology of Aging (ERIBA), University Medical Center Groningen (UMCG), Groningen, the Netherlands for providing photos of the 3MR mice. Some of the photomicrographs were generated from projects funded by the University of the Philippines (OVPAA-BPhD-2016-08, MCV) and the U.S. National Institutes of Health (NIH)/National Institute on Aging (NIA) (K99AG041221, MCV).

REFERENCES

1. Hayflick, Leonard, and Paul Sidney Moorhead. 1961. The serial cultivation of human diploid cell strains. *Experimental Cell Research* 25: 585–621.
2. Pignolo, Robert J., Mitch O. Rotenberg, and Vincent J. Cristofalo. 1994. Alterations in contact and density-dependent arrest state in senescent WI-38 Cells. *In Vitro Cellular & Developmental Biology—Animal* 30 (7): 471–476. doi: 10.1007/BF02631316.
3. Marcotte, Richard, Chantale Lacelle, and Eugenia Wang. 2004. Senescent fibroblasts resist apoptosis by downregulating caspase-3. *Mechanisms of Ageing and Development* 125 (10): 777–783. doi: 10.1016/j.mad.2004.07.007.
4. Hampel, Barbara, Mechthild Wagner, David Teis, Werner Zwerschke, Lukas A. Huber, and Pidder Jansen-Dürr. 2005. Apoptosis resistance of senescent human fibroblasts is correlated with the absence of nuclear IGFBP-3. *Aging Cell* 4 (6): 325–330. doi: 10.1111/j.1474-9726.2005.00180.x.
5. Campisi, Judith, and Fabrizio d'Adda di Fagagna. 2007. Cellular senescence: When bad things happen to good cells. *Nature Reviews Molecular Cell Biology* 8 (9): 729–740. doi: 10.1038/nrm2233.
6. Hayflick, Leonard. 1998. How and why we age. *Experimental Gerontology* 33 (7–8): 639–653.
7. Hayflick, Leonard. 1974. The longevity of cultured human cells. *Journal of the American Geriatrics Society* 22 (1): 1–12. doi: 10.1111/j.1532-5415.1974.tb02152.x.
8. Rheinwald, James G., and Howard Green. 1975. Serial cultivation of strains of human epidermal keratinocytes: The formation keratinizing colonies from single cells. *Cell* 6 (3): 331–343. doi: 10.1016/S0092-8674(75)80001-8.
9. Bierman, Edwin L. 1978. The effect of donor age on the in vitro life span of cultured human arterial smooth-muscle cells. *In Vitro* 14 (11): 951–955. doi: 10.1007/BF02616126.
10. Tice, Raymond R., Edward L. Schneider, David Kram, and Phil Thorne. 1979. Cytokinetic analysis of the impaired proliferative response of peripheral lymphocytes from aged humans to phytohemagglutinin. *Journal of Experimental Medicine* 149 (5): 1029–1041. doi: 10.1084/jem.149.5.1029.
11. Tassin, J., E. Malaise, and Y. Courtois. 1979. Human lens cells have an in vitro proliferative capacity inversely proportional to the donor age. *Experimental Cell Research* 123 (2): 388–392. doi: 10.1016/0014-4827(79)90483-X.
12. Mueller, Stephen, Eliot M. Rosen, and Eliot M. Levine. 1980. Cellular senescence in a cloned strain of bovine fetal aortic endothelial cells. *Science* 207 (4433): 889 891. doi: 10.1126/science.7355268.
13. Wagner, Wolfgang, Patrick Horn, Mirco Castoldi, Anke Diehlmann, Simone Bork, Rainer Saffrich, Vladimir Benes, et al. 2008. Replicative senescence of mesenchymal stem cells: A continuous and organized process. *PLoS ONE* 3 (5): e2213. doi: 10.1371/journal.pone.0002213.
14. Feng, Qiang, Shi-Jiang Lu, Irina Klimanskaya, Ignatius Gomes, Dohoon Kim, Young Chung, George R. Honig, Kwang-Soo Kim, and Robert Lanza. 2010. Hemangioblastic derivatives from human induced pluripotent stem cells exhibit limited expansion and early senescence. *Stem Cells* 28 (4): 704–712. doi: 10.1002/stem.321.

15. Park, In-Kyung, Sean J. Morrison, and Michael F. Clarke. 2004. Bmi1, stem cells, and senescence regulation. *Journal of Clinical Investigation* 113 (2): 175–179. doi: 10.1172/JCI200420800.

16. Jacobs, Jacqueline J. L., Karin Kieboom, Silvia Marino, Ronald A DePinho, and Maarten van Lohuizen. 1999. The oncogene and polycomb-group gene bmi-1 regulates cell proliferation and senescence through the ink4a locus. *Nature* 397 (6715): 164–168. doi: 10.1038/16476.

17. Park, In-Kyung, Dalong Qian, Mark Kiel, Michael W. Becker, Michael Pihalja, Irving L. Weissman, Sean J. Morrison, and Michael F. Clarke. 2003. Bmi-1 is required for maintenance of adult self-renewing haematopoietic stem cells. *Nature* 423 (6937): 302–305. doi: 10.1038/nature01587.

18. Molofsky, Anna V., Shenghui He, Mohammad Bydon, Sean J. Morrison, and Ricardo Pardal. 2005. Bmi-1 promotes neural stem cell self-renewal and neural development but not mouse growth and survival by repressing the p16Ink4a and p19 Arf senescence pathways. *Genes and Development* 19 (12): 1432–1437. doi: 10.1101/gad.1299505.

19. Gonos, Efstathios S., Anastasia Derventzi, Marie Kveiborg, Georgia Agiostratidou, Mustapha Kassem, Brian F. C. Clark, Parmjit S. Jat, and Suresh I. S. Rattan. 1998. Cloning and identification of genes that associate with mammalian replicative senescence. *Experimental Cell Research* 240 (1): 66–74. doi: 10.1006/excr.1998.3948.

20. Narita, Masako Masashi, Sabrina Núñez, Edith Heard, Masako Masashi Narita, Athena W. Lin, Stephen A. Hearn, David L. Spector, Gregory J. Hannon, and Scott W. Lowe. 2003. Rb-mediated heterochromatin formation and silencing of E2F target genes during cellular senescence. *Cell* 113 (6): 703–716. doi: 10.1016/S0092-8674(03)00401-X.

21. Zhang, Hong, Kuang-Hung Pan, and Stanley N. Cohen. 2003. Senescence-specific gene expression fingerprints reveal cell-type-dependent physical clustering of up-regulated chromosomal loci. *Proceedings of the National Academy of Sciences* 100 (6): 3251–3256. doi: 10.1073/pnas.2627983100.

22. Freund, Adam, Remi-Martin Laberge, Marco Demaria, and Judith Campisi. 2012. Lamin B1 loss is a senescence-associated biomarker. *Molecular Biology of the Cell* 23 (11): 2066–2075. doi: 10.1091/mbc. E11-10-0884.

23. Davalos, Albert R., Misako Kawahara, Gautam K. Malhotra, Nicholas Schaum, Jiahao Huang, Urvi Ved, Christian M. Beausejour, Jean-Philippe Coppé, Francis Rodoer, and Judith Campisi. 2013. p53-dependent release of alarmin HMGB1 is a central mediator of senescent phenotypes. *Journal of Cell Biology* 201 (4): 613–629. doi: 10.1083/jcb.201206006.

24. Coppé, Jean-Philippe, Christopher K. Patil, Francis Rodier, Yu Sun, Denise P. Muñoz, Joshua Goldstein, Peter S. Nelson, Pierre-Yves Desprez, and Judith Campisi. 2008. Senescence-associated secretory phenotypes reveal cell-nonautonomous functions of oncogenic RAS and the p53 tumor suppressor. *PLoS Biology* 6 (12): e301. doi: 10.1371/journal.pbio.0060301.

25. Kuilman, Thomas, and Daniel S. Peeper. 2009. Senescence-messaging secretome: SMS-Ing cellular stress. *Nature Reviews Cancer* 9 (2): 81–94. doi: 10.1038/nrc2560.

26. Rayess, Hani, Marilene B. Wang, and Eri S. Srivatsan. 2012. Cellular senescence and tumor suppressor gene p16. *International Journal of Cancer* 130 (8): 1715–1725. doi: 10.1002/ijc.27316.

27. Sherwood, S W, D. Rush, J. L. Ellsworth, and R. T Schimke. 1988. Defining cellular senescence in IMR-90 cells: A flow cytometric analysis. *Proceedings of the National Academy of Sciences of the United States of America* 85 (23): 9086–9090.

28. Mao, Zhiyong, Zhonghe Ke, Vera Gorbunova, and Andrei Seluanov. 2012. Replicatively senescent cells are arrested in G1 and G2 phases. *Aging (Albany NY)* 4 (6): 431–435. doi: 10.18632/aging. 100467.

29. d'Adda di Fagagna, Fabrizio, Philip M. Reaper, Lorena Clay-Farrace, Heike Fiegler, Philippa Carr, Thomas von Zglinicki, Gabriele Saretzki, et al. 2003. A DNA damage checkpoint response in telomere-initiated senescence. *Nature* 426 (6963): 194–198. doi: 10.1038/nature02118.

30. Waldman, Todd, Kenneth W Kinzler, and Bert Vogelstein. 1995. p21 Is necessary for the p53-mediated G1 arrest in human cancer cells. *Cancer Research* 55 (22): 5187–5190.

31. Gire, Véronique, and Vjekoslav Dulić. 2015. Senescence from G2 arrest, revisited. *Cell Cycle* 14 (3): 297–304. doi: 10.1080/15384101.2014.1000134.

32. Waldman, Todd, Christoph Lengauer, Kenneth W. Kinzler, and Bert Vogelstein. 1996. Uncoupling of S phase and mitosis induced by anticancer agents in cells lacking p21. *Nature* 381 (6584): 713–716. doi: 10.1038/381713a0.

33. Duque, Alvaro, and Pasko Rakic. 2011. Different effects of bromodeoxyuridine and [3H]thymidine incorporation into DNA on cell proliferation, position, and fate. *Journal of Neuroscience* 31 (42): 15205–15217. doi: 10.1523/JNEUROSCI.3092-11.2011.

34. Gratzner, Howard G., A. Pollack, D. J. Ingram, and R. C. Leif. 1976. Deoxyribonucleic acid replication in single cells and chromosomes by immunologic techniques. *The Journal of Histochemistry and Cytochemistry* 24 (1): 34–39.

35. Nowakowski, R. S., S. B. Lewin, and M. W. Miller. 1989. Bromodeoxyuridine immunohistochemical determination of the lengths of the cell cycle and the DNA-synthetic phase for an anatomically defined population. *Journal of Neurocytology* 18 (3): 311–318. doi: 10.1111/j.1467-6486.2009.00883.x.

36. Gratzner, Howard G. 1982. Monoclonal antibody to 5-bromo- and 5-iododeoxyuridine: A new reagent for detection of DNA replication placental mononuclear phagocytes as a source of interleukin-1. *Science* 218: 474–475. doi: 10.1126/science.7123245.

37. Shapiro, Howard M. 2003. *Practical Flow Cytometry.* Hoboken, NJ, USA: John Wiley & Sons, Inc. doi: 10.1002/0471722731.

38. Darzynkiewicz, Zbigniew. 2010. Critical aspects in analysis of cellular DNA content. In *Current Protocols in Cytometry*, Chapter 7: 1–9, Unit 7.2. Hoboken, NJ, USA: John Wiley & Sons, Inc. doi: 10.1002/0471142956.cy0702s52.

39. Schmid, Ingrid, Christel H. Uittenbogaart, and Janis V. Giorgi. 1991. A gentle fixation and permeabilization method for combined cell surface and intracellular staining with improved precision in DNA quantification. *Cytometry* 12 (3): 279–285. doi: 10.1002/cyto.990120312.

40. Jackson, James G., and Olivia M. Pereira-Smith. 2006. p53 is preferentially recruited to the promoters of growth arrest genes p21 and GADD45 during replicative senescence of normal human fibroblasts. *Cancer Research* 66 (17): 8356–8560. doi: 10.1158/0008-5472.CAN-06-1752.

41. Trougakos, Ioannis P., Aggeliki Saridaki, George Panayotou, and Efstathios S. Gonos. 2006. Identification of differentially expressed proteins in senescent human embryonic fibroblasts. *Mechanisms of Ageing and Development* 127 (1): 88–92. doi: 10.1016/j.mad.2005.08.009.

42. Stein, Gretchen H., Linda F. Drullinger, Alexandre Soulard, and Vjekoslav Dulić. 1999. Differential roles for cyclin-dependent kinase inhibitors p21 and p16 in the mechanisms of senescence and differentiation in human fibroblasts. *Molecular and Cellular Biology* 19 (3): 2109–2117. doi: 10.1128/MCB.19.3.2109.

43. Espinosa, Joaquín M., Ramiro E. Verdun, and Beverly M. Emerson. 2003. p53 functions through stress- and promoter-specific recruitment of transcription initiation components before and after DNA damage. *Molecular Cell* 12 (4): 1015–1027.

44. Alcorta, D. A., Y. Xiong, D. Phelps, G. Hannon, D. Beach, and J. C. Barrett. 1996. Involvement of the cyclin-dependent kinase inhibitor p16 (INK4a) in replicative senescence of normal human fibroblasts. *Proceedings of the National Academy of Sciences of the United States of America* 93 (24): 13742–13747. doi: 10.1073/pnas.93.24.13742.

45. Serrano, Manuel, Athena W. Lin, Mila E. McCurrach, David Beach, and Scott W. Lowe. 1997. Oncogenic Ras provokes premature cell senescence associated with accumulation of p53 and p16INK4a. *Cell* 88 (5): 593–602. doi: 10.1016/S0092-8674(00)81902-9.

46. Portefaix, Jean Michel, Cristina Fanutti, Claude Granier, Evelyne Crapez, Richard Perham, Jean Grenier, Bernard Pau, and Maguy Del Rio. 2002. Detection of anti-p53 antibodies by ELISA using p53 synthetic or phage-displayed peptides. *Journal of Immunological Methods* 259 (1): 65–75. doi: 10.1016/S0022-1759(01)00494-X.

47. Herbig, Utz, Wendy A. Jobling, Benjamin P. C. Chen, David J. Chen, and John M. Sedivy. 2004. Telomere shortening triggers senescence of human cells through a pathway involving ATM, p53, and p21CIP1, but not p16INK4a. *Molecular Cell* 14 (4): 501–513. doi: 10.1016/S1097-2765(04)00256-4.

48. Schneider, Sylke, Kazumi Uchida, Dennis Salonga, Ji Min Yochim, Kathleen D. Danenberg, and Peter V. Danenberg. 2004. Quantitative determination of p16 gene expression by RT-PCR. In *Checkpoint Controls and Cancer: Volume 2:Activation and Regulation Protocols*, 281: 91–103. NJ: Humana Press. doi: 10.1385/1-59259-811-0:091.

49. Cherneva, R. V., O. B. Georgiev, D. B. Petrov, I. I. Dimova, and D. I. Toncheva. 2014. Expression levels of p53 messenger RNA detected by real time PCR in tumor tissue, lymph nodes and peripheral blood of patients with non-small cell lung cancer—New perspectives for clinicopathological application. *Biotechnology & Biotechnological Equipment* 23 (2): 1247–1249, Taylor & Francis. doi: 10.1080/13102818.2009.10817647.

50. Dimri, G., X. Lee, G. Basile, M. Acosta, G. Scott, C. Roskelley, E. E. Medrano, et al. 1995. A biomarker that identifies senescent human cells in culture and in aging skin in vivo. *Proceedings of the National Academy of Sciences of the United States of America* 92 (20): 9363–9367. doi: 10.1073/pnas.92.20.9363.

51. Itahana, Koji, Judith Campisi, and Goberdhan P. Dimri. 2007. Methods to detect biomarkers of cellular senescence: The senescence-associated beta-galactosidase assay. *Methods in Molecular Biology (Clifton, N.J.)* 371 (1): 21–31. doi: 10.1007/978-1-59745-361-5_3.

52. Debacq-Chainiaux, Florence, Jorge D. Erusalimsky, Judith Campisi, and Olivier Toussaint. 2009. Protocols to detect senescence-associated beta-galactosidase (SA-Betagal) activity, a biomarker of senescent cells in culture and in vivo. *Nature Protocols* 4 (12): 1798–1806. doi: 10.1038/nprot.2009.191.

53. Kurz, David J., Stephanie Decary, Ying Hong, and Jorge D. Erusalimsky. 2000. Senescence-associated (beta)-galactosidase reflects an increase in lysosomal mass during replicative ageing of human endothelial cells. *Journal of Cell Science* 113 (20): 3613–3622.

54. Lee, Bo Yun, Jung A. Han, Jun Sub Im, Amelia Morrone, Kimberly Johung, Edward C. Goodwin, Wim J. Kleijer, Daniel DiMaio, and Eun Seong Hwang. 2006. Senescence-associated β-galactosidase is lysosomal β-galactosidase. *Aging Cell* 5 (2): 187–195. doi: 10.1111/j.1474-9726.2006.00199.x.

55. Yegorov, Yegor E., Sergey S. Akimov, Ralf Hass, Alexander V. Zelenin, and Igor A. Prudovsky. 1998. Endogenous beta-galactosidase activity in continuously nonproliferating cells. *Experimental Cell Research* 243 (1): 207 211. doi: 10.1006/excr.1998.4169.

56. Severino, Joseph, R. G. Allen, S Balin, Arthur Balin, and Vincent J. Cristofalo. 2000. Is beta-galactosidase staining a marker of senescence in vitro and in vivo? *Experimental Cell Research* 257 (1): 162–171. doi: 10.1006/excr.2000.4875.

57. Untergasser, G., R. Gander, H. Rumpold, E. Heinrich, E. Plas, and P. Berger. 2003. TGF-beta cytokines increase senescence-associated beta-galactosidase activity in human prostate basal cells by supporting differentiation processes, but not cellular senescence. *Experimental Gerontology* 38 (10): 1179–1188. doi: 10.1016/j.exger.2003.08.008.

58. Nuss, Jonathan E., Kashyap B. Choksi, James H. DeFord, and John Papaconstantinou. 2008. Decreased enzyme activities of chaperones PDI and BiP in aged mouse livers. *Biochemical and Biophysical Research Communications* 365 (2): 355–361. doi: 10.1016/j.bbrc.2007.10.194.

59. Kaufman, Randal J. 2002. Orchestrating the unfolded protein response in health and disease. *Journal of Clinical Investigation* 110 (10): 1389–1398. doi: 10.1172/JCI200216886.

60. Cho, Joon-Ho, Deepak Kumar Saini, W. K. Ajith Karunarathne, Vani Kalyanaraman, and N. Gautam. 2011. Alteration of Golgi structure in senescent cells and its regulation by a g protein γ subunit. *Cellular Signalling* 23 (5): 785–793. doi: 10.1016/j.cellsig.2011.01.001.

61. Guilly, Y Le, M. Simon, P. Lenoir, and M. Bourel. 1973. Long-term culture of human adult liver cells: Morphological changes related to in vitro senescence and effect of donor's age on growth potential. *Gerontologia* 19 (5): 303–313.

62. Stanulis-Praeger, Betzabé M. 1987. Cellular senescence revisited: A review. *Mechanisms of Ageing and Development* 38 (1): 1–48. doi: 10.1016/0047-6374(87)90109-6.

63. Matsumura, T. 1980. Multinucleation and polyploidization of aging human cells in culture. *Advances in Experimental Medicine and Biology* 129: 31–38.

64. Wang, Eugenia, and Doris Gundersen. 1984. Increased organization of cytoskeleton accompanying the aging of human fibroblasts in vitro. *Experimental Cell Research* 154 (1): 191–202. doi: 10.1016/0014-4827(84)90679-7.

65. Chen, Qin M., Victoria C. Tu, Jeffrey Catania, Maggi Burton, Olivier Toussaint, and Tarrah Dilley. 2000. Involvement of Rb family proteins, focal adhesion proteins and protein synthesis in senescent morphogenesis induced by hydrogen peroxide. *Journal of Cell Science* 113: 4087–4097.

66. von Zglinicki, T., E. Nilsson, W. D. Döcke, and U. T. Brunk. 1995. Lipofuscin accumulation and ageing of fibroblasts. *Gerontology* 41 (Suppl 2): 95–108.

67. Georgakopoulou, E. A., K. Tsimaratou, K. Evangelou, Marcos-P. J. Fernandez, V. Zoumpourlis, I. P. Trougakos, D. Kletsas, J. Bartek, M. Serrano, and V. G. Gorgoulis. 2012. Specific lipofuscin staining as a novel biomarker to detect replicative and stress-induced senescence. A method applicable in cryopreserved and archival tissues. *Aging* 5 (1): 37–50. doi: 10.18632/aging.100527.

68. Stewart, Grant S., Bin Wang, Colin R. Bignell, A. Malcolm R. Taylor, and Stephen J. Elledge. 2003. MDC1 is a mediator of the mammalian DNA damage checkpoint. *Nature* 421 (6926): 961–966. doi: 10.1038/nature01446.

69. Rogakou, Emmy P., Duane R. Pilch, Ann H. Orr, Vessela S. Ivanova, and William M. Bonner. 1998. DNA double-stranded breaks induce histone H2AX phosphorylation on serine 139. *Journal of Biological Chemistry* 273 (10): 5858–5868. doi: 10.1074/jbc.273.10.5858.

70. Schultz, Linda B., Nabil H. Chehab, Asra Malikzay, and Thanos D. Halazonetis. 2000. p53 binding protein 1 (53BP1) is an early participant in the cellular response to DNA double-strand breaks. *The Journal of Cell Biology* 151 (7): 1381–1390. doi: 10.1083/jcb.151.7.1381.

71. Paull, Tanya T., Emmy P. Rogakou, Vikky Yamazaki, Cordula U. Kirchgessner, Martin Gellert, and William M. Bonner. 2000. A critical role for histone H2AX in recruitment of repair factors to nuclear foci after DNA damage. *Current Biology* 10 (15): 886–895. doi: 10.1016/S0960-9822(00)00610-2.

72. Stiff, Tom, Mark O'Driscoll, Nicole Rief, Kuniyoshi Iwabuchi, Markus Löbrich, and Penny A. Jeggo. 2004. ATM and DNA-PK function redundantly to phosphorylate H2AX after exposure to ionizing radiation. *Cancer Research* 64 (7): 2390–2396. doi: 10.1158/0008-5472.CAN-03-3207.

73. Kuo, Linda J., and Li-Xi Yang. 2008. Gamma-H2AX—A novel biomarker for DNA double-strand breaks. *In Vivo (Athens, Greece)* 22 (3): 305–309.

74. Di Micco, R., Fumagalli, M., Cicalese, A., Piccinin, S., Gasparini, P., Luise, C. et al. d'Adda di Fagagna, F. 2006. Oncogene-induced senescence is a DNA damage response triggered by DNA hyper-replication. *Nature* 444 (7119): 638–642. https://doi: 10.1038/nature05327.

75. Takai, Hiroyuki, Agata Smogorzewska, Titia de Lange, R. A. Marciniak, F. B. Johnson, L. Guarente, R. S. Maser, et al. 2003. DNA damage foci at dysfunctional telomeres. *Current Biology* 13 (17): 1549–1556. doi: 10.1016/s0960-9822(03)00542-6.

76. Rodier, Francis, Jean-Philippe Coppé, Christopher K. Patil, Wieteke A. M. Hoeijmakers, Denise P. Muñoz, Saba R. Raza, Adam Freund, Eric Campeau, Albert R. Davalos, and Judith Campisi. 2009. Persistent DNA damage signalling triggers senescence-associated inflammatory cytokine secretion. *Nature Cell Biology* 11 (8): 973–979. doi: 10.1038/ncb1909.

77. Rodier, Francis, D. P. Munoz, Robert Teachenor, Victoria Chu, Oanh Le, Dipa Bhaumik, J.-P. Coppe, et al. 2011. DNA-SCARS: Distinct nuclear structures that sustain damage-induced senescence growth arrest and inflammatory cytokine secretion. *Journal of Cell Science* 124 (1): 68–81. doi: 10.1242/jcs.071340.

78. Higgs, Douglas R., Douglas Vernimmen, Jim Hughes, and Richard Gibbons. 2007. Using genomics to study how chromatin influences gene expression. *Annual Review of Genomics and Human Genetics* 8 (1): 299–325.

79. Grewal, Shiv I. S., and Sarah C. R. Elgin. 2002. Heterochromatin: New possibilities for the inheritance of structure. *Current Opinion in Genetics and Development* 12 (2): 178–187. doi: 10.1016/S0959-437X(02)00284-8.

80. Howard, Bruce H. 1996. Replicative senescence: Considerations relating to the stability of heterochromatin domains. *Experimental Gerontology* 31 (1): 281–293. doi: 10.1016/0531-5565(95)00022-4.

81. Zhang, Rugang, Maxim V. Poustovoitov, Xiaofen Ye, Hidelita A. Santos, Wei Chen, Sally M. Daganzo, Jan P. Erzberger, et al. 2005. Formation of MacroH2A-containing senescence-associated heterochromatin foci and senescence driven by ASF1a and HIRA. *Developmental Cell* 8 (1): 19–30. doi: 10.1016/j.devcel.2004.10.019.

82. Maison, Christèle, and Geneviève Almouzni. 2004. HP1 and the dynamics of heterochromatin maintenance. *Nature Reviews. Molecular Cell Biology* 5 (4): 296–304. doi: 10.1038/nrm1355.

83. Zhang, Rugang, Song-tao Liu, Wei Chen, Michael Bonner, John Pehrson, Timothy J. Yen, and Peter D. Adams. 2007. HP1 proteins are essential for a dynamic nuclear response that rescues the function of perturbed heterochromatin in primary human cells. *Molecular and Cellular Biology* 27 (3): 949–962. doi: 10.1128/MCB.01639-06.

84. Ye, Xiaofen, Brad Zerlanko, Rugang Zhang, Neeta Somaiah, Marc Lipinski, Paolo Salomoni, and Peter D. Adams. 2007. Definition of pRB- and p53-dependent and -independent steps in HIRA/ASF1a-mediated formation of senescence-associated heterochromatin foci. *Molecular and Cellular Biology* 27 (7): 2452–2465. doi: 10.1128/MCB.01592-06.

85. Krohne, Georg, and Ricardo Benavente. 1986. The nuclear lamins. A multigene family of proteins in evolution and differentiation. *Experimental Cell Research* 162 (1): 1–10. doi: 10.1016/0014-4827(86)90421-0.

86. Dechat, T., K. Pfleghaar, K. Sengupta, T. Shimi, D. K. Shumaker, L Solimando, and R. D. Goldman. 2008. Nuclear lamins: Major factors in the structural organization and function of the nucleus and chromatin. *Genes & Development* 22 (7): 832–853. doi: 10.1101/gad.1652708.

87. Goldman, Robert D., Yosef Gruenbaum, Robert D. Moir, Dale K. Shumaker, and Timothy P. Spann. 2002. Nuclear lamins: Building blocks of nuclear architecture. *Genes and Development* 16 (5): 533–547. doi: 10.1101/gad.960502.

88. Stewart, Colin L., and Brian Burke. 1987. Teratocarcinoma stem cells and early mouse embryos contain only a single major lamin polypeptide closely resembling lamin B. *Cell* 51 (3): 383–392. doi: 10.1016/0092-8674(87)90634-9.

89. Lin, Feng, and Howard J. Worman. 1995. Structural organization of the human gene (LMNB1) encoding nuclear lamin B1. *Genomics* 27 (2): 230–236. doi: 10.1006/geno.1995.1036.

90. Tilli, C. M. L. J., F. C. S. Ramaekers, J. L. V. Broers, C. J. Hutchison, and H. A. M. Neumann. 2003. Lamin expression in normal human skin, actinic keratosis, squamous cell carcinoma and basal cell carcinoma. *The British Journal of Dermatology* 148 (1): 102–109. doi: 10.1046/j.1365-2133.2003.05026.x.

91. Broers, J. L. V., Barbie M. Machiels, H. J. H. Kuijpers, Frank Smedts, Ronald Van Den Kieboom, Yves Raymond, and F. C. S. Ramaekers. 1997. A- and B-type lamins are differentially expressed in normal human tissues. *Histochemistry and Cell Biology* 107 (6): 505–517. doi: 10.1007/s004180050138.

92. Rober, Ruth-Ariane, Klaus Weber, and Mary Osborn. 1989. Differential timing of nuclear lamin A/C expression in the various organs of the mouse embryo and the young animal: A developmental study. *Development* 105: 365–378.

93. Dreesen, Oliver, Alexandre Chojnowski, Peh Fern Ong, Tian Yun Zhao, John E. Common, Declan Lunny, E. Birgitte Lane, et al. 2013. Lamin B1 fluctuations have differential effects on cellular proliferation and senescence. *Journal of Cell Biology* 200 (5): 605–617. doi: 10.1083/jcb.201206121.

94. Reddy, K. L., J. M. Zullo, E. Bertolino, and H. Singh. 2008. Transcriptional repression mediated by repositioning of genes to the nuclear lamina. *Nature* 452 (7184): 243–247. doi: 10.1038/nature06727.

95. Sadaie, Mahito, Rafik Salama, Thomas Carroll, Kosuke Tomimatsu, Tamir Chandra, Andrew R. J. Young, Masako Narita, et al. 2013. Redistribution of the lamin B1 genomic binding profile affects rearrangement of heterochromatic domains and SAHF formation during senescence. *Genes and Development* 27 (16): 1800–1808. doi: 10.1101/gad.217281.113.

96. Shimi, Takeshi, Veronika Butin-Israeli, Stephen A. Adam, Robert B. Hamanaka, Anne E. Goldman, Catherine A. Lucas, Dale K. Shumaker, Steven T. Kosak, Navdeep S. Chandel, and Robert D. Goldman. 2011. The role of nuclear lamin B1 in cell proliferation and senescence. *Genes and Development* 25 (24): 2579–2593. doi: 10.1101/gad.179515.111.

97. Grosschedl, Rudolf, Klaus Giese, and John Pagel. 1994. HMG domain proteins: Architectural elements in the assembly of nucleoprotein structures. *Trends in Genetics* 10 (3): 94–100. doi: 10.1016/0168-9525(94)90232-1.

98. Bianchi, Marco E., and Alessandra Agresti. 2005. HMG proteins: Dynamic players in gene regulation and differentiation. *Current Opinion in Genetics and Development* 15 (5): 496–506. doi: 10.1016/j.gde.2005.08.007.

99. Reeves, Raymond, and Jennifer E. Adair. 2005. Role of high mobility group (HMG) chromatin proteins in DNA repair. *DNA Repair* 4 (8): 926–938. doi: 10.1016/j.dnarep.2005.04.010.

100. Pil, P. M., and S. J. Lippard. 1992. Specific binding of chromosomal protein HMG1 to DNA damaged by the anticancer drug cisplatin. *Science* 256 (5054): 234–237. doi: 10.1126/science.1566071.

101. Lange, Sabine S., Madhava C. Reddy, and Karen M. Vasquez. 2009. Human HMGB1 directly facilitates interactions between nucleotide excision repair proteins on triplex-directed psoralen interstrand cross-links. *DNA Repair* 8 (7): 865–872. doi: 10.1016/j.dnarep.2009.04.001.

102. Pasheva, Evdokia A., Iliya G. Pashev, and Alain Favre. 1998. Preferential binding of high mobility group 1 protein to UV-damaged DNA: Role of the COOH-terminal domain. *Journal of Biological Chemistry* 273 (38): 24730–24736. doi: 10.1074/jbc.273.38.24730.

103. Lotze, Michael T., and Kevin J. Tracey. 2005. High-mobility group box 1 protein (HMGB1): Nuclear weapon in the immune arsenal. *Nature Reviews Immunology* 5 (4): 331–342. doi: 10.1038/nri1594.

104. Bianchi, Marco E. 2007. DAMPs, PAMPs and alarmins: All we need to know about danger. *Journal of Leukocyte Biology* 81 (1): 1–5. doi: 10.1189/jlb.0306164.

105. Yamada, Shingo, and Ikuro Maruyama. 2007. HMGB1, a novel inflammatory cytokine. *Clinica Chimica Acta* 375 (1): 36–42. doi: 10.1016/j.cca.2006.07.019.

106. Wiley, Christopher D., Michael C. Velarde, Pacome Lecot, Su Liu, Ethan A. Sarnoski, Adam Freund, Kotaro Shirakawa, et al. 2016. Mitochondrial dysfunction induces senescence with a distinct secretory phenotype. *Cell Metabolism 23* (2): 303–314. doi: 10.1016/j.cmet.2015.11.011.

107. Coppé, Jean-Philippe, Pierre-Yves Desprez, Ana Krtolica, and Judith Campisi. 2010. The senescence-associated secretory phenotype: The dark side of tumor suppression. *Annual Review of Pathological Mechanical Disease* 5 (1): 99–118. doi: 10.1146/annurev-pathol-121808-102144.

108. Campisi, Judith. 2011. Cellular senescence: Putting the paradoxes in perspective. *Current Opinion in Genetics and Development* 21 (1), 107–12. doi: 10.1016/j.gde.2010.10.005.

109. Laberge, Remi-Martin, Yu Sun, Arturo V. Orjalo, Christopher K. Patil, Adam Freund, Lili Zhou, Samuel C. Curran, et al. 2015. MTOR regulates the pro-tumorigenic senescence-associated secretory phenotype by promoting IL1A translation. *Nature Cell Biology* 17 (8): 1049–1061. doi: 10.1038/ncb3195.

110. Coppé, Jean-Philippe, Katalin Kauser, Judith Campisi, and Christian M. Beauséjour. 2006. Secretion of vascular endothelial growth factor by primary human fibroblasts at senescence. *The Journal of Biological Chemistry* 281 (40): 29568–29574. doi: 10.1074/jbc.M603307200.

111. Freund, Adam, Arturo V. Orjalo, Pierre Yves Desprez, and Judith Campisi. 2010. Inflammatory networks during cellular senescence: Causes and consequences. *Trends in Molecular Medicine* 16 (5): 238–246. doi: 10.1016/j.molmed.2010.03.003.

112. Moiseeva, Olga, Xavier Deschênes-Simard, Emmanuelle St-Germain, Sebastian Igelmann, Geneviève Huot, Alexandra E. Cadar, Véronique Bourdeau, Michael N. Pollak, and Gerardo Ferbeyre. 2013. Metformin inhibits the senescence-associated secretory phenotype by interfering with IKK/NF-κB activation. *Aging Cell* 12 (3): 489–498. doi: 10.1111/acel.12075.

113. Acosta, Juan Carlos, Ana O'Loghlen, Ana Banito, Maria V. Guijarro, Arnaud Augert, Selina Raguz, Marzia Fumagalli, et al. 2008. Chemokine signaling via the CXCR2 receptor reinforces senescence. *Cell* 133 (6): 1006–1018. doi: 10.1016/j.cell.2008.03.038.

114. Acosta, Juan Carlos, Ana Banito, Torsten Wuestefeld, Athena Georgilis, Peggy Janich, Jennifer P. Morton, Dimitris Athineos, et al. 2013. A complex secretory program orchestrated by the inflammasome controls paracrine senescence. *Nature Cell Biology* 15 (8): 978–990. doi: 10.1038/ncb2784.

115. Nelson, Glyn, James Wordsworth, Chunfang Wang, Diana Jurk, Conor Lawless, Carmen Martin-Ruiz, and Thomas von Zglinicki. 2012. A senescent cell bystander effect: Senescence-induced senescence. *Aging Cell* 11 (2): 345–349. doi: 10.1111/j.1474-9726.2012.00795.x.

116. Kuilman, Thomas, Chrysiis Michaloglou, Liesbeth C. W. Vredeveld, Sirith Douma, Remco van Doorn, Christophe J Desmet, Lucien A. Aarden, Wolter J. Mooi, and Daniel S. Peeper. 2008. Oncogene-induced senescence relayed by an interleukin-dependent inflammatory network. *Cell* 133 (6): 1019–1031. doi: 10.1016/j.cell.2008.03.039.

117. Zhu, Yi, Jacqueline L. Armstrong, Tamara Tchkonia, and James L. Kirkland. 2014. Cellular senescence and the senescent secretory phenotype in age-related chronic diseases. *Current Opinion in Clinical Nutrition and Metabolic Care* 17 (4): 324–328. doi: 10.1097/MCO.0000000000000065.

118. Harley, Calvin B., A. Bruce Futcher, and Carol W. Greider. 1990. Telomeres shorten during ageing of human fibroblasts. *Nature* 345 (6274): 458–460. doi: 10.1038/345458a0.

119. Olovnikov, A. M. 1973. A theory of marginotomy. The incomplete copying of template margin in enzymic synthesis of polynucleotides and biological significance of the phenomenon. *Journal of Theoretical Biology* 41 (1): 181–190. doi: 10.1016/0022-5193(73)90198-7.

120. Campisi, Judith. 1997. The biology of replicative senescence. *European Journal of Cancer* 33 (5): 703–709. doi: 10.1016/S0959-8049(96)00058-5.

121. Kaufmann, W. K., and R. S. Paules. 1996. DNA damage and cell cycle checkpoints. *The FASEB Journal* 10 (2): 238–247.

122. Lukas, Jiri, Claudia Lukas, and Jiri Bartek. 2011. More than just a focus: The chromatin response to DNA damage and its role in genome integrity maintenance. *Nature Cell Biology* 13 (10): 1161–1169. doi: 10.1038/ncb2344.

123. d'Adda di Fagagna, Fabrizio. 2008. Living on a break: Cellular senescence as a DNA-damage response. *Nature Reviews Cancer* 8 (7): 512–522. doi: 10.1038/nrc2440.

124. Fielden, E. M., B. D. Michael, and K. M. Prise. 1997. Radiation damage to DNA: Techniques, quantitation and mechanisms. *Radiation Research* 148 (5): 481. doi: 10.2307/3579326.

125. Rastogi, Rajesh P., Richa, Ashok Kumar, Madhu B. Tyagi, and Rajeshwar P. Sinha. 2010. Molecular mechanisms of ultraviolet radiation-induced DNA damage and repair. *Journal of Nucleic Acids* 2010: 592980. doi: 10.4061/2010/592980.

126. Prasad, R., W. A. Beard, V. K. Batra, Y. Liu, D. D. Shock, and S. H. Wilson. 2011. A review of recent experiments on step-to-step 'hand-off' of the DNA intermediates in mammalian base excision repair pathways. *Molecular Biology* 45 (4): 536–550. doi: 10.1134/S0026893311040091.

127. Roninson, Igor B. 2003. Tumor cell senescence in cancer treatment. *Cancer Research* 63 (11): 2705–2715. doi: 10.1016/j.mce.2006.03.017.

128. Gewirtz, David A., Shawn E. Holt, and Lynne W. Elmore. 2008. Accelerated senescence: An emerging role in tumor cell response to chemotherapy and radiation. *Biochemical Pharmacology* 76 (8): 947–957. doi: 10.1016/j.bcp.2008.06.024.

129. Ewald, Jonathan A., Joshua A. Desotelle, George Wilding, and David F. Jarrard. 2010. Therapy-induced senescence in cancer. *Journal of the National Cancer Institute* 102 (20): 1536–1546. doi: 10.1093/jnci/djq364.

130. Chang, Bey-Dih, Eugenia V. Broude, Milos Dokmanovic, Hongming Zhu, Adam Ruth, Yongzhi Xuan, Eugene S. Kandel, Ekkehart Lausch, Konstantin Christov, and Igor B. Roninson. 1999. A senescence-like phenotype distinguishes tumor cells that undergo terminal proliferation arrest after exposure to anticancer agents. *Cancer Research* 59 (15): 3761–3767. doi: 10.1038/nrc2961.

131. Chang, Jianhui, Yingying Wang, Lijian Shao, Remi-Martin Laberge, Marco Demaria, Judith Campisi, Krishnamurthy Janakiraman, et al. 2015. Clearance of senescent cells by ABT263 rejuvenates aged hematopoietic stem cells in mice. *Nature Medicine* 22 (1): 78–83. doi: 10.1038/nm.4010.

132. Ewald, Jonathan A., Noel Peters, Joshua A. Desotelle, F. Michael Hoffmann, and David F. Jarrard. 2009. A high-throughput method to identify novel senescence-inducing compounds. *Journal of Biomolecular Screening* 14 (7): 853–858. doi: 10.1177/1087057109340314.

133. Han, Z., W. Wei, S. Dunaway, J. W. Darnowski, P. Calabresi, J. Sedivy, E. A. Hendrickson, K. V. Balan, P. Pantazis, and J. H. Wyche. 2002. Role of p21 in apoptosis and senescence of human colon cancer cells treated with camptothecin. *Journal of Biological Chemistry* 277 (19): 17154–17160. doi: 10.1074/jbc. M112401200.

134. Linge, Annett, Karina Weinhold, Robert Bläsche, Michael Kasper, and Kathrin Barth. 2007. Downregulation of caveolin-1 affects bleomycin-induced growth arrest and cellular senescence in A549 cells. *International Journal of Biochemistry and Cell Biology* 39 (10): 1964–1974. doi: 10.1016/j. biocel.2007.05.018.

135. te Poele, Robert H, Andrei L. Okorokov, Lesley Jardine, Jeffrey Cummings, and Simon P. Joel. 2002. DNA damage is able to induce senescence in tumor cells in vitro and in vivo. *Cancer Research* 62 (6): 1876–1883. doi: 10.1016/0014-4827(65)90211-9.

136. Majumder, Pradip K., Chiara Grisanzio, Fionnuala O'Connell, Marc Barry, Joseph M. Brito, Qing Xu, Isil Guney, et al. 2008. A prostatic intraepithelial neoplasia-dependent p27 kip1 checkpoint induces senescence and inhibits cell proliferation and cancer progression. *Cancer Cell* 14 (2): 146–155. doi: 10.1016/j.ccr.2008.06.002.

137. Davies, Kelvin. 1999. The broad spectrum of responses to oxidants in proliferating cells: A new paradigm for oxidative stress. *IUBMB Life* 48 (1): 41–47. doi: 10.1080/713803463.

138. Yoshikawa, Toshikazu, and Yuji Naito. 2002. What is oxidative stress? *Japan Medical Association* 124 (11): 1549–1553. doi: 10.1016/S0026-0495(00)80077-3.

139. Burdon, R. H., and Catherine Rice-Evans. 1989. Free radicals and the regulation of mammalian cell proliferation. *Free Radical Research Communications* 6 (6): 345–358, Taylor & Francis. doi: 10.3109/10715768909087918.

140. Wiese, A. G., R. E. Pacifici, and K. J. Davies. 1995. Transient adaptation of oxidative stress in mammalian cells. *Archives of Biochemistry and Biophysics* 318 (1): 231–240. doi: 10.1006/abbi.1995.1225.

141. Kurz, David J., Stephanie Decary, Ying Hong, Elisabeth Trivier, Alexander Akhmedov, and Jorge D. Erusalimsky. 2004. Chronic oxidative stress compromises telomere integrity and accelerates the onset of senescence in human endothelial cells. *Journal of Cell Science* 117 (11): 2417–2426. doi: 10.1242/jcs.01097.

142. Dasari, Arvind, Janine N. Bartholomew, Daniela Volonte, and Ferruccio Galbiati. 2006. Oxidative stress induces premature senescence by stimulating caveolin-1 gene transcription through p38 mitogen-activated protein kinase/Sp1-mediated activation of two GC-rich promoter elements. *Cancer Research* 66 (22): 10805–10814. doi: 10.1158/0008-5472.CAN-06-1236.

143. Galbiati, Ferruccio, Daniela Volonte, J. Liu, F. Capozza, P. G. Frank, L. Zhu, R. G. Pestell, and Michael P Lisanti. 2001. Caveolin-1 expression negatively regulates cell cycle progression by inducing G(0)/G(1) arrest via a p53/p21(WAF1/Cip1)-dependent mechanism. *Molecular Biology of the Cell* 12 (8): 2229–2244.

144. Volonte, Daniela, Kun Zhang, Michael P. Lisanti, and Ferruccio Galbiati. 2002. Expression of caveolin-1 induces premature cellular senescence in primary cultures of murine fibroblasts. *Molecular Biology of the Cell* 13 (7): 2502–2517. doi: 10.1091/mbc.01-11-0529.

145. Aird, Katherine M., Gao Zhang, Hua Li, Zhigang Tu, Benjamin G. Bitler, Azat Garipov, Hong Wu, et al. 2013. Suppression of nucleotide metabolism underlies the establishment and maintenance of oncogene-induced senescence. *Cell Reports* 3 (4): 1252–1265. doi: 10.1016/j.celrep.2013.03.004.

146. Chandler, Hollie, and Gordon Peters. 2013. Stressing the cell cycle in senescence and aging. *Current Opinion in Cell Biology* 25 (6): 765–771. doi: 10.1016/j.ceb.2013.07.005.

147. Gil, Jesús, and Gordon Peters. 2006. Regulation of the INK4b–ARF–INK4a tumour suppressor locus: All for one or one for all. *Nature Reviews Molecular Cell Biology* 7 (9): 667–677. doi: 10.1038/nrm1987.

148. Sage, Julien, George J. Mulligan, Laura D. Attardi, Abigail Miller, Siqi Chen, Bart Williams, Elias Theodorou, and Tyler Jacks. 2000. Targeted disruption of the three rb-related genes leads to loss of G1 control and immortalization. *Genes and Development* 14 (23): 3037–3050. doi: 10.1101/gad.843200.

149. Dannenberg, J. H., A. van Rossum, L. Schuijff, and H. te Riele. 2000. Ablation of the retinoblastoma gene family deregulates G(1) control causing immortalization and increased cell turnover under growth-restricting conditions. *Genes & Development* 14 (23): 3051–3064. doi: 10.1101/gad.847700.

150. Dai, Charlotte Y., and Greg H. Enders. 2000. p16INK4a can initiate an autonomous senescence program. *Oncogene* 19 (13): 1613. doi: 10.1038/sj.onc.1203438.

151. Beauséjour, Christian M., Ana Krtolica, Francesco Galimi, Masashi Narita, Scott W. Lowe, Paul Yaswen, and Judith Campisi. 2003. Reversal of human cellular senescence: Roles of the p53 and p16 pathways. *EMBO Journal* 22 (16): 4212–4222. doi: 10.1093/emboj/cdg417.

152. Takahashi, Akiko, Naoko Ohtani, Kimi Yamakoshi, Shin-ichi Iida, Hidetoshi Tahara, Keiko Nakayama, Keiichi I. Nakayama, Toshinori Ide, Hideyuki Saya, and Eiji Hara. 2006. Mitogenic signalling and the p16INK4a-Rb pathway cooperate to enforce irreversible cellular senescence. *Nature Cell Biology* 8 (11): 1291–1297. doi: 10.1038/ncb1491.

153. Imai, Yoshinori, Akiko Takahashi, Aki Hanyu, Satoshi Hori, Seidai Sato, Kazuhito Naka, Atsushi Hirao, Naoko Ohtani, and Eiji Hara. 2014. Crosstalk between the Rb pathway and AKT signaling forms a quiescence-senescence switch. *Cell Reports* 7 (1): 194–207. doi: 10.1016/j.celrep.2014.03.006.

154. Moiseeva, Olga, Véronique Bourdeau, Antoine Roux, Xavier Deschênes-Simard, and Gerardo Ferbeyre. 2009. Mitochondrial dysfunction contributes to oncogene-induced senescence. *Molecular and Cellular Biology* 29 (16): 4495–4507. doi: 10.1128/MCB.01868-08.

155. Kang, S., J.-P. Louboutin, P. Datta, C. P. Landel, D. Martinez, A. S. Zervos, D. S. Strayer, T. Fernandes-Alnemri, and E. S. Alnemri. 2013. Loss of HtrA2/Omi activity in non-neuronal tissues of adult mice causes premature aging. *Cell Death and Differentiation* 20 (2): 259–269. doi: 10.1038/cdd.2012.117.

156. Davidovic, Mladen, Goran Sevo, Petar Svorcan, Dragoslav P. Milosevic, Nebojsa Despotovic, and Predrag Erceg. 2010. Old age as a privilege of the 'selfish ones.' *Aging and Disease* 1 (2): 139–146.

157. Chen, Jian Hua, C. Nicholes Hales, and Susan E. Ozanne. 2007. DNA damage, cellular senescence and organismal ageing: Causal or correlative? *Nucleic Acids Research* 35 (22): 7417–7428. doi: 10.1093/nar/gkm681.

158. Campisi, Judith. 2005. Senescent cells, tumor suppression, and organismal aging: good citizens, bad neighbors. *Cell* 120 (4): 513–522. doi: 10.1016/j.cell.2005.02.003.

159. Jeyapalan, Jessie C., Mark Ferreira, John M. Sedivy, and Utz Herbig. 2007. Accumulation of senescent cells in mitotic tissue of aging primates. *Mechanisms of Ageing and Development* 128 (1): 36–44. doi: 10.1016/j.mad.2006.11.008.

160. Krishnamurthy, Janakiraman, Chad Torrice, Matthew R. Ramsey, Grigoriy I. Kovalev, Khalid Al-Regaiey, Lishan Su, and Norman E. Sharpless. 2004. Ink4a/Arf expression is a biomarker of aging. *The Journal of Clinical Investigation* 114 (9): 1299–1307. doi: 10.1172/JCI22475.

161. Geng, Yi Qun, Ji Tian Guan, Xiao Hu Xu, and Yu Cai Fu. 2010. Senescence-associated beta-galactosidase activity expression in aging hippocampal neurons. *Biochemical and Biophysical Research Communications* 396 (4): 866–869. doi: 10.1016/j.bbrc.2010.05.011.

162. Velarde, Michael C., James M. Flynn, Nicholas U. Day, Simon Melov, and Judith Campisi. 2012. Mitochondrial oxidative stress caused by Sod2 deficiency promotes cellular senescence and aging phenotypes in the skin. *Aging* 4 (1): 3–12. doi: 10.18632/aging.100423.

163. Wang, Chunfang, Diana Jurk, Mandy Maddick, Glyn Nelson, Carmen Martin-Ruiz, and Thomas Von Zglinicki. 2009. DNA damage response and cellular senescence in tissues of aging mice. *Aging Cell* 8 (3): 311–323. doi: 10.1111/j.1474-9726.2009.00481.x.

164. Jaskelioff, Mariela, Florian L. Muller, Ji-Hye Hye Paik, Emily Thomas, Shan Jiang, Andrew C. Adams, Ergun Sahin, et al. 2011. Telomerase reactivation reverses tissue degeneration in aged telomerase-deficient mice. *Nature* 469 (7328): 102–106. doi: 10.1038/nature09603.

165. Baker, Darren J., Carmen Perez-Terzic, Fang Jin, Kevin S. Pitel, Kevin S. Pitel, Nicolas J. Niederländer, Karthik Jeganathan, et al. 2008. Opposing roles for p16Ink4a and p19Arf in senescence and ageing caused by bubR1 insufficiency. *Nature Cell Biology* 10 (7): 825–836. doi: 10.1038/ncb1744.

166. Baker, Darren J., Tobias Wijshake, Tamar Tchkonia, Nathan K. LeBrasseur, Bennett G. Childs, Bart van de Sluis, James L. Kirkland, and Jan M. van Deursen. 2011. Clearance of p16Ink4a-positive senescent cells delays ageing-associated disorders. *Nature* 479 (7372): 232–236. doi: 10.1038/nature10600.

167. Baker, Darren J., Bennett G. Childs, Matej Durik, Melinde E. Wijers, Cynthia J. Sicbcn, Jian Zhong, Rachel A. Saltness, et al. 2016. Naturally occurring p16 ink4a—Positive cells shorten healthy lifespan. *Nature* 530 (7589): 1–5. doi: 10.1038/nature16932.

168. Demaria, Marco, Naoko Ohtani, Sameh A. Youssef, Francis Rodier, Wendy Toussaint, James R. Mitchell, Remi-Martin Laberge, et al. 2014. An essential role for senescent cells in optimal wound healing through secretion of PDGF-AA. *Developmental Cell* 31 (6): 722–733. doi: 10.1016/j.devcel.2014.11.012.

169. Jun, Joon-Il, and Lester F. Lau. 2010. The matricellular protein CCN1 induces fibroblast senescence and restricts fibrosis in cutaneous wound healing. *Nature Cell Biology* 12 (7): 676–685. doi: 10.1038/ncb2070.

170. Zlokovic, Berislav V. 2005. Neurovascular mechanisms of Alzheimer's neurodegeneration. *Trends in Neurosciences* 28 (4): 202–208. doi: 10.1016/j.tins.2005.02.001.

171. Katzman, R., and T. Saitoh. 1991. Advances in Alzheimer's disease. *FASEB Journal* 5 (3): 278–286.

172. Mega, M. S., J. L. Cummings, T. Fiorello, and J. Gornbein. 1996. The spectrum of behavioral changes in Alzheimer's disease. *Neurology* 46 (1): 130–135. doi: 10.1212/WNL.46.1.130.

173. Bhat, Rekha, Elizabeth P. Crowe, Alessandro Bitto, Michelle Moh, Christos D. Katsetos, Fernando U. Garcia, Frederick Bradley Johnson, John Q. Trojanowski, Christian Sell, and Claudio Torres. 2012. Astrocyte senescence as a component of Alzheimer's disease.. *PLoS ONE* 7 (9): e45069. doi: 10.1371/journal.pone.0045069.

174. Onyango, Isaac G., Jameel Dennis, and Shaharyah M. Khan. 2016. Mitochondrial dysfunction in Alzheimer's disease and the rationale for bioenergetics based therapies. *Aging and Disease* 7 (2): 201–214. doi: 10.14336/ad.2015.1007.

175. Chartier-Harlin, M. C., Fiona Crawford, Henry Houlden, Andrew Warren, David Hughes, Liana Fidani, Alison Goate, Martin Rossor, Penelope Roques, and John Hardy. 1991. Early-onset Alzheimer's disease caused by mutations at codon 717 of the beta-amyloid precursor protein gene. *Nature* 353 (6347): 844–846. doi: 10.1038/353844a0.

176. Lott, Ira T., and Elizabeth Head. 2001. Down syndrome and Alzheimer's disease: A link between development and aging. *Mental Retardation and Developmental Disabilities Research Reviews* 7 (3): 172–178. doi: 10.1002/mrdd.1025.

177. He, N., W.-L. Jin, K.-H. Lok, Y. Wang, M. Yin, and Z.-J. Wang. 2013. Amyloid-$\beta 1$–42 oligomer accelerates senescence in adult hippocampal neural stem/progenitor cells via formylpeptide receptor 2. *Cell Death and Disease* 4 (11): e924. doi: 10.1038/cddis.2013.437.

178. McShea, A., P. L. Harris, K. R. Webster, A. F. Wahl, and M. A. Smith. 1997. Abnormal expression of the cell cycle regulators P16 and CDK4 in Alzheimer's disease. *The American Journal of Pathology* 150 (6): 1933–1939.

179. Dunnett, Stephen B., and A. Björklund. 1999. Prospects for new restorative and neuroprotective treatments in Parkinson's disease. *Nature* 399 (6738 Suppl): A32–A39. doi: 10.1038/399a032.

180. Shahed, Joohi, and Joseph Jankovic. 2007. Motor symptoms in Parkinson's disease. *Handbook of Clinical Neurology* 83: 329–342. doi: 10.1016/S0072-9752(07)83013-2.

181. Chinta, S. J. Lieu, C. A., M. Demaria, R.-M. M. Laberge, J. Campisi, and J. K. Andersen. 2013. Environmental stress, ageing and glial cell senescence: A novel mechanistic link to Parkinson's disease? *Journal of Internal Medicine* 273 (5): 429–436. doi: 10.1111/joim.12029.

182. Riddle, David R., and Robin J. Lichtenwalner. 2007. Neurogenesis in the adult and aging brain. In *Brain Aging: Models, Methods, and Mechanisms*. Boca Raton, FL: CRC Press/Taylor & Francis.

183. Parker, W. Davis, Janice K. Parks, and Russell H. Swerdlow. 2008. Complex I deficiency in Parkinson's disease frontal cortex. *Brain Research* 1189 (1): 215–218. doi: 10.1016/j.brainres.2007.10.061.

184. Schapira, A. H. V., J. M. Cooper, D. Dexter, J. B. Clark, P. Jenner, and C. D. Marsden. 1990. Mitochondrial complex I deficiency in Parkinson's disease. *Journal of Neurochemistry* 54 (3): 823–827. doi: 10.1111/j.1471-4159.1990.tb02325.x.

185. Keeney, Paula M., Jing Xie, Roderick A. Capaldi, and James P. Bennett. 2006. Parkinson's disease brain mitochondrial complex I has oxidatively damaged subunits and is functionally impaired and misassembled. *The Journal of Neuroscience* 26 (19): 5256–5264. doi: 10.1523/JNEUROSCI.0984-06.2006.

186. Chinta, S. J., G. Woods, M. Demaria, J. Campisi, and J. K. Andersen. 2016. Cellular senescence induced by paraquat drives neuropathology associated with Parkinson's disease. *Movement Disorders* 31: S252.

187. American Diabetes Association. 2010. Standards of medical care in diabetes–2010. *Diabetes Care* 33 (Suppl 1): S11–61. doi: 10.2337/dc10-S011.

188. Dey, Lucy, Anoja S. Attele, and Chun-Su Yuan. 2002. Alternative therapies for type 2 diabetes. *Alternative Medicine Review* 7 (1): 45–58.

189. Hu, Frank B. 2011. Globalization of diabetes: The role of diet, lifestyle, and genes. *Diabetes Care*, 34: 1249–1257. doi: 10.2337/dc11-0442.

190. Tchkonia, Tamara, Dean E. Morbeck, Thomas Von Zglinicki, Jan Van Deursen, Joseph Lustgarten, Heidi Scrable, Sundeep Khosla, Michael D. Jensen, and James L. Kirkland. 2010. Fat tissue, aging, and cellular senescence. *Aging Cell* 9 (5): 667–684. doi: 10.1111/j.1474-9726.2010.00608.x.

191. Sone, H., and Y. Kagawa. 2005. Pancreatic beta cell senescence contributes to the pathogenesis of type 2 diabetes in high-fat diet-induced diabetic mice. *Diabetologia* 48 (1): 58–67. doi: 10.1007/s00125-004-1605-2.

192. Minamino, Tohru, Masayuki Orimo, Ippei Shimizu, Takeshige Kunieda, Masataka Yokoyama, Takashi Ito, Aika Nojima, et al. 2009. A crucial role for adipose tissue p53 in the regulation of insulin resistance. *Nature Medicine* 15 (9): 1082–1087. doi: 10.1038/nm.2014.

193. Fu, Qiuli, Zhenwei Qin, Jiexin Yu, Yinhui Yu, Qiaomei Tang, Danni Lyu, Lifang Zhang, Zhijian Chen, and Ke Yao. 2016. Effects of senescent lens epithelial cells on the severity of age-related cortical cataract in humans: A case-control study. *Medicine* 95 (25): e3869. doi: 10.1097/MD.0000000000003869.

194. Baker, Darren J., Karthik B. Jeganathan, J. Douglas Cameron, Michael Thompson, Subhash Juneja, Alena Kopecka, Rajiv Kumar, et al. 2004. BubR1 insufficiency causes early onset of aging-associated phenotypes and infertility in mice. *Nature Genetics* 36 (7): 744–749. doi: 10.1038/ng1382.

195. Bourne, Rupert R. A., Gretchen A. Stevens, Richard A. White, Jennifer L. Smith, Seth R. Flaxman, Holly Price, Jost B. Jonas, et al. 2013. Causes of vision loss worldwide, 1990–2010: A systematic analysis. *The Lancet Global Health* 1 (6): e339–e349. doi: 10.1016/S2214-109X(13)70113-X.

196. Young, Richard W. 1987. Pathophysiology of age-related macular degeneration. *Survey of Ophthalmology* 31 (5): 291–306. doi: 10.1016/0039-6257(87)90115-9.

197. Williams, Rebecca A., B. L. Brody, R. G. Thomas, R. M. Kaplan, and S. I. Brown. 1998. The psychosocial impact of macular degeneration. *Archives of Ophthalmology* 116 (4): 514–520. doi: 10.1001/archopht.116.4.514.

198. Bird, A. C., N. M. Bressler, S. B. Bressler, I. H. Chisholm, G. Coscas, M. D. Davis, P. T. V. M. de Jong, et al. 1995. An international classification and grading system for age-related maculopathy and age-related macular degeneration. *Survey of Ophthalmology* 39 (5): 367–374. doi: 10.1016/S0039-6257(05)80092-X.

199. Lim, Laurence S., Paul Mitchell, Johanna M. Seddon, Frank G. Holz, and Tien Y. Wong. 2012. Age-related macular degeneration. *The Lancet* 379 (9827): 1728–1738. doi: 10.1016/S0140-6736(12)60282-7.

200. Green, W. R., P. J. McDonnell, and J. H. Yeo. 1985. Pathologic features of senile macular degeneration. *Ophthalmology* 92 (5): 615–627. doi: 10.1097/00006982-200507001-00011.

201. Applegate, Raymond A., Anthony J. Adams, John C. Cavender, and Frank Zisman. 1987. Early color vision changes in age-related maculopathy. *Applied Optics* 26 (8): 1458. doi: 10.1364/AO.26.001458.

202. Beatty, Stephen, Hui-Hiang Koh, M. Phil, David Henson, and Michael Boulton. 2000. The role of oxidative stress in the pathogenesis of age-related macular degeneration. *Survey of Ophthalmology* 45 (2): 115–134. doi: 10.1016/S0039-6257(00)00140-5.

203. Ding, Xiaoyan, Mrinali Patel, and Chi-Chao Chan. 2009. Molecular pathology of age-related macular degeneration. *Progress in Retinal and Eye Research* 28 (1): 1–18. doi: 10.1016/j.preteyeres.2008.10.001.

204. Ambati, Jayakrishna, Akshay Anand, Stefan Fernandez, Eiji Sakurai, Bert C. Lynn, William A. Kuziel, Barrett J. Rollins, and Balamurali K. Ambati. 2003. An animal model of age-related macular degeneration in senescent Ccl-2- or Ccr-2-deficient mice. *Nature Medicine* 9 (11): 1390–1397. doi: 10.1038/nm950.

205. Kozlowski, Michael R. 2012. RPE cell senescence: A key contributor to age-related macular degeneration. *Medical Hypotheses* 78 (4): 505–510. doi: 10.1016/j.mehy.2012.01.018.

206. Marazita, Mariela C., Andrea Dugour, Melisa D. Marquioni-Ramella, Juan M. Figueroa, and Angela M. Suburo. 2016. Oxidative stress-induced premature senescence dysregulates VEGF and CFH expression in retinal pigment epithelial cells: Implications for age-related macular degeneration. *Redox Biology* 7: 78–87. doi: 10.1016/j.redox.2015.11.011.

207. Mendis, Shanthi, Pekka Puska, and B. Norrving. 2011. *Global Atlas on Cardiovascular Disease Prevention and Control*. Geneva: World Health Organization.

208. Institute of Medicine. 2010. Promoting cardiovascular health in the developing world. In *Promoting Cardiovascular Health in the Developing World: A Critical Challenge to Achieve Global Health*, Gregory R. Bock and Jamie A. Goode, eds. Washington, DC: National Academies Press. doi: 10.17226/12815.

209. Micha, Renata, Georgios Michas, and Dariush Mozaffarian. 2012. Unprocessed red and processed meats and risk of coronary artery disease and type 2 diabetes—An updated review of the evidence. *Current Atherosclerosis Reports* 14 (6): 515–524. doi: 10.1007/s11883-012-0282-8.

210. Finks, Shannon W., Anita Airee, Sheryl L. Chow, Tracy E. Macaulay, Michael P. Moranville, Kelly C. Rogers, and Toby C. Trujillo. 2012. Key articles of dietary interventions that influence cardiovascular mortality. *Pharmacotherapy: The Journal of Human Pharmacology and Drug Therapy* 32 (4): e54–e87. doi: 10.1002/j.1875-9114.2011.01087.x.

211. Naylor, R. M., D. J. Baker, and J. M. van Deursen. 2013. Senescent cells: A novel therapeutic target for aging and age-related diseases. *Clinical Pharmacology & Therapeutics* 93 (1): 105–116. doi: 10.1038/clpt.2012.193.

212. Jani, B., and C. Rajkumar. 2006. Ageing and vascular ageing. *Postgraduate Medical Journal* 82 (968): 357–362. doi: 10.1136/pgmj.2005.036053.

213. Hansson, Göran K., Anna-Karin L. Robertson, and Cecilia Söderberg-Nauclér. 2006. Inflammation and atherosclerosis. *Annual Review of Pathology* 1 (1): 297–329. doi: 10.1146/annurev.pathol.1.110304.100100.

214. Tsirpanlis, George. 2009. Cellular senescence and inflammation: A noteworthy link. *Blood Purification* 28 (1): 12–14. doi: 10.1159/000210032.

215. Tsirpanlis, George. 2005. Inflammation in atherosclerosis and other conditions: A response to danger. *Kidney and Blood Pressure Research* 28 (4): 211–217. doi: 10.1159/000087121.

216. Wang, Julie C., and Martin Bennett. 2012. Aging and atherosclerosis: Mechanisms, functional consequences, and potential therapeutics for cellular senescence. *Circulation Research* 111 (2): 245–259. doi: 10.1161/CIRCRESAHA.111.261388.

217. Matthews, Charles, Isabelle Gorenne, Stephen Scott, Nicola Figg, Peter Kirkpatrick, Andrew Ritchie, Martin Goddard, and Martin Bennett. 2006. Vascular smooth muscle cells undergo telomere-based senescence in human atherosclerosis: Effects of telomerase and oxidative stress. *Circulation Research* 99 (2): 156–164. doi: 10.1161/01.RES.0000233315.38086.bc.

218. Jeon, Yun Kyung, Bo Hyun Kim, and In Joo Kim. 2016. The diagnosis of osteoporosis. *Journal of the Korean Medical Association* 59 (11): 842. doi: 10.5124/jkma.2016.59.11.842.

219. Ginaldi, Lia, Maria Cristina Di Benedetto, and Massimo De Martinis. 2005. Osteoporosis, inflammation and ageing. *Immunity & Ageing* 2 (1): 14. doi: 10.1186/1742-4933-2-14.

220. Manolagas, Stavros C., Teresita Bellido, and Robert L. Jilka. 2007. Sex steroids, cytokines and the bone marrow: New concepts on the pathogenesis of osteoporosis. In *Ciba Foundation Symposium 191-Non-Reproductive Actions of Sex Steroids*, 187–202. John Wiley & Sons, Ltd. doi: 10.1002/9780470514757.ch11.

221. Ginaldi, Lia, Francesca Ciccarelli, and Massimo De Martinis. 2013. Effect of senescence on bone remodeling: The role of inflammageing. *Rejuvenation Research* 16 (1): 22–23.

222. Franceschi, Claudio, and Judith Campisi. 2014. Chronic inflammation (inflammaging) and its potential contribution to age-associated Diseases. *Journals of Gerontology—Series A Biological Sciences and Medical Sciences* 69 (1 Suppl): S4–S9. doi: 10.1093/gerona/glu057.

223. Yun, A. Joon, and Patrick Y. Lee. 2004. Maladaptation of the link between inflammation and bone turnover may be a key determinant of osteoporosis. *Medical Hypotheses* 63 (3): 532–537. doi: 10.1016/S0306-9877(03)00326-8.

224. Ishihara, Katsuhiko, and Toshio Hirano. 2002. IL-6 in autoimmune disease and chronic inflammatory proliferative disease. *Cytokine and Growth Factor Reviews* 13 (4-5): 357–368. doi: 10.1016/S1359-6101(02)00027-8.

225. Moschen, A. R., A. Kaser, B. Enrich, O. Ludwiczek, M. Gabriel, P. Obrist, A. M. Wolf, and H. Tilg. 2005. The RANKL/OPG system is activated in inflammatory bowel disease and relates to the state of bone loss. *Gut* 54 (4): 479–487. doi: 10.1136/gut.2004.044370.

226. Roodman, G. D. 1993. Role of cytokines in the regulation of bone resorption. *Calcified Tissue International* 53: S94–S98.

227. Manolagas, Stavros C. 1995. Role of cytokines in bone resorption. *Bone* 17 (2): 63S–67S. doi: 10.1016/8756-3282(95)00180-L.

228. Schett, Georg. 2011. Effects of inflammatory and anti-inflammatory cytokines on the bone. *European Journal of Clinical Investigation* 41 (12): 1361–1366. doi: 10.1111/j.1365-2362.2011.02545.x.

229. Aw, Danielle, Alberto B. Silva, and Donald B. Palmer. 2007. Immunosenescence: Emerging challenges for an ageing population. *Immunology* 120 (4): 435–446. doi: 10.1111/j.1365-2567.2007.02555.x.

230. Chou, Jennifer P., and Rita B. Effros. 2013. T Cell replicative senescence in human aging. *Current Pharmaceutical Design* 19 (9): 1680–1698. doi: 10.2174/1381612811319090016.

231. Doherty, Timothy J. 2003. Invited review: Aging and sarcopenia. *Journal of Applied Physiology* 95 (4): 1717–1727. doi: 10.1152/japplphysiol.00347.2003.

232. Balagopal, P., Olav E. Rooyackers, Debrorah B. Adey, Philip A. Ades, and K. Sreekumaran Nair. 1997. Effects of aging on in vivo synthesis of skeletal muscle myosin heavy-chain and sarcoplasmic protein in humans. *American Journal of Physiology-Endocrinology and Metabolism* 273 (4): E790–E800.

233. Rowan, Sharon L., Karolina Rygiel, Fennigje M. Purves-Smith, Nathan M. Solbak, Douglas M. Turnbull, and Russell T. Hepple. 2012. Denervation causes fiber atrophy and myosin heavy chain co-expression in senescent skeletal muscle. *PLoS ONE* 7 (1): e29082. doi: 10.1371/journal.pone.0029082.

234. Marzetti, Emanuele, Riccardo Calvani, Matteo Cesari, Thomas W. Buford, Maria Lorenzi, Bradley J. Behnke, and Christiaan Leeuwenburgh. 2013. Mitochondrial dysfunction and sarcopenia of aging: From signaling pathways to clinical trials. *The International Journal of Biochemistry & Cell Biology* 45 (10): 2288–2301.

235. Rygiel, Karolina A., Martin Picard, and Doug M. Turnbull. 2016. The ageing neuromuscular system and sarcopenia–A mitochondrial perspective. *The Journal of Physiology* 594 (16): 1–14. doi: 10.1113/JP271212.

236. Fulle, Stefania, Feliciano Protasi, Guglielmo Di Tano, Tiziana Pietrangelo, Andrea Beltramin, Simona Boncompagni, Leonardo Vecchiet, and Giorgio Fanò. 2004. The contribution of reactive oxygen species to sarcopenia and muscle Ageing. *Experimental Gerontology* 39 (1): 17–24. doi: 10.1007/978-88-470-0376-7_6.

237. Sousa-Victor, Pedro, Susana Gutarra, Laura García-Prat, Javier Rodriguez-Ubreva, Laura Ortet, Vanessa Ruiz-Bonilla, Mercè Jardí, et al. 2014. Geriatric muscle stem cells switch reversible quiescence into senescence. *Nature* 506 (7488): 316–321. doi: 10.1038/nature13013.

238. Nature America. 2000. The aging populations of developed countries are likely to present a growing market for arthritis therapies. *Nature Biotechnology* 18: IT12–IT14. doi: 10.1038/80045.

239. Oliver, J. E., and A. J. Silman. 2006. Risk factors for the development of rheumatoid arthritis. *Scandinavian Journal of Rheumatology* 35 (3): 169–174. doi: 10.1080/03009740600718080.

240. Firestein, G. S. 2003. Evolving concepts of rheumatoid arthritis. *Nature* 423 (6937): 356–361. doi: 10.1038/nature01661.

241. Price, Jo S., Jasmine G. Waters, Clare Darrah, Caroline Pennington, Dylan R. Edwards, Simon T. Donell, and Ian M. Clark. 2002. The role of chondrocyte senescence in osteoarthritis. *Aging Cell* 1 (1): 57–65.

242. Zhu, Yi, Tamara Tchkonia, Tamar Pirtskhalava, Adam C. Gower, Husheng Ding, Nino Giorgadze, Allyson K. Palmer, et al. 2015. The Achilles' heel of senescent cells: From transcriptome to senolytic drugs. *Aging Cell* 14 (4): 644–658. doi: 10.1111/acel.12344.

243. Zhu, Yi, Tamara Tchkonia, Heike Fuhrmann-Stroissnigg, Haiming M. Dai, Yuanyuan Y. Ling, Michael B. Stout, Tamar Pirtskhalava, et al. 2016. Identification of a novel senolytic agent, navitoclax, targeting the Bcl-2 family of anti-apoptotic factors. *Aging Cell* 15 (3): 428–435. doi: 10.1111/acel.12445.

244. Roos, Carolyn M., Bin Zhang, Allyson K. Palmer, Mikolaj B. Ogrodnik, Tamar Pirtskhalava, Nassir M. Thalji, Michael Hagler, et al. 2016. Chronic senolytic treatment alleviates established vasomotor dysfunction in aged or atherosclerotic mice. *Aging Cell* 15 (5): 973–977. doi: 10.1111/acel.12458.

245. Naghavi, M., H. Wang, R. Lozano, A. Davis, X. Liang, M. Zhou, et al. 2015. Global, regional, and national age-sex specific all-cause and cause-specific mortality for 240 causes of death, 1990–2013: Asystematic analysis for the Global Burden of Disease Study 2013. *Lancet* 385 (9963): 117–171.

246. Mayo Clinic. 2017. Senescence, frailty, and mesenchymal stem cell functionality in chronic kidney disease: Effect of senolytic agents. Accessed January 22, 2018. Available from: https://clinicaltrials.gov/ct2/show/NCT02848131

247. Unity Biotechnology. Unity Biotechnology [Internet]. Accessed January 18, 2017. Available from: http://unitybiotechnology.com/

248. Velarde, Michael C., and Marco Demaria. 2016. Targeting senescent cells: Possible implications for delaying skin aging: A mini-review. *Gerontology* 62 (5): 513–518. doi: 10.1159/000444877.

249. Pitozzi, Vanessa, Alessandra Mocali, Anna Laurenzana, Elisa Giannoni, Ingrid Cifola, Cristina Battaglia, Paola Chiarugi, Piero Dolara, and Lisa Giovannelli. 2013. Chronic resveratrol treatment ameliorates cell adhesion and mitigates the inflammatory phenotype in senescent human fibroblasts. *Journals of Gerontology—Series A Biological Sciences and Medical Sciences* 68 (4): 371–381. doi: 10.1093/gerona/gls183.

250. Lim, Hyun, Haeil Park, and Hyun Pyo Kim. 2015. Effects of flavonoids on senescence-associated secretory phenotype formation from bleomycin-induced senescence in BJ fibroblasts. *Biochemical Pharmacology* 96 (4): 337–348. doi: 10.1016/j.bcp.2015.06.013.

251. Ido, Yasuo, Albert Duranton, Fan Lan, Karen A. Weikel, Lionel Breton, and Neil B. Ruderman. 2015. Resveratrol prevents oxidative stress-induced senescence and proliferative dysfunction by activating the AMPK-FOXO3 cascade in cultured primary human keratinocytes. *PLoS ONE* 10 (2): e0115341. doi: 10.1371/journal.pone.0115341.

252. Holmes-McNary, M., and A. S. Baldwin. 2000. Chemopreventive properties of trans-resveratrol are associated with inhibition of activation of the IkappaB kinase. *Cancer Research* 60 (13): 3477–3483. doi: 10.1126/science.275.5297.218.

17 Osteoimmunology in Aging

Lia Ginaldi, Daniela Di Silvestre, Maria Maddalena Sirufo, and Massimo De Martinis

CONTENTS

INTRODUCTION

An unexpected relationship between bone and immune cells has recently emerged, resulting in the establishment of osteoimmunology, an innovative discipline that investigates interactions between the skeleton and immune system, the so-called immunoskeletal interface. Recent discoveries in this research field highlighted a new landscape of osteoporosis, and enabled the development of innovative therapies for its treatment. From this perspective, osteoporosis could be regarded as an age-related chronic immune-mediated disease, sharing with other age-related disorders a common inflammatory background [1].

Bone and immune cells are strictly interconnected: the immune system is able to influence bone remodeling, predominantly through inflammatory mediators, and even the bone is actively involved in regulating the immune system. Moreover, the immunoskeletal interface is also involved in the regulation of important body functions beyond bone remodeling and plays a central role in a variety of pathophysiological conditions [2–4]. There is an intricate communication among bone cells. Moreover, bone influences the activity of other organs and is also influenced by cells of other systems of the body. Bone is a metabolically active tissue, subjected to numerous stresses during life. The crosstalk between immune and bone cells drive the various phases of bone formation and resorption that coexist in a dynamic equilibrium throughout life. The normal bone remodeling process is essential for the maintenance of bone homeostasis. An altered remodeling can occur as the consequence of several factors and can result in an incomplete coupling between bone resorption and bone formation. Abnormal increase in osteoclastogenesis and osteoclast activity results in bone diseases such as osteoporosis, where resorption exceeds formation causing decreased bone density and increased risk of fractures [5]. Several pathological conditions impact on bone turnover leading to skeletal damage: in inflammatory arthritis, abnormal osteoclast activation results in periarticular erosions, bone metastases cause painful osteolytic lesions, in periodontitis, as a consequence of bacterial proliferation, inflammatory cells produce osteoclastogenic cytokines stimulating alveolar bone resorption [2,5–8]. On the other hand, an excessive accumulation of bone mass occurs in osteopetrosis, a rare disease affecting bone turnover [9,10].

Osteoporosis is a major cause of morbidity and mortality in older people. Several risk factors implicated in the development of senile osteoporosis exert their effects through immunologically mediated modulation of bone remodeling. In the complex scenario of osteoimmunology, senile osteoporosis is an example of the central role of immune-mediated inflammation in determining bone resorption. The pathogenesis of osteoporosis in the context of immunosenescence is therefore mediated by inflammatory mechanisms that include hyperproduction of proinflammatory and osteoclastogenic cytokines by senescent immune cells, as well as age-related remodeling of transcription factors and receptors known to influence osteoclasts and osteoblasts. Inflammaging itself therefore plays a role in bone remodeling changes observed during aging through an age-related remodeling of the immunoskeletal interface [11].

PHYSIOLOGICAL BONE TURNOVER: CELLS AND SIGNALING PATHWAYS

Bone is a mineralized connective tissue, continuously remodeled during life through the concerted actions of bone cells. It is therefore a plastic tissue which undergoes continuous adaptation during lifetime to preserve the structure of the skeleton and to control mineral homeostasis [10]. The bone remodeling process is under the control of several local and systemic factors, such as cytokines, chemokines, growth factors, hormones, biomechanical stimulations, etc. that all together contribute to bone homeostasis. An imbalance between bone resorption and formation results in bone diseases including osteoporosis.

Bone tissue is formed by a protein and mineral salt matrix in which are embedded the bone cells: osteoblasts, osteocytes, osteoclasts, and bone lining cells. In addition to bone cells, many other types of cells take part in bone composition, including immune cells, mesenchymal and hematopoietic stem cells, stromal and cartilage cells, all linked by a dense network of signals [12]. For example, the phagocytosis of apoptotic cells by osteal macrophages, key mediators of fracture repair, is a critical process in both clearing dead cells and replacement of progenitor cells to maintain bone homeostasis [13].

Bone turnover requires two coordinated processes: bone formation, driven by osteoblasts, and bone resorption, mediated by osteoclasts [14]. During bone remodeling, there is an intricate communication among bone cells. Moreover, there is a complex communication between bone cells and other organs, indicating the dynamic and multifunctional nature of bone tissue.

Osteoblasts, the precursors of osteocytes, originate from mesenchymal stem cells that can also give rise to adipocytes, marrow stromal cells, and chondrocytes. There are multiple subpopulations of mesenchymal stem cells [15–18]. The specific stem cell that maintains the postnatal skeleton is an osteochondroreticular stem cell that generates osteoblasts, chondrocytes, and reticular marrow stromal cells, but not adipocytes. The osteoblast precursor cells increase the expression of the osteopontin receptor (CD44) and the receptor for stromal cell-derived factor 1-SDF1 (CXCR4) and become mature osteoblasts. They migrate along chemotactic gradients into regions of bone formation, attracted by vascular endothelial cells expressing SDF1 [19]. The commitment of mesenchymal stem cell toward the osteoprogenitor lineage involves the expression of specific genes, following timely programmed steps, including the synthesis of bone morphogenetic proteins (BMPs) and members of the Wingless (Wnt) pathway [20]. By action of peroxisome proliferative-activated receptor gamma (PPARγ), the principal regulator of adipogenesis, the mesenchymal stem cell (MSC) differentiates into adipocytes rather than into OB. In contrast, the expression of runt-related transcription factor 2 (Runx2), associated with OB differentiation, and Osterix, an OB zinc finger-containing transcription factor, shift the equilibrium toward the osteoblastogenesis [21,22]. In addition, through the induction of c-fos, the PPARγ promotes osteoclastogenesis. Mechanical loading promotes OB differentiation and inhibits adipogenesis by downregulating PPARγ or by stimulating a durable β-catenin signal [23]. The expressions of runt-related transcription factors 2 (Runx2), distal-less homeobox 5 (Dlx5), osterix (Osx), fibroblast growth factor (FGF), microRNAs, and connexin are also crucial for osteoblast differentiation [24–28]. Other osteoblast-related genes, such as alpha-1 type I collagen (ColIA1), alkaline phosphatase (ALP), bone sialoprotein (BSP), bone gamma-carboxyglutamate protein (BGLAP), and osteocalcin (OCN), are in turn upregulated by Runx2 [24,29].

Osteocytes are cells with sensing and signal transport functions. They represent the structural cells in the bone and synthesize the bone matrix proteins which, along with the mineral component, determine the quality of the bone [2]. They also produce factors that influence osteoblast and osteo-clast activities and exhert a strict control of the bone remodeling process by acting as mechano-sensors, leading to the activation of repair processes. Osteocytes derive from osteoblasts: at the end of a bone formation cycle, a subpopulation of osteoblasts downregulate osteoblastic receptors and express osteocyte markers including dentine matrix protein 1 (DMP1) and sclerostin [30–32] becoming osteocytes. At the same time, they are entrapped within the mineralized bone matrix where their cytoplasmic processes cross bone canaliculi connect to other neighboring osteocytes processes, as well as to osteoblasts and bone lining cells [33,34]. Osteocytes are able to translate mechanical stimuli into biochemical signals [35]. Upon mechanical stimulation, osteocytes produce several secondary messengers, for example, adenosine triphosphate (ATP), nitric oxide (NO), Ca^{2+}, and prostaglandins (PGE_2 and PGI_2) which influence bone physiology [30]. Moreover, osteocyte apoptosis constitutes a chemotactic signal to osteoclastic bone resorption [36–38]. In this way, the osteocytes act as orchestrators of bone remodeling [33]. RANKL secreted by osteocytes stimu-lates osteoclastogenesis, while prostaglandin E_2 (PGE_2), NO, and insulin-like growth factor (IGF) stimulate osteoblast activity. Conversely, osteocytes produce osteoprotegerin (OPG) that inhibits osteoclastogenesis; moreover, osteocytes produce sclerostin and dickkopf WNT signaling path-way inhibitor (DKK-1) that decrease osteoblast activity. Increased expression of RANKL, vascu-lar endothelial growth factor (VEGF), monocyte chemoattractant protein-1 (CCL2), high mobility group box protein 1 (HMGB1), and macrophage colony-stimulating factor (M-CSF) in osteocytes promote local osteoclastogenesis [38–42].

Bone lining cells are quiescent flat-shaped osteoblasts covering bone surfaces where neither resorption nor formation occurs. However, these cells can adopt a cuboidal appearance acquiring secretory activity. These cells prevent direct interaction between osteoclasts and the bone matrix, when bone resorption should not occur. Moreover, they are able to produce OPG and RANKL, participating in osteoclast differentiation and function [43].

Osteoclasts, derived from bone marrow precursors which give rise also to professional antigen presenting cells (APC) (dendritic cells and macrophages), are multinucleated myeloid cells, special-ized to resorb bone by removing mineralized bone matrix through the production of lysosomal enzymes (tartrate-resistant acid phosphatase [TRAP] and catepsin k). Their differentiation from myeloid lineage precursors is driven by osteoblasts through the production of M-CSF, receptor activator of nuclear factor-kB (NF-kB) ligand (RANKL), and other co-stimulatory factors [44]. Osteoclasts are not only bone resorbing cells, but also a source of cytokines that influence the activ-ity of other cells. For example, they modulate the bone remodeling cycle by secreting clastokines that control osteoblast activity, and directly regulate the hematopoietic stem cell niche [45].

A complex receptor network mediates bone remodeling as well as the crosstalk between the bone and immune system. The main activation signal for bone resorption is the stimulation of the receptor RANK on osteoclasts and their precursors by its specific ligand RANKL expressed by osteoblasts and stromal cells. A central role in this system is also played by the ligand OPG, a decoy receptor of RANKL, which prevents bone resorption by acting as a competitive inhibitory receptor of RANKL [46]. The binding of RANKL to its receptor RANK on osteoclasts and their precursors is the main activation signal for bone resorption. The osteoblast-derived M-CSF links to its receptor c-fms on the surface of osteoclast cell precursors enabling the RANK/RANKL signal. Through the adapter protein tumor necrosis factor receptor-associated factor 6 (TRAF6), RANK receptor, expressed on osteoclast, activates NF-kB as well as other transcription factors, such as mitogen-activated protein kinases (MAPKs), c-fos, activator protein 1 (AP1), up to nuclear factor of activated T cells (NFATc1). Under the influence of the RANKL/RANK interaction, NFATc1 also induces the expression of DC-STAMP, which is crucial for the fusion of osteoclast precursors [47,48]. NFATc1 is the hub of various signaling pathways. Together with other transcription factors, it induces osteo-clast differentiation and proliferation, leading to the activation of genes codifying for calcitonin receptor, cathepsin k, and TRAP [49,50]. Both integrins and CD44 facilitate the attachment of the osteoclast podosomes to the bone surface [51,52] and mediate the formation and maintenance of the ruffled border, which is essential for osteoclast activity, enabling the trafficking of lysosomal and endosomal components, such as TRAP, cathepsin K, and matrix metalloproteinase-9 (MMP-9). Cathepsin K expressed in osteoclasts is involved in bone matrix type I collagen degradation, in dendritic cell activation through the toll-like receptor (TLR) 9, and supports the secretion of interleukin (IL)-6 and IL-23 [53]. TRAP and MMP-9 play essential roles in bone matrix demineralization and degradation.

Many other receptor pathways, most of which are shared by immune cells, interact with RANK, some co-stimulators and others inhibitors. The inhibitor receptor system ephrin (Eph) B2/B4 allows the passage of signals bidirectionally between osteoclasts and osteoblasts [54]. It inhibits osteoclast differentiation by blocking c-fos and the NFATc1 transcriptional cascade in osteoclast cell lineage and contemporaneously favors the coupling of bone formation and resorption through the induction of osteogenetic regulatory genes in osteoblasts [55]. The ephrinB2/ephrinB4 binding therefore functions as a coupling factor in bone remodeling process [56]. Other coupling factors are semaphorins, glycoproteins involved in several biological processes such as immune response, tumor progression, and bone remodeling, among others [57–59]. Semaphorin4D (Sema4D) expressed in osteoclasts binds to its osteoblast receptor (Plexin-B1) inhibiting IGF-1 pathway, essential for osteoblast differentiation [60], whereas Sema3A in osteoblasts is an inhibitor of osteoclastogenesis [61]. During bone remodeling osteoclasts inhibit bone formation by expressing Sema4D, in order to initiate bone resorption, whereas osteoblasts express Sema3A that suppresses bone resorption, prior to bone formation.

The canonical Wnt/beta (β) catenin pathway stimulates osteoblasts to differentiate, promotes their mineralization activity, and inhibits their apoptosis in response mainly to mechanical load. It comprises a family of proteins and transmembrane receptors consisting of the assembly of subunits, namely Frizzled (Fz) proteins and low density lipoprotein-related receptors (LRP-5, LRP-6). The Wnt ligand is a secreted glycoprotein that binds to Frizzled receptors, leading to the formation of a larger cell surface complex with LRP5/6. Activation of the Wnt receptor complex triggers the stabilization of β-catenin that migrates into the nucleus where it regulates the transcription of target genes inducing the differentiation of mesenchymal stem cells toward osteoblast differentiation and bone formation. WNT signaling also inhibits progenitor commitment toward adipogenic and chondrogenic lineages and indirectly inhibits osteoclastogenesis and bone resorption by increasing OPG secretion in osteoblasts and steocytes [62]. The BMP pathway is a Wnt associative stimulator signal [18], whereas the Wnt/β-catenin signaling pathway inhibitor Dickkopf Homolog-1 (DKK-1), the secreted frizzled-related protein (sFRP) and sclerostin, the product of the SOST gene which binds to LRP5/6 receptors, are natural inhibitors of the Wnt signaling pathway [62]. The inflammatory cytokine tumor necrosis factor-alpha (TNF-α) stimulates DKK-1 and Sost expression inducing osteoblast apoptosis. Glucocorticoids induce DKK-1 and sFRP expression, thus inducing osteoporosis by inhibiting the Wnt/β-catenin signal on osteoblasts [63].

Bone matrix provides support for bone cells but also exerts an essential role in bone homeostasis by releasing several molecules, including osteopontin, fibronectin, collagen, BSPs, and adhesion molecules, variously involved in the interaction between the bone cells and bone matrix [64]. Osteopontin increases bone resorption by inducing the expression of the osteoclastic immune receptor CD44, essential for cell migration, and by directly enhancing osteoclast attachment to bone extracellular matrix, required for the activation of osteoclast precursors. As a consequence of bone resorption, more osteoponin is further released from the extracellular matrix into the bone microenvironment and into the blood, thus amplifying local and systemic osteoclastogenesis [65].

IMMUNE REGULATION OF BONE REMODELING: THE CYTOKINE NETWORK

Both local and systemic factors influence bone homeostasis. Autocrine and paracrine molecules such as factors of the bone matrix released during bone resorption, as well as cytokines, growth factors, and prostaglandins produced by bone cells [66], variously interfere with each other in the control of bone turnover. Parathyroid hormone (PTH), PTH-related protein (PTHrP), calcitonin, 1,25-dihydroxyvitamin D3 (calcitriol), glucocorticoids, androgens, and estrogens are among systemic factors that influence bone remodeling. IGFs, transforming growth factor β (TGF-β), BMPs, FGF, and platelet-derived growth factor (PDGF), stored in the bone matrix and released after osteoclast bone resorption [67], act as soluble coupling factors in bone remodeling [68].

Interestingly, the cells of the immune system, mainly activated T and B lymphocytes and dendritic cells, express RANKL, that stimulates bone resorption through NFATc1, which is also a crucial factor in immune system regulation [4,44]. Other examples of shared receptor signals are the immunoglobulin (Ig)-like receptors which amplify the NFATc1 signal. The TLRs, stimulated by pathogen-associated molecular patterns (PAMP), utilize TRAF6 in their cascade signaling [69]. TLR are able to activate both the synthesis and release of proinflammatory and osteoclastogenic cytokines from immune cells. The osteoclast-associated receptor (OSCAR), which mediates interactions between osteoblasts and osteoclasts [70], is also involved in the regulation of both the adaptive and innate immunity [71–73]. TNF induces the expression of OSCAR and other receptors important for osteoclast differentiation on the surface of peripheral blood monocytes [74]. RANKL and CD40L expressed on T cells, APC, stromal cells, and osteoblasts, activate the cognate receptors RANK and CD40 in osteoclast precursors and osteoblasts, respectively. CD40 activation in B cells promotes their OPG secretion thereby decreasing bone resorption. CD40L also increases the commitment of MSC to the osteoblastic lineage. Through CD80/86 signaling in osteoclast progenitors, T cells suppress osteoclast differentiation [75]. CD80/CD86 downregulates osteoclastogenesis by binding to CTLA4, which is highly expressed on Treg surface [76].

Inflammation and bone remodeling share same mediators, such as cytokines and transcription factors and same signaling pathways drive both inflammatory processes and bone turnover. T lymphocytes in the bone marrow regulate bone remodeling in physiological as well as in pathological conditions. During inflammatory diseases or in conditions characterized by low-grade systemic inflammation, such as menopause and aging, bone resorption is driven by inflammatory cytokines produced by activated T lymphocytes. However, bone marrow T cells also support bone formation by binding of T cell co-stimulatory molecules to their receptors on bone cells and releasing cytokines and factors able to activate Wnt signaling in osteoblasts. The final effect of T lymphocytes on bone therefore depends on their activation state and phenotype: activated central memory CD8+ lymphocytes, abundant in postmenopausal women with osteoporotic fractures, secrete high levels of the osteoclastogenic cytokine TNF-α; Th17 lymphocytes stimulates the release of RANKL by osteoblasts and osteocytes and upregulate RANK expression on osteoclasts; instead, T regulatory (Treg) cells exert anti-osteoclastogenic activity by producing suppressor cytokines, including IL-4, IL-10, and TGF-β [75].

Inflammatory cytokines, including TNF-α, IL-1, IL-6, and IL-17, are crucial mediators of acute and chronic inflammation as well as strong inducers of bone resorption. An excessive or abnormal immune activation can induce osteoporosis, as for example in autoimmune and infection diseases. Also, in senile and postmenopausal osteoporosis, an increased inflammatory background and the presence of RANKL producing activated T cells can induce skeletal fragility. However, there are also cytokines able to counteract bone resorption and exert osteoblastogenic properties, resulting in a complex bone remodeling cytokine network [77]. For example, IFN-γ exerts anti-osteoclastogenic effects in physiological bone remodeling, by binding to specific osteoclast receptors and inducing TRAF6 proteosomal degradation with consequent inhibition of the transduction signal mediated by RANKL. However, since IFN-γ is a powerful stimulator of class II major histocompatibility complex (MHCII) antigen expression on APC, the final effect of IFN-γ is skewed toward bone resorption in postmenopausal osteoporosis, inflammation, or infections, through T lymphocyte activation and RANKL expression and osteoclastogenic proinflammatory cytokine production [78–82]. Homing, differentiation, and activation of osteoclast precursors are driven by a cascade of cytokines. Furthermore, these cells are also capable of producing proinflammatory cytokines and chemokines amplifying inflammatory processes by recruiting and activating CD8 T lymphocytes.

Moreover, the osteoclast precursors can also enhance the expression of the suppressor of cytokine signaling (SOCS) counteracting inflammation [83].

Finally, the same bone cells are able to influence or also even perform many immune functions, such as cytokine production and antigen presentation [84–86]. Cytokines secreted by bone cells drive naive T cell differentiation into several lineages, leading to the expansion of mature T cell populations that further regulate bone homeostasis. Osteoclasts selectively recruit and activate CD8+ T cells expressing CD25 and FoxP3, creating a regulatory loop: osteoclasts induce Treg, and then Treg blunt osteoclastic bone resorption by decreasing inflammatory/osteoclastogenic cytokine production and stimulating bone formation [75]. Skeletal stem cells recruit and activate neutrophils via the release of interleukin IL-6 and IL-8, IFN-β, GM-CSF, and macrophage migration inhibitory factor (MIF) [87]. A subset of mesenchymal cells expressing Osterix, a marker of bone precursors, regulate the maturation of early B lymphoid precursors through IGF-1 production [88] and shift the differentiation of T lymphocytes toward functional regulatory cells. In addition, skeletal stem cells express active TLRs, through which they sense bone microenvironment [87], induce macrophages to switch from classically activated proinflammatory (M1) to alternatively activated anti-inflammatory (M2) phenotype and inhibit mast cell degranulation attenuating allergic reactions [89].

THE IMPACT OF AGING ON BONE TURNOVER

Bone remodeling is variously influenced by the aging process [90]. Advancing age and loss of bone mass and strength are closely linked. During aging, the amount of bone resorbed by osteoclasts is not fully restored with bone deposited by osteoblasts and this imbalance leads to bone loss. The decline in bone strength is mainly due to reduction in trabecular and cortical bone density, decreased cortical thickness, and increase in cortical porosity [91].

Several interconnected factors contribute to senile osteoporosis progression. Physical activity stimulates bone formation and a decrease in mobility with age contributes to the loss of bone mass. An increase in the production of endogenous glucocorticoids with age, as well as enhanced sensitivity of bone cells to glucocorticoids represents another age-associated pathogenetic mechanism of senile osteoporosis [92]. Endogenous glucocorticoids are strong inhibitors of bone formation by stimulating osteoblast apoptosis and decreasing osteoblastogenesis. The adverse role of estrogen deficiency on bone mass after menopause contributes to the acceleration of skeletal involution in aged women. Elevated osteoblast and osteocyte apoptosis and decreased osteoblast numbers characterize age-related skeletal changes. Increased apoptosis and defective proliferation and differentiation of stem cell populations, as well as diversion of these progenitor cells toward the adipocyte lineage [93–95], contribute to defective osteoblast numbers in the aging skeleton.

Similar to other tissues, oxidative stress increases in bone with age, contributing to the effects of the aging process on bone and its cellular constituents. In the elderly, both lipid oxidation mediated by ROS increased production and Wnt signaling suppression, contribute to bone formation decline. Reactive oxygen species (ROS) decrease both number and synthetic capacity of osteocytes, which are the orchestrators of bone resorption and formation via production of RANKL and sclerostin in response to hormonal and mechanical stimuli, respectively [96]. The elevated osteocyte apoptosis with age is associated with diminished levels of RANKL and sclerostin protein in bone. RANKL stimulates ROS production that, in turn, is required for osteoclastogenesis. Oxidative stress decreases osteoblast and osteocyte life span. In addition, oxidized polyunsaturated fatty acids induce the association of PPAR with β-catenin and promote its degradation, attenuating osteoblastogenesis through a downregulation of the Wnt signaling. Oxidized lipids also potentiate oxidative stress, enhance osteoblast apoptosis, and inhibit BMP-2 induced osteoblast differentiation. Via these mechanisms, lipid oxidation potentiates oxidative stress and contributes to the decay in bone formation that occurs with aging. Moreover, ROS can determine changes in bone matrix proteins contributing to skeletal fragility and increased risk of fractures. Resveratrol is an antioxidant agent able to both decrease NF-k activation induced by RANKL and promote osteogenesis in mesenchymal stem cells via the Sirt1/FoxO3 axis stimulation [97].

Transcriptional activity of NF-kB is increased in a variety of tissues with aging and is associated with numerous age-related degenerative and inflammatory diseases, including osteoporosis [98]: NF-kB regulates expression of cytokines, growth factors, and genes that drive apoptosis, cell cycle progression, cell senescence, and inflammation. With aging, there is a loss of tissue homeostasis leading to an impaired ability to respond to genotoxic, inflammatory, and oxidative stresses which all stimulate the NF-kB transcription factor [99]. An aberrant NF-kB signaling in aging plays a central role in determining senile osteoporosis. NF-kB is upregulated in response to accumulated DNA damage which drives tissue degeneration. NF-κB signaling and inflammatory cytokine secretion are upregulated in osteoporosis. Caloric restriction, which is able to extend longevity, inhibits NF-κB signaling and downregulates expression of several inflammatory genes, functioning via inflammatory suppression.

Several signaling pathways implicated in aging, including IGF-1, mTOR, SIRT1, and p53, induce cellular senescence, a tumor-suppressive mechanism that impacts on cell growth, immune responses, apoptosis, and metabolism. The altered transcriptional phenotype of senescent cells, consisting of increased expression of IL-6, IL-8, IL-7, MCP-2, MIP-3, intercellular adhesion molecule (ICAM), Il-1α, and Il-β amplify bone resorption. In particular, IL-6 and TNFα levels are markers of increased risk of developing osteoporosis [100].

The tumor suppressor protein p53 is a critical cellular stress sensor whose activation increases in the bone with advancing age [15,55]. Depending on the degree of activation, it promotes growth arrest and repair, apoptosis or cellular senescence, thus contributing to the appearance of aging phenotypes, including osteoporosis [70,101,102]. The negative effect of p53 on osteoblast generation is mediated by the repression of Runx2 and Osterix expression [97,103]. In addition, p53 activity in osteoblasts attenuates osteoclast generation by decreasing M-CSF expression. Defects in telomere maintenance, similar to p53 activation, impair osteoblast differentiation and promote osteoporosis [14]. The progressive loss of telomere function during aging triggers chronic activation of p53 [104]. Telomeres are DNA and nucleoprotein complexes at chromosome ends that function to preserve chromosomal integrity.

Age-related metabolic changes strongly impact on bone homeostasis. Advanced glycation end products (AGEs) increase with aging. AGEs are a heterogeneous group of bioactive molecules that are formed by the nonenzymatic glycation of proteins, lipids, and nucleic acids and are implicated in the pathophysiological processes of many inflammatory and dysmetabolic age-associated diseases. In particular, the interaction between AGE and its receptor receptor for advanced glycation end products (RAGE) is involved in the development and progression of both senile and diabetes-associated osteoporosis by inhibiting proliferation and inducing osteoblast apoptosis. The binding of AGE to organic bone matrix may also increase the fragility of bones [105].

Both normal and pathological aging are often associated with a reduced autophagic potential [106]. Autophagy, upregulated in stressful conditions, is a metabolic process by which eukaryotic cells degrade and recover damaged macromolecules and organelles into autophagosomes. Excessive autophagy is harmful to cells and leads to damage or massive cell death. However, autophagy deficiency increases oxidative stress levels in osteoblasts, decreases bone mineralization, and induces RANKL secretion.

A common feature of aging is the alteration in tissue distribution and composition. The redistribution of adipose tissue toward visceral and ectopic sites has dramatic effects on metabolic and immune functions. Dysfunctional adipocytes are less able to store fat. During senescence, the increased number of dysfunctional adipocytets contributes to impaired bone health and fragility fractures [107–110]. The pathway of differentiation of preadipocytes to mature adipocytes becomes impaired in the elderly. Increased ectopic adiposity is linked to insulin resistance leading to metabolic syndrome and diabetes. In addition, the risk of developing cardiovascular disease is increased with elevated visceral adipose distribution. Beyond increased ectopic adiposity, the effect of impaired adipose tissue function is an elevation in systemic free fatty acids, a common feature of many metabolic disorders. In particular, saturated fatty acids are capable of inducing insulin resistance and inflammation through lipid mediators such as ceramide, which can increase the risk of developing atherosclerosis and may represent a driving factor for the metabolic syndrome, characterized by increased insulin resistance, cardiovascular disease risk, and inflammation in older adults [111].

IMMUNOSENESCENCE AND BONE REMODELING

The amplification of inflammatory reactions leading to bone resorption and skeletal fragility is the hallmark of senescence and inflammaging that is the condition of chronic inflammation characterizing aging. Inflammaging results in immunoskeletal interface disturbances causing osteoporosis [112,113]. This condition also represents the background of a broad range of age-related diseases with an inflammatory pathogenesis [101]. Many of these age-related diseases are associated to osteoporosis. However, also in the absence of overt inflammatory diseases, the heightened catabolic signals induced by inflammation enhance apoptosis of osteoblasts and muscle cells, causing both osteoporosis and sarcopenia.

Several hormones, variously modified during aging, affect skeletal metabolism through immune-mediated mechanisms. PTH is an endocrine regulator of calcium and phosphorus metabolism. Secondary hyperparathyroidism is involved in the pathogenesis of senile osteoporosis [114]. The catabolic effect of PTH is mostly mediated by enhanced production of RANKL and decreased production of OPG by osteoblasts and stromal cells. The PTH anabolic effect is mediated by Wnt signaling activation by increasing β-catenin levels, promoting LRP6 signaling, and decreasing sclerostin production [75].

Estrogen plays crucial roles in bone homeostasis: the decrease in estrogen level at menopause is the main cause of bone loss and osteoporosis in postmenopausal females. Osteoimmunology highlights the immune modulation of bone homeostasis, leading to a shift in the concept of postmenopausal osteoporosis, which represents a clear example of the mutual influence between the immune system, bone, and endocrine system. Bone changes typical of the menopausal estrogen decline are currently considered the consequence of an inflammatory condition [115,116]. Both immune and bone cells express estrogen receptors. Menopausal estrogen decline leads to lymphocyte activation and hyperproduction of inflammatory cytokines resulting in the chronic osteoclast stimulation responsible for bone loss and increased fracture risk in postmenopausal women [117,118]. Bone cells express estrogen receptors and are direct targets for estrogens. Estrogen preserves bone homeostasis by inhibiting osteoblast and osteocyte apoptosis and suppressing osteoclast formation and activity through the inhibition of RANKL production. Estrogens also stimulate the OPG production and reduce levels of inflammatory osteoclastogenic cytokines, including IL-1, IL-6, IL-11, TNF-α, TNF-β, and M-CSF, further inhibiting osteoclastogenesis.

The lifelong exposure to oxidative stress and chronic antigenic load leads to the loss of the regulatory process which counteracts macrophage and T cell activation [102]. During senescence, besides the impaired Treg function, the number of effector memory cells is increased [103]. These are senescent cells with proinflammatory properties, secreting several inflammatory cytokines able to influence bone remodeling. Interestingly, this immunological pattern (accumulation of activated cells and memory/effector lymphocytes secreting proinflammatory cytokines) characterizing immunosenescence, is also peculiar of other immunological conditions associated with osteoporosis, such as rheumatoid arthritis, AIDS, chronic viral infections, etc. [119].

The increased production of proinflammatory cytokines with aging derives from a chronic hyperactivation of macrophages and dendritic cells, as well as memory and senescent T cells. These cytokines induce expansion of osteoclast precursors which in turn may contribute to the maintenance of inflammation through their capability to produce proinflammatory cytokines themselves and recruit other inflammatory cells, rendering the inflammation chronic (Figure 17.1). Osteoclastogenesis and inflammation are directly proportional to osteoclast precursor levels in peripheral blood.

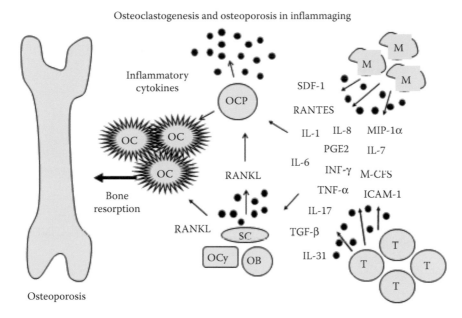

FIGURE 17.1 Osteoclastogenesis and osteoporosis in inflammaging. Cells that induce bone resorption are activated lymphocytes and macrophages that produce cytokines and inflammatory mediators that stimulate differentiation and activation of osteoclasts, regulated by a variety of cytokines. Release of cells into the circulation from the bone marrow and homing from the blood stream to peripheral tissues where the immature osteoclast precursor differentiation into mature bone resorbing osteoclasts is a complex process involving growth factors, adhesion molecules, cytokines, and chemokines. In addition, they themselves produce a great amount of inflammatory mediators, thus supporting and amplifying inflammatory reactions: OC (osteoclasts), OCP (osteoclast precursors), SC (stromal cells), OCy (osteocytes), OB (osteoblasts), M (macrophages), and T (T lymphocytes).

Thymic T cell production declines rapidly with advancing age, conditioning the immune phenotype of elderly people and subjects with senile osteoporosis. Moreover, multiple mechanisms, including antigen-driven clonal expansion impose replicative stress on T cells and induce the biological program of cellular senescence with characteristic phenotypic changes. T cell immunosenescence is associated with profound changes in T cell functional profile and leads to accumulation of CD4+ T cells which have lost CD28 but have gained killer immunoglobulin-like receptors (KIRs), markers of natural killer cells. They also exhibit cytolytic capability and produce large amounts of proinflammatory cytokines. Characteristic of an aged immune profile is the accumulation of activated memory cells expressing RANKL, preferentially resident in the bone and secreting osteoclastogenic proinflammatory cytokines. During aging, the number of effector memory CD45RA positive CD8+ T cells is increased. They are mainly senescent and proinflammatory cells, able to secrete large amounts of proinflammatory cytokines involved in the regulation of bone turnover.

Both number and function of Treg cells decline with age. Tregs antagonize immune-mediated bone resorption by suppressing inflammation via cell contact dependent mechanisms and through the production of soluble factors, including IL-10 and TGF-β. FoxP3 positive regulatory T lymphocytes expressing CTLA-4 balance IL-17 induced bone resorption by inhibiting osteoclast activity [104,120]. Aging compromises Treg generation and function, facilitating inflammatory processes and osteoporosis.

Changes in the cytokine milieu are major characteristics of the aging process as well as of age-related diseases [9]. Cytokines which function as stimulators or inhibitors of bone resorption are variously modified during aging and may elicit their effects directly, by acting on the osteoclast precursors or mature cells or indirectly, mainly modulating RANKL/OPG expression. TNF-α may

TABLE 17.1

Main Osteoclastogenic Cytokines Involved in Bone Remodeling during Aging

IL-1	• TRAF6 expression stimulation, NF-kB induction and MAPKs activation
	• OC maturation induction
	• PGE$_2$ and RANKL expression induction on OB
	• Enhancement of RANKL induced osteoclastogenesis
	• Amplification of RANKL secretion by stromal cells induced by TNF-α
IL-6	• Inhibition of SHP2/MEK2/ERK and SHP2/PI3 K/Akt2 mediated OB differentiation
	• RANKL production by synovial and bone marrow stromal cells
	• Synergy with IL-1, TNF-α and PGE
	• Generalized inflammatory effects
IL-7	• T and B cell activation
	• Promotion of RANKL and osteoclastogenic cytokine production by T cells
	• Synergy with IL-1, TNF-α and IFN-γ
IL-8	• RANKL increase
IL-11	• Increase of RANKL/OPG ratio
IL-17	• Induction of RANKL expression on OB, synovial cells and fibroblasts
	• Synthesis induction of bone matrix degrading enzymes (MMP)
	• Induction of TNF-α and IL6 production by inflammatory cells
	• Synergy with TNF-α, IL-1 and PGE
IL-23	• OC activation in inflammatory pathologies
	• TH17 increase
	• IL-17 and RANKL induction
IL-31	• Stimulation of proinflammatory cytokine, chemokine, and matrix metalloproteinase secretion (TNF-α, IL-1β, IL-8, MMP)
IL-32	• Proinflammatory cytokine production stimulation
	• OCP differentiation induction
TNF-α	• M-CSF and RANKL expression induction and inflammation
	• OC transcription factor (TRAF2, NF-kB) activation through TNF-R1
	• OBP maturation inhibition, OB activity inhibition, and OB apoptosis induction
	• Induction of negative regulators of the Wnt pathway expression
	• Inhibition of genes involved in bone formation (APH, alkaline phosphatase, 1,25-(OH)2 D3 receptor, PTH receptor)
TGF-β	• Promotion of RANKL induced osteoclastogenesis
INF-γ	• MHCII antigen expression induction on APC
	• Antigen presentation enhancement and in T cell activation
	• Osteoclastogenic cytokine production enhancement
	• Synergy with IL-12
M-CSF	• RANK expression upregulation on OCP
	• OCP pool expansion
	• OC formation (in synergy with RANKL)
	• Monocyte/macrophage lineage differentiation
G-CSF	• OB apoptosis
	• MSC apoptosis

Abbreviations: GM-CSF, granulocyte–macrophage colony-stimulating factor; M-CSF, monocyte/macrophage colony-stimulating factor; IFN, interferon; IL, interleukin; RANK, receptor activator of nuclear factor-κB; RANKL, receptor activator of nuclear factor-κB ligand; TH, T-helper; TNF, tumor necrosis factor; TGF, transforming growth factor; OC, osteoclast; OCP, osteoclast precursor; OB, osteoblast; OBP, osteoblast precursor; OPG, osteoprotegerin; TRAF, TNF receptor-associated factor; PTH, parathyroid hormone; PG, prostaglandin; MHCII, major histocompatibility complex-class II; APC, antigen presenting cells; 1,25-(OH)2 D3, dihydroxyvitamin D3; APH, alkaline phosphatase; MMP, matrix metalloproteinase; NF-κB, nuclear factor kappa-light-chain-enhancer of activated B cells; TNF-R, TNF-receptor; MAPKs, mitogen-activated protein kinases; ERK, extracellular signal-regulated kinases; PI3 K, phosphatidylinositol-3-kinase; Wnt, Wingless pathway.

act directly on the osteoclast precursors to promote osteoclast differentiation or alternatively may promote osteoclastogenesis indirectly through the induction of the expression of RANKL and colony-stimulating factor-1 (CSF-1) in bone marrow stromal cells and bone lining cells. TNF-α may also function to increase the CD11b+ osteoclast precursor cell population. The proinflammatory cytokine IL-1 can act by promoting both the fusion of mononuclear osteoclast precursors to form osteoclasts and the survival and function of mature osteoclasts. Table 17.1 lists the main osteoclastogenic cytokines involved in bone remodeling during aging.

Adipose tissue becomes dysfunctional in aging and produces abnormal levels of proinflammatory mediators, thus contributing to low-grade systemic inflammation commonly observed in the elderly. On the other hand, proinflammatory cytokines are capable of inhibiting adipocyte differentiation, leading to cachexia. Several cell populations within the fat tissue can exacerbate the development of the chronic, low-grade inflammation associated with aging. Various immunocytes infiltrate fat tissue and communicate closely with adipocytes. Visceral fat tissue infiltrating macrophages have a proinflammatory phenotype which exacerbates the chronic inflammation of aging [92]. Visceral adipose tissue increases in the elderly. It contains dysfunctional-activated macrophages producing inflammatory cytokines, including IL-1, IL-6, and TNF-α, that can alter the quality of the bone, making it more fragile [111].

Adipose tissue produced proinflammatory cytokines and adipokines modulate the activity of OC and OB. Fat tissue, mainly visceral fat tissue, may increase bone resorption through the production of inflammatory cytokines such as IL-6 and TNF-α, which stimulate osteoclast activity through the regulation of the RANKL/RANK/OPG pathway [121]. Leptin and adiponectin act on the bone through different signaling pathways with contrasting effects [107]. Leptin signaling regulates osteogenesis by skeletal stem cells. It promotes adipogenesis and inhibits osteogenesis by activating Jak2/Stat3 signaling [16,17]. Both adipocytes and osteoblasts express RANKL and their modulation is influenced by inflammatory mediators. Adipokines produced by fat cells are involved in the regulation of bone metabolism and insulin action, neurodegenerative diseases, and cancer cell proliferative signaling. Osteoblast lineage cells express receptors for adiponectin, leptin, angiotensin II, insulin, and IGF-1, able to influence the RANK/RANKL/OPG signaling pathway [65]. Adipokines can be divided into pro- and anti-inflammatory subgroups. Leptin is a proinflammatory adipokine that induces TNF-α and IL-6 production by monocytes, chemokines production by macrophages, and Th1 cytokine production from polarized CD4+ T cells. Leptin also supports proliferation of activated T cells. Adiponectin is an anti-inflammatory adipokine whose plasma levels are strongly correlated with insulin sensitivity and glucose tolerance. It can directly interfere with inflammatory cytokine production in macrophages and can induce the expression of the anti-inflammatory cytokine IL-10. TNF-α and IL-6 inhibit adiponectin production in adipocytes [92]. With aging, there is an imbalance in the adipokine profile and function resulting in an enhanced proinflammatory and bone resorption activity [122].

CONCLUSION

Osteoimmunology therefore allows a new approach to senile osteoporosis that is now considered an inflammatory condition. The immunoskeletal interface represents a systemic model of integrated signaling pathways and cytokines working in a cooperative fashion. The immune and skeletal systems interact with each other both in physiological and pathological conditions. Aging influences the complex crosstalk between bone and immune cells which occurs in the dynamic bone microenvironment. The great majority of the biological mechanisms implicated in the aging process, such as cell senescence, proinflammatory immune profile, apoptosis, and metabolism imbalance, are also implicated in bone remodeling. A central role in this age-related crosstalk remodulation is played by proinflammatory and osteoclastogenic cytokines which drive bone resorption.

REFERENCES

1. Ginaldi L, De Martinis M. 2016. Osteoimmunology and beyond. *Curr Med Chem*, 23(33):3754–3774.
2. Takayanagi H. 2007. Osteoimmunology: Shared mechanisms and crosstalk between the immune and bone systems. *Nat Immunol*, 7:292–304.
3. Greenblatt MB, Shim J-H. 2013. Osteoimmunology: A brief introduction. *Immune Netw*, 13:111–115.
4. Takayanagi, H. 2015. Osteoimmunology in 2014: Two-faced immunology-from osteogenesis to bone resorption. *Nat Rev Rheumatol*, 11:74–76.
5. Feng X, McDonald JM. 2011. Disorders of bone remodeling. *Annu Rev Pathol Mech Dis*, 6:121–145.
6. Longhini R, de Oliveira PA, Sasso-Cerri E, Cerri PS. 2014. Cimetidine reduces alveolar bone loss in induced periodontitis in rat molars. *J Periodontol*, 85(8):1115–1125.
7. Jain N, Jain GK, Javed S et al. 2008. Recent approaches for the treatment of periodontitis. *Drug Discov Today*, 13(21–22):932–943.
8. Seeman E, Delmas PD. 2006. Bone quality—The material and structural basis of bone strength and fragility. *N Engl J Med*, 354(21):2250–2261.
9. Sobacchi C, Schulz A, Coxon FP, Villa A, Helfrich MH. 2013. Osteopetrosis: Genetics, treatment and new insights into osteoclast function. *Nat Rev Endocrinol*, 9(9):522–536.
10. Florencio-Silva R, da Silva Sasso GR, Sasso-Cerri E, Simões MJ, Cerri PS. 2015. Biology of bone tissue: Structure, function, and factors that influence bone cells. *Biomed Res Int*, Article ID 421746, 17 Review Article, 5–7.
11. De Martinis M, Di Benedetto MC, Mengoli LP, Ginaldi L. 2006. Senile osteoporosis: Is it an immune-mediated disease? *Inflamm Res*, 55:399–404.
12. Chan CK, Seo EY, Chen JY, Lo D, McArdle A et al. 2015. Identification and specification of the mouse skeletal stem cell. *Cell*, 160:285–298.
13. Sinder BP, Pettit AR, McCauley LK. 2015. Macrophages: Their emerging roles in bone. *J Bone Miner Res*, 30(12):2140–2149.
14. Mensah KA, Li J, Schwarz EM. 2010. The emerging field of osteoimmunology. *Immunol Res*, 45(2-3):100–113.
15. Birbrair A, Frenette PS. 2016. Niche heterogeneity in the bone marrow. *Ann N Y Acad Sci*, 1370:82.
16. Morrison SJ, Scadden DT. 2014. The bone marrow niche for haematopoietic stem cells. *Nature*, 505:327–334.
17. Yue R, Zhou BO, Shimada IS, Zhao Z, Morrison SJ. 2016. Leptin receptor promotes adipogenesis and reduces osteogenesis by regulating mesenchymal stromal cells in adult bone marrow. *Cell Stem Cell*, 18:782–796.
18. Worthley DL, Churchill M, Compton JT, Tailor Y et al. 2015. Gremlin 1 identifies a skeletal stem cell with bone, cartilage, and reticular stromal potential. *Cell*, 160:269–284.
19. Rauner M, Sipos W, Thiele S, Pietschmann P. 2013. Advances in osteoimmunology: Pathophysiologic concepts and treatment opportunities. *Int Arch Allergy Immunol*, 160:114–125.
20. Grigoriadis AE, Heersche JNM, Aubin JE. 1988. Differentiation of muscle, fat, cartilage, and bone from progenitor cells present in a bone-derived clonal cell population: Effect of dexamethasone. *J Cell Biol*, 106(6):2139–2151.
21. Matsubara T, Kida K, Yamaguchi A et al. 2008. BMP2 regulates Osterix through Msx2 and Runx2 during osteoblast differentiation. *J Biol Chem*, 283(43):29119–29125.
22. Cao Y, Zhou Z, de Crombrugghe B et al. 2005. Osterix, a transcription factor for osteoblast differentiation, mediates antitumor activity in murine osteosarcoma. *Cancer Res*, 65(4):1124–1128.
23. Monroe DG, McGee-Lawrence ME, Oursler MJ, Westendorf JJ. 2012. Update on Wnt signaling in bone cell biology and bone disease. *Gene*, 492(1):1–18.
24. Capulli M, Paone R, Rucci N. 2014. Osteoblast and osteocyte: Games without frontiers. *Arch Biochem Biophys*, 561:3–12.
25. Kapinas K, Kessler C, Ricks T, Gronowicz G, Delany AM. 2010. MiR-29 modulates Wnt signaling in human osteoblasts through a positive feedback loop. *J Biol Chem*, 285(33):25221–25231.
26. Zhang Y, Xie R-L, Croce CM. et al. 2011. A program of microRNAs controls osteogenic lineage progression by targeting transcription factor Runx2. *Proc Natl Acad Sci U S A*, 108(24):9863–9868.
27. Montero AY, Okada Y, Tomita M et al. 2000. Disruption of the fibroblast growth factor-2 gene results in decreased bone mass and bone formation. *J Clin Invest*, 105(8):1085–1093.
28. Buo AM, Stains JP. 2014. Gap junctional regulation of signal transduction in bone cells. *FEBS Lett*, 588(8):1315–1321.

29. Fakhry M, Hamade E, Badran B, Buchet R, Magne D. 2013. Molecular mechanisms of mesenchymal stem cell differentiation towards osteoblasts. *World J Stem Cells*, 5(4):136–148.

30. Bonewald LF. 2011. The amazing osteocyte. *J Bone Miner Res*, 26(2):229–238.

31. Mikuni-Takagaki Y, Kakai Y, Satoyoshi M et al. 1195. Matrix mineralization and the differentiation of osteocyte-like cells in culture. *J Bone Miner Res*, 10(2):231–242.

32. Poole KES, van Bezooijen RL, Loveridge N et al. 2005. Sclerostin is a delayed secreted product of osteocytes that inhibits bone formation. *FASEB J*, 19(13):1842–1844.

33. Dallas SL, Prideaux M, Bonewald LF. 2013. The osteocyte: An endocrine cell … and more. *Endocr Rev*, 34(5):658–690.

34. Civitelli R, Lecanda F, Jørgensen NR, Steinberg TH. 2002. Intercellular junctions and cell-cell communication in bone. In *Principles of Bone Biology*, Bilezikan JP, Raisz L, and Rodan GA (eds), Academic Press, San Diego, CA, pp. 287–302.

35. Rochefort GY, Pallu S, Benhamou CL. 2010. Osteocyte: The unrecognized side of bone tissue. *Osteoporos Int*, 21(9):1457–1469.

36. Aguirre JI, Plotkin LI, Stewart SA et al. 2006. Osteocyte apoptosis is induced by weightlessness in mice and precedes osteoclast recruitment and bone loss. *J Bone Miner Res*, 21(4):605–615.

37. Plotkin LI. 2014. Apoptotic osteocytes and the control of targeted bone resorption. *Curr Osteoporos Rep*, 12(1):121–126.

38. Bellido T. 2014. Osteocyte-driven bone remodeling. *Calcif Tissue Int*, 94(1):25–34.

39. Nakashima T, Hayashi M, Fukunaga T et al. 2011. Evidence for osteocyte regulation of bone homeostasis through RANKL expression. *Nat Med*, 17(10):1231–1234.

40. Wu AC, Morrison NA, Kelly WL, Forwood MR. 2013. MCP-1 expression is specifically regulated during activation of skeletal repair and remodeling. *Calcif Tissue Int*, 92(6):566–575.

41. Zhou Z, Han J-Y, Xi C-X et al. 2008. HMGB1 regulates RANKL-induced osteoclastogenesis in a manner dependent on RAGE. *J Bone Miner Res*, 23(7), 1084–1096.

42. Harris SE, MacDougall M, Horn D et al. 2012. Meox2Cre-mediated disruption of CSF-1 leads to osteopetrosis and osteocyte defects. *Bone*, 50(1):42–53.

43. Andersen TL, Sondergaard TE, Skorzynska KE et al. 2009. A physical mechanism for coupling bone resorption and formation in adult human bone. *Am J Pathol*, 174(1):239–247.

44. Teitelbaum SL. 2000. Bone resorption by osteoclasts. *Science*, 289:1504–1508.

45. Charles JF, Aliprantis AO. 2014. Osteoclasts: More than 'bone eaters'. *Trends Mol Med*, 20(8): 449–459.

46. Brown JP, Roux C, Ho PR et al. 2014. Denosumab significantly increases bone mineral density and reduces bone turnover compared with monthly oral ibandronate and risedronate in postmenopausal women who remained at higher risk for fracture despite previous suboptimal treatment with an oral bisphosphonate. *Osteoporos Int*, 25:1953–1961.

47. Miyamoto T. 2006. The dendritic cell-specific transmembrane protein DC-STAMP is essential for osteoclast fusion and osteoclast bone-resorbing activity. *Mod Rheumatol*, 16(6):341–342.

48. Kobayashi Y, Udagawa N, Takahashi N. 2009. Action of RANKL and OPG for osteoclastogenesis. *Crit Rev Eukaryot Gene Expr*, 19(1):61–72.

49. Gohda J, Akiyama T, Koga T, Takayanagi H, Tanaka S, Inoue J-I. 2005. RANK-mediated amplification of TRAF6 signaling leads to NFATc1 induction during osteoclastogenesis. *EMBO J*, 24:790–799.

50. Li S, Miller CH, Giannopoulou E, Hu X, Ivashkiv LB, Zhao B. 2014. RBP-J imposes a requirement for ITAM-mediated costimulation of osteoclastogenesis. *J Clin Invest*, 124(11):5057–5073.

51. Ljusberg J, Wang Y, Lång P et al. 2005. Proteolytic excision of a repressive loop domain in tartrate-resistant acid phosphatase by cathepsin K in osteoclasts. *J Biol Chem*, 280(31):28370–28381.

52. de Souza Faloni, AP, Sasso-Cerri, E, Rocha FRG, Katchburian E, Cerri PS. 2012. Structural and functional changes in the alveolar bone osteoclasts of estrogen-treated rats. *J Anat*, 220(1):77–85.

53. Yu H. 2016. Sphingosine-1-phosphate receptor 2 regulates proinflammatory cytokine production and osteoclastogenesis. *PLoS One*, 11(5):e0156303.

54. Edwards CM, Mundy GR. 2008. Eph receptors and ephrin signaling pathways: A role in bone homeostasis. *Int J Med Sci*, 5(5):263–272.

55. Matsuo K. 2010. Eph and ephrin interactions in bone. *Adv Exp Med Biol*, 658:95–103.

56. Zhao C, Irie N, Takada Y et al. 2006. Bidirectional ephrinB2–EphB4 signaling controls bone homeostasis. *Cell Metab*, 4(2):111–121.

57. Negishi-Koga T, Shinohara M, Komatsu N et al. 2011. Suppression of bone formation by osteoclastic expression of semaphorin 4D. *Nat Med*, 17(11):1473–1480.

58. Sutton ALM, Zhang X, Dowd DR, Kharode YP, Komm BS, MacDonald PN. 2008. Semaphorin 3B is a 1,25-dihydroxyvitamin D3-induced gene in osteoblasts that promotes osteoclastogenesis and induces osteopenia in mice. *Mol Endocrinol*, 22(6):1370–1381.

59. Hughes A, Kleine-Albers J, Helfrich MH, Ralston SH, Rogers MJ. 2012. A class III semaphorin (Sema3e) inhibits mouse osteoblast migration and decreases osteoclast formation in vitro. *Calcif Tissue Int*, 90(2):151–162.

60. Yang YH, Buhamrah A, Schneider A, Lin YL, Zhou H, Bugshan A, Basile JR. 2016. Semaphorin 4D promotes skeletal metastasis in breast cancer. *PLoS One*, 11(2):e0150151.

61. Hayashi M, Nakashima T, Taniguchi M, Kodama T, Kumanogoh A, Takayanagi H. 2012. Osteoprotection by semaphorin 3A. *Nature*, 485(7396):69–74.

62. Deal C. 2012. Bone loss in rheumatoid arthritis: Systemic, periarticular, and focal. *Curr Rheumatol Rep*, 14:231–237.

63. Klontzas M, Kenanidis E, MacFarlane R et al. 2016. Investigational drugs for fracture healing: Preclinical and clinical data. *Expert Opin Investig Drugs*, 25(5):585–596.

64. Helfrich MH, Stenbeck G, Nesbitt MA et al. 2008. Integrins and adhesion molecules. In *Principles of Bone Biology*, Bilezikan, JP, Raisz LG, Martin TJ (eds), Academic Press, Elsevier, San Diego, CA, pp. 385–424.

65. Musso G, Paschetta E, Gambino R, Cassader M, Molinaro F. 2013. Interactions among bone, liver, and adipose tissue predisposing to diabesity and fatty liver. *Trends Mol Med*, 19(9):522–535.

66. Manolagas SC. 2000. Birth and death of bone cells: Basic regulatory mechanisms and implications for the pathogenesis and treatment of osteoporosis. *Endocr Rev*, 21(2):115–137.

67. Howard GA, Bottemiller BL, Turner RT, Rader JI, Baylink DJ. 1981. Parathyroid hormone stimulates bone formation and resorption in organ culture: Evidence for a coupling mechanism. *Proc Natl Acad Sci U S A*, 78(5):3204–3208.

68. Linkhart TA, Mohan S, Baylink DJ. 1996. Growth factors for bone growth and repair: IGF, TGF beta and BMP. *Bone*, 19(1):1S–12S.

69. Kassem A, Henning P, Kindlund B, Lindholm C, Lerner UH. 2015. TLR5, a novel mediator of innate immunity-induced osteoclastogenesis and bone loss. *Faseb J*, 29(11):4449–4460.

70. Barrow AD, Raynal N, Andersen TL et al. 2011. OSCAR is a collagen receptor that costimulates osteo-clastogenesis in DAP12-deficient humans and mice. *J Clin Invest*, 121(9):3505–3516.

71. Koga T, Inui M, Inoue K et al. 2004. Costimulatory signals mediated by the ITAM motif cooperate with RANKL for bone homeostasis. *Nature*, 428(6984):758–763.

72. Humphrey MB, Nakamura MC. 2015. A comprehensive review of immunoreceptor regulation of osteo-clasts. *Clin Rev Allergy Immunol*. doi:10.1007/s12016-015.8521-8.

73. Kasagi S, Chen W. 2013. TGF-beta1 on osteoimmunology and the bone component cells. *Cell Biosci*, 3:4.

74. Goettsch C, Helas S, Jessberger R, Lorenz C et al. 2011. The role of osteoclast-associated receptor in osteoimmunology. *J Immunol*, 186:13–18.

75. Pacifici R. 2016. T cells, osteoblasts, and osteocytes: Interacting lineages key for the bone anabolic and catabolic activities of parathyroid hormone. *Ann N Y Acad Sci*, 1364:11–24.

76. Schett G. 2009. Osteoimmunology in rheumatic diseases. *Arthritis Res Ther*, 11:210. doi:10.1186/ar2571.

77. Khosla S. 2013. Pathogenesis of age-related bone loss in humans. *J Gerontol A Biol Sci Med Sci*, 68:1226–1235.

78. Ginaldi L, De Martinis M, Ciccarelli F, Saitta S, Imbesi S, Mannucci C, Gangemi S. 2015. Increased levels of interleukin 31 (IL-31) in osteoporosis. *BMC Immunol*, 16:60. doi: 10.1186/s12865-015-0125-9.43,47,48.

79. Ogawa H, Mukai K, Kawano Y, Minegishi Y, Karasuyama H. 2012. Th2-inducing cytokines IL-4, and IL-33 synergistically elicit the expression of transmembrane TNF-α on macrophages through the auto-crine action of IL-6. *Biochem Biophys Res Commun*, 420(1):114–118.

80. Wang Y, Wu NN, Mou YQ, Chen L, Deng ZL. 2013. Inhibitory effects of recombinant IL-4 and recom-binant IL-13 on UHMWPE-induced bone destruction in the murine air pouch model. *J Surg Res*, 180(2):e73–81.

81. Salem S, Gao C, Li A, Wang H, Nguyen Yamamoto L, Goltzman D, Henderson JE, Gros P. 2014. A novel role for interferon regulatory factor 1 (IRF1) in regulation of bone metabolism. *J Cell Mol Med*, 18(8):1588–1598.

82. Raychaudhuri SP, Raychaudhuri SK. 2016. IL-23/IL-17 axis in spondyloarthritis-bench to bedside. *Clin Rheumatol*, 35(6):1437–1441.

83. Zhang X, Alnaeeli M, Singh B, Teng YT. 2009. Involvement of SOCS3 in regulation of CD11c+ den-dritic cell-derived osteoclastogenesis and severe alveolar bone loss. *Infect Immun*, 77(5):2000–2009.

84. Liu W, Zhang X. 2015. Receptor activator of nuclear factor-κB ligand (RANKL)/RANK/osteoprote-gerin system in bone and other tissues. *Mol Med Rep*, 11(5):3212–3218.

85. Charles JF, Aliprantis AO. 2014. Osteoclasts: More than 'bone eaters'. *Trends Mol Med*, 20(8):449–459.
86. Dallas SL, Prideaux M, Bonewald LF. 2013. The osteocyte: An endocrine cell ... and more. *Endocr. Rev*, 34, 658–690.
87. Kim SH, Das A, Chai JC et al. 2016. Transcriptome sequencing wide functional analysis of human mesenchymal stem cells in response to TLR4 ligand. *Sci Rep*, 6:30311. doi: 10.1038.
88. Yu VW, Lymperi S, Oki T, Jones A et al. 2016. Distinctive mesenchymal–parenchymal cell pairings govern B cell differentiation in the bone marrow. *Stem Cell Rep*, 7(2):220–235.
89. Uccelli A, de Rosbo NK. 2015. The immunomodulatory function of mesenchymal stem cells: Mode of action and pathways. *Ann N Y Acad Sci*, 1351:114–126.
90. Almeida A. 2012. Aging mechanisms in bone. *BoneKEy Rep*, 1, Article number: 102. doi: 10.1038/bonekey.2012.102.
91. Zebaze RM, Ghasem-Zadeh A, Bohte A, Iuliano-Burns S, Mirams M, Price RI et al. 2010. Intracortical remodelling and porosity in the distal radius and post-mortem femurs of women: A cross-sectional study. *Lancet*, 375:1729–1736.
92. DiSpirito JR, Mathis D. 2015. Immunological contributions to adipose tissue homeostasis. *Semin Immunol*, 27(5):315–321.
93. Kassem M, Marie PJ. 2011. Senescence-associated intrinsic mechanisms of osteoblast dysfunctions. *Aging Cell*, 10:191–197.
94. Almeida M, Han L, Martin-Millan M, Plotkin LI, Stewart SA, Roberson PK et al. 2007. Skeletal involution by age-associated oxidative stress and its acceleration by loss of sex steroids. *J Biol Chem*, 282:27285–27297.
95. Syed FA, Modder UI, Roforth M, Hensen I, Fraser DG, Peterson JM et al. 2010. Effects of chronic estrogen treatment on modulating age-related bone loss in female mice. *J Bone Miner Res*, 25:2438–2446.
96. Xiong J, O'Brien CA. 2012. Osteocyte RANKL: New insights into the control of bone remodeling. *J Bone Miner Res*, 27:499–505.
97. Tseng PC, Hou SM, Chen RJ, Peng HW, Hsieh CF, Kuo ML, Yen ML. 2011. Resveratrol promotes osteogenesis of human mesenchymal stem cells by upregulating RUNX2 gene expression via the SIRT1/FOXO3A axis. *J Bone Miner Res*, 26(10):2552–2563.
98. Clauson CL, Niedernhofer LJ, and Robbins PD. 2011. *Aging Dis*, 2(6):449–465.
99. Adler A, Sinha, S, Kawahara, TLA, Zhang, JY, Segal, E, Chang, HY. 2007. Motif module map reveals enforcement of aging by continual NF-kB activity. *Genes Dev*, 21(24):3244–3257.
100. Ginaldi L, Di Benedetto MC, De Martinis M. 2005. Osteoporosis, inflammation and aging. *Immun Aging*, 2:14.
101. Ginaldi L, Mengoli LP, De Martinis M. 2009. Osteoporosis, inflammation and aging. In *Handbook on Immunosenescence: Basic Understanding and Clinical Applications*. Fulop T, Franceschi C, Hirokawa K, Pawelec G (eds), The Netherlands: Springer-Verlag Press, pp. 1329–1352.
102. Föger-Samwald U, Vekszler G, Hörz-Schuch E, Salem S, Wipperich M, Ritschl P, Mousavi M, Pietschmann P. 2016. Molecular mechanisms of osteoporotic hip fractures in elderly women. *Exp Gerontol*, 73:49–58.
103. Takayanagi H. 2015. SnapShot: Osteoimmunology. *Cell Metab*, 21:502 el.
104. Buchwald ZS, Kiesel JR, DiPaolo R, Pagadala MS, Aurora R. 2012. Osteoclast activated FoxP3+ CD8+ T-cells suppress bone resorption in vitro. *PLoS One*, 7(6):e38199.
105. Semba RD, Nicklett EJ, Ferrucci L. 2010. Does accumulation of advanced glycation end products contribute to the aging phenotype? *J Gerontol A Biol Sci Med Sci*, 65A(9):963–975.
106. Rubinsztein DC, Marino G, Kroemer G. 2011. Autophagy and aging. *Cell*, 146:682–695.
107. Madeira E, Mafort TT, Madeira M et al. 2014. Lean mass as a predictor of bone density and microarchitecture in adult obese individuals with metabolic syndrome. *Bone*, 59:89–92.
108. Sun K, Liu J, Lu N, Sun H, Ning G. 2014. Association between metabolic syndrome and bone fractures: A meta-analysis of observational studies. *BMC Endocr Disord*, 14:13.
109. Kishida K, Tohru Funahashi T, Shimomura I. 2014. Adiponectin as a routine clinical biomarker. *Best Pract Res Clin Endocrinol Metab*, 28:119–130.
110. Cildir G, Akıncılar SC, Tergaonkar V. 2013. Chronic adipose tissue inflammation: All immune cells on the stage. *Trends Mol Med*, 19(8):487–500.
111. Pararasa C, Bailey CJ, Griffiths HR. 2015. Ageing, adipose tissue, fatty acids and inflammation. *Biogerontology*, 16(2):235–248.
112. D'Amelio P, Sassi F. 2016. Osteoimmunology: From mice to humans. *Bonekey Rep*, 5:802.
113. Schwarz P, Jørgensen NR, Abrahamsen B. 2014. Status of drug development for the prevention and treatment of osteoporosis. *Expert Opin Drug Discov*, 9:1–9.

114. Li JY, Walker LD, Tyagi AM, Adams J, Weitzmann MN, Pacifici R. 2014. The sclerostin-independent bone anabolic activity of intermittent PTH treatment is mediated by T-cell-produced Wnt10b. *J Bone Miner Res*, 29(1):43–54.
115. Weitzmann MN. 2013. The role of inflammatory cytokines, the RANKL/OPG axis, and the immuno-skeletal interface in physiological bone turnover and osteoporosis. *Scientifica*, 2013:125705.
116. Weitzmann MN. 2014. T-cells and B-cells in osteoporosis. *Curr Opin Endocrinol Diabetes Obes*, 21(6):461–467.
117. Mansoori MN, Tyagi AM, Shukla P et al. 2016. Methoxyisoflavones formononetin and isoformononetin inhibit the differentiation of Th17 cells and B-cell lymphopoesis to promote osteogenesis in estrogen-deficient bone loss conditions. *Menopause*, 23(5):565–576.
118. Li H, Lu Y, Qian J, Zheng Y, Zhang M, Bi E, He J, Liu Z, Xu J, Gao JY, Yi Q. 2014. Human osteoclasts are inducible immunosuppressive cells in response to T cell-derived IFN-γ and CD40 ligand in vitro. *J Bone Miner Res*, 29(12):2666–2675.
119. Mallon PW. 2014. Aging with HIV: Osteoporosis and fractures. *Curr Opin HIV AIDS*, 9:428–435.
120. Sućur A, Katavić V, Kelava T, Jajić Z, Kovačić N, Grčević D. 2014. Induction of osteoclast progenitors in inflammatory conditions: Key to bone destruction in arthritis. *Int Orthop*, 38(9):1893–1903.
121. Sigl V, Penninger JM. 2014. RANKL/RANK—From bone physiology to breast cancer. *Cytokine Growth Factor Rev*, 25(2):205–214.
122. Arai Y, Takayama M, Abe Y, Hirose N. 2011. Adipokines and aging. *J Atheroscler Thromb*, 18(7):545–550.

18 Manipulating Aging to Treat Age-Related Disease

V. Mallikarjun and J. Swift

CONTENTS

INTRODUCTION

Age is one of the highest risk factors for neurodegenerative, metabolic and inflammatory diseases, and cancer, as well as general frailty and infirmity that, while not currently classed as a disease, certainly impair quality of life. While aging may seem inexorable and unavoidable, over a century of biomedical research has suggested quite the opposite.[1–3] Such work has shown aspects of aging to be tractable, malleable processes that can be manipulated to prevent or reverse age-related disease progression. Similar to hypertension—a key risk factor for cardiovascular disease that is benign by itself—treating aging directly has the potential to improve both the quality and quantity of life in elderly individuals. This chapter will present some of the most promising attempts to prevent or reverse age-related disease including pharmacological manipulation of aging pathways and cell-based regenerative medicine (RM) therapies to reverse the progression of the various degenerative diseases of aging.

DRUGS ACTING ON BIOCHEMICAL PATHWAYS INVOLVED IN AGING

Several experimental models have shown that starvation without malnutrition (commonly known as dietary/caloric restriction: DR) can extend the life span in several model organisms.[4–6] However, the severest, typically most longevity-extending forms of DR are associated with several drawbacks, such as impaired wound healing and immune function,[7] that make it impractical for a person to follow, and undesirable due to negative effects on mood and physical activity levels. In mice, DR-mediated longevity extension is associated with lower morbidity and tumor incidence at older ages.[6] Importantly, a report on nonhuman primates suggested that DR had no positive effect on longevity.[8] However, a recent study suggested that the control primates in the Mattison et al.[8] study may have also been subjected to some level of DR and went on to show that DR did indeed have a positive effect on primate longevity.[4] Even before results from nonhuman primate studies were available, considerable efforts have been made to develop drugs that mimic the beneficial effects of DR while avoiding detrimental side effects. Notably, given the exorbitant cost of bring a new drug to market, there is considerable interest in repurposing existing clinically available drugs that have since been found to affect age-related pathways in addition to their intended mechanism (reviewed in Reference 9).

RAPAMYCIN AND mTOR INHIBITION

Key to the longevity-enhancing effects of DR is the inhibition of "growth-promoting" pathways such as insulin-like signaling and the mechanistic target of rapamycin (mTOR) pathways.[10–12] Inhibition of mTOR signaling upregulates cellular quality control mechanisms such as autophagy and proteasome-mediated protein turnover and is thought to improve protein and organelle homeostasis.[13,14] Rapamycin has been repeatedly shown to extend the murine life span[12] and is currently approved for clinical use as an immunosuppressant following organ transplantation, and recent evidence has shown that RAD001—a rapamycin analog—can actually improve immune function (as measured by the improved vaccination response) in the elderly.[15] Furthermore, inhibition of mTOR has been shown to alleviate loss of neurogenesis in mouse models of Parkinson's and Alzheimer's diseases.[16,17] However, rapamycin has been shown to have negative effects in obese diabetic mice,[18] demonstrating how the effects of mTOR inhibition can be modulated by genetic background, similar to those of DR itself.[19,20] Also, high-dose rapamycin feeding in a genetically heterogeneous mouse cohort resulted in significant testicular atrophy and cataracts.[21] Recently, the Dog Aging Project (DAP; dogagingproject.com) has been set up to assess how generalizable the effects of rapamycin treatment are in a genetically heterogeneous population in a variety of environments. DAP aims to investigate if rapamycin treatment has a beneficial effect on longevity in domestic dogs, reasoning that these dogs share the same environment as their human owners and so a positive effect observed in dogs would prove promising for applications to human longevity, or at least stimulate discussion regarding demographic effects on rapamycin-enhanced longevity. However, given the multifaceted response to DR and the subtly different response in longevity invoked by mTOR inhibition,[22] it is unlikely that the full beneficial response to DR can be completely recapitulated by only mTOR inhibition.

METFORMIN

Metformin is an antidiabetic drug that has been shown to have pleiotropic downstream effects that extend the murine life span.[23] Metformin is known to modulate glucose and energy homeostasis through regulation of $5'$ adenosine monophosphate-activated protein kinase (AMPK) activity.[24] However, others have shown that metformin can also affect many other pathways that can contribute to its antiaging and longevity-enhancing effects.[25,26] Recently, metformin has become the subject of a first-of-its-kind clinical trial (funded by the American Federation of Aging Research) to see if metformin can delay the progression of age-related diseases.[27] Importantly, the TAME (Targeting Aging with Metformin) study will set a benchmark in trial design and data analysis for any future potential antiaging drugs. Obviously, measuring life spans and constructing Kaplan–Meier survival curves is not possible in human cohorts within a practical time frame. However, as co-morbidities appear with increasing frequency in aging human populations, the TAME study aims to measure any effect on the aging process by measuring the time between appearance of new comorbidities in adults aged over 65 years. This approach could still require upwards of 10 years to measure significant effects. It remains to be seen if aging can be quantified through the development of more efficient biomarker-based approaches (see References 9, 28 for reviews covering the use of biomarkers for aging research).

SENOLYTICS (ABLATION OF SENESCENT CELLS)

It has recently been shown that ablation of senescent cells in aged mice can restore youthful function.[29] Senescence represents a specific cell fate wherein cells stop proliferation. The classic indicators of senescence include an enlarged, flat morphology, positive staining for senescence-induced β-galactosidase activity and the secretion of a panel of proinflammatory, extracellular-matrix (ECM) remodeling factors collectively known as SASP (senescence-associated secreted proteome).[30] Senescent cells are typically cleared by the immune system; however, they can persist in immunocompromised individuals such as the elderly where they become one of the major drivers of the chronic, sterile low-grade inflammatory syndrome that is characteristic of advancing age.

Senescence can be induced through telomere shortening following many replication cycles, DNA damage, oncogene activation, or mitochondrial dysfunction.[31,32] Baker et al.[29] have shown that global clearance of senescent cells from old mice can restore youthful function in some tissues, notably cardiac function, where ablation of senescent cells prevented cardiac hypertrophy in 18-month-old mice following treatment with isoproterenol, and also protected kidney and adipose tissue function. However, the authors note that removal of senescent cells did not reverse or prevent age-related decline in the liver and colon, suggesting that these tissues may be refractory to senescent cell clearance. Furthermore, Baker et al.[29] also showed that ablation of senescent cells impaired wound healing, in line with the established role of senescent cells and SASP in cell migration during wound healing. Furthermore, correct timing of appearance and clearance of senescent cells has been associated with the repeated flawless regeneration of amputated limbs in salamanders,[33] as well as wound healing in mammals.[34] These results suggest that greater increases in longevity could be achieved with more temporal control of senescent cell ablation. Other potential complementary strategies have been proposed which include manipulating SASP itself to provide fine-grain control over the beneficial short-term (cancer prevention and wound healing) and negative long-term (inflammation and impaired regeneration) effects of senescence.[35] Considerable effort is currently being imposed in discovering therapeutic strategies, such as senolytic drugs, to translate senescent cell ablation into the clinic. Interestingly, senescent cells have been shown to rely on specific anti-apoptotic BCL2 family member proteins to allow them to persist in aged tissues long after they would be cleared in a younger individual. Specific inhibition of BCL-W and BCL-XL by the small molecule ABT-737 specifically ablated senescent cells *in vitro* and *in vivo*.[36] These results demonstrate that senescent cell ablation *in vivo* on unmodified individuals is a feasible, promising strategy to prevent and reverse some age-related diseases.

ADDRESSING AGE-RELATED DISEASE WITH REGENERATIVE MEDICINE

RM involves the use of cells or tissues grown *in vitro* and implanted into a patient to replace cells lost due to aging or disease. Applying RM, specifically in stem cell (SC) therapies and tissue engineering, to the diseases of aging is an attractive strategy for two reasons: first, RM could restore function to aged/degenerated tissues either by restoring the ability of that tissue to repair itself or by replacing the tissue entirely. These goals can be achieved by manipulating endogenous SC populations,[37] or by introducing new SCs or even replacement tissues synthesized *in vitro* (tissue engineering).[38] Second, pharmacological approaches to tackle aging require sufficient knowledge to find target proteins and develop drugs to act on them. In contrast, RM is able to bypass much of this molecular and metabolic complexity by simply restoring/replacing tissues as they are lost, rather than trying to intervene in the complex mechanisms that cause them to be lost in the first place. The exact causes of aging are still debated, and completely retarding their progression would require a massive overhaul of human metabolism. This is well beyond currently available technology and of limited use to those already suffering from late-stage age-related diseases.

The ideal for therapeutic tissue engineering is to use SCs derived from the patient's own cells to craft autologous replacement tissues. However, for treating age-related disease, tissues will have to be derived from old individuals, which may present several methodological problems. A major determinant of physiological age of tissues is the chemical and mechanical signals cells within those tissues (particularly SCs) receive from their growth substrate.[39] Recent evidence suggests that signals cells receive from an aged environment can impair the efficacy of autologous RM in aged individuals.[38,40] The ECM provides structural support to which cells adhere and also forms the primary means through which cells receive these signals. Degeneration of the ECM structure and function is a feature of many age-related diseases, notably cardiovascular disease and sarcopenia (age-related loss of skeletal muscle mass), where skeletal and cardiac muscles, along with smooth arterial wall muscles, stiffen and become less elastic with age,[41] leading to structural weakness and increased chance of injury. Interestingly, Blau and colleagues were able to increase the diminished proliferative capacity of *ex vivo* muscle SCs derived from aged mice via use of biomaterials tailored to mimic the *in vivo* stiffness of muscle tissue and inhibition of p38α/β MAPK signaling.[38] This increased the ability of aged muscle SCs to restore function to an aged model of muscle injury, demonstrating that *ex vivo* correction of erroneous *in vivo* signaling events can facilitate the generation of physiologically youthful tissues for use in RM. Nonetheless, it is debatable whether cells that are metabolically compromised due to aging can be used to generate physiologically young tissues. For instance, type II diabetes is a common metabolic disease whose incidence increases with age. Peterson and colleagues have shown that mesenchymal SCs derived from diabetic mice and cultured *in vitro*, outside their original diabetic environment, proved therapeutically ineffective compared to SCs derived from control animals,[42] suggesting that a compromised metabolism may effectively "scar" cells, rendering them therapeutically nonviable.

While data from Blau and colleagues[38] suggest that defects resulting from incorrect signaling can be corrected *ex vivo*, correcting defects whose accumulation is a consequence of normal metabolic processes may be necessary in some cell types before SCs derived from these cells can be used to generate healthy tissues. For example, another potential pitfall, particularly when manufacturing larger tissues for use in humans, is the need to expand SC populations *in vitro* prior to implantation. Extensive *in vitro* expansion risks exhaustion of telomeres, especially in aged tissues where telomeres may already be significantly shortened and where relatively few SCs may be responsive to *ex vivo* corrective therapies (as was the case observed with inhibition of p38α/β MAPK signaling by Blau and colleagues[38]). Shortening of telomeres *in vitro* could therefore limit the capacity of SC therapies to replace dysfunctional tissues. Alternatively, use of induced pluripotent SCs (iPSCs)[43] could address issues with replicative capacity, but these cells have reported problems with *in vivo* tumorigenicity and teratogenicity. Another possibility would be to transiently induce telomerase (the enzyme responsible for extending telomeres, not normally expressed in somatic tissues) expression in culture during SC expansion prior to implantation.[44,45] Despite these concerns, the possibility of correcting both extrinsic and intrinsic defects through tailored biomaterials, genetics, and pharmacology remains promising, and is significantly simpler to engineer *ex vivo* than *in vivo*.

Applying RM to age-related degenerative diseases is, in principle, no different from applying it to other degenerative diseases. However, special concerns when creating autologous tissues from elderly individuals include addressing: (1) the dysfunction of cells isolated from aged tissues prior to tissue manufacture, and (2) of the aged niche that subsequent engineered tissues will inhabit. Aging is a systemic phenotype and any engineered tissues that are introduced into an old individual will have to exist within a systemically aged environment. Studies involving heterochronic parabiosis (a technique wherein two mice of different ages are made to share a circulatory system) have highlighted circulating factors, such as various members of the TGFβ family,[37,46,47] that can reverse or enhance the progression of aging through their promotion or inhibition of SC niche maintenance. Induction or repression of such factors could be employed to ensure that engineered tissues are not prematurely aged from introduction into an aged individual.

CONCLUSION

Currently, the most promising pharmacological approaches to treating aging and age-related disease target many features of aging, such as rapamycin and metformin acting to modulate energy, organelle and protein homeostasis, and senescent cell ablation seeking to alter intercellular signaling pathways by decreasing SASP-mediated signaling. The multifaceted, pleiotropic nature of the most promising treatments illustrates the multifaceted nature of aging itself. Furthermore, integrating information about many different pathways to produce drugs for treating aging directly with minimal side effects and optimizing RM to work better in aged individuals demonstrates the need for sophisticated systems biology to parse the complexity of aging systems. Finally, the complexity of treating aging on an individual basis demonstrates the acute need for individualized medicine[48] in cases where long-term medication may be necessary, as is the case with preventing or reversing age-related disease.

REFERENCES

1. Kenyon, C., Chang, J., Gensch, E., Rudner, A. and Tabtiang, R. A *C. elegans* mutant that lives twice as long as wild type. *Nature* **366**, 461–4, 1993.
2. Kirkwood, T. B. L. Understanding the odd science of aging. *Cell* **120**, 437–47, 2005.
3. López-Otín, C., Blasco, M. A., Partridge, L., Serrano, M. and Kroemer, G. The hallmarks of aging. *Cell* **153**, 1194–217, 2013.
4. Colman, R. J. et al. Caloric restriction reduces age-related and all-cause mortality in rhesus monkeys. *Nat. Commun.* **5**, 3557, 2014.
5. Lin, S.-J. et al. Calorie restriction extends *Saccharomyces cerevisiae* lifespan by increasing respiration. *Nature* **418**, 344–8, 2002.
6. Lee, C. K., Klopp, R. G., Weindruch, R. and Prolla, T. A. Gene expression profile of aging and its retardation by caloric restriction. *Science* **285**, 1390–3, 1999.
7. Kristan, D. M. Calorie restriction and susceptibility to intact pathogens. *Age (Omaha).* **30**, 147–56, 2008.
8. Mattison, J. A. et al. Impact of caloric restriction on health and survival in rhesus monkeys from the NIA study. *Nature* **489**, 318–21, 2012.
9. Mallikarjun, V. and Swift, J. Therapeutic manipulation of ageing: Repurposing old dogs and discovering new tricks. *EBioMedicine* **14**, 24–31, 2016.
10. Wang, P.-Y. et al. Long-lived Indy and calorie restriction interact to extend life span. *Proc. Natl. Acad. Sci. USA* **106**, 9262–7, 2009.
11. Bartke, A. and Brown-Borg, H. Life extension in the dwarf mouse. *Curr. Top. Dev. Biol.* **63**, 189–225, 2004.
12. Harrison, D. E. et al. Rapamycin fed late in life extends lifespan in genetically heterogeneous mice. *Nature.* **460**, 392–5, 2009.
13. Hill, C. et al. mTORC1 links protein quality and quantity control by sensing chaperone availability. *J. Biol. Chem.* **285**, 27385–95, 2010.
14. Yu, L. et al. Termination of autophagy and reformation of lysosomes regulated by mTOR. *Nature* **465**, 942–6, 2010.
15. Mannick, J. B. et al. mTOR inhibition improves immune function in the elderly. *Sci. Transl. Med.* **6**, 268ra179, 2014.
16. Malagelada, C., Jin, Z. H., Jackson-Lewis, V., Przedborski, S. and Greene, L. A. Rapamycin protects against neuron death in *in vitro* and *in vivo* models of Parkinson's disease. *J. Neurosci.* **30**, 1166–75, 2010.
17. Richardson, A., Galvan, V., Lin, A. L. and Oddo, S. How longevity research can lead to therapies for Alzheimer's disease: The rapamycin story. *Exp. Gerontol.* **68**, 51–8, 2015.
18. Sataranatarajan, K. et al. Rapamycin increases mortality in db/db mice, a mouse model of Type 2 diabetes. *J. Gerontol. A. Biol. Sci. Med. Sci.* **71**, 850–7, 2016.
19. Liao, C. Y., Rikke, B. A., Johnson, T. E., Diaz, V. and Nelson, J. F. Genetic variation in the murine lifespan response to dietary restriction: From life extension to life shortening. *Aging Cell.* **9**, 92–5, 2010.
20. Mulvey, L., Sinclair, A. and Selman, C. Lifespan modulation in mice and the confounding effects of genetic background. *J. Genet. Genomics.* **41**, 497–503, 2014.
21. Wilkinson, J. E. et al. Rapamycin slows aging in mice. *Aging Cell.* **11**, 675–82, 2012.
22. Garratt, M., Nakagawa, S. and Simons, M. J. P. Comparative idiosyncrasies in life extension by reduced mTOR signalling and its distinctiveness from dietary restriction. *Aging Cell.* **15**, 737–43, 2016.
23. Anisimov, V. N. et al. If started early in life, metformin treatment increases life span and postpones tumors in female SHR mice. *Aging (Albany. NY).* **3**, 148–57, 2011.
24. Shaw, R. J. et al. The kinase LKB1 mediates glucose homeostasis in liver and therapeutic effects of metformin. *Science* **310**, 1642–6, 2005.

25. Foretz, M. et al. Metformin inhibits hepatic gluconeogenesis in mice independently of the LKB1/AMPK pathway via a decrease in hepatic energy state. *J. Clin. Invest.* **120**, 2355–69, 2010.

26. Wheaton, W. W. et al. Metformin inhibits mitochondrial complex I of cancer cells to reduce tumorigenesis. *Elife* **2014**, 1–18, 2014.

27. Barzilai, N., Crandall, J. P., Kritchevsky, S. B. and Espeland, M. A. Metformin as a tool to target aging. *Cell. Metab.* **23**, 1060–65, 2016.

28. Martin-Ruiz, C. and von Zglinicki, T. Biomarkers of healthy ageing expectations and validation. *Proc. Nutr. Soc.* **73**, 422–9, 2014.

29. Baker, D. J. et al. Naturally occurring p16Ink4a-positive cells shorten healthy lifespan. *Nature* 1–20, 2016. doi:10.1038/nature16932.

30. Coppé, J.-P. et al. Senescence-associated secretory phenotypes reveal cell-nonautonomous functions of oncogenic RAS and the p53 tumor suppressor. *PLoS Biol.* **6**, 2853–68, 2008.

31. Wiley, C. D. et al. Mitochondrial dysfunction induces senescence with a distinct secretory phenotype. *Cell Metab.* **23**, 303–14, 2015.

32. Moiseeva, O., Bourdeau, V., Roux, A., Deschênes-Simard, X. and Ferbeyre, G. Mitochondrial dysfunction contributes to oncogene-induced senescence. *Mol. Cell. Biol.* **29**, 4495–507, 2009.

33. Yun, M. H., Davaapil, H. and Brockes, J. P. Recurrent turnover of senescent cells during regeneration of a complex structure. *Elife* **4**, 1–16, 2015.

34. Ritschka, B. et al. The senescence-associated secretory phenotype induces cellular plasticity and tissue regeneration. *Genes. Dev.* 1–12, 2017. Doi:10.1101/gad.290635.116

35. Malaquin, N., Martinez, A. and Rodier, F. Keeping the senescence secretome under control: Molecular reins on the senescence-associated secretory phenotype. *Exp. Gerontol.* **82**, 39–49, 2016.

36. Yosef, R. et al. Directed elimination of senescent cells by inhibition of BCL-W and BCL-XL. *Nat. Commun.* **7**, 11190, 2016.

37. Sinha, M. et al. Restoring systemic GDF11 levels reverses age-related dysfunction in mouse skeletal muscle. *Science* **344**, 649–52, 2014.

38. Cosgrove, B. D. et al. Rejuvenation of the muscle stem cell population restores strength to injured aged muscles. *Nat. Med.* **20**, 255–64, 2014.

39. Sun, Y. et al. Rescuing replication and osteogenesis of aged mesenchymal stem cells by exposure to a young extracellular matrix. *FASEB J.* **25**, 1474–85, 2011.

40. Fan, M. et al. The effect of age on the efficacy of human mesenchymal stem cell transplantation after a myocardial infarction. *Rejuvenation. Res.* **13**, 429–38, 2010.

41. Phillip, J. M., Aifuwa, I., Walston, J. and Wirtz, D. The mechanobiology of aging. *Annu. Rev. Biomed. Eng.* **17**, 113–41, 2015.

42. Shin, L. and Peterson, D. A. Impaired therapeutic capacity of autologous stem cells in a model of Type 2 diabetes. *Stem. Cells. Transl. Med.* **1**, 125–35, 2012.

43. Takahashi, K. and Yamanaka, S. Induction of pluripotent stem cells from mouse embryonic and adult fibroblast cultures by defined factors. *Cell.* **126**, 663–76, 2006.

44. Harley, C. B. et al. A natural product telomerase activator as part of a health maintenance program. *Rejuvenation Res.* **14**, 45–56, 2011.

45. Sahin, E. and Depinho, R. A. Linking functional decline of telomeres, mitochondria and stem cells during ageing. *Nature* **464**, 520–8, 2010.

46. Katsimpardi, L. et al. Vascular and neurogenic rejuvenation of the aging mouse brain by young systemic factors. *Science* **344**, 630–4, 2014.

47. Yousef, H. et al. Systemic attenuation of the TGF-β pathway by a single drug simultaneously rejuvenates hippocampal neurogenesis and myogenesis in the same old mammal. *Oncotarget.* **6**, 11959–78, 2015.

48. Topol, E. J. Individualized medicine from prewomb to tomb. *Cell.* **157**, 241–53, 2014.

19 Therapeutic Options to Enhance Poststroke Recovery in Aged Humans

Aurel Popa-Wagner, Dumbrava Danut, Roxana Surugiu, Eugen Petcu, Daniela-Gabriela Glavan, Denissa-Greta Olaru, and Raluca Sandu Elena

CONTENTS

INTRODUCTION

Stroke represents the second most common cause of death in Europe, and the third most common cause of death in the United States and Canada [1,2]. In demographically developed countries, the average age at which stroke occurs is around 73 years, reflecting the older age structure of these countries. The probability of a first stroke and first transitory ischemic attack is around 1.6 per 1000 and 0.42 per 1000, respectively. In less developed regions, the average age of stroke will be younger due to the different population age structure resulting from higher mortality rates and competing causes of death.

Stroke patients are at highest risk of death in the first weeks after the event, and between 20% and 50% die within the first month depending on type, severity, age, comorbidity, and effectiveness of treatment of complications. Patients who survive may be left with no disability or with mild, moderate, or severe disability. Considerable spontaneous recovery occurs up to about 6 months [3]. However, patients with a history of stroke are at risk of a subsequent event of around 10% in the first year and 5% per year thereafter [4].

The negative consequences of stroke extend well beyond the victims themselves, ultimately including families, caregivers, social networks, and employers. The proportion of patients achieving independence in self-care by 1 year after a stroke ranges from around 60% to 83%. This wide variation relates to whether the studies are community based or hospital based, which activities are considered in estimating independence, and the methods used to rate ability. In established marked economies (EMEs), depending on the organization of hospital services, between 10% and 15% of survivors are resident in an institution at 1 year [5].

AGE IS THE PRINCIPAL RISK FACTOR FOR STROKE

The incidence of stroke increases significantly with age both in men and women with incidence rates accelerating exponentially above 70 years [2]. However, there are gender differences in the incidence by age subgroups. Men aged up to 75 years are more likely to be hit by stroke than women. The risk of having a stroke then becomes higher in women than men aged 85 years or older [2]. This may be attributed to sex-related differences in life expectancy of women and the development of age-related atherosclerosis. It should be noted that the age-associated decline in functional reserve is most pronounced after the age of 85, and implies an impaired response to stressors and illnesses. Importantly, age-associated variability in functional reserve is due, in large part, to genetic and long-term lifestyle factors [6–8].

STROKE MODELS USING AGED ANIMALS ARE CLINICALLY MORE RELEVANT

Since stroke afflicts mostly the elderly comorbid patients, it is highly desirable to test the efficacy of cell therapies in an appropriate animal stroke model. Animal models of stroke often ignore age and comorbidities frequently associated with senescence, and this could be one of the explanations for unsuccessful bench-to-bedside translation of neuroprotective strategies.

Worldwide, stroke is increasing in parallel with modernization, changes in lifestyle, and the growing elderly population. In particular, rates in Eastern Europe have been increasing such that currently the highest rates are found in countries such as Bulgaria, Romania, and Hungary. Among women and men, individuals with a low-risk lifestyle (smoking, exercising daily, consuming a prudent diet including moderate alcohol, and having a healthy weight during midlife) had a significantly lower risk of stroke than individuals without a low-risk lifestyle. Therefore, the relatively high incidence of stroke may be due in part to the impact of numerous known risk factors: arterial hypertension, diabetes, high cholesterol, smoking, alcoholism, obesity, stress, and a sedentary lifestyle [9].

Comorbidities like diabetes and arterial hypertension, or comorbidity factors such as hypercholesterolemia, are common in elderly persons and are associated with a higher risk of stroke, increased mortality, and disability [10]. Moreover, the simultaneous presence of vascular diabetic complications associated comorbidities like hypertension and chronic diabetes significantly increased the level of ischemic damage [11].

Currently, there are several different rodent models with comorbidities such as the spontaneously hypertensive rat (SHR) model, spontaneously hypertensive rat stroke-prone (SHRSP), the Streptozocin rat model for diabetes, and the high-fat diet or high-sugar diet Sprague Dawley® rats. Diffusion (DWI)- and perfusion (PWI)-weighted magnetic resonance imaging of temporal changes in the ischemic penumbra in SHRSP showed increased infarct size early after stroke as compared to the normotensive WKY rats. Moreover, the infarct volume after 60 min MCAO was greater in SHRSP (36% ± 4% of hemisphere volume) than in SHR (19% ± 5%) or WKY normotensive rats (5% ± 2%) [12.

High blood pressure is a major risk factor for stroke. Large clinical trials have shown that ACE inhibitors reduce the incidence of stroke by up to 43% [13]. However, since normotensive patients also benefit from ACE inhibition, it has been suggested that these effects may also be independent of the blood pressure-lowering effects of ACE inhibition [14]. Indeed, neither short (7 days) nor long term (42 days) prior to stroke administration of ACE inhibitors to SHR reduced the infarct size despite lowering the blood pressure while WKY normotensive rats paradoxically showed marked reductions in infarct volume [15].

By MRI, hyperglycemia was also shown to accelerate infarct progression in cortical areas [16]. However, the mechanisms of hyperglycemia-associated infarct progression remain unclear. It could be that hyperglycemia-associated neuroinflammation aggravates brain infarction by hemorrhagic transformation leading to blood–brain barrier (BBB) disruption and neuronal cell death [17,18].

Our knowledge about the molecular and cellular mechanisms underlying accelerated infarct progression in subjects with metabolic syndrome is still poor. Some studies report a strong connection between nutrition and body weight, on the one hand, and increased oxidative stress or proinflammatory changes in the brain, which promotes neural imbalance and glucose level elevation, on the other [19]. Zhang et al. [20] suggested that metabolic inflammatory changes in the brain are linked to the IKK/NF-kB signaling pathway.

Observational studies have shown a strong correlation between blood lipid levels and stroke [21]. In animal models, it could be shown that VEGF-induced angiogenesis is compromised by hyperlipidemia and provided an explanation of poor efficacy of angiogenic therapies in patients suffering from hyperlipidemia [22].

Overnutrition and hypercholesterolemia may not only be responsible for metabolic inflammation of the brain but can also induce mitochondrial dysfunction and increased oxidative stress (NADPH oxidase dependent), which may promote metabolic syndrome and related diseases [19,23]. Cellular stress and hyperglycemia are known to accelerate the aging process. In this light, a better understanding of molecular factors and signaling pathways underlying the metabolic syndrome as well as the contribution of comorbidities to stroke-induced sequelae may be translated into more successful treatments or prevention therapies against age-associated diseases, which in turn would improve the life span and quality of life.

Effects of age and gender on stroke incidence, functional recovery, and mortality have not only been shown in humans but also in animal models [24,25]. Indeed, the age-dependent increase in the evolution of ischemic tissue into infarction strongly suggests that age is a biological marker for the variability in tissue outcome in acute human stroke [26].

Over the past 10 years, a variety of models of middle cerebral artery occlusion (MCAO) have been established in rodents [27]. MCAO in aged rodents has been produced with permanent or transient occlusion for 30–120 min using (i) MCA ligation after craniectomy [28]; (ii) intraluminal thread occlusion [29]; using a hook attached to a micromanipulator [30]; cauterization [31,32]; photothrombosis [33]; endothelin injection [34,35], injection of a thrombus via ECA [36], or intraluminal thrombus formation by thrombin injection (using occlusion of distal branches of the middle cerebral artery (MCA).

Since focal cerebral ischemia is technically difficult to perform in very old rats and since, based on epidemiological studies, human stroke occurs more often in late middle-aged (60–70 years old) subjects [37], it is advisable to use middle-aged instead of very old animals for stroke research [38].

SPONTANEOUS STROKE RECOVERY IN AGED PATIENTS AND ANIMALS

In clinical practice, physical therapy is widely used for stimulating poststroke brain recovery [39–41]. Recovery is thought to occur via recruitment of neighboring neuronal circuitries [42].

Stroke patients regain some of their lost neurological functions during the first weeks or months after the stroke. In contrast, in animal models of stroke, complete spontaneous recovery may occur in young rats, depending on the size and location of the ischemic lesion. However, stroke recovery is delayed and often incomplete in aged rats. Under normal conditions, young rats begin to show improvements in neurological function starting by day 2 post stroke, whereas in aged rats, neurological recovery is hardly detectable before days 4–5, with about 75% of the functional improvement observed in young rats [43].

It has been hypothesized that older brains may be more vulnerable in part because of decreased rates of compensatory oligodendrogenesis due to an age-related decline in cyclic AMP response element-binding protein (CREB)-mediated oligodendrogenesis after brain injury [44].

Housing experimental animals in an enriched environment enhances the recovery from brain damage both in young and aged animals [43]. When aged rats were allowed to recover in an enriched environment, the delay period was shortened and behavioral performance was significantly improved. The improvement in task performance positively correlated with slower infarct development, fewer proliferating astrocytes, and smaller size of the glial scar [43]. Even more effective rehabilitation of the contralateral forelimb could be achieved by Corbett and colleagues by combining the enriched environment with training [45].

Spontaneous recovery is common if the infarct is located in the striatum, a subcortical structure that exhibits activity-dependent plasticity and is important for controlling movement and motor learning. The enhanced recovery was associated with structural and synaptic plasticity in the contralesional striatum [46] (Figure 19.1). This may explain why patients with subcortical lacunar stroke are more likely to have early functional recovery after stroke [47,48]. Other studies suggest that the beneficial effect could be due to *in situ* secretion of neuroprotective factors by the transplanted cells. For example, human-derived inducible pluripotent cells (iPSCs) implanted into the striatum of young animals at 1 week after MCAO protected substantia nigra from atrophy, probably through a trophic effect [49].

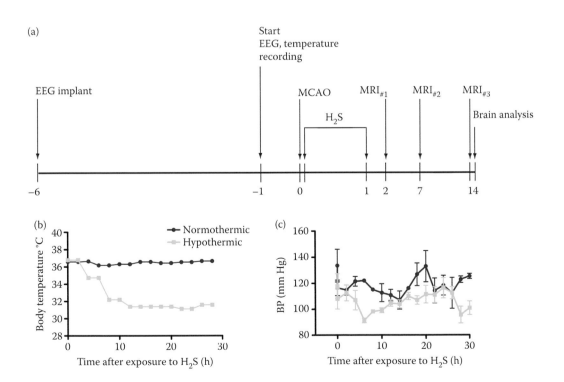

FIGURE 19.1 Stroke therapy by H_2S-induced hypothermia. (a) Schematic overview of the experimental design. (b) Time course of whole body cooling after stroke in aged rats immersed in an atmosphere containing 40 ppm H_2S. (c) Time course of arterial blood pressure (BP) in rats immersed in an atmosphere containing 40 ppm H_2S. BP dropped by 30% and reached a minimum after 8 h poststroke. The ups and downs of blood pressure in normothermic animals followed the circadian rhythmicity.

NON-PHARMACEUTICAL APPROACHES TO STROKE TREATMENT: GASEOUS HYPOTHERMIA

To minimize the incapacitating sequelae of stroke, a promising focus of research is on long-term neuroprotective strategies that minimize functional impairment by preventing the death of neurons, which continues for days to weeks following focal cerebral ischemia. A viable alternative to conventional drug-based therapies is physical cooling, or hypothermia, either confined to the head or including the entire body [50–53].

The feasibility of hypothermia (either by surface or endovascular cooling) has been addressed by several studies in stroke patients. Stroke patients were exposed for 6–24 h to mild hypothermia (in the range 33–35.5°C). Hypothermia was well tolerated but its clinical benefits are limited, especially in the long term as measured by the NIHSS (National Institutes of Health Stroke Scale) scores [54–56]. However, one study reported that the infarct volume was lower in hypothermia patients than in non-hypothermia patients [57]. More recently, it was shown that patients with ischemic stroke, who underwent mild hypothermia after recanalization, had less cerebral edema and showed improved clinical outcome [58,59].

The feasibility and clinical outcome after longer exposure to hypothermia (24 h) is ongoing in a large multicenter phase 3 randomized trial funded by the European Union (EuroHYP-1). In this ongoing study, researchers are attempting to mirror the significant improvement in clinical outcomes noted after reanimation in patients with cardiac arrest. EuroHYP-1 participants are randomly assigned to either hypothermia and medical treatment or best medical treatment alone for acute ischemic stroke [60].

Methods to achieve surface cooling include water mattresses, alcohol bathing, and ice packing or, more recently, using convective air blankets. Although feasible, maintenance of a constant body temperature using these approaches is difficult mainly because of temperature-dependent redistribution of blood flow due to vasoconstriction of superficial vasculature. Therefore, a simple method to pharmacologically induce long-term, regulated lowering of whole body temperature is highly desirable.

The concept of drug-induced cooling comes from the phenomenon of hibernation in some mammalian species which hibernate in a self-created hydrogen sulfide (H_2S) environment, which results in lowered body temperature and slower metabolism [61]. Experimentally, H_2S, a weak, reversible inhibitor of oxidative phosphorylation, induces a suspended animation-like state in mice by lowering body temperature to 15°C during an exposure time of 6 h at an ambient temperature of 13°C [61]. Using a modified version of this procedure, we previously showed that poststroke exposure of aged rats to H_2S-induced hypothermia for 48 h resulted in a 50% reduction in infarct volume without causing obvious neurological or physiological side effects [62,63].

Therapeutic hypothermia has yielded inconsistent results with regard to the relationship between depth of cooling and its effectiveness in reducing infarct volume and improving behavioral recovery. Thus, a systematic study showed that cooling at 34°C over 4 h poststroke was superior to other temperatures in the range 32–37°C [52,64,65]. Therefore, prolonged hypothermia (at least 24 h) and a longer survival time (at least 2 weeks) is a better study design to test the efficacy of hypothermia in experimental models and clinical trials. Along this line, prolonged cooling (33°C for 24 h and then 35°C for 24 h) started at 2.5 h after reperfusion prevented the contralateral limb impairment in food pellet retrieval and reduced the infarct volume by 40% in an experimental model [66].

Despite encouraging results from experiments with young animals, human stroke trials of neuroprotective factors which could be indicated after thrombolysis to limit infarct expansion and promote tissue recovery have not yielded satisfactory results [67]. One possible explanation for this discrepancy between laboratory and clinical outcomes is the role that age plays in the recovery of the brain from insult [68]. In this regard, the aged post-acute animal model is clinically most relevant to stroke rehabilitation [43,69–71].

Efficient neuroprotection requires a long-term regulated lowering of whole body temperature. Previously, we reported that 48 h poststroke exposures to H_2S effectively lowers whole body temperature and confers neuroprotection in aged animals [62,63]. Since the duration of hypothermia in most clinical trials was between 24 and 48 h, and in order to avoid the complications associated with longer exposure to hypothermia often seen in humans, we asked if a 24 h exposure to gaseous hypothermia, within the optimal hypothermia depth of 32°C, has the same neuroprotective efficacy as a 48 h exposure.

TWENTY-FOUR HOUR HYPOTHERMIA HAS TEMPORARY EFFICACY IN REDUCING BRAIN INFARCTION AND INFLAMMATION IN AGED RATS

Hypothermia is one of the very few therapeutic options for stroke. In animal studies of focal ischemia, short-term hypothermia often reduces infarct size. As an alternative approach to surface and endovascular cooling, gaseous hypothermia enables precise temperature control that is achieved by the simple inspiration of a mixture of air and H_2S. The disagreeable odor of H_2S can be circumvented by using injectable formulations that yield therapeutic doses of the gas; a number of these procedures are already in clinical trials [72]. In addition, a number of drugs that induce hypothermia have been reported recently [73].

In clinical trials, hypothermia has been applied for 2–48 h with various beneficial effects [74,75]. Therefore, the optimal time range of exposure to hypothermia yielding the most efficacious neuroprotection is not clear. While most authors agree that mild hypothermia (i.e., hypothermia range 32–34°C) provides optimal neuroprotection [73,76], the ideal hypothermic therapy in terms of depth and duration of mild hypothermia has to be established yet. For example, brief exposure (2–3 h) to postischemic hypothermia may appear quite effective with short survival times (less than 24 h). In an attempt to find an optimal therapeutic window in aged rats, animals were exposed for 24 h to an atmosphere containing 80 ppm hydrogen sulfide (H_2S) and 19.5% O_2, achieved by mixing room air with 5000 ppm H_2S-balanced nitrogen at a flow rate of 3 L/min. After 2 h, the concentration of H_2S was reduced to 40 ppm (the toxicity threshold for H_2S is 80 ppm) (Figure 19.1a). The temperature outside the experimental box was maintained at 21°C in a well-ventilated room. Water was available during this period, although no appetitive activity was observed. Carbon dioxide, oxygen, and H_2S were measured continuously using appropriate gas detectors (GfG, Dortmund, Germany) placed directly in the cage. Exposure of aged rats to an atmosphere containing a low dose of H_2S after stroke led to a gradual decrease in whole body temperature, which stabilized at 32 ± 0.5°C after 12 h (Figure 19.1b, filled circles). After 24 h, animals were returned to normal atmospheric conditions. The animals recovered within minutes and did not show any signs of neurological or physiological deficits related to H_2S-induced hypothermia.

During the first 24 h of exposure to hypothermia, the arterial blood pressure dropped by 30% and reached a minimum after 7 h poststroke and abolished the circadian rhythmicity seen in control animals (Figure 19.1c). After 7 h, the blood pressure started to normalize and reached control levels after 24 h poststroke. After 24 h, animals were returned to normal atmospheric conditions. However, 24 h of exposure to H_2S-induced hypothermia did not result in robust neuroprotection after stroke. On the contrary, after returning to normothermia, the inflammatory reaction resumes at an even higher pace so that by day 7, the infarct/edema volume exceeded that of controls, suggesting that neuronal death was simply postponed (Figure 19.2c vs. d). A longer exposure (48 h) was found to be more efficacious and resulted in a consistent reduction in infarct size and phagocytic activity at 2 and 14 d poststroke (Figure 19.3a vs. b; c vs. d).

Behaviorally, hypothermia had a limited beneficial effect on the bilateral sensorimotor coordination (inclined plane) and asymmetry of sensorimotor deficit (cylinder test) (Figure 19.4). Body weight regulation was significantly improved in hypothermic animals poststroke compared to animals kept at room temperature (Figure 19.4d). Finally, we noted a decreased number of phagocytic cells in hypothermia animals (Figure 19.5b vs. A). At day 7 after stroke, 79% of cells expressing

Normothermia   Hypothermia

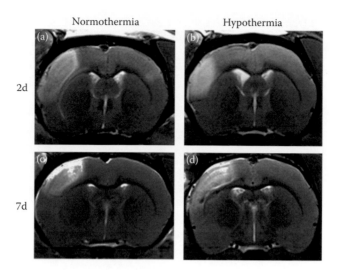

FIGURE 19.2 A short exposure (24 h) to mild hypothermia had a limited efficacy on the infarct volume in cooled animals. After a temporary reduction at day 2 after stroke (a vs. b), the infarct volume at day 7 was higher than in normothermic animals (d vs. c).

the polymorphonuclear (PMN) cells antigen also expressed ANXA1 (Figure 19.5c and d). At 14 d, poststroke ANXAA1 and ED1 co-localized in 47% of cells, in the infarct area (Figure 19.5e and f). After removing H_2S, the number of ANXA1-positive cells increased steadily (Figure 19.5c through f, h).

MECHANISMS OF HYPOTHERMIA-INDUCED PROTECTION AGAINST ISCHEMIC INJURY

A prerequisite for the successful translation of neuroprotective treatments from animal studies to clinical use is a more detailed understanding of the cellular and molecular mechanisms underlying the beneficial effects of long-term hypothermia for stroke treatment.

In spite of the obvious functional benefit and neuroprotective effect of reperfusion under mild hypothermic conditions, the mechanism of this treatment is not yet fully understood. The beneficial effects of hypothermia have been attributed to diminished excitotoxicity, neuroinflammation, apoptosis, free radical production, seizure activity, BBB disruption, blood vessel leakage, and/or cerebral thermopooling [77]. Other neuroprotective mechanisms could include increased neurogenesis and vascular density after stroke. Recently, we reported that the microenvironment of the injured, aged brain is fully capable of neurorestorative processes [78,79]. However, whether stroke stimulates endogenous neurogenesis is still debated, especially in aged subjects in whom neurogenesis is normally decreased [39,80].

Normal aging is characterized by a chronic, low-grade, proinflammatory state with an overexpression of systemic inflammatory factors, including proinflammatory cytokines [81,82]. Age-associated inflammation in the brain manifests primarily by the chronic activation of perivascular and parenchymal macrophages/microglia expressing proinflammatory cytokines and an increased number of astrocytes [83,84]. Therefore, it is not surprising that the age-associated proinflammatory state may contribute to the fulminant phagocytic activity of brain macrophages in the first 3 days poststroke associated with the rapid development of the infarct in aged animals [69].

However, 48 h of exposure to mild hypothermia resulted in drastic downregulation of annexin A1 (ANXA1), a major proinflammatory protein that is upregulated after stroke and is expressed by PMN cells in the perilesional area of the aged rat brain (Figure 19.5a vs b).

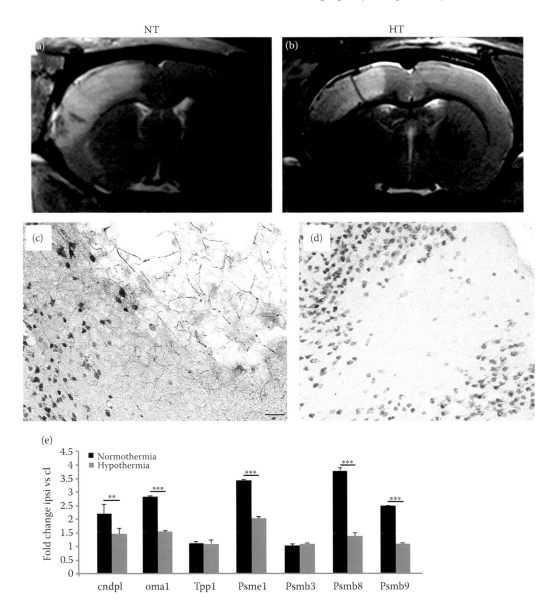

FIGURE 19.3 Forty-eight hours of exposure to mild hypothermia resulted in a consistent reduction in infarct size at 14 d poststroke. Thus, the cortical lesion, as defined by the region of T2 hyperintensity, was reduced by the hypothermic treatment (b) as compared with the normothermic group (a), by ~42%. Similar results have been achieved by immunocytochemistry (c vs d) showing tissue preservation in cooled aniamls (d). Tissue preservation in cooled animals was associated with large decreases (fold changes >2) in the transcripts coding for proteasome activator complex subunit 1 (Psme), proteasome subunit beta type-8 and type 9 (Psmb8, Psmb9) as well as significant decreases in the mRNA coding for Oma1, a mitochondrial metalloendopeptidase, and, to a lesser extent, Cndp1, a member of the M20 metalloprotease family, early after stroke (c).

At tissue level one, a clear effect of hypothermia was the preservation of the infarct core, suggesting that the phagocytic activity of microglia was diminished in the animals kept under hypothermic conditions in the first 2 days poststroke. At the transcription level, hypothermia caused a reduction in the mRNA coding for caspase 12, NF-kappa B, and grp78 in the peri-infarcted region, suggesting an overall decrease in the transcriptional activity related to inflammation and apoptosis [62].

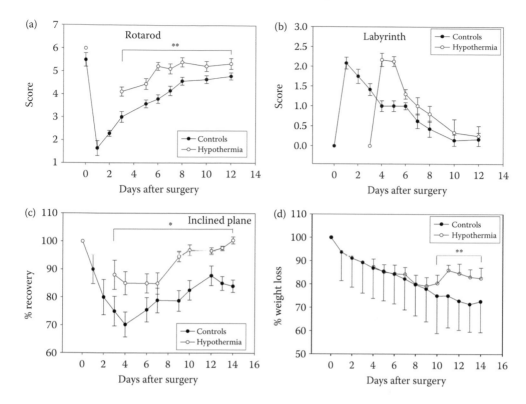

FIGURE 19.4   Behaviorally, the recovery of complex motor function (Rotarod) was significantly improved ($p < 0.03$) in animals kept under deep hypothermia for 48 h (a). In the labyrinth test, recovery of spatial learning based on positive reinforcement, as well as working and reference memory, did not differ significantly between the two groups (b). On the inclined plane, the beneficial effect (15% increase; $p < 0.05$) of hypothermia extended over the entire study period (c). Body weight regulation was significantly improved in hypothermic animals poststroke compared to animals kept at room temperature ($p < 0.025$) (d).

HYPOTHERMIA DIMINISHES THE EXPRESSION OF GENES CODING FOR PROTEASES

An exacerbated upregulation of genes coding for proteases in the aged brain after stroke may be one reason for the severity of the damage as well as the brain's resistance to neuroprotective therapies in the elderly. Prolonged exposure to H_2S-induced hypothermia reduced infarct volume after stroke. As assessed by NeuN immunohistochemistry, the infarct core was better preserved in aged hypothermic animals compared to the tissue from aged normothermic animals; we hypothesized that the expression of genes coding for proteases might be attenuated by hypothermia. Indeed, proteasome inhibition has been shown to induce long-term neuroprotection after cerebral ischemia that is associated with reduced infarct size, improved functional neurological deficits, decreased BBB breakdown, and enhanced angioneurogenesis [33,85–88]. Proteasome subunit beta type-1, 8, and 9 are essential proteins that contribute to the complete assembly of the 20S proteasome complex that recognizes degradable proteins, including stroke-damaged proteins, for protein quality control purposes.

We hypothesized that long-term exposure to hypothermia may attenuate the expression of genes encoding proteases and improve the odds of tissue preservation and functional rehabilitation in poststroke aged rats. To verify this hypothesis, poststroke aged rats were exposed to 2 days of hypothermia and the expression of several proteasome-related genes assessed by qRT-PCR. Indeed, it was found that two days of H_2S-induced hypothermia significantly lowers the expression of genes

Normothermia   Hypothermia

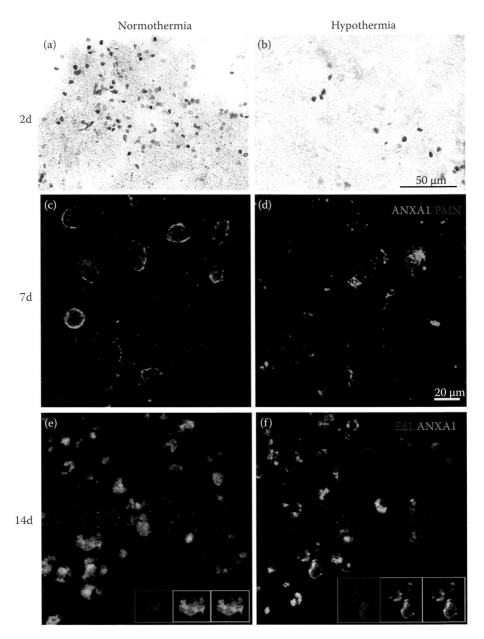

FIGURE 19.5   Forty-eight hours of exposure to H$_2$S-induced hypothermia decreased inflammatory responses to stroke and resulted in a consistent reduction in phagocytic activity at 2 d and 14 d poststroke. In the control group, approximately equal numbers of cells displayed activated microglia-like and macrophage-like phenotypes (a, inset), while in the hypothermia group, the activated microglia-like phenotype predominated (b, inset). The inset in (a) represents the phagocytic (macrophage-like) phenotype, and the inset in (b) represents the activated microglia phenotype. However, 24 hrs of exposure to H2S-induced hypothermia led to a decreased number of activated microglia (Iba1-positive) but an increased number of phagocytic cells (ANXA1 and ED1-positive cells) in the infarct area of hypothermic animals. Twenty-four h exposure to an H$_2$S atmosphere reduced the early poststroke expression of two inflammatory markers, annexin a1 (ANXA1) and ED1 (a, b). However, ANXA1- and ED1-positive cells did not overlap early after stroke (a, b, insets). At day 7 after stroke, 79% of cells expressing the PMN cells antigen also expressed ANXA1 (c, d). At 14 d poststroke, ANXAA1 and ED1 co-localized in 47% of cells in the infarct area (e, f). After removing H$_2$S, the number of ANXA1-positive cells increased steadily (c–f).

coding for proteasomal proteins early after stroke, enhances microvascular density, and improves indices of functional recovery at 2 weeks after stroke in aged rats [89] (Figure 19.3e).

HYPOTHERMIA ENHANCES POSTSTROKE ANGIOGENESIS IN AGED RATS

Mild hypothermia led to increased angiogenesis after focal ischemia in young-adult rats [90,91]. However, in this study, neovascularization was detected using a non-specific marker, fluorescein isothiocyanate–dextran, which does not distinguish between old and newly formed blood vessels. Using an anti-collagen IV specific marker, it could be shown that hypothermia is also effective in enhancing the density of the newly formed vascular network in the formerly infarcted area of aged rats during the recovery phase after stroke (Figure 6a, b, d).

HYPOTHERMIA DOES NOT STIMULATE NEUROGENESIS IN POSTSTROKE AGED RATS

Recent research has revealed that young-adult animals exposed to mild hypothermia show an increased endogenous repair capacity in the brain. In these subjects, the number of newly formed neurons in the dentate gyrus is higher compared to normothermic animals [92–94]. Likewise, the number of newly born striatal neurons in aged rodents after stroke was similar to that in young-adult rodents [95,96] despite a 50% decline in neurogenesis in the SVZ of older animals compared with young-adult animals [97]. There are very few studies relating hypothermia to neurogenesis. One study of adult rats showed that a short exposure to hypothermia (45 min) had no effect on neurogenesis in a rat model of forebrain ischemia [98].

Recently, we reported that the microenvironment of the injured, aged brain is fully capable of neurorestorative processes [78,79]. Furthermore, the relatively large number of DCX + positive cells amid large numbers of Iba1-immunoreactive-activated microglial cells in the SVZ and the injured area of aged rat brains suggests that activated microglia may not be detrimental to neurogenesis [99]. Hypothermia was not, however, beneficial to neurogenesis as indicated by the number of DCX$^+$ positive cells in the SVZ and infarct area (Figure 19.6c), suggesting that factors other than inflammation may block the beneficial effect of hypothermia on poststroke neurogenesis.

POSTSTROKE EEG AFTER HYPOTHERMIA IN AGED RATS

Although the first description of the hypometabolic state induced by H_2S was referred to as a state of "suspended animation," the effects on electrocortical activity were not investigated [61]. Thus, we found that following exposure to H_2S, the normal sleep-wake oscillations are replaced by a low-amplitude EEG dominated by a 4 Hz rhythmic activity, reminiscent of EEG recordings in hibernating animals [100]. Intriguingly, an increase in cortical oscillations close to 4 Hz was recently found to precede epileptic seizures in a rat model of absence epilepsy [101], raising the possibility that neocortical power-down by H_2S in our study arrested the development of seizures with an additional neuroprotective effect.

Early after MCAO in rats, several EEG abnormalities were identified including nonconvulsive seizures, periodic epileptiform discharges, and intermittent rhythmic delta activity [102]. From 7 to 14 days after MCAO, we did not find evidence of abnormalities in the background EEG, although we did not carry out high density recordings to explore changes localized to the affected hemisphere. We did find, however, a persistent alteration in the sleep cycles post MCAO, particularly those related to their circadian rhythm. This is not entirely new. Few previous studies reported various sleep alterations occurring during the first weeks following experimental MCAO in rodents [103,104]. Intriguingly, one of these studies reported altered circadian variability of sleep patterns [104] similar to postischemic impairment of other circadian rhythms reported in stroke patients [105] and experimental models [106]. The observed increase in wakefulness during the nocturnal

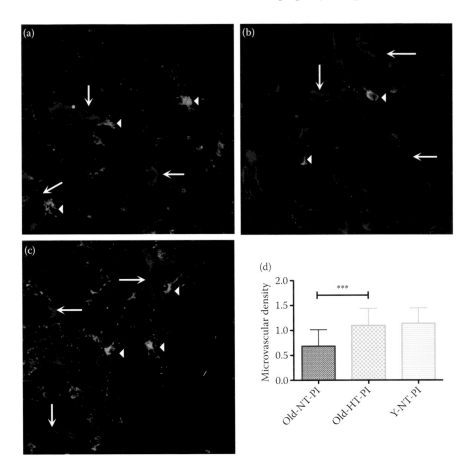

FIGURE 19.6 Hypothermia enhanced microvascular density but not neurogenesis in the infarcted area of aged rats. Note the scattered DCX-positive cells in the perilesional area of the normothermic rats (a, arrowheads) in the vicinity of new blood vessels (a, arrows, green). Hypothermia did not increase the number of DCX-positive cells (b, arrowheads and e) to the number shown by the young animals (e). However, in the perilesional cortex of aged rats subjected to hypothermia, we noted a significant increase (d; 1.8-fold, $p = 0.023$) in the microvascular density highlighted with anti-collagen IV antibodies (b, arrows). Remarkably, microvascular density in hypothermic aged rats was similar to that in normothermic young rats (c, d).

period after MCAO did not seem to recover from 7 to 14 days after stroke. This indicated that the impairment reflected a persistent injury, rather than a transient functional perturbation. Although post MCAO sleep changes were reportedly a poor indicator of infarct size [104], it is tempting to speculate that the sleep impairment reflected impaired cognitive functioning, which becomes especially relevant during the nocturnal period when rats are more active. Nevertheless, our data suggest that H_2S hypothermia did not prevent the occurrence of post MCAO circadian sleep impairment.

GRANULOCYTE-COLONY STIMULATING FACTOR AFTER STROKE IN AGED RATS

The hematopoietic factor G-CSF (granulocyte colony stimulating factor) effectively reduces infarct size and improves functional outcome after various types of experimental stroke [107–110]. Under ischemic conditions, G-CSF inhibits programmed neuronal cell death [111] and stimulates neural progenitor cell differentiation. These mechanisms and others, including immunomodulation and blood vessel plasticity, are currently thought to be responsible for infarct size reduction and improved functional outcome in young-adult rodent stroke models [112]. G-CSF exerts a wide range of potential effects [112] and can reduce the number of fatal hemorrhages after experimental thrombolysis [113]. It has been suggested that G-CSF exerts a therapeutic effect after stroke by anti-apoptotic properties and by reducing excitotoxicity-driven penumbral apoptosis [114]. The latter effect was considered strong enough to reduce the lesion size in young animals [115]. However, the aging brain is in need of increased glutamate signaling, which is reflected by the abundant expression of Na^+-dependent membrane glutamate transporters, particularly in white matter areas. This, in turn, renders the aging brain highly susceptible to ischemic excitotoxicity, swiftly exhausting mitochondrial capacities [116].

The broad preclinical body of evidence of G-CSF's efficacy in stroke studies is reflected by a close alignment along the STAIR recommendations. In fact, G-CSF is currently viewed to be one of the best preclinical studied candidate stroke drugs in recent years that was translated into clinical development [117]. One potential weakness of the preclinical dataset is, however, the lack of proof in aged subjects. It is in fact a general drawback of preclinical evaluations of candidate stroke drugs that, owing to cost-effectiveness and practicability, most studies were done in young animals. A lack of data from aged subjects in preclinical studies may, at least in part, explain the failure of candidate neuroprotective drugs in clinical trials. The aged brain has, compared to the young brain, an enhanced susceptibility to stroke and displays a limited recovery from an ischemic injury [30,32,43,69].

G-CSF treatment was recently shown to induce substantially more neural progenitor cells and immature neurons in subcortical regions adjacent to the infarcted area. G-CSF also increased neurogenesis in the dentate gyrus of the hippocampus [114]. This cell-regenerative effect in young-adult animals could be reconfirmed to some extent for the aged rats that have been treated with G-CSF for 14 days after stroke [118]. Although G-CSF treatment in aged rats increased the number of proliferating cells in the dentate gyrus and in the subventricular zone (SVZ), in aged rats there were more newborn neurons only in the SVZ of the damaged hemisphere. Likewise, G-CSF treatment in aged rats after stroke enhances survival, functional neurological recovery [118].

Despite some positive impact of G-CSF on the aged brain, apart from anti-excitotoxicity, this situation may have limited G-CSF monotherapy benefits, [118] although a remaining benefit was clearly shown. We tested the hypothesis that treating poststroke aged rats with the combination of bone marrow-derived mononuclear cells (BM MNC) and G-CSF improves the long-term (56 days) functional outcome by compensating the delay before G-CSF comes to full effect. To this end, 1×10^6 syngeneic BM MNC per kg bodyweight (BW) in combination with G-CSF (50 μg/kg, intraperitoneal application, continued for 28 days) were administered via the jugular vein to aged Sprague Dawley rats at 6 h poststroke (Figure 19.7a). Infarct volume was measured by magnetic resonance imaging at 3 and 48 days poststroke and additionally by immunohistochemistry at day 56 (Figure 19.7a). Functional recovery was tested during the entire poststroke survival period. Daily G-CSF treatment led to robust and consistent improvement of neurological function, but did not alter final infarct volumes. This result was unexpected since benefits of G-CSF and BM MNC treatment paradigms in stroke, independently from each other, have been repeatedly reported by independent experiments and groups and were hypothesized to work synergistically, especially in the aged, stroke-lesioned brain. The combination of G-CSF and BM MNC did not further improve poststroke recovery. The lack of an additional benefit may be due to a hitherto not well investigated interaction between both approaches and, to a minor extent, to the insensitivity of the aged brains to regenerative mechanisms. Also, considering recent findings on other tandem approaches involving G-CSF in animal models featuring relevant comorbidities, we conclude that such combination therapies are not the optimal approach to treat the acutely injured aged brain.

Current knowledge suggests that administered BM MNC provide indirect neuroprotection leading to infarct size reduction after ischemic damage in a time window of up to 1 month [119]. G-CSF, in turn, induces BM MNC mobilization while the SDF-1/CXCR4 system causes BM MNC to invade

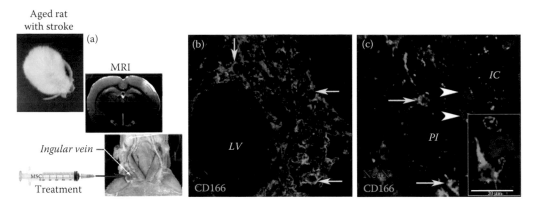

FIGURE 19.7 Cell therapy of stroke testing the hypothesis that treating poststroke aged rats with the combination of bone marrow-derived mononuclear cells (BM MNC) and G-CSF improves the long-term (56 days) functional outcome by compensating the delay before G-CSF comes to full effect. To this end, 1×10^6 syngeneic BM MNC per kg bodyweight (BW) in combination with G-CSF (50 μg/kg, intraperitoneal application, continued for 28 days) were administered via the jugular vein to aged Sprague Dawley® rats at 6 h poststroke. Note that some of the injected cells reached the lateral ventricle (b) and peri-infarcted area (c, inset).

the ischemic brain [109,120], where they are believed to exert therapeutic effects. However, the initiation of this potentially beneficial effect may take simply too much time. Although a granulocyte boost is seen after about 48 h, peaking of G-CSF-based mobilization can take up to 9 days [121], which is beyond the therapeutic time window for BM MNC. Since endogenous G-CSF is not available in sufficient concentrations directly after the ischemic event [115], a combination therapy providing (i) G-CSF in sufficient amounts to act as a neuroprotective agent and (ii) exogenous BM MNC early enough to bridge the time gap until G-CSF-based endogenous BM MNC mobilization comes to full effect seemed promising but failed to fulfill expectations.

One may assume that either the lesioned and aged rat brain environment was insensitive to regenerative mechanisms by BM MNC or cell treatment has been mainly ineffective. Indeed, the aggravated impact of ischemic damage on the aged brain is well known while potential detrimental effects of ageing on BM MNC have been anticipated [38,68,122]. Moreover, technical complications may come into play as well: a limiting influence of long-term cryopreservation on the therapeutic efficacy of umbilical cord blood MNC, a population very similar to that of BM MNC, has been discussed recently [123]. However, deriving syngeneic cells from young animals and limiting cryopreservation to no more than 4 weeks in our experiment might have limited such aging and cryopreservation effects on the donor side. Although a remaining impact cannot be excluded per se, a complete failure of the BM MNC treatment seems unlikely. An alternative explanation for the reduced efficacy of the combination treatment could be interference between both treatment regimens. A recent study in hypertensive animals demonstrated that intravenously administered BM MNC occupies splenic granulocyte clearance capacities for apoptotic cells [124]. This clearance system usually removes apoptotic granulocytes from the circulation, which represents an important anti-inflammatory mechanism. Being already compromised by externally administered BM MNC, the swift and early granulocyte boost from the BM by G-CSF may have completely exhausted the clearance system in our treatment scenario. This detrimental interaction may have caused a sustained systemic and central proinflammatory bias, leading to subtle additional damage, not enhancing but partly reducing the neuroprotective G-CSF effect. It remains for further investigation to determine whether this interaction is even more relevant in the aged brain.

False-negative results are a common phenomenon when selected sample sizes are too small to reveal small-scale treatment effects with the G-CSF and G-CSF + BM MNC group outcomes just differing randomly from each other. To prevent such scenarios, we chose relatively large samples (n = 21 animals per group). This group strength is large enough to detect an effect size of 20%, which is recommended for experimental stroke therapies [125]. We therefore consider that insufficient study power is not very likely to have "masked" a positive effect of G-CSF + BM MNC.

FUNCTIONAL NEUROLOGICAL RECOVERY AND TISSUE REPAIR AFTER NEURAL TISSUE TRANSPLANTATION

Although rehabilitation is important for improving functional recovery in the early stages after stroke, it does not provide a replacement of lost tissue. However, if and how the aged brain responds to grafted neural tissue is largely unknown. For example, mouse fetal hippocampal NSCs implanted into the injured hippocampus of 24-month-old rats exhibited limited neuronal plasticity, robust astrocytic differentiation, and impaired migration [126].

Most clinical studies conducted so far used neural cells derived from human fetal donors. The techniques to achieve effective survival and growth of neuronal tissues transplanted into the CNS are meanwhile well established [127]. Even though effective, neural grafting has, however, not become a standard treatment for several reasons, including the limited supply of fetal tissue of human origin, and the beneficial effects have been controversial [128]. Of the various options, stem cell therapy presents us with a viable alternative [129].

In order to enable the replacement of lost tissue, cell replacement strategies were used in human stroke patients [130,131]. These early clinical studies lacked appropriate control groups.

STEM CELL THERAPY IN SUBCORTICAL STROKE: ROLE OF ENDOGENOUS NEUROGENESIS AND AGING

The ultimate goal of stroke treatment is restoration of neurological function. Stroke is a heavily undertreated disease demanding a vigorous search for new therapies.

Despite improving knowledge about stroke pathology, therapeutical benefits for stroke patients are limited. Distinguished by a necrotic core surrounded by the ischemic area (penumbra), stroke is still the largest cause of disability in stroke survivors. Crucial for the recovery phase are the first days and weeks after stroke. Previous studies have shown that neuroplasticity allows for brain "remodeling" even years after stroke. However, despite the recent progression in stroke research, the major problems to be solved for stroke survivors remain the restorative process.

Spontaneous recovery is common whenever the infarct is located in the striatum, a subcortical structure that exhibits activity-dependent plasticity and is important for controlling movement and motor learning. Neurological recovery is associated with structural dendritic and synaptic plasticity in the contralesional striatum [46] and axonal plasticity in the contralesional motor cortex [132], which may explain why patients with subcortical stroke are likely to exhibit functional neurological recovery [47,48].

Cell-based therapy augments this endogenous response. Thus, human iPSCs implanted into the striatum of young-adult animals at 1 week after MCAO protected substantia nigra from atrophy, probably through a trophic effect via the release of survival-promoting growth factors [49]. However, how cells are transplanted and where they are placed after stroke are important issues in graft survival and efficacy in promoting behavioral recovery. Data from many groups have shown that stroke increases proliferation of neuronal progenitors in the ipsilateral subventricular region of young-adult rodents with a maximum at 1–2 weeks, and the newly generated neuroblasts migrate to the damaged area in the peri-infarcted striatum over a period of several months. Eventually, the neuroblasts differentiate into medium-sized spiny neurons and may become part of the neuronal network [133–137]. It seems that the injected cells themselves can also stimulate neurogenesis in the SVZ [33,136].

Here, we demonstrated that intracerebral transplantation of NSI-566RSC, a spinal cord-derived NSC line, at two sites within the striatum reduced behavioral deficits associated with ischemic stroke. Significant improvements in both motor and neurological tests were detected in the NSI-566RSC-treated stroke animals. In addition, the results revealed significant dose-dependent differences in the behavioral improvement across treatment groups at post-transplantation periods with the highest NSI-566RSC dose showing the most significant improvement in both motor and neurological tests.

These results demonstrated safety and efficacy of NSI-566RSC in a subacute model of ischemic stroke in rats [138].

However, the proportion of surviving neurons is discouragingly low [133,139,140]. In animal models, the number of new striatal neurons in aged rodents after stroke was similar to that in young animals [95,96], despite a 50% decline in neurogenesis in the SVZ of elderly rodents compared to young-adult animals [97,141].

Similar findings have been reported in humans [135,142–144]. Earlier studies on postmortem human brains provided evidence that there might be SVZ cell proliferation and neuroblast formation after stroke even in aged patients [135,143,145]. The finding that new neurons are continuously added in the adult human striatum [146], along with the presence of an increased number of putative neuroblasts in the human striatum after stroke, lends support to this hypothesis [143]. However, whether endogenous neurogenesis contributes to spontaneous recovery after stroke has not yet been established. In addition, age, comorbidities, physical condition of the patient, and severity of disease could substantially influence these steps and, therefore, the outcome of the healing process.

The establishment of induced pluripotent stem cells (iPSCs) offers new prospects for stroke treatment. iPSCs can be generated from a patient, avoiding both ethical problems and immune rejection and a limited differentiation potential of adult stem cells [147]. However, if and how the aged brain responds to grafted cells is largely unknown. The experiments done so far have yielded conflicting results. For example, mouse fetal hippocampal NSCs implanted into the injured hippocampus of 24-month-old rats exhibited limited neuronal plasticity, robust astrocytic differentiation, and impaired migration. Yet another study using NSCs transplanted into young-adult (3-month-old) and aged (24-month-old) rat brains at 1 day after stroke reportedly reduced ischemic brain injury in aged rats [148]. In stroke models, hiPSC-lt-NES cells derived from a young-adult male have the potential to survive, differentiate into immature and mature neurons, and migrate to the peri-infarct area of aged rats. The treated aged rats showed improved behavioral recovery after implantation into the stroke-injured striatum and cortex of adult rats [149,150]. In a recent study, it could be shown that human iPSC survived and differentiated into neurons after intracortical transplantation in aged rats with cortical stroke and also improved functional recovery in cylinder tests at 4 and 7 weeks [79].

Recent studies indicate that iPSCs can also be generated from aged humans and differentiate into specific cell types [79,151,152]. Moreover, it seems that the re-differentiation efficiency of human fibroblasts via iPSCs into functional motor neurons is the same as in 29–82-year-old individuals [153].

CONCLUSIONS

It thus appears that H_2S-induced hypothermia has a pleiotropic effect by: (i) reducing the metabolic rate, (ii) inducing a hibernation-like state; (iii) reducing the epileptic forms of EEG activity; and (iv) inducing a sleep-deprivation state. Therefore, the ability of ANXA1 to control and contain inflammation may play a pivotal role in postischemic recovery.

Our results suggest that H_2S-induced hypothermia, via targeting multiple points of intervention, could have a higher probability of success in treating stroke as compared to other monotherapies. Therefore, a better understanding of the pathophysiology of the ischemic injury processes on which hypothermia and H_2S act will serve to further promote the use of this promising method to reduce mortality and morbidity caused by stroke. However, the optimal conditions for therapeutic hypothermia, such as temperature and the initiation and duration of cooling, must be individualized [154]. We noted that the potential for neurogenesis is also preserved in aged, stroke-injured brains and the environment of the aged brain is not hostile to cell therapies. There remain significant developmental and translational issues that remain to be resolved in future studies such as (i) understanding the differentiation into specific phenotypes. Upon transplantation, the differentiated cells often de-differentiate [155]; (ii) tumorigenesis remains a significant concern [156]; (iii) anti-neuroinflammatory therapies are a potential target to promote regeneration and repair in diverse injury and neurodegenerative conditions by stem cell therapy; (iv) efficacy of cell therapy can be enhanced by physical rehabilitation [127]. We recommend that, in a real clinical practice involving older poststroke patients, successful regenerative therapies would have to be carried out for a much longer time. The BM MSC therapy in aged rodents warrants further investigation including repeated administrations of therapeutic cells at several time points after stroke and using various combinations with G-CSF or other relevant growth factors/cytokines.

Finally, a better understanding of potential risks of stem cell therapies in stroke shall make the translation of cell therapies safer. Likewise, awareness of potential risks of stem cell therapies may help improve their efficacy in order to achieve therapeutic goals [157].

REFERENCES

1. Lloyd-Jones DM. Cardiovascular risk prediction: Basic concepts, current status, and future directions. *Circulation.* 2010;121(15):1768–77.
2. Roger VL, Go AS, Lloyd-Jones DM, Benjamin EJ, Berry JD, Borden WB et al. Heart disease and stroke statistics—2012 update: A report from the American Heart Association. *Circulation.* 2012;125(1):e2–e220.
3. Bonita R, Beaglehole R. Recovery of motor function after stroke. *Stroke.* 1988;19(12):1497–500.
4. Burn J, Dennis M, Bamford J, Sandercock P, Wade D, Warlow C. Long-term risk of recurrent stroke after a first-ever stroke. *The Oxfordshire Community Stroke Project. Stroke.* 1994;25(2):333–7.
5. Appelros P, Nydevik I, Viitanen M. Poor outcome after first-ever stroke: Predictors for death, dependency, and recurrent stroke within the first year. *Stroke.* 2003;34(1):122–6.
6. Tacutu R, Budovsky A, Fraifeld VE. The NetAge database: A compendium of networks for longevity, age-related diseases and associated processes. *Biogerontology.* 2010;11(4):513–22.
7. Tacutu R, Budovsky A, Yanai H, Fraifeld VE. Molecular links between cellular senescence, longevity and age-related diseases—A systems biology perspective. *Aging (Albany NY).* 2011;3(12):1178–91.
8. Wolfson M, Budovsky A, Tacutu R, Fraifeld V. The signaling hubs at the crossroad of longevity and age-related disease networks. *Int J Biochem Cell Biol.* 2009;41(3):516–20.
9. Donnan GA, Davis SM. Breaking the 3 h barrier for treatment of acute ischaemic stroke. *Lancet Neurol.* 2008;7(11):981–2.
10. Goldstein LB, Bushnell CD, Adams RJ, Appel LJ, Braun LT, Chaturvedi S et al. Guidelines for the primary prevention of stroke: A guideline for healthcare professionals from the American Heart Association/American Stroke Association. *Stroke.* 2011;42(2):517–84.
11. Rewell SS, Fernandez JA, Cox SF, Spratt NJ, Hogan L, Aleksoska E et al. Inducing stroke in aged, hypertensive, diabetic rats. *J Cereb Blood Flow Metab.* 2010;30(4):729–33.
12. McCabe C, Gallagher L, Gsell W, Graham D, Dominiczak AF, Macrae IM. Differences in the evolution of the ischemic penumbra in stroke-prone spontaneously hypertensive and Wistar-Kyoto rats. *Stroke.* 2009;40(12):3864–8.
13. Yusuf S. After the HOPE Study. ACE inhibitor now for every diabetic patient? Interview by Dr. Dirk Einecke. *MMW Fortschr Med.* 2000;142(44):10.
14. Sleight P. The role of angiotensin-converting enzyme inhibitors in the treatment of hypertension. *Curr Cardiol Rep.* 2001;3(6):511–8.
15. Porritt MJ, Chen M, Rewell SS, Dean RG, Burrell LM, Howells DW. ACE inhibition reduces infarction in normotensive but not hypertensive rats: Correlation with cortical ACE activity. *J Cereb Blood Flow Metab.* 2010;30(8):1520–6.
16. Martin A, Rojas S, Chamorro A, Falcon C, Bargallo N, Planas AM. Why does acute hyperglycemia worsen the outcome of transient focal cerebral ischemia? Role of corticosteroids, inflammation, and protein O-glycosylation. *Stroke.* 2006;37(5):1288–95.
17. Kumari S, Anderson L, Farmer S, Mehta SL, Li PA. Hyperglycemia alters mitochondrial fission and fusion proteins in mice subjected to cerebral ischemia and reperfusion. *Transl Stroke Res.* 2012;3(2):296–304.

18. Soejima H, Ogawa H, Morimoto T, Nakayama M, Okada S, Sakuma M et al. Aspirin possibly reduces cerebrovascular events in type 2 diabetic patients with higher C-reactive protein level: Subanalysis from the JPAD trial. *J Cardiol*. 2013;62(3):165–70.

19. Cai D, Liu T. Inflammatory cause of metabolic syndrome via brain stress and NF-kappaB. *Aging (Albany NY)*. 2012;4(2):98–115.

20. Zhang X, Zhang G, Zhang H, Karin M, Bai H, Cai D. Hypothalamic IKKbeta/NF-kappaB and ER stress link overnutrition to energy imbalance and obesity. *Cell*. 2008;135(1):61–73.

21. Iso H, Jacobs DR, Jr., Wentworth D, Neaton JD, Cohen JD. Serum cholesterol levels and six-year mortality from stroke in 350,977 men screened for the multiple risk factor intervention trial. *N Engl J Med*. 1989;320(14):904–10.

22. Zechariah A, ElAli A, Doeppner TR, Jin F, Hasan MR, Helfrich I et al. Vascular endothelial growth factor promotes pericyte coverage of brain capillaries, improves cerebral blood flow during subsequent focal cerebral ischemia, and preserves the metabolic penumbra. *Stroke*. 2013;44(6):1690–7.

23. Herz J, Hagen SI, Bergmuller E, Sabellek P, Gothert JR, Buer J et al. Exacerbation of ischemic brain injury in hypercholesterolemic mice is associated with pronounced changes in peripheral and cerebral immune responses. *Neurobiol Dis*. 2014;62:456–68.

24. Bergerat A, Decano J, Wu CJ, Choi H, Nesvizhskii AI, Moran AM et al. Prestroke proteomic changes in cerebral microvessels in stroke-prone, transgenic[hCETP]-hyperlipidemic, Dahl salt-sensitive hypertensive rats. *Mol Med*. 2011;17(7–8):588–98.

25. Gokcay F, Arsava EM, Baykaner T, Vangel M, Garg P, Wu O et al. Age-dependent susceptibility to infarct growth in women. *Stroke*. 2011;42(4):947–51.

26. Ay H, Koroshetz WJ, Vangel M, Benner T, Melinosky C, Zhu M et al. Conversion of ischemic brain tissue into infarction increases with age. *Stroke*. 2005;36(12):2632–6.

27. Bacigaluppi M, Pluchino S, Peruzzotti-Jametti L, Kilic E, Kilic U, Salani G et al. Delayed post-ischaemic neuroprotection following systemic neural stem cell transplantation involves multiple mechanisms. *Brain*. 2009;132(Pt 8):2239–51.

28. Wang LC, Futrell N, Wang DZ, Chen FJ, Zhai QH, Schultz LR. A reproducible model of middle cerebral infarcts, compatible with long-term survival, in aged rats. *Stroke*. 1995;26(11):2087–90.

29. Sutherland GR, Dix GA, Auer RN. Effect of age in rodent models of focal and forebrain ischemia. *Stroke*. 1996;27(9):1663–7;discussion 8.

30. Popa-Wagner A, Schroder E, Walker LC, Kessler C. Beta-amyloid precursor protein and ss-amyloid peptide immunoreactivity in the rat brain after middle cerebral artery occlusion: Effect of age. *Stroke*. 1998;29(10):2196–202.

31. Katsman D, Zheng J, Spinelli K, Carmichael ST. Tissue microenvironments within functional cortical subdivisions adjacent to focal stroke. *J Cereb Blood Flow Metab*. 2003;23(9):997–1009.

32. Rosen CL, Dinapoli VA, Nagamine T, Crocco T. Influence of age on stroke outcome following transient focal ischemia. *J Neurosurg*. 2005;103(4):687–94.

33. Zhang RL, Chopp M, Roberts C, Jia L, Wei M, Lu M et al. Ascl1 lineage cells contribute to ischemia-induced neurogenesis and oligodendrogenesis. *J Cereb Blood Flow Metab*. 2011;31(2):614–25.

34. Soleman S, Yip P, Leasure JL, Moon L. Sustained sensorimotor impairments after endothelin-1 induced focal cerebral ischemia (stroke) in aged rats. *Exp Neurol*. 2010;222(1):13–24.

35. Trueman RC, Harrison DJ, Dwyer DM, Dunnett SB, Hoehn M, Farr TD. A critical re-examination of the intraluminal filament MCAO model: Impact of external carotid artery transection. *Transl Stroke Res*. 2011;2(4):651–61.

36. DiNapoli VA, Huber JD, Houser K, Li X, Rosen CL. Early disruptions of the blood–brain barrier may contribute to exacerbated neuronal damage and prolonged functional recovery following stroke in aged rats. *Neurobiol Aging*. 2008;29(5):753–64.

37. Feigin V, Anderson N, Gunn A, Rodgers A, Anderson C. The emerging role of therapeutic hypothermia in acute stroke. *Lancet Neurol*. 2003;2(9):529.

38. Popa-Wagner A, Carmichael ST, Kokaia Z, Kessler C, Walker LC. The response of the aged brain to stroke: Too much, too soon? *Curr Neurovasc Res*. 2007;4(3):216–27.

39. Hermann DM, Chopp M. Promoting brain remodelling and plasticity for stroke recovery: Therapeutic promise and potential pitfalls of clinical translation. *Lancet Neurol*. 2012;11(4):369–80.

40. Honmou O, Onodera R, Sasaki M, Waxman SG, Kocsis JD. Mesenchymal stem cells: Therapeutic outlook for stroke. *Trends Mol Med*. 2012;18(5):292–7.

41. Liepert J, Hamzei F, Weiller C. Lesion-induced and training-induced brain reorganization. *Restor Neurol Neurosci*. 2004;22(3–5):269–77.

42. Hallett M. Plasticity of the human motor cortex and recovery from stroke. *Brain Res Brain Res Rev.* 2001;36(2–3):169–74.

43. Buchhold B, Mogoanta L, Suofu Y, Hamm A, Walker L, Kessler C et al. Environmental enrichment improves functional and neuropathological indices following stroke in young and aged rats. *Restor Neurol Neurosci.* 2007;25(5–6):467–84.

44. Miyamoto N, Pham LD, Hayakawa K, Matsuzaki T, Seo JH, Magnain C et al. Age-related decline in oligodendrogenesis retards white matter repair in mice. *Stroke.* 2013;44(9):2573–8.

45. Hicks AU, Hewlett K, Windle V, Chernenko G, Ploughman M, Jolkkonen J et al. Enriched environment enhances transplanted subventricular zone stem cell migration and functional recovery after stroke. *Neuroscience.* 2007;146(1):31–40.

46. Qin L, Jing D, Parauda S, Carmel J, Ratan RR, Lee FS et al. An adaptive role for BDNF Val66Met polymorphism in motor recovery in chronic stroke. *J Neurosci.* 2014;34(7):2493–502.

47. Bejot Y, Catteau A, Caillier M, Rouaud O, Durier J, Marie C et al. Trends in incidence, risk factors, and survival in symptomatic lacunar stroke in Dijon, France, from 1989 to 2006: A population-based study. *Stroke.* 2008;39(7):1945–51.

48. Rothrock JF, Clark WM, Lyden PD. Spontaneous early improvement following ischemic stroke. *Stroke.* 1995;26(8):1358–60.

49. Polentes J, Jendelova P, Cailleret M, Braun H, Romanyuk N, Tropel P et al. Human induced pluripotent stem cells improve stroke outcome and reduce secondary degeneration in the recipient brain. *Cell Transplant.* 2012;21(12):2587–602.

50. Esposito E, Ebner M, Ziemann U, Poli S. In cold blood: Intraarterial cold infusions for selective brain cooling in stroke. *J Cereb Blood Flow Metab.* 2014;34(5):743–52.

51. Hennerici MG, Kern R, Szabo K. Non-pharmacological strategies for the treatment of acute ischaemic stroke. *Lancet Neurol.* 2013;12(6):572–84.

52. Kollmar R, Blank T, Han JL, Georgiadis D, Schwab S. Different degrees of hypothermia after experimental stroke: Short- and long-term outcome. *Stroke.* 2007;38(5):1585–9.

53. Wu TC, Grotta JC. Hypothermia for acute ischaemic stroke. *Lancet Neurol.* 2013;12(3):275–84.

54. Hemmen TM, Raman R, Guluma KZ, Meyer BC, Gomes JA, Cruz-Flores S et al. Intravenous thrombolysis plus hypothermia for acute treatment of ischemic stroke (ICTuS-L): Final results. *Stroke.* 2010;41(10):2265–70.

55. Kammersgaard LP, Rasmussen BH, Jorgensen HS, Reith J, Weber U, Olsen TS. Feasibility and safety of inducing modest hypothermia in awake patients with acute stroke through surface cooling: A case-control study: The Copenhagen Stroke Study. *Stroke.* 2000;31(9):2251–6.

56. Wan YH, Nie C, Wang HL, Huang CY. Therapeutic hypothermia (different depths, durations, and rewarming speeds) for acute ischemic stroke: A meta-analysis. *J Stroke Cerebrovasc Dis.* 2014;23(10):2736–47.

57. De Georgia MA, Krieger DW, Abou-Chebl A, Devlin TG, Jauss M, Davis SM et al. Cooling for acute ischemic brain damage (COOL AID): A feasibility trial of endovascular cooling. *Neurology.* 2004;63(2):312–7.

58. Hong JM, Lee JS, Song HJ, Jeong HS, Choi HA, Lee K. Therapeutic hypothermia after recanalization in patients with acute ischemic stroke. *Stroke.* 2014;45(1):134–40.

59. Piironen K, Tiainen M, Mustanoja S, Kaukonen KM, Meretoja A, Tatlisumak T et al. Mild hypothermia after intravenous thrombolysis in patients with acute stroke: A randomized controlled trial. *Stroke.* 2014;45(2):486–91.

60. van der Worp HB, Macleod MR, Bath PM, Demotes J, Durand-Zaleski I, Gebhardt B et al. EuroHYP-1: European multicenter, randomized, phase III clinical trial of therapeutic hypothermia plus best medical treatment vs. best medical treatment alone for acute ischemic stroke. *Int J Stroke.* 2014;9(5):642–5.

61. Blackstone E, Morrison M, Roth MB. H2S induces a suspended animation-like state in mice. *Science.* 2005;308(5721):518.

62. Florian B, Vintilescu R, Balseanu AT, Buga AM, Grisk O, Walker LC et al. Long-term hypothermia reduces infarct volume in aged rats after focal ischemia. *Neurosci Lett.* 2008;438(2):180–5.

63. Joseph C, Buga AM, Vintilescu R, Balseanu AT, Moldovan M, Junker H et al. Prolonged gaseous hypothermia prevents the upregulation of phagocytosis-specific protein annexin 1 and causes low-amplitude EEG activity in the aged rat brain after cerebral ischemia. *J Cereb Blood Flow Metab.* 2012;32(8):1632–42.

64. Huh PW, Belayev L, Zhao W, Koch S, Busto R, Ginsberg MD. Comparative neuroprotective efficacy of prolonged moderate intraischemic and postischemic hypothermia in focal cerebral ischemia. *J Neurosurg.* 2000;92(1):91–9.

65. Maier CM, Ahern K, Cheng ML, Lee JE, Yenari MA, Steinberg GK. Optimal depth and duration of mild hypothermia in a focal model of transient cerebral ischemia: Effects on neurologic outcome, infarct size, apoptosis, and inflammation. *Stroke*. 1998;29(10):2171–80.

66. Colbourne F, Corbett D, Zhao Z, Yang J, Buchan AM. Prolonged but delayed postischemic hypothermia: A long-term outcome study in the rat middle cerebral artery occlusion model. *J Cereb Blood Flow Metab*. 2000;20(12):1702–8.

67. Sacco RL, Chong JY, Prabhakaran S, Elkind MS. Experimental treatments for acute ischaemic stroke. *Lancet*. 2007;369(9558):331–41.

68. Popa-Wagner A, Buga AM, Kokaia Z. Perturbed cellular response to brain injury during aging. *Ageing Res Rev*. 2011;10(1):71–9.

69. Badan I, Buchhold B, Hamm A, Gratz M, Walker LC, Platt D et al. Accelerated glial reactivity to stroke in aged rats correlates with reduced functional recovery. *J Cereb Blood Flow Metab*. 2003;23(7):845–54.

70. Lucke-Wold BP, Logsdon AF, Turner RC, Rosen CL, Huber JD. Aging, the metabolic syndrome, and ischemic stroke: Redefining the approach for studying the blood–brain barrier in a complex neurological disease. *Adv Pharmacol*. 2014;71:411–49.

71. Petcu EB, Smith RA, Miroiu RI, Opris MM. Angiogenesis in old-aged subjects after ischemic stroke: A cautionary note for investigators. *J Angiogenes Res*. 2010;2:26.

72. Jha S, Calvert JW, Duranski MR, Ramachandran A, Lefer DJ. Hydrogen sulfide attenuates hepatic ischemia-reperfusion injury: Role of antioxidant and antiapoptotic signaling. *Am J Physiol Heart Circ Physiol*. 2008;295(2):H801–6.

73. Johansen FF, Hasseldam H, Rasmussen RS, Bisgaard AS, Bonfils PK, Poulsen SS et al. Drug-induced hypothermia as beneficial treatment before and after cerebral ischemia. *Pathobiology*. 2014;81(1):42–52.

74. Clark DL, Penner M, Wowk S, Orellana-Jordan I, Colbourne F. Treatments (12 and 48 h) with systemic and brain-selective hypothermia techniques after permanent focal cerebral ischemia in rat. *Exp Neurol*. 2009;220(2):391–9.

75. Wei G, Hartings JA, Yang X, Tortella FC, Lu XC. Extraluminal cooling of bilateral common carotid arteries as a method to achieve selective brain cooling for neuroprotection. *J Neurotrauma*. 2008;25(5):549–59.

76. Goossens J, Hachimi-Idrissi S. Combination of therapeutic hypothermia and other neuroprotective strategies after an ischemic cerebral insult. *Curr Neuropharmacol*. 2014;12(5):399–412.

77. Chopp M, Knight R, Tidwell CD, Helpern JA, Brown E, Welch KM. The metabolic effects of mild hypothermia on global cerebral ischemia and recirculation in the cat: Comparison to normothermia and hyperthermia. *J Cereb Blood Flow Metab*. 1989;9(2):141–8.

78. Balseanu AT, Buga AM, Catalin B, Wagner DC, Boltze J, Zagrean AM et al. Multimodal approaches for regenerative stroke therapies: Combination of granulocyte colony-stimulating factor with bone marrow mesenchymal stem cells is not superior to G-CSF alone. *Front Aging Neurosci*. 2014;6:130.

79. Tatarishvili J, Oki K, Monni E, Koch P, Memanishvili T, Buga AM et al. Human induced pluripotent stem cells improve recovery in stroke-injured aged rats. *Restor Neurol Neurosci*. 2014;32(4):547–58.

80. Jinno S. Decline in adult neurogenesis during aging follows a topographic pattern in the mouse hippocampus. *J Comp Neurol*. 2011;519(3):451–66.

81. Buga AM, Di Napoli M, Popa-Wagner A. Preclinical models of stroke in aged animals with or without comorbidities: Role of neuroinflammation. *Biogerontology*. 2013;14(6):651–62.

82. Lucin KM, Wyss-Coray T. Immune activation in brain aging and neurodegeneration: Too much or too little? *Neuron*. 2009;64(1):110–22.

83. Akiyama H, Chaboissier MC, Martin JF, Schedl A, de Crombrugghe B. The transcription factor Sox9 has essential roles in successive steps of the chondrocyte differentiation pathway and is required for expression of Sox5 and Sox6. *Genes Dev*. 2002;16(21):2813–28.

84. Ye SM, Johnson RW. Increased interleukin-6 expression by microglia from brain of aged mice. *J Neuroimmunol*. 1999;93(1–2):139–48.

85. Buchan AM, Li H, Blackburn B. Neuroprotection achieved with a novel proteasome inhibitor which blocks NF-kappaB activation. *Neuroreport*. 2000;11(2):427–30.

86. Doeppner TR, Mlynarczuk-Bialy I, Kuckelkorn U, Kaltwasser B, Herz J, Hasan MR et al. The novel proteasome inhibitor BSc2118 protects against cerebral ischaemia through HIF1A accumulation and enhanced angioneurogenesis. *Brain*. 2012;135(Pt 11):3282–97.

87. Henninger N, Sicard KM, Bouley J, Fisher M, Stagliano NE. The proteasome inhibitor VELCADE reduces infarction in rat models of focal cerebral ischemia. *Neurosci Lett*. 2006;398(3):300–5.

88. Zhang SC. Neural subtype specification from embryonic stem cells. *Brain Pathol*. 2006;16(2):132–42.

89. Sandu RE, Uzoni A, Ciobanu O, Moldovan M, Anghel A, Radu E et al. Post-stroke gaseous hypothermia increases vascular density but not neurogenesis in the ischemic penumbra of aged rats. *Restor Neurol Neurosci.* 2016 Feb 24;34(3):401–14. doi: 10.3233/RNN-150600.

90. Xie YC, Li CY, Li T, Nie DY, Ye F. Effect of mild hypothermia on angiogenesis in rats with focal cerebral ischemia. *Neurosci Lett.* 2007;422(2):87–90.

91. Yenari MA, Han HS. Neuroprotective mechanisms of hypothermia in brain ischaemia. *Nat Rev Neurosci.* 2012;13(4):267–78.

92. Bregy A, Nixon R, Lotocki G, Alonso OF, Atkins CM, Tsoulfas P et al. Posttraumatic hypothermia increases doublecortin expressing neurons in the dentate gyrus after traumatic brain injury in the rat. *Exp Neurol.* 2012;233(2):821–8.

93. Silasi G, Colbourne F. Therapeutic hypothermia influences cell genesis and survival in the rat hippocampus following global ischemia. *J Cereb Blood Flow Metab.* 2011;31(8):1725–35.

94. Silasi G, Klahr AC, Hackett MJ, Auriat AM, Nichol H, Colbourne F. Prolonged therapeutic hypothermia does not adversely impact neuroplasticity after global ischemia in rats. *J Cereb Blood Flow Metab.* 2012;32(8):1525–34.

95. Ahlenius H, Visan V, Kokaia M, Lindvall O, Kokaia Z. Neural stem and progenitor cells retain their potential for proliferation and differentiation into functional neurons despite lower number in aged brain. *J Neurosci.* 2009;29(14):4408–19.

96. Darsalia V, Heldmann U, Lindvall O, Kokaia Z. Stroke-induced neurogenesis in aged brain. *Stroke.* 2005;36(8):1790–5.

97. Enwere E, Shingo T, Gregg C, Fujikawa H, Ohta S, Weiss S. Aging results in reduced epidermal growth factor receptor signaling, diminished olfactory neurogenesis, and deficits in fine olfactory discrimination. *J Neurosci.* 2004;24(38):8354–65.

98. Lasarzik I, Winkelheide U, Thal SC, Benz N, Lorscher M, Jahn-Eimermacher A et al. Mild hypothermia has no long-term impact on postischemic neurogenesis in rats. *Anesth Analg.* 2009;109(5):1632–9.

99. Tobin MK, Bonds JA, Minshall RD, Pelligrino DA, Testai FD, Lazarov O. Neurogenesis and inflammation after ischemic stroke: What is known and where we go from here. *J Cereb Blood Flow Metab.* 2014;34(10):1573–84.

100. Deboer T. Brain temperature dependent changes in the electroencephalogram power spectrum of humans and animals. *J Sleep Res.* 1998;7(4):254–62.

101. Sitnikova E, van Luijtelaar G. Electroencephalographic precursors of spike-wave discharges in a genetic rat model of absence epilepsy: Power spectrum and coherence EEG analyses. *Epilepsy Res.* 2009;84(2–3):159–71.

102. Hartings JA, Williams AJ, Tortella FC. Occurrence of nonconvulsive seizures, periodic epileptiform discharges, and intermittent rhythmic delta activity in rat focal ischemia. *Exp Neurol.* 2003;179(2):139–49.

103. Baumann CR, Kilic E, Petit B, Werth E, Hermann DM, Tafti M et al. Sleep EEG changes after middle cerebral artery infarcts in mice: Different effects of striatal and cortical lesions. *Sleep.* 2006;29(10):1339–44.

104. Meng H, Liu T, Borjigin J, Wang MM. Ischemic stroke destabilizes circadian rhythms. *J Circadian Rhythms.* 2008;6:9.

105. Takekawa H, Miyamoto M, Miyamoto T, Hirata K. Circadian rhythm abnormalities in the acute phase of cerebral infarction correlate with poor prognosis in the chronic phase. *Auton Neurosci.* 2007;131(1–2):131–6.

106. Mortola JP. Hypoxia and circadian patterns. *Respir Physiol Neurobiol.* 2007;158(2–3):274–9.

107. Han JL, Blank T, Schwab S, Kollmar R. Inhibited glutamate release by granulocyte-colony stimulating factor after experimental stroke. *Neurosci Lett.* 2008;432(3):167–9.

108. Lee ST, Chu K, Jung KH, Ko SY, Kim EH, Sinn DI et al. Granulocyte colony-stimulating factor enhances angiogenesis after focal cerebral ischemia. *Brain Res.* 2005;1058(1–2):120–8.

109. Shyu WC, Lin SZ, Yang HI, Tzeng YS, Pang CY, Yen PS et al. Functional recovery of stroke rats induced by granulocyte colony-stimulating factor-stimulated stem cells. *Circulation.* 2004;110(13):1847–54.

110. Xiao BG, Lu CZ, Link H. Cell biology and clinical promise of G-CSF: Immunomodulation and neuroprotection. *J Cell Mol Med.* 2007;11(6):1272–90.

111. Komine-Kobayashi M, Zhang N, Liu M, Tanaka R, Hara H, Osaka A et al. Neuroprotective effect of recombinant human granulocyte colony-stimulating factor in transient focal ischemia of mice. *J Cereb Blood Flow Metab.* 2006;26(3):402–13.

112. Minnerup J, Heidrich J, Wellmann J, Rogalewski A, Schneider A, Schabitz WR. Meta-analysis of the efficacy of granulocyte-colony stimulating factor in animal models of focal cerebral ischemia. *Stroke.* 2008;39(6):1855–61.

113. dela Pena IC, Yoo A, Tajiri N, Acosta SA, Ji X, Kaneko Y et al. Granulocyte colony-stimulating factor attenuates delayed tPA-induced hemorrhagic transformation in ischemic stroke rats by enhancing angiogenesis and vasculogenesis. *J Cereb Blood Flow Metab.* 2015;35(2):338–46.

114. Schneider A, Kruger C, Steigleder T, Weber D, Pitzer C, Laage R et al. The hematopoietic factor G-CSF is a neuronal ligand that counteracts programmed cell death and drives neurogenesis. *J Clin Invest.* 2005;115(8):2083–98.

115. Bratane BT, Bouley J, Schneider A, Bastan B, Henninger N, Fisher M. Granulocyte-colony stimulating factor delays PWI/DWI mismatch evolution and reduces final infarct volume in permanent-suture and embolic focal cerebral ischemia models in the rat. *Stroke.* 2009;40(9):3102–6.

116. Baltan S. Excitotoxicity and mitochondrial dysfunction underlie age-dependent ischemic white matter injury. *Adv Neurobiol.* 2014;11:151–70.

117. Philip M, Benatar M, Fisher M, Savitz SI. Methodological quality of animal studies of neuroprotective agents currently in phase II/III acute ischemic stroke trials. *Stroke.* 2009;40(2):577–81.

118. Popa-Wagner A, Stocker K, Balseanu AT, Rogalewski A, Diederich K, Minnerup J et al. Effects of granulocyte-colony stimulating factor after stroke in aged rats. *Stroke.* 2010;41(5):1027–31.

119. Komatsu K, Honmou O, Suzuki J, Houkin K, Hamada H, Kocsis JD. Therapeutic time window of mesenchymal stem cells derived from bone marrow after cerebral ischemia. *Brain Res.* 2010;1334:84–92.

120. Sharma S, Yang B, Strong R, Xi X, Brenneman M, Grotta JC et al. Bone marrow mononuclear cells protect neurons and modulate microglia in cell culture models of ischemic stroke. *J Neurosci Res.* 2010;88(13):2869–76.

121. Hill QA, Buxton D, Pearce R, Gesinde MO, Smith GM, Cook G. An analysis of the optimal timing of peripheral blood stem cell harvesting following priming with cyclophosphamide and G-CSF. *Bone Marrow Transplant.* 2007;40(10):925–30.

122. Wagner DC, Bojko M, Peters M, Lorenz M, Voigt C, Kaminski A et al. Impact of age on the efficacy of bone marrow mononuclear cell transplantation in experimental stroke. *Exp Transl Stroke Med.* 2012;4(1):17.

123. Weise G, Lorenz M, Posel C, Maria Riegelsberger U, Storbeck V, Kamprad M et al. Transplantation of cryopreserved human umbilical cord blood mononuclear cells does not induce sustained recovery after experimental stroke in spontaneously hypertensive rats. *J Cereb Blood Flow Metab.* 2014;34(1):e1–9.

124. Posel C, Scheibe J, Kranz A, Bothe V, Quente E, Frohlich W et al. Bone marrow cell transplantation time-dependently abolishes efficacy of granulocyte colony-stimulating factor after stroke in hypertensive rats. *Stroke.* 2014;45(8):2431–7.

125. Macleod MR, van der Worp HB, Sena ES, Howells DW, Dirnagl U, Donnan GA. Evidence for the efficacy of NXY-059 in experimental focal cerebral ischaemia is confounded by study quality. *Stroke.* 2008;39(10):2824–9.

126. Shetty AK, Hattiangady B, Rao MS. Vulnerability of hippocampal GABA-ergic interneurons to kainate-induced excitotoxic injury during old age. *J Cell Mol Med.* 2009;13(8B):2408–23.

127. Dunnett SB. Neural tissue transplantation, repair, and rehabilitation. *Handb Clin Neurol.* 2013;110:43–59.

128. Morizane A, Li JY, Brundin P. From bench to bed: The potential of stem cells for the treatment of Parkinson's disease. *Cell Tissue Res.* 2008;331(1):323–36.

129. Stoll EA. Advances toward regenerative medicine in the central nervous system: Challenges in making stem cell therapy a viable clinical strategy. *Mol Cell Ther.* 2014;2:12.

130. Bang OY, Lee JS, Lee PH, Lee G. Autologous mesenchymal stem cell transplantation in stroke patients. *Ann Neurol.* 2005;57(6):874–82.

131. Kondziolka D, Wechsler L, Goldstein S, Meltzer C, Thulborn KR, Gebel J et al. Transplantation of cultured human neuronal cells for patients with stroke. *Neurology.* 2000;55(4):565–9.

132. Reitmeir R, Kilic E, Reinboth BS, Guo Z, ElAli A, Zechariah A et al. Vascular endothelial growth factor induces contralesional corticobulbar plasticity and functional neurological recovery in the ischemic brain. *Acta Neuropathol.* 2012;123(2):273–84.

133. Arvidsson A, Collin T, Kirik D, Kokaia Z, Lindvall O. Neuronal replacement from endogenous precursors in the adult brain after stroke. *Nat Med.* 2002;8(9):963–70.

134. Hou SW, Wang YQ, Xu M, Shen DH, Wang JJ, Huang F et al. Functional integration of newly generated neurons into striatum after cerebral ischemia in the adult rat brain. *Stroke.* 2008;39(10):2837–44.

135. Jin K, Wang X, Xie L, Mao XO, Zhu W, Wang Y et al. Evidence for stroke-induced neurogenesis in the human brain. *Proc Natl Acad Sci U S A.* 2006;103(35):13198–202.

136. Mine Y, Tatarishvili J, Oki K, Monni E, Kokaia Z, Lindvall O. Grafted human neural stem cells enhance several steps of endogenous neurogenesis and improve behavioral recovery after middle cerebral artery occlusion in rats. *Neurobiol Dis.* 2013;52:191–203.

137. Thored P, Arvidsson A, Cacci E, Ahlenius H, Kallur T, Darsalia V et al. Persistent production of neurons from adult brain stem cells during recovery after stroke. *Stem Cells.* 2006;24(3):739–47.
138. Tajiri N, Quach DM, Kaneko Y, Wu S, Lee D, Lam T et al. Behavioral and histopathological assessment of adult ischemic rat brains after intracerebral transplantation of NSI-566RSC cell lines. *PLoS One.* 2014;9(3):e91408.
139. Lindvall O, Kokaia Z. Neurogenesis following stroke affecting the adult brain. *Cold Spring Harb Perspect Biol.* 2015;7(11).
140. Parent JM, Vexler ZS, Gong C, Derugin N, Ferriero DM. Rat forebrain neurogenesis and striatal neuron replacement after focal stroke. *Ann Neurol.* 2002;52(6):802–13.
141. Tropepe V, Craig CG, Morshead CM, van der Kooy D. Transforming growth factor-alpha null and senescent mice show decreased neural progenitor cell proliferation in the forebrain subependyma. *J Neurosci.* 1997;17(20):7850–9.
142. Kojima T, Hirota Y, Ema M, Takahashi S, Miyoshi I, Okano H et al. Subventricular zone-derived neural progenitor cells migrate along a blood vessel scaffold toward the post-stroke striatum. *Stem Cells.* 2010;28(3):545–54.
143. Macas J, Nern C, Plate KH, Momma S. Increased generation of neuronal progenitors after ischemic injury in the aged adult human forebrain. *J Neurosci.* 2006;26(50):13114–9.
144. Marti-Fabregas J, Romaguera-Ros M, Gomez-Pinedo U, Martinez-Ramirez S, Jimenez-Xarrie E, Marin R et al. Proliferation in the human ipsilateral subventricular zone after ischemic stroke. *Neurology.* 2010;74(5):357–65.
145. Minger SL, Ekonomou A, Carta EM, Chinoy A, Perry RH, Ballard CG. Endogenous neurogenesis in the human brain following cerebral infarction. *Regen Med.* 2007;2(1):69–74.
146. Ernst A, Alkass K, Bernard S, Salehpour M, Perl S, Tisdale J et al. Neurogenesis in the striatum of the adult human brain. *Cell.* 2014;156(5):1072–83.
147. Yuan T, Liao W, Feng NH, Lou YL, Niu X, Zhang AJ et al. Human induced pluripotent stem cell-derived neural stem cells survive, migrate, differentiate, and improve neurologic function in a rat model of middle cerebral artery occlusion. *Stem Cell Res Ther.* 2013;4(3):73.
148. Liu X, Ye R, Yan T, Yu SP, Wei L, Xu G et al. Cell based therapies for ischemic stroke: From basic science to bedside. *Prog Neurobiol.* 2014;115:92–115.
149. Oki K, Tatarishvili J, Wood J, Koch P, Wattananit S, Mine Y et al. Human-induced pluripotent stem cells form functional neurons and improve recovery after grafting in stroke-damaged brain. *Stem Cells.* 2012;30(6):1120–33.
150. Tornero D, Wattananit S, Gronning Madsen M, Koch P, Wood J, Tatarishvili J et al. Human induced pluripotent stem cell-derived cortical neurons integrate in stroke-injured cortex and improve functional recovery. *Brain.* 2013;136(Pt 12):3561–77.
151. Mohamad O, Drury-Stewart D, Song M, Faulkner B, Chen D, Yu SP et al. Vector-free and transgene-free human iPS cells differentiate into functional neurons and enhance functional recovery after ischemic stroke in mice. *PLoS One.* 2013;8(5):e64160.
152. Phanthong P, Raveh-Amit H, Li T, Kitiyanant Y, Dinnyes A. Is aging a barrier to reprogramming? Lessons from induced pluripotent stem cells. *Biogerontology.* 2013;14(6):591–602.
153. Boulting GL, Kiskinis E, Croft GF, Amoroso MW, Oakley DH, Wainger BJ et al. A functionally characterized test set of human induced pluripotent stem cells. *Nat Biotechnol.* 2011;29(3):279–86.
154. Kim JH, Seo M, Han HS, Park J, Suk K. The neurovascular protection afforded by delayed local hypothermia after transient middle cerebral artery occlusion. *Curr Neurovasc Res.* 2013;10(2):134–43.
155. Kalladka D, Muir KW. Brain repair: Cell therapy in stroke. *Stem Cells Cloning.* 2014;7:31–44.
156. Riess P, Molcanyi M, Bentz K, Maegele M, Simanski C, Carlitscheck C et al. Embryonic stem cell transplantation after experimental traumatic brain injury dramatically improves neurological outcome, but may cause tumors. *J Neurotrauma.* 2007;24(1):216–25.
157. Boltze J, Arnold A, Walczak P, Jolkkonen J, Cui L, Wagner DC. The dark side of the force—constraints and complications of cell therapies for stroke. *Front Neurol.* 2015;6:155.

Section IV

Mechanisms of Aging

20 Increase in Mitochondrial DNA Fragments inside Nuclear DNA during the Lifetime of an Individual as a Mechanism of Aging

Gustavo Barja

CONTENTS

INTRODUCTION

The updated mitochondrial free-radical theory of aging (MFRTA), first proposed by Denham Harman [1], can be responsible for a substantial fraction of the aging rate. Reactive oxygen species (ROS) are deleterious substances continuously produced in our aerobic tissue cells, especially in mitochondria. Ironically oxygen, one of the most fundamental substances for animal life, can indirectly be responsible for our progressive degradation during our lifetime.

Free radicals attack cellular macromolecules, proteins, lipids, or DNA. The cells can support such free-radical flux because they have a large and diverse set of enzymatic and nonenzymatic antioxidants. Due to this, many researchers have tried to increase animal longevity by boosting antioxidant levels, either increasing them in the diet or through overexpression of antioxidant-codifying genes. However, these manipulations have been unsuccessful, especially in mammals. It seems that the problem is not so simple and that investigators have been focusing on the wrong place.

It is well known that long-lived animal species, such as cow, humans, or many birds, produce small amounts of ROS per unit time in the mitochondrial respiratory chain compared to short-lived mammals, such as rats and mice [2–6]. They have a low rate of mitROSp (mitochondrial ROS production) which can contribute to explain their smaller degree of oxidative damage at mtDNA [7] and their slower rate of aging. In addition dietary calorie restriction (DR), the most robust experimental manipulation that increases mammalian longevity, also decreases mitROSp and oxidative damage to mtDNA [8]. The same is true concerning the other known manipulations that increase mammalian longevity, methionine restriction (MetR), or rapamycin treatment [9–11]. The close vicinity between mtDNA and the mitROS generator, even their physical contact, can explain the small capacity of antioxidants to intercept ROS before they can damage mtDNA. It seems that mother nature has "designed" this to avoid large modifications in aging rate after antioxidant ingestion. The ultimate reason likely is that such modification in aging rate would be catastrophic for the group or species.

The kind of oxidative damage critical for aging depends on the rate at which mitochondria generate ROS. Even if these ROS attack cellular or mitochondrial proteins and lipids, they can be repaired. However, when mtDNA is subjected to free radical attack, in addition to suffering point mutations, it can break down also generating big deletions and mtDNA fragments. In the past it was thought that the point mutations and big deletions in mtDNA could be a fundamental cause of aging. However, due to the high copy number of mtDNA per cell, today it does not seem that those mutations can reach levels high enough in tissues of old animals to cause aging. In this chapter, I focus on a more recent alternative possibility: that the mtDNA fragments generated by mitROS insert inside nuclear DNA (nDNA) and significantly contribute to cause aging.

MtDNA DELETIONS AND POINT MUTATIONS DURING AGING

Oxidative damage to mtDNA bases, like 8-oxodG (8-oxo-7,8-dihydro-2′deoxyguanosine), is known to be repaired in the mitochondria. But mitROS, in addition to DNA base and sugar oxidative modifications, have the capacity to produce double-strand breaks (DSBs) in DNA in general, and especially in the very nearby situated mtDNA. Fragmentation of mtDNA through DSBs by the nearby generated mitROS could most probably be one cause of the accumulation of mtDNA mutations, including large mtDNA deletions, with age [12]. Recently, it has been proposed that mtDNA mutations can also be due to errors during DNA replication and repair, rather than to mitROSp. However, whereas those random errors can contribute to accumulated damage during the life span of a single individual, they cannot be responsible for the strongly different longevity of the different species nor for the change in longevity induced by the different kinds of DRs, because these longevities are genetically, instead of randomly, controlled. In other words, there are no plausible mechanisms that would lead rats to commit 30-fold more errors than humans during mtDNA replication or repair. Replication and repair as source of mtDNA mutations suffers the same limitation that many other wrong proposals based on random processes (e.g., wear and tear theories of aging). Instead, the longevity of a species, or fine tuning of longevity to a new level in DR, is determined by the genome. Therefore, it must be due to the existence of genetically programmed processes residing in the cell nucleus which respond to environmental nutrient availability (during DRs) to finally increase animal life span.

Large mtDNA deletions increase with age in mammalian tissues and have been proposed among the end-point detrimental effects causing aging. Since mtDNA is highly compacted, without introns, the large deletions detected in old tissues would lead to the lack of many genes coding for electron transport chain (ETC) or mitochondrial ribosome subunits in a single mtDNA circle molecule. However it is now clear that, with the exception of a few tissues, the level of these deletions does not reach the threshold needed, in homoplasmy (all the mtDNA copies of a cell are similar), to be of deleterious functional consequences in most tissues of old animals. The high level of heteroplasmia (both WT and mutated mtDNA present inside a cell) of mtDNA, due to the presence of thousands of mitochondria per cell, and various mtDNA copies per mitochondrion, strongly protects against the direct functional effects of mtDNA mutations. Many copies of mtDNA are present in each cell. Only if a large majority of these mtDNA copies are mutated, would mitochondrial ATP production be compromised. Cells essentially homoplasmic for deleted mtDNA are abundant in a few areas like substantia nigra in humans [13,14], but in the brain in general and in other vital tissues of old individuals their percentage is too low (<2%) to be a cause of aging.

MtDNA FRAGMENTS INCREASE INSIDE nDNA DURING THE LIFETIME OF THE INDIVIDUAL

DSBs in mtDNA can produce mtDNA deletions. But they can also generate mtDNA fragments, the missing overlooked segments deleted from mtDNA. In agreement with earlier proposals [15,16] those fragments that can escape from mitochondria at least through the permeability transition pore, have been observed in rodent brain cytosolic fractions [17,18], and are also present inside the nucleus [19,20]. It was proposed that this could randomly change nuclear gene information and thereby contribute to cause cancer and aging [15]. It has been simultaneously demonstrated that mtDNA fragments accumulate inside nDNA with age (Figure 20.1) both in yeast [19] and in rat liver and brain [20], and that such accumulation causes damage and accelerates aging in yeast [21]. Mouse liver also accumulates mtDNA fragments inserted inside nDNA with age [11]. The lack of detrimental consequences of mtDNA point mutations or deletions in tissues of old animals due to heteroplasmia of mutated mtDNA inside cells was a strong problem for MFRTA and the mitochondrial theory of aging in general. The lack of end-point damage left the theory without a mechanism to generate aging. However, this can be solved with the phenomenon of mtDNA fragments accumulation with age inside nDNA. While a main cause of aging resides in the mitochondria—the rate of mitROSp—the location seems to be different concerning its main consequence for aging. Such consequence is situated at the nucleus where mtDNA fragments accumulate causing havoc. Concerning the final consequences for aging, the end point of mitochondrial damage was previously not found because it is likely that most researchers were looking in the wrong place. As regards the main cause of aging we must look at the mitochondria: mitROSp breaking down mtDNA into fragments. And as regards mitROSp consequences we must look at the nucleus: mtDNA fragments insertion inside nDNA, at the chromosomes. The insertion of mtDNA fragments inside nDNA is most important because almost all the cellular structural genes and their regulatory regions are situated inside the nucleus, not in the mitochondria. Therefore, it is logical to expect that, even if the cause of damage is situated in the mitochondria, the final target will be the nucleus. And the number of cells affected by mtDNA fragment accumulation in the nucleus could be extensive.

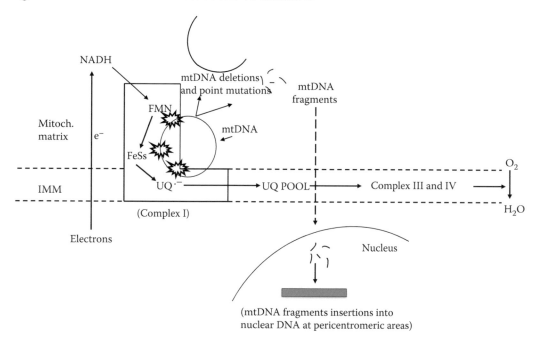

(mtDNA fragments insertions into
nuclear DNA at pericentromeric areas)

FIGURE 20.1 MtDNA fragments insert inside nuclear DNA during aging. Mitochondrial ROS production generates DSBs at mtDNA. The mtDNA fragments exit mitochondria and enter the nucleus, where they insert into nDNA. Such insertions increase with age in yeast and in rat and mouse tissues, and increase chronological aging in yeast. They are mostly localized at pericentromeric areas in rodents. Those insertions can promote aging by various different mechanisms. Therefore, although mitROSp is a main effector of aging, the main final target irreversibly damaged can be the nucleus. IMM = inner mitochondrial membrane; e^- = electrons.

Such mtDNA fragment accumulation means that aging is generated by collaboration between the mitochondria and nucleus. They collaborate because aging is an adaptive function useful for the species carrying it. This can only be understood by changing the mind set from the outdated selfish gene hypothesis and from natural selection acting at the individual level only, to considering aging "bad" for the individual (decreasing its biological fitness) but "good" for the group. A change of mind set from mainstream theories that postulate, without evidence, that aging is due to many random processes (like in the case of inert matter) to programmed aging theories fully compatible with the fact that longevity is widely different in many different species, is needed. Longevity is genetically determined, and there is a gene expression tissue-specific response to the different DRs and rapamycin which is programmed inside the nucleus. There are many dozens of known genes that promote aging (and much less antiaging genes), and hundreds of them are waiting to be discovered inside the nucleus, those that compose the aging program (AP). "Wear and tear"-based theories of aging are wrong because they overlook the fact that animals are not inanimate object like cars or chairs. Animals are alive, their longevity is determined in their genome, and they actively regulate their functions, including aging, which is also a physiological species-specific function. This was widely overlooked for decades due to thinking about aging in just a single species: mouse only, or human only. On the contrary, when different species with different longevities are compared, the concept of *aging rate* as a fundamental one for aging quickly arises. And the search for the causes of the different aging rates of the different species arises too. The different animal species generate aging *endogenously*, from inside, at a pace compatible with their population number and density in the wild.

An increase in mtDNA fragments inside nDNA has been also observed during reprogramming of somatic cells to pluripotent cells and during differentiation [22]. Such accumulation seems a highly conserved mechanism linked to aging from yeast to mammals. Strong gene loss in mtDNA during evolution took place due to transfer of most mtDNA sequences to the nucleus. Although only 13 polypeptide coding genes remain in human mtDNA, the total amount of mtDNA per cell is quite large. The copy number of mtDNA per diploid nuclear genome in human tissues is around 5000 [23], and this number is expected to be even higher in mice due to their higher weight-specific metabolic rate compared to humans. Even a conservative calculation of 5000 mtDNA copies would correspond to around 12.5% of the total amount of nDNA in a mouse hepatocyte. This is 1.4-fold (40%) higher than the DNA amount present in all the nDNA structural genes, assuming that around 1% of total nDNA codes for proteins. Therefore, the amount of mtDNA fragments than can be produced for insertion into nDNA can be quantitatively important compared to the nDNA amount. There is more than enough mtDNA to cause significant damage to nDNA. This is even more true with the realization that mtDNA replicates independently of nDNA between successive cell divisions, whereas in the absence of pathology or stress, hepatic cells do not replicate their nDNA. On the other hand, it is well known that oxidative damage to mtDNA, measured as 8-oxodG, is around one order of magnitude higher than in nDNA. Since there is no known well-established ROS generator inside the nucleus, it is possible that the small amount of 8-oxodG detected in nDNA, or at least a substantial part of it, is not generated inside the nuclear compartment. Instead, mtDNA guanines are oxidized by ROS generators in the mitochondria and travel from the mitochondria to the nucleus with the mtDNA fragments containing them, justifying the presence of 8-oxodG in the nucleus at least in part.

Rapamycin is the only known drug that increases mammalian longevity. Strikingly, dietary treatment with rapamycin during 7 weeks, in addition to decreasing mitROSp, totally reversed mtDNA fragments accumulation in nDNA [11], and decreased lipofuscin (non-autophagocytosed materials) in middle-aged mice. This indicates that the accumulation of mtDNA fragments inserted inside nDNA during aging is not a random process. Instead, it strongly suggests that there is a flux of these fragments through nDNA that has a functional role likely related to aging. The decrease in lipofuscin might be due to a decrease of damaged mitochondria, and/or to the increase in autophagy also induced by the drug [11]. The changes observed for both mitROSp and mtDNA fragments inserted in nDNA are strikingly similar because full reversion of age-related increased values occurred in both cases. This could be due to a cause–effect relationship between these two parameters, because ROS also have strong capacity to produce DSBs and then DNA fragmentation.

Therefore, part of the increase in longevity of rapamycin-treated mice can be due to: (i) a decrease in mitROSp, (ii) a decrease in the insertion of mtDNA fragments inside nDNA, and (iii) an increase in autophagy. Interestingly, in a parallel experiment we did not detect significant changes in mitROSp in the heart after the rapamycin treatment. Therefore, the reason why rapamycin increases life span to a lower extent than DR can be due to the impact of DR on more longevity signaling pathways than those involving mTOR. But it can also be due to the possibility that rapamycin decreases mitROSp, or increases autophagy, in some but not in all organs [11].

Rapamycin dietary treatment in mice decreased mitROSp and FRL (free-radical leak) in mouse liver, but it also lowered the amount of complex I, which contains the ROS generator relevant for aging. That was associated with an increase (instead of decrease) in mitochondrial biogenesis [11]. A possible explanation of this apparently paradoxical result is that rapamycin could selectively induce mitochondrial biogenesis from the more youthful pool of liver mitochondria. These are expected to show lower FRL than the more damaged ones. That selection could be another reason why rapamycin decreased mitROSp. It has been observed that rapamycin increases autophagy and mitochondrial biogenesis in mouse heart suggesting that damaged mitochondria are replaced by newly synthesized ones to rejuvenate mitochondrial homeostasis. Treatment with rapamycin also restores PGC1α-TFEB signaling and abrogates impairment of mitochondrial quality control and neurodegenerative features, while TFEB induction restores mitochondrial function and cell viability [24]. Likely related to replacement of old mitochondria with younger ones, it has been observed that DR (which also inhibits mTOR) and rapamycin both lower mitROSp and FRL and decrease the amount of the "matrix domain-only" of complex I in the mitochondria of mouse liver [25]. That matrix domain contains the mitROS generator at which mitROSp decreases both during the rapamycin treatment [11] and during DR [25–30].

mtDNA fragments leave the mitochondria toward the nucleus inserting into nDNA during the lifetime of the individual, both in yeast and in vital mitotic and postmitotic mammalian tissues (11; Figure 20.1). They are visualized under the microscope using fluorescent *in situ* hybridization techniques heavily concentrated at the pericentromeric regions, which are rich in tandem repeats (satellite DNA and TEs—transposable elements), and co-localize with MaSat (major satellites) in mice. MaSat and MiSat show high sequence conservation across the centromeric domain of all mouse chromosomes. In addition to MaSat and MiSat, two new satellite sequences, MS3 and MS4, have been found in the mouse centromeric and pericentromeric regions [31]. Thus, the mouse genome contains at least four satDNA types: AT-rich (MaSat and MiSat) and CG-rich (MS3 and MS4). On metaphase chromosomes MS3 and MS4 are located at the centromeric region. FISH analysis during the cell cycle showed that all mapped satDNA fragments, except MaSat, belong to the outer layer of the chromocenters in the G0/G1 phase. MS3 is likely involved in centromere formation. The centromeric region consists of a small block of MiSat and MS3 followed by a pericentromeric block of MaSat with MS4 [32]. Inside the block of the long-range cluster, MaSat repeats intermingle mostly with MS4, while MiSat intermingle with MS3. It is possible to find MS3 fragments in the MaSat array and MS4 fragments in the MiSat array. In each satDNA fragment only part of the DNA is methylated. MS3 and MS4 are heavily methylated, being GC-rich. Pericentromeric satellite DNA fragments are more methylated than centromeric ones. MS4 is the most methylated of the four families of satellite DNA while MiSat is methylated only to a minimal extent. The range of the probes used in those studies did not cover the whole centromeric region and the existence of unknown sequences in the mouse centromere is likely. Interestingly, the subtelomeric regions are also rich in tandem repeats including TEs. Could they perhaps be also involved in the insertion of mtDNA fragments into nDNA mechanism of aging? If that were true it could affect, like in the case of the pericentromeric fragments, not only to mitotic, but also the postmitotic tissue cells. Thus, telomere involvement in aging could be due to this mechanism instead of the classic one implying the loss of terminal telomeric repeated sequences. Such loss only occurs in mitotic cells.

The large majority of DNA of the early endosymbiotic α-proteobacteria or cyanobacteria that collaborated to create the eukaryotic animal and plant organelles has been transferred to the nucleus during evolutionary time, leading to the present small genomes of mitochondria and plastids. Interestingly, both plastid and mitochondrial DNA fragments from evolutionary origin present in nDNA are also observed at pericentromeric regions in animals and plants. These fragments suffer vigorous shuffling at the pericentromeric regions, exit them perhaps aided by transposable elements, and are observed dispersed at much lower concentrations across the length of most or all chromosomes depending on the species [33,34]. Thus, the pericentromeric regions seem to be the "entry doors" for the access of mitochondrial (and plastid) DNA fragments into the chromosomes. There is a continuous flux of mtDNA fragments entering and leaving the pericentromeric regions. Perhaps the mtDNA fragments leaving the pericentromeric areas are transferred with the help of TEs to structural genes or their regulatory regions where the mtDNA fragments would modify them.

Rapamycin treatment in mice decreases mitROSp (like the different kinds of DRs) and therefore the breaking of mtDNA into fragments would be lowered. The decreased transfer rate of mtDNA fragments to the pericentromeric regions, together with a lack of change (initially at least) in their rate of removal from MaSat, can explain the observed decreased levels of mtDNA fragments present at the pericentromeric area of the chromosomes in rapamycin-treated animals [11]. On the other hand, the steady-state amount of mtDNA fragments at pericentromers is higher in middle-aged mice than in young adult mice, because the rate of mitROSp, and then the flux of mtDNA fragments insertion at the pericentromeric areas, is also higher in middle-aged mice.

MtDNA FRAGMENTS INSIDE nDNA: CONSEQUENCES FOR AGING

The accumulation of mtDNA fragments inside nDNA with age can cause aging through various different mechanisms. Movement of mtDNA fragments from the pericentromeric areas to other chromosome regions, after *in situ* reshuffling aided by local TEs, would not be strange because retroelements or their fragments are essential components of the normal centromeres [35], tandemly repeated DNA families are present at pericentromeric MaSat, and transposable elements related superfamilies and TE-related tandem repeats are also located along all the chromosomes. These include retroviruses like MTA transposons and tandemly repeated LINE-1 (long interspersed nuclear elements; Reference 36). Chromosomes contain host fragments inserted via retroposons. Retroposons or their fragments may be necessary components of normal centromeres in higher eukaryotes [35]. The human centromere is enriched in LINE elements [37] and the kangaroo centromere contains ERV class retroposons together with the typical satellite DNA repeats [38]. LINE-1 of 2 Kb in size has been recently localized at chromocenters in mouse and human heterochromatin [39]. Ectopic recombination with flanking TEs might explain mtDNA movement from MaSat pericentromeric region to other specific genomic locations because mitochondrial sequences have been localized, at smaller concentrations than at the pericentromeric regions, all along the chromosome length of all chromosomes in various species [33,34]. If this transfer of mtDNA fragments to other genomic regions also occurs during aging it would have the potential to cause many different kinds of genetic damage, genomic instability, and chromosomal rearrangements.

Insertions of mtDNA sequences in genomic DNA can cause genomic instability, and mtDNA fragments have been observed inserted into oncogenes in cancer patients [40]. Extrachromosomal DNA circles have been also observed in pluripotent cells harboring mtDNA fragments [22]. Tens of thousands of extrachromosomal circular DNAs (microDNAs) related to mismatch repair pathways and transcriptional changes have been observed in all kinds of normal tissues—including the brain—of old individuals, and their generation leaves behind deletions in different genomic loci [41,42]. Strikingly, mtDNA transfer to the nuclear genome has been related to decreased nucleosome occupancy, loss of histones leading to gene induction, DNA strand breaks, large-scale chromosomal alterations, translocations, and retrotranspositions at least during replicative aging in yeast [43]. Such strongly deleterious changes could help to explain the genomic instability typically observed in aging.

In yeast replicative aging is due to replication slowing down or stalling at many chromosomal sites including centromeres, long-terminal repeats (LTRs), and telomeric and also subtelomeric regions. It is also associated with alterations in chromatin induced by loss of core histones H3 and H4 (whereas their overexpression extends replicative life span), and with increases in specific gene transcription [44]. Many of the involved sites were identified in chromosome translocations and are fragile sites [45–46] that break more often [44]. Mutation rates, DSBs in nDNA, mitotic recombination rates, chromosome translocations [47], deletions, and duplications increase during aging not only in replicating yeast [43,44,48,49], but also in prematurely aging mice and perhaps also in mammals aging normally [50].

During chronological aging in yeast, the accumulation of mtDNA fragments in the nucleus also occurs [19,21]. Interestingly, the chromosome rearrangements described above have been related to mtDNA fragments and rDNAs. Fragmentation of mtDNA by mitROS (through DSBs) would also generate these two kinds of DNAs, mtDNA fragments containing the 13 genes codifying for ETC components, and those containing the mitochondrial genes codifying for mitochondrial ribosomal proteins. These two kinds of mtDNA fragments seem to be used to damage the nuclear genome at a rate controlled by the rate of mitROSp at complex I, fast in rats and slow in humans. Thus, the rate of mitROSp determines the rate of mtDNA fragments insertion into nDNA, and then the chromosome rearrangements and cell death or malfunction. This last kind of cell fate would likely produce additional damage because cell malfunction in postmitotic tissues would perturb many of their surrounding neighbor cells that were initially healthy. The damaged but still alive cells could consume much higher than normal substrate amounts and could produce much higher quantities of or abnormal metabolic products. This will damage or kill the neighboring cells through starvation or waste product intoxication, and strongly amplify the initial damage compared to the loss of a single cell. This will add to subcellular positive feedback vicious cycles that help to explain the exponential increase in mortality rate with age of the individual observed in rodents, humans, and similar animals. In the case of the mitotic tissues, cell malfunction would also amplify the damage since it can increase cancer incidence.

Another possibility is that mtDNA fragments insertion in the pericentromeric region has detrimental effects by affecting the nearby situated centromere. This DNA region binds the modified histone CEN H3 (Cen PA) which in turn binds to microtubules. Therefore, chromatid separation during mitosis could be also affected, also generating chromosome abnormalities like aneuploidy which can be lethal to cells due to chromosome loss. If cell division is finally impeded, this could also impose functional although restricted limitations, as in the case of telomere shortening, to mitotic tissues. On the other hand, pericentromeric areas also influence chromosome binding to the inside of the nuclear envelope favoring chromosome movement, rearrangements, and local recombination. And MaSat, where the mtDNA insertions have been localized, is also involved in heterochromatin formation and sister chromatid cohesion [51].

At the least, a third kind of mechanism is possible. The two mechanisms described above are simple and likely. Both state that mtDNA fragments accumulation in the nucleus finally cause chromosomal damage and instability, unequal chromosome segregation, or cell division impairment, as the main aging irreversible end points leading to cell death or malfunction. But there is also the possibility that mtDNA fragments insertion into nDNA can also speed up aging by affecting regulatory DNA sequences or structural genes. Regions harboring tandemly repeated sequences like the pericentromeric regions are thought to be involved in gene regulatory functions. The mtDNA fragments inserted at MaSat, after local reshuffling and TE-aided movement to other genomic locations, could be perhaps distributed to specific chromosome locations containing gene regulatory regions or structural genes and would detrimentally modify them finally promoting both cancer and aging. Whatever the mechanism involved, the initial fully theoretical proposal concerning mtDNA fragments insertion inside nDNA [15] would only be wrong as regards the suggested randomness of the process. Random insertion of mtDNA fragments inside nDNA would cause rather limited damage because the structural genes represent only around 1% of total nDNA. Instead, the flux of mtDNA fragments through nDNA seems to be under the control of regulated processes which respond to longevity extensors like rapamycin, and is aided by facilitated local insertion, reshuffling, removal, and transfer to other, perhaps specific, chromosome regions.

Evidence concerning the possibility that the mtDNA fragments insertion damages structural genes already exists. The Pallister–Hall syndrome, a condition usually inherited in an autosomal dominant fashion was due, in a studied patient, to the novo nucleic acid transfer from the mitochondrial to the nuclear genome [52]. This 72-bp insertion into exon 14 of the GL13 gene, creates a premature stop codon and predicted a truncated protein product. This case demonstrates that *de novo* mitochondrial-nuclear transfer of nucleic acid is a novel mechanism of human inherited disease. In another investigation the breakpoint junctions of a familial constitutional reciprocal translocation were cloned, sequenced, and analyzed [53]. Within the 10-kb region flanking the breakpoints, chromosome 11 had 25% repeat elements whereas chromosome 9 had 98% repeats, 95% of which were L1-type LINE elements. At the breakpoint junction of derivative chromosome 9, an unusually large 41-bp insertion was found, which showed full identity to mitochondrial DNA sequence between nucleotides 896 and 936. Analysis of the human genome failed to show the preexistence of the inserted sequence anywhere in the genome, indicating that the insertion was derived from human mtDNA and captured into the junction during the DSB repair process.

mtDNA TRANSFER TO THE NUCLEUS IN EVOLUTION AND AGING

The transfer of mtDNA fragments from mitochondria to nucleus during aging is strongly reminiscent of what happened during evolution after the symbiogenesis event that created the eukaryotic cell around 2 billion years ago. During such evolution most genes of the initially free-living α-proteobacteria were transferred to what now is the nDNA of the eukaryotic cell. This process generated the well-known NUMTs and NUPTs (nuclear mitochondrial or nuclear plastid DNA) classically considered "pseudogenes" assumed to be nonfunctional, which abound in the eukaryotic cell nucleus [40]. However, now we know that mtDNA transfer to the nucleus also occurs during the life time of the individual from young to aged. The same kind of process, occurring at such different time scales, constitutes another example of the old observation that in various cases "ontogeny recapitulates phylogeny," in this case applied to programed aging as a continuation of development from egg to adult. Evolution is often opportunistic. Perhaps ROS were used by the early pre-symbiotic free-living eubacteria to generate mutations that could increase diversity for fast adaptation to changing environments. Most mutations are detrimental, but high mutation rates combined with lateral gene transfer could have speeded up evolution in the community of free-living bacteria. Nowadays, however, the mitROSp function has changed to promote aging of the multicellular animal. Thus, the archaeon gave to the α-proteobacteria a secure and comfortable place to live, through collaboration instead of competition. And the α-proteobacteria which entered the archaeon to live inside it for ever after, supplied it with much more ATP per gram of glucose, and perhaps later donated to the archaeon or its multicellular descendants a valuable "gift," a mechanism of aging. This was important for most multicellular animals because aging, like sex, increases diversity (variability). And diversity is the raw material necessary for the action of Darwinian natural selection. Without variability there is nothing to select. In this way, aging and sex speeded up the rate of evolution and thus the chances of survival in an ever changing terrestrial environment. Perhaps this is also why most multicellular animals have adopted and positively selected aging, incorporating it into their genome and adult life. Sex is among the numerous features of complex cells that originated during the evolutionary transition toward eukaryotes. Sexual reproduction is a nearly universal feature of eukaryotic organisms, and sex is thought to have arisen once in the last common ancestor to all eukaryotes [54]. The evolution of sex has been traditionally analyzed in terms of costs and benefits of chromosome segregation and recombination, paying little attention to historical circumstances such as the process of eukaryogenesis and mitochondrial endosymbiosis. Cell fusion, for example, may be viewed as a result of proto-mitochondrial manipulations allowing endosymbionts to spread through the population without the rigors of a free-living stage, at the same time allowing for the reciprocal recombination stabilizing the newly emerging nuclear genome and restoring favorable conditions for growth and replication [55]. Despite numerous apparent disadvantages, such as breaking up co-adapted gene combinations or the need to find a partner, sex is almost universal among the eukaryotic species. Sex uncovers the hidden genetic variation, allows selection to "see" individual genes and gene combinations, and thus increases the efficacy of selection at least in finite populations subject to genetic drift [56]. Whole-cell fusion and reciprocal sex are almost unknown in bacteria that tend to share genetic information exclusively through lateral gene transfer, although some modern archaeans may exhibit more eukaryote-like mating. Studies of modern eukaryotes suggest that seemingly fully developed sexual life cycles were very likely present in the common ancestor of all eukaryotes [54]. Sexual life cycles suggest several important connections to mitochondrial endosymbiosis. First, the fluidity of membranes that permits whole-cell fusion might have also permitted symbiont acquisition. Second, whole-cell fusion can permit exchange of intracellular symbionts without the attendant rigors of a free-living stage of the life cycle. Additionally, it has been suggested [57] that the evolution of sex could have been driven by high mutation rates and genome instability during the phase of intense bombardment of the nuclear genome by mitochondrial genes and introns.

Like in the case of chromosome rearrangements, the mtDNA fragments insertion into nDNA seems also to be opportunistically used both during evolution and during aging. ROS could have been used by ancestor free-living bacteria to evolve (because they cause mutations), and were used later by multicellular eukaryotes to age at a determined rate in each species. Thus, looking to what happens nowadays when mtDNA fragments accumulate during the lifetime of the individual can help us to understand what happened during evolution, and vice versa. The mitROSp at complex I-dependent mtDNA fragments insertion in nDNA would be a further example illustrating Theodosius Dobzhanský's famous sentence that *Nothing in Biology Makes Sense Except in the Light of Evolution*. Perhaps such a kind of damage mechanism is the missing piece that gerontology has been looking for through a whole century.

In summary, the lack of increase to phenotypic threshold (around 70% deleted) of mtDNA deletions due to the very high copy number of mtDNA in heteroplasmia per cell was a strong problem for the validity of MFRTA, similar to the main origin of mtDNA point mutations from replication/repair instead of from mitROSp as deduced from the relative frequency of base transversions versus transitions. However, now there is evidence that mtDNA fragments accumulate inside nDNA with age in yeast and mammals, and that this promotes aging [21]. Furthermore, the longevity increasing drug rapamycin reverses such age-related accumulation in mouse liver strongly paralleling what happens for complex I mitROSp [11]. All that means that MFRTA cannot be considered "dead" any more except by those not aware of the recent breakthrough discoveries of mtDNA transfer from mitochondria to the nucleus during the lifetime of the individual. On the contrary MFRTA, like the phoenix, emerges again alive, well, and all pristine from its ashes. Perhaps the failure to detect mtDNA mutations relevant for aging has been due to looking in the wrong place. The ROS related to aging should be looked for in mitochondria. However, what could be more important for aging would not be the (remaining) deleted mtDNA, but what is lacking in mtDNA after the deletion: the mtDNA fragments. The consequences that are likely to be relevant for aging: the mtDNA fragments found inserted into nDNA inside the nucleus. Therefore, we should look both at the mitochondria and to the nucleus to understand aging.

CONCLUSIONS

MitROS are constantly produced throughout life at a different rate in each species in agreement with their longevity. Such mitROSp leads to the generation of oxidative damage in mtDNA (e.g., 8-oxodG), which is repaired, and can also lead to point mutations in the process. But mitROS, substances produced by the organism that have capacity to break covalent bonds, also generate single and DSBs in mtDNA. This leads to irreversible forms of damage (mutations) like mtDNA deletions, and mtDNA fragments which enter the chromosomes through the pericentromeric regions and accumulate in nDNA during aging. The steady-state level of 8-oxodG in mtDNA is a marker of the flow of ROS-dependent damage generation and repair through mtDNA. Its measurement is a useful marker of the rate of generation of mtDNA deletions and mtDNA fragments. Mutations can also arise due to processes unrelated to oxidative stress like mtDNA replication and repair. However, it is highly unlikely that these last mechanisms of damage generation are related to longevity, because their random nature cannot explain the determination of longevity during DRs and in different animal species. It has been argued that the types of base mutations (transitions or transversions) present mainly in mtDNA indicate that they come mainly from mtDNA replication and repair. This has been taken as evidence against MFRTA. But this applies only to base substitutions and not to mitROS-induced DNA strand breaks leading to mtDNA large deletions, and to mtDNA fragments insertion inside nDNA. When irreversibly damaged mtDNA reaches a high threshold level in a cell, approaching homoplasmy of mutated mtDNA, mitochondrial ATP generation through oxidative phosphorylation is decreased to levels great enough to contribute to aging. But it seems that the level of mtDNA deletions in homoplasmy in tissue cells from old individuals do not reach such a threshold level to contribute to explain aging.

However, it is now known that mtDNA fragments accumulate during aging inside nDNA in yeast, rat liver and brain, and mouse liver [11,19–21], causing an increase in chronological aging at least in yeast [19]. Recent investigations show that such accumulation, as well as increases in complex I mitROSp and FRL with age in mouse liver, is fully reversed by rapamycin dietary treatment [11]. This is accompanied by rapamycin-induced strong increases in autophagy fully reverting to young levels, and by partial reversion of lipofuscin accumulation with age [11]. Interestingly, recent studies indicate that the mtDNA fragments do not enter nDNA randomly. Instead they are directed to the pericentromeric regions as a main "entry door" to the nucleus. From there, they can potentially disseminate to other chromosome regions, including perhaps regulatory regions, thus contributing to aging and to promoting mortality at old age. Further research is urgently needed to clarify this possibility. Alternatively, the mtDNA fragments can potentially alter the information coded in nDNA. They could be directed specifically to structural genes or their regulatory regions, thus promoting cell death, cell malfunction, or cellular malignant transformation, and thus cancer and aging.

REFERENCES

1. Harman D. The biological clock: The mitochondria? *J Am Geriatr Soc* 1972; 20: 145–7.
2. Ku HH, Brunk UT, Sohal RS. Relationship between mitochondrial superoxide and hydrogen peroxide production and longevity of mammalian species. *Free Rad Biol Med* 1993; 15: 621–7.
3. Lambert A, Boysen H, Buckingham JA, Yang T, Podlutsky A, Austad SN et al. Low rates of hydrogen peroxide production by isolated heart mitochondria associate with long maximum lifespan in vertebrate homeotherms. *Aging Cell* 2007; 6: 607–18.
4. Barja G, Cadenas S, Rojas C, Pérez-Campo R, López-Torres M. Low mitochondrial free radical production per unit O_2 consumption can explain the simultaneous presence of high longevity and high metabolic rates in birds. *Free Rad Res* 1994; 21: 317–28.
5. Herrero A, Barja G. H_2O_2 production of heart mitochondria and aging rate are slower in canaries and parakeets than in mice: Sites of free radical generation and mechanisms involved. *Mech Ageing Dev* 1998; 103: 133–46.
6. Barja G. Updating the mitochondrial free radical theory of aging: An integrated view, key aspects and confounding concepts. *Antiox Redox Signaling* 2013; 19: 1420–45.
7. Barja G, Herrero A. Oxidative damage to mitochondrial DNA is inversely related to maximum life span in the heart and brain of mammals. *FASEB J* 2000; 14: 312–18.
8. Gredilla R, Barja G. The role of oxidative stress in relation to caloric restriction and longevity. *Endocrinol* 2005; 146: 3713–7.

9. Sanz A, Caro P, Ayala V, Portero-Otin M, Pamplona R, Barja G. Methionine restriction decreases mitochondrial oxygen radical generation and leak as well as oxidative damage to mitochondrial DNA and proteins. *FASEB J* 2006; 20: 1064–73.

10. Sanchez-Roman I, Barja G. Regulation of longevity and oxidative stress by nutritional interventions: Role of methionine restriction. *Exper Gerontol* 2013; 48: 1030–42.

11. Martínez-Cisuelo V, Gómez J, García-Junceda I, Naudí A, Cabré R, Mota-Martorell N et al. Rapamycin reverses age-related increases in mitochondrial ROS production at complex I, oxidative stress, accumulation of mtDNA fragments inside nuclear DNA, and lipofuscin level in liver of middle-aged mice. *Exper Gerontol* 2016; 83: 130–8.

12. Sato A, Nakada K, Shitara H, Kasahara A, Yonekawa H, Hayashi JI. Deletion-mutant mtDNA increases in somatic tissues but decreases in female germ cells with age. *Genetics* 2007; 177: 2031–7.

13. Kraytsberg Y, Kudryavtseva E, McKee AC, Geula C, Kowall NW, Khrapko K. Mitochondrial DNA deletions are abundant and cause functional impairment in aged human substantia nigra neurons. *Nat Genet* 2006; 38; 518–20.

14. Bender A, Kishnan K, Morris CM, Taylor GA, Reeve AK, Perry RH et al. High levels of mitochondrial DNA deletions in substantia nigra neurons in aging and Parkinson disease. *Nat Genet* 2006; 38: 515–7.

15. Richter C. Do mitochondrial DNA fragments promote cancer and aging? *FEBS Lett* 1988; 241: 1–5.

16. Suter M, Richter C. Fragmented mitochondrial DNA is the predominant carrier of oxidized DNA bases. *Biochemistry* 1999; 38: 459–64.

17. Patrushev M, Kasymov V, Patrusheva V, Ushakova T, Gogvadze V, Gaziev A. Mitochondrial permeability transition triggers the release of mtDNA fragments. *Cell Mol Life Sci* 2004; 61: 3100–3.

18. Patrushev M, Kasymov V, Patrusheva V, Ushakova T, Gogvadze V, Gaziev AI. Release of mitochondrial DNA fragments from brain mitochondria of irradiated mice. *Mitochondrion* 2006; 6: 43–7.

19. Cheng X, Ivessa AS. The migration of mitochondrial DNA fragments to the nucleus affects the chronological aging process of Saccharomyces cerevisiae. *Aging Cell* 2010; 9: 919–23.

20. Caro P, Gómez J, Arduini A, González-Sánchez M, González García M, Borrás C et al. Mitochondrial DNA sequences are present inside nuclear DNA in rat tissues and increase with age. *Mitochondrion* 2010; 10: 479–86.

21. Cheng X, Ivessa AS. Accumulation of linear mitochondrial DNA fragments in the nucleus shortens the chronological life span of yeast. *Eur J Cell Biol* 2012; 91: 782–8.

22. Schneider JS, Cheng X, Zhao Q, Underbayev C, Gonzalez JP, Raveche ES, Fraidenraich D, Ivessa AS. Reversible mitochondrial DNA accumulation in nuclei of pluripotent stem cells. *Stem Cells Dev* 2014;23: 2712–9.

23. Miller FJ, Rosenfeldt FL, Zhang C, Linnane AW, Nagley P. Precise determination of mitochondrial DNA copy number in human skeletal and cardiac muscle by a PCR-based assay: Lack of change of copy number with age. *Nucleic Acids Res* 2003; 31:e61.

24. Siddiqui A, Bhaumik D, Chinta SJ, Rane A, Rajagopalan S, Lieu CA, Lithgow GJ, Andersen JK. Mitochondrial quality control via PGC1α-TFEB signaling pathway is compromised by parkin Q311X mutation but independently restored by rapamycin. *J Neurosci* 2015; 35: 12833–44.

25. Miwa S, Jow H, Baty K, Johnson A, Czapiewski R, Saretzki G et al. Low abundance of the matrix arm of complex I in mitochondria predicts longevity in mice. *Nat Commun* 2014; May 12; 5: 3837. doi: 10.1038/ncomms4837.

26. Genova ML, Ventura B, Giuliano G, Bovina C, Formiggini G, Parenti Castelli G et al. The state of production of superoxide radical in mitochondrial Complex I is not a bound semiquinone but presumably iron-sulphur cluster N2. *FEBS Lett* 2001; 505: 364–68.

27. Kushnareva Y, Murphy A, Andreyev A. Complex I-mediated reactive oxygen species generation: Modulation by cytochrome c and NAD(P+) oxidation-reduction state. *Biochem J* 2002; 368: 545–53.

28. Herrero A, Barja G. Sites and mechanisms responsible for the low rate of free radical production of heart mitochondria in the long-lived pigeon. *Mech Ageing Dev* 1997; 98: 95–111.

29. Herrero A, Barja G. Localization of the site of oxygen radical generation inside the Complex I of heart and non-synaptic brain mammalian mitochondria. *J Bioenerg Biomembr* 2000; 32: 609–15.

30. Gredilla R, Sanz A, Lopez-Torres M, Barja G. Caloric restriction decreases mitochondrial free radical generation at Complex I and lowers oxidative damage to mitochondrial DNA in the rat heart. *FASEB J* 2001; 15: 1589–91.

31. Kuznetsova IS, Prusov AN, Enukashvily NI, Podgornaya OI. New types of mouse centromeric satellite DNAs. *Chromosome Res* 2005; 13: 9–25.

32. Kuznetsova I, Podgornaya O, Ferguson-Smith MA. High-resolution organization of mouse centromeric and pericentromeric DNA. *Cytogenet Genome Res* 2006; 112: 248–55.

33. Matsuo M, Ito Y, Yamuchi R, Obokata J. The rice nuclear genome continuously integrates, shuffles, and eliminates the chloroplast genome to cause chloroplast-nuclear DNA flux. *The Plant Cell* 2005; 17: 665–75.

34. Michalovova M, Vyskot B, Keinovsky E. Analysis of plastid and mitochondrial insertions in the nucleus (NUPTs and NUMTs) of six plant species: Size, relative age and chromosomal localization. *Heredity* 2013; 111: 314–20.

35. Komissarov AS, Kuznetsova IS, Podgornaia OI. Mouse centromeric tandem repeats in silico and *in situ*. *Russian J Genet* 2010; 46: 1080–83.

36. Komissarov AS, Gavrilova EV, Demin SJ, Ishov AM, Podgornaya OI. Tandemly repeated DNA families in the mouse genome. *BMC Genomics* 2011;12: 531. doi: 10.1186/1471-2164-12-531.

37. Chueh AC, Northrop EL, Brettingham-Moore KH, Choo KH, Wong LH. LINE retrotransposon RNA is an essential structural and functional epigenetic component of a core neocentromeric chromatin. *PLoS Genet* 2009; e1000354. doi: 0.1371/journal.pgen.1000354.

38. Carone DM, Longo MS, Ferreri GC, Hall L, Harris M, Shook N, Bulazel KV, Carone BR, Obergfell C, O'Neill MJ, O'Neill RJ. A new class of retroviral and satellite encoded small RNAs emanates from mammalian centromeres. *Chromosoma* 2009; 118: 113–25. doi: 10.1007/s00412-008-0181-5.

39. Kuznetsova IS, Ostromyshenskii DI, Komissarov AS, Prusov AN, Waisertreiger IS, Gorbunova AV, Trifonov VA, Ferguson-Smith MA, Podgornaya OI. LINE-related component of mouse heterochromatin and complex chromocenters's composition. *Chromosome Res* 2016; 24: 309–23.

40. Gaziev AI and Shaikhaev GO. Nuclear mitochondrial pseudogenes. *Mol Biol (Russia)* 2010; 44: 358–68.

41. Shibata Y, Kumar P, Layer R, Willcox S, Gagan JR, Griffith JD, Dutta A. Extrachromosomal microDNAs and chromosomal microdeletions in normal tissues. *Science* 2012; 336: 82–6.

42. Dillon LW, Kumar P, Shibata Y, Wang YH, Willcox S, Griffith JD, Pommier Y, Takeda S, Dutta A. Production of extrachromosomal MicroDNAs is linked to mismatch repair pathways and transcriptional activity. *Cell Rep* 2015; 11: 1749–59.

43. Hu Z, Chen K, Xia Z, Chavez M, Pal S, Seol JH, Chen CC, Li W, Tyler JK. Nucleosome loss leads to global transcriptional up-regulation and genomic instability during yeast aging. *Genes Dev* 2014; 28: 396–408.

44. Cabral M, Cheng X, Singh S, Ivessa AS. Absence of non-histone protein complexes at natural chromosome pause sites results in reduced replication pausing in aging yeast cells. *Cell Reports* 2016; 17: 1747–54.

45. Raveendranathan M, Chattopadhyay S, Bolon YT, Haworth J, Clarke DJ, Bielinsky AK. Genome wide replication profiles of S-phase checkpoint mutants reveal fragile sites in yeast. *EMBO J* 2006; 25: 3627–39.

46. Song W, Dominska M, Greenwell PW, Petes TD. Genome-wide high resolution mapping of chromosomal fragile sites in *Saccharomyces cerevisiae*. *Proc Natl Acad Sci USA* 2014; 111: E2210–18.

47. Lemoine FJ, Degtyareva NP, Lobachev K, Petes TD. Chromosomal translocations in yeast induced by low levels of DNA polymerase a model for chromosomal fragile sites. *Cell* 2005; 120: 587–98.

48. McMurray MA and Gottschling DE. An age-induced switch to a hyper-recombinational state. *Science* 2003; 301: 1908–11.

49. Lindstrom DL, Leverich CK, Henderson KA, Gottschling DE. Replicative age induces mitotic recombination in the ribosomal RNA gene cluster of *Saccharomyces cerevisiae*. *PLoS Genet* 2011; 7(3):e1002015. doi: 10.1371/journal.pgen.1002015.

50. Lopez-Otin C, Blasco MA, Partridge L, Serrano M, Kroemer G. The hallmarks of aging. *Cell* 2013; 153: 1194–1217.

51. Guenatri M, Bailly D, Maison C, Almouzni G. Mouse centric and pericentric satellite repeats form distinct functional heterochromatin. *J Cell Biol* 2004; 166: 493–505.

52. Turner C, Killoran C, Thomas NS, Rosenberg M, Chuzhanova NA, Johnston J et al. Human genetic disease caused by *de novo* mitochondrial-nuclear DNA transfer. *Hum Genet* 2003; 112: 303–9.

53. Willett-Brozick JE, Savul SA, Richey LE, Baysal BE. Germ line insertion of mtDNA at the breakpoint junction of a reciprocal constitutional translocation. *Hum Genet* 2001; 109: 216–23.

54. Goodenough U, Heitman J. Origins of eukaryotic sexual reproduction. *Cold Spring Harb Perspect Biol* 2014; 6. pii: a016154. doi:10.1101/cshperspect.a016154.

55. Blackstone NW, Green DR. The evolution of a mechanism of cell suicide. *Bioessays* 1999; 21: 84–8.

56. Radzvilavicius AL, Blackstone NW. Conflict and cooperation in eukaryogenesis: Implications for the timing of endosymbiosis and the evolution of sex. *J R Soc Interface* 2015; 12: 0584. doi: 10.1098/rsif.2015.0584.

57. Lane N. Energetics and genetics across the prokaryote–eukaryote divide. *Biol Direct* 2011; 6: 35. doi: 10.1186/1745-6150-6-35.

21 Mitochondrial Oxidative Stress in Aging and Healthspan

Dao-Fu Dai

CONTENTS

INTRODUCTION

The roles of free radicals in aging have a long, complicated, and controversial history. The free radical hypothesis of aging was first proposed by Denham Harman (1956), in which he proposed that free radical-induced macromolecular damage accumulated with age and was the primary factor in aging as well as an essential determinant of life expectancy.[1] This hypothesis was alluring and simple and it has been one of the most broadly tested theories of aging. Several studies have confirmed the increase in reactive oxygen species (ROS)-associated macromolecular damages with aging. However, the causal roles of ROS in aging remained controversial. These controversies came from mixed results from experimental and clinical studies supplementing antioxidants, which albeit reducing ROS, did not consistently show antiaging effects. In the *Caenorhabditis elegans* model of aging, several strains deficient in respiratory proteins that produce excess ROS are in fact longer lived than wild type (WT) controls.[2] In mammals, mice with deletion of various antioxidant enzymes had little effect on life span[3–6] and, even more importantly, overexpression of various antioxidants including superoxide dismutase (SOD) and peroxisomal catalase (pCAT) failed to extend the life span in mice.[4–6] Furthermore, naked mole rats that have a very long life span (10–30 years) compared to mice (2–3 years) provide ther evidence against the free radical theory of aging. Despite more extensive oxidative damage found in naked mole rats over their lifetime, naked mole rats have substantially longer life span compared to other rodents, suggesting that ROS may not necessarily play a causative role in aging.[7]

MITOCHONDRIAL FREE RADICAL THEORY OF AGING

Harman further refined his theory in 1972 to specify that mitochondria are the main sites of ROS production and also the main targets of ROS-induced damage.[8] This mitochondrial version of the free radical theory of aging postulates that mitochondria are generating ROS from the electron transport chain during oxidative phosphorylation, which may attack several mitochondrial macromolecules to induce oxidative damage. The mitochondrial version of the free radical theory also suggests that previous experiments that failed to validate the general free radical theory might be explained by the lack of experimental interventions to target mitochondrial ROS. Based on this, although general antioxidants have been failed, antioxidants targeted to mitochondria are expected to extend longevity and healthspan.

Direct evidence of mitochondrial free radical of aging are supported by studies applying transgenic mouse overexpressing catalase targeted mitochondria (mCAT), compared with mice overexpressing catalase targeted to the endogenously targeted site of peroxisome (pCAT) and targeted to nucleus (nCAT). Comparison of catalase overexpression targeted to various sites (nucleus, peroxisomes, or mitochondria) allows a direct assessment of the relative importance of ROS (especially H_2O_2) in nucleus, cytoplasmic versus those in mitochondrial compartments. There are several advantages of using catalase as antioxidant enzyme in this context. First, catalase does not directly consume adenosine tri-phosphate (ATP) or alter glutathione or NAD(P)H redox balance when it detoxifies H_2O_2. Second, catalase converts two H_2O_2 molecules to water and oxygen only at relatively high H_2O_2 levels. The Km of the catalytic activity of catalase for conversion of H_2O_2 to water is >10 mM. At lower levels, when a second H_2O_2 may be in limited supply, catalase indeed acts as a peroxidase and oxidizes a variety of substrates,[9] so it is less likely to interfere with low level of H_2O_2 concentrations that may be involved in signaling or hormesis (see below).[10,11] Thus, this strategy can be utilized to investigate the effects of reduced H_2O_2 on longevity and healthspan. The initial report demonstrated that of the three mouse longevity cohorts of catalase overexpression, the mCAT enhanced both maximal and median life span extension by approximately 20%.[12] The pCAT showed a modest and overall nonsignificant extension of life span, and contradictory to our initial hypothesis, the nCAT had no positive effect on life span.[12] Of note, the life span extension effects of mCAT are not observed when the ethical censoring rate for "end-of-life" pathology was very high. Since the definition of "end-of-life" varies in different experiments and may include several organ systems, investigations of healthspans may provide further insights into the study of aging.

As increase in maximal life span is suggestive of a protection of the underlying aging processes, mCAT mice which had mitochondrial catalase expression highest in the heart and skeletal muscle, displayed reduced cardiac pathology, arteriosclerosis, and myopathy phenotypes. This reduced cardiac pathology was associated with reduction in H_2O_2 release from isolated cardiac mitochondria, decreased accumulation of oxidized DNA, and decreased susceptibility of mitochondrial aconitase to H_2O_2-induced damage.[12] All of these beneficial effects may be explained by enhanced antioxidant defenses in mitochondria.[13]

Several follow-up studies (discussed in section Mitochondrial Oxidative Stress in Healthspan and Potential Interventions) have demonstrated vicious cycle of ROS generation within mitochondria that involved electron transport chains, mitochondrial DNA, and endogenous antioxidants (Figure 21.1).

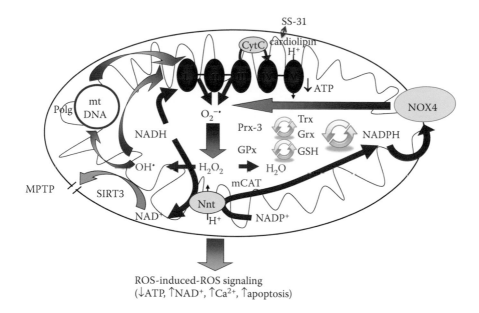

FIGURE 21.1 ROS amplification within mitochondria.

Mitochondrial ROS is produced by respiratory complexes during oxidative phosphorylation and the ROS may induce oxidative damage to respiratory complexes. These oxidatively damaged respiratory complex proteins or defective respiratory protein subunits produced by mutant mtDNA may further increase ROS production because of inefficient electron transport. Increase in mitochondrial ROS production may lead to changes in the mitochondrial redox balance, which is regulated by glutathione and NAD(P)H redox system (see Figure 21.1). The reductants NAD(P)H are utilized to regenerate glutathione peroxidase (GPx) and peroxiredoxin (Prx), the primary intrinsic mitochondrial antioxidant enzymes that detoxify H_2O_2, preventing its conversion to the highly reactive hydroxyl or peroxynitrite radicals. Furthermore, the reductive potential of nicotineamide adenine dinucleotide phosphate (NADPH) is in balance with that of NADH, via the activity of nicotinamide nucleotide transferase (NNT); the balance of nicotinamide adenine dinucleotide (NAD)/NADH regulates sirtuin histone deacetylases, including mitochondrial SIRT3.[14] This mechanism links the redox homeostasis with NAD^+-dependent epigenetic regulation of diverse metabolic consequences (section Proteostasis, Autophagy, and Mitochondrial Dysfunction in Aging and Longevity and Mitochondrial Oxidative Stress in Healthspan). Many studies using mCAT mouse models have provided evidence supporting the ROS vicious cycle in mitochondria and are consistent with mitochondrial free radical theory.

MITOCHONDRIAL HORMESIS

In recent years, some studies have shown that not all of the effects of ROS are necessarily harmful. Increasing evidence has shown the important roles of ROS in normal physiological signaling and stress response. Dietary restriction, especially glucose restriction, has been shown to preferentially induce mitochondrial metabolism to extend life span in various model organisms, including *Saccharomyces cerevisae*,[15] *Drosophila melanogaster*,[16] and *C. elegans*.[17] Increased mitochondrial metabolism and respiration is expected to increase the production of ROS. For example, glucose restriction in *C. elegans* extended life span by inducing mitochondrial respiration and increasing oxidative stress, and this adenosine monophosphate activated kinase (AMPK)-dependent life span extension was abolished by pretreatment of antioxidant N-acetyl cysteine (NAC), indicating that ROS is crucial in the signaling to induce stress resistance.[17] This concept of mitochondrial hormesis (mitohormesis) proposes that at low levels of ROS, induced by caloric restriction, exercise,[18] or other stimuli, have beneficial effect in promoting health and longevity by inducing endogenous protective mechanisms that subsequently enhancing stress resistance[19] and overall reduction of chronic oxidative damage.[19]

The studies in *C. elegans* showed that increase ROS from mitochondria can significantly increase life span through the ROS-mediated activation of HIF-1.[20] Furthermore, metformin has been shown to extend life span in *C. elegans* by blocking respiratory complex I, which increased ROS, leading to the activation of peroxiredoxin, mitogen-activated protein kinase cascade, and subsequent stress-resistance pathways.[21] Mice with heterozygous deletion of MCLK1 (which encodes an enzyme in ubiquinone biosynthesis) demonstrated life span extension despite an increase in mitochondrial ROS resulting from reduction of mitochondrial electron transport, tricarboxylic acid cycle, and ATP synthesis. Surprisingly, the mitochondrial dysfunction in Mclk1[+/−] mice was accompanied by a decrease in oxidative damage to cytosolic proteins as well as by a decrease in plasma isoprostanes, a systemic biomarker of oxidative stress and aging.[22] These findings support the beneficial effect of mitochondrial ROS in mitohormesis.

Mitohormesis represents a retrograde mitochondrial stress signaling to the nucleus. Stresses that cause mitochondrial perturbation may include increased ROS, decline in energetic substrates, misfolded/damaged proteins, or mitochondrial toxins. Several endogenous protective mechanisms have been reported. Low levels of ROS induce upregulation of endogenous antioxidants, which reduce chronic oxidative damage and subsequently improve healthspan.[19] Unfolded peptides induce activation of mitochondrial unfolded protein response (mtUPR) to improve proteostasis. Damaged proteins and organelles activate autophagy to remove harmful proteins and organelles. Decline in energetics substrates may induce mitochondrial biogenesis and metabolic adaptations (see review[23]).

As discussed in section Mitochondrial Free Radical Theory of Aging and Mitochondrial Oxidative Stress in Healthspan, the majority of publications using mitochondria-targeted catalase (mCAT) mice highlight its protective effects and potential therapeutic uses of mitochondrial antioxidants. How can this evidence supporting mitochondrial free radical theory be reconciled with mitohormesis? These seemingly contradictory findings might be explained using a model of continuum of ROS. Low levels of ROS serve important physiological signaling roles that are critical for metabolism, protein turnover, cellular differentiation, stress response (i.e., mitohormesis), and apoptosis[24,25] (Figure 21.2). Low level ROS is also required for the renewal of stem cells as over-suppression of ROS significantly limits renewal of stem cells and reduces neurogenesis in mice.[26] On the other end of the spectrum, rapid rise of pathological or high levels of ROS may react with proteins, DNA, and lipids to damage important components of the cell. In aging and various disease models, increasing levels of ROS and oxidative damage are widely documented and supported by the protective effects of mCAT, which are consistent with "mitochondrial free radical theory"[27] (Figure 21.2). While mCAT suppression of pathological levels of ROS is protective in aging and diseases, mCAT demonstrates adverse effect on young healthy mice, such as impaired

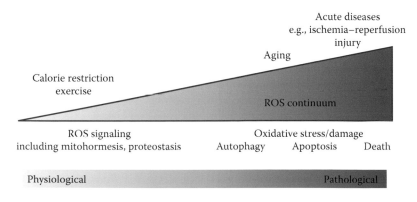

FIGURE 21.2   ROS continuum.

bactericidal activity.[28] A recent study reported that mCAT overexpression in young mouse hearts exhibited cardiac proteome profiles resembling that found in old wild-type mouse hearts. These suggest an adverse impact of removing the more beneficial effects of mitochondrial ROS in signaling.[29] Furthermore, while moderate overexpression of mCAT normalized mitochondrial ROS, preserved mitochondrial function, and attenuated impairment of Mitofusin 2 null hearts, in contrast, super-suppression of mitochondrial ROS by high mCAT overexpression impaired compensatory autophagy and failed to improve either mitochondrial fitness or cardiomyopathy in these mice.[30]

Taken together, the studies discussed here suggest that an ideal antioxidant therapy might be one that prevents oxidative damage induced under pathological conditions without interfering with ROS needed for physiological ROS signaling or hormesis. Thus, mitochondrial antioxidant may not be universally beneficial, and the beneficial effects are observed only in a setting when "pathological" oxidative stress or a high burst of ROS is anticipated.

PROTEOSTASIS, AUTOPHAGY, AND MITOCHONDRIAL DYSFUNCTION IN AGING AND LONGEVITY

Proteostasis or protein homeostasis is the equilibrium state between protein synthesis and degradation. Proteostasis is a fundamental mechanism to maintain cellular and organismal wellbeing. Previous studies demonstrate that aging and age-dependent degenerative diseases are associated with abnormal accumulation of defective proteins and biomolecules as a result of oxidative damage and inefficient removal of damaged proteins.[31] Examples of such proteins include neurofibrillary tangles (NFTs) and beta amyloid proteins in Alzheimer's disease (AD), alpha synuclein in Parkinson's disease (PD),[32] lipofuscin and senile amyloid in cardiac and skeletal muscle ageing,[33,34–36] and damaged crystallins in cataracts.[37] The defective proteostasis has been observed prior to development of various degenerative diseases,[38] emphasizing critical roles of maintaining protein quality control.

Several cellular mechanisms are involved in proteostasis and protein quality control. Autophagy is one of the main mechanisms degrading the vast majority of proteins. The other mechanism includes ubiquitin–proteasome pathway. Autophagic degradation involves lysosomes, which contains several digestive enzymes.[39] Three major pathways have been described based on the delivery mechanisms of macromolecules to the lysosome: microautophagy, macroautophagy, and chaperone-mediated autophagy (CMA).[40] Microautophagy acts through invaginations of the lysosomal membrane directly engulfing cytoplasmic macromolecules into the lysosome followed by enzymatic degradation.[40,41] Macroautophagy involves formation of autophagosomes, which are double-membrane vesicles formed by phagophores engulfing cytosolic proteins and organelles. These autophagosomes fuse with lysosomes, leading to the degradation of the sequestered cellular contents by lysosomal enzymes.[42,43] CMA is a targeted degradation, in which cytosolic proteins containing a pentapeptide KFERQ motif are targeted and translocated across the lysosomal membrane, resulting in the degradation of specific proteins.[44] These three mechanisms of autophagy may occur simultaneously in various cell types.[39,45,46] Macroautophagy is the most extensively studied of these three mechanisms and will be the focus of the following discussion.

Autophagy has been shown to be an important determinant of longevity, as shown by life span extension observed in organisms with enhanced autophagy, such as brain-specific overexpression of Atg8a in *Drosophila*[47] and ubiquitous overexpression of Atg5 in mice.[48] Overexpression of these autophagosome component molecules extends life span and maintains youthful phenotypes in old organisms. The protective effect is mediated by enhanced cleaning of toxic protein aggregates and damaged mitochondria in aging, including those containing harmful levels of ROS. These harmful molecules or organelles are removed by autophagy in normal young tissues (Figure 21.3a). In aging, the capacity of autophagy is exhausted by excessive ROS, oxidative damage, and misfolded proteins, leading to accumulation of these harmful molecules, which might cause cell death, organ dysfunction, and eventually lead to organism mortality (Figure 21.3b). This is consistent with the mitochondrial free radical theory of aging discussed above. Since low level of ROS is known to be

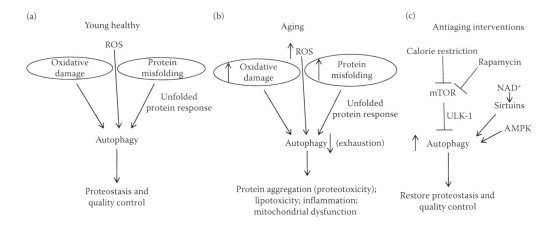

FIGURE 21.3 The relationship between ROS-induced damage and autophagy in aging and antiaging interventions. (a) Autophagy is induced by ROS, oxidatively damaged macromolecules, and misfolded proteins (via unfolded protein response) as a part of protein homeostasis and quality control in young healthy tissue. (b) In aged tissues, there is a substantial increase in ROS, oxidative damage, and misfolded proteins, which exhausts the capacity of autophagy and results in protein aggregation, lipotoxicity, mitochondrial, and other organelle dysfunction, leading to increased inflammation and organ dysfunction. (c) Several antiaging interventions have been shown to enhance autophagy and restore proteostasis.

one of the main signaling mechanisms inducing autophagy to remove damaged organelles, this is also consistent with the concept of mitochondrial hormesis discussed above.[19]

Additional evidence to support the association between autophagy and aging is shown by antiaging intervention studies (section Potential Interventions). Many interventions that have been shown to extend life span, including mammalian target of rapamycin (mTOR) inhibition by rapamycin[49] or calorie restriction (CR)[50] and *Sirtuin* activators[51] are mediated through enhanced autophagy (Figure 21.3c).[32,45,52–54] Similar to CR, inhibition of mTORC1 by rapamycin is well documented to extend life span in invertebrate models of aging, including flies,[55] worms,[56,57] and yeast,[58,59] as well as in mice.[49,60] In yeast and *Drosophila*, the life span extension by rapamycin was inhibited by deletion or silencing of Atg1, Atg7, or Atg5,[53,54] indicating that enhanced autophagy is required for the life span extension benefit of mTORC1 inhibition. Sirtuins are class III histone deacetylases that deacetylate (activate) numerous transcription factors, cofactors, histones, and enzymes in response to metabolic stress using NAD$^+$ as a cofactor. Dependent on cytoplasmic or mitochondrial NAD$^+$, sirtuins act as a "sensor" protein for nutrition and energy balance.[61,62] Sirtuins deacetylations of downstream pathways counteract the pathogenic mechanisms underlying aging and longevity as well as several age-related diseases, including diabetes, cardiovascular disease, cancer, inflammatory diseases, and neurodegenerative conditions.[63] For example, Sirt1 has been shown to affect regulators of autophagy and many autophagy-related genes, including Atg5, Atg7, and Atg8.[64] Sirt1 life span extension effect is also dependent on autophagy, as deletion of Beclin-1 (Atg6) which suppressed the induction of autophagy by Sirt, abolish its life span extension effect.[52] Furthermore, several other interventions improving protein quality control have also been shown to improve health and aging in invertebrate and mammalian models.[32,39,45,46]

UBIQUITIN–PROTEASOME SYSTEM IN PROTEOSTASIS

The ubiquitin–proteasome system (UPS) is the other major protein degradation pathway independent of lysosome. In contrast to the generalized and relatively nonspecific degradation in autophagy, UPS degradation is more specific to particular proteins. While autophagy degrades a complex mixture of cytoplasmic biomolecules within the vesicles using lysosomal hydrolases, the UPS specifically targets ubiquitinated proteins (poly-ubiquitin tagging of proteins to be eliminated) and utilizes proteasome for degradation.[39,45,65] This targeted degradation is regulated by a sophisticated mechanism having a high spatial and temporal precision (see reviews).[32,39,45,65,66] The roles of UPS in degrading other macromolecules or organelles are still unknown.

Similar to the perturbation of autophagy, the disruption of UPS function can also cause the accumulation of abnormal protein inclusions, leading to severe toxicity and cell death.[31,66] Genetic depletion of proteasome subunits in mouse brains was shown to induce neuronal protein inclusions leading to neurodegenerative diseases.[31] The role of UPS in aging, however, remains unclear. Studies in *C. elegans* and *Drosophila* demonstrate the action of UPS on specific longevity pathways.[45,66,67] In *C. elegans*, the ubiquitin ligase RLE-1 targets a key component in the insulin/ insulin growth factor (IGF) pathway, daf-16, and leads to its degradation by the proteasome.[67] Thus, its inhibition extends life span in worms. In *Drosophila*, overexpression of parkin, a ubiquitin ligase which promotes mitophagy, has been shown to extend life span.[68] These are examples of UPS effects that promote or inhibit specific longevity pathways. Although previous studies show that proteasomal activity decline with age and are restored in long-lived animals under CR,[45,66] the effect of manipulation of overall UPS function on aging is still unknown.

In summary, the perturbation of proteostasis plays an important role in aging and restoration of protein homeostasis such as autophagy, and is protective against aging and age-related disease (Figure 21.3c). Both autophagy and the UPS as two major mechanisms controlling proteostasis are expected to work in synchrony to regulate protein degradation. However, the interactions between various mechanisms of autophagy and UPSs remain largely unknown. Further studies are needed to investigate the mechanisms underlying the protective effects of enhancing proteostasis by autophagy and UPS on life span and healthspan benefits.

MITOCHONDRIAL DYNAMICS, MITOPHAGY, AND AGING

Mitochondria are dynamic organelles continuously undergoing fusion and fission in order to maintain normal shape, number, and function.[69] Mitochondrial dynamics have been implicated in aging. Prior studies reported significant morphological changes of mitochondria in aging and cardiomyopathy,[70,71] implying that mitochondrial fusion and fission may play critical roles in mitochondrial maintenance and quality control in aging. As a quality control mechanism, mitochondrial fission has been shown to generate mitochondrial fragments with lower oxidative phosphorylation activity, resulting in lower membrane potential, and these mitochondrial fragments will be targeted for degradation by ubiquitination through Pink and Parkin-mediated pathway.[72]

Mitofusin1 (Mfn1), mitofusins2 (Mfn2), and OPA1 are proteins involved in mitochondrial fusion. Abnormality of these proteins may cause abnormal mitochondrial structure and function, resulting in inefficient cellular respiration in many tissues,[73–75] including the aging heart.[76,77] Disruption of both Mfn1 and Mfn2 is embryonic lethal. Deletion of Mfn1 may cause mitochondrial fragmentation, which lead to cardiac hypertrophy and dysfunction.[77] In the heart, deficiency of Mfn2 is associated with the disruption of cell cycle progression, cardiac hypertrophy, reduced oxidative metabolism, and altered mitochondrial permeability transition, which may cause systolic dysfunction.[77] Consistent with this, downregulation of Mfn2 has been reported in several experimental models of heart failure.[78] Suppression of Mfn2 and OPA1 in cardiomyocytes exhibits altered mitochondrial morphology, resulting in large pleomorphic and irregular mitochondria with disrupted cristae structure.[77,79] Mitochondrial fusion is an essential mechanism to maintain mitochondrial integrity and function. Partial deletion of both Mfn1 and 2 in mice has been shown to cause mitochondrial fragmentation, impaired mitochondrial respiration, and cardiomyopathy.[80] The changes in mitochondrial dynamics and their role in aging warrant further study.

Mitophagy is a special form of autophagy that degrades damaged mitochondria as a part of quality control (see Reference 81). One of the best studied mechanisms of mitophagy involves phosphate and tensin homolog (PTEN)-induced putative kinase 1 (PINK1)-Parkin-mitofusin 2 (Mfn2) complex, which marks a depolarized mitochondria to be engulfed by autophagosomes through an LC3-receptor-dependent mechanism.[82] PINK1 induces phosphorylation of ubiquitin, which recruits LC3-receptor proteins and other autophagy factors to mitochondria.[83] Deletion of Parkin led to the accumulation of disorganized mitochondria in aged mouse cardiomyocytes.[84] This study emphasizes the essential role of Parkin to mediate autophagy in cardiac aging.[85] In summary, mitochondrial protein quality declined with age because of increase oxidative damage and impairment in many of these quality control mechanisms in old age.

MITOCHONDRIAL OXIDATIVE STRESS IN HEALTHSPAN

The roles of oxidative stress and ROS-mediated signaling are well documented in several age-related diseases. As discussed above, the applications of mCAT model provide direct evidence of mitochondrial ROS (H_2O_2) involvement in these diseases.

CARDIOMETABOLIC DISEASES

Metabolic Syndrome and Atherosclerosis

Metabolic syndrome is a constellation of metabolic disorders including abdominal obesity, impaired fasting glucose and insulin resistance, elevated blood pressure, as well as abnormal cholesterol and triglyceride levels. Metabolic syndrome increased the risk of diabetes and atherosclerotic cardiovascular diseases independent of obesity.[86] Metabolic syndrome has been associated with proinflammatory and prothrombotic states as well as dyslipidemia,[87,88] which may promote atherogenesis.

Oxidative stresses play a critical role in the pathogenesis of metabolic syndrome.[89] In skeletal muscles of both rodents and humans, a high fat diet increases the mitochondrial potential to emit H_2O_2, shifts the cellular redox environment to a more oxidized state, and decreases the redox-buffering capacity, independent of mitochondrial respiratory dysfunction.[89] Mechanistically, obese skeletal muscle demonstrates reduced capacity of fatty acid oxidation, leading to lipid accumulation within skeletal muscles which subsequently result in accumulation of fatty acyl-CoA's and other proinflammatory metabolites. This in turn impairs β-cell function and dysregulate insulin signing,[89,90] which further leads to increased hydrogen peroxide emission in mitochondria. The overexpression of mCAT exhibits a preservation of mitochondrial function, energy metabolism, and insulin sensitivity in skeletal muscle fed a high fat diet[89] as well as in aged skeletal muscles.[91] The role of oxidative stresses in metabolic syndrome has also been supported in clinical studies. An observational longitudinal clinical study in young nondiabetic adults reported associations between oxidative stress markers, including serum F_2-isoprostanes and oxidized low density lipoprotein, and insulin resistance measured by homeostasis model assessment (HOMA-IR).[92]

The roles of mitochondrial ROS in atherosclerosis are supported by amelioration of atherosclerosis by mCAT. Overexpression of mCAT in macrophages reduces the burden of atherosclerotic lesions in low density lipoprotein receptor$^{-/-}$ mice fed with a high fat diet.[93] The mCAT has also been shown to suppress the proinflammatory pathway, including NF-κB-mediated entry of monocytes into atherosclerotic lesions.[93] Furthermore, mCAT reduces insulin resistance in skeletal muscle and the improvement of insulin sensitivity may be beneficial in slowing atherosclerosis. These findings suggest that mitochondrial protection represents a viable strategy to attenuate progression of atherosclerosis.

Cardiac Aging and Heart Failure

ROS has been implicated in the pathogenesis of several heart diseases, including hypertension, cardiac hypertrophy in aging and pressure overload, ischemia–reperfusion injury, as well as heart failure. Several sources of ROS have been documented, including mitochondria, NADPH oxidases (NOX), xanthine oxidase, monoamine oxidase, and nitric oxide synthase. Free radicals generated by these sources are maintained at the physiological levels by several endogenous antioxidant systems, including SOD, catalase, thioredoxin (TRX), glutaredoxin (GRX), GPxs, and glutathione reductase (GR). At physiological levels, ROS acts as signaling molecules (such as in mitohormesis), but at pathological levels in response to noxious stimuli, increase in ROS may activate autophagy as a defense mechanism to prevent propagation of ROS in damaged mitochondria. Substantial burst in ROS may incite opening of the mitochondrial permeability transition pore (mPTP) leading to cytochrome c release and activation of apoptosis.

In aged hearts, there is an age-dependent increase in mitochondrial ROS production and impaired ROS detoxification.[94–96] Impaired electron transport function may directly lead to electron leakage and subsequent generation of mtROS. Since the heart is a highly metabolic active organ dependent on ATP generated by mitochondria, it is particularly susceptible to mitochondrial oxidative damage. Consistent with this, impairment of mitochondrial energetics has been documented in human and experimental animals with heart failure.[97] The molecular mechanisms of this may include mitochondrial biogenesis that does not keep up with the increasing demand,[98] mitochondrial uncoupling and decreased substrate availability,[99] and increased mitochondrial DNA deletions.[100]

Previous studies reported significant increase in mitochondrial oxidative damage in aged or failing mouse hearts. These include increase in oxidatively damaged protein shown by increased carbonylation, increase in mitochondrial DNA point mutation and deletion frequencies, presumably due to oxidative stress. Increased mitochondrial damage is expected to stimulate signaling for mitochondrial biogenesis, indicated by upregulation of the master regulator PPAR-γ Coactivator-1-α (PGC-1α) and its downstream transcription factors as well as increase in mtDNA copy number in ageing hearts.[101] Overexpression of mCAT attenuated mitochondrial oxidative damage in aged hearts, concomitant with significant amelioration of cardiac aging phenotypes. These protective effects are indicated by attenuation of age-dependent left ventricular hypertrophy and diastolic dysfunction, and improvement of overall myocardial performance.[101]

The critical role of mitochondria in cardiac aging and heart failure is further reinforced by mice with proofreading-deficient homozygous mutation of mitochondrial polymerase gamma (Polg$^{D257A/D257A}$ designated as Polg$^{m/m}$).[102,103] These mice demonstrated shortened life span and several "accelerated aging-like" phenotypes, including kyphosis, graying and loss of hair, anemia, osteoporosis, and age-dependent cardiomyopathy.[102,104] They display mtDNA point mutations and deletions that substantially increase with age. Concomitant mCAT overexpression in double transgenic mice partially rescue mitochondrial damage and cardiomyopathy in these Polg$^{m/m}$ mice. These findings suggest that mitochondrial ROS and mitochondrial DNA damage are part of a vicious cycle of ROS-induced ROS release (Figure 21.1).[104] The mechanism of mtROS amplification may explain the observations that damaged mitochondria from aged or failing mouse hearts produce more ROS than healthy mitochondria in young hearts. Interestingly, endurance exercise has been shown to enhance the performance of skeletal muscle and attenuate some phenotypes of cardiac dysfunction in these Polg$^{m/m}$ mice.[105] The proposed mechanisms were augmentation of mitochondrial biogenesis (possibly via exercise-induced ROS as seen in mitohormesis), which may compensate the mitochondrial dysfunction in these mice.

Accumulation of mtDNA deletions in old age has also been documented in various tissues in humans, including the heart.[106,107] Genetic mutation of mitochondrial enzymes may manifest as idiopathic hypertrophic and dilated cardiomyopathies in human patients.[108] Likewise, mitochondrial DNA and protein oxidative damage have been reported in various experimental models of cardiac hypertrophy and heart failure,[109] such as in chronic infusion of angiotensin II and Gαq overexpression in mice.[100] By comparing the effect of mCAT and pCAT, we have demonstrated that mCAT, but not pCAT, are protective against cardiac hypertrophy, fibrosis and diastolic dysfunction induced by angiotensin II, as well as heart failure phenotypes in Gαq overexpressing mice.[100] These findings emphasize the central role of mitochondrial ROS in cardiac hypertrophy and heart failure.

These findings also support the mechanism of ROS amplification within mitochondria. Angiotensin II binds to ATR1, a Gαq coupled-receptor, and then activates NOX2 via protein kinase C.[110] ROS generated from NOX2 at the cell membrane and/or from NOX4 at the mitochondrial membrane can stimulate electron leakage from respiratory complexes and induce further mtROS production.[100,111,112] Mechanisms of mtROS amplification may include ROS-induced ROS release as well as a ROS-mtDNA damage vicious cycle (Figure 21.1). The latter is supported by the observations that primary damage to mtDNA, either in Polg$^{m/m}$ or by administration of azidothymidine (AZT), is sufficient to increase ROS, causing cardiac hypertrophy leading to heart failure.[100,101,113] AZT is an anti-HIV nucleoside analog which works through inhibition of retroviral reverse transcriptase. It also inhibits mitochondrial polymerase gamma which may cause several side effects. In experimental animals, AZT-induced cardiomyopathy is comparable to that seen in the Polg$^{m/m}$ mouse, and it is attenuated by mCAT.[113] These findings suggest that primary damage to mitochondrial DNA in aged Polg$^{m/m}$ or in response to AZT lead to cardiomyopathy. The fact that mCAT partially rescue the cardiomyopathy phenotype support the mtROS–mtDNA damage vicious cycle. Thus, breaking the ROS vicious cycle within mitochondria by mCAT or mitochondrial-targeted antioxidants is effective in attenuating both cardiac hypertrophy and failure (Figure 21.1).

Recent studies have shown that NOX4 localized to the mitochondrial membrane. This supports the importance of mitochondrial ROS in relation to NAD metabolism in models of cardiac hypertrophy and failure. NOX4 activation consumes NADPH and directly generates superoxide anions leading to mitochondrial oxidative damage.[114,115] Within mitochondria, superoxide anions are converted by SOD to become hydrogen peroxide, which is physiologically detoxified by peroxiredoxin-3 (Prx-3) and GPx. After their oxidation by hydrogen peroxide, these enzymes are replenished using the reductive power of NADPH. However, the consumption and exhaustion of NADPH by NOX4 further aggravate the mitochondrial vicious cycle (Figure 21.1). In contrast, enhancing mitochondrial antioxidants using mCAT or other small molecules approach may break this vicious cycle by removing superoxide or hydrogen peroxide without exhausting glutathione or NADPH. NADPH can itself be regenerated from NADP$^+$ by electron exchange with NADH, catalyzed by nicotinamide nucleotide transhydrogenase (Nnt). Thus, cardiomyocyte mitochondrial redox status is closely interrelated with NAD metabolism. This is further linked to sirtuins (sensors of the ratio of NAD$^+$/NADH), particularly mitochondrial SIRT3. Sirtuins play a key role in the epigenetic control of cardiac response to aging and stress.

Ischemia–Reperfusion Injury

Ischemia–reperfusion injury is well known to involve ROS. ROS accumulate during ischemia,[116] causing mitochondrial respiratory complex dysfunction, which further produces a burst of ROS during reperfusion. Ischemia also induced lactic acidosis as a result of anaerobic glycolysis, and this will enhance the conversion of the superoxide and hydrogen peroxide to the highly reactive hydroxyl free radicals. Many of these conditions associated with ischemia–reperfusion, including accumulation of ROS, acidic pH, and rapid rise in [Ca$^{2+}_i$], may open the mPTP, leading to more mitochondrial ROS release. This ROS-induced ROS release is one of the mechanism of ROS amplification within mitochondria, as discussed above.[117] In consistence with these, several mitochondrial protective molecules have shown protective effects in various experimental models and in a small clinical trial (e.g., SS31, cyclosporine,[118] see section Potential Interventions).

NEURODEGENERATIVE DISEASES

Aging is associated with progressive decline in nervous system functional performance. This intrinsic nervous system aging may include slower reaction times, degeneration of sensory and motor function, and decline in cognitive and memory performance, which are found in AD, PD, and many other neurodegenerative diseases.

Age-Related Sensorineural Hearing Loss

The prevalence of age-dependent sensorineural hearing loss (presbycusis) significantly increases in the elderly population, estimated to be 30%–35% in the population aged 65–75 years and 40%–50%

in those older than 75 years of age.[119] It is characterized by preferential hearing loss for high pitched sound, causing difficulty in understanding speech in the elderly. The pathology exhibits gradual loss of sensory hair cells, spiral ganglion neurons, and stria vascularis cells in the inner ear cochlea. The involvement of ROS in presbycusis is evidenced by the observations that ROS induces expression of the mitochondrial pro-apoptotic gene Bak in primary cochlear cells, while mCAT suppresses Bak expression, reduces cochlear cell death, and prevents presbycusis. Furthermore, deletion of Bak attenuates age-dependent apoptotic cell deaths and prevents presbycusis in mice.[120] These findings support a central role of mitochondrial ROS-induced apoptotic pathway in presbycusis.[120] In addition, CR prevents presbycusis via reduction of oxidative damage by upregulation of mitochondrial deactylase SIRT3. In response to CR, SIRT3 directly deacetylates and activates mitochondrial isocitrate dehydrogenase 2, leading to increased NADPH levels and an increased ratio of reduced-to-oxidized glutathione in mitochondria, thereby enhancing the mitochondrial glutathione antioxidant defense system.[121]

Retinitis Pigmentosa

Although retinitis pigmentosa (RP) is not a disease of old age, it is a group of inherited retinal degenerative disease causing death of rod and/or cone photoreceptor cells. The clinical manifestations of RP depend on the type of photoreceptors initially affected by the disease. Rod cells are more often affected before cone cells. Since rod cells can function in less intense light than cone cells, night blindness is usually the earliest symptoms of RP. As the disease deteriorates, there is progressive loss of color vision, visual acuity, and sight in the central visual field due to the degeneration of cone cells. Several genes have been associated with RP, such as the rhodopsin gene, the mutation of which accounts for 15%–25% of human RP cases. Rhodopsin mutation may cause protein misfolding leading to the progressive loss of rod cells. Because rod cells are the most numerous photoreceptors in the retina and they are highly metabolic active, the death of rod cells in the outer retina would increase oxidative stress. The propagation of these ROS may lead to cone cell death as the disease progress. As such, antioxidant treatment with various vitamins and SOD mimetics have been shown to delay photoreceptor cell death in mouse models of RP,[122] both in early onset *rd1/rd1* and late onset *rd10/rd10* mice (see review for further detail of mouse model of retinal degeneration[123]). In *rd10/rd10* mouse model of RP, inducible expression of both SOD2 and mCAT, but not either alone, significantly reduce protein oxidative damage and ameliorate cone cell death. Interestingly, overexpression of SOD1 alone increased oxidative stress and accelerated cone cell death.[124] This study emphasizes the crucial roles of mitochondrial oxidative stress in the progression of cone cell death and suggests potential therapeutic application of mitochondrial antioxidants in RP, which has no effective treatment to date.

Alzheimer's Disease

AD is the one of the most prevalent neurodegenerative disease, affecting approximately five million Americans. The clinical manifestation of AD is memory impairment and dementia. Deficits in other cognitive functions may appear later, after the development of memory impairment.

The pathology of AD involves two hallmark lesions: NFT and the amyloid plaque.[125] The NFTs are made up of accumulations of abnormally phosphorylated tau within the cytoplasm of certain susceptible neurons. The amyloid plaques arise from β-amyloid peptide (Aβ) through proteolytic processing of amyloid precursor protein (APP) by β-secretase and γ-secretase (presenilin1/2). Both NFT and Aβ have a characteristic distribution pattern. In fact, the hierarchical and temporal pattern of NFTs accumulation among brain regions is so consistent that the topography of these lesions has been clinically used in the staging of AD pathology.[126] Most AD cases are sporadic and manifest in old age. In contrast, familial autosomal dominant AD cases have an early onset and comprise less than 1% of all AD cases. Genes implicated in the early onset AD include β-APP, presenilin 1 and presenilin 2.[127] Mutation of the APP gene affects the cleavage of APP by β-secretase or γ-secretase to generate various forms of Aβ. The Aβ peptides have a tendency to form oligomer aggregates and become toxic, especially the long form, $A\beta_{1-42}$. Presenilins are integral membrane proteins that function as the proteolytic components of γ-secretase. Mutations of presenilins result in increased production of $A\beta_{1-42}$.

The central roles of mitochondria in AD have been recently reviewed.[128] Many of the abnormal protein aggregates in AD, including APP and presenilin, have been isolated in mitochondrial fractions.[129,130] As a component of amyloid plaque, $A\beta$ is imported into the mitochondrial cristae through the translocase of the outer mitochondrial membrane complex.[131] Furthermore, increased mitochondrial oxidative stress and damage to mitochondrial enzyme complexes have been observed in early AD.[132–135] One of these mechanisms involves $A\beta$, which inhibits mitochondrial function by the inhibition of the electron transport chain, particularly complex III and IV, which further leads to increased ROS production, decreased ATP production, and facilitation of cytochrome c release.[134,136,137] Additional insults to mitochondria may include altered Ca^{2+} homeostasis,[138] and increased mitochondrial DNA mutations and deletions.[132] Concomitant overexpression of mCAT in $A\beta$ precursor protein (PP) overexpressing mice have been shown to decrease neuronal $A\beta$ toxicity and oxidative injury, and extends the life span of $A\beta$ PP overexpressing mice. This data provide direct evidence that mitochondrial oxidative stress plays a primary role in AD pathology, and supports the possibility that mitochondria-targeted antioxidants might be beneficial to treat patients with AD.

In addition to oxidative stress, alterations of mitochondrial dynamics have also been implicated in AD.[139–141] It has been shown that S-nitrosylation of Drp1, a mitochondrial fission protein, mediates $A\beta$-related mitochondrial fission and neuronal injury.[139] Increased production of $A\beta$ interacts with Drp1, which is a critical factor in mitochondrial fragmentation, abnormal mitochondrial dynamics, and synaptic damage.[141]

Parkinson's Disease

PD is the second most common neurodegenerative disorder. The clinical manifestations include resting tremor, bradykinesia, rigidity, and postural instability. The characteristics pathology of PD is a gradual loss of pigmented dopaminergic neurons in the substantia nigra pars compacta, and accumulation of Lewy bodies in catecholaminergic neurons of the brainstem in the substantia nigra and locus ceruleus. Lewy bodies are abnormal aggregates of proteins composed predominantly of α-synuclein and ubiquitin.

There are substantial evidences of the central role of mitochondria in the pathogenesis of PD. A few genetic loci have been mapped in rare familial PD cases, and are sequentially named PARK1 to PARK11 (see Reference 142). Several genes associated with familial PD have been identified in these loci, and the majority of them are related to mitochondria. PARK1 gene encodes α-synuclein, which has been implicated in the maintenance of mitochondrial membranes.[143] Increased amount of α-synuclein binding to mitochondria inhibits mitochondrial fusion and thereby triggers PD pathology, which can be rescued by PINK1, Parkin, and DJ-1.[143] PARK8 encodes the leucine-rich repeat kinase 2 (LRRK2) and its mutations have been associated with mitochondrial oxidative phosphorylation dysfunction.[144] LRRK2 regulates mitochondrial dynamics by a direct interaction with DLP1, a mitochondrial fission protein.[145] PARK7 encodes DJ-1, the mutation of which is associated with complex I defects, increased mitochondrial ROS, reduced mitochondrial membrane potential, and altered mitochondrial morphology and dynamics.[146–148] PARK2 and PARK6 encode Parkin and PTEN-induced kinase 1 (PINK1), which is involved in mitochondrial dynamics (fusion/fission) and turnover by mitophagy.[149–151]

Several rodent models have been used to recapitulate the pathology of PD, including mice with genetic manipulation of many of the genes mentioned above and rodents treated with environmental toxins (see Reference 152). Indeed, the majority of the environmental toxins causing PD-like phenotypes are mitochondrial complex I inhibitors, such as 1-methyl-4-phenyl-1,2,5,6-tetrahydropyridine (MPTP), paraquat, or rotenone. Inhibition of complex I is associated with impaired mitochondrial respiration leading to increased mitochondrial oxidative stress to proteins, lipids, and DNA, which may further activate mitochondrial-dependent apoptotic pathways and cause dopaminergic neuronal cell death.

Direct evidence was shown by the protective effect of mCAT mouse brains against MPTP-induced mitochondrial ROS production and subsequent dopaminergic neuron degeneration in substantia nigra pars compacta.[153] In contrast, mice with defective mitochondria are more susceptible to MPTP-induced dopaminergic neuronal cell death. For example, harlequin mice with partial deficiency of apoptosis inducing factor, which is required for maintenance of mitochondrial complex I, display increased susceptibility to MPTP. This is reversed by the antioxidant tempol (SOD-mimetic).[153] Another study demonstrated that mitochondrial protective peptide SS31 (Bendavia) protects dopaminergic neurons and preserves striatal dopamine levels in mice treated with MPTP (see section Potential Interventions).[154]

SKELETAL MUSCLE PATHOLOGY

Sarcopenia, or the decline of skeletal muscle tissue with age, is one of the most important causes of functional decline and loss of independence in older adults. Skeletal muscles, like the heart and brain, are highly metabolic active and dependent on ATP production from mitochondria for muscle contraction. Different from the brain and heart, the skeletal muscles demonstrate a broad dynamic range of energetic demand differing over an order of magnitude between resting and intense muscle contraction.[155,156] Hence, skeletal muscles demonstrate long periods of low metabolic flux at rest, interspersed by shorter periods of high metabolic flux during sustained muscle contraction. At rest, mitochondria produce relatively more ROS, while during contraction, the majority of ROS are generated from NADPH and xanthine oxidase.[157] These non-mitochondrial ROS during muscle contraction play an important role in muscle adaptation to exercise.[157,158]

Mitochondrial oxidative stress has been implicated in skeletal muscle dysfunction, atrophy, and sarcopenia[159–163] with aging and pathological stress. One characteristic of aging skeletal muscle is a decline in the quality of the mitochondria. A study using the *extensor digitorum longus* muscle from old age C57Bl/6 mice revealed disrupted stoichiometry of the electron transport system, and particularly increased expression of respiratory complex I proteins relative to other complex subunits. This disrupted respiratory complexes[164] reduced the respiratory efficiency measured by flux per unit mitochondria. Overexpression of mCAT attenuated this age-related decline in mitochondrial respiratory efficiency and better preserved mitochondrial function, leading to enhanced skeletal muscle physiology in aging. Interestingly, preservation of mitochondrial function with age by mCAT also improved age-dependent decline in insulin sensitivity, glucose metabolism, and increased accumulation of intramyocellular lipid in aged skeletal muscle. Similar protection of skeletal muscle insulin sensitivity by mCAT was observed in mice fed a high fat diet (see section Cardiometabolic Diseases).[165]

The molecular physiology of aged mouse skeletal muscles demonstrate leaky calcium release channel ryanodine receptor 1 (RyR1) located on the sarcoplasmic reticulum (SR), in association with muscle atrophy, weakness, and reduced exercise tolerance. The mechanisms include oxidative modifications that destabilize the interaction between RyR1 and calstabin1, which lead to increased calcium leak.[166] This calcium leak reduces SR calcium load and hence reduces calcium release in response to muscle activation, resulting in reduced force production. Increased calcium leak also increases mitochondrial calcium uptake. In young normal muscles, rapid increase in mitochondrial calcium uptake may enhance tricarboxylic acid (TCA) cycle and increase mitochondrial ATP production.[167,168] However, under chronic stress increase in mitochondrial calcium can increase mitochondrial ROS production.[167,169] Thus, the elevated calcium leak in aged skeletal muscle can initiate a vicious cycle of mitochondrial ROS amplification and further RyR1 calcium leak. The mCAT prevented this vicious cycle and stabilized the RyR1 and calstabin1 interaction.[166,170,171] Aged mCAT skeletal muscles had improved specific force, increased calcium release amplitude, reduced calcium leak, and increased SR loading compared with age-matched wild-type skeletal muscles.

Reducing mtH_2O_2 by mCAT also reduces muscle weakness and atrophy with chronic disease. Muscle weakness and wasting in cancer patients receiving chemotherapy is associated with increased fatigue and frailty, resulting in reduced quality of life of cancer survivors.[172,173] Both cancer and chemotherapeutic agents (e.g., anthracycline)[160,174,175] independently contribute to skeletal muscle dysfunction. In this context, mitochondrial oxidative stress has been shown as a causative factor in muscle atrophy and weakness following exposure to anthracyclines.[160,175] Mice inoculated with breast cancer cells treated with a single dose of doxorubicin displayed skeletal muscle dysfunction in parallel with mitochondrial oxidative damage and dysfunction. The mCAT reduced mtH_2O_2 production, better preserved mitochondrial respiration in permeabilized muscle fibers, and this was associated with reduced protein oxidative damage and preservation of muscle mass and force production. Consistent with this, other studies reported protection of skeletal muscle dysfunction following doxorubicin treatment using mitochondrial-targeted small molecules to reduce oxidative stress.[160]

The efficacy of mCAT in ameliorating skeletal muscle pathology in various models emphasizes important roles of mitochondrial H_2O_2 as a mediator of muscle pathology and suggests a potential strategy targeting mitochondrial oxidative stress to prevent muscle dysfunction and frailty.

CANCER

Oxidative damage to nucleic acids and proteins is widely documented in carcinogenesis. Mitochondria have also been implicated in carcinogenesis. For example, in human patients with ulcerative colitis, loss of mitochondrial cytochrome oxidase has been shown to associate with the development of colonic dysplasia (precancerous state), linking mitochondrial damage to carcinogenesis in humans.[176]

The role of mitochondrial ROS in carcinogenesis and cancer progression is supported by the use of mCAT mouse models. In a mouse end of-life pathology study, overexpression of mCAT had reduced non-hematopoietic tumor burden.[177] The mCAT expression has also been shown to ameliorate metastatic breast cancer in the PyMT mouse model of breast cancer. In this study, mCAT reduced invasive primary breast tumor and had a 30% lower incidence of pulmonary metastasis. Both tumor cells and lung fibroblasts in mCAT/PyMT double transgenic mice demonstrated reduced ROS and enhanced resistance to H_2O_2-induced cell death, which may confer the protective effects in mCAT mice.[178]

The role of mitochondrial ROS in carcinogenesis was also shown in ataxia telangiectasia mutated null mice ($ATM^{-/-}$). ATM kinase is recruited during the DNA damage response and redox sensing by the phosphorylation of many key proteins that initiate activation of the DNA damage checkpoint, leading to cell cycle arrest, DNA repair, or apoptosis. Patients with ATM mutation may present with severe ataxia due to cerebellar degeneration, immune defects, as well as increased risk of lymphomas and leukemias.[179] Mice with mutated null $ATM^{-/-}$ had thymic lymphomas and neurodegenerative phenotypes. It has been shown that reducing mitochondrial ROS by mCAT in $ATM^{-/-}$ mice reduced the propensity to develop thymic lymphoma, improved bone marrow hematopoiesis and macrophage differentiation *in vitro*, and partially rescued memory T-cell development.[179]

POTENTIAL INTERVENTIONS

CALORIE RESTRICTION

CR is the most reproducible intervention shown to extend both healthspan and life span across different model organisms from yeast to mice.[180,181] The life span extension effects of CR in nonhuman primates are still debatable. In the Wisconsin National Primate Research Center Studies, a long term approximately 30% restricted diet since young adulthood significantly improves age-related and all cause survival,[182,183] while a 22%–24% calorie reduction below control levels implemented in young and older age rhesus monkeys at the National Institute on Aging has not improved survival outcomes.[184,185]

Inhibition of mTORC1 is shown to mediate the beneficial effects of CR, although CR has also been shown to modulate other signaling pathways. One proposed mechanism of CR is that limited nutrient availability drives cells from a proliferative and energetic state to a somatic maintenance state, preserving limited resources for the most critical processes, including mitochondrial energy production. In *Drosophila,* dietary restriction has been shown to preferentially maintain mitochondrial protein translation while suppressing the overall protein translation, mediated through d4EBP, one of the main downstream signalings of mTORC1. Consistent with this, CR in rodents preferentially maintains mitochondrial biogenesis while suppressing overall DNA synthesis.[186,187] CR reduces mitochondrial oxidative damage through enhancement of endogenous antioxidant systems and reduction of ROS production from NOX.[188,189]

In the heart, CR reduces age-associated apoptosis through protection from DNA damage, enhanced DNA repair, and alterations in apoptosis-related gene expression.[190,191] The cardioprotective effects of CR in aged hearts are consistent with those found in mCAT mice, including attenuation of mitochondrial oxidative damage, better preservation of youthful mitochondrial proteome, in parallel with amelioration of age-dependent cardiac hypertrophy and diastolic dysfunction. CR also protects against myocardial ischemia and reduces vascular inflammatory markers.[191,192] CR affects expression of genes involved in fibrosis, extracellular matrix maintenance, inflammation, and fatty acid metabolism.[191] CR increased markers of autophagy (LC3-II/I ratio, Beclin-1) and reduced phosphorylation of mTOR as well as Akt/glycogen synthase kinase-3β. This was associated with better preservation of cardiac structure and function.[181] In addition to rodents, dietary restriction has been shown to reduce the aging-associated decline in cardiac function, and ameliorates cardiac hypertrophy and features of cardiomyopathy in human and nonhuman primates.[182,190,193–196]

THE mTORC1 PATHWAY AND RAPAMYCIN

The mTORC1 is an important regulator of cell growth and size, which plays a critical role in cell growth, function, and protein homeostasis in aging.[197,198] The target of rapamycin (TOR) is a signaling pathway modulated by a wide variety of environmental signal input, including nutrients, amino acids, hormones, and mitogens, and then exerts adaptive responses within the cell. These responses may include regulation of transcription and translation, autophagy, apoptosis, mitochondrial biogenesis, lipid metabolism, glycolysis, and inflammation. The mTOR is a serine/threonine kinase in the PI3K family, forming two distinct complexes—mTORC1 and mTORC2. The mTORC1complex, downstream of the Akt and PI3K pathways, includes the raptor and is well known to be inhibited by rapamycin. The mTORC2 complex, activated by the RAS and RAF signaling cascade,[199] includes the rictor protein. It has been shown that chronic treatment with rapamycin can also inhibit mTORC2 in a cell-specific manner.[200]

Important pathways downstream of TORC1 include regulation of cap-dependent initiation of translation via 4EBP1, control of ribosomal protein biosynthesis via S6K, and regulation of autophagy via Ulk1. Through inhibition of mTORC1, rapamycin, similar to CR, has been well documented to extend life span in invertebrate models of aging, including flies,[55] worms,[56,57] and yeast,[58,59] as well as in mice.[49,60] The mTORC1 signaling is suppressed in stress conditions such as low ATP concentration and low nutrient availability, as seen in CR. Rapamycin is the best documented agent that is believed to function as a CR mimetic by inhibiting mTORC1. Similar to CR, chronic rapamycin treatment has been shown to extend life span, even if initiated later in life after middle age in mice.[49]

Modulation of mTORC1 has been shown to improve cardiac structure and function. In *Drosophila*, upregulation of d4eBP was sufficient to mitigate the age-related decline in cardiac function.[201] It is presumed that both insulin/IGF and TOR signaling prevent some aging-related cardiac dysfunction by modulating 4eBP.[202] Downstream of mTORC1, 4eBP binds eIF4E, inhibiting cap-dependent initiation of translation.[203] In mice, shorter term of rapamycin for 10–12 weeks initiated at middle age (24–25 months) significantly reversed age-associated cardiac hypertrophy, diastolic dysfunction and inflammation,[195,204] as well as attenuated age-related cardiac proteomic remodeling, comparable to but weaker than the effects of CR.[195] While the total proteome turnover of the aging mouse heart was not significantly different from young controls, the cardiac proteome had a significantly increased half-life after CR or rapamycin treatment, concurrent with a reduction of detectable protein oxidation and ubiquitination.

Since rapamycin has been shown to inhibit ULK phosphorylation and induces autophagy, it is expected that autophagic markers should be elevated after rapamycin treatment. However, a recent study demonstrated that in old mice treated with rapamycin, autophagic markers increased in aged hearts only during the first week of rapamycin treatment, and then return to baseline since the second week of rapamycin treatment.[205] There was a concordant increase in mitochondrial biogenesis markers over the first 2 weeks of rapamycin treatment and these markers returned to baseline thereafter. These data suggest that rapamycin temporarily induces autophagy, presumably to remove damaged mitochondria, and then induces mitochondrial biogenesis to replenish and rejuvenate mitochondrial homeostasis. The phenotype of age-dependent diastolic dysfunction started to improve 2–4 weeks after treatment, and was maintained throughout the 10-week course of rapamycin treatment. The above studies suggest that the better preserved mitochondrial proteome, especially the pathways associated with better and younger mitochondrial functional profile (electron transport chain, TCA cycle, fatty acid metabolism) is among the mechanisms underlying the beneficial effects of short-term CR or rapamycin treatment.[195]

Rapamycin has tremendous effects on the proteome remodeling across multiple aged tissues throughout the body, including the heart, liver, adrenal glands, skeletal muscles, and bone marrow.[206,207] While the effect of rapamycin on proteome remodeling in aged hearts closely resembles the effect of CR on aged hearts, in contrast, the effect of rapamycin on aged liver is quite different from that seen in CR treated aged liver.[208] This study suggests that rapamycin may act as CR mimetic on some organs, but not on others.

ANTIOXIDANTS

The roles of ROS in aging and healthspan have been extensively discussed above. Based on these observations, several clinical studies have been performed to test the therapeutic potential of dietary antioxidant supplement in a wide range of diseases including cancer, gastrointestinal, neurological, rheumatoid, endocrine, and cardiovascular diseases. The majority of these studies have shown little to no efficacy. In a large clinical meta-analysis of 232,606 participants, vitamin antioxidant supplement (beta carotene, vitamins A, C, and E, and selenium) did not show any benefit on mortality or disease prevention. Indeed, some ingredients have shown slightly but significantly increased mortality (34).

The beneficial effects of mCAT expression in the protection of multiple disease mouse models implicate a critical role of mitochondrial oxidative stress in the pathogenesis of multiple diseases and provide strong scientific basis for the development of mitochondrial antioxidant drugs. The potential benefits of pharmacological mitochondrial antioxidants, however, have not been tested to the same extent as general antioxidants.

Two main approaches have been used to deliver pharmacologic compounds to mitochondria. The first approach is by conjugating redox agents to triphenylphosphonium ion (TPP$^+$) and taking advantage of the potential gradient across the inner mitochondrial membrane. Mito Q and SkQl are TPP$^+$ conjugated to ubiquinone and plastoquinone, respectively; and they utilize the mitochondrial membrane potential across the inner mitochondrial membrane to deliver these redox-active compounds into the mitochondrial matrix.[209,210] The second approach is by utilizing the affinity to a mitochondrial component to target the mitochondria without relying on mitochondrial potential.

The Szeto-Schiller (SS) compounds are tetrapeptides with an alternating aromatic-cationic amino acids motif that selectively bind to cardiolipin (CL) on the inner mitochondrial membrane[211–213] to target delivery to the mitochondria. The effects of these pharmacologic interventions on longevity and healthspan are discussed below.

TPP[+]-Conjugated Mitochondrial Antioxidants

MitoQ (10-(6′-ubiquinonyl) decyltriphenylphosphonium bromide) selectively concentrates in the mitochondria and prevents mitochondrial oxidative damage.[210] Several studies have demonstrated the beneficial effects of MitoQ in aging and age-dependent diseases. In *Drosophila*, MitoQ prolongs life span of SOD-deficient flies but not normal WT flies and improves pathology associated with antioxidant deficiency.[214] In a *C. elegans* model of AD, MitoQ extends life span and improves healthspan without reducing ROS production, protein carbonyl content, and mtDNA damage burden in this transgenic *C. elegans* with muscle-specific expression of human Aβ.[215] The protective effect of MitoQ against Aβ-toxicity has also been demonstrated in primary neurons from amyloid-beta PP transgenic mice and neuroblastoma cells incubated with Aβ.[216] In a triple transgenic mouse model of AD (3xTg-AD), MitoQ treatment for 5 months prevents cognitive decline and AD-like neuropathology, supporting the therapeutic potential of MitoQ in AD.[217] MitoQ treatment has also been shown to confer neuroprotection in cell culture and mouse models of PD.[218]

In addition to its neuroprotective effects, MitoQ treatment has been shown to confer cardioprotection in multiple models,[219–222] including cardiac ischemia–reperfusion,[219] and spontaneous hypertensive rats, in which MitoQ treatment for 8 weeks reduced systolic blood pressure and attenuated cardiac hypertrophy.[221] A recent study showed that administration of MitoQ to the storage solution of donor hearts prevents ischemia–reperfusion-related injury after heart transplantation.[220]

SkQ1, another TPP[+] conjugated mitochondrial-targeted antioxidant, has been shown to prolong life span of *Podospora*, *Ceriodaphnia*, *Drosophila*, and female outbred spontaneous hypertensive rats.[223] A recent study reported that SkQ1 can also extend the life span of male BALB/c and C57Bl/6 mice.[224] Like MitoQ, SkQ1has been shown to have beneficial effects in models of cardiac ischemia–reperfusion.[225] Moreover, it is also protective against renal and brain ischemic injuries.[225] Interestingly, a study showed that the administration of SkQ1 via the diet can prevent age-induced cataract and retinopathies in senescence-accelerated OXYS rats, and SkQ1 eye drops can reverse cataract in middle-aged OXYS rats and Wistar rats.[226]

One limitation of TPP[+] conjugated antioxidants is their dependence on mitochondrial membrane potential to penetrate the mitochondria, given that mitochondrial membrane potential is often compromised in pathological conditions. Moreover, MitoQ and SkQ have also been shown to inhibit respiration and disrupt mitochondrial membrane potential at concentrations above 5 to 25 μM.[209,210] Because MitoQ and SkQ are both quinone derivatives with potential pro-oxidant properties, optimal dosages that exert beneficial antioxidant effect but not excessive pro-oxidant activities must be carefully evaluated before using these interventions.

MITOCHONDRIAL PROTECTIVE STRATEGIES

The SS peptides are tetrapeptides with an alternating aromatic-cationic amino acids motif that were incidentally found to concentrate in the inner mitochondrial membrane over 1000-fold compared with the cytosolic concentration.[225,227,228] While the positively charged TPP[+]-based compounds (MitoQ and SkQ1) target to mitochondria because of potential gradient, the mitochondrial uptake of SS peptides is independent of mitochondrial membrane potential since they can also concentrate within depolarized mitochondria.[227,228] The SS-31 peptide (H-D-Arg-Dmt-Lys-Phe-NH2), also known as Bendavia, Elamipretide, or MTP-131, is the best-characterized of these peptides. SS-31 was initially thought to exert its protective effect by the ROS-scavenging activity of the dimethyltyrosine residue.[221] However, more recent studies showed that SS-31 selectively interacts with CL, which is enriched within mitochondrial inner membranes.[229] The binding of SS-31 to CL on the inner mitochondrial membrane alters the interaction of CL with cytochrome c.[213] This altered interaction preserves Met80-heme ligation of cytochrome c and favors cytochrome c electron carrier activity while inhibiting its peroxidase activity.[212,213] In a renal ischemia–reperfusion model, SS-31 treatment can increase ATP production and reduce ROS generation, preventing CL peroxidation and preserving cristae membrane integrity.[213] These findings suggest that ROS-independent mechanisms may contribute to the protective effects of SS-31, with reduced ROS production as a secondary benefit. This mechanism may explain how SS-31 protects mitochondrial cristae architecture, prevents mitochondrial swelling, and attenuates renal dysfunction after ischemia or IR.[211,230] In other words, SS-31 protects mitochondria by preserving mitochondrial cristae ultrastructure and enhancing the efficiency of electron transfer by cytochrome c, thereby reducing free radical production.

Several studies have shown the protective effects of SS-31 in various models of age-related diseases, including AD.[216,231] Similar to MitoQ, SS-31 prevents Aβ-toxicity in primary neurons from Aβ PP transgenic mice and neuroblastoma cells incubated with Aβ.[216] SS-31 treatment rescues the impairment of mitochondrial dynamics and antegrade transport and prevents synaptic degeneration caused by Aβ-toxicity.[231] The neuroprotective effect of SS-31 was also demonstrated in a mouse model of PD. SS-31 protects dopaminergic neurons in a dose-dependent manner and preserves striatal dopamine levels in mice treated with 1-methyl-4-phenyl-1,2,3,6-tetrahydropyridine (MPTP).[154] SS-20, a version of SS peptide without dimethyltyrosine residue, also showed neuroprotection on dopaminergic neurons of MPTP-treated mice. The same study also showed that SS-31 and SS-20 prevented the inhibition of oxygen consumption, ATP production, and mitochondrial swelling by MPTP treatment in isolated mitochondria. These findings suggest that the ROS-scavenging activity of SS-31 is not necessary for neuroprotection and that preservation of mitochondrial ATP production and inhibition of mitochondrial permeability transition may mediate the neuroprotective effects of SS peptides.[154]

Beside neuroprotection, SS peptides also exert cardioprotection in multiple disease models. SS-31 has been demonstrated to offer cardioprotective effects in cardiac ischemia–reperfusion injury, reperfusion arrhythmia, and myocardial infarction models.[232–237] SS-31 also protect against cardiac hypertrophy and heart failure. In a similar manner to mCAT expression, SS-31 prevents Ang II-induced cardiac hypertrophy and preserves cardiac function in transverse aortic constriction (TAC)-induced heart failure model.[238–240] Electron microscopy analysis showed that SS-31 preserves cardiac mitochondria from TAC-induced abnormalities and proteomic analysis showed that SS-31 attenuates TAC-induced proteomic remodeling, more prominently preserving mitochondrial proteins, and to a lesser extent also attenuating TAC-induced changes in non-mitochondrial proteins.[239] In a large animal model of heart failure, administration of SS-31 for three months significantly improved systolic function in dogs with microembolization-induced heart failure. SS-31 treated group had normalized levels of plasma biomarkers, and preserved mitochondrial function and bioenergetics in the myocardium of dogs with advanced heart failure.[241]

Similar to the heart, skeletal muscle has a high-energy demand and is highly dependent on mitochondrial energy production to function. SS-31 has been shown to acutely reverse age-related impairment in mitochondrial energetics in and lead to improve muscle performance.[159] The effect of longer-term SS-31 treatment on skeletal muscle aging is under further investigation. In a hindlimb immobilization model and a casting model, SS-31 has been shown to prevent disuse skeletal muscle atrophy by attenuating ROS production and protease activation.[161,242] Similar protective effects of SS-31 have also been demonstrated in an inactivity-induced diaphragm dysfunction.[243] Doxorubicin is an anthracycline cancer chemotherapy drug that has been shown to cause cardiac and skeletal muscle myopathy. SS-31 protects cardiac and skeletal muscles from Doxorubicin-induced mitochondrial ROS production and prevents Doxorubicin-induced atrophy and dysfunction.[160] These findings implicate the role of mtROS in muscle weakness and dysfunction and support the potential of mitochondrial targeting antioxidants as therapies.

The promising results of preclinical studies of mitochondrial-targeted antioxidants and mitochondrial protective peptides discussed above have led to clinical trials on neurodegenerative, cardiac, and renal ischemia–reperfusion injury, skeletal muscle and ocular diseases, heart failure as well as mitochondrial diseases.[244,245] The roles of these interventions on aging and aging-associated diseases warrant further clinical investigations as several preclinical studies have demonstrated their great therapeutic potential in various diseases.

CONCLUSION

Convincing evidence has supported the roles of mitochondrial ROS and oxidative damage in aging and several age-related diseases, as shown by the protective effects of mice overexpressing mitochondrial catalase (mCAT) and these are consistent with the mitochondrial free radical theory of aging. However, the application of antioxidants may be beneficial or detrimental, depending on the context whether a pathological burst of ROS is anticipated. In the absence of a large amount of ROS, antioxidant supplements may be harmful, as physiological low levels of ROS are crucial signaling molecules that are required to induce beneficial stress response defense. These mechanisms include endogenous antioxidant systems, proteostasis, and autophagy that protect against ROS-induced damage in young tissues. In aged tissues, because of substantial increase in ROS, oxidative damage, and misfolded proteins, coupled with the impairment of proteostasis and autopaghy, there is an accumulation of protein aggregation and lipotoxicity, leading to mitochondrial and other organelle dysfunction, subsequently causing organ dysfunction. Several antiaging interventions, such as CR, rapamycin, and sirtuins activators, have been shown to enhance autophagy and restore proteostasis, thereby better preserving mitochondrial functions. Mitochondrial antioxidants and mitochondrial protective SS peptides protect mitochondria against ROS-induced damage and ameliorate several age-related degenerative diseases.

REFERENCES

1. Harman D. Aging: A theory based on free radical and radiation chemistry. *J Gerontol.* July 1956;11(3):298–300.
2. Van Raamsdonk JM, Hekimi S. Deletion of the mitochondrial superoxide dismutase sod-2 extends lifespan in *Caenorhabditis elegans*. *PLoS Genet.* February 2009;5(2):e1000361.
3. Zhang Y, Ikeno Y, Qi W et al. Mice deficient in both Mn superoxide dismutase and glutathione peroxidase-1 have increased oxidative damage and a greater incidence of pathology but no reduction in longevity. *J Gerontol A Biol Sci Med Sci.* 2009; September 23:1–9.
4. Perez VI, Van Remmen H, Bokov A, Epstein CJ, Vijg J, Richardson A. The overexpression of major antioxidant enzymes does not extend the lifespan of mice. *Aging Cell.* February 2009;8(1):73–75.
5. Jang YC, Perez VI, Song W et al. Overexpression of Mn superoxide dismutase does not increase life span in mice. *J Gerontol A Biol Sci Med Sci.* November 2009;64(11):1114–1125.
6. Huang TT, Carlson EJ, Gillespie AM, Shi Y, Epstein CJ. Ubiquitous overexpression of CuZn superoxide dismutase does not extend life span in mice. *J Gerontol A Biol Sci Med Sci.* January 2000;55(1):B5–9.
7. Lewis KN, Andziak B, Yang T, Buffenstein R. The naked mole-rat response to oxidative stress: Just deal with it. *Antioxid Redox Signal.* October 2013;19(12):1388–1399.
8. Harman D. The biologic clock: The mitochondria? *J Am Geriatr Soc.* April 1972;20(4):145–147.
9. Percy ME. Catalase: An old enzyme with a new role? *Can J Biochem Cell Biol.* October 1984;62(10):1006–1014.
10. Abe K, Makino N, Anan FK. pH dependency of kinetic parameters and reaction mechanism of beef liver catalase. *J Biochem.* February 1979;85(2):473–479.
11. Agar NS, Sadrzadeh SM, Hallaway PE, Eaton JW. Erythrocyte catalase: A somatic oxidant defense? *J Clin Invest.* January 1986;77(1):319–321.
12. Schriner SE, Linford NJ, Martin GM et al. Extension of murine life span by overexpression of catalase targeted to mitochondria. *Science.* June 24, 2005;308(5730):1909–1911.
13. Linford NJ, Schriner SE, Rabinovitch PS. Oxidative damage and aging: Spotlight on mitochondria. *Cancer Res.* March 1, 2006;66(5):2497–2499.

14. Bause AS, Haigis MC. SIRT3 regulation of mitochondrial oxidative stress. *Exp Gerontol*. July 2013;48(7):634–639.
15. Lin SJ, Kaeberlein M, Andalis AA et al. Calorie restriction extends *Saccharomyces cerevisiae* lifespan by increasing respiration. *Nature*. July 18, 2002;418(6895):344–348.
16. Zid BM, Rogers AN, Katewa SD et al. 4E-BP extends lifespan upon dietary restriction by enhancing mitochondrial activity in *Drosophila. Cell*. October 2, 2009;139(1):149–160.
17. Schulz TJ, Zarse K, Voigt A, Urban N, Birringer M, Ristow M. Glucose restriction extends *Caenorhabditis elegans* life span by inducing mitochondrial respiration and increasing oxidative stress. *Cell Metab*. October 2007;6(4):280–293.
18. Ristow M, Zarse K, Oberbach A et al. Antioxidants prevent health-promoting effects of physical exercise in humans. *Proc Natl Acad Sci U S A*. May 26, 2009;106(21):8665–8670.
19. Ristow M, Zarse K. How increased oxidative stress promotes longevity and metabolic health: The concept of mitochondrial hormesis (mitohormesis). *Exp Gerontol*. June 2010;45(6):410–418.
20. Lee SJ, Hwang AB, Kenyon C. Inhibition of respiration extends *C. elegans* life span via reactive oxygen species that increase HIF-1 activity. *Curr Biol*. December 7, 2010;20(23):2131–2136.
21. De Haes W, Frooninckx L, Van Assche R et al. Metformin promotes lifespan through mitohormesis via the peroxiredoxin PRDX-2. *Proc Natl Acad Sci U S A*. June 17, 2014;111(24):E2501–2509.
22. Lapointe J, Stepanyan Z, Bigras E, Hekimi S. Reversal of the mitochondrial phenotype and slow development of oxidative biomarkers of aging in long-lived Mclk1$^{+/-}$ mice. *J Biol Chem*. July 24, 2009;284(30):20364–20374.
23. Yun J, Finkel T. Mitohormesis. *Cell Metab*. May 6, 2014;19(5):757–766.
24. Starkov AA. The role of mitochondria in reactive oxygen species metabolism and signaling. *Ann N Y Acad Sci*. December 2008;1147:37–52.
25. Reczek CR, Chandel NS. ROS-dependent signal transduction. *Curr Opin Cell Biol*. April 2015;33:8–13.
26. Le Belle JE, Orozco NM, Paucar AA et al. Proliferative neural stem cells have high endogenous ROS levels that regulate self-renewal and neurogenesis in a PI3K/Akt-dependant manner. *Cell Stem Cell*. January 2011;8(1):59–71.
27. Shi Y, Buffenstein R, Pulliam DA, Van Remmen H. Comparative studies of oxidative stress and mitochondrial function in aging. *Integr Comp Biol*. November 2010;50(5):869–879.
28. West AP, Brodsky IE, Rahner C et al. TLR signalling augments macrophage bactericidal activity through mitochondrial ROS. *Nature*. April 2011;472(7344):476–480.
29. Basisty N, Dai DF, Gagnidze A et al. Mitochondrial-targeted catalase is good for the old mouse proteome, but not for the young: "Reverse" antagonistic pleiotropy? *Aging Cell*. August 2016;15(4):634–645.
30. Song M, Chen Y, Gong G, Murphy E, Rabinovitch PS, Dorn GW. Super-suppression of mitochondrial reactive oxygen species signaling impairs compensatory autophagy in primary mitophagic cardiomyopathy. *Circ Res*. July 2014;115(3):348–353.
31. Bedford L, Hay D, Devoy A et al. Depletion of 26S proteasomes in mouse brain neurons causes neurodegeneration and Lewy-like inclusions resembling human pale bodies. *J Neurosci*. August 2008;28(33):8189–8198.
32. Douglas PM, Dillin A. Protein homeostasis and aging in neurodegeneration. *J Cell Biol*. September 2010;190(5):719–729.
33. Christians ES, Benjamin IJ. Proteostasis and REDOX state in the heart. *Am J Physiol Heart Circ Physiol*. January 2012;302(1):H24–37.
34. Hedhli N, Pelat M, Depre C. Protein turnover in cardiac cell growth and survival. *Cardiovasc Res*. November 2005;68(2):186–196.
35. Vinciguerra M, Musaro A, Rosenthal N. Regulation of muscle atrophy in aging and disease. *Adv Exp Med Biol*. 2010;694:211–233.
36. Marzetti E, Calvani R, Bernabei R, Leeuwenburgh C. Apoptosis in skeletal myocytes: A potential target for interventions against sarcopenia and physical frailty—A mini-review. *Gerontology*. 2012;58(2):99–106.
37. Surguchev A, Surguchov A. Conformational diseases: Looking into the eyes. *Brain Res Bull*. January 2010;81(1):12–24.
38. de Magalhães JP. From cells to ageing: A review of models and mechanisms of cellular senescence and their impact on human ageing. *Exp Cell Res*. October 2004;300(1):1–10.
39. Morimoto RI, Cuervo AM. Protein homeostasis and aging: Taking care of proteins from the cradle to the grave. *J Gerontol A Biol Sci Med Sci*. February 2009;64(2):167–170.
40. Schneider JL, Cuervo AM. Autophagy and human disease: Emerging themes. *Curr Opin Genet Dev*. June 2014;26:16–23.

41. Mijaljica D, Prescott M, Devenish RJ. Different fates of mitochondria: Alternative ways for degradation? *Autophagy*. January–February 2007;3(1):4–9.
42. Levine B, Kroemer G. Autophagy in the pathogenesis of disease. *Cell*. January 11, 2008;132(1):27–42.
43. Gatica D, Chiong M, Lavandero S, Klionsky DJ. Molecular mechanisms of autophagy in the cardiovascular system. *Circ Res*. January 30, 2015;116(3):456–467.
44. Dice JF, Terlecky SR, Chiang HL et al. A selective pathway for degradation of cytosolic proteins by lysosomes. *Semin Cell Biol*. December 1990;1(6):449–455.
45. Koga H, Kaushik S, Cuervo AM. Protein homeostasis and aging: The importance of exquisite quality control. *Ageing Res Rev*. April 2011;10(2):205–215.
46. Madeo F, Tavernarakis N, Kroemer G. Can autophagy promote longevity? *Nat Cell Biol*. Sep 2010;12(9):842–846.
47. Simonsen A, Cumming RC, Brech A, Isakson P, Schubert DR, Finley KD. Promoting basal levels of autophagy in the nervous system enhances longevity and oxidant resistance in adult *Drosophila*. *Autophagy*. February 2008;4(2):176–184.
48. Pyo JO, Yoo SM, Ahn HH et al. Overexpression of Atg5 in mice activates autophagy and extends lifespan. *Nat Commun*. 2013;4:2300.
49. Harrison DE, Strong R, Sharp ZD et al. Rapamycin fed late in life extends lifespan in genetically heterogeneous mice. *Nature*. July 16, 2009;460(7253):392–395.
50. Wohlgemuth SE, Julian D, Akin DE et al. Autophagy in the heart and liver during normal aging and calorie restriction. *Rejuvenation Res*. September 2007;10(3):281–292.
51. Hsu CP, Odewale I, Alcendor RR, Sadoshima J. Sirt1 protects the heart from aging and stress. *Biol Chem*. March 2008;389(3):221–231.
52. Morselli E, Maiuri MC, Markaki M et al. Caloric restriction and resveratrol promote longevity through the Sirtuin-1-dependent induction of autophagy. *Cell Death Dis*. 2010;1:e10.
53. Bjedov I, Toivonen JM, Kerr F et al. Mechanisms of life span extension by rapamycin in the fruit fly *Drosophila melanogaster*. *Cell Metab*. January 2010;11(1):35–46.
54. Alvers AL, Wood MS, Hu D, Kaywell AC, Dunn WA, Jr., Aris JP. Autophagy is required for extension of yeast chronological life span by rapamycin. *Autophagy*. August 2009;5(6):847–849.
55. Kapahi P, Zid BM, Harper T, Koslover D, Sapin V, Benzer S. Regulation of lifespan in *Drosophila* by modulation of genes in the TOR signaling pathway. *Curr Biol*. May 25, 2004;14(10):885–890.
56. Vellai T, Takacs-Vellai K, Zhang Y, Kovacs AL, Orosz L, Muller F. Genetics: Influence of TOR kinase on lifespan in *C. elegans*. *Nature*. December 11, 2003;426(6967):620.
57. Jia K, Chen D, Riddle DL. The TOR pathway interacts with the insulin signaling pathway to regulate *C. elegans* larval development, metabolism and life span. *Development*. August 2004;131(16):3897–3906.
58. Kaeberlein M, Powers RW, 3rd, Steffen KK et al. Regulation of yeast replicative life span by TOR and Sch9 in response to nutrients. *Science*. November 18, 2005;310(5751):1193–1196.
59. Powers RW, 3rd, Kaeberlein M, Caldwell SD, Kennedy BK, Fields S. Extension of chronological life span in yeast by decreased TOR pathway signaling. *Genes Dev*. January 15, 2006;20(2):174–184.
60. Miller RA, Harrison DE, Astle CM et al. Rapamycin, but not resveratrol or simvastatin, extends life span of genetically heterogeneous mice. *J Gerontol A Biol Sci Med Sci*. February 2011;66(2):191–201.
61. Ozden O, Park SH, Kim HS et al. Acetylation of MnSOD directs enzymatic activity responding to cellular nutrient status or oxidative stress. *Aging (Albany NY)*. February 2011;3(2):102–107.
62. Choudhary C, Kumar C, Gnad F et al. Lysine acetylation targets protein complexes and co-regulates major cellular functions. *Science*. August 14, 2009;325(5942):834–840.
63. Chalkiadaki A, Guarente L. Sirtuins mediate mammalian metabolic responses to nutrient availability. *Nat Rev Endocrinol*. May 2012;8(5):287–296.
64. Lee IH, Cao L, Mostoslavsky R et al. A role for the NAD-dependent deacetylase Sirt1 in the regulation of autophagy. *Proc Natl Acad Sci U S A*. March 4, 2008;105(9):3374–3379.
65. Wong E, Cuervo AM. Integration of clearance mechanisms: The proteasome and autophagy. *Cold Spring Harb Perspect Biol*. December 2010;2(12):a006734.
66. Jana NR. Protein homeostasis and aging: Role of ubiquitin protein ligases. *Neurochem Int*. April 2012;60(5):443–447.
67. Li W, Gao B, Lee SM, Bennett K, Fang D. RLE-1, an E3 ubiquitin ligase, regulates *C. elegans* aging by catalyzing DAF-16 polyubiquitination. *Dev Cell*. February 2007;12(2):235–246.
68. Rana A, Rera M, Walker DW. Parkin overexpression during aging reduces proteotoxicity, alters mitochondrial dynamics, and extends lifespan. *Proc Natl Acad Sci U S A*. May 2013;110(21):8638–8643.
69. Bereiter-Hahn J, Voth M. Dynamics of mitochondria in living cells: Shape changes, dislocations, fusion, and fission of mitochondria. *Microsc Res Tech*. February 15, 1994;27(3):198–219.

70. Hom J, Sheu SS. Morphological dynamics of mitochondria—A special emphasis on cardiac muscle cells. *J Mol Cell Cardiol*. June 2009;46(6):811–820.

71. Ong SB, Hausenloy DJ. Mitochondrial morphology and cardiovascular disease. *Cardiovasc Res*. October 1, 2010;88(1):16–29.

72. Matsuda N, Sato S, Shiba K et al. PINK1 stabilized by mitochondrial depolarization recruits Parkin to damaged mitochondria and activates latent Parkin for mitophagy. *J Cell Biol*. April 19, 2010;189(2):211–221.

73. Chen H, Chomyn A, Chan DC. Disruption of fusion results in mitochondrial heterogeneity and dysfunction. *J Biol Chem*. July 15, 2005;280(28):26185–26192.

74. Szabadkai G, Simoni AM, Chami M, Wieckowski MR, Youle RJ, Rizzuto R. Drp-1-dependent division of the mitochondrial network blocks intraorganellar Ca^{2+} waves and protects against Ca^{2+}-mediated apoptosis. *Mol Cell*. October 8, 2004;16(1):59–68.

75. Westermann B. Mitochondrial fusion and fission in cell life and death. *Nat Rev Mol Cell Biol*. December 2010;11(12):872–884.

76. Bossy-Wetzel E, Barsoum MJ, Godzik A, Schwarzenbacher R, Lipton SA. Mitochondrial fission in apoptosis, neurodegeneration and aging. *Curr Opin Cell Biol*. December 2003;15(6):706–716.

77. Papanicolaou KN, Khairallah RJ, Ngoh GA et al. Mitofusin-2 maintains mitochondrial structure and contributes to stress-induced permeability transition in cardiac myocytes. *Mol Cell Biol*. March 2011;31(6):1309–1328.

78. Fang L, Moore XL, Gao XM, Dart AM, Lim YL, Du XJ. Down-regulation of mitofusin-2 expression in cardiac hypertrophy in vitro and in vivo. *Life Sci*. May 16, 2007;80(23):2154–2160.

79. Piquereau J, Caffin F, Novotova M et al. Down-regulation of OPA1 alters mouse mitochondrial morphology, PTP function, and cardiac adaptation to pressure overload. *Cardiovasc Res*. June 1, 2012;94(3):408–417.

80. Chen Y, Liu Y, Dorn GW, 2nd. Mitochondrial fusion is essential for organelle function and cardiac homeostasis. *Circ Res*. December 9, 2011;109(12):1327–1331.

81. Saito T, Sadoshima J. Molecular mechanisms of mitochondrial autophagy/mitophagy in the heart. *Circ Res*. April 10, 2015;116(8):1477–1490.

82. Chen Y, Dorn GW, 2nd. PINK1-phosphorylated mitofusin 2 is a Parkin receptor for culling damaged mitochondria. *Science*. April 26, 2013;340(6131):471–475.

83. Lazarou M, Sliter DA, Kane LA et al. The ubiquitin kinase PINK1 recruits autophagy receptors to induce mitophagy. *Nature*. August 20, 2015;524(7565):309–314.

84. Kubli DA, Quinsay MN, Gustafsson AB. Parkin deficiency results in accumulation of abnormal mitochondria in aging myocytes. *Commun Integr Biol*. July 1, 2013;6(4):e24511.

85. Leon LJ, Gustafsson AB. Staying young at heart: Autophagy and adaptation to cardiac aging. *J Mol Cell Cardiol*. June 2016;95:78–85.

86. Vassallo P, Driver SL, Stone NJ. Metabolic syndrome: An evolving clinical construct. *Prog Cardiovasc Dis*. 2016;59(2):172–177.

87. Executive summary of the third report of the national cholesterol education program (NCEP) expert panel on detection, evaluation, and treatment of high blood cholesterol in adults (adult treatment panel III). *JAMA*. May 16, 2001;285(19):2486–2497.

88. Kaur J. A comprehensive review on metabolic syndrome. *Cardiol Res Pract*. 2014;2014:943162.

89. Anderson EJ, Lustig ME, Boyle KE et al. Mitochondrial H2O2 emission and cellular redox state link excess fat intake to insulin resistance in both rodents and humans. *J Clin Invest*. March 2009;119(3):573–581.

90. Kahn SE, Hull RL, Utzschneider KM. Mechanisms linking obesity to insulin resistance and type 2 diabetes. *Nature*. December 14, 2006;444(7121):840–846.

91. Lee HY, Choi CS, Birkenfeld AL et al. Targeted expression of catalase to mitochondria prevents age-associated reductions in mitochondrial function and insulin resistance. *Cell Metab*. December 1, 2010;12(6):668–674.

92. Park K, Gross M, Lee DH et al. Oxidative stress and insulin resistance: The coronary artery risk development in young adults study. *Diabetes Care*. July 2009;32(7):1302–1307.

93. Wang Y, Wang GZ, Rabinovitch PS, Tabas I. Macrophage mitochondrial oxidative stress promotes atherosclerosis and nuclear factor-kappaB-mediated inflammation in macrophages. *Circ Res*. January 31, 2014;114(3):421–433.

94. Mammucari C, Rizzuto R. Signaling pathways in mitochondrial dysfunction and aging. *Mech Ageing Dev*. July–August 2010;131(7–8):536–543.

95. Trifunovic A, Larsson NG. Mitochondrial dysfunction as a cause of ageing. *J Intern Med.* February 2008;263(2):167–178.
96. Terzioglu M, Larsson NG. Mitochondrial dysfunction in mammalian ageing. *Novartis Found Symp.* 2007;287:197–208; discussion 208-113.
97. Ventura-Clapier R, Garnier A, Veksler V. Transcriptional control of mitochondrial biogenesis: The central role of PGC-1alpha. *Cardiovasc Res.* July 15, 2008;79(2):208–217.
98. Goffart S, von Kleist-Retzow J-C, Wiesner RJ. Regulation of mitochondrial proliferation in the heart: Power-plant failure contributes to cardiac failure in hypertrophy. *Cardiovasc Res.* 2004;64(2):198–207.
99. Murray AJ, Anderson RE, Watson GC, Radda GK, Clarke K. Uncoupling proteins in human heart. *Lancet.* November 13–19, 2004;364(9447):1786–1788.
100. Dai DF, Johnson SC, Villarin JJ et al. Mitochondrial oxidative stress mediates angiotensin II-induced cardiac hypertrophy and G{alpha}q overexpression-induced heart failure. *Circ Res.* April 1, 2011;108(7):837–846.
101. Dai DF, Santana LF, Vermulst M et al. Overexpression of catalase targeted to mitochondria attenuates murine cardiac aging. *Circulation.* June 2, 2009;119(21):2789–2797.
102. Trifunovic A, Wredenberg A, Falkenberg M et al. Premature ageing in mice expressing defective mitochondrial DNA polymerase. *Nature.* May 27, 2004;429(6990):417–423.
103. Kujoth GC, Hiona A, Pugh TD et al. Mitochondrial DNA mutations, oxidative stress, and apoptosis in mammalian aging. *Science.* July 15, 2005;309(5733):481–484.
104. Dai DF, Chen T, Wanagat J et al. Age-dependent cardiomyopathy in mitochondrial mutator mice is attenuated by overexpression of catalase targeted to mitochondria. *Aging Cell.* August 2010;9(4):536–544.
105. Safdar A, Bourgeois JM, Ogborn DI et al. Endurance exercise rescues progeroid aging and induces systemic mitochondrial rejuvenation in mtDNA mutator mice. *Proc Natl Acad Sci U S A.* 2011;108(10):4135–40.
106. Corral-Debrinski M, Stepien G, Shoffner JM, Lott MT, Kanter K, Wallace DC. Hypoxemia is associated with mitochondrial DNA damage and gene induction. Implications for cardiac disease. *JAMA.* October 2, 1991;266(13):1812–1816.
107. Zhang C, Bills M, Quigley A, Maxwell RJ, Linnane AW, Nagley P. Varied prevalence of age-associated mitochondrial DNA deletions in different species and tissues: A comparison between human and rat. *Biochem Biophys Res Commun.* January 23, 1997;230(3):630–635.
108. DiMauro S, Schon EA. Mitochondrial respiratory-chain diseases. *N Engl J Med.* June 26, 2003;348(26):2656–2668.
109. Marin-Garcia J, Goldenthal MJ, Moe GW. Abnormal cardiac and skeletal muscle mitochondrial function in pacing-induced cardiac failure. *Cardiovasc Res.* October 2001;52(1):103–110.
110. Mollnau H, Wendt M, Szocs K et al. Effects of angiotensin II infusion on the expression and function of NAD(P)H oxidase and components of nitric oxide/cGMP signaling. *Circ Res.* March 8, 2002;90(4):E58–E65.
111. Doughan AK, Harrison DG, Dikalov SI. Molecular mechanisms of angiotensin II-mediated mitochondrial dysfunction: Linking mitochondrial oxidative damage and vascular endothelial dysfunction. *Circ Res.* February 29, 2008;102(4):488–496.
112. Kimura S, Zhang GX, Nishiyama A et al. Mitochondria-derived reactive oxygen species and vascular MAP kinases: Comparison of angiotensin II and diazoxide. *Hypertension.* March 2005;45(3):438–444.
113. Kohler JJ, Cucoranu I, Fields E et al. Transgenic mitochondrial superoxide dismutase and mitochondrially targeted catalase prevent antiretroviral-induced oxidative stress and cardiomyopathy. *Lab Invest.* July 2009;89(7):782–790.
114. Ago T, Kuroda J, Pain J, Fu C, Li H, Sadoshima J. Upregulation of Nox4 by hypertrophic stimuli promotes apoptosis and mitochondrial dysfunction in cardiac myocytes. *Circ Res.* April 16, 2010;106(7):1253–1264.
115. Kuroda J, Ago T, Matsushima S, Zhai P, Schneider MD, Sadoshima J. NADPH oxidase 4 (Nox4) is a major source of oxidative stress in the failing heart. *Proc Natl Acad Sci U S A.* August 31, 2010;107(35):15565–15570.
116. Becker LB, vanden Hoek TL, Shao ZH, Li CQ, Schumacker PT. Generation of superoxide in cardiomyocytes during ischemia before reperfusion. *Am J Physiol.* December 1999;277(6 Pt 2):H2240–H2246.
117. Zorov DB, Juhaszova M, Sollott SJ. Mitochondrial ROS-induced ROS release: An update and review. *Biochim Biophys Acta.* 2006;1757(5–6):509–517.
118. Piot C, Croisille P, Staat P et al. Effect of cyclosporine on reperfusion injury in acute myocardial infarction. *N Engl J Med.* July 31, 2008;359(5):473–481.

119. Yueh B, Shapiro N, MacLean CH, Shekelle PG. Screening and management of adult hearing loss in primary care: Scientific review. *JAMA*. April 16, 2003;289(15):1976–1985.
120. Someya S, Xu J, Kondo K et al. Age-related hearing loss in C57BL/6J mice is mediated by Bak-dependent mitochondrial apoptosis. *Proc Natl Acad Sci U S A*. November 17, 2009;106(46):19432–19437.
121. Someya S, Yu W, Hallows WC et al. Sirt3 mediates reduction of oxidative damage and prevention of age-related hearing loss under caloric restriction. *Cell*. November 24, 2010;143(5):802–812.
122. Komeima K, Rogers BS, Campochiaro PA. Antioxidants slow photoreceptor cell death in mouse models of retinitis pigmentosa. *J Cell Physiol*. December 2007;213(3):809–815.
123. Veleri S, Lazar CH, Chang B, Sieving PA, Banin E, Swaroop A. Biology and therapy of inherited retinal degenerative disease: Insights from mouse models. *Dis Model Mech*. February 2015;8(2):109–129.
124. Usui S, Komeima K, Lee SY et al. Increased expression of catalase and superoxide dismutase 2 reduces cone cell death in retinitis pigmentosa. *Mol Ther*. May 2009;17(5):778–786.
125. Serrano-Pozo A, Frosch MP, Masliah E, Hyman BT. Neuropathological alterations in Alzheimer disease. *Cold Spring Harb Perspect Med*. September 2011;1(1):a006189.
126. Braak H, Braak E. Neuropathological stageing of Alzheimer-related changes. *Acta Neuropathol*. 1991;82(4):239–259.
127. Tanzi RE. The genetics of Alzheimer disease. *Cold Spring Harb Perspect Med*. 2012 Oct 1;2(10). pii: a006296.
128. Adiele RC, Adiele CA. Mitochondrial regulatory pathways in the pathogenesis of Alzheimer's disease. *J Alzheimers Dis*. July 6, 2016;53(4):1257–1270.
129. Hansson CA, Frykman S, Farmery MR et al. Nicastrin, presenilin, APH-1, and PEN-2 form active gamma-secretase complexes in mitochondria. *J Biol Chem*. December 3, 2004;279(49):51654–51660.
130. Walls KC, Coskun P, Gallegos-Perez JL et al. Swedish Alzheimer mutation induces mitochondrial dysfunction mediated by HSP60 mislocalization of amyloid precursor protein (APP) and beta-amyloid. *J Biol Chem*. August 31 , 2012;287(36):30317–30327.
131. Hansson Petersen CA, Alikhani N, Behbahani H et al. The amyloid beta-peptide is imported into mitochondria via the TOM import machinery and localized to mitochondrial cristae. *Proc Natl Acad Sci U S A*. September 2, 2008;105(35):13145–13150.
132. Coskun PE, Beal MF, Wallace DC. Alzheimer's brains harbor somatic mtDNA control-region mutations that suppress mitochondrial transcription and replication. *Proc Natl Acad Sci U S A*. July 20, 2004;101(29):10726–10731.
133. Massaad CA, Amin SK, Hu L, Mei Y, Klann E, Pautler RG. Mitochondrial superoxide contributes to blood flow and axonal transport deficits in the Tg2576 mouse model of Alzheimer's disease. *PLoS One*. 2010;5(5):e10561.
134. Terni B, Boada J, Portero-Otin M, Pamplona R, Ferrer I. Mitochondrial ATP-synthase in the entorhinal cortex is a target of oxidative stress at stages I/II of Alzheimer's disease pathology. *Brain Pathol*. January 2010;20(1):222–233.
135. Lustbader JW, Cirilli M, Lin C et al. ABAD directly links Abeta to mitochondrial toxicity in Alzheimer's disease. *Science*. April 16, 2004;304(5669):448–452.
136. Crouch PJ, Blake R, Duce JA et al. Copper-dependent inhibition of human cytochrome c oxidase by a dimeric conformer of amyloid-beta1-42. *J Neurosci*. January 19, 2005;25(3):672–679.
137. Atamna H, Boyle K. Amyloid-beta peptide binds with heme to form a peroxidase: Relationship to the cytopathologies of Alzheimer's disease. *Proc Natl Acad Sci U S A*. February 28, 2006;103(9):3381–3386.
138. Ferreiro E, Oliveira CR, Pereira CM. The release of calcium from the endoplasmic reticulum induced by amyloid-beta and prion peptides activates the mitochondrial apoptotic pathway. *Neurobiol Dis*. June 2008;30(3):331–342.
139. Cho DH, Nakamura T, Fang J et al. S-nitrosylation of Drp1 mediates beta-amyloid-related mitochondrial fission and neuronal injury. *Science*. April 3, 2009;324(5923):102–105.
140. Itoh K, Nakamura K, Iijima M, Sesaki H. Mitochondrial dynamics in neurodegeneration. *Trends Cell Biol*. February 2013;23(2):64–71.
141. Manczak M, Calkins MJ, Reddy PH. Impaired mitochondrial dynamics and abnormal interaction of amyloid beta with mitochondrial protein Drp1 in neurons from patients with Alzheimer's disease: Implications for neuronal damage. *Hum Mol Genet*. July 1, 2011;20(13):2495–2509.
142. Thomas B, Beal MF. Parkinson's disease. *Hum Mol Genet*. October 15, 2007;16(Spec No. 2):R183–194.
143. Kamp F, Exner N, Lutz AK et al. Inhibition of mitochondrial fusion by alpha-synuclein is rescued by PINK1, Parkin and DJ-1. *Embo J*. October 20, 2010;29(20):3571–3589.
144. Mortiboys H, Johansen KK, Aasly JO, Bandmann O. Mitochondrial impairment in patients with Parkinson disease with the G2019S mutation in LRRK2. *Neurology*. November 30, 2010;75(22):2017–2020.

145. Wang X, Yan MH, Fujioka H et al. LRRK2 regulates mitochondrial dynamics and function through direct interaction with DLP1. *Hum Mol Genet*. May 1, 2012;21(9):1931–1944.
146. McCoy MK, Cookson MR. DJ-1 regulation of mitochondrial function and autophagy through oxidative stress. *Autophagy*. May 2011;7(5):531–532.
147. Wang X, Petrie TG, Liu Y, Liu J, Fujioka H, Zhu X. Parkinson's disease-associated DJ-1 mutations impair mitochondrial dynamics and cause mitochondrial dysfunction. *J Neurochem*. June 2012;121(5):830–839.
148. Cookson MR. Parkinsonism due to mutations in PINK1, parkin, and DJ-1 and oxidative stress and mitochondrial pathways. *Cold Spring Harb Perspect Med*. September 2012;2(9):a009415.
149. Geisler S, Holmstrom KM, Skujat D et al. PINK1/Parkin-mediated mitophagy is dependent on VDAC1 and p62/SQSTM1. *Nat Cell Biol*. February 2010;12(2):119–131.
150. Youle RJ, Narendra DP. Mechanisms of mitophagy. *Nat Rev Mol Cell Biol*. January 2011;12(1):9–14.
151. Yu W, Sun Y, Guo S, Lu B. The PINK1/Parkin pathway regulates mitochondrial dynamics and function in mammalian hippocampal and dopaminergic neurons. *Hum Mol Genet*. August 15, 2011;20(16):3227–3240.
152. Melrose HL, Lincoln SJ, Tyndall GM, Farrer MJ. Parkinson's disease: A rethink of rodent models. *Exp Brain Res*. August 2006;173(2):196–204.
153. Perier C, Bove J, Dehay B et al. Apoptosis-inducing factor deficiency sensitizes dopaminergic neurons to parkinsonian neurotoxins. *Ann Neurol*. August 2010;68(2):184–192.
154. Yang L, Zhao K, Calingasan NY, Luo G, Szeto HH, Beal MF. Mitochondria targeted peptides protect against 1-methyl-4-phenyl-1,2,3,6-tetrahydropyridine neurotoxicity. *Antioxid Redox Signal*. September 2009;11(9):2095–2104.
155. Conley KE. Cellular energetics during exercise. In: Jones JH, ed. *Comparative Vertebrate Exercise Physiology: Unifying Physiological Principles*. Vol. 38A. San Diego, CA: Academic Press, Inc; 1994: 1–39.
156. Conley KE, Jubrias SA, Esselman PC. Oxidative capacity and ageing in human muscle. *J Physiol*. July 1, 2000;526(Pt 1):203–210.
157. Powers SK, Jackson MJ. Exercise-induced oxidative stress: Cellular mechanisms and impact on muscle force production. *Physiol Rev*. October 2008;88(4):1243–1276.
158. Ristow M, Zarse K, Oberbach A et al. Antioxidants prevent health-promoting effects of physical exercise in humans. *Proc Natl Acad Sci U S A*. 2009 May 26;106(21):8665–70.
159. Siegel MP, Kruse SE, Percival JM et al. Mitochondrial-targeted peptide rapidly improves mitochondrial energetics and skeletal muscle performance in aged mice. *Aging Cell*. October 2013;12(5):763–771.
160. Min K, Kwon OS, Smuder AJ et al. Increased mitochondrial emission of reactive oxygen species and calpain activation are required for doxorubicin-induced cardiac and skeletal muscle myopathy. *J Physiol*. April 15, 2015;593(8):2017–2036.
161. Min K, Smuder AJ, Kwon OS, Kavazis AN, Szeto HH, Powers SK. Mitochondrial-targeted antioxidants protect skeletal muscle against immobilization-induced muscle atrophy. *J Appl Physiol (1985)*. November 2011;111(5):1459–1466.
162. McClung JM, Kavazis AN, Whidden MA et al. Antioxidant administration attenuates mechanical ventilation-induced rat diaphragm muscle atrophy independent of protein kinase B (PKB Akt) signalling. *J Physiol*. November 15, 2007;585(Pt 1):203–215.
163. Laitano O, Ahn B, Patel N et al. Pharmacological targeting of mitochondrial reactive oxygen species counteracts diaphragm weakness in chronic heart failure. *J Appl Physiol (1985)*. April 1, 2016;120(7):733–742.
164. Kruse SE, Karunadharma PP, Basisty N et al. Age modifies respiratory complex I and protein homeostasis in a muscle type-specific manner. *Aging Cell*. 2016; Feb;15(1):89–99.
165. Anderson EJ, Lustig ME, Boyle KE et al. Mitochondrial H2O2 emission and cellular redox state link excess fat intake to insulin resistance in both rodents and humans. *J Clin Invest*. 2009 Mar; 119(3):573–81.
166. Andersson DC, Betzenhauser MJ, Reiken S et al. Ryanodine receptor oxidation causes intracellular calcium leak and muscle weakness in aging. *Cell Metab*. August 3, 2011;14(2):196–207.
167. Brookes PS, Darley-Usmar VM. Role of calcium and superoxide dismutase in sensitizing mitochondria to peroxynitrite-induced permeability transition. *Am J Physiol Heart Circ Physiol*. January 2004;286(1):H39–46.
168. Kavanaugh NI, Ainscow EK, Brand MD. Calcium regulation of oxidative phosphorylation in rat skeletal muscle mitochondria. *Biochem Biophys Acta*. 2000;1457:57–70.
169. Gutierrez-Perez A, Cortes-Rojo C, Noriega-Cisneros R et al. Protective effects of resveratrol on calcium-induced oxidative stress in rat heart mitochondria. *J Bioenerg Biomembr*. April 2011;43(2):101–107.
170. Andersson DC, Marks AR. Fixing ryanodine receptor Ca leak—A novel therapeutic strategy for contractile failure in heart and skeletal muscle. *Drug Discov Today Dis Mech*. Summer 2010;7(2):e151–e157.

171. Andersson DC, Meli AC, Reiken S et al. Leaky ryanodine receptors in beta-sarcoglycan deficient mice: A potential common defect in muscular dystrophy. *Skelet Muscle.* 2012;2(1):9.

172. Butt Z, Rosenbloom SK, Abernethy AP et al. Fatigue is the most important symptom for advanced cancer patients who have had chemotherapy. *J Natl Compr Canc Netw.* May 2008;6(5):448–455.

173. Bower JE, Ganz PA, Desmond KA, Rowland JH, Meyerowitz BE, Belin TR. Fatigue in breast cancer survivors: Occurrence, correlates, and impact on quality of life. *J Clin Oncol.* February 2000;18(4):743–753.

174. Gilliam LA, Lark DS, Reese LR et al. Targeted overexpression of mitochondrial catalase protects against cancer chemotherapy-induced skeletal muscle dysfunction. *Am J Physiol Endocrinol Metab.* August 1, 2016;311(2):E293–E301.

175. Gilliam LA, Fisher-Wellman KH, Lin CT, Maples JM, Cathey BL, Neufer PD. The anticancer agent doxorubicin disrupts mitochondrial energy metabolism and redox balance in skeletal muscle. *Free Radic Biol Med.* December 2013;65:988–996.

176. Ussakli CH, Ebaee A, Binkley J et al. Mitochondria and tumor progression in ulcerative colitis. *J Natl Cancer Inst.* August 21, 2013;105(16):1239–1248.

177. Treuting PM, Linford NJ, Knoblaugh SE et al. Reduction of age-associated pathology in old mice by overexpression of catalase in mitochondria. *J Gerontol A Biol Sci Med Sci.* August 2008;63(8):813–822.

178. Goh J, Enns L, Fatemie S et al. Mitochondrial targeted catalase suppresses invasive breast cancer in mice. *BMC Cancer.* 2011;11:191.

179. D'Souza AD, Parish IA, Krause DS, Kaech SM, Shadel GS. Reducing mitochondrial ROS improves disease-related pathology in a mouse model of ataxia-telangiectasia. *Mol Ther.* January 2013;21(1):42–48.

180. Speakman JR, Mitchell SE. Caloric restriction. *Mol Aspects Med.* June 2011;32(3):159–221.

181. Han X, Turdi S, Hu N, Guo R, Zhang Y, Ren J. Influence of long-term caloric restriction on myocardial and cardiomyocyte contractile function and autophagy in mice. *J Nutr Biochem.* December 2012;23(12):1592–1599.

182. Colman RJ, Anderson RM, Johnson SC et al. Caloric restriction delays disease onset and mortality in rhesus monkeys. *Science.* July 10, 2009;325(5937):201–204.

183. Colman RJ, Beasley TM, Kemnitz JW, Johnson SC, Weindruch R, Anderson RM. Caloric restriction reduces age-related and all-cause mortality in rhesus monkeys. *Nat Commun.* April 1, 2014;5:3557.

184. Mattison JA, Roth GS, Beasley TM et al. Impact of caloric restriction on health and survival in rhesus monkeys from the NIA study. *Nature.* September 13, 2012;489(7415):318–321.

185. Cava E, Fontana L. Will calorie restriction work in humans? *Aging (Albany NY).* July 2013;5(7):507–514.

186. Drake JC, Peelor FF, 3rd, Biela LM et al. Assessment of mitochondrial biogenesis and mTORC1 signaling during chronic rapamycin feeding in male and female mice. *J Gerontol A Biol Sci Med Sci.* December 2013;68(12):1493–1501.

187. Miller BF, Robinson MM, Bruss MD, Hellerstein M, Hamilton KL. A comprehensive assessment of mitochondrial protein synthesis and cellular proliferation with age and caloric restriction. *Aging Cell.* February 2012;11(1):150–161.

188. Gredilla R, Sanz A, Lopez-Torres M, Barja G. Caloric restriction decreases mitochondrial free radical generation at complex I and lowers oxidative damage to mitochondrial DNA in the rat heart. *Faseb J.* July 2001;15(9):1589–1591.

189. Csiszar A, de Cabo R, Ungvari Z. Caloric restriction and cardiovascular disease. In *Calorie Restriction, Aging and Longevity* 2010: 263–277.

190. Maeda H, Gleiser CA, Masoro EJ, Murata I, McMahan CA, Yu BP. Nutritional influences on aging of Fischer 344 rats: II. Pathology. *J Gerontol.* November 1985;40(6):671–688.

191. Dhahbi JM, Tsuchiya T, Kim HJ, Mote PL, Spindler SR. Gene expression and physiologic responses of the heart to the initiation and withdrawal of caloric restriction. *J Gerontol A Biol Sci Med Sci.* March 2006;61(3):218–231.

192. Spaulding CC, Walford RL, Effros RB. Calorie restriction inhibits the age-related dysregulation of the cytokines TNF-alpha and IL-6 in C3B10RF1 mice. *Mech Ageing Dev.* February 1997;93(1–3):87–94.

193. Taffet GE, Pham TT, Hartley CJ. The age-associated alterations in late diastolic function in mice are improved by caloric restriction. *J Gerontol A Biol Sci Med Sci.* November 1997;52(6):B285–B290.

194. Niemann B, Chen Y, Issa H, Silber RE, Rohrbach S. Caloric restriction delays cardiac ageing in rats: Role of mitochondria. *Cardiovasc Res.* November 1, 2010;88(2):267–276.

195. Dai DF, Karunadharma PP, Chiao YA et al. Altered proteome turnover and remodeling by short-term caloric restriction or rapamycin rejuvenate the aging heart. *Aging Cell.* June 2014;13(3):529–539.

196. Shinmura K, Tamaki K, Sano M et al. Impact of long-term caloric restriction on cardiac senescence: Caloric restriction ameliorates cardiac diastolic dysfunction associated with aging. *J Mol Cell Cardiol.* January 2011;50(1):117–127.

197. Kurdi M, Booz GW. Three 4-letter words of hypertension-related cardiac hypertrophy: TRPC, mTOR, and HDAC. *J Mol Cell Cardiol.* June 2011;50(6):964–971.

198. Jung CH, Ro SH, Cao J, Otto NM, Kim DH. mTOR regulation of autophagy. *FEBS Lett.* April 2, 2010;584(7):1287–1295.

199. Dobashi Y, Watanabe Y, Miwa C, Suzuki S, Koyama S. Mammalian target of rapamycin: A central node of complex signaling cascades. *Int J Clin Exp Pathol.* June 20, 2011;4(5):476–495.

200. Lamming DW, Ye L, Katajisto P et al. Rapamycin-induced insulin resistance is mediated by mTORC2 loss and uncoupled from longevity. *Science.* March 30, 2012;335(6076):1638–1643.

201. Wessells R, Fitzgerald E, Piazza N et al. d4eBP acts downstream of both dTOR and dFoxo to modulate cardiac functional aging in *Drosophila. Aging Cell.* September 2009;8(5):542–552.

202. Partridge L, Alic N, Bjedov I, Piper MD. Ageing in *Drosophila*: The role of the insulin/Igf and TOR signalling network. *Exp Gerontol.* May 2011;46(5):376–381.

203. Sonenberg N, Hinnebusch AG. Regulation of translation initiation in eukaryotes: Mechanisms and biological targets. *Cell.* February 20, 2009;136(4):731–745.

204. Flynn JM, O'Leary MN, Zambataro CA et al. Late-life rapamycin treatment reverses age-related heart dysfunction. *Aging cell.* October 2013;12(5):851–862.

205. Chiao YA, Kolwicz SC, Basisty N et al. Rapamycin transiently induces mitochondrial remodeling to reprogram energy metabolism in old hearts. *Aging (Albany NY).* February 2016;8(2):314–327.

206. Wilkinson JE, Burmeister L, Brooks SV et al. Rapamycin slows aging in mice. *Aging Cell.* August 2012;11(4):675–682.

207. Chen C, Liu Y, Zheng P. mTOR regulation and therapeutic rejuvenation of aging hematopoietic stem cells. *Sci Signal.* 2009;2(98):ra75.

208. Karunadharma PP, Basisty N, Dai DF et al. Subacute calorie restriction and rapamycin discordantly alter mouse liver proteome homeostasis and reverse aging effects. *Aging Cell.* August 2015;14(4): 547–557.

209. Antonenko YN, Avetisyan AV, Bakeeva LE et al. Mitochondria-targeted plastoquinone derivatives as tools to interrupt execution of the aging program. 1. Cationic plastoquinone derivatives: Synthesis and in vitro studies. *Biochemistry (Mosc).* December 2008;73(12):1273–1287.

210. Kelso GF, Porteous CM, Coulter CV et al. Selective targeting of a redox-active ubiquinone to mitochondria within cells: Antioxidant and antiapoptotic properties. *J Biol Chem.* February 16, 2001;276(7):4588–4596.

211. Birk AV, Liu S, Soong Y et al. The mitochondrial-targeted compound SS-31 re-energizes ischemic mitochondria by interacting with cardiolipin. *J Am Soc Nephrol.* July 2013;24(8):1250–1261.

212. Szeto HH. First-in-class cardiolipin therapeutic to restore mitochondrial bioenergetics. *Br J Pharmacol.* 2014 Apr;171(8):2029–50.

213. Birk AV, Chao WM, Bracken C, Warren JD, Szeto HH. Targeting mitochondrial cardiolipin and the cytochrome c/cardiolipin complex to promote electron transport and optimize mitochondrial ATP synthesis. *Br J Pharmacol.* 2014 Apr;171(8):2017–28.

214. Magwere T, West M, Riyahi K, Murphy MP, Smith RA, Partridge L. The effects of exogenous antioxidants on lifespan and oxidative stress resistance in *Drosophila melanogaster. Mech Ageing Dev.* April 2006;127(4):356–370.

215. Ng LF, Gruber J, Cheah IK et al. The mitochondria-targeted antioxidant MitoQ extends lifespan and improves healthspan of a transgenic *Caenorhabditis elegans* model of Alzheimer disease. *Free Radic Biol Med.* June 2014;71:390–401.

216. Manczak M, Mao P, Calkins MJ et al. Mitochondria-targeted antioxidants protect against amyloid-beta toxicity in Alzheimer's disease neurons. *J Alzheimers Dis.* 2010;20(Suppl 2):S609–631.

217. McManus MJ, Murphy MP, Franklin JL. The mitochondria-targeted antioxidant MitoQ prevents loss of spatial memory retention and early neuropathology in a transgenic mouse model of Alzheimer's disease. *J Neurosci.* November 2, 2011;31(44):15703–15715.

218. Ghosh A, Chandran K, Kalivendi SV et al. Neuroprotection by a mitochondria-targeted drug in a Parkinson's disease model. *Free Radic Biol Med.* December 1, 2010;49(11):1674–1684.

219. Adlam VJ, Harrison JC, Porteous CM et al. Targeting an antioxidant to mitochondria decreases cardiac ischemia-reperfusion injury. *FASEB J.* July 2005;19(9):1088–1095.

220. Dare AJ, Logan A, Prime TA et al. The mitochondria-targeted anti-oxidant MitoQ decreases ischemia-reperfusion injury in a murine syngeneic heart transplant model. *J Heart Lung Transplant.* November 2015;34(11):1471–1480.

221. Graham D, Huynh NN, Hamilton CA et al. Mitochondria-targeted antioxidant MitoQ10 improves endothelial function and attenuates cardiac hypertrophy. *Hypertension.* August 2009;54(2):322–328.

222. Supinski GS, Murphy MP, Callahan LA. MitoQ administration prevents endotoxin-induced cardiac dysfunction. *Am J Physiol Regul Integr Comp Physiol*. October 2009;297(4):R1095–1102.
223. Anisimov VN, Bakeeva LE, Egormin PA et al. Mitochondria-targeted plastoquinone derivatives as tools to interrupt execution of the aging program. 5. SkQ1 prolongs lifespan and prevents development of traits of senescence. *Biochemistry (Mosc)*. December 2008;73(12):1329–1342.
224. Anisimov VN, Egorov MV, Krasilshchikova MS et al. Effects of the mitochondria-targeted antioxidant SkQ1 on lifespan of rodents. *Aging (Albany NY)*. November 2011;3(11):1110–1119.
225. Bakeeva LE, Barskov IV, Egorov MV et al. Mitochondria-targeted plastoquinone derivatives as tools to interrupt execution of the aging program. 2. Treatment of some ROS- and age-related diseases (heart arrhythmia, heart infarctions, kidney ischemia, and stroke). *Biochemistry (Mosc)*. December 2008;73(12):1288–1299.
226. Neroev VV, Archipova MM, Bakeeva LE et al. Mitochondria-targeted plastoquinone derivatives as tools to interrupt execution of the aging program. 4. Age-related eye disease. SkQ1 returns vision to blind animals. *Biochemistry (Mosc)*. December 2008;73(12):1317–1328.
227. Zhao K, Zhao GM, Wu D et al. Cell-permeable peptide antioxidants targeted to inner mitochondrial membrane inhibit mitochondrial swelling, oxidative cell death, and reperfusion injury. *J Biol Chem*. August 13, 2004;279(33):34682–34690.
228. Doughan AK, Dikalov SI. Mitochondrial redox cycling of mitoquinone leads to superoxide production and cellular apoptosis. *Antioxid Redox Signal*. November 2007;9(11):1825–1836.
229. Birk AV, Chao WM, Bracken C, Warren JD, Szeto HH. Targeting mitochondrial cardiolipin and the cytochrome c/cardiolipin complex to promote electron transport and optimize mitochondrial ATP synthesis. *Br J Pharmacol*. April 2014;171(8):2017–2028.
230. Szeto HH, Liu S, Soong Y et al. Mitochondria-targeted peptide accelerates ATP recovery and reduces ischemic kidney injury. *J Am Soc Nephrol*. June 2011;22(6):1041–1052.
231. Calkins MJ, Manczak M, Mao P, Shirendeb U, Reddy PH. Impaired mitochondrial biogenesis, defective axonal transport of mitochondria, abnormal mitochondrial dynamics and synaptic degeneration in a mouse model of Alzheimer's disease. *Hum Mol Genet*. December 1, 2011;20(23):4515–4529.
232. Brown DA, Hale SL, Baines CP et al. Reduction of early reperfusion injury with the mitochondria-targeting peptide bendavia. *J Cardiovasc Pharmacol Ther*. January 2014;19(1):121–132.
233. Cho J, Won K, Wu D et al. Potent mitochondria-targeted peptides reduce myocardial infarction in rats. *Coron Artery Dis*. May 2007;18(3):215–220.
234. Dai W, Shi J, Gupta RC, Sabbah HN, Hale SL, Kloner RA. Bendavia, a mitochondria-targeting peptide, improves postinfarction cardiac function, prevents adverse left ventricular remodeling, and restores mitochondria-related gene expression in rats. *J Cardiovasc Pharmacol*. December 2014;64(6):543–553.
235. Kloner RA, Hale SL, Dai W et al. Reduction of ischemia/reperfusion injury with bendavia, a mitochondria-targeting cytoprotective Peptide. *J Am Heart Assoc*. June 2012;1(3):e001644.
236. Shi J, Dai W, Hale SL et al. Bendavia restores mitochondrial energy metabolism gene expression and suppresses cardiac fibrosis in the border zone of the infarcted heart. *Life Sci*. November 15, 2015;141:170–178.
237. Szeto HH. Mitochondria-targeted cytoprotective peptides for ischemia-reperfusion injury. *Antioxid Redox Signal*. March 2008;10(3):601–619.
238. Dai DF, Chen T, Szeto H et al. Mitochondrial targeted antioxidant peptide ameliorates hypertensive cardiomyopathy. *J Am Coll Cardiol*. June 28, 2011;58(1):73–82.
239. Dai DF, Hsieh EJ, Chen T et al. Global proteomics and pathway analysis of pressure-overload-induced heart failure and its attenuation by mitochondrial-targeted peptides. *Circ Heart Fail*. September 1, 2013;6(5):1067–1076.
240. Dai DF, Hsieh EJ, Liu Y et al. Mitochondrial proteome remodelling in pressure overload-induced heart failure: The role of mitochondrial oxidative stress. *Cardiovasc Res*. January 1, 2012;93(1):79–88.
241. Sabbah HN, Gupta RC, Kohli S, Wang M, Hachem S, Zhang K. Chronic Therapy with elamipretide (MTP-131), a novel mitochondria-targeting peptide, improves left ventricular and mitochondrial function in dogs with advanced heart failure. *Circ Heart Fail*. February 2016;9(2):e002206.
242. Talbert EE, Smuder AJ, Min K, Kwon OS, Szeto HH, Powers SK. Immobilization-induced activation of key proteolytic systems in skeletal muscles is prevented by a mitochondria-targeted antioxidant. *J Appl Physiol (1985)*. August 15, 2013;115(4):529–538.
243. Powers SK, Hudson MB, Nelson WB et al. Mitochondria-targeted antioxidants protect against mechanical ventilation-induced diaphragm weakness. *Crit Care Med*. July 2011;39(7):1749–1759.
244. Szeto HH. First-in-class cardiolipin-protective compound as a therapeutic agent to restore mitochondrial bioenergetics. *Br J Pharmacol*. April 2014;171(8):2029–2050.
245. Dai DF, Chiao YA, Marcinek DJ, Szeto HH, Rabinovitch PS. Mitochondrial oxidative stress in aging and healthspan. *Longev Healthspan*. 2014;3:6.

22 Immunology of Aging and Cancer Development

T. Fulop, J. M. Witkowski, G. Dupuis,
A. Le Page, A. Larbi, and G. Pawelec

CONTENTS

INTRODUCTION

Aging is clearly a very complex process [1]. Many theories have been put forward to explain it, but no overarching theory explaining the entire process has been accepted thus far. This is because we still do not have a complete, global view of all changes occurring with aging; a systems approach is required to explain the whole process. In any organism, the extensively studied immune system is likely to be involved at least in part in the aging process [2]. Immune response affects many physiological and pathophysiological processes which are changed with aging. Moreover, the immune system interacts with many others, such as the endocrinological and neurological systems; this notion and the perceived importance of these interactions both in physiology and pathology resulted in the concept of neuro-endocrine–immune interactions being of prime importance in controlling many physiological processes [3]. It is tempting to propose a key role for the immune changes occurring through aging in the occurrence of various age-related major diseases including cancer [4]. In this chapter, we will review the changes in the immune response in aging and especially their potential contribution to cancer development.

IMMUNITY AS THE MASTER FOR PROTECTION AGAINST CHALLENGES

Every living organism continuously faces internal and external challenges to its integrity, arising from microorganisms, metabolic stress, altered macromolecules, or cell transformation, irradiation etc. Thus, every living organism from plants to vertebrates evolved sophisticated defense systems of different degrees of complexity. In every organisms except vertebrates, the defense system relies on innate immunity, while in all vertebrates (both with and without jaw) for unknown reasons, a more sophisticated immune system evolved with a far greater range of receptor specificity and endowed with adaptive memory—hence its designation "the adaptive immune system" [5]. Thus, in all vertebrates, the immune system is composed of two distinct, but closely interacting arms, the innate and adaptive arms, with distinct roles in the maintenance of organismal integrity. The innate arm responds immediately to a limited constellation of molecules present in microorganisms whereas adaptive immunity responds more slowly in a specific manner to universal types of diverse antigens and can develop specific memory permitting more rapid and stronger responses to later recurrence of the same challenge [5].

IMMUNE CHANGES WITH AGING

It is often assumed that immune changes with age must be detrimental and thus contribute to or even cause many diseases for which incidence is increased with age and which are considered as age-related diseases, such as cancer, autoimmune disorders, and chronic inflammatory diseases [6]. This concept was conceptualized by the denominating age-associated immune changes as immunosenescence, but this is only acceptable when those particular immune changes have indeed been shown to be linked with detrimental clinical outcomes [7,8]. However, this term was and still is commonly associated with age-related immunodeficiency without essential data being available. Clearly, although some parts of the immune system are less efficient in the elderly than in the young, efficient protection from many challenges is retained in older adults [7]. Thus, notably the adaptive immune system and to some degree also innate immunity undergoes remodeling processes which may permit survival for extended periods—may be up to 120 years in humans.

In the following we will describe the changes, their putative causes, and how in our opinion they should be interpreted in view of the present conceptualization of the aging process. This way of approaching age-associated immune changes permits their consideration not only as solely detrimental, but as potentially beneficial adaptation which may redirect the focus of interest in these topics elsewhere than originally and traditionally thought.

INNATE IMMUNE SYSTEM

The cells of the innate immune system are the first to respond to immune challenges of any sort, such as pathogens, cell debris, or tumor cells. This first interaction needs to be well orchestrated as its efficacy helps determine whether an adaptive immune response will be solicited. Furthermore, the innate response must be tightly controlled in order to avoid uncontrolled inflammation impacting on the adaptive immune system and the appearance of age-related diseases [9,10].

The innate immune system is composed of circulating molecules such as the complement system proteins and cells including neutrophils, macrophages, natural killer cells (NK cells), and the main antigen-presenting cells and dendritic cells (DCs). In most instances, this armamentarium is sufficient for possible elimination of the challenge and restores homeostasis.

Neutrophils in the circulation have a very short life span and die by apoptosis or a process known as NETosis [11]. They are the first cells attracted to the site of infection and initiate a complex cascade of functions which may culminate in phagocytosis of microorganisms and their intracellular destruction. They also kill targets extracellularly either by extruding neutrophil extracellular traps (NETs) or by antibody-dependent cellular cytotoxicity (ADCC). They produce multiple inflammatory mediators such as cytokines and chemokines. There is increasing experimental evidence that neutrophil phenotypes and the above-mentioned functions are all changed with aging [12–16]. It is of interest that the tendency is not only to decrease but also to increase certain parameters. The phenotypic changes are not yet well characterized, but a recent study described that circulating TNF levels and mitochondrial DNA content are the major determinant of neutrophil phenotypes in the elderly [17].

These soluble factors in the serum are typically increased during inflammation and also in the elderly as part of the inflammaging phenomenon [18]. Neutrophils from elderly donors are morphologically more immature, have higher levels of intracellular reactive oxygen species (ROS), and higher expression of the activation markers CD11b and HLA-DR [19–21]. This is consistent with the well-established notion that aging is accompanied by a basal activation of neutrophils (and, as we will see later, of the other innate cells). This is an important characteristic of the aging innate immune system. In contrast, receptor-driven activation, such as that resulting from ligation of receptors for GM-CSF, FMLP, and FCγ is decreased by aging [22]. Among these functions the most important effector functions, changed with aging, are chemotaxis, phagocytosis, intracellular killing, and apoptosis. The causes responsible for these altered effector functions are associated with the increased basal activation and resulting decreased ability for further increases in receptor signaling [23].

It has been shown that the numbers of receptors on the cell surface do not change with age, but signaling initiated by these receptors does. The most important signaling pathways for the initiation of the effector functions are the JAK/STAT, Akt/PI3 K, and MAPK pathways [24,25]. It was shown that activation of these signaling pathways after specific stimulation with GM-CSF, or via Fc receptors, CR3 receptors, or FMLP receptors is altered in neutrophils from the elderly. The most dramatic changes occur in the akt/PI3K pathway, which appears to be a master controller of the other signaling pathways [26]. As mentioned above, these signaling molecules are already in an activated state prior to stimulation, so that additional stimulation is difficult. Thus, it seems that neutrophils are already activated at the "quiescent" state in aging.

Other cells of the innate immune system are also affected by aging. Monocytes/macrophages have different phenotypes [27], also changing in the direction of more inflammatory phenotypes. The intermediate CD14^{high}CD16^{high} monocytes subpopulation is increased at the expense of the classical phenotype which normally represents 80% of the circulating monocyte population [28–31]. This is also in accordance with the presence of a more inflammatory milieu. The differentiation of monocytes to macrophages is also perturbed in some way, but a clear general picture is difficult to draw because of variation from tissue to tissue and because the state of differentiation is related to pathological phenomena [32,33]. Macrophage phenotypes exhibit a great plasticity as many surface markers can be present simultaneously on Type 1 and Type 2 macrophages. Conservatively, it can be suggested that in the resting state most of the macrophages resemble more an M1, so being of inflammatory type, whereas on specific stimulation they mostly differentiate into M2 anti-inflammatory cells, possibly contributing to an inadequate acute response otherwise expected in such circumstances. Thus, the functions of monocytes/macrophages are altered with aging, and in particular the secretion of pro-inflammatory cytokines mainly at the basal state is increased [19]. Moreover, their most important effector functions, including chemotaxis, phagocytosis, intracellular killing, and secretion of bactericidal products are also altered with aging [9,34,35]. These changes may contribute to the increased frequency and severity of infections observed with advancing age.

Along the same lines, when considering the causes of these functional alterations it was shown that certain types of receptors such as TLR1/2 are increased with aging, while others did not change [15,36]. In contrast, the function of the TLR is decreased with aging [15,37]. Also, as in the case of neutrophils, alterations in signal transduction were described in monocytes with aging. Among these changes the excessive activation of NFkB is striking [23].

One other very important type of cell of the innate system which has also been extensively studied is the DCs, which are crucial for antigen presentation to the adaptive immune system. Moreover, DCs determine the type of adaptive immune response by the production of different constellations of cytokines. Until recently, most available data indicated few changes in DCs with aging, but newer evidence shows many subtle changes that could have crucial effects on adaptive immune responses in the elderly [38–40]. Among these, and similar to the situation in other innate cells as outlined above, the most important are the increased basal production of pro-inflammatory cytokines, in addition to DC-specific decreased antigen processing and presentation to T cells. These changes may have dramatic consequences for the engagement of the adaptive immune response toward a specific antigen in the elderly.

Finally, NK cells have also been extensively studied in the context of aging [41–43]. Their characteristic surface markers include CD56, which defines two distinct populations of CD56^{high} and

CD56^{dim} cells with different functions; CD56^{high} NK cells are more cytotoxic while CD56^{dim} cells are more secretory. With aging, there is a decrease of the CD56^{high} and an increase of the CD56^{dim} subpopulation [9]. The functional consequences of this are that aging individuals are less able to control virus-infected or transformed cells, which may contribute to increased infections and cancers in the elderly [4]. It is of note that the number of NK cells is commonly increased with aging, which may be a compensatory mechanism due to the decreased functionality on a per cell basis and again represents another adaptation to aging in the immune response, even in the innate immune system.

Taken together, the majority of studies document that the innate immune system does show significant changes with age both at the phenotypic and functional levels. The most striking change is the increased basal activation state which may assure a certain level of pre-activated readiness to deal with assorted challenges, but which may also be associated with the kind of "immunoparalysis" for responses to additional specific activation, which could potentially contribute to age-related diseases such as cancer. This state may be explained by the continuous challenges during the immune life history of elderly persons including infections, cancers, and metabolic stresses over extended periods of time [44–46]. It may be suggested that these challenges act through the trained memory of innate immune cells [19]. By virtue of the accumulation of different innate cell types resulting from the individual's history of exposures, they will have a kind of memory assisting a more efficacious response to re-challenge [47]. However, this system may become maladapted with two possible major consequences either manifesting in immunoparalysis or in a hyperinflammatory state [44]. With aging, we can see both phenomena [19]. The underlying mechanism is an epigenetic adaptation to the challenges which will cause a metabolic shift favoring these innovative ways of functioning of the innate cells. This again may represent an adaptation through a physiological process to maintain a certain readiness of the system to cross the threshold of an otherwise failed activation, but not always being fully capable of further reactivity. Nevertheless, this divergent maladaptation may be targetable in future by metabolic manipulation in specific circumstances such as infections and cancer [48–52].

Adaptive Immune Changes

There are marked differences in measures of adaptive immunity in younger and older subjects, but despite extensive efforts, it is still not known what real changes are caused by aging, what the exact causes and the clinical/pathological consequences are [7,53–57]. It is largely accepted that the proportion and numbers of naïve T cells decrease with aging and that effector antigen-committed T cells (mainly in the CD8$^+$ compartment) increase. These cells have a limited antigen repertoire and are more inflammatory, with a decreased clonal expansion capacity. There are multiple phenotypic differences discernible by employing surface markers, such as CD28, CD27, CCR7, and CD45RA [58–60]. The most widely reported is CD28 expression, the absence of which on CD8$^+$ T cells is commonly considered to define a state of immune aging, as well as being further increased in much age-related pathologies, such as atherosclerosis, rheumatoid arthritis, or cardiac diseases [61]. Moreover, CD8$^+$ CD28$^-$ T cells are often considered to be senescent due to parallels with replicative senescence in fibroblast cultures, such as decreased proliferation, increased pro-inflammatory cytokine production, and shortened telomere length [62–64]. Some of these cells may indeed be senescent but the majority may in fact represent a normal late-stage differentiated phenotype of effector memory cells. Nonetheless, at least some may exhibit the so-called "senescence-associated secretory phenotype" (SASP) typical for other senescent cell types which could contribute to the inflammaging phenomenon [65–66]. However, these cells are likely to be beneficial in that many are specific for CMV, accumulating mostly in CMV-seropositive individuals and probably crucial for controlling latent infection [67–74].

The question may also be raised as to whether the decrease of the naïve cells is necessarily as detrimental as perceived earlier. The only source of naïve T cells is thymic-processed T cell precursors from the bone marrow. However, hematopoietic output changes with age, and more importantly for T cell production, thymic involution beginning even before puberty greatly reduces the release of naïve T cells to the periphery. Nonetheless, a greatly reduced production of naïve T cells continues in most people until quite late in life, possibly even up to 70 years of age. Additionally, the phenomenon of homeostatic proliferation driven by cytokines such as IL-7 and IL-15 expands peripheral naïve T cells as a possible compensatory mechanism for reduced thymic output [75,76]. It is unclear why thymic involution occurs so radically in humans, but presumably represents an evolutionarily advantageous adaptive process to avoid diverting resources from the maintenance of defense against known antigens to the costly generation of eventually unnecessary naïve T cells. It is of note that studies showed that the elderly exhibiting low numbers of naïve CD8[+] T cells did not present an increased mortality over a longitudinal follow up of 2–9 years under certain circumstances [77].

Given that T cell differentiation may be flexible, like many biological phenomena in aging, as with macrophages where there is constant reprogramming from one macrophage type to the other and vice versa, it can also be proposed that the same applies to T cells, because T cell differentiation is dependent on the metabolic programming of the cells. Thus, a certain plasticity would be expected, which would permit T cells to differentiate and re-differentiate between the central memory and effector memory and possibly even late-stage TEMRA phenotypes and functions [57,78]. The same idea may also be proposed for the phenotypes and functional characteristics of T cells in exhausted or senescent states [62]. This would help to explain why not all the elderly suffer from infections or cancer and would permit the modulation of some T cell states such as exhaustion in certain cancers [79]. However, currently there is no experimental evidence for this level of plasticity in T cell differentiation pathways. Still, it could be proposed that the immune metabolic pathways active through mTOR influence the immune phenotype and lineage skewing as a means of imbuing these cells with a degree of plasticity [80–83]. The epigenetic determination of the predominance of ox-phos metabolism or anaerobic glycolysis (Warburg effect) impacts markedly on T cell lineage commitment [84,85]. These phenomena are not well understood within the context of aging; however, some data suggest that there is a predominance of the ox-phos pathway for energy production [61]. It is of note that the metabolic needs of the naïve, memory, and effector T cells are quite different, and this could change with age [61]. An imbalance between two enzymes regulating glucose metabolism in T cells, 6-phosphofructo-2-kinase/fructose 2,6-biphosphate 3 (PFKFB3), and glucose-6-phosphate dehydrogenase (G6PD) emerges with aging [86,87]. The activity of the former is decreased while that of the latter increased resulting in altered pyruvate production. This in turn results in energy deprivation and leads to anabolism as well as to a decreased oxidative state impairing appropriate proliferation and lineage differentiation which may be shifted toward immunosuppressive phenotypes such as Tregs [88]. By virtue of also participating in the autophagy process, these enzymes induce alterations impeding energy metabolism, and favor the overexpression of CD39 on T cells, especially on CD4[+] T cells. This renders them more susceptible to apoptosis, contributing to decreased effector CD4[+] T cell survival with aging [89]. These changes would also influence signal transduction in T cells [23].

In addition to phenotypic changes, there are parallel functional changes in T cells with age, such as clonal expansion and cytokine/chemokine production [7]. These alterations may partly be explained by altered receptor signaling in T cells at each step of the cascade [23], but are most strikingly related to very early signaling events [89–92].

The formation of the immune synapse between the APC and T cell is altered as the membrane of old T cells becomes more rigid, resulting in decreased lipid raft coalescence, which is indispensable for the correct functioning of the immune synapse [93]. This contributes to a decreased signaling in T cells by impairing activation of the essential tyrosine kinase, Lck [94–96] due to decreased activation of ZAP70. The primary mechanism responsible for decreased Lck activation is the inability to modulate phosphatase inhibitory activity [96]. *In vitro* repression of the SHP-1 phosphatase can restore the proliferation and IL-2 secretion of T cells in the aged [96], as was also demonstrated for several other phosphatases of the DUSP family [97,98].

Not only T cells but also B cells are clearly different in younger and older individuals [99–101]. With increasing age, like with the T cells, there is a shift from naïve to effector B cells. Secreted antibodies have lower affinity for specific antigens as well as less diversity, but auto-antibody production is increased.

WHAT COULD BE THE CONSEQUENCES OF THE IMMUNE CHANGES WITH AGING?

Whether these age-related changes have any clinical or pathological impact has been the subject of intensive research. It is proposed that the increased prevalence of age-associated disease including cancer is a major consequence of age-associated low-grade chronic inflammation and the presence of immunosenescence. However, direct relationships between these age-related diseases either with the low-grade inflammation or immunosenescence *per se* has never been clearly demonstrated.

Low-Grade Inflammation as an Adaptation

Aging is commonly associated with the presence of low-grade inflammation first identified by Franceschi et al. [18] and then refined since that time [102,103]. Such "inflammaging" may at least partly reflect unbalanced pro- and anti-inflammatory activities in favor of the pro-inflammatory state due to over-activity of innate immunity. However, increases in levels of pro-inflammatory cytokines such as IL-1, TNF-α, and IL-6 with aging may also originate from damaged or senescent nonimmune cells showing the SASP. Increased immune pro-inflammatory activity may also be an adaptive mechanism [19,104]. Inflammation is necessary for maintaining longevity, as demonstrated for semi-supercentenarians [105]. In that study, inflammatory status correlated positively with longevity better than any of the other bio-markers studied. This was also shown in centenarians where the importance of the balance between the pro-inflammatory and anti-inflammatory parameters was crucial [106,107]. Even with higher inflammation in aging an effective anti-inflammatory response may be efficacious to decrease its pathological consequences. At this point it is important to consider the entire phenomenon besides evaluating only one aspect [108]. So perhaps the problem is not aging per se but the mechanisms which drive the low-grade inflammation to an unbalanced state resulting in hyperinflammation.

Several mechanisms have been proposed as causing the low-grade inflammation. It may be either by exogenous stimulation resulting from infections with microorganisms especially chronic infectious agents (predominantly cytomegalovirus or CMV), or endogenous agents such as modified macromolecules (proteins and DNA), cancer cells, or the microbiota. As suggested above, trained innate memory could contribute to the low-grade inflammation with aging. It is thus likely that this phenomenon may be adaptive, but reflects the usual immune pathology accompanying immune responses.

Low-Grade Inflammation as a Determinant of the Inflammatory Diseases of the Elderly

As mentioned above, it is the dysregulation of an adaptive inflammatory response occurring with aging under various stimuli that may induce most of the apparently different diseases of the elderly [6,104]. Probably during the development of these diseases, the equilibrium between pro-inflammatory and anti-inflammatory signals is disrupted and the above-mentioned "hyperinflammation" will pave the way for the pathological consequences. Excessive inflammation during the inflammaging process increases susceptibility to various pathologies; diseases as diverse as diabetes mellitus, cardiovascular diseases, Alzheimer's disease, cancer, and sarcopenia may all be related to the inflammatory process observed in elderly subjects. Of course, the relationship between inflammaging and these diseases is complex and occurs through many different intertwined pathways [102] largely influencing each other. In the remainder of this chapter, we will specifically examine the relationship between inflammaging and immunosenescence, and the development of cancer.

CANCER DEVELOPMENT AND IMMUNE CHANGES WITH AGING

The incidence and prevalence of various cancers increase with age until at least the ninth decade [4]. The finding of clinically detectable cancer can be a lengthy process. The transformation of a pre-cancerous cell to an overt cancer depends on many different physiological and pathological mechanisms. The primary events of DNA damage by free radicals, ionizing radiation, carcinogenic chemicals, and pathogens which fail to be repaired or trigger apoptosis occur decades before the clinical appearance of a cancer and are enhanced during low-grade chronic inflammation such as that occurring during aging [109]. Thus inflammaging and oxidative stresses induce genetic and epigenetic modifications which will alter cytokine expressions, oncogenes, and tumor suppressor gene transcription, and thus ultimately contribute to carcinogenesis [110]. The progression of these pathological processes resulting in clinically manifest cancer development is influenced by the efficacious functionality of the immune system [111]. One of the most important components of the adaptive immune system is the cytotoxic CD8+ T cells, crucial at every step of immunosurveillance against cancer development. However, immune escape variants of cancer cells resistant to CD8+ T cell killing eventually emerge, resulting in tumor progression, which may even subvert the immune system to facilitate cancer growth. The fading of the immunosurveillance and the killing capabilities as occurring with aging may contribute to cancer development; however, there are other changes which may facilitate and contribute to carcinogenesis in elderly individuals [4].

Thus, all the above-mentioned systemic immune phenotypic and functional changes occurring with aging in both the innate and adaptive immune systems may contribute to carcinogenesis [4,72,102,103,110]. Changes in the effector functions, specifically in the cytotoxicity of NK cells and CD8+ T cells are likely to be important in this context. Their decrease in aging may contribute to the survival of tumor cells and their dissemination throughout the body. Generalized alterations in TLR functions in innate immune cells also favor carcinogenesis by decreasing the capacity to eliminate transformed cells either directly or by antibody-dependent cytotoxicity. Increased and uncontrolled secretion of immunosuppressive substances such as indoleamine 2,3-dioxygenase which strongly inhibits T cell activation and clonal expansion is an important immunosuppressive mechanism in aging. The age-associated failure of DCs to process and present tumor antigens to T cells also contributes to the development and progression of cancer. It is of note that, as with tumor progression the malignant cells become less antigenic and hide from the immune response, therefore the DCs processing of the recognizable antigens become more important for fighting the cancer. However, the functions of the DCs also decrease with age.

Not only do the tumors become more immunologically inert, the decreased proportion of naïve T cells may also impede the response to novel tumor-related antigens. Finally, there are also increased immunosuppressive regulatory cells in aging such as Tregs and myeloid-derived suppressor cells (MDSCs). With aging, the number of Tregs increase in the lymphoid tissues, which may help to suppress the development of anti-tumor T cell responses. Also MDSCs have been shown to increase during aging, either systemically or in the tumor microenvironment, directly contributing to the decreased anti-tumor T cell response as well as that of the innate immune response [4].

Furthermore, another important aspect of the immune surveillance of cancers is the tumor microenvironment. It is a very complex setting composed of different immune cells (tumor-associated neutrophils (TANs), tumor-associated macrophages (TAMs), MDSC, Tregs, and cytotoxic T cells) and of other cells like fibroblasts, as well as the extracellular matrix. The interactions among these various cells will influence whether they will fight or favor further tumor growth and as such affect the eventual clinical outcome [112]. In response to tumor-derived factors, the tumor-infiltrating macrophages (TAM) preferentially release anti-inflammatory mediators such as IL-4 and TGFβ favoring the development of M2 macrophages, which may promote tumor growth in elderly individuals. On the other hand, inflammatory infiltration by tumor-associated pro-inflammatory neutrophils (TAN), which may be also perverted to an anti-inflammatory phenotype, may also influence tumor development and growth. IL-6, as a pleiotropic cytokine known to increase with aging may be considered as instrumental in this activity through its role as a growth factor and an anti-apoptotic factor. TNFα may also participate in carcinogenesis as it triggers DNA damage, angiogenesis, invasion, and metastasis. TNFα, as one of the best known pro-inflammatory cytokines, supports the aging-associated low-grade inflammation (inflammaging) and as such, either directly or through inflammation, participates in the changes leading to carcinogenesis. Therefore, as the anti-inflammatory innate cells start to dominate and this increases the anti-inflammatory cytokines such as IL-10 and TGF-β, the cytokine balance will be tipped toward the anti-inflammatory outcome. Thus, this age-related imbalance among pro- and anti-inflammatory cytokines as well as the subversion of the targeted cytokine actions in the tumor microenvironment may favor the tumor growth instead of combating it. It is of note that the role of the various cytokines is dependent on the balance of the cells in the tumor microenvironment, the tumor progression phase, and the various metabolic conditions prevailing in the tumor microenvironment. Together in elderly subjects systemically as well as locally, the cytokines no matter what their original function could favor tumor development.

One fundamental question is whether these mostly carcinogenesis favoring changes in the immune response occurring with aging would impact on therapy in the era of immunomodulatory antibody therapy when applied in elderly patients. Blockade of immune-inhibitory checkpoints by antibodies targeting CTLA-4 and PDL-1 surface molecules has recently demonstrated spectacular successes in the treatment of melanoma, non-small cell lung carcinoma, and other solid tumors [113]. Therefore, considering changes to immunity with age, the question arises of whether these therapies would be equally effective in the elderly? Thus far, there are only a few reports on this outcome but those which deal with this question reported clinical results as favorable in the elderly as in young subjects [114,115]. This could imply that at least the exhausted T cells of the elderly when relieved from checkpoint inhibition may mediate anticancer activities as efficiently as in young subjects. However, whether this would be sufficient to maintain this favorable effect in view of the other changes with aging in the immune-inflammatory environment requires further investigation. Nevertheless, these are encouraging results to be confirmed, but suggest that some type of modulation of the immune functions in aging may have beneficial clinical effects.

CONCLUSION

The so-called "immunosenescence" together with inflammaging may be considered an adaptive mechanism occurring over the life span due to continuous antigenic challenges aimed to maintain protection against most already-encountered pathogens. In a balanced genetic and environmental milieu, these changes may contribute to longevity. This is underlined by the fact that in real life the elderly are doing much better than would be predicted from current paradigms about age-associated immune changes. However, dysregulation of this adaptation may result in pathological conditions due to immune paralysis of the innate and adaptive immune system concomitant to hyperinflammation as predicted by the trained innate memory and decreased adaptive immune response state (Figure 22.1). Alterations in both the innate and adaptive immune responses are likely contributing concomitantly to the deregulated inflammaging process. One of the diseases affecting particularly the elderly is cancer and its increased occurrence may be related to decreased immune response with aging either systemically or locally. However, immunotherapy in elderly patients thus far seems to be as efficient as in young patients. To assess globally the effect of immune changes in aging we should study this in a global manner using systems biology approaches. Taking into account these considerations we should be very careful in our interventions to "rejuvenate" the elderly immune system before determining when, where, and how to intervene.

ACKNOWLEDGMENTS

This work is partly supported by grants from the Canadian Institutes of Health Research (No. 106634 and No. 106701), the Université de Sherbrooke, and the Research Center on Aging; Polish

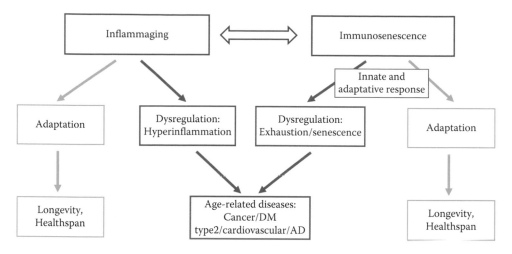

FIGURE 22.1 Schematic representation of age-related immune changes leading to increased cancer susceptibility development.

Ministry of Science and Higher Education statutory grant 02-0058/07/262 to JMW; Agency for Science Technology and Research (A*STAR) to AL.

REFERENCES

1. Fulop T and Robert L (Eds). Facts and theories. *Interdiscip Top Gerontol.* 2014;39 pp Karger, Basel.
2. Walford RL. The immunologic theory of aging. *Gerontologist.* 1964; 4:195–197.
3. Fulop T, Witkowski JM, Pawelec G, Cohen A, Larbi A. On the immunological theory of aging. *Interdiscip Top Gerontol.* 2014;39, 163–76.
4. Fulop T, Larbi A, Witkowski JM, Kotb R, Hirokawa K, Pawelec G. Immunosenescence and cancer. *Crit Rev Oncog.* 2013;18(6):489–513.
5. Müller L, Fülöp T, Pawelec G. Immunosenescence in vertebrates and invertebrates. *Immun Ageing.* 2013;10(1):12.
6. Fülöp T, Dupuis G, Witkowski JM, Larbi A. The role of immunosenescence in the development of age-related diseases. *Rev Invest Clin.* 2016;68(2):84–91.
7. Muller L, Pawelec G. The aging immune system: Dysregulation, compensatory mechanisms and prospects for intervention. In *Handbook of the Biology of Aging.* Keberlein M and Martin GM (Eds) Eleesevier, Amsterdam, 8th Edition, pp. 407–431.
8. Pawelec G. Hallmarks of human "immunosenescence": Adaptation or dysregulation? *Immun Ageing.* 2012;9(1):15.
9. Solana R, Tarazona R, Gayoso I, Lesur O, Dupuis G, Fulop T. Innate immunosenescence: Effect of aging on cells and receptors of the innate immune system in humans. *Semin Immunol.* 2012;24(5):331–41.
10. Fortin CF, McDonald PP, Lesur O, Fülöp T Jr. Aging and neutrophils: There is still much to do. *Rejuvenation Res.* 2008;11(5):873–82.
11. Sørensen OE, Borregaard N. Neutrophil extracellular traps—The dark side of neutrophils. *J Clin Invest.* 2016;126(5):1612–20.
12. Montgomery RR, Shaw AC. Paradoxical changes in innate immunity in aging: Recent progress and new directions. *J Leukoc Biol.* 2015;98(6):937–43.
13. Tseng CW, Liu GY. Expanding roles of neutrophils in aging hosts. *Curr Opin Immunol.* 2014; 29:43–8.
14. Wessels I, Jansen J, Rink L, Uciechowski P. Immunosenescence of polymorphonuclear neutrophils. *Sci World J.* 2010;10:145–60.
15. Fulop T, Larbi A, Douziech N, Fortin C, Guérard KP, Lesur O, Khalil A, Dupuis G. Signal transduction and functional changes in neutrophils with aging. *Aging Cell.* 2004;3(4):217–26.
16. Wenisch C, Patruta S, Daxböck F, Krause R, Hörl W. Effect of age on human neutrophil function. *J Leukoc Biol.* 2000;67(1):40–5.

17. Verschoor CP, Loukov D, Naidoo A, Puchta A, Johnstone J, Millar J, Lelic A et al. Circulating TNF and mitochondrial DNA are major determinants of neutrophil phenotype in the advanced-age, frail elderly. *Mol Immunol*. 2015;65(1):148–56.

18. Franceschi C, Bonafè M, Valensin S, Olivieri F, De Luca M, Ottaviani E, De Benedictis G. Inflamm-aging. An evolutionary perspective on immunosenescence. *Ann N Y Acad Sci*. 2000;908:244–54.

19. Fulop T, Dupuis G, Baehl S, Le Page A, Bourgade K, Frost E, Witkowski JM, Pawelec G, Larbi A, Cunnane S. From inflamm-aging to immune-paralysis: A slippery slope during aging for immune-adaptation. *Biogerontology*. 2016;17(1):147–57.

20. Nogueira-Neto J, Cardoso AS, Monteiro HP, Fonseca FL, Ramos LR, Junqueira VB, Simon KA. Basal neutrophil function in human aging: Implications in endothelial cell adhesion. *Cell Biol Int*. 2016;40(7):796–802.

21. Ogawa K, Suzuki K, Okutsu M, Yamazaki K, Shinkai S. The association of elevated reactive oxygen species levels from neutrophils with low-grade inflammation in the elderly. *Immun Ageing*. 2008;5:13.

22. Bandaranayake T, Shaw AC. Host resistance and immune aging. *Clin Geriatr Med*. 2016;32(3):415–32.

23. Fulop T, Le Page A, Fortin C, Witkowski JM, Dupuis G, Larbi A. Cellular signaling in the aging immune system. *Curr Opin Immunol*. 2014;29:105–11.

24. Fortin CF, Larbi A, Dupuis G, Lesur O, Fülöp T Jr. GM-CSF activates the Jak/STAT pathway to rescue polymorphonuclear neutrophils from spontaneous apoptosis in young but not elderly individuals. *Biogerontology*. 2007;8(2):173–87.

25. Fortin CF, Larbi A, Lesur O, Douziech N, Fulop T Jr. Impairment of SHP-1 down-regulation in the lipid rafts of human neutrophils under GM-CSF stimulation contributes to their age-related, altered functions. *J Leukoc Biol*. 2006;79(5):1061–72.

26. Sapey E, Greenwood H, Walton G, Mann E, Love A, Aaronson N, Insall RH, Stockley RA, Lord JM. Phosphoinositide 3-kinase inhibition restores neutrophil accuracy in the elderly: Toward targeted treatments for immunosenescence. *Blood*. 2014;123(2):239–48.

27. Dalton HJ, Armaiz-Pena GN, Gonzalez-Villasana V, Lopez-Berestein G, Bar-Eli M, Sood AK. Monocyte subpopulations in angiogenesis. *Cancer Res*. 2014;74(5):1287–93.

28. Puchta A, Naidoo A, Verschoor CP, Loukov D, Thevaranjan N, Mandur TS, Nguyen PS et al. TNF drives monocyte dysfunction with age and results in impaired anti-pneumococcal immunity. *PLoS Pathog*. 2016;12(1):e1005368.

29. Pararasa C, Ikwuobe J, Shigdar S, Boukouvalas A, Nabney IT, Brown JE, Devitt A, Bailey CJ, Bennett SJ, Griffiths HR. Age-associated changes in long-chain fatty acid profile during healthy aging promote pro-inflammatory monocyte polarization via PPARγ. *Aging Cell*. 2016;15(1):128–39.

30. Baëhl S, Garneau H, Lorrain D, Viens I, Svotelis A, Lord JM, Cabana F, Larbi A, Dupuis G, Fülöp T. Alterations in monocyte phenotypes and functions after a hip fracture in elderly individuals: A 6-month longitudinal study. *Gerontology*. 2016;62(5):477–90.

31. de Pablo-Bernal RS, Cañizares J, Rosado I, Galvá MI, Alvarez-Ríos AI, Carrillo-Vico A, Ferrando-Martínez S et al. Monocyte phenotype and polyfunctionality are associated with elevated soluble inflammatory markers, cytomegalovirus infection, and functional and cognitive decline in elderly adults. *J Gerontol A Biol Sci Med Sci*. 2016;71(5):610–8.

32. Wang Y, Wehling-Henricks M, Samengo G, Tidball JG. Increases of M2a macrophages and fibrosis in aging muscle are influenced by bone marrow aging and negatively regulated by muscle-derived nitric oxide. *Aging Cell*. 2015;14(4):678–88.

33. Verschoor CP, Johnstone J, Loeb M, Bramson JL, Bowdish DM. Anti-pneumococcal deficits of monocyte-derived macrophages from the advanced-age, frail elderly and related impairments in PI3K-AKT signaling. *Hum Immunol*. 2014;75(12):1192–6.

34. Goronzy JJ, Weyand CM. Immune aging and autoimmunity. *Cell Mol Life Sci*. 2012;69(10):1615–23.

35. Lloberas J, Celada A. Effect of aging on macrophage function. *Exp Gerontol*. 2002;37(12):1325–31.

36. Alvarez-Rodriguez L, Lopez-Hoyos M, Garcia-Unzueta M, Amado JA, Cacho PM, Martinez-Taboada VM. Age and low levels of circulating vitamin D are associated with impaired innate immune function. *J Leukoc Biol*. 2012;91(5):829–38.

37. Weinberger B, Grubeck-Loebenstein B. Vaccines for the elderly. *Clin Microbiol Infect*. 2012;18(Suppl 5):100–8.

38. Magrone T, Jirillo E. Disorders of innate immunity in human ageing and effects of nutraceutical administration. *Endocr Metab Immune Disord Drug Targets*. 2014;14(4):272–82.

39. Qian F, Wang X, Zhang L, Lin A, Zhao H, Fikrig E, Montgomery RR. Impaired interferon signaling in dendritic cells from older donors infected *in vitro* with West Nile virus. *J Infect Dis*. 2011;203:1415–1424.

40. Linton PJ, Thoman ML. Immunosenescence in monocytes, macrophages, and dendritic cells: Lessons learned from the lung and heart. *Immunol Lett.* 2014;162(1 Pt B):290–7.

41. Tarazona R, Campos C, Pera A, Sanchez-Correa B, Solana R. Flow cytometry analysis of NK cell phenotype and function in aging. *Methods Mol Biol.* 2015;1343:9–18.

42. Pera A, Campos C, López N, Hassouneh F, Alonso C, Tarazona R, Solana R. Immunosenescence: Implications for response to infection and vaccination in older people. *Maturitas.* 2015;82(1):50–5.

43. Isitman G, Tremblay-McLean A, Lisovsky I, Bruneau J, Lebouché B, Routy JP, Bernard NF. NK cells expressing the inhibitory killer immunoglobulin-like receptors (iKIR) KIR2DL1, KIR2DL3 and KIR3DL1 are less likely to be CD16+ than their iKIR negative counterparts. *PLoS One.* 2016;11(10):e0164517.

44. Netea MG, Joosten LA, Latz E, Mills KH, Natoli G, Stunnenberg HG, O'Neill LA, Xavier RJ. Trained immunity: A program of innate immune memory in health and disease. *Science.* 2016;352, 2(6284):aaf1098.

45. Kleinnijenhuis J, Quintin J, Preijers F, Joosten LA, Ifrim DC, Saeed S, Jacobs C et al. Bacillus Calmette–Guerin induces NOD2-dependent nonspecific protection from reinfection via epigenetic reprogramming of monocytes. *Proc Natl Acad Sci USA* 2012;109:17537–42.

46. Quintin J, Saeed S, Martens JH, Giamarellos-Bourboulis EJ, Ifrim DC, Logie C, Jacobs L et al. *Candida albicans* infection affords protection against reinfection via functional reprogramming of monocytes. *Cell Host Microbe* 2012;12:223–32.

47. Bekkering S, Blok BA, Joosten LA, Riksen NP, van Crevel R, Netea MG. In vitro experimental model of trained innate immunity in human primary monocytes. *Clin Vaccine Immunol.* 2016 Dec 5;23(12):926–933.

48. Oppermann U. Why is epigenetics important in understanding the pathogenesis of inflammatory musculoskeletal diseases? *Arthritis Res Ther.* 2013;15:209.

49. Kyburz D, Karouzakis E, Ospelt C. Epigenetic changes: The missing link. *Best Pract Res Clin Rheumatol.* 2014;28:577–87.

50. Ostuni R, Piccolo V, Barozzi I, Polletti S, Termanini A, Bonifacio S, Curina A, Prosperini E, Ghisletti S, Natoli G. Latent enhancers activated by stimulation in differentiated cells. *Cell.* 2013;152:157–71.

51. Cheng SC, Quintin J, Cramer RA, Shepardson KM, Saeed S, Kumar V, Giamarellos- et al. mTOR- and HIF-1α-mediated aerobic glycolysis as metabolic basis for trained immunity. *Science.* 2014;345:1250684.

52. Saeed S, Quintin J, Kerstens HH, Rao NA, Aghajanirefah A, Matarese F, Cheng SC et al. Epigenetic programming of monocyte-to-macrophage differentiation and trained innate immunity. *Science.* 2014;345:1251086.

53. Yanes RE, Gustafson CE, Weyand CM, Goronzy JJ. Lymphocyte generation and population homeostasis throughout life. *Semin Hematol.* 2017;54(1):33–38.

54. Kim C, Fang F, Weyand CM, Goronzy JJ. The life cycle of a T cell after vaccination—Where does immune ageing strike? *Clin Exp Immunol.* 2017;187(1):71–81.

55. Tu W, Rao S. Mechanisms underlying T cell immunosenescence: Aging and cytomegalovirus infection. *Front Microbiol.* 2016 Dec 27;7:2111.

56. Kennedy RB, Ovsyannikova IG, Haralambieva IH, Oberg AL, Zimmermann MT, Grill DE, Poland GA. Immunosenescence-related transcriptomic and immunologic changes in older individuals following influenza vaccination. *Front Immunol.* 2016 Nov 2;7:450.

57. Larbi A, Fulop T. From "truly naïve" to "exhausted senescent" T cells: When markers predict functionality. *Cytometry A.* 2014;85(1):25–35.

58. Onyema OO, Njemini R, Forti LN, Bautmans I, Aerts JL, De Waele M, Mets T. Aging-associated subpopulations of human CD8+ T-lymphocytes identified by their CD28 and CD57 phenotypes. *Arch Gerontol Geriatr.* 2015;61,(3):494–502.

59. Sallusto F, Lenig D, Förster R, Lipp M, Lanzavecchia A. Pillars article: Two subsets of memory T lymphocytes with distinct homing potentials and effector functions. *Nature.* 1999. 401: 708–712. *J Immunol.* 2014 Feb 1;192(3):840–4.

60. Romero P, Zippelius A, Kurth I, Pittet MJ, Touvrey C, Iancu EM, Corthesy P et al. Four functionally distinct populations of human effector-memory CD8+ T lymphocytes. *J Immunol.* 2007;178(7):4112–9.

61. Weyand CM, Goronzy JJ. Aging of the immune system. Mechanisms and therapeutic targets. *Ann Am Thorac Soc.* 2016;13(Supplement_5):S422–S428.

62. Wherry EJ, Kurachi M. Molecular and cellular insights into T cell exhaustion. *Nat Rev Immunol.* 2015;15(8):486–99.

63. Onyema OO, Njemini R, Bautmans I, Renmans W, De Waele M, Mets T. Cellular aging and senescence characteristics of human T-lymphocytes. *Biogerontology.* 2012;13(2):169–81.

64. Effros RB. Replicative senescence of CD8 T cells: Effect on human ageing. *Exp Gerontol.* 2004;39(4):517–24.

65. Loaiza N, Demaria M. Cellular senescence and tumor promotion: Is aging the key? *Biochim Biophys Acta.* 2016;1865(2):155–67.

66. Tchkonia T, Zhu Y, van Deursen J, Campisi J, Kirkland JL. Cellular senescence and the senescent secretory phenotype: Therapeutic opportunities. *J Clin Invest.* 2013;123(3):966–72.

67. Pawelec G. Immunosenescence: Role of cytomegalovirus. *Exp Gerontol.* 2014;54:1–5.

68. Pawelec G, Akbar A, Caruso C, Solana R, Grubeck-Loebenstein B, Wikby A. Human immunosenescence: Is it infectious? *Immunol Rev.* 2005;205:257–68.

69. Solana R, Tarazona R, Aiello AE, Akbar AN, Appay V, Beswick M, Bosch JA et al. CMV and Immunosenescence: From basics to clinics. *Immun Ageing.* 2012;9:23.

70. Pawelec G, McElhaney JE, Aiello AE, Derhovanessian E. The impact of CMV infection on survival in older humans. *Curr Opin Immunol.* 2012;24:507–11.

71. van Baarle D, Tsegaye A, Miedema F, Akbar A. Significance of senescence for virus-specific memory T cell responses: Rapid ageing during chronic stimulation of the immune system. *Immunol Lett.* 2005;97:19–29.

72. Bauer ME, Fuente Mde L. The role of oxidative and inflammatory stress and persistent viral infections in immunosenescence. *Mech Ageing Dev.* 2016;158:27–37.

73. Weltevrede M, Eilers R, de Melker HE, van Baarle D. Cytomegalovirus persistence and T-cell immunosenescence in people aged fifty and older: A systematic review. *Exp Gerontol.* 2016;77:87–95.

74. Söderberg-Nauclér C, Fornara O, Rahbar A. Cytomegalovirus driven immunosenescence-An immune phenotype with or without clinical impact? *Mech Ageing Dev.* 2016;158:3–13.

75. Qi Q, Zhang DW, Weyand CM, Goronzy JJ. Mechanisms shaping the naïve T cell repertoire in the elderly—Thymic involution or peripheral homeostatic proliferation? *Exp Gerontol.* 2014; 54:71–4.

76. Appay V, Sauce D. Naive T cells: The crux of cellular immune aging? *Exp Gerontol.* 2014; 54:90–3.

77. Derhovanessian E, Maier AB, Hähnel K, Zelba H, de Craen AJ, Roelofs H, Slagboom EP et al. Lower proportion of naïve peripheral CD8+ T cells and an unopposed pro-inflammatory response to human Cytomegalovirus proteins *in vitro* are associated with longer survival in very elderly people. *Age (Dordr).* 2013;35(4):1387–99.

78. Stervbo U, Bozzetti C, Baron U, Jürchott K, Meier S, Mälzer JN, Nienen M et al. Effects of aging on human leukocytes (part II): Immunophenotyping of adaptive immune B and T cell subsets. *Age (Dordr).* 2015;37(5):93.

79. Pawelec G. Immunosenescence and cancer. *Biogerontology.* 2017;18(4):717–21.

80. Fernández-Ramos AA, Poindessous V, Marchetti-Laurent C, Pallet N, Loriot MA. The effect of immunosuppressive molecules on T-cell metabolic reprogramming. *Biochimie.* 2016;127:23–36.

81. Caza T, Landas S. Functional and phenotypic plasticity of CD4(+) T cell subsets. *Biomed Res Int.* 2015;2015:521957.

82. Delgoffe GM, Kole TP, Zheng Y, Zarek PE, Matthews KL, Xiao B, Worley PF, Kozma SC, Powell JD. The mTOR kinase differentially regulates effector and regulatory T cell lineage commitment. *Immunity.* 2009 Jun 19;30(6):832–44.

83. Kouidhi S, Elgaaied AB, Chouaib S. Impact of metabolism on T-cell differentiation and function and cross talk with tumor microenvironment. *Front Immunol.* 2017 Mar 13;8:270.

84. Leoni C, Vincenzetti L, Emming S, Monticelli S. Epigenetics of T lymphocytes in health and disease. *Swiss Med Wkly.* 2015;145:w14191.

85. Peng M, Yin N, Chhangawala S, Xu K, Leslie CS, Li MO. Aerobic glycolysis promotes T helper 1 cell differentiation through an epigenetic mechanism. *Science.* 2016;3546311:481–484.

86. Yang Z, Shen Y, Oishi H, Matteson EL, Tian L, Goronzy JJ, Weyand CM. Restoring oxidant signaling suppresses proarthritogenic T cell effector functions in rheumatoid arthritis. *Sci Transl Med.* 2016;8(331):331ra38.

87. Yang Z, Fujii H, Mohan SV, Goronzy JJ, Weyand CM. Phosphofructokinase deficiency impairs ATP generation, autophagy, and redox balance in rheumatoid arthritis T cells. *J Exp Med.* 2013;210(10):2119–34.

88. Wen Z, Shimojima Y, Shirai T, Li Y, Ju J, Yang Z, Tian L et al. NADPH oxidase deficiency underlies dysfunction of aged CD8+ Tregs. *J Clin Invest.* 2016;126(5):1953–67.

89. Fang F, Yu M, Cavanagh MM, Hutter Saunders J, Qi Q, Ye Z, Le Saux S et al. Expression of CD39 on activated T cells impairs their survival in older individuals. *Cell Rep.* 2016 Feb 9;14(5):1218–31.

90. Varin A, Larbi A, Dedoussis GV, Kanoni S, Jajte J, Rink L, Monti D et al. In vitro and *in vivo* effects of zinc on cytokine signalling in human T cells. *Exp Gerontol.* 2008;43(5):472–82.

91. Larbi A, Pawelec G, Wong SC, Goldeck D, Tai JJ, Fülöp T. Impact of age on T cell signaling: A general defect or specific alterations? *Ageing Res Rev.* 2011;10:370–378.

92. Goronzy JJ, Li G, Yu M, Weyand CM. Signaling pathways in aged T cells—a reflection of T cell differentiation, cell senescence and host environment. *Semin Immunol.* 2012;24:365–372.

93. Larbi A, Dupuis G, Khalil A, Douziech N, Fortin C, Fülöp T Jr. Differential role of lipid rafts in the functions of CD4+ and CD8+ human T lymphocytes with aging. *Cell Signal.* 2006;18(7):1017–30.

94. Yeo GC, Tarakanova A, Baldock C, Wise SG, Buehler MJ, Weiss AS. Subtle balance of tropoelastin molecular shape and flexibility regulates dynamics and hierarchical assembly. *Sci Adv.* 2016;2(2):e1501145.

95. Le Saux S, Weyand CM, Goronzy JJ. Mechanisms of immunosenescence: Lessons from models of accelerated immune aging. *Ann N Y Acad Sci.* 2012;1247:69–82.

96. Le Page A, Fortin C, Garneau H, Allard N, Tsvetkova K, Tan CT, Larbi A, Dupuis G, Fülöp T. Downregulation of inhibitory SRC homology 2 domain-containing phosphatase-1 (SHP-1) leads to recovery of T cell responses in elderly. *Cell Commun Signal.* 2014 Jan 9;12:2.

97. Yu M, Li G, Lee WW, Yuan M, Cui D, Weyand CM, Goronzy JJ. Signal inhibition by the dual-specific phosphatase 4 impairs T cell-dependent B-cell responses with age. *Proc Natl Acad Sci U S A.* 2012;109(15):E879–88.

98. Li G, Yu M, Lee WW, Tsang M, Krishnan E, Weyand CM, Goronzy JJ. Decline in miR-181a expression with age impairs T cell receptor sensitivity by increasing DUSP6 activity. *Nat Med.* 2012;18(10):1518–24.

99. Leandro MJ. B-cell subpopulations in humans and their differential susceptibility to depletion with anti-CD20 monoclonal antibodies. *Arthritis Res Ther.* 2013;15(Suppl 1):S3.

100. Boyd SD, Liu Y, Wang C, Martin V, Dunn-Walters DK. Human lymphocyte repertoires in ageing. *Curr Opin Immunol.* 2013;25(4):511–5.

101. Frasca D, Diaz A, Romero M, Blomberg BB. The generation of memory B cells is maintained, but the antibody response is not, in the elderly after repeated influenza immunizations. *Vaccine.* 2016;34(25):2834–40.

102. Xia S, Zhang X, Zheng S, Khanabdali R, Kalionis B, Wu J, Wan W et al. An update on inflamm-aging: Mechanisms, prevention, and treatment. *J Immunol Res.* 2016;2016:8426874.

103. Fougère B, Boulanger E, Nourhashémi F, Guyonnet S, Cesari M. Chronic inflammation: Accelerator of biological aging. *J Gerontol A Biol Sci Med Sci.* 2016. [Epub ahead of print]

104. Monti D, Ostan R, Borelli V, Castellani G, Franceschi C. Inflammaging and human longevity in the omics era. *Mech Ageing Dev.* 2016. [Epub ahead of print]

105. Arai Y, Martin-Ruiz CM, Takayama M, Abe Y, Takebayashi T, Koyasu S, Suematsu M et al. Inflammation, but not telomere length, predicts successful ageing at extreme old age: A longitudinal study of semi-supercentenarians. *EBioMedicine.* 2015;2(10):1549–58.

106. Giuliani C, Pirazzini C, Delledonne M, Xumerle L, Descombes P, Marquis J, Mengozzi G et al. Centenarians as extreme phenotypes: An ecological perspective to get insight into the relationship between the genetics of longevity and age-associated diseases. *Mech Ageing Dev.* 2017 Feb 27. [Epub ahead of print]

107. Minciullo PL, Catalano A, Mandraffino G, Casciaro M, Crucitti A, Maltese G, Morabito N et al. Inflammaging and anti-inflammaging: The role of cytokines in extreme longevity. *Arch Immunol Ther Exp (Warsz).* 2016;64(2):111–26.

108. Morrisette-Thomas V, Cohen AA, Fülöp T, Riesco É, Legault V, Li Q, Milot E et al. Inflamm-aging does not simply reflect increases in pro-inflammatory markers. *Mech Ageing Dev.* 2014;139:49–57.

109. Coussens LM, Werb Z. Inflammation and cancer. *Nature.* 2002;420(6917):860–7.

110. Franceschi C, Campisi J. Chronic inflammation (inflammaging) and its potential contribution to age-associated diseases. *J Gerontol A Biol Sci Med Sci.* 2014;69(Suppl 1):S4–9.

111. Zitvogel L, Tesniere A, Kroemer G. Cancer despite immunosurveillance: Immunoselection and immunosubversion. *Nat Rev Immunol.* 2006;6(10):715–27.

112. Renner K, Singer K, Koehl GE, Geissler EK, Peter K, Siska PJ, Kreutz M. Metabolic hallmarks of tumor and immune cells in the tumor microenvironment. *Front Immunol.* 2017 Mar 8;8:248.

113. Park J, Kwon M, Shin EC. Immune checkpoint inhibitors for cancer treatment. *Arch Pharm Res.* 2016;39(11):1577–1587.

114. Helissey C, Vicier C, Champiat S. The development of immunotherapy in older adults: New treatments, new toxicities? *J Geriatr Oncol.* 2016;7(5):325–33.

115. Nishijima TF, Muss HB, Shachar SS, Moschos SJ. Comparison of efficacy of immune checkpoint inhibitors (ICIs) between younger and older patients: A systematic review and meta-analysis. *Cancer Treat Rev.* 2016;45:30–7.

23 Oxidation of Ion Channels in the Aging Process

Federico Sesti

CONTENTS

INTRODUCTION

Aging is a multifactorial phenomenon that involves numerous changes at the level of the cell. Well-established mechanisms of aging are telomere shortening [1,2], DNA methylation [3,4] and damage [5], and release of reactive oxygen species (ROS) in the cell [6,7]. ROS can target virtually any cellular component including lipids, ribonucleic and deoxyribonucleic acids, and proteins. The ion channels, integral membrane proteins that allow the passive diffusion of ions across the plasma membrane, are a common substrate of ROS [8]. Ion channels are primarily responsible for nerve and muscle excitability and are implicated in a variety of biological functions that range from sensory transduction, to control of processes as varied as blood volume homeostasis, hormone secretion, epithelial transport of nutrients and ions, and T-cell activation [9].

A large number of diseases are associated with ion-channel dysfunction including epilepsy, congenital deafness, cardiac arrhythmia, and hypersecretion of insulin [10–16]. Typically, ion channels result from the assembly of several subunits. Such assemblies involve a circular arrangement of four identical or related pore-forming or alpha-subunits, packed around a water-filled pore and accessory or beta-subunits that modulate basic channel's attributes such as permeation, gating, trafficking, abundance on the plasma membrane, and pharmacology [17,18]. The genomes of *Homo sapiens*, *Mus musculus* (mouse), *Gallus gallus* (chicken), and *Fugu rupribes* (puffer fish) contain more than 200 ion channel genes each and invertebrate organisms including *Caenorhabditis elegans* (nematode), *Drosophila melanogaster* (fly), and *Anopheles gambiae* (mosquito) have similar numbers of ion-channel genes in their genomes [19]. Further heterogeneity arises from heteromeric ion channel complexes formed by the combination of different gene products.

The movement of ions through pores is controlled by gates, which may be opened or closed in response to electrical or chemical signals, temperature, or mechanical force. Some of the major families of ion channels are briefly described below:

- *Voltage-gated, sodium (Nav), calcium (Cav), and potassium (Kv)—selective, channels.* These channels open or close in response to changes in the voltage across the plasma membrane and preferably conduct (selectivity) a single type of ion from which they take their name. The human families of voltage-gated Na^+ and Ca^{2+} channels have more than 10 members each, whereas the family of voltage-gated K^+ channels contains at least 40 members, which are divided into 12 subfamilies. The pore-forming (or α-) subunits of Na^+ and Ca^{2+} channels are large polypeptides containing up to 4000 amino acids. These are composed of four homologous repeat domains of six transmembrane spanning domains (TMD)s each, that assemble around a symmetry axis to form the ion-conducting pathway (pore) and selectivity filter of the channel. In contrast, the pore-forming subunits of K^+ channels have a single-domain homologous of the pore-forming subunits of Na^+ and Ca^{2+} channels and functional K^+ channels are tetramers resulting from the assembly of four pore-forming subunits. Kv heterogeneity is further enhanced by the ability to form channels composed by different pore-forming subunits and to assemble with at least four distinct types of accessory (or β-) subunits (Kvβs, KCNEs, KChIPs, and DPPLs).
- *Calcium-activated K^+ (KCa) channels.* There are eight members in the human family, divided into three groups: large (BK), intermediate (IK), and small (SK) conductance. BK and IK channels can also be activated by voltage, whereas SK channels are voltage insensitive and can only be opened by intracellular calcium. KCa pore-forming subunits can have six or seven TMDs and make homo- and hetero–tetrameric complexes. In addition, they generally form complexes with four different kinds of the accessory β-subunit (β1–β4) which can be either inhibitory or excitatory.
- *Two-pores (2-P) and inward rectifier (Kir) K^+ channels.* Together they make 30 members in the human family. 2-P channels contain two tandem K_V-like domains fused on one polypeptide chain. Kir may exist in the membrane as homo- or hetero-oligomers and each monomer possesses between 2 and 4 TMDs. K_{ir} channels require phosphatidylinositol 4,5 bisphosphate (PIP$_2$) for activation. Kir6.x genes provide the pore-forming subunits of ATP-sensitive (KATP) potassium channels.
- *Cyclic nucleotide-gated channels.* These channels are nonselective for cations and are opened by binding of intracellular cyclic adenosine monophosphate (cAMP) and/or cyclic guanosine monophosphate (cGMP). They are heteromeric complexes containing pore-forming and multiple β-subunits.
- *Hyperpolarization-activated cyclic nucleotide-activated channels.* These channels are opened by hyperpolarization and are also sensitive to the cyclic nucleotides cAMP and cGMP which alter their voltage sensitivity.
- *Transient receptor potential (TRP) channels.* This family contains at least 28 members. TRP channels can be gated by voltage, calcium, pH, redox state, osmolality, mechanical force, and some members seem to be constitutively open. Some TRP channels are nonselective cation channels; others appear to be selective for calcium and magnesium.
- *Chloride channels.* This family consists of approximately 13 members. These channels are poorly selective for small anions.

Changes in the expression and/or functional properties of ion channels during aging have been observed in a variety of organisms and tissues and are summarized in Table 23.1. Here I review cases in which these modifications have been mechanistically linked to physiological impairment. These cases mostly revolve around Ca^{2+} and K^+ channels operating in the muscles and nervous systems of both vertebrate and invertebrate organisms. ROS are the most common causative agents behind

TABLE 23.1

Oxidative Modifications of Ion Channels in Aging and Disease

Tissue	Channel	Organism	References
Invertebrates			
Nervous system	MEC-4, EGL-19; KVS-1, KCNQ, Nav	Worm (*C. elegans*), fly (*D. melanogaster*)	[20–23]
Muscle	SUR; KCNQ, UNC-103	Worm (*C. elegans*), fly (*D. melanogaster*)	[24–27]
Vertebrates			
Skeletal muscle	SK3, RyR1, Cav1.1, Cavβ1, K(ATP)	Rat; mouse	[28–40]
Cardiac muscle	Cav1.x; RyR2, Nav1.5, K(ATP); Kv4.2, Kv1.2	Rat, mouse, dog, rabbit	[41–59]
Smooth muscle	Cav1.x; RyR2, Orai1, TRPC, BK, BKβ1; K(ATP);	Rat, mouse, *H. sapiens*	[60–76]
Endothelium	IK1; SK4	Mouse	[77,78]
Cochlea	Kir4.1; Kv7.1; Kv1.1; Kv3.1	Rat; mouse	[79,80]
Sino-atrial node	RyR2; Cav1.2, Kv1.5; HCN1, Nav1.1, Nav1.5, Nav1.6, Navβ1, Navβ3	Rat; mouse	[81–84]
Central nervous system	RyR2, RyR3, Cav1.x, Cav2.2, Nav, Nav1.1, Nav1.2, Kv4.2; Kv4.3, Kv2.1, SK3; BK, TRPM2, TRPV1,	Rat; mouse	[85–109]
Peripheral nervous system	Nav1.8, HCN	Mouse, *H. sapiens*	[110,111]
Glia	Kv; Kv1.3; Kir4.1, AQP4	Mouse	[112–114]
Connective tissue	Kv1.1; BKβ1, RyR	*H. sapiens*	[115,116]
Blood (red cells)	Cav1.x, SK3; IK4	*H. sapiens*	[117,118]

MEC-4 = MEChanosensory abnormality 4 (amiloride-sensitive Na+ channel); EGL-19 = EGg laying defective 19 (voltage-gated Ca^{2+} channel); KVS-1 = K (potassium) voltage-sensitive channel subunit 1; UNC-103 = UNCoordinated (ether-a-go-go-related K$^+$ channel); Orai1 = calcium release-activated calcium channel protein 1; AQP4 = aquaporin 4. When the specific gene product is not known, currents are indicated without numbers (e.g., Nav for voltage-dependent sodium currents; HCN for hyperpolarization and cyclic nucleotide-activated currents, etc.).

those alterations of ion channels' function. In summary, the status of current studies indicates that oxidation of ion channels by ROS is a general and widespread mechanism of aging vulnerability.

MUSCLE

VOLTAGE-GATED CA^{2+} CHANNELS AND RYANODINE RECEPTORS IN SKELETAL MUSCLE

The process that converts an electrical stimulus into a contraction of skeletal muscle is called excitation–contraction coupling (E–C coupling) [119–122]. Membrane depolarization triggers diffusion into the cytoplasm of calcium stored in the sarcoplasmic reticulum that by binding to the actin filaments of the sarcomere, allows them to slide over the myosin filaments. In skeletal muscle fibers, voltage-dependent L-type Cav1.1 calcium channels (VDCCs; also called dihydropyridine receptors or DHPRs), are coupled to the gatekeepers of calcium stores, the ryanodine receptors type 1 (RyR1) so that they act as switches that in response to membrane depolarization, undergo conformational changes that allosterically open the RyR1 [9]. During aging, multiple processes converge to impair calcium handling by the VDCCs and the RyR1s, leading to loss of muscle fitness. Studies in rodents [28–30,123,124] and humans [41,125] have revealed a progressive E–C uncoupling in aging fibers primarily caused by disorganization and/or loss of VDCCs and their accessory subunits [31].

The molecular basis of VDCC downregulation have not been completely elucidated but both exercise and caloric restriction partially restore the loss of VDCCs implicating the IGF insulin-like pathway. Indeed, overexpression of the IGF-like receptor in mice recovers the expression of VDCCs in aging fibers and consequently restores muscle strength [42,43,126,127]. The idea that the diet can affect E–C uncoupling is further supported by the observation that partial removal of cholesterol from the membranes of skeletal muscle fibers alters the levels of the VDCCs via caveolin-3. It is as triad fractions from aged rats have similar levels of RyR1, but lower caveolin-3 protein levels compared from triad fractions from young rats [128] (concomitantly the aged animals exhibit E–C coupling impairment without significant changes in the integrity of the sarcoplasmic reticulum). Calorie restriction and the IGF-like pathway decrease oxidative stress—a condition characterized by an imbalance between pro- and antioxidants—in mice [129–131]. Oxidative stress is also responsible for the dysregulated leakage of intracellular calcium by the RyR1 of old skeletal muscle fibers. The RyR1 leaks calcium because it is oxidized at thiol groups (a single RyR1 subunit contains roughly 100 cysteine residues, many of which are redox-sensitive [132–135]) and depleted of stabilizing accessory subunit calstabin1 [32]. Indeed, overexpressing antioxidant enzyme catalase in mitochondria regains muscle strength in old fibers by diminishing the oxidation of the RyR1 [136]. Furthermore, the RyR1 is modulated by calmodulin MaPKII kinase (CaMKII), that becomes activated during E–C coupling and acts to oppose the opening of the RyR1 [33]. In aging fibers, methionine residues in CaMKII, including met109 and met124 are subjected to oxidation [137]. The functional results of these modifications primarily weaken the inhibitory action of CaMKII thereby contributing to the abnormal release of calcium through the RyR1.

VOLTAGE-GATED CA^{2+} CHANNELS AND RYANODINE RECEPTORS IN CARDIAC MUSCLE

The mechanisms through which Ca^{2+} channels cause age-dependent impairment in skeletal muscle are preserved to a considerable extent in cardiac muscle. In fact, the channels are conserved (Cav1.2 and RyR type 2), but unlike skeletal muscle, in cardiac muscle they are not physically coupled. The handling of calcium in cardiac muscle depends primarily on a mechanism known as calcium-induced calcium release (CICR) in which the discharge of calcium from the intracellular stores is triggered by an influx of the same ions into the cell through voltage-dependent opening of the VDCCs [9,122]. Likewise in skeletal muscle, age-associated increase of mitochondrial ROS [138,139] leads to oxidation of the RyR2, which in turn leaks calcium [44–46,140]. Studies have further underscored an important role for RyR2 stabilizing subunit Calstabin 2 in cardiac aging [141]. ROS also act to modulate CaMKII thereby affecting its regulation of the RyR2 [46,47]. Changes in the VDCC, which affect the muscular action potential, have been observed in aging cardiomyocytes, but it is not clear whether they have an impact on CICR [48–50,142–144]. Overall, oxidant-dependent dysregulated RyR2 activity contributes to the loss of strength of the muscle and promotes pulmonary vein arrhythmogenesis during aging [60] and when these modifications are suppressed by expressing catalase, muscle strength is restored.

VOLTAGE-GATED CA^{2+} CHANNELS AND RYANODINE RECEPTORS IN VASCULAR SMOOTH MUSCLE

Like skeletal and cardiac muscle, smooth muscle also relies on calcium to achieve contraction. Unlike the other two muscle types though, smooth muscle contractions are initiated by the calcium-activated phosphorylation of myosin (via the calcium–calmodulin–myosin light–chain kinase complex) rather than calcium binding to the troponin complex [122]. Most of the studies on the role of ion channels in the aging process of smooth muscle have focused on the vasculature. Smooth muscle produces vasodilation and vasoconstriction that are regulated by multiple mechanisms including release of paracrine agents from endothelial cells (nitric oxide, NO), as well as by the autonomic nervous system and catecholamine (norepinephrine and epinephrine) secreted by adrenal glands [122,145]. Early studies showed that aging reduces the response of small arteries to the vasodilatory effect of calcium channel blockers implicating these proteins in the aging process of smooth cells [146–149]. This hypothesis has further been supported by the finding that L-type VDCCs are down-regulated in the large [61] and small arteries [62] of old rats thereby explaining the differential response to blockers. There is also circumstantial evidence that suggests that defects in either RyRs or inositol 1,4,5-trisphosphate IP_3 receptors (IP_3R) may contribute to defective calcium handling in aging vascular smooth muscle cells [150] but the details of these mechanisms await further elucidation.

ATP-Sensitive K$^+$ Channels in Skeletal and Cardiac Muscle

The ATP-sensitive K$^+$, K(ATP), channels are sensors of cell metabolism, that respond to changes in ADP/ATP (adenosine triphosphate and adenosine diphosphate) levels. In the heart—where they were first discovered by Noma [151]—sarcoplasmic (sarc) K(ATP) channels are predominantly closed under normal conditions and thus do not contribute to myocyte excitability. They become activated in response to metabolic stresses such as hypoxia and ischemia, where the hyperpolarization of the membrane that follows their opening exerts cardioprotection by slowing down calcium influx. SarcK(ATP) current is markedly diminished in aging cardiac muscle, a commonly believed major contributing factor to ischemia vulnerability in the elderly [51–54,152]. The impairment of K(ATP) current is primarily due to a decrease of the open probability (p_o), and also to sex-specific downregulation [51,54]. What causes p_o decrease in cardiac K(ATP) channels is not known, but in K(ATP) channels of aging skeletal muscle, oxidation of thiol groups by ROS leads to a robust p_o decrease [34] and it seems therefore reasonable to hypothesize that cardiac K(ATP) channels undergo similar modifications. K(ATP) channels are heteromeric complexes formed by an inward rectifier K_{ir} pore-forming subunit and a sulfonylurea receptor (SUR) subunit, along with additional components [153]. Studies in aging rats have shown that sarcSUR2A is downregulated in female but not male hearts [51,53]. Together, the decrease in p_o and the lack of sarcSUR2A contribute to limit the availability of functional sarcK(ATP) channels leading to increased sensitivity to metabolic injury in old myocytes. Notably, the fly *Drosophila melanogaster* possesses a primitive heart, named the dorsal vessel, that expresses a K(ATP) channel [24]. The sulfonylurea SUR subunit of *Drosophila* K(ATP) is downregulated in aging flies and exposure of older animals to pinacidil, a K(ATP) agonist, decreases pacing-induced failure rate, underscoring a significant conservation in the mechanisms that regulate the expression of K(ATP) channels during aging.

Ca^{2+}-Activated K$^+$ Channels in Vascular Smooth Muscle and Endothelium

Vasodilation results from the hyperpolarization of smooth muscle cells that form the walls of blood vessels. As the membrane potential of the cell becomes negative, mainly following opening of KCa channels, E–C coupling and consequently, contraction, are inhibited [146,154,155]. As age progresses, the coronary arteries of mammalian hearts become stiffer, thicker, and have higher spontaneous contractile activity, even in the absence of atherosclerosis. This is due, in part, to marked downregulation of both the pore-forming subunit and the β1 accessory subunit of large conductance Ca^{2+}-activated K$^+$ (BK) channels [63,64,156]. In humans, the decrease of the protein levels of the BK pore-forming subunit and indirectly of the amounts of BK current, has been estimated to be 50% in elderly (60–70 year old) subjects compared with normal adults (20–40 year old) and similar decreases have been reported to occur in rats [63]. Physical inactivity, which is considered a risk factor for cardiovascular disease in humans, may contribute to BK channel downregulation during aging, as studies have shown that low-intensity exercise in rats ameliorated the BK-mediated decline in the vasodilatory properties of coronary arteries [65]. In addition sphingolipid composition also induces KCa upregulation [66]. Alterations in BK channel expression and the beneficial effects of exercise appear to be a general theme in the microvasculature including in superior epigastric arteries (amid conflicting reports of upregulation [67,68] or downregulation [69]), mesenteric arteries

[70], soleus and gastrocnemius arterioles [157], and cerebrovascular myocytes [158] of old rodents. As mentioned before, the endothelium contributes to the control of vasodilation through the release of endothelium-derived vasodilators such as NO. In addition, the endothelium initiates membrane hyperpolarization that spreads to the surrounding smooth muscle cells through myoendothelial junctions. Early studies found that endothelial-mediated vasodilation was impaired in aged mesenteric arteries and blockade of KCa channels differently affected vasodilation in young and old rats [159–161]. It was later shown that in superior epigastric arteries of old mice and in coronary arterioles from human patients undergoing cardiac surgery, electrical conduction along the endothelium was hampered due to hyperactivation of small and intermediate-conductance KCa^+ (SK and IK) channels [77,162]. Those studies further implicated ROS as the driving force behind SK and IK hyperactivity, as catalase improved electrical conduction in old mice whereas hydrogen peroxide (H_2O_2) specifically activated those channels in young but not in old animals [77]. The mechanisms leading to endothelial H_2O_2 increase during aging are not known, but they do not seem to originate from defective NO synthase activity. Rather, it was found that in primary mouse aortic endothelial cells, age-related oxidative stress caused superoxide dismutase upregulation and downregulation of catalase and glutathione peroxidase 1 [78] suggesting that a likely cause for the generation of more H_2O_2 is the dysregulated expression of antioxidant defenses.

NERVOUS SYSTEM

VOLTAGE-GATED CA^{2+} CHANNELS AND RYANODINE RECEPTORS IN THE CENTRAL NERVOUS SYSTEM

Neurons in the brain use calcium to achieve learning and memory formation through mechanisms that are largely similar to those used by muscle cells to achieve contraction. It is therefore not coincidental that calcium handling is impaired in aging neurons through dysregulation of the same mechanisms and types of channels. Most of the studies that have examined the role of calcium channels in the aging process of the brain have focused on long-term potentiation (LTP) in the hippocampus [163–165]. LTP is a process in which patterns of presynaptic activity result in long-lasting strengthening of synapses following calcium-induced phosphorylation of the α-amino-3-hydroxy-5-methyl-4-isoxazolepropionic acid receptor (AMPAR). Calcium enters into the cell primarily via N-methyl-D-aspartate receptors (NMDAR) and is simultaneously released from intracellular stores through CICR mechanisms that involve either VDCCs or RyR type 2 and 3 (RyR2/3) or IP_3Rs activated by metabotropic glutamate receptors [166–174]. During periods of intense neuronal activity, calcium enters with each action potentially faster than the cell can remove it. Consequently, the level of calcium progressively rises until it reaches the threshold for activation of SK channels whose current gives rise to prolonged periods of hyperpolarization (slow after hyperpolarization, sAHP) during which time neural activity is suppressed. In aged hippocampal neurons the levels of intracellular calcium are higher than normal [175]. This facilitates the activation of SK channels and therefore decreases susceptibility to LTP induction by enhancing the sAHP [85,176–178].

Early studies focused on VDCCs as the most likely culprits of increased intracellular calcium levels in aged neurons. It was found that both the protein [86,87,179,180] and current [88,89] amounts of the VDCCs increased in the aging hippocampus in part due to phosphorylation [90] and tumor necrosis factor (TNF) signaling [181]. Studies using VDCCs' antagonists [178,182,183] calcium chelators [175] or Ca^{2+}-regulating hormones [91] showed that VDCC inhibition improved cognitive decline [184–186] and LTP [178]. Other studies pointed to a role of the CICR mechanisms via dysregulated activity of the RyRs [88,184–189] or of the IP_3Rs [167], in the increase of free intracellular calcium in aging hippocampal neurons. Martini et al. [190] reported a marked reduction in the density of IP_3Rs along with impairment of Ca^{2+}-dependent protein kinase C phosphorylation in microsomes derived from the cerebellum of aging rats. Whether IP3R signaling plays a role in LTP impairment during aging still awaits elucidation even though recent evidence indicates that superoxide anion modifies thiol groups within the IP_3R leading to sensitization of calcium release [167].

Kumar and Foster [185] and Gant et al. [92] independently implicated the RyR in the impairment of LTP during aging. In particular, Gant et al. showed that pharmacological blockade of the RyR reduced or eliminated differences in the excitability of young versus aging neurons by slowing calcium rise during repetitive synaptic stimulation—thereby shortening the sAHP [92,191].

The crucial role of the RyR in the learning process of the hippocampus during aging is now well established as treatments that inhibit the CICR mechanism have a marked effect on rescuing learning defects in aging animals [93,185,187,192]. For example, expression of RyR-accessory subunit calstabin2 (FK506-binding protein 12.6/1b), a homolog of muscle calstabin1, declines in the hippocampus of aged rats and its overexpression results in enhancement of spatial memory along with reduction of the sAHP [193,194]. Likewise, in skeletal and cardiac muscle, the cause for RyR dysregulation in aging neurons is oxidation by ROS. Dithiothreitol (DTT), a reducing agent, decreases the sAHP in old but not young rats, and blockade of RyRs or depletion of intracellular calcium stores suppresses DTT effects, whereas DTT-mediated decrease in sAHP is not affected by inhibition of other Ca^{2+} pathways including VDCC-mediated pathways [93,192,195]. Using single RyRs from rat brain cortex incorporated into lipid bilayers, Bull et al. showed that the activation of the RyR by calcium depends on the redox state of the receptor [196,197]. They further showed that the RyR2 and RyR3 from the cortex of rat brains subjected to cerebral ischemia opened more readily and thus leaked more calcium than normal, due to S-glutathionylation of unidentified cysteine residues [198].

VOLTAGE-GATED K$^+$ CHANNELS IN THE CENTRAL NERVOUS SYSTEM

The voltage-gated delayed rectifier K^+ channel Kv2.1 (KCNB1 gene) carries a major somatodendritic current in the cortex and hippocampus [199,200]. Loss of function due to mutations in KCNB1 have been linked to early infantile epileptic encephalopathy (EIEE) [201–203] and KCNB1 knock out in mice can cause hippocampal hyperexcitability and seizures [204]. Kv2.1 is susceptible to redox through oxidation of cysteines that cross-link Kv2.1 subunits to each other giving rise to oligomers [94]. Kv2.1 oligomers do not conduct current and accumulate in the plasma membrane from where they activate Src and downstream C-Jun N-terminal kinases (JNKs) that act to destabilize mitochondria leading to ROS leakage and apoptosis [205,206]. Recently, it was shown that the apoptotic signal originates from integrins, which form macromolecular complexes with Kv2.1 channels [207]. The initial stimulus is transduced to Fyn and possibly other Src family members by Focal Adhesion Kinase (FAK). The mechanism by which Kv2.1 oligomers activate integrin signaling has not been elucidated yet but it has been proposed that Kv2.1 oxidation may favor integrin clustering, thereby facilitating the recruitment and activation of FAK and Src/Fyn kinases. A cysteine residue located in the N-terminus of the channel, cys73, mediates Kv2.1 oligomerization. When cys73 is mutated to alanine (C73A), the mutant channel retains wild-type characteristics except those which do not oligomerize and consequently do not induce apoptosis. Kv2.1 oligomers have provided a good biochemical marker of the oxidative status of the channel and indirectly, of the cell, because they can be easily detected in a Western blot. Indeed, Kv2.1 oligomers are present in the brains of aging mice and in larger amounts in injured brains and in the brains of the 3xTg-AD mouse model of Alzheimers disease [208–214], where they promote hyperexcitability and apoptosis [215]. Yu et al. constructed transgenic mice expressing a human C73A Kv2.1 mutant in the cortex and hippocampus to test the effects of diminished Kv2.1 oxidation in the brain in a live animal [207]. To induce oxidative stress under controlled and reproducible experimental conditions (animals of the same age subject to constant levels of oxidative stress only in the cortex and hippocampus) the brain was injured using the lateral fluid percussion (LFP) method. The decrease of Kv2.1 oxidation in the transgenic C73A (Tg-C73A) injured brain led to reduced inflammation and cell death and improved cognitive outcome after the injury. The researchers further showed that use of a FDA-approved drug, dasatinib, to inhibit the apoptotic pathway activated in response to Kv2.1 oxidation, improved cellular damage and behavioral outcome. Conditions of oxidative stress have also been shown to accelerate trafficking of Kv2.1 channels to the plasma membrane *in vitro* [216]. This mechanism is intrinsically dependent on the cytosolic levels of Zn^{2+} and Ca^{2+} which increase when the cell is undergoing apoptosis. Liberated Zn^{2+} and Ca^{2+} promote phosphorylation of Kv2.1 by several protein kinases, including apoptosis signal-regulating kinase 1 (ASK1), p38 MAPK-dependent kinase, c-Src tyrosine kinase, and Ca(2+)/calmodulin-dependent protein kinase II (CaMKII) [217–221]. In turn, phosphorylation facilitates the interaction of Kv2.1 with syntaxin favoring the surge of channels to the plasma membrane. The increase of the Kv2.1 current that follows is thought to be proapoptotic; however, the accumulation of Kv2.1 channels alone in the membrane is sufficient to induce cell death [206].

K+ CHANNELS IN MUSCLE AND NERVOUS SYSTEM OF INVERTEBRATES

Invertebrate models offer many advantages for the experimental study of aging, including short life span and powerful genetics. Indeed, they have significantly contributed to advance our understanding of the role of ion channels in the aging process. *C. elegans* has provided the first experimental evidence that oxidation of a K+ channel as a side effect of aging constitutes a mechanism of loss of neuronal function. *C. elegans* expresses a voltage-gated K+ channel, named KVS-1, which is the homolog of Kv2.1 (it was this homology that later led to the identification of Kv2.1 as a substrate for ROS in the brain) [222,223]. The KVS-1 current can be described as A-type (rapidly activating–inactivating) by virtue of a "ball-and-chain" N-inactivation domain located in the first 40 amino acids of the channel [224]. When the N-inactivation domain is deleted, the KVS-1 current becomes non-inactivating delayed rectifier, like the Kv2.1 current. Oxidants convert the KVS-1 current from A-type to delayed rectifier type [20]. Cysteine 73 is conserved in KVS-1 at position 113 (taking into account the extra 40 amino acids of the N-inactivation domain of KVS-1). When cys113 is mutated to serine (C113S), the KVS-1 current becomes redox-insensitive, indicating that oxidation of this cysteine impairs the N-inactivation mechanism, although it remains to be determined whether oxidation of this residue leads to the formation of disulfide-bonded cysteines and thus of oligomers or alternatively, sulfinic or sulfonic acid. Genetic (knock in worms expressing non-oxidizable C113S KVS-1 mutant) and electrophysiological evidence underscores significant oxidation of cys113 during aging which in turn, impacts the excitability of the neurons where the channel operates, causing progressive decline of sensory function [20]. Oxidative stress is a hallmark of aging and *C. elegans* has provided a compelling example of a physiological process through which voltage-gated K+ channels lead to sensorial decline. A homolog of the mammalian voltage-gated KCNQ1 K+ channel, which provides the pore-forming subunit of the slow component of the delayed repolarizing current in the human ventricle (I_{Ks}), is also expressed in the dorsal vessel and nervous system of *D. melanogaster* [21,25]. *Drosophila* KCNQ conducts a slowly activating/inactivating voltage-gated current that can be suppressed by muscarinic acetylcholine receptor agonists. RT-PCR evidence has shown that *Drosophila* KCNQ expression declines with age leading to impairment of the dorsal vessel and nervous function. Specifically, in the dorsal vessel, *Drosophila* KCNQ downregulation is associated with increased "arrhythmogenesis" while in the nervous system, it is linked to associative memory decline.

CONCLUSIONS

The status of current studies underscores a primary role of ion channels in the aging process across species. Most importantly, there appears to be significant conservation of mechanism in different tissues including the nervous system, cardiovascular system and musculature. ROS appear to represent the primary driving force behind changes in the function of ion channels during aging. Thus, oxidation of RyRs promotes calcium leakage in muscle fibers and neurons, leading to impaired E–C coupling and LTP. Similarly, oxidation of K+ channels affects endothelial signaling and increases apoptosis and ischemia vulnerability. Overall, these oxidative modifications contribute to loss of muscle fitness, cognitive impairment, and increased vulnerability to cardiovascular disease in the elderly.

ACKNOWLEDGMENTS

We thank Shuang Liu for critically reading the manuscript. This work was supported by a NSF grant (1456675) and a NIH grant (1R21NS096619–01) to Federico Sesti.

REFERENCES

1. Hayflick, L. 1965. The limited in vitro lifetime of human diploid cell strains. *Exp Cell Res* 37, 614–636.
2. Hayflick, L. 1987. Origins of longevity. in *Modern Biological Theories of Aging* (Warner, H. R., Butler, R. N., Sprott, R. L., and Schneider, E. L. eds.), Raven Press, New York. pp 21–34.
3. Berdyshev, G. D., Korotaev, G. K., Boiarskikh, G. V., and Vaniushin, B. F. 1967. Nucleotide composition of DNA and RNA from somatic tissues of humpback and its changes during spawning. *Biokhimiia* 32, 988–993.
4. Marioni, R. E., Shah, S., McRae, A. F., Chen, B. H., Colicino, E., Harris, S. E., Gibson, J. et al. 2015. DNA methylation age of blood predicts all-cause mortality in later life. *Genome Biol* 16, 25.
5. Freitas, A. A., and de Magalhaes, J. P. 2011. A review and appraisal of the DNA damage theory of aging. *Mutat Res* 728, 12–22.
6. Harman, D. 1956. Aging: A theory based on free radical and radiation chemistry. *J Gerontol* 11, 298–300.
7. Harman, D. 1972. The biologic clock: The mitochondria? *J Am Geriatr Soc* 20, 145–147.
8. Sesti, F., Liu, S., and Cai, S. Q. 2010. Oxidation of potassium channels by ROS: A general mechanism of aging and neurodegeneration? *Trends Cell Biol* 20, 45–51.
9. Hille, B. 2001. *Ionic Channels of Excitable Membranes*, 3rd ed., Sinauer Associates., Sunderland, MA.
10. Sing, N., Charlier, C., Stauffer, D., DuPont, B., Leach, R., Melis, R., Ronen, G. et al. 1998. A novel potassium channel gene, KCNQ2, is mutated in an inherited epilepsy of newborns. *Nat Genet* 18, 25–29.
11. Tyson, J., Tranebjaerg, L., Bellman, S., Wren, C., Taylor, J., Bathen, J., Aslaksen, B. et al. 1997. IsK and KvLQT1: Mutations in either of the two subunits of the slow component of the delayed rectifier potassium channel can cause Jervell and Lange–Nielsen syndrome. *Hum Mol Genet* 6, 2179–2185.
12. Browne, D., Gancher, S., Nutt, J., Brunt, E., Smith, E., Kramer, P., and Litt, M. 1994. Episodic ataxia/myokymia syndrome is associated with point mutations in the human potassium channel gene, KCNA1. *Nat Genet* 8, 136–140.
13. Jurkiewicz, N. K., Wang, J., Fermini, B., Sanguinetti, M. C., and Salata, J. J. 1996. Mechanism of action potential prolongation by RP 58866 and its active enantiomer, terikalant. Block of the rapidly activating delayed rectifier K+ current, IKr. *Circulation* 94, 2938–2946.
14. Nestorowicz, A., Inagaki, N., Gonoi, T., Schoor, K., Wilson, B., Glaser, B., Landau, H. et al. 1997. A nonsense mutation in the inward rectifier potassium channel gene, Kir6.2, is associated with familial hyperinsulinism. *Diabetes* 46, 1743–1748.
15. Simon, D. B., Karet, F. E., Rodriguez-Soriano, J., Hamdan, J. H., DiPietro, A., Trachtman, H., Sanjad, S. A. et al. 1996. Genetic heterogeneity of Bartter's syndrome revealed by mutations in the K+ channel, ROMK. *Nat Genet* 14, 152–156.
16. Ackerman, M. 1998. The long QT syndrome: Ion channel diseases of the heart. *Mayo Clin Proc* 73, 250–269.
17. MacKinnon, R. 1991. Determination of the subunit stoichiometry of a voltage-activated potassium channel. *Nature* 350, 232–235.
18. McCrossan, Z. A., and Abbott, G. W. 2004. The MinK-related peptides. *Neuropharmacology* 47, 787–821.
19. Jegla, T. J., Zmasek, C. M., Batalov, S., and Nayak, S. K. 2009. Evolution of the human ion channel set. *Comb Chem High Throughput Screen* 12, 2–23.
20. Cai, S. Q., and Sesti, F. 2009. Oxidation of a potassium channel causes progressive sensory function loss during aging. *Nat Neurosci* 12, 611–617.
21. Cavaliere, S., Malik, B. R., and Hodge, J. J. 2013. KCNQ channels regulate age-related memory impairment. *PLoS One* 8, e62445.
22. Jiang, H. C., Hsu, J. M., Yen, C. P., Chao, C. C., Chen, R. H., and Pan, C. L. 2015. Neural activity and CaMKII protect mitochondria from fragmentation in aging *Caenorhabditis elegans* neurons. *Proc Natl Acad Sci U S A* 112, 8768–8773.
23. Reenan, R. A., and Rogina, B. 2008. Acquired temperature-sensitive paralysis as a biomarker of declining neuronal function in aging Drosophila. *Aging Cell* 7, 179–186.

24. Akasaka, T., Klinedinst, S., Ocorr, K., Bustamante, E. L., Kim, S. K., and Bodmer, R. 2006. The ATP-sensitive potassium (KATP) channel-encoded dSUR gene is required for Drosophila heart function and is regulated by tinman. *Proc Natl Acad Sci U S A* 103, 11999–12004.

25. Ocorr, K., Reeves, N. L., Wessells, R. J., Fink, M., Chen, H. S., Akasaka, T., Yasuda, S. et al. 2007. KCNQ potassium channel mutations cause cardiac arrhythmias in Drosophila that mimic the effects of aging. *Proc Natl Acad Sci U S A* 104, 3943–3948.

26. Guo, X., Navetta, A., Gualberto, D. G., and Garcia, L. R. 2012. Behavioral decay in aging male *C. elegans* correlates with increased cell excitability. *Neurobiol Aging* 33, 1483 e1485–e1423.

27. Swiatkowski, P., and Sesti, F. 2013. Delayed pharyngeal repolarization promotes abnormal calcium buildup in aging muscle. *Biochem Biophys Res Commun* 433, 354–357.

28. Renganathan, M., Messi, M. L., and Delbono, O. 1997. Dihydropyridine receptor–ryanodine receptor uncoupling in aged skeletal muscle. *J Membr Biol* 157, 247–253.

29. Wang, Z. M., Messi, M. L., and Delbono, O. 2000. L-Type Ca(2+) channel charge movement and intracellular Ca(2+) in skeletal muscle fibers from aging mice. *Biophys J* 78, 1947–1954.

30. O'Connell, K., Gannon, J., Doran, P., and Ohlendieck, K. 2008. Reduced expression of sarcalumenin and related Ca2+ -regulatory proteins in aged rat skeletal muscle. *Exp Gerontol* 43, 958–961.

31. Taylor, J. R., Zheng, Z., Wang, Z. M., Payne, A. M., Messi, M. L., and Delbono, O. 2009. Increased CaVbeta1A expression with aging contributes to skeletal muscle weakness. *Aging Cell* 8, 584–594.

32. Andersson, D. C., Betzenhauser, M. J., Reiken, S., Meli, A. C., Umanskaya, A., Xie, W., Shiomi, T. et al. 2011. Ryanodine receptor oxidation causes intracellular calcium leak and muscle weakness in aging. *Cell Metab* 14, 196–207.

33. Boschek, C. B., Jones, T. E., Smallwood, H. S., Squier, T. C., and Bigelow, D. J. 2008. Loss of the calmodulin-dependent inhibition of the RyR1 calcium release channel upon oxidation of methionines in calmodulin. *Biochemistry* 47, 131–142.

34. Tricarico, D., and Camerino, D. C. 1994. ATP-sensitive K+ channels of skeletal muscle fibers from young adult and aged rats: Possible involvement of thiol-dependent redox mechanisms in the age-related modifications of their biophysical and pharmacological properties. *Mol Pharmacol* 46, 754–761.

35. Tricarico, D., Petruzzi, R., and Camerino, D. C. 1997. Changes of the biophysical properties of calcium-activated potassium channels of rat skeletal muscle fibres during aging. *Pflugers Arch* 434, 822–829.

36. Vergara, C., and Ramirez, B. U. 1997. Age-dependent expression of the apamin-sensitive calcium-activated K+ channel in fast and slow rat skeletal muscle. *Exp Neurol* 146, 282–285.

37. Hwang, C. Y., Kim, K., Choi, J. Y., Bahn, Y. J., Lee, S. M., Kim, Y. K., Lee, C. et al. 2014. Quantitative proteome analysis of age-related changes in mouse gastrocnemius muscle using mTRAQ. *Proteomics* 14, 121–132.

38. Ferrington, D. A., Krainev, A. G., and Bigelow, D. J. 1998. Altered turnover of calcium regulatory proteins of the sarcoplasmic reticulum in aged skeletal muscle. *J Biol Chem* 273, 5885–5891.

39. Damiani, E., Larsson, L., and Margreth, A. 1996. Age-related abnormalities in regulation of the ryanodine receptor in rat fast-twitch muscle. *Cell Calcium* 19, 15–27.

40. Luin, E., and Ruzzier, F. 2007. The role of L- and T-type Ca2+ currents during the *in vitro* aging of murine myogenic (i28) cells in culture. *Cell Calcium* 41, 479–489.

41. Delbono, O., O'Rourke, K. S., and Ettinger, W. H. 1995. Excitation-calcium release uncoupling in aged single human skeletal muscle fibers. *J Membr Biol* 148, 211–222.

42. Renganathan, M., and Delbono, O. 1998. Caloric restriction prevents age-related decline in skeletal muscle dihydropyridine receptor and ryanodine receptor expression. *FEBS Lett* 434, 346–350.

43. Renganathan, M., Messi, M. L., and Delbono, O. 1998. Overexpression of IGF-1 exclusively in skeletal muscle prevents age-related decline in the number of dihydropyridine receptors. *J Biol Chem* 273, 28845–28851.

44. Cooper, L. L., Li, W., Lu, Y., Centracchio, J., Terentyeva, R., Koren, G., and Terentyev, D. 2013. Redox modification of ryanodine receptors by mitochondria-derived reactive oxygen species contributes to aberrant Ca2+ handling in ageing rabbit hearts. *J Physiol* 591, 5895–5911.

45. Ren, J., Li, Q., Wu, S., Li, S. Y., and Babcock, S. A. 2007. Cardiac overexpression of antioxidant catalase attenuates aging-induced cardiomyocyte relaxation dysfunction. *Mech Ageing Dev* 128, 276–285.

46. Guo, X., Yuan, S., Liu, Z., and Fang, Q. 2014. Oxidation- and CaMKII-mediated sarcoplasmic reticulum Ca(2+) leak triggers atrial fibrillation in aging. *J Cardiovasc Electrophysiol* 25, 645–652.

47. Xu, A., and Narayanan, N. 1998. Effects of aging on sarcoplasmic reticulum Ca2+-cycling proteins and their phosphorylation in rat myocardium. *Am J Physiol* 275, H2087–2094.

48. Walker, K. E., Lakatta, E. G., and Houser, S. R. 1993. Age associated changes in membrane currents in rat ventricular myocytes. *Cardiovasc Res* 27, 1968–1977.

49. Gan, T. Y., Qiao, W., Xu, G. J., Zhou, X. H., Tang, B. P., Song, J. G., Li, Y. D. et al. 2013. Aging-associated changes in L-type calcium channels in the left atria of dogs. *Exp Ther Med* 6, 919–924.

50. Xu, G. J., Gan, T. Y., Tang, B. P., Chen, Z. H., Mahemuti, A., Jiang, T., Song, J. G. et al. 2013. Alterations in the expression of atrial calpains in electrical and structural remodeling during aging and atrial fibrillation. *Mol Med Rep* 8, 1343–1352.

51. Ranki, H. J., Crawford, R. M., Budas, G. R., and Jovanovic, A. 2002. Ageing is associated with a decrease in the number of sarcolemmal ATP-sensitive K+ channels in a gender-dependent manner. *Mech Ageing Dev* 123, 695–705.

52. Raveaud, S., Verdetti, J., and Faury, G. 2009. Nicorandil protects ATP-sensitive potassium channels against oxidation-induced dysfunction in cardiomyocytes of aging rats. *Biogerontology* 10, 537–547.

53. Sudhir, R., Sukhodub, A., Du, Q., Jovanovic, S., and Jovanovic, A. 2011. Ageing-induced decline in physical endurance in mice is associated with decrease in cardiac SUR2A and increase in cardiac susceptibility to metabolic stress: Therapeutic prospects for up-regulation of SUR2A. *Biogerontology* 12, 147–155.

54. Bao, L., Taskin, E., Foster, M., Ray, B., Rosario, R., Ananthakrishnan, R., Howlett, S. E. et al. 2013. Alterations in ventricular K(ATP) channel properties during aging. *Aging Cell* 12, 167–176.

55. Dun, W., Yagi, T., Rosen, M. R., and Boyden, P. A. 2003. Calcium and potassium currents in cells from adult and aged canine right atria. *Cardiovasc Res* 58, 526–534.

56. Liu, S. J., Wyeth, R. P., Melchert, R. B., and Kennedy, R. H. 2000. Aging-associated changes in whole cell K(+) and L-type Ca(2+) currents in rat ventricular myocytes. *Am J Physiol Heart Circ Physiol* 279, H889–900.

57. Signore, S., Sorrentino, A., Borghetti, G., Cannata, A., Meo, M., Zhou, Y., Kannappan, R. et al. 2015. Late Na(+) current and protracted electrical recovery are critical determinants of the aging myopathy. *Nat Commun* 6, 8803.

58. Baba, S., Dun, W., Hirose, M., and Boyden, P. A. 2006. Sodium current function in adult and aged canine atrial cells. *Am J Physiol Heart Circ Physiol* 291, H756–761.

59. Wu, C. C., Su, M. J., Chi, J. F., Wu, M. H., and Lee, Y. T. 1997. Comparison of aging and hypercholesterolemic effects on the sodium inward currents in cardiac myocytes. *Life Sci* 61, 1539–1551.

60. Wongcharoen, W., Chen, Y. C., Chen, Y. J., Chen, S. Y., Yeh, H. I., Lin, C. I., and Chen, S. A. 2007. Aging increases pulmonary veins arrhythmogenesis and susceptibility to calcium regulation agents. *Heart Rhythm* 4, 1338–1349.

61. Fukuda, T., Kuroda, T., Kono, M., Miyamoto, T., Tanaka, M., and Matsui, T. 2014. Attenuation of L-type Ca(2)(+) channel expression and vasomotor response in the aorta with age in both Wistar–Kyoto and spontaneously hypertensive rats. *PLoS One* 9, e88975.

62. Albarwani, S. A., Mansour, F., Khan, A. A., Al-Lawati, I., Al-Kaabi, A., Al-Busaidi, A. M., Al-Hadhrami, S. et al. 2016. Aging reduces L-type calcium channel current and the vasodilatory response of small mesenteric arteries to calcium channel blockers. *Front Physiol* 7, 171.

63. Marijic, J., Li, Q., Song, M., Nishimaru, K., Stefani, E., and Toro, L. 2001. Decreased expression of voltage- and Ca(2+)-activated K(+) channels in coronary smooth muscle during aging. *Circ Res* 88, 210–216.

64. Nishimaru, K., Eghbali, M., Lu, R., Marijic, J., Stefani, E., and Toro, L. 2004. Functional and molecular evidence of MaxiK channel beta1 subunit decrease with coronary artery ageing in the rat. *J Physiol* 559, 849–862.

65. Albarwani, S., Al-Siyabi, S., Baomar, H., and Hassan, M. O. 2010. Exercise training attenuates ageing-induced BKCa channel downregulation in rat coronary arteries. *Exp Physiol* 95, 746–755.

66. Choi, S., Kim, J. A., Kim, T. H., Li, H. Y., Shin, K. O., Lee, Y. M., Oh, S. et al. 2015. Altering sphingolipid composition with aging induces contractile dysfunction of gastric smooth muscle via K(Ca) 1.1 upregulation. *Aging Cell* 14, 982–994.

67. Hayoz, S., Bradley, V., Boerman, E. M., Nourian, Z., Segal, S. S., and Jackson, W. F. 2014. Aging increases capacitance and spontaneous transient outward current amplitude of smooth muscle cells from murine superior epigastric arteries. *Am J Physiol Heart Circ Physiol* 306, H1512–1524.

68. Hu, Z., Ma, A., Tian, H., Xi, Y., Fan, L., and Wang, T. 2013. Effects of age on expression of BKca channel in vascular smooth muscle cells from mesenteric arteries of spontaneously hypertensive rats. *J Physiol Biochem* 69, 945–955.

69. Shi, L., Liu, X., Li, N., Liu, B., and Liu, Y. 2013. Aging decreases the contribution of MaxiK channel in regulating vascular tone in mesenteric artery by unparallel downregulation of alpha- and beta1-subunit expression. *Mech Ageing Dev* 134, 416–425.

70. Shi, L., Liu, B., Zhang, Y., Xue, Z., Liu, Y., and Chen, Y. 2014. Exercise training reverses unparallel downregulation of MaxiK channel alpha- and beta1-subunit to enhance vascular function in aging mesenteric arteries. *J Gerontol A Biol Sci Med Sci* 69, 1462–1473.

71. Kawano, T., Tanaka, K., Chi, H., Kimura, M., Kawano, H., Eguchi, S., and Oshita, S. 2010. Effects of aging on isoflurane-induced and protein kinase A-mediated activation of ATP-sensitive potassium channels in cultured rat aortic vascular smooth muscle cells. *J Cardiovasc Pharmacol* 56, 676–685.

72. Feleder, E. C., Peredo, H. A., Mendizabal, V. E., and Adler-Graschinsky, E. 1998. Effects of aging on ATP sensitive K(+) channels and on prostanoid production in the rat mesenteric bed. *Age (Omaha)* 21, 183–188.

73. Zhao, C., Ikeda, S., Arai, T., Naka-Mieno, M., Sato, N., Muramatsu, M., and Sawabe, M. 2014. Association of the RYR3 gene polymorphisms with atherosclerosis in elderly Japanese population. *BMC Cardiovasc Disord* 14, 6.

74. Yang, Y., Zhu, J., Wang, X., Xue, N., Du, J., Meng, X., and Shen, B. 2015. Contrasting patterns of agonist-induced store-operated Ca2+ entry and vasoconstriction in mesenteric arteries and aorta with aging. *J Cardiovasc Pharmacol* 65, 571–578.

75. Toth, P., Csiszar, A., Tucsek, Z., Sosnowska, D., Gautam, T., Koller, A., Schwartzman, M. L. et al. 2013. Role of 20-HETE, TRPC channels, and BKCa in dysregulation of pressure-induced Ca2+ signaling and myogenic constriction of cerebral arteries in aged hypertensive mice. *Am J Physiol Heart Circ Physiol* 305, H1698–1708.

76. Du, Q., Jovanovic, S., Tulic, L., Sljivancanin, D., Jack, D. W., Zizic, V., Abdul, K. S. et al. 2013. KATP channels are up-regulated with increasing age in human myometrium. *Mech Ageing Dev* 134, 98–102.

77. Behringer, E. J., Shaw, R. L., Westcott, E. B., Socha, M. J., and Segal, S. S. 2013. Aging impairs electrical conduction along endothelium of resistance arteries through enhanced Ca2+-activated K+ channel activation. *Arterioscler Thromb Vasc Biol* 33, 1892–1901.

78. Choi, S., Kim, J. A., Li, H. Y., Shin, K. O., Oh, G. T., Lee, Y. M., Oh, S. et al. 2016. KCa 3.1 upregulation preserves endothelium-dependent vasorelaxation during aging and oxidative stress. *Aging Cell* 15, 801–810.

79. Yang, H., Xiong, H., Huang, Q., Pang, J., Zheng, X., Chen, L., Yu, R. et al. 2013. Compromised potassium recycling in the cochlea contributes to conservation of endocochlear potential in a mouse model of age-related hearing loss. *Neurosci Lett* 555, 97–101.

80. Jung, D. K., Lee, S. Y., Kim, D., Joo, K. M., Cha, C. I., Yang, H. S., Lee, W. B. et al. 2005. Age-related changes in the distribution of Kv1.1 and Kv3.1 in rat cochlear nuclei. *Neurol Res* 27, 436–440.

81. Tellez, J. O., McZewski, M., Yanni, J., Sutyagin, P., Mackiewicz, U., Atkinson, A., Inada, S. et al. 2011. Ageing-dependent remodelling of ion channel and Ca2+ clock genes underlying sino-atrial node pacemaking. *Exp Physiol* 96, 1163–1178.

82. Liu, J., Sirenko, S., Juhaszova, M., Sollott, S. J., Shukla, S., Yaniv, Y., and Lakatta, E. G. 2014. Age-associated abnormalities of intrinsic automaticity of sinoatrial nodal cells are linked to deficient cAMP-PKA-Ca(2+) signaling. *Am J Physiol Heart Circ Physiol* 306, H1385–1397.

83. Huang, X., Yang, P., Yang, Z., Zhang, H., and Ma, A. 2016. Age-associated expression of HCN channel isoforms in rat sinoatrial node. *Exp Biol Med (Maywood)* 241, 331–339.

84. Huang, X., Du, Y., Yang, P., Lin, S., Xi, Y., Yang, Z., and Ma, A. 2015. Age-dependent alterations of voltage-gated Na(+) channel isoforms in rat sinoatrial node. *Mech Ageing Dev* 152, 80–90.

85. Blank, T., Nijholt, I., Kye, M. J., Radulovic, J., and Spiess, J. 2003. Small-conductance, Ca2+-activated K+ channel SK3 generates age-related memory and LTP deficits. *Nat Neurosci* 6, 911–912.

86. Herman, J. P., Chen, K. C., Booze, R., and Landfield, P. W. 1998. Up-regulation of alpha1D Ca2+ channel subunit mRNA expression in the hippocampus of aged F344 rats. *Neurobiol Aging* 19, 581–587.

87. Nunez-Santana, F. L., Oh, M. M., Antion, M. D., Lee, A., Hell, J. W., and Disterhoft, J. F. 2014. Surface L-type Ca2+ channel expression levels are increased in aged hippocampus. *Aging Cell* 13, 111–120.

88. Thibault, O., and Landfield, P. W. 1996. Increase in single L-type calcium channels in hippocampal neurons during aging. *Science* 272, 1017–1020.

89. Moyer, J. R., Jr., and Disterhoft, J. F. 1994. Nimodipine decreases calcium action potentials in rabbit hippocampal CA1 neurons in an age-dependent and concentration-dependent manner. *Hippocampus* 4, 11–17.

90. Davare, M. A., and Hell, J. W. 2003. Increased phosphorylation of the neuronal L-type Ca(2+) channel Ca(v)1.2 during aging. *Proc Natl Acad Sci U S A* 100, 16018–16023.

91. Brewer, L. D., Porter, N. M., Kerr, D. S., Landfield, P. W., and Thibault, O. 2006. Chronic 1alpha,25-(OH)2 vitamin D3 treatment reduces Ca2+-mediated hippocampal biomarkers of aging. *Cell Calcium* 40, 277–286.

92. Gant, J. C., Sama, M. M., Landfield, P. W., and Thibault, O. 2006. Early and simultaneous emergence of multiple hippocampal biomarkers of aging is mediated by Ca2+-induced Ca2+ release. *J Neurosci* 26, 3482–3490.

93. Hopp, S. C., D'Angelo, H. M., Royer, S. E., Kaercher, R. M., Adzovic, L., and Wenk, G. L. 2014. Differential rescue of spatial memory deficits in aged rats by L-type voltage-dependent calcium channel and ryanodine receptor antagonism. *Neuroscience* 280, 10–18.

94. Cotella, D., Hernandez-Enriquez, B., Wu, X., Li, R., Pan, Z., Leveille, J., Link, C. D. et al. 2012. Toxic role of K+ channel oxidation in mammalian brain. *J Neurosci* 32, 4133–4144.

95. Simkin, D., Hattori, S., Ybarra, N., Musial, T. F., Buss, E. W., Richter, H., Oh, M. M., et al. 2015. Aging-related hyperexcitability in CA3 pyramidal neurons is mediated by enhanced A-Type K+ channel function and expression. *J Neurosci* 35, 13206–13218.

96. Power, J. M., Wu, W. W., Sametsky, E., Oh, M. M., and Disterhoft, J. F. 2002. Age-related enhancement of the slow outward calcium-activated potassium current in hippocampal CA1 pyramidal neurons *in vitro*. *J Neurosci Off J Soc Neurosci* 22, 7234–7243.

97. Balaban, H., Naziroglu, M., Demirci, K., and Ovey, I. S. 2017. The protective role of selenium on scopolamine-induced memory impairment, oxidative stress, and apoptosis in aged rats: The involvement of TRPM2 and TRPV1 channels. *Mol Neurobiol* 54, 2852–2868.

98. Gong, L., Gao, T. M., Huang, H., and Tong, Z. 2000. Redox modulation of large conductance calcium-activated potassium channels in CA1 pyramidal neurons from adult rat hippocampus. *Neurosci Lett* 286, 191–194.

99. Kelly, K. M., Kume, A., Albin, R. L., and Macdonald, R. L. 2001. Autoradiography of L-type and N-type calcium channels in aged rat hippocampus, entorhinal cortex, and neocortex. *Neurobiol Aging* 22, 17–23.

100. Behringer, E. J., Vanterpool, C. K., Pearce, W. J., Wilson, S. M., and Buchholz, J. N. 2009. Advancing age alters the contribution of calcium release from smooth endoplasmic reticulum stores in superior cervical ganglion cells. *J Gerontol A Biol Sci Med Sci* 64, 34–44.

101. Gokulrangan, G., Zaidi, A., Michaelis, M. L., and Schoneich, C. 2007. Proteomic analysis of protein nitration in rat cerebellum: Effect of biological aging. *J Neurochem* 100, 1494–1504.

102. Vanterpool, C. K., Vanterpool, E. A., Pearce, W. J., and Buchholz, J. N. 2006. Advancing age alters the expression of the ryanodine receptor 3 isoform in adult rat superior cervical ganglia. *J Appl Physiol 1985* 101, 392–400.

103. Dunia, R., Buckwalter, G., Defazio, T., Villar, F. D., McNeill, T. H., and Walsh, J. P. 1996. Decreased duration of Ca(2+)-mediated plateau potentials in striatal neurons from aged rats. *J Neurophysiol* 76, 2353–2363.

104. Murchison, D., and Griffith, W. H. 1996. High-voltage-activated calcium currents in basal forebrain neurons during aging. *J Neurophysiol* 76, 158–174.

105. Randall, A. D., Booth, C., and Brown, J. T. 2012. Age-related changes to Na+ channel gating contribute to modified intrinsic neuronal excitability. *Neurobiol Aging* 33, 2715–2720.

106. Chung, Y. H., Joo, K. M., Kim, M. J., and Cha, C. I. 2003. Age-related changes in the distribution of Na(v)1.1 and Na(v)1.2 in rat cerebellum. *Neuroreport* 14, 841–845.

107. Ovey, I. S., and Naziroglu, M. 2015. Homocysteine and cytosolic GSH depletion induce apoptosis and oxidative toxicity through cytosolic calcium overload in the hippocampus of aged mice: Involvement of TRPM2 and TRPV1 channels. *Neuroscience* 284, 225–233.

108. Belrose, J. C., Xie, Y. F., Gierszewski, L. J., MacDonald, J. F., and Jackson, M. F. 2012. Loss of glutathione homeostasis associated with neuronal senescence facilitates TRPM2 channel activation in cultured hippocampal pyramidal neurons. *Mol Brain* 5, 11.

109. Farajnia, S., Meijer, J. H., and Michel, S. 2015. Age-related changes in large-conductance calcium-activated potassium channels in mammalian circadian clock neurons. *Neurobiol Aging* 36, 2176–2183.

110. Jankelowitz, S. K., McNulty, P. A., and Burke, D. 2007. Changes in measures of motor axon excitability with age. *Clin Neurophysiol* 118, 1397–1404.

111. Moldovan, M., Rosberg, M. R., Alvarez, S., Klein, D., Martini, R., and Krarup, C. 2016. Aging-associated changes in motor axon voltage-gated Na(+) channel function in mice. *Neurobiol Aging* 39, 128–139.

112. Schilling, T., and Eder, C. 2015. Microglial K(+) channel expression in young adult and aged mice. *Glia* 63, 664–672.

113. Charolidi, N., Schilling, T., and Eder, C. 2015. Microglial Kv1.3 channels and P2Y12 receptors differentially regulate cytokine and chemokine release from brain slices of young adult and aged mice. *PLoS One* 10, e0128463.

114. Gupta, R. K., and Kanungo, M. 2013. Glial molecular alterations with mouse brain development and aging: Up-regulation of the Kir4.1 and aquaporin-4. *Age (Dordr)* 35, 59–67.

115. Zironi, I., Gaibani, P., Remondini, D., Salvioli, S., Altilia, S., Pierini, M., Aicardi, G. et al. 2010. Molecular remodeling of potassium channels in fibroblasts from centenarians: A marker of longevity? *Mech Ageing Dev* 131, 674–681.

116. Huang, M. S., Adebanjo, O., Moonga, B. S., Goldstein, S., Lai, F. A., Lipschitz, D. A., and Zaidi, M. 1998. Upregulation of functional ryanodine receptors during *in vitro* aging of human diploid fibroblasts. *Biochem Biophys Res Commun* 245, 50–52.

117. Tiffert, T., Daw, N., Etzion, Z., Bookchin, R. M., and Lew, V. L. 2007. Age decline in the activity of the Ca2+-sensitive K+ channel of human red blood cells. *J Gen Physiol* 129, 429–436.

118. Romero, P. J., Romero, E. A., Mateu, D., Hernandez, C., and Fernandez, I. 2006. Voltage-dependent calcium channels in young and old human red cells. *Cell Biochem Biophys* 46, 265–276.

119. Sandow, A. 1952. Excitation-contraction coupling in muscular response. *Yale J Biol Med* 25, 176–201.

120. Huxley, A. F., and Niedergerke, R. 1954. Structural changes in muscle during contraction; interference microscopy of living muscle fibres. *Nature* 173, 971–973.

121. Huxley, H., and Hanson, J. 1954. Changes in the cross-striations of muscle during contraction and stretch and their structural interpretation. *Nature* 173, 973–976.

122. Widmaier, P., Raff, H., and Strang, T. 2010. *Vander's Human Physiology: The Mechanisms of Body Function*, 12th ed., McGraw-Hill, New York.

123. Plant, D. R., and Lynch, G. S. 2002. Excitation-contraction coupling and sarcoplasmic reticulum function in mechanically skinned fibres from fast skeletal muscles of aged mice. *J Physiol* 543, 169–176.

124. Weisleder, N., Brotto, M., Komazaki, S., Pan, Z., Zhao, X., Nosek, T., Parness, J. et al. 2006. Muscle aging is associated with compromised Ca2+ spark signaling and segregated intracellular Ca2+ release. *J Cell Biol* 174, 639–645.

125. Boncompagni, S., d'Amelio, L., Fulle, S., Fano, G., and Protasi, F. 2006. Progressive disorganization of the excitation-contraction coupling apparatus in aging human skeletal muscle as revealed by electron microscopy: A possible role in the decline of muscle performance. *J Gerontol A Biol Sci Med Sci* 61, 995–1008.

126. Mayhew, M., Renganathan, M., and Delbono, O. 1998. Effectiveness of caloric restriction in preventing age-related changes in rat skeletal muscle. *Biochem Biophys Res Commun* 251, 95–99.

127. Zheng, Z., Messi, M. L., and Delbono, O. 2001. Age-dependent IGF-1 regulation of gene transcription of Ca2+ channels in skeletal muscle. *Mech Ageing Dev* 122, 373–384.

128. Barrientos, G., Llanos, P., Hidalgo, J., Bolanos, P., Caputo, C., Riquelme, A., Sanchez, G. et al. 2015. Cholesterol removal from adult skeletal muscle impairs excitation-contraction coupling and aging reduces caveolin-3 and alters the expression of other triadic proteins. *Front Physiol* 6, 105.

129. Holzenberger, M., Dupont, J., Ducos, B., Leneuve, P., Geloen, A., Even, P. C., Cervera, P. et al. 2003. IGF-1 receptor regulates lifespan and resistance to oxidative stress in mice. *Nature* 421, 182–187.

130. Dubey, A., Forster, M. J., Lal, H., and Sohal, R. S. 1996. Effect of age and caloric intake on protein oxidation in different brain regions and on behavioral functions of the mouse. *Arch Biochem Biophys* 333, 189–197.

131. Sohal, R. S., and Weindruch, R. 1996. Oxidative stress, caloric restriction, and aging. *Science* 273, 59–63.

132. Lanner, J. T., Georgiou, D. K., Joshi, A. D., and Hamilton, S. L. 2010. Ryanodine receptors: Structure, expression, molecular details, and function in calcium release. *Cold Spring Harb Perspect Biol* 2, a003996.

133. Voss, A. A., Lango, J., Ernst-Russell, M., Morin, D., and Pessah, I. N. 2004. Identification of hyperreactive cysteines within ryanodine receptor type 1 by mass spectrometry. *J Biol Chem* 279, 34514–34520.

134. Aracena, P., Aguirre, P., Munoz, P., and Nunez, M. T. 2006. Iron and glutathione at the crossroad of redox metabolism in neurons. *Biol Res* 39, 157–165.

135. Aracena-Parks, P., Goonasekera, S. A., Gilman, C. P., Dirksen, R. T., Hidalgo, C., and Hamilton, S. L. 2006. Identification of cysteines involved in S-nitrosylation, S-glutathionylation, and oxidation to disulfides in ryanodine receptor type 1. *J Biol Chem* 281, 40354–40368.

136. Umanskaya, A., Santulli, G., Xie, W., Andersson, D. C., Reiken, S. R., and Marks, A. R. 2014. Genetically enhancing mitochondrial antioxidant activity improves muscle function in aging. *Proc Natl Acad Sci U S A* 111, 15250–15255.

137. McCarthy, M. R., Thompson, A. R., Nitu, F., Moen, R. J., Olenek, M. J., Klein, J. C., and Thomas, D. D. 2015. Impact of methionine oxidation on calmodulin structural dynamics. *Biochem Biophys Res Commun* 456, 567–572.

138. Fernandez-Sanz, C., Ruiz-Meana, M., Miro-Casas, E., Nunez, E., Castellano, J., Loureiro, M., Barba, I. et al. 2014. Defective sarcoplasmic reticulum-mitochondria calcium exchange in aged mouse myocardium. *Cell Death Dis* 5, e1573.

139. Jahangir, A., Ozcan, C., Holmuhamedov, E. L., and Terzic, A. 2001. Increased calcium vulnerability of senescent cardiac mitochondria: Protective role for a mitochondrial potassium channel opener. *Mech Ageing Dev* 122, 1073–1086.

140. Zhu, X., Altschafl, B. A., Hajjar, R. J., Valdivia, H. H., and Schmidt, U. 2005. Altered Ca2+ sparks and gating properties of ryanodine receptors in aging cardiomyocytes. *Cell Calcium* 37, 583–591.
141. Yuan, Q., Chen, Z., Santulli, G., Gu, L., Yang, Z. G., Yuan, Z. Q., Zhao, Y. T. et al. 2014. Functional role of Calstabin2 in age-related cardiac alterations. *Sci Rep* 4, 7425.
142. Roberts, J., Mortimer, M. L., Ryan, P. J., Johnson, M. D., and Tumer, N. 1990. Role of calcium in adrenergic neurochemical transmission in the aging heart. *J Pharmacol Exp Ther* 253, 957–964.
143. Xu, G. J., Gan, T. Y., Tang, B. P., Chen, Z. H., Jiang, T., Song, J. G., Guo, X. et al. 2013. Age-related changes in cellular electrophysiology and calcium handling for atrial fibrillation. *J Cell Mol Med* 17, 1109–1118.
144. Rosen, M. R., Reder, R. F., Hordof, A. J., Davies, M., and Danilo, P., Jr. 1978. Age-related changes in Purkinje fiber action potentials of adult dogs. *Circ Res* 43, 931–938.
145. Guyton, A., Hall, J. 2006. Chapter 17: Local and humoral control of blood flow by the tissues. in *Textbook of Medical Physiology* (Gruliow, R. B. t. e. ed), pp. 195–203. Elsevier Inc., Philadelphia, Pennsylvania.
146. Amenta, F., Ferrante, F., Mancini, M., Sabbatini, M., Vega, J. A., and Zaccheo, D. 1995. Effect of long-term treatment with the dihydropyridine-type calcium channel blocker darodipine (PY 108-068) on the cerebral capillary network in aged rats. *Mech Ageing Dev* 78, 27–37.
147. Yu, H. J., Wein, A. J., and Levin, R. M. 1996. Age-related differential susceptibility to calcium channel blocker and low calcium medium in rat detrusor muscle: Response to field stimulation. *Neurourol Urodyn* 15, 563–576.
148. Wanstall, J. C., and O'Donnell, S. R. 1989. Age influences responses of rat isolated aorta and pulmonary artery to the calcium channel agonist, Bay K 8664, and to potassium and calcium. *J Cardiovasc Pharmacol* 13, 709–714.
149. Wanstall, J. C., and O'Donnell, S. R. 1989. Influence of age on calcium entry blocking drugs in rat aorta is spasmogen-dependent. *Eur J Pharmacol* 159, 241–246.
150. del Corsso, C., Ostrovskaya, O., McAllister, C. E., Murray, K., Hatton, W. J., Gurney, A. M. et al. 2006. Effects of aging on Ca^{2+} signaling in murine mesenteric arterial smooth muscle cells. *Mech Ageing Dev* 127, 315–323.
151. Noma, A. 1983. ATP-regulated K+ channels in cardiac muscle. *Nature* 305, 147–148.
152. Schulman, D., Latchman, D. S., and Yellon, D. M. 2001. Effect of aging on the ability of preconditioning to protect rat hearts from ischemia-reperfusion injury. *Am J Physiol Heart Circ Physiol* 281, H1630–1636.
153. Inagaki, N., Gonoi, T., Clement, J. P. T., Namba, N., Inazawa, J., Gonzalez, G., Aguilar-Bryan, L., Seino, S., and Bryan, J. 1995. Reconstitution of IKATP: An inward rectifier subunit plus the sulfonylurea receptor. *Science* 270, 1166–1170.
154. Busse, R., Edwards, G., Feletou, M., Fleming, I., Vanhoutte, P. M., and Weston, A. H. 2002. EDHF: Bringing the concepts together. *Trends Pharmacol Sci* 23, 374–380.
155. Ledoux, J., Werner, M. E., Brayden, J. E., and Nelson, M. T. 2006. Calcium-activated potassium channels and the regulation of vascular tone. *Physiology (Bethesda)* 21, 69–78.
156. Giulumian, A. D., Clark, S. G., and Fuchs, L. C. 1999. Effect of behavioral stress on coronary artery relaxation altered with aging in BHR. *Am J Physiol* 276, R435–440.
157. Kang, L. S., Kim, S., Dominguez, J. M.2nd, Sindler, A. L., Dick, G. M., and Muller-Delp, J. M. 2009. Aging and muscle fiber type alter K+ channel contributions to the myogenic response in skeletal muscle arterioles. *J Appl Physiol 1985* 107, 389–398.
158. Li, N., Liu, B., Xiang, S., and Shi, L. 2016. Similar enhancement of BK(Ca) channel function despite different aerobic exercise frequency in aging cerebrovascular myocytes. *Physiol Res* 65, 447–459.
159. Zhou, E., Qing, D., and Li, J. 2010. Age-associated endothelial dysfunction in rat mesenteric arteries: Roles of calcium-activated K(+) channels (K(ca)). *Physiol Res* 59, 499–508.
160. Mantelli, L., Amerini, S., and Ledda, F. 1995. Roles of nitric oxide and endothelium-derived hyperpolarizing factor in vasorelaxant effect of acetylcholine as influenced by aging and hypertension. *J Cardiovasc Pharmacol* 25, 595–602.
161. Fujii, K., Ohmori, S., Tominaga, M., Abe, I., Takata, Y., Ohya, Y., Kobayashi, K. et al. 1993. Age-related changes in endothelium-dependent hyperpolarization in the rat mesenteric artery. *Am J Physiol* 265, H509–516.
162. Feher, A., Broskova, Z., and Bagi, Z. 2014. Age-related impairment of conducted dilation in human coronary arterioles. *Am J Physiol. Heart Circ Physiol* 306, H1595–1601.
163. McNaughton, B. L., Barnes, C. A., Rao, G., Baldwin, J., and Rasmussen, M. 1986. Long-term enhancement of hippocampal synaptic transmission and the acquisition of spatial information. *J Neurosci: Off J Soc Neurosci* 6, 563–571.

164. Morris, R. G., Anderson, E., Lynch, G. S., and Baudry, M. 1986. Selective impairment of learning and blockade of long-term potentiation by an N-methyl-D-aspartate receptor antagonist, AP5. *Nature* 319, 774–776.

165. Bliss, T. V., and Collingridge, G. L. 1993. A synaptic model of memory: Long-term potentiation in the hippocampus. *Nature* 361, 31–39.

166. Malenka, R. C., and Bear, M. F. 2004. LTP and LTD: An embarrassment of riches. *Neuron* 44, 5–21.

167. Bansaghi, S., Golenar, T., Madesh, M., Csordas, G., RamachandraRao, S., Sharma, K., Yule, D. I. et al. 2014. Isoform- and species-specific control of inositol 1,4,5-trisphosphate (IP3) receptors by reactive oxygen species. *J Biol Chem* 289, 8170–8181.

168. Johnston, D., Williams, S., Jaffe, D., and Gray, R. 1992. NMDA-receptor-independent long-term potentiation. *Ann Rev Physiol* 54, 489–505.

169. Futatsugi, A., Kato, K., Ogura, H., Li, S. T., Nagata, E., Kuwajima, G., Tanaka, K. et al. 1999. Facilitation of NMDAR-independent LTP and spatial learning in mutant mice lacking ryanodine receptor type 3. *Neuron* 24, 701–713.

170. Wu, J., Rowan, M. J., and Anwyl, R. 2004. An NMDAR-independent LTP mediated by group II metabotropic glutamate receptors and p42/44 MAP kinase in the dentate gyrus *in vitro*. *Neuropharmacology* 46, 311–317.

171. Moosmang, S., Haider, N., Klugbauer, N., Adelsberger, H., Langwieser, N., Muller, J., Stiess, M. et al. 2005. Role of hippocampal Cav1.2 Ca2+ channels in NMDA receptor-independent synaptic plasticity and spatial memory. *J Neurosci: Off J Soc Neurosci* 25, 9883–9892.

172. Berridge, M. J. 1998. Neuronal calcium signaling. *Neuron* 21, 13–26.

173. Verkhratsky, A. 2002. The endoplasmic reticulum and neuronal calcium signalling. *Cell Calcium* 32, 393–404.

174. Verkhratsky, A., and Petersen, O. H. 2002. The endoplasmic reticulum as an integrating signalling organelle: From neuronal signalling to neuronal death. *Eur J Pharmacol* 447, 141–154.

175. Ouanounou, A., Zhang, L., Charlton, M. P., and Carlen, P. L. 1999. Differential modulation of synaptic transmission by calcium chelators in young and aged hippocampal CA1 neurons: Evidence for altered calcium homeostasis in aging. *J Neurosci* 19, 906–915.

176. Faber, E. S. 2010. Functional interplay between NMDA receptors, SK channels and voltage-gated Ca2+ channels regulates synaptic excitability in the medial prefrontal cortex. *J Physiol* 588, 1281–1292.

177. Power, J. M., Wu, W. W., Sametsky, E., Oh, M. M., and Disterhoft, J. F. 2002. Age-related enhancement of the slow outward calcium-activated potassium current in hippocampal CA1 pyramidal neurons *in vitro*. *J Neurosci* 22, 7234–7243.

178. Norris, C. M., Halpain, S., and Foster, T. C. 1998. Reversal of age-related alterations in synaptic plasticity by blockade of L-type Ca2+ channels. *J Neurosci* 18, 3171–3179.

179. Porter, N. M., Thibault, O., Thibault, V., Chen, K. C., and Landfield, P. W. 1997. Calcium channel density and hippocampal cell death with age in long-term culture. *J Neurosci* 17, 5629–5639.

180. Bangalore, R., and Triggle, D. J. 1995. Age-dependent changes in voltage-gated calcium channels and ATP-dependent potassium channels in Fischer 344 rats. *Gen Pharmacol* 26, 1237–1242.

181. Sama, D. M., Mohmmad Abdul, H., Furman, J. L., Artiushin, I. A., Szymkowski, D. E., Scheff, S. W., and Norris, C. M. 2012. Inhibition of soluble tumor necrosis factor ameliorates synaptic alterations and Ca2+ dysregulation in aged rats. *PLoS One* 7, e38170.

182. Rose, G. M., Ong, V. S., and Woodruff-Pak, D. S. 2007. Efficacy of MEM 1003, a novel calcium channel blocker, in delay and trace eyeblink conditioning in older rabbits. *Neurobiol Aging* 28, 766–773.

183. Batuecas, A., Pereira, R., Centeno, C., Pulido, J. A., Hernandez, M., Bollati, A., Bogonez, E. et al. 1998. Effects of chronic nimodipine on working memory of old rats in relation to defects in synaptosomal calcium homeostasis. *Eur J Pharmacol* 350, 141–150.

184. Bodhinathan, K., Kumar, A., and Foster, T. C. 2010. Intracellular redox state alters NMDA receptor response during aging through Ca2+/calmodulin-dependent protein kinase II. *J Neurosci: Off J Soc Neurosci* 30, 1914–1924.

185. Kumar, A., and Foster, T. C. 2004. Enhanced long-term potentiation during aging is masked by processes involving intracellular calcium stores. *J Neurophysiol* 91, 2437–2444.

186. Norris, C. M., Halpain, S., and Foster, T. C. 1998. Reversal of age-related alterations in synaptic plasticity by blockade of L-type Ca2+ channels. *J Neurosci : Off J Soc Neurosci* 18, 3171–3179.

187. Bodhinathan, K., Kumar, A., and Foster, T. C. 2010. Redox sensitive calcium stores underlie enhanced after hyperpolarization of aged neurons: Role for ryanodine receptor mediated calcium signaling. *J Neurophysiol* 104, 2586–2593.

188. Kumar, A., Bodhinathan, K., and Foster, T. C. 2009. Susceptibility to calcium dysregulation during brain aging. *Front Aging Neurosci* 1, 2.

189. Oh, M. M., Oliveira, F. A., and Disterhoft, J. F. 2010. Learning and aging related changes in intrinsic neuronal excitability. *Front Aging Neurosci* 2, 2.

190. Martini, A., Battaini, F., Govoni, S., and Volpe, P. 1994. Inositol 1,4,5-trisphosphate receptor and ryanodine receptor in the aging brain of Wistar rats. *Neurobiol Aging* 15, 203–206.

191. Kumar, A., and Foster, T. C. 2004. Enhanced long-term potentiation during aging is masked by processes involving intracellular calcium stores. *J Neurophysiol* 91, 2437–2444.

192. Paula-Lima, A. C., Adasme, T., and Hidalgo, C. 2014. Contribution of Ca2+ release channels to hippocampal synaptic plasticity and spatial memory: Potential redox modulation. *Antioxid Redox Signal* 21, 892–914.

193. Gant, J. C., Chen, K. C., Kadish, I., Blalock, E. M., Thibault, O., Porter, N. M., and Landfield, P. W. 2015. Reversal of aging-related neuronal Ca2+ dysregulation and cognitive impairment by delivery of a transgene encoding FK506-binding protein 12.6/1b to the Hippocampus. *J Neurosci* 35, 10878–10887.

194. Gant, J. C., Chen, K. C., Norris, C. M., Kadish, I., Thibault, O., Blalock, E. M., Porter, N. M. et al. 2011. Disrupting function of FK506-binding protein 1b/12.6 induces the Ca(2)+-dysregulation aging phenotype in hippocampal neurons. *J Neurosci* 31, 1693–1703.

195. Bodhinathan, K., Kumar, A., and Foster, T. C. 2010. Redox sensitive calcium stores underlie enhanced after hyperpolarization of aged neurons: Role for ryanodine receptor mediated calcium signaling. *J Neurophysiol* 104, 2586–2593.

196. Bull, R., Finkelstein, J. P., Humeres, A., Behrens, M. I., and Hidalgo, C. 2007. Effects of ATP, Mg2+, and redox agents on the Ca2+ dependence of RyR channels from rat brain cortex. *Am J Physiol. Cell Physiol* 293, C162–171.

197. Bull, R., Marengo, J. J., Finkelstein, J. P., Behrens, M. I., and Alvarez, O. 2003. SH oxidation coordinates subunits of rat brain ryanodine receptor channels activated by calcium and ATP. *Am J Physiol. Cell Physiol* 285, C119–128.

198. Bull, R., Finkelstein, J. P., Galvez, J., Sanchez, G., Donoso, P., Behrens, M. I., and Hidalgo, C. 2008. Ischemia enhances activation by Ca2+ and redox modification of ryanodine receptor channels from rat brain cortex. *J Neurosci: Off J Soc Neurosci* 28, 9463–9472.

199. Murakoshi, H., and Trimmer, J. S. 1999. Identification of the Kv2.1 K+ channel as a major component of the delayed rectifier K+ current in rat hippocampal neurons. *J Neurosci* 19, 1728–1735.

200. Trimmer, J. S. 1993. Expression of Kv2.1 delayed rectifier K+ channel isoforms in the developing rat brain. *FEBS Lett* 324, 205–210.

201. Thiffault, I., Speca, D. J., Austin, D. C., Cobb, M. M., Eum, K. S., Safina, N. P., Grote, L. et al. 2015. A novel epileptic encephalopathy mutation in KCNB1 disrupts Kv2.1 ion selectivity, expression, and localization. *J Gen Physiol* 146, 399–410.

202. Torkamani, A., Bersell, K., Jorge, B. S., Bjork, R. L., Jr., Friedman, J. R., Bloss, C. S., Cohen, J. et al. 2014. De novo KCNB1 mutations in epileptic encephalopathy. *Ann Neurol* 76, 529–540.

203. Saitsu, H., Akita, T., Tohyama, J., Goldberg-Stern, H., Kobayashi, Y., Cohen, R., Kato, M. et al. 2015. De novo KCNB1 mutations in infantile epilepsy inhibit repetitive neuronal firing. *Sci Rep* 5, 15199.

204. Speca, D. J., Ogata, G., Mandikian, D., Bishop, H. I., Wiler, S. W., Eum, K., Wenzel, H. J. et al. 2014. Deletion of the Kv2.1 delayed rectifier potassium channel leads to neuronal and behavioral hyperexcitability. *Genes Brain Behav* 13, 394–408.

205. Wu, X., Hernandez-Enriquez, B., Banas, M., Xu, R., and Sesti, F. 2013. Molecular mechanisms underlying the apoptotic effect of KCNB1 K+ channel oxidation. *J Biol Chem* 288, 4128–4134.

206. Yu, W., Parakramaweera, R., Teng, S., Gowda, M., Sharad, Y., Thakker-Varia, S., Alder, J., and Sesti, F. 2016. Oxidation of KCNB1 potassium channels causes neurotoxicity and cognitive impairment in a mouse model of traumatic brain injury. *J Neurosci* 36, 11084–11096.

207. Yu, W., Gowda, M., Sharad, Y., Singh, S. A., and Sesti, F. 2017. Oxidation of KCNB1 potassium channels triggers apoptotic integrin signaling in the brain. *Cell Death Dis* 8, e2737.

208. Cotella, D., Hernandez-Enriquez, B., Wu, X., Li, R., Pan, Z., Leveille, J., Link, C. D. et al. 2012. Toxic role of k+ channel oxidation in mammalian brain. *J Neurosci : Off J Soc Neurosci* 32, 4133–4144.

209. Oddo, S., Caccamo, A., Shepherd, J. D., Murphy, M. P., Golde, T. E., Kayed, R., Metherate, R. et al. 2003. Triple-transgenic model of Alzheimer's disease with plaques and tangles: Intracellular Abeta and synaptic dysfunction. *Neuron* 39, 409–421.

210. Smith, I. F., Hitt, B., Green, K. N., Oddo, S., and LaFerla, F. M. 2005. Enhanced caffeine-induced Ca2+ release in the 3xTg-AD mouse model of Alzheimer's disease. *J Neurochem* 94, 1711–1718.

211. Sensi, S. L., Rapposelli, I. G., Frazzini, V., and Mascetra, N. 2008. Altered oxidant-mediated intraneuronal zinc mobilization in a triple transgenic mouse model of Alzheimer's disease. *Exp Gerontol* 43, 488–492.
212. Yao, J., Irwin, R. W., Zhao, L., Nilsen, J., Hamilton, R. T., and Brinton, R. D. 2009. Mitochondrial bioenergetic deficit precedes Alzheimer's pathology in female mouse model of Alzheimer's disease. *Proc Natl Acad Sci USA* 106, 14670–14675.
213. Chou, J. L., Shenoy, D. V., Thomas, N., Choudhary, P. K., Laferla, F. M., Goodman, S. R., and Breen, G. A. 2011. Early dysregulation of the mitochondrial proteome in a mouse model of Alzheimer's disease. *J Proteomics* 74, 466–479.
214. McManus, M. J., Murphy, M. P., and Franklin, J. L. 2011. The mitochondria-targeted antioxidant MitoQ prevents loss of spatial memory retention and early neuropathology in a transgenic mouse model of Alzheimer's disease. *J Neurosci: Off J Soc Neurosci* 31, 15703–15715.
215. Frazzini, V., Guarnieri, S., Bomba, M., Navarra, R., Morabito, C., Mariggio, M. A., and Sensi, S. L. 2016. Altered Kv2.1 functioning promotes increased excitability in hippocampal neurons of an Alzheimer's disease mouse model. *Cell Death Dis* 7, e2100.
216. Pal, S. K., Takimoto, K., Aizenman, E., and Levitan, E. S. 2006. Apoptotic surface delivery of K+ channels. *Cell Death Differ* 13, 661–667.
217. Aras, M. A., and Aizenman, E. 2005. Obligatory role of ASK1 in the apoptotic surge of K+ currents. *Neurosci Lett* 387, 136–140.
218. Redman, P. T., Hartnett, K. A., Aras, M. A., Levitan, E. S., and Aizenman, E. 2009. Regulation of apoptotic potassium currents by coordinated zinc-dependent signalling. *J Physiol* 587, 4393–4404.
219. Redman, P. T., He, K., Hartnett, K. A., Jefferson, B. S., Hu, L., Rosenberg, P. A., Levitan, E. S. et al. 2007. Apoptotic surge of potassium currents is mediated by p38 phosphorylation of Kv2.1. *Proc Natl Acad Sci USA* 104, 3568–3573.
220. McCord, M. C., and Aizenman, E. 2013. Convergent Ca2+ and Zn2+ signaling regulates apoptotic Kv2.1 K+ currents. *Proc Natl Acad Sci USA* 110, 13988–13993.
221. Norris, C. A., He, K., Springer, M. G., Hartnett, K. A., Horn, J. P., and Aizenman, E. 2012. Regulation of neuronal proapoptotic potassium currents by the hepatitis C virus nonstructural protein 5A. *J Neurosci: Off J Soc Neurosci* 32, 8865–8870.
222. Bianchi, L., Kwok, S. M., Driscoll, M., and Sesti, F. 2003. A potassium channel-MiRP complex controls neurosensory function in *Caenorhabditis elegans. J Biol Chem* 278, 12415–12424.
223. Park, K. H., Hernandez, L., Cai, S. Q., Wang, Y., and Sesti, F. 2005. A family of K+ channel ancillary subunits regulate taste sensitivity in *Caenorhabditis elegans. J Biol Chem* 280, 21893–21899.
224. Cai, S. Q., and Sesti, F. 2007. A new mode of regulation of N-type inactivation in a *Caenorhabditis elegans* voltage-gated potassium channel. *J Biol Chem* 282, 18597–18601.

24 Lipid Raft Alteration and Functional Impairment in Aged Neuronal Membranes

Julie Colin, Lynn Gregory-Pauron, Frances T. Yen,
Thierry Oster, and Catherine Malaplate-Armand

CONTENTS

INTRODUCTION

Life expectancy has increased in most developed countries, which has led to a higher proportion of elderly people in the world's population. However, this increase is not accompanied by a lengthening of the health span, since aging is characterized by progressive deterioration in cellular and organ functions. The brain is particularly vulnerable to aging processes, and this is clearly demonstrated by the onset of age-related neurodegenerative diseases.

The mechanisms involved in aging, particularly in the brain, remain for the most part unclear. A key challenge in neurosciences is the identification of age-sensitive parameters and aging-promoting factors, as well as that of their impact on cognitive functions. Recently, neuronal membranes gained increased interest when considered as a cellular compartment with a fine micro-organization that can influence most brain functions. This led to the emergence of a very exciting research field to explore and understand the mechanisms that control the cross-talk involved in the reciprocal segregation of membrane lipids and associated proteins allowing the numerous processes residing in the neuronal membranes. This could also explain the complex and ambivalent properties of lipids, including dietary fatty acids (FA) and cholesterol, well known to impact neuronal susceptibility to cellular stress and thereby to aging-related diseases such as Alzheimer's disease (AD).

INFLUENCE OF LIPIDS ON BRAIN AGING

METABOLIC DISORDERS IN AGING

Lipid metabolism changes significantly with age, with a clearly defined impact on many age-related diseases, especially metabolic, cardiovascular, and neurodegenerative diseases. In both men and women, plasma lipid levels are modified similarly during aging, with some age-dependent differences. Indeed, plasma cholesterol and triglycerides (TG) increase until 50 years in men and 60 years in women to stabilize in a more or less durable way. After 70 years, plasma TG go through a decline phase. Low-density lipoprotein (LDL)-associated cholesterol levels undergo identical changes to total cholesterol, with an earlier plateau in men (50–60 years old) than in women (60–70 years old). High-density lipoprotein (HDL)-associated cholesterol decreases slightly until 50 years and finally stabilizes in men, while this decrease continues throughout the life for women [1,2]. A study on rodents described similar metabolic changes in humans (increased body weight and body fat), as well as impaired free fatty acid (FFA) metabolism. This alteration was reflected by a decrease in gene expression involved in the FA oxidation in muscle, liver, and adipose tissue and was accompanied by an increase in plasma concentrations of total FA, especially stearic acid (C18:0) and palmitic acid (C16:0), which could result from disturbances in FA transport, uptake, and storage [3].

OBESITY, METABOLIC DISORDERS, AND COGNITIVE FUNCTIONS

Metabolic syndrome is defined by a series of related disorders, including obesity, insulin resistance, glucose intolerance, and hyperlipidemia (for a review, see Reference [4]). Obesity is a major factor not only for the development of metabolic syndrome but also for the development of cardiovascular diseases and type 2 diabetes (for a review, see Reference [5]). Much attention has been paid to obesity as a risk factor for AD. Increasing evidence suggests an association between juvenile or adulthood obesity and the risk of dementia in later life [6–9]. For example, an epidemiological study found that adult obese participants (body mass index, BMI $\geq$ 30) had a 35% higher risk of dementia than those of normal weight (BMI 18.6–24.9) [10]. A positive association was also previously shown between Aβ peptide plasma levels, BMI, and fat mass [11]. However, the impact of obesity at the end of life remains controversial [12–14].

A study in 70–88-year-old women showed that the risk of developing AD from 70 years of age increases by 36% when BMI increases by 1 kg/m^2 [15]. In contrast, other studies reported that the risk of dementia increases in people with low BMI or those who have suffered significant weight loss before diagnosis [16–18]. The relationship between obesity and AD may depend on the age of onset of obesity and the weight loss that preceded the disease [12–14,19].

Obesity may also play an essential role in cognitive decline during normal aging, but neurobiological damage caused by obesity is poorly understood. Obesity has been reported to be associated with lower brain volumes, suggesting that it may increase the risk of neurodegeneration [20,21]. In addition, obesity was associated with structural abnormalities in the brain. Indeed, several studies demonstrated by diffusion tensor imaging that integrity of white matter was adversely affected in individuals with a high BMI [22,23]. Finally, recent studies indicate that the link between obesity and dysfunction includes oxidative stress and neuroinflammation.

In rodents, diet-induced obesity (DIO) leads to increased inflammatory factors and immune cells (astrocytes and microglia) in peripheral tissues as well as in critical brain regions dedicated to the maintenance of energy balance [24]. Another study showed that obesity in animals caused an elevation in reactive oxygen species (ROS) generation and mitochondrial dysfunction, both in plasma and brain tissues [9]. By using experimental models of insulin resistance and obesity, many studies found that chronic obesity and insulin resistance conditions led to cognitive impairments and synaptic dysfunction [25–27]. Some of them demonstrated that a brief occurrence of obesity and insulin resistance during childhood/adolescence might induce irreversible epigenetic modifications in the brain that persist following restoration of normal metabolic homeostasis, leading to synaptic dysfunction during aging [27,28].

Epidemiological and clinical studies demonstrated that the deleterious effects of obesity on higher cortical function were exacerbated in the elderly population (for a review, see Reference [29]). Studies comparing young (7-month-old) and aged (24-month-old) obese C57Bl/6 mice placed on a high-fat diet (HFD) found that aging exacerbated the obesity-induced decline in microvascular density, both in the hippocampus and cortex. The extent of hippocampal microvascular rarefaction and impairment of hippocampal-dependent cognitive function was positively correlated. Aging exacerbated obesity-induced loss of pericyte coverage on cerebral microvessels and alters the gene expression signature of hippocampal angiogenesis (e.g., increasing the expression vascular endothelial growth factor A, which likely contributes to microvascular rarefaction). Aging was also found to exacerbate obesity-induced oxidative stress and induction of nicotinamide adenine dinucleotide phosphate (NADPH) oxidase, as well as impaired cerebral blood flow responses to whisker stimulation [30]. Aging also worsened obesity-induced systemic inflammation and blood–brain barrier (BBB) disruption, as indicated by the increase in circulating levels of proinflammatory cytokines and by that of extravasated immunoglobulin G in the hippocampus, respectively [31].

Metabolic abnormalities associated with obesity include components of lipoprotein metabolism, such as plasma cholesterol and TG transport. Thus, obesity is often accompanied by dyslipidemia. The characteristic abnormalities of obesity are hypertriglyceridemia and reduced plasma HDL-cholesterol level. These same abnormalities are observed in the elderly [2]. In the common form of obesity, leptin resistance, resulting from decreased availability of leptin in the hypothalamus and/or impaired leptin activity during aging, may accentuate lipid storage. Leptin is a hormone predominantly secreted by adipose tissues and circulating in plasma, which regulates the body's fat reserves and food intake by controlling the sensation of satiety [32]. Leptin receptors are widely expressed in the brain and many studies link obesity to peripheral hyperleptinemia, decreased leptin transport through BBB, and signaling alteration [33–35]. Moreover, studies showed that dysregulation and/or leptin deficiency, frequently found in obese patients, were related to cognitive deficits associated with neurodegenerative diseases [33,36]. For example, leptin was shown to lower the β-secretase-mediated amyloidogenic cleavage of neuronal amyloid precursor protein (APP) by a mechanism that might involve changes in membrane lipid composition [35,36]. Moreover, leptin administration was shown to decrease Aβ peptide levels and Tau phosphorylation in the brain of AD transgenic mice, while improving the learning abilities and memory performance of these animals [33,35–37].

Finally, age is an unmodifiable risk factor for pathological aging. Environmental factors such as metabolic disorders could promote the deleterious processes associated with aging. This suggests that food intake and especially the HFDs are major modifiable risk factors for pathological aging.

Deleterious Effects of HFDs on Brain Function

Diets rich in saturated fat, refined sugars, or cholesterol induce metabolic conditions leading to obesity, type 2 diabetes, and hypercholesterolemia. These chronic pathologies are not only known for their impact on cardiovascular risk but are also described as capable of promoting brain aging [38]. Indeed, the incidence of AD is higher in populations consuming hypercaloric diets with a high content of saturated fats [39,40]. For instance, the reported higher saturated fat and refined sugar intake were associated with cognitive decline in aging and increased incidence of AD [40–44]. Other examples show a correlation between excessive intake of *trans* fatty acids (TFA) (intakes exceeding 2% of total energy intake) and an increased risk of coronary heart disease. These adverse effects have been associated with changes in the serum lipoprotein profile, plasma TG, and inflammatory markers such as C-reactive protein, including an increase in LDL cholesterol and a decrease in HDL cholesterol [45–47].

An epidemiological study found a relative risk of AD 5-fold higher in people over 65 who consume about 4.8 g/day of TFA compared to those who eat only 1.8 g/day [43]. Animal research has similarly demonstrated impairment of brain function following maintenance on a high saturated fat and refined sugar diet. For example, rat studies have shown that short exposure (about 20 days) to diets rich in saturated fat or sugars alters memory capacity and is associated with the development of neuroinflammation and oxidative stress in the hippocampus [48]. In wild-type mice or AD model mice (Tg2576 transgenic mice or after intracerebral injection of Aβ peptide), diets rich in saturated fat, sugar, or cholesterol favor amyloidogenesis as well as memory deficits [8,49–53]. In contrast, a study of AD transgenic mice showed that high TFA consumption modulated brain FA profiles, including increased monounsaturated FA and decreased polyunsaturated fatty acids (PUFA), but had no significant effect on the main neuropathological features of AD, such as Aβ production [46]. These impairments correspond to neurophysiological changes, such as changes in proteins expression, which support synaptic plasticity in the brain and are essential for learning and memory (Table 24.1).

Importantly, the aging brain is considered to be particularly sensitive to these aspects of HFD exposure [67–69] and provides a vulnerable environment to which a HFD could cause more damage. A study investigated the effects of a high-fat/high cholesterol (HFHC) diet on cognitive performance, neuroinflammation markers, and phosphorylated Tau (p-Tau) pathological markers in the hippocampus of young (4-month-old) versus aged (14-month-old) male rats [70]. Young and aged rats fed the HFHC diet exhibited worse performance on a spatial working memory task. They also exhibited a significant reduction in NeuN and calbindin-D28k immunoreactivity as well as an increased activation of microglial cells in the hippocampal formation. Western blot analysis showed higher levels of hippocampal p-Tau in aged HFHC rats, suggesting abnormal phosphorylation of the Tau protein following the HFHC diet exposure. These results demonstrate that this type of diet impairs the cognitive capacities in aged rodents, indicating that the aging process gives rise to a higher vulnerability to diet-induced alterations in hippocampal function [70]. Other studies have not shown this exacerbation, but considered that adult mice were more susceptible to the physiological and anxiety-like effects of HFD consumption than aged mice, while aged mice displayed deficits in spatial cognition regardless of dietary influence. Indeed, normal aging was shown to affect spatial memory abilities in rodents. Other studies reported significant effects of HFD on spatial reference memory in young rodents. Nonetheless, it was clear that the aged rats on HFHC diet constituted the group exhibiting the worst performance on both working memory and reference memory aspects of this task, confirming data from clinical studies that HFHC diets predispose individuals to cognitive impairment with aging [69–72].

Disruptions of lipid metabolism and homeostasis could therefore be considered as risk factors for aging. Interestingly, particular dietary lipids such as *n*-3 PUFA and especially one of them, docosahexaenoic acid (DHA, C22:6), can also provide significant protection, not only by correcting dyslipidemia but also by preserving cognitive functions. Their incorporation in neuronal membranes is increasingly documented to lower the changes related to aging, which could therefore lead to strategies aimed at preserving brain function.

MEMBRANES AND BRAIN FUNCTION

The characterization of biological membranes has considerably evolved over the past 20 years. Today, it is widely accepted that the membrane is composed of two coexisting liquid phases: liquid-disordered and liquid-ordered domains, the latter constituting the microdomains or lipid rafts.

MEMBRANE NON-RAFT DOMAINS: COMPOSITION AND PROPERTIES

Lipids are more than inert structural components delimitating the extra- and intracellular compartments. By the numerous properties that they display, they can influence and modify many cellular functions: the FA composition of phospholipids (PL) determines the biophysical properties of cell membranes, and protein functionalities or localizations [73]. Membrane non-raft domains are composed mainly of the PL containing PUFA whose physical properties explain their relative abundance in neuronal membranes. These domains constitute a liquid-disordered environment that influences the membrane architecture and organization. Changes in the composition of these non-raft domains

TABLE 24.1
HFD Effects on Brain Function

Diet Duration	Investigated Tissue/ Region	Observed Effects	References
		Neurotransmission	
60% HFD 12 weeks	Mouse brain	Dopamine dysregulation	[54,55]
HSFD 8 weeks	Rat caudate putamen	Higher serotonin 5-HT2A receptor	[56]
	Rat mammillary nucleus	Lower serotonin 5-HT2A and 5-HT2C receptor	[40]
75% HSFD 8 weeks	Rat hippocampus	Higher neuronal calcium sensor 1 Lower guanine nucleotide-binding protein G(o) subunit alpha Lower glutamate receptor, ionotropic, AMPA 3	[40]
60% HFD 5 months	Rat hippocampus	Lower glutamate-to-glutamine ratio	[57]
		Membrane Fluidity	
TFA-rich diet lifelong	Rat striatum	Lower Na+/K+ ATPase activity Higher protein carbonyl	[58]
		Cellular Assembly and Organization	
75% HSFD 8 weeks	Rat hippocampus	Higher profilin 1 Lower tropomyosin 1, alpha	[40]
HFD 8 weeks	Rat hippocampus	Lower MAP-2	[59]
		Neuronal Proliferation and Differentiation	
75% HSFD 8 weeks	Rat hippocampus	Lower tubulin beta-3 Lower protein tyrosine kinase 2	[40]
		Synaptic Plasticity and Long-Term Potentiation	
76% HSFD 8 weeks	Rat hippocampus	Lower Ca^{2+}/calmodulin-dependent protein kinase	[40]
		Neuro-Inflammation and Oxidative Stress	
Palmitic acidincubation 12 h	Hypothalamic cell line (mHypoA-CLU192)	Lower mitofusin 2 (MTF2) Higher glucose-regulated protein 78/immunoglobulin heavy chain-binding protein (GRP78/BIP)	[60]
Palmitic acid incubation 1–24 h	Mouse cortical neurons	Higher GRP78	[61]
Palmitic acid incubation 6–24 h	SH-SY5Y cells	Higher spliced X-box binding protein mRNA (XBP-1) Higher BiP mRNA	[62]
45% HFD 13 weeks	Mouse arcuate nucleus of hypothalamus	Lower MTF2	[60]
35% HFD 9 months	3xTg-AD mice cortex	Higher GFAP	[50]
40% HFD 3–12 months	Mice brain	Higher GFAP	[63]
HFD 5 months	Rat cortex	Higher prostaglandin E2 Higher cyclooxygenase-1 and -2	[64]

(Continued)

TABLE 24.1 (*Continued*)
HFD Effects on Brain Function

Diet Duration	Investigated Tissue/ Region	Observed Effects	References
		Apoptosis	
Palmitic acid incubation 6–24 h	SH-SY5Y cells	Higher caspase-3 Higher caspase-9	[65]
		BBB Dysfunction	
HSFD 6 months	Rat hippocampus	Lower SMI-71 (an antibody specific to the rat endothelial barrier protein)	[66]
40% HFD 3–12 months		Lower occludin	[63]
		Metabolic and Bioenergetics Function	
60% HFD 5 months	Rat hippocampus	Lower total creatine Higher N-acetylaspartylglutamic acid Lower myoinositol Lower serine	[57]

AD: Alzheimer's disease; BBB: blood–brain barrier; GFAP: glial fibrillary acidic protein; HFD: high-fat diet; HSFD: high saturated fat diet; TFA: *trans* fatty acid.

create a more or less fluid environment that can alter the properties of the raft domains [74]. The chains of the PUFA predominantly found in these regions have high disorder rates of molecular orientation. They adopt complex spatial conformations imposed by their unsaturation. The flexibility of the chain results from rapid interconversion between the conformational states, which provides a rough fluctuation surface [75]. Many molecular dynamics simulations have been carried out to determine the DHA conformation in space and its influence on the membrane structure. On the basis of energy minimization simulations, a model reveals DHA according to a dynamic structure that can alternately adopt four conformations in space [76]. Thus, membranes rich in long-chain PUFA are distorted due to the steric hindrance and rapid reorientation of the hydrocarbon chains, which leads to a transition from an l_o phase to an l_d phase [77].

The conclusions remain controversial on the PUFA effect on the lipid bilayer due to the variability of the biological membranes. However, data show that DHA enrichment affects the fluidity, permeability, melting, and elasticity of membranes and the transport of transmembrane (TM) proteins, as summarized in Table 24.2. The raft domains could also be modulated by FAs and DHA [78]. DHA-containing PL-enriched domains are the opposite of rafts and represent the extreme case of non-raft domains, which are low cholesterol and highly disordered domains [74]. Modulation of DHA membrane content could alter the lipid environment of rafts and disrupt their organization and size in membranes *in vitro* and *in vivo* due to the incompatibility and low affinity between PUFA and cholesterol. Indeed, this poor affinity could lead to mixing of the PL chains, and consequently to phase separation of the membrane constituents [79].

PLs can also undergo rapid deacylation and reacylation processes involving the presence of phospholipase A_2 and acyltransferases that ensure the permanent exchange of FA between the different PL classes. The deacylation–reacylation cycle is an important mechanism responsible for the introduction of PUFA into neural membrane glycerophospholipids [80]. This phenomenon makes it possible to preserve the neuronal membrane by controlling the balance between free and esterified FA. Indeed, high levels of free unsaturated FAs induce cell membrane destabilization, oxidative stress, and possibly cell death, whereas the reacylation of these FA in PLs restores membrane integrity and ensures the cell membrane survival [81].

TABLE 24.2

Effects of PUFA Supplementation on Membrane Properties

Membrane Properties	Investigated Tissue Target	Effects	References
Fluidity	Human erythrocyte membrane	Higher unsaturation in membrane Lower viscosity Higher fluidity	[82,83]
	Rat lymph node lymphocytes	Higher DHA in membrane PLs Higher fluidity	[84]
	Human retinal pigment epithelium cells	Higher membrane DHA Higher fluidity	[85]
Permeability	T84, human colonic adenocarcinoma cell line	Higher DHA in membrane PLs Lower permeability	[86]
	ECV304, human vascular endothelial cell line	Higher occludin Lower permeability	[87]
	Mitochondria from cardiomyopathic hamsters	Lower permeability Lower apoptosis	[88]
Fusion	dipalmitoyl-PC (DPPC) vesicles	Higher fusogenic rate of liposomes	[89]
	Frog cilia-derived sensory organelles	Higher colocalization of effector proteins of vesicular fusion	[90]
Elasticity	NMR and X-ray diffraction studies	Higher compressibility of of unsaturated chains	[91]
TM proteins/ neurotransmission	B82, murine fibroblasts cell line	Inhibited depolarization K^+ channels	[92]
	Rat cardiomyocytes	Inhibited depolarization Na^+ channels	[93]
	Rat brain	Inhibited depolarization Na^+ channels	[94]
	Rat cardiomyocytes	Inhibited depolarization $Ca2^+$ channels	[95]
	Rat cerebral cortex	Inhibited depolarization $Ca2^+$ channels	[96]
	Rat forebrain	Higher glutamate receptor	[97]
	Rat substantia nigra neuron	Lower γ-amino-butyric acid (GABA) receptor	[98]
	Rat endothelial cells of brain	Higher GLUT1 transporter	[99]
	Rat brain	Differences depending on the brain area for dopaminergic and serotoninergic system	[100]
Rafts domains	Human lymphocytes	PLD1 activation by exclusion from rafts: immunosuppressor effect	[101]
	Mouse lymphocytes	Lower rafts grouping by cross-linking of GM1	[102]
	JY human B-lymphoblasts	Moved major histocompatibility complex (MHC) from non-raft to raft-like domains	[103]

MEMBRANE RAFT DOMAINS

In 1997, Simons and Ikonen were the first to propose a molecular mechanism for lateral heterogeneity of biological membranes. According to these authors, the molecular basis for the raft model lay in the preferential lateral association of long-chain saturated sphingolipids and cholesterol within the exo-leaflet where cholesterol filled the voids between sphingolipid molecules [104].

Composition of Rafts

The rafts constitute preformed entities in the cell membrane and are present in different parts of the lipid bilayer [105]. These microdomains are asymmetric and contain 30%–50% cholesterol, 3–5 levels higher than other membrane regions [106–108], and up to 70% of cellular sphingomyelins (SM) [106,109]. Their external sheet is particularly rich in sphingoglycolipids. 10%–15% of the membrane sphingolipids and 10%–20% of the membrane glycolipids are concentrated in the rafts. Their cytoplasmic sheet is rich in glycerolipids. If the plasma membrane is very rich in PL, including phosphatidylethanolamine (PE) and phosphatidylcholine (PC), less than 30% of the lipids contained in the rafts are of this type [106–108,110]. The rafts are also very rich in gangliosides, in particular the GM1 which is almost exclusively located there [109]. Within these rafts, the hydrocarbon chains of the FAs are mainly saturated and in conformation stretched to form with the cholesterol a phase less fluid than the rest of the plasma membrane (Figure 24.1). The presence of raft-specific stabilizing proteins, such as flotillins or caveolins, is also necessary to maintain a stable conformation [105,110,111].

Localization and Organization

The necessary condition for raft formation in cell membranes is the presence of sufficient cholesterol and sphingolipids. There is an increasing gradient of cholesterol from the endoplasmic reticulum (ER) to the plasma membrane. As for the biosynthesis of the sphingolipids, it takes place entirely in the Golgi, suggesting that the rafts form in this intracellular compartment before being incorporated into the vesicles to be addressed to the plasma membrane [104,112–114]. Thus, although rafts are predominantly localized in the plasma membrane of cells, they have also been identified in other organelles such as endosomes, caveosomes, and phagosomes, organelles connected to the Golgi and/or plasma membrane [115–118]. Once incorporated into the membrane, the rafts perpetually undergo endo- and exocytosis phenomena under the influence of certain parameters such as cholesterol, proof that cholesterol plays a decisive role in the formation and renewal of rafts [119]. It has been shown that a high cholesterol depletion of the cells leads to the rupture of the rafts and to the diffusion of their constituents along the plasma membrane [108].

The mechanisms that control the biogenesis, size, and lifetime of rafts are still poorly understood, but it seems that microtubules and actin microfilaments are privileged partners of rafts. Indeed, rafts isolated from primary cortical neurons co-immunostain with microtubule proteins such as microtubule-associated protein 2 (MAP-2), β-tubulin II, or Tau. The exposure of neurons to a cholesterol-depleting membrane (MβCD), or to an agent inhibiting the metabolism of sphingoglycolipids (D-PDMP, D-threo-1-Phenyl-2-decanoylamino-3-morpholino-1-propanol), leads to neuronal retraction before neuronal death. These effects would be due to a significant disturbance of the rafts' association with the microtubules proteins [120]. The cytoskeleton could play a role in the rafts' stabilization required in particular to facilitate the recruitment of the partner proteins [121,122].

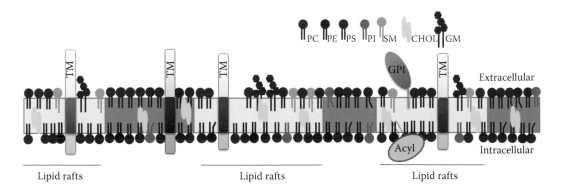

FIGURE 24.1 A simplified model of lipid rafts in cell membranes. The PL (PC in blue, PE in red, PS in purple, and phosphatidylinositol (PI) in orange) and cholesterol (CHOL) are distributed in both the leaflets, whereas sphingolipids (SM) and gangliosides (GM) are enriched in the outer leaflet of the bilayer. Rafts are specialized membrane domains containing high concentrations in cholesterol and sphingolipids. The acyl chains in lipids rafts are generally long and saturated (straight line in lipid tails), whereas those in non-raft domains are shorter and contain singly or multiply unsaturated acyl chains (oblique line in lipid tails). Raft domains contain concentrations of acylated (Acyl) and GPI-anchored (GPI) proteins, whereas TM proteins are associated with both non-raft and raft domains.

RAFT-SPECIFIC PROTEINS AND ASSOCIATED BRAIN FUNCTIONS

Caveolin and Flotillin

Caveolae were the first membrane microdomains identified and the only ones identifiable by their morphology and observable by microscopy. Caveolae are membrane invaginations, with a diameter of 25–150 nm, present in the Trans-Golgian network, exocytosis vesicles, ER, and plasma membrane [123]. Morphologically, they are abundant in the endothelium, muscle cells, adipocytes, and pulmonary epithelial cells [123–125]. Studies have also revealed that these structures are present in the central nervous system (CNS) [126,127]. Caveolae are structures rich in sphingolipids and cholesterol, as well as caveolin and cavine [107,108,128,129]. These caveolae are in particular involved in the transcytosis of molecules in the endothelial cells and in the endocytosis of bacterial toxins [104]. In the brain, the restrictive nature of the BBB requires cellular machinery to transfer and deliver macromolecules to the brain, which involved transcytosis for which the caveolae could play a role. Indeed, they interact in particular in the endocytosis of the plasma membrane receptors and their specific ligand such as insulin, transferrin, or lipoproteins [130], and in a small proportion, leptin [131].

Cells without caveolae have low density and detergent-resistant membrane domains [132]. These domains are characterized by the presence of specific proteins, the flotillins, which could represent a functional analog of caveolin [117]. These proteins were initially described as proteins of axonal regeneration. They exist under two very conserved isoforms, flotillin-1 and -2, suggesting an important cellular function [133]. Flotillins can insert into the plasma membrane and in endomembrane, which can also be favored upon palmitoylation or myristoylation depending on which isoform is modified [133,134]. Flotillin-1 has been described to associate with the membrane compartments of endocytes, phagosomes, Golgi, and nucleus [116–118]. The N-terminal domain of flotillin-1 contains two putative hydrophobic membrane binding sites allowing its association while the C-terminal domain contains an α helical region which may be involved in oligomerization or protein–protein interactions [117,133].

The rafts are structured and stabilized by the formation of flotillin homo- and oligomers which are sporadically distributed along the plasma membranes [118,135]. The flotillins are ubiquitously expressed and it has been reported that flotillin-1 is particularly present in the CNS, adipose and muscle tissues, and erythrocytes [117]. These proteins can be used as raft biomarkers [118]. They have shown that rafts have different characteristics and distributions depending on the cell type and stage of development [136]. For example, in mature neurons, rafts primarily accumulate in the membranes of the cell bodies and the postsynaptic axon [135,137]. Flotillins are associated not only with rafts but also with GPI proteins in the neurons [135,138], suggesting a role in signal transduction [107,133,135,138].

Role of Rafts in Neuronal Membranes

On the basis of their capacity to provide the proper environment to segregate functional groups of proteins, rafts are considered as key players in many cellular processes including endocytosis and signal transduction, being involved in particular in the endocytosis of the interleukin-2, high-affinity IgE, and insulin receptors [139]. The role of rafts in exocytosis has also been extensively reported. For example, the concentration of certain SNARE complex proteins [140] was reported to be up to 25 times higher in rafts [111]. Furthermore, rafts are thought to play an important role in synaptic signaling, as demonstrated by the enrichment of synaptic proteins, such as the synaptosomal-associated protein (SNAP) [141] and the postsynaptic density (PSD) in rafts of rat forebrain synaptic membranes and pheochromocytoma PC12 cells [142]. Cerebral and, in particular, neuronal function can also be affected due to raft-dependent neurotransmitter transporter activity and trafficking, as is the case for choline and serotonin in cells stably expressing the respective transporters [143,144]. Through lateral segregation of membrane-associated proteins, rafts also play a role in the regulation of protein–protein interactions. On the one hand, clustering of proteins within rafts statistically favors their interaction. On the other hand, inclusion of a particular protein within rafts prevents its interaction with proteins located outside of rafts or in distinct subpopulations of rafts. This latter situation results in the inhibition of the signaling complex assembly and subsequent activation of cascade events [145,146].

Recruitment of receptors into rafts can be necessary for activation of specific signaling pathways, as in the case of the activation of the TrkA receptor by nerve growth factor (NGF), or TrkB receptor by brain-derived neurotrophic factor (BDNF) [147–149]. Other examples point toward ligand-induced relocation of membrane receptors in various cell types. Indeed, the epidermal growth factor (EGF) receptor locates outside of rafts when activated by its ligand, as shown in b82 and NR6 mouse fibroblasts [150], whereas the insulin growth factor (IGF) receptor is recruited into rafts upon treatment of human hepatoma HuH7 cells by insulin [151]. Moreover, the NGF-activated (phosphorylated) and palmitoylated TrkA and p75^{NTR} receptors are clustered in the caveolae-like domains of the plasma membrane in mouse 3T3 fibroblasts and in PC12 cells treated with their ligand, but not after caveolae disruption by filipin [152]. Recently, SH-SY5Y cells were used as dopaminergic neurons to demonstrate that the neuroprotective effect of *glial cell line-derived neurotrophic factor* (GDNF) is influenced by the relocation of NCAM-140 into rafts, which can be suppressed by palmitoylation inhibition [153].

In summary, rafts are involved in the regulation of numerous cellular processes, such as the activation of receptors, signal transduction, and signaling pathways. For this reason, rafts are sometimes described as signalosomes, and thus as recruitment, or interaction, platforms for receptors activated by ligand binding [154,155]. Considering the central roles played by rafts in neuronal signaling processes, particularly in synaptic function, modifications in the properties of rafts can thus be expected to adversely affect neuronal function.

IMPACT OF AGING ON NEURONAL MEMBRANES

Cerebral aging is generally acknowledged as a complex process involving multiple factors (for a review, see References [156–158]). In particular, long-term exposure to nutritional risk factors can cause variations in molecular composition and concentration of cerebral lipid membranes, which in turn lead to neuronal vulnerability and the development of pathological conditions. On the basis of our previous work and the literature, we have developed the hypothesis that these alterations affect the organization of the cerebral membrane and particularly the raft functionalities. For example, the membrane lipid composition of the frontal cortex was studied in 20–100-year-old human subjects with no reported neurological disorders or psychiatric disease at the time of death [159]. The authors reported a progressive loss of cerebral membrane lipid after the age of 20 years and an accentuated deficit after the age of 80 years. In individuals aged over 20 years, total PL, cholesterol, cerebrosides, and sulfatides decrease in a curvilinear manner as a function of time. Ganglioside ratios change as well, with a decrease in GM1 and an increase in GM3 levels [159]. More recently, numerous lipid compositional analyses focused on different zones of human and rodent brain, as well as primary cultures of neurons, and confirmed lipid alterations associated with aging (Table 24.3). In light of the impact of lipid composition on phase separation behavior of bilayer membranes, it appears evident that age-associated variations in lipid composition would lead to modifications of the physicochemical properties of membranes and provide a mechanistic explanation for age-related disturbances of raft functionality, particularly in synaptic membranes, potentially contributing to neuropathological events [160].

The capacity of synaptic vesicles to fuse with the synaptic membrane is a determinant parameter for neurotransmitter release. It requires significant structural changes both in vesicle and presynaptic membranes. Implication of particular fusogenic PL, such as phosphatidylinositol-4,5-bisphosphate and phosphatidic acid, was reported at the different steps (site selection, priming, and fusion) of the exocytotic process by affecting membrane topology [172].

Cholesterol plays an important role in the fusion process for defining exocytosis sites. While cholesterol promotes negative membrane curvature, its depletion leads to dose-dependent inhibition of vesicle fusion kinetics and Ca^{2+} sensitivity [173,174], as well as to lower extent of regulated

TABLE 24.3

Lipid Changes in Human Brain during Aging

Lipid	Effect of Aging	References
Cholesterol	Decreased	[161–163]
Ceramide	Increased	[164]
Ganglioside	Lower GM1, higher GM3	[163,165]
Sphingosine	Higher C20:C18	[166,167]
Phospholipids	Decreased	[97,161,163]
Saturated FA	Decreased	[168]
Monounsaturated FA	Increased	[168]
Polyunsaturated FA	Decreased	[97,168–171]

dopamine exocytosis from PC12 cells [141]. Cholesterol depletion not only limits the ability of synaptic membranes to fuse [175], but also impacts synaptic transmission and cerebral function [174]. Likewise, sphingolipids are enriched in presynaptic membranes and influence the fusion kinetics of synaptic vesicles. In addition, an appropriate SM-to-cholesterol ratio appears to be important for the vesicular fusion process, since an increase in SM dramatically decreases fusion efficiency [176,177]. On the basis of liposome models with various lipids including PL (dioleoyl forms of PC, PE, and phosphatidylserine [PS]), SM, and cholesterol (each of them playing a part in optimizing the membrane for fusion), it was reported that the natural lipid composition of synaptic vesicles is very close to the optimal one with respect to fusogenicity [176], although this process could also be influenced by membrane-associated proteins such as the SNARE proteins. Accordingly, SM concentration is observed to increase in aging, which affects the fusion capacity of the synaptic membrane [178]. In addition, the structure and hydrophilic properties of gangliosides provide a strong membrane curvature capability that plays a vital role in neurotransmitter release from synaptosomes [179]. Therefore, the decrease in brain concentrations of both SM and gangliosides, generally observed over the course of aging, leads to a progressive decline in curvature and fusion capacities of synaptic membranes [180]. Moreover, a local increase in cholesterol, associated with caveolin-1 oligomerization, was observed in high-density rafts of brains from old rats [181]. Also, by studying protein distribution in the raft fractions of rat brain, it was demonstrated that protein concentration in rafts undergoes significant modification during the process of aging [182], although the overall protein concentration in the membrane remains unaltered. Furthermore, the aged tissues displayed a reduced antioxidant activity and lower levels of proteins involved in neurotransmitter secretion, such as synapsin, as well as in cytoskeleton-bound synaptic proteins [110].

The assembly of the SNARE complex was demonstrated to be negatively impacted as a result of aging-related decrease of PUFA in non-raft domains of the synaptic membrane [175,183]. Conceivably, this decrease could be explained by higher peroxidation and lower incorporation of PUFA, which can lead to subsequently disturbed synaptic transmission as a consequence, as suggested by the correlation between lipid peroxidation and AD [184]. Peroxidation can modify the physicochemical properties of membranes, leading not only to an increase in membrane permeability, TM diffusion, and loss of leaflet asymmetry, but also to modified phase separation, which could account for the alteration of raft formation and properties. In turn, these effects are likely to interfere with membrane-associated protein function, which would ultimately influence neuron-specific pathways and promote brain aging.

CONCLUSIONS

Lipids are the fundamental components of biological membranes. A wide variety of alterations in brain lipid composition have been correlated with normal aging process as with pathological aging, and it is still difficult to explain and even to establish a clear causal link between these numerous and complex modifications and cognitive decline. One of the mechanisms to explain these effects is the influence of lipids on the architecture and fluidity of neuronal membranes. The lipid composition of brain membranes is therefore a crucial parameter, especially regarding *n-3* PUFA levels. As the predominant PUFA in neuronal membranes, DHA plays a major role in neurotransmission, synaptic plasticity, neurotrophic factor response, cognitive functions, and neuronal survival. There is also evidence that DHA can influence the physicochemical properties of the lipid bilayer as well as on the lateral segregation of proteins in the lipid rafts. This suggests that maintaining the integrity of membrane microdomains may represent a strategy for preventing or slowing neuronal dysfunctions and brain vulnerability associated with aging.

REFERENCES

1. A. Ferrara, E. Barrett-Connor, J. Shan, Total, LDL, and HDL cholesterol decrease with age in older men and women. The Rancho Bernardo Study 1984–1994, *Circulation*. 96, 1997, 37–43.
2. M. Kuzuya, F. Ando, A. Iguchi, H. Shimokata, Changes in serum lipid levels during a 10 year period in a large Japanese population. A cross-sectional and longitudinal study, *Atherosclerosis*. 163, 2002, 313–320.
3. R.H. Houtkooper, C. Argmann, S.M. Houten, C. Cantó, E.H. Jeninga, P.A. Andreux, C. Thomas, R. Doenlen, K. Schoonjans, J. Auwerx, The metabolic footprint of aging in mice, *Sci. Rep.* 1, 2011, 134. doi:10.1038/srep00134.
4. H.N. Ginsberg, Y.-L. Zhang, A. Hernandez-Ono, Metabolic syndrome: Focus on dyslipidemia, *Obes. Silver Spring Md.* 14(Suppl 1), 2006, 41S–49S. doi:10.1038/oby.2006.281.
5. J.-P. Després, I. Lemieux, Abdominal obesity and metabolic syndrome, *Nature*. 444, 2006, 881–887. doi:10.1038/nature05488.
6. G.M. Pasinetti, J.A. Eberstein, Metabolic syndrome and the role of dietary lifestyles in Alzheimer's disease, *J. Neurochem.* 106, 2008, 1503–1514. doi:10.1111/j.1471-4159.2008.05454.x.
7. C. Ballard, S. Gauthier, A. Corbett, C. Brayne, D. Aarsland, E. Jones, Alzheimer's disease, *Lancet*. 377, 2011, 1019–1031. doi:10.1016/S0140-6736(10)61349-9.
8. M. Maesako, K. Uemura, M. Kubota, A. Kuzuya, K. Sasaki, N. Hayashida, M. Asada-Utsugi et al. Exercise is more effective than diet control in preventing high fat diet-induced β-amyloid deposition and memory deficit in amyloid precursor protein transgenic mice, *J. Biol. Chem.* 287, 2012, 23024–23033. doi:10.1074/jbc.M112.367011.
9. W. Ma, L. Yuan, H. Yu, Y. Xi, R. Xiao, Mitochondrial dysfunction and oxidative damage in the brain of diet-induced obese rats but not in diet-resistant rats, *Life Sci*. 110, 2014, 53–60. doi:10.1016/j.lfs.2014.07.018.
10. R.A. Whitmer, The epidemiology of adiposity and dementia, *Curr. Alzheimer Res.* 4, 2007, 117–122.
11. K. Balakrishnan, G. Verdile, P.D. Mehta, J. Beilby, D. Nolan, D.A. Galvão, R. Newton, S.E. Gandy, R.N. Martins, Plasma Abeta42 correlates positively with increased body fat in healthy individuals, *J. Alzheimers Dis. JAD.* 8, 2005, 269–282.
12. D. Cao, H. Lu, T.L. Lewis, L. Li, Intake of sucrose-sweetened water induces insulin resistance and exacerbates memory deficits and amyloidosis in a transgenic mouse model of Alzheimer disease, *J. Biol. Chem.* 282, 2007, 36275–36282. doi:10.1074/jbc.M703561200.
13. C. Boitard, A. Cavaroc, J. Sauvant, A. Aubert, N. Castanon, S. Layé, G. Ferreira, Impairment of hippocampal-dependent memory induced by juvenile high-fat diet intake is associated with enhanced hippocampal inflammation in rats, *Brain. Behav. Immun.* 40, 2014, 9–17. doi:10.1016/j.bbi.2014.03.005.
14. E.M. Knight, I.V. Martins, S. Gümüşgöz, S.M. Allan, C.B. Lawrence, High-fat diet-induced memory impairment in triple-transgenic Alzheimer's disease (3xTgAD) mice is independent of changes in amyloid and tau pathology, *Neurobiol. Aging*. 35, 2014, 1821–1832. doi:10.1016/j.neurobiolaging.2014.02.010.
15. D. Gustafson, E. Rothenberg, K. Blennow, B. Steen, I. Skoog, An 18-year follow-up of overweight and risk of Alzheimer disease, *Arch. Intern. Med.* 163, 2003, 1524–1528. doi:10.1001/archinte.163.13.1524.
16. D.K. Johnson, C.H. Wilkins, J.C. Morris, Accelerated weight loss may precede diagnosis in Alzheimer disease, *Arch. Neurol.* 63, 2006, 1312–1317. doi:10.1001/archneur.63.9.1312.
17. J.A. Luchsinger, B. Patel, M.-X. Tang, N. Schupf, R. Mayeux, Measures of adiposity and dementia risk in elderly persons, *Arch. Neurol.* 64, 2007, 392–398. doi:10.1001/archneur.64.3.392.

18. A.R. Atti, K. Palmer, S. Volpato, B. Winblad, D. De Ronchi, L. Fratiglioni, Late-life body mass index and dementia incidence: Nine-year follow-up data from the Kungsholmen Project, *J. Am. Geriatr. Soc.* 56, 2008, 111–116. doi:10.1111/j.1532-5415.2007.01458.x.

19. A.L. Fitzpatrick, L.H. Kuller, O.L. Lopez, P. Diehr, E.S. O'Meara, W.T. Longstreth, J.A. Luchsinger, Midlife and late-life obesity and the risk of dementia: Cardiovascular health study, *Arch. Neurol.* 66, 2009, 336–342. doi:10.1001/archneurol.2008.582.

20. B.G. Windham, S.T. Lirette, M. Fornage, E.J. Benjamin, K.G. Parker, S.T. Turner, C.R. Jack, M.E. Griswold, T.H. Mosley, Associations of brain structure with adiposity and changes in adiposity in a middle-aged and older biracial population, *J. Gerontol. A. Biol. Sci. Med. Sci.* 2016. doi:10.1093/gerona/glw239.

21. L. Ronan, A.F. Alexander-Bloch, K. Wagstyl, S. Farooqi, C. Brayne, L.K. Tyler, Cam-CAN, P.C. Fletcher, Obesity associated with increased brain age from midlife, *Neurobiol. Aging.* 47, 2016, 63–70. doi:10.1016/j.neurobiolaging.2016.07.010.

22. B.M. Bettcher, C.M. Walsh, C. Watson, J.W. Miller, R. Green, N. Patel, B.L. Miller, J. Neuhaus, K. Yaffe, J.H. Kramer, Body mass and white matter integrity: The influence of vascular and inflammatory markers, *PloS One.* 8, 2013, e77741. doi:10.1371/journal.pone.0077741.

23. K.M. Stanek, S.M. Grieve, A.M. Brickman, M.S. Korgaonkar, R.H. Paul, R.A. Cohen, J.J. Gunstad, Obesity is associated with reduced white matter integrity in otherwise healthy adults, *Obes. Silver Spring Md.* 19, 2011, 500–504. doi:10.1038/oby.2010.312.

24. A.D. de Kloet, D.J. Pioquinto, D. Nguyen, L. Wang, J.A. Smith, H. Hiller, C. Sumners, Obesity induces neuroinflammation mediated by altered expression of the renin-angiotensin system in mouse forebrain nuclei, *Physiol. Behav.* 136, 2014, 31–38. doi:10.1016/j.physbeh.2014.01.016.

25. N.Z. Gerges, A.M. Aleisa, K.A. Alkadhi, Impaired long-term potentiation in obese zucker rats: Possible involvement of presynaptic mechanism, *Neuroscience.* 120, 2003, 535–539.

26. A.M. Stranahan, E.D. Norman, K. Lee, R.G. Cutler, R.S. Telljohann, J.M. Egan, M.P. Mattson, Diet-induced insulin resistance impairs hippocampal synaptic plasticity and cognition in middle-aged rats, *Hippocampus.* 18, 2008, 1085–1088. doi:10.1002/hipo.20470.

27. J. Wang, D. Freire, L. Knable, W. Zhao, B. Gong, P. Mazzola, L. Ho, S. Levine, G.M. Pasinetti, Childhood and adolescent obesity and long-term cognitive consequences during aging, *J. Comp. Neurol.* 523, 2015, 757–768. doi:10.1002/cne.23708.

28. E. Velkoska, T.J. Cole, M.J. Morris, Early dietary intervention: Long-term effects on blood pressure, brain neuropeptide Y, and adiposity markers, *Am. J. Physiol. Endocrinol. Metab.* 288, 2005, E1236–E1243. doi:10.1152/ajpendo.00505.2004.

29. G.N. Bischof, D.C. Park, Obesity and aging: Consequences for cognition, brain structure, and brain function, *Psychosom. Med.* 77, 2015, 697–709. doi:10.1097/PSY.0000000000000212.

30. Z. Tucsek, P. Toth, S. Tarantini, D. Sosnowska, T. Gautam, J.P. Warrington, C.B. Giles et al., Aging exacerbates obesity-induced cerebromicrovascular rarefaction, neurovascular uncoupling, and cognitive decline in mice, *J. Gerontol. A. Biol. Sci. Med. Sci.* 69, 2014, 1339–1352. doi:10.1093/gerona/glu080.

31. Z. Tucsek, P. Toth, D. Sosnowska, T. Gautam, M. Mitschelen, A. Koller, G. Szalai, W.E. Sonntag, Z. Ungvari, A. Csiszar, Obesity in aging exacerbates blood-brain barrier disruption, neuroinflammation, and oxidative stress in the mouse hippocampus: Effects on expression of genes involved in beta-amyloid generation and Alzheimer's disease, *J. Gerontol. A. Biol. Sci. Med. Sci.* 69, 2014, 1212–1226. doi:10.1093/gerona/glt177.

32. M. Zamboni, A.P. Rossi, F. Fantin, G. Zamboni, S. Chirumbolo, E. Zoico, G. Mazzali, Adipose tissue, diet and aging, *Mech. Ageing Dev.* 2013. doi:10.1016/j.mad.2013.11.008.

33. J. Harvey, Leptin regulation of neuronal excitability and cognitive function, *Curr. Opin. Pharmacol.* 7, 2007, 643–647. doi:10.1016/j.coph.2007.10.006.

34. B.L. Tang, Leptin as a neuroprotective agent, *Biochem. Biophys. Res. Commun.* 368, 2008, 181–185. doi:10.1016/j.bbrc.2008.01.063.

35. C.D. Morrison, Leptin signaling in brain: A link between nutrition and cognition? *Biochim. Biophys. Acta.* 1792, 2009, 401–408. doi:10.1016/j.bbadis.2008.12.004.

36. S.J. Greco, S. Sarkar, J.M. Johnston, X. Zhu, B. Su, G. Casadesus, J.W. Ashford, M.A. Smith, N. Tezapsidis, Leptin reduces Alzheimer's disease-related tau phosphorylation in neuronal cells, *Biochem. Biophys. Res. Commun.* 376, 2008, 536–541. doi:10.1016/j.bbrc.2008.09.026.

37. D.C. Fewlass, K. Noboa, F.X. Pi-Sunyer, J.M. Johnston, S.D. Yan, N. Tezapsidis, Obesity-related leptin regulates Alzheimer's Abeta, *FASEB J. Off. Publ. Fed. Am. Soc. Exp. Biol.* 18, 2004, 1870–1878. doi:10.1096/fj.04-2572com.

38. B. Beck, G. Pourié, Ghrelin, neuropeptide Y, and other feeding-regulatory peptides active in the hippocampus: Role in learning and memory, *Nutr. Rev.* 71, 2013, 541–561. doi:10.1111/nure.12045.

39. L.M. Refolo, B. Malester, J. LaFrancois, T. Bryant-Thomas, R. Wang, G.S. Tint, K. Sambamurti, K. Duff, M.A. Pappolla, Hypercholesterolemia accelerates the Alzheimer's amyloid pathology in a transgenic mouse model, *Neurobiol. Dis.* 7, 2000, 321–331. doi:10.1006/nbdi.2000.0304.

40. H.M. Francis, M. Mirzaei, M.C. Pardey, P.A. Haynes, J.L. Cornish, Proteomic analysis of the dorsal and ventral hippocampus of rats maintained on a high fat and refined sugar diet, *Proteomics.* 13, 2013, 3076–3091. doi:10.1002/pmic.201300124.

41. M.H. Eskelinen, T. Ngandu, E.-L. Helkala, J. Tuomilehto, A. Nissinen, H. Soininen, M. Kivipelto, Fat intake at midlife and cognitive impairment later in life: A population-based CAIDE study, *Int. J. Geriatr. Psychiatry.* 23, 2008, 741–747. doi:10.1002/gps.1969.

42. D. Knopman, L.L. Boland, T. Mosley, G. Howard, D. Liao, M. Szklo, P. McGovern, A.R. Folsom, Atherosclerosis Risk in Communities (ARIC) Study Investigators. Cardiovascular risk factors and cognitive decline in middle-aged adults, *Neurology.* 56, 2001, 42–48.

43. M.C. Morris, D.A. Evans, J.L. Bienias, C.C. Tangney, D.A. Bennett, N. Aggarwal, J. Schneider, R.S. Wilson, Dietary fats and the risk of incident Alzheimer disease, *Arch. Neurol.* 60, 2003, 194–200.

44. W.B. Grant, A. Campbell, R.F. Itzhaki, J. Savory, The significance of environmental factors in the etiology of Alzheimer's disease, *J. Alzheimers Dis. JAD.* 4, 2002, 179–189.

45. S.-L. Niu, D.C. Mitchell, B.J. Litman, Trans fatty acid derived phospholipids show increased membrane cholesterol and reduced receptor activation as compared to their cis analogs, *Biochemistry (Mosc.).* 44, 2005, 4458–4465. doi:10.1021/bi048319+.

46. A. Phivilay, C. Julien, C. Tremblay, L. Berthiaume, P. Julien, Y. Giguère, F. Calon, High dietary consumption of trans fatty acids decreases brain docosahexaenoic acid but does not alter amyloid-beta and tau pathologies in the 3xTg-AD model of Alzheimer's disease, *Neuroscience.* 159, 2009, 296–307. doi:10.1016/j.neuroscience.2008.12.006.

47. I.A. Brouwer, A.J. Wanders, M.B. Katan, Effect of animal and industrial trans fatty acids on HDL and LDL cholesterol levels in humans—A quantitative review, *PLoS ONE.* 5, 2010. doi:10.1371/journal.pone.0009434.

48. J.E. Beilharz, J. Maniam, M.J. Morris, Short exposure to a diet rich in both fat and sugar or sugar alone impairs place, but not object recognition memory in rats, *Brain. Behav. Immun.* 37, 2014, 134–141. doi:10.1016/j.bbi.2013.11.016.

49. L. Li, D. Cao, D.W. Garber, H. Kim, K. Fukuchi, Association of aortic atherosclerosis with cerebral beta-amyloidosis and learning deficits in a mouse model of Alzheimer's disease, *Am. J. Pathol.* 163, 2003, 2155–2164.

50. C. Julien, C. Tremblay, A. Phivilay, L. Berthiaume, V. Emond, P. Julien, F. Calon, High-fat diet aggravates amyloid-beta and tau pathologies in the 3xTg-AD mouse model, *Neurobiol. Aging.* 31, 2010, 1516–1531. doi:10.1016/j.neurobiolaging.2008.08.022.

51. T.L. Davidson, S.L. Hargrave, S.E. Swithers, C.H. Sample, X. Fu, K.P. Kinzig, W. Zheng, Inter-relationships among diet, obesity and hippocampal-dependent cognitive function, *Neuroscience.* 253, 2013, 110–122. doi:10.1016/j.neuroscience.2013.08.044.

52. S.H. Park, J.H. Kim, K.H. Choi, Y.J. Jang, S.S. Bae, B.T. Choi, H.K. Shin, Hypercholesterolemia accelerates amyloid β-induced cognitive deficits, *Int. J. Mol. Med.* 31, 2013, 577–582. doi:10.3892/ijmm.2013.1233.

53. E.L.G. Moreira, J. de Oliveira, D.F. Engel, R. Walz, A.F. de Bem, M. Farina, R.D.S. Prediger, Hypercholesterolemia induces short-term spatial memory impairments in mice: Up-regulation of acetylcholinesterase activity as an early and causal event?, *J. Neural Transm. Vienna Austria 1996.* 121, 2014, 415–426. doi:10.1007/s00702-013-1107-9.

54. Z. Vucetic, J.L. Carlin, K. Totoki, T.M. Reyes, Epigenetic dysregulation of the dopamine system in diet-induced obesity, *J. Neurochem.* 120, 2012, 891–898. doi:10.1111/j.1471-4159.2012.07649.x.

55. J. Carlin, T.E. Hill-Smith, I. Lucki, T.M. Reyes, Reversal of dopamine system dysfunction in response to high-fat diet, *Obes. Silver Spring Md.* 21, 2013, 2513–2521. doi:10.1002/oby.20374.

56. T.M. du Bois, C. Deng, W. Bell, X.-F. Huang, Fatty acids differentially affect serotonin receptor and transporter binding in the rat brain, *Neuroscience.* 139, 2006, 1397–1403. doi:10.1016/j.neuroscience.2006.02.068.

57. K. Raider, D. Ma, J.L. Harris, I. Fuentes, R.S. Rogers, J.L. Wheatley, P.C. Geiger et al., A high fat diet alters metabolic and bioenergetic function in the brain: A magnetic resonance spectroscopy study, *Neurochem. Int.* 97, 2016, 172–180. doi:10.1016/j.neuint.2016.04.008.

58. V.T. Dias, F. Trevizol, R.C.S. Barcelos, F.T. Kunh, K. Roversi, K. Roversi, A.J. Schuster et al., Lifelong consumption of trans fatty acids promotes striatal impairments on Na(+)/K(+) ATPase

activity and BDNF mRNA expression in an animal model of mania, *Brain Res. Bull.* 118, 2015, 78–81. doi:10.1016/j.brainresbull.2015.09.005.

59. A.-C. Granholm, H.A. Bimonte-Nelson, A.B. Moore, M.E. Nelson, L.R. Freeman, K. Sambamurti, Effects of a saturated fat and high cholesterol diet on memory and hippocampal morphology in the middle-aged rat, *J. Alzheimers Dis. JAD.* 14, 2008, 133–145.

60. B. Diaz, L. Fuentes-Mera, A. Tovar, T. Montiel, L. Massieu, H.G. Martínez-Rodríguez, A. Camacho, Saturated lipids decrease mitofusin 2 leading to endoplasmic reticulum stress activation and insulin resistance in hypothalamic cells, *Brain Res.* 1627, 2015, 80–89. doi:10.1016/j.brainres.2015.09.014.

61. A. Camacho, S. Rodriguez-Cuenca, M. Blount, X. Prieur, N. Barbarroja, M. Fuller, G.E. Hardingham, A. Vidal-Puig, Ablation of PGC1 beta prevents mTOR dependent endoplasmic reticulum stress response, *Exp. Neurol.* 237, 2012, 396–406. doi:10.1016/j.expneurol.2012.06.031.

62. H.-W. Kim, J.S. Rao, S.I. Rapoport, M. Igarashi, Regulation of rat brain polyunsaturated fatty acid (PUFA) metabolism during graded dietary n-3 PUFA deprivation, *Prostaglandins Leukot. Essent. Fatty Acids.* 85, 2011, 361–368. doi:10.1016/j.plefa.2011.08.002.

63. R. Takechi, M.M. Pallebage-Gamarallage, V. Lam, C. Giles, J.C. Mamo, Aging-related changes in blood-brain barrier integrity and the effect of dietary fat, *Neurodegener. Dis.* 12, 2013, 125–135. doi:10.1159/000343211.

64. X. Zhang, F. Dong, J. Ren, M.J. Driscoll, B. Culver, High dietary fat induces NADPH oxidase-associated oxidative stress and inflammation in rat cerebral cortex, *Exp. Neurol.* 191, 2005, 318–325. doi:10.1016/j.expneurol.2004.10.011.

65. J. Kim, Y.-J. Park, Y. Jang, Y.H. Kwon, AMPK activation inhibits apoptosis and tau hyperphosphorylation mediated by palmitate in SH-SY5Y cells, *Brain Res.* 1418, 2011, 42–51. doi:10.1016/j.brainres.2011.08.059.

66. L.R. Freeman, V. Haley-Zitlin, C. Stevens, A.-C. Granholm, Diet-induced effects on neuronal and glial elements in the middle-aged rat hippocampus, *Nutr. Neurosci.* 14, 2011, 32–44. doi:10.1179/1743132 11X12966635733358.

67. C.D. Morrison, P.J. Pistell, D.K. Ingram, W.D. Johnson, Y. Liu, S.O. Fernandez-Kim, C.L. White, M.N. Purpera, R.M. Uranga, A.J. Bruce-Keller, J.N. Keller, High fat diet increases hippocampal oxidative stress and cognitive impairment in aged mice: Implications for decreased Nrf2 signaling, *J. Neurochem.* 114, 2010, 1581–1589. doi:10.1111/j.1471-4159.2010.06865.x.

68. R.M. Uranga, A.J. Bruce-Keller, C.D. Morrison, S.O. Fernandez-Kim, P.J. Ebenezer, L. Zhang, K. Dasuri, J.N. Keller, Intersection between metabolic dysfunction, high fat diet consumption, and brain aging, *J. Neurochem.* 114, 2010, 344–361. doi:10.1111/j.1471-4159.2010.06803.x.

69. J.P. Kesby, J.J. Kim, M. Scadeng, G. Woods, D.M. Kado, J.M. Olefsky, D.V. Jeste, C.L. Achim, S. Semenova, Spatial cognition in adult and aged mice exposed to high-fat diet, *PloS One.* 10, 2015, e0140034. doi:10.1371/journal.pone.0140034.

70. A. Ledreux, X. Wang, M. Schultzberg, A.-C. Granholm, L.R. Freeman, Detrimental effects of a high fat/high cholesterol diet on memory and hippocampal markers in aged rats, *Behav. Brain Res.* 312, 2016, 294–304. doi:10.1016/j.bbr.2016.06.012.

71. A. Gonzalo-Ruiz, J.L. Pérez, J.M. Sanz, C. Geula, J. Arévalo, Effects of lipids and aging on the neurotoxicity and neuronal loss caused by intracerebral injections of the amyloid-beta peptide in the rat, *Exp. Neurol.* 197, 2006, 41–55. doi:10.1016/j.expneurol.2005.06.008.

72. T. Pancani, K.L. Anderson, L.D. Brewer, I. Kadish, C. DeMoll, P.W. Landfield, E.M. Blalock, N.M. Porter, O. Thibault, Effect of high-fat diet on metabolic indices, cognition, and neuronal physiology in aging F344 rats, *Neurobiol. Aging.* 34, 2013, 1977–1987. doi:10.1016/j.neurobiolaging.2013.02.019.

73. B. Langelier, A. Linard, C. Bordat, M. Lavialle, C. Heberden, Long chain-polyunsaturated fatty acids modulate membrane phospholipid composition and protein localization in lipid rafts of neural stem cell cultures, *J. Cell. Biochem.* 110, 2010, 1356–1364. doi:10.1002/jcb.22652.

74. S.R. Wassall, W. Stillwell, Polyunsaturated fatty acid-cholesterol interactions: Domain formation in membranes, *Biochim. Biophys. Acta.* 1788, 2009, 24–32. doi:10.1016/j.bbamem.2008.10.011.

75. W. Stillwell, S.R. Wassall, Docosahexaenoic acid: Membrane properties of a unique fatty acid, *Chem. Phys. Lipids.* 126, 2003, 1–27.

76. S.E. Feller, K. Gawrisch, A.D. MacKerell, Polyunsaturated fatty acids in lipid bilayers: Intrinsic and environmental contributions to their unique physical properties, *J. Am. Chem. Soc.* 124, 2002, 318–326.

77. D.C. Mitchell, B.J. Litman, Molecular order and dynamics in bilayers consisting of highly polyunsaturated phospholipids, *Biophys. J.* 74, 1998, 879–891. doi:10.1016/S0006-3495(98)74011-1.

78. P. Yaqoob, S.R. Shaikh, The nutritional and clinical significance of lipid rafts, *Curr. Opin. Clin. Nutr. Metab. Care.* 13, 2010, 156–166. doi:10.1097/MCO.0b013e328335725b.

79. M.R. Brzustowicz, V. Cherezov, M. Caffrey, W. Stillwell, S.R. Wassall, Molecular organization of cholesterol in polyunsaturated membranes: Microdomain formation, *Biophys. J.* 82, 2002, 285–298. doi:10.1016/S0006-3495(02)75394-0.

80. A.A. Farooqui, L.A. Horrocks, T. Farooqui, Deacylation and reacylation of neural membrane glycerophospholipids, *J. Mol. Neurosci. MN.* 14, 2000, 123–135.

81. A.A. Farooqui, W.-Y. Ong, L.A. Horrocks, Biochemical aspects of neurodegeneration in human brain: Involvement of neural membrane phospholipids and phospholipases A2, *Neurochem. Res.* 29, 2004, 1961–1977.

82. T. Kamada, T. Yamashita, Y. Baba, M. Kai, S. Setoyama, Y. Chuman, S. Otsuji, Dietary sardine oil increases erythrocyte membrane fluidity in diabetic patients, *Diabetes.* 35, 1986, 604–611.

83. C. Popp-Snijders, J.A. Schouten, J. van der Meer, E.A. van der Veen, Fatty fish-induced changes in membrane lipid composition and viscosity of human erythrocyte suspensions, *Scand. J. Clin. Lab. Invest.* 46, 1986, 253–258.

84. P.C. Calder, P. Yaqoob, D.J. Harvey, A. Watts, E.A. Newsholme, Incorporation of fatty acids by concanavalin A-stimulated lymphocytes and the effect on fatty acid composition and membrane fluidity, *Biochem. J.* 300(Pt 2), 1994, 509–518.

85. T. Said, J. Tremblay-Mercier, H. Berrougui, P. Rat, A. Khalil, Effects of vegetable oils on biochemical and biophysical properties of membrane retinal pigment epithelium cells, *Can. J. Physiol. Pharmacol.* 91, 2013, 812–817. doi:10.1139/cjpp-2013-0036.

86. L.E.M. Willemsen, M.A. Koetsier, M. Balvers, C. Beermann, B. Stahl, E.A.F. van Tol, Polyunsaturated fatty acids support epithelial barrier integrity and reduce IL-4 mediated permeability *in vitro*, *Eur. J. Nutr.* 47, 2008, 183–191. doi:10.1007/s00394-008-0712-0.

87. W.G. Jiang, R.P. Bryce, D.F. Horrobin, R.E. Mansel, Regulation of tight junction permeability and occludin expression by polyunsaturated fatty acids, *Biochem. Biophys. Res. Commun.* 244, 1998, 414–420. doi:10.1006/bbrc.1998.8288.

88. T.F. Galvao, R.J. Khairallah, E.R. Dabkowski, B.H. Brown, P.A. Hecker, K.A. O'Connell, K.M. O'Shea et al., Marine n3 polyunsaturated fatty acids enhance resistance to mitochondrial permeability transition in heart failure but do not improve survival, *Am. J. Physiol. Heart Circ. Physiol.* 304, 2013, H12–H21. doi:10.1152/ajpheart.00657.2012.

89. W. Ehringer, D. Belcher, S.R. Wassall, W. Stillwell, A comparison of the effects of linolenic (18:3 omega 3) and docosahexaenoic (22:6 omega 3) acids on phospholipid bilayers, *Chem. Phys. Lipids.* 54, 1990, 79–88.

90. J. Mazelova, N. Ransom, L. Astuto-Gribble, M.C. Wilson, D. Deretic, Syntaxin 3 and SNAP-25 pairing, regulated by omega-3 docosahexaenoic acid, controls the delivery of rhodopsin for the biogenesis of cilia-derived sensory organelles, the rod outer segments, *J. Cell Sci.* 122, 2009, 2003–2013. doi:10.1242/jcs.039982.

91. B.W. Koenig, H.H. Strey, K. Gawrisch, Membrane lateral compressibility determined by NMR and x-ray diffraction: Effect of acyl chain polyunsaturation. *Biophys. J.* 73, 1997, 1954–1966.

92. J.S. Poling, S. Vicini, M.A. Rogawski, N. Salem, Docosahexaenoic acid block of neuronal voltage-gated K+ channels: Subunit selective antagonism by zinc, *Neuropharmacology.* 35, 1996, 969–982.

93. J.X. Kang, A. Leaf, Evidence that free polyunsaturated fatty acids modify Na+ channels by directly binding to the channel proteins, *Proc. Natl. Acad. Sci. U. S. A.* 93, 1996, 3542–3546.

94. T.A. Kumosani, S.S. Moselhy, Modulatory effect of cod-liver oil on Na(+)-K(+) ATPase in rats' brain, *Hum. Exp. Toxicol.* 30, 2011, 267–274. doi:10.1177/0960327110371699.

95. Y.F. Xiao, A.M. Gomez, J.P. Morgan, W.J. Lederer, A. Leaf, Suppression of voltage-gated L-type Ca2+ currents by polyunsaturated fatty acids in adult and neonatal rat ventricular myocytes, *Proc. Natl. Acad. Sci. U. S. A.* 94, 1997, 4182–4187.

96. S.D. Kearns, M. Haag, The effect of omega-3 fatty acids on Ca-ATPase in rat cerebral cortex, *Fatty Acids.* 67, 2002, 303–308.

97. S.C. Dyall, G.J. Michael, R. Whelpton, A.G. Scott, A.T. Michael-Titus, Dietary enrichment with omega-3 polyunsaturated fatty acids reverses age-related decreases in the GluR2 and NR2B glutamate receptor subunits in rat forebrain, *Neurobiol. Aging.* 28, 2007, 424–439. doi:10.1016/j.neurobiolaging.2006.01.002.

98. H. Hamano, J. Nabekura, M. Nishikawa, T. Ogawa, Docosahexaenoic acid reduces GABA response in substantia nigra neuron of rat, *J. Neurophysiol.* 75, 1996, 1264–1270.

99. F. Pifferi, M. Jouin, J.M. Alessandri, U. Haedke, F. Roux, N. Perrière, I. Denis, M. Lavialle, P. Guesnet, n-3 Fatty acids modulate brain glucose transport in endothelial cells of the blood-brain barrier, *Prostaglandins Leukot. Essent. Fatty Acids.* 77, 2007, 279–286. doi:10.1016/j.plefa.2007.10.011.

100. S. Chalon, Omega-3 fatty acids and monoamine neurotransmission, *Prostaglandins Leukot. Essent. Fatty Acids.* 75, 2006, 259–269. doi:10.1016/j.plefa.2006.07.005.

101. M.L. Diaz, N. Fabelo, R. Marín, Genotype-induced changes in biophysical properties of frontal cortex lipid raft from APP/PS1 transgenic mice, *Front. Physiol.* 3, 2012, 454. doi:10.3389/fphys.2012.00454.

102. B.D. Rockett, A. Franklin, M. Harris, H. Teague, A. Rockett, S.R. Shaikh, Membrane raft organization is more sensitive to disruption by (n-3) PUFA than nonraft organization in EL4 and B cells, *J. Nutr.* 141, 2011, 1041–1048. doi:10.3945/jn.111.138750.

103. S.R. Shaikh, M. Edidin, Immunosuppressive effects of polyunsaturated fatty acids on antigen presentation by human leukocyte antigen class I molecules, *J. Lipid Res.* 48, 2007, 127–138. doi:10.1194/jlr.M600365-JLR200.

104. K. Simons, E. Ikonen, Functional rafts in cell membranes, *Nature.* 387, 1997, 569–572. doi:10.1038/42408.

105. C. Mencarelli, P. Martinez-Martinez, Ceramide function in the brain: When a slight tilt is enough, *Cell. Mol. Life Sci. CMLS.* 70, 2013, 181–203. doi:10.1007/s00018-012-1038-x.

106. A. Prinetti, V. Chigorno, G. Tettamanti, S. Sonnino, Sphingolipid-enriched membrane domains from rat cerebellar granule cells differentiated in culture. A compositional study, *J. Biol. Chem.* 275, 2000, 11658–11665.

107. L.J. Pike, Lipid rafts: Bringing order to chaos, *J. Lipid Res.* 44, 2003, 655–667. doi:10.1194/jlr.R200021-JLR200.

108. L.J. Pike, Lipid rafts: Heterogeneity on the high seas, *Biochem. J.* 378, 2004, 281–292. doi:10.1042/BJ20031672.

109. A.E. Cremesti, F.M. Goni, R. Kolesnick, Role of sphingomyelinase and ceramide in modulating rafts: Do biophysical properties determine biologic outcome?, *FEBS Lett.* 531, 2002, 47–53.

110. W.F.D. Bennett, D.P. Tieleman, Computer simulations of lipid membrane domains, *Biochim. Biophys. Acta.* 1828, 2013, 1765–1776. doi:10.1016/j.bbamem.2013.03.004.

111. M.F. Hanzal-Bayer, J.F. Hancock, Lipid rafts and membrane traffic, *FEBS Lett.* 581, 2007, 2098–2104. doi:10.1016/j.febslet.2007.03.019.

112. D.A. Brown, E. London, Structure and function of sphingolipid- and cholesterol-rich membrane rafts, *J. Biol. Chem.* 275, 2000, 17221–17224. doi:10.1074/jbc.R000005200.

113. F.R. Maxfield, D. Wüstner, Intracellular cholesterol transport, *J. Clin. Invest.* 110, 2002, 891–898. doi:10.1172/JCI16500.

114. K. Simons, J.L. Sampaio, Membrane organization and lipid rafts, *Cold Spring Harb. Perspect. Biol.* 3, 2011, a004697. doi:10.1101/cshperspect.a004697.

115. D.A. Brown, E. London, Structure of detergent-resistant membrane domains: Does phase separation occur in biological membranes?, *Biochem. Biophys. Res. Commun.* 240, 1997, 1–7. doi:10.1006/bbrc.1997.7575.

116. H. Kokubo, J.B. Helms, Y. Ohno-Iwashita, Y. Shimada, Y. Horikoshi, H. Yamaguchi, Ultrastructural localization of flotillin-1 to cholesterol-rich membrane microdomains, rafts, in rat brain tissue, *Brain Res.* 965, 2003, 83–90.

117. I.C. Morrow, R.G. Parton, Flotillins and the PHB domain protein family: Rafts, worms and anaesthetics, *Traffic Cph. Den.* 6, 2005, 725–740. doi:10.1111/j.1600-0854.2005.00318.x.

118. C.A.O. Stuermer, Microdomain-forming proteins and the role of the reggies/flotillins during axon regeneration in zebrafish, *Biochim. Biophys. Acta.* 1812, 2011, 415–422. doi:10.1016/j.bbadis.2010.12.004.

119. S. Mukherjee, F.R. Maxfield, Role of membrane organization and membrane domains in endocytic lipid trafficking, *Traffic Cph. Den.* 1, 2000, 203–211.

120. S.N. Whitehead, S. Gangaraju, A. Aylsworth, S.T. Hou, Membrane raft disruption results in neuritic retraction prior to neuronal death in cortical neurons, *Biosci. Trends.* 6, 2012, 183–191.

121. N. Föger, R. Marhaba, M. Zöller, Involvement of CD44 in cytoskeleton rearrangement and raft reorganization in T cells, *J. Cell Sci.* 114, 2001, 1169–1178.

122. K. Itoh, M. Sakakibara, S. Yamasaki, A. Takeuchi, H. Arase, M. Miyazaki, N. Nakajima, M. Okada, T. Saito, Cutting edge: Negative regulation of immune synapse formation by anchoring lipid raft to cytoskeleton through Cbp-EBP50-ERM assembly, *J. Immunol. Baltim. Md 1950.* 168, 2002, 541–544.

123. T. Fujimoto, H. Hagiwara, T. Aoki, H. Kogo, R. Nomura, Caveolae: From a morphological point of view, *J. Electron Microsc. (Tokyo).* 47, 1998, 451–460.

124. E.J. Smart, G.A. Graf, M.A. McNiven, W.C. Sessa, J.A. Engelman, P.E. Scherer, T. Okamoto, M.P. Lisanti, Caveolins, liquid-ordered domains, and signal transduction, *Mol. Cell. Biol.* 19, 1999, 7289–7304.

125. B.P. Head, H.H. Patel, P.A. Insel, Interaction of membrane/lipid rafts with the cytoskeleton: Impact on signaling and function: Membrane/lipid rafts, mediators of cytoskeletal arrangement and cell signaling, *Biochim. Biophys. Acta BBA Biomembr.* 1838, 2014, 532–545. doi:10.1016/j.bbamem.2013.07.018.

126. P.L. Cameron, J.W. Ruffin, R. Bollag, H. Rasmussen, R.S. Cameron, Identification of caveolin and caveolin-related proteins in the brain, *J. Neurosci. Off. J. Soc. Neurosci.* 17, 1997, 9520–9535.

127. T. Ikezu, H. Ueda, B.D. Trapp, K. Nishiyama, J.F. Sha, D. Volonte, F. Galbiati et al., Affinity-purification and characterization of caveolins from the brain: Differential expression of caveolin-1, -2, and -3 in brain endothelial and astroglial cell types, *Brain Res.* 804, 1998, 177–192.

128. M. Sargiacomo, P.E. Scherer, Z. Tang, E. Kübler, K.S. Song, M.C. Sanders, M.P. Lisanti, Oligomeric structure of caveolin: Implications for caveolae membrane organization, *Proc. Natl. Acad. Sci. U. S. A.* 92, 1995, 9407–9411.

129. C.M. Stary, Y.M. Tsutsumi, P.M. Patel, B.P. Head, H.H. Patel, D.M. Roth, Caveolins: Targeting pro-survival signaling in the heart and brain, *Front. Physiol.* 3, 2012, 393. doi:10.3389/fphys.2012.00393.

130. J.E. Preston, N. Joan Abbott, D.J. Begley, Transcytosis of macromolecules at the blood-brain barrier, *Adv. Pharmacol San Diego Calif.* 71, 2014, 147–163. doi:10.1016/bs.apha.2014.06.001.

131. H. Tu, H. Hsuchou, A.J. Kastin, X. Wu, W. Pan, Unique leptin trafficking by a tailless receptor, *FASEB J. Off. Publ. Fed. Am. Soc. Exp. Biol.* 24, 2010, 2281–2291. doi:10.1096/fj.09-143487.

132. L.J. Pike, X. Han, K.-N. Chung, R.W. Gross, Lipid rafts are enriched in arachidonic acid and plasmenyl-ethanolamine and their composition is independent of caveolin-1 expression: A quantitative electrospray ionization/mass spectrometric analysis, *Biochemistry (Mosc.)*. 41, 2002, 2075–2088.

133. C. Neumann-Giesen, B. Falkenbach, P. Beicht, S. Claasen, G. Lüers, C.A.O. Stuermer, V. Herzog, R. Tikkanen, Membrane and raft association of reggie-1/flotillin-2: Role of myristoylation, palmitoylation and oligomerization and induction of filopodia by overexpression, *Biochem. J.* 378, 2004, 509–518. doi:10.1042/BJ20031100.

134. L. Rajendran, M. Masilamani, S. Solomon, R. Tikkanen, C.A.O. Stuermer, H. Plattner, H. Illges, Asymmetric localization of flotillins/reggies in preassembled platforms confers inherent polarity to hematopoietic cells, *Proc. Natl. Acad. Sci. U. S. A.* 100, 2003, 8241–8246. doi:10.1073/pnas.1331629100.

135. C.A. Stuermer, D.M. Lang, F. Kirsch, M. Wiechers, S.O. Deininger, H. Plattner, Glycosylphosphatidyl inositol-anchored proteins and fyn kinase assemble in noncaveolar plasma membrane microdomains defined by reggie-1 and -2, *Mol. Biol. Cell.* 12, 2001, 3031–3045.

136. L. Colombaioni, M. Garcia-Gil, Sphingolipid metabolites in neural signalling and function, *Brain Res. Brain Res. Rev.* 46, 2004, 328–355. doi:10.1016/j.brainresrev.2004.07.014.

137. M. Edidin, The state of lipid rafts: From model membranes to cells, *Annu. Rev. Biophys. Biomol. Struct.* 32, 2003, 257–283. doi:10.1146/annurev.biophys.32.110601.142439.

138. L. Rajendran, K. Simons, Lipid rafts and membrane dynamics, *J. Cell Sci.* 118, 2005, 1099–1102. doi:10.1242/jcs.01681.

139. C. Lamaze, A. Dujeancourt, T. Baba, C.G. Lo, A. Benmerah, A. Dautry-Varsat, Interleukin 2 receptors and detergent-resistant membrane domains define a clathrin-independent endocytic pathway, *Mol. Cell.* 7, 2001, 661–671.

140. R. Jahn, R.H. Scheller, SNAREs—Engines for membrane fusion, *Nat. Rev. Mol. Cell Biol.* 7, 2006, 631–643. doi:10.1038/nrm2002.

141. C. Salaün, G.W. Gould, L.H. Chamberlain, Lipid raft association of SNARE proteins regulates exocytosis in PC12 cells, *J. Biol. Chem.* 280, 2005, 19449–19453. doi:10.1074/jbc.M501923200.

142. T. Suzuki, J. Zhang, S. Miyazawa, Q. Liu, M.R. Farzan, W.-D. Yao, Association of membrane rafts and postsynaptic density: Proteomics, biochemical, and ultrastructural analyses, *J. Neurochem.* 119, 2011, 64–77. doi:10.1111/j.1471-4159.2011.07404.x.

143. L.K. Cuddy, W. Winick-Ng, R.J. Rylett, Regulation of the high-affinity choline transporter activity and trafficking by its association with cholesterol-rich lipid rafts, *J. Neurochem.* 128, 2014, 725–740. doi:10.1111/jnc.12490.

144. F. Magnani, C.G. Tate, S. Wynne, C. Williams, J. Haase, Partitioning of the serotonin transporter into lipid microdomains modulates transport of serotonin, *J. Biol. Chem.* 279, 2004, 38770–38778. doi:10.1074/jbc.M400831200.

145. S. Sonnino, M. Aureli, S. Grassi, L. Mauri, S. Prioni, A. Prinetti, Lipid rafts in neurodegeneration and neuroprotection, *Mol. Neurobiol.* 2013. doi:10.1007/s12035-013-8614-4.

146. V.A.M. Villar, S. Cuevas, X. Zheng, P.A. Jose, Localization and signaling of GPCRs in lipid rafts, *Methods Cell Biol.* 132, 2016, 3–23. doi:10.1016/bs.mcb.2015.11.008.

147. S. Suzuki, T. Numakawa, K. Shimazu, H. Koshimizu, T. Hara, H. Hatanaka, L. Mei, B. Lu, M. Kojima, BDNF-induced recruitment of TrkB receptor into neuronal lipid rafts: Roles in synaptic modulation, *J. Cell Biol.* 167, 2004, 1205–1215. doi:10.1083/jcb.200404106.

148. D.B. Pereira, M.V. Chao, The tyrosine kinase Fyn determines the localization of TrkB receptors in lipid rafts, *J. Neurosci. Off. J. Soc. Neurosci.* 27, 2007, 4859–4869. doi:10.1523/JNEUROSCI.4587-06.2007.

149. S. Pryor, G. McCaffrey, L.R. Young, M.L. Grimes, NGF causes TrkA to specifically attract microtubules to lipid rafts, *PloS One*. 7, 2012, e35163. doi:10.1371/journal.pone.0035163.

150. C. Mineo, G.N. Gill, R.G. Anderson, Regulated migration of epidermal growth factor receptor from caveolae, *J. Biol. Chem*. 274, 1999, 30636–30643.

151. S. Vainio, S. Heino, J.-E. Mansson, P. Fredman, E. Kuismanen, O. Vaarala, E. Ikonen, Dynamic association of human insulin receptor with lipid rafts in cells lacking caveolae, *EMBO Rep*. 3, 2002, 95–100. doi:10.1093/embo-reports/kvf010.

152. C.S. Huang, J. Zhou, A.K. Feng, C.C. Lynch, J. Klumperman, S.J. DeArmond, W.C. Mobley, Nerve growth factor signaling in caveolae-like domains at the plasma membrane, *J. Biol. Chem*. 274, 1999, 36707–36714.

153. L. Li, H. Chen, M. Wang, F. Chen, J. Gao, S. Sun, Y. Li, D. Gao, NCAM-140 Translocation into lipid rafts mediates the neuroprotective effects of GDNF, *Mol. Neurobiol*. 2016. doi:10.1007/s12035-016-9749-x.

154. A.M. Sebastião, M. Colino-Oliveira, N. Assaife-Lopes, R.B. Dias, J.A. Ribeiro, Lipid rafts, synaptic transmission and plasticity: Impact in age-related neurodegenerative diseases, *Neuropharmacology*. 64, 2013, 97–107. doi:10.1016/j.neuropharm.2012.06.053.

155. J.H. Lorent, I. Levental, Structural determinants of protein partitioning into ordered membrane domains and lipid rafts, *Chem. Phys. Lipids*. 192, 2015, 23–32. doi:10.1016/j.chemphyslip.2015.07.022.

156. J. Zhang, Q. Liu, Cholesterol metabolism and homeostasis in the brain, *Protein Cell*. 6, 2015, 254–264. doi:10.1007/s13238-014-0131-3.

157. Z. Xue-shan, P. juan, W. Qi, R. Zhong, P. Li-hong, T. Zhi-han, J. Zhi-sheng, W. Gui-xue, L. Lu-shan, Imbalanced cholesterol metabolism in Alzheimer's disease, *Clin. Chim. Acta*. 456, 2016, 107–114. doi:10.1016/j.cca.2016.02.024.

158. H.-L. Wang, Y.-Y. Wang, X.-G. Liu, S.-H. Kuo, N. Liu, Q.-Y. Song, M.-W. Wang, Cholesterol, 24-Hydroxycholesterol, and 27-Hydroxycholesterol as surrogate biomarkers in cerebrospinal fluid in mild cognitive impairment and Alzheimer's disease: A meta-analysis, *J. Alzheimers Dis. JAD*. 51, 2016, 45–55. doi:10.3233/JAD-150734.

159. J.A. Conquer, M.C. Tierney, J. Zecevic, W.J. Bettger, R.H. Fisher, Fatty acid analysis of blood plasma of patients with Alzheimer's disease, other types of dementia, and cognitive impairment, *Lipids*. 35, 2000, 1305–1312.

160. J. Thomas, C.J. Thomas, J. Radcliffe, C. Itsiopoulos, Omega-3 fatty acids in early prevention of inflammatory neurodegenerative disease: A focus on Alzheimer's disease, *BioMed Res. Int*. 2015, 2015, 172801. doi:10.1155/2015/172801.

161. M. Söderberg, C. Edlund, K. Kristensson, G. Dallner, Lipid compositions of different regions of the human brain during aging, *J. Neurochem*. 54, 1990, 415–423.

162. M.G. Martin, S. Perga, L. Trovò, A. Rasola, P. Holm, T. Rantamäki, T. Harkany, E. Castrén, F. Chiara, C.G. Dotti, Cholesterol loss enhances TrkB signaling in hippocampal neurons aging *in vitro*, *Mol. Biol. Cell*. 19, 2008, 2101–2112. doi:10.1091/mbc.E07-09-0897.

163. L. Svennerholm, Designation and schematic structure of gangliosides and allied glycosphingolipids, *Prog. Brain Res*. 101, 1994, XI–XIV. doi:10.1016/S0079-6123(08)61935-4.

164. R.G. Cutler, J. Kelly, K. Storie, W.A. Pedersen, A. Tammara, K. Hatanpaa, J.C. Troncoso, M.P. Mattson, Involvement of oxidative stress-induced abnormalities in ceramide and cholesterol metabolism in brain aging and Alzheimer's disease, *Proc. Natl. Acad. Sci. U. S. A*. 101, 2004, 2070–2075. doi:10.1073/pnas.0305799101.

165. E. Di Pasquale, J. Fantini, H. Chahinian, M. Maresca, N. Taïeb, N. Yahi, Altered ion channel formation by the Parkinson's-disease-linked E46K mutant of alpha-synuclein is corrected by GM3 but not by GM1 gangliosides, *J. Mol. Biol*. 397, 2010, 202–218. doi:10.1016/j.jmb.2010.01.046.

166. J.-E. Mansson, M.-T. Vanier, L. Svennerholm, Changes in the fatty acid and sphingosine composition of the major gangliosides of human brain with age, *J. Neurochem*. 30, 1978, 273–275. doi:10.1111/j.1471-4159.1978.tb07064.x.

167. S. Sonnino, V. Chigorno, Ganglioside molecular species containing C18- and C20-sphingosine in mammalian nervous tissues and neuronal cell cultures, *Biochim. Biophys. Acta*. 1469, 2000, 63–77.

168. R.K. McNamara, Y. Liu, R. Jandacek, T. Rider, P. Tso, The aging human orbitofrontal cortex: Decreasing polyunsaturated fatty acid composition and associated increases in lipogenic gene expression and stearoyl-CoA desaturase activity, *Prostaglandins Leukot. Essent. Fatty Acids*. 78, 2008, 293–304. doi:10.1016/j.plefa.2008.04.001.

169. A. Létondor, B. Buaud, C. Vaysse, L. Fonseca, C. Herrouin, B. Servat, S. Layé, V. Pallet, S. Alfos, Erythrocyte DHA level as a biomarker of DHA status in specific brain regions of n-3 long-chain PUFA-supplemented aged rats, *Br. J. Nutr*. 112, 2014, 1805–1818. doi:10.1017/S0007114514002529.

170. I. Denis, B. Potier, C. Heberden, S. Vancassel, Omega-3 polyunsaturated fatty acids and brain aging, *Curr. Opin. Clin. Nutr. Metab. Care.* 18, 2015, 139–146. doi:10.1097/MCO.0000000000000141.

171. D. Cutuli, Functional and structural benefits induced by omega-3 polyunsaturated fatty acids during aging, *Curr. Neuropharmacol.* 15, 2017, 534–542.

172. M.R. Ammar, N. Kassas, S. Chasserot-Golaz, M.-F. Bader, N. Vitale, Lipids in regulated exocytosis: What are they doing?, *Front. Endocrinol.* 4, 2013. doi:10.3389/fendo.2013.00125.

173. M.A. Churchward, T. Rogasevskaia, J. Höfgen, J. Bau, J.R. Coorssen, Cholesterol facilitates the native mechanism of Ca2+-triggered membrane fusion, *J. Cell Sci.* 118, 2005, 4833–4848. doi:10.1242/jcs.02601.

174. A. Kumar, D. Baycin-Hizal, Y. Zhang, M.A. Bowen, M.J. Betenbaugh, Cellular traffic cops: The interplay between lipids and proteins regulates vesicular formation, trafficking, and signaling in mammalian cells, *Curr. Opin. Biotechnol.* 36, 2015, 215–221. doi:10.1016/j.copbio.2015.09.006.

175. M.D. Ledesma, M.G. Martin, C.G. Dotti, Lipid changes in the aged brain: Effect on synaptic function and neuronal survival, *Prog. Lipid Res.* 51, 2012, 23–35. doi:10.1016/j.plipres.2011.11.004.

176. M.E. Haque, T.J. McIntosh, B.R. Lentz, Influence of lipid composition on physical properties and pegmediated fusion of curved and uncurved model membrane vesicles: "nature's own" fusogenic lipid bilayer, *Biochemistry (Mosc.).* 40, 2001, 4340–4348.

177. T. Rogasevskaia, J.R. Coorssen, Sphingomyelin-enriched microdomains define the efficiency of native Ca(2+)-triggered membrane fusion, *J. Cell Sci.* 119, 2006, 2688–2694. doi:10.1242/jcs.03007.

178. L. Trovò, P.P. Van Veldhoven, M.G. Martín, C.G. Dotti, Sphingomyelin upregulation in mature neurons contributes to TrkB activity by Rac1 endocytosis, *J. Cell Sci.* 124, 2011, 1308–1315. doi:10.1242/jcs.078766.

179. S. Sonnino, L. Mauri, V. Chigorno, A. Prinetti, Gangliosides as components of lipid membrane domains, *Glycobiology.* 17, 2007, 1R–13R. doi:10.1093/glycob/cwl052.

180. E. Posse de Chaves, S. Sipione, Sphingolipids and gangliosides of the nervous system in membrane function and dysfunction, *FEBS Lett.* 584, 2010, 1748–1759. doi:10.1016/j.febslet.2009.12.010.

181. P. Marquet-de Rougé, C. Clamagirand, P. Facchinetti, C. Rose, F. Sargueil, C. Guihenneuc-Jouyaux, L. Cynober, C. Moinard, B. Allinquant, Citrulline diet supplementation improves specific age-related raft changes in wild-type rodent hippocampus, *Age Dordr. Neth.* 35, 2013, 1589–1606. doi:10.1007/s11357-012-9462-2.

182. L. Jiang, J. Fang, D.S. Moore, N.V. Gogichaeva, N.A. Galeva, M.L. Michaelis, A. Zaidi, Age-associated changes in synaptic lipid raft proteins revealed by two-dimensional fluorescence difference gel electrophoresis, *Neurobiol. Aging.* 31, 2010, 2146–2159. doi:10.1016/j.neurobiolaging.2008.11.005.

183. J. Schumann, A. Leichtle, J. Thiery, H. Fuhrmann, Fatty acid and peptide profiles in plasma membrane and membrane rafts of PUFA supplemented RAW264.7 macrophages, *PloS One.* 6, 2011, e24066. doi:10.1371/journal.pone.0024066.

184. R. Volinsky, P.K.J. Kinnunen, Oxidized phosphatidylcholines in membrane-level cellular signaling: From biophysics to physiology and molecular pathology, *FEBS J.* 280, 2013, 2806–2816. doi:10.1111/febs.12247.

25 Autophagy: The Way to Death or Immortality? Activators and Inhibitors of Autophagy as Possible Modulators of the Aging Process

Galina V. Morgunova, Alexander A. Klebanov, and Alexander N. Khokhlov

CONTENTS

INTRODUCTION

Autophagy is translated from Greek as "self-eating," which fully reflects the essence of the process: various cytoplasmic substrates of the cell are delivered to the lysosomes, in which they are then destroyed. Three main types of autophagy can be distinguished; they differ in the volume of sequestered substrates and the mechanism of delivery of cytoplasmic components to the lysosomes. The most studied type is *macroautophagy*, which leads to the degradation of cell organelles and large macromolecules via the formation of special membrane structures, autophagolysosomes. At first, a portion of the cytoplasm with damaged substrates is sequestered in a two-membrane structure, phagophore, which then closes and becomes the autophagosome. The fusion of the latter with the lysosome forms the autophagolysosome. With the help of macroautophagy, the cell can renew the intracellular nonnuclear material (Rubinsztein et al. 2011), destroying the old structures and creating new ones from the building blocks obtained as a result of "digestion." The "digestion" of mitochondria (mitophagy) is of particularly great importance (Yen and Klionsky 2008; Gottlieb and Carreira 2010; Morgunova et al. 2016b), because the quality control of these organelles is essential for the long-term existence of the cell. In another type of autophagy, *microautophagy*, lysosome regions form invaginations, thus sequestering small structures and macromolecules without forming autophagosomes. Cells use microautophagy when they experience energy deficiency. This is the way (digestion of small portions of the cytoplasm) which is used for obtaining energy by yeast (Vicencio et al. 2008). Finally, *chaperone-mediated autophagy* does not require lysosomal membrane rearrangement: the "faulty" structures are transported to the lysosome with the involvement of chaperone proteins (Yang et al. 2005; Yen and Klionsky 2008). Only the soluble cytosolic proteins can be transported this way; and to get inside lysosomes, they should be unfolded (Massey et al. 2006). The activity of macroautophagy and chaperone-mediated autophagy dramatically increases in stress, which helps cells to adapt to the environment (Mortimore et al. 1988).

Over the past 10 years, the interest of researchers (including gerontologists) in autophagy has significantly increased. Apparently, it is this process which helps cells to eliminate the "faulty" organelles emerging with aging. In addition, a number of facts indicate the impact of autophagy on life span and aging. It is known that its activity decreases with age and that many age-related pathologies are associated with the disturbance of this process (Yen and Klionsky 2008; Rubinsztein et al. 2011). Stimulation of autophagy may have an antiaging effect (Madeo et al. 2010). It is also known that the effect of calorie restriction on life span is largely realized through autophagy (Levine et al. 2011; Rubinsztein et al. 2011). Nevertheless, in our opinion, autophagy has a limited effect on life span. Cells can obtain nutrients as a result of digestion of their own cytoplasm and can eliminate damaged organelles and macromolecules and renew them; however, if damage occurs in DNA, the main template of the cell, autophagy becomes useless.

In this chapter, we would like to consider the data on the relationship between autophagy (primarily macroautophagy) and aging of organisms and cell cultures, as well as to present our views on both the interpretation of data obtained in the studies of autophagy in cytogerontological experiments and the methodology of such studies. In addition, we will try to analyze some problems related to the prospects of pharmaceutical modulation of autophagy and to the possible effect of these interventions on aging and longevity.

AUTOPHAGY AND AGING

According to the definition to which we adhere, aging is the set of age-related changes in the organism leading to an increase in the probability of death (Khokhlov 2010, 2014b; Khokhlov et al. 2012, 2014, 2017; Khokhlov and Morgunova 2017). Over time, the ability of the organism to withstand environmental impacts decreases, the ability to resist infections is reduced, and the risk of development of age-related diseases increases. Apparently, the decline in the activity of autophagy contributes to all of these changes.

The positive effect of autophagy on life span was shown on model organisms. Pharmacological manipulations associated with the inhibition of the TOR (target of rapamycin) complex, which, similarly to calorie restriction, increase the life span of *Caenorhabditis elegans* and *Drosophila melanogaster*, activate autophagy (Bjedov et al. 2010; Morselli et al. 2010). Genetic inhibition of autophagy in *C. elegans* shows that dietary restriction does not increase the life span (Meléndez et al. 2003). *D. melanogaster* individuals carrying mutations in autophagy-regulating genes *Atg7* and *Atg8a* (*Atg* is short for "autophagy-related genes") are hypersensitive to oxidative stress and have a shorter life span than the control flies (Juhász et al. 2007; Simonsen et al. 2008). Enhancing the *Atg8a* gene expression, conversely, increases their average life span and resistance to oxidative stress (Simonsen et al. 2008). In the mutants for the *Atg7* gene, the development of neurodegenerative pathologies is observed (Juhász et al. 2007). In addition, an age-related decline in autophagy gene expression was observed in the nervous tissue of *D. melanogaster*, which entails the accumulation of markers of neurodegenerative diseases (Simonsen et al. 2008). Knockout of *Atg7* and *Atg5* genes in the brain leads to the formation of inclusion bodies in the cytoplasm of neurons in mice; such inclusions were shown to accumulate in the brain with aging (Hara et al. 2006; Komatsu et al. 2006). Moreover, it was demonstrated that these genes are required for normal functioning of the central nervous system: *Atg5* deficiency leads to disturbances in motor function (Hara et al. 2006), and *Atg7* deficiency causes disturbances in coordination and massive loss of neurons in the cerebellar cortex and cerebral hemispheres in mice (Komatsu et al. 2006). In another study, it was shown that the *Atg7* gene knockout in skeletal muscles of mice leads to atrophy of muscle fibers and development of degradation therein. In these animals, there is an accumulation of protein aggregates, damaged mitochondria and membrane structures, sarcoplasmic reticulum distension, vacuolization of cytoplasm, and apoptosis in myocytes (Masiero et al. 2009). Thus, maintaining the activity of autophagy at a certain level is an absolute prerequisite for normal physiological functioning of skeletal muscles and neurons.

In aging cells, damaged macromolecules and organelles are accumulated (Khokhlov 2013a, 2013b). This is ballast material, and it is difficult for the cell to eliminate it. Apparently, one of the factors that cause the accumulation of such "waste" is the decline in autophagy (Cuervo et al. 2005). It must be emphasized that the renewal of cellular material via macroautophagy is essential for postmitotic cells, which cannot divide and thus renew their contents. Cardiovascular and neurodegenerative diseases develop with age, just because cardiomyocytes and neurons accumulate "waste" proteins and damaged organelles throughout life. "Cleaning" cells via autophagy can help them to function longer.

The role of autophagy in the vital activity of postmitotic cells is well illustrated by the processes associated with the functioning of skeletal muscle fibers and neurons. Since the muscles constitute a significant part of the body weight of an individual, their contribution to the metabolism of the entire body is very significant. Skeletal muscles are a "storehouse" of amino acids that the body can use during fasting, they also utilize a considerable portion of glucose, etc. (Sandri 2010; Brook et al. 2016). With age, muscle fibers lose cells and undergo atrophy due to an imbalance between synthesis and degradation of proteins (Cuervo and Dice 1998; Sanchez et al. 2012; Brook et al. 2016). Fasting, disturbed innervation, or injuries, similarly to aging, lead to hyperactivation of catabolic pathways, which leads to the loss of muscle mass. In view of this, it would be logical to assume that autophagy more likely damages muscle fibers rather than protects them. Besides, the TOR complex activity is necessary for the synthesis of muscle fibers. However, data described in the above-mentioned study by Masiero et al. (Masiero et al. 2009; Masiero and Sandri 2010) contradict this assumption. Violation of the mechanism of autophagy leads to abnormal development of muscles and atrophy of fibers.

Neurons are even more prone to accumulating degraded proteins with age than muscle fibers (Cuervo and Dice 1998). The main age-related neurodegenerative human diseases—Alzheimer's disease, Parkinson's disease, and Huntington's chorea—are associated with the accumulation of waste in nerve cells (Rubinsztein et al. 2011; Stefanova and Kolosova 2016). It was found that, in Alzheimer's disease, β-amyloid accumulates in autophagosomes, but autophagosomes do not fuse with lysosomes, as a result of which the peptide is not destroyed, as should be occurring in normal nerve cells. Experiments on mice with neurodegenerative lesions and on Drosophila, for which a model of Huntington's disease was developed, showed that the induction of autophagy decreased the number of aggregates of the mutant protein huntingtin (Ravikumar et al. 2004). The "faulty" organelles and proteins are eliminated from nerve cells via both macroautophagy and chaperone-mediated autophagy. Maintaining these types of autophagy in the "old" nerve cells at the level characteristic of the "young" cells would probably help to prevent the development of neurodegenerative diseases.

Nevertheless, there is a standpoint according to which autophagy plays a negative role in the life activity of the organism. Cancer cells can use this process to survive (e.g., after chemotherapy) (Gewirtz 2013). It was established that the treatment of breast tumor cells and colon cancer cells with anticancer drugs activates autophagy (Goehe et al. 2012). At the same time, autophagy inhibitors are used to fight against certain tumors (Degenhardt et al. 2006). In our opinion, this proves once again the viewpoint that autophagy is a universal process that facilitates the survival of any cells. As mentioned in our previous work (Morgunova et al. 2016a), it is important to distinguish the effect of a factor on individual cells within the organism and the effect on the organism itself. For example, the transition of a cell to the senescent state, on the one hand, is beneficial for the organism, because the cell does not become cancerous, but, on the other hand, the number of senescent cells becomes too high and they disturb the functioning of tissues and organs. If a malignant tumor has already formed, autophagy cannot help the organism. However, this process can also be used by the normal cells; it is particularly necessary for neurons, cardiomyocytes, and long-lived immune memory cells (Gottlieb and Carreira 2010). We still tend to believe that autophagy is a way to protect cells from adverse conditions rather than a tool to trigger their death.

AUTOPHAGY, CELLULAR SENESCENCE, AND
CELL PROLIFERATION RESTRICTION

In cytogerontology, there are two basic models of cellular aging—the Hayflick model (replicative aging) (Hayflick and Moorhead 1961; Hayflick 1965) and the "stationary phase aging" of cell cultures (Akimov and Khokhlov 1998; Khokhlov 2013a, 2013d; Khokhlov et al. 2015; Khokhlov and Morgunova 2015). In the classic replicative aging "according to Hayflick," the number of cell divisions is counted. After a certain number of passages, normal cells cease to divide and become senescent, because their telomeres become highly shortened. To date, a model is often used in which cells are made "senescent" by damaging them, so that they become unable to divide. This is the so-called stress-induced premature senescence (SIPS) (Toussaint et al. 2000; Jeyapalan and Sedivy 2008; Khokhlov 2013d). Such a formulation of the problem makes it possible to use not only normal but also transformed cells in experiments. It should be noted that the proponents of this model believe that both autophagy and SIPS protect the organism from cancer (White and Lowe 2009). There is also a standpoint that autophagy can serve as a trigger that activates DNA damage and, thereby, initiates SIPS (Young et al. 2009). We believe that autophagy normally should not trigger this process and that this effect is probably associated with excessive activation of autophagy, which is caused by creating nonphysiological conditions (e.g., using chemotherapy).

Yeasts, unicellular eukaryotic organisms, are often used as a model object to study autophagy. Such experiments are performed using primarily the chronological aging model, when cells reach the stationary growth phase, after which their proliferation stops (Fabrizio and Longo 2003; Khokhlov 2016). Activation of macroautophagy prolongs the life of "chronologically aging" yeast cells (Kaeberlein 2007; Alvers et al. 2009), whereas its inhibition leads to premature cell death (Herman 2002). Restriction of the amount of amino acids in the nutrient medium leads to an increased life span of *Saccharomyces cerevisiae*; however, this effect disappears if the process of autophagy is disturbed (Matecic et al. 2010). Thus, the effect of calorie restriction in yeast is likely to be mediated via autophagy. It was also demonstrated that the short-lived *S. cerevisiae* mutants have mutations in the *Atg* genes (Matecic et al. 2010).

Bacteria have a mechanism that is similar in principle to the autophagy of eukaryotes, which is manifested in a decrease in cell size ("dwarfing"), since part of the cytoplasm is "digested." This process is also activated in the stationary growth phase of culture (Nyström 2004) and facilitates the survival of cells under nutrient deficiency conditions.

The "stationary phase aging" of cultured cells derived from multicellular organisms is similar to the chronological aging in yeast (Khokhlov 2016; Morgunova et al. 2017). Cells become "old" in the stationary phase of growth, when the culture reaches a monolayer, after which proliferation stops as a result of contact inhibition. This model system is based on the concept that cell proliferation restriction leads to the accumulation of macromolecular lesions in cultured cells, similar to the lesions that accumulate with age in the postmitotic cells of a multicellular organism (Khokhlov 2013a–c; Morgunova et al. 2015). In our experiments, we have repeatedly observed vacuolation of the cytoplasm of "stationary phase aged" transformed Chinese hamster cells and normal human fibroblasts. This indicates an active digestion of the cytoplasmic material, when organelles and macromolecules in cells acquire numerous defects. However, the degradation process continues in cells that are unable to divide, and, sooner or later, autophagy ceases to "rescue" them. If the nuclear structures (most importantly, DNA) rather than the cytoplasmic structures in the postmitotic cell are damaged, then the negative consequences of such disturbances for cells can only be prevented by the DNA repair system. Nevertheless, even this system cannot completely eliminate all errors. However, if the cytoplasm contains no damaged mitochondria, then reactive oxygen species are not generated in excess and, therefore, the risk of DNA damage is reduced. At the same time, it should be emphasized that there are organisms which made another choice: they just dump the cells that have accumulated defects and replace them with new ones. It is this way which is used by the freshwater hydra (Khokhlov 2014a). Due to the high proliferation rate, damaged cells in the organism of the hydra are constantly "diluted" with new ones. For higher organisms and, especially, for humans, this approach is not suitable, due to the presence of highly differentiated cell populations that ensures the normal functioning of individuals. For this reason, such organisms have to search for a way to eliminate the defects in postmitotic cells. However, autophagy plays an important role even in the hydra's life, helping the polyp to survive during starvation (Chera et al. 2009), which suppresses the renewal of cells of the organism.

We think that it is better to trigger macroautophagy in a natural way rather than by using various calorie restriction mimetics (CRM)—rapamycin, resveratrol, etc., because addiction to them may develop over time (Alayev et al. 2015; Morgunova et al. 2016b). In such a natural way, this process is activated in our "stationary phase aging" model.

INHIBITORS AND ACTIVATORS OF AUTOPHAGY

Since autophagy is a complex multi-stage process, it is not entirely true to talk about specific activators and inhibitors of this process. There are substances that mimic the effect of dietary restriction; they are the factors that are peculiar activators of autophagy.

Perhaps the most popular among the substances was and remains rapamycin, obtained in 1970 from a strain of *Streptomyces hygroscopicus* (Wullschleger et al. 2006). Initially, the metabolite was used as a substance that suppressed the proliferation of mammalian cells and immunity, but today rapamycin is considered a promising medicine against cancer, cardiovascular pathologies, metabolic syndrome, and many other age-related diseases (Blagosklonny 2006; Wullschleger et al. 2006). Small doses of rapamycin (sufficient to slow the cell division but to not completely block this process) increase the life span of yeast *S. cerevisiae* in the model of chronological aging (Powers et al. 2006; Alvers et al. 2009); as it was shown in one of these papers, this effect is mediated exactly through the activation of macroautophagy (Alvers et al. 2009).

Rapamycin was one of the drugs that actually prolonged the life of mammals (mice), with both the average and maximum life expectancy of animals increased—in females by 14%, in males by 9% (Harrison et al. 2009). Later, additional studies were conducted in which the average life span was increased by as much as 23% in males and 26% in females (Miller et al. 2014). It should be noted that in these experiments the dose of the substance was three times higher than in the previous work.

Several years ago, in studies of cancer-prone male mice C57BL/6J Rj, it was shown that although rapamycin prolonged the life of animals, it almost did not affect the phenotypic manifestations of old age (Neff et al. 2013). Rapamycin in capsules was given to mice for 12 months starting at the age of 4, 13, or 20–22 months. Control animals received empty capsules. After that, a number of behavioral and physiological indicators were evaluated. The age dynamics of many of these indices in mice receiving rapamycin turned out to be the same as in control animals. The authors emphasize that the extent of the improvement under the influence of the drug of some indicators, for example, memory and the spatial learning, depended little on the age of the animals. Thus, it is assumed that the reason for the increase in the life span of mice under the influence of rapamycin lies in its ability to slow down the development of tumors, which are the main cause of death in these animals.

As regards cancer cells, they are really sensitive to rapamycin, which made the substance a promising antitumor agent; under the influence of rapamycin the growth of cancer cells stops in the G_1 phase (Hashemolhosseini et al. 1998; Hidalgo and Rowinsky 2000). It is also likely that rapamycin can induce apoptosis in cancer cells (Hosoi et al. 1999; Noh et al. 2004). Since the PI3K/TORC1 (phosphatidylinositol-4,5-bisphosphate 3-kinase/mammalian TOR complex 1) signaling pathway is hyperactivated in a variety of tumors (Alayev et al. 2015), rapamycin can really help to prevent tumor development. However, over time, addiction to the drug develops, due to the appearance of a feedback loop and the activation of Akt (protein kinase B). In this regard, the solution of the problem considered by some researchers is the use of combinations of an mTOR inhibitor with a PI3K or Akt inhibitor (Sun et al. 2005; Alayev et al. 2015).

Complex interactions of pathways that are somehow associated with TOR are likely to prevent rapamycin from being used as a universal pill against all sorts of age-related diseases, although it is obvious that studying these pathways will make it possible to understand many regularities of development of these pathologies. To date, in addition to the rapamycin itself, a number of its analogs exist, the so-called rapologs (Yang et al. 2013). Among them are temsirolimus, everolimus, and ridaforolimus.

The aforementioned vegetable product resveratrol, activating sirtuins, is also considered as CRM (Alarcón de la Lastra and Villegas 2005; Yen and Klionsky 2008; Kaeberlein 2010). It is the resveratrol that some researchers propose to use in combination with rapamycin (Alayev et al. 2015) exactly in the cases when rapamycin is addictive. However, it seems to us that this does not solve the problem, addiction can arise again, and the endless supplements of new inhibitors can eventually block quite useful metabolic pathways.

Equally popular is metformin, a drug long used in the treatment of diabetes (Anisimov et al. 2008; Smith et al. 2010; Martin-Montalvo et al. 2013). The choice of this drug as a CRM is completely logical since diabetes and obesity (usually, a result of overnutrition) are related issues, and therefore, the medicine against the first should, to some extent, also help from the second. Metformin and other biguanides (buformin and phenformin) probably can indeed activate autophagy, but data on the effect of these compounds on healthy (nondiabetic) animals is still insufficient (Anisimov 2010, 2013). The mechanism of action of biguanides is associated with the activation of AMPK (AMP-activated protein kinase) (Zhou et al. 2001). AMPK is a sensor that reacts to a lack of energy, which is expressed in an increase in the ratio of AMP/ATP. Metformin activates AMPK, and it, in turn, inhibits TOR and takes the cell into an "economical" mode of existence.

One of the first CRM was the analog of the usual glucose-2-deoxyglucose (Lane et al. 1998; Roth et al. 2001). Although the use of 2-deoxyglucose did mimic the effect of dietary restriction, the same group of researchers later found out that this analog has a toxic effect—it causes vacuolization of myocytes in rats and increases the mortality of these animals (Minor et al. 2010).

An activator of autophagy is also trehalose, a disaccharide that protects cells from environmental stresses (Sarkar et al. 2007). Trehalose is produced by cells of many species (non-mammals) and it activates autophagy via a TOR-independent pathway. Importantly, the autophagy induced by trehalose improves the "digestion" of mutant huntingtin and α-synuclein (Sarkar et al. 2007). It also reduces the endogenous tau protein content and its aggregation (Krüger et al. 2012). All these data once again underscore the extremely important role of autophagy in protecting against neurodegenerative diseases.

There are some other drugs that may also mimic the effects of dietary restriction, but so far their effect has not been proven in convincing experiments.

Specific inhibitors of autophagy do not exist, although there are a number of compounds that can suppress this process. Almost all of them are not effective enough (Wu et al. 2013). In this regard, active searching for more efficient derivatives is conducted. Autophagy can be suppressed by manipulation with *Atg* genes, but, as described in section Autophagy and Aging this does not lead to anything good.

One of the most popular inhibitors of autophagy is 3-MA (3-methyladenine), which inhibits the rapamycin-sensitive signaling pathway (Shigemitsu et al. 1999; Yang et al. 2013). Its effect is associated with PI3K suppression—that is, it is not worth to talk about the suppression of autophagy itself. However, this substance is poorly soluble at room temperature and is effective only at high concentrations. Attempts have been made to create a number of 3-MA derivatives that are devoid of these shortcomings (Wu et al. 2013).

Other popular inhibitors of autophagy are chloroquine and hydroxychloroquine. They are used in medicine primarily as substances for the fight against malaria, and also as immunosuppressants in the treatment of rheumatoid arthritis. The use of chloroquine and its derivatives makes it possible to increase the effectiveness of treatment of tumors in preclinical studies. Among chloroquine derivatives the most effective one is Lys05 (McAfee et al. 2012).

However, 3-MA, chloroquine, and other nonspecific inhibitors are still not the best pharmaceutical preparations, since they suppress not only autophagy but also other processes in the cell. In addition, they are all poorly understood, so their use can lead to the appearance of side effects with time. Nevertheless, the search for new inhibitors of autophagy still looks reasonable, precisely because cancer cells use autophagy for survival.

The above data allow us to draw some conclusions. First, addiction is likely to develop both to CRM and to inhibitors of autophagy, which is quite natural, since there are various alternative ways of implementation of the considered signal cascades in the cell. Second, some of the CRM (such as deoxyglucose or metformin at high doses) may have a toxic effect, and what is the most dangerous is that this effect may be deferred. In addition, CRM, perhaps, are struggling only with the consequences of aging—age-related diseases. The best way to increase longevity is precisely the physiological calorie restriction. So far, this method has allowed researchers to significantly extend the life of rodents (McCay et al. 1935; Masoro 2005) and retard age-related changes in primates (Weindruch 1996; Roth et al. 1999, 2001). Potentially, this approach will make it possible to prolong the active life of people.

CONCLUSIONS

1. Autophagy activity decreases with age and many age-related pathologies are associated with the disturbance of this process. Maintaining macroautophagy and chaperone-mediated autophagy in the "old" postmitotic cells at the level characteristic of the "young" cells would probably help to prevent the development of some diseases.
2. Various experiments show the positive effect of autophagy on life span of model organisms and negative effect of inhibition of this process on health and longevity laboratory animals and cell culture.
3. Autophagy has a limited effect on life span if damage occurs in DNA, the main template of the cell.
4. Specific activators and inhibitors of autophagy do not exist because this process is extremely complex. It is better to trigger autophagy in a natural way rather than by using various CRM.
5. Cancer cells also can use autophagy to survive; this fact proves the viewpoint that this process is a universal self-defense mechanism. Due to this, it is necessary to continue searching for new ways to not only activate but also suppress autophagy.

Despite the existence of the standpoint that autophagy is a negative regulator of vital functions, there is more evidence of its beneficial effects (including gerontology). All types of autophagy in one way or another extend life span. Since autophagy is more important for postmitotic cells, the study of the role of this process in the determination of life span and regulation of aging of organisms in the model of "stationary phase aging" of cell cultures, in which cell proliferation is stopped due to contact inhibition, can be considered the most appropriate. It looks like the life-extending effect of calorie restriction does not need to be mimicked; people need to really eat less.

REFERENCES

Akimov, S.S., and A.N. Khokhlov. 1998. Study of "stationary phase aging" of cultured cells under various types of proliferation restriction. *Ann NY Acad Sci* 854:520.

Alarcón de la Lastra, C., and I. Villegas. 2005. Resveratrol as an anti-inflammatory and anti-aging agent: Mechanisms and clinical implications. *Mol Nutr Food Res* 49:405–430.

Alayev, A., Berger, S.M., Kramer, M.Y., Schwartz, N.S. and M.K. Holz. 2015. The combination of rapamycin and resveratrol blocks autophagy and induces apoptosis in breast cancer cells. *J Cell Biochem* 116:450–457.

Alvers, A.L., Wood, M.S., Hu, D., Kaywell, A.C., Dunn, W.A., Jr. and J.P. Aris. 2009. Autophagy is required for extension of yeast chronological life span by rapamycin. *Autophagy* 5:847–849.

Anisimov, V.N. 2010. Metformin for aging and cancer prevention. *Aging (Albany NY)* 2:76–774.

Anisimov, V.N. 2013. Metformin: Do we finally have an anti-aging drug? *Cell Cycle* 12:3483–3489.

Anisimov, V.N., Berstein, L.M., Egormin, P.A. et al. 2008. Metformin slows down aging and extends life span of female SHR mice. *Cell Cycle* 7:2769–2773.

Bjedov, I., Toivonen, J.M., Kerr, F. et al. 2010. Mechanisms of life span extension by rapamycin in the fruit fly *Drosophila melanogaster*. *Cell Metab* 11:35–46.

Blagosklonny, M.V. 2006. Aging and immortality: Quasi-programmed senescence and its pharmacologic inhibition. *Cell Cycle* 5:2087–2102.

Brook, M.S., Wilkinson, D.J., Phillips, B.E. et al. 2016. Skeletal muscle homeostasis and plasticity in youth and ageing: Impact of nutrition and exercise. *Acta Physiol (Oxf)* 216:15–41.

Chera, S., Buzgariu, W., Ghila, L. and B. Galliot. 2009. Autophagy in Hydra: A response to starvation and stress in early animal evolution. *Biochim Biophys Acta* 1793:1432–1443.

Cuervo, A.M., and J.F. Dice. 1998. How do intracellular proteolytic systems change with age? *Front Biosci* 3:d25–d43.

Cuervo, A.M., Bergamini, E., Brunk, U.T., Droge, W., Ffrench, M. and A. Terman. 2005. Autophagy and aging: The importance of maintaining "clean" cells. *Autophagy* 1:131–140.

Degenhardt, K., Mathew, R., Beaudoin, B. et al. 2006. Autophagy promotes tumor cell survival and restricts necrosis, inflammation, and tumorigenesis. *Cancer Cell* 10:51–64.

Fabrizio, P., and V.D. Longo. 2003. The chronological lifespan of *Saccharomyces cerevisiae*. *Aging Cell* 2:73–81.

Gewirtz, D.A. 2013. Autophagy and senescence. A partnership in search of definition. *Autophagy* 9:808–812.

Goehe, R.W., Di, X., Sharma, K. et al. 2012. The autophagy-senescence connection in chemotherapy: Must tumor cells (self) eat before they sleep? *J Pharmacol Exp Ther* 343:763–778.

Gottlieb, R.A., and R.S. Carreira. 2010. Autophagy in health and disease. 5. Mitophagy as a way of life. *Am J Physiol Cell Physiol* 299:C203–C210.

Hara, T., Nakamura, K., Matsui, M. et al. 2006. Suppression of basal autophagy in neural cells causes neuro-degenerative disease in mice. *Nature* 441:885–889.

Harrison, D.E., Strong, R., Sharp, Z.D. et al. 2009. Rapamycin fed late in life extends lifespan in genetically hctcrogeneous mice. *Nature* 460:392–395.

Hashemolhosseini, S., Nagamine, Y., Morley, S.J., Desrivières, S., Mercep, L. and S. Ferrari. 1998. Rapamycin inhibition of the G_1 to S transition is mediated by effects on cyclin D1 mRNA and protein stability. *J Biol Chem* 273:14424–14429.

Hayflick, L., and P.S. Moorhead. 1961. The serial cultivation of human diploid cell strains. *Exp Cell Res* 25:585–621.

Hayflick, L. 1965. The limited *in vitro* lifetime of human diploid cell strains. *Exp Cell Res* 37:614–636.

Herman, P.K. 2002. Stationary phase in yeast. *Curr Opin Microbiol* 5:602–607.

Hidalgo, M., and E.K. Rowinsky. 2000. The rapamycin-sensitive signal transduction pathway as a target for cancer therapy. *Oncogene* 19:6680–6686.

Hosoi, H., Dilling, M.B., Shikata, T. et al. 1999. Rapamycin causes poorly reversible inhibition of mTOR and induces p53-independent apoptosis in human rhabdomyosarcoma cells. *Cancer Res* 59:886–894.

Jeyapalan, J.C., and J.M. Sedivy. 2008. Cellular senescence and organismal aging. *Mech Aging Dev* 129:467–474.

Juhász, G., Érdi, B., Sass, M. and T.P. Neufeld. 2007. Atg7-dependent autophagy promotes neuronal health, stress tolerance, and longevity but is dispensable for metamorphosis in Drosophila. *Genes Dev* 21:3061–3066.

Kaeberlein, M. 2010. Resveratrol and rapamycin: Are they anti-aging drugs? *Bioessays* 32:96–99.

Kaeberlein, M., Burtner, C.R. and B.K. Kennedy. 2007. Recent dcvclopments in yeast aging. *PLoS Genet* 3:e84.

Khokhlov, A.N. 2010. Does aging need an own program or the existing development program is more than enough? *Russ J Gen Chem* 80:1507–1513.

Khokhlov, A.N. 2013a. Does aging need its own program, or is the program of development is the program of development quite sufficient for it? Stationary cell cultures as a tool to search for anti-aging factors. *Curr Aging Sci* 6:14–20.

Khokhlov, A.N. 2013b. Impairment of regeneration in aging: Appropriateness or stochastics? *Biogerontology* 14:703–708.

Khokhlov, A.N. 2013c. Decline in regeneration during aging: Appropriateness or stochastics? *Russ J Dev Biol* 44:336–341.

Khokhlov, A.N. 2013d. Evolution of the term "cellular senescence" and its impact on the current cytogerontological research. *Moscow Univ Biol Sci Bull* 68:158–161.

Khokhlov, A.N. 2014a. On the immortal hydra. Again. *Moscow Univ Biol Sci Bull* 69:153–157.

Khokhlov, A.N. 2014b. What will happen to molecular and cellular biomarkers of aging in case its program is canceled (provided such a program does exist)? *Adv Gerontol* 4:150–154.

Khokhlov, A.N. 2016. Which aging in yeast is "true?" *Moscow Univ Biol Sci Bull* 71:11–13.

Khokhlov, A.N., and G.V. Morgunova. 2015. On the constructing of survival curves for cultured cells in cytogerontological experiments: A brief note with three hierarchy diagrams. *Moscow Univ Biol Sci Bull* 70:67–71.

Khokhlov, A.N., and G.V. Morgunova. 2017. Testing of geroprotectors in experiments on cell cultures: Pros and cons. In *Anti-Aging Drugs: From Basic Research to Clinical Practice*, ed. A.M. Vaiserman, 53–74. London: Royal Society of Chemistry.

Khokhlov, A.N., Klebanov, A.A., Karmushakov, A.F., Shilovsky, G.A., Nasonov, M.M. and G.V. Morgunova. 2014. Testing of geroprotectors in experiments on cell cultures: Choosing the correct model system. *Moscow Univ Biol Sci Bull* 69:10–14.

Khokhlov, A.N., Klebanov, A.A. and G.V. Morgunova. 2017. Anti-aging drug discovery in experimental gerontological studies: From organism to cell and back. In *This Book*, ed. S. Ahmad, 575–592. Boca Raton: Taylor & Francis.

Khokhlov, A.N., Morgunova, G.V., Ryndina, T.S. and F. Coll. 2015. Pilot study of a potential geroprotector, "Quinton Marine Plasma," in experiments on cultured cells. *Moscow Univ Biol Sci Bull* 70:7–11.

Khokhlov, A.N., Wei, L., Li, Y. and J. He. 2012. Teaching cytogernotology in Russia and China. *Adv Gerontol* 25:513–516.

Komatsu, M., Waguri, S., Chiba, T. et al. 2006. Loss of autophagy in the central nervous system causes neurodegeneration in mice. *Nature* 441:880–884.

Krüger, U., Wang, Y., Kumar, S. and E.M. Mandelkow. 2012. Autophagic degradation of tau in primary neurons and its enhancement by trehalose. *Neurobiol Aging* 33:2291–2305.

Lane, M.A., Ingram, D.K. and G.S. Roth. 1998. 2-Deoxy-D-glucose feeding in rats mimics physiologic effects of calorie restriction. *J Anti-Aging Med* 1:327–337.

Levine, B., Mizushima, N. and H.W. Virgin. 2011. Autophagy in immunity and inflammation. *Nature* 469:323–335.

Madeo, F., Tavernarakis, N. and G. Kroemer. 2010. Can autophagy promote longevity? *Nat Cell Biol* 12:842–846.

Martin-Montalvo, A., Mercken, E.M., Mitchell, S.J. et al. 2013. Metformin improves healthspan and lifespan in mice. *Nat Commun* 4:2192.

Masiero, E., Agatea, L., Mammucari, C. et al. 2009. Autophagy is required to maintain muscle mass. *Cell Metabolism* 10:507–515.

Masiero, E., and M. Sandri. 2010. Autophagy inhibition induces atrophy and myopathy in adult skeletal muscles. *Autophagy* 6:307–309.

Masoro, E.J. 2005. Overview of caloric restriction and ageing. *Mech Ageing Dev* 126:913–922.

Massey, A.C., Kiffin, R. and A.M. Cuervo. 2006. Autophagic defects in aging. Looking for an "emergency exit?" *Cell Cycle* 5:1292–1296.

Matecic, M., Smith, D.L., Jr., Pan, X. et al. 2010. A microarray-based genetic screen for yeast chronological aging factors. *PLoS Genet* 6:e1000921.

McAfee, Q., Zhang, Z., Samanta, A. et al. 2012. Autophagy inhibitor Lys05 has single-agent antitumor activity and reproduces the phenotype of a genetic autophagy deficiency. *Proc Natl Acad Sci U S A* 109:8253–8258.

McCay, C.M., Crowell, M.F. and L.A. Maynard. 1935. The effect of retarded growth upon the length of life span and upon the ultimate body size one figure. *J Nutr* 10:63–79.

Meléndez, A., Tallóczy, Z., Seaman, M., Eskelinen, E.L., Hall, D.H. and B. Levine. 2003. Autophagy genes are essential for dauer development and life-span extension in *C. elegans*. *Science* 301:1387–1391.

Miller, R.A., Harrison, D.E., Astle, C.M. et al. 2014. Rapamycin-mediated lifespan increase in mice is dose and sex dependent and metabolically distinct from dietary restriction. *Aging Cell* 13: 468–477.

Minor, R.K., Smith, D.L. Jr., Sossong, A.M. et al. 2010. Chronic ingestion of 2-deoxy-D-glucose induces cardiac vacuolization and increases mortality in rats. *Toxicol Appl Pharmacol* 243:332–339.

Morgunova, G.V., Klebanov, A.A. and A.N. Khokhlov. 2016a. Interpretation of data about the impact of biologically active compounds on viability of cultured cells of various origin from a gerontological point of view. *Moscow Univ Biol Sci Bull* 71:67–70.

Morgunova, G.V., Klebanov, A.A. and A.N. Khokhlov. 2016b. Some remarks on the relationship between autophagy, cell aging, and cell proliferation restriction. *Moscow Univ Biol Sci Bull* 71:207–211.

Morgunova, G.V., Klebanov, A.A., Marotta, F. and A.N. Khokhlov. 2017. Culture medium pH and stationary phase/chronological aging of different cells. *Moscow Univ Biol Sci Bull* 72:47–51.

Morgunova, G.V., Kolesnikov, A.V., Klebanov, A.A. and A.N. Khokhlov. 2015. Senescence-associated β-galactosidase—A biomarker of aging, DNA damage, or cell proliferation restriction? *Moscow Univ Biol Sci Bull* 70:165–167.

Morselli, E., Maiuri, M.C., Markaki, M. et al. 2010. Caloric restriction and resveratrol promote longevity through the Sirtuin-1-dependent induction of autophagy. *Cell Death Dis* 1:e10.

Mortimore, G.E., Lardeux, B.R. and C.E. Adams. 1988. Regulation of microautophagy and basal protein turnover in rat liver. Effects of short-term starvation. *J Biol Chem* 263:2506–2512.

Neff, F., Flores-Dominguez, D., Ryan, D.P. et al. 2013. Rapamycin extends murine lifespan but has limited effects on aging. *J Clin Invest* 123: 3272–3291.

Noh, W.C., Mondesire, W.H., Peng, J. et al. 2004. Determinants of rapamycin sensitivity in breast cancer cells. *Clin Cancer Res* 10:1013–1023.

Nyström, T. 2004. Stationary-phase physiology. *Annu Rev Microbiol* 58:161–181.

Powers, R.W. III, Kaeberlein, M., Caldwell, S.D., Kennedy, B.K. and S. Fields. 2006. Extension of chronological life span in yeast by decreased TOR pathway signaling. *Genes Dev* 20:174–184.

Ravikumar, B., Vacher, C., Berger, Z. et al. 2004. Inhibition of mTOR induces autophagy and reduces toxicity of polyglutamine expansions in fly and mouse models of Huntington disease. *Nat Genet* 36:585–595.

Roth, G.S., Ingram, D.K. and M.A. Lane. 1999. Calorie restriction in primates: Will it work and how will we know? *J Am Geriatr Soc* 47:896–903.

Roth, G.S., Ingram, D.K. and M.A. Lane. 2001. Caloric restriction in primates and relevance to humans. *Ann NY Acad Sci* 928:305–315.

Rubinsztein, D.C., Mariño, G. and G. Kroemer. 2011. Autophagy and aging. *Cell* 146:682–695.

Sanchez, A. M., Csibi, A., Raibon, A. et al. 2012. AMPK promotes skeletal muscle autophagy through activation of forkhead FoxO3a and interaction with Ulk1. *J Cell Biochem* 113:695–710.

Sandri, M. 2010. Autophagy in skeletal muscle. *FEBS Lett* 584:1411–1416.

Sarkar, S., Davies, J.E., Huang, Z., Tunnacliffe, A. and D.C. Rubinsztein. 2007. Trehalose, a novel mTOR-independent autophagy enhancer, accelerates the clearance of mutant huntingtin and α-synuclein. *J Biol Chem* 282:5641–5652.

Shigemitsu, K., Tsujishita, Y., Hara, K., Nanahoshi, M., Avruch, J. and K. Yonezawa. 1999. Regulation of translational effectors by amino acid and mammalian target of rapamycin signaling pathways. Possible involvement of autophagy in cultured hepatoma cells. *J Biol Chem* 274:1058–1065.

Simonsen, A., Cumming, R.C., Brech, A., Isakson, P., Schubert, D.R. and K.D. Finley. 2008. Promoting basal levels of autophagy in the nervous system enhances longevity and oxidant resistance in adult *Drosophila*. *Autophagy* 4:176–184.

Smith, D.L., Elam, C.F., Mattison, J.A. et al. 2010. Metformin supplementation and life span in Fischer-344 rats. *J Gerontol A Biol Sci Med Sci* 65A:468–474.

Stefanova, N.A., and N.G. Kolosova. 2016. Evolution of Alzheimer's disease pathogenesis conception. *Moscow Univ Biol Sci Bull* 71:4–10.

Sun, S.Y., Rosenberg, L.M., Wang, X. et al. 2005. Activation of Akt and eIF4E survival pathways by rapamycin-mediated mammalian target of rapamycin inhibition. *Cancer Res* 65:7052–7058.

Toussaint, O., Medrano, E.E. and T. Von Zglinicki. 2000. Cellular and molecular mechanisms of stress-induced premature senescence (SIPS) of human diploid fibroblasts and melanocytes. *Exp Gerontol* 35:927–945.

Vicencio, J.M., Galluzzi, L., Tajeddine, N. et al. 2008. Senescence, apoptosis or autophagy? When a damaged cell must decide its path—A mini-review. *Gerontology* 54:92–99.

Weindruch, R. 1996. The retardation of aging by caloric restriction: Studies in rodents and primates. *Toxicol Pathol* 24:742–745.

White, E., and S.W. Lowe. 2009. Eating to exit: Autophagy-enabled senescence revealed. *Genes Dev* 23:784–787.

Wu, Y., Wang, X., Guo, H. et al. 2013. Synthesis and screening of 3-MA derivatives for autophagy inhibitors. *Autophagy* 9:595–603.

Wullschleger, S., Loewith, R. and M.N. Hall. 2006. TOR signaling in growth and metabolism. *Cell* 124:471–484.

Yang, Y.P., Hu, L.F., Zheng, H.F. et al. 2013. Application and interpretation of current autophagy inhibitors and activators. *Acta Pharmacol Sin* 34:625–635.

Yang, Y.P., Liang, Z.Q., Gu, Z.L. and Z.H. Qin. 2005. Molecular mechanism and regulation of autophagy. *Acta Pharmacol Sin* 26:1421–1434.

Yen, W.-L., and D.J. Klionsky. 2008. How to live long and prosper: Autophagy, mitochondria, and aging. *Physiology (Bethesda)* 23:248–262.

Young, A.R., Narita, M., Ferreira, M. et al. 2009. Autophagy mediates the mitotic senescence transition. *Genes Dev* 23:798–803.

Zhou, G., Myers, R., Li, Y. et al. 2001. Role of AMP-activated protein kinase in mechanism of metformin action. *J Clin Invest* 108:1167–1174.

Section V

Treatments in Aging

26 Aging
Grounds and Determents

Sreeja Lakshmi and Preetham Elumalai

CONTENTS

INTRODUCTION

Aging is normally considered as a difficult stage in the life of an individual where we come across diseases and disabilities, physically and emotionally. It puts one to the ramparts of health issues where healthcare costs skyrocket. Aging brings a stage of entropy. Though conventional strategies and beliefs related to aging still exist, the time has come with talks on "healthy aging" which give a good sense of feeling that aged people can stay healthy and active in their years ahead. According to the National Institute of Aging, the average life expectancy at birth in the United States in 1970 was 70.8 years, which increased to 78 years in 2008 and as per the U.S. Census Bureau, it is assumed to reach 79.5 years by 2020.

GROUNDS OF AGING

The past two decades have seen a fruitful period of investigations regarding the biological basis of aging. Aging is an essential, inevitable physiological phenomenon and a collective consequence of genetic, environmental, and lifestyle factors (Figure 26.1) [1]. Even the correct definition of the aging mechanism is still a complicated issue, increase in the risk of death by decreasing the ability to survive owing to the accumulation of deleterious molecular damages in cells and tissues characterize aging [2].

Aging is associated with a myriad of diseases and functional deficits like cancer, cardiovascular diseases, neurodegenerative disorders, arteriosclerosis, arthritis and osteoporosis, diabetes, and dementia, to name a few, and many kinds of "non-disease" aging manifestations, like age-related blood vessel stiffening, high blood pressure, decreasing elasticity of lung fiber, falling vital capacity caused by gradual collagen cross-linking, skin slacking, eyesight degeneration, joint stiffening, lipofuscin formation, etc. [3]. Even if the causes and consequences of aging goes in a chain, some important facts underlying aging are discussed below:

Genomic instability   Intercellular communication   Loss of proteostasis

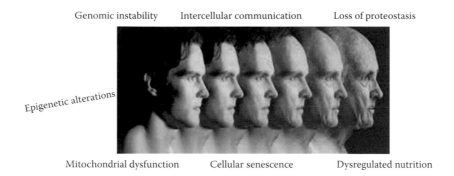

Epigenetic alterations

Mitochondrial dysfunction   Cellular senescence   Dysregulated nutrition

FIGURE 26.1 Schematic representation of major contributors to aging. (Modified from C. Lopez-Otin et al., 2013.)

1. Impairment in intercellular communication

   Alteration at the intercellular communication level is one among the causatives of aging under which inflammatory responses play an important role. These changes can affect various systems like the nervous system and endocrinal systems [4,5]. Failure of the immune system to evade pathogens, secretion of proinflammatory cytokines, and activation of nuclear factor kappa B (NFκB) results in increased production of interferons, interleukins, and tumor necrosis factor-α (TNF-α) [5,6]. As age progresses, senescent cells impair normal functioning of tissues by excessively accumulating in them. By secreting mediators of inflammation, they induce aging in neighboring cells (contagious aging) and cause inflammation as well as support tumor progression. This is accompanied by increased DNA damage as a result of inefficient DNA repair system(s) [7]. Inflammation is associated with diseased conditions like atherosclerosis, type-2 diabetes, and obesity [8,9]. It was also found that inflammatory responses activate NFκB in the hypothalamus and reduce the synthesis of gonadotropin-releasing hormone (GnRH) [10], which in turn enhances age associated changes like reduced neurogenesis, bone and muscle weakness, etc.

2. Instability of genetic harmony

   Increased accumulation of damaged DNA is another hallmark of aging which is associated with various exogenous physical, chemical, and biological agents and reactive oxygen species (ROS) like endogenous factors [11,12]. Genome integrity is adversely affected by damage in nuclear and mitochondrial DNA (mt DNA). Studies in mice and humans have shown that impairment in DNA repair mechanisms underpin numerous premature aging diseases like Werner's syndrome, Bloom syndrome, etc. [13]. Like nuclear DNA, mt DNA equally contributes to a large extent to aging [14]. Studies with mice devoid of mt DNA polymerase γ showed premature aging and reduced lifetime owing to impaired mitochondrial function [15]. Premature aging syndromes are also contributed by laminopathies—a defective nuclear architecture [16]. Abnormalities in nuclear lamina cause aging syndromes like the Hutchinson–Gilford syndrome as a result of mutations in genes coding proteins of lamina or alterations in the laminar structure [17]. Telomere aberrations are also potential contributors toward aging [18].

3. Cellular senescence

   Accumulation of senescent cells increases with aging. Normally, tissues possess a cell replacement mechanism that enables efficient removal of senescent cells and regeneration of new ones. However, in aged organisms, the system becomes inefficient so that senescent cells start accumulating, resulting in aging [1]. P16^{INK4a}/Rb and p19^{ARF}/p53 pathways rank at the top in implementing aging through oncogenic or mitogenic alterations [19]. Stem cell decline, a cumulative consequence of various forms of tissue damage, also contributes as one of the major characteristics of aging [20].

4. Mitochondrial dysfunction

Mitochondrial dysfunction and aging have been suggested to be closely related and researches underpinning the background of this synergy keep on highly challenging. Adenosine triphosphate (ATP) production in mitochondria reduces as the efficacy of the electron transport chain (ETC) diminishes [6]. Oxygen is central to life, but as a by-product of oxidative metabolisms of certain bio-compounds, ROS such as hydrogen peroxide, hydroxyl radicals, and superoxide anions are generated which are highly toxic to cells called oxidative stress [21,22]. Against these ROS, biological systems have evolved the production of antioxidants which can maintain a balance between the production of antioxidation and ROS. Catalase, superoxide dismutase (SOD), glutathione peroxidase, glutathione reductase, thioredoxin, and thioredoxin reductase are the important enzymes maintaining this homeostasis. Any imbalance between generation of ROS and antioxidants can lead to undue oxidation of biomolecules by ROS (Figure 26.2). Besides being a promising player in major cases of life-threatening disorders, oxidative stress followed by inflammatory responses leading to pathological changes is a key factor in aging of organisms. Oxidative stress and impaired cholinergic function result in cognitive decline and memory deficits in aged people [23]. Destruction of protein structure and induction of certain functional mutations in DNA, membrane dysfunction, and cell lysis are among the consequences of oxidative stress. Formation of ROS in mitochondria also adversely affects different components of the electron transport chain (ETC), which in turn can cause the depletion in ATP formation and subsequent damage in mitochondria and other cellular organelles [21,22,24].

5. Dysregulated nutrition

The importance of diet restriction in increasing life span has been gaining importance as a prime topic of debates, which highlights the relevance of diet, exercise, and lifestyle in our daily routine. Pathways like the "insulin and IGF-1 signaling (IIS) pathway," which participate in glucose sensing, and pathways like mammalian target of Rapamycin

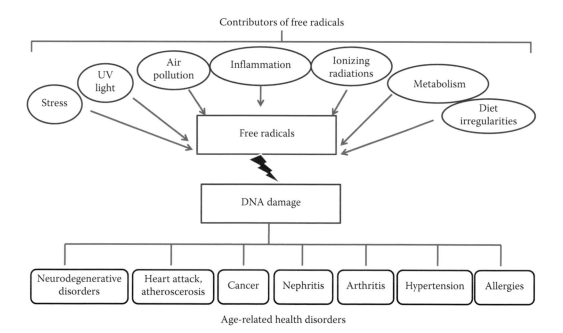

FIGURE 26.2 Causes and consequences of free radicals. Free radicals generated from a series of processes damage DNA, followed by organ damage, resulting in manifestations of aging highlighted with age-related disorders.

(mTOR), AMP-activated protein kinase (AMPK), and sirtuins in sensing high amino acid concentration, high adenosine mono phosphate (AMP) level, and NAD^+, respectively, are all having importance in the process of aging [24]. Also, studies have shown that aging is accelerated by metabolic signaling [25]. Metabolic pathways especially mitochondrial activity is considered as the most promising measures of longevity [25].

6. Loss of proteostasis

   Various forms of endogenous as well as exogenous stress, like oxidative stress, heatshock, etc., lead to improper folding of proteins during protein synthesis. Unfolded proteins have different fates: either refolded or get destroyed by the ubiquitin–proteasome pathway or autophagy. As aging progresses, the ubiquitin–proteasome pathway and lysosomal pathway decline and their failure leads to the accumulation of damaged proteins resulting in proteotoxic effects and aging [1]. Age-related neurodegenerative disorders like Alzheimer's disease (AD) and Parkinson's diseases (PDs) have been associated with the aggregation of misfolded or unfolded proteins [26].

7. Alterations in epigenetic factors

   Several studies have come up with the influence of epigenetic alterations like alterations in DNA methylation, posttranslational modification of histones, and chromatin remodeling in aging. They can also result in progeria like syndromes [27]. Mice models and human patients suffering from the progeroid syndrome have been identified in exhibiting greater DNA methylation and histone modifications than in normal aging [28]. Aging is also characterized by regulation of telomere length and increase in transcriptional changes encoding mitochondrial, lysosomal, and inflammatory systems [29,30].

DETERMENTS OF AGING

Scientific research now is heading toward biomedical interventions and natural strategies to reduce the disabilities associated with aging. Lifestyle factors like diet, physical activity, and remedies from natural resources, such as herbal and marine compounds, are on board to delay or even prevent age-associated disorders and increase the life longevity. A general description and the role of some important factors in creating healthy aging is described below.

Caloric restriction is found to be the most studied way for life longevity [31]. It refers to the diet with a specific amount of calories lower than in normal diet, but with essential nutrients. The effect of calorie restriction (CR) has been studied employing various organisms, from unicellular to mammals [31]. Physiological changes associated with CR are elicited by hormonal response which causes activation of intercellular signaling cascades. Cells are susceptible to decreased risk of cancerous transformation owing to increased efficiency of antioxidant systems. Caloric restriction sometimes leads to lack of energy due to lack of nutrients. On the other hand, it induces mitochondrial respiration and causes hyperpolarization of mitochondria, leading to excess production of ROS. In this case, uncouplers of oxidative phosphorylation perform as a geroprotective agent [32,33]. Anionic uncoupler, 2,4-dinitrophenol is the most debated uncoupler along with lipophilic compounds [31].

Sirtuins are another promising agent playing a role in slowing down the aging process, thus prolonging life span with improvement in health [7]. Sirtuins were discovered in *Saccharomyces cerevisiae* and further studies have shown their presence from bacteria to humans [34]. In humans, the sirtuin family comprises seven members—SIRT1-7. Although sirtuins belong to the histone deacetylases class III family, they also play roles other than protein deacetylation [7]. The effect of CR increases the level of all sirtuins except SIRT-4 [35]. While SIRT-1 and SIRT-2 serve as markers of either senescence or some diseases such as cancer, diabetes, cardiovascular, and neurodegenerative disorders, SIRT-3 was found to be the only sirtuin found to influence life longevity in humans [7]. Studies have shown that SIRT-3 decreases the level of ROS in cells and increases the activity of enzymes involved in CR [36,7]. Studies have also reported the importance of SIRT-6 and 7 in life span extension. An increase in the level of SIRT-6 resulted in reduction of the insulin-like growth factor (IGF)-1 pathway, which in turn caused facilitated glucose tolerance and reduced fat accumulation as well as resulted in extended life span in mice [37].

Premature aging and progeria like symptoms were observed in SIRT7-/- mice [38]. The mechanism by which sirtuins modulate senescence is highly complex. Aging creates genomic instability followed by DNA damage. Unrepairable DNA damage results in senescence. Sirtuins sustain genomic integrity by repairing DNA damage, which results in a slowing down of the aging process [7]. SIRT-1 is also found to activate a certain repair protein synthesis to repair damaged DNA [39]. Furthermore, SIRT-3 plays an important role in the antioxidative defense mechanism [7] and SIRT-1 plays a pivotal role in epigenetic modifications and exercise-induced beneficial effects to improve the quality of life in which, as studies have shown that sirtuins get activated by mild to moderate exercise [40,41].

Curcumin extracted from the turmeric plant root was found to have an antiaging effect through sirtuin activation. It activates SIRT1 and attenuates mitochondrial oxidative damage elicited by myocardial ischemia–reperfusion injury [42]. Bisdemethoxycurcumin, a curcuminoid, was found to reverse premature senescence induced by oxidative stress in W138 fibroblast by activation of the SIRT1/AMPK signaling pathway. Studies also demonstrated a significant reduction in arterial disinfection and oxidative stress through curcumin supplementation [43,44]. Curcumin exerts beneficiary effects at lower concentrations but if concentration increases, it leads to adverse effects [7].

Resveratrol (3,5,4′-trihydroxy-*trans*-stilbene), a natural polyphenol primarily found in grapes and hence in red wine, has antiaging and antioxidant properties which makes it useful for common age-related diseases, such as obesity, diabetes, cancer, neurodegenerative, and cardiovascular diseases [45]. Resveratrol has a wide range of unique antiaging properties, including cardiovascular benefits via increased nitric oxide production, downregulation of vasoactive peptides, lowered levels of oxidized low-density lipoprotein, and cyclooxygenase inhibition; possible benefits on ad by breakdown of beta-amyloid and direct effects on neural tissues; phytohormonal actions; anticancer properties via modulation of signal transduction, which translates into anti-initiation, anti-promotion, and anti-progression effects; antimicrobial effects; and sirtuin activation, which is believed to be involved in the caloric restriction-longevity effect [46]. Applying a low dose of resveratrol in mice was found to inhibit certain parameters of aging mimicking caloric restriction. Thus, resveratrol can be applied in clinical trials on humans to check if it has any effect on the aging process [47].

HORMONES IN ANTIAGING

Human Growth Hormone

Studies have shown that human growth hormone (hGH) has beneficial effects in elderly people and its supplementation is considered to increase muscle strength, improve the immune system, and increase libido [48]. The side effects of hGH, such as weight gain, increased blood pressure, and induction if diabetes, suggest that it has failed to live up to expectations [49]. However, mice studies have shown that a low dose of growth hormone (GH) treatment can increase the life span in aged mice [50], whereas genetically modified mice producing a high concentration of GH live less longer than controls while mice producing less GH live longer [51]. Studies in humans with a deficiency in GH signaling due to a defect in the GH receptor indicate a strong cancer protection due to decreased GH signaling [52]. Human studies also hint at a supernova effect: hGH makes patients feel better but might actually diminish their life span [52].

Insulin-Like Growth Factor 1

IGF-1 is another hormone that may play a role in aging and can be purchased as a supplement. Endogenous production of IGF-1 is induced by GH and, like GH, the levelof IGF-1 declines with age. In mice, low levels of IGF-1 appear to correlate with longevity; mutations in mice having lower IGF-1 seem to extend life span [53]. Like hGH, IGF-1 injections could be counterproductive.

Low levels of IGF-1 is associated with long life span as proved with multiple animal models. Although IGF-1 does appear to play a role in the aging process, whether it can be used as an antiaging factor is not clear at this stage. Clearly, IGF-1 injections are unlikely to extend life span and, like hGH, may even be harmful [53].

Other hormones whose production decreases with age include dehydroepiandrosterone (DHEA) and melatonin. DHEA has been reported to improve the wellbeing of the elderly in a variety of ways: improved memory, immune system, muscle mass, sexual appetite, and benefits to the skin. Protection against cancer has also been argued but there is no strong scientific evidence for this. Minor side effects such as acne have also been reported. One clinical trial in elderly women found no evidence of benefits from DHEA [54].

Melatonin

It is a hormone mostly involved in sleep and circadian rhythms, the latter hypothesized by some to be associated with aging and life-extension [55]. Melatonin appears to have antioxidant functions and may have some beneficial effects in elderly patients in particular in terms of sleep [56]. Some of its proponents claim that it delays the aging process and many age-related diseases, though this is not proven yet. In mice, melatonin can increase life span but also appears to increase cancer incidence [57]. In humans, there are no data to determine whether melatonin extends longevity, though it might have benefits in some patients [58]. Although it can be used for jet lag and some sleep disorders, it may also cause sleep disorders such as nightmares. One study claimed that melatonin levels do not decrease with age, except maybe at night, although owing to diseases or drugs elderly persons can have low levels of melatonin [59].

Estrogen

For women, estrogen is a popular antiaging therapy. This hormone is generally used in conjunction with others such as in hormone replacement therapy. It does appear to reduce some of the effects of menopause by protecting against heart disease and osteoporosis. On the other hand, it can increase the risk of breast cancer and lead to weight gain and thrombosis as side effects. Although there is a vast amount of literature on the advantages and disadvantages of hormone replacement therapy, in the context of aging, there is no evidence that estrogen plays any role as an antiaging agent. For men, testosterone has also been considered as an antiaging agent, but there is no evidence even if it might have some benefits like increased sexual function and muscle mass [60].

Marine Reservoirs as Natural Remedies

Marine environment is a hot spot owing to its distinctive evolutionary background and extreme biodiversity. A significant amount of progress has been achieved by the scientific research and pharmacological fields in exploring the role of specific and potent bioactive components from the sea creatures. Much awareness has been gained recently regarding the correlation between lifestyle factors associated with diet and therapeutics from natural resources, and health.

The extreme cost of therapeutic practices, such as chemotherapy and drug treatments and their side effects, seems to be big ramparts in treatments against diseases like cancer and cardiovascular and neurodegenerative disorders, to list a few. This created much attention in deriving bioactive compounds from natural resources, including the marine ecosystem. Marine lipids, especially polyunsaturated fatty acids (PUFAs), are proving to be one of the best nutraceuticals in modern biology as well as in pharmaceutical applications owing to their potential effects in preventing health disorders and in promoting health. Awareness of development of products from marine sources with upgraded nutritional importance to the human especially for those who are suffering with aforementioned chronic illnesses, is of extreme importance in todays' lifestyle.

Edible marine organisms such as fish and shellfish, are rich in polyunsaturated fatty acids (PUFAs), particularly, ω-3 fatty acids (ω-3FA) (eicosapentanoic acid (EPA) and docosahexanoic acid (DHA). Apart from PUFAs, other lipid components like monogalactosyldiacylglycerols

(MGDGs), Leucettamol A, diacylglyceryltrimethylhomoserines, and glycero-phospholipids (GPLs) from various marine reservoirs also possess health beneficiaries [61]. EPA and DHA constitute 70%–75% of total PUFA. EPA and DHA can exert hypolipidemic activity by decreasing cholesterol, triglycerides, low-density lipoprotein (LDL), and very low-density lipoprotein (VLDL)-cholesterol in the systemic circulation. Also, they are capable of elevating high-density lipoprotein (HDL)-cholesterol, which lowers the rate of coronary heart diseases and reduces the risk of atherosclerosis and stroke. They also have an influence on kidney function by modulating the retention of water and removal of excess sodium. Dietary consumption of EPA and DHA is beneficial in alleviating a wide range of cancers. Intake of ω-3 fatty acid (FA) also reduces the activities of cartilage destroying enzymes, which are responsible for joint destruction in rheumatoid arthritis. Diet containing ω-3 FA will allow tissue to more efficiently absorb and metabolize glucose in the absence of insulin. Diseases such as asthma, diabetes, psoriasis, thyrotoxicosis, multiple sclerosis, etc. can be moderated by ω-3 FAs. DHA is critical to normal eye and vision development [62]. Along with another FA called linoleic acid, it makes 1/3rd of fatty acids in the human brain and retina. DHA also increases memory power [62]. ω-3 FAs have much beneficiaries on the age-associated disorders in reverting or delaying the disease stage. Age-related macular degeneration (AMD) is a disease associated with aging, resulting in losing central vision. Reports have shown that DHA delays the progression of dementia and AMD [63]. Application of ω-3 FA as therapeutic tools has been studied in treating psychiatric conditions like depression and bipolar disorders, etc. [64]. DHA and EPA are widely prescribed for cancer patients as well. ω-3 FAs are shown to have anticancer effects, along with their role in ameliorating the secondary complications of cancer [65]. Investigations indicate that supplementation with fish oil (43 g/day) or EPA/DHA (41 g EPA and 40.8 g DHA per day), which is associated with positive clinical outcomes such as ω-3 Fas, has gained much attention in reducing certain types of cancers, including breast and colon cancers [65]. Unlimited usages of ω-3 FAs have taken them on par in scientific research and therapeutic applications. Yet the mechanisms by which both marine-derived n-3 PUFAs and other fish oil-derived compounds mediate their effects are yet to be fully explored. Apart from PUFAs, MGDGs-3 and -4 produced from marine microalgae were found to downregulate inducible nitric oxide synthase protein and thereby suppress nitric oxide production [66]. Yet another marine lipid leucettamol isolated from sponge was also found to be an anticancer agent which upregulates the activity of the tumor suppressor p53 protein [67].

CONCLUSION

Aging is an inevitable, continuous, complex, and dynamic process accompanied by many deteriorative and degenerative changes, which include several changes from the cellular to the organism level. Aging is influenced by many factors and threatening disorders such as neurodegeneration. Delaying the aging process with healthy life can provide enormous social benefits. Further research on aging will aim for the early detection, prevention, treatment, and reversal of age-related dysfunction, disorders, and diseases. It is a healthcare model intended to promote innovative science and research to prolong the healthy life span in humans. Support for the use of antiaging chemicals is surging recently, but in parallel their hazardous side effects also cause fear. Hence, natural treatment practices are on board to provide a novel gateway for healthy aging. Even if the topic of healthy aging seems to be arable and escalating in the scientific research community, more investigations are needed to ascertain the efficacy and benefits of antiaging agents.

ACKNOWLEDGMENT

The authors deeply acknowledge Department of Health Research, ICMR, New Delhi, India, for the encouragement and financial assistance (No.R.12013/04/2017-HR) towards HRD fellowship-Women Scientist with break in career to SL. The authors would also like to thank Dr. T. V. Sankar for helpful discussions and extend appreciation for continued support.

REFERENCES

1. Lopez-Otin C, Blasco MA, Partridge L, Serrano M and Kroemer G 2013. The Hallmarks of aging. *Cell* 153:1194–1217.
2. Rajawat YS and Bossis I 2008. Autophagy in aging and in neurodegenerative disorders. *Hormones (Athens)* 7(1):46–61.
3. Cutler RG and Mattson M P 2006. The adversities of aging. *Ageing Res. Rev.* 5:221–238.
4. Russel SJ and Kahn CR 2007. Endocrine regulation of ageing. *Nat. Rev. Mol. Cell Biol.* 8:681–691.
5. Salminen A, Kaamiranta K and Kauppinen A 2012. Inflammaging: Disturbed interplay between autophagy and inflammasomes. *Aging (Albany NY)* 4:166–175.
6. Green DR, Galluzi L and Kroemer G 2011. Mitochondria and the autophagy-inflammation-cell death axis in organismal aging. *Science* 333:1109–1112.
7. Grabowska W, Sikora E and Bielak-Zmijewska A 2017. Sirtuins, a promising target in slowing down the ageing process. *Biogerontology* 18:447–476. doi: 10.1007/s10522-017-9685
8. Barzilai N, Huffmann DH, Muzumdar RH and Bartke A 2012. The critical role of metabolic pathways in aging. *Diabetes* 61:1315–1322.
9. Tabas I 2010. Macrophage death and defective inflammation resolution in atherosclerosis. *Nat. Rev. Immunol.* 10:36–46.
10. Zhang G, Li J et al. 2013. Hypothalamic programming of systemic ageing involving IKK-B, NFkappa B and GnRH. *Nature* 497:211–216.
11. Hoeijmakers JH 2009. DNA damage, aging and cancer. *N. Engl. J. Med.* 361:1475–1485.
12. Moskalev AA, Shaposhnikov MV et al. 2012. The role of DNA damage and repair in aging through the prism of Koch-like criteria. *Aging Res. Rev.* 12:661–684.
13. Gregg SQ, Gutierrez V et al. 2012. A mouse model of accelerated liver aging caused by a defect in DNA repair. *Hepatologz* 55:609–611.
14. Park CB and Larsson NG 2011. Mitochondrial DNA mutations in disease and aging. *J. Cell. Biol.* 193:809–818.
15. Vermulst M, Wanagat J et al. 2008. DNA deletions and clonal mutations drive premature aging in mitochondrial mutator mice. *Nat. Genet.* 40:392–394.
16. Worman HJ 2012. Nuclear lamins and laminopathies. *J. Pathol.* 226:316–325.
17. De sandre-Giovannoli A, Bernard R. et al. 2003. Lamin, a truncation in Hutchinson–Gilford progeria. *Science* 300:2055.
18. Blackburn EH, Greider CW and Szostak JW 2006. Telomeres and telomerase: The path from maize, Tetrahymnea and yeast to human cancer and aging. *Nat. Med.* 12:1133–1138.
19. Serrano M, Lin AW, McCurrach ME, Beach D and Lowe SW 1997. Oncogenic ras provokes premature cell senescence associated with accumulation of p53 and p16INK4a. *Cell* 88:593–602.
20. Rando TA and Chang HY 2012. Aging, rejuvenation and epigenetic reprogramming: Resetting the aging clock. *Cell* 148:46–57.
21. Lin MT and Beal MF 2006. Mitochondrial dysfunction and oxidative stress in neurodegenerative diseases. *Nature* 443:787–795.
22. Andersen JK 2004. Oxidative stress in neurodegeneration: Cause or consequence? *Nat. Rev. Neurosci.* 5:S18–S25.
23. Sohal RS, Mockett RJ and Orr WC 2002. Mechanisms of aging: An appraisal of the oxidative stress hypothesis. *Free Radic. Biol. Med.* 33:575–586.
24. Lakshmi S and Elumalai P 2014. Dietary polyphenols and brain health. In *Food and Brain Health*, M. Essa, M. A. Memon, and M. Akbar (eds.). Nova Science Publishers. Inc, New York, pp. 123–132.
25. Houtkooper RH, Williams RW and Auwrex J 2010. Metabolic networks of longevity. *Cell* 142:9–14.
26. Fontana L, Patridge L and Longo VD 2010. Extending healthy life span—From yeast to humans. *Science* 328:321–326.
27. Powers ET, Marimoto RI, Dillin A, Kelly JW and Balch WE 2009. Biological and chemical approaches to diseases of proteostasis deficiency. *Annu. Rev. Biochem.* 78:959–991.
28. Fraga MF and Esteller M 2007. Epigenetics and aging: The targets and the marks. *Trends Genet.* 23:413–418.
29. Osorio FG, Varela I et al. 2010. Nuclear envelope alterations generate in aging-like epigenetic pattern in microdeficient in Zmpsle24 metaloprotease. *Aging Cell* 9:947–957.
30. deMagalhaes JP, Curada J and Church GM. 2009. Meta analysis of age-related gene expression profiles identifies common signatures of aging. *Bioinformatics* 25:871–881.
31. Knorre DA and Severin FF. 2016. Uncouplers of oxidation and phosphorylation as antiaging compounds. *Biochemistry (Moscow)* 81:1438–1444.

32. Korshunov SS, Skulachev VP and Starkov AA 1997. High protonic potential actuates a mechanism of production of reactive oxygen species in mitochondria. *FEBS Lett.* 416:15–18.
33. Skulachev VP. 1996. Role of uncoupled and non-coupled oxidations in maintenance of safely low levels of oxygen and its one-electron reductants. *Quart. Rev. Biophys.* 29:169–202.
34. Vaquero A. 2009. The conserved role of sirtuins in chromatin regulation. *Int. J. Dev. Biol.* 53:303–322.
35. Watroba M and Szukiewicz D 2016. The role of sirtuins in aging and age-related diseases. *Adv. Med. Sci.* 61:52–62.
36. Jing E, Emanuelli B, Hirschey MD, Boucher J, Lee KY, Lom-bard D, Verdin EM and Kahn CR 2011. Sirtuin-3 (Sirt3) regulates skeletal muscle metabolism and insulin signaling via altered mitochondrial oxidation and reactive oxygen species production. *Proc. Natl. Acad. Sci. USA* 108:14608–14613.
37. Kanfi Y, Naiman S, Amir G, Peshti V, Zinman G, Nahum L, Bar-Joseph Z and Cohen HY 2012. The sirtuin SIRT6 regulates lifespan in male mice. *Nature* 483:218–221.
38. Vakhrusheva O, Smolka C, Gajawada P, Kostin S, Boettger T, Kubin T, Braun T and Bober E 2008. Sirt7 increases stress resistance of cardiomyocytes and prevents apoptosis and inflammatory cardiomyopathy in mice. *Circ. Res.* 102(6):703–710.
39. Chung S, Yao H, Caito S, Hwang JW, Arunachalam G and Rahman I 2010. Regulation of SIRT1 in cellular functions: Role of polyphenols. *Arch. Biochem. Biophys.* 501(1):79–90.
40. Bayod S, del Valle J, Lalanza JF, Sanchez-Roige S, de Luxan-Delgado B, Coto-Montes A, Canudas AM, Camins A, Escorihuela RM and Pallas M 2012. Long-term physical exercise induces changes in sirtuin 1 pathway and oxidative parameters in adult rat tissues. *Exp. Gerontol.* 47:925–935.
41. Csiszar A, Labinskyy N, Jimenez R, Pinto JT, Ballabh P, Losonczy G, Pearson KJ, de Cabo R and Ungvari Z 2009. Anti-oxidative and anti-inflammatory vasoprotective effects of caloric restriction in aging: Role of circulating factors and SIRT1. *Mech. Ageing Dev.* 130:518–527.
42. Yang Y, Duan W, Lin Y, Yi W, Liang Z, Yan J, Wang N, Deng C, Zhang S, Li Y, Chen W, Yu S, Yi D and Jin Z 2013. SIRT1 activation by curcumin pretreatment attenuates mitochondrial oxidative damage induced by myocardial ischemia reperfusion injury. *Free Radic. Biol. Med.* 65:667–679.
43. Kitani K, Osawa T and Yokozawa T 2007. The effects of tetrahydrocurcumin and green tea polyphenol on the survival of male C57BL/6 mice. *Biogerontology* 8:567–573.
44. Fleenor BS, Sindler AL, Marvi NK, Howell KL, Zigler ML, Yoshizawa M and Seals DR 2013. Curcumin ameliorates arterial dysfunction and oxidative stress with aging. *Exp. Gerontol.* 48:269–276.
45. Baur JA and Sinclair DA. 2006. Therapeutic potential of resveratrol: The *in vivo* evidence. *Nat. Rev. Drug Discov.* 5:493–506.
46. Baxter RA 2008. Anti-aging properties of resveratrol: Review and report of a potent new antioxidant skin care formulation. *J. Cosmet. Dermatol.* 7:2–7.
47. Barger JL, Kayo T, Vann JM, Arias EB, Wang J, Hacker TA, Wang Y, Raederstorff D, Morrow JD, Leeuwenburgh C, Allison DB, Saupe KW, Cartee GD, Weindruch R and Prolla TA 2008. A low dose of dietary resveratrol partially mimics caloric restriction and retards aging parameters in mice. *PLoS ONE* 4(3):e2264. doi: 10.1371/journal.pone.0002264.
48. Blackman MR, Sorkin JD et al. 2002. Growth hormone and sex steroid administration in healthy aged women and men: A randomized controlled trial. *JAMA* 288(18):2282–2292.
49. Liu H, Bravata DM, Olkin I, Nayak S, Roberts B, Garber AM and Hoffman AR 2007. Systematic review: The safety and efficacy of growth hormone in the healthy elderly. *Ann. Intern. Med.* 146(2):104–115.
50. Khansari DN and Gustad T 1991. Effects of long-term, low-dose growth hormone therapy on immune function and life expectancy of mice. *Mech. Ageing Dev.* 57(1):87–100.
51. Laron Z 2005. Do deficiencies in growth hormone and insulin-like growth factor-1 (IGF-1) shorten or prolong longevity? *Mech. Ageing Dev.* 126(2):305–307.
52. Guevara-Aguirre J, Balasubramanian P et al. 2011. Growth hormone receptor deficiency is associated with a major reduction in pro-aging signaling, cancer, and diabetes in humans. *Sci. Transl. Med.* 3(70):70ra13.
53. Miller RA 2005. Genetic approaches to the study of aging. *J. Am. Geriatr. Soc.* 53(9 Suppl):S284–S286.
54. Nair KS, Rizza RA et al. 2006. DHEA in elderly women and DHEA or testosterone in elderly men. *N. Engl. J. Med.* 355(16):1647–1659.
55. Froy O and Miskin R 2007. The interrelations among feeding, circadian rhythms and ageing. *Prog. Neurobiol.* 82(3):142–150.
56. Poeggeler B 2005. Melatonin, aging, and age-related diseases: Perspectives for prevention, intervention, and therapy. *Endocrine* 27(2):201–212.
57. Anisimov VN, Zavarzina NY et al. 2001. Melatonin increases both life span and tumor incidence in female CBA mice. *J. Gerontol. A Biol. Sci. Med. Sci.* 56(7):B311–B323.

58. Karasek M 2004. Melatonin, human aging, and age-related diseases. *Exp. Gerontol.* 39(11–12):1723–1729.
59. Zhao ZY, Xie Y, Fu YR, Bogdan A and Touitou Y 2002. Aging and the circadian rhythm of melatonin: A cross-sectional study of Chinese subjects 30-110 yr of age. *Chronobiol. Int.* 19(6):1171–1182.
60. Dominguez LJ, Barbagallo M and Morley JE 2009. Anti-aging medicine: Pitfalls and hopes. *Aging Male* 12(1):13–20.
61. Stonik VA and Fedorov SN 2014. Marine low molecular weight natural products as potential cancer preventive compounds. *Mar. Drugs* 12:636–671.
62. Luchtman DW and Song C 2013. Cognitive enhancement by omega-3 fatty acids from child-hood to old age: Findings from animal and clinical studies. *Neuropharmacology* 64:550–565.
63. Johnson EJ and Schaefer EJ 2006. Potential role of dietary n-3 fatty acids in the prevention of dementia and macular degeneration. *Am. J. Clin. Nutr.* 83(suppl): 1494S–1498S.
64. Prior PL and Galduroz JC 2012. N-3 Fatty acids: Molecular role and clinical uses in psychiatric disorders. *Adv. Nutr.* 3:257–265.
65. Vaughan VC, Hassing MR and Lewandowski PA 2013. Marine polysaturated fatty acids and cancer therapy. *Br. J. Cancer* 108:486–492.
66. Banskota AH, Gallant P, Stefanova R, Melanson R and O'Leary SJB 2013. Monogalactosyldiacylglycerols, potent nitric oxide inhibitors from the marine microalga *Tetraselmis chui. Nat. Prod. Res.* 27:1084–1090.
67. Tsukamoto S, Takeuchi T, Rotinsulu H, Mangindaan REP, van Soest RWM, Ukai K, Kobayashi H, Namikoshi M, Ohta T, Yokosawa H 2008. Leucettamol A: A new inhibitor of Ubc13-Uev1A interaction isolated from a marine sponge, Leucetta aff. microrhaphis. *Bioorg. Med. Chem. Lett.* 18:6319–6320.

27 Skin Aging Clock and Its Resetting by Light-Emitting Diode Low-Level Light Therapy

R. Glen Calderhead

CONTENTS

INTRODUCTION

The aging process in the skin is a combination of chronological events occurring at cellular and tissue levels, together with other damaging processes induced by exogenous environmental stimuli, the most common being exposure to terrestrial levels of sunlight on the skin. Chronological aging is a natural process that takes place at all levels in the skin, from the epidermis down through the dermis and its extracellular matrix (ECM), and just as we age, so does the skin as its active components become less active [1]. Extrinsic aging, associated with a variety of exogenous environmental stimuli, accelerates the degradation of the fibrous structure of the ECM in particular through induction of ultraviolet (UV)-mediated damage [2], and the aging skin becomes even more tired and older-looking.

Good skin care with appropriate topical moisturizing and nutritive preparations, for example, creams, gels, and sera, together with a rigorous daily UVA/B sunscreen regimen, can slow down both the intrinsic chronological process and help protect against the extrinsic environmental insult, but no amount of nonsurgical skin care can stop the process completely and nothing can really turn it back, apart from the resort to surgical interventions of varying degrees of invasiveness: even then, while the structure of the skin and underlying tissues may be repaired and reconstructed to give the appearance of a return to youth, the surgery has not actually altered the aging process itself and indeed, in the older patient, repair will take longer because of the destructive effects of that aging process on the metabolism of the dermis, and the epidermis above it [3].

What we need is actually a way to stop and, if possible, turn back the chronological clock governing the aging of skin cells and structures, and to help to redress at the cellular and subcellular levels the insidious damage wrought by the extrinsic environmental skin stressors. There is a possible way, and that is to use the endogenous healing and reparative powers of our body. One way to achieve this is by deliberately creating some form of controlled damage to induce the wound healing process through the three stages of inflammation, proliferation, and remodeling. However, another more interesting way is through totally noninvasive interventions with low levels of incident photon energy—light: how elegant it would be to be able to combat the ravages of light in the skin, particularly UV damage, with carefully selected light therapy—phototherapy. A direct energy transfer occurs between the incident light energy and the target cells absorbing that energy, without energy loss through the creation of heat (athermal reaction) and without any damage (atraumatic reaction), and that energy revitalizes the energy pool of the cells and in consequence, the organism. This is the basis of phototherapy (as distinct to photosurgery), also known as photobiomodulation (PBM) or low-level light therapy (LLLT). LLLT, at appropriate wavelengths and parameters and from appropriate sources, can rejuvenate the skin's cellular components. LLLT could help to refresh the skin in its entirety, including the ECM and its vascular supply, indeed to affect beneficially all the layers and components of the dermis and the epidermis: in effect, noninvasively resetting the skin aging clock.

INTRINSIC AGING OF THE SKIN

The skin is the largest living organ of our bodies and is in a constant state of renewal. The skin provides the first stage of protection against external environmental insults while maintaining the homeostasis and hydration of the tissues beneath it. It acts as part of our body's thermostat system through dermal blood flow regulation, and helping to keep us cool through sweating, with at any given time around 20% of our total blood volume is pulsing through the dermis. However, like any system which is in constant use, it requires some form of maintenance. If there is damage, it needs to be repaired, Most of all, as in any machine that is in constant use every minute of every day of every year, despite good maintenance, the skin inevitably gets tired and ages. Skin ages in two ways, intrinsically from the inside, and extrinsically from outside influences.

Intrinsic or chronological aging occurs in every organ in the body. Chronological aging of the skin mirrors this natural process whereby as we age a number of things happen. The thicker epidermis of youth becomes notably thinner with the passage of time as the mother keratinocytes in the basal layer of the epidermis slow down and produce fewer and poorer-quality daughter keratinocytes which make up the prickle cell layer, the thickest epidermal layer. Tight cell–cell adhesion in the stratum spinosum deteriorates. The characteristic wave-like formation of the dermoepidermal junction with the epidermal rete peg formations and dermal papillary processes gradually flattens out. This reduces the area of contact between the dermis and the epidermis where the former oxygenates and feeds the latter through the basement membrane, with adverse effects on the living cells in the basal layer. Plump collagen bundles in the dermal ECM become less abundant and cross-linking between fibers breaks down. As collagen is responsible for the shear strength of the skin, its ability to withstand stretching without tearing, the skin gets weaker. Elastic fibers lose their ability to reform deformed skin as elastin quality decreases and, particularly around the muscles of mimicry, folds appear in the skin which start as fine lines and enlarge to form wrinkles which are visible even with the skin at rest. Areas of subdermal fat begin to migrate downward to form fat bags under the eyes and jowling at the lower mandible. Reparative processes slow down so wound repair takes longer. The pool of ECM mesenchymal cells which can differentiate into new, active fibroblasts is depleted, and the quality and energy of the fibroblasts, the absolute key cell which maintains homeostasis of the entire ECM, deteriorates. In turn, the main type of collagen in the ECM, type I, begins to be replaced by the less-tough type III, and elastotic changes to the skin occur with many interstitial spaces in the ECM seen on hematoxylin and eosin staining [4]. Even with the best maintenance program using topical cosmetics and cosmeceuticals, chronological aging is inevitable: without any such program, it simply develops faster.

PHOTOAGING OF THE SKIN

Apart from the intrinsic endogenous aging process summarized above, the extrinsic environmental processes compound the senescence of cells and fibers which comprise the epidermis and dermis, and the major villain is the effect of terrestrial solar radiation on the skin, in particular the UV component consisting of ultraviolet (UVA) and much smaller quantities of UVB. Figure 27.1 shows a typical graphic representation of terrestrial solar radiation. Of the light that reaches the Earth's surface, on average, infrared radiation makes up just over 49% while visible light provides just over 42%. Although UV radiation makes up only around 8% of the total amount of solar radiation, it is the most potentially damaging agent.

Light, comprising minute particles of pure energy without mass known as photons, is classed by its wavelength measured in nanometers (1 nm = 1 × 10^{-9} m). Figure 27.2a illustrates a typical sinusoid waveform of a beam of light, and the wavelength (λ in the figure) is measured from the beginning to the end of one complete cycle. In Figure 27.2b there are several waveforms with different wavelengths. It can be noted that, as the wavelength gets shorter, so the waveform alters over time on the x-axis with more cycles per given unit of time. This is referred to as the frequency of light, and is measured in cycles per second, or hertz (Hz). One cycle per second is 1 Hz, 10 cycles is 10 Hz and so on, and it can be noted in Figure 27.2b that the frequency increases in direct proportion to the decreasing wavelength: in other words, shorter wavelengths have higher frequencies [5,6]. Because of this, UV radiation, and to a lesser extent shorter wavelengths of visible light, are more damaging to the skin than the rest of the visible and infrared energy [7]. This is because of the energy level of the photons which make up a beam of light, so light can be said to comprise photons propagating linearly in a waveform path (Figure 27.2), and each photon has its own photon energy.

When a photon is absorbed by, for example, a skin cell, the energy from the photon is transferred to the energy pool of the cell with, in the optimum case, beneficial results. However, the energy level of photons is directly related to the frequency of the light, and the higher the frequency, the greater the energy level of the photons [8]. Thus light at shorter wavelengths consists of photons with higher energy levels than those at longer wavelengths. There comes a point, a critical threshold, at which that level of energy is intrinsically harmful to the cell which cannot cope with the sudden influx of energy: increasing levels of energy beyond that threshold can cause disruption to the molecules and atoms making up the cell, and if the energy is high enough, can dissociate the atoms, literally

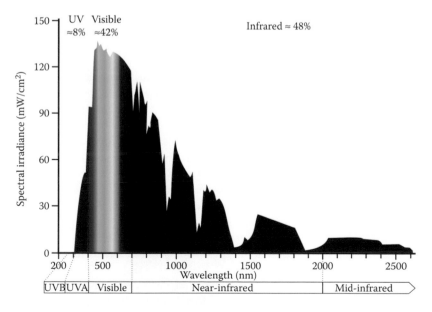

FIGURE 27.1 Graphic illustration of the waveband of sunlight at the surface of the Earth showing the proportion of UV energy and its component parts to visible light and infrared energy and its components, in addition to the relative spectral irradiance in mW/cm^2.

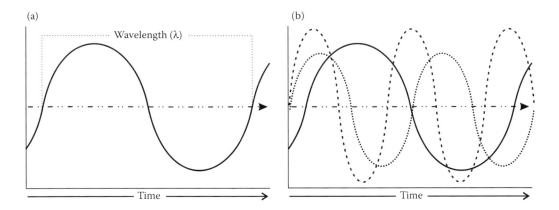

FIGURE 27.2   Concepts of wavelength and frequency illustrated. (a) The wavelength (λ), of a photon (light) is measured as one complete cycle as the photon propagates linearly in a sinusoidal wave. (b) As the wavelength of light energy decreases, the frequency (cycles per second, Hz) increases in direct proportion.

bursting them apart, releasing components as free-electron particles called ions. x-ray energy is a good example of ionizing radiation.

Although not as powerful as x-ray energy, solar UV energy occupies the waveband of 200–400 nm, and is divided into UVC (200–280 nm), UVB (280–320 nm), and UVA (320–400 nm), all of which have potentially damaging photon energy levels. UVC is frankly carcinogenic, but does not reach the Earth at all because of the stratospheric ozone layer which absorbs it. The terrestrial levels of UV therefore comprise UVB and UVA. The ozone layer also absorbs most of the incident UVB radiation, but the very small fraction which reaches the Earth is considered to be the major source of the more severe solar radiation-induced damage to the epidermis and dermis, including the induction of skin cancer [9]. UVA makes up the major part of solar UV radiation reaching the Earth. Considering the adverse effect of high photon energy levels, the inherent damage potential of terrestrial UVA–B is from 3.5 to 5 times that of the near-infrared waveband and even visible blue light has almost three times the damage potential. Bearing these facts in mind, we should remember that all terrestrial UV radiation which is most damaging to the cellular system, especially damage to DNA, can lead to carcinogenicity [9,10].

UVA radiation, which was previously considered to be much less harmful than UVB, is now known to have the potential to react strongly with organic oxygen (O_3) found in tissues and cells to create singlet oxygen $^1O_2$. This free radical can attack cells and tissues through the process called oxidative stress. This is basically an over-production of free radicals beyond the ability of the cells to counteract their harmful effects (detoxification) through antioxidants.

The fibrous component of the dermal ECM, collagen fibers and elastic fibers, gives skin its inherent strength. As mentioned above, chronological aging weakens these fibers and thus the overall integrity of the skin. Oxidative stress rapidly accelerates the deleterious effects of chronological aging on collagen and elastin [11,12], creating even worse skin laxity.

Excessive oxidative stress can also induce apoptosis in skin cells [13], a programmed cell death which takes a toll on the pool of dermal matrix mesenchymal cells and fibrocytes (quiescent fibroblasts) which differentiate when required to replace aging fibroblasts as they approach the end of their approximately 8-week life span.

Although intrinsic aging plays an important role in the deterioration of the epidermis and dermis as we naturally age, it is the extrinsic UV radiation together with short-wave visible blue light which can cause the greater amount of concomitant and synergistic damage. The best solution therefore appeared to be to try to redress the ravages of both time- and particularly light-associated damage, in other words to rejuvenate the age-damaged skin.

REPAIRING TIME- AND LIGHT-MEDIATED SKIN DAMAGE IN AGING SKIN

Surgical intervention was the first reparative strategy. Surgical approaches seek to repair the appearance of the aging face by reconstructing tissues to simulate the look of youth, and the skilled surgeon can certainly achieve that. However, this approach does nothing for the skin aging clock which keeps ticking in the simulated youthful face. The skin may look tight, but it's still "old." I have argued that true rejuvenation of the aging skin can only be achieved by stimulating the body's natural wound repair process, and early interventions other than plastic surgery did just that in an invasive way by creating damage. Mechanical dermabrasion removed the epidermis with a rapidly rotating burr, rather like a sander on wood [14]. Although it enjoyed some popularity, it was messy, extremely painful, and the side-effect potential was extremely high with long patient downtime [15]. Wound healing was certainly induced, but the incidence of textural and pigmentary changes was unacceptably high and it fell into disuse. The development of chemical peeling techniques took over from mechanical dermabrasion, where varying strengths and formulae of acid-based compounds were used to create damage in the skin to a controlled depth followed by shedding of the damaged skin accompanied by regeneration. Peels are still a popular method of skin rejuvenation [16], and in the hands of an expert, chemical peels can still give very good results: the occasional adverse side effect can still, however, prevail [17].

PHOTOSURGICAL SKIN REJUVENATION

The laser entered the skin rejuvenation arena in the late 1990s and continues very successfully in various forms until today, so light, the "L" in "laser" (light amplification by stimulated emission of radiation), began to be used to repair time-and light-damaged skin. Skin contains around 85% water, so two laser types were identified which had high absorption in water to use that as their target, or chromophore, namely the carbon dioxide (CO_2) and erbium yttrium aluminum garnet (Er:YAG) lasers (Figure 27.3) [18,19]. At appropriate incident power densities, the laser energy was rapidly absorbed in the water in the target tissue resulting in an explosive ablative effect as the tissue water was vaporized, hence the technique was described as ablative laser resurfacing. In the hands of the expert, the laser could ablate the epidermis cleanly off the dermis, leaving some residual thermal damage (RTD) in the dermis, basically tissue coagulation with varying degrees of necrosis. RTD was the holy grail of ablative resurfacing to kick-start the body's wound healing process, namely inflammation, proliferation, and remodeling. The inflammatory phase immediately after actual wounding leads to the proliferative or regenerative phase by which mostly the dermal fibroblasts are stimulated to produce new collagen and elastin to renew and tighten the skin together with reepithelization with a renewed epidermis. Ultimately, all the new fibers and structures are tightened and firmed up in the months-long remodeling process.

Various ablative laser resurfacing techniques were championed with varying amounts of success, but in general, the approach at its best became, and remains, the gold standard for the resurfacing of severely aged skin [15,20,21]. However, side effects could be severe with crusting and oozing making the patient's participation in social activities very difficult for at least a week, and prolonged erythema, over weeks or even months, was also a problem. Both pigmentary and textural changes were potential unwanted side effects, even with high levels of clinical technique and laser technology. Full-face ablative resurfacing was therefore a "no pain, no gain" approach.

Patients started to demand something as or nearly as effective, but with less downtime and unwanted sequelae and so there was a brief phase of new laser devices that aimed to create controlled coagulation in the dermis under an unharmed and chilled epidermis, so called "nonablative laser rejuvenation" [22]. The theory was good, however the practice was not, and patient satisfaction with this approach was universally low [23]. Although good neocollagenesis was recognized in the dermis, the epidermis remained as the "same old epidermis" and that was what the patients saw in the mirror, not the nice new dermis.

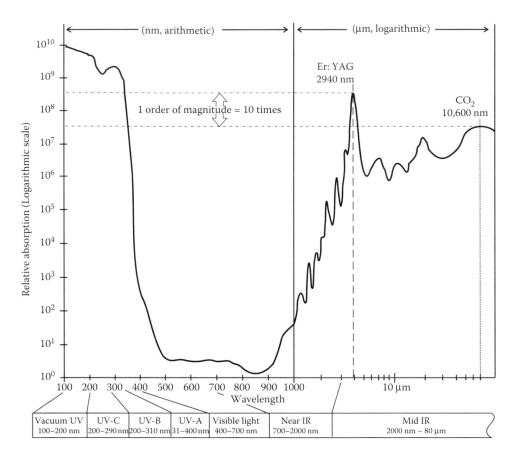

FIGURE 27.3 Water absorption curve showing superior absorption for Er:YAG compared with CO_2. (Adapted from Calderhead RG: Wavelength and penetration. In Calderhead RG, *Photobiological Basics of Photosurgery and Phototherapy*. 2011, Hanmi Medical, Seoul, South Korea. pp. 55–57. Used with permission.)

In an effort to achieve the efficacy of ablative resurfacing while minimizing patient downtime and sequelae, the next phase was fractionation technology [24], where the laser beam was fractionated into pulses of multiple μm-wide microbeams, leaving a larger ratio of untreated to treated skin to accelerate wound healing, and minimize downtime. Fractional laser resurfacing went through a nonablative phase, which initially suffered the same lack of patient satisfaction as nonablative skin resurfacing with more or less the "same old epidermis" syndrome as the earlier strategy, and finally ablative fractional resurfacing triumphed, once again with the CO_2 and Er:YAG lasers [24–26]. The rejuvenation cycle had gone through 360°. While not achieving the safe effects in severely aged skin as the fully ablative approach, the fractional ablative approach continues to deliver excellent results in the rejuvenation of aging skin with much less downtime and significantly fewer sequelae. Patients, however, demanded less and less invasive techniques for their rejuvenation procedures, as efficacious as possible but with no pain and side effects, so another approach was actively sought after.

PHOTOTHERAPEUTIC SKIN REJUVENATION

The fractional nonablative and ablative laser and energy-based technologies for the rejuvenation of aged skin are all invasive, although minimally invasive compared with the fully ablative approach, and are extremely effective particularly for very severely photoaged and damaged skin. It was considered ideal to capture as much of this efficacy as possible in a totally noninvasive way, taking into account that ever-increasing patient-driven demand for effective rejuvenation without side effects or downtime.

Early on in the development of lasers in surgery and medicine it was noticed incidentally that the "L" of laser, light, had a very beneficial effect on target cells and tissue at very low incident photon intensities [27], and indeed could accelerate wound healing in a totally noninvasive manner as well as alleviating pain [28]. This was the birth of phototherapy and the medical laser took a new direction diametrically opposite to the concept of using photothermal damage to achieve clinical efficacy, such as in the photorejuvenation of aging skin. Instead, very low photon intensities were used to achieve a level of reaction in target cells below their damage threshold, but that could produce a useful clinical effect in target tissue, without heat (athermal) and damage (atraumatic). Low level laser therapy (LLLT) was the term coined by Ohshiro and Calderhead to describe this new arm of the medical laser [28], with the principle illustrated in Figure 27.4. Phototherapy, or LLLT, is therefore defined as the use of low incident levels of photon energy to induce some form of clinical effect at subcellular, cellular, and tissue levels but without heat or damage.

Dedicated small laser systems were quickly developed based on laser diode (LD) technology and excited a great deal of interest in LLLT. Controlled studies appeared in the literature showing excellent healing of normal or compromised wounds with LLLT [29–32], and LLLT was clearly shown to accelerate wound healing in both animal and human studies [33]. So, what does wound healing actually accomplish? The answer must include good collagen renewal with remodeling, restoration of the architecture, elasticity and hydration of the skin, improved blood supply, and enhanced cellularity of both the dermis and the epidermis. These are the exact mechanisms which also underpin the rejuvenation of aging skin. Since LLLT could heal both traumatic and iatrogenic wounds, focus moved to the use of LD-LLLT for skin rejuvenation with the face as an obvious main target. However, the face is quite a large area of tissue, and lasers by their point source nature can irradiate

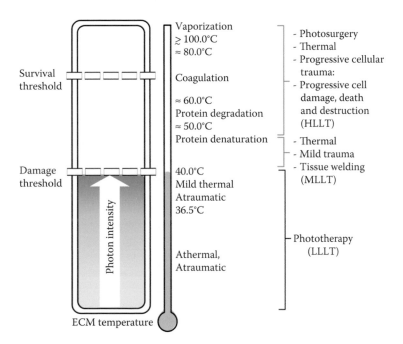

FIGURE 27.4 Schematic of a cell showing its arbitrary damage and survival thresholds. Increasing intensities of light, that is, photon energy, are absorbed in the cell, and as long as the level of reaction is below the photothermal damage threshold, the result is a photoactivated cell, in other words a phototherapeutic reaction. Further increases in photon intensity lead to higher temperatures in the cell and surrounding tissues, leading eventually to photodestruction.

only one time-consuming small area of skin at a time, even when defocused, and tend to be rather expensive. Some method for treating the entire face in a more convenient manner was thus required.

The thoughts of the aesthetic community turned to light-emitting diodes (LEDs), a non-laser cousin of the LD, as a potential phototherapy light source. At the time, LEDs were inexpensive and very bright, but were unfortunately also inappropriate for clinical indications having very divergent beams with low and unstable output powers and a very large spectral output. That meant they could not be used to target the specific chromophores identified as required for efficient wound healing, and by extension, photorejuvenation. Fortunately, the Space Medicine Laboratory at the USA National Aeronautical and Space Administration developed a new generation LED, the NASA LED, which was orders of magnitude more powerful than previous LEDs, had much narrower divergence, and probably most importantly, offered quasimonochromaticity [34]. Because the emitted wavelength was only a few nanometers either side of the rated value, these could be used effectively in LLLT applications including wound healing, as very soon thereafter was demonstrated by the developers of the NASA LED [35]. A leading US photobiologist proposed that LEDs were part of the phototherapy spectrum, and suggested that the acronym LLLT should now cover both laser- and light-based sources under the revised term "low level *light* therapy" [36,37]. LED-LLLT very soon garnered, and continues to garner, a very large number of research and clinical articles on wound healing in animals and man, showing accelerated healing in traumatic and iatrogenic wounds, in up to half the historically anticipated time [38,39], with prophylaxis shown for LED-LLLT against hypertrophic scar formation [40]. It was therefore justifiably argued that LED-LLLT should be very interesting for skin rejuvenation, and indeed an increasing body of solid scientific evidence for the efficacy of LED-LLLT for skin rejuvenation has been building up over the past few years [41–49].

With this increased interest in LED-LLLT for skin rejuvenation, a brief look at what LEDs actually are would therefore be appropriate. Like LDs, LEDs are very efficient light-producing semiconductor chips with the advantage of being significantly less expensive than LDs. Large numbers of LEDs can also be mounted in adjustable planar arrays to irradiate entire areas of the body in a hands-free manner, making them ideal for rejuvenation of, for example, the face or décolleté. Unlike LDs, LEDs emit noncoherent light, but in a very narrow band centered around the rated wavelength, with more than 98% of the photons at that wavelength, for example, 830 ± 5 nm. This is called quasimonochromaticity, and is important because, as will be examined below, the skin cells and other structures which are the targets for LED-LLLT rejuvenation respond best to light of specific wavelengths.

Like LDs, LEDs emit a divergent beam of light, but unlike LDs, this light cannot be collimated (i.e., made into a parallel beam) and focused because it is noncoherent and not like laser energy. However, LEDs offer an interesting phenomenon whereby the intensity of light at the target can be higher that at the LEDs themselves because of the way the LED beams interact with each other, known as photon interference. Although full collimation is impossible, energy from an LED array can be "squeezed" into a slightly more narrow beam, thereby increasing the intensity at the target and enhancing the potential for effective skin rejuvenation (Figure 27.5). This is why LED-LLLT for photorejuvenation is best applied from as large an active treatment head as possible, and with the treatment head at a distance from the target tissue rather than right next to it. Photon interference is one of the mysteries of quantum physics, but this phenomenon in LED-LLLT practice has been demonstrated and reported through actual measurements with power meters in a working system [50].

In summary, LEDs therefore have inherent and powerful advantages for applications in skin photorejuvenation.

- The solid state nature, as described above, means that LEDs require no flashlamps and no energy is lost through heat generation.
- LEDs are very efficient, needing only a low-voltage current to produce a great deal of light energy.

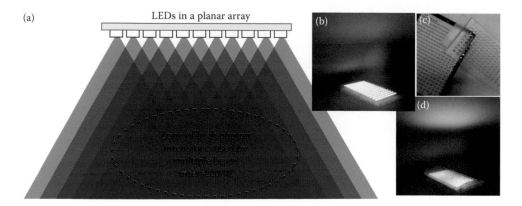

FIGURE 27.5 The principle of photon interference and semicollimation illustrated. (a) LEDs in an array showing multiple intersection of the beams creating the zone of high photon intensity due to photon interference between the individual LED beams. (b) 830 nm LED array in action captured with an infrared camera showing the intensity of light reaching the target in the upper part of the photograph. Note the light energy lost to lateral scatter. (c) The LED array on the left with an optically transparent panel of semicollimating lenses (right) precisely designed to fit over each LED in the panel. (d) The same LED array as in 5b at the same output power, but fitted with the semicollimating lens panel. Note the significantly higher photon intensity of the energy reaching the target and the minimized loss through scatter around the LED array. (5b, c and d courtesy of the R&D Department, Lutronic Corporation, Goyang, South Korea.)

- LEDs are as capable as LDs of photoactivating the skin rejuvenation cells with proven quasimonochromaticity.
- LEDs can be mounted in large planar arrays, and when these panels are articulated, the irradiator or treatment head can be adjusted to deliver even levels of energy to large contoured areas of the body, such as the entire face, in a hands-free manner.
- Because they are not emitting laser energy, LED-LLLT systems are inherently safe and can be legally operated by trained assistants without requiring a physician operator thus freeing up clinicians to attend to other patients.
- LED-LLLT treatment is comfortable, pain-free, and side-effect free, and is well-tolerated by patients of all ages from infants to centenarians.
- Finally, LEDs are less expensive that even laser diodes, so this enables manufacturers to keep the costs down for both clinicians and patients.

These advantages make LED-based systems extremely attractive for LED-LLLT in skin photorejuvenation as a totally noninvasive way to make skin look younger and tighter, and not only to appear younger, but actually to be younger, from the inside out, in a noninvasive, athermal, and atraumatic manner.

LED-LLLT FOR SKIN REJUVENATION: WAVELENGTH IS THE KEY

Apart from the large body of evidence available on the efficacy of LED-LLLT for wound healing [27–33,35–40], as noted above a number of publications specifically on skin rejuvenation with LEDs have appeared [41–49]. The wavelengths reported in these studies are in the near infrared and visible red of 830 nm and 633 nm, respectively. It is clear that wavelength is most important to consider when experimenting with irradiating living biological tissue, as Karu has suggested on identifying the targets for LLLT-LED, namely the photoreceptors [51,52]. One study reported on pulsed visible yellow LED-LLLT at 595 nm [53], but the results in that study failed to be replicated and use of this wavelength in monotherapy was hence discontinued.

The first law of photobiology states that absorption needs to occur before a reaction can take place, and wavelength is the primary factor in photorejuvenation which determines not only what the targets are, but how deeply light can penetrate to reach these targets. Figure 27.6 shows a photospectrogram of the penetration of polychromatic white light through a human hand *in vivo* measured by the author. The system determines the optical density (OD) of the intervening tissue, and thus how deeply the light can penetrate at specific wavelengths. Best penetration was seen at the waveband from 630 to 840 nm, thus illustrating the useful depth of penetration at 633 and 830 nm, the two wavelengths which have been reported as the most interesting ones for skin rejuvenation. When LED energy cannot reach the target cells, absorption obviously cannot take place and thus both 633 and 830 nm with their superior penetration have found good applications in LED-LLLT photorejuvenation, with 830 nm showing more favorable results [54,55].

Referring to the above studies, one article which clearly shows the efficacy of LED-LLLT in skin rejuvenation is the outstanding study by Lee et al., which deserves a more detailed explanation to show the real potential of LED-LLLT in skin photorejuvenation [43]. As its central theme, the study sought to compare the efficacy of LED-LLLT photorejuvenation among 830 nm alone (N = 21), 633 nm alone (N = 18), and 830 nm followed by 633 nm (N = 22) with a sham-irradiated control group (N = 17). The numbers in each group were those patients who completed the treatments (8 sessions over 4 weeks) and the follow-up period (12 weeks from the final treatment session), thus representing the potential for good statistical strength.

In the independently-assessed clinician global assessment based on the clinical photography, all the treatment groups scored significantly higher than the control group. There was no significant difference seen among the three treated groups. In terms of patient satisfaction, no patient was dissatisfied with the result in any of the treatment groups, but the drop-out rate was very high in the placebo sham-irradiated group. When examining those subjects in the LED-treated groups who scored their result as excellent at the 12-week assessment, the lowest number was in the 633 nm group followed by the 830 + 633 nm group, and the highest number was in the 830 nm group. The 830 nm group also showed the fastest subjective response (Figure 27.7).

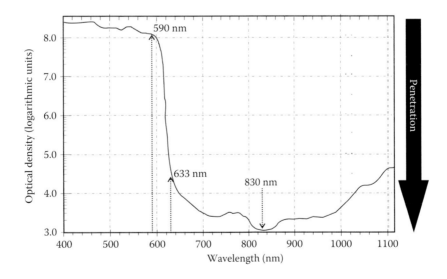

FIGURE 27.6 Transmission photospectrogram of the author's hand using "white light" (400–1100 nm) shown on the x-axis, and the OD of the tissue on the y-axis in logarithmic units. Penetration in indicated on the z-axis. The higher the OD value, the poorer the penetration. Penetration at 590 nm (yellow) is poor, at 633 nm (red) it is almost 4 orders of magnitude better, and at 830 nm (near-IR), it is 5 orders of magnitude better.

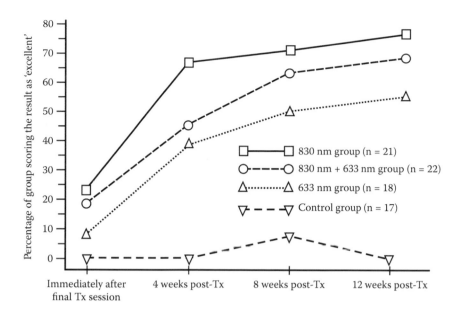

FIGURE 27.7 Percentage of subjects in each group scoring their skin rejuvenation result as "excellent" immediately after the final treatment session, then at 4, 8, and 12 weeks after the final session. (Drawn based on patient subjective satisfaction data in Lee SY et al. A prospective, randomized, placebo-controlled, double-blinded, and split-face clinical study on LED phototherapy for skin rejuvenation: Clinical, profilometric, histologic, ultrastructural, and biochemical evaluations and comparison of three different treatment settings. *J Photochem Photobiol B* 2007; 88: 51–67.)

A Cutometer was used to measure skin elasticity objectively. All treatment groups showed increasing improvement during the 12-week follow-up, with the best improvement seen in the 830 nm group. The steady increase seen in both patient satisfaction and skin elasticity during the 12-week assessment can be explained by the remodeling process improving the new collagen and elastin laid down during the treatment period, which could be correlated with the proliferative phase of wound healing induced by the LED-LLLT. The Cutometer readings additionally clearly showed wrinkle reduction in the LED-LLLT groups, particularly the 830 nm group, with improvement noted as the 12-week follow-up period progressed in a similar manner to those of the increasing skin elasticity and patient satisfaction values.

The histological specimens showed much improved collagen and elastin deposition in all the treatment groups. Ultrastructural examination of specimens with transmission electron microscopy backed up the histological findings, consistently showing young-looking, plump, and fibroplasic fibroblasts in the treated groups surrounded by well-organized collagen bundles compared with the untreated groups in which fibroblasts were rather thin and surrounded by disorganized collagen fibers. This showed replacement of the older and tired fibroblasts by young and active ones, an essential component for resetting the skin aging clock.

Immunohistochemistry gave an interesting insight into the potential photoprotective effect being conferred on the new collagen fibers by tissue inhibitors of matrix metalloproteinases (TIMPs) following LED-LLLT. Additionally, real-time polymerase chain reaction (RT-PCR) assessments illustrated upregulation of connexin 43, suggesting better cell–cell adhesion of the daughter keratinocytes in the stratum spinosum which correlated with the increased thickness and cellularity of the epidermis as seen in the histological photomicrographs.

In short, the Lee study showed, both from gross photography and detailed tissue assessments at cellular and subcellular levels, that LED-LLLT had successfully and noninvasively turned back the skin aging clock in all the treated group patients, with 830 nm LED-LLLT showing the best results.

Levels of Action of LED-LLLT Skin Rejuvenation

Taking all of the evidence clearly shown in Lee et al.'s article discussed above and in other articles exploring LED-LLLT at tissue, cellular and subcellular levels, it can be argued that LED-LLLT intervention has been proven to be capable of rejuvenating intrinsically aged and photodamaged skin in a noninvasive, athermal and atraumatic manner: that it can also in fact turn back the skin aging clock through the interaction of LED-LLLT and all levels of cutaneous targets, consisting of the cells and components of the epidermis and ECM of the dermis. Although LED-LLLT of necessity takes time and cannot target the deepest of wrinkles and roughest of skin, the results are tangible for the mild-to-moderately photoaged patient who does not want downtime and aggressive intervention, and who is prepared to wait, because LED-LLLT is working from the inside out rather than the outside in, through targeting specific cells. It could be said that the body is being stimulated to rejuvenate aging skin by itself, in a completely natural manner.

First of all, LED-LLLT works at the systemic level benefiting areas of the body not directly targeted. This has been well demonstrated, particularly in a study on the indirect healing of standardized burns in the rat and mouse model using the systemic effect [56]. On post-burn day 7, all of the indirectly irradiated wounds were at least 70% healed, and some 40% were fully healed compared with the control wounds. It is believed that photoproducts created by light–cell interaction in the irradiated tissue are carried by the vascular or lymphatic system throughout the body, thereby accelerating the healing of the wounds in the treated animals even though they were not directly irradiated. The blood vascular system is another good candidate for the systemic effect. In a study on the axial pattern flap in the rat model, treatment group animals were treated with 830 nm LLLT aimed at the iliolumbar feeder artery for the flap, while control animals were treated in exactly the same way but without actual irradiation [57]. At 90 min after a single treatment, speckle Doppler flowmetry showed increased full-flap perfusion in the treated animals but not in the control animals, which was matched by statistically significantly better flap survival in the treated group.

At the cellular level, the mast cell is another potential target in LED-LLLT photorejuvenation. In the Lee study discussed above, proliferation of both collagen and elastin was clearly seen as the second regenerative phase of wound healing. It has been demonstrated that treatment of normal skin *in vivo* in one arm of human subjects acted on the mast cells to induce degranulation, accompanied by recruitment of more mast cells, macrophages, and neutrophils 48 hours after a single 830 nm LED-LLLT treatment compared with the contralateral unirradiated arm [58]. Although a very small and nonsignificant increase in mast cells and macrophages, but not neutrophils, was seen in the contralateral arm, no degranulation of mast cells occurred, indicating a mild form of systemic effect after a single 830 nm irradiation. These findings suggested that active degranulation of mast cells, normally only seen in wounds or an allergic response, created a mild inflammatory response in the 830 nm LED-LLLT irradiated skin but without any physical wound, in fact, it created what I would term a "quasi-wound." In the study, the irradiated arms had none of the traditional signs of inflammation, but in fact the ultrastructural TEM findings showed that an inflammatory response had been created which I suggest would lead to the proliferative stage of wound healing with enhanced collagen and elastin deposition, just as seen in practice in the Lee study analyzed above. During degranulation of mast cells, one of the components released into the ECM during the final stage of degranulation is superoxide dismutase (SOD), an extremely powerful exogenous antioxidant. The presence of SOD in the ECM could well be an added protective factor in LED-LLLT photorejuvenated skin against oxidative stress, just as the TIMPs seen 48 hr after LED-LLLT in the Lee study could exert a photoprotective influence on the new collagen.

As for other cell types associated with photorejuvenation and turning back the skin aging clock, studies have shown that macrophages, when treated with LED-LLLT *in vitro*, not only work harder and faster in phagocytosing targets, but also release manyfold greater amounts of fibroblast growth factor (FGF), a beneficial trophic factor for fibroblasts in their proliferating phase [59,60]. Pooled human neutrophils treated with 830 nm demonstrated increased chemotaxis toward their target and enhanced opsonic activity through oxidative burst activity [61,62], in addition to synthesis of trophic factors.

The fibroblast is an essential target for rejuvenating the aging skin, not only for collagen deposition but also for maintaining ECM homeostasis. Already mentioned above was the histological evidence for LED-LLLT-mediated enhanced fibroblast and elastic fiber deposition and ultrastructural evidence of renewal of youthful and fibroplasic fibroblasts in aged skin following LED-LLLT [43]. In a recent real-time polymerase chain reaction study it was shown that near-IR LED-LLLT activated the elastin (ELN) gene in human fibroblasts, the gene responsible for elastin protein production for neoelastinogenesis and human fibroblast functionality [63]. Most interestingly, the older the subject from whom the fibroblasts were pooled, the higher was the ELN upregulation.

An ultrastructural study of human skin following a course of *in vivo* LED-LLLT irradiations showed rapid transformation of intracellular vimentin granules in the fibroblast cytosol into vimentin fibrils with development of significant cellular fibroplasia [64]. Vimentin is copolymerized with desmin in the Golgi complex to form procollagen filaments, the soluble precursor of extracellular tropocollagen, which in turn form collagen fibers and bundles. Unirradiated skin showed no such changes.

The keratinocyte has also the nickname of the "cytocyte" because it is capable of synthesis of a large number of different cytokines, many of which are important in the skin rejuvenation process. Some studies have shown that, when keratinocytes are cultured in medium and irradiated with LED-LLLT, fibroblasts proliferated and synthesized collagen remarkably faster in the keratinocyte-conditioned medium both *in vitro* and *in vivo* than in the control normal medium [65,66]. The keratinocyte is therefore not to be ignored in the skin photorejuvenation program.

CONCLUSIONS

Phototherapy, particularly with LEDs as LED-LLLT, has the strong potential to redress the damage in the skin caused by both time and external factors, and not only that, to turn back the skin aging clock rather than simply reconstructing or ablating older skin to make it look younger. It has to be said that LED-LLLT does not result in a rapid transformation. On the other hand, LED-LLLT as a monotherapy does produce good results in mild-to-moderate skin damage with a series of treatments followed by some top-up sessions after a of a few months.

LED-LLLT rejuvenates skin by first inducing the inflammatory response athermally, with an atraumatic quasi-wound. LED-LLLT then goes on to enhance the proliferation phase of wound healing, and to tighten skin through a photoenhanced remodeling phase. Moreover, LED-LLLT addresses both the epidermis and the dermis, very important when we consider it is the epidermis that people see in the mirror.

This can be taken one stage further and it could be suggested that, while monotherapy LED-LLLT does work, and works well, it might be even more exciting to combine LED-LLLT as an adjunct to any other form of rejuvenation treatment, from one as mild as a powder dermabrasion, through nonablative and ablative fractional laser procedures, to frank surgical approaches.

We could go one stage even further and LED-LLLT could start to be applied in subjects in their late teens before the first signs of aging have begun to appear, to act as a noninvasive prophylaxis against both intrinsic and extrinsic aging, to help dramatically slow down the skin aging clock and postpone or even help avoid the use of the more drastic forms of skin rejuvenation. This could be applied in combination with good creams and sera, and of course a solid UVA/B sunscreen regimen. That would be true "photoantiaging"!.

The Scottish poet, Robert Burns, wrote that "Nae man can tether time nor tide" ... but perhaps we can slow it down or even turn it back in the skin from the inside out with the noninvasive power of LLLT, particularly using LEDs.

REFERENCES

1. Makrantonaki E, Zouboulis CC.The skin as a mirror of the aging process in the human organism: State of the art and results of the aging research in the German National Genome Research Network 2 (NGFN-2). *Exp Gerontol* 2007; 42: 879–886.

2. Farage MA, Miller KW, Elsner P, Maibach HI. Intrinsic and extrinsic factors in skin ageing: A review. *Int J Cosmet Sci* 2008; 30: 87–95.

3. Ashcroft GS, Horan MA, Ferguson MWJ. Aging is associated with reduced deposition of specific extracellular matrix components, an upregulation of angiogenesis, and an altered inflammatory response in a murine incisional wound healing model. *J Invest Derm* 1997; 4: 430–437.

4. Friedman O. Changes associated with the aging face. *Facial Plast Surg Clin North Am* 2005; 13: 371–380.

5. http://www.dermatology.ca/skin-hair-nails/skin/photoaging/what-is-photoaging/ (Accessed 02/27/2017).

6. http://naturalfrequency.com/wiki/solar-radiation (Accessed 03/10/2017).

7. Sjerobabski Masnec I, Poduje S. Photoaging. *Coll Antropol* 2008; 32(Suppl 2): 177–180.

8. Calderhead RG. Wavelength and frequency. In Calderhead RG (ed.), *Photobiological Basics of Photosurgery and Phototherapy.* 2011; Hanmi Medical, Seoul, Korea. pp 7–10.

9. Jugé R, Breugnot J, Da Silva C, Bordes S, Closs B, Aouacheria A. Quantification and characterization of UVB-Induced mitochondrial fragmentation in normal primary human keratinocytes. *Sci Rep.* 2016 Oct 12; epub ahead of print. Published online October 12, 2016. doi: 10.1038/srep35065

10. Moura Valejo Coelho M, Matos TR, Apetato M. The dark side of the light: Mechanisms of photocarcinogenesis. *Clin Dermatol.* 2016; 34: 563–570.

11. McAdam E, Brem R, Karran P. Oxidative stress-induced protein damage inhibits DNA repair and determines mutation risk and therapeutic efficacy. *Mol Cancer Res* 2016; 14: 612–622.

12. Liu C, Yang Q, Fang G, Li BS, Wu DB et al. Collagen metabolic disorder induced by oxidative stress in human uterosacral ligament-derived fibroblasts: A possible pathophysiological mechanism in pelvic organ prolapse. *Mol Med Rep* 2016; 13: 2999–3008.

13. Bayir H, Kagan V. Bench-to-bedside review: Mitochondrial injury, oxidative stress and apoptosis—There is nothing more practical than a good theory. *Crit Care* 2008; 12: 206.

14. Mandy SH. Dermabrasion. *Semin Cutan Med Surg* 1996; 15: 162–169.

15. Kirkland EB, Gladstone HB, Hantash BM. What's new in skin resurfacing and rejuvenation? *G Ital Dermatol Venereol* 2010; 145: 583–596.

16. Cortez EA, Fedok FG, Mangat DS. Chemical peels: Panel discussion. *Facial Plast Surg Clin North Am* 2014; 22: 1–23.

17. Truchuelo M, Cerdá P, Fernández LF. Chemical peeling: A useful tool in the office. *Actas Dermosifiliogr.* 2016; 108: 315–322.

18. Fitzpatrick RE. Maximizing benefits and minimizing risk with CO_2 laser resurfacing. *Dermatol Clin* 2002; 20: 77–86.

19. Alster TS, Lupton JR. Erbium:YAG cutaneous laser resurfacing. *Dermatol Clin* 2001; 19: 453–466.

20. Ross EV, Miller C, Meehan K, McKinlay J, Sajben P et al. One-pass CO_2 versus multiple-pass Er:YAG laser resurfacing in the treatment of rhytides: A comparison side-by-side study of pulsed CO_2 and Er:YAG lasers. *Dermatol Surg.* 2001; 27(8): 709–715.

21. Trelles MA, Allones I, Luna R. One-pass resurfacing with a combined-mode erbium: YAG/CO_2 laser system: A study in 102 patients. *Br J Dermatol* 2002; 146: 473–480.

22. Levy JL, Trelles M, Lagarde JM, Borrel MT, Mordon S. Treatment of wrinkles with the nonablative 1,320-nm Nd:YAG laser. *Ann Plast Surg* 2001; 47: 482–488.

23. Trelles MA, Allones I, Luna R. Facial rejuvenation with a nonablative 1320 nm Nd:YAG laser: A preliminary clinical and histologic evaluation. *Dermatol Surg.* 2001 Feb; 27(2): 111–116.

24. Brightman LA1, Brauer JA, Anolik R, Weiss E, Karen J et al. Ablative and fractional ablative lasers. *Dermatol Clin* 2009; 27: 4479–4489.

25. Clementoni MT, Gilardino P, Muti GF, Beretta D, Schianchi R. Non-sequential fractional ultrapulsed CO_2 resurfacing of photoaged facial skin: Preliminary clinical report. *J Cosmet Laser Ther* 2007; 9: 218–225.

26. Trelles MA, Mordon S, Velez M, Urdiales F, Levy JL. Results of fractional ablative facial skin resurfacing with the erbium:yttrium-aluminium-garnet laser 1 week and 2 months after one single treatment in 30 patients. *Lasers Med Sci* 2009; 24: 186–194.

27. Mester E, Jászsági-Nagy E. Biological effects of laser radiation. *Radiobiol Radiother (Berl).* 1971; 12: 377–385.

28. Ohshiro T, Calderhead RG. *Low Level Laser Therapy: A Practical Introduction*. 1988, John Wiley & Sons, Chichester.

29. Reddy GK, Stehno-Bittel L, Enwemeka CS. Laser photostimulation of collagen production in healing rabbit Achilles tendons. *Lasers Surg Med* 1998; 22: 281–287.

30. Byrnes KR, Barna L, Chenault VM, Waynant RW, Ilev IK et al. Photobiomodulation improves cutaneous wound healing in an animal model of type II diabetes. *Photomed Laser Surg* 2004; 22: 281–290.

31. Chung TY, Peplow PV, Baxter GD. Laser photobiomodulation of wound healing in diabetic and non-diabetic mice: Effects in splinted and unsplinted wounds. *Photomed Laser Surg* 2010; 28: 251–261.

32. Magri AM, Fernandes KR, Assis L, Mendes NA, da Silva Santos AL et al. Photobiomodulation and bone healing in diabetic rats: Evaluation of bone response using a tibial defect experimental model. *Lasers Med Sci* 2015; 30: 1949–1957.

33. Ohshiro T. *Low Level Laser Therapy: Practical Application*. 1992, John Wiley & Sons, Chichester, UK.

34. Whelan HT, Houle JM, Whelan NT, Donohoe DL et al. The NASA light-emitting diode medical program: Progress in space flight and terrestrial applications. *Space Tech. App. Int'l. Forum* 2000; 504: 37–43.

35. Whelan HT, Smits RL Jr, Buchman EV, Whelan NT et al. Effect of NASA light-emitting diode (LED) irradiation on wound healing. *J Clin Laser Med Surg* 2001; 19: 305–314.

36. Smith KC. Laser (and LED) therapy is phototherapy. *Photomed Laser Surg* 2005; 23: 78–80.

37. Smith KC. Laser and LED photobiology. *Laser Ther* 2010; 19: 72–78.

38. Trelles MA, Allones I, Mayo E. Combined visible light and infrared light-emitting diode (LED) therapy enhances wound healing after laser ablative resurfacing of photodamaged facial skin. *Med Laser App* 2006; 28: 165–175.

39. Min PK, Goo BCL. 830 nm light-emitting diode low level light therapy (LED-LLLT) enhances wound healing: A preliminary study. *Laser Ther* 2013; 22: 43–49.

40. Park YJ, Kim SJ, Song HS, Kim SK, Lee JH et al. Prevention of thyroidectomy scars in Asian adults with low-level light therapy. *Derm Surg* 2015; 42: 526–534.

41. Russell BA, Kellett N, Reilly LR. A study to determine the efficacy of combination LED light therapy (633 nm and 830 nm) in facial skin rejuvenation. *J Cosmet Laser Ther* 2005; 7: 196–200.

42. Trelles MA. Phototherapy in anti-aging and its photobiologic basics: A new approach to skin rejuvenation. *J Cosmet Dermatol* 2006; 5: 87–91.

43. Lee SY, Park KH, Choi JW, Kwon JK, Lee DR et al. A prospective, randomized, placebo-controlled, double-blinded, and split-face clinical study on LED phototherapy for skin rejuvenation: Clinical, profilometric, histologic, ultrastructural, and biochemical evaluations and comparison of three different treatment settings. *J Photochem Photobiol B* 2007; 88: 51–67.

44. Baez F, Reilly LR. The use of light-emitting diode therapy in the treatment of photoaged skin. *J Cosmet Dermatol* 2007; 6: 189–194.

45. Sadick NS. A study to determine the efficacy of a novel handheld light-emitting diode device in the treatment of photoaged skin. *J Cosmet Dermatol* 2008; 7: 263–267.

46. Boulos PR, Kelley JM, Falcão MF, Tremblay JF, Davis RB et al. In the eye of the beholder: Skin rejuvenation using a light-emitting diode photomodulation device. *Dermatol Surg* 2009; 35: 229–329.

47. Calderhead RG, Vasily DB. Low level light therapy with light-emitting diodes for the aging face. *Clin Plast Surg* 2016; 43: 541–550.

48. Barolet D, Roberge CJ, Auger FA, Boucher A, Germain L. Regulation of skin collagen metabolism *in vitro* using a pulsed 660 nm LED light source: Clinical correlation with a single-blinded study. *J Invest Dermatol* 2009; 129: 2751–9275.

49. Gold MH. Light-emitting diode. *Curr Probl Dermatol* 2011; 42: 173–180.

50. Park MK, Kim BJ, Kim NM, Mun SK Hong HK. The measurement of optimal power distance in LEDs. *Korean J Derm* 2011; 49: 125–130 (in Korean, abstract in English).

51. Karu TI, Kolyakov SF. Exact action spectra for cellular responses relevant to phototherapy. *Photomed Laser Surg* 2005; 23: 355–361.

52. Karu, T. Identification of the photoreceptors. In Karu, T. *Ten Lectures on Basic Science of Laser Phototherapy*. 2007, Prima Books AB, Grangesberg, Sweden. pp. 115–142.

53. Weiss RA, McDaniel DH, Geronemus RG, Weiss MA, Beasley KL et al. Clinical experience with light-emitting diode (LED) photomodulation. *Dermatol Surg* 2005; 31(9 Pt 2): 1199–1205.

54. Kim WS, Calderhead RG. Is light-emitting diode phototherapy (LED-LLLT) really effective? *Laser Ther* 2011; 20, 206–215.

55. Calderhead RG, Kim WS, Ohshiro T, Trelles MA, Vasily DB. Adjunctive 830 nm light-emitting diode therapy can improve the results following aesthetic procedures. *Laser Ther.* 2015; 24: 277–289.

56. Lee GY, Kim WS. The systemic effect of 830-nm LED phototherapy on the wound healing of burn injuries: A controlled study in mouse and rat models. *J Cosmet Laser Ther* 2012; 14: 107–110.

57. Kubota J. Effects of diode laser therapy on blood flow in axial pattern flaps in the rat model. *Lasers Med Sci* 2002; 17: 146–153.

58. Calderhead RG, Kubota J, Trelles MA, Ohshiro T. One mechanism behind LED phototherapy for wound healing and skin rejuvenation: Key role of the mast cell. *Laser Ther* 2008; 17: 141–148.

59. Young S, Bolton P, Dyson M, Harvey W, Diamantopoulos C. Macrophage responsiveness to light therapy. *Lasers Surg Med* 1989; 9: 497–505.

60. Bolton PA, Dyson M, Young S. Macrophage responsiveness to light therapy: A dose-response study. *Laser Ther* 2004; 14(Pilot Issue): 23–28.

61. Osanai T, Shiroto C, Mikami Y, Kudou E et al. Measurement of GaAlAs diode laser action on phagocytic activity of human neutrophils as a possible therapeutic dosimetry determinant. *Laser Ther* 1990; 2: 123–134.

62. Dima VF, Suzuki K, Liu Q, Koie T et al. Laser and neutrophil serum opsonic activity. *Roum Arch Microbiol Immunol* 1996; 55: 277–283.

63. Roncati M, Lauritano D, Cura F, Carinci F. Evaluation of light-emitting diode (LED-835 nm) application over human gingival fibroblast: An *in vitro* study. *J Biol Regul Homeost Agents* 2016; 30(2 Suppl 1): 161–167.

64. Takezaki S, Omi T, Sato S, Kawana S. Ultrastructural observations of human skin following irradiation with visible red light-emitting diodes (LEDs): A preliminary *in vivo* report. *Laser Ther* 2005; 14: 153–160.

65. Samoilova KA, Bogacheva ON, Obolenskaya KD, Blinova MI et al. Enhancement of the blood growth promoting activity after exposure of volunteers to visible and infrared polarized light. Part I: Stimulation of human keratinocyte proliferation *in vitro*. *Photochem Photobiol Sci*. 2004 Jan; 3(1): 96–101.

66. Tian YS1, Kim NH, Lee AY. Antiphotoaging effects of light-emitting diode irradiation on narrow-band ultraviolet B-exposed cultured human skin cells. *Dermatol Surg* 2012; 38: 1695–1703.

Section VI

Healthy and Successful Aging

28 Social Structure and Healthy Aging
Case Studies*

Jong In Kim and Gukbin Kim

CONTENTS

INTRODUCTION

The survival probability of becoming centenarians (SPBC) has greatly increased over the past decades. However, it is difficult to determine whether the natural life span of humans has increased overall [1]. Thus, the SPBC may have been affected by socioeconomic indicators. The analysis of the socioeconomic factors in a social structure perspective may help identify the factors associated with healthy aging.

The population of centenarians has been steadily increasing globally [2]. However, over the past half-century, the average number of centenarians has dramatically increased [3]. The determinants associated with the increase in the production of centenarians include the cohort size at chosen ages, along with the probability of survival at chosen ages [4]. Therefore, the SPBC is defined as an output of the survival productivity of centenarians by a population. The SPBC (70) in this case study is the SPBC for those who are aged 70.

In a social structure perspective, healthy aging is an important concept for the health of the older adult. It seeks for older adults to have access to improved sanitation facilities (SF), public expenditure on health (PEH), satisfaction with their level of income, and using mobile phones as the standard of living, which are essential as well. Therefore, this chapter seeks to determine the association between such socioeconomic factors and the SPBC (70).

Healthy aging is influenced by biological, environmental, and psychosocial factors in a social structure perspective [5]. In such elements, socioeconomic factors of modifiable risk factors have been researched in some studies. Previous studies have shown that health environmental factors such as drinking water and sanitation can predict death in incidences of disease [6,7]. The association between the use of telephones and the improvement of health has also been studied [8,9]. Finally, previous reports supported that the associations between national economic and relative physical fitness among older adults are positive [10]. However, it is unknown as to whether such associations are applicable to the SPBC (70) [11,12].

---

* This case study is reedited with regard to the authors' following article: "Kim JI, & Kim G: Factors affecting the survival probability of becoming a centenarian for those aged 70, based on the human mortality database: Income, health expenditure, telephone, and sanitation. *BMC Geriatrics*, 2014; 14(1):113." Its copyright is retained by the authors.

Therefore, this case study examines the associations of independent socioeconomic factors with the SPBC (70). In additional, this case study estimates the influences and correlates between the SPBC (70) and national income (NI), health expenditure, mobile telephones, and sanitation in a social structure perspective.

FRAMEWORK OF SOCIOECONOMIC FACTORS

The proposed framework of this case study represents the socioeconomic factors for the SPBC (70) (Figure 28.1). Centenarians have been influenced by socioeconomic factors in this study period. Healthy aging is also influenced by social environmental factors in a social structure perspective [5,13]. In additional, healthy aging indicates physical activity with the absence of diseases for at least 100 years, characterized by the preservation of functional capacity, and well-being for active life in society [14–19]. Thus, the SPBC (70) of this case study has excluded the effect by hereditary factors.

The indirectly affecting environmental health factors can be found in sanitation and water [6,7]. In additional, lack of SF can be a greater challenger of good health against waterborne infections [20,21] among older adults in a social structure perspective. On the other hand, the association between the power of mobile phones and health has also been studied [8,9,18]. The hypothesis of this case study is that the associations between TS and the SPBC (70) may differ between developed and undeveloped countries [18]. Although the associations between health expenditures and an older adult are correlated [10,14,18], it is unknown whether such associations are applicable to centenarians with PEH and NI in this case study [18,19].

Hence we set up the hypothesis for input, progress, and output. Inputs are the age of the older adults, and the progress of the long term is SF, TS, PHE, and NI, and outputs are the difference of the SPBC (70). These factors of social structure (SF, TS, PHE, and NI) could be explained by

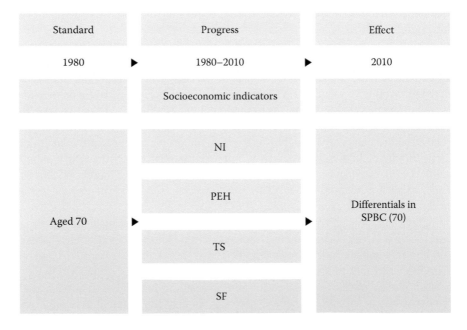

FIGURE 28.1 Description of theoretical framework. SPBC (70): survival probability of becoming centenarians for those aged 70 (per 10,000); NI: national income, per capita (constant 2005 international US$) (1990–2010); PEH: public expenditure on health (% of GDP) (2000–2010); TS: telephone subscribers (per 100 people) (1980–1990); SF: sanitation facilities (%), urban (2005–2010). (From Kim JI, Kim, G: *BMC Geriatr* 2014; 14(1):113.)

successful aging, which may have been affected when becoming 100 years old in 2010, for people aged 70 in 1980.

ESTIMATION OF THE SPBC (70)

This study utilized the human mortality database (HMD) for calculating the survival probability of centenarians [12,18]. The SPBC (70) indicates that the number of those aged 100 in the year 2010, divided by the size of the matching cohort at the age of 70 in 1980 [4,14,18,19]. This calculation to the production of centenarians indicates the number of people aged 70 in 1980 who reached 100 in 2010 [18]. The SPBC (70) consist of cohorts who were born in 1910, who are is expected to live until becoming centenarians in 2010 and thus, the SPBC is an indicator for assessing the production of centenarians (70).

PREDICTION FACTORS FOR SURVIVAL PROBABILITY

To investigate relationships between the socioeconomic factors associated with healthy aging, we conducted a multiple regression analysis for the SPBC (70). Tables 28.1, 28.2, and Figure 28.2 present an analysis of the socioeconomic factors related to the SPBC (70). Significant positive correlation coefficients have been found between the SPBC (70) and socioeconomic indicators (see Table 28.1). A log scale was also used for all explanatory variables (Figure 28.2). Finally, the SPBC (70) of male and female predictors were higher NI and TS, as well as higher PEH and SF as socioeconomic factors ($R^2 = 0.422$, $P < 0.004$) (Table 28.2).

SOCIAL STRUCTURE ON HEALTHY AGING

This case study has investigated the socioeconomic factors associated with the SPBC (70) of healthy aging from a social structure perspective. The factors are as follows:

First, SF of the environmental health indicators has been found as a significant factor in the SPBC (70) [6,7,20–22]. In the case study, Japan, Canada, Switzerland, and the United Kingdom were the highest among all 32 countries; whereas Russia was the lowest. In

TABLE 28.1
Correlations of the Indications with SPBC (70)

Variable		Correlations Coefficient	P-Value
SPBC (70) MF	NI	0.555	0.0001
	PEH	0.583	0.0001
	TS	0.611	0.0001
	SF	0.382	0.0311
SPBC (70) M	NI	0.421	0.0017
	PEH	0.457	0.0081
	TS	0.521	0.0021
	SF	0.351	0.0491
SPBC (70) F	NI	0.568	0.0001
	PEH	0.594	0.0001
	TS	0.623	0.0001
	SF	0.391	0.0281

Source: From Kim JI, Kim, G: *BMC Geriatr* 2014; 14(1):113.

TABLE 28.2

Multiple Regression Models for Predicting SPBC (70)

Model 1

(1) $Y_1 = -50.959 + 2.450\text{E-5 } X_1 + 9.692 X_2 + 1.268 X_3 + 0.327 X_4 + \text{En}$

$R^2 = 0.422$,   F-Value $= 4.933$,   P $= 0.004$

Model 2

(2) $Y_2 = -30.612 + 0.00001 X_1 + 1.984 X_2 + 0.586 X_3 + 0.412 X_4 + \text{En}$

$R^2 = 0.303$,   F-Value $= 2.939$,   P $= 0.039$

Model 3

(3) $Y_3 = -89.981 + 0.00001 X_1 + 15.034 X_2 + 1.974 X_3 + 0.512 X_4 + \text{En}$

$R^2 = 0.439$,   F-Value $= 5.278$,   P $= 0.0031$

Model 1, 2, 3: Income level (+) Expenditure of health (+) Using telephone
(+) Improved sanitation

$Y_1 =$ SPBC (70) Male & Female
$Y_2 =$ SPBC (70) Male
$Y_3 =$ SPBC (70) Female

$X_1 =$ NI
$X_2 =$ PEH
$X_3 =$ TS
$X_4 =$ SF

Source: From Kim JI, Kim, G: *BMC Geriatr* 2014; 14(1):113.

---

countries with low SF, this can be a major contributing factor for the SPBC (70). The death rate stemming from the ISF is also an important public health concern. Older adults who are susceptible to the risk of SF might not have an equal SPBC (70). In Russia the occurrence of the typhus epidemic increased especially in the region where older adults were living under poor sanitation conditions [23]. Russia has the lowest SF because it has insufficient government investment and infrastructure for health. The SPBC (70) in this case study has been indirectly affected by environment health factors (SF). Therefore, the lower the ISF, the lower the promotion of health in preventing infectious diseases and poor hygiene among older adults [20–22]. Concurrently, the policy of using SF may help to maintain the requisite health environment for the SPBC (70). Thus, the correlations between the SPBC (70) and SF are significantly positive, indicating that SF is a crucial contributor to the SPBC (70).

Second, TS reflects the quality of life, the standard of living, and the health promotion aspect by accessing health-related information [8,9,14,18,19,24,25].

We suggest that it is a significant contributory factor to healthy aging because decreases in TS lead to decrease in the SPBC (70) from a social structure perspective. In this case study, it was found that the TS in Poland was the lowest; whereas, in Sweden TS was the highest. TS can be a crucial contributing factor to a higher SPBC (70). In this case study it was found that there is a high correlation between TS and the SPBC (70) because TS reflects the high-level of the quality of life in developed countries and the public investment in telecommunications network infrastructure [14,18]. In additional, telephone counseling can improve mental health and also maintains social relationships in older adults [26,27]. Thus, such counseling appears to have a latent effect on death in late life [27]. Telemedicine health services are complementary for improving health status in the older adults [28]. Therefore, TS is an important contributor to the SPBC (70).

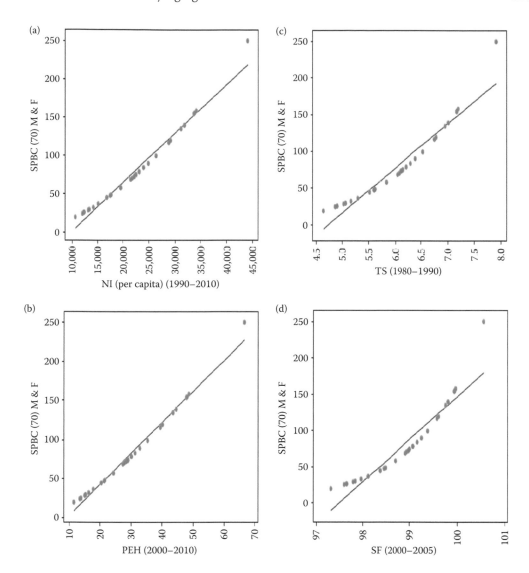

FIGURE 28.2 Association of SPBC (70) male and female with (a) NI; (b) PEH (%); (c) TS (%); and (d) SF (%), 10,000 per. (From Kim JI, Kim, G: *BMC Geriatr* 2014; 14(1):113.)

Finally, PEH and NI reflect the standard of living as economic factors relating to the SPBC (70) [10,14]. In the current case study, PEH in Russia and NI in the Ukraine were the lowest; whereas, the PEH in France and NI in Luxembourg were the highest [18]. We suggest that they are significant contributory factors to healthy aging because decreases in PEH and NI lead to a decrease in the SPBC (70) from a social structure perspective.

CONCLUSIONS

Socioeconomic factors appear to have a key latent effect on the SPBC (70). This case study has identified four socioeconomic indicators associated with the SPBC (70) from the social structure perspective: higher overall NI, PEH, TS, and SF make for healthy aging. Thus, the socioeconomic level seems to affect the SPBC (70) importantly.

REFERENCES

1. Väinö K: On the survival of centenarians and the span of life. *Popul Stud: J Demogr* 1988; 42(3):389–406.
2. Byass P: Towards a global agenda on ageing. *Global Health Action* 2008; 1. DOI: 10.3402/gha. v1i0.1908
3. Berr C, Balard F, Blain H, Robine JM: How to define old age: Successful aging and/or longevity. *Med Sci* 2012; 28(3):281–7.
4. Robine, J-M, Paccaud, F: Nonagenarians and centenarians in Switzerland, 1860–2001: A demographic analysis. *J Epidemiol Community Health* 2005; 59:31–7.
5. Candore G, Balistreri CR, Listì F, Grimaldi MP, Vasto S, Colonna-Romano G, Franceschi C et al.: Immunogenetics, gender, and longevity. *Ann N Y Acad Sci* 2006; 1089:516–37.
6. Cairncross S, Hunt C, Boisson S, Bostoen K, Curtis V, Fung IC, Schmidt WP: Water, sanitation and hygiene for the prevention of diarrhoea. *Int J Epidemiol* 2010; 39(Suppl 1):i193–205.
7. Lin HJ, Sung TI, Chen CY, Guo HR: Arsenic levels in drinking water and mortality of liver cancer in Taiwan. *J Hazard Mater* 2013; 262:1132–8.
8. Kaplan WA: Can the ubiquitous power of mobile phones be used to improve health outcomes in developing countries? *Global Health* 2006; 2:9.
9. McBride CM, Rimer BK: Using the telephone to improve health behavior and health service delivery. *Patient Educ Couns* 1999; 37(1):3–18.
10. Theou O, Brothers TD, Rockwood MR, Haardt D, Mitnitski A, Rockwood K: Exploring the relationship between national economic indicators and relative fitness and frailty in middle-aged and older Europeans. *Age Ageing* 2013; 42(5):614–9.
11. Rosero-Bixby L: The exceptionally high life expectancy of Costa Rican nonagenarians. *Demography* 2008; 45(3):673–91.
12. University of California, Berkeley (USA), and Max Planck Institute for Demographic Research (Germany). Human Mortality Database. Available at www.mortality.org, accessed on June 02, 2014.
13. Mossakowska M, Barcikowska M, Broczek K, Grodzicki T, Klich-Raczka A, Kupisz-Urbanska M, Podsiadly-Marczykowska T et al.: Polish centenarians programme. Multidisciplinary studies of successful ageing: Aims, methods, and preliminary results. *Exp Gerontol* 2008; 43(3):238–44.
14. Kim JI: Social factors associated with centenarian rate (CR) in 32 OECD countries. *BMC Int Health Hum Rights* 2013; 13(1):16.
15. Baker J, Meisner BA, Logan AJ, Kungl AM, Weir P: Physical activity and successful aging in Canadian older adults. *J Aging Phys Act* 2009; 17(2):223–35.
16. Rowe JW, Kahn RL: Human aging: Usual and successful. *Science* 1987; 237(4811):143–9.
17. Hsu HC: Impact of morbidity and life events on successful aging. *Asia Pac J Public Health* 2011; 23(4):458–69.
18. Kim JI, Kim, G: Factors affecting the survival probability of becoming a centenarian for those aged 70, based on the human mortality database: Income, health expenditure, telephone, and sanitation. *BMC Geriatr* 2014; 14(1):113.
19. Kim JI, Kim G: Social structural influences on healthy aging: Community-level socioeconomic conditions and survival probability of becoming a centenarian for those aged 65 to 69 in South Korea. *Int J Aging Hum Dev* 2015; 81(4):241–59.
20. Steyer A, Torkar KG, Gutiérrez-Aguirre I, Poljšak-Prijatelj M: High prevalence of enteric viruses in untreated individual drinking water sources and surface water in Slovenia. *Int J Hyg Environ Health* 2011; 214(5):392–8.
21. Wright JA, Yang H, Rivett U, Gundry SW: Public perception of drinking water safety in South Africa 2002–2009: A repeated cross-sectional study. *BMC Public Health* 2012; 12(1):556.
22. Tumwebaze IK, Orach CG, Niwagaba C, Luthi C, Mosler HJ: Sanitation facilities in Kampala slums, Uganda: Users' satisfaction and determinant factors. *Int J Environ Health Res* 2013; 23(3):191–204.
23. Tarasevich I, Rydkina E, Raoult D: Outbreak of epidemic typhus in Russia. *Lancet* 1998; 352(9134):1151.
24. Kim JI, Kim G: Labor force participation and secondary education of gender inequality index (GII) associated with healthy life expectancy (HLE) at birth. *Int J Equity in Health* 2014; 13(1):106.
25. Kim JI, Kim G: Country-level socioeconomic indicators associated with healthy life expectancy: Income, urbanization, schooling, and internet users: 2000–2012. *Social Ind Res* 2016; 129(1): 391–402.

26. Ailshire JA, Crimmins EM: Psychosocial factors associated with longevity in the United States: Age differences between the old and oldest-old in the health and retirement study. *J Aging Res* 2011; 2011:530534. PMCID: PMC3199053.
27. Wu JY, Leung WY, Chang S, Lee B, Zee B, Tong PC, Chan JC: Effectiveness of telephone counselling by a pharmacist in reducing mortality in patients receiving polypharmacy: Randomised controlled trial. *Br Med J* 2006; 333(7567):522.
28. Wang F: Economic evaluations of the effects of longevity on telemedicine and conventional healthcare provision. *Telemed J E Health* 2011; 17(6):431–4.

29 Successful Aging
Role of Cognition, Personality, and Whole-Person Wellness

Peter Martin, Leonard W. Poon, Kyuho Lee, Yousun Baek, and Jennifer A. Margrett

CONTENTS

INTRODUCTION

Few topics on aging have stimulated so much interest among gerontologists as has the concept of successful aging. Havighurst (1961) first introduced the topic in the first issue of *The Gerontologist*. Since then, articles and monographs by Rowe and Kahn (1997) and by Baltes and Baltes (1990a) have encouraged generations of researchers to debate, test, and evaluate psychosocial concepts of successful aging.

This chapter is consistent with other articles that highlight new trends and perspectives on this topic (Martin et al., 2015). We will first summarize "traditional" and well-established perspectives on successful aging. In our second segment, we will focus primarily on successful cognition and functioning in late life, and a third section links personality as an important topic for a deeper understanding of the construct of successful aging. Finally, we will attempt a linkage between successful aging and "whole-person" wellness.

EARLY THEORIES ON SUCCESSFUL AGING

Since frailty, impairment, and dependency continue to be prevalent beliefs of normal aging, researchers and practitioners alike have come up with alternative perspectives of aging. One perspective is the model of successful aging (Cheng et al., 2015). Successful aging has been at the forefront of a debate between two extremes: disengagement theory (Cumming and Henry, 1961) and activity theory (Havighurst, 1961, 1963). Disengagement theory assumes that successful aging is the desire of older adults for withdrawal from active life in order to prepare them for the end of life. Activity theory, on the contrary, suggests that successful aging involves continued activities and attitudes from midlife onward in order to retain a positive sense about oneself. Williams and Wirths (1965) provided a compromise model by including both engagement in activities and the ability to disengage as important factors of successful aging. Adding to a more individualized perspective, Havighurst (1961, 1963) emphasized the subjectivity of successful aging demonstrating the importance of feelings of happiness and satisfaction.

ROWE AND KAHN'S MACARTHUR MODEL OF SUCCESSFUL AGING

Criticizing prevalent research that emphasized "normal aging" (Bell, Rose, and Damon, 1972; Shock et al., 1984; Palmore, Nowlin, Busse, Siegler, and Maddox, 1985), Rowe and Kahn (1987, 1997) first presented the components of successful aging about three decades ago. These components included (i) low probability of disease and disease-related disability, (ii) high cognitive and functional capacity, and (iii) active engagement with life. The model notes that successful aging is achieved if all three components are realized. This model has been attractive to researchers because of its easy application for evaluating older adults' "success" in aging. The model became popular, inspiring gerontologists to formulate more than 100 variations of the original model (Rowe and Kahn, 2015).

However, this model also has raised criticism, and one obvious criticism was that according to this model only very few older adults were able to maintain the high levels of functioning that would label them as "successful." Cho, Martin, and Poon (2012), for example, found that only 15% of octogenarians fell into the successful aging group, and none of the centenarians in their sample satisfied all three components, highlighting the irony that older adults who survived up to 100 years belong to the less successful group when compared to octogenarians. Willcox and colleagues (2006) also found that only 11% of adults 85 years and older were considered successful.

A second criticism of the approach was that the criteria of successful aging were regarded as fixed end points rather than emphasizing a process where individuals strive for personal goals throughout their life span (Baltes and Carstensen, 1996; Pearlin and McKean Skaff, 1996). In addition, missing elements such as biological components (Masoro, 2001), spirituality (Crowther, Parker, Achenbaum, Larimore, and Koenig, 2002), social structure (Riley, 1998), and subjective constructions (Bowling and Dieppe, 2005) were mentioned. Finally, some proponents argued that this model should be abandoned altogether because the term "successful aging" can create stigma and discriminate older adults with disabilities and impairment (von Faber et al., 2001).

MacArthur Model 2.0

Acknowledging the criticisms of the MacArthur Model since its inception, Rowe and Kahn (2015) recently proposed the MacArthur Model 2.0, adding societal factors to the model of successful aging. They suggested three main goals for scholars: First, they argued that researchers and practitioners should develop more strategies, programs, and policies to serve the aging population, not just focusing on individual needs such as education, work, retirement, and housing. Second, changes in individuals' life course ought to be taken into account relative to changes in longevity: lengthened life expectancy brought change in the tradition of age-segregated roles such as education for youth, work for midlife, and leisure for later life. Rowe and Kahn argued that later life should allow for more varied roles. Last, they suggested that societies should encourage the aging population's human capital: since current older people are fully capable of participating in productive activities, older adults should be involved in paid work and civil engagement. Notwithstanding Rowe and Kahn's remedy, Tesch-Römer and Wahl (2017) pointed out the limitations of the MacArthur Model 2.0. They argued that Rowe and Kahn did not consider that prolonged longevity could also result in longer periods of poor health. Therefore, Tesch-Römer and Wahl suggested that we should consider functional loss, frailty, dying, and death as part of successful aging, so that the concept of successful aging acknowledges concurrent or consecutive periods of health, disease, and disabilities.

SELECTIVE OPTIMIZATION WITH COMPENSATION

Baltes and Baltes (1990a, b) acknowledged aging-related losses (particularly as related to physical and biological domains) and emphasized plasticity that can assist older adults in experiencing more gains than losses. The basic assumption of the selective optimization with compensation (SOC) model is that people use three components of adaptation when facing opportunities (e.g., education, social network) and limitations in resources (e.g., time, energy, and decline in physical health): selection, optimization, and compensation. "Selection" refers to older adults' restricted involvement in activities when they lose functional capacity. Freund and Baltes (1998) explained that this selection process is a loss-based selection, in that people select behaviors in response to their losses. "Optimization" is an effort to maximize what one still has or can do in order to maintain activities (e.g., exercising; Baltes and Carstensen, 1999). Finally, "compensation" refers to efforts to achieve goals with alternative means (e.g., using assistive devices, changing exercise types).

The SOC model overcame many of the limitations of Rowe and Kahn's (1987) successful aging model. Unlike Rowe and Kahn's model, SOC assumes that successful aging is not a homogeneous process but has many different routes and outcomes rather than a final state. In fact, Baltes and Baltes (1990a, b) suggested that multiple subjective and objective criteria are equally important to define successful aging. In addition, SOC acknowledges individual and cultural variations.

Like Rowe and Kahn's model, researchers have used the SOC model in empirical studies. Freund and Baltes (1998) used the Berlin Aging Study data to investigate the empirical value of the SOC model. They reported that people who adopted SOC-related strategies had higher scores in successful aging measures, such as subjective well-being and positive emotions, and an absence of loneliness.

PROACTIVITY MODEL

In contrast to Rowe and Kahn's (1987, 1997) model, the proactivity model (Kahana and Kahana, 1996) was proposed emphasizing strategies to effectively deal with stressors associated with aging in order to avoid negative outcomes. The proactivity model explores pathways to maintain psychological well-being and engagement in social activities even for older adults with illness, frailty, and social losses. Therefore, the main approach of the research applying this model is to find proactive coping strategies with which people prevent future stressors or minimize the negative effects of the stressors. This model is unique in that it places emphasis on the values of human agency in the face of stressors (Kahana, Kelley-Moore, and Kahana, 2012).

According to the proactive theory, older adults can take advantage of proactive coping in order to ensure the preservation of quality of life. When individuals deal with stressors at an early stage, the negative influence of the stressor may be less significant when compared to stressors that fully emerge (Aspinwall and Taylor, 1997). Three types of actions, both corrective and preventive, make up the elements of proactive behaviors and occur prior to the stressor: health-promoting behaviors, planning, and helping others to receive social support in difficult situations.

A number of studies have provided empirical support for this theory. Kahana et al. (2002) reported that older adults experienced an increase in the quality of life eight years after engaging in preventive health behaviors, such as exercise, or refraining from smoking. Kahana et al. (2012) also employed this model in a longitudinal study and reported that proactive adaptations of marshaling supports and planning for the future played a mediating role in the relationship between internal (i.e., active coping and religious coping) and external (i.e., financial and social) resources and quality of life.

Although the proactivity model has faced some criticism that it was overly simplifying the process of coping strategies by neglecting older adults' intra- and interindividual differences (Ouwehand, Ridder, and Bensing, 2007), several studies have demonstrated a potential of this perspective for practical applications. For example, Bode, de Ridder, and Bensing (2006) evaluated an aging preparation education program employing proactive coping. The investigators asked participants to anticipate what they would regret in five years if they did not use proactive coping strategies. Subsequently, they built up strategies to cope with the "warning" signals and mentally simulated the problem solving activities to imagine the attained status. Then they practiced the strategies in the real world and received feedback from other group members. The results of the program revealed that participants experienced a significant increase in proactive coping competencies after the program.

OTHER MODELS

RYFF'S SIX-FACTOR MODEL

Carol Ryff (1989) pointed out four limitations of existing theories of successful aging. The first limitation refers to the absence of theoretical frameworks of well-being. The second shortcoming suggests that mostly negative measurements of successful aging such as anxiety, depression, worry, and illness are included. Next, there is little attention on the possibilities of continued growth and development in older age. Fourth, there is insufficient recognition that well-being is a cultural construction varying by time and region. Ryff tried to address these limitations with a new model including six dimensions of successful aging: (1) self-acceptance; (2) positive relations with others; (3) autonomy; (4) control over one's environment; (5) purpose in life; and (6) personal growth. Many studies have included Ryff's scale of psychological well-being (RPWB). There is a discussion about the number of dimensions contained in the scale (Kafka and Kozma, 2002; Ryff and Singer, 2006; Springer, Hauser, and Freese, 2006; van Dierendonck, Díaz, Rodríguez-Carvajal, Blanco, and Moreno-Jiménez, 2008). This model contributed to the literature by suggesting important dimensions of successful aging and emphasizing positivity rather than just an absence of negative experiences (van Dierendonck et al., 2008).

It is interesting to note that each model of successful aging includes a unique combination of factors. For example, Phelan and Larson (2002) summarized seven major elements based on eleven models of successful aging. These are: life satisfaction, longevity, freedom from disability, mastery/growth, active engagement with life, high/independent functioning, and positive adaptation. Some researchers underscore the importance of including individuals' subjective perceptions and judgment. Understandably, different models predict different levels of successful aging. For example, Strawbridge, Wallhagen, and Richard (2002) demonstrated that 50.3% of older adults responded that they were aging successfully, whereas only 18.8% of older adults were in the category of successful aging as defined by Rowe and Kahn's (1987) criteria.

Strawbridge et al. (2002) also suggested including self-ratings, or self-perceptions, when assessing successful aging. How does one reconcile if there are differences between objective measures outlined by different researcher-based models compared to individual self-related level of success or satisfaction? This is an important and yet unanswered question that has opened a new level of scrutiny and discussion among researchers and policy makers alike.

Although many models contain physical health components, several alternative models emphasize the importance of psychosocial domains of successful aging. In order to further conceptualize a model based on self-ratings of successful aging, it is important to take into account the cognitive capacity, as well as personality traits and coping strategies, filters through which we view the world.

SUCCESSFUL AGING AND COGNITION

This section addresses the questions (i) how much cognition individuals need to reliably rate their level of successful aging, (ii) how does cognition influence one's quality of life, and (iii) how do cognitive processes change in normal and pathological aging. On the basis of our current knowledge, we have a significant amount of information on normative and pathological cognitive changes in aging. However, answers to the first two questions are equivocal, and we hope our chapter will garner new thinking and research directions for these questions.

There is no argument that cognition is important in everyday functioning. Cognition is our ability to perceive, learn, remember, think, plan, and make decisions—all the necessary tools that make us humans. It is the central processing system from which humans navigate and survive in the environment. It is easy to imagine that without the ability to cognate there would not be any quality of life or successful aging. Functional capacity, on the other hand, is the ability to navigate in our environment in everyday life. Some of these functions are more physical in nature, such as eating, bathing, dressing, toileting, walking or transferring, and continence. Others are more instrumental in nature that would allow a person to live independently, such as housework, preparing meals, taking medications, and managing money. Cognition and functioning are intrinsically interrelated. Cognition governs all everyday functions, some more so than others.

A question arises on what happens when cognition and functional capacities fall from normal to abnormal levels and how these processes impact the level of success in one's aging. The answers to the question relate in an understanding of normative aging changes in cognition and function followed by an understanding of how these systems fail in aging that could significantly compromise the quality of life and level of success in aging.

Normative Changes

Among publications on age-related changes on psychological processes, many have focused on cognitive differences and changes. The proportion of these publications indicates the relative importance of cognition in the aging of individuals.

Cognition and aging have been studied from different perspectives in experimental psychology (e.g., Salthouse, 1991, 2017a, 2017b), psychometrics (e.g., Schaie, 2013), clinical diagnosis and treatment (e.g., Poon, 1986) as well as cognition in everyday contexts (Poon, Rueben, and Wilson, 1992). Different approaches have employed different cognitive models and measures. We can draw four major conclusions in normal aging of cognitive processes (Schaie, 2013). One, most cognitive abilities peak at about 45–50 years and then decline in some fashion until death. Two, some abilities such as the learning and retention of rote information (crystallized intelligence) decline very little in normal aging, while the learning of new information (fluid intelligence) tends to decline over the adult life span. Three, when inspecting the change in a constellation of abilities, most individuals may show some decline in one type of ability, while almost none decline in all types of abilities in normal aging. Four, although studies were able to identify normative or average changes in different types of cognition at different age decades, there is a significant amount of differences across individuals. *This is a major reason why it is difficult to make differential cognitive or dementia diagnosis, especially in early stages of dysfunction.* Later on in this chapter, we will further elaborate this important aspect of individual variability in cognitive functions and differential diagnosis of dysfunction. These four sets of findings have demonstrated that older adults in the normal time course of aging and without pathologies can retain their cognitive abilities to strive to age successfully.

As noted earlier, cognitive abilities and changes in physical and mental functions are associated with functional capacity in everyday life or the ability to navigate in the daily environment. Intuitively, unimpaired older adults should be able to function well in their everyday environment. Data do confirm this hypothesis with four conclusions. One, performance ability plateaus at about age 50 years when it may gradually decline (Schaie, 2013). Two, great stability is found of functional performances over time *within* individuals (Finlayson et al., 2005). Three, performance errors, while small in healthy aging, tend to occur more frequently with more complex compared to easy tasks (Finlayson et al., 2005). Finally, there are three types of functional trends *between* individuals; they are associated with minimal changes, and with early onset and late onset of functional impairment. The latter two trends often characterize individuals with physical or mental health challenges (Liang et al., 2003).

ABNORMAL AND PATHOLOGICAL CHANGES

Owing to the importance of cognition in everyday functioning, a primary concern among community-dwelling older adults is losing one's memory (Lowenthal and Berkman, 1967). Dementia is a general term that describes a constellation of symptoms that include the gradual or sudden losses in memory, language, judgment, motor skills, and other intellectual functions. One or several diseases, such as Alzheimer's disease (AD), can cause dementia. Some dementias are reversible, such as those caused by vitamin B12 deficiency, while others, such as Alzheimer's, are progressive declining gradually over time. The clinical symptoms and time course of dementia vary depending on the etiology of causal diseases, and different types of dementia may present different pathology, location, and structure of the brain. Unfortunately, there are no viable, proven interventions to delay or eliminate dementias at present.

According to the 2016 facts and figures distributed by the Alzheimer's Association, of the 5.4 million Americans with AD, an estimated 5.2 million people are 65 and older, and approximately 200,000 are under age 65 with early onset dementias. The prevalence of individuals aged 71 and older was 13.9%. Dementia prevalence increased with age, from 5% of those in the seventh decade to 37.4% among those 90 and older (Plassman, Langa et al., 2007). It is interesting to note that reporting of dementia prevalence among centenarians varied between 27% and 75% with a mean of about 60% (Poon et al., 2012).

A study of a population-based sample of centenarians by a team of psychologists, neuropsychologists, neuropathologists, geneticists, and clinicians showed that a quarter of centenarians had no signs of dementia. In addition, about a quarter had signs of transient confusion, and about half showed classical behavioral signs of dementia varying in severity. Finally, only 5% were in the most severe stage of dementia (Poon et al., 2012). Further, this study found that the criterion used for the definition of dementia in the same sample of centenarians could influence the prevalence from 52% to 78%. In addition, the demographic distribution characteristics of the sample (e.g., gender, race, living arrangements, and education) could influence prevalence as these concomitant characteristics influence interindividual differences in cognition.

As noted earlier in this chapter, interindividual differences in cognition have presented difficulties in clinical diagnosis of dementia, especially in early stages of cognitive dysfunction (Poon, 1993, 1994, 1997; Poon et al., 1986). As a rule of thumb, clinical diagnosis of behavioral symptoms of dementia at the moderate to severe stages is comparatively easy to document. The clinical challenge is to diagnose whether the etiology is reversible or progressive and clinicians can use a battery of diagnostic tests excluding potential causal factors and including longitudinal history and repeat testing (Poon et al., 1986). Early cases of dysfunction present challenges to clinical diagnosis as variations on normal cognitive functioning between individuals are large; furthermore, variations within a person from time to time could also be large owing to environment and affect influences. Hence, if a diagnosis is questionable at early detection of changes, a prudent approach is to wait and test again later to either confirm or reject the first impression.

Two questions arise on whether and how abnormal changes of cognition impact functional capacity or physical frailty, and also whether cognition and functional capacity are independent processes. Frailty is a clinical syndrome defined by three or more of the following: unintentional weight loss, exhaustion, weakness, slow walking speed, and low physical activities (Fried et al., 2001). It also refers to the inability to perform physical and instrumental tasks in navigating everyday activities.

It is interesting to note that some older demented adults can still be physically strong and active, though they may be having difficulty with performing complex activities. The literature is supportive of the positive relationships between dementia and cognitive impairments with decrement in activities of physical (ADL) and instrumental daily living (IADL; Mortiz, Kasl, and Berkman, 1995; Blaum, Ofstedal, and Liang, 2002). This finding led Liang and his colleagues (2003) to examine patterns of change in a large longitudinal sample of older adults in Japan. They found that over 60% of the sample had little impairment and functional limitation before 75 years with gradual limitations to age 90. About 20% of the sample, an early onset group, had more initial functional limitation at age 60 and then either stayed at that level or slightly accelerated in limitation. Finally, another 13% of the sample had a late onset of functional limitations at age 80 and then remained stable in the intervening years. Furthermore, the study provided empirical evidence that self-rated health and cognitive impairment are associated with the three trajectories. Liang et al. found that greater cognitive impairment was associated with early onset trajectories while poorer self-related health was associated with late onset. Both self-rated poor health and cognitive impairment at baseline were associated with the probability of dying. These findings showed that the impact of cognitive impairment and chronic diseases can significantly affect individuals' functioning and, in turn, influence the degree of successful aging.

THEORY TO CONSOLIDATE NORMAL AND PATHOLOGICAL COGNITIVE AGING CHANGES

Aging directly affects the central nervous system (CNS), the central processing unit of both physical and mental functioning. One of the primary outcomes of aging is the slowing of the CNS. Reaction time and cognitive processing time are common measures of reaction time (Cerella, Poon, and Williams, 1980). Furthermore, patterns of functional cognitive processing times of simple to complex cognitive functions from young to middle-aged to older adults can estimate the magnitude of aging impact on the CNS (Cerella et al., 1980; Poon, 1993, 1997). This line of research showed slowed CNS in normal aging by a constant factor in the processing of simple to complex information. We observe pathological changes that further depressed the rate of processing compared to normative cohorts. The finding of aging on CNS functioning has been translated to an information processing approach in the study of aging and for clinical memory assessment (Erickson, Poon, Walsh-Sweeney, 1980; Kazniak, Poon, and Riege, 1986). Hence, changes in the CNS by patterns of observed processing or performance times can explain the complexity of aging phenomena, especially in the cognitive and functional domains.

Investigators have not utilized this line of research to its full potential, and we hope this chapter could ignite new interests to its applications. Specifically, we do know that there are specific strengths and deficits in cognitive processes in normal aging, and these deficits show in pathological aging. Perhaps CNS decline can measure the rate of change, and vice versa. A logical next step is to query at what junction of processing rate decline makes self-rating of successful aging not reliable. The answers would then provide new insights into how different levels of cognitive functioning can contribute to successful aging.

SUCCESSFUL AGING AND PERSONALITY

Cognitive abilities and functional capacity are undoubtedly important for older adults to manage successfully their everyday life. However, even when health and functions decline, a number of old and very old persons continue to enjoy life and feel they are still doing well. How can we explain such discrepancies?

Physical health, cognitive functioning, and social engagement are only momentary snapshots of a person who brings many decades of life to a current situation. An important component of successful aging, often overlooked, includes an individual's personality. We define personality as unique characteristics that account for one's patterns of behavior and experience. Another analogy is that personality is the prism, or lens, that an individual reacts to in his or her surroundings. Different individuals react differently with variation in styles. Such patterns are usually established by genetics, environmental, and social factors (Friedman and Kern, 2014), which determine individuals' personality. Since personality tends to be relatively stable, we can describe and explain the underlying causes of individual differences. Some researchers, on the other hand, have argued that there is evidence that personality is changeable and flexible. Baltes (1987) suggested that the developmental concept of plasticity makes clear that the potential flexibility of personality relies on external experiences. A genetic potential for high self-regulation (i.e., conscientiousness), for example, may not be maintained throughout the life span unless it is reinforced by many years of experience with self-controlling behavior (McAdams and Adler, 2006). However, it is noticeable that personality plasticity is unlimited and that personality does not change much, especially in adulthood. Personality traits do not change easily even when older adults experience challenging situations. Although there may be individual differences in stability, it is an important fact that personality is what causes a person to exhibit consistency in behavior and distinguishes one person from another.

There are a number of ways in which one can view personality as a form of successful aging: through personality traits, resilience, and life stories or biographies. In the following sections, we will address all three components of personality.

PERSONALITY TRAITS AND SUCCESSFUL AGING

The study of personality traits has a long history in the psychological sciences. Among the most popular approaches are the Big-5 dimensions outlined by Costa and McCrae (1985). There appears to be consensus that the basic structure of personality is composed of five basic tendencies: neuroticism (i.e., reversed emotional stability), extraversion, openness to experience, conscientiousness, and agreeableness. Since personality provides information on how individuals cope with the world and adapt to their situations, much gerontological research has noted the important association between personality traits and outcomes related to successful aging. All five-personality traits relate to physical and cognitive health, as well as social engagement (Lodi-Smith and Roberts, 2012; Luchetti, Terracciano, Stephan, and Sutin, 2015; Morack, Infurna, Ram, and Gerstorf, 2013).

Neuroticism—a tendency to experience negative affect such as anxiety, hostility, vulnerability, and depression—is the primary factor often negatively related to health outcomes and longevity. People who are anxious and worried are more likely to get sick, experience cognitive decline, and feel depressed and lonely (Tetzner and Schuty, 2016). This trait therefore promotes less favorable aging trajectories. Neuroticism is also associated with lower functioning in instrumental activities of daily living and worse quality of life in later years (Chapman, Duberstein, and Lyness, 2007). This is because neurotic individuals tend to lead their lives in a way that increases the likelihood to experience negative events (e.g., sensitive to their failures), and they are less likely to use problem-focused approaches to cope with these events (Lahey, 2009). In summary, neuroticism generally predicts shorter, less happy, and less successful lives (Lahey, 2009). However, it is not always a negative factor in all circumstances. People with high levels of neuroticism may stay healthy, unless they feel overwhelmed by stress. They are often anxious about their health, and thus they would more frequently seek medical advice and treatment (Friedman, 2000).

Extraverted people, on the other hand, are more likely to do well as they become older. Extraversion, which contains assertiveness, gregariousness, activity, and positive affect, relates to social activities and relationships. Extraverted people seek more help, find solutions, and support through their social relationships. Since they tend to show high levels of positive emotion and sociability (McCrae and Costa, 1999), they may also more frequently participate in productive activities, such as volunteering, which will lead them to remain active into late life.

Much like extraverted people, individuals who are open to new experiences also appear to be better off as they age. The openness components (e.g., imaginative, creative, and intellectual curiosity) suggest that people high on this trait show more willingness to try out new things and enjoy new experiences, which will ultimately help adjust to new life situations. Open individuals, who may have a strong intellectual curiosity, perform better in cognitive tests (Sharp, Reynolds, Pedersen, and Gatz, 2010).

Another strong association between a personality trait and a health outcome is the trait of conscientiousness. Conscientious individuals tend to be very organized, responsible, and goal-directed. They have integrity and exercise self-control. Therefore, they are more likely to engage in effective health behaviors such as exercising, good nutritional behavior, and going to a regular check-up. Conscientious people are also less likely to smoke (Bogg and Roberts, 2004). The consequences of conscientiousness are related to better IADL functioning and health-related quality of life (Chapman et al., 2007), as well as better cognitive functioning (Wilson et al., 2015). The influence of conscientiousness on adaptation may relate to childhood experiences. Children who were rated as conscientious in childhood were more likely to survive longer and stay in better health (Friedman et al., 1993).

Finally, agreeableness—a tendency to be altruistic, friendly, and cooperative with others—is also an important trait for successful aging. Similar to extraversion, agreeableness includes traits focusing on interpersonal relationships and social interactions. Yet what separates agreeableness from extraversion is that agreeableness is conceptually linked to "motives for maintaining positive relations with others" (Jensen-Campbell and Graziano, 2001, p. 327). Thus, people who are more agreeable are more likely to adhere to a treatment plan to avoid conflicts when they are sick. Individuals high in agreeableness give and receive greater social support (Bowling, Beehr, and Swader, 2005). Their prosocial behaviors may make it easier to get involved in social activities, which in turn lead them to age successfully. They are also more likely to accept challenges rather than fighting them.

There is now ample evidence that personality traits directly relate to health outcomes and this would argue for an important role that each trait plays in adjusting to adverse circumstances. The aging literature is now ready for new and more integrative approaches to understand the causality of the link between personality traits and successful aging. Past research has focused on the different combinations or configurations of these traits. There seems to be a consensus that relatively low values on neuroticism and relatively high scores on extraversion, openness, conscientiousness, and agreeableness in combination may be the best predictor of health outcomes and successful aging (Martin, Baenziger, MacDonald, Siegler, and Poon, 2009; Baek, Martin, Siegler, Davey and Poon, 2016). Crowe, Andel, Pedersen, Fratiglioni, and Gatz (2006) also provided evidence that the combination of personality is part of optimal functioning late in life. In their longitudinal study, Crowe et al. demonstrated that the grouping of neuroticism and low extraversion placed individuals more at risk for cognitive impairment.

Next to including the Big-5 dimensions of personality, successful aging may also relate to the personality components of self-esteem and optimism. People high in self-esteem may feel a sense of agency that allows them to take fate in their own hands and control even difficult situations. Optimistic people may not allow themselves to view challenging experiences negatively. All of this may relate directly to aspects of resilience.

RESILIENCE AND SUCCESSFUL AGING

Although there is no broad consensus on resilience, some have defined this disposition as an important characteristic when facing difficult situations (Masten, 2001). Resilient people "bounce back" from adversity and view their life as changeable. They also draw on some forms of positive emotions in times of stress and utilize them to recover from difficulties (Tugade and Fredrickson, 2004). Resilience can be a person-centered variable (Masten, 2001) or a variable in a larger network of predictors and outcomes that determines how well a person adjusts to adversity. Although resilience is often a reaction in response to stressors, it may play an important role in maintaining well-being in later life, when individuals may lose functioning (Baltes and Baltes, 1990). Particularly among the oldest old, resilience is an important survivorship characteristic. If individuals experience many challenges but maintain good mental health, they show relatively high levels of resilience (Martin, MacDonald, Margrett, and Poon, 2010). In older populations, higher levels of resilience protect against an increase in activity of daily living limitations (Manning et al., 2016).

Life Stories and Successful Aging

Next to personality traits and resilience perspectives, some have argued that aging studies need to consider a broader framework for personality, integrating models of personality that emphasize structures and processes with different foci of personality (Hooker and McAdams, 2003). This structure would include life stories and biographies that have become a major part of old and very old people. Self-narratives become part of a stable personality structure that includes parallel process constructs relevant to a successful life. Individual life stories become an integral part of personality and identity in late life (McAdams and Perls, 2006). Wong and Watt (1991) found that older adults who remember their past lives as worthwhile, learn from experiences, and use them to solve problems tend to age successfully. Not much research links life stories or biographies to personality and successful aging but integrating the entire life span into the fabric of a successfully led life is a plausible approach to the study of successful aging.

SUCCESSFUL AGING AND WHOLE-PERSON WELLNESS (WPW)

Several important considerations for future research, practice, and policy arise from the previous discussion. Addressing these shortcomings by incorporating a whole-person wellness (WPW) perspective can lead us to new directions in the field.

The first area for consideration is reconciliation of the disconnect between objective and subjective indicators of "successful" aging. Models employing a "yes/no" dichotomy to assess successful aging may not accurately portray older adults' real-world functioning. For example, a simple assessment of the presence or absence of disease does not take into account the severity of a physical condition. Compared to the past, more conditions have moved to a designation of "chronic" and individuals can manage them effectively. In addition, lack of diagnosis does not guarantee absence of disease and wellness. For example, it may be more beneficial for an older adult to be diagnosed and treated than remain undiagnosed; additionally, some treatments may have positive effects on conjunctive systems (e.g., cardiovascular treatment and cognitive health; Margrett et al., 2017). As a result, researchers must look deeper and carefully vet indicators used to represent the concept of successful aging.

Diversity increases with age, resulting in differential values, skills, and resources. However, many approaches to the study of successful aging overlook individual differences in older adults' priorities, the diversity of methods employed by older adults to address personal priorities, and the connectedness of well-being and functional domains. The result is neglect of older adults' everyday realities—including what is relevant and important to each person as well as how individual goals and motivations are achieved. For example, following conventional definitions of successful aging, older adults who prioritize (or select) areas of well-being such as spirituality may be considered less successful compared to other older adults who focus on maintaining a high level of physical functioning. To achieve their priorities, older adults can demonstrate profound resilience and adjust to experienced adversity throughout their life course. Resilience, compensation, and adaptation rely on multiple functional domains and resources.

Unfortunately, while numerous scholars investigating successful aging have defined it as a multidimensional paradigm, to date, largely disparate models illustrate "discrete" successful aging functional domains and its outcomes. While individual functional domains (e.g., physical health, cognition, interpersonal relationships) supporting successful aging are important, it is prudent to consider the "whole," which comprises interacting and synergistic components. Various functional domains, indirectly or directly, can affect one another and may share common underlying mechanisms. An example of such a multifaceted relationship is the association between cognitive and sensory functioning (e.g., Baltes and Lindenberger, 1997; Pichora-Fuller, 2014). In addition, functional domains may also support one another and challenges may compensate for deficits in one domain.

WPW AND ITS RELEVANCE TO SUCCESSFUL AGING

By considering one domain in isolation, we are likely to miss complex, interactive effects that enhance understanding of successful aging and provide data for prevention and intervention efforts. Integrative models provide a productive method to understanding and promoting successful aging across functional domains (Srivastava and Das, 2013). We can conceptualize such models as incorporating a WPW perspective.

WPW incorporates six domains including physical, social, emotional, intellectual, occupational, and spiritual (Hettler, 1976; Kang, Russ, and Ryu, 2008). In his original conceptualization, Hettler noted WPW application "as a pathway to optimal living...." He emphasized the importance of programs and approaches which "help people achieve their full potential" and "affirm and mobilize people's positive qualities and strengths (p. 2)." Indeed, the descriptions of the wellness domains provide a direct connection to successful aging. The outcome of "wellness" within each domain is important as well as the process used to achieve wellness. For instance, Hettler described occupational wellness as utilizing skills and talents to achieve "personal satisfaction and enrichment in one's life through work." Physical wellness is accomplished through personal responsibility and care (e.g., physical activity, nutrition, avoidance of unhealthy behaviors; Hettler). Emotional wellness refers to the ability to regulate emotions, accept limitations, and cope with stress (Hettler). Spiritual wellness denotes "a deep appreciation for the depth and expanse of life and natural forces" and ultimately, development of "actions... consistent with your beliefs and values" (Hettler). In his description of social wellness, Hettler described our interconnectedness with the environment and one another and development of curiosity as we strive for intellectual wellness. Hettler's descriptions of wellness domains emphasize the need for being an active agent in one's wellness including reflection and life choices.

Adoption of a whole-person approach to the investigation and application of successful aging can be advantageous. First, we can enhance ecological validity by measuring domains that are important to and selected by older adults, thereby, reflecting individuals' everyday contexts. We can also increase ecological validity at a broader level through incorporation of multiple domains that provide a more accurate assessment of successful aging as it is relevant and experienced within the context of specific life stages and cohorts (Cherry, Marks, Benedetto, Sullivan, and Barker, 2013; Nosraty, Jylha, Raittila, and Lumme-Sandt, 2015). Another benefit of considering the whole person and adopting a more comprehensive approach is to better understand complex findings related to gene–environment interactions. A third benefit is the utility/value added to prevention and intervention efforts aimed at fostering successful aging. Unfortunately, wellness activities are still likely to be fragmented and not indicative of WPW (e.g., Russ, 2012). Integration of activities that benefit multiple components of WPW may increase both programmatic efficacy and efficiency.

APPLICATION OF WPW TO SUCCESSFUL AGING

Peisah's (2016) application of successful aging to aging within a professional context provides an illustrative example from which we can extrapolate several crucial principles. Foremost, underlying her application is the cornerstone of biopsychosocial functioning required to achieve personal success. She describes successful aging as a process in which adults are required to consciously assess their individual strengths, challenges, and changes in order to set goals and initiate a "renewable personal ageing plan." Peisah reminds us of the importance of reserves and their delicate balance with contextual demands and the role of "proactive adaption."

Just as the synergy between functional domains is critical to understanding factors enhancing successful aging, so too are the contexts in which older adults function. For instance, context is critical as we consider older adults' motivation. Individuals' goals, as well as definitions of "success," are likely to vary across context. Context also shapes experiences and resources. Bioecological and systems approaches allow us to consider various levels of influence on older adults' successful aging across time (White and Klein, 2008). Levels of influence run from close to far and include the individual, close others and systems such as families and work (micro level), communities (exo level), and society and culture (macro level). On the micro level, the workplace can be its own microcosm reflecting successful aging, and the outcomes in the workplace certainly affect aging outside of the work environment (Robson, Hansson, Abalos, and Booth, 2006). At the micro and exo levels, we will be remiss to neglect the impact of technology on successful aging. A developing area of research is the availability of online environments and the ability of these communities and supports to enhance multidimensional aspects of wellness (e.g., Olson, 2013). One way to demonstrate the ability of technology to influence successful aging is through utilization of smart home and gerontechnology (e.g., Demiris and Hensel, 2008). From a provider perspective, adoption of an approach that embodies multidisciplinary teamwork, whole-person and person-centered principles to health and wellness can enhance delivery of care and ultimately serve to promote successful aging (Menard et al., 2015; Doohan, Coutinho, Lochner, Wohler, and DeVoe, 2016). At the broadest level, communities and macro-level policies and resources shape our learning and living environments. Proponents of successful aging are capitalizing on these influences through efforts to promote age-friendly universities (Association for Gerontology in Higher Education, 2017) and cities (Chatterjee and King, 2014).

In terms of considering an individual's development toward successful aging over time, it is important to consider life span influences on successful aging. We do not arrive at older adulthood and subsequently demonstrate "success"—it is a process borne out over time given the context of an individual's micro- and macro-level influences. As described by Brandt, Deindl, and Hank (2012), selected "origins" of successful aging may be evident very early in life (e.g., childhood health and family socioeconomic status). Such findings carry important significance for policy and practice.

CONCLUSION

As evidenced by recent reviews, the debate continues surrounding the definition and evaluation of successful aging. Some argue that subjective and objective criteria should be used (Kusumastuti et al., 2016). Others have emphasized the importance of taking a multidimensional approach, moving toward less emphasis on biomedical and more psychosocial approaches (Cosco, Prina, Perales, Stephan, and Brayne, 2014).

Perhaps we are moving beyond what are necessary and sufficient components and segmented approaches of successful aging and more to a cohesive life span model of successful aging. Adults are active agents in their successful aging—identifying priorities, building resources, and taking actions to facilitate their own development within their individual context. Connectedness occurs across the life span as well as across dimensions of functioning and well-being.

In this way, we consider more than individual constituent components, such as a minimum level of physical and cognitive functioning, and integrate domains of functioning and well-being important to the person. Such a model can also include components that help to facilitate successful aging such as personality (individual level) and resources (family, community, and societal levels). A multidimensional approach focused on WPW provides integration and avenues for understanding complex relationships between functional domains, which bodes well for prevention and intervention efforts.

REFERENCES

Aspinwall, L. G., and Taylor, S. E. 1997. A stitch in time: Self-regulation and proactive coping. *Psychological Bulletin*, 121, 417–436.

Association for Gerontology in Higher Education. 2017. *AGHE Joins Age-Friendly University Initiative.* Retrieved from https://www.geron.org/press-room/press-releases/2016-press-releases/637-aghe-joins-age-friendly-university-initiative.

Baek, Y., Martin, P., Siegler, I. C., Davey, A., and Poon, L. W. 2016. Personality traits and successful aging: Findings from the Georgia Centenarian Study. *International Journal of Aging and Human Development*, 80, 1–22. doi:10.1177/0091415016652404.

Baltes, M. M., and Carstensen, L. L. 1996. The process of successful ageing. *Ageing and Society*, 16, 397–422.

Baltes, M. M., and Carstensen, L. L. 1999. Social-psychological theories and their applications to aging: From individual to collective. In V. L. Bengtson and K. W. Schaie (Eds.), *Handbook of Theories of Aging* (pp. 209–226). New York: Springer.

Baltes, P. B. 1987. Theoretical propositions of life-span developmental psychology: On the dynamics between growth and decline. *Developmental Psychology*, 23, 611–626.

Baltes, P. B., and Baltes, M. M. 1990a. Psychological perspectives on successful aging: The model of selective optimization with compensation. In P. B. Baltes and M. M. Baltes (Eds.), *Successful Aging: Perspectives from the Behavioral Sciences* (pp. 1–34). Cambridge, UK: Cambridge University Press. doi:10.1017/cbo9780511665684.003.

Baltes, P. B., and Baltes, M. M. (Eds.). *Successful Aging: Perspectives from the Behavioral Sciences*, Cambridge, UK: Cambridge University Press. doi:10.1017/CBO9780511665684.

Baltes, P. B., and Lindenberger, U. 1997. Emergence of a powerful connection between sensory and cognitive functions across the adult life span: A new window to the study of cognitive aging? *Psychology and Aging*, 12(1), 12–21.

Bell, B., Rose, C. L., and Damon, A. 1972. The normative aging study: An interdisciplinary and longitudinal study of health and aging. *The International Journal of Aging and Human Development*, 3(1), 5–17. doi:10.2190/ggvp-xlb5-pc3n-ef0g.

Bode, C., de Ridder, D. T. D., and Bensing, J. M. 2006. Preparing for aging: Development, feasibility and preliminary results of an educational program for midlife and older based on proactive coping theory. *Patient Education and Counseling*, 61(2), 272–278. doi:10.1016/j.pec.2005.04.006.

Bogg, T., and Roberts, B. W. 2004. Conscientiousness and health-related behaviors: A meta-analysis of the leading behavioral contributors to mortality. *Psychological Bulletin*, 130(6), 887–919.

Bowling, N. A., Beehr, T. A., and Swader, W. M. 2005. Giving and receiving social support at work: The roles of personality and reciprocity. *Journal of Vocational Behavior*, 67(3), 476–489. doi:10.1016/j.jvb.2004.08.004.

Bowling, A., and Dieppe, P. 2005. What is successful ageing and who should define it? *BMJ*, 331, 1548–1551. doi:10.1136/bmj.331.7531.1548.

Blaum C. S, Ofstedal M. B, and Liang J. (2002) Low cognitive performance, comorbid disease, and task-specific disability: findings from a nationally representative survey. *Journal of Gerontology, Series A: Biological Sciences and Medical Sciences*, 57(8), M523–531.

Brandt, M., Deindl, C., and Hank, K. 2012. Tracing the origins of successful aging: The role of childhood conditions and social inequality in explaining later life health. *Social Science and Medicine*, 74(9), 1418–1425.

Cerella, J., Poon, L. W., and Williams, D. 1980. Age and the complexity hypothesis. In L. W. Poon (Ed.), *Aging in the 1980s: Psychological Issues* (pp. 332–340). Washington, DC: American Psychological Association.

Chapman, B., Duberstein, P., and Lyness, J. M. 2007. Personality traits, education, and health-related quality of life among older adults primary care patients. *Journals of Gerontology, Series B: Psychological Sciences and Social Sciences*, 62B, P343–P352.

Chatterjee, A., and King, J. 2014. *Best Cities for Successful Aging.* Santa Monica, CA: Milken Institute. Retrieved from http://successfulaging.milkeninstitute.org/2014/best-cities-for-successful-aging-report-2014.pdf.

Cheng, S. T., Fung, H. H., Li, L. W., Li, T., Woo, J., and Chi, I. 2015. Successful aging: Concepts, reflections and its relevance to Asia. In S. Cheng, H. H. Fung, I. Chi, L. W. Li., and J. Woo (Eds.), *Successful Aging* (pp. 1–18). Dordrecht, Netherlands: Springer.

Cherry, K. E., Marks, L. D., Benedetto, T., Sullivan, M. C., and Barker, A. 2013. Perceptions of longevity and successful aging in very old adults. *Journal of Religion, Spirituality and Aging*, 25(4), 288–310.

Cho, J., Martin, P., and Poon, L. W. 2012. The older they are, the less successful they become? Findings from the Georgia Centenarian Study. *Journal of Aging Research*, 1–8. doi: 10.1155/2012/695854.

Cosco, T. D., Prina, A. M., Perales, J., Stephan, B. C., and Brayne, C. 2014. Operational definitions of successful aging: A systematic review. *International Psychogeriatrics*, 26(3), 373–381.

Costa, P. T. Jr., and McCrae, R. R. 1985. *The NEO Personality Inventory Manual*. Odessa, FL: Psychological Assessment Resources.

Crowe, M., Andel, R., Pedersen, N. L., Fratiglioni, L., and Gatz, M. 2006. Personality and risk of cognitive impairment 25 years later. *Psychology and Aging*, 21, 573–580.

Crowther, M. R., Parker, M. W., Achenbaum, W. A., Larimore, W. L., and Koenig, H. G. 2002. Rowe and Kahn's model of successful aging revisited positive spirituality—The forgotten factor. *The Gerontologist*, 42(5), 613–620. doi:10.1093/geront/42.5.613.

Cumming, E., and Henry, W. E. 1961. *Growing Old: The Process of Disengagement*. New York: Basic Books.

Demiris, G., and Hensel, B. K. 2008. Technologies for an aging society: A systematic review of "smart home" applications. *IMIA Yearbook of Medical Informatics*, 3, 33–40.

Doohan, N., Coutinho, A. J., Lochner, J., Wohler, D., and DeVoe, J. 2016. "A paradox persists when the paradigm is wrong": Pisacano Scholars' reflections from the inaugural Starfield Summit. *The Journal of the American Board of Family Medicine*, 29(6), 793–804.

Eaton, N. R., Krueger, R. F., South, S. C., Gruenewald, T. L., Seeman, T. E., and Roberts, B. W. 2012. Genes, environments, personality, and successful aging: Toward a comprehensive developmental model in later life. *The Journals of Gerontology Series A: Biological Sciences and Medical Sciences*, 67, 480–488. doi:10.1093/Gerona/gls090.

Erickson, R. C., Poon, L. W., and Walsh-Sweeney, L. 1980. Clinical memory testing of the elderly. In L. W. Poon, J. L. Fozard, L. S. Cermak, D. Arenberg, and L. W. Thompson (Eds.), *New Directions in Memory and Aging: Proceedings of the George A Talland Memorial Conference* (pp. 379–402). Hillsdale, NJ: Lawrence Erlbaum Associates.

Finlayson, M., Mallison, T., and Barbosa, V.M. 2005. Activities of daily living (ADL) and instrumental activities of daily living (IADL) items were stable over time in a longitudinal study on aging. *Journal of Clinical Epidemiology*, 58 (4): 338–349.

Freund, A. M., and Baltes, P. B. 1998. Selection, optimization, and compensation as strategies of life management: Correlations with subjective indicators of successful aging. *Psychology and Aging*, 13(4), 531–543. doi:10.1037/0882-7974.13.4.531.

Fried, L.P., Tangen, C. M., Walston, J., Newman, A. B., Hirsch, C., Gottdiener, J., Seeman, T., Tracy, R., Kop, W. J., Burke, G., and Mcburnie, M. A. 2001. Frailty in older adults: Evidence for a phenotype. *Journal of Gerontology A: Biological Sciences & Medical Sciences*, 56, M146–M156.

Friedman, H. S., Tucker, J. S., Tomlinson-Keasey, C., Schwartz, J. E., Wingaard, D. L., and Criqui, M. H. 1993. Does childhood personality predict longevity? *Journal of Personality and Social Psychology*, 65, 176–185.

Friedman, H. S., and Kern, M. L. 2014. Personality, well-being, and health. *Annual Review of Psychology*, 65, 719–742.

Friedman, H. S. 2000. Long-term relations of personality and health: Dynamisms, mechanisms, tropisms. *Journal of Personality*, 68(6), 1089–1107.

Havighurst, R. J. 1961. Successful aging. *The Gerontologist*, 1, 8–13. doi:10.1093/geront/1.1.8.

Havighurst, R. J. 1963. Successful aging. In R. H. Williams, C. Tibbitts, and W. Donahue (Eds.), *Processes of Aging* (pp. 299–320). New York: Atherton Press.

Hettler, B. 1976. The Six Dimensions of Wellness Model [Handout]. Retrieved from the National Wellness Institute website: http://www.nationalwellness.org.

Hooker, K., and McAdams, D. P. 2003. Personality reconsidered: A new agenda for aging research. *Journals of Gerontology, Series B: Psychological Sciences and Social Sciences*, 58B, P96–P304.

Jensen-Campbell, L. A., and Graziano, W. G. 2001. Agreeableness as a moderator of interpersonal conflict. *Journal of Personality*, 69(2), 323–362.

Kafka, G. J., and Kozma, A. 2002. The construct validity of Ryff's scales of psychological well-being (SPWB) and their relationship to measures of subjective well-being. *Social Indicators Research*, 57(2), 171–190.

Kahana, E., and Kahana, B. 1996. Conceptual and empirical advances in understanding aging well through proactive adaptation. In V. Bengtson (Ed.), *Adulthood and Aging: Research on Continuities and Discontinuities* (pp. 18–40). New York: Springer.

Kahana, E., Kelley-Moore, J., and Kahana, B. 2012. Proactive aging: A longitudinal study of stress, resources, agency, and well-being in late life. *Aging and Mental Health*, 16(4), 438–451.

Kang, M., Russ, R. R., and Ryu, J. S. 2008. *Wellness for Older Adults in Daily Life*. Division of Agricultural Sciences and Natural Resources, Stillwater, OK: Oklahoma State University.

Kaszniak, A. W., Poon, L. W., and Riege, W. H. 1986. Assessing memory deficits: An information-processing approach. In L. W. Poon, T. Crook et al. (Eds.), *Handbook for Clinical Memory Assessment of Older Adults* (pp. 168–188). Washington, DC: American Psychological Association.

Kusumastuti, S., Derks, M. G., Tellier, S., Di Nucci, E., Lund, R., Mortensen, E. L., and Westendorp, R. G. 2016. Successful ageing: A study of the literature using citation network analysis. *Maturitas*, 93, 4–12. doi:10.1016/j.maturitas.2016.04.010.

Lahey, B. B. 2009. Public health significance of neuroticism. *American Psychologist*, 64, 241–256.

Liang, J., Shaw, B. A., and Krause, N. 2003. Changes in functional status among older adults in Japan: Successful and usual aging. *Psychology and Aging*, 18(4), 684–695.

Lodi-Smith, J., and Roberts, B. W. 2012. Concurrent and prospective relationships between social engagement and personality traits in older adulthood. *Psychology and Aging*, 27(3), 720–727. doi: 10.1037/a0027044.

Lowenthal, M. F., and Berkman, P. L. 1967. *Aging and Mental Disorder in San Francisco*. San Francisco, CA: Jossey-Bass.

Luchetti, M., Terracciano, A., Stephan, Y., and Sutin, A. R. 2015. Personality and cognitive decline in older adults: Data from a longitudinal sample and meta-analysis. *Journals of Gerontology Series B: Psychological Sciences and Social Sciences*, 71(4), 591–601. doi:10.1093/geronb/gbu184.

Manning, L. K., Carr, D. C., and Kail, B. C. 2016. Do higher levels of resilience buffer the deleterious impact of chronic disability in later life? *The Gerontologist*, 56, 514–524. doi:10.1083/geront/gnu068.

Margrett, J. A., Schofield, T., Martin, P., Poon, L. W., Masaki, K., and Willcox, B. J. 2017. *Predictors of Cognitive Maintenance and Decline among Older Japanese-American Men: The Kuakini Honolulu Heart Program/Honolulu-Asia Aging Study*. Manuscript submitted for publication.

Martin, P., Baenziger, J., MacDonald, M., Siegler, I., and Poon, L. W. 2009. Engaged lifestyle, personality, and mental status among centenarians. *Journal of Adult Development*, 16, 199–208.

Martin, P., Kelly, N., Kahana, B., Kahana, E., Willcox, B., Willcox, D. C., and Poon, L. W. 2015. Definitions of successful aging: A tangible or elusive concept? *The Gerontologist*, 55, 14–25. doi:10.1093/geront/gnu044.

Martin, P., MacDonald, M., Margrett, J., and Poon, L. W. 2010. Resilience and longevity: Expert survivorship of centenarians. In P. Fry and C. Keyes (Eds.), *New Frontiers in Resilient Aging: Life-Strengths and Well-Being in Late Life* (pp. 213–238). New York, NY: Cambridge University Press.

Masoro, E. J. 2001. "Successful aging:" Useful or misleading concept? *The Gerontologist*, 41, 415–418.

Masten, A. S. 2001. Ordinary magic. *American Psychologist*, 56, 227–238. doi:10.1037//0003-066X.56.3.227.

McCrae, R. R., and Costa P. T., Jr. 1999. A five-factor theory of personality. In L. A. Pervin and O. P. John (Eds.) *Handbook of Personality: Theory and Research* (Vol. 2, pp. 139–153). New York: Guilford Press.

McAdams, D. P., and Pals, J. L. 2006. The new Big Five: Fundamental principles for an integrative science of personality. *American Psychologist*, 61, 204–217.

McAdams, D. P., and Adler, J. M. 2006. How does personality develop? In D. K. Mroczek, and T. D. Little (Eds.) *Handbook of Personality Development* (pp. 469–492). Mahwah, NJ: Lawrence Erlbaum Associates.

Menard, M. B., Weeks, J., Anderson, B., Meeker, W., Calabrese, C., O'Bryon, D., and Cramer, G. D. 2015. Consensus recommendations to NCCIH from research faculty in a transdisciplinary academic consortium for complementary and integrative health and medicine. *The Journal of Alternative and Complementary Medicine*, 21(7), 386–394.

Moritz, D. J., Kasl, S. V., and Berkman, L. F. (1995). Cognitive functioning and the incidence of limitations in activities of daily living in an elderly community sample. *American Journal of Epidemiology*, 141(1), 41–49.

Morack, J., Infurna, F. J., Ram, N., and Gerstorf, D. 2013. Trajectories and personality correlates of change in perception of physical and mental health across adulthood and old age. *International Journal of Behavioral Development*, 37(6), 475–484. doi:10. 1177/0165025413992605.

Nosraty, L., Jylhä, M., Raittila, T., and Lumme-Sandt, K. 2015. Perceptions by the oldest old of successful aging, Vitality 90+ Study. *Journal of Aging Studies*, 32, 50–58.

Olson, K. 2013. Transcending place and transforming aging through online and offline experiences. *Generations*, 37(4), 84.

Ouwehand, C., de Ridder, D. T., and Bensing, J. M. 2007. A review of successful aging models: Proposing proactive coping as an important additional strategy. *Clinical Psychology Review*, 27(8), 873–884.

Palmore, E., Nowlin, J., Busse, E., Siegler, I., and Maddox, G. 1985. *Normal Aging III*. Durham, NC: Duke University Press.

Pearlin, L. I., & Skaff, M. M. 1996. Stress and the life course: A paradigmatic alliance. *The Gerontologist*, 36(2), 239–247. doi:10.1093/geront/36.2.239.

Peisah, C. 2016. Successful ageing for psychiatrists. *Australasian Psychiatry*, 24(2), 126–130.

Phelan, E. A., and Larson, E. B. 2002. "Successful Aging"—Where next? *Journal of the American Geriatrics Society*, 50(7), 1306–1308. doi:10.1046/j.1532-5415.2002.t01-1-50324.x.

Pichora-Fuller, M. K. 2014. A successful aging perspective on the links between hearing and cognition. SIG 6 Perspectives on hearing and hearing disorders. *Research and Diagnostics*, 18(2), 53–59.

Plassman B. L., Langa, K. M., Fisher, G. G., Heeringa, S. G., Weir, D. R., Ofstedal, M. B., Burke, J. R., Hurd, M. D., Potter, G. G., Rodgers, W. L., Steffens, D. C., Willis, R. J., and Wallace, R. B. 2007. Prevalence of dementia in the United States: The aging, demographics, and memory study. *Neuroepidemiology*, 29(1–2), 125–132.

Poon, L. W., Clayton, G. M., Martin, P., Johnson, M. A., Courtenay, B. C., Sweaney, A. L., Merriam, S. B., Pless, B. S., and Thielman, S. B. 1992. The Georgia Centenarian Study. *International Journal of Aging and Human Development*, 34, 1–17.

Poon, L. W. 1993. Assessing neuropsychological changes in pharmacological trials. *Clinical Neuropharmacology*, 16, S31–S38.

Poon, L. W. 1994. Commentary: On the paradox of improving sensitivity of ADL scales for the detection of behavioral changes in early dementia. *International Psychogeriatrics*, 6(2), 171–177.

Poon, L. W. 1997. Cognitive slowing with age: Implications for the clinical diagnosis of pathology. *The Annual Report of Educational Psychology in Japan*, 36, 178–183.

Poon, L. W. (Ed.) 1986. *Handbook for Clinical Memory Assessment of Older Adults*. Washington, DC: American Psychological Association.

Poon, L. W., Woodard, J. L., Miller, L. S., Green, R., Gearing, M., Davey, A., Arnold, J., Martin, P., Siegler, I. C., Nahapetyan, L., Kim, Y. S., and Markesbery, W. 2012. Understanding dementia prevalence among centenarians. *Journal of Gerontology: Series A: Biological Sciences and Medical Sciences*, 67A(4), 358–365. doi:10.1093/gerona/glr250.

Riley M. W. 1998. Letter to the editor. *The Gerontologist*, 38,151.

Robson, S. M., Hansson, R. O., Abalos, A., and Booth, M. 2006. Successful aging criteria for aging well in the workplace. *Journal of Career Development*, 33(2), 156–177.

Rowe, J. W., and Kahn, R. L. 1987. Human aging: Usual and successful. *Science*, 237, 143–149. doi:10.1126/science.3299702.

Rowe, J. W., and Kahn, R. L. 2015. Successful aging 2.0: Conceptual expansions for the 21st century. *The Journals of Gerontology Series B: Psychological Sciences and Social Sciences*, 70(4), 593–596.

Rowe, J. W. and Kahn, R. L. 1997. Successful aging. *The Gerontologist*, 37(4), 433–440. doi:10.1093/geront/37.4.433.

Russ, R. 2012. Wellness activities of rural older adults in the Great Plains. *Journal of Rural Research and Policy*, 7(1), 1–14.

Ryff, C. D. 1989. Beyond Ponce de Leon and life satisfaction: New directions in quest of successful ageing. *International Journal of Behavioural Development*, 12, 35–55.

Ryff, C. D., and Singer, B. H. 2006. Best news yet on the six-factor model of well-being. *Social Science Research*, 35(4), 1103–1119. doi:10.1016/j.ssresearch.2006.01.002.

Salthouse, T.A. 1991. *Mechanisms of Age-cognition Relations in Adulthood*. Hillsdale, NJ: Lawrence Erlbaum Associates.

Salthouse, T.A. 2017a. Contributions of the individual differences approach to cognitive aging. *Journal of Gerontology, Series B: Psychological Sciences and Social Sciences*, 72, 7–15. doi: 10.1093/geronb/gbw069.

Salthouse, T.A. 2017b. Shared and unique influences on age-related cognitive change. *Neuropsychology*, 31, 11–19. doi: 10.1037/neu0000330.

Schaie, K. W. 2013. *Developmental Influences on Adult Intelligence: The Seattle Longitudinal Study* (2nd ed.). New York: Oxford University Press.

Sharp, E. S., Reynolds, C. A., Pedersen, N. L., and Gatz, M. 2010. Cognitive engagement and cognitive aging: is openness protective? *Psychology and Aging*, 25(1), 60–73.

Shock, N. W. Greulich, R. C., Andres, R., Arenberg, D., Costa, P. T., Jr., Lakatta, E. G., and Tobin, J. D. 1984. *Normal Human Aging: The Baltimore Longitudinal Study of Aging (NIH publication No. 84–2450)*. Bethesda, MD: National Institutes of Health.

Springer, K. W., Hauser, R. M., and Freese, J. 2006. Bad news indeed for Ryff's six-factor model of well-being. *Social Science Research*, 35(4), 1120–1131.

Srivastava, K., and Das, R. C. 2013. Personality pathways of successful ageing. *Industrial Psychiatry Journal*, 22(1), 1–3.

Strawbridge, W. J., Wallhagen, M. I., and Cohen, R. D. 2002. Successful aging and well-being self-rated compared with Rowe and Kahn. *The Gerontologist*, 42(6), 727–733.

Strawbridge, W. J., Wallhagen, M. I., and Cohen, R. D. 2002. Successful aging and well-being self-rated compared with Rowe and Kahn. *The Gerontologist*, 42(6), 727–733. doi:10.1093/geront/42.6.727.

Tesch-Römer, C., and Wahl, H. W. 2017. Toward a more comprehensive concept of successful aging: Disability and care needs. *The Journals of Gerontology Series B: Psychological Sciences and Social Sciences*, 72, 310–318. doi:https://doi.org/10.1093/geronb/gbw162.

Tetzner, J., and Schuth, M. 2016. Anxiety in late adulthood: Associates with gender, education, and physical and cognitive function. *Psychology and Aging*, 31, 532–544. doi:10.1037/pag0000118.

Tugade, M. M., and Fredrickson, B. L. 2004. Resilient individuals use positive emotions to bounce back from negative emotional experiences. *Journal of Personality and Social Psychology*, 86(2), 320–333.

van Dierendonck, D., Díaz, D., Rodríguez-Carvajal, R., Blanco, A., and Moreno-Jiménez, B. 2008. Ryff's six-factor model of psychological well-being, a Spanish exploration. *Social Indicators Research*, 87(3), 473–479.

von Faber, M., Bootsma-van der Wiel, A., van Exel, E., Gussekloo, J., Lagaay, A. M., van Dongen, E., Knook, D. L., van der Geest, S., and Westendorp, R. G. 2001. Successful aging in the oldest old: Who can be characterized as successfully aged? *Archives of Internal Medicine*, 161, 2694–2700. doi:10.1001/archinte.161.22.2694.

White, J. M., and Klein, D. M. 2008. *Family Theories*. Thousand Oaks, CA: Sage Publications.

Williams, R. H., and Wirths, C. G. 1965. *Lives Through the Years: Styles of Life and Successful Aging*. New York: Atherton Press.

Wilson, R. S., Boyle, P. A., Yu, L., Segawa, E., Sythma, J., and Bennett, D. A. 2015. Conscientiousness, dementia-related pathology, and trajectories of cognitive aging. *Psychology and Aging*, 30, 74–82. doi:10.1037/pag0000013.

Wong, P. T., and Watt, L. M. 1991. What types of reminiscence are associated with successful aging? *Psychology and Aging*, 6(2), 272–279.

30 Physical Activity in Prevention of Glucocorticoid Myopathy and Sarcopenia in Aging

Teet Seene and Priit Kaasik

CONTENTS

INTRODUCTION

Aging and a reduced physical level are mainly responsible for the progressive decline in several physiological capacities in the elderly. Aging is a physiological process that includes a gradual decrease in skeletal muscle mass, strength, and endurance coupled with an ineffective response to tissue damage [1]. Aging is not a disease, though aging does come with an increasing susceptibility to disease. Among the aging population, muscle wasting is mainly a result of the decrease of fast-twitch (FT) muscle fibers and atrophy of these fibers [2]. Reduced muscle elasticity, increased tone, and stiffness with a concomitant decrease in cytoskeletal proteins titin and nebulin and contractile protein myosin content are accompanied by muscle atrophy [3]. Both sarcopenic and glucocorticoid-caused myopathic muscles have impaired locomotion, general weakness, and diminished capacity for regeneration [4]. In both cases the decrease of myofibrillar proteins synthesis rate has been shown [5,6]. The catabolic action of glucocorticoids and aging on skeletal muscle also depends on the functional activity of muscle [7,8]. Increased functional activity leads to interaction between signaling pathways in mitochondrial and myofibrillar compartments and causes muscle hypertrophy and maintains endurance capacity in the elderly and myopathic population [9]. Endurance type of activity improves capillary blood supply, increases mitochondrial biogenesis, muscle oxidative capacity, and causes qualitative remodeling of type I and IIA fibers. Resistance type of muscular activity causes anabolic and anticatabolic effect in skeletal muscle [9].

This chapter will focus on the development of sarcopenia and the role of sarcopenia in the etiology of disability in the elderly and glucocorticoid-caused myopathic patients from the functional activity aspect and how different types of physical activity programs may prevent/decelerate sarcopenia and disability.

AGING SKELETAL MUSCLE

SARCOPENIA

Sarcopenia has been considered to be a minor modifiable risk factor for health outcomes, and it plays a significant role in the etiology of disability [10,11]. In essence, sarcopenia is an imbalance between protein synthesis and degradation rate [9]. Aging-induced sarcopenia (Figure 30.1) is a result of decreased synthesis and increased degradation of myofibrillar proteins, which leads to the slower turnover rate of muscle proteins, especially contractile proteins, and this, in turn, leads to the decrease in muscle strength [12,13]. Decrease in the protein synthesis rate is affected by the translational process occurring in older human skeletal muscle whereas the transcriptional process appears to be unaltered when compared with those in younger men [14]. Skeletal muscle fibers have a remarkable capacity to regenerate [15] and this depends on the number of satellite cells (Scs) under the basal lamina of fibers and their oxidative capacity [16]. A decrease in the number of Scs has been shown in the FT muscle fibers of elderly subjects [17]. In sarcopenic muscle, the decrease in the Scs pool and the length of telomeres might explain the higher prevalence of muscle injuries and delayed muscle regeneration [18]. Functionally heterogeneous Scs with different properties may be recruited for different tasks, for example, muscle regeneration [19,20]. After severe

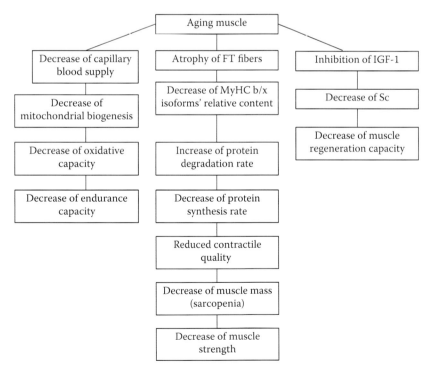

FIGURE 30.1   Morpho-functional changes in aging skeletal muscle. *FT*, fast-twitch; *IGF*, insulin-like growth factor; *Sc*, satellite cell.

damage, muscles in old rodents did not regenerate as well as muscles in adults [4]. The decreased regeneration capacity of muscles has been shown to be due to extrinsic causes rather than because of an intrinsic limitation of muscles, but it is a combination of both extrinsic and intrinsic factors that contribute to reduced skeletal muscle regeneration [21,22]. A contraction-induced muscle injury to weight bearing muscles in old rodents causes deficits in muscle mass and force [23]. The degradation rate of contractile proteins in rat skeletal muscle during aging increased about two times, and muscle strength and motor activity decreased at the same time [24]. Increasing dietary protein intake in combination with the use of anabolic agents attenuates muscle loss.

AGING AND INACTIVITY

Aging and inactivity or disuse of organs are associated with a decline in muscle mass, structure, and strength [2]. A sedentary lifestyle, bed rest, spaceflight, and hindlimb suspension lead the skeletal muscle to microcirculatory disturbances, atrophy, protein loss, changes in contractile properties, and fiber-type switching [25,26]. In both young and aged skeletal muscle, oxidative stress increases in response to inactivity [27] and may have an important role in mediating muscle atrophy. Inactivity results in a decrease in the number of myonuclei and an increase in the number of apoptotic myonuclei in skeletal muscle [28]. Heat-shock protein (HSP) 70 inhibits caspase-dependent and caspase-independent apoptotic pathways and may function in the regulation of muscle size via the inhibition of necrotic muscle fiber distribution and apoptosis in aged muscle [29]. The decline in elderly muscle mass primarily results from type II fiber atrophy and loss in the number of these muscle fibers. Increased variability in fiber size, accumulation of nongrouping, scattered, and angulated fibers, and the expansion of extracellular space are characteristics of muscle atrophy [30]. Beyond the loss of muscle size due to reduced fiber number and myofibrillar proteins that underlie muscle weakness in the elderly [31], impairments in neural activation have been found, as well as potential alterations in other muscular properties that may reduce contractile quality defined as a reduction in involuntary force production per unit muscle size [32]. The functional and structural decline of the neuromuscular system is a recognized cause of decreased strength, impaired performance of daily activities, and loss of independence in the elderly [33]. Loss of muscle strength in older adults is weakly associated with the loss of lean body mass [34]. It means that muscle weakness in older adults is more related to impairments in neural activation and/ or reductions in the intrinsic force generating capacity of skeletal muscle. Effective physical activity in the elderly increases both muscle oxidative capacity and contractile property, enhancing their life quality by improving muscle functional capacity and plasticity.

GLUCOCORTICOID-CAUSED MYOPATHY

Glucocorticoids have an anti-inflammatory effect, but there are also many side effects. It is well known that glucocorticoid-caused myopathy as well as Cushing disease lead to a marked reduction in muscle mass, wasting of muscle, loss of strength, and selective atrophy of FT muscle fibers. The catabolic effect of glucocorticoids depends on individual tissues. The catabolic action of glucocorticoids in skeletal muscle depends on the type of muscle fibers [7].

CATABOLIC ACTION OF GLUCOCORTICOIDS IN MUSCLE FIBERS

FT muscles, particularly type IIB fibers, which have the lowest oxidative capacity among the striated muscle fibers, are most sensitive to the catabolic action of glucocorticoids (Figure 30.2). Myofibrils of type IIB muscle fibers are thinner, mainly because of splitting of myofibrils. Disarray of myosin filaments begins from the periphery of myofibrils, spreads to the central part of the sarcomere near the H-zone, and is distributed over all the A-band [5]. In these fibers the activity of non-lysosomal proteases increases significantly after glucocorticoid administration [7]. Large doses of glucocorticoids depress testosterone and insulin levels [35,36], although the inhibition of protein synthesis

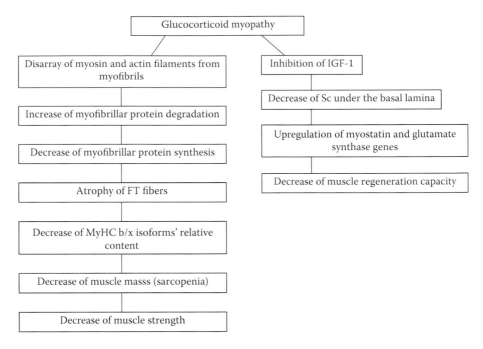

FIGURE 30.2 Changes in glucocorticoid-caused myopathic muscle. *FT*, fast-twitch; *Sc*, satellite cell; *IGF-1*, insulin-like growth factor; *MyHC*, myosin heavy chain.

after glucocorticoid administration plays a lesser role in the atrophy of muscle fibers than accelerated protein catabolism [37]. As the content of lysosomes in type IIB/X muscle fibers is low, the role of lysosomal proteases in the development of corticosteroid myopathy is not significant. It has been shown that the process of atrophy starts from myosin filaments. Myosin filaments lyzed after separation [5], while actin filaments are more resistant to the catabolic action of glucocorticoids.

Muscle weakness in corticosteroid myopathy is mainly the result of destruction and atrophy of the myofibrillar compartment of skeletal muscle [4]. In agreement with Kelly et al. [38], who about three decades ago stated that the terms catabolic and myopathic should be used carefully as they can be misleading descriptions of the action of glucocorticoids on the level of the organism as well as on the level of skeletal muscle. Disappearance of about 20% of myosin filaments from myofibrils of muscle fibers with low oxidative capacity and decrease of myosin heavy chain (MyHC) IIb isoform relative content [4] are the explanation for decreased muscle strength and motor activity in the case of corticosteroid myopathy. This is qualitative remodeling of the myofibrillar compartment of myopathic type II B/X fibers but not whole skeletal muscle. The higher the degree of atrophy, the lower the muscle elasticity and the higher the tone. Muscle tone is dependent on changes in innervation. It has been shown that the neuromuscular synapses of glucocorticoid myopathic FT muscles are destroyed [39]. A decrease of titin and myosin [40] and of the ratio of nebulin and MyHC in myopathic muscle [3], shows that these changes in contractile and elastic proteins are the result of elevated catabolism of the above-mentioned proteins in skeletal muscle. This is the reason for reduced elasticity and generation of tension in glucocorticoid-caused myopathic muscle.

CATABOLIC ACTION OF GLUCOCORTICOIDS IN CONTRACTILE APPARATUS

Muscle fibers and myofibrils of glucocorticoid-caused myopathic glycolytic muscle are thinner in comparison with the control group and disappeared completely from one fifth of the area of myofibrils of myopathic muscle myosin filaments [4]. The intensive destruction of myofibrils and degradation of contractile proteins, including MyHC IIb isoform [41], are the main reasons for reduced muscle strength, motor activity, and weakness in glucocorticoid-caused myopathic rats [42]. Destruction of myofibrils (Figure 30.2) starts from the periphery of glycolytic muscle fibers, from myosin filaments, and spreads all over the myofibrillar compartment [5]. The second reason why myofibrils of myopathic muscle fibers are thinner is the slower myofibrillar protein synthesis rate and assembly of filaments. The decrease of the MyHC IIb isoform relative content and the increase of the MyHC IId isoform show that the quantitative changes in myofibrils are significantly related to the qualitative remodeling of thick myofilaments in myopathic glycolytic muscle fibers [42]. Changes in the myofibril ultrastructure of myopathic muscle fibers are also related to the functional modification of glycolytic muscle fibers. These modifications were not observed in muscle fibers with higher oxidative capacity in myopathic rodents [42]. Glucocorticoids caused wasting in senescent and young rodents as a result of the loss of FT fibers, their myofibrils, contractile proteins, and conversion of muscle fibers with low oxidative capacity into higher oxidative capacity. The myosin HC and actin synthesis rate in aging rats decreased by about 20%–30% and contractile proteins turned over in old subjects very slowly; the same tendency continued after glucocorticoid treatment [37].

CATABOLIC ACTION OF GLUCOCORTICOIDS IN AGING MUSCLE

During aging muscle strength decreases significantly; for example, the hindlimb grip strength in old rats decreases by about 50% due to sarcopenia [24]. Glucocorticoid treatment decreases muscle strength in both young and old groups. In the old group the decrease was more significant than in the young. Aging is accompanied by general weakness and impaired locomotion [6]. Motor activity in old rats decreased in comparison with the young group and glucocorticoid treatment reduced it in both age groups [24]. An excess of glucocorticoids decreases the skeletal muscle regeneration capacity in the young and old groups and is in correlation with a decrease of Scs under the basal lamina.

Old glucocorticoid-caused myopathic rats have only one half the number of Scs that young ones do. Despite that skeletal muscle regeneration, although the adipose tissue content was significantly higher in old myopathic rats and muscle, the strength decrease was about 50% [24]. In glucocorticoid-treated aged rats muscle wasting was more rapid than in young ones and recovery of muscle mass took twice as long as in the young [43]. The reason is the decrease of the stimulatory effect of insulin and IGF-1 in the skeletal muscle of old rats, which is twice as severe in the young [44].

Aging-caused sarcopenia and glucocorticoid-caused myopathy both develop as a result of the decrease and damage of Scs [45]. In old myopathic muscle, degradation of contractile proteins doubles [24]. Many intrinsic changes in cells accompany aging: accumulation of oxidative damage, decline in genome maintenance, and diminished mitochondrial function [46,47]. These changes may induce muscle destruction and therefore delay the regeneration of the myofibrillar compartment. Increased functional activity of muscle tissue causes fast recovery of muscle contractile structures and strength and depends on the oxidative capacity of muscle [2]. Contractile proteins' synthesis rate is more intensive in muscle fibers with a high oxidative capacity [48,49]. A decrease in adenosine monophosphate-activated protein kinase (AMPK) activity in the elderly is the reason for reduced mitochondrial function. AMPK is activated in response to moderate functional activity and this may be an effective measure in the prevention of disability and diseases in the elderly [9].

ROLE OF PHYSICAL ACTIVITY IN RETARDING MUSCLE ATROPHY

More than four decades ago, the preventive role of exercise in the development of muscle atrophy during glucocorticoid administration was shown [50]. From the historical view point, endurance type of physical activity has been found to be an effective measure in retarding skeletal muscle atrophy associated with the administration of glucocorticoids [7,51].

On many occasions, specific physical activity programs are tailored for rehabilitation needs. For example, in case of glucocorticoid-caused myopathy, both endurance and strength exercise training have been shown to play a preventive role in the development of muscle atrophy, but a combination of both with different frequency, intensity, and duration seems to be more effective. From the contraction nature, four model systems have given the desired effect: endurance exercise, strength exercise, muscle functional overload, and *in vitro* cell culture stimulation [52]. Later intensive short-lasting exercise training has shown to have an anticatabolic effect on the contractile apparatus and the ECM of skeletal muscle [53]. Glucocorticoids increased myofibrillar protein degradation in FT muscles, while fibril- and network-forming collagen specific mRNA levels decreased at the same time in FT and ST muscles [54]. Both the myofibrillar apparatus and the ECM play a crucial role in changes of muscle strength during glucocorticoid administration and following muscle loading [55].

ANTICATABOLIC EFFECT OF PHYSICAL ACTIVITY IN SKELETAL MUSCLE

About half of century ago it was shown that the catabolic action of corticosteroids depends on the state of functional activity of skeletal muscle [50]. Moderate endurance-type exercise has been shown to be effective in retarding muscle atrophy [8] and protecting against wasting [51]. The effect of moderate physical activity (Figure 30.3) in inducing a less pronounced catabolic effect of glucocorticoid is caused by the elevation of anticatabolic activity of this type of physical activity and related with the endogenous action of androgens in the stimulation of anticatabolic activity in type IIB/X muscle fibers [7]. Skeletal muscles with higher oxidative capacity, particularly ST fibers, are less sensitive to the catabolic action of glucocorticoids, this phenomenon was explained by the lesser elevation of proteolytic activity in these muscle fibers, but in muscle fibers where oxidative capacity

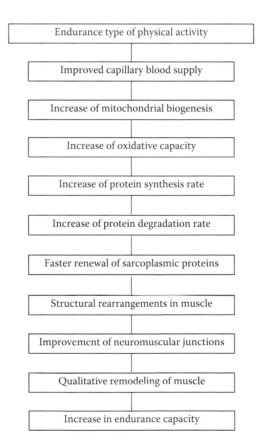

FIGURE 30.3   Effect of endurance type of physical activity in skeletal muscle.

is low, catabolic activity was more pronounced [5]. As ST muscle fibers are involved in the maintenance of static body posture, in slow repetitive movements, and being functionally active even when FT fibers are passive, it may also explain why there is no significant catabolic action of glucocorticoids and atrophy of ST muscle fibers [5]. Atrophy of muscle fibers with low oxidative capacity is the result of inhibition of insulin-like growth factor-1 (IGF-1) [56] and upregulation of two genes, myostatin and glutamate synthase [57]. Increased functional activity in the elderly improves glucose intolerance and insulin signaling, reduces tumor necrosis factor (TNF)-α, increases adiponectin and IGF-1 concentrations, and reduces total and abdominal visceral fat [58]. Increase of muscle activity increases the rate of synthesis of myofibrillar proteins [59] via a mammalian target of rapamycin (mTOR)-activating proteins within the nitrogen-activated protein kinase signaling [60]. The recovery from the last exercise session, particularly from intensive exercise, is faster in the young than in the elderly [61].

EFFECT OF RESISTANCE TYPE OF PHYSICAL ACTIVITY

Muscle atrophy contributes to, but does not completely explain the decrease in force in the elderly. The age-related decrease in muscle mass and strength is a consequence of the complete loss of fibers associated with the decrease in the number of motor units and fiber atrophy [23]. In recent years, resistance exercise has become one of the fastest growing forms of physical activity for different purposes: improving athletic performance, enhancing general health and fitness, rehabilitation after surgery or an injury, or just for the fun of it. Resistance exercise has shown to be an effective measure in the elderly, improving glucose intolerance, including improvements in insulin signaling defects, reduction in TNF-α, increases in adiponectin and IGF-1 concentrations, and reductions in total and abdominal visceral fat [58]. Resistance exercise improves skeletal muscle metabolism and through it muscle function in the elderly and their life quality [2]. Resistance exercise (Figure 30.4)

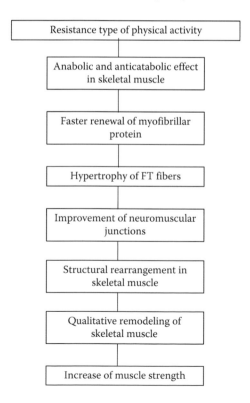

FIGURE 30.4   Effect of resistance type of physical activity in skeletal muscle. *FT,* fast-twitch.

enhances the synthesis rate of myofibrillar proteins but not that of sarcoplasmic proteins [59] and this is related to mTOR by activating proteins within the nitrogen-activated protein kinase signaling [60]. A significant difference was observed between previously trained young and old participants in recovery from resistance training (RT) [62]. These results suggest a more rapid recovery in the young group. It seems that recovery from more damaging resistance exercise is slower as a result of age, whereas there are no age-related differences in recovery from less damaging metabolic fatigue [61]. It has been shown that RT, during which the power of exercise increased less than 5% per session, caused hypertrophy of both FT and ST muscle fibers, an increase of myonuclear number via fusion of Scs with damaged fibers or the formation of new muscle fibers as a result of myoblasts' fusion in order to maintain myonuclear domain size [63]. It has been shown that contractile proteins turned over faster in type I and IIA fibers than in IIB fibers and the turnover rate of skeletal muscle proteins in skeletal muscle depends on the functional activity of the muscle [49]. The turnover rate of myofibrillar proteins in aging skeletal muscle is related to the changes in MyHC isoforms' composition [2].

The effect of RT on the increase of the turnover rate of skeletal muscle contractile proteins in old age is relatively small [2]. Adaptational changes first appeared in newly formed or regenerating fibers and these changes lead to the remodeling of the contractile apparatus and an increase in the strength generating capability of muscle. These changes are more visible in muscle fibers with higher oxidative capacity. The recovery of locomotory activity after unloading is as fast as the recovery of muscle strength. It is related to the regeneration of muscle structure from disuse atrophy [64]. This fact suggests the presence of functionally immature muscle fibers during the recovery process following disuse atrophy [64]. So, the recovery of skeletal muscle mechanical properties depends on the structural and metabolic peculiarities of skeletal muscle [49]. As a complex of factors contributes to the development of muscle wasting and weakness in the elderly, skeletal muscle unloading and glucocorticoid-caused myopathy, it is complicated to find one certain measure for rehabilitation. As lack of strength is one of the central reasons for muscle weakness, it seems to be most realistic to use resistance type of activity/RT for this purpose in the elderly. RT is a strong stimulus for muscle metabolism in the elderly, particularly for the contractile machinery of muscle.

HYPERTROPHY OF MUSCLE FIBERS

RT increases the cross-sectional area (CSA) of the whole muscle and individual muscle fibers, and increases myofibrillar size and number (Figure 30.4). The hypertrophy response to RT is related to the activation of satellite cells (Scs) in the early stage of training [65]. RT causes fiber hypertrophy in two ways: damaged fibers regenerate as a result of the fusion with Sc [66] as it is proved by the incorporation of ^{3}H thymidine into the nucleus of the muscle fiber [63], and via Sc activation under the basal lamina, division and after that myosymplasts fuse with each other and form myotubes [65]. RT also causes other morphological adaptations, such as hyperplasia, changes in muscle fine architecture, in myofilament density, and in the structures of connective tissue [65]. RT mainly causes an increase in the CSA of IIX/IIB and IIA fibers. Structural changes in skeletal muscle during RT are fiber specific. RT enhances the synthesis rate of myofibrillar proteins, not of sarcoplasmic proteins, and this is related to the mTOR complex by activating proteins with mitogen-activated protein kinase (MAPK) signaling [60]. Recovery from intensive RT-caused damages is slower as a result of age, whereas there are no age-related differences in recovery from less damaging metabolic fatigue [61]. Recovery from RT, during which the power of exercise increases less than 5% per session, causes hypertrophy of both FT and ST muscle fibers and an increase in the myonuclear number. This is achieved via Sc fusion with damaged fibers or the formation process of new muscle fibers as a result of myoblasts' fusion in order to maintain myonuclear domain size [63]. RT increases the level of IGF-I and mechano-growth factor (MGF) in skeletal muscle and these factors support faster recovery of muscle tissue [67].

EFFECT OF ENDURANCE TYPE OF PHYSICAL ACTIVITY

As oxidative capacity of skeletal muscle decreases in the elderly, endurance type of activity/ endurance training (ET) is effective in stimulating mitochondrial biogenesis and improving their functional parameters [68]. In combination with RT, the oxidative capacity and subsequently the turnover rate of contractile proteins in elderly skeletal muscle increases. This increase of the turn-over rate of muscle proteins leads to the increase in skeletal muscle plasticity. It has recently been shown that that the plasticity of individual development in the elderly makes it possible to modify the age-associated decline even in maximal physical performance [69]. Another positive influence of ET in the elderly is related to an increase in the ability of cardiovascular factors and to a lesser extent, to an increase in muscle mitochondrial concentration and capacity [70]. The increase in muscle oxidative capacity and contractile property is an effective measure for enhancing life quality in the elderly by improving skeletal muscle functional capacity and plasticity. It has recently been shown that the individual development of muscle plasticity in the elderly makes it possible to modify the age-associated decline even in maximal physical performance at least for some time [69]. The higher aerobic capacity in trained elderly people is related to an increase in the abilities of the cardiovascular system and to the lesser extent to an increase in muscle mitochondrial concentration [70]. It means that regular aerobic activity provides a foundation for an increase in muscle oxidative capacity in the elderly. It is useful to repeat the viewpoint of Suominen [69] that adequate physical performance is an essential element of a healthy and productive life among the elderly. Netz [71] studied the effect of physical activity on the moderating role of fitness improvement and mode of exercise to the potential mechanisms for explaining the physical activity affect relationship and found that neither improved fitness or exercise modality serve as moderators of physical activity effect on affect. However, with increasing age, managing everyday activities becomes less self-evident, although there are gender differences in physical functioning [72]. Functional limitation is an objective measure of the consequences of disease and impairment [73]. It seems that the turnover rate of contractile proteins provides the mechanism by which the effect of exercise causes changes in accordance with the needs of the contractile apparatus. As the contractile protein turn-over rate depends on the oxidative capacity of muscle and muscle oxidative capacity decreases in the elderly, it is obvious that endurance exercise stimulates an increase in the oxidative capacity of skeletal muscle by an increase in mitochondrial biogeneses and supports faster protein turnover during RT in order to increase muscle function. It has been shown that the aging-associated reduction in AMPK activity may be a factor in reduced mitochondrial function [74]. In response to contractile activity, AMPK activation was registered only in aging FT muscles [75]. It is known that AMPK is activated in response to ET [76] and related to the metabolic adaptation of skeletal muscle. Later it has been shown that $\alpha 1$ isoform of AMPK is the regulator of skeletal muscle growth, but not of metabolic adaptation [77]. As factors such as health, physical function, and independence constitute components of the quality of life in the elderly, physiological functioning of skeletal muscle in the elderly has significance in determining the ability to maintain independence and an active interaction with the environment [2,78].

CHANGES IN MUSCLE METABOLISM

According to Kramer and Erickson [79], successful aging is guaranteed when elderly people use widespread participation in low-cost and low-tech exercise for further improving their fitness and reducing the risk of disability. Type I muscle fibers contain a large number of myonuclei and Sc compared with FT IIB fibers. Fast to slow fiber transition has been shown to be associated with increases in Sc activation, content, and fusion to transforming fibers, especially within the IIB fibers [80]. The number of Sc in very different stages of development under the basal lamina of type I and FT type IIA muscle fibers increases during endurance type of activity [81,82] AMPK is activated in response to endurance type of physical activity [76] and related to the metabolic adaptation of skeletal muscle in both the young and old. AMPK function includes glucose transport, glycogen metabolism, fatty acid oxidation, and transcriptional regulation of structural muscle genes [83]. α1 isoform of AMPK is the regulator of skeletal muscle growth and α2 isoform regulates metabolic adaptation [77]. Increased mitochondrial biogeneses via AMPK is accompanied by the suppression of myofibrillar protein synthesis through pathways mediated by MAPK, nuclear factor kappa B (NF-κB) mTOR, and tuberous sclerosis complex (TSC) [84]. IGF-I expression is higher in ST fibers [85] and myostatin in fibers with higher oxidative capacity (type I and IIA) [86]. The components of the degradation system of muscle proteins, such as ubiquitin ligases muscle atrophy F-box (MAFbx) and muscle ring finger (MuRF) are about twofold higher in fibers with higher oxidative capacity [84] also in the elderly. It was shown that the number of Sc in rat skeletal muscle increased about 3.5 times during endurance type of physical activity [87]. Both oxidative capacity and Sc number in muscle fibers, which determine muscle regenerative capacity, are higher in young than in old muscle. Protein turnover in skeletal muscle is relatively slow, especially contractile proteins and endurance type of physical activity stimulates protein turnover [88]. The turnover rate of myosin heavy chain (MyHC) and myosin light chain (MyLC) isoforms provides a mechanism by which the type and amount of protein changes in accordance with the needs of the contractile machinery during adaptation to endurance type of physical activity [89]. Endurance type of activity mainly increases the number of Sc under the basal lamina of type I and IIA muscle fibers and increases the regeneration capacity of these fibers [49]. The mechanism associated with activity-induced shifts in myosin expression is the key to understanding the plasticity of skeletal muscle as the hypertrophied muscle fiber has adapted to a chronic overload via an alteration in its phenotype [90]. The mechanisms involved in regulating changes in the myosin expression and in the muscle mass may have different sensitivities to mechanical load [91].

EFFECT OF CONCURRENT STRENGTH AND ENDURANCE TYPE OF PHYSICAL ACTIVITY

Concurrent physical activity for strength and endurance/concurrent strength and ET has shown to decrease the gain in muscle mass in comparison with activity for strength alone [92]. This effect was explained by AMPK blocking the activation of mammalian target of rapamycin complex-1 (TORC 1) by phosphorylating and activating the tuberous sclerosis complex-2 (TSC 2) [93]. This interference in skeletal muscle strength development was also explained by alterations in the protein synthesis induced by the high volume of endurance type of activity or by frequent physical activity sessions [94] or was related to impairment of neural adaptations [95]. Concurrent strength and endurance type of activity in elderly men has shown that strength gain was similar to that observed with strength type of activity alone, although the volume was half of that [96]. Using lower physical activity volumes in concurrent training in older men [97] in comparison with endurance and resistance activity alone leads to similar strength enhancement with absence of interference in this population [98]. In the elderly population, improvement in both strength and cardiorespiratory fitness is important and concurrent activity is the best strategy to enhance cardiorespiratory fitness as it has widely been shown in the literature [95]. Concurrent resistance and endurance type of physical activity combination is the fastest way to prevent muscle atrophy in the elderly due to the anabolic and anticatabolic effect of resistance type of physical activity and at the same time, endurance type of activity increases the oxidative capacity of skeletal muscle [99]. The qualitative remodeling of muscle fibers as the result of increased physical activity also prevents the development of muscle weakness. Physical activity has the ability to influence the function of muscle fibers modifying their structure and metabolism and promoting the release of growth factors and other signaling molecules, such as nitric oxide, which work through the paracrine system to activate Sc. Physical activity not only supports muscle regeneration capacity in top athletes, aging populations, and in patients suffering with different genesis muscle atrophy, but is the fastest way of regeneration of damaged muscle fibers [100].

CONCLUSIONS

Aging-induced sarcopenia is a result of decreased synthesis and increased degradation rate of myofibrillar proteins, which leads to slower turnover rate of contractile proteins and to a significant decrease of muscle strength. Dexamethasone treatment in both the young and old led to quite similar results, but these changes are more significant in the aging group. Aging- and dexamethasone-induced sarcopenic muscles have diminished regenerative capacity. Sarcopenia is the risk factor for health outcomes and plays a significant role in the etiology of disability in the elderly. As a complex of factors contributes to the development of muscle wasting and weakness in the elderly, it is difficult to find any specific measure for rehabilitation. As the lack of strength is one main reason for muscle weakness, it seems to be most realistic to use resistance type of physical activity for this purpose in the elderly. Resistance type of activity is a strong stimulus for muscle metabolism in the elderly, particularly for the contractile machinery of muscle. The contractile proteins turnover rate provides a mechanism by which the effect of exercise-caused changes can be assessed in accordance with the needs of the contractile apparatus. As the contractile proteins turnover rate depends on the oxidative capacity of muscle and muscle oxidative capacity decreases in the elderly, it is obvious that endurance exercise stimulates an increase in the oxidative capacity of skeletal muscle by an increase in mitochondrial biogenesis and supports faster protein turnover during RT, as a result muscle function, and thereby the quality of life, in the elderly improves. The regeneration of skeletal muscle from the damage caused by exercise is faster in muscles with higher oxidative capacity. Using both resistance and endurance type of physical activity (concurrent resistance and ET) in the elderly and glucocorticoid-caused myopathic patients makes it possible to modify the age-associated decline in muscle function and decelerate the development of muscle weakness.

ACKNOWLEDGMENTS

This study was supported by the funds of the Ministry of Education and Research of the Republic of Estonia, research project number TMVSF14058I.

CONFLICT OF INTEREST

No conflict of interest.

REFERENCES

1. Degenes H, Alway SE. 2006. Control of muscle size during disuse, disease, and aging. *Int J Sports Med*, 27, 94–99. http://dx.doi.org/10.1055/s-2005-837571
2. Seene T, Kaasik P, Riso EM. 2012. Review on aging, unloading and reloading: Changes in skeletal muscle quantity and quality. *Arch Gerontol Geriatr*, 54, 374–380.http://dx.doi.org/10.1016/j.archger.2011.05.002
3. Aru M, Alev K, Gapeyeva H, Vain A, Puhke R et al. 2013. Glucocorticoid-induced alterations in titin, nebulin, myosin heavy chain isoform content and viscoelastic properties of rat skeletal muscle. *Adv Biol Chem*, 3, 70–75. [CrossRef]
4. Kaasik P, Aru M, Alev K, Seene T. 2012. Aging and regenerative capacity of skeletal muscle in rats. *Curr Aging Sci*, 5, 126–130. http://dx.doi.org/10.2174/1874609811205020126
5. Seene T, Umnova M, Alev K, Pehme A. 1988. Effect of glucocorticoids on contractile apparatus of rat skeletal muscle. *J Steroid Biochem*, 29, 313–317. [CrossRef]
6. Attaix D, Mosoni L, Dardevet D, Combaret L, Patureau Mirand P et al. 2005. Altered responses in skeletal muscle protein turnover during aging in anabolic and catabolic periods. *Int J Biochem Cell Biol*, 37, 1962–1973. [CrossRef] [PubMed]
7. Seene T, Viru A. 1982. The catabolic effect of glucocorticoids on different types of skeletal muscle fibres and its dependence upon muscle activity and interaction with anabolic steroids. *J Steroid Biochem*, 16, 349–352. [CrossRef]
8. Hickson R, Davis J. 1981. Partial prevention of glucocorticoid induced muscle atrophy by endurance training. *Am. J. Physiol*, 241, E226–E232. [PubMed]

9. Seene T, Kaasik P. 2012. Role of exercise therapy in prevention of decline in aging muscle function: Glucocorticoid myopathy and unloading. *J. Aging Res*, 2012, 172492. [CrossRef] [PubMed]

10. Lauretani F, Russo C.R, Bandinelli S, Bartali B, Cavazzini C et al. 2003. Age-associated changes in skeletal muscles and their effect on mobility: An operation diagnosis of sarcopenia. *J Appl Physiol*, 9, 1851–1860. http://dx.doi.org/10.1152/japplphysiol.00246.2003

11. Clark BC, Manini TM. 2010. Functional consequences of sarcopenia and dynapenia in the elderly. *Curr Opin Clin Nutr Metab Care*, 13, 271–276. http://dx.doi.org/10.1097/MCO.0b013e328337819e.

12. Evans WJ. 2010. Skeletal muscle loss: Cachexia, sarcopenia, and inactivity. *Am J Clin Nutr*, 91, 1123S–1127S. http://dx.doi.org/10.3945/ajcn.2010.28608A

13. Evans WJ, Paolisso G, Abbatecola AM, Corsonello, Bustacchini S et al. 2010. Frailty and muscle metabolism dysregulation in the elderly. *Biogerontology*, 11, 527–536. http://dx.doi.org/10.1007/s10522-010-9297-0.

14. Roberts MD, Kerksick CM, Dalbo VJ, Hassell SE, Tucker PS et al. 2010. Molecular attributes of human skeletal muscle at rest and after unaccustomed exercise: An age comparison. *J Strength Cond Res*, 24, 1161–1168. http://dx.doi.org/10.1519/JSC.0b013e3181da786f

15. Bassaglia Y, Gautron J. 1995. Fast and slow rat muscles degenerate and regenerate differently a after crush injury. *J Muscle Res Cell Motil*, 16, 420–429. http://dx.doi.org/10.1007/BF00114507

16. Shultz E, Darr K. 1990. The role of satellite cells in adaptive or induced fiber transformations. In: *The Dynamic State of Muscle Fibers*, D. Pette, Ed., W. de Gruyter, Berlin, 667–681.

17. Verney J, Kadi, Charifi N, Feasson L, Saafi MA et al. 2008. Effects of combined lower body endurance and upper body resistance training on the satellite cell pool in elderly subjects. *Muscle Nerve*, 38, 1147–1154. http://dx.doi.org/10.1002/mus.21054

18. Kadi F, Ponsot E. 2010. The biology of satellite cells and telomeres in human skeletal muscle: Effects of aging and physical activity. *Scand J Med Sci Sports*, 20, 39–48. http://dx.doi.org/10.1111/j.1600-0838.2009.00966.x

19. Malatesta M, Perdoni F, Muller S, Pellicciari C, Zancanaro C. 2010. Pre-mRNA processing is partially impaired in satellite cell nuclei from aged muscles. *J Biomed Biotehnol*, 2010, Article ID: 410405.

20. Tatsumi R. 2010. Mechano-biology of skeletal muscle hypertrophy and regeneration: Possible mechanism of stretch-induced activation of resident myogenic stem cells. *Anim Sci J*, 81, 11–20. http://dx.doi.org/10.1111/j.1740-0929.2009.00712.x

21. Carlson BM, Dedkov EI, Borisov AB, Faulkner JA. 2001. Skeletal muscle regeneration in very old rats. *J Gerontol. Series A, Biol Sci Med Sci*, 56, B224–B233.http://dx.doi.org/10.1093/gerona/56.5.b224

22. Conboy IM, Conboy MJ, Wagers AJ, Girma ER, Weissman IL et al. 2005. Rejuvenation of aged progenitor cells by exposure to a young systemic environment. *Nature*, 433, 760–764. http://dx.doi.org/10.1038/nature03260

23. Rader EP, Faulkner JA. 2006. Recovery from contraction-induced injury is impaired in weight-bearing muscles of old male mice. *J Appl Physiol*, 100, 656–661. http://dx.doi.org/10.1152/japplphysiol.00663.2005

24. Kaasik P, Umnova M, Pehme A, Alev K, Aru M et al. 2007. Ageing and dexamethasone associated sarcopenia: Peculiarities of regeneration. *J Steroid Biochem Mol Biol*, 105, 85–89. http://dx.doi.org/10.1016/j.jsbmb.2006.11.024

25. Haus JM, Carrithers JA, Trappe SW, Trappe TA. 2007. Collagen, cross-linking, and advanced glycation end products in aging human skeletal muscle. *J Appl Physiol*, 103, 2068–2076. http://dx.doi.org/10.1152/japplphysiol.00670.2007

26. Pasiakos SM, Vislocky LM, Carbone JW, Altieri N, Konopelski K et al. 2010. Acute energy deprivation affects skeletal muscle protein synthesis associated intracellular signaling proteins in physically active adults. *J Nutr*, 140, 745–751. http://dx.doi.org/10.3945/jn.109.118372

27. Siu PM, Pistilli EE, Alway SE. 2008. Age-dependent increase in oxidative stress in gastrocnemius muscle with unloading. *J Appl Physiol*, 105, 1695–1705. http://dx.doi.org/10.1152/japplphysiol.90800.2008

28. Leeuwenburgh C, Gurley CM, Strotman BA, Dupont-Versteegden EE. 2005. Age-related differences in apoptosis with disuse atrophy in soleus muscle. *Am J Physiol. Regul Integr Comp Physiol*, 288, R1288–R1296. http://dx.doi.org/10.1152/ajpregu.00576.2004

29. Ogata T, Machida S, Oishi Y, Higuchi M, Muraoka I. 2009. Differential cell death regulation between adult-unloaded and aged rat soleus muscle. *Mech Ageing Dev*, 130, 328–336. http://dx.doi.org/10.1016/j.mad.2009.02.001

30. Buford TW, Anton SD, Judge AR, Marzctti E, Wohlgemuth SE et al. 2010. Models of accelerated sarcopenia: Critical pieces for solving the puzzle of age-related muscle atrophy. *Ageing Res Rev*, 9, 369–383. http://dx.doi.org/10.1016/j.arr.2010.04.004

31. Clark BC, Manini TM. 2008. Sarcopenia ≠ dynapenia. *J Gerontol Series A, Biol Sci Med Sci*, 63, 829–834. http://dx.doi.org/10.1093/gerona/63.8.829

32. Gonzales E, Messi ML, Delbono O. 2000. The specific force of single intact extensor digitorum longus and soleus mouse muscle fibers declines with aging. *J Membr Biol*, 178, 175–183. http://dx.doi.org/10.1007/s002320010025

33. Perrey S, Rupp T. 2009. Altitude-induced changes in muscle contractile properties. *High Alt Med Biol*, 10, 175–182. http://dx.doi.org/10.1089/ham.2008.1093

34. Gandevia SC. 2001. Spinal and supraspinal factors in human muscle fatigue. *Physiol Rev*, 81, 1725–1789

35. Tomas F, Munro H, Young V. 1979. Effect of glucocorticoid administration on the rate of muscle protein breakdown *in vivo* in rats, as measured by urinary excretion on N-methylhistidine. *Biochem J*, 178, 139–146. [CrossRef] [PubMed]

36. Smals A, Kloppenborg P, Benroad T. 1977. Plasma testosterone profiles in Cushing's syndrome. *J Clin Endocrinol Metab*, 45, 240–245. [CrossRef] [PubMed]

37. Seene T. 1994. Turnover of skeletal muscle contractile proteins in glucocorticoid myopathy. *J Steroid Biochem Mol Biol*, 50, 1–4. [CrossRef]

38. Kelly F, McGrath J, Goldspink D, Gullen M. 1986. A morphological (biochemical) study on the action of corticosteroids on rat skeletal muscle. *Muscle Nerve*, 9, 1–10. [CrossRef] [PubMed]

39. Seene T, Umnova M and Kaasik P. 1999. The exercise myopathy. In: *Overload, Performance Incompetence, and Regeneration in Sport*, Lehmann, M., Foster, C., Gastmann, U., Keizer, H., Steinacker, J., Eds., Kluwer Academic/Plenum Publishers: New York, NY, USA; Boston, MA, USA; Dordrecht, The Netherlands; London, UK; Moscow, Russia, pp. 119–130.

40. Hayashi K, Tada O, Higuchi K and Ohtsuka A. 2000. Effects of corticosterone on connection content and protein breakdown in rat skeletal muscle. *Biosci Biotechnol Biochem*, 64, 2686–2688. [CrossRef]

41. Fernandez-Rodriques, E, Stewart P and Cooper M. 2009. The pituitary-adrenal axis and body composition. *Pituitary*, 12, 105–115. [CrossRef] [PubMed]

42. Seene T, Kaasik P, Pehme A, Alev K and Riso EM. 2003. The effect of glucocorticoids on the myosin heavy chain isoforms' turnover in skeletal muscle. *J Steroid Biochem Mol Biol*, 86, 201–206. [CrossRef] [PubMed]

43. Dardevet D, Sornet C, Taillandier D, Savary I, Attaix D and Grizard J. 1995. Sensitivity and protein turn overresponse to glucocorticoids are different in skeletal muscle from adult and old rats. Lack of regulation of the ubiquitin-proteasome proteolytic pathway in ageing. *J Clin Investig*, 96, 2113–2119. [PubMed]

44. Dardevet D, Sornet C, Savary I, Debras E, Patureau Mirand P and Grizard J. 1998. Glucocorticoid effects on insulin and IGF-1-regulated muscle protein metabolism during aging. *J Endocrinol*, 156, 83–89. [CrossRef] [PubMed]

45. Machida S and Booth F. 2004. Increased nuclear proteins in muscle satellite cells in aged animals as compared to young growing animals. *Exp Gerontol*, 39, 1521–1525. [CrossRef] [PubMed]

46. Hasty P, Campisi J, Hoeijmakers J, Van Steeg H and Vijg J. 2003. Aging and genome maintenance: Lessons from the mouse? *Science*, 299, 1355–1359. [CrossRef] [PubMed]

47. Ames B. 2004. Alpha lipoic acid's role in delaying the mitochondrial decay of aging. *Ann N Y Acad Sci*, 1019, 406–411. [CrossRef] [PubMed]

48. Seene T, Alev K, Kaasik P and Pehme A. 2007. Changes in fast-twitch muscle oxidative capacity and myosin isoforms modulation during endurance training. *J Sports Med Phys Fitness*, 47, 124–132. [PubMed]

49. Seene T, Kaasik P and Umnova M. 2009. Structural rearrangements in contractile apparatus and resulting skeletal muscle remodelling: Effect of exercise training. *J Sports Med Phys Fitness*, 49, 410–423. [PubMed]

50. Goldberg A. 1969. Protein turnover in skeletal muscle in experimental corticosteroid myopathy. *J Biol Chem*, 244, 3217–3222. [PubMed].

51. Czerwinski S, Kurowski T, O'Neill T and Hickson R. 1987. Initiating regular exercise protects against muscle atrophy from glucocorticoids. *J Appl Physiol*, 63, 1504–1510. [PubMed]

52. Czerwinski SM and Hickson R.C. 1990. Glucocorticoid receptor activation during exercise in muscle. *J Appl Physiol*, 68, 1615–1620.

53. Riso EM, Ahtikoski AM, Umnova M, Kaasik, P., Alev, K et al. 2003. Partial prevention of muscle atrophy in excessive level of glucocorticoids by exercise: Effect on contractile proteins and extracellular matrix. *Balt J Lab Anim Sci*, 13, 5–12.

54. Riso EM, Ahtikoski AM, Alev K, Kaasik P, Pehme A and Seene T. 2008. Relationship between extracellular matrix, contractile apparatus, muscle mass and strength in case of glucocorticoid myopathy. *J Steroid Biochem Mol Biol*, 108, 117–120.

55. Riso EM, Ahtikoski AM, Takala TES and Seene T. 2010. The effect of unloading and reloading on the extracellular matrix in skeletal muscle: Changes in muscle strength and motor activity. *Biol Sport*, 27, 89–94.

56. Singleton J, Baker B and Thorburn A. 2000. Dexamethasone inhibits insulin-like growth factor signaling and potentiates myoblast apoptosis. *Endocrinology*, 14, 2945–2950. [CrossRef] [PubMed]

57. Carballo-Jane E, Pandit S, Santoro S, Freund C, Luell S, Harris G, Forrest M, Sitlani A. 2004. Skeletal muscle: A dual system to measure glucocorticoid—Dependent transactivation and transrepression of gene regulation. *J Steroid Biochem Mol Biol*, 88, 191–201. [CrossRef] [PubMed]

58. Flack K. Davy K, Hulver M, Winett R, Friasrd M and Davy B. 2011. Aging, resistance training, and diabetes prevention. *J Aging Res*, 2011, Article ID 127315, 12. [CrossRef] [PubMed]

59. Moore D, Tang J, Burd N, Rerecich T, Tarnopolsky M and Phillips S. 2009. Differential stimulation of myofibrillar and sarcoplasmic protein synthesis with protein ingestion at rest and after resistance exercise. *J Physiol*, 587, 897–904. [CrossRef] [PubMed]

60. Moore D, Atherton P, Rennie M, Tarnopolsky M and Phillips S. 2011. Resistance exercise enhances mTOR and MAPK signalling in human muscle over that seen at rest after bolus protein ingestion. *Acta Physiol (Oxf.)*, 201, 365–372. [CrossRef] [PubMed]

61. Fell J and Williams D. 2008. The effect of aging on skeletal-muscle recovery from exercise: Possible implications foraging athletes. *J Aging Phys Act*, 16, 97–115. [PubMed]

62. McLester JR, Bishop PA, Smith J et al. 2003. A series of studies—A practical protocol for testing muscular endurance recovery. *J Strength Cond Res*, 17, 259–273.

63. Seene T, Pehme A, Alev K, Kaasik P, Umnova M and Aru M, 2010. Effects of resistance training on fast- and slow-twitch muscles in rats. *Biol Sport*, 27, 221–229.

64. Itai Y, Kariya Y and Hoshino Y. 2004. Morphological changes in rat hindlimb muscle fibres during recovery from disuse atrophy. *Acta Physiol Scand*, 181, 217–224.

65. Folland JP and Williams AG. 2007. The adaptations to strength training: Morphological and neurological contributions to increased strength. *Sports Med*, 37, 145–168. http://dx.doi.org/10.2165/00007256-200737020-00004

66. Allen DL, Roy RR and Edgerton VR. 1999. Myonuclear domains in muscle adaptation and disease. *Muscle Nerve*, 22, 1350–1360. http://dx.doi.org/10.1002/(SICI)1097-4598(199910)22:10<1350::AID-MUS3>3.0.CO;2-8

67. Seene T, Umnova M, Kaasik P, Alev K and Pehme A. 2008. Overtraining injuries in athletic population. In: *Skeletal Muscle Damage and Repair*, P. Tiidus, Ed., Human Kinetics Books, Champaign, IL, USA. 173–184.

68. Ljubicic V, Joseph AM, Saleem A, Uguccioni G., Collu-Marchese, M., Lai, RY, Nguyen LM. and Hood DA. 2010. Transcriptional and post-transcriptional regulation of mitochondrial biogenesis in skeletal muscle: Effects of exercise and aging. *Biochim Biophys Acta*, 1800, 223–234. http://dx.doi.org/10.1016/j.bbagen.2009.07.031

69. Suominen H. 2011. Ageing and maximal physical performance. *Eur Rev Aging Phys Act*, 8, 37–42.

70. Sagiv M, Goldhammer E, Ben-Sira D and Amir R. 2010. Factors defining oxygen uptake at peak exercise in aged people. *Eur Rev Aging Phys Act*, 7, 1–2. http://dx.doi.org/10.1007/s11556-010-0061-x

71. Netz Y. 2009. Type of activity and fitness benefits as moderators of the effect of physical activity on affect in advanced age: A review. *Eur Rev Aging Phys Act*, 6, 19–27.

72. Kuh D, Bassey EJ, Butterworth S, Hardy R and Wadsworth MEJ. 2005. Grip strength, postural control, and functional leg power in a representative cohort of British men and women: Associations with physical activity, health status, and socioeconomic conditions. *J Gerontol A*, 60, 224–231.

73. Guralnik JM, and Ferrucci L. 2003. Assessing the building blocks of function: Utilizing measures of functional limitation. *Am J Prev Med*, 25, 112–121.

74. Reznick RM, Zong HJ, Li J et al. 2007. Aging-associated reductions in AMP-activated protein kinase activity and mitochondrial biogenesis. *Cell Metab*, 5, 151–156.

75. Thomson D M, Brown JD, Fillmore N et al. 2009. AMP activated protein kinase response to contractions and treatment with the AMPK activator AICAR in young adult and old skeletal muscle. *J Physiol*, 587, 2077–2086.

76. Winder WW and Hardie DG. 1996. Inactivation of acetyl-CoA carboxylase and activation of AMP-activated protein kinase in muscle during exercise. *Am J Physiol*, 270, E299–E304.

77. Mcgee SL, Mustard KJ, Hardie DG and Baar K. 2008. Normal hypertrophy accompanied by phosphorylation and activation of AMP-activated protein kinase $\alpha 1$ following overload in LKB1 knockout mice. *J Physiol*, 586, 1731–1741.

78. Spirduso WW and DL Cronin DL. 2001. Exercise dose-response effects on quality of life and independent living in older adults. *Med Sci Sports Exerc*, 33(suppl 6), S598–S608.

79. Kramer AF, and Erickson KI. 2007. Capitalizing on cortical plasticity: Influence of physical activity on cognition and brain function. *Trends Cogn Sci*, 11, 342–348.

80. Putman CT, Düsterhöft S and Pette D. 2001. Satellite cell proliferation in low frequency-stimulated fast muscle of hypothyroid rat. *Am J Physiol. Cell Physiol*, 279, C682–C690.

81. Magaudda L, Di Mauro D, Trimarchi F and Anastasi G. 2004. Effects of physical exercise on skeletal muscle fiber: Ultrastructural and molecular aspects. *Basic Appl Myol*, 14, 17–21.

82. Umnova M and Seene T. 1991. The effect of increased functional load on the activation of satellite cells in the skeletal muscle of adult rats. *Int J Sports Med*, 12, 501–504. http://dx.doi.org/10.1055/s-2007-1024723

83. Hardie DG and Sakamoto K. 2006. AMPK: A key sensor of fuel and energy status in skeletal muscle. *Physiology*, 21, 48–60. http://dx.doi.org/10.1152/physiol.00044.2005

84. van Wessel T, de Haan A, van der Laarse WJ and Jaspers RT. 2010. The muscle fiber type-fiber size paradox: Hypertrophy or oxidative metabolism? *Eur J Appl Physiol*, 110, 665–694. http://dx.doi.org/10.1007/s00421-010-1545-0

85. Stitt TN, Drujan D, Clarke BA, Panaro F, Timofeyva Y, Kline WO, Gonzalez M, Yancopoulos GD and Glass DJ. 2004. The IGF-1/PI3 K/Akt pathway prevents expression of muscle atrophy-induced ubiquitin ligases by inhibiting FOXO transcription factors. *Mol Cell*, 14, 395–403. http://dx.doi.org/10.1016/S1097-2765(04)00211-4

86. van der Vusse GJ, Glatz JF, Stam HC and Reneman RS. 1992. Fatty acid homeostasis in the normoxic and ischemic heart. *Physiol Rev*, 72, 881–940.

87. Seene TL and Umnova M. 1992. Relations between the changes in the turnover rate of contractile proteins, activation of satellite cells and ultra-structural response of neuromuscular junctions in the fast-oxidative-glycolytic muscle fibres in endurance trained rats. *Basic Appl Myol*, 2, 39–46.

88. Seene T, Kaasik P and Alev K. 2011. Muscle protein turnover in endurance training: A review. *Int J Sports Med*, 32, 905–911. http://dx.doi.org/10.1055/s-0031-1284339

89. Alev K, Kaasik P, Pehme A, Aru M, Parring A.-M, Selart A and Seene T. 2009. Physiological role of myosin light and heavy chain isoforms in fast- and slow-twitch muscles: Effect of exercise. *Biol Sport*, 26, 215–234. http://dx.doi.org/10.5604/20831862.894654

90. Pette D. 2001. Historical perspectives: Plasticity of mammalian skeletal muscle. *J Appl Physiol*, 90, 1119–1124.

91. Hernandez JM, Fedele MJ and Farrell PA. 2000. Time course evaluation of protein synthesis and glucose uptake after acute resistance exercise in rats. *J Appl Physiol*, 88, 1142–1149.

92. Hickson RC. 1980. Interference of strength development by simultaneously training for strength and endurance. *Eur J Appl Physiol Occup Physiol*, 45, 255–263.

93. Inoki K, Li Y, Zhu T, Wu J and Guan KJ. 2002. TSC2 is phosphorylated and inhibited by Akt and suppresses mTOR signalling. *Nat Cell Biol*, 4, 648–657.

94. Nader GA. 2006. Concurrent strength and endurance training: From molecules to man. *Med Sci Sports Exerc*, 38, 1965–1970.

95. Cadore EL, Pinto RS, Lhullier FL, R et al. 2010. Physiological effects of concurrent training in elderly men. *Int J Sports Med*, 31, 689–697.

96. Wood RH, Reyes R, Welsch MA et al. 2001. Concurrent cardiovascular and resistance training in healthy older adults. *Med Sci Sports Exerc*, 33, 1751–1758.

97. Izquierdo M, Ibanez J, Häkkinen K, Kraemer WJ, Larrion JL and Gorostiaga EM. 2004. Once weekly combined resistance and cardiovascular training in healthy older men. *Med Sci Sports Exerc*, 36, 435–443.

98. Karavirta L, Tulppo MP, Laaksonen DE et al. 2009. Heart rate dynamics after combined endurance and strength training in older men. *Med Sci Sports Exerc*, 41, 1436–1443.

99. Seene T, Kaasik P. 2016. Role of myofibrillar protein catabolism in development of glucocorticoid myopathy: Aging and functional activity aspects. *Metabolites*, 6, 15; doi:3390/metabo6020015.

100. Seene T, Kaasik P. 2013. Muscle damage and regeneration: Response to exercise training. *Health*, 5:136–145.

31 Reductionism versus Systems Thinking in Aging Research

Marios Kyriazis

CONTENTS

INTRODUCTION

Simple but relevant definitions of "aging" depend on the context where the term is used. In biology, I define aging as "the imperfect repair of time-related damage." In clinical situations, aging is simply "time-related dysfunction." A reductionist approach in dealing with age-related degeneration is the cornerstone of the existing paradigm of aging research. In reductionism we decompose a concept, we simplify a problem by some orders of magnitude so that to make it easily solvable. We take a complex issue, we then select a central, isolated, and perhaps limited component which we study in detail [1].

This paradigm is saturated with practical obstacles, immense theoretical shortcomings, clinical naivetés, and concepts that terminate the thread of progress. Notwithstanding this, many eminent and less eminent researchers seem unable or unwilling to even discuss alternative modes of research [2]. Nevertheless, there exist such alternative models, which, if properly researched, can lead to a widespread reduction, or even elimination, of age-related degeneration in humans. Here I will consider such models, which are based on a more inclusive worldview that depends on the interactions of humans with their environment. This environment is increasingly becoming more technological, leading to the perhaps inevitable notion of human–machine hybridization, where the line between the biological and the virtual is increasingly blurred. These models are based on a "systems thinking" approach: *the understanding of the relationships between the parts of the system in relation to the whole, and the interdependency of such structures of dynamic systems* [3].

Humans have been trying to eliminate death by aging (colloquially, find the "elixir of youth") for several centuries [4], even millennia, and this race to "solve aging" has intensified in the past century [5]. Anybody who has even a passing interest in eliminating death by aging is aware of stories, myths, and fables about alchemists, pseudo-doctors, snake oil salesmen, and also scientists, philosophers, and legitimate academics who tried, unsuccessfully, to find an answer. Even in today's high-tech society, this search for a physical therapy that can make us immune to the ravages of time continues, albeit employing a more refined terminology:

- "rejuvenation biotechnologies" [6]
- "stem cell interventions" [7]
- "anti-senescence drugs" [8,9]
- "parabiosis" [10]
- And other related interventions [11].

All of these efforts, in their entirety, aim to deal with aging by discovering and offering a physical therapy or intervention, an item or an object that can be given by the healer to the patient and simply make aging "go away"; in other words, "solve aging" [12]. This physical therapy is inherently reductionist and mechanistic, and refers to anything such as a tablet, an injection, a liquid or an elixir, a physical medical procedure or anything else using objects that exist physically [13]. Although a physical, item-based therapy may be of some use in alleviating certain clinical manifestations of age-related degeneration (for instance, tablets to improve memory in dementia, injections to improve joint function in osteoarthritis), this methodology has hitherto been proven insufficient with regard to treatments aimed at the underlying process of aging itself [14,15].

It is important at this stage to emphasize that I consider two elements in aging (Figure 31.1): The first is the process of aging itself (time-related damage *and* the inability to repair this damage). This leads to dysfunction in the clinical sense. The second is the diseases and clinical conditions that are manifestations of this time-related dysfunction. It is well accepted that medical and biomedical treatments or therapies exist for a wide variety of age-related conditions (e.g., Parkinson's disease, age-related macular degeneration, dementia, arthritis, cardiovascular disease, and skin aging), but currently there is no (and not likely to ever be) physical therapy that can diminish the process of

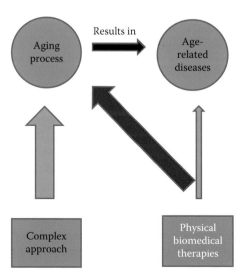

FIGURE 31.1 The relationship between the basic phenomenon of aging, age-related diseases, and the therapeutic concepts used against each. By using a complex "systems thinking" approach (which I discuss in some detail below), we diminish the impact of aging and thus reduce age-related diseases, in contrast to the current biomedical paradigm that, despite trying to influence the aging process, in reality remains limited and it is only somewhat effective against certain age-related conditions. Green = "it is effective against." Red = "it is ineffective against." Thickness of arrows indicates strength of efficacy.

aging, stop age-related degeneration, and thus achieve a state where the age-related mortality of humanity tends to zero (negligible senescence). In simple terms, there are currently no effective ways to radically extend human life span [16].

PHYSICAL ITEM-INSPIRED THERAPIES

I will now discuss in more depth some current concepts regarding treatments or therapies that are based on reductionism: therapies that can be given by the doctor to the patient in an attempt to control, reduce, or eliminate aging and achieve an age-related rate of mortality which tends to zero. These concepts are inspired from First or Second Phase Science models (see below). Examples in this respect include drugs (Table 31.1), physical interventions with stem cell therapies, or genetic manipulations.

However, a more close analysis of the general principle (i.e., offering a drug by a healer to a patient) results in several problems. I have discussed elsewhere problems associated with compliance, unpredictable effects, and translational issues [13,24] and I quote [13]:

> The fact that an effective therapy against ageing may theoretically be developed in the laboratory, does not necessarily align with the actual use of this therapy by the public. It is well known that non-compliance is a widespread problem in medicine [25] and, even in life-threatening conditions; the use of life-saving therapies can be suboptimal [26]. For example, it was shown that only 75% of coronary heart disease patients take sufficient medicine for it to be effective [27]. An example comparable to therapies for ageing rejuvenation is that of antiretroviral medication which can save a patient from certain death from AIDS. A relevant study shows that suboptimal adherence may be a problem in nearly 50% of some patients [28]. Worsening compliance is proportional to the number of drugs taken [29]. It is also inversely proportional to the number of times a patient has to take the therapy each day. If a patient has to take the medication once a day, the average compliance rate is 80%. This drops to 50% for those who have to take their medication four times a day [30]. However, there are several other predictors of non-compliance, all of which add unknown variables to the problem (Table 31.2) [2].

There are several other obstacles associated with this model, obstacles that are physical, translational, and conceptual. Rejuvenation biotechnologies aim to develop fault prediction models and then lessen the impact of these biomedical faults by devising molecular or cellular repairing strategies. With regard to human aging, however, this approach simply cannot work because it does not take into account the complex, dynamic, and nonlinear processes that define a human being. Bell et al. [37] have suggested that a failing organism may be better repaired through a process of autonomous self-repair, which can be enhanced by non-biomedical means.

I am now suggesting that it is time to abandon a large element of our enthusiasm for item-based therapies and embrace a wider worldview, a new methodology of thinking about aging. The complexity of aging cannot simply be addressed by using reductionist concepts. To paraphrase Ashby's

---

TABLE 31.1

Drugs or Remedies That Can Be Used in Aging Control

- Senolytics (to selectively induce death of senescence cells) such as Quercetin, Dasatinib, Navitoclax, or combinations [17]
- Rapamycin [18]
- Life span-extending herbs [19]
- Metformin [20,21]
- Telomere and telomerase modulators [22]
- Sirtuins [23]

TABLE 31.2

Major Predictors of Nonadherence to Medication

- Psychological problems—particularly depression—and also cognitive impairment, which is common in older people who would be candidates for treatment with rejuvenation biotechnologies [32]
- Asymptomatic disease—which would be an issue in younger healthy people who enroll for putative treatment [33]
- Side effects of the medication—likely to be several, as a wide range of therapeutic interventions will have to be deployed [34]
- Administrative issues—such as inadequate follow-up or discharge planning, poor relationship between patient and provider, missed appointments, lack of health insurance, and cost [35]
- Complexity of the treatment itself [36]

Source: Barat, I. et al. 2001. Drug therapy in the elderly: What doctors believe and patients actually do. *Br J Clin Pharmacol*, 51(6): 615–622.

Note: The totality of these problems significantly reduces the likelihood that a drug can be taken appropriately by any given patient.

---

law [38]: *the complexity of the treatment must match the complexity of the process we want to treat.* We should end our obsession with *magnification* (i.e., the study of magnified cells, tissues, enzymes, DNA, etc.), which is essentially pure reductionism, and consider how matters look in *miniaturization* (in other words, "zoom out"). We need to consider how an aging human organism interacts with other such organisms, the entire society and even the planet. Consider cultural, technological, and philosophical elements of what it is to be an aging human.

HUMAN AGING AND A STEPWISE INCREASE IN SCIENTIFIC SOPHISTICATION

While many current approaches to dealing with aging are based upon an "observer-object" model, new approaches consider the interaction of humans with their environment and the relationships that originate from this interaction. The environment plays a crucial role in influencing our phenotype. Although the gene pool is set, the way of interpreting the genetic information is subjected to influences from the environment. The information contained on our genes has been compared to a cookbook which provides precise directions for the preparation of a meal, but the execution (i.e., the phenotype in this example) may vary according to the environment, in which the cook is. Although the genes are more or less stable, the transcription of the information and the final results are different [39]. In order to be able to suggest novel ways of looking at the complex phenomenon of aging, we now need to discuss a general framework of scientific enquiry, based on three different models (Figure 31.2).

The reductionist Cartesian [40] model of the so-called "First Phase Science" was (and still is, to a large extent) the cornerstone of scientific inquiry. In this instance, we have a clear separation between the observer and the object of the observation [41] (Figure 31.2a). In First Phase Science, there is a distinction between a detached observer (such as a clinician, a therapist, etc.) and the observed (the patient, the illness, the interventional therapy). However, real-life situations are alien to this type of inquiry. Real-life everyday situations are extremely complex and depend on a wide diversity of interacting variables. Attempts aimed at remediating this shortcoming resulted in the emergence of the Second Phase Science model, where the observer and the observed are intermingled (Figure 31.2b). The immersion of the observer in the process of observation creates real-life variables which force the system under observation to react accordingly (Table 31.3).

With regard to aging research, Second Phase Science models predict that it may still be possible to design biomedical interventions that can involve both the patient and the clinician in a mutual manner. However, this is still a rather theoretical prediction based on reductionism and on a mechanistic model of inquiry, both of which have no real value in this context. This, and the previous

(a)

(b)

(c)

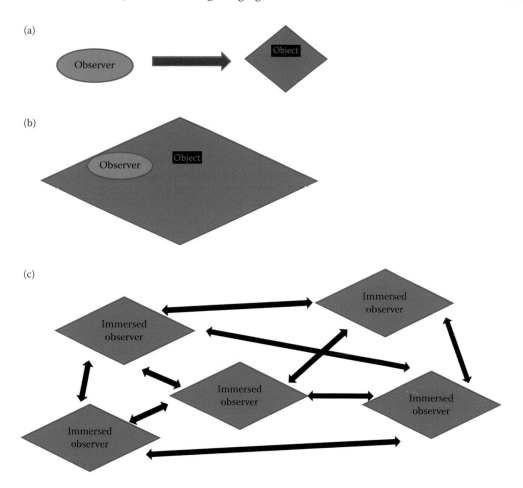

FIGURE 31.2 (a) First Phase Science: The "detached observer." (b) Second Phase Science: The "immersed observer." (c) Third Phase Science: multiple immersed observers with complex interacting perspectives.

model (First Phase Science), fail to capture the real intricacies of the aging process, which also depend on cultural, social, and technological interactions, not just medicine or biology [43]. For instance, we know [44] that degeneracy in biological systems suggests that one gene does not equate to one function. Degeneracy is *the ability of elements that are structurally different to perform the same function or yield the same output* [45]. Aiming to repair one gene by using rejuvenation biotechnologies may not give the expected result due to the fact that other elements would have been involved in its function, elements which would now be "disorganized" if the original gene is repaired. There are multiple backup systems which tend to respond to the damage, and the repair of one specific gene would change the dynamical architecture of the system.

TABLE 31.3

Clinical Examples of Second Phase Science Models (Where the Observer and the Object of the Observation Are "Immersed" in Each Other)

- Unblinded clinical trials, where both the observer (clinician) and object of observation (patient) freely exchange information
- The relationship then can develop during psychoanalysis, between the therapist and the patient
- Real-time adaptable therapeutic procedures which change according to feedback received from the patient [42]

The therapy would also depend on the phenomenon of "Degeneracy Lifting": *when systems display different behaviors which depend on the environment they are in at any given time*. If we assume that a certain treatment has been developed against aging, the effect of this therapy will depend on the environment where the treatment is being applied, and it would not give a consistently valid result each time. Hyperconnectivity (enhanced connections between the different elements of a system, such as cells in the human organism) results in degeneracy lifting and this modifies the function of the epigenome [46]. The treated system may recruit multiple pathways in order to achieve a function, and thus the artificial intervention may disrupt this. There is no room for adaptability in the case of unforeseen environmental perturbations. The original damage is no longer relevant, and repairing it would not give the expected results. The entire process depends on environmental cues that can, for example, involve microRNAs [47], itself an extremely complex issue. Therefore, we must take into account the multiple interacting influences between the observer and the object, that is, not only how the therapy interacts with a target gene/cell/organ, but also how multiple such targets interact between them and their environment. In this case, the model of Third Phase Science could provide ample room for explaining these complex interactions [41,48] (Figure 32.2C).

I quote from [2]:

> *In Third Phase science, there are different interacting viewpoints which create more complexity and multi-dimensionality. The influences and viewpoints of the observer interact with those of other observers and result in a diversity of viewpoints which may form a more comprehensive overall picture (Figure 31.2C). In this model, the result is more dynamic, complex and integrative, representing a situation which is closer to a real-life situation. This model may describe the complexity of the human organism in much more accurate terms compared to the two other models of science. The model describes variable range correlations, when interactions at a local level may exhibit global effects, and also when global patterns affect local elements, a situation we typically encounter when we attempt to treat a complex organism in real life.*

MOLECULAR PATHOLOGICAL EPIDEMIOLOGY: ANOTHER NAIL IN THE COFFIN OF REJUVENATION BIOTECHNOLOGIES

In an attempt to bridge the gap between these first two models of aging research (reductionist approach) and the third one which defines a more comprehensive, multidimensional view of aging (a "systems thinking" approach), I will now discuss the concept of Molecular Pathological Epidemiology (MPE), and how this may share elements with cybernetic and "wholistic" (as in "whole") views on aging. A reductionist First or Second Phase Science model is at odds with a concept developed by Shuji Ogino, that of MPE [49]. MPE combines elements from epidemiology and pathology in order to study the heterogeneity and etiology of disease and degeneration. In this case, the focus is both at a microscopic (molecular and cellular) and macroscopic (organismal and population) levels. Therefore MPE may act as a link between reductionism and systems thinking perspectives. In this way, it is possible to form a more inclusive view of the different disease processes. In essence, MPE helps us study the influence of the environment on genetic factors, and it shares many frontiers with Social Genomics, although it is also focused on epidemiological matters. However, if we are using only a simplistic model to study aging, we may encounter problems such as an uncontrollable and unpredictable multiplication of random errors in the repair process which a theoretical antiaging drug/therapy may be associated with [50]. On the basis of principles developed within MPE, we may find that there are several inherently heterogeneous interacting pathological processes which may lead to disease. We cannot adequately predict, or even study, these multiple processes which can cause significant variations of diseases even in similar patients [51]. The net result is that in order for a therapy to be effective for any given patient, it is necessary to develop tailor-made interventions for each patient, and not rely on generic treatments. This may prove to be an impossible task [52]. It has been termed the "Precision Medicine Concept" or the "Unique Disease Principle," which suggests that each individual patient exhibit unique disease profiles based on an interaction of genetic, epigenetic, and other factors such as cell interactions, nutrition, lifestyle, microbial exposure, social and racial elements, age differences, etc. All of these factors need to be taken into account in the designing and application of any putative anti-aging therapy. In other words, both the biological and the social/environmental elements need to be taken together, whereas currently the reductionist model of one "fit-all" therapy takes into account only the biological element.

In addition, there are issues with the repair process in general. Drawing inspiration from the realm of engineering of self-repairing systems, Bell et al. [37] conclude that: (before attempting a repair) … *The diagnosis must be confirmed, to avoid undesirable events such as 'good' components being unnecessarily removed or routed around.* This is a basic principle, in accordance with MPE, and is clearly at odds with those who hold that it is not necessary to know the diagnosis, but just repair all apparent damage preemptively.

A HORMETIC–CYBERNETIC APPROACH TO AGING RESEARCH

Another possible bridge between reductionism and systems thinking may be found in a cybernetic view of aging, one that also considers a human organism as a "whole," that is, an autonomous entity that interacts and depends on its environment in a "wholistic" manner (Figure 31.3). In this respect, it is necessary to consider the concept of "homeodynamic space" developed by biogerontologist Suresh Rattan. This virtual space indicates the limits of the survival ability and buffering capacity of any dynamic biological system. Aging is associated with shrinkage of, or dysfunctions within, this space [53]. Associated with this concept is the general concept of hormesis and the cellular stress response. Hormesis is a process defined by a nonlinear, "U"-shaped, "low dose stimulation, high dose inhibition" principle (see Reference 54), whereby exposure to a weak stimulus may positively challenge the organism by upregulating the stress response and result in health benefits, whereas an excessive, suboptimal, or prolonged exposure can result in damage and disease. The

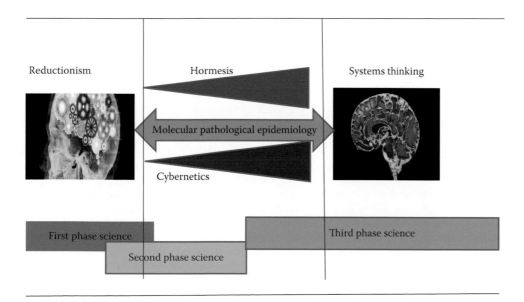

FIGURE 31.3 Bridging the gap between *Reductionism* and *Systems Thinking*: Both the concept of hormesis and the use of cybernetic principles touch on reductionism but are mainly applicable closer to the systems thinking domain. MPE provides a more stable bond between the two domains. First and Second Phase Science describe mainly the Reductionist model, whereas the Third Phase Science concept is more applicable in the systems thinking realm.

concept of hormesis is based on mild exposures to new environmental information, a continual though not excessive state of "novelty" which modulates the homoeostatic mechanisms and redefines the homeodynamic space. The hormetic response is triggered by exposure to any physical, chemical, biological, mental, or other challenges, which may disturb the cellular or organismic homeostatic mechanisms. Hormesis is invoked when one operates at the "outer edge of their comfort zone," when the stimulus confronting them is challenging but "doable." This, translated into clinical, practical terms, should mean "positive stress": frustrating but in a pleasant, positive way, which provokes rather than annoys.

The stress response is an innate mechanism which senses any external or internal challenges or disturbances, and then initiates maintenance and remodeling mechanisms that lead to adaptation and survival. Demirovic and Rattan [53] state:

> *... An effective strategy, which makes use of (stress response) for achieving healthy aging and extending the healthspan, is that of strengthening the homeodynamics through repeated mild stress-induced hormesis by physical, biological and nutritional hormetins...*

Returning to the cybernetic interpretation of aging, it is necessary to explain that, in that context, a human organism is seen as a "Complex Adaptive System" and is thus subjected to cybernetic laws and principles, which aim to define regulation of damage control [38]. A complex adaptive system (CAS) is a collection or a network of connected structures ("agents," such as cells, for instance) which form an overall "meta-structure." These interconnected agents help the overall structure adapt and survive [55]. A system is a CAS when its local elements cannot easily be decoupled from systemic patterns. Its individual structures (e.g., cells) allow for a diversity of properties and actions, and each such property or action can be achieved through multiple diverse ways (i.e., degeneracy, as described above). All of these characteristics are found in the human organism, where different structures achieve similar functions and where local actions may have global effects.

Both our individual components within our body, and us, as whole organisms within society, must obey the edicts of network theory, evolution, adaptation, anti-fragility, complexity, and other characteristics of dynamic systems [56]. The cyberneticist Francis Heylighen has described, in general terms, the buffering mechanisms that aim to improve damage control during aging. He quotes [56]:

> *Buffering can be optimized by providing sufficient rest together with plenty of nutrients: amino acids, antioxidants, methyl donors, vitamins, minerals, etc. Knowledge and the range of action can be extended by subjecting the organism to an as large as possible variety of challenges. These challenges are ideally brief so as not to deplete resources and produce irreversible damage. However, they should be sufficiently intense and unpredictable to induce an overshoot in the mobilization of resources for damage repair, and to stimulate the organism to build stronger capabilities for tackling future challenges. This allows them to override the trade-offs and limitations that evolution has built into the organism's repair processes in order to conserve potentially scarce resources. Such acute, "hormetic" stressors strengthen the organism in part via the "order from noise" mechanism that destroys dysfunctional structures by subjecting them to strong, random variations. They include heat and cold, physical exertion, exposure, stretching, vibration, fasting, food toxins, micro-organisms, environmental enrichment and psychological challenges. The proposed buffering-challenging strategy may be able to extend life indefinitely, by forcing a periodic rebuilding and extension of capabilities, while using the Internet as an endless source of new knowledge about how to deal with disturbances.*

In this example, there is a strict bond between damage control and hormetic stressors which, via complex mechanisms [57], improve overall function. This example also provides a bridge between reductionism (individual interventions) and systems thinking (whole body effects).

Another example where a CAS must obey the cybernetic laws is the case of the "Law of Requisite Action" and the "Engagement Axiom" [58]. The first asserts that the capacity of a system to implement a plan of action effectively depends strongly on the true engagement of the individual components of that system, and the part they play in designing (that plan of action). The Engagement

Axiom [59] underlines the importance of engaging individual participants in designing action plans for complex social systems [60], but it can also be applied in clinical medicine: If we disregard the input and feedback from the patients, then our treatment plans are bound to fail. Reductionist biomedical rejuvenation technologies disregard this principle by offering treatment to a group of patients who just passively accept the treatment offered, without true opportunities to participate in the process: this is, in effect, an example of the First Phase Science approach.

Last but not least, a relevant example is the cybernetic principle of Downward Causation which states that "all processes at the lower level of a hierarchy are restrained by, and act in conformity to the laws of the higher level" [61]. In this case, rejuvenation biotechnologies that repair genes, molecules, or cells (i.e., individual components of a system, at the lower level of hierarchy) will in effect be offering unpredictable therapies because they disregard the cumulative effects at a higher level of the hierarchy.

SYSTEMS THINKING IN AGING

These simple, scientifically naïve models of aging do not account for many other phenomena that are associated with complexity, such as emergence, or "Non-additive Determinism" [62]. Non-additive determinism specifies that analyzing individual components of a system does not clearly determine the overall properties of a system, or, to put it more simply, a system is not defined by the sum of all of its single components, but there are other emergent properties that add another layer of complexity in the system's behavior [63]. An additional relevant principle here is that of "Reciprocal Determinism" [63], whereby individual biological and individual social factors influence each other in a complex manner, leading to results which may be unnoticeable when only the biological or only the social aspect is considered in isolation. Both the biological and the sociocultural environment of the target patient must be taken into account [64]. I quote from my paper [2]:

> .... the separation of the observer from the object of the observation creates problems which become irrelevant when the object of the observation is examined within the context of its environment. In a classic experiment, the effects of the administration of amphetamine or placebo to nonhuman primates were examined and assessed [65]. The assessment of each individual primate did not show any significant effects, but when each primate's position in the social hierarchy was considered, there were emergent effects. Amphetamine administration assessed in the context of the primates' social environment showed that there was amplified dominant behaviour in those primates who occupied high ranks in their society, and increased submissive behaviour in primates who were in the lower ranks of the social hierarchy.

So, in this discussion, we are now moving one step higher, from examining human components and human organisms, to study humans as a part of society. The science of social genomics provides supporting evidence within the general concept of a gene–society interaction, and it suggests that biological functions are inherently and mutually dependent upon the social environment [66]. Let me briefly discuss social genomics because this is becoming increasingly relevant in the study of the complex interactions between humans and their society [67]. Social genomics study how the various social factors and cultural processes affect the activity of our genes [68]. Socially originating challenges, information, or stimuli may invoke biological responses such as increased expression of hundreds of gene transcripts [69]. Cole [69] states that:

> Systems-level capabilities emerge from groups of individual, socially sensitive genomes ... and transcriptional biofeedback (that) empirically optimizes individual well-being in the context of the unique genetic, geographic, historical, developmental, and social contexts ... Studies of human social genomics are now clarifying which specific types of human genes are subject to social regulation and mapping the social signal transduction pathways that mediate these effects. The results of these analyses are shedding new light on the molecular basis for social influences on individual heath, the genomic basis for human thriving, and the metagenomic capabilities that emerge from networked communities of socially sensitive genomes ...

The science of social genomics and the notion of gene–culture coevolution share common elements [70] and both are now becoming more relevant in formulating a complex view of aging. The gene–culture coevolution concept explains the changing interactions between genetic evolution and cultural evolution (also known as the Dual Inheritance theory). An extension of the concept of gene–culture coevolution [71,72] defines influences from our technological culture upon our own health and, at the same time, how we influence the evolution of the environment itself [73], characteristics which define the Third Phase Science model. This amalgam between sociotechnical/cultural elements with biological structures and processes is also studied through principles developed within social genomics, and more details are discussed below.

AN EXAMPLE OF SYSTEMS THINKING IN AGING RESEARCH: THE INDISPENSABLE SOMA HYPOTHESIS

Above, I have discussed some of the shortcomings of a reductionist, item-based research model and how this is inadequate in research concerning human aging [74]. First and Second Phase Science models need to give way to a different approach, one that is based on the general concept of Third Phase Science [75]. The basis of this model is that, as humans progressively interact with technology and with each other, this leads to new realities concerning both the human organism and the society we are in [76]. These realities are heavily and necessarily dependent upon epigenetic mechanisms which remodel our phenotypic landscape [77].

Therefore, the time is now ripe to present a model of aging which incorporates all elements of systems thinking discussed above (such as hormesis, cybernetics, MPE, social genomics, and others (Figure 31.3). I have already discussed in detail, in Chapter 4, I will describe in detail a hypothesis which is based on the above concepts, and concerns the continual battle for survival between the somatic neuron and the germ line. Just to summarize this, by virtue of us living in a high quality, information-rich technological ecosystem, there is a hormetic-style activation of neuronal stress response (and the DNA damage response mechanisms) which, via a variety of sensor molecules (Table 31.4), may both enhance the function of the neuron AND downregulate the function of the germline [78]. In the final analysis, we may encounter a situation whereby a human–computer hybrid entity is both biologically and technologically robust, and able to survive without any age-related degeneration, due to a shift of repair resources from the germ line to the soma [79].

I quote from my paper [6]

> ..... within a high-level cognitive ecosystem, we initiate a series of upregulation sequences and involvement of certain evolutionary functions such as degeneracy [84], exaptation [85] including exaptation of transposable elements [86], adaptive response (an appropriate reaction to an environmental demand) [87,88] and others, which cause our body to repair the ageing damage and thus completely avoid chronic degenerative diseases such as arthritis, heart disease, senile dementia, Alzheimer Parkinson etc. In this way, the life span will increase dramatically [89].

TABLE 31.4
Stress Response Sensor Molecules

Hormetic information and data from our technological, digital environment are integrated in the amygdala and prefrontal cortex, among others, and upregulate the stress response in neurons. These neurons generate stress response sensor molecules such as:

IRE-1 (Inositol-requiring enzyme 1) [80]

PERK (Protein kinase RNA-like endoplasmic reticulum kinase [81]

ATF6 (activating-transcription-factor-6) [82]

JNK (c-Jun N-terminal kinase 1) [8]

Thus, the Indispensable Soma hypothesis discusses the details of an increasing integration of humans with a highly techno-cognitive environment, hormetic cognitive stresses, and energetic trade-offs as described by the traditional disposable soma theory [90]. For example, research shows that increased and meaningful engagement with online social platforms is correlated with increased life span [91]. This study compared 12 million Facebook users to nonusers and found that those users who accept more online friends have a lower mortality risk. Although many questions remain, it is suggested here that online social interactions (i.e., technologically aided exchange of meaningful social information) may have a certain value in maintaining a low mortality risk, through hitherto poorly studied mechanisms. Cognitive training is known to produce mental benefits both in healthy older people and in Alzheimer's disease patients [92].

This line of thought is in accordance with a general "systems thinking" approach to aging and sees humans as indispensable organisms within a technological ecosystem. It is possible to examine some consequences of the reduction or elimination of age-related degeneration (due to upregulated somatic repairs) and a prolonged human life span. In this instance, it is worth highlighting an apparently strange argument: that the longer we live, the more chances we have for improved repairs. Time is a resource in aging. Living longer may be a blessing in the sense that it allows for more time to achieve cellular repair. Fast-living animals do not have sufficient time to detect and consequently repair cellular damage. In contrast, slow-living organisms have sufficient time. Lorenzini states [93]:

> *The implication of these results is that longer-lived species delay cell cycle progression to a greater degree than shorter-lived species, allowing for higher fidelity repair. We suggest that the ability to devote longer periods of time to repair and maintenance is a key feature of longer-lived species, and that evolutionary pressure to complete repair and resume cell division is a determinant of species life span. Thus, time is a resource that must be managed by the organism to attempt to maximize the fidelity of repair.*

Support for this argument may come from the general concept of "Slower is Faster" [94]. The Slower is Faster (SIF) phenomenon characterizes situations when the individual components of a system operate at a moderate rate and this leads to an overall better performance of the system (as opposed to when the components work really efficiently themselves, but the overall effect is that the system operates less efficiently). This is equivalent to the "Faster is Slower" effect, found in many situations such as traffic and traffic control systems, social situations, ecological and financial dynamics, etc. [95]. This concept can be applied in the case of aging when aged and damaged systems operate better when their individual components do not function at a peak capacity (but nevertheless still function efficiently), an effect reminiscent of hormesis where low-level stimulation is beneficial, whereas higher stimulation may be detrimental.

Concepts based on hormesis, cybernetics, and social genomics may help us understand some of the mechanisms that may lead to an increased life span. When many millions of humans interact constructively through technology and provide instant feedback, the information load to the brain is magnified accordingly, and several mechanisms may act in order to enhance neuronal repair:

1. *Biological amplification*: Activation of one process locally can have beneficial repercussions elsewhere in the body. General examples of biological amplification are: deregulation of luteinizing hormone may result in an associated increase in Alzheimer's disease [96]. Testosterone, a sexual hormone, increases bioavailability of antioxidants, and also improves immune responsiveness [97]. Physical exercise not only improves somatic health but also mental health [98].
2. *Challenge propagation*: In cybernetic terms, a challenge (for instance, a cognitive encounter with digital technology) will lead to a change in the environment of the neuron (an adaptation to the stimulus), and this change will then act as a new challenge for other

neurons to respond to. Therefore, the adaptation diffuses through the network and results in an improved performance of the entire network. The same is true in the case of humans who may be able to influence others and collectively improve the function of the human society

3. *Selective reinforcement*: This is a general concept in cybernetics, cognitive sciences, and other disciplines. It is based upon behavior modification which seeks to increase the occurrence of certain desired events, and decrease the occurrence of certain undesirable events, with the result that only positive events are retained: the reinforcing response is selective and not random. This increases the degrees of freedom, adaptation, and information content of the evolutionary process.

4. *Exaptation*: A character or property that was evolved for one purpose, facilitating a new function in adaptation of an organism or a system; existing structure gaining a new function. Once a beneficial function becomes operational in a biological system, and a change of context occurs (when the environmental information becomes increasingly more cognitive), there will be fast emergence of radical new functions. In this case, the biological mechanisms necessary for this new function are already present but they are just adapting to a new function. In exaptation, there is the facility to develop a new function just by reusing a mechanism that is already present.

CONCLUSIONS

In this paper, I have tried to highlight the importance of shifting our attention from a well-defined reductionist thinking to a wider, apparently vague "systems thinking" model with regard to aging. However, there are examples where mechanisms may share common frontiers with both approaches. Although this discussion may appear to be anti-reductionist and pro "systems thinking," in effect it recognizes the value of both. Chen [99] quotes:

> *Reductionism proposes that an intervention program can be broken into crucial components for rigorous analyses; system thinking views an intervention program as dynamic and complex, requiring a holistic examination. In spite of their contributions, reductionism and systems thinking represent the extreme ends of a theoretical spectrum; many real-world programs, however, may fall in the middle.*

The aging process is a mind-numbingly complex phenomenon (in fact, a collection of phenomena). We cannot hope to interfere in a significant and positive manner if we just rely on simplistic concepts origination from outdated models of reductionist research. We need to move away from intellectually and clinically naïve concepts involving physical, item-based interventions, and instead consider a much more complex worldview. The phenomenon of aging can be totally eliminated from certain groups of humans who are playing an indispensable role in maintaining the adaptation and evolution of the entire complex human–computer ecosystem [100]. I have already discussed in detail, in Chapter 4 about the Indispensable Soma hypothesis, an example of a concept based on "systems thinking" in aging research.

ACKNOWLEDGMENTS

I would like to thank John Stewart, Michael Singer, and Yiannis Laouris for general discussions during the preparation of this concept.

REFERENCES

1. Kaiser, M.I. 2011. The limits of reductionism in the life sciences. *Hist Philos Life Sci* 33(4):453–76.
2. Kyriazis, M. 2017. Third phase science: Defining a novel model of research into human ageing. *Front Biosci (Landmark Ed)* 22:982–90.

3. Cabrera, D., Cabrera, L. 2015 *Systems Thinking Made Simple: New Hope for Solving Wicked Problems.* Odyssean, Ithaca, NY. ISBN 978-0996349307.
4. Demontis, L. 2015. Comparative analysis of medieval and modern scientific research on ageing reveals many conceptual similarities. *Gerontol Geriatric Res* 4/3:216:1–4.
5. Stambler, I. 2014. *A History of Life-Extensionism in the Twentieth Century.* Rison Lezion, Israel.
6. de Grey, A.D. 2016. Rejuvenation biotechnology: The industry emerges, but short-termism looms. *Rejuvenation Res* 19(3):193–4.
7. Conboy, I.M., Conboy, M.J., Rebo, J. 2015. Systemic problems: A perspective on stem cell aging and rejuvenation. *Aging (Albany NY)* 7(10):754–65.
8. Zhu, Y., Tchkonia, T., Pirtskhalava, T. 2015. The Achilles' heel of senescent cells: From transcriptome to senolytic drugs. *Aging Cell* 14(4):644–58.
9. Roos, C.M., Zhang, B., Palmer, A.K. et al. 2016. Chronic senolytic treatment alleviates established vasomotor dysfunction in aged or atherosclerotic mice. *Aging Cell* 15(5):973–7.
10. Baht, G.S., Silkstone, D., Vi, L., Nadesan, P. et al. 2015. Exposure to a youthful circulation rejuvenates bone repair through modulation of β-catenin. *Nat Commun* 6:7131.
11. Huffman, D.M., Schafer, M.J., LeBrasseur, N.K. 2016. Energetic interventions for healthspan and resiliency with aging. *Exp Gerontol* pii: S0531-5565(16)30147-4.
12. Anonymous. 2016. The Renaissance of Rejuvenation Biotechnology, August 9 http://www.longevityreporter.org/blog/2016/8/8/the-renaissance-of-rejuvenation-biotechnology (accessed January 15, 2017).
13. Kyriazis, M., Apostolides, A. 2015. The fallacy of the longevity elixir: Negligible senescence may be achieved, but not by using something physical. *Curr Aging Sci* 8(3):227–34.
14. Kyriazis, M. 2014. The impracticality of biomedical rejuvenation therapies: Translational and pharmacological barriers. *Rejuvenation Res* 17(4):390–6.
15. Carnes, B., Olshansky, S.J., Hayflick, L. 2013. Can human biology allow most of us to become centenarians? *J Gerontol A Biol Sci Med Sci* 68(2):136–42.
16. Guerin, J. 2004. Emerging area of aging research: Long-lived animals with negligible senescence. *Ann N Y Acad Sci* 1019:518–20.
17. Zhu, Y., Tchkonia, T., Fuhrmann-Stroissnigg, H. et al. 2016. Identification of a novel senolytic agent, navitoclax, targeting the Bcl-2 family of anti-apoptotic factors. *Aging Cell* 15(3):428–35.
18. Arriola ApeloI, S.I., Lamming, D.W. 2016. An InhibiTOR of aging emerges from the soil of Easter Island. *J Gerontol A Biol Sci Med Sci* 71(7):841–9.
19. Ratho, L., Pant, A., Awasthi, H. et al. 2017. An antidiabetic polyherbal phytomedicine confers stress resistance and extends lifespan in *Caenorhabditis elegans*. *Biogerontology* 18(1):131–147.
20. Barzilai, N., Crandall, J.P., Kritchevsky, S.B. et al. 2016. Metformin as a tool to target aging. *Cell Metab* 23(6):1060–5.
21. Niedernhofer, L.J., Kirkland, J.L., Ladiges, W. 2017. Molecular pathology endpoints useful for aging studies. *Ageing Res Rev* 35:241–249.
22. Strazhesko, I.D., Tkacheva, O.N., Akasheva, D.U. et al. 2016. Atorvastatin therapy modulates telomerase activity in patients free of atherosclerotic cardiovascular diseases. *Front Pharmacol* 7:347.
23. Camins, A., Sureda, F.X., Junyent, F. et al. 2010. Sirtuin activators: Designing molecules to extend life span. *Biochim Biophys Acta* 1799(10–12):740–9.
24. Kyriazis, M. 2015. Translating laboratory anti-aging biotechnology into applied clinical practice: Problems and obstacles. *World J Transl Med* 4(2): 51–4.
25. Choudhry, N.K., Fischer, M.K., Avorn, J. et al. 2011. The implications of therapeutic complexity on adherence to cardiovascular medications. *Arch Intern Med* 171:814–22.
26. Gellad, W.F., Grenard, K., McGlynn, E.A. 2009. A review of barriers to medication adherence: A framework for driving policy options. RAND. http://www.rand.org/content/dam/rand/pubs/technical_reports/2009/RAND_TR765.pdf (accessed January 4, 2017).
27. Jackevicius, C.A., Li, P., Tu, J.V. 2008. Prevalence, predictors, and outcomes of primary non-adherence after acute myocardial infarction. *Circulation* 117:1028–36.
28. Huan, Z., Fuzhi, W., Lu, L. et al. 2016. Comparisons of adherence to antiretroviral therapy in a high-risk population in China: A systematic review and meta-analysis. *PLoS One* 11(1):e0146659.
29. Chapman, R.H., Benner, J.S., Petrilla, A.A., et al. 2005. Predictors of adherence with antihypertensive and lipid-lowering therapy. *Arch Intern Med* 163:1147–52.
30. Claxton, A.J., Cramer, J., Pierce, C. 2012. A systematic review of the associations between dose regimens and medication compliance. *Clin Ther* 23:1296–310.
31. Barat, I., Andreasen, F., Damsgaard, E.M.S. 2001. Drug therapy in the elderly: What doctors believe and patients actually do. *Br J Clin Pharmacol* 51(6):615–22.

32. Lenze, E.J., Miller, M.D., Dew, M.A. et al. 2001. Subjective health measures and acute treatment outcomes in geriatric depression. *Int J Geriatr Psychiatry* 16(12):1149–55.
33. Donadiki, E.M., Jiménez-García, R., Hernández-Barrera, V. et al. 2014. Health belief model applied to non-compliance with HPV vaccine among female university students. *Public Health* 128(3):268–73.
34. Malhotra, S., Karan, R.S., Pandhi, P. 2001. Drug related medical emergencies in the elderly: Role of adverse drug reactions and non-compliance. *Postgrad Med* 77(913):703–7.
35. Shepherd, J.G., Locke, E., Zhang, Q. et al. 2014. Health services use and prescription access among uninsured patients managing chronic diseases. *J Community Health* 39(3):572–83.
36. Cochrane, G.M., Horne, R., Chanez, P. 1999. Compliance in asthma. *Respir Med* 93(11):763–9.
37. Bell, C., McWilliam, R., Purvis, A., Tiwari, A. 2013. Concepts of self-repairing systems. *Meas Control* 46(6):176–9.
38. Ashby, W.R. 1958. Requisite variety and its implications for the control of complex systems. *Cybernetica (Namur)* 1(2):1–17.
39. Knox S. 2015. Gene x environment interactions as dynamical systems: Clinical implications. *AIMS Mol Sci* 3(1):1–11.
40. Harre, R. 2002. *Cognitive Science: A Philosophical Introduction.* Sage Publications, London.
41. Bausch, K.C. 2014. The theory and practice of third phase science, social systems and design. *Trans Syst Sci* 1:129–45.
42. Hall, A.M., Ferreira, P.H., Maher, C.G. et al. 2010. The influence of the therapist-patient relationship on treatment outcome in physical rehabilitation. *Phys Ther* 90:1099–110.
43. Gershenson, C., Fernández, N. 2012. Complexity and information: Measuring emergence, self-organization, and homeostasis at multiple scales. *Complexity* 18(2):29–44.
44. Mason, P. H. 2015. Degeneracy: Demystifying and destigmatizing a core concept in systems biology. *Complexity* 20:12–21.
45. Gally, J. A. 2001. Degeneracy and complexity in biological systems. *Proc Natl Acad Sci USA.* 98(24):13763–8.
46. Subramaniam, A.R., Pan, T., Cluzel, P. 2013. Environmental perturbations lift the degeneracy of the genetic code. *Proc Natl Acad Sci USA* 110(6):2419–24.
47. Voskarides, K. 2017. Plasticity vs mutation. The role of microRNAs in human adaptation. *Mech Ageing Dev* 163:36–39.
48. Metcalf, G.S. 2014. (Ed). *Social Systems and Design.* Springer, Ashland, USA.
49. Ogino, S., Campbell, P.T., Nishihara, R. et al. 2015. Proceedings of the second international molecular pathological epidemiology (MPE) meeting. *Cancer Causes Control* 26(7):959–72.
50. Stratton, M.R., Campbell, P.J. 2009. The cancer genome. *Nature* 458(7239):719–24.
51. Ogino, S., Nishihara, R., VanderWeele, T.J. et al. 2016. Review article: The role of molecular pathological epidemiology in the study of neoplastic and non-neoplastic diseases in the era of precision medicine. *Epidemiology* 27(4):602–11.
52. Nishi, A., Milner, D.A., Giovannucci, E.L. et al. 2016. Integration of molecular pathology, epidemiology and social science for global precision medicine. *Expert Rev Mol Diagn* 16(1):11–23.
53. Demirovic, D., Rattan, S.I. 2013. Establishing cellular stress response profiles as biomarkers of homeodynamics, health and hormesis. *Exp Gerontol* 48(1):94–8.
54. Kyriazis, M. 2016. Hormesis and Adaptation. In: *Challenging Ageing: The Anti-Senescence Effects of Hormesis, Environmental Enrichment, and Information Exposure.* Bentham Science Publishers, UAE, pp. 3–37.
55. Mitleton-Kelly, E. 2003. Ten principles of complexity and enabling infrastructures. In: Mitleton-Kelly, E. (Ed). *Complex Systems and Evolutionary Perspectives of Organisations: The Application of Complexity Theory to Organisations.* Elsevier, Oxford, UK. pp. 23–50. ISBN 9780080439570.
56. Heylighen, F. 2014. Cybernetic principles of aging and rejuvenation: The buffering- challenging strategy for life extension. *Curr Aging Sci* 7(1):60–75.
57. Kyriazis, M. 2016. Engagement with a technological environment for ongoing homoeostasis maintenance. In: *Challenging Ageing: The Anti-Senescence Effects of Hormesis, Environmental Enrichment, and Information Exposure 2016.* Bentham Science Publishers, UAE.
58. Christakis, A., Laouris, Y. 2007. Harnessing the wisdom of the people. The tree of social action. *Workshop Organized in Proceedings of the 3rd International Conference of the Hellenic Society of Systems Sciences*, Pireus, 26–28 May.
59. Özbekhan, H. 1970. On some of the fundamental problems in planning. *Technol Forecast* 1(3):235–40.
60. Laouris, Y., Laouri, R., Christakis, A. 2008. Communication praxis for ethical accountability: The ethics of the tree of action: Dialogue and breaking down the wall in Cyprus. *Syst Res Behav Sci* 25:1–16.

61. Campbell, D.T. 1974. Downward causation in hierarchically organised biological systems. In: F.J. Ayala, T. Dobzhansky (Eds). *Studies in the Philosophy of Biology*. Macmillan Publishers, UK. pp. 179–86.

62. Necka, E.A., Cacioppo, S., Cacioppo, J.T. 2015. *Social Neuroscience of the Twenty-First Century*. The University of Chicago, Chicago, IL, USA.

63. Cacioppo, J.T., Berntson, G.G. 1992. Social psychological contributions to the decade of the brain: Doctrine of multilevel analysis. *Am Psychol* 47, 1019–28.

64. Kyriazis, M. 2016. Opinion paper: A cognitive-cultural segregation and the three stages of aging. *Curr Aging Sci* 9(2):81–6.

65. Haber, S.N., Barchas, P.R. 1983. The regulatory effect of social rank on behavior after amphetamine administration. In: Barchas, P.R. (Ed). *Social Hierarchies: Essays Toward a Socio-Physiological Perspective*. Greenwood, Westport, CT, pp. 119–32.

66. Cole, S.W. 2013. Social regulation of human gene expression: Mechanisms and implications for public health. *Am J Public Health* 103(S1):S84–92.

67. Richerson, P.J., Boyd, R., Henrich, J. 2010. Gene–culture coevolution in the age of genomics. In: *The Light of Evolution*. Volume IV. The Human Condition National Academies Press, OpenBook, Washington DC.

68. Grewenb, K.M., Coffeya, K.A., Algoea, S.B. et al. 2013. A functional genomic perspective on human well-being. *PNAS* 110:33.

69. Cole, S.W. 2014. Human social genomics. *PLoS Genet* 10(8):e1004601. doi:10.1371/journal.pgen.1004601.

70. Gintisn, H. 2003. The Hitchhiker's guide to altruism: Gene–culture coevolution, and the internalization of norms. *J. Theor Biol* 220:407–18.

71. Feldman, M.W., Laland, K.N. 1996. Gene–culture coevolutionary theory. *Trends Ecol Evol* 11:453–7.

72. Chudek, M., Henrich, J. 2011. Culture-gene coevolution, norm-psychology and the emergence of human prosociality. *Trends Cogn Sci* 15(5):218–26.

73. Kyriazis, M. 2014. Technological integration and hyper-connectivity: Tools for promoting extreme human lifespans. *Complexity* 20:15–24.

74. Kyriazis, M. 2014. Editorial: Novel approaches to an old problem: Insights, theory and practice for eliminating aging. *Curr Aging Sci* 7(1):1–2.

75. Helbing, D. 2011. FuturICT-New Science and Technology to Manage Our Complex, Strongly Connected World. http://arxiv.org/abs/1108.6131 (accessed January 16, 2017).

76. Kyriazis, M. 2015. Systems neuroscience in focus: From the human brain to the global brain? *Front Syst Neurosci* 9:7.

77. Kyriazis, M. 2014. Information-Sharing, Adaptive Epigenetics and Human Longevity. http://arxiv.org/abs/1407.6030 (accessed January 12, 2017).

78. Kyriazis, M. 2016. Clinical effects of a "Human-Computer" interaction. *SSRN* June 21. http://ssrn.com/abstract=2798529.

79. Kyriazis, M. 2014. Reversal of informational entropy and the acquisition of germ-like immortality by somatic cells. *Curr Aging Sci* 7(1):9–16.

80. Levi-Ferber, M., Gian, H., Dudkevich, R. et al. 2015. Transdifferentiation mediated tumor suppression by the endoplasmic reticulum stress sensor IRE-1 in *C. elegans*. *Elife* Jul 20:4. doi: 10.7554/eLife.08005.

81. Chavez-Valdez, R., Flock, D.L., Martin, L.J. et al. 2016. Endoplasmic reticulum pathology and stress response in neurons precede programmed necrosis after neonatal hypoxia-ischemia. *Int J Dev Neurosci* 48:58–70.

82. Yoshikawa, A., Kamide, T., Hashida, K. et al. 2015. Deletion of Atf6α impairs astroglial activation and enhances neuronal death following brain ischemia in mice. *J Neurochem* 132(3):342–53.

83. Riches, J.J., Reynolds, K. 2014. Jnk1 activity is indispensable for appropriate cortical interneuron migration in the developing cerebral cortex. *J Neurosci* 34(43):14165–6.

84. Seifert, L., Komar, J., Araujo, D. et al. 2016. Neurobiological degeneracy: A key property for functional adaptations of perception and action to constraints. *Neurosci Biobehav Rev* 69:159–165.

85. Glinsky, G.V. 2016. Mechanistically distinct pathways of divergent regulatory DNA creation contribute to evolution of human-specific genomic regulatory networks driving phenotypic divergence of Homo sapiens. *Genome Biol Evol* 8(9):2774–88.

86. de Souza, F.S., Franchini, L.F., Rubinstein, M. 2013. Exaptation of transposable elements into novel cis-regulatory elements: Is the evidence always strong? *Mol Biol Evo* 30(6):1239–51.

87. Pfeuty, B., Thommen, Q. 2016. Adaptive benefits of storage strategy and dual AMPK/TOR signaling in metabolic stress response. *PLoS One* 11(8):e0160247.

88. Tosato, V., Sims, J., West, N. et al. 2017. Post-translocational adaptation drives evolution through genetic selection and transcriptional shift in *Saccharomyces cerevisiae*. *Curr Genet* 63(2):281–292.

89. Kyriazis, M. 2016. The Indispensable Soma Hypothesis https://figshare.com/articles/New_draft_item_ The_Indispensable_Soma_Hypothesis/3079732 (accessed January 17, 2017).

90. Kirkwood, T.B. 1977. Evolution of ageing. *Nature* 270(5635):301–14.

91. Hobbs, W.R., Burke, M., Christakis, N.A., Fowler, J.H. 2016. Online social integration is associated with reduced mortality risk. *PNAS* 113(46):12980–4.

92. Giuli, C., Papa, R., Lattanzio, F., Postacchini, D. 2016. The effects of cognitive training for elderly: Results from My Mind Project. *Rejuv Res* 19(6):485–94. doi:10.1089/rej.2015.1791.

93. Lorenzini, A., Stamato, T., Sell, C. 2011. The disposable soma theory revisited: Time as a resource in the theories of aging. *Cell Cycle.* 10(22):3853–6.

94. Gershenson, C., Helbing, D. 2015. When slower is faster. *Complexity* 21:9–15. doi:10.1002/cplx.2136.

95. Suzuno, K., Tomoeda, A., Ueyama, D. 2013. Analytical investigation of the faster-is-slower effect with a simplified phenomenological model. *Phys Rev E Stat Nonlin Soft Matter Phys* 88(5):052813.

96. Palm, R., Chang, J., Blair, J. et al. 2014. Down-regulation of serum gonadotropins but not estrogen replacement improves cognition in aged-ovariectomized 3xTg AD female mice. *J Neurochem* 130(1), 115–25. http://doi.org/10.1111/jnc.12706.

97. Fanaei, H., Karimian, S.M., Sadeghipour, H.R. et al. 2014. Testosterone enhances functional recovery after stroke through promotion of antioxidant defenses, BDNF levels and neurogenesis in male rats. *Brain Res* 1558:74–83. doi: 10.1016/j.brainres.2014.02.028.

98. Hogan, C.L., Mata, J., Carstensen, L.L. 2013. Exercise holds immediate benefits for affect and cognition in younger and older adults. *Psychol Aging* 28(2):587–94. doi: 10.1037/a0032634.

99. Chen, H.T. 2016. Interfacing theories of program with theories of evaluation for advancing evaluation practice: Reductionism, systems thinking, and pragmatic synthesis. *Eval Program Plann* 59:109–18.

100. Kyriazis, M. (Ed). 2017. *Re-Thinking Ageing: A Cross-Disciplinary Perspective (Special Issue)*. Mech Ageing Dev 163:1.

Section VII

Anti-Aging Drugs

32 Anti-Aging Drug Discovery in Experimental Gerontological Studies

From Organism to Cell and Back

Alexander N. Khokhlov, Alexander A. Klebanov, and Galina V. Morgunova

CONTENTS

INTRODUCTION

There is the issue of what we call "the problem of reductionism" in gerontology. In the majority of gerontological theories proposed in the past few decades, the mechanisms of both "normal" and accelerated or retarded aging of multicellular organisms are reduced to certain macromolecular changes (no matter stochastic or programmed) in their constituent cells. As a consequence, numerous model systems have been developed to study "age-related" changes in the cells relieved from "organismal noise" associated with the functioning of the neurohumoral system. Such reductionism in experimental gerontology (everything depends on adverse changes in individual cells) has played its role, particularly in the development of various cytogerontological models including the Hayflick model and also some models used in our laboratory, such as the stationary phase aging model, the cell kinetics model for testing of geroprotectors (anti-aging compounds or physical factors) and geropromoters (any factors which accelerate aging), and the model based on evaluation of cell colony-forming capacity.

Starting from the Weismann's and Carrel's pioneer works, this reductionist approach to experimental gerontological studies has been increasing its popularity for more than 100 years. However, it appears that today the construction of the survival curves of the test animal/human cohorts is still the most reliable way to estimate the efficiency of interventions in the aging process, though unfortunately this method is inefficient in terms of labor, time, and finance expenditures.

In this chapter, we are going to substantiate why we consider it feasible now to go back from experimental gerontological investigations on isolated cells to the "classic" gerontological research, that is, to both experiments in animals and clinical trials.

DEFINITIONS AND SOME GENERAL REMARKS

According to the classic definition of aging we share, it is a combination of changes in an organism leading to an increase in the probability of its death (rate of mortality) (Comfort 1964; Khokhlov 2010a; Khokhlov et al. 2012; Wei et al. 2012). It should be noted that the data on increasing or decreasing the life span affected by various factors are often interpreted in the studies as a modification of the aging process per se. However, aging and life span are not necessarily interrelated. If people did not age at all, they would not live eternally anyway. People would die because of random reasons, and the life expectancy would be increased "only" up to 700–800 years (Comfort 1964; Khokhlov 2013a). We guess that Steven Austad was right when noting (http://www.salon.com/2000/03/30/immortal/), in response to the question if there is any theoretical limit that would keep increased longevity from becoming immortality, "The only limit is that there is no such thing as immortality because accidents still happen. The theoretical limit is human behaviour, not human physiology. If teenagers didn't drive cars like crazy people, that would probably have more effect on life expectancy than curing cancer."

It is known that there exist both aging and non-aging organisms. The former can be distinguished from the latter simply by the shape of the survival curves of respective cohorts (Khokhlov 2010b, 2013a). The aging organisms die "according to Gompertz law," whereas the non-aging ones die "exponentially." In the very rare cases of the complete absence of death, for example, in the case of freshwater hydra populations under certain conditions (Martínez 1998; Martínez and Bridge 2012; Jones et al. 2014; Khokhlov 2014a), the survival curve is simply a horizontal line.

The conclusion whether or not a given factor affects the aging process is made on the basis of the pattern of modification of such curves under the influence of this factor. It can be assumed that a "true" geroprotector (any physical or chemical factor retarding the increase in the probability of death with age) should cause a rightward shift in the survival curve without changing its shape (i.e., it should increase both the average and maximum life span). In addition, the survival curve must not be exponential. However, a hypothetical "immortalizer" that makes the survival curve horizontal (i.e., virtually abolishes the death of the members of the cohort) could also be regarded as a geroprotector—in this case, an "ideal" one. It should also be emphasized that, in our opinion, it is not very important in this approach whether aging is a programmed process or whether it is only a "by-product" of the program of development and is determined by the stochastic processes triggered after the completion of the program (Minot 1908; Bidder 1932; Dilman 1971; Austad 1999, 2004; Hayflick 2007; Holliday 2007; Khokhlov 2010a, 2013a, 2013b, 2013c, 2014b).

The factors that increase the life span of the non-aging organisms apparently cannot be considered geroprotectors, because they do not affect the process of increasing the probability of death with age. Regarding the drugs that are used to combat the age-related diseases, formally, with their help we can slow down (or postpone) the age-associated increase in the probability of death but hardly can affect the maximum (species-specific) life span. If such drugs are also regarded as geroprotectors, then this group should include almost everything that ensures the normal existence of an organism (water, food, vitamins, trace elements, etc.). We share the point of view that the age-related diseases are *the result* of the aging process and not vice versa.

The growing interest in experimental gerontological research during recent years has unfortunately resulted in a paradoxical situation: although the number of publications in this field is increasing, only a minor part of them is actually devoted to the mechanisms of aging. In our opinion, this is due, among others, to the following methodological problems:

1. As a rule, the authors ignore the aforementioned classical definition of aging as a complex of age-related changes that increase the probability of death.
2. The emphasis in such studies is on increase or decrease in life span, although this often, as previously said, has no relation to modification of the aging process (in particular, it is possible to prolong the life span of non-aging organisms, while the fact of aging itself is not necessarily indicative of low longevity).

3. The control group often consists of animals with certain abnormalities or genetic disorders, so that any favorable influence on the corresponding pathological processes results in life span prolongation (e.g., Kolosova et al. 2006, 2013; Shabalina et al. 2017).

4. Too much significance is assigned to increase or decrease in *the average* life span, which is largely determined by factors unrelated to aging.

5. An increasing number of gerontological experiments are performed on model systems providing only indirect information on the mechanisms of aging, and its interpretation largely depends on the basic concept supported by a given research team. In particular, this concerns the usage of the term "cell/cellular senescence," which was originally introduced to designate a complex of various adverse changes occurring in normal cells due to the exhaustion of their proliferative potential (Swim and Parker 1957; Hayflick and Moorhead 1961; Hayflick 1965, 1979b). Today, however, many authors apply it to the phenomenon of suppression of proliferative activity in cells (including transformed cells) under the effect of various DNA damaging factors, which is accompanied by a certain cascade of intracellular events (Campisi 2011, 2013, Sikora et al. 2011).

There are also some extra problems we will address in the next sections concerning various approaches to testing of geroprotectors in experiments on cultured cells.

EXPERIMENTAL GERONTOLOGICAL MODELS USING ISOLATED CELLS

Cytogerontology deals with the analysis of aging mechanisms on cultured cells (Hayflick 1979a; Khokhlov 1988, 2002, 2013a, 2013c). It is the cytogerontological approach that is increasingly often used to test potential geroprotectors. It should be emphasized that cytogerontology as a branch of gerontology cannot successfully develop in the absence of correct general gerontological concepts and definitions described in Section "Definitions and Some General Remarks". Consequently, we will review various approaches to testing of geroprotectors in experiments on cultured cells keeping all our general considerations in mind.

What is not often remembered is that the foundations of this science were laid by August Weismann (1885, 1892) as early as in the late nineteenth century. As for the term "cytogerontology," it was introduced by Leonard Hayflick (1979a, 1996) to describe research on aging *in vitro*, that is, "age-related" changes in cultures of normal cells that have exhausted their mitotic potential (in fact, it is this replicative senescence that was subsequently named the "Hayflick phenomenon"). The term "cytogerontology" has then been extended to any studies on the mechanisms of aging in experiments on cell cultures (Kirkwood and Cremer 1982; Khokhlov 1992a, 2002, 2003; Alinkina et al. 2012).

A. Weismann was the first to emphasize the essential distinction between germ line cells, whose population is basically immortal, and somatic cells, which age and die (Weismann 1885, 1892). Thus, the cornerstone of his concept is that there exist the mortal soma and the immortal "germ plasm" (*Keimplasma*). However, Weismann failed to give a clear definition of what cell aging/senescence is, and this probably accounted for the findings and conclusions made by Alexis Carrel (1912, 1913), who laid experimental foundations of cytogerontology in the early twentieth century.

Carrel was interested to test whether somatic cells isolated from higher animals would "senesce" and die instead of propagating indefinitely. To this end, he developed a procedure for culturing epithelial or fibroblast-like cells in special flasks, which is still used today with only minor modifications. However, the results of his experiments did not fit the "mortal soma" concept: some cell strains derived from chicken embryos could be maintained in culture almost indefinitely, without showing any signs of degradation. This is why gerontologists in the twentieth century for almost 50 years considered somatic cells to be capable of unlimited replication, until the experiments performed in the 1950s and 1960s by Swim and Parker (1957) and, subsequently, by Hayflick (Hayflick and Moorhead 1961; Hayflick 1965) showed that the results obtained by Carrel were apparently artifactual. In fact, almost all normal animal cells have proved to have a limited proliferation potential, being capable of no more than 100–120 divisions in culture (about 50 cell population doublings).

Unfortunately, the model based on the Hayflick limit concept (aging *in vitro*) is apparently not directly related to the mechanisms of aging, as has been repeatedly noted previously (Olovnikov 2007a, 2007b; Khokhlov 2010a, 2010b; Macieira-Coelho 2011; Wei et al. 2012; Khokhlov et al. 2012, 2013a). In other words, we cannot conclusively explain why we age by relying solely on the phenomenon of limited mitotic potential of normal cells, which is practically never fully utilized *in vivo*. However, owing to A.M. Olovnikov's theory of marginotomy (Olovnikov 1971, 1973, 1996), we at least know today how this phenomenon is realized in the cells.

It is not out of the question that if the human life span were to be extended several fold, some cell populations would eventually exhaust their mitotic potential (thereby reaching the Hayflick limit), which could have resulted in the "second wave" of aging, but this has not been seen to occur so far. It should be noted, however, that some researchers still hold to the opinion that the shortening of telomeres in the cells is the key mechanism of aging. In particular, according to V.M. Mikhelson (Mikhelson 2001; Mikhelson and Gamaley 2012), certain "mosaicism" in the proliferative parameters, observed in a highly organized multicellular organism, allows the shortening of telomeres to be considered as an important factor in aging and longevity.

The body of evidence for the gerontological value of the Hayflick phenomenon is based only on the series of *correlations* (Khokhlov 2003, 2010b) like reduced mitotic potential of fibroblasts from the patients with progeria, direct relationship of this parameter to the species life span, or its inverse relationship to the age of the cell donor, etc. When demonstrating that Hayflick's model is appropriate for studying the aging mechanisms, it is usually emphasized that various changes, similar to those in the cells of an aging organism, take place in normal cultured cells as the number of cell population doublings increases (Hayflick and Moorhead 1961; Hayflick 1965, 1979b, 1991; Khokhlov 1988). In other words, cells either accumulate or lose something during aging *in vitro* in the same way as during aging *in vivo*. Therefore, again it is the case of *correlation*; this time it is correlation of the changes of certain biomarkers of aging (BA).

Despite its "correlativity," Hayflick's model has been widely used. On the bsais of this model, a large amount of data was obtained, which explained many problematic aspects in the life activity of organisms. In particular, it concerned the mechanisms of development and malignant transformation. However, the study of aging *in vitro*, in our opinion, practically did not help gerontologists to understand the fundamental mechanisms of aging and longevity.

Keeping in mind the main topic of the review, we should note that, when testing geroprotectors on the model system, researchers have followed either (1) the proliferative potential of the cells studied, or (2) the process of the accumulation of various BA.

We developed another "correlative" model for testing of geroprotectors and geropromoters—the "cell kinetics model" (Chirkova et al. 1984; Khokhlov et al. 1985b; Khokhlov 1992a). It is based on the well-known inverse correlation between the "age" of cultured cells (i.e., age of their donor) and their saturation density (Schneider and Smith 1981). This term is used for the maximum density (number of cells per square unit) of a cell culture in the stationary phase of growth when the cells stop propagating due to the contact inhibition. It was assumed that the higher the saturation density is, the "younger" the studied cells are. The model allowed us to perform preliminary testing (Khokhlov et al. 1985c, 1987b; Khokhlov 1988, 1992a) of a lot of different compounds and factors, such as gamma irradiation, DNA-alkylating agent thiophosphamide, low frequency electromagnetic field, antioxidants 2-ethyl-6-methyl-3-hydroxypyridine chlorohydrate and butylated hydroxytoluene, etc., that are interesting from a gerontological point of view, but, ironically, it also revealed no information about the real mechanisms of aging or its modulation.

Unlike the Hayflick model and the cell kinetics model, which are based on a series of correlations, our model of stationary phase aging (accumulation of "age-dependent" injuries in cultured cells whose proliferation is restricted in a certain way, preferably by contact inhibition) is a "*gist*" model based on the assumption that processes taking place in this model system are essentially similar to those in an aging multicellular organism (Vilenchik et al. 1981; Khokhlov 1988, 1992b, 1998, 2003; Akimov and Khokhlov 1998; Alinkina et al. 2012; Yablonskaya et al. 2013). In fact, this assumption directly issues from our concept that the restriction of cell proliferation is the main mechanism providing for the accumulation of macromolecular defects in cells of aging multicellular organisms (Khokhlov 1988, 2010a, 2010b, 2013a).

In this connection, it should be mentioned that we have always considered it to be natural that any person adhering to a certain gerontological theory, after giving a clear definition of aging, must also attempt to answer, among others, the following questions formulated by one of us some years ago (Khokhlov 2003, 2004):

1. How universal is the aging process?
2. Why do the non-aging species, concomitantly with the aging organisms (whose death probability increases with age), also exist (some of the coelenterate, fishes, and others)?
3. Why do different species have different life spans?
4. Why are tumor cells "immortal," that is, possess unlimited mitotic (cell population doubling) potential, as distinguished from normal cells? If one does not recognize that the Hayflick phenomenon plays a role in aging *in vivo*, then this question can be ignored. Also, it should not be forgotten—"mortal" tumor cells and "immortal" normal cells do exist and this fact should also be taken into consideration (Macieira-Coelho 1999).
5. How are the processes that determine the aging of organisms discriminated for dividing and non-dividing (e.g., such as neurons) cells? It could be asked how both those cell-types age, but to do that, it should be first defined what it is—the cell aging.
6. By what means does the germ line, that is, the population of germ cells that provides the transfer of genetic information through a practically endless series of generations, "escape" aging? And what is the purpose of the cytoplasmic sex determinant?
7. How do plants, bacteria, fungi, Protozoa, mycoplasmas, etc., age (if it could be said at all that they age)?
8. In what way could the evolution of the aging process be imagined in terms of the concept being formulated (let us remember ideas of the "early" and "late" August Weismann (Kirkwood and Cremer 1982)?

To answer these questions without accepting that what is defined as a "system approach to the problem" is, in our opinion, impossible. It looks like answering them would be real within the framework of our conception of cell proliferation restriction as the primary cause for the accumulation of macromolecular defects that, in turn, leads to the aging of the organism (Khokhlov 1998, 2010a, 2010b, 2013a, 2013b).

Our numerous experiments provide evidence that changes in the cells occurring in our cyto-gerontological model system of stationary phase aging are indeed similar to those in the cells of aging multicellular organisms. They include accumulation of DNA single-strand breaks and DNA-protein crosslinks, DNA demethylation, changes in the level of spontaneous sister chromatid exchanges, structural defects in the cell nucleus, alterations in the plasma membrane, retardation of mitogen-stimulated proliferation, impairment of colony-forming capacity, changes in dealkylase activity of cytochrome P450, accumulation of 8-oxo-2'-deoxyguanosine (a known biomarker of aging and oxidative stress) in the DNA, increase in the number of cells with senescence-associated beta-galactosidase (SA-β-Gal) activity, inhibition of poly(ADP-ribosyl)ation of chromatin proteins, etc. (Khokhlov et al. 1984a, 1985a, 1986, 1987a, 1988; Khokhlov 1988, 2013a; Prokhorov et al. 1994; Shram et al. 2006; Esipov et al. 2008; Vladimirova et al. 2012).

It should be emphasized that such experiments can be performed with cells of different origin, including human and animal cells, bacteria (Nyström 2002; Pletnev et al. 2015), yeasts (currently most widely used in experiments on stationary phase aging; (Khokhlov 2016), plant cells (Khokhlov 1988), microalgae (Ushakov et al. 1992), mycoplasmas (Khokhlov et al. 1984b;

Kapitanov and Aksenov 1990), etc. This provides a basis for the evolutionary approach to the analysis of experimental results (Khokhlov 1994). Moreover, the "age-related" changes in cells of stationary cultures can be revealed within a relatively short time: as a rule in 2–3 weeks after the start of the experiment.

The stationary phase aging of yeasts is called "chronological aging" and most frequently studied in *Saccharomyces cerevisiae*. It is observed in a population of yeast cells in the stationary phase of growth when their proliferation is stopped in one way or another (Fabrizio and Longo 2003). In this case, the viability of cells is usually estimated by their ability to form colonies in a fresh growth medium (Breitenbach et al. 2012). The chronological aging of yeasts should be distinguished from their so-called "replicative aging." The latter is based on the phenomenon of a limited number of daughter cells that can be generated by one mother cell. This model is very similar to the Hayflick model. However, unlike the cultured human and animal cells, the daughter cell of the yeast *S. cerevisiae*, which is typically much smaller than the mother cell, is formed as a result of asymmetric budding. In this case, the mother cell loses its ability for such budding after a certain number of divisions and then undergoes degradation and lysis, and the daughter cells "are born very young." This process is similar to the aging of the stem cell pool in higher organisms (Laun et al. 2007). It should be noted that, for the yeast *Schizosaccharomyces pombe*, in which two identical daughter cells are formed as a result of symmetrical division (fission) of one mother cell, only the chronological aging model can be used (Roux et al. 2006).

There is a point of view that the chronological aging of yeast and the stationary phase aging of cultured animal and human cells are a consequence of growth medium acidification (Burtner et al. 2009; Leontieva and Blagosklonny 2011; Murakami et al. 2011; Kaeberlein and Kennedy 2012). However, a number of recent publications indicate that the process influences, to a certain extent, the rate of "aging" of cells in the stationary phase of growth but does not determine it completely. Apparently, the key factor in this case is the cell proliferation restriction which leads to "aging" of the cells even under physiologically optimal conditions (Morgunova et al. 2017).

During chronological aging of yeasts and stationary phase aging of mammalian cells, the medium is getting acidified to pH < 4. Preventing the medium acidification could make it possible to increase the culture life span (Burhans and Weinberger 2009; Burtner et al. 2009; Murakami et al. 2011), but the cells will still die out, albeit at a slower rate. Effects of the medium acidification observed during chronological aging and stationary phase aging can be explained by the activation of highly conserved growth signaling pathways leading to the oxidative stress development (Burhans and Weinberger 2009; Yucel et al. 2014); these processes, in turn, can be involved in aging of multicellular organisms and play a role in their age-related diseases (Burhans and Weinberger 2009; Fabrizio and Wei 2011).

A while ago, we studied the effect of buffer capacity of growth medium on stationary phase aging of transformed Chinese hamster cells. We found that HEPES at 20 mM had no effect on the cell growth, and the growth curves reached the plateau level on the same day. However, the cells grown with HEPES, on the one hand, reached lower saturation density than the control ones (i.e., were "older" in terms of the gerontological cell kinetics model), and, on the other, underwent stationary phase aging at a much slower rate (though still they were "getting older"). It can be assumed that extracellular pH, which, by the way, is well correlated with intracellular pH (Akatov et al. 1985; Kurkdjian and Guern 1989), is very important (I.A. Arshavsky's concept [Arshavsky 1982] on a role of the acidic alteration in aging) but not the key factor determining survival of cells in a stationary culture.

Finally, it is important to note that, in studies on the Hayflick model, it is fairly difficult to correctly perform repeated experiments with the same cell strain, because the cells continuously change from passage to passage ("no man ever steps into the same river twice"), whereas the stationary phase aging model, as already mentioned above, allows for experimentation with transformed (or normal but immortalized) animal and human cells with an unlimited mitotic potential, so that multiple replication of an experiment is no longer a problem (Khokhlov 2012).

EVALUATION OF CULTURED CELLS DYING OUT IN CYTOGERONTOLOGICAL EXPERIMENTS

During many years of research on the stationary phase aging model, our premise was that cultured cells whose proliferation is restricted in some way (preferably by contact inhibition) accumulate "age-related" defects similar to those in cells of aging multicellular organisms (and geroprotectors should postpone/retard the accumulation), with the kinetics of cell death in this model system remaining behind the scene. Our subsequent studies have shown that mammalian cells in this model die out in accordance with the Gompertz law; that is, they age in the true sense (Khokhlov 2010b; Khokhlov et al. 2014; Khokhlov and Morgunova 2017). In other words, the probability of their death increases exponentially with age, as in aging animals and humans. Incidentally, similar results were obtained with the suspension cultures of *Acholeplasma laidlawii* (Kapitanov and Aksenov 1990), and our previous experiments with this mycoplasma showed that its stationary phase aging could be successfully delayed by treatment with a geroprotective antioxidant 2-ethyl-6-methyl-3-hydroxypyridine chlorohydrate (Khokhlov et al. 1984b).

It should be noted that most of the cell survival curves in our studies were obtained with transformed animal and human cells. Under appropriate conditions, most cancer cells are capable of proliferating indefinitely, with a given cell line (but not individual cells!) being "immortal." For example, the well-known HeLa cell line has been maintained in hundreds of laboratories for over 60 years. However, when the growth of such a culture is restricted by certain physiological means (not causing cell death), various defects at different structural and functional levels begin to accumulate in the cells, and the probability of their death increases; that is, the cells, as already mentioned, age in the true sense (Khokhlov 2012). At the same time, with regard to the reliability theory, it should be taken into account that an aging multicellular organism should not necessarily consist of senescing cells: the cells can simply die out "by exponent" (i.e., without senescence), as in the case of radioactive decay.

Usually, no special methods of cell viability assessment were used in such our experiments with human and animal cell cultures, and the proportion of cells survived by a given moment of time was determined visually, simply by counting the cells under a light microscope. Hence, the question has arisen how adequately the viability of an individual cell is evaluated using such an approach. This aspect is especially important for correctly constructing the survival curve's right tail, where the scattering of data points reaches a maximum because of significant reduction in the absolute number of the cell population (Yablonskaya et al. 2013).

It should be noted that the correct assessment of cell viability is a problem for all specialists working with cell cultures, but it is especially acute in case of cytogerontological experiments, where attention is focused on the temporal dynamics of the live/dead cell ratio in culture. It is such a parameter that should be determined in the first place in studies on cell aging both in the Hayflick model and in our model of stationary phase aging. However, this task is not as simple as it may seem at first glance. First, the cells may divide, thereby disrupting the integrity of the cell cohort; second, it is fairly difficult to correctly determine the time of death of a particular cell: the period of dying may be commensurate with cell life span, and it is tough to tell what stage in this long process is the point of no return (Kroemer et al. 2009), after which the cell can be certainly considered dead.

Now a variety of probes are available for assessing cell viability (Khokhlov and Morgunova 2015), but the results obtained with different probes unfortunately differ from each other. This is not surprising, because the rationales for using certain probes are based on different concepts of what exactly is the main criterion of cell viability (the integrity of the plasmalemma, the ability to synthesize ATP, the level of dehydrogenase activity, cell respiration rate, etc.). In other words, this is a fairly common situation when a given cell is classified as live in one test and as dead in another. The accuracy of analysis may be improved by using standard reagent kits containing several molecular probes each (Stoddart 2011), but this does not solve the problem in general.

In our experiments, we repeatedly determined the proportion of dead cells in the same "stationary aged" culture (not subcultured for 2–3 weeks) by directly examining the cells under a microscope and by taking digital images of the culture and taking cell counts on a computer display. In both cases, the cells were examined either "as is" (without any special treatment) or after adding dyes/probes commonly used for differential staining of live and dead cells (in particular, trypan blue, methylene blue, neutral red, and MTT). In many cases, the dead cell ratio detected by these methods proved to differ significantly, which casts doubt on the efficiency of such an approach to cell viability assessment in cytogerontological research. It should also be noted that some popular dyes have a number of side effects, which researchers often fail to mention. In particular, this concerns tetrazolium salts (MTT, XTT, etc.), which are inexpensive and can be used in experiments with cells of different origin, from bacteria to mammalian cells. However, some specialists consider that these probes are not optimal for assessing cell viability, even though they allow correct estimation of metabolic activity (Berridge et al. 2005). In particular, cell metabolic activity may change due to a variety of factors, even when the number of live cells in the population remains unchanged (Keepers et al. 1991).

Three groups of approaches to assessing the viability of cultured cells were diagrammatically represented in one of our papers (Khokhlov and Morgunova 2015). The paper does not cover all possible variants of live/dead cell tests but provides an idea of how broad the spectrum of such approaches can be. All methods have certain advantages and drawbacks. For example, the occurrence of holes in the plasma membrane is not necessarily fatal for the cell, since sometimes the membrane can be repaired (Reddy et al. 2001).

Meanwhile, there is one method that usually gives a correct answer to the question about the proportion of dead cells in the test culture under study, in which the viability of cells is estimated from their colony-forming efficiency (CFE) (Khokhlov et al. 1991, Smith et al. 2002; Maier et al. 2008). This method was widely introduced in cytogerontological experimentation in the 1970s, with the development of studies on the Hayflick phenomenon, that is, aging *in vitro* (Good and Smith 1974; Smith et al. 1977). In particular, this was due to the fact that the proportion of colonies consisting of at least 64 cells (in some studies, at least 16 cells) proved to be a good indicator of the "biological age" of normal cell culture, well correlated with the cell population doubling level. The CFE assay is also actively used to test the mitogenic or cytotoxic activity of various compounds (Kuczek and Axelrod 1987; Khokhlov et al. 1991).

This assay is usually performed by plating 100–200 cells from the test culture into Petri dishes and evaluating the number and size of colonies grown after several days. The same approach is suitable for determining the CFE of cells from donors of different ages in studies on their aging *in vivo*, but it is inapplicable to postmitotic or very slowly dividing cells composing organs which are critically important for the aging process (neurons, cardiomyocytes, hepatocytes, etc.). Unfortunately, the viability of such cells can only be assessed using the aforementioned probes for measuring a certain functional parameter. However, the choice of such a parameter largely depends on what concepts of aging are supported by the researcher, while the idea that if a cell divides, then this is certainly a live cell is evident to all gerontologists. This is why, when possible, it is most expedient to rely on measurements of CFE as the best indicator of cell viability in the population studied. Unfortunately, this method in its routine variant often fails to reveal subtle modifications of CFE manifested as changes in the distribution of colonies by size rather than in their number. Hence, it is often difficult to compare histograms of size distribution for colonies formed in different Petri dishes, for example, in experiments on the effect on cell cultures of certain biologically active substances (in particular, potential geroprotectors or geropromoters). To facilitate this procedure, it would be desirable to have a certain numerical parameter providing an integrated characteristic of each histogram, which allows simple statistical analysis of the data.

To this end, we have modified the method of CFE assay by distributing the colonies grown in a dish into size classes and calculating the average weighted number of the class (AWNC) for each distribution (Alinkina et al. 2012; Yablonskaya et al. 2013). It has been assumed that an increase in

AWNC (a shift of the distribution to a larger colony size) is indicative of improvement in the functional status of the test culture (i.e., a reduction of its "biological age"), while a decrease in this parameter is evidence that the culture is "getting older." This approach may be ineffective in some cases, since it is theoretically possible that AWNC remains the same while the shape of the histogram changes, but we have never observed such a situation in our experiments. An additional advantage of this approach is that it provides for a lower scattering of results obtained by different researchers for the same Petri dishes with cell colonies. Therefore, more researchers may be involved in the tedious process of cell counting in all the colonies in order to accelerate it, without any significant increase in the contribution of subjectivity to the dispersion of the results. To plot the distribution of colonies by size, we divide them into 17 classes with regard to the number of cells per colony (1–15, 16–31, 32–47 ... 240–255, 256 and greater). Thus, all classes except the first and the last are of the same size (16 cells).

The above approach to the testing of biologically active compounds on cell cultures is well illustrated by the results of our study on the effect of hydrated C_{60}-fullerene on the CFE of transformed Chinese hamster cells (Yablonskaya et al. 2013). These results confirm the geropromoter activity of the test agent, which has been revealed in previous experiments with the stationary phase aging model. As found in this study, the calculation of CFE and, especially, AWNC markedly simplifies the interpretation of experimental data and practically eliminates the problem of subjectivity in taking colony cell counts. To date, we have performed a number of cell culture experiments with various potential geroprotectors and geropromoters, and the results obtained provide conclusive evidence for the expediency of the proposed approach for rapid testing of such agents (Alinkina et al. 2012; Khokhlov et al. 2015).

To be objective, it should be noted that analysis of stationary phase cell aging by CFE assay may be complicated by the fact that cultured cells should be first removed from the growth substrate by treatment with special agents (usually a mixture of trypsin and Versene solutions) that disrupts the calcium-protein bridges attaching the cells to the surface and then plated at a very low density into Petri dishes or culture plates with fresh medium. This procedure is fairly traumatic and may even be fatal, especially for "elderly" cells, and the scattering of data on their CFE may sharply increase at the late stages of cell survival in this model system (Yablonskaya et al. 2013).

EXEGESIS OF EFFECT OF ANTI-AGING AGENTS ON VIABILITY OF CULTURED CELLS IN CYTOGERONTOLOGICAL STUDIES

On the basis of the data reviewed in the former section, it could be assumed that the solution of problems related to evaluating the viability of cultured cells in cytogerontological experiments, with special emphasis placed on the problems associated with construction of the survival curves for cultured cells in the stationary phase aging model, should ensure successful testing of potential geroprotectors in experiments based on this model as well as on some other cytogerontological model systems. However, the following questions (in addition to some items formulated in Section "Definitions and Some General Remarks") regarding the interpretation of data obtained in such studies in application to humans, whose aging is of primary interest to us, remain open:

1. Whether the factors (chemical or physical) that improve the viability of cultured cells should always slow down the aging of a multicellular organism, and vice versa?
2. How important is it which criteria of cell viability are used in testing geroprotectors in cytogerontological experiments?
3. How can the interpretation of results obtained in a study depend on the origin of cells that were used in this study?

We will try to answer these questions below.

When studying potential geroprotectors in cytogerontological experiments (i.e., in experiments on cell cultures), we usually evaluate their effect on cell viability. However, the criteria of this viability, as already mentioned, may be fundamentally different depending on the theory of aging to which a particular researcher adheres. In particular, the concept according to which the aging of a multicellular organism is caused by the limited mitotic potential of the normal cells constituting this organism has been very popular for many years. For this reason, the compounds that increase the proliferative potential (the Hayflick limit) of such cells *in vitro* were automatically regarded as geroprotectors (it should be emphasized that we are talking about the proliferative *potential* of cells but not about their proliferative *activity*; unfortunately, these parameters are very often confused in the cytogerontological literature). At the same time, data according to which the aging of an organism is largely determined by its postmitotic or very slowly proliferating cells (neurons, cardiomyocytes, hepatocytes, egg cells, etc.), which never have enough time to realize even the "normal" proliferative potential during the lifetime of the "host," have been ignored (Cristofalo et al. 1998; Khokhlov 2010b). The majority of human cells do not proliferate or proliferate very slowly because they *should not* do so rather than because they *cannot* do so. Therefore, the induction of telomerase activity in the normal cells, leading to a significant increase in their mitotic potential (possibly even making it unlimited), cannot be realized in these cells. Also, for the cells of the organism that already have telomerase (stem and germ line cells), this induction is even more useless.

If a test compound has a positive effect on the proliferative *activity* of cells (which is manifested, e.g., in increasing their CFE), the effects of this drug on the organism can be dual type. On the one hand, for some cells (e.g., those involved in the regeneration processes), such stimulation can be useful. On the other hand, this effect can, firstly, stimulate the proliferation of those cells that, as mentioned above, should not divide, and secondly, increase the probability of rapid propagation of the precancerous (or even cancerous) cells present in the organism. An increase in the incidence of benign tumors also cannot be ruled out. However, it should be emphasized that the evaluation of the CFE of cells is one of the few methods that provides data on the characteristics of *individual* cells rather than the cell population in general (Khokhlov et al. 1991). In the latter case, information about the possible subpopulations of cells that may differently respond to the test compound is lost due to averaging. For example, under the influence of a test factor, the content of 8-oxo-2′-deoxyguanosine, a popular aging biomarker (Esipov et al. 2008), in DNA of different cells may increase, decrease, or remain unchanged. As a result, the estimation of the content of 8-oxo-2′-deoxyguanosine *on average* can lead to the conclusion about the absence of changes in this parameter.

Certain parameters used to assess cell viability in cytogerontological experiments can be purely "correlative" (Khokhlov 2003), so that their interpretation becomes even more complicated. For example, this applies to the saturating density of a cell culture. It is known that, for normal diploid cells, this parameter is inversely well correlated with the age of the cell donor (in this case, the cause–effect relationships remain unclear). It was this index we used in our cell kinetics model (see Section "Experimental Gerontological Models Using Isolated Cells") to assess potential geroprotectors. It was assumed that the factors that increase the saturating density of the culture and thereby reduce the "biological age" of cells should have a positive effect on the viability and aging of a multicellular organism. However, in this case, we may face the same problems as in interpreting the data of experiments on CFE. It is not obvious that an improved ability of cells to reach a high saturating density in culture will slow down the aging of a multicellular organism in all cases. It cannot be ruled out that it may have no effect on the aging process at all or may even accelerate it.

It is very important which cell types are used in cytogerontological experiments on testing potential geroprotectors—normal or transformed cells of multicellular organisms, unicellular eukaryotic or prokaryotic organisms, etc. As noted above, differences in interpreting the results of geroprotector testing that were obtained on normal and transformed human or animal cells can become quite apparent when these results are extrapolated to humans, many of whom die from cancer. In particular, the biologically active compounds that reduce the viability of cultured cancer cells can extend the life span of humans and experimental animals, similar to the agents that increase the viability of normal cultured cells (Morgunova et al. 2016a). The use of unicellular organisms, such as bacteria or yeast, makes it possible to estimate the effect of various agents on the cells that represent independent organisms. However, a bacterium, for example, is so dramatically different from a mammalian cell that the same compound can kill the former but have hardly any effect on the viability of the latter (e.g., this refers to antibiotics).

In our opinion, the use of the stationary phase aging model in many cases makes it possible to avoid many of these problems, because the key factor that triggers the "aging" of all cells used in experiments is the restriction of cell proliferation with the help of various quite physiological impacts. A classic example is the chronological aging of yeast (Breitenbach et al. 2012; Khokhlov 2016), the results of studies of which are often pretty successfully used for studying the mechanisms of aging of humans and animals. In particular, experiments with the yeast *S. cerevisiae* showed that rapamycin, a well-known TOR inhibitor, in small doses that are sufficient for slowing down the proliferation of ycast cells but do not completely block this process, increases the culture life span in the chronological aging model (Powers et al. 2006; Alvers et al. 2009). Later, this compound was shown to extend the life span of experimental animals—mice (Harrison et al. 2009; Miller et al. 2014) and fruit flies (Bjedov et al. 2010). It should be noted that, according to the ideas of some researchers (Alvers et al. 2009; Rubinsztein et al. 2011; Morgunova et al. 2016b), the positive "gerontological" effect of rapamycin may be associated with the activation of autophagy. It also cannot be ruled out that the beneficial effect of rapamycin on the life span of animals may be due to its ability to suppress the emergence and development of malignant tumors (Blagosklonny 2006; Neff et al. 2013). As already mentioned above, in this case, it can hardly be considered a geroprotector. In addition, it is interesting to note that animals may develop tolerance to rapamycin over time. For this reason, some authors suggest that this drug should be used in combination with other active compounds, such as resveratrol (Alayev et al. 2015). Unfortunately, such problems are unlikely to be "caught" in cytogerontological studies.

SOME REMARKS ON BIOMARKERS OF CELL AGING/SENESCENCE

As already mentioned above, today, the construction of the survival curves of the test animal/human cohorts is the most reliable way to estimate the efficiency of interventions in the aging process. Unfortunately, this method is inefficient in terms of labor, time, and finance expenditures. Consequently, overeager gerontologists currently rely mainly on so-called BA. Space limitations do not allow us to dwell on the essence of this term, but this is not necessary, since the relevant literature is available to any reader. It should only be noted that the researchers who use this term usually have in mind not so much the markers of aging itself as the markers of biological age. In other words, the markers (parameters) are well correlated with the chronological age of the test organisms but not with aging, that is, the time-dependent increase in the probability of death.

An illustrative example to this issue is the situation with human hair turning gray: the relative amount of gray hairs is well correlated with age but shows practically no correlation with mortality. Thus, relevant parameters in gerontology are those related to the basic mechanisms of aging, preferably in a cause-and-effect mode, and the majority of gerontologists consider that these are cellular or molecular mechanisms. The batteries of tests for determining the biological age (in other words, the degree of senescence) based on evaluation of various physiological parameters, which have been used on a wide scale, gradually recede in the past, giving way to studies with emphasis on "fundamental" BA—that is, on certain cellular or molecular characteristics. Moreover, these parameters are currently usually tied in with the phenomenon named cell/cellular senescence, which is central in cytogerontology.

It was initially considered that cell senescence takes place "by itself," that is, it is driven by an intrinsic mechanism, and all subsequent changes in the cells are mere *consequences* of this process. In fact, this fully applies to the mechanism of telomere shortening with every cell division, discovered by Alexey Olovnikov (Olovnikov 1971). In the 1980s, one of us formulated the concept of aging (Khokhlov 1988) according to which the restriction of cell proliferation imposed during development (due to the formation of populations of highly differentiated postmitotic or very slowly dividing cells) is the main cause of age-dependent accumulation of various macromolecular defects (mainly DNA damage) in the cells. This concept provides a simple explanation to "age-dependent" changes in senescent cell cultures: since cell proliferation at later passages is retarded, spontaneous DNA injuries are no longer "diluted" among newly emerging cells and their frequency in the population as a whole increases. The population aspect is very important, since some cells fully retain the ability to divide, but their proportion decreases with passaging, so that cell senescence is manifested at the level of the whole *cell population*. In essence, our stationary phase aging model (Vilenchik et al. 1981; Khokhlov 1988, 1992b, 1998, 2013a; Akimov and Khokhlov 1998) was based on 100% suppression of cell proliferation in culture by contact inhibition or some other physiological factor, with consequent accumulation of "age-dependent" defects in the cell population. In this case as well, we first made the cells "senesce" and only then analyzed them for certain biomarkers of *in vivo* aging (e.g., DNA breaks). Thus, in the "classic" approach, it was assumed that cell senescence is driven by a certain intrinsic mechanism, which leads to the emergence of various macromolecular defects (first of all, DNA damage) in the cells.

In recent years, however, cell senescence is understood primarily as the appearance or accumulation in the cells (most often, transformed cells not prone to replicative senescence) of certain "BA" (this time in quotation marks, because the situation is by no means related to real aging) under the impact of various external factors causing DNA damage (oxidative stress, H_2O_2, mitomycin C, doxorubicin, ethanol, ionizing radiation, etc.) (Jeyapalan and Sedivy 2008; Kuilman et al. 2010; Lawless et al. 2010; Giaimo and D'Adda di Fagagna 2012; Campisi 2013). This phenomenon is referred to as DNA damage response (DDR). Within this definition, the "senescence" of cells takes place under the impact of DNA-damaging agents rather than on itself. It is also called "stress-induced premature senescence" (Toussaint et al. 2002). The aforementioned BA include SA-β-Gal activity, expression of p53 and p21 proteins as well as of regulators of inflammation such as IL-6 or IL-8, activation of oncogenes, etc. Therefore, cell "senescence" in the context of the above definition occurs not by itself but *because* of the impact of DNA-damaging agents. In our opinion, such an approach is very important for defining the strategy of cancer control but, yet again, leads away from the study of actual mechanisms of organismal aging (Khokhlov 2013d). A similar view was expressed by the famous gerontologist Denham Harman in his brief comment published in the journal *Biogerontology* (Harman 2009). It should be emphasized that, in our stationary phase aging model (Khokhlov 1988, 1992b, 1998, 2013a), we also observe certain BA in cell cultures, but in this case they appear due to restriction of proliferation by contact inhibition, that is, by a physiological factor that itself causes no damage to the cells. This situation is closely similar to what takes place in a multicellular organism.

The most popular biomarker of cellular senescence is SA-β-Gal (pH 6 β-galactosidase). The enzyme β-galactosidase, a lysosomal hydrolase, cleaves off the terminal β-galactose from the compounds containing it (lactose, keratin sulfates, sphingolipids, etc.). It is involved in some "minor" metabolic reactions and is present in almost all tissues. This enzyme exhibits maximum activity at pH 4.0; however, the difference in this index between the "old" and "young" cells can be better detected by certain biochemical methods at pH 6.0. The feasibility of using SA-β-Gal activity as a BA was first postulated by Dimri et al. (1995), who demonstrated that the expression of this enzyme increases with aging both *in vitro* and *in vivo*. In subsequent years, this BA was widely used in cytogerontological experiments to assess the "age" of cells and is currently the most common in the studies (Debacq-Chainiaux et al. 2008; Sikora et al. 2011) based on the definition of cellular senescence that we do not accept. However, in parallel, several studies were published whose authors emphasized that SA-β-Gal activity in cells is not so good a BA, because, in many cases, it depends not so much on age (both *in vivo* and *in vitro*) as on the method of research and/or the presence of certain pathologies as well as, what is most important, on the proliferative status of the cells (Yegorov et al. 1998; Krishna et al. 1999; Choi et al. 2000; Severino et al. 2000; Untergasser et al. 2003; Kang et al. 2004; Cristofalo 2005). It seems that cell proliferation restriction, for whatever reason (differentiation, contact inhibition, DDR, some diseases, etc.), is the factor that causes stimulation of SA-β-Gal expression. In other words, SA-β-Gal appears even in the "young" cells if their proliferation is suppressed. Not long ago, we showed (Vladimirova et al. 2012) that in the stationary phase culture of transformed Chinese hamster cells, the proportion of cells in which SA-β-Gal is detected by the method of Dimri et al. increases with time, and this is accompanied, on the one hand, by an increase in the level of poly(ADP-ribose) in the cells and, on the other hand, by a decline in their capacity to synthesize poly(ADP-ribose) in response to DNA damage induced by hydrogen peroxide (H_2O_2). Such data, in our opinion, provide further evidence of the viability of our concept of aging, which postulates the crucial role of cell proliferation restriction in the accumulation in cells of various macromolecular defects (the most important of which are DNA lesions), which in turn lead to a deterioration in the functioning of organs and tissues and further increase in the probability of death of macroorganisms (Khokhlov 2010b, 2013a, 2014b).

It is also interesting to note that, in the experiments designed to compare the effects of "stationary-phase" or "stress-induced" (exposure to 4% ethanol for 2 h per day for 5 days) aging on the transformed Chinese hamster cells, we showed that the percentage of cells stained for SA-β-Gal by the method of Dimri et al. in a 14-day-old "stationary-phase-old" culture was much higher than in the "young" (7-day-old) control culture but comparable to that detected in 7-day-old cells incubated with ethanol (Morgunova et al. 2015).

Finally, we would like to mention another study by Lee et al. (2006), where it was shown that, both in the "stress-induced premature senescence" and in the replicative senescence, "according to Hayflick," SA-β-Gal does not accumulate if the expression of the *GLB1* gene, which encodes the lysosomal β-galactosidase, is disrupted.

CONCLUSIONS

1. We think that any "true" geroprotector should retard the age-related increase in the probability of death of aging organisms causing a rightward shift in the survival curve and increasing both the average and maximum life span.

2. We do not think that the drugs which are used to combat the age-related diseases could be considered geroprotectors, as well as the factors that increase the life span of the non-aging organisms.

3. At present, there are several cytogerontological models which are used for testing of potential geroprotectors. The most popular among them are the Hayflick model, the stationary phase (chronological) aging model, and the cell kinetics model. In our opinion, the least number of problems associated with interpreting the results of testing potential geroprotectors in cytogerontological experiments arises when such studies are performed using the model of stationary phase aging (which is based on the concept of cell proliferation restriction as the main cause of accumulation of macromolecular lesions in cells of multicellular organisms with age, leading to the deterioration of the functioning of tissues and organs and, as a result, an increase in the probability of death) of normal cells. However, even this approach will not give the *final* answer to the question of whether or not the studied factor is a "true" geroprotector. Answering this question will inevitably require both experiments in animals and clinical trials.

4. The cytogerontological models mentioned can be effectively used to test various agents (drugs) or their combinations for their potential ability to accelerate or retard aging only if their effect is realized at the cell level. Unfortunately, we have recently got the impression that even the data obtained with "gist" cell culture models cannot be directly extrapolated to the organism as a whole. Our cytogerontological tests of various geroprotectors on the models of stationary phase aging, cell kinetics, and cell CFE have shown that these factors fairly often have no favorable effect on the viability of cultured cells, even though they prolong the life span of experimental animals and improve the state of human health. This fact suggests that the effect of a geroprotector in many cases manifests itself only at the organismal level (probably due to activation/suppression of certain biochemical or

neurophysiological processes) and is not limited to the improvement of viability of individual cells. Apparently, the same is also true of geropromoters. Thus, it was probably a serious mistake to perform experiments with cell cultures so as to exclude the influence of the endocrine and central nervous systems (which actually was the main purpose of gerontologists, beginning from studies by Alexis Carrel 1912, 1913). By all accounts, the results of cytogerontological experiments should be thoroughly verified in studies on laboratory animals and even in clinical trials (provided this complies with ethical principles of human subject research). Of course, this will lessen our chances of an early breakthrough in studies aimed at retarding the process of aging, but the reliability of the obtained data will be significantly higher.

5. Regarding the approaches to cell viability testing in cytogerontological experiments, the choice of methods to this end depends mainly, apparently, on the researchers' ideas about molecular and cellular mechanisms of aging. The most appropriate method, the evaluation of CFE, though optimal for cell viability assessment, is not applicable to postmitotic or very slowly propagating cells. Unfortunately, many problems encountered when using popular molecular probes designed for live/dead cell viability assays remain open.

6. When interpreting the results of geroprotector testing in experiments on cell cultures, the conclusions strongly depend on which cell types are used (see Section "Exegesis of Effect of Anti-aging Agents on Viability of Cultured Cells in Cytogerontological Studies"). In particular, they could vary greatly for normal and transformed animal cells.

7. Instead of analyzing the effect of a potential anti-aging factor on the proliferative potential of cultured cells or their stationary phase life span, we can follow some BA during cell aging/senescence *in vitro* or stationary phase (chronological) aging. If all that was said in Section "Some Remarks on Biomarkers of Cell Aging/Senescence" is true, then it may well be that canceling the aging process will not necessarily cause any significant changes in the age-dependent dynamics of those BA (regardless of whether they accumulate or disappear) that are directly connected with the proliferative status of cells forming organs and tissues. This should apply at least to those BA that are not directly involved in the mechanisms responsible for the age-dependent increase in the probability of death. If certain BA are "gist" markers, that is, the aging process cannot be retarded without affecting these BA, then the postulated mechanism of aging canceling should provide an explanation as to how these BA will be continuously removed from postmitotic or very slowly proliferating cells.

Finally, here is the main conclusion. As already noted above, it appears that gerontologists analyzing the possibilities for retarding or even blocking the aging process currently have no fully adequate alternative to the construction of survival curves for the cohorts of animals or humans, even though this approach is highly expensive and requires great labor expenditures. Apparently, all the cytogerontological models reviewed provide only preliminary testing of potential anti-aging factors.

REFERENCES

Akatov, V.S., Lezhnev, E.I., Vexler, A.M. and L.N. Kublik. 1985. Low pH value of pericellular medium as a factor limiting cell proliferation in dense cultures. *Exp Cell Res* 160:412–418.

Akimov, S.S. and A.N. Khokhlov. 1998. Study of "stationary phase aging" of cultured cells under various types of proliferation restriction. *Ann NY Acad Sci* 854:520.

Alayev, A., Berger, S.M., Kramer, M.Y., Schwartz, N.S. and M.K. Holz. 2015. The combination of rapamycin and resveratrol blocks autophagy and induces apoptosis in breast cancer cells. *J Cell Biochem* 116:450–457.

Alinkina, E.S., Vorobyova, A.K., Misharina, T.A., Fatkullina, L.D., Burlakova, E.B. and A.N. Khokhlov. 2012. Cytogerontological studies of biological activity of oregano essential oil. *Moscow Univ Biol Sci Bull* 67:52–57.

Alvers, A.L., Wood, M.S., Hu, D., Kaywell, A.C. Dunn, W.A. Jr. and J.P. Aris. 2009. Autophagy is required for extension of yeast chronological life span by rapamycin. *Autophagy* 5:847–849.

Arshavsky, I.A. 1982. *Physiological Mechanisms and Regularities of Individual Development.* Moscow: Nauka (in Russian).

Austad, S.N. 1999. *Why We Age: What Science is Discovering About The Body's Journey Through Life.* 1st edition. Chichester: John Wiley & Sons, Inc.

Austad, S.N. 2004. Rebuttal to Bredesen: 'The non-existent aging program: How does it work?' *Aging Cell* 3:253–254.

Berridge, M.V., Herst, P.M. and A.S. Tan. 2005. Tetrazolium dyes as tools in cell biology: New insights into their cellular reduction. *Biotechnol Annu Rev* 11:127–152.

Bidder, G.P. 1932. Senescence. *Br Med J* 2:583–585.

Bjedov, I., Toivonen, J.M., Kerr, F. et al. 2010. Mechanisms of life span extension by rapamycin in the fruit fly *Drosophila melanogaster. Cell Metab* 11:35–46.

Blagosklonny, M.V. 2006. Aging and immortality: Quasi-programmed senescence and its pharmacologic inhibition. *Cell Cycle* 5:2087–2102.

Breitenbach, M., Jazwinski, S.M. and P. Laun, Eds. 2012. *Aging Research in Yeast: Subcell. Biochem. Vol. 57.* Netherlands: Springer.

Burhans, W.C. and M. Weinberger. 2009. Acetic acid effects on aging in budding yeast: Are they relevant to aging in higher eukaryotes? *Cell Cycle* 8:2300–2302.

Burtner, C.R., Murakami, C.J., Kennedy, B.K. and M. Kaeberlein. 2009. A molecular mechanism of chronological aging in yeast. *Cell Cycle* 8:1256–1270.

Campisi, J. 2011. Cellular senescence: Putting the paradoxes in perspective. *Curr Opin Genet Dev* 21:107–112.

Campisi, J. 2013. Aging, cellular senescence, and cancer. *Annu Rev Physiol* 75:685–705.

Carrel, A. 1912. Artificial activation of the growth *in vitro* of connective tissue. *J Exp Med* 17:14–19.

Carrel, A. 1913. Contributions to the study of the mechanism of the growth of connective tissue. *J Exp Med* 18:287–299.

Chirkova, E.Iu., Golovina, M.E., Nadzharian, T.L. and A.N. Khokhlov. 1984. Cellular kinetic model for studying geroprotectors and geropromotors. *Dokl Akad Nauk SSSR* 278:1474–1476.

Choi, J., Shendrik, I., Peacocke, M. et al. 2000. Expression of senescence associated beta-galactosidase in enlarged prostates from men with benign prostatic hyperplasia. *Urology* 56:160–166.

Comfort, A. 1964. *Ageing: The Biology of Senescence.* 2nd edition. London: Routledge & Kegan Paul.

Cristofalo, V.J. 2005. SA beta Gal staining: Biomarker or delusion. *Exp Gerontol* 40:836–838.

Cristofalo, V.J., Allen, R.G., Pignolo, R.J., Martin, B.G. and J.C. Beck. 1998. Relationship between donor age and the replicative lifespan of human cells in culture: A reevaluation. *Proc Natl Acad Sci USA* 95:10614–10619.

Debacq-Chainiaux, F., Pascal, T., Boilan, E., Bastin, C., Bauwens, E. and O. Toussaint. 2008. Screening of senescence-associated genes with specific DNA array reveals the role of IGFBP-3 in premature senescence of human diploid fibroblasts. *Free Radic Biol Med* 44:1817–1832.

Dilman, V.M. 1971. Age-associated elevation of hypothalamic threshold to feedback control, and its role in development, ageing, and disease. *Lancet* 297:1211–1219.

Dimri, G.P., Lee, X., Basile, G. et al. 1995. A biomarker that identifies senescent human cell in culture and in aging skin *in vivo. Proc Natl Acad Sci USA* 92:9363–9367.

Esipov, D.S., Gorbacheva, T.A., Khairullina, G.A., Klebanov, A.A., Nguyen, T.N. and A.N. Khokhlov. 2008. 8-oxo-2'-deoxyguanosine accumulation in DNA from "stationary phase aging" cultured cells. *Adv Gerontol* 21:485–487.

Fabrizio, P. and V.D. Longo. 2003. The chronological life span of *Saccharomyces cerevisiae. Aging Cell* 2:73–81.

Fabrizio, P. and M. Wei. 2011. Conserved role of medium acidification in chronological senescence of yeast and mammalian cells. *Aging (Albany NY)* 3:1127–1129.

Giaimo, S. and F. D'Adda di Fagagna. 2012. Is cellular senescence an example of antagonistic pleiotropy? *Aging Cell* 11:378–383.

Good, P.I. and J.R. Smith. 1974. Age distribution of human diploid fibroblasts: A stochastic model for *in vitro* aging. *Biophys J* 14:811–823.

Harman, D. 2009. About "Origin and evolution of the free radical theory of aging: A brief personal history, 1954–2009". *Biogerontology* 10:783.

Harrison, D.E., Strong, R., Sharp, Z.D. et al. 2009. Rapamycin fed late in life extends lifespan in genetically heterogeneous mice. *Nature* 460:392–395.

Hayflick, L. 1965. The limited *in vitro* lifetime of human diploid cell strains. *Exp Cell Res* 37:614–636.

Hayflick, L. 1979a. Progress in cytogerontology. *Mech Ageing Dev.* 9:393–408.

Hayflick, L. 1979b. The cell biology of aging. *J Invest Dermatol* 73:8–14.

Hayflick, L. 1991. Aging under glass. *Mutat Res/DNAging* 256:69–80.

Hayflick, L. 1996. *How and Why We Age*. New York: Ballantine Books.

Hayflick, L. 2007. Entropy explains aging, genetic determinism explains longevity, and undefined terminology explains misunderstanding both. *PLoS Genet* 3:e220.

Hayflick, L. and P.S. Moorhead. 1961. The serial cultivation of human diploid cell strains. *Exp Cell Res* 25:585–621.

Holliday, R. 2007. *Aging: The Paradox of Life. Why We Age*. Dordrecht: Springer.

Jeyapalan, J.C. and J.M. Sedivy. 2008. Cellular senescence and organism aging. *Mech Aging Dev* 129:467–474.

Jones, O.R., Scheuerlein, A., Salguero-Gómes, R. et al. 2014. Diversity of ageing across the tree of life. *Nature* 505:169–173.

Kaeberlein, M. and B.K. Kennedy. 2012. A new chronological survival assay in mammalian cell culture. *Cell Cycle* 11:201–202.

Kang, H.T., Lee, C.J., Seo, E.J., Bahn, Y.J., Kim, H.J. and E.S. Hwang. 2004. Transition to an irreversible state of senescence in HeLa cells arrested by repression of HPV E6 and E7 genes. *Mech Ageing Dev* 125:31–40.

Kapitanov, A.B. and M.Y. Aksenov. 1990. Ageing of prokaryotes. *Acholeplasma laidlawii* as an object for cell ageing studies: A brief note. *Mech Ageing Dev* 54:249–258.

Keepers, Y.P., Pizao, P.E., Peters, G.J., van Ark-Otte, J., Winograd, B. and H.M. Pinedo. 1991. Comparison of the sulforhodamine B protein and tetrazolium (MTT) assays for *in vitro* chemosensitivity testing. *Eur J Cancer* 27:897–900.

Khokhlov, A.N. 1988. *Proliferatsiya i Starenie (Cell Proliferation and Aging), Itogi Nauki i Tekhniki VINITI AN SSSR. Ser. Obshchie Problemy Fiziko-Khimicheskoi Biologii (Advances in Science and Technology, VINITI Akad. Sci. USSR, Ser. General Problems of Physicochemical Biology)* [*Cell Proliferation and Aging. Advances in Science and Technology, VINITI Akad. Sci. USSR, Ser. General Problems of Physicochemical Biology*]. Moscow: VINITI (in Russian).

Khokhlov, A.N. 1992a. The cell kinetics model for determination of organism biological age and for geroprotectors or geropromoters studies. In *Biomarkers of Aging: Expression and Regulation. Proceeding*, ed. F. Licastro, and C.M. Caldarera. Bologna: CLUEB, pp. 209–216.

Khokhlov, A.N. 1992b. Stationary cell cultures as a tool for gerontological studies. *Ann NY Acad Sci* 663:475–476.

Khokhlov, A.N. 1994. Evolutionary cytogerontology as a new branch of experimental gerontology. *Age* 17:159.

Khokhlov, A.N. 1998. Cell proliferation restriction: Is it the primary cause of aging? *Ann NY Acad Sci* 854:519.

Khokhlov, A.N. 2002. Results and perspectives of cytogerontologic studies in modern time. *Tsitologiia* 44:1143–1148.

Khokhlov, A.N. 2003. Cytogerontology at the beginning of the third millennium: From "correlative" to "gist" models. *Russ J Dev Biol* 34:321–326.

Khokhlov, A.N. 2004. In search of "gist" cytogerontological models. In *Longevity, Aging and Degradation Models in Reliability, Public Health, Medicine and Biology*, ed. V. Antonov, C. Huber, M. Nikulin, and V. Polischook. St. Petersburg: St. Petersburg State Polytechnical University, vol. 1, pp. 84–92.

Khokhlov, A.N. 2010a. Does aging need an own program or the existing development program is more than enough? *Russ J Gen Chem* 80:1507–1513.

Khokhlov, A.N. 2010b. From Carrel to Hayflick and back, or what we got from the 100-year cytogerontological studies. *Biophysics* 55:859–864.

Khokhlov, A.N. 2012. Can cancer cells age? Stationary cell culture approach to the problem solution. In *Visualizing of Senescent Cells in Vitro and in Vivo. Programme and Abstracts*, Warsaw, Poland, December 15–16, p. 49.

Khokhlov, A.N. 2013a. Does aging need its own program, or is the program of development quite sufficient for it? Stationary cell cultures as a tool to search for anti-aging factors. *Curr Aging Sci* 6:14–20.

Khokhlov, A.N. 2013b. Impairment of regeneration in aging: Appropriateness or stochastics? *Biogerontology* 14:703–708.

Khokhlov, A.N. 2013c. Decline in regeneration during aging: Appropriateness or stochastics? *Russ J Dev Biol* 44:336–341.

Khokhlov, A.N. 2013d. Evolution of the term "cellular senescence" and its impact on the current cytogerontological research. *Moscow Univ Biol Sci Bull* 68:158–161.

Khokhlov, A.N. 2014a. On the immortal hydra. Again. *Moscow Univ Biol Sci Bull* 69:153–157.

Khokhlov, A.N. 2014b. What will happen to molecular and cellular biomarkers of aging in case its program is canceled (provided such a program does exist)? *Adv Gerontol* 4:150–154.

Khokhlov, A.N. 2016. Which aging in yeast is "true"? *Moscow Univ Biol Sci Bull* 71:11–13.

Khokhlov, A.N., Chirkova, E.Iu. and A.N. Chebotarev. 1985a. Changes in the level of sister chromatid exchanges in cultured Chinese hamster cells undergoing limited proliferation. *Tsitol Genet* 19:90–92.

Khokhlov, A.N., Chirkova, E.Iu. and A.N. Chebotarev. 1987a. Changes in the number of sister chromatid exchanges in cultured Chinese hamster cells with limited proliferation. Additional studies. *Tsitol Genet* 21:186–190.

Khokhlov, A.N., Chirkova, E.Yu. and A.I. Gorin. 1986. Strengthening of the DNA–protein complex during stationary phase aging of cell cultures. *Bull Exp Biol Med* 101:437–440.

Khokhlov, A.N., Chirkova, E.Iu. and T.L. Nadzharian. 1984a. DNA degradation in resting cultured Chinese hamster cells. *Tsitologiia* 26:965–968.

Khokhlov, A.N., Golovina, M.E., Chirkova, E.Iu. and T.L. Nadzharian. 1985b. Analysis of the kinetic growth patterns of cultured cells. I. The model. *Tsitologiia* 27:960–965.

Khokhlov, A.N., Golovina, M.E., Chirkova, E.Iu. and T.L. Nadzharian. 1985c. Analysis of the kinetic growth patterns of cultured cells. II. The action of ionizing radiation, an alkylating agent and a low-frequency electromagnetic field. *Tsitologiia* 27:1070–1075.

Khokhlov, A.N., Golovina, M.E., Chirkova, E.Iu. and T.L. Nadzharian. 1987b. Analysis of the kinetic growth patterns of cultured cells. III. The effect of inoculation density, of a geroprotector-antioxidant and of stationary-phase aging. *Tsitologiia* 29:353–357.

Khokhlov, A.N., Kirnos, M.D. and B.F. Vaniushin. 1988. The level of DNA methylation and "stationary phase aging" in cultured cells. *Izv Akad Nauk SSSR Ser Biol* 3:476–478.

Khokhlov, A.N., Klebanov, A.A., Karmushakov, A.F., Shilovsky, G.A., Nasonov, M.M. and G.V. Morgunova. 2014. Testing of geroprotectors in experiments on cell cultures: Choosing the correct model system. *Moscow Univ Biol Sci Bull* 69:10–14.

Khokhlov, A.N. and G.V. Morgunova. 2015. On the constructing of survival curves for cultured cells in cytogeron-tological experiments: A brief note with three hierarchy diagrams. *Moscow Univ Biol Sci Bull* 70:67–71.

Khokhlov, A.N. and G.V. Morgunova. 2017. Testing of geroprotectors in experiments on cell cultures: Pros and cons. In *Anti-Aging Drugs: From Basic Research to Clinical Practice*, ed. A.M. Vaiserman. Royal Society of Chemistry, pp. 53–74.

Khokhlov, A.N., Morgunova, G.V., Ryndina, T.S. and F. Coll. 2015. Pilot study of a potential geroprotector, "Quinton Marine Plasma," in experiments on cultured cells. *Moscow Univ Biol Sci Bull* 70:7–11.

Khokhlov, A.N., Prokhorov, L.Yu., Ivanov, A.S. and A.I. Archakov. 1991. Effects of cholesterol- or 7-keto-choles-terol-containing liposomes on colony-forming ability of cultured cells. *FEBS Lett* 290:171–172.

Khokhlov, A.N., Ushakov, V.L., Kapitanov, A.B. and T.L. Nadzharian. 1984b. Effect of the geroprotector 2-ethyl-6-methyl-3-hydroxypyridine chlorohydrate on the cell proliferation of *Acholeplasma laidlawii*. *Dokl Akad Nauk SSSR* 274:930–933.

Khokhlov, A.N., Wei, L., Li, Y. and J. He. 2012. Teaching cytogernotology in Russia and China. *Adv Gerontol* 25:513–516.

Kirkwood, T.B. and T. Cremer. 1982. Cytogerontology since 1881: A reappraisal of August Weismann and a review of modern progress. *Hum Genet* 60:101–121.

Kolosova, N.G., Shcheglova, T.V., Sergeeva, S.V. and L.V. Loskutova. 2006. Long-term antioxidant supple-mentation attenuates oxidative stress markers and cognitive deficits in senescent-accelerated OXYS rats. *Neurobiol Aging* 27:1289–1297.

Kolosova, N.G., Vitovtov, A.O., Muraleva, N.A., Akulov, A.E., Stefanova, N.A. and M.V. Blagosklonny. 2013. Rapamycin suppresses brain aging in senescence-accelerated OXYS rats. *Aging (Albany NY)* 5:474–484.

Krishna, D.R., Sperker, B., Fritz, P. and U. Klotz. 1999. Does pH 6 beta-galactosidase activity indicate cell senescence? *Mech Ageing Dev* 109:113–123.

Kroemer, G., Galluzzi, L., Vandenabeele, P. et al. 2009. Classification of cell death: Recommendations of the Nomenclature Committee on Cell Death 2009. *Cell Death Differ* 16:3–11.

Kuczek, T. and D.E. Axelrod. 1987. Tumor cell heterogeneity: Divided-colony assay for measuring drug response. *Proc Natl Acad Sci USA* 84:4490–4494.

Kuilman, T., Michaloglou, C., Mooi, W.J. and D.S. Peeper. 2010. The essence of senescence. *Genes Dev* 24:2463–2479.

Kurkdjian, A. and J. Guern. 1989. Intracellular pH: Measurement and importance in cell activity. *Annu Rev Plant Physiol Plant Mol Biol* 40:271–303.

Laun, P., Bruschi, C.V., Dickinson, J.R. et al. 2007. Yeast mother cell-specific ageing, genetic (in)stability, and the somatic mutation theory of ageing. *Nucleic Acid Res* 35:7514–7526.

Lawless, C., Wang, C., Jurk, D., Merz, A., von Zglinicki, T. and J.F. Passos. 2010. Quantitative assessment of markers for cell senescence. *Exp Gerontol* 45:772–778.

Lee, B.Y., Han, J.A., Im, J.S. et al. 2006. Senescence-associated β-galactosidase is lysosomal-galactosidase. *Aging Cell* 5:187–195.

Leontieva, O.V. and M.V. Blagosklonny. 2011. Yeast-like chronological senescence in mammalian cells: Phenomenon, mechanism and pharmacological suppression. *Aging (Albany NY)* 3:1078–1091.

Macieira-Coelho, A. 1999. Comparative biology of cell immortalization. Cell Immortalization. In: *Cell Immortalization (Progress in Molecular and Subcellular Biology, vol. 24)*, ed. A. Macieira-Coelho. Berlin-Heidelberg: Springer-Verlag, pp. 51–80.

Macieira-Coelho, A. 2011. Cell division and aging of the organism. *Biogerontology* 12:503–515.

Maier, A.B., Maier, I.L., van Heemst, D. and R.G.J. Westendorp. 2008. Colony formation and colony size do not reflect the onset of replicative senescence in human fibroblasts. *J Gerontol A Biol Sci Med Sci* 63:655–659.

Martínez, D. 1998. Mortality patterns suggest lack of senescence in hydra. *Exp Gerontol* 33:217–225.

Martínez, D.E. and D. Bridge. 2012. Hydra, the everlasting embryo, confronts aging. *Int J Dev Biol* 56:479–487.

Mikhelson, V.M. 2001. Replicative mosaicism might explain the seeming contradictions in the telomere theory of aging. *Mech Ageing Dev* 122:1361–1365.

Mikhelson, V.M. and I.A. Gamaley. 2012. Telomere shortening is a sole mechanism of aging in mammals. *Curr Aging Sci* 5:203–208.

Miller, R.A., Harrison, D.E., Astle, C.M. et al. 2014. Rapamycin-mediated lifespan increase in mice is dose and sex dependent and metabolically distinct from dietary restriction. *Aging Cell* 13:468–477.

Minot, C.S. 1908. *The Problem of Age, Growth, and Death; a Study of Cytomorphosis, Based on the Lectures at the Lowell Institute, March, 1907*. New York–London: G. P. Patnam's Sons.

Morgunova, G.V., Klebanov, A.A. and A.N. Khokhlov. 2016a. Interpretation of data about the impact of biologically active compounds on viability of cultured cells of various origin from a gerontological point of view. *Moscow Univ Biol Sci Bull* 71:67–70.

Morgunova, G.V., Klebanov, A.A. and A.N. Khokhlov. 2016b. Some remarks on the relationship between autophagy, cell aging, and cell proliferation restriction. *Moscow Univ Biol Sci Bull* 71:207–211.

Morgunova, G.V., Klebanov, A.A., Marotta, F. and A.N. Khokhlov. 2017. Culture medium pH and stationary phase/chronological aging of different cells. *Moscow Univ Biol Sci Bull* 72:47–51.

Morgunova, G.V., Kolesnikov, A.V., Klebanov, A.A. and A.N. Khokhlov. 2015. Senescence-associated β-galactosidase—A biomarker of aging, DNA damage, or cell proliferation restriction? *Moscow Univ Biol Sci Bull* 70:165–197.

Murakami, C.J., Wall, V., Basisty, N. and M. Kaeberlein. 2011. Composition and acidification of the culture medium influences chronological aging similarly in vineyard and laboratory yeast. *PloS One* 6:e24530.

Neff, F., Flores-Dominguez, D., Ryan, D.P. et al. 2013. Rapamycin extends murine lifespan but has limited effects on aging. *J Clin Invest* 123:3272–3291.

Nyström, T. 2002. Aging in bacteria. *Curr Opin Microbiol* 5:596–601.

Olovnikov, A.M. 1971. Principle of marginotomy in the template synthesis of polynucleotides. *Dokl Akad Nauk SSSR* 201:1496–1499.

Olovnikov, A.M. 1973. A theory of marginotomy. The incomplete copying of template margin in enzymic synthesis of polynucleotides and biological significance of the phenomenon. *J Theor Biol* 41:181–190.

Olovnikov, A.M. 1996. Telomeres, telomerase and aging; origin of the theory. *Exp Gerontol* 31:443–448.

Olovnikov, A.M. 2007a. Hypothesis: Lifespan is regulated by chronomere DNA of the hypothalamus. *J Alzheimer's Dis* 11:241–252.

Olovnikov, A.M. 2007b. Role of paragenome in development. *Russ J Dev Biol* 38:104–123.

Pletnev, P., Osterman, I., Sergiev, P., Bogdanov, A. and O. Dontsova. 2015. Survival guide: *Escherichia coli* in the stationary phase. *Acta Naturae* 7:22.

Powers III, R.W., Kaeberlein, M., Caldwell, S.D., Kennedy, B.K. and S. Fields. 2006. Extension of chronological life span in yeast by decreased TOR pathway signaling. *Genes Dev* 20:174–184.

Prokhorov, L.Yu., Petushkova, N.A. and A.N. Khokhlov. 1994. Cytochrome P-450 and "stationary phase aging" of cultured cells. *Age* 17:162.

Reddy, A., Caler, E.V. and N.W. Andrews. 2001. Plasma membrane repair is mediated by Ca^{2+}-regulated exocytosis of lysosomes. *Cell* 106:157–169.

Roux, A.E., Quissac, A., Chartrand, P., Ferbeyre, G. and L.A. Rokeach. 2006. Regulation of chronological aging in *Schizosaccharomyces pombe* by the protein kinases Pka1 and Sck2. *Aging Cell* 5:345–357.

Rubinsztein, D.C., Mariño, G. and G. Kroemer. 2011. Autophagy and aging. *Cell* 146:682–695.

Schneider, E.L. and J.R. Smith. 1981. The relationship of *in vitro* studies to *in vivo* human aging. *Int Rev Cytol* 69:261–270.

Severino, J., Allen, R.G., Balin, S., Balin, A. and V.J. Cristofalo. 2000. Is beta-galactosidase staining a marker of senescence *in vitro* and *in vivo*? *Exp Cell Res* 257:162–171.

Shabalina, I.G., Vyssokikh, M.Y., Gibanova, N. et al. 2017. Improved health-span and lifespan in mtDNA mutator mice treated with the mitochondrially targeted antioxidant SkQ1. *Aging (Albany NY)* 9:315–339.

Shram, S.I., Shilovskii, G.A. and A.N. Khokhlov. 2006. Poly(ADP-ribose)-polymerase-1 and aging: Experimental study of possible relationship on stationary cell cultures. *Bull Exp Biol Med* 141:628–632.

Sikora, E., Arendt, T., Bennett, M. and M. Narita. 2011. Impact of cellular senescence signature on ageing research. *Ageing Res Rev* 10:146–152.

Smith, J.R., Pereira-Smith, O. and P.I. Good. 1977. Colony size distribution as a measure of age in cultured human cells. A brief note. *Mech Ageing Dev* 6:283–286.

Smith, J.R., Venable, S., Roberts, T.W., Metter, E.J., Monticone, R. and E.L. Schneider. 2002. Relationship between *in vivo* age and *in vitro* aging: Assessment of 669 cell cultures derived from members of the Baltimore Longitudinal Study of Aging. *J Gerontol A Biol Sci Med Sci* 57:B239–B246.

Stoddart, M.J. 2011. Cell viability assays: Introduction. *Methods Mol Biol* 740:1–6.

Swim, H.E. and R.F. Parker. 1957. Culture characteristics of human fibroblasts propagated serially. *Am J Hyg* 66:235–243.

Toussaint, O., Dumont, P., Remacle, J. et al. 2002. Stress-induced premature senescence or stress-induced senescence-like phenotype: One *in vivo* reality, two possible definitions? *Sci World J* 2:230–247.

Untergasser, G., Gander, R., Rumpold, K., Heinrich, E., Plas, E. and P. Berger. 2003. TGF-beta cytokines increase senescence-associated beta-galactosidase activity in human prostate basal cells by supporting differentiation processes, but not cellular senescence. *Exp Gerontol* 38:1179–1188.

Ushakov, V.L., Gusev, M.V. and A.N. Khokhlov. 1992. Is it worth to study mechanisms of the cellular aging in experiments with the blue-green algae? A critical review, part 1. *Vestn Mosk Univ Ser 16 Biol* 1:3–15 (in Russian).

Vilenchik, M.M., Khokhlov, A.N. and K.N. Grinberg. 1981. Study of spontaneous DNA lesions and DNA repair in human diploid fibroblasts aged *in vitro* and *in vivo*. *Stud Biophys* 85:53–54.

Vladimirova, I.V., Shilovsky, G.A., Khokhlov, A.N. and S.I. Shram. 2012. "Age-related" changes of the poly(ADPribosyl)ation system in cultured Chinese hamster cells. In *Visualizing of Senescent Cells in Vitro and in Vivo, Programme and Abstracts*, Warsaw, Poland, December 15–16, p. 108.

Wei, L., Li, Y., He, J. and A.N. Khokhlov. 2012. Teaching the cell biology of aging at the Harbin Institute of Technology and Moscow State University. *Moscow Univ Biol Sci Bull* 67:13–16.

Weismann, A. 1885. *Die Kontinuitat des Keimplasmas als Grundlage einer Theorie der Vererbung (The Continuity of the Germ Plasm as the Basis of a Theory of Heredity)*. Jena: G. Fisher Ferlag.

Weismann, A. 1892. *Das Keimplasma. Eine Theorie der Vererbung (The Germ Plasm. A Theory of Heredity)*. Jena: G. Fisher Ferlag.

Yablonskaya, O.I., Ryndina, T.S., Voeikov, V.L. and A.N. Khokhlov. 2013. A paradoxical effect of hydrated C_{60}-fullerene at an ultralow concentration on the viability and aging of cultured Chinese hamster cells. *Moscow Univ Biol Sci Bull* 68:63–68.

Yegorov, Y.E., Akimov, S.S., Hass, R., Zelenina, V. and I.A. Prudovsky. 1998. Endogenous beta-galactosidase activity in continuously nonproliferating cells. *Exp Cell Res* 243:207–211.

Yucel, E.B., Eraslan, S. and K.O. Ulgen. 2014. The impact of medium acidity on the chronological life span of *Saccharomyces cerevisiae*—lipids, signaling cascades, mitochondrial and vacuolar functions. *FEBS J* 281:1281–1303.

Section VIII

Aging in Caenorhabditis elegans

33 *Caenorhabditis elegans* Aging is Associated with a Decline in Proteostasis

Elise A. Kikis

CONTENTS

CAENORHABDITIS ELEGANS AS A MODEL SYSTEM TO STUDY AGING

C. elegans has an average life span of 21 days under typical laboratory conditions, and has for many decades been an important model system for the study of genetic and environmental factors that affect aging. In 1976, Klass and Hirch published a paper in *Nature* describing an alternative developmental program in *C. elegans* called the dauer stage that allows for nearly indefinite longevity with the absence of normal aging biomarkers [1]. However, this dauer stage, while fascinating, is a worm-specific phenomenon and arguably not relevant to human aging.

Nonetheless, the study of *C. elegans* aging ballooned in the early 1990s due to a series of papers suggesting that conserved genetic pathways shorten the life span of wild-type animals. One such landmark paper was published in 1993 by Cynthia Kenyon and fittingly titled, "A *C. elegans* mutant that lives twice as long as wild type" [2]. In that paper, Kenyon et al. showed that disrupting the *C. elegans* ortholog of the insulin receptor (daf-2) significantly lengthened life span. This, and other studies showing similar effects when components of the insulin/IGF (insulin-like growth factor) signaling pathway were disrupted [3], provided the first evidence that life span is genetically controlled. Specifically, when the kinase activity of IP3K (encoded by *age-1* in *C. elegans*) is activated in wild-type animals via the downregulation of DAF-2 (Dauer Formation), the resultant signaling cascade ultimately represses the forkhead transcription factor DAF-16 [4,5]. In animals harboring either *age-1* or *daf-2* mutations, DAF-16 is held in a constitutively de-repressed state leading to life span extension. Thus, mutations in *daf-16* itself shorten life span relative to that of wild-type *C. elegans*. The life span extending insulin/IGF signaling pathway is shown (Figure 33.1 and Reference [6]).

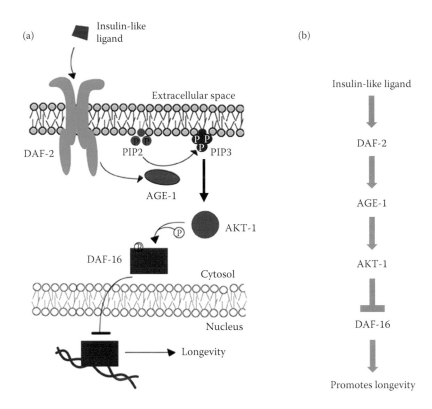

FIGURE 33.1 Longevity and the *C. elegans* insulin signaling pathway. (a) Cartoon representation of the insulin signaling pathway. The insulin-like ligand triggers a signal transduction cascade upon binding the receptor daf-2. The cascade culminates with the phosphorylation and inhibition of the transcription factor daf-16 that would otherwise induce life span-extending gene expression. (b) Schematic representation of the *C. elegans* insulin signaling pathway.

NEURODEGENERATIVE DISEASES AND AGING

Human aging is often marked by the late onset of neurodegenerative diseases. Most have misfolding at their core and are thereby referred to as diseases of protein conformation [7]. Such diseases include Parkinson's disease (PD), amyotrophic lateral sclerosis (ALS), prion disease, Alzheimer's disease (AD) and polyglutamine-expansion diseases including Huntington's disease (HD), Kennedy's disease, and spinocerebellar ataxias (SCAs). All of the above are invariably characterized by the deposition of misfolded protein in the form of aggregates, inclusions bodies, or plaques. Such protein misfolding is toxic, leading to "toxic gain-of-function" phenotypes [8,9]. Furthermore, each of these diseases is progressive, age-dependent, and usually fatal.

Support for protein misfolding being the molecular underpinning of these diseases comes from a substantial body of work employing transgenic model systems such as *Saccharomyces cerevisiae*, *C. elegans*, and *Drosophila* [10–15], in which many molecular, cellular, and behavioral phenotypes associated with neurodegenerative disease have been recapitulated and studied. These invertebrate models for protein folding diseases have led to the discovery of components of a proteostasis network that modulates protein folding under normal conditions and in the presence of disease-associated proteins. However, the focus of this chapter will be on the work with *C. elegans* that has demonstrated that the ability of cells and organisms to maintain proteostasis is significantly compromised by aging and other processes that increase the overall load of misfolded protein.

C. ELEGANS AS A MODEL SYSTEM TO STUDY PROTEIN MISFOLDING

C. elegans has a fully sequenced genome, a short lifecycle, relatively simple body plan, and straight-forward transgenic technology, making it a powerful model system to address a wide array of sci-entific questions. As a testament to the power of its genetics, several recent Nobel Prizes have been awarded based on work done with *C. elegans*. Brenner, Sulston, and Horvitz were awarded a Nobel Prize in 2002 for their work on apoptosis in *C. elegans* [16,17]. Fire and Mello were awarded a Nobel Prize in 2006 for their work in *C. elegans* on the mechanisms of gene knockdown by RNAi [18]. Finally, Chalfie, who did much work with *C. elegans* was awarded a Nobel Prize in 2008 for Green Fluorescent Protein (GFP) [19].

While making mutants of *C. elegans* is straightforward using conventional methods such as treatment with chemical mutagens like ethyl methanesulfonate (EMS), gene knockdown by RNAi is extraordinarily easy and effective, often making it the method of choice, especially for the testing of candidate genes. A genome-wide RNAi library is commercially available and includes cloned fragments of every gene transformed into *Escherichia coli* for feeding. *C. elegans* will knockdown target gene expression globally when exposed to double-stranded RNA even when that RNA is simply ingested. Thus, *C. elegans,* which typically feed on bacteria, will knockdown the expression of a target gene when the ingested bacteria express the corresponding double-stranded RNA [20].

Expressing transgenes in *C. elegans* is done by injecting gene constructs into the gonads of young adults and identifying any offspring carrying the transgene. These transgenes are usually not inte-grated into the genome until gamma irradiation-induced random integration, although "extrachro-mosomal arrays" can be reliably inherited. *C. elegans* studies in which human disease-associated proteins were expressed in different *C. elegans* tissues have allowed cell-type specificity with respect to aggregation and toxicity to be examined genetically as described later in this chapter.

MODELING POLYGLUTAMINE PROTEIN AGGREGATION AND TOXICITY IN C. ELEGANS

Polyglutamine (polyQ) disorders are a family of nine different neurodegenerative disorders all of which are caused by an expansion of a polyQ tract, albeit in different proteins, each encoded by a cytosine-adenine-guanine (CAG) trinucleotide repeat in the corresponding gene. As such, these are heritable genetic disorders and include HD, Machado–Joseph disease (MJD), spinobulbar muscular atrophy (SBMA), dentatorubral–pallidoluysian atrophy (DRPLA), and five SCAs. For all of these diseases, the polyQ expansion destabilizes the affected protein thereby disrupting the thermody-namics of folding and causing protein misfolding and aggregation. Outside of the polyQ tract itself, each of the affected proteins in each of these diseases share no common sequences or functions. The key to disease is thus the polyQ tract, such that the age of disease onset and symptom severity is inversely proportional to the length of the polyQ tract [21–23].

In 1999, the laboratory of Ann Hart published the first *C. elegans* polyQ model [11]. They expressed a polyQ-containing N-terminal fragment of the human htt protein in sensory neurons. To model polyQ length-dependent effects, they generated either wild-type (2 or 23) polyQ repeat lengths or disease-associated (95 or 150) polyQ lengths and found that the longer polyQ tracts caused neurodegeneration [11].

The fact that expanded polyQ tracts seem to be toxic in many different protein contexts led to the hypothesis that they may be inherently toxic and therefore sufficient to cause disease even on their own. To test this, the laboratory of Richard Morimoto generated naked polyQ tracts fused to the N-terminus of yellow fluorescent protein (YFP) for visualization and expressed them in body wall muscle cells [24] or neurons [25] under the control of tissue-specific promoters. They found that polyQ alone aggregates in a length-dependent manner, with the threshold for aggregation being almost identical to that observed in patients with HD. Aggregation correlated with a marked decline in motility, indicating that expanded polyQ is toxic to *C. elegans*. In a parallel study, these same

TABLE 33.1

C. elegans **PolyQ Models**

Expressed Protein	Tissue/Cell Type	Phenotypes	References
PolyQ-YFP	Body wall muscle cells	PolyQ length-dependent and age-dependent aggregation and toxicity	Morley et al. [24]
PolyQ-YFP	All neurons	PolyQ length-dependent aggregation and toxicity. No age-dependent changes observed	Brignull et al. [25]
PolyQ-YFP	Intestines	Disrupted proteostasis during infection	Mohri-Shiomi et al. [26]
Htt171	Sensory neurons	PolyQ length-dependent aggregation and toxicity. Age-dependent aggregation of Q150	Faber et al. [11]
GFP-Htt exon 1	Body wall muscle cells	PolyQ length-dependent aggregation and toxicity	Wang et al. [29]
Full-length ataxin-3	Neurons	PolyQ length-dependent and age-dependent aggregation and toxicity	Teixeira-Castro et al. [28]
Ataxin-3 CT	Neurons	PolyQ length-dependent and age-depdendent aggregation and toxicity	Teixeira-Castro et al. [28]
Ataxin-3 CT	Body wall muscle cells	PolyQ length-dependent aggregation and toxicity. No age-dependent changes observed	Christie et al. [27]
Htt513	Body wall muscle cells	PolyQ length-dependent aggregation and toxicity. No age-dependent changes observed	Lee et al. [94]

polyQ-YFP peptides were expressed in *C. elegans* intestinal cells where they aggregated in a polyQ length-dependent manner, with aggregation being exacerbated in response to infection. This indicated that infection likely causes proteotoxic stress to which polyQ folding is especially sensitive [26]. Together, these models provide evidence that expanded polyQ tracts are toxic in several cellular contexts.

That is not to say, however, that the protein context in which a polyQ tract is embedded has no effect on aggregation and toxicity. In fact, a polyQ-containing fragment of the ataxin-3 protein (ataxin-3 CT) expressed in body wall muscle cells [27] or neurons [28], resulted in aggregation and toxicity. However, age-dependent changes in aggregation were observed for some models [24,25] but not others [27]. A list of polyQ proteins expressed in *C. elegans* is shown (Table 33.1).

MODELING Aβ AGGREGATION AND TOXICITY IN *C. ELEGANS*

Thirty percent of the human population can expect to suffer from AD, assuming a life span of greater than 85 years. This makes AD the number one leading cause of dementia around the world. The genetics underlying AD are not as simple and straightforward as the monogenic polyQ disorders described above. For one thing, AD is polygenic, being characterized by several different misfolded proteins, including the amyloid β (Aβ) peptide and tau. Aβ is a small peptide that is derived from the larger amyloid precursor protein (APP) following certain proteolytic cleavages. APP is a single-pass transmembrane protein that has a large domain on the extracellular side of the plasma membrane. Despite all the research done to date on AD, not much is known about the normal function of APP, although some evidence suggests that it may be involved in a gene regulatory pathway similar to that of *Notch* [30].

Under normal, non-disease, conditions, APP is sequentially cleaved by two proteases, α-secretase and γ-secretase at the cell surface. This usually leads to the formation of a nontoxic peptide. However, certain mutant forms of APP exist that undergo noncanonical cleavage. These APP mutants are incorporated into vesicles containing both γ-secretase and a different secretase known as β-secretase or BACE1. Together, they cleave APP in such a way as to produce the

TABLE 33.2

C. elegans Aβ Models

Expressed Protein	Tissue/Cell Type	Phenotypes	References
$A\beta_{3-42}$	Body wall muscle cells	Paralysis	Link et al. [38]
$A\beta_{3-42}$	Neurons	Paralysis	Wu et al. [39]
$A\beta_{1-42}$	Body wall muscle cells	Toxic only with elevated temperature	McColl et al. [40]

disease-associated $A\beta_{1-42}$ peptide. $A\beta_{1-42}$ is then secreted and forms amyloid plaques in the extracellular space, characteristic of AD [31,32].

$A\beta_{1-42}$ aggregates in a concentration-dependent manner. We see the effect of this in a pronounced way in patients with Down syndrome. The trisomy 21 that is the genetic cause of Down syndrome leads to APP overexpression. In turn, Aβ deposits occur earlier in Down syndrome patients than those with only two copies of chromosome 21 [33–35]. Similarly, fragile X syndrome (a trinucleotide repeat disorder) causes APP gene expression to be upregulated, leading to increased Aβ production and therefore impaired neurodevelopment [36].

From a genetic standpoint, those otherwise healthy individuals who are most at risk for developing AD at some point in life are those carrying the APOE4 allele of the gene that encodes apolipoprotein isoform 4. This protein isoform leads to an increased amount of extracellular Aβ deposition, likely by initiating an altered gene expression program [37].

As shown in Table 33.2, two different forms of the Aβ peptide have been expressed in *C. elegans*. Namely, an N-terminally truncated $A\beta_{3-42}$ was expressed in a heat-inducible manner in body wall muscle cells [38] and the same peptide was also expressed in neurons [39]. A full length disease-associated $A\beta_{1-42}$ peptide was more recently expressed in *C. elegans* body wall muscle cells where it was shown to be highly toxic only under high heat [40]. The fact that phenotypes were seen only with elevated temperature seems to suggest that temperature and $A\beta_{1-42}$ together may drain cells of the same resources, such as molecular chaperones, setting up too great a competition. Alternatively, elevated temperature may affect the folding of $A\beta_{1-42}$ peptide such that it adopts an especially toxic conformation.

MODELING TAU, A-SYN, SUPEROXIDE DISMUTASE 1, AND TDP-43 AGGREGATION AND TOXICITY IN *C. ELEGANS*

Hyperphosphorylated tau is found in the brains of AD patients as well as those suffering from other tauopathies such as FTDL-17, which is an autosomal dominant form of frontotemporal lobar degeneration (FTLD) caused by any one of a number of mutations in the tau protein that is encoded by a gene on chromosome 17 [41,42]. In its normal conformation, tau interacts with the lateral edges of microtubules thereby preventing two microtubules from packing too tightly with each other. While mutant tau is unable to carry out this function, FTDL-17 is inherited as an autosomal dominant disorder thought to be the result of toxic properties associated with the mutant tau rather than from loss of function [43]. This is supported by evidence that overexpression of tau also leads to FTDL-17.

In summary, mutant tau misfolds, becomes hyperphosphorylated, and aggregates in the human brain. This ultimately results in the formation of the characteristic neurofibrillary tangles [43]. To model tau aggregation and toxicity, the human tau protein was expressed in *C. elegans* neurons where it led to neurodegeneration and impaired neurotransmission [44].

Unlike FTDL-17, other types of FTDL are sporadic rather than heritable. Such is the case of the most common form of FTDL called FTDL-U. FTDL-U patients present with ubiquitinated aggregates [45] comprising misfolded TDP-43 [46]. TDP-43 contains two RNA binding motifs that may play a role in RNA splicing [47]. This is very similar to what is seen in patients with sporadic ALS

TABLE 33.3
Tau, TDP-43, SOD1, α-Synculein

Expressed Protein	Tissue/Cell Type	Phenotypes	References
Tau	Neurons	Neurodegeneation and impaired neurotransmission	Kraemer et al. [44]
TDP-43	Neurons	Impaired motility	Ash et al. [48]
SOD1	Body wall muscle cells	Aggregation and mild motor defects	Gidalevitz et al. [49]
α-Synuclein	Body wall muscle cells	Neuronal loss, motor defects	Lakso et al. [53]
α-Synuclein	Neurons	Age-dependent aggregation	van Ham et al. [54]

[46], suggesting that sporadic ALS and FTDL-U may have nearly identical underlying pathologies. Human TDP-43 was expressed in *C. elegans* neurons and shown to be neurotoxic [48].

Not all forms of ALS are sporadic. Rare familial forms of ALS are caused by any one of a number of mutations in superoxide dismutase 1 (SOD1). When various SOD1 mutants were expressed in *C. elegans* body wall muscle cells, they aggregated and disrupted muscle cell function, with some mutation-specific differences with respect to interaction with the cellular environment [49].

PD is also caused by protein misfolding. It is a progressive neurodegenerative disorder in which patients suffer from tremors, rigidity, and loss of dopaminergic neurons in the substantia nigra brain region. At least three different mutations in the human α-synuclein gene are known to be genetic determinants for PD. α-synuclein is a synaptic protein whose mutant forms aggregate and comprise the Lewy bodies that are typically found in PD brains [50–52]. To model α-synuclein aggregation and toxicity, it was expressed in the neurons [53] and body wall muscle cells [54] of *C. elegans*, resulting in neuronal dysfunction and motor loss in both tissues [53,54] (Table 33.3).

Interestingly, α-synuclein has recently been shown to migrate between cells in a manner reminiscent of infectious prion proteins. In one study, normal neurons were grafted to PD brains. After a period of time, the grafted neurons picked up α-synuclein, which in turn formed the characteristic Lewy body type of aggregates [55–57].

MODELING PRION PROTEINS IN *C. ELEGANS*

In certain rare instances, neurodegenerative disease can be caused by infectious proteins known as prion proteins (PrP). For example, Creutzfeldt–Jacob disease is a prion disorder also known as a transmissible spongiform encephalopathy (TSE). When PrP adopts an alternative, self-propagating conformation, PrP^{Sc}, disease symptoms may eventually appear [58–60]. Interestingly, like the neurodegenerative diseases described above, aging is a significant risk factor for TSEs, meaning that exposure to PrP^{Sc} may not lead to symptoms until late in life, usually after 60 years of age. Most cases of TSE are sporadic, meaning that a normal PrP-encoding gene mutates into a form that encodes the PrP^{Sc} variant during aging. Other cases are familial, such that a faulty PrP^{Sc} gene is inherited. Other cases of TSE in people occur following the ingestion of beef from cattle exposed to bovine spongiform encephalopathy, also known as mad cow disease. Once the PrP^{Sc} prion conformation forms, PrP^{Sc} itself is able to seed the misfolding of normal PrP protein into the PrP^{Sc} conformation and nucleate protein aggregation.

Research conducted over the past several years has revealed that the mechanism of PrP^{Sc} action, specifically its infectious nature, may not be as different from the other misfolded and aggregation-prone proteins as was originally thought [61,62]. For example, misfolded α-synuclein from the Lewy bodies of PD brains can move between tissues in much the same way as has been described for PrP^{Sc}. Likewise, misfolded Aβ can trigger further Aβ aggregation via a seeding mechanism. When Aβ was injected into mice overexpressing APP, those mice readily formed plaques typically

TABLE 33.4
Prion Proteins

Expressed Protein	Tissue/Cell Type	Phenotypes	References
Sup35	Body wall muscle cells	Aggregation and toxicity	Nussbaum-Krammer et al. [65]
MoPrPSC	Body wall muscle cells	Phenotypes ranged from completely normal to complete paralysis	Park et al. [66]

associated with AD [63]. Even aggregates of mutant htt protein associated with HD can be taken up by tissue culture cells where they seed the aggregation of normal, endogenous, htt protein [64].

Much of what we know about prion proteins comes from work done on yeast cells that have their own endogenous prion proteins. The most well studied of the yeast prions is a domain of the Sup35 protein [60]. This Sup35 prion domain has been expressed in *C. elegans* body wall muscle cells to model its misfolding and toxicity in a transgenetic system, and, importantly, in a multicellular organism. Like the human disease-associated proteins that are toxic not only to the human brain, but also to *C. elegans* neurons and body wall muscle cells, the Sup35 prion domain aggregated in *C. elegans* body wall muscle cells and was toxic, as determined by motility defects in the transgenic animals [65]. The mouse ortholog of the human PrP (MoPrP) was likewise expressed in *C. elegans* body wall muscle cells where it sometimes resulted in total paralysis of the affected animal [66]. Table 33.4 lists the *C. elegans* prion models and describes the resultant phenotypes with respect to aggregation and toxicity.

ENDOGENOUS PROTEOSTASIS SENSORS

The minimalistic approach of expressing disease-associated proteins in *C. elegans* has its advantages, but is arguably a poor way to study disease as too many factors typically associated with neurodegeneration are missing in model invertebrates. Thus, the models so far described in this chapter cannot be considered to be disease models per se. Instead, they are models of protein folding. They can even be used as sensors of the folding environment because when overall misfolding increases, aggregation of these proteins, such as polyQ alone, increases and can be easily visualized [67]. Such folding sensors have made it possible to probe the protein folding environment under a variety of biological conditions, thereby providing insight into the contexts in which proteins fold and the large-scale challenges cells face to combat stress placed on protein folding. Other important folding sensors include metastable proteins that have been used to obtain constant feedback on how much stress the *C. elegans* protein folding machinery is under. Such sensors include temperature-sensitive variants of certain endogenous proteins. Under conditions of little to no folding stress, the temperature-sensitive proteins fold normally at lower, permissive, temperatures, but misfold and result in mutant phenotypes at higher, restrictive, temperatures. However, as protein folding stress increases, misfolding occurs even at permissive temperatures, resulting in mutant phenotypes [67].

Interestingly, the expression of expanded polyQ in the genetic background is a source of sufficient folding stress to cause temperature-sensitive proteins to misfold [49,67]. Such misfolding also occurred in a wild-type genetic background early in aging, indicating that aging is accompanied by an increase in proteotoxic stress [68].

USING *C. ELEGANS* TO IDENTIFY COMPONENTS OF THE PROTEOSTASIS NETWORK

The proteostasis network is defined as the total complement of cellular factors that are required to maintain proteostasis, or more specifically, to maintain a healthy balance between protein synthesis, folding, and degradation to ensure a healthy proteome. In the category of protein folding regulators, molecular chaperones play a crucial role in that the heat shock response (HSR) and the unfolded protein response (UPR) protect cells from heat-induced protein damage [69,70]. The HSR is mediated by the transcription factor heat shock factor 1 (HSF1), which triggers a rapid increase in the transcription and translation of molecular chaperones. In parallel, global translation rates decline, allowing the heat-shocked cells to switch away from protein synthesis and toward protein refolding or degradation [71]. While we typically think about the molecular chaperones acting under conditions of acute proteotoxic stress, they also have a housekeeping function in that HSF1 [72,73], UPR signaling components [74–76], and some heat-inducible molecular chaperones are actually encoded by essential genes that function early in development, rendering mutants embryonic lethal. As such, these stress-inducible pathways must be required to maintain proteostasis under normal conditions as well as conditions of acute stress.

Nonetheless, when molecular chaperones are overexpressed, cell growth is negatively impacted [77,78] and can sometimes contribute to cancer [79–81]. Therefore, it seems that the proteostasis network must be precisely regulated to ensure the health of the organism. Consistent with this, a significant body of evidence suggests that dysregulation of proteostasis during aging is one of the main causes of cellular dysfunction and the onset of age-related neurodegenerative disease.

Clearly, identifying the full complement of proteins that together represent the proteostasis network will lead to a better understand of how proteostasis is regulated. Many of the models of protein misfolding described in this chapter have been used as tools for this very purpose.

Because of the aforementioned availability of genome-wide RNAi libraries and the ease at which such libraries can be screened for gene knockdown phenotypes in *C. elegans*, this has usually been the method of choice for the identification of proteostasis network components. One such study involved screening the library for gene inactivations that led to increased or early aggregation of polyQ-YFP in body wall muscle cells [82]. Not surprisingly, this screen revealed that molecular chaperones are required to mitigate polyQ protein misfolding. Interestingly, not all molecular chaperones were identified. Instead, the chaperonin containing TCP-1 (CCT) molecular chaperone complex seems to play a predominant role in this process. Other genes were identified that function at various points along the protein maturation pathway. Specifically, 186 proteins in total were identified, and they included transcription factors, splicing factors, protein elongation factors, and even some proteins involved in protein degradation.

To expand on this preliminary 186 member proteostasis network, a complementary screen was performed taking the opposite approach of identifying genes that when knocked down by RNAi a caused a *decrease* in both polyQ and mutant SOD1 aggregation [83]. As such, the first screen identified suppressors of aggregation and this second screen identified enhancers. Such enhancers included additional proteins involved in metabolism and RNA processing [83].

A third genome-wide screen was performed, this time taking an HSR-centric approach, as the HSR seems to be central to proteostasis regulation. A reporter construct was employed in which GFP expression was placed under the control of a heat shock promoter. Gene inactivations leading to either an HSR in the absence of heat shock or a lack of HSR under heat shock conditions were identified [84]. Interestingly, several of the same proteins identified in previous screens were also identified here, including the CCT complex, and certain proteasomal subunits [82,84]. Together, these data provide support for the hypothesis that these two multiprotein complexes play outsized roles in proteostasis regulation.

In addition to genome-wide screens, a handful of targeted RNAi studies have been performed. For example, a candidate screen for molecular chaperons was performed on temperature-sensitive mutants as folding sensors. Molecular chaperones were identified that are specifically required for *C. elegans* muscle homeostasis [85]. Specifically, Hsp90 and its co-chaperones Sti-1, Aha1, and Cep23 were identified [85].

Similarly, another targeted study was performed to identify the entire complement of molecular chaperones that are required to maintain proteostasis even when the misfolded protein load is abnormally high due to the presence of neurodegenerative disease-associated proteins. All 332 known or presumptive molecular chaperones were targeted for gene knock down in *C. elegans* by RNAi. This was done both in $A\beta_{3\text{-}42}$ animals and in animals expressing expanded polyQ-YFP in body wall muscle cells. As evidence that the collection of proteostasis regulators identified in screens described here are starting to approach saturation, the CCT chaperonin complex as well as Hsp90, two Hsp90 co-chaperones (sti-1 and cdc-37), and two DnaJ chaperones (dnj-8 and dnj-12) that likely act as Hsp70 co-chaperones were identified [86]. Nonetheless, additional targeted screens continue to be performed, mostly to determine whether specific disease-associated proteins interact with unique subsets of the proteostasis machinery.

Furthermore, proteomics studies have been performed as an important complement to the genetic studies described above. For example, proteins that physically interact with polyQ-YFP were isolated by immunoprecipitation and identified by mass spectroscopy [87]. Glutamine and asparagine-rich proteins known as pqn proteins resemble polyQ proteins and seem to bind to polyQ aggregates with very high affinity. In addition, a non-pqn protein, CRMP3-associated molecule (CRAM-1), seems to be a primitive molecular chaperone that binds with slightly lower affinity to polyQ aggregates but appears to have an important role in blocking the turnover of polyQ aggregates [87].

Together, the screens described here have significantly expanded our understanding of how deep into the cell the proteostasis network extends. The proteostasis network seems to operate at all the points along the pathway for protein synthesis and maturation. Specifically, it includes transcriptional regulators, splicing factors, the protein synthesis machinery including ribosomal subunits, molecular chaperones, and factors involved in protein turnover (Figure 33.2).

HOW GENETIC VARIATION IMPACTS PROTEOSTASIS

Early studies of HD revealed that different individuals responded to the same HD-associated mutation in different ways. While a longer polyQ tract is predictive of an earlier age of onset of disease, a significant variation in actual age of onset was observed for people with the same polyQ lengths [23]. This provided some of the first evidence that genetic background significantly impacts the toxicity of the mutant htt protein. To more thoroughly study this phenomenon, a large cohort of HD sufferers in Venezuela was examined, leading to the finding that a range of ages at which disease symptoms first appear is possible for the same polyQ tract lengths (Figure 33.3).

Similar effects of genetic background have been described in *C. elegans* models of protein folding. The temperature-sensitive alleles, described above as sensors for the folding environment, harbor mutations in endogenous genes. As such, they can be considered part of the spectrum of possible genetic variation in these otherwise inbred laboratory animals. Because the temperature-sensitive alleles encode proteins that are metastable with respect to their folding, we can think of their genetic background as contributing more misfolded protein that needs to be dealt with by the proteostasis network. PolyQ was more prone to aggregate in this genetic background than it was in the wild-type background [67]. The data suggested that polyQ misfolding in the temperature-sensitive background was a direct consequence of the proteostasis network becoming overwhelmed by the total amount of misfolded protein. This was consistent with a later study examining the effect of temperature-sensitive mutations on *C. elegans* expressing SOD1 in body wall muscle cells [49,67]. Together, these data provide strong evidence that the proteostasis network is not very robust [88].

The possibility that these findings were the result of an artifact of temperature-sensitive mutations, rather than indicative of genetic background having such a strong influence on proteostasis, needed to be ruled out. To address this, four different wild isolates of *C. elegans* were obtained and the polyQ-YFP transgene was introduced via introgression to produce a series of recombinant inbred (RI) lines [89]. Consistent with the findings from temperature-sensitive alleles, changing even the wild-type background profoundly affected the aggregation propensity of polyQ-YFP.

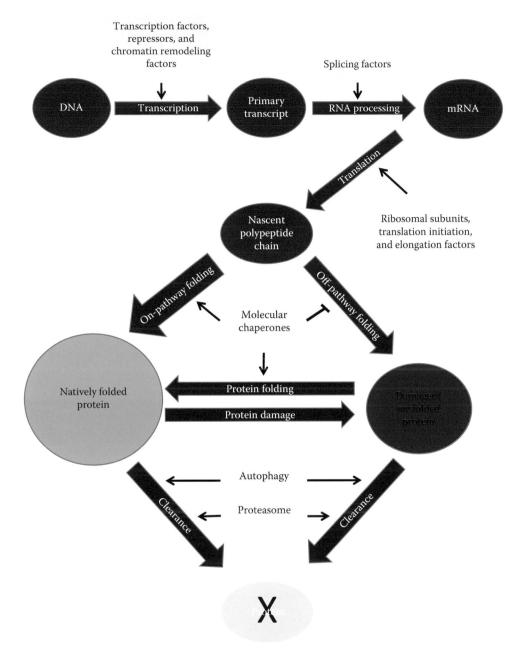

FIGURE 33.2 The proteostasis network. The pathways involved in protein metabolism are depicted. Cellular processes are indicated with blue arrows and metabolic intermediates are indicated in purple circles. Functional categories comprising the proteostasis network are shown in boxes, with → representing positive regulation and —| representing negative regulation. The thick arrow for on-pathway folding indicates that under normal conditions on-pathway folding predominates.

This confirmed that the proteostasis network is indeed not very robust, which may have profound consequence not only for age-related neurodegenerative disease, but also for normal healthy aging. Specifically, as the overall load of misfolded proteins increases, perhaps due to accumulated damage during aging, the ability to maintain proteostasis declines—and this is magnified when disease-associated proteins are present.

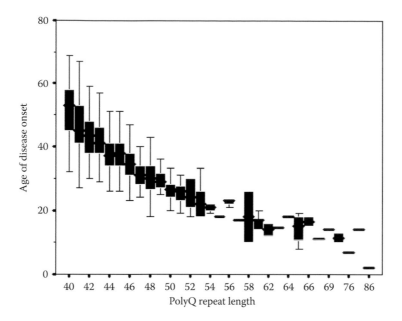

FIGURE 33.3 Age of Huntington disease onset as a function of polyQ repeat length. (Adapted from Wexler et al. 2004. *Proc. Natl. Acad. Sci.*)

CELLS LOSE THEIR ABILITY TO MAINTAIN PROTEOSTASIS DURING AGING

Some of the first evidence that an aging proteome impacts the overall protein folding environment and the ability of cells and organisms to maintain proteostasis came from an important *C. elegans* study [68]. The temperature-sensitive mutations were used to detect age-dependent changes in proteostasis. Even in an otherwise wild-type background, these temperature-sensitive mutant proteins only folded correctly at the permissive temperature in young animals. As animals reached even the earliest stages of adulthood (animals were still within their reproductive prime), the temperature-sensitive proteins started to misfold, leading to their characteristic phenotypes even at what is usually a permissive temperature [68]. This means that something, such as accumulated protein damage, reached a critical point right at the transition to adulthood.

To determine what exactly was happening during aging that led to a proteostasis decline, the specific physiological processes associated with the transition to aging were examined more closely. Consistent with the onset of reproduction triggering proteostasis decline in *C. elegans*, germ line stem cell arrest restored proteostasis to its preadult level of robustness [90]. Because this rescue was shown to depend on stress responsive signaling pathways such as the HSR and the insulin-like signaling pathway, the authors concluded that as germ line stem cells proliferate, they send a signal to the rest of the body that in turn triggers the suppression of stress responses and thus a decline in the buffering capacity of the proteostasis network of somatic cells [90]. This means that the ability of the proteostasis network to buffer against protein misfolding is so limited that even normal physiological processes can be overwhelming.

Age-dependent changes in the integrity of the proteome were shown more directly using a proteomics approach in which misfolded proteins were identified over an aging time course in wild-type *C. elegans* [91]. Older animals had more misfolded proteins than younger animals. A decline in proteostasis triggered by accumulated protein damage during aging seems to be a trigger for age-dependent onset of neurodegenerative disease. Therefore, the ability of metastable, but otherwise wild type, *C. elegans* proteins to trigger aggregation of polyQ-YFP, is consistent with our understanding of the mechanism of disease progression. Likewise, when some of the proteins identified by proteomics were overexpressed in the polyQ-YFP background, polyQ-YFP aggregation increased [91].

Later studies further supported these earlier findings. Specifically, another large-scale proteomics study involving an analysis of 5000 proteins revealed that at least one-third of them change in abundance during aging, resulting in a stoichiometric disruption for many crucial multiprotein complexes such as the ribosome [92]. Those proteins that were upregulated during aging were often found to aggregate, especially in older animals [92]. In addition, the HSR, which, as discussed above, forms the core of the proteostasis network, was shown to undergo an age-dependent decline [68,93].

CONCLUSIONS

C. elegans models of protein misfolding have contributed greatly to our current understanding of the proteostasis network, its buffering capacity, and its regulation. Many *C. elegans* models of protein misfolding have been developed, and have contributed to our understanding that the buffering capacity of the proteostasis network is limited, likely contributing to the onset of neurodegenerative disease symptoms. Furthermore, this limitation seems to be most pronounced during aging when accumulated protein damage, and perhaps even signaling events associated with age-dependent physiological events, triggers a disruption in core components of the proteostasis network. The normal decline in proteostasis that may be a part of what is usually considered to be healthy aging likely exacerbates neurodegenerative disease-associated proteotoxicity, leading to an age-dependent onset of symptoms.

REFERENCES

1. Klass M, Hirsh D. 1976. Non-ageing developmental variant of *Caenorhabditis elegans*. *Nature* 260: 523–525.
2. Kenyon C, Chang J, Gensch E, Rudner A, Tabtiang R. 1993. A *C. elegans* mutant that lives twice as long as wild type. *Nature* 366: 461–464.
3. Johnson TE. 1990. Increased life-span of age-1 mutants in *Caenorhabditis elegans* and lower Gompertz rate of aging. *Science* 249: 908–912.
4. Morris JZ, Tissenbaum HA, Ruvkun G. 1996. A phosphatidylinositol-3-OH kinase family member regulating longevity and diapause in *Caenorhabditis elegans*. *Nature* 382: 536–539.
5. Lin K, Dorman JB, Rodan A, Kenyon C. 1997. daf-16: An HNF-3/forkhead family member that can function to double the life-span of *Caenorhabditis elegans*. *Science* 278: 1319–1322.
6. Scerbak C, Vayndorf EM, Parker JA, Neri C, Driscoll M et al. 2014. Insulin signaling in the aging of healthy and proteotoxically stressed mechanosensory neurons. *Front Genet* 5: 212.
7. Evans DL, Marshall CJ, Christey PB, Carrell RW. 1995. Heparin binding site, conformational change, and activation of antithrombin. *Biochemistry* 34: 3478.
8. Kakizuka A. 1998. Protein precipitation: A common etiology in neurodegenerative disorders? *Trends Genet* 14: 396–402.
9. Chiti F, Dobson CM. 2006. Protein misfolding, functional amyloid, and human disease. *Annu Rev Biochem* 75: 333–366.
10. Warrick JM, Paulson HL, Gray-Board GL, Bui QT, Fischbeck KH et al. 1998. Expanded polyglutamine protein forms nuclear inclusions and causes neural degeneration in *Drosophila*. *Cell* 93: 939–949.
11. Faber PW, Alter JR, MacDonald ME, Hart AC. 1999. Polyglutamine-mediated dysfunction and apoptotic death of a *Caenorhabditis elegans* sensory neuron. *Proc Natl Acad Sci USA* 96: 179–184.
12. Feany MB, Bender WW. 2000. A *Drosophila* model of Parkinson's disease. *Nature* 404: 394–398.
13. Krobitsch S, Lindquist S. 2000. Aggregation of huntingtin in yeast varies with the length of the polyglutamine expansion and the expression of chaperone proteins. *Proc Natl Acad Sci USA* 97: 1589–1594.
14. Satyal SH, Schmidt E, Kitagawa K, Sondheimer N, Lindquist S et al. 2000. Polyglutamine aggregates alter protein folding homeostasis in *Caenorhabditis elegans*. *Proc Natl Acad Sci USA* 97: 5750–5755.
15. Parker JA, Connolly JB, Wellington C, Hayden M, Dausset J et al. 2001. Expanded polyglutamines in *Caenorhabditis elegans* cause axonal abnormalities and severe dysfunction of PLM mechanosensory neurons without cell death. *Proc Natl Acad Sci USA* 98: 13318–13323.

16. Sulston JE. 2003. *Caenorhabditis elegans*: The cell lineage and beyond (Nobel lecture). *Chembiochem* 4: 688–696.
17. Horvitz HR. 2003. Worms, life, and death (Nobel lecture). *Chembiochem* 4: 697–711.
18. Mello CC. 2007. Return to the RNAi world: Rethinking gene expression and evolution. *Cell Death Differ* 14: 2013–2020.
19. Chalfie M. 2010. The 2009 Lindau Nobel Laureate meeting: Martin Chalfie, Chemistry 2008. *J Vis Exp* 35.
20. Timmons L, Court DL, Fire A. 2001. Ingestion of bacterially expressed dsRNAs can produce specific and potent genetic interference in *Caenorhabditis elegans*. *Gene* 263: 103–112.
21. Andrew SE, Goldberg YP, Kremer B, Telenius H, Theilmann J et al. 1993. The relationship between trinucleotide (CAG) repeat length and clinical features of Huntington's disease. *Nat Genet* 4: 398–403.
22. Duyao M, Ambrose C, Myers R, Novelletto A, Persichetti F et al. 1993. Trinucleotide repeat length instability and age of onset in Huntington's disease. *Nat Genet* 4: 387–392.
23. Snell RG, MacMillan JC, Cheadle JP, Fenton I, Lazarou LP et al. 1993. Relationship between trinucleotide repeat expansion and phenotypic variation in Huntington's disease. *Nat Genet* 4: 393–397.
24. Morley JF, Brignull HR, Weyers JJ, Morimoto RI. 2002. The threshold for polyglutamine-expansion protein aggregation and cellular toxicity is dynamic and influenced by aging in *Caenorhabditis elegans*. *Proc Natl Acad Sci USA* 99: 10417–10422.
25. Brignull HR, Moore FE, Tang SJ, Morimoto RI. 2006. Polyglutamine proteins at the pathogenic threshold display neuron-specific aggregation in a pan-neuronal *Caenorhabditis elegans* model. *J Neurosci* 26: 7597–7606.
26. Mohri-Shiomi A, Garsin DA. 2008. Insulin signaling and the heat shock response modulate protein homeostasis in the *Caenorhabditis elegans* intestine during infection. *J Biol Chem* 283: 194–201.
27. Christie NT, Lee AL, Fay HG, Gray AA, Kikis EA. 2014. Novel polyglutamine model uncouples proteotoxicity from aging. *PLoS One* 9: e96835.
28. Teixeira-Castro A, Ailion M, Jalles A, Brignull HR, Vilaca JL et al. 2011. Neuron-specific proteotoxicity of mutant ataxin-3 in *C. elegans*: Rescue by the DAF-16 and HSF-1 pathways. *Hum Mol Genet* 20: 2996–3009.
29. Wang H, Lim PJ, Yin C, Rieckher M, Vogel BE et al. 2006. Suppression of polyglutamine-induced toxicity in cell and animal models of Huntington's disease by ubiquilin. *Hum Mol Genet* 15: 1025–1041.
30. Konietzko U, Goodger ZV, Meyer M, Kohli BM, Bosset J et al. 2010. Co-localization of the amyloid precursor protein and Notch intracellular domains in nuclear transcription factories. *Neurobiol Aging* 31: 58–73.
31. Citron M, Oltersdorf T, Haass C, McConlogue L, Hung AY et al. 1992. Mutation of the beta-amyloid precursor protein in familial Alzheimer's disease increases beta-protein production. *Nature* 360: 672–674.
32. Goate A, Chartier-Harlin MC, Mullan M, Brown J, Crawford F et al. 1991. Segregation of a missense mutation in the amyloid precursor protein gene with familial Alzheimer's disease. *Nature* 349: 704–706.
33. Teller JK, Russo C, DeBusk LM, Angelini G, Zaccheo D et al. 1996. Presence of soluble amyloid beta-peptide precedes amyloid plaque formation in Down's syndrome. *Nat Med* 2: 93–95.
34. Armstrong RA. 1994. Differences in beta-amyloid (beta/A4) deposition in human patients with Down's syndrome and sporadic Alzheimer's disease. *Neurosci Lett* 169: 133–136.
35. Beyreuther K, Pollwein P, Multhaup G, Monning U, Konig G et al. 1993. Regulation and expression of the Alzheimer's beta/A4 amyloid protein precursor in health, disease, and Down's syndrome. *Ann N Y Acad Sci* 695: 91–102.
36. Malter JS, Ray BC, Westmark PR, Westmark CJ. 2010. Fragile X syndrome and Alzheimer's disease: Another story about APP and beta-amyloid. *Curr Alzheimer Res* 7: 200–206.
37. Theendakara V, Peters-Libeu CA, Spilman P, Poksay KS, Bredesen DE et al. 2016. Direct transcriptional effects of apolipoprotein E. *J Neurosci* 36: 685–700.
38. Link CD, Taft A, Kapulkin V, Duke K, Kim S et al. 2003. Gene expression analysis in a transgenic *Caenorhabditis elegans* Alzheimer's disease model. *Neurobiol Aging* 24: 397–413.
39. Wu Y, Wu Z, Butko P, Christen Y, Lambert MP et al. 2006. Amyloid-beta-induced pathological behaviors are suppressed by Ginkgo biloba extract EGb 761 and ginkgolides in transgenic *Caenorhabditis elegans*. *J Neurosci* 26: 13102–13113.
40. McColl G, Roberts BR, Pukala TL, Kenche VB, Roberts CM et al. 2012. Utility of an improved model of amyloid-beta (Abeta(1)(-)(4)(2)) toxicity in *Caenorhabditis elegans* for drug screening for Alzheimer's disease. *Mol Neurodegener* 7: 57.
41. Hutton M, Lendon CL, Rizzu P, Baker M, Froelich S et al. 1998. Association of missense and 5'-splice-site mutations in tau with the inherited dementia FTDP-17. *Nature* 393: 702–705.

42. Hong M, Zhukareva V, Vogelsberg-Ragaglia V, Wszolek Z, Reed L et al. 1998. Mutation-specific functional impairments in distinct tau isoforms of hereditary FTDP-17. *Science* 282: 1914–1917.

43. Goedert M, Jakes R. 2005. Mutations causing neurodegenerative tauopathies. *Biochim Biophys Acta* 1739: 240–250.

44. Kraemer BC, Zhang B, Leverenz JB, Thomas JH, Trojanowski JQ et al. 2003. Neurodegeneration and defective neurotransmission in a *Caenorhabditis elegans* model of tauopathy. *Proc Natl Acad Sci USA* 100: 9980–9985.

45. Vaccaro A, Tauffenberger A, Aggad D, Rouleau G, Drapeau P et al. 2012. Mutant TDP-43 and FUS cause age-dependent paralysis and neurodegeneration in *C. elegans*. *PLoS One* 7: e31321.

46. Neumann M, Sampathu DM, Kwong LK, Truax AC, Micsenyi MC et al. 2006. Ubiquitinated TDP-43 in frontotemporal lobar degeneration and amyotrophic lateral sclerosis. *Science* 314: 130–133.

47. Buratti E, Dork T, Zuccato E, Pagani F, Romano M et al. 2001. Nuclear factor TDP-43 and SR proteins promote in vitro and in vivo CFTR exon 9 skipping. *EMBO J* 20: 1774–1784.

48. Ash PE, Zhang YJ, Roberts CM, Saldi T, Hutter H et al. 2010. Neurotoxic effects of TDP-43 overexpression in *C. elegans*. *Hum Mol Genet* 19: 3206–3218.

49. Gidalevitz T, Krupinski T, Garcia S, Morimoto RI. 2009. Destabilizing protein polymorphisms in the genetic background direct phenotypic expression of mutant SOD1 toxicity. *PLoS Genet* 5: e1000399.

50. Polymeropoulos MH, Lavedan C, Leroy E, Ide SE, Dehejia A et al. 1997. Mutation in the alpha-synuclein gene identified in families with Parkinson's disease. *Science* 276: 2045–2047.

51. Kruger R, Kuhn W, Muller T, Woitalla D, Graeber M et al. 1998. Ala30Pro mutation in the gene encoding alpha-synuclein in Parkinson's disease. *Nat Genet* 18: 106–108.

52. Zarranz JJ, Alegre J, Gomez-Esteban JC, Lezcano E, Ros R et al. 2004. The new mutation, E46K, of alpha-synuclein causes Parkinson and Lewy body dementia. *Ann Neurol* 55: 164–173.

53. Lakso M, Vartiainen S, Moilanen AM, Sirvio J, Thomas JH et al. 2003. Dopaminergic neuronal loss and motor deficits in *Caenorhabditis elegans* overexpressing human alpha-synuclein. *J Neurochem* 86: 165–172.

54. van Ham TJ, Thijssen KL, Breitling R, Hofstra RM, Plasterk RH et al. 2008. *C. elegans* model identifies genetic modifiers of alpha-synuclein inclusion formation during aging. *PLoS Genet* 4: e1000027.

55. Angot E, Steiner JA, Lema Tome CM, Ekstrom P, Mattsson B et al. 2012. Alpha-synuclein cell-to-cell transfer and seeding in grafted dopaminergic neurons in vivo. *PLoS One* 7: e39465.

56. Kordower JH, Dodiya HB, Kordower AM, Terpstra B, Paumier K et al. 2011. Transfer of host-derived alpha synuclein to grafted dopaminergic neurons in rat. *Neurobiol Dis* 43: 552–557.

57. Li JY, Englund E, Holton JL, Soulet D, Hagell P et al. 2008. Lewy bodies in grafted neurons in subjects with Parkinson's disease suggest host-to-graft disease propagation. *Nat Med* 14: 501–503.

58. Ironside JW. 1998. Prion diseases in man. *J Pathol* 186: 227–234.

59. Lindquist S. 1997. Mad cows meet psi-chotic yeast: The expansion of the prion hypothesis. *Cell* 89: 495–498.

60. Lindquist S. 1996. Mad cows meet mad yeast: The prion hypothesis. *Mol Psychiatry* 1: 376–379.

61. Soto C, Estrada L, Castilla J. 2006. Amyloids, prions and the inherent infectious nature of misfolded protein aggregates. *Trends Biochem Sci* 31: 150–155.

62. Brundin P, Melki R, Kopito R. 2010. Prion-like transmission of protein aggregates in neurodegenerative diseases. *Nat Rev Mol Cell Biol* 11: 301–307.

63. Meyer-Luehmann M, Coomaraswamy J, Bolmont T, Kaeser S, Schaefer C et al. 2006. Exogenous induction of cerebral beta-amyloidogenesis is governed by agent and host. *Science* 313: 1781–1784.

64. Yang W, Dunlap JR, Andrews RB, Wetzel R. 2002. Aggregated polyglutamine peptides delivered to nuclei are toxic to mammalian cells. *Hum Mol Genet* 11: 2905–2917.

65. Nussbaum-Krammer CI, Park KW, Li L, Melki R, Morimoto RI. 2013. Spreading of a prion domain from cell-to-cell by vesicular transport in *Caenorhabditis elegans*. *PLoS Genet* 9: e1003351.

66. Park KW, Li L. 2008. Cytoplasmic expression of mouse prion protein causes severe toxicity in *Caenorhabditis elegans*. *Biochem Biophys Res Commun* 372: 697–702.

67. Gidalevitz T, Ben-Zvi A, Ho KH, Brignull HR, Morimoto RI. 2006. Progressive disruption of cellular protein folding in models of polyglutamine diseases. *Science* 311: 1471–1474.

68. Ben-Zvi A, Miller EA, Morimoto RI. 2009. Collapse of proteostasis represents an early molecular event in *Caenorhabditis elegans* aging. *Proc Natl Acad Sci USA* 106: 14914–14919.

69. Westerheide SD, Morimoto RI. 2005. Heat shock response modulators as therapeutic tools for diseases of protein conformation. *J Biol Chem* 280: 33097–33100.

70. Ron D, Walter P. 2007. Signal integration in the endoplasmic reticulum unfolded protein response. *Nat Rev Mol Cell Biol* 8: 519–529.

71. Banerji SS, Theodorakis NG, Morimoto RI. 1984. Heat shock-induced translational control of HSP70 and globin synthesis in chicken reticulocytes. *Mol Cell Biol* 4: 2437–2448.

72. Xiao X, Zuo X, Davis AA, McMillan DR, Curry BB et al. 1999. HSF1 is required for extra-embryonic development, postnatal growth and protection during inflammatory responses in mice. *EMBO J* 18: 5943–5952.

73. Santos SD, Saraiva MJ. 2004. Enlarged ventricles, astrogliosis and neurodegeneration in heat shock factor 1 null mouse brain. *Neuroscience* 126: 657–663.

74. Zhang P, McGrath B, Li S, Frank A, Zambito F et al. 2002. The PERK eukaryotic initiation factor 2 alpha kinase is required for the development of the skeletal system, postnatal growth, and the function and viability of the pancreas. *Mol Cell Biol* 22: 3864–3874.

75. Reimold AM, Iwakoshi NN, Manis J, Vallabhajosyula P, Szomolanyi-Tsuda E et al. 2001. Plasma cell differentiation requires the transcription factor XBP-1. *Nature* 412: 300–307.

76. Zhang K, Wong HN, Song B, Miller CN, Scheuner D et al. 2005. The unfolded protein response sensor IRE1alpha is required at 2 distinct steps in B cell lymphopoiesis. *J Clin Invest* 115: 268–281.

77. Elefant F, Palter KB. 1999. Tissue-specific expression of dominant negative mutant *Drosophila* HSC70 causes developmental defects and lethality. *Mol Biol Cell* 10: 2101–2117.

78. Feder JH, Rossi JM, Solomon J, Solomon N, Lindquist S. 1992. The consequences of expressing hsp70 in *Drosophila* cells at normal temperatures. *Genes Dev* 6: 1402–1413.

79. Nylandsted J, Rohde M, Brand K, Bastholm L, Elling F et al. 2000. Selective depletion of heat shock protein 70 (Hsp70) activates a tumor-specific death program that is independent of caspases and bypasses Bcl-2. *Proc Natl Acad Sci USA* 97: 7871–7876.

80. Whitesell L, Mimnaugh EG, De Costa B, Myers CE, Neckers LM. 1994. Inhibition of heat shock protein HSP90-pp60v-src heteroprotein complex formation by benzoquinone ansamycins: Essential role for stress proteins in oncogenic transformation. *Proc Natl Acad Sci USA* 91: 8324–8328.

81. Whitesell L, Lindquist SL. 2005. HSP90 and the chaperoning of cancer. *Nat Rev Cancer* 5: 761–772.

82. Nollen EA, Garcia SM, van Haaften G, Kim S, Chavez A et al. 2004. Genome-wide RNA interference screen identifies previously undescribed regulators of polyglutamine aggregation. *Proc Natl Acad Sci USA* 101: 6403–6408.

83. Silva MC, Fox S, Beam M, Thakkar H, Amaral MD et al. 2011. A genetic screening strategy identifies novel regulators of the proteostasis network. *PLoS Genet* 7: e1002438.

84. Guisbert E, Czyz DM, Richter K, McMullen PD, Morimoto RI. 2013. Identification of a tissue-selective heat shock response regulatory network. *PLoS Genet* 9: e1003466.

85. Frumkin A, Dror S, Pokrzywa W, Bar-Lavan Y, Karady I et al. 2014. Challenging muscle homeostasis uncovers novel chaperone interactions in *Caenorhabditis elegans*. *Front Mol Biosci* 1: 21.

86. Brehme M, Voisine C, Rolland T, Wachi S, Soper JH et al. 2014. A chaperome subnetwork safeguards proteostasis in aging and neurodegenerative disease. *Cell Rep* 9: 1135–1150.

87. Ayyadevara S, Balasubramaniam M, Gao Y, Yu LR, Alla R et al. 2015. Proteins in aggregates functionally impact multiple neurodegenerative disease models by forming proteasome-blocking complexes. *Aging Cell* 14: 35–48.

88. Gidalevitz T, Kikis EA, Morimoto RI. 2010. A cellular perspective on conformational disease: The role of genetic background and proteostasis networks. *Curr Opin Struct Biol* 20: 23–32.

89. Gidalevitz T, Wang N, Deravaj T, Alexander-Floyd J, Morimoto RI. 2013. Natural genetic variation determines susceptibility to aggregation or toxicity in a *C. elegans* model for polyglutamine disease. *BMC Biol* 11: 100.

90. Shemesh N, Shai N, Ben-Zvi A. 2013. Germline stem cell arrest inhibits the collapse of somatic proteostasis early in *Caenorhabditis elegans* adulthood. *Aging Cell* 12: 814–822.

91. David DC, Ollikainen N, Trinidad JC, Cary MP, Burlingame AL et al. 2010. Widespread protein aggregation as an inherent part of aging in *C. elegans*. *PLoS Biol* 8: e1000450.

92. Walther DM, Kasturi P, Zheng M, Pinkert S, Vecchi G et al. 2015. Widespread proteome remodeling and aggregation in aging *C. elegans*. *Cell* 161: 919–932.

93. Liang V, Ullrich M, Lam H, Chew YL, Banister S et al. 2014. Altered proteostasis in aging and heat shock response in *C. elegans* revealed by analysis of the global and de novo synthesized proteome. *Cell Mol Life Sci* 71: 3339–3361.

94. Lee AL, Ung HM, Sands LP, Kikis EA. 2017. A new *Caenorhabditis elegans* model of human huntingtin513 aggregation and toxicity in body wall muscles. *PLoS ONE* 12(3): e0173644.

Section IX

Hibernation and Aging

34 Hibernation and Aging
Molecular Mechanisms of Mammalian Hypometabolism and Its Links to Longevity

Cheng-Wei Wu and Kenneth B. Storey

CONTENTS

MAMMALIAN HIBERNATION

When challenged with unfavorable environmental conditions, organisms must adapt to their surroundings in order to survive, this is especially crucial for animals that encounter the cold and harsh conditions of winter. For many wintering mammals, the solution is hibernation, an adaptive physiological state characterized by the extreme depression of basal metabolic rate and body temperature. Mammals that undergo hibernation include monotremes, rodents, marsupials, bats, shrews, hedgehogs, black bear, and lemurs (Dausmann et al. 2004, Storey 2010). When hibernators enter torpor, they first undergo reductions in metabolic rate followed by a subsequent drop in body temperature. Once animals enter full torpor, their body temperature drops to near ambient, with heart rates reduced to approximately 5 beats per minute from 200 to 300, and respiration reduced to 4–6 breath per minute from 100 to 200. This dormant state can last between 5 and 15 days, after which torpor is interrupted by spontaneous arousal back to euthermia for 1 day, then followed by reentry into deep torpor. This torpor cycle is repeated for 10–20 bouts until the end of hibernation season (Carey, Andrews, and Martin 2003). This type of metabolic depression is also observed in other non-mammalian species as an adaptive response to stressful environments, with examples including freeze tolerance in frogs, diapause/dauer in nematodes, anoxia tolerance in turtles, and a estivation in snails (Churchill and Storey 1993, Fielenbach and Antebi 2008, Krivoruchko and Storey 2010, Wu, Biggar, and Storey 2013). Mammalian hypometabolism is of special interest as many hibernators must also deal with the threat of hypothermia development as they reduce their core body temperature to near ambient (~5°C) during torpor. To coordinate and endure such drastic physiological changes, hibernators must essentially reprogram their whole body metabolism in order to maintain this dormant state while also activating appropriate cytoprotective responses to avoid cellular damage. In this chapter, we will briefly discuss metabolic regulations of hibernation and cellular processes that are regulated to help establish and maintain this hypometabolic state.

METABOLIC REPROGRAMMING AND ENERGY RESERVES

Many hibernators do not cache food in their burrows during the hibernation season, and must rely on their internal reserves as the primary source of metabolic fuel. For animals like the ground squirrel, they undergo periods of hyperphagia in the fall as preparation for the hibernation season, a process that allows the animal to increase their body mass by up to 50%, mainly through an upsurge in the triglycerides content in their white adipose tissues (Dark 2005, Storey 2010). To take advantage of the increase in lipid contents, hibernators reduce their carbohydrate oxidation to initiate a switch toward a reliance on oxidation of fat during hibernation. This metabolic switch is in part catalyzed by the upregulation of *pyruvate dehydrogenase kinase* isoenzyme 4 (PDK4) mRNA in multiple tissues during hibernation (Buck, Squire, and Andrews 2002). PDK4 phosphorylates the pyruvate dehydrogenase complex (PDC), a step that inactivates PDC-catalyzed conversion of pyruvate from glycolysis into acetyl-CoA, and inhibits its subsequent oxidative reactions in the mitochondria. Enzymatic activities of pyruvate dehydrogenase are shown to be significantly reduced across different tissues during hibernation, and this reduction is observed across multiple hibernating species (Heldmaier et al. 1999). The inactivation of carbohydrate metabolism is also evident by a change in the respiratory quotient (a measurement of ratio between CO_2 production to O_2 consumption) to 0.7 in animals during hibernation, suggesting a reliance on lipid as the primary source of metabolic energy. Genes that are directly involved in fatty acid catabolism also significantly increase expression during hibernation, among these include *carnitine palmitoyltransferase I, acetyl-CoA acetyltransferase,* and *fatty acid binding proteins (fabp)* (Williams et al. 2011). FABPs are a class of small proteins that function as lipid chaperones by binding reversibly to hydrophobic ligands such as saturated and unsaturated long chain fatty acids, bile acids, and other lipids (Furuhashi and Hotamisligil 2008). Due to its role in fatty acid trafficking, FABPs are important in coordinating lipid metabolisms in cells, and are crucial in hibernation as lipids are the primary metabolic fuel utilized during torpor. In addition to being upregulated in hibernating animals, FABPs in hibernators are also found to function over a broad range of temperatures. Compared to rats, FABPs in the liver of hibernating ground squirrels display temperature insensitive dissociating constant (K_d) for substrates oleate and *cis*-parinarate when measured at temperatures of 5°C, 25°C, and 37°C (Stewart, English, and Storey 1998). These unique properties would ensure proper lipid transport is maintained independently of the fluctuating body temperature experienced by hibernators during the torpor arousal cycle.

While this switch toward fat oxidation allows hibernators to take advantage of their internal lipid storage as metabolic fuel during torpor, other metabolic processes must be appropriately coordinated and regulated to reduce nonessential energy expenditure and ensure sufficient fuel reserves are in place to survive the full hibernation season. In general, virtually all cellular processes are suppressed compared to euthermia during hibernation, as the reduction of body temperature alone decreases the rate of all biochemical reactions. Studies to date have identified several molecular mechanisms for which key cellular processes are shut off in a regulated and reversible manner, so that upon arousal from hibernation, these same cellular processes can be rapidly resumed to support euthermia. In the following sections, we will highlight the molecular mechanisms that regulate RNA transcription and protein synthesis, which arc two major biological processes reversibly suppressed during hibernation.

TRANSCRIPTIONAL AND TRANSLATIONAL REGULATION DURING GROUND SQUIRREL HIBERNATION

When hibernators are in full torpor, their metabolic rates can be reduced to 1%–3% of their basal rates in normothermia. In this torpid state, most cellular and metabolic functions are suppressed in order to conserve energy Adenosine triphosphate (ATP) usage during torpor, meanwhile energy is prioritized to ensure cell preservation mechanisms are activated to protect vital tissue organs and maintain homeostasis. Although the complete metabolic reprogramming that takes place during hibernation is not fully known, considerable progress has been made to identify many key cellular processes that are regulated during torpor. These include inhibitions of energetically costly processes such as global transcription and protein synthesis (Frerichs et al. 1998, Carey, Andrews, and Martin 2003, Morin and Storey 2006).

Active mRNA transcriptional machinery in mammalian cells can account for 1%–10% of its basal metabolic rate, an energetic cost that is not available during hibernation (Rolfe and Brown 1997). In hibernating ground squirrels, the rate of RNA synthesis is reduced at both the initiation and the elongation stage, resulting in a state of transcriptional arrest during torpor (van Breukelen and Martin 2002). The rate of RNA transcription can also be influenced by the accessibility of the DNA template, which are complexed within histone proteins that are subject to posttranslational modifications that determine whether the chromatin is in a transcriptionally active or repressive state (Zhang and Reinberg 2001). The histone tails can undergo a variety of modifications that include acetylation, phosphorylation, methylation, ubiquitination, and more recently known SUMOylation and ADP-ribosylation (Bannister and Kouzarides 2011). When histone acetyltransferase (HAT) proteins acetylate histone tails, chromatins generally exhibit a relaxed and open state and transcription is activated by the increase in accessibility of DNA to interacting proteins. Conversely, when histone tails are deacetylated by histone deacetylase (HDAC) proteins, transcription is suppressed by the formation of the highly repressive heterochromatin structures (Eberharter and Becker 2002). In the skeletal muscle of hibernating ground squirrels, the acetylation of histone H3 was found to be significantly reduced by 25% compared to euthermic squirrels, and this is also accompanied by a 1.8-fold increase in HDAC activity and 1.2–1.5 fold increase in levels of HDAC 1 and 4 proteins (Morin and Storey 2006). Furthermore, activity of RNA polymerase II was also found to be significantly reduced by 43% during hibernation (Morin and Storey 2006). In addition to histone modifications, gene expression can also be controlled via DNA methylation. In general, methylation of DNA is associated with the blocking of gene expression, via changes to the interaction between the DNA and interacting proteins. When methylation takes place at the CpG islands of the promoter, it recruits the binding of methyl-CpG-binding domain proteins (MBDs) that function to block transcription. In the brown adipose tissue, the HDAC proteins are also shown to be upregulated during hibernation, in addition, the global levels of methylated DNA were increased by 1.7-fold, and this increase was coupled with the upregulation of MBD1 protein by 1.9-fold (Biggar and Storey 2014). These results suggest the potential suppression of transcription via increased binding of MBD1 proteins to methylated DNA. Although global transcription effectively ceases during hibernation (van Breukelen and Martin 2002, Morin and Storey 2006), several transcriptomic studies have shown that a cohort of transcripts are actually upregulated during torpor. Genes that are upregulated during hibernation include those involved in stress responses, metabolic regulations, lipid metabolism, and vesicle transports, implicating the importance of these cellular processes in establishing the hibernation phenotype (Hampton et al. 2011, Williams et al. 2011, Schwartz, Hampton, and Andrews 2013).

The epigenetic mechanisms of transcriptional regulation are well conserved across species, and appear to be an important mechanism utilized by hibernators to switch off transcription during torpor to prevent the synthesis of unwarranted genes during torpor (Figure 34.1a). As global transcription is arrested during hibernation, it is not surprising to find that similar modes of inhibition were also found for protein biosynthesis. Although essential to sustain mammalian life, a near complete arrest of protein synthesis takes place in ground squirrels during torpor. In the brains of hibernating squirrels, activity of protein synthesis was reduced to 0.04% of its euthermic values as determined by autoradiographic detection, with similar inhibitions also observed in the liver (0.09%) and heart (0.10%) (Frerichs et al. 1998). When the rate of protein synthesis was measured in hibernator's cell free lysates at 37°C, activity of protein translation in the brain was still reduced

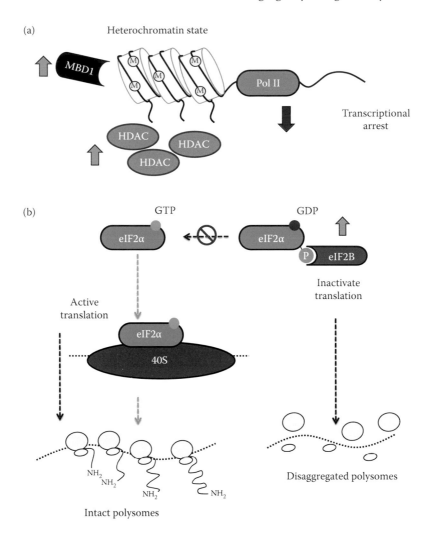

FIGURE 34.1 Transcriptional and translational control during hibernation. (a) In hibernators, transcriptional arrest is accomplished by a combination of increase in HDAC proteins, DNA methylation, and methyl binding proteins (MBD), along with a decrease in the activity of RNA polymerase II. (b) The eIF2α protein is a rate-limiting factor in the assembly of the initiation complex for protein translation; in hibernators, phosphorylation of eIF2α is significantly increased, a process that inhibits the GDP to GTP recycling of eIF2α that is required for its subsequent function with the ribosome.

by more than three-fold, suggesting that translational arrest was a product of both hypothermia and active mechanisms of inhibition (Frerichs et al. 1998). One of the rate-limiting steps of protein synthesis is the formation of the translation initiation complex regulated by the eukaryotic initiation factors-2 (eIF2), which is controlled and inhibited via phosphorylation at serine-51 on the alpha subunit of the protein. To initiate mRNA translation, eIF2 binds to guanosine-5′-triphosphate (GTP) and transfers Met-tRNAi to the ribosomal subunit, where GTP is hydrolyzed to guanosine diphosphate (GDP) to release eIF2 from the ribosome. When eIF2 is phosphorylated at the alpha subunit, the GDP bound eIF2 cannot be recycled and exchanged to GTP, a reaction catalyzed by eIF2B, thereby preventing the subsequent formation of the eIF2-GTP-Met-tRNAi complex that is required for translation initiation (Figure 34.1b) (Kimball 1999). In hibernating ground squirrels, although the total protein levels of eIF2α remained unchanged, the levels of phosphorylation at serine-51 was highly elevated in the brain, with 13% of eIF2α phosphorylated during hibernation compared to 2%

in active animals (Frerichs et al. 1998). These disruptions to the translation initiation machinery is also supported by evidence of polysome disaggregation, with multiple tissues that include the brain, liver, and kidney all showing a decrease in polysome formation and increase in monosome levels in animals under torpor (Frerichs et al. 1998, Knight et al. 2000, Hittel and Storey 2002). As formation of polysomes is a positive indicator of active protein translation, these results demonstrate that the amount of translation ready ribosomes is significantly reduced and protein synthesis is inhibited during hibernation.

Although only briefly illustrated here, hibernators undergo significant metabolic reprogramming during torpor. The switch toward the reliance on lipids as the primary source of metabolic fuel allows the hibernators to take advantage of their fat reserves stored during the fall, which reduces the need to forage for food during the unfavorable conditions of winter. To make the most of their fuel reserves, hibernators reduce their metabolic cost by inhibiting both transcription and translation (Figure 34.1), a process that helps to establish and maintain a hypometabolic state that is aimed at maximizing survival during winter.

HIBERNATION IN THE CONTEXT OF AGING

At first glance, there are numerous medical benefits in understanding molecular mechanisms that regulate hibernation, specifically, how it can be useful in the context of organ cryopreservation. Mammals that undergo hibernation can escape cellular damages of hypothermia even when their body temperature is reduced to near zero, a process that basically functions to naturally preserve their internal organs at a low temperature (Storey 2010). Hibernators are also viewed as a potential model to study ischemia and reperfusion related injuries, as they are able to tolerate rapid changes in blood flow experienced between torpor and arousal periods without extensive damages to oxygen sensitive tissues such as the heart and brain. However, as we further our understanding of the biochemical mechanisms that underlie the metabolic controls of hibernation, it appears that the regulation patterns of many signaling pathways appears to be similar to those seen in other animal models that have extended longevity. Interestingly, it was recently observed that bats that can hibernate are also extremely long-lived, with the Brandt's bat (*Myotis brandtii*) having a maximum life span of 38–41 years (Podlutsky et al. 2005). In addition to bats, other hibernators such as squirrels, hamsters, marmots, and mouse lemurs also display an extended life span compared to other mammals of similar size (Wu and Storey 2016). In Turkish hamsters, it was shown that there is a positive correlation between the amount of time the animal spent in hibernation and their maximal life span, suggesting that hibernation could function to slow down the aging process (Lyman et al. 1981). One simple way to interpret these results would be that the increase in longevity is caused by a passive effect where animals do not age at a normal rate while under hibernation. Another interpretation would be that animals that hibernate must possess specific features that enable them to adapt and enter a state of torpor, and that some of these features that help to establish hibernation also can have antiaging effects. Regardless, both interpretations do suggest that when animals hibernate, the process of aging is slowed. In the following sections, we will discuss aspects of hibernation that show a positive connection between torpor and longevity, and the potential of using hibernators as a model to studying the complex phenomenon of aging.

TEMPERATURE AND AGING

Temperature can influence the rate of biological and chemical reactions, and is a major factor that determines life span in many organisms. In both poikilotherms and homeotherms, a general inverse correlation exists between temperature and life span, such that reductions in ambient temperature of poikilotherms, or core temperature in homeotherms, can be associated with different levels of life span extension (Conti et al. 2006, Conti 2008). In poikilotherms like *Drosophila melanogaster* and *Caenorhabditis elegans*, effects of temperature on life span can be easily studied by reducing the temperature of the environment at which the animal is maintained. In fruit flies, life span can be doubled by lowering the ambient temperature from 27°C to 21°C (Miquel et al. 1976); similarly, worms cultivated at a low temperature of 10°C live nearly four times longer than those at 25°C (Van Voorhies and Ward 1999). In mammals, however, the effects of temperature on life span are more complex; most mammals maintain a narrow range of temperature homeostasis that is strictly regulated, with hypothermia induced when the core body temperature is reduced by more than 2°C. Temperature is also a major factor that drives the induction of hibernation; ground squirrels are able to cope with the cold by abandoning euthermia and enter hibernation, by reducing their core body temperature to near ambient, and can do so without experiencing the common hypothermia-related damages seen in other mammals. Since ground squirrels spend a significant portion of their life hibernating at a lowered body temperature, it can be hypothesized that this unique poikilothermia allows hibernators to benefit from the positive effect that reduced temperatures have on life span (as seen in flies and worms). Studies in rodents have provided some interesting findings that may support this hypothesis, with examples of different long-lived mice also showing slight reductions in their core body temperature. The Ames dwarf mice, which is homozygous for the Prop1^{df} (prophet of pituitary factor 1) loss-of-function mutation live about 1 year longer (350 days in males and 470 days in females) than their wild-type siblings. The mutation to Prop1^{df} hinders the normal development of the pituitary that results in deficiencies in levels of growth hormone, prolactin, and thyroid stimulating hormone (Brown-Borg et al. 1996). More interestingly, the Ames dwarf mice show a reduction of approximately 1.6°C in core body temperature compared to the wild-type mice. Similar reductions in core body temperature are also shown in the long-lived Snell mice, which a carry mutation in the *Pit-1* (pituitary-specific positive transcription factor 1) locus and are phenotypically alike to Ames dwarf mice with similar hormone deficiencies (Bartke and Brown-Borg 2004). Mice with knockout to the growth hormone receptor (GHR) protein also show a significant increase in their life span by 38%–55% along with a slight reduction in core body temperature (Hauck et al. 2001). These mutant mice demonstrate the role hormone regulations have on life span, but more interesting, provide a potential link between temperature and longevity in mammals. One possible explanation for the temperature reduction in these long-lived mice is that the deficiencies in levels of growth hormone could function to reduce their overall metabolic rate, thereby affecting the core body temperature of the animal.

Another key hormone that is linked to longevity is insulin, which is significantly reduced in several long-lived mice, and a knockout mutation to the insulin-like growth factor-1 (IGF) receptor protein significantly increases the life span of the mice (Holzenberger et al. 2003, Selman et al. 2008). The effects of IGF-1 on life span are also conserved in other species, with disruptions to this signaling axis shown to positively influence life span and delay aging in flies and worms (Kenyon et al. 1993, Clancy et al. 2001). Studies have also identified the IGF-1 signaling pathway as an important regulator of hibernation in multiple species (Buck, Squire, and Andrews 2002, Carey, Andrews, and Martin 2003, Wu and Storey 2012). In the next section, we will discuss the regulatory mechanisms of insulin signaling in different hibernating mammals, and how it could be linked to the long-lived phenotype.

INSULIN SIGNALING CONTROL IN HIBERNATION

Insulin is a key metabolic hormone that is responsible for regulating whole body glucose homeostasis, and is essential for the development and growth of many tissues. The levels of serum insulin has been shown to fluctuate at different stages of hibernation, with the highest insulin levels observed during the early periods of torpor (September–October), and with the lowest levels observed while the animals are in full torpor (December–January) (Buck, Squire, and Andrews 2002). It has been proposed that the decrease in serum insulin levels during hibernation could function to promote PDK4 expression to stimulate lipid metabolism (Andrews 2007), as a previous study showed that insulin can suppress the expression of PDK4 in the rat skeletal muscle (Kim et al. 2006). The reduction in serum insulin levels during hibernation would also suggest that the signal cascades downstream of insulin are also affected during hibernation. The action of insulin is initiated upon binding to the insulin receptor, which triggers a downstream phosphorylation cascade that relays the signal transduction through multiple kinases that include PI3-kinase (phosphatidylinositol-4,5-bisphosphate 3-kinase), Akt (also known as protein kinase B), and mammalian target of rapamycin (mTOR) (Figure 34.2a). Activation of Akt by PI3-kinase through insulin signaling has been shown to promote and stimulate the translocation of the principal glucose transporter GLUT4

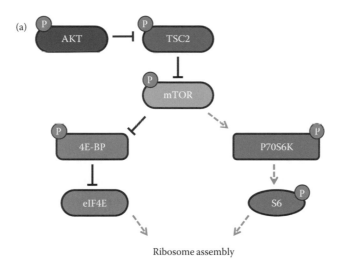

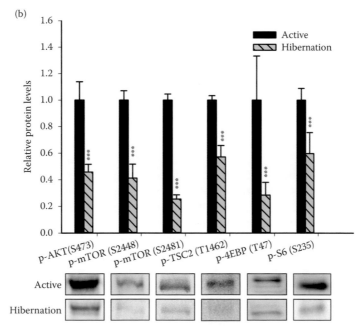

FIGURE 34.2 The Akt signaling cascade. (a) When Akt is activated via phosphorylation, it subsequently phosphorylates and inactivates TSC2 which is a negative regulator of mTOR. Activation of mTOR mediates protein translation via its regulatory roles on 4E-BP and P70S6K; phosphorylation of 4E-BP by mTOR inhibits its negative effect on eIF4E, while phosphorylation of P70S6 kinase by mTOR promotes the subsequent phosphorylation of the S6 ribosomal protein, both of which are involved in the assembly of the ribosome translational complex. (b). Phosphorylation pattern of Akt, mTOR, TSC2, 4E-BP, and S6 ribosomal proteins in the skeletal muscle of the hibernating ground squirrel *Ictidomys tridecemlineatus*. (Data replotted from Wu, C. W., and K. B. Storey. 2012. *J Exp Biol* 215 (Pt 10):1720–7. doi: 10.1242/jeb.066225.)

to the cell surface, where it mediates the removal of glucose from circulation (Huang and Czech 2007). In ground squirrels, the phosphorylation levels of Akt at its activation residue (Ser-473) are significantly reduced in the brain, liver, and skeletal muscle during hibernation; this decrease in Akt phosphorylation is also associated with a reduction in the Akt kinase activity, with 42% decrease observed in the brain and 60%–65% decrease observed in the liver and skeletal muscle (Cai et al. 2004, Abnous, Dieni, and Storey 2008). Although Akt phosphorylates multiple downstream effectors, one of the most characterized downstream target is mTOR, a nutrient sensing cytoplasmic kinases that regulates cell growth and development. The mTOR protein forms two distinct complexes, mTORC1 and mTORC2, each regulating different downstream networks. mTORC1, which responds to nutrient, energy, and oxygen levels, has been shown to be an important and conserved regulator of aging in multiple species. When Akt is active, it promotes the activation of mTORC1 by directly phosphorylating and inactivating two negative regulators of mTOR, proline-rich Akt substrate 40kDa (PRAS40) and tuberous sclerosis protein 2 (TSC2) (Figure 34.2a) (Nascimento et al. 2010). Genetic inhibitions of TORC1 in *C. elegans* have been shown to extend life span and promote an increase resistance toward oxidative stress, through mechanisms that require the transcription factors SKN-1/Nrf-2 and DAF-16/FOXO. Pharmacological inhibition of mTOR through rapamycin feeding in mice has also been shown to extend life span, with an increase of 14% observed in females and 9% in males (Harrison et al. 2009). These results suggest an evolutionary conserved role where suppression of the TOR pathway can positively influence life span. In ground squirrels, the phosphorylation of mTOR at its activation residue (Ser-2448) is significantly reduced by >50% during hibernation in the skeletal muscle (Figure 34.2b) (Wu and Storey 2012). This decrease in mTOR activation during hibernation is evident by a similar decrease in the phosphorylation of its downstream targets eIF4E binding protein (4E-BP) and S6 ribosomal protein, which together regulate the initiation of cap-dependent protein translation (Figure 34.2b). Although studies in *C. elegans* have shown that transcription factors like SKN-1/Nrf-2 and DAF-16/FOXO play an essential role in regulating TOR mediate extension of life span, the direct TOR downstream targets of 4E-BP and S6 kinase (S6K) have also been shown to influence aging through their regulatory roles on protein translation. Mutation in the S6K gene alone have been shown to extend life span in multiple organisms (Johnson, Rabinovitch, and Kaeberlein 2013), and genetic knockdown of translation initiation factors can also extend life span in *C. elegans* (Pan et al. 2007). In *Drosophila*, overexpression of an activated allele of 4E-BP, a protein that inhibits protein translation at the initiation stage by binding to eIF4E, can extend life span by 11% in male and 22% in female flies (Kapahi et al. 2004). Although 4E-BP is regulated by phosphorylation during hibernation, it has also been shown that the protein levels of 4E-BP undergo drastic seasonal changes in the ground squirrel. In the summer active squirrels, levels 4E-BP protein in the liver is essentially undetectable compared to the winter hibernating squirrels, suggesting that 4E-BP is abundantly overexpressed when animals are in hibernation (van Breukelen, Sonenberg, and Martin 2004).

The regulatory pattern of insulin signaling observed during ground squirrel hibernation, which include the reduction of serum insulin level, decrease in Akt and mTOR activity, and inhibition of protein synthesis through 4E-BP and S6 protein, although currently viewed as a strategy to reduce energy expenditure during hibernation, also mirrors mechanisms that have been shown to significantly increase life span in other species. The involvement of the insulin signaling pathway in longevity and hibernation was also recently shown to be influenced by genetic factors, with the genome sequencing of a long-lived hibernating Brandt's bat *M. brandtii*. In the early 2000s, scientists captured a male Brandt's bat in the wild that was originally tagged and released in the 1960s, with this particular bat calculated to be approximately 41 years old (Podlutsky et al. 2005). By the longevity quotient standard, this bat lives 9.8-times longer than expected given its body size, a number that was easily the highest for any mammal known to date measured either in the wild or in captivity. Although both hibernating and non-hibernating bats are considered long-lived, bats that hibernate live an average of six years longer than their non-hibernating equals (Wilkinson and South 2002). When the genome of the Brandt's bat was sequenced, unique evolutionary adaptations were found in forms of distinct amino acid substitutions for several genes, these include changes to genes that are involved in the regulation of hibernation, echolocation, vision, and reproduction (Seim et al. 2013). At the transcriptomic level, bats show many similar molecular changes during hibernation when compared to ground squirrels, these include the downregulation of genes involved in protein synthesis, and the upregulation of genes involved in lipid metabolism, oxidative stress, and adaptation to starvation (Seim et al. 2013). A the genetic level, unique nucleotide substitutions in the bat include changes to the SLC45A2 transporter protein that is involved in sound absorption, and CNGB3 protein that regulates sensory transduction of cone receptors. Most interestingly, unique amino acid substitutions were also found for the GHR and IGF-1 receptor (IGF-1R) proteins in the Brandt's bat, at amino acids residues within the highly conserved transmembrane domain of each protein. These unique mutations to the GHR/IGF-1R proteins are only present in the *Vespertilionoidea* superfamily, and are absent in other mammals like horse, cat, human, and mouse (Seim et al. 2013). As mentioned earlier, mutations to either GHR and IGF-1 proteins can lead to an extended life span in mice, and its potential role in hibernator's longevity is supported by an increase in the expression of several insulin associated genes in the Brandt's bat, at a level that is comparable to the upregulation observe between GHR mutant and wild-type mice. Furthermore, Brandt's bats also show similar expression changes in 11 out of 15 genes differentially expressed between long-lived and wild-type mice (Seim et al. 2013). Taken together, these results suggest an important overlapping role of the insulin signaling pathway in contributing to hibernation as well as promoting a long-lived phenotype.

Transcription Factors Downstream of Insulin Signaling during Hibernation

To date, the role of insulin signaling on life span is well established, with the downstream process of protein synthesis playing a crucial role in mediating the long-lived phenotype in many species. However, studies in *C. elegans* have shown that the insulin signaling pathway also functions to regulate two key proteins that are prominently involved in aging, which are the SKN-1/Nrf-2 and DAF-16/FOXO transcription factors. The insulin receptor DAF-2 in *C. elegans* directly inhibits the longevity-promoting transcription factors SKN-1 and DAF-16 in parallel pathways, and mutations to the DAF-2 protein that attenuates insulin signaling and extends life span is in part dependent on the relief of its inhibition on these two transcription factors (Lin et al. 2001, Murphy et al. 2003, Tullet et al. 2008).

The Nrf-2 transcription factor, homologous to cap 'n' collar (CNC) in *D. melanogaster* and SKN-1 in *C. elegans*, is best characterized by its evolutionary conserved role in oxidative stress. Nrf-2 is activated upon elevated levels of oxidative stress or xenobiotics, and responds by translocating into the nucleus and induces expression of detoxification genes through binding to the antioxidant response elements (ARE) of the target promoters (Sykiotis and Bohmann 2008, Sykiotis et al. 2011). Elevated protein levels of Nrf-2 have been observed in fibroblast cells of the long-lived Snell dwarf mice (Leiser and Miller 2010), and also in the extremely long-lived naked mole rats (Lewis et al. 2015). Naked mole rats like the Brandt's bat is another example of extreme longevity, with a maximum life span of 31 years in captivity, five times longer than the expected life span based on its body size (Lewis et al. 2015). In the naked mole rats, mRNA and protein levels of Nrf-2 are both highly elevated compared to wild-type mice, and this is combined with a decrease in mRNA and protein levels of the Nrf-2 inhibitor protein Keap1. The link between Nrf-2 activity and life span is not just limited to the naked mole rat, as the Nrf-2 ARE binding activity is positively correlated with the maximum life span potential across 10 different species (Lewis et al. 2015). Conversely, Nrf-2 deficient mice are hypersensitive to a wide range of stressors, and *C. elegans* carrying a loss-of-function mutation to the *skn-1* gene show significant reductions in life span (Tullet et al. 2008). In ground squirrels, the protein levels of Nrf-2 are upregulated by two to three fold in multiple tissues that include the heart, liver, skeletal muscle, and brown adipose tissue during hibernation (Morin et al. 2008, Xu et al. 2013). This increase in Nrf-2 protein expression is also accompanied by an increase in its transcriptional activation, evident by the increase in mRNA level of its downstream target gene *heme oxygenase-1* (Ni and Storey 2010). This upregulation of Nrf-2 in ground squirrels is thought to serve as a mechanism to elevate its antioxidant defense during torpor, however, studies in other species would suggest that the increase in Nrf-2 protein levels could also positively promote longevity of the animal.

Another class of transcription factor that is prominently involved in the insulin signaling pathway and longevity are the forkhead box O (FOXO) proteins, which in mammals is comprised of a family of four members, FOXO1, 3, 4, and 6. The FOXO transcription factors are directly phosphorylated and negatively regulated by insulin signaling cascade through the Akt kinase, and similar to Nrf-2, the FOXO proteins can be activated in the presence of stress to induce downstream genes involved in cellular protective responses that include cell cycle arrest, DNA repair, redox balance, apoptosis, and many more (Figure 34.3a) (Eijkelenboom and Burgering 2013, Morris et al. 2015). In genome-wide association studies in human set out to identify "long-lived" genes, it was shown that genetic variations and single-nucleotide polymorphisms within the *foxo3a* gene was significantly associated with the longevity of centenarians (Willcox et al. 2008, Flachsbart et al. 2009). Activation and overexpression of FOXO protein in the fat body of *Drosophila* can increase life span by 15.5% in males and 19.4% in females (Hwangbo et al. 2004). Conversely in *C. elegans*, a deletion to the FOXO ortholog DAF-16 significantly reduces the worm's life span compared to the wild type, and loss of DAF-16 largely suppress the long-lived phenotype of the DAF-2 insulin receptor mutant (Lin et al. 2001). And in mice, protein levels of FOXO3 and FOXO4 have been shown to decline by 25% and 18% respectively with age (Furuyama et al. 2002). These findings suggest that like Nrf-2, an increase in activity and expression of the FOXO proteins may have beneficial effects on life span. In ground squirrels, the total levels FOXO3a protein is overexpressed is by 3.6-fold in the skeletal muscle during hibernation, and this is accompanied by an increase in its nuclear localization and transcriptional activation of its downstream genes that include cell cycle inhibitors p27 and cyclin G2, and the antioxidant enzyme catalase (Figure 34.3b–d) (Wu and Storey 2014).

Although it has been known for over a decade the positive role whereby FOXO confers longevity, the genes downstream of FOXO that are responsible for this phenotype remain largely elusive. A popular hypothesis is that FOXO mediates longevity by activating expression of genes involved in damage response to reactive oxygen species (ROS) such as *super oxide dismutases* (*sod*), forming an elevated protective system that limits cellular damage and increases life span. However, recent studies have shown that worms with deletion to all of its *sod* isoforms (*sod-1* to *5*) have a normal life span, and deletion mutations to the mitochondrial *sod-2* isoform alone can actually increase longevity (Van Raamsdonk and Hekimi 2009, 2012). In *C. elegans*, knockdown of the DAF-16 protein alters the expression of over 3000 genes, these include regulation of genes involved in multiple stress responses (oxidative, heat, and pathogen), proteasome mediated protein degradation, lipid metabolism, and ribosome processing (Murphy et al. 2003). Although some of the downstream genes can mimic the longevity phenotypes of DAF-16, no single target gene alone can recapitulate the full effect of DAF-16 on life span, which upon deletion cause worms to be short-lived and abolishes the life span extension of the *daf-2* mutants (Lin et al. 2001, Murphy et al. 2003). These studies have concluded that activation of DAF-16/FOXO likely functions to promote longevity by exerting the cell toward protections from a variety of extracellular stressors while promoting metabolic adjustments to optimize energy metabolism (Tullet 2015), forming a cellular environment much like the one established during hibernation.

LONGEVITY IN OTHER METABOLIC STRESS MODELS

Over the past few decades, multiple breakthrough studies in model organisms have identified genes and pathways that have a remarkable influence on life span. While it is tempting to try and find a single regulator of longevity, the majority of studies suggest that although a single gene can confer

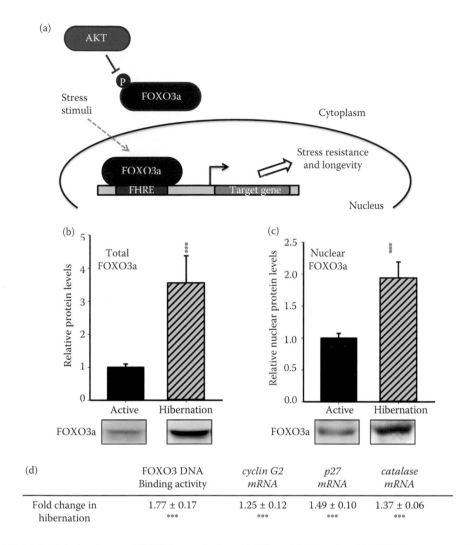

FIGURE 34.3 Mechanisms of FOXO3a regulation. (a) When Akt is active, FOXO3a is sequestered from the nucleus via an inhibitory phosphorylation. However, upon stimulation by stress, FOXO3a translocates into the nucleus to induce the expression of downstream genes involved in the regulation of stress resistance and longevity. (b) Total protein levels of FOXO3a in the skeletal muscle of the hibernating ground squirrel, ***$P<0.05$. (c) Nuclear protein levels of FOXO3a in the skeletal muscle of the hibernating ground squirrel, ***$P<0.05$. (d) FOXO3a DNA binding activity and downstream target gene expression in the skeletal muscle of the hibernating ground squirrel, ***$P < 0.05$. (Data replotted from Wu, C.W., and K.B. Storey. *Mol Cell Biochem.* 390(1–2):185–95. doi: 10.1007/s11010-014-1969-7.)

a long-lived phenotype, it exerts its effect through global changes to multiple downstream cellular processes that ranges from stress resistance, energy metabolism, to proteome homeostasis. Similarly, in hibernation, studies to date have not identified a single gene or pathway that alone regulate this hypometabolic adaptation, but rather suggest that the torpor phenotype is a product of systemic change at the cellular level that is coordinated by multiple biological processes. While the long-lived phenotype and hibernation may undergo a number of unique cellular changes, there are commonalities in terms of how cellular protection and energy metabolism are prominently involved in both phenotypes. This then puts forward the question of, are there links between longevity and other hypometabolic adaptations beside hibernation?

ANOXIA-TOLERANT TURTLES, A MODEL OF NEGLIGIBLE SENESCENCE

Negligible senescence, is a term coined to describe organisms that show very slow or negligible signs of aging. Animals that display negligible senescence have a long life span, and also encompass unique resistance toward common age-related declines that include decrease in reproduction with age, decline in physiological capacity, and the development of diseases (Finch 2009). Freshwater turtles, which include the *Chyrsemys* and *Trachemys* genera, are well-known models of negligible senescence that can live upward of 60 years without signs of aging (Congdon et al. 2003). Freshwater turtles are also extraordinary in their ability to endure anoxia, and can survive for weeks at a time without oxygen (Krivoruchko and Storey 2010). Much like hibernation in ground squirrels, anoxia stress induces profound metabolic depression in turtles that can reduce their metabolic rate to 10%–20% of their aerobic resting rate (Krivoruchko and Storey 2010). The shift toward anaerobiosis in turtles requires complex biochemical changes that include an increase in glycogen storage to fuel glycolysis, and utilization of skeletal muscle and exoskeletal shell to manage lactate accumulation to avoid acidosis. At the molecular level, anoxia-tolerant turtles undergo many cellular changes that are also seen in hibernators, which include global decreases in ATP turnover, ion channel arrest, complete suppression of protein synthesis, and induction of stress responsive genes (Hochachka et al. 1996). Turtles also have an exceptional antioxidant defense, as they can experience toxic levels of ROS generated when the oxygen supply is restored after prolong periods of anoxia. To protect against reperfusion-related ROS damages during re-oxygenation, turtle brains have been shown to have two to three times the amount of cortex level ascorbic acid compared to mammals (Rice, Lee, and Choy 1995), and turtles also show constitutively high activities of key antioxidant enzymes that include catalase, SOD, and alkyl hydroperoxide reductase (Willmore and Storey 1997). Much like hibernators, turtles show similarities in mechanisms of metabolic depression when faced with anoxia, suggesting this may be a conserved response in many species when confronted with unfavorable environmental conditions. These studies have suggested that molecular mechanisms regulating anoxia tolerance in turtles could activate protective signaling pathways that may be key in rescuing cellular declines implicated in aging, and contribute to the longevity of turtles.

HYPOXIA TOLERANCE IN NAKED MOLE RATS

Naked mole rats, *Heterocephalus glaber*, as mentioned in brief earlier, are mouse-sized rodents that in captivity show an extremely long-lived phenotype of up to 28 years. Similar to hibernators, the naked mole rat's body temperature can fluctuate in accordance to their ambient surroundings, an exceptionally rare trait of poikilothermia in mammals, and compared to other rodents, the metabolic rate of naked mole rats is approximately 70% of that in mice (Buffenstein 2005). In their natural subterranean environment of East Africa, naked mole rats live in poorly ventilated burrows with densely packed soil. As such, they are naturally hypoxic tolerant and can survive in as low as 3% oxygen for several hours. When exposed to a hypoxic condition of 7% oxygen for 1 hour, naked mole rat's metabolic rate is reduced to approximately 30% of its basal value measured at 21% oxygen (Pamenter, Dzal, and Milsom 2015). Although severe reduction in metabolic rates has been observed in mammals such as hibernators, they are aided by the profound decrease in their body temperature, which passively can reduce the metabolic demand of the animal. Naked mole rats, however, retain a constant body temperature of approximately 32°C during hypoxia exposure, suggesting that major cellular changes likely take place at the molecular level to compensate and support such drastic change in metabolic energetics. In this respect, the naked mole rat is an extremely attractive model to study mechanisms of metabolic depression independent of temperature, which may yield interesting insights compared to hibernators that experience a reduction in body temperature under metabolic depression. Next to the metabolic flexibility, naked mole rats are also extremely stress tolerant, with their fibroblast showing elevated resistance toward exposure to the heavy metal cadmium, ROS generator paraquat, heat, and the chemical mutagen methyl methane sulfonate compared to mice fibroblast (Salmon et al. 2008). Although there are limited molecular data available detailing regulation of signaling pathways in the naked mole rats, they do show an approximately 50-fold increase in the expression of tumor suppressive p53 protein in their fibroblast compared to mice, which is a key transcription factor that regulates cellular responses to multiple genotoxic stressors that include DNA damage, hypoxia, nutrient deprivation, oxidative stress, and more (Bieging and Attardi 2012). This extreme tolerance toward a wide range of stressors suggest that the mole rats likely possess enhanced mechanisms to protect against cellular damage, a trait that may contribute to its extended longevity by maintaining a state of cellular integrity and genomic stability.

It is unlikely a coincidence that animals that display enhanced stress resistance also show signs of extended longevity. Although the view on the free radical theory of aging remains debated, it is recognized that exposure to harmful stressors can cause damages at the macromolecular level, and when in excess can lead to loss of cellular integrity and genomic instability that can contribute to aging and the development of age-related diseases. As such, naturally stress-resistant models like fresh water turtles and naked mole rats are a valuable resource to study the biology of aging, and can provide some unique perspectives to understand the relationship between stress resistance and aging.

PERSPECTIVE AND CONCLUSION

At the surface, the link between hibernation and longevity appears to be casual, where the extended life span of many hibernators can be interpreted as a result of an increase in duration of inactivity, a period where the animals experience near negligible senescence. However, closer examination reveals similarities at the molecular level of how these two phenotypes can be related, and how genes and pathways that can extend life span in some organisms are also important in the establishment of a hibernation phenotype. Strategic changes to the insulin signaling pathway appears to be a central and conserved regulator of longevity extension in many species, and in hibernators, this pathway functions as an important regulator of cellular energetics by its control over protein synthesis. The importance of this pathway was also demonstrated in the hibernating Brandt's bat, where unique amino acid substitutions were found within the GHR and IGF-1R proteins that are linked to changes within the insulin axis in a manner similar to those observed in long-lived Ames and Snell mice. Hibernators also show significant upregulation of many key proteins (Nrf-2, FOXO3a) that promote stress resistance during torpor, and overexpression of these same proteins have been shown to significantly increase life span in other animal models like *C. elegans* and *D. melanogaster*. The similarities between the long-lived phenotypes may not be idiosyncratic to hibernators, and may be also associated with other animal models that undergo hypometabolism during exposure to different environmental stresses. As briefly discussed, freshwater turtles and naked mole rats are two species that are considered to be extremely long-lived, and are both capable of undergoing metabolic depression when exposed to unfavorable conditions of oxygen deprivation. Establishing a potential relationship between longevity not only to hibernation, but also hypometabolism in general would be valuable, as this will assist researchers in identifying common mechanisms regulating metabolic depression among different species that may be associated with the long-lived phenotype.

In conclusion, in our pursuit to understand the biology of aging, there may be valuable lessons in taking a comparative approach to understand how environmental stress responses like hibernation are regulated at the molecular level, which may provide valuable insights on how these metabolic adaptations can positively antagonize aging and promote longevity.

ACKNOWLEDGMENTS

Thanks go to J.M. Storey for editorial review of the manuscript. Research in the Storey laboratory is funded by a Discovery grant from the Natural Sciences and Engineering Research Council (NSERC) of Canada to K.B.S. (Grant #6793). C.W.W. is supported by a NSERC postdoctoral fellowship.

REFERENCES

Abnous, K., C. A. Dieni, and K. B. Storey. 2008. Regulation of Akt during hibernation in Richardson's ground squirrels. *Biochim Biophys Acta* 1780 (2):185–93. doi: 10.1016/j.bbagen.2007.10.009.

Andrews, M. T. 2007. Advances in molecular biology of hibernation in mammals. *Bioessays* 29 (5):431–40. doi: 10.1002/bies.20560.

Bannister, A. J., and T. Kouzarides. 2011. Regulation of chromatin by histone modifications. *Cell Res* 21 (3):381–95. doi: 10.1038/cr.2011.22.

Bartke, A., and H. Brown-Borg. 2004. Life extension in the dwarf mouse. *Curr Top Dev Biol* 63:189–225. doi: 10.1016/S0070-2153(04)63006-7.

Bieging, K. T., and L. D. Attardi. 2012. Deconstructing p53 transcriptional networks in tumor suppression. *Trends Cell Biol* 22 (2):97–106. doi: 10.1016/j.tcb.2011.10.006.

Biggar, Y., and K. B. Storey. 2014. Global DNA modifications suppress transcription in brown adipose tissue during hibernation. *Cryobiology* 69 (2):333–8. doi: 10.1016/j.cryobiol.2014.08.008.

Brown-Borg, H. M., K. E. Borg, C. J. Meliska, and A. Bartke. 1996. Dwarf mice and the ageing process. *Nature* 384 (6604):33. doi: 10.1038/384033a0.

Buck, M. J., T. L. Squire, and M. T. Andrews. 2002. Coordinate expression of the PDK4 gene: A means of regulating fuel selection in a hibernating mammal. *Physiol Genomics* 8 (1):5–13. doi: 10.1152/physiolgenomics.00076.2001.

Buffenstein, R. 2005. The naked mole-rat: A new long-living model for human aging research. *J Gerontol A Biol Sci Med Sci* 60 (11):1369–77.

Cai, D., R. M. McCarron, E. Z. Yu, Y. Li, and J. Hallenbeck. 2004. Akt phosphorylation and kinase activity are down-regulated during hibernation in the 13-lined ground squirrel. *Brain Res* 1014 (1–2):14–21. doi: 10.1016/j.brainres.2004.04.008.

Carey, H. V., M. T. Andrews, and S. L. Martin. 2003. Mammalian hibernation: Cellular and molecular responses to depressed metabolism and low temperature. *Physiol Rev* 83 (4):1153–81. doi: 10.1152/physrev.00008.2003.

Churchill, T. A., and K. B. Storey. 1993. Dehydration tolerance in wood frogs: A new perspective on development of amphibian freeze tolerance. *Am J Physiol* 265 (6 Pt 2):R1324–32.

Clancy, D. J., D. Gems, L. G. Harshman, S. Oldham, H. Stocker, E. Hafen, S. J. Leevers, and L. Partridge. 2001. Extension of life-span by loss of CHICO, a *Drosophila* insulin receptor substrate protein. *Science* 292 (5514):104–6. doi: 10.1126/science.1057991.

Congdon, J. D., R. D. Nagle, O. M. Kinney, R. C. van Loben Sels, T. Quinter, and D. W. Tinkle. 2003. Testing hypotheses of aging in long-lived painted turtles (*Chrysemys picta*). *Exp Gerontol* 38 (7):765–72.

Conti, B. 2008. Considerations on temperature, longevity and aging. *Cell Mol Life Sci* 65 (11):1626–30. doi: 10.1007/s00018-008-7536-1.

Conti, B., M. Sanchez-Alavez, R. Winsky-Sommerer, M. C. Morale, J. Lucero, S. Brownell, V. Fabre et al. 2006. Transgenic mice with a reduced core body temperature have an increased life span. *Science* 314 (5800):825–8. doi: 10.1126/science.1132191.

Dark, J. 2005. Annual lipid cycles in hibernators: Integration of physiology and behavior. *Annu Rev Nutr* 25:469–97. doi: 10.1146/annurev.nutr.25.050304.092514.

Dausmann, K. H., J. Glos, J. U. Ganzhorn, and G. Heldmaier. 2004. Physiology: Hibernation in a tropical primate. *Nature* 429 (6994):825–6. doi: 10.1038/429825a.

Eberharter, A., and P. B. Becker. 2002. Histone acetylation: A switch between repressive and permissive chromatin. Second in review series on chromatin dynamics. *EMBO Rep* 3 (3):224–9. doi: 10.1093/embo-reports/kvf053.

Eijkelenboom, A., and B. M. Burgering. 2013. FOXOs: Signalling integrators for homeostasis maintenance. *Nat Rev Mol Cell Biol* 14 (2):83–97. doi: 10.1038/nrm3507.

Fielenbach, N., and A. Antebi. 2008. *C. elegans* dauer formation and the molecular basis of plasticity. *Genes Dev* 22 (16):2149–65. doi: 10.1101/gad.1701508.

Finch, C. E. 2009. Update on slow aging and negligible senescence—A mini-review. *Gerontology* 55 (3):307–13. doi: 10.1159/000215589.

Flachsbart, F., A. Caliebe, R. Kleindorp, H. Blanché, H. von Eller-Eberstein, S. Nikolaus, S. Schreiber, and A. Nebel. 2009. Association of FOXO3A variation with human longevity confirmed in German centenarians. *Proc Natl Acad Sci U S A* 106 (8):2700–5. doi: 10.1073/pnas.0809594106.

Frerichs, K. U., C. B. Smith, M. Brenner, D. J. DeGracia, G. S. Krause, L. Marrone, T. E. Dever, and J. M. Hallenbeck. 1998. Suppression of protein synthesis in brain during hibernation involves inhibition of protein initiation and elongation. *Proc Natl Acad Sci U S A* 95 (24):14511–6.

Furuhashi, M., and G. S. Hotamisligil. 2008. Fatty acid-binding proteins: Role in metabolic diseases and potential as drug targets. *Nat Rev Drug Discov* 7 (6):489–503. doi: 10.1038/nrd2589.

Furuyama, T., H. Yamashita, K. Kitayama, Y. Higami, I. Shimokawa, and N. Mori. 2002. Effects of aging and caloric restriction on the gene expression of Foxo1, 3, and 4 (FKHR, FKHRL1, and AFX) in the rat skeletal muscles. *Microsc Res Tech* 59 (4):331–4. doi: 10.1002/jemt.10213.

Hampton, M., R. G. Melvin, A. H. Kendall, B. R. Kirkpatrick, N. Peterson, and M. T. Andrews. 2011. Deep sequencing the transcriptome reveals seasonal adaptive mechanisms in a hibernating mammal. *PLoS One* 6 (10):e27021. doi: 10.1371/journal.pone.0027021.

Harrison, D. E., R. Strong, Z. D. Sharp, J. F. Nelson, C. M. Astle, K. Flurkey, N. L. Nadon et al. 2009. Rapamycin fed late in life extends lifespan in genetically heterogeneous mice. *Nature* 460 (7253):392–5. doi: 10.1038/nature08221.

Hauck, S. J., W. S. Hunter, N. Danilovich, J. J. Kopchick, and A. Bartke. 2001. Reduced levels of thyroid hormones, insulin, and glucose, and lower body core temperature in the growth hormone receptor/binding protein knockout mouse. *Exp Biol Med (Maywood)* 226 (6):552–8.

Heldmaier, G., M. Klingenspor, M. Werneyer, B. J. Lampi, S. P. Brooks, and K. B. Storey. 1999. Metabolic adjustments during daily torpor in the Djungarian hamster. *Am J Physiol* 276 (5 Pt 1):E896–906.

Hittel, D., and K. B. Storey. 2002. The translation state of differentially expressed mRNAs in the hibernating 13-lined ground squirrel (*Spermophilus tridecemlineatus*). *Arch Biochem Biophys* 401 (2):244–54. doi: 10.1016/S0003-9861(02)00048-6.

Hochachka, P. W., L. T. Buck, C. J. Doll, and S. C. Land. 1996. Unifying theory of hypoxia tolerance: Molecular/metabolic defense and rescue mechanisms for surviving oxygen lack. *Proc Natl Acad Sci U S A* 93 (18):9493–8.

Holzenberger, M., J. Dupont, B. Ducos, P. Leneuve, A. Géloën, P. C. Even, P. Cervera, and Y. Le Bouc. 2003. IGF-1 receptor regulates lifespan and resistance to oxidative stress in mice. *Nature* 421 (6919):182–7. doi: 10.1038/nature01298.

Huang, S., and M. P. Czech. 2007. The GLUT4 glucose transporter. *Cell Metab* 5 (4):237–52. doi: 10.1016/j.cmet.2007.03.006.

Hwangbo, D. S., B. Gershman, M. P. Tu, M. Palmer, and M. Tatar. 2004. *Drosophila* dFOXO controls lifespan and regulates insulin signalling in brain and fat body. *Nature* 429 (6991):562–6. doi: 10.1038/nature02549.

Johnson, S. C., P. S. Rabinovitch, and M. Kaeberlein. 2013. mTOR is a key modulator of ageing and age-related disease. *Nature* 493 (7432):338–45. doi: 10.1038/nature11861.

Kapahi, P., B. M. Zid, T. Harper, D. Koslover, V. Sapin, and S. Benzer. 2004. Regulation of lifespan in *Drosophila* by modulation of genes in the TOR signaling pathway. *Curr Biol* 14 (10):885–90. doi: 10.1016/j.cub.2004.03.059.

Kenyon, C., J. Chang, E. Gensch, A. Rudner, and R. Tabtiang. 1993. A *C. elegans* mutant that lives twice as long as wild type. *Nature* 366 (6454):461–4. doi: 10.1038/366461a0.

Kim, Y. I., F. N. Lee, W. S. Choi, S. Lee, and J. H. Youn. 2006. Insulin regulation of skeletal muscle PDK4 mRNA expression is impaired in acute insulin-resistant states. *Diabetes* 55 (8):2311–7. doi: 10.2337/db05-1606.

Kimball, S. R. 1999. Eukaryotic initiation factor eIF2. *Int J Biochem Cell Biol* 31 (1):25–9.

Knight, J. E., E. N. Narus, S. L. Martin, A. Jacobson, B. M. Barnes, and B. B. Boyer. 2000. mRNA stability and polysome loss in hibernating Arctic ground squirrels (*Spermophilus parryii*). *Mol Cell Biol* 20 (17):6374–9.

Krivoruchko, A., and K. B. Storey. 2010. Forever young: Mechanisms of natural anoxia tolerance and potential links to longevity. *Oxid Med Cell Longev* 3 (3):186–98. doi: 10.4161/oxim.3.3.12356.

Leiser, S. F., and R. A. Miller. 2010. Nrf2 signaling, a mechanism for cellular stress resistance in long-lived mice. *Mol Cell Biol* 30 (3):871–84. doi: 10.1128/MCB.01145-09.

Lewis, K. N., E. Wason, Y. H. Edrey, D. M. Kristan, E. Nevo, and R. Buffenstein. 2015. Regulation of Nrf2 signaling and longevity in naturally long-lived rodents. *Proc Natl Acad Sci U S A* 112 (12):3722–7. doi: 10.1073/pnas.1417566112.

Lin, K., H. Hsin, N. Libina, and C. Kenyon. 2001. Regulation of the *Caenorhabditis elegans* longevity protein DAF-16 by insulin/IGF-1 and germline signaling. *Nat Genet* 28 (2):139–45. doi: 10.1038/88850.

Lyman, C. P., R. C. O'Brien, G. C. Greene, and E. D. Papafrangos. 1981. Hibernation and longevity in the Turkish hamster *Mesocricetus brandti*. *Science* 212 (4495):668–70.

Miquel, J., P. R. Lundgren, K. G. Bensch, and H. Atlan. 1976. Effects of temperature on the life span, vitality and fine structure of *Drosophila melanogaster*. *Mech Ageing Dev* 5 (5):347–70.

Morin, P., Jr., Z. Ni, D. C. McMullen, and K. B. Storey. 2008. Expression of Nrf2 and its downstream gene targets in hibernating 13-lined ground squirrels, *Spermophilus tridecemlineatus*. *Mol Cell Biochem* 312 (1–2):121–9. doi: 10.1007/s11010-008-9727-3.

Morin, P., Jr., and K. B. Storey. 2006. Evidence for a reduced transcriptional state during hibernation in ground squirrels. *Cryobiology* 53 (3):310–8. doi: 10.1016/j.cryobiol.2006.08.002.

Morris, B. J., D. C. Willcox, T. A. Donlon, and B. J. Willcox. 2015. FOXO3: A major gene for human longevity—A mini-review. *Gerontology* 61 (6):515–25. doi: 10.1159/000375235.

Murphy, C. T., S. A. McCarroll, C. I. Bargmann, A. Fraser, R. S. Kamath, J. Ahringer, H. Li, and C. Kenyon. 2003. Genes that act downstream of DAF-16 to influence the lifespan of *Caenorhabditis elegans*. *Nature* 424 (6946):277–83. doi: 10.1038/nature01789.

Nascimento, E. B., M. Snel, B. Guigas, G. C. van der Zon, J. Kriek, J. A. Maassen, I. M. Jazet, M. Diamant, and D. M. Ouwens. 2010. Phosphorylation of PRAS40 on Thr246 by PKB/AKT facilitates efficient phosphorylation of Ser183 by mTORC1. *Cell Signal* 22 (6):961–7. doi: 10.1016/j.cellsig.2010.02.002.

Ni, Z., and K. B. Storey. 2010. Heme oxygenase expression and Nrf2 signaling during hibernation in ground squirrels. *Can J Physiol Pharmacol* 88 (3):379–87. doi: 10.1139/Y10-017.

Pamenter, M. E., Y. A. Dzal, and W. K. Milsom. 2015. Adenosine receptors mediate the hypoxic ventilatory response but not the hypoxic metabolic response in the naked mole rat during acute hypoxia. *Proc Biol Sci* 282 (1800):20141722. doi: 10.1098/rspb.2014.1722.

Pan, K. Z., J. E. Palter, A. N. Rogers, A. Olsen, D. Chen, G. J. Lithgow, and P. Kapahi. 2007. Inhibition of mRNA translation extends lifespan in *Caenorhabditis elegans*. *Aging Cell* 6 (1):111–9. doi: 10.1111/j.1474-9726.2006.00266.x.

Podlutsky, A. J., A. M. Khritankov, N. D. Ovodov, and S. N. Austad. 2005. A new field record for bat longevity. *J Gerontol A Biol Sci Med Sci* 60 (11):1366–8.

Rice, M. E., E. J. Lee, and Y. Choy. 1995. High levels of ascorbic acid, not glutathione, in the CNS of anoxia-tolerant reptiles contrasted with levels in anoxia-intolerant species. *J Neurochem* 64 (4):1790–9.

Rolfe, D. F., and G. C. Brown. 1997. Cellular energy utilization and molecular origin of standard metabolic rate in mammals. *Physiol Rev* 77 (3):731–58.

Salmon, A. B., A. A. Sadighi Akha, R. Buffenstein, and R. A. Miller. 2008. Fibroblasts from naked mole-rats are resistant to multiple forms of cell injury, but sensitive to peroxide, ultraviolet light, and endoplasmic reticulum stress. *J Gerontol A Biol Sci Med Sci* 63 (3):232–41.

Schwartz, C., M. Hampton, and M. T. Andrews. 2013. Seasonal and regional differences in gene expression in the brain of a hibernating mammal. *PLoS One* 8 (3):e58427. doi: 10.1371/journal.pone.0058427.

Seim, I., X. Fang, Z. Xiong, A. V. Lobanov, Z. Huang, S. Ma, Y. Feng et al. 2013. Genome analysis reveals insights into physiology and longevity of the Brandt's bat *Myotis brandtii*. *Nat Commun* 4:2212. doi: 10.1038/ncomms3212.

Selman, C., S. Lingard, A. I. Choudhury, R. L. Batterham, M. Claret, M. Clements, F. Ramadani et al. 2008. Evidence for lifespan extension and delayed age-related biomarkers in insulin receptor substrate 1 null mice. *FASEB J* 22 (3):807–18. doi: 10.1096/fj.07-9261com.

Stewart, J. M., T. E. English, and K. B. Storey. 1998. Comparisons of the effects of temperature on the liver fatty acid binding proteins from hibernator and nonhibernator mammals. *Biochem Cell Biol* 76 (4):593–9.

Storey, K. B. 2010. Out cold: Biochemical regulation of mammalian hibernation—A mini-review. *Gerontology* 56 (2):220–30. doi: 10.1159/000228829.

Sykiotis, G. P., and D. Bohmann. 2008. Keap1/Nrf2 signaling regulates oxidative stress tolerance and lifespan in *Drosophila*. *Dev Cell* 14 (1):76–85. doi: 10.1016/j.devcel.2007.12.002.

Sykiotis, G. P., I. G. Habeos, A. V. Samuelson, and D. Bohmann. 2011. The role of the antioxidant and longevity-promoting Nrf2 pathway in metabolic regulation. *Curr Opin Clin Nutr Metab Care* 14 (1):41–8. doi: 10.1097/MCO.0b013e32834136f2.

Tullet, J. M. 2015. DAF-16 target identification in *C. elegans*: Past, present and future. *Biogerontology* 16 (2):221–34. doi: 10.1007/s10522-014-9527-y.

Tullet, J. M., M. Hertweck, J. H. An, J. Baker, J. Y. Hwang, S. Liu, R. P. Oliveira, R. Baumeister, and T. K. Blackwell. 2008. Direct inhibition of the longevity-promoting factor SKN-1 by insulin-like signaling in *C. elegans*. *Cell* 132 (6):1025–38. doi: 10.1016/j.cell.2008.01.030.

van Breukelen, F., and S. L. Martin. 2002. Reversible depression of transcription during hibernation. *J Comp Physiol B* 172 (5):355–61. doi: 10.1007/s00360-002-0256-1.

van Breukelen, F., N. Sonenberg, and S. L. Martin. 2004. Seasonal and state-dependent changes of eIF4E and 4E-BP1 during mammalian hibernation: Implications for the control of translation during torpor. *Am J Physiol Regul Integr Comp Physiol* 287 (2):R349–53. doi: 10.1152/ajpregu.00728.2003.

Van Raamsdonk, J. M., and S. Hekimi. 2009. Deletion of the mitochondrial superoxide dismutase sod-2 extends lifespan in *Caenorhabditis elegans*. *PLoS Genet* 5 (2):e1000361. doi: 10.1371/journal.pgen.1000361.

Van Rakamsdonk, J. M., and S. Hekimi. 2012. Superoxide dismutase is dispensable for normal animal lifespan. *Proc Natl Acad Sci U S A* 109 (15):5785–90. doi: 10.1073/pnas.1116158109.

Van Voorhies, W. A., and S. Ward. 1999. Genetic and environmental conditions that increase longevity in *Caenorhabditis elegans* decrease metabolic rate. *Proc Natl Acad Sci U S A* 96 (20):11399–403.

Wilkinson, G. S., and J. M. South. 2002. Life history, ecology and longevity in bats. *Aging Cell* 1 (2):124–31.

Willcox, B. J., T. A. Donlon, Q. He, R. Chen, J. S. Grove, K. Yano, K. H. Masaki, D. C. Willcox, B. Rodriguez, and J. D. Curb. 2008. FOXO3A genotype is strongly associated with human longevity. *Proc Natl Acad Sci U S A* 105 (37):13987–92. doi: 10.1073/pnas.0801030105.

Williams, C. T., A. V. Goropashnaya, C. L. Buck, V. B. Fedorov, F. Kohl, T. N. Lee, and B. M. Barnes. 2011. Hibernating above the permafrost: Effects of ambient temperature and season on expression of metabolic genes in liver and brown adipose tissue of arctic ground squirrels. *J Exp Biol* 214 (Pt 8):1300–6. doi: 10.1242/jeb.052159.

Willmore, W. G., and K. B. Storey. 1997. Antioxidant systems and anoxia tolerance in a freshwater turtle *Trachemys scripta elegans*. *Mol Cell Biochem* 170 (1–2):177–85.

Wu, C. W., K. K. Biggar, and K. B. Storey. 2013. Dehydration mediated microRNA response in the African clawed frog *Xenopus laevis*. *Gene* 529 (2):269–75. doi: 10.1016/j.gene.2013.07.064.

Wu, C. W., and K. B. Storey. 2012. Regulation of the mTOR signaling network in hibernating thirteen-lined ground squirrels. *J Exp Biol* 215 (Pt 10):1720–7. doi: 10.1242/jeb.066225.

Wu, C. W., and K. B. Storey. 2014. FoxO3a-mediated activation of stress responsive genes during early torpor in a mammalian hibernator. *Mol Cell Biochem* 390 (1–2):185–95. doi: 10.1007/s11010-014-1969-7.

Wu, C. W., and K. B. Storey. 2016. Life in the cold: Links between mammalian hibernation and longevity. *Biomol Concepts* 7 (1):41–52. doi: 10.1515/bmc-2015-0032.

Xu, R., E. Andres-Mateos, R. Mejias, E. M. MacDonald, L. A. Leinwand, D. K. Merriman, R. H. Fink, and R. D. Cohn. 2013. Hibernating squirrel muscle activates the endurance exercise pathway despite prolonged immobilization. *Exp Neurol* 247:392–401. doi: 10.1016/j.expneurol.2013.01.005.

Zhang, Y., and D. Reinberg. 2001. Transcription regulation by histone methylation: Interplay between different covalent modifications of the core histone tails. *Genes Dev* 15 (18):2343–60. doi: 10.1101/gad.927301.

Section X

Mathematical Modeling of Aging

35 The Role of Mathematical Modeling in Understanding Aging

Mark Mc Auley, Amy Morgan, and Kathleen Mooney

CONTENTS

INTRODUCTION

The process of growing old and aging has intrigued and troubled scholars, philosophers, and scientists from the beginning of civilization. The question of why we age is challenging, however thanks to the contributions of many eminent scientists this conundrum has largely been resolved. Mathematical models have been to the forefront in helping to unravel this puzzle and in this chapter we discuss the contribution they have made to biogerontology [1,2]. In order for mathematical models to be successfully applied to unraveling why aging occurs it was necessary to understand aging from an evolutionary perspective. The first scientist to seriously consider aging through the lens of evolution was the brilliant German biologist August Weismann. In 1881, a mere 22 years after Darwin published *On The Origin of Species by Means of Natural Selection* [3], Weisman outlined a purely adaptive theory to explain aging [4]. He reasoned that senescence is the result of an evolved limitation to the division potential of somatic cells and argued this limitation evolved because it is beneficial to eliminate old individuals from a population, thus freeing up resources for younger individuals. This theory resonated with much of the evolutionary and social demographic thinking of the nineteenth century as Weismann hypothesized senescence or "programmed death" as it became known was an evolved trait and its utility was to remove decrepit individuals from a population. As a theory it remains controversial for a number of reasons; first, if a trait was to be selected for by evolution, for example, "programmed aging/death" it would have been necessary for it to manifest itself in such a way as to have a bearing on the survival of the organism; and as aging is rarely witnessed in the wild due to high rates of extrinsic mortality, there would have been limited selection pressure for "programmed death" to have evolved. It is uncertain if Weismann recognized this limitation but he did go on to propose a second nonadaptive theory that considered old individuals as neutral and the evolution of aging as a "panmixia," where neutral characters decline during evolution [5]. Thus, Weismann's original idea of "programmed aging" has been in the main consigned to the annals of history. Although the idea of "programmed aging" still has its advocacy recently, Mitteldorf argued for a form of programmed aging by postulating that aging is regulated by an epigenetic clock [6].

Weisman did not have the mathematical machinery at his disposal to formalize his ideas. However, following Weismann the mathematical foundations to fully explore aging from an evolutionary perspective were laid in the 1920s by the statistical geneticist R.A. Fisher and the polymath J.B.S. Haldane. Fisher and Haldane both developed mathematical models which utilized a Malthusian parameter as a key determinant of fitness [7,8]. Although their models did not explicitly address the question of why aging occurs, the introduction of a Malthusian parameter was a seminal event as it laid the foundations for the development of future mathematical models of senescence. Moreover, the theoretical musings of Fisher and Haldane were utilized by Sir Peter Medawar in 1952 when he formulated the next significant theory, which attempted to explain why we age [9]. Even though Medawar did not use mathematics, his arguments were strongly influenced by the mathematical logic introduced by Fisher and Haldane. The essence of Medawars' hypothesis is that the force of natural selection decreases with age. Based on this premise, it is argued that if a fatal trait is expressed before reproductive age there will be strong selection pressure to remove it. Therefore, the majority of organisms bearing the trait will die before passing it on to their offspring. Moreover, if a harmful trait is expressed after reproductive age many individuals would have died throughout the process of evolution before the trait had an opportunity to be expressed. Thus, the focal point of this theory is that aging can be defined by the accumulation of late-acting mutations. This idea is now known as the mutation accumulation theory. In 1957 George C. Williams published a theory which is also based on the decline of natural selection with age [10]. Williams proposed that genes exist with antagonistic effects. According to Williams these genes have beneficial effects early in life and detrimental effects in older age. For example, if a gene conferred a beneficial effect on reproductive ability it would be selected for, despite any adverse effects of it later in life. This theory is now referred to as the "antagonistic pleiotropy theory." In contrast to Medawar and Williams, in 1966 Hamilton did utilize mathematics in an attempt to understand the evolution of senescence [11]. Like Fisher and Haldane before him, Hamilton utilized a Malthusian parameter embroidered within mathematical equations to represent organismal fitness. Hamilton's mathematical model was able to corroborate the arguments of Medawar and Williams that the force of natural selection diminishes with age.

The 1970s were a halcyon period for biogerontology as mathematical models significantly enhanced our fundamental understanding of aging. For instance, Brian Charlesworth continued the mathematical treatment of aging with the aid of Malthusian parameters, as originally conceptualized by Haldane, Fisher, and Hamilton [12,13]. In addition to the work of Charlesworth, an alternative evolutionary theory of aging encased within a mathematical framework was introduced by Thomas Kirkwood [14]. Known as the disposable soma theory; the kernel of this idea is that: it is necessary for energy to be divided between investment in reproduction and somatic maintenance. As extrinsic mortality is high in the wild, the optimal strategy for the allocation of this energy is based on the life expectancy of the organism. If maintenance is set too high, energy will be wasted when the organism dies. If maintenance is set too low this will result in premature intrinsic mortality. Thus, according to Kirkwood organisms have evolved so the amount of energy invested in somatic maintenance is sufficient to reach reproductive years but less than needed to achieve indefinite survival.

It can be argued the elegance of the disposable soma theory is that it provides a natural fusion between evolutionary theory and the cellular maintenance and repair mechanisms which have been identified as key to how aging unfolds. In the last few decades biogerontology has made remarkable progress in understanding these processes [1]. It is now known that aging is underscored by an array of cellular processes [15]. Moreover, it is beyond doubt that the collective dysregulation of these processes over time eventually renders an organism increasingly likely to die. However, despite the burgeoning knowledge of these processes as individual entities, to date we only have a rudimentary understanding of how they interact with one another to shape the trajectory of aging and onset of disease. This is an issue of huge significance to biogerontology because aspects of the aging processes are malleable and gaining a deeper understanding of the crosstalk which exists between these mechanisms could enable this malleability to be therapeutically exploited. However, its realization will not be straightforward. The dynamic behavior which underpins the biological processes associated with aging are not trivial and overcoming this problem will be challenging. Fortunately, as with unraveling the evolution of aging, mathematical models are beginning to come to the forefront in helping to solve this problem. Moreover, systems biology; a paradigm which has at its core a non-reductionist view of biology is helping to further embed mathematical models within bioscience [16,17].

Mathematical models are at the centre of systems biology due to their ability to handle the detailed information and complex dynamics associated with biological systems and models now routinely sit alongside conventional laboratory experiments. Thus, mathematical modeling presents a solution to the challenges associated with studying the inherent complexity associated with aging. In this chapter, we will, (1) outline what mathematical modeling is, (2) discuss the main theoretical paradigms used to build models, (3) elucidate why mechanistic mathematical models are a necessity for studying aging, (4) provide the details of the software and modeling exchange formats which are used to model biological systems, and (5) discuss existing mathematical models of aging.

WHAT ARE MATHEMATICAL MODELS?

As outlined in the introduction, the application of mathematical models to the study of aging has been around for a considerable period of time. However, in order to fully appreciate the utility of a model it is necessary to understand what it is and what it aims to do. In general terms a model can be defined as a process whereby a biological system is described in a precise manner using mathematics. When Fisher and Haldane developed their mathematical models computers were unavailable, however nowadays computers are routinely used to solve the equations which underpin models; something which is becoming increasingly important as models of biological systems are becoming progressively grand in scale, and thus are only tractable with the aid of a computer. However, regardless of its size, a worthwhile model should aim to provide an alternative view of the mechanisms which underpin the biology of aging. In essence something that would otherwise be unlikely or impossible to do by using conventional means. Consequently, when a model is configured correctly it provides several advantages over experimental approaches. The advantages mathematical modeling provides for improving the understanding of aging are outlined in Table 35.1.

Steps Involved in Building a Model

In order to build a mathematical model a number of steps are required (see References 18–22]. The first element of model building involves identifying previous models. This is a pivotal step as the model may have been built previously or a model may exist which can be adapted. If no suitable model exists, then it is necessary to assemble a new one. The next step of model building involves clearly defining a hypothesis. This step should be informed by clarity of thought and it is unadvisable to begin model building in the absence of a well-defined hypothesis. Moreover, a cogent and cohesive hypothesis is imperative as it will establish the utility of the model while also dictating its boundary points. An ill-defined model which does not have a clear aim can rapidly evolve into a cumbersome description of the biology with limited utility. Unfortunately, there are many recent

TABLE 35.1

Benefits of Mathematical Modeling to Understanding Aging

Advantage	Explanation
Representing the complexity of aging	Aging is phenomenally complex and is underscored by a network of overlapping processes. Modeling can help to represent this biological detail cohesively so the collective dynamics of these process can be explored in a holistic fashion.
Generation of novel insights about aging	Modeling can improve our understanding of the underlying biology associated with aging.
Identification of areas for further experimental research	A model can pin-point gaps in our current knowledge, and isolate a particular aspect of the biology that merits further experimental investigation.
A model may generate counterintuitive explanations about aging	Explanations and unusual predictions about aging can come to light that would otherwise be unapparent if the system was not studied in an integrated manner.
Ethical constrains	There are many experiments which cannot be conducted due to ethical constraints. For example, modeling provides a means of examining the effects of a radical dietary regime such as caloric restriction (CR).
The time spans involved in studying aging	Conventional approaches are limited when it comes to studying aging over extended timeframes. Mathematical models offer a solution to this as model simulations can take place over relatively short periods of time
Analysing intrinsic aging under a wide range of conditions	Mathematical models provide an efficient way to examine the aging process under an array of conditions. For example, analyzing the effects of an array of dietary components together with lifestyle factors such as physical activity.
Examining differing paradigms	There are many different ideas about aging; thus mathematical models can be utilized to compare and contrast different ideas and hypotheses.
Keeping up to date with the rapidly evolving field of aging	Biogerontology is a field which is continually evolving. Thus, a model can be easily updated as our knowledge of the biology of aging changes.

examples of this type of model, which is overly descriptive and does not have a clear purpose. When a satisfactory hypothesis is formulated the next step involves determining what variables to include in the model and deciding how these variables interact. Evidently, this process of abstraction can be a challenging aspect of model building, particularly if the boundary points of the model are nebulous due to an ill-defined hypothesis. When the list of variables and their relationships are characterized it is commonplace to then translate them into a diagram which encapsulates the essence of the model. This part of the assembly process captures the essence of the model and facilitates its communication. The diagram is akin to an electrical wiring diagram which serves as an unambiguous template for entering the details of the model into a computer.

There are many software tools which can be used to facilitate model building and a discussion of them is beyond the scope of this chapter [23]. Key to this part of the process is that the software tool which is selected is conducive to the level of expertise of the modeler [20]. Certain software tools have a steep learning curve for those unfamiliar with them. Fortunately, an array of intuitive computational systems biology software tools now exists for those new to modeling [23]. When a suitable tool is identified, the model is assembled and the mathematics underpinning the model solved so that it can be analyzed, explored, and interpreted. If the output from the model is deemed acceptable and in line with the behavior of its biological counterpart, then the idea or hypothesis, which the model was designed to answer, can be evaluated. However, if a model is not aligned with the biology, this can indicate one of two things: firstly, the model may not be a realistic representation of the biological system. In this case it is necessary to adjust the model, its parameters or both. Alternatively, the model could be "correct" and it has revealed something counterintuitive or a previously unknown aspect of the biological system. In this case this would open the opportunity for some novel laboratory experiments to explore this finding in greater depth. Either way these points

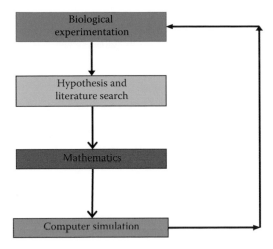

FIGURE 35.1 Steps involved in building a mathematical model.

serve to illustrate that modeling is a cyclical process and should be closely coupled with experimentation (Figure 35.1).

MATHEMATICAL MODELING FRAMEWORKS

There are two main mathematical paradigms which are generally utilized to assemble a model of a biological system. The leading paradigm is based on viewing a biological system as deterministic in nature. This means that randomness is assumed to have a negligible effect on the system of interest. Deterministic models are generally constructed with ordinary differential equations (ODEs). ODEs depend on one independent variable (time) and they emerge when there is a necessity to mathematically describe interdependent variables; for example, the rate of change in concentration of a certain cellular entity, for example, within the context of an aging cell this could be fluctuating levels of reactive oxygen species (ROS). If ROS are used as an example, it can be assumed ROS levels are influenced by their rates of production and neutralization by antioxidants. Thus, this relationship can be followed over time by using an ODE. As a first step to doing this it is possible to represent the relationship between the two entities using biochemical reactions:

$$\varnothing \xrightarrow{k_1} \text{ROS} \quad \text{(Production)}$$

$$\text{ROS} \xrightarrow{k_2} \varnothing \quad \text{(Neutralization)}$$

Based on these biochemical reactions the following ODE can be derived:

$$\frac{d\text{ROS}}{dt} = k_1\text{ROS} - k_2\text{ROS}$$

The left-hand side of the equation encapsulates the production rate of ROS while the right-hand side of the equation represents the neutralization of ROS by antioxidants. Using this approach, we can assemble a model of the variables and their interactions. The differential equations can then be solved by a computer which uses numerical integration. Figure 35.2 presents the output from the model following numerical integration.

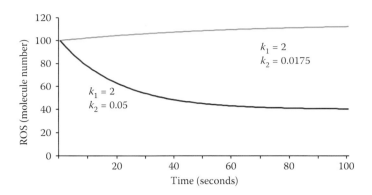

FIGURE 35.2 Example of deterministic simulations of the model of ROS generation and neutralization. The green line represents ROS when antioxidant levels are low, while the blue line represents a situation where the rate of antioxidant neutralization is sufficient to cause a drop in ROS.

The second mathematical framework employed to capture the dynamics of biological systems are stochastic models. In contrast to deterministic models stochastic models are founded on the premise that it is necessary to take into account the intrinsic mechanical fluctuations in the binding and diffusion dynamics of the molecules involved in the reactions [24,25]. This becomes particularly important when considering small populations of molecules rather than concentrations. For example, in a microenvironment, defined by low-molecular populations, molecules react at discrete time-points via random collisions between individual particles. This theoretical framework has become particularly important for modeling aging as random molecular behavior and the accumulation of damage by chance is widely regarded as key to how intracellular aging progresses [26]. Therefore, this scenario necessitates an alternative mathematical paradigm to the continuous deterministic approach of ODEs. Stochastic models attempt to capture the discrete random collisions between molecules, and are built using propensity functions. The key difference between a stochastic and a deterministic model is that given the same set of initial conditions and parameters a deterministic model will produce the same output; however, a stochastic model will produce a different solution with each simulation. This key difference is highlighted in Figure 35.3.

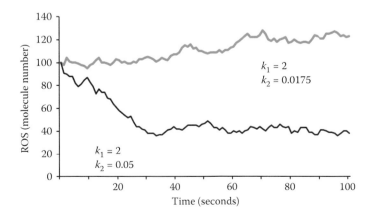

FIGURE 35.3 Example of stochastic simulations of the simple model of ROS generation and neutralization. The green line represents ROS when antioxidant levels are low, while the blue line represents a situation where the rate of antioxidant neutralization is sufficient to cause a drop in ROS levels. Note the different profile of the stochastic simulations compared to the deterministic simulations.

MODELING SOFTWARE, STANDARDS, AND REPOSITORIES

Over the years the computational systems biology community has produced an extensive range of software which can be used to assemble a mathematical model for aging research. A worthwhile place to find a tool is the Systems Biology Markup Language (SBML) [27] software matrix (http://sbml.org/SBML_Software_Guide/SBML_Software_Matrix).

At present, SBML is the leading schema for exchanging computational systems biology models. The website contains descriptions of a wide range of tools; however, CellDesigner [28] and Copasi [29] are two software tools worth highlighting, because they possess intuitive user interfaces and are straightforward to use. The systems biology fraternity has also developed BioModels [30], an online repository for archiving models which have been coded in SBML and which have been accepted for publication. BioModels provide each deposited model with a unique identification number and it is categorized as either curated or non-curated. BioModels administrators also validate a model to ensure its output matches the results in its corresponding publication.

WHY STUDIES ON AGING NEED INTEGRATED MATHEMATICAL MODELS

The aging process is characterized by its global nature where interactions occur over a range of spatial and temporal scales (Figure 35.4). Until recently it has been commonplace to investigate aging in a reductionist manner and efforts have focused on unraveling the dynamics of one particular aspect of aging in isolation. However, in order to fully understand aging and its impact on healthspan it is necessary to view these processes in an integrated manner. Consider cellular senescence, which takes place when differentiating cells stop dividing [31]. This process is closely associated with telomere attrition [32]. Moreover, both telomere erosion and cell senescence have been shown to be modulated by oxidative stress. For instance, as demonstrated by Kruz et al. ROS have been shown to increase the rate of telomere loss and accelerate the onset of senescence [33]. In addition, it has been postulated that the beneficial effects of CR are mediated via a reduction in the production of ROS [34]. Moreover, a range of other processes have been implicated in cell aging. For instance, the accumulation of unrepaired DNA damage has long been regarded as a significant underlying

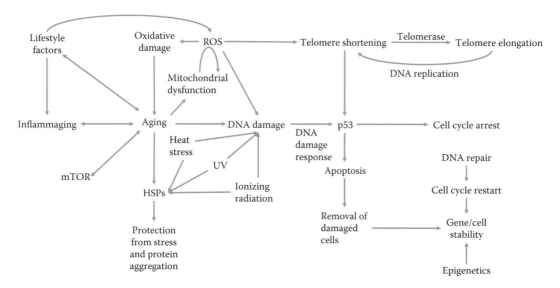

FIGURE 35.4 An integrated view of aging. The aging process is defined by many individual mechanisms interacting in a collective manner. An improved understanding of aging and healthspan will only be gained by studying the dynamics of this complex phenomenon in an integrated fashion. Mathematical modeling provides an ideal framework for doing this.

component of aging. DNA damage is caused by extrinsic factors such as environmental pollutants and UV light, or intrinsic factors such as ROS [35]. Furthermore, ROS also damage proteins, some of which can be corrected by molecular chaperones [36]. However, overtime the efficiency of the chaperone system diminishes with age and its dysregulation results in the further accumulation of protein damage [37]. In addition to these molecular processes, the pathways defined by the mammalian target of rapamycin (mTOR), which is underpinned by two complexes (mTORC1 and mTORC2), could have a significant role to play in aging and healthspan [38]. This is because they regulate a plethora of nutrient and hormonal signals, which underpin processes such as cell growth, cell size regulation, the metabolic state of the cell, mitochondrial homeostasis, and autophagy. It is widely accepted that the effects of CR could in part be modulated by mTOR or a component of one of the many pathways it overlaps with. In addition to metabolic pathways, such as that defined by mTOR, growing findings from the field of epigenetics also emphasize the integrated nature of aging. Most notably, a wealth of experimental evidence now connects DNA methylation status to healthspan and aging [39]. For example, DNA methylation patterns have been shown to alter as a result of exposure to ROS [40].

Evidently, given the findings we have presented, it is apparent that the processes associated with aging are deeply integrated. Unfortunately, to date mathematical models have mainly focused on individual aspects of cellular aging and are somewhat narrowly defined in terms of their scope. However, despite their limited nature, they have contributed significantly to our understanding of aging [1]. The next section will discuss these models and their findings. Furthermore, the goal of this section will be to stress the growing need to integrate these seemingly disparate models to create an overall dynamic picture of aging.

Modeling ROS and Mitochondria

Harmans' free-radical theory postulates that aging is the result of the accumulation of free-radical damage over time [41,42]. Produced mainly in the mitochondria, free radicals induce oxidative damage, which in turn leads to mitochondrial dysfunction and further ROS production. To explore the relationship ROS have with aging Kirkwood and Kowald used a mathematical model to show that increased amounts of antioxidants reduced mitochondrial DNA damage [43]. Kirkwood and Kowald also used mathematical modeling to explore the accumulation of mtDNA deletion mutants with age [44]. They proposed the presence of a product inhibition loop that reduces the rate of transcription when the product is in excess, and if deletion events occur at crucial genes, then mtDNA deletion mutants have the ability to replicate at a significantly increased rate compared to wild-type mtDNA. Using single-cell simulations it was shown that a single mutant can outgrow the wild-type population and produce a COX-negative phenotype in just a few months [44]. Notably, Miwa et al. investigated the proposed relationship CR has with mitochondrial energy metabolism to suggest that CR is associated with a reduction in ROS [34,45]. Moreover, Jacob et al. used mathematical modeling to determine that mitochondrial outer membrane permeabilization, a central process to apoptosis, was dependent on both spatiotemporal tBid and ROS signaling. Interestingly, models of each singular process failed to completely match experimental data, while an integrated model of both processes was able to accurately reproduce experimental findings [46]. This finding is key as it makes it apparent that ROS intersect with many other elements of aging. In recognition of this we recently suggested that ROS could be used as the "adhesive" to "glue" together the seemingly disparate aspects of aging into a holistic framework. In addition we suggested that mathematical modeling could act as the conduit for exploring the integrated picture of aging [47].

Telomeres

Telomeres are protective caps found at the end of chromosomes which consist of a repeated TTAGGG sequence. Telomeres shorten with each cell division and in the 1990s this phenomenon was inextricably connected to cellular aging by Harley et al. [48]. Following this key finding, mathematical models have significantly helped to unravel the relationship telomeres have with cellular aging. Using mathematical modeling, Buijs et al. discovered that telomere length is regulated by negative feedback [49]. Moreover, Rubelj and Vondraček demonstrated through mathematical modeling that abrupt telomere shorting by recombination or nuclease digestion was responsible for the onset of cell senescence [50]. Mathematical modeling was also used by Proctor and Kirkwood to demonstrate that telomere attrition can be increased by excessive oxidative stress, and is also coupled to reduced replicative life span [51]. Using this finding as its foundation, the model by Proctor and Kirkwood was subsequently updated to explore the hypothesis that cell-cycle arrest is due to telomere uncapping, as a result of telomere shortening, rather than telomere shortening alone. Results from the updated model were more in line with experimental data, than the findings from the previous model, thus supporting the hypothesis that cell senescence is regulated by telomere uncapping [52]. Furthermore, Hirt et al. demonstrated by using a deterministic model that low concentrations of the potential anticancer drug RHPS4, a G-quadruplex ligand, can reduce telomere length, while high concentrations can destabilize the system and induce senescence and apoptosis [53].

mTOR

Encoded by the *MTOR* gene, mTOR is the catalytic subunit of the five subunit complex mTORC1 and six subunit complex mTORC2. mTORC1 plays an important role in cell growth and proliferation through protein, lipid, and organelle synthesis. This protein is less well understood, but is thought to be involved in cell survival, metabolism, proliferation, and cytoskeleton organization [54]. The mTOR pathway is complex due to the multiple interactions with upstream regulators and downstream targets, with one comprehensive systems biology graphical notation diagram of the mTOR signaling network containing 964 species and 777 reactions [55]. Activation of the mTOR pathway is observed in age-related diseases such as cancer, insulin resistance, and type 2 diabetes mellitus, while inhibition of the pathway has been shown to protect model organisms from such conditions, and extend life span [38]. Using a mathematical model, Kriete et al. showed that a reduced sensitivity of mTOR to low ATP had a significant effect on life span, and reduced the rate of oxidized protein and damaged mitochondria accumulation [56]. Mathematical modeling was also used by Dalle et al. to show that mTOR inhibition could reduce mitochondrial mass after exposure to cell senescence inducing irradiation [57]. Many mathematical models have focused on the role of the mTOR pathway in carcinogenesis [58]. For instance, Mosca et al. compared metabolic states in proliferating cancer and normal cells, which were created by alterations to the level of signaling through the PI3 K/Akt/mTOR pathway [59]. Intriguingly, Wang and Krueger suggest that mathematical modeling of carcinogenesis could provide insights into potential therapeutic strategies for this disease [60]. Moreover, Thobe and Siebert mathematically modeled the crosstalk between the mitogen-activated protein kinases (MAPK) and mTOR signaling pathways to test chemotherapeutic drugs, and found that Nexavar and Selumetinib, either used individually, or in combination, were unable to induce cell apoptosis as had been observed by previous experimental work [61]. More recently, from an integrated aging perspective we proposed creating a mathematical model which would be capable of investigating crosstalk between mTOR and silent information regulator proteins—sirtuins [62]. Sirtuins are a group of enzymes heavily implicated with aging that operate as NAD+-dependent deacetylases [63]. Our logic was underpinned by the fact that sirtuins take part in a broad range of cellular activities, from nutrient sensing to DNA damage and their activities overlap with the mTOR pathway [64]. Thus, it is logical to integrate these two key cellular processes.

FOLATE METABOLISM

By acting as a carbon donor in folate-mediated one carbon metabolism (FOCM), folate plays a vital role in DNA methylation and nucleotide synthesis. This pathway and its relationship with health have been extensively mathematically modeled (reviewed by Mc Auley et al. [65]). As DNA methylation aberrations are associated with healthspan [66–68], it is imperative that the mechanisms underpinning FOCM and aging are well characterized. Using mathematical modeling, Reed et al. made several observations of FOCM, including that low-folate concentrations can amplify the inverse relationship between folate and homocysteine [69]. Furthermore, the model was used to investigate vitamin B12 deficiency. This model was later utilized to explore the effect of genetic polymorphisms on carcinogenesis-relevant mechanisms. It was determined that the model was robust to single genetic polymorphisms; however, the effect of the genetic polymorphism was amplified in the presence of low-folate status [70]. The model was again utilized in 2011 to analyze the effect of excess intracellular folate on the rate of methylation and purine deoxynucleosides and thymidylate synthesis. It was found that while excess intracellular folate produced a modest increase in DNA methylation capacity and purine synthesis, excessive folate resulted in an almost fivefold increase thymidylate synthesis, which may be associated with an increased risk for preneoplastic lesions or cancer [71]. This is significant as, although there is substantial evidence that folate has an important role to play in the prevention of cancer, very high levels of folate may promote carcinogenesis [72]. Not only has mathematical modeling been used to study how aberrations to FOCM may impact cancer pathogenesis, Morrison and Allegra used modeling to explore folate metabolism within human breast carcinoma (MCF-7) cells [73]. More recently, and although not specifically directed at aging, a model of folate metabolism was developed by Salcedo-Sora and Mc Auley [74]. This model was able to isolate specific targets embedded within folate metabolism, which congruently dovetailed with our current understanding of how antifolates behave *in vivo*. Moreover, output from the model reflected emerging experimental findings which suggest a key component of folate metabolism known as the folinic acid substrate cycle is a decisive mechanism deployed during active cell growth. Interestingly, despite the importance of DNA methylation and the folate cycle to aging and health, to date no mathematical model has successfully modeled the full scope of their relationship. Thus, in order to improve the understanding of this association we recently proposed a rationale for the development of a combined model which captures the full extent of crosstalk between DNA methylation and the folate cycle and their intersection with aging [65].

THE DNA DAMAGE RESPONSE

Genome instability, and the accumulation of DNA damage has been associated with aging and age-related diseases such as cancer [75]. The tumor suppressor protein p53 plays a key role in the DNA damage response, by initiating DNA repair, growth arrest, senescence, and apoptosis [76]. There are several mathematical models of p53 and the DNA damage response. For example, Ma et al. created a model that showed double stranded DNA breaks, induced by ionizing radiation, form complexes with DNA repair proteins, which are detected by ataxia telangiectasia mutated (ATM), and in turn activate p53, following its phosphorylation. The model found that when stochasticity is assumed for the number of DNA breaks and repair process, increasing ionizing radiation induced an increase in the coordinated dynamics of p53 and Mdm2, a negative regulator of p53 [77]. p53 stabilization can be brought about by both p14^{ARF} and ATM. Proctor and Grey created two stochastic models of these mechanisms in the p53/Mdm2 system; the ARF model predicted that a threshold level of DNA damage, which accumulates with age, is required for the development of oscillations, while the ATM model oscillations are more varied, which may account for the variation seen in the experimental data of ARF negative cells [78]. Building on the model by Ma et al., Sun et al. created a stochastic delay model to analyze p53 dynamics under stressed and non-stressed conditions. The model showed that in the absence of stress, p53 pulses could be activated by DNA double strand breaks, something which can occur during normal cell cycling, with the first pulse associated with the G1 phase [79]. Chong et al. expanded the model by Sun et al. to define p53 excitable dynamics as type II excitability. The authors went on to discuss the elevation of Mdm2, MdmX and Wip1, inhibitors of p53, in cancer cells, and showed that suppression of these overexpressed feedback regulators can reactivate the default oscillatory behavior of p53 [80]. Dolan et al. also analyzed the effect of anticancer therapeutics by combining two preexisting models [81,82], and Dolan et al. were able to study the effect of radiation therapy where it was shown that repeated low-dose irradiation caused 55% of the cells to undergo early senescence in comparison to 38% of cells for a single high dose [83].

CHAPERONES

Heat-shock proteins (HSPs) are highly conserved chaperones that have several important roles, including the regulation of protein folding, protein transport, autophagy, and cell protection from stress [84]. Overexpression of these molecular chaperones is not only associated with heat shock, but also other stressors including; cold temperatures [85], heavy metals [86], infection [87], UV light [88], and ischemia [89]. When stressors are absent, HSPs are bound to heat-shock factors (HSFs). Stress causes the dissociation of HSFs leading to induction of transcription and translation of further HSPs, which in turn bind to the newly synthesized HSPs to inhibit further synthesis [90]. Interestingly, Butov et al. showed that elevation of HSPs can result in an increase in longevity, through a process called hormesis, as seen in the nematode worm *Caenorhabditis elegans*, and a corresponding stochastic model [91]. With age however, the heat-shock response is attenuated, resulting in the accumulation of aberrant proteins which is associated with the pathogenesis of diseases including Alzheimer, Parkinson, and Huntington disease [92]. Using stochastic simulations, Proctor et al., showed HSP90 synthesis is upregulated to maintain homeostasis when there is an increase in denatured proteins as a result of elevated stress. Importantly, it was demonstrated that chronic stress led to an insufficient HSP90 production, resulting in protein aggregation, with chaperone damage and ATP decline exacerbating aggregation [93]. This model was later updated to include HSP70. Based on this updated model, three principles were uncovered: (1) the main trigger for protein homeostasis failure is due to oxidative stress, thus the reduction of oxidative stress is significant, (2) HSP action and cell death pathways are vital, particularly in the long-term to prevent protein aggregation and disease progression, and (3) although the inhibition of cell death pathways can lead to prolonged cell survival, it cannot prevent protein aggregation [94].

MATHEMATICALLY MODELING HEALTHSPAN AND ITS INTERSECTION WITH AGING

The goal of our group is to utilize the potential of mathematical modeling to explore the dysregulation of whole-body systems as a result of intrinsic aging. Evidently this has implications for the onset of diseases whose prevalence increases with age such as cardiovascular disease (CVD), and dementia. One area of intense focus for these investigations is the intersection between the hypothalamic–pituitary–adrenal axis and brain aging. In particular, we reasoned a mathematical model would be capable of investigating how stress induced fluctuations in cortisol levels impinge on hippocampal atrophy [95]. The findings from the *in silico* system reflected the dynamics of its *in vivo* counterpart. Moreover, the mathematical model was able to suggest that both acute and chronic levels of cortisol accelerated age-associated hippocampal atrophy with concomitant loss of hippocampal neuronal activity. The second and primary area of research focus for our group is the intersection between lipid metabolism, aging, and healthspan [96–101]. The logic which underscores our approach to mathematically modeling this system is to represent it in an integrated fashion because as we recently documented aging effects every component of this system [102]. The first mathematical model we developed of this system brings together all the key elements of cholesterol balance in the body [103,104]. The model included mechanistic details of cholesterol ingestion, excretion, synthesis, low-density lipoprotein receptor (LDLr) turnover, and reverse cholesterol transport. The model incorporated intrinsic aging by adopting key findings from the experimental literature. The model was utilized to explore the impact of age-related alterations to cholesterol metabolism on LDL-C concentrations, which is the gold standard risk factor for CVD. The model found that for each 10% increase in the rate of cholesterol absorption, there was a 12.5 mg/dL increase in LDL cholesterol. In addition, the model showed that when cholesterol absorption was increased by 30% between the age of 20 and 60 years this caused a 34 mg/dL increase in plasma LDL-C. However, the cardinal finding of the model centered on how age-associated changes to the number of hepatic LDL receptors (LDLr) impact plasma LDL-C. It was found that when the number of hepatic LDLr declined by 50% by 65 years of age this caused LDL-C to increase significantly. The model was encased in SBML and archived in BioModels (http://www.ebi.ac.uk/biomodels-main/BIOMD0000000434). This means our model is straightforward to update and is available to the community as a whole. In 2016 our group updated the model to include further regulatory mechanisms and an array of additional features including intestinal microfloral enzymes and the detailed steps involved in both cholesterol biosynthesis and bile acid synthesis [105]. Using the new model, it was found the parameters associated with the less characterized aspects of cholesterol metabolism are exceptionally sensitive to change. Moreover, the model found that the cholesterol biosynthesis pathway is particularly robust. The updated model was also used to explore the effect different cholesteryl ester transfer protein (CETP) genotypes had on cholesterol metabolism. The rationale for doing this was that mutations in the *CETP* gene have been associated with exceptional longevity. When this investigation was conducted the *in silico* model revealed that when with intrinsic aging was combined with a genotype indicative of low-CETP activity, this resulted in a significant increase in LDL-C, when compared to simulations which only included intrinsic aging. These findings emphasized the utility of mathematical models for facilitating our understanding of cholesterol metabolism and aging [106].

CONCLUSIONS

It is apparent from our discussion of aging and the mathematical models which have been developed to represent it, that there is a growing need to create models which are capable of integrating the various aspects of the biology aging into a unified framework. It is possible this journey will commence when some of the models discussed in this chapter are merged together. If the models are encoded in an exchange format such as SBML in theory this will make the task significantly more straightforward. Once integrated the unified model would offer the real possibility of being able to explore the synergies which exist between the mechanisms underpinning aging. A deeper understanding of these can only benefit our understanding of healthspan and will help to identify lifestyle interventions which will alleviate the pernicious effects of aging.

REFERENCES

1. Mc Auley MT, Guimera AM, Hodgson D, McDonald N, Mooney KM, Morgan AE et al. Modeling the molecular mechanisms of ageing. *Biosci Rep.* 2017.
2. Mc Auley MT, Mooney KM. Computational systems biology for aging research. *Interdiscip Top Gerontol.* 2015;40:35–48.
3. Darwin C. *The Origin of Species, 150th Anniversary Edition.* ed. Signet Classic, Penguin Publishing Group, New York; 2003.
4. Weismann A. The duration of life. In: *Essays Upon Heredity and Kindred Biological Problems.* Vol I. Clarendon Press, Oxford; 1881.
5. Kirkwood TB, Cremer T. Cytogerontology since 1881: A reappraisal of August Weismann and a review of modern progress. *Hum Genet.* 1982;60(2):101–21.
6. Mitteldorf J. An epigenetic clock controls aging. *Biogerontology.* 2016;17(1):257–65.
7. Fisher RA. *The Genetical Theory of Natural Selection: A Complete Variorum Edition.* Oxford University Press. 1930:7–15.
8. Haldane JBS. A mathematical theory of natural and artificial selection, part V: Selection and mutation. *Mathematical Proceedings of the Cambridge Philosophical Society.* Vol. 23. No. 07. Cambridge University Press. 1927.

9. Medawar PB. An Unsolved Problem of Biology. An Inaugural Lecture Delivered at University College, London, 6 December, 1951. 1952.

10. Williams GC. Pleiotropy, natural selection and the evolution of senescence. *Evolution.* 1957;11:398–411.

11. Hamilton WD. The moulding of senescence by natural selection. *J Theor Biol.* 1966;12(1):12–45.

12. Charlesworth B. Selection in populations with overlapping generations. I. The use of Malthusian parameters in population genetics. *Theor Popul Biol.* 1970;1(3):352–70.

13. Charlesworth B. Selection in populations with overlapping generations. VI. Rates of change of gene frequency and population growth rate. *Theor Popul Biol.* 1974;6(1):108–33.

14. Kirkwood TB. Evolution of ageing. *Nature.* 1977;270(5635):301–4.

15. Passos JF, Kirkwood TB. Biological and physiological aspects of aging. *Textbook Geriatr Dentistr.* 2015;7.

16. Klipp E, Liebermeister W, Wierling C, Kowald A, Herwig R. *Systems Biology: A Textbook.* John Wiley & Sons, Hoboken, New Jersey; 2016.

17. Palsson B, Palsson BØ. *Systems Biology.* Cambridge University Press; 2015.

18. Choi H, Mc Auley MT, Lawrence DA. Prenatal exposures and exposomics of asthma. *AIMS Environ Sci.* 2015;2(1):87–109.

19. Mc Auley MT, Choi H, Mooney K, Paul E, Miller VM. Systems biology and synthetic biology: A new epoch for toxicology research. *Adv. Toxicol.* 2015;2015:14.

20. Mc Auley MT, Proctor CJ, Corfe BM, Cuskelly CJ, Mooney KM. Nutrition research and the impact of computational systems biology. *J Comput Sci Syst Biol.* 2013;6:271–85.

21. Stockton, D.J., Schilstra, M., Khalil, R. and McAuley, M. Biological control processes and their application to manufacturing planning. Advances in manufacturing technology-XXI. *Proceedings of the 5th international conference on manufacturing research (ICMR2007),* 2001: 11th–13th September 2007. pp. 259–264.

22. Stockton DJ, Schilstra M, Khalil R, McAuley M. Applying biological control approaches to finite capacity scheduling. *Flex Autom Intell Manuf.* 2008;2:853–9.

23. Bartocci E, Lió P. Computational modeling, formal analysis, and tools for systems biology. *PLoS Comput Biol.* 2016;12(1):e1004591.

24. Wilkinson DJ. Stochastic modeling for quantitative description of heterogeneous biological systems. *Nat Rev Genet.* 2009;10(2):122–33.

25. Paulsson J, Berg OG, Ehrenberg M. Stochastic focusing: Fluctuation-enhanced sensitivity of intracellular regulation. *Proc Natl Acad Sci.* 2000;97(13):7148–53.

26. Garinis GA, Van der Horst GT, Vijg J, Hoeijmakers JH. DNA damage and ageing: New-age ideas for an age-old problem. *Nat Cell Biol.* 2008;10(11):1241–7.

27. Hucka M, Finney A, Bornstein BJ, Keating SM, Shapiro BE, Matthews J et al. Evolving a lingua franca and associated software infrastructure for computational systems biology: The Systems Biology Markup Language (SBML) project. *Syst Biol (Stevenage).* 2004;1(1):41–53.

28. Funahashi A, Matsuoka Y, Jouraku A, Kitano H, Kikuchi N, editors. CellDesigner: A modeling tool for biochemical networks. *Proceedings of the 38th Conference on Winter Simulation;* 2006: Winter Simulation Conference, Monterey, California.

29. Hoops S, Sahle S, Gauges R, Lee C, Pahle J, Simus N et al. COPASI—A complex pathway simulator. *Bioinformatics.* 2006;22(24):3067–74.

30. Li C, Donizelli M, Rodriguez N, Dharuri H, Endler L, Chelliah V et al. BioModels database: An enhanced, curated and annotated resource for published quantitative kinetic models. *BMC Syst Biol.* 2010;4:92.

31. Shay JW, Wright WE. Hayflick, his limit, and cellular ageing. *Nat Rev Mol Cell Biol.* 2000;1(1):72–6.

32. Herbig U, Jobling WA, Chen BP, Chen DJ, Sedivy JM. Telomere shortening triggers senescence of human cells through a pathway involving ATM, p53, and p21 CIP1, but not p16 INK4a. *Mol Cell.* 2004;14(4):501–13.

33. Kurz DJ, Decary S, Hong Y, Trivier E, Akhmedov A, Erusalimsky JD. Chronic oxidative stress compromises telomere integrity and accelerates the onset of senescence in human endothelial cells. *J Cell Sci.* 2004;117(11):2417–26.

34. Qiu X, Brown K, Hirschey MD, Verdin E, Chen D. Calorie restriction reduces oxidative stress by SIRT3-mediated SOD2 activation. *Cell Metab.* 2010;12(6):662–7.

35. Finkel T, Holbrook NJ. Oxidants, oxidative stress and the biology of ageing. *Nature.* 2000;408(6809): 239–47.

36. Goldberg AL. Protein degradation and protection against misfolded or damaged proteins. *Nature.* 2003;426(6968):895–9.

37.Ső C, Csermely P. Aging and molecular chaperones. *Exp Gerontol*. 2003;38(10):1037–40.
38. Johnson SC, Rabinovitch PS, Kaeberlein M. mTOR is a key modulator of ageing and age-related disease. *Nature*. 2013;493(7432):338–45.
39. Calvanese V, Lara E, Kahn A, Fraga MF. The role of epigenetics in aging and age-related diseases. *Ageing Res Rev*. 2009;8(4):268–76.
40. Franco R, Schoneveld O, Georgakilas AG, Panayiotidis MI. Oxidative stress, DNA methylation and carcinogenesis. *Cancer Lett*. 2008;266(1):6–11.
41. Harman D. The Biologic clock: The mitochondria? *J Am Geriatr Soc*. 1972;20(4):145–7.
42. Harman D. Aging: A theory based on free radical and radiation chemistry. 1955.
43. Kirkwood TBL, Kowald A. The free-radical theory of ageing—Older, wiser and still alive. *BioEssays*. 2012;34(8):692–700.
44. Kowald A, Kirkwood TBL. Transcription could be the key to the selection advantage of mitochondrial deletion mutants in aging. *Proc Natl Acad Sci USA*. 2014;111(8):2972–7.
45. Miwa S, Lawless C, von Zglinicki T. Mitochondrial turnover in liver is fast *in vivo* and is accelerated by dietary restriction: Application of a simple dynamic model. *Aging Cell*. 2008;7(6):920–3.
46. Jacob SF, Würstle ML, Delgado ME, Rehm M. An Analysis of the truncated Bid- and ROS-dependent spatial propagation of mitochondrial permeabilization waves during apoptosis. *J Biol Chem*. 2015; 291:4603–4613.
47. Mooney KM, Morgan AE, Mc Auley MT. Aging and computational systems biology. *Wiley Interdiscip Rev: Syst Biol Med*. 2016;8(2):123–39.
48. Harley CB, Futcher AB, Greider CW. Telomeres shorten during ageing of human fibroblasts. *Nature*. 1990;345(6274):458.
49. Buijs Jod, van den Bosch PPJ, Musters MWJM, van Riel NAW. Mathematical modeling confirms the length-dependency of telomere shortening. *Mech Ageing Dev*. 2004;125(6):437–44.
50. Rubelj I, Vondracek Z. Stochastic mechanism of cellular aging—Abrupt telomere shortening as a model for stochastic nature of cellular aging. *J Theor Biol*. 1999;197(4):425–38.
51. Proctor CJ, Kirkwood TBL. Modeling telomere shortening and the role of oxidative stress. *Mech Ageing Dev* 2002;123(4):351–63.
52. Proctor CJ, Kirkwood TBL. Modeling cellular senescence as a result of telomere state. *Aging Cell*. 2003;2(3):151–7.
53. Hirt BV, Wattis JAD, Preston SP. Modeling the regulation of telomere length: The effects of telomerase and G-quadruplex stabilising drugs. *J Math Biol*. 2014;68(6):1521–52.
54. Laplante M, Sabatini DM. mTOR signaling at a glance. *J Cell Sci*. 2009;122(20):3589–94.
55. Caron E, Ghosh S, Matsuoka Y, Ashton-Beaucage D, Therrien M, Lemieux S et al. A comprehensive map of the mTOR signaling network. *Mol Syst Biol*. 2010;6:453.
56. Kriete A, Bosl WJ, Booker G. Rule-based cell systems model of aging using feedback loop motifs mediated by stress responses. *PLOS J Comput Biol*. 2010;6(6):e1000820.
57. Dalle Pezze P, Nelson G, Otten EG, Korolchuk VI, Kirkwood TBL, von Zglinicki T et al. Dynamic Modeling of Pathways to cellular senescence reveals strategies for targeted interventions. *PLOS Comput Biol*. 2014;10(8):e1003728.
58. Li T, Wang G. Computer-aided targeting of the PI3K/Akt/mTOR pathway: Toxicity reduction and therapeutic opportunities. *Int J Mol Sci*. 2014;15(10):18856–91.
59. Mosca E, Alfieri R, Maj C, Bevilacqua A, Canti G, Milanesi L. Computational modeling of the metabolic states regulated by the kinase akt. *Front Physiol*. 2012;3:418.
60. Wang G, Krueger GRF. Computational analysis of mTOR signaling pathway: Bifurcation, carcinogenesis, and drug discovery. *Anticancer Res*. 2010;30(7):2683–8.
61. Thobe K, Siebert H, editors. Discrete modeling of the mapk-mtor pathway connection in cancer. *11th Bioinformatics Research and Education Workshop*, Berlin, Germany; 2013.
62. Mc Auley MT, Mooney KM, Angell PJ, Wilkinson SJ. Mathematical modeling of metabolic regulation in aging. *Metabolites*. 2015;5(2):232–51.
63. Guarente L. Sirtuins, aging, and medicine. *N Engl J Med*. 2011;364(23):2235–44.
64. Houtkooper RH, Pirinen E, Auwerx J. Sirtuins as regulators of metabolism and healthspan. *Nat Rev Mol Cell Biol*. 2012;13(4):225–38.
65. Mc Auley MT, Mooney KM, Salcedo-Sora JE. Computational modeling folate metabolism and DNA methylation: Implications for understanding health and ageing. *Brief Bioinform*. 2016;bbw116.
66. Liu AY, Scherer D, Poole E, Potter JD, Curtin K, Makar K et al. Gene-diet-interactions in folate-mediated one-carbon metabolism modify colon cancer risk. *Mol Nutr Food Res*. 2013;57(4). doi: 10.1002/mnfr.201200180.

67. Chen P, Li C, Li X, Li J, Chu R, Wang H. Higher dietary folate intake reduces the breast cancer risk: A systematic review and meta-analysis. *Br J Cancer.* 2014;110(9):2327–38.

68. De S, Shaknovich R, Riester M, Elemento O, Geng H, Kormaksson M et al. Aberration in DNA methylation in B-cell lymphomas has a complex origin and increases with disease severity. *PLOS Genet.* 2013;9(1):e1003137.

69. Reed MC, Nijhout HF, Neuhouser ML, Gregory JF, Shane B, James SJ et al. A mathematical model gives insights into nutritional and genetic aspects of folate-mediated one-carbon metabolism. *J Nutr.* 2006;136(10):2653–61.

70. Ulrich CM, Neuhouser M, Liu AY, Boynton A, Gregory JF, Shane B et al. Mathematical modeling of folate metabolism: Predicted effects of genetic polymorphisms on mechanisms and biomarkers relevant to carcinogenesis. *Cancer Epidemiol Biomarkers Prev.* A publication of the American Association for Cancer Research, Cosponsored by the American Society of Preventive Oncology. 2008;17(7):1822–31.

71. Neuhouser ML, Nijhout HF, Gregory JF, Reed MC, James SJ, Liu A et al. Mathematical modeling predicts the effect of folate deficiency and excess on cancer related biomarkers. *Cancer Epidemiol Biomarkers Prev.* American Association for Cancer Research, Cosponsored by the American Society of Preventive Oncology. 2011;20(9):1912–7.

72. Ulrich CM. Folate and cancer prevention: A closer look at a complex picture. *Am J Clin Nutr.* 2007;86(2):271–3.

73. Morrison PF, Allegra CJ. Folate cycle kinetics in human breast cancer cells. *J Biol Chem.* 1989;264(18): 10552–66.

74. Enrique Salcedo-Sora J, Mc Auley MT. A mathematical model of microbial folate biosynthesis and utilisation: Implications for antifolate development. *Mol Biosyst.* 2016;12(3):923–33.

75. Vijg J, Suh Y. Genome instability and aging. *Annu Rev Physiol.* 2013;75:645–68.

76. Williams AB, Schumacher B. p53 in the DNA damage repair process. *Cold Spring Harb Perspect Med.* 2016;6(5). doi:10.1101/cshperspect.a026070 a.

77. Ma L, Wagner J, Rice JJ, Hu W, Levine AJ, Stolovitzky GA. A plausible model for the digital response of p53 to DNA damage. *Proc Natl Acad Sci USA.* 2005;102(40):14266–71.

78. Proctor CJ, Gray DA. Explaining oscillations and variability in the p53-Mdm2 system. *BMC Syst Biol.* 2008;2:75.

79. Sun T, Yang W, Liu J, Shen P. Modeling the basal dynamics of P53 system. *PLOS ONE.* 2011;6(11):e27882.

80. Chong KH, Samarasinghe S, Kulasiri D. Mathematical modeling of p53 basal dynamics and DNA damage response. *Math Biosci.* 2015;259:27–42.

81. Dolan D, Nelson G, Zupanic A, Smith G, Shanley D. Systems modeling of NHEJ reveals the importance of redox regulation of Ku70/80 in the dynamics of DNA damage foci. *PLOS ONE.* 2013;8(2):e55190.

82. Passos JF, Nelson G, Wang C, Richter T, Simillion C, Proctor CJ et al. Feedback between p21 and reactive oxygen production is necessary for cell senescence. *Mol Syst Biol.* 2010;6(1):1–14.

83. Dolan DWP, Zupanic A, Nelson G, Hall P, Miwa S, Kirkwood TBL et al. Integrated stochastic model of DNA damage repair by non-homologous end joining and p53/p21-mediated early senescence signaling. *PLOS Comput Biol.* 2015;11(5):e1004246.

84. Li Z, Srivastava P. *Heat-Shock Proteins. Current Protocols in Immunology.* John Wiley & Sons, Inc., Hoboken, New Jersey; 2001.

85. Laios E, Rebeyka IM, Prody CA. Characterization of cold-induced heat shock protein expression in neonatal rat cardiomyocytes. *Mol Cell Biochem.* 1997;173(1–2):153–9.

86. Wagner M, Hermanns I, Bittinger F, Kirkpatrick CJ. Induction of stress proteins in human endothelial cells by heavy metal ions and heat shock. *Am J Physiol.* 1999;277(5 Pt 1):L1026–33.

87. Varano della Vergiliana JF, Lansley SM, Porcel JM, Bielsa S, Brown JS, Creaney J et al. Bacterial infection elicits heat shock protein 72 release from pleural mesothelial cells. *PLOS ONE.* 2013;8(5):e63873.

88. Simon MM, Reikerstorfer A, Schwarz A, Krone C, Luger TA, Jäättelä M et al. Heat shock protein 70 overexpression affects the response to ultraviolet light in murine fibroblasts. Evidence for increased cell viability and suppression of cytokine release. *J Clin Investig* 1995;95(3):926–33.

89. Zhang PL, Lun M, Schworer CM, Blasick TM, Masker KK, Jones JB et al. Heat shock protein expression is highly sensitive to ischemia-reperfusion injury in rat kidneys. *Ann Clin Lab Sci.* 2008;38(1):57–64.

90. Kiang JG, Tsokos GC. Heat shock protein 70 kDa: Molecular biology, biochemistry, and physiology. *Pharmacol Ther.* 1998;80(2):183–201.

91. Butov A, Johnson T, Cypser J, Sannikov I, Volkov M, Sehl M et al. Hormesis and debilitation effects in stress experiments using the nematode worm *Caenorhabditis elegans*: The model of balance between cell damage and HSP levels. *Exp Gerontol.* 2001;37(1):57–66.

92. Wyttenbach A, Arrigo AP. *The Role of Heat Shock Proteins during Neurodegeneration in Alzheimer's, Parkinson's and Huntington's Disease. Heat Shock Proteins in Neural Cells.* Springer, London; 2009, p. 81–99.

93. Proctor CJ, Sőti C, Boys RJ, Gillespie CS, Shanley DP, Wilkinson DJ et al. Modeling the actions of chaperones and their role in ageing. *Mech Ageing Dev.* 2005;126(1):119–31.

94. Proctor CJ, Lorimer IAJ. Modeling the role of the Hsp70/Hsp90 system in the maintenance of protein homeostasis. *PLOS ONE.* 2011;6(7):e22038.

95. Mc Auley MT, Kenny RA, Kirkwood TB, Wilkinson DJ, Jones JJ, Miller VM. A mathematical model of aging-related and cortisol induced hippocampal dysfunction. *BMC Neurosci.* 2009;10:26.

96. Mc Auley MT, Mooney KM. Computationally modeling lipid metabolism and aging: A mini-review. *Comput Struct Biotechnol J.* 2015;13:38–46.

97. Mc Auley MT, Mooney KM. Lipid metabolism and hormonal interactions: Impact on cardiovascular disease and healthy aging. *Expert Rev Endocrinol Metab.* 2014;9(4):357–67.

98. Mooney KM, Mc Auley MT. Cardiovascular disease and healthy ageing. *J Integr Cardiol.* 2015;1(4):76–8.

99. Kilner J, Corfe BM, McAuley MT, Wilkinson SJ. A deterministic oscillatory model of microtubule growth and shrinkage for differential actions of short chain fatty acids. *Mol Biosyst.* 2016;12(1):93–101.

100. Morgan AE, Mooney KM, Wilkinson SJ, Pickles NA, Mc Auley MT. Investigating cholesterol metabolism and ageing using a systems biology approach. *Proc Nutr Soc.* 2016;1–14.

101. Morgan A, Mooney K, Mc Auley M. Obesity and the dysregulation of fatty acid metabolism: Implications for healthy aging. *Expert Rev Endocrinol Metab.* 2016;11(6):501–10.

102. Morgan AE, Mooney KM, Wilkinson SJ, Pickles NA, Mc Auley MT. Cholesterol metabolism: A review of how ageing disrupts the biological mechanisms responsible for its regulation. *Ageing Res Rev.* 2016;27:108–24.

103. Mc Auley MT, Wilkinson DJ, Jones JJ, Kirkwood TB. A whole-body mathematical model of cholesterol metabolism and its age-associated dysregulation. *BMC Syst Biol.* 2012;6:130.

104. Mc Auley M, Jones J, Wilkinson D, Kirkwood T. Modeling lipid metabolism to improve healthy ageing. *BMC Bioinformatics.* 2005;6(3):1.

105. Morgan AE, Mooney KM, Wilkinson SJ, Pickles NA, Mc Auley MT. Mathematically modeling the dynamics of cholesterol metabolism and ageing. *Biosystems.* 2016;145:19–32.

106. Mc Aule MT, Mooney KM. LDL-C levels in older people: Cholesterol homeostasis and the free radical theory of ageing converge. *Med Hypotheses.* 2017;104:15–19.

Index

A

Accelerated aging, 14, 197, 241, 263, 266, 403
ACE inhibitors, 97, 102, 106, 358, 359
Adaptive immunity, 64, 424, 426
Adaptive phenotype, 26, 33
Adaptive programmed aging, 33
Age related genes, 47, 198
Age-related degeneration, 12, 559
Age-related diseases, 13, 47, 129, 305–321, 342, 351–354, 401, 412, 424, 456
Age-related macular degeneration, 91, 100, 318, 495
Aging brain, 10, 186, 194, 279, 369, 458
Aging clock, 124, 499–511
Aging genes, 126–127, 196–197
Aging related diseases (ARDs), 45, 455
Aging, 3–14, 23–38
ALT-711 as antiaging, 13
Alzheimer's disease, 91, 101, 122, 182, 282, 287, 317, 352, 405, 428, 455, 477, 492, 530, 569, 600
AMP-activated protein kinase, 491
Amyloid precursor protein (APP), 317, 405, 457, 602
Angiogenic pericytes, 274
Antagonistic pleiotropy theory, 25, 33, 45, 65, 638
Anthropocene, 150–151, 153, 155, 156
Antiaging strategies, 129
Antiaging therapies, 13, 15, 199
Anti-neuroinflammatory therapies, 373
Antioxidant, 13, 396, 410–411, 641
Apoptosis, 76, 81, 144, 321, 408, 424, 443, 545
Apoptotic protease activating factor, 1, 76
Ataxia telangiectasia, xix, 408, 646
Autophagolysosomes, 475
Autophagy, 34, 265, 285, 286, 341, 352, 386, 399–402, 427, 475–481, 492, 587, 647

B

Baller–Gerold syndrome (BGS), 216
Barrier-to-autointegration factor (BAF), 80
Base excision repair (BER), 176, 214
Bats, 153, 165, 617, 621, 625
Biodiversity intactness index (BII), 154
Biological age, 47, 51, 52, 183, 584, 585, 586
Biomarkers, 51–52, 81, 263–264, 312, 587–588
Biotechnologies, 561, 564–565
BLM, 210, 219, 220
Blood-brain barrier, 271–294, 359, 457
Bloom syndrome, 210, 490
BM MNC, 369, 370
Body mass index (BMI), 97, 104, 182, 456
Brain aging, 456–458, 647
Brain, 100–104, 287, 293, 457–458, 458–464

C

C. elegans, 77, 79, 124, 125, 176, 184, 187, 188, 190, 191, 195, 599–610, 625, 626

Calorie restriction (CR), 187, 400, 408–409, 476, 492
Calorie restriction mimetics, 479
Calorie restriction, 187, 383, 400, 408–409, 440, 476, 492
Cancer, 212, 237–242, 248–251, 265, 313–314, 408, 423–430
Carcinogenesis, 408, 646
Cardiometabolic disease, 402–404
Cardiovascular disease (CVD), 45, 96–97, 100–102, 105, 122, 135, 183, 197, 210, 238, 239, 259, 261, 262, 264, 265, 319, 341–342, 351, 354, 400, 402, 410, 428, 441, 456, 647
Catalase, 396, 442, 491
Cataract, 210, 238, 317, 318
CD4T cells, 261, 263
Cell cultures, 119, 312, 585
Cell proliferation restriction, 478–479, 581
Cell turnover, 95, 100
Cellular senescence, 26, 34, 44, 175, 306–307, 311, 313–315, 315–320, 478–479, 490, 579, 588, 643
Cellular signaling, 13, 315
Central nervous system (CNS), 96, 211, 272, 442–444, 463, 476, 531, 590
Cerebral circulation, 289, 290
Chaperone-mediated autophagy, 475, 477
Chlamydomonas reinhardtii, 136
Cholesterol, 464–465
Chromatin remodeling, 46, 61, 126
Chromatin, 28, 47, 61, 123–125, 311
Chronological aging, 178, 389, 499, 500, 502, 582
Cockayne syndrome, 231–235
Coenzyme Q_{10}, 13
Co-evolution, 29, 76
Cognition, 529–531
Co-morbidities, 352
Corticosteroid myopathy, 546
CpG island, 179, 619
C-reactive protein (CRP), 263, 457
CSA, 233, 234, 550
CSB, 233, 234, 235
Culture, 29, 148–150, 161–170, 308
Curcumin, 493
Cyclin dependent kinase inhibitor, 308
Cytogerontology, 478, 579
Cytokines, 319, 340, 343, 344

D

Death probability, 581
Deep neural networks (DNs), 52
Deleterious mutations, 25, 26, 30
Dementia with Lewy bodies (DLB), 98, 102–103
Demographic models of aging, 45, 60
Deterministic model, 196, 641–642, 645
Developmental programs, 32–36, 37, 140–143, 171
Diabetes, 37, 95, 96, 105, 265, 318
Dietary caloric restriction (DR), 383
Digital aging atlas, 6
Disposable soma theories, 57